Smart Cities Policies and Financing

Approaches and Solutions

Smart Cities Policies and Financing

Approaches and Solutions

Edited by
John R. Vacca

ELSEVIER

Elsevier
Radarweg 29, PO Box 211, 1000 AE Amsterdam, Netherlands
The Boulevard, Langford Lane, Kidlington, Oxford OX5 1GB, United Kingdom
50 Hampshire Street, 5th Floor, Cambridge, MA 02139, United States

Library of Congress Cataloging-in-Publication Data
A catalog record for this book is available from the Library of Congress

British Library Cataloguing-in-Publication Data
A catalogue record for this book is available from the British Library

ISBN: 978-0-12-819130-9

For information on all Elsevier publications visit our website at
https://www.elsevier.com/books-and-journals

Publisher: Joe Hayton
Acquisitions Editor: Kathryn Eryilmaz
Editorial Project Manager: Barbara Makinster
Production Project Manager: Punithavathy Govindaradjane
Cover Designer: Matthew Limbert

Typeset by TNQ Technologies

In memory of Ronald D. Spangler (1946–2019)

Contents

Contributors xxiii
About the editor xxv
Foreword xxvii
Preface xxix
Acknowledgments xxxv

Part I
Overview of urbanization policies and financial trends: a comprehensive introduction

1. Introduction to industry 4.0: smart cities policies and financing

Amir Hooshang Fakhimi, Mojtaba Nazarzadeh Dehbozorgi and Javad Majrouhi Sardroud

1.1 Introduction 3
1.2 Smart city policy 3
1.3 Industry revolutions 6
1.4 Smart city generations development 6
1.5 Summary 12
1.6 Chapter review questions/exercises 12
True/false 12
Multiple choice 12
Exercise 13
Problem 13
Hands-on projects 13
Project 13
Case projects 13
Problem 13
Optional team case project 13
Problem 13
References 13

Part II
Implementing smart infrastructure: some key challenges and science, technology and innovation-driven policy and financial instruments

2. The need to localize smart infrastructure policies and financing

Louis Burns

2.1 Introduction 17
2.2 Context, culture, and economics for local smart cities 17
2.2.1 Context and smart cities 17
2.2.2 Culture and smart cities 18
2.2.3 Economics and smart cities 19
2.3 Holistic problem-solving for smart cities 20
2.3.1 Universal frameworks 20
2.3.2 Universal design principles 20
2.3.3 The need for civic engagement and involvement 21
2.3.4 Local public participation processes 21
2.3.5 Local metrics for success 22
2.4 Locally relevant policies and financing concepts 22
2.4.1 Local balance sheets 23
2.4.2 Local budgeting 23
2.4.3 Local project selection criteria 23
2.4.4 Local incremental development 24
2.5 Key civic policy and financing instruments 25
2.5.1 Alternative engagement models to ensure the viability of IT initiatives 25
2.5.2 Charrettes 25
2.5.3 Civic crowdfunding 26
2.5.4 Civic hackathons 26
2.5.5 Commons-based organizations 26
2.5.6 Community land trust 26
2.5.7 Co-ops 26
2.5.8 Crowdsourcing 26
2.5.9 Dashboards 27
2.5.10 Economic development corporations 27
2.5.11 Local currencies 27
2.5.12 Local stock exchanges 27
2.5.13 Municipal annexation 27
2.5.14 Municipal bonds 28
2.5.15 Municipal Service Coverage 28
2.5.16 Public asset corporation 28
2.5.17 Public participation methods 28
2.5.18 Public–private partnerships (PPP) 28
2.5.19 Tactical Urbanism 28
2.5.20 Tax increment financing (TIF) 29

2.5.21 Upgrading to smart city infrastructure during routine maintenance 29
2.5.22 Zoning as a policy instrument 29
2.6 Summary 29
2.7 Chapter review questions/exercises 30
True/false 30
Multiple choice 30
Exercise 30
Problem 30
Hands-on projects 30
Project 30
Case projects 30
Problem 30
Optional team case project 30
Problem 30
References 30

3. Revolutionizing the localization of smart city infrastructure policies and financing

Michela Longo and Wahiba Yaici

3.1 Introduction 33
3.2 Mobility trend 33
3.2.1 Vehicle to everything—V2X 33
3.2.2 Mobility as a service (MaaS) 33
3.2.3 Autonomous vehicle 34
3.2.4 Electrification or the switch to electrical technology 34
3.3 Smart mobility and important factors 34
3.4 Mobility drivers and trends 35
3.4.1 Policies to reduce local and global emissions 35
3.4.2 The decrease in costs for the purchase of electric vehicles 35
3.4.3 Improved battery performance 35
3.4.4 Less expensive and more performing vehicles 35
3.4.5 Possibility of selling energy on the grid (vehicle-to-grid—V2G) 35
3.4.6 Reduction of emissions and fight against climate change 36
3.5 New actions for to improve the e-mobility 36
3.6 Summary 37
3.7 Chapter review questions/exercises 37
True/false 37
Multiple choice 37
Exercise 38
Problem 38
Hands-on projects 38
Project 38
Case projects 38
Problem 38
Optional team case project 38
Problem 38
References 38

4. Harnessing the local innovation system: an overarching solution for policy and financing instruments for promoting the localization of smart infrastructure

Michela Longo and Wahiba Yaici

4.1 Introduction 41
4.2 Digital player 41
4.3 How mobility is changing 42
4.4 Charging infrastructure 42
4.5 Charging stations for public enviroment 43
4.5.1 Scenario 1: extended stop 44
4.5.2 Scenario 2: short stop 44
4.5.3 Scenario 3: stop for charging 44
4.6 Cost of charging an electric vehicle 45
4.7 Summary 45
4.8 Chapter review questions/exercises 46
True/false 46
Multiple choice 46
Exercise 46
Problem 46
Hands-on projects 46
Project 46
Case projects 47
Problem 47
Optional team case project 47
Problem 47
References 47

5. Smart counties: technologies, considerations, characteristics, challenges, policies, and theoretical concerns

Evan Evangelopoulos

5.1 Introduction 49
5.2 Smart technologies 49
5.2.1 The internet 50
5.2.2 Sensors and the internet of things 50
5.2.3 User-generated content capabilities 51
5.2.4 Cloud computing 51
5.2.5 Smartphones and smartphone applications 51
5.2.6 GIS 52

5.2.7 Analytics 53
5.2.8 Augmented and virtual reality 54
5.2.9 Artificial intelligence (AI) 54
5.3 The smartification equation: from collection to application 55
5.4 Smartification considerations 56
5.4.1 Overall considerations of smartification applications 56
5.4.2 Instrumental considerations in smartification applications 56
5.4.3 Value-based considerations in smartification applications 57
5.5 Smartification in counties 57
5.5.1 The nature of county—city differences—unique county characteristics 57
5.5.2 Unique county services 58
5.5.3 County departments 58
5.5.4 Smart technology applications by county department 59
5.5.5 County characteristics, challenges, smartification, and departments 60
5.5.6 County planning department, priorities, applications, and considerations 63
5.5.7 County priorities 64
5.5.8 Policies for smart counties 64
5.6 General smartification concerns and benefits 68
5.6.1 Technology-related general concerns 68
5.6.2 Major participant-related general concerns 68
5.6.3 Spectrum of desired smartness and overall benefits 69
5.7 Theoretical concerns 69
5.8 Smartification and smart growth 70
5.9 Summary 71
5.9.1 Technologies 71
5.9.2 Process 72
5.9.3 In counties 72
5.9.4 Benefits, concerns, and theory 72
5.10 Chapter review questions/exercises 73
True/false 73
Multiple choice 73
Exercise 73
Problem 73
Hands-on projects 73
Project 73
Case projects 73
Problem 73
Optional team case project 74
Problem 74
References 74
Further reading 78

6. From community plan to clean energy infrastructure: policy, partnership, and investment

Chris Castro and R. Bruce Stephenson

6.1 Introduction 79
6.2 Activist informed policy 80
6.3 The roadmap to carbon-free energy 82
6.3.1 Conduct a municipal and community-wide forecast 82
6.3.2 Prepare a renewable energy and optimization analysis 82
6.3.3 Establish solar and electric vehicle design guidelines 82
6.3.4 Create a ranking matrix for solar siting 82
6.3.5 Develop a clean energy strategy toolkit 83
6.3.6 Utilize computer modeling 83
6.3.7 Determine current and future load forecasts 83
6.3.8 Invest in pilot projects 84
6.4 The future takes shape: shutting down coal and envisioning equity 84
6.5 Summary 86
6.6 Chapter review questions/exercises 86
True/false 86
Multiple choice 86
Exercise 87
Problem 87
Hands-on projects 87
Project 87
Case projects 87
Problem 87
Optional team case project 87
Problem 87
References 87

7. Energy efficiency in urban planning for smart cities in the developing world

Doaa M. El-Sherif

7.1 Introduction 89
7.1.1 The importance of efficient energy planning for cities 89
7.1.2 The sustainable energy efficient path for cities 89
7.2 Guidelines for energy efficiency in urban planning 89
7.2.1 Energy efficiency potential in the building sector 90
7.2.2 Why energy efficiency for urban planning 90
7.2.3 Principles of energy efficiency for urban planning 90

7.3 Developing energy-efficient urban plan for cities 90
- 7.3.1 Step 1: designating a lead office, finding champions 90
- 7.3.2 Step 2: establishing partnerships 91
- 7.3.3 Step 3: finding the hooks 91
- 7.3.4 Step 4: conducting an energy and greenhouse gas (GHG) emissions audit 91
- 7.3.5 Step 5: analyzing the collected data and developing a draft plan 91
- 7.3.6 Step 6: building public and internal support: the participation process 91
- 7.3.7 Step 7: finalizing the plan 91
- 7.3.8 Step 8: implementing and financing the plan 92
- 7.3.9 Step 9: monitoring and evaluating the plan 92
- 7.3.10 Step 10: publicizing the benefits 92

7.4 Developing an strategic urban plan for cities in Egypt 92
7.5 Responsibilities 92
7.6 Different steps 92
- 7.6.1 Start-u 93
- 7.6.2 Data collection 93
- 7.6.3 Data Analysis 93
- 7.6.4 City consultation 93
- 7.6.5 Strategy formulation 93
- 7.6.6 Approval of strategic urban plans 94
- 7.6.7 Evaluation of plans 94

7.7 Summary 94
7.8 Chapter review questions/exercises 95
- True/false 95
- Multiple choice 95

Exercise 95
- Problem 95

Hands-on projects 95
- Project 95

Case projects 95
- Problem 95

Optional team case project 95
- Problem 95

References 95

8. Transportation policies for connected and automated mobility in smart cities

Mizanur Rahman, Steven Polunsky and Steven Jones

8.1 Introduction 97
8.2 Connected and automated mobility 98
- 8.2.1 Connected and automated vehicles 98
- 8.2.2 Connected and automated infrastructure 98
- 8.2.3 Transportation data 99

8.3 Safety guidance for automated vehicles 99
8.4 Communication for connected mobility 101
8.5 Cybersecurity and privacy 101
8.6 Cooperative driving automation 105
8.7 Navigation of CAVs 106
8.8 Policies for connected and automated mobility services 108
- 8.8.1 Connected and automated mobility 108
- 8.8.2 Field report: Connecticut Department of Transportation, Connecticut, US 108
- 8.8.3 Field report: Florida, US 108
- 8.8.4 Field report: case studies and pilot projects by university of Oregon, Oregon, US 109
- 8.8.5 Mobility as a service (MaaS) 109
- 8.8.6 Private sector involvement 109

8.9 Parking and land use 109
8.10 Legislation 110
8.11 Public perception 111
8.12 Summary 111
8.13 Chapter review questions/exercises 112
- True/false 112
- Multiple choice 112

Exercise 112
- Problem 112

Hands-on projects 112
- Project 112

Case projects 112
- Problem 112

Optional team case project 112
- Problem 112

References 113
Further reading 116

9. Public–private-partnerships (PPP) enabled smart city funding and financing

Ali Mohammad Mirzaee and Javad Majrouhi Sardroud

9.1 Introduction 117
9.2 Literature review 118
9.3 Smart city definitions 119
9.4 The potential funding options for smart cities 121

9.5 PPP definition, its theory, and advantages 123
9.6 Components of PPP-Based funding and financing 123
9.7 Funding and financing strategies in PPP-Based smart cities 124
9.8 Components of PPP-Based smart cities funding and financing 125
9.8.1 Pillars and pilot of smart cities 125
9.8.2 Optimization and distribution level in capital structure 126
9.8.3 Multisector partnerships 126
9.9 Summary 128
9.10 Chapter review questions/exercises 128
True/false 128
Multiple choice 128
Exercise 129
Problem 129
Hands-on projects 129
Project 129
Case projects 129
Problem 129
Optional team case project 129
Problem 129
References 129
Further reading 131

Part III
Human resources skills gaps

10. Preparing students for interdisciplinary work: green infrastructure curricula at Toronto Metropolitan University, Toronto, Ontario, Canada

Sheila Boudreau, Gareth Gransaull, Nina-Marie Lister and Gary Pritchard

10.1 Introduction 135
10.2 Smart cities and green infrastructure 135
10.2.1 Considerations for practice 136
10.3 Educating professionals 137
10.3.1 Social science theory for interdisciplinary practice 138
10.3.2 Supportive models for interdisciplinary practice 141
10.4 Summary 150
10.4.1 Class materials: questions for class discussion 151
10.5 Chapter review questions/exercises 152
True/false 152
Multiple choice 152
Exercise 152
Problem 152
Hands-on Projects 152
Project 152
Case projects 152
Problem 152
Optional team case project 153
Problem 153
References 153
Further reading 153

11. Teaching sustainability in metro Orlando: the evolution of the pragmatic liberal arts at Rollins college

R. Bruce Stephenson

11.1 Introduction 155
11.2 Common sense education and an ideal campus: Hamilton Holt's legacy 155
11.3 Bike trails, nature preserves, and green communities 157
11.4 Winter park: sacrosanct space and active transportation 162
11.5 Greenworks action plan: making Orlando the greenest city in the Southeast 164
11.6 Sustainability's challenge: affordable housing and homelessness 165
11.7 Summary 166
11.8 Chapter review questions/exercises 166
True/false 166
Multiple choice 167
Exercise 167
Problem 167
Hands-on projects 167
Project 167
Case projects 167
Problem 167
Optional team case project 167
Problem 167
References 167

12. Partnering with technology firms to train smart city workforces

Neena Pahuja

12.1 Introduction 169
12.2 Why skill 170
12.3 Technology-based skilling verticals for a smart city 171
12.3.1 Telecommunication connectivity and infrastructure creation and maintenance 171
12.3.2 Traffic management 171

12.3.3 Data analytics and monetization 172
12.3.4 Healthcare 172
12.3.5 Disaster management training 172
12.3.6 Sales and marketing skills to promote local 172
12.3.7 Smart asset management 172
12.3.8 Industry 4.0 and digital manufacturing 172
12.3.9 Usage of 3D printing for designing 172
12.3.10 Smart energy: renewable energy and energy conservation 173
12.3.11 eCommerce and supply chain management 173
12.3.12 Information security 173
12.3.13 Supporting specially enabled persons and elders 173
12.3.14 Entrepreneurial skill 173
12.3.15 Change management 173

12.4 Smart city training framework 173

12.5 Upskilling in creating a smart city 175

12.6 Technology horizontals for skilling 175

12.6.1 Information and communication technology 175
12.6.2 Internet of things 176
12.6.3 Sensors 176
12.6.4 Information security and safety 176
12.6.5 Geospatial technology 176
12.6.6 Artificial intelligence (AI) 177
12.6.7 Blockchain 177

12.7 Preparing for the future technologies 177

12.7.1 Supporting Internet of transportation and logistics 4.0 177
12.7.2 Connected city drones 178
12.7.3 Connected healthcare 178

12.8 Expectation from technology vendors 178

12.8.1 Supporting innovation and digital city 178
12.8.2 Digital workplace and centers of excellence 179

12.9 Summary 179

12.10 Chapter review questions/exercises 179

True/false 179
Multiple choice 180

Exercise 180

Problem 180

Hands-on projects 180

Project 180

Case projects 180

Problem 180

Optional team case project 180

Problem 180

References 180

13. Training professional smart city workforces through a partnership with technology firms

Tuncay Ercan and Mahir Kutay

13.1 Introduction 181

13.2 Smart city workforces 182

13.2.1 New features of workforce 182
13.2.2 Training capabilities and opportunities 183
13.2.3 Training capabilities and opportunities government trends 183
13.2.4 Status in the private sector 183

13.3 Skills gap 184

13.3.1 Skills analysis for smart city workforce 184
13.3.2 Assessing the skills needed 185
13.3.3 Act on collected data and create a skillset inventory 186
13.3.4 Methods for skills development 186
13.3.5 Compensation of skill shortages 186
13.3.6 Policymaking for skills gap 187

13.4 Relations between education and areas of expertise 187

13.4.1 Partnering with technology firms to train smart city workforces 188
13.4.2 Government effect 188
13.4.3 IT effect in training 189
13.4.4 Automation 190
13.4.5 In-service training 190
13.4.6 Internship 190
13.4.7 University—industry cooperation 190

13.5 Partnership with technology firms 191

13.5.1 Drive technology and services 192
13.5.2 Labor transition 192
13.5.3 Relations with technology partners 192
13.5.4 Role of e-training and distance education 193

13.6 Conclusion 193

13.6.1 Recommendations and solutions for the future 193

13.7	**Summary**	194
13.8	**Chapter review questions/exercises**	194
	True/false	194
	Multiple choice	194
Exercise		194
	Problem	194
Hands-on projects		195
	Project	195
Case projects		195
	Problem	195
Optional team case project		195
	Problem	195
References		195
Further reading		196

14. Training smart city workforces with a wide range of electronic and digital technologies

Doaa M. El-Sherif and Essam E. Khalil

14.1	**Introduction**	197
14.2	**Characteristics**	198
14.3	**Frameworks**	198
	14.3.1 Technology framework	198
	14.3.2 Human framework	199
	14.3.3 Institutional framework	199
	14.3.4 Energy framework	200
	14.3.5 Data management framework	200
14.4	**Criticism**	200
14.5	**Summary**	200
14.6	**Chapter review questions/exercises**	201
	True/false	201
	Multiple choice	201
Exercise		201
	Problem	201
Hands-on projects		201
	Project	201
Case projects		201
	Problem	201
Optional team case project		201
	Problem	201
References		202

Part IV
Lack of financing and well-developed business models

15. Digitization, automation, operation, and monetization: the changing management of sidewalk and kerb 2000–25

Alanna Coombes and Bern Grush

15.1	**Introduction**	207
	15.1.1 A century of changes at the kerb	207
15.2	**Twenty-first century footpaths, sidewalks, or pavements**	210
	15.2.1 Heading toward the autonomous future	211
	15.2.2 Prioritizing space	213
15.3	**Rights to the footway and kerb**	214
	15.3.1 Property rights	214
	15.3.2 Kerb rights	215
	15.3.3 Right to the smart city	215
	15.3.4 Is smart, smart?	215
15.4	**Summary**	216
15.5	**Chapter review questions/exercises**	216
	True/false	216
	Multiple choice	216
Exercise		217
	Problem	217
Hands-on projects		217
	Project	217
Case projects		217
	Problem	217
Optional team case project		217
	Problem	217
References		217

16. Digitization, automation, operation, and monetization: standardizing the management of sidewalk and kerb 2025–50

Bern Grush and Alanna Coombes

16.1	**Introduction**	219
16.2	**Delivery robot technology will out-disrupt the passenger robotaxi**	220
	16.2.1 The safety barrier for delivery robots is much lower than for robotaxis	220
	16.2.2 The fear barrier for delivery robots is much lower than for robotaxis	220
	16.2.3 Concerns for job loss from delivery robots are lower	221
	16.2.4 Delivery robots will have fewer enemies and more friends than robotaxis	221
	16.2.5 Development and deployment costs for delivery robots is far lower	221
	16.2.6 The regulatory barrier to delivery robot deployment is lower	221
	16.2.7 The travel environment for pathway robots is a more complex	221

16.2.8 Psychotechnical barriers around full autonomy favor delivery robots 222
16.2.9 Delivery robots evoke fewer perceived privacy issues 222
16.2.10 Delivery robots gives rise to fewer perceived security issues 223
16.2.11 The total risk equation for delivery robots is a magnitude lower 223
16.2.12 The risk of nonreadiness to cities and towns 223

16.3 An international standard to manage pathway robots 223
16.3.1 ISO 4448 223
16.3.2 The context for a standard for ground-based automated mobility 224
16.3.3 Five intentions for standardizing kerbside and pathway automation 225
16.3.4 Standard components 225
16.3.5 Robotic road vehicles for passengers or goods 226
16.3.6 Service robots on pedestrian pathways 227
16.3.7 Certification for use: kerbside and pathway certification for automation 231
16.3.8 Looking forward 232

16.4 Pathway robots and business improvement areas 232
16.4.1 The context of a *business improvement district* for delivery robots 232
16.4.2 Using pathway robots for consumer deliveries 232
16.4.3 Business Improvement Areas 233
16.4.4 Are pathway robots a good idea? 234
16.4.5 Are pathway robots a bad idea? 234
16.4.6 Business change and societal issues 235
16.4.7 Can local thinking sustain local growth? 235
16.4.8 Potential municipal and BIA responses to pathway robots 236

16.5 Summary 236
16.5.1 Afterword 237

16.6 Chapter review questions/exercises 237
True/false 237
Multiple choice 237

Exercise 238
Problem 238

Hands-on projects 238
Project 238

Case projects 238
Problem 238

Optional team case project 238
Problem 238

References 238

17. Innovative smart applications for solid waste management

Essam E. Khalil and Doaa M. El-Sherif

17.1 Introduction 239
17.1.1 Better environment 239

17.2 Various methods of waste disposal 239
17.2.1 Landfills 239
17.2.2 Incineration/combustion 240
17.2.3 Recovery and recycling 240
17.2.4 Plasma gasification 240
17.2.5 Composting 240
17.2.6 Waste to energy (recover energy) 240
17.2.7 Special waste disposal 240
17.2.8 Avoidance/waste minimization 240

17.3 The best methods of waste management 241
17.3.1 Bottom line 241
17.3.2 Functional elements of smart solid waste management 242
17.3.3 Integrated solid waste management 243

17.4 Recycling 244
17.4.1 Composting of organic solid wastes 244

17.5 Energy from incineration 245

17.6 Summary 245

17.7 Chapter review questions/exercises 246
True/false 246
Multiple choice 246

Exercise 246
Problem 246

Hands-on projects 247
Project 247

Case projects 247
Problem 247

Optional team case project 247
Problem 247

References 247

18. Digital infrastructure policies for data security and privacy in smart cities

Sabrina De Capitani di Vimercati, Sara Foresti, Giovanni Livraga and Pierangela Samarati

18.1 Introduction 249
18.2 Data anonymity 250
18.2.1 *k*-anonymity 252
18.2.2 Differential privacy 254
18.3 Data encryption 255
18.3.1 Enabling selective access 255
18.3.2 Enabling fine-grained access 256
18.4 Collaborative computations 257
18.5 Summary 258
18.6 Chapter review questions/exercises 259
True/false 259
Multiple choice 259
Exercise 259
Problem 259
Hands-on projects 259
Project 259
Case projects 259
Problem 259
Optional team case project 259
Problem 259
Acknowledgments 260
References 260

19. Financing smart cities—an Indian case study

Prasun Kamal Agrawal and Bipin Pradeep Kumar

19.1 Introduction 263
19.2 Convergence of missions 264
19.3 Six differentiators 265
19.3.1 Competition 265
19.3.2 Different starting points 265
19.3.3 Special-Purpose Vehicles (SPVs) 266
19.3.4 Citizen engagement 266
19.3.5 ABD (area-based development) + Pan City strategy 266
19.3.6 Convergence 267
19.4 Various sources of finance for an smart city 267
19.5 The SPV—an Agra case study 267
19.5.1 Structure of the SPV in ensuring financial control 268
19.5.2 Structure of the SPV in ensuring accountability 268
19.6 ICCC project of Agra smart city limited 268
19.7 The tender evaluation process by cities 269
19.7.1 Stage 1: prequalification 269
19.7.2 Stage 2: technical evaluation 269
19.7.3 Stage 3: commercial evaluation 269
19.7.4 Stage 4: final score calculation through QCBS 270
19.8 Innovative quality cum based selection (QCBS) methodology 270
19.9 Payment terms and schedule 270
19.10 Conclusion 271
19.11 Summary 271
19.12 Chapter review questions/exercises 272
True/false 272
Multiple choice 272
Exercise 272
Problem 272
Hands-on projects 272
Project 272
Case projects 272
Problem 272
Optional team case project 272
Problem 272
Acknowledgments 272
References 273

Part V
Governance: balancing top-down and bottom-up policy approaches

20. Policy instruments for facilitating smart city governance by promoting platforms for bottom-up participatory governance

Lidia Mierzejewska and Magdalena Wdowicka

20.1 Introduction 277
20.2 Smart city governance—understanding, aspects and the role of ICT 278
20.3 Basic policies for smart city governance 279
20.3.1 Public participation 279
20.3.2 Technology 281
20.3.3 Policy of open government data (OGD) 282
20.4 Platforms supporting bottom-up participatory governance 283
20.4.1 Coproduction 284
20.4.2 Cocreation of value 290
20.5 Summary 291
20.6 Chapter review questions/exercises 292

True/false 292
Multiple choice 292
Exercises 293
Problem 293
Hands-on projects 293
Case projects 293
Problem 293
Optional team case project 293
Problem 293
References 293

21. Promoting platforms for bottom-up participatory governance: a policy instrument approach through the facilitation of strategic smart city governance

Carles Agustí i Hernàndez

21.1 Introduction 297
21.1.1 Politics 297
21.1.2 Smart cities 298
21.1.3 Governance 298
21.1.4 Platforms 298
21.1.5 Bottom up 299
21.1.6 Participation 299
21.2 The cities of the future 299
21.2.1 Mobility 299
21.2.2 Urbanism 300
21.2.3 Green spaces 300
21.3 Governance of smart cities 300
21.4 Instruments of governance and cultural change 301
21.4.1 Listening 301
21.4.2 Observing 301
21.4.3 Capture 301
21.4.4 The creation of channels 301
21.4.5 Internalization 301
21.4.6 The answer 301
21.4.7 Share 302
21.4.8 From the outside in and from the top down 302
21.4.9 Citizen participation 302
21.4.10 Transparency 302
21.4.11 Open data 302
21.4.12 Open government 302
21.4.13 Collective intelligence 303
21.4.14 Cocreation spaces 303
21.4.15 Participatory platforms 303
21.5 Summary 303
21.6 Chapter review questions/exercises 303
True/false 303
Multiple choice 303
Exercise 304
Problem 304
Hands-on projects 304
Project 304
Case projects 304
Problem 304
Optional team case project 304
Problem 304
References 304

22. Policy instruments for facilitating smart city governance

Doaa M. El-Sherif and Essam E. Khalil

22.1 Introduction 305
22.2 Literature review: methods and the corpus 305
22.3 The components of smart governance 307
22.3.1 Governmental organization 307
22.3.2 Citizen participation 307
22.3.3 Use of technology 308
22.3.4 What are the outcomes of smart governance? 309
22.4 Summary 312
22.5 Chapter review questions/exercises 313
True/false 313
Multiple choice 314
Exercise 314
Problem 314
Hands-on projects 314
Project 314
Case projects 314
Problem 314
Optional team case project 314
Problem 314
References 314
Further reading 317

Part VI
Designing inclusive smart cities policy instruments

23. Provide affordable smart infrastructure policy applications for informal sectors

V. Padmavathi and K. Aruna

23.1 Introduction 321
23.1.1 Connectivity 321
23.1.2 Data processing 321
23.1.3 User interface 321
23.2 Smart cities and infrastructure 321
23.2.1 Smart city 322
23.2.2 Smart infrastructure 322
23.3 Energy 322
23.3.1 Key principles 322
23.4 Transport 322
23.5 Traffic management 323

23.5.1 Connected cars 323
23.5.2 Vehicle tracking systems 323
23.5.3 Public transport management 323
23.5.4 IoT technology for connected public transport systems 323
23.6 Health care 323
23.6.1 How an IoT-enabled care device works 324
23.6.2 Applications of IoT in healthcare 325
23.7 Market growth 325
23.7.1 Industrial internet revolution will be a key market trend 325
23.7.2 Networking technology to dominate the IIoT market 326
23.7.3 Industrial IoT 326
23.8 Implementation 327
23.8.1 PIR sensor's working principle 328
23.8.2 LDR sensor 328
23.8.3 Working principle of light dependent resistor 328
23.9 Results and discussions 329
23.9.1 Circuit connections 329
23.9.2 Prototype model front view 330
23.9.3 Arduino IDE output 330
23.9.4 Blynk cloud representation 330
23.9.5 Threshold alert 330
23.9.6 Motion reorganization 332
23.10 Summary 332
23.11 Chapter review questions/exercises 333
True/false 333
Multiple choice 333
Exercise 333
Problem 333
Hands-on projects 333
Project 333
Case projects 333
Problem 333
Optional team case project 333
Problem 333
References 333

24. Customizable service level agreement (SLA) generator platform using FCAPS management to enhance Quality of Experience (QoE) on Internet of Things (IoT)

Cathryn Peoples, Adrian Moore, Mohammad Zoualfaghari, Parag Kulkarni and Mamun Abu-Tair

24.1 Introduction 335
24.2 Research problem 336
24.3 Prior art 337
24.4 Research proposal 338
24.5 High-level ontology design 342
24.6 SLA assignment process 343
24.7 Summary 349
24.8 Chapter review questions/exercises 350
True/false 350
Multiple choice 350
Exercise 350
Problem 350
Hands-on projects 350
Project 350
Case projects 350
Problem 350
Optional team case project 351
Problem 351
References 351

25. Effective value capturing of mobility and public transportation through efficient urban planning

Ritu Mohanty-Padora and Bipin Pradeep Kumar

25.1 Introduction 353
25.2 Urban policies in India 354
25.3 Beginning with a strategic vision for urban planning and development 355
25.3.1 Case study 1: SPA for new smart business district in Mumbai 355
25.3.2 Case study 2: SPV for smart city with Tender SURE and mobility in Bengaluru 356
25.4 Capturing value by implementing unique financing mechanisms 357
25.4.1 Case study 1: Mumbai's new smart business district at Wadala 357
25.4.2 Case study 2: Tender SURE and mobility hubs at Bangalore Smart Cities 358
25.5 Global lessons and way forward 359
25.6 Conclusion 360
25.7 Summary 361
25.8 Chapter review questions/exercises 361
True/false 361
Multiple choice 361
Exercise 362
Problem 362
Hands-on projects 362
Project 362
Case projects 362
Problem 362
Optional team case project 362
Problem 362
Acknowledgments 362
References 362

Part VII
Smart infrastructure design principles and financial policy approaches

26. Urban design strategies and the smart city paradigm

Mallory B.E. Baches

26.1 Introduction 365
26.2 Unpacking the smart city paradigm 365
26.2.1 Criticism of the smart city paradigm and its impacts on urbanism 366
26.3 The role of urban design in shaping cities 367
26.3.1 Urban design as a framework for understanding cities 368
26.4 Smart city technologies and urban design challenges 369
26.4.1 Use of digital technologies in public engagement and participatory design 370
26.4.2 Urban design for smart(er) cities 370
26.4.3 The practice of urban design with digital technologies 371
26.5 The future design of cities 372
26.6 Summary 373
26.7 Chapter review questions/exercises 373
True/false 373
Multiple choice 374
Exercise 374
Problem 374
Hands-on projects 374
Project 374
Case projects 374
Problem 374
Optional team case project 374
Problem 374
References 374

27. Smart City Technologies plus Nature-Based Solutions: viable and valuable resources for urban resilience

Ernesto Philibert Petit

27.1 Introduction 377
27.2 Urban resilience policies in the context of sustainability: the continuing battle for sustainability will be held in the cities 378
27.2.1 Resilience is the emergency room of sustainability 378
27.2.2 Urban resilience policies, their foci, and the threats they are devised to face 379
27.3 Nature-Based Solutions, the ultimate resource for urban resilience 380
27.3.1 Defined concepts of NBS 380
27.3.2 Benefits and costs of NBS 383
27.3.3 Implementation of NBS 384
27.3.4 Assessment of NBS 385
27.4 Smart City Technologies applied to NBS 386
27.4.1 A Smart City Technologies *plus* Nature-Based Solutions conceptual chart 386
27.4.2 Glossary of defined concepts of Smart City Technologies 387
27.5 Viability of NBS 389
27.5.1 Policy compliance 389
27.5.2 Fundability and solvency 390
27.5.3 Dynamic efficiency: a substantial positive benefit—cost ratio over time 390
27.5.4 Efficiency and dynamic efficiency 390
27.6 Added value of SCT *plus* NBS 391
27.6.1 Effectiveness 391
27.6.2 Stakeholder engagement and social equity 391
27.6.3 Synergism with other urban infrastructures 391
27.6.4 Synergism with other urban systems 392
27.7 Propositions for an upcoming agenda: nature-based smart urban projects 393
27.7.1 Network approach 393
27.7.2 Integration of disciplines 393
27.7.3 Integration to urban networks 393
27.7.4 Research by design 393
27.8 Summary 393
27.9 Chapter review questions/exercises 394
True/false 394
Multiple choice 394
Exercise 394
Problem 394
Hands-on projects 394
Project 394
Case projects 394
Problem 394
Optional team case project 394
Problem 394

References 395
Further reading 397

28. Financial and technical planning of decarbonized hybrid energy systems for modern cities

Hossam A. Gabbar, Muhammad R. Abdussami and Md Ibrahim Adham

28.1 Introduction 399
28.2 Nuclear-Renewable Integrated Energy System (N-R IES) 399
28.3 A case study: modeling of system architecture 400
28.3.1 Electric and thermal demand 400
28.3.2 Microreactors 402
28.3.3 Solar PV panels 402
28.3.4 Wind turbines 403
28.3.5 Battery Energy Storage (BES) 404
28.3.6 Converter 404
28.3.7 Thermal Load Controller 404
28.3.8 Boiler 404
28.3.9 Microgrid controller 404
28.4 Optimization problem 404
28.5 Simulation results 406
28.6 Sensitivity assessment 407
28.7 Future work 412
28.8 Summary 412
28.9 Chapter review questions/exercises 412
True/false 412
Multiple choice 412
Exercise 413
Problem 413
Hands-on projects 413
Project 413
Case projects 413
Problem 413
Optional team case project 413
Problem 413
References 413

29. Reactive power compensation for voltage variation mitigation in grid with renewable energy

Om Prakash Mahela, Ashok Gocher, Baseem Khan, Sunil Agarwal, Akhil Ranjan Garg and Hassan Haes Alhelou

29.1 Introduction 415
29.1.1 Literature review 416
29.1.2 Contribution of the research work 417
29.1.3 Organization of the manuscript 417
29.2 Overview of utility grid 417
29.3 Methodology 417
29.3.1 Problem formulation 418
29.3.2 Base transmission system 418
29.3.3 Planning of reactive compensation to mitigate voltage variations 418
29.4 Results and discussion 420
29.4.1 Load flow study to check feasibility of the additional reactive compensation 421
29.4.2 Proposed cases of study 421
29.4.3 Observations of study 421
29.4.4 Effect of TCR on voltage under the condition of RE generation variation 422
29.5 Summary 423
29.6 Chapter review questions/exercises 424
True/false 424
Multiple choice 424
Exercise 425
Problem 425
Hands-on projects 425
Project 425
Case projects 425
Problem 425
Optional team case project 425
Problem 425
References 425

30. DC smart micro grid protection system

Endeshaw Solomon, Baseem Khan, Esayas Gidey, Mesfin Fanuel, Om Prakash Mahela and Hassan Haes Alhelou

30.1 Introduction 427
30.2 DC micro grids 427
30.2.1 DC micro grid design 427
30.2.2 DC micro grid control strategy 427
30.3 DC protection system 428
30.3.1 Protection devices 428
30.3.2 Grounding options 428
30.4 Principle of protection system 429
30.4.1 Nonunit protection system 429
30.4.2 Unit protection 429
30.4.3 Other protection problems 431
30.5 Research prospect 432
30.5.1 Grounding and protection system 432

30.5.2 Control and protection of coordinated strategy 432
30.5.3 Comparison of optimal power flow methods 432
30.6 Summary 433
30.7 Chapter review questions/exercises 434
True/false 434
Multiple choice 434
Exercise 434
Problem 434
Hands-on projects 434
Project 434
Case projects 434
Problem 434
Optional team case project 434
Problem 434
References 435

31. Identification and classification of faults using Stockwell transform and decision rule

Om Prakash Mahela, Vishnu Dutt Sharma, Baseem Khan, Sunil Agarwal and Hassan Haes Alhelou

31.1 Introduction 439
31.1.1 Related work 439
31.1.2 Contribution of the research work 440
31.1.3 Organization of the chapter 440
31.2 Methodology and test system 441
31.2.1 Problem formulation 441
31.2.2 Test system used for the study 441
31.2.3 Current supported algorithm used for identification of fault conditions 441
31.3 Result and discussion 442
31.3.1 Faults on phase-A and involvement of ground 442
31.3.2 Fault on phases-A and B and without involvement of ground 443
31.3.3 Fault on phases-A and B and involvement of ground 444
31.3.4 Fault on all the three phases and without involvement of ground 445
31.3.5 Fault on all the three phases and involvement of ground 446
31.3.6 Impact of fault incidence angle 446
31.3.7 Impact of variations in the fault impedance 449
31.3.8 Effect variations in fault location 451
31.3.9 Fault classification 453
31.4 Summary 453
31.5 Chapter review questions/exercises 453
True/false 453
Multiple choice 454
Exercise 454
Problem 454
Hands-on projects 454
Project 454
Case projects 454
Problem 454
Optional team case project 454
Problem 454
References 454
Further reading 455

32. Evaluation of fuel production technologies by using renewable energy for smart cities

Celal Hakan Canbaz, Orhan Ekren, Banu Y. Ekren and Vikas Kumar

32.1 Introduction 457
32.1.1 Energy management and planning in smart cities 458
32.2 Renewable fuel production technologies 459
32.3 Fuel production by using solar energy 460
32.3.1 Thermolysis of water 460
32.3.2 Thermochemical cycles 460
32.3.3 Gaseous hydrocarbon cracking 461
32.3.4 Gasification by solar energy 461
32.3.5 Solar energy fuel production in smart cities 463
32.4 Fuel production by using wind energy 465
32.4.1 Wind energy fuel production in smart cities 467
32.5 Summary 468
32.6 Chapter review questions/exercises 468
True/false 468
Multiple choice 469
Exercise 469
Problem 469
Hands-on projects 469
Project 469
Case projects 469
Problem 469
Optional team case project 469
Problem 469
References 469

33. Solving traffic congestion and parking scarcity with narrow track vehicles

Richard W. Woodbury

33.1	Introduction: the traffic challenge	471
33.2	Potential solutions	472
33.3	Another way of thinking	474
33.4	Disrupting disruption	477
33.5	What are the alternatives?	477
33.6	Potential financial benefits to cities	477
33.7	Summary	478
33.7.1	Commuters	478
33.7.2	Municipalities	478
33.7.3	Communities in general	479
33.8	Chapter review questions/exercises	479
	True/false	479
	Multiple choice	479
Exercise		480
	Problem	480
Hands-on projects		480
	Project	480
Case projects		480
	Problem	480
Optional team case project		480
	Problem	480
References		480
Further reading		480

34. Modeling and simulation for connected and automated vehicle (CAV) deployment and performance evaluation

Hossam A. Gabbar, Jing Ren and Patrick C.K. Hung

34.1	Introduction	481
34.2	CAV use cases	482
34.2.1	Vehicle platooning	482
34.2.2	Lane change	483
34.2.3	Intersection traffic	483
34.2.4	Obstacle avoidance	483
34.3	CAV integrated modeling and simulation	483
34.3.1	CAV functional modeling	484
34.3.2	Integrated CAV simulation	487
34.4	Control strategies with deep learning	494
34.4.1	Overview	494
34.4.2	System designs and descriptions	495
34.4.3	Case study	495
34.5	Communication and cybersecurity layers for CAV systems	497
34.6	CAV test platform	499
34.6.1	CAVTE design requirements	499
34.6.2	Test functions	499
34.6.3	Function test analysis	501
34.6.4	CAVTE demonstration	501
34.7	Robot-assisted CAV systems	504
34.8	CAV deployment in smart cities	505
34.9	Summary	506
34.10	Chapter review questions/exercises	507
	True/false	507
	Multiple choice	507
Exercise		507
	Problem	507
Hands-on projects		507
	Project	507
Case projects		507
	Problem	507
Optional team case project		507
	Problem	507
Acknowledgment		508
References		508
Further reading		509

35. VUCA: volatility, uncertainty, complexity, and ambiguity in our smart city and tech finance future

John Sanphillippo

35.1	Introduction	511
35.2	What is new is old again	511
35.3	Flows of sewage, streams of revenue	513
35.4	All drivers are above average	514
35.5	Repaving paradise, potholes in parking lots	515
35.6	Complexity meets complications	516
35.7	Unintended consequences	517
35.8	The onerous ship	517
35.9	Summary	518
35.10	Chapter review questions/exercises	518
	True/false	518
	Multiple choice	518
Exercise		519
	Problem	519
Hands-on projects		519
	Project	519
Case projects		519
	Problem	519
Optional team case project		519
	Problem	519
References		519

Part VIII
Appendices

Appendix A: List of top smart cities and critical infrastructure policies and financing implementation and deployment companies 525

Appendix B: List of smart cities and critical infrastructure policies and financing products 541

Appendix C: List of smart cities and critical infrastructure policies and financing standards 545

Appendix D: List of miscellaneous smart cities and critical infrastructure policies and financing resources 551

Appendix E: Smart cities and critical infrastructure policies and financing frequently asked questions 553

Appendix F: List of smart cities and critical infrastructure policies and financing case studies 557

Appendix G: Answers to review questions/exercises, hands-on projects, case projects, and optimal team case project by chapter 561

Appendix H: Glossary 589

Index 591

Contributors

Muhammad R. Abdussami, Faculty of Energy Systems and Nuclear Science, Ontario Tech University, Oshawa, ON, Canada

Mamun Abu-Tair, Ulster University, Antrim, United Kingdom

Md Ibrahim Adham, Faculty of Energy Systems and Nuclear Science, Ontario Tech University, Oshawa, ON, Canada

Sunil Agarwal, Department of Electrical Engineering, Apex Institute of Engineering and Technology, Jaipur, Rajasthan, India

Prasun Kamal Agrawal, India Japan Technology Leader, Mumbai, Maharashtra, India

Carles Agustí i Hernàndez, Open Government Area of Presidency, Barcelona, Spain

Hassan Haes Alhelou, Tishreen University, Latakia, Syria

K. Aruna, A.V.C. College of Engineering, Mayiladuthurai, Tamilnadu, India

Mallory B.E. Baches, Congress for the New Urbanism, Savannah, GA, United States

Sheila Boudreau, Toronto Metropolitan University, Toronto, ON, Canada

Louis Burns, Urban Prosperity Network, Richardson, TX, United States

Celal Hakan Canbaz, Ege University, Izmir, Bornova, Turkey

Chris Castro, Office of Sustainability and Resilience, Orlando, FL, United States

Alanna Coombes, University of Southern California, Los Angeles, CA, United States

Sabrina De Capitani di Vimercati, Università degli Studi di Milano, Milan, Italy

Mojtaba Nazarzadeh Dehbozorgi, Central Tehran Branch, Islamic Azad University, Tehran, Iran

Orhan Ekren, Ege University, Izmir, Bornova, Turkey

Banu Y. Ekren, Yasar University, Izmir, Turkey

Doaa M. El-Sherif, Housing & Building National Research Center (HBRC), Ministry of Housing, Utilities & Urban Communities, Cairo, Egypt

Tuncay Ercan, Yasar University, Izmir, Turkey

Evan Evangelopoulos, Pinal County Community Development Department, Florence, AZ, United States

Amir Hooshang Fakhimi, Kashan University, Islamic Azad University, Kashan, Iran

Mesfin Fanuel, Department of Electrical and Computer Engineering, Hawassa University, Hawassa, Ethiopia

Sara Foresti, Università degli Studi di Milano, Milan, Italy

Hossam A. Gabbar, Faculty of Energy Systems and Nuclear Science, Faculty of Engineering and Applied Science, Ontario Tech University, Oshawa, ON, Canada

Akhil Ranjan Garg, Department of Electrical Engineering, Faculty of Engineering and Architecture, J.N.V. University, Jodhpur, Rajasthan, India

Esayas Gidey, Department of Electrical and Computer Engineering, Hawassa University, Hawassa, Ethiopia

Ashok Gocher, Department of Electrical Engineering, Apex Institute of Engineering and Technology, Jaipur, Rajasthan, India

Gareth Gransaull, University of Western Ontario, Toronto, ON, Canada

Bern Grush, Harmonize Mobility Inc., Toronto, ON, Canada

Patrick C.K. Hung, Faculty of Business and IT, Ontario Tech University, Oshawa, ON, Canada

Steven Jones, The University of Alabama, Tuscaloosa, AL, United States

Essam E. Khalil, Cairo University, Cairo, Egypt

Baseem Khan, Department of Electrical and Computer Engineering, Hawassa University, Hawassa, Ethiopia

Parag Kulkarni, United Arab Emirates University, Al Ain, Abu Dhabi Emirate, United Arab Emirates

Bipin Pradeep Kumar, Systems Biology Institute, Tokyo, Japan

Vikas Kumar, University of the West of England, Bristol, United Kingdom

Mahir Kutay, Yasar University, Izmir, Turkey

Nina-Marie Lister, Toronto Metropolitan University, Toronto, ON, Canada

Giovanni Livraga, Università degli Studi di Milano, Milan, Italy

Michela Longo, Politecnico di Milano, Milan, Italy

Om Prakash Mahela, Power System Study Division, Rajasthan Rajya Vidyut Prasarn Nigam Ltd., Jaipur, Rajasthan, India

Lidia Mierzejewska, Adam Mickiewicz University, Poznań, Poland

Ali Mohammad Mirzaee, Department of Civil Engineering, Central Tehran Branch, Islamic Azad University, Tehran, Iran

Ritu Mohanty-Padora, Urban Planning and Design Consultant, Mumbai, Maharashtra, India

Adrian Moore, Ulster University, Antrim, United Kingdom

V. Padmavathi, A.V.C. College of Engineering, Mayiladuthurai, Tamilnadu, India

Neena Pahuja, ERNET India, Ministry of Electronics and Information Technology, New Delhi, Delhi, India

Cathryn Peoples, Ulster University, Londonderry, United Kingdom

Ernesto Philibert Petit, Environmental Structure Research Group, Queretaro, Mexico

Steven Polunsky, The University of Alabama, Tuscaloosa, AL, United States

Gary Pritchard, University of Guelph, Guelph, ON, Canada; Carleton University, Ottawa, ON, Canada

Mizanur Rahman, The University of Alabama, Tuscaloosa, AL, United States

Jing Ren, Faculty of Engineering and Applied Science, Ontario Tech University, Oshawa, ON, Canada

Pierangela Samarati, Università degli Studi di Milano, Milan, Italy

John Sanphillippo, Granola Shotgun, San Francisco, CA, United States

Javad Majrouhi Sardroud, Department of Civil Engineering, Central Tehran Branch, Islamic Azad University, Tehran, Iran

Vishnu Dutt Sharma, Department of Electrical Engineering, Apex Institute of Engineering and Technology, Jaipur, Rajasthan, India

Endeshaw Solomon, Department of Electrical and Computer Engineering, Hawassa University, Hawassa, Ethiopia

R. Bruce Stephenson, Environmental Studies, Rollins College, Winter Park, FL, United States

John R. Vacca, TechWrite, Pomeroy, OH, United States

Magdalena Wdowicka, Adam Mickiewicz University, Poznań, Poland

Richard W. Woodbury, Commuter Cars Corporation, Spokane, WA, United States

Wahiba Yaici, CanmetENERGY Research Centre, Ottawa, ON, Canada

Mohammad Zoualfaghari, BT Group Plc, Suffolk, United Kingdom

About the editor

John R. Vacca is an information technology consultant, researcher, professional writer, editor, reviewer, and internationally known, best-selling author based in Pomeroy, Ohio. Since 1982, John has authored/edited 84 books (some of his most recent books include):

- Cloud Computing Security: Foundations and Challenges, Second Edition [*Publisher*: CRC Press (an imprint of Taylor & Francis Group, LLC) (November 2, 2020)]
- Solving Urban Infrastructure Problems Using Smart City Technologies: Handbook on Planning, Design, Development, and Regulation, First Edition [*Publisher*: Elsevier Inc.) (September 30, 2020)]
- Online Terrorist Propaganda, Recruitment, and Radicalization, First Edition [*Publisher:* CRC Press (an imprint of Taylor & Francis Group, LLC) (August 9, 2019)]
- Nanoscale Networking and Communications Handbook, First Edition [*Publisher:* CRC Press (an imprint of Taylor & Francis Group, LLC) (July 22, 2019)]
- Computer and Information Security Handbook, 3E [*Publisher:* Morgan Kaufmann (an imprint of Elsevier Inc.) (June 10, 2017)]
- Cloud Computing Security: Foundations and Challenges [*Publisher:* CRC Press (an imprint of Taylor & Francis Group, LLC) (September 14, 2016)]
- Security in the Private Cloud [*Publisher:* CRC Press (an imprint of Taylor & Francis Group, LLC) (August 26, 2016)]
- Handbook of Sensor Networking: Advanced Technologies and Applications [*Publisher:* CRC Press (an imprint of Taylor & Francis Group, LLC) (January 14, 2015)]
- Network and System Security, Second Edition, 2E [*Publisher:* Syngress (an imprint of Elsevier Inc.) (September 23, 2013)]
- Cyber Security and IT Infrastructure Protection [*Publisher:* Syngress (an imprint of Elsevier Inc.) (September 23, 2013)]
- Managing Information Security, Second Edition, 2E [*Publisher:* Syngress (an imprint of Elsevier Inc.) (September 23, 2013)]
- Computer and Information Security Handbook, 2E [*Publisher:* Morgan Kaufmann (an imprint of Elsevier Inc.) (May 31, 2013)]
- Identity Theft (Cybersafety) [*Publisher:* Chelsea House Pub (April 1, 2012)]

and, more than 600 articles in the areas of advanced storage, computer security, and aerospace technology (copies of articles and books are available upon request).

John was also a configuration management specialist, computer specialist, and the computer security official (CSO) for NASA's space station program (Freedom) and the International Space Station Program, from 1988 until his retirement from NASA in 1995.

In addition, John is also an independent online book reviewer. Finally, John was one of the security consultants for the MGM movie titled: "AntiTrust," which was released on January 12, 2001. A detailed copy of his author bio can be viewed at URL: http://www.johnvacca.com. John can be reached at: john2164@windstream.net.

Foreword

John and I met while working at NASA 30 years ago. I was immediately taken with his sincere passion for making this world a better place for humankind. Because of this humble integrity and solid character, I introduced him to a coworker who shared identical values. They fell in love and married. I feel selfishly responsible for laying a piece of the foundation, which allows John to prolifically write excellent books under the supportive strength from his wonderful wife, Bee.

Over the years, John moved his vision from astronomical space to a more immediate mission to improve planet earth. He now champions improving the world's infrastructure through ideas that stimulate technology advancements and policy. His extensive expertise in Information Technology is his core base to derive connections among people, businesses, cities, and commercial industry.

John is uniquely qualified to provide insightful concepts and strategies to harness science, technology, and innovation to today's stagnated infrastructure. In this book, he collaborates with the world's foremost experts and harnesses them together to offer a cogent approach for smart city infrastructures.

This book is smartly organized by addressing the infusion needs across all stakeholders, including engineers, policymakers, managers, and government. With all of John's books, there's a summarized set of recommendations and solutions. This will leave the reader an on-ramp to further discuss, challenge, and advance this significant effort.

It remains my privilege to have been a forward part of this discourse.

David D. Lee, BSEE, MBA
Aerospace Engineering Manager
Houston, Texas

Preface

This comprehensive handbook serves as a professional reference, as well as a practitioner's guide to the smart cities need of an integrated approach to harness the full potential of infrastructure policy making and financial planning. It offers an in-depth coverage of how integrated approaches are effective tools for capturing the dynamic relations between people, policies, financing, and environments; and where they are most often useful and effective when coproduced by scientists from multiple disciplines, policymakers, financial planners, managers, civil society representatives, and all relevant stakeholders.

A functional approach has been adopted regarding the key role of science, technology, and innovation (STI), including information and communications technologies (ICTs), in the design, development, and management of smart cities and infrastructure policies and financing. This approach explores practical problems and solutions in identifying the challenges encountered in the implementation of smart infrastructure policies and financing and the role of STI in addressing them.

This handbook also examines a new urban agenda that focuses on policies, financing, and strategies that can result in effectively harnessing the power and forces behind urbanization. Smart cities are expected to play an important role in this new urban agenda. In other words, the implementation of smart cities infrastructure policies and financing concepts, especially in developing countries, faces numerous challenges. This handbook discusses all of these challenges and the role STI urban communities can play in overcoming them, including some policy and financing instruments driven by STI that could help to address each challenge.

With the preceding background, this handbook invites contributed submissions by experts in the field that propose state-of-the-art solutions for smart cities policies and financing problems. It brings together experts from academia, government, and industry to give their views on how governments can allocate research funds to smart city projects and provide incentives to make such projects a priority within their STI communities, which can be used to improve the lives of billions of people living in towns and cities around the globe. A wide range of topics are covered providing exclusive insights into how we can transform our urban areas to make investments or create local adaptations, that they can join with other cities confronting similar developmental challenges, as well as with technology partners, to conceptualize, finance, implement, and exploit complementary policy competences and share lessons learned.

The primary audience for this handbook consists of researchers and practitioners in industry and academia, as well as Executives; Policy-makers; Regulators; City planners; Utility network specialists; Industry experts; Startuppers; Chief Information Officers; Heads of Economic Development; Heads of Sustainability; Heads of Planning; Heads of Transport; Heads of Innovation; e-government leads; Heads of Inward Investment; Financial investors and entrepreneurs, Heads of Growth and Regeneration; Councilors; Local Authority officials; Community managers; Development Managers; Directors; Senior Managers; Strategic Directors; Heads of Planning; Urban planners; Project and Program Managers and Heads of Research and Innovation that will be drawn from Government; Government leaders involved with new urban economic growth, designing new cities, and urban creative regions or zones; Engineers and designers; Local decision-makers; Local Authorities; Local Enterprise Partnerships; Higher Education, Health and Social Care; Transport; Business unit leaders; and Local Businesses and the Voluntary and Private Sector. This comprehensive reference and practitioner's guide will also be of value to a secondary audience, which is comprised of students in upper-division undergraduate and graduate-level courses in the engineering of smart cities and infrastructure technologies.

1.1 Organization of this book

The book is organized into eight parts composed of 35 contributed chapters by leading experts in their fields, as well as eight appendices, including an extensive glossary of smart cities and critical infrastructure terms and acronyms.

1.1.1 Part I: Overview of urbanization policies and financial trends: a comprehensive introduction

Part I discusses key urbanization policies and financial trends and their links to the 2030 Agenda for Sustainable Development. For instance:

Chapter 1, "Introduction to Industry 4.0: Smart Cities Policies and Financing," addresses the evolution of smart cities platforms, along with industrial revolutions, and the key points of policies applicable in the field of smart cities using industry 4.0 (city 4.0). Simultaneous attention is given to the financial effects and smart cities policies in the fourth generation of the different industrial revolution and is the key highlight that can be considered by policymakers in creating or developing smart cities.

1.1.2 Part II: Implementing smart infrastructure: some key challenges and science, technology, and innovation-driven policy and financial instruments

Part II describes how the implementation of smart infrastructure concepts, especially in developing countries, faces numerous challenges. This part discusses some of these challenges and the role science, technology, and innovation (STI) communities can play in overcoming them, including some policy instruments driven by STI that could help to address each challenge. For instance:

Chapter 2, "The Need to Localize Smart Infrastructure Policies and Financing," presents an overview of how smart city infrastructure policies and financing concepts need to be responsive to the local population and political economy.

Chapter 3, "Revolutionizing the Localization of Smart City Infrastructure Policies and Financing," presents a general discussion of the diffusion of E-mobility. It includes a brief introduction and particular attention barriers and actions needed to enable large-scale deployment of electric vehicles and associated charging stations.

Chapter 4, "Harnessing the Local Innovation System: An Overarching Solution for Policy and Financing Instruments for Promoting the Localization of Smart Infrastructure," presents a general discussion of the criteria that administrators have to use for the diffusion of charging stations (CS). It includes a brief introduction of and current developments regarding the new players and the new mode to the live concept of mobility; the different types of charging stations (private and public) currently available according to their characteristics; and the different scenarios that are necessary to consider for the diffusion of CS and the usage costs.

Chapter 5, "Smart Counties: Technologies, Considerations, Characteristics, Challenges, Policies, and Theoretical Concerns," explores the technologies that enabled the terms smart cities, counties, or regions. It presents the different parts of the smartification process; the major considerations; the specific applications to counties; and, the nature of the challenges, along with the theoretical implications of smart technology applications.

Chapter 6, "From Community Plan to Clean Energy Infrastructure: Policy, Partnership, and Investment," presents a general discussion of walkable urbanism and clean energy; and how they are essential to ending dependence on fossil fuels. It also looks at how Orlando, Florida's investment in sustainability, illustrates how coupling political will and a definitive vision can empower citizens and initiate change.

Chapter 7, "Energy Efficiency in Urban Planning for Smart Cities in the Developing World," focuses on the participatory approaches that are vital to the planning processes at the cities' level; energy-efficient design and urban planning that is the most cost effective way to reduce energy consumption of buildings; energy aspects that should be incorporated into the planning process, as a cross-cutting issue, which intersects with all other aspects and sectors; a comparison between the Egyptian strategic urban planning process and the international energy-efficient urban planning process; and how the Egyptian process has a high inclusion of the participatory process and all local actors.

Chapter 8, "Transportation Policies for Connected and Automated Mobility in Smart Cities," first presents an overview of connected and automated mobility systems, including connected and automated vehicles, transportation infrastructure, transportation data, and several brief field reports. Second, policies and guidance on safety; communication; cooperative driving automation; cybersecurity and privacy; and, navigation systems (GPS/GNSS) for connected and automated mobility are covered. Third, the chapter covers policies for connected and automated mobility services and parking and land use. Finally, legislation and public perception are presented for improving user acceptance for new technologies, such as connected and automated mobility systems.

Chapter 9, "Public–Private-Partnerships (PPP) Enabled Smart City Funding and Financing," consolidates the main strategies and key components of the PPP-based smart city in a proposed funding and financing framework.

1.1.3 Part III: Human resources skills gaps

Part III covers how human resources skills need to be available to ensure that all the different facets of the city are adequately and efficiently addressed in order for a smart city to be successful in its endeavors. For example, a digital or data layer must be added to all relevant operations, and more technology vendors and the integrated functioning of different departments must be dealt with. Human resources skills include planning and design; digital citizenship; data literacy; and policy and financing implementation and management. Investing in smart people, not only smart technology, is essential. Currently there is little research, especially in developing countries, that quantifies the exact skills deficit. Thus, to begin with, cities should conduct their own analyses of skills deficits. Some key policy and financing instruments are also discussed in this part. For instance:

Chapter 10, "Preparing Students for Interdisciplinary Work: Green Infrastructure Curricula at Toronto Metropolitan University, Toronto, Ontario, Canada," discusses the role of interdisciplinary approaches to green infrastructure work in the education of future planning professionals. Using a range of theories from complex systems science, Indigenous Knowledge, social psychology, and other fields, the chapter demonstrates how designers and planners must transcend disciplinary boundaries if they are to address the complex socioecological challenges that smart cities will invariably face. Through applied case studies at Toronto Metropolitan University, the chapter demonstrates the importance of creating an intersection between theory and practice, and identifies how social science techniques can be used in real-world settings to build relationships and promote interpersonal understanding in cross-sectoral, multistakeholder infrastructure projects.

Chapter 11, "Teaching Sustainability in Metro Orlando: The Evolution of the Pragmatic Liberal Arts at Rollins College," outlines how Orlando, Florida, adopted a sustainability plan, and how faculty and students worked in concert with the Sustainability and Resilience Office to illustrate investments in sustainability in the areas of livability, transportation, solid waste, local food, and water resources. The chapter also looks at how such partnerships are essential not only for educating the campus and the community, but by accelerating the pace for achieving sustainability. It also discusses how in 2021, Rollins College made sustainability a strategic priority, and how faculty versed in project-based learning are leading the effort to infuse sustainability into the curriculum. Finally, this chapter describes how this initiative not only documents the evolution of project-based learning, but also marks the generation that will envision and build smart cities.

Chapter 12, "Partnering with Technology Firms to Train Smart City Workforces," presents the need for training by collating some of the new skilling requirements of Smart Cities. It then gets into details about new methodologies that can be used for skilling; and how some of the technology companies can partner in handling this. The chapter also brings in some new thoughts to encourage the companies to spend and partner for reskilling in the Public—Private Partnership (PPP) mode, to ensure continuity in this activity. The chapter also suggests the possibility of creating a "Resource Pool" for a city for getting a better return on investment (ROI)—that is required for creating a smarter workforce. Finally, it also goes into some suggestions on how this smarter workforce can lead to a "Community Led Innovation" environment.

Chapter 13, "Training Professional Smart City Workforces Through a Partnership with Technology Firms," proposes new workforce qualifications and training requirements in smart cities. The chapter also discusses the perspectives of government and the private sector on this subject. Next, it lists the competencies required for the qualifications of the workforce in smart cities and provides information on policies and how to meet the needs in this regard. Also, while giving information about training and expertise areas for the required workforce; the impact of government; Information Technologies; and in-service training; the importance of internship and university—industry cooperation are explained in detail. Finally, this chapter provides information on labor transitions, relations with technology companies, e-education, distance education, company collaborations, and workplace support.

Chapter 14, "Training Smart City Workforces with a Wide Range of Electronic and Digital Technologies," examines the smart city as an urban area that uses different types of electronic methods and sensors to collect data. Next, the chapter looks at insights gained from data that are used to manage assets, resources, and services efficiently. It also looks at data collected from citizens, devices, buildings, and assets, which is then processed and analyzed to monitor and manage traffic and transportation systems, power plants, utilities, water supply networks, waste, crime detection, information systems, schools, libraries, hospitals, and other community services. Finally, the chapter describes the smart city concept; integrates information and communication technology (ICT); and various physical devices connected to the IoT (Internet of things) network to optimize the efficiency of city operations and services that connect to citizens.

1.1.4 Part IV: Lack of financing and well-developed business models

Part IV discusses how smart infrastructure projects require pooling public and private resources through creative financing and public—private partnership models. Policies, taxation, and regulatory certainty also play an important role in this process. Governments need to address such issues in order to encourage the private sector to grow and innovate in new, thoughtful, and increasingly strategic ways to invest in smart city projects. Strategically funding smart city infrastructure and technology investments is critical to the realization of smarter cities. Smart city projects are often complex undertakings, involving long time horizons, multiple stakeholders, and risks. Citizens must be made aware of the costs involved, the associated benefits, and the prices they will be charged before the commencement of projects. Some key policy instruments are also discussed in this part. For instance:

Chapter 15, "Digitization, Automation, Operation, and Monetization: The Changing Management of Sidewalk and Kerb 2000–25," covers the need for community, business and political agreement on who has rights to the kerb and footways. The chapter also discusses how these rights need to be turned into clearly defined priorities that meet the needs of citizens, including those traditionally excluded, and businesses. Next, it also explains how public space, including the kerb and footways, should be designed for community, artistic expression, and livability. Finally, the chapter describes how these vital public spaces (like the city centers in which they exist) need to adapt to the needs of current and future generations, by addressing their economic, social, community, and well-being.

Chapter 16, "Digitization, Automation, Operation, and Monetization: Standardizing the Management of Sidewalk and Kerb 2025–50," describes why autonomous, mobile robots will become pervasive long before fully automated passenger vehicles will. The chapter outlines an international standard for *Ground-based Automated Mobility*, currently being drafted, inclusive of allocation and reservation systems for these robots, as well as goods and passenger road vehicles. Finally, it describes a scenario for the operation of this technology within a community, including its implied benefits and drawbacks.

Chapter 17, "Innovative Smart Applications for Solid Waste Management," discusses the different types of solid waste and their common sources. This chapter also covers the stages in solid waste management and appropriate disposal methods.

Chapter 18, "Digital Infrastructure Policies for Data Security and Privacy in Smart Cities," addresses the problem of protecting large data collections in the context of smart cities; and illustrates possible approaches for effectively anonymizing data and encrypting them—while performing computations.

Chapter 19, "Financing Smart Cities—An Indian Case Study," is centered on the financial framework of the Smart Cities Mission. It outlines the smart city landscape along with other urban transformation missions in India. The key differentiating drivers; the various sources of financing; and the mechanism, called: Convergence are further discussed. Finally, the Agra Smart City Limited example elaborated on in this chapter illustrates the governance structure in place for implementing and monitoring the mission.

1.1.5 Part V: Governance: balancing top-down and bottom-up policy approaches

Part V discusses why Smart cities call for new governance models. Effective smart city management needs to balance both top-down and bottom-up governance policy approaches. On the one hand, collating the information generated by smart sensors deployed in different smart infrastructures and taking policy actions, especially during emergencies, may require strong top-level leadership and top-down execution processes. On the other hand, bottom-up governance policy approaches, including citizen-driven innovations and cocreation, have been the defining characteristic of much of smart city infrastructure. Managing a good balance between these two policy approaches is therefore important. Achieving such a balance helps city governments harness the synergy between various participants (such as universities, the private sector, civil society, and local and municipal governments). Furthermore, an efficient governance model for smart cities requires breaking down silos across different government departments. Information islands act as the greatest barrier to resource integration in the course of development, at both the technical and management levels of smart cities. Governance models need to be reformed in order that the data from smart infrastructure may be made available and used effectively in decision-making processes. City government administrators thus face the challenge of devising new forms of governance processes that adequately place citizen needs at the core of the governance process by adequately balancing top-down and bottom-up governance policy approaches. For instance:

Chapter 20, "Policy Instruments for Facilitating Smart City Governance by Promoting Platforms for Bottom-Up Participatory Governance," addresses the implementation of smart city governance and the adoption of appropriate development policies (to increase public participation, improve access to ICT technologies, and the open data policy), which are conducive to the development of platforms intended for stimulating bottom-up participatory. Attention is also paid to platforms that serve the phenomena called coproduction (including geoparticipation, crowdsourcing, crowdfunding, hackathons, Living Labs, gamification) and cocreation of value (e.g., urban hacktivism and BarCamps).

Chapter 21, "Promoting Platforms for Bottom-Up Participatory Governance: A Policy Instrument Approach Through the Facilitation of Strategic Smart City Governance," reviews what participatory democracy has been and its evolution—especially in recent years. This chapter will also look at how these advances and transformations have been made possible, in order

to generate participatory governance techniques and tools that are applied to Smart Cities and their characteristics, urban planning, mobility, and the role of citizens. Finally, the chapter addresses the governance of cities, and more specifically, analyzes the instruments of governance and participation that we have and how to implement them.
Chapter 22, "Policy Instruments for Facilitating Smart City Governance," reviews smart governance, in the sense of ICT-enabled government—citizen collaboration to advance urban sustainability. Next, the chapter looks at why the evidence that smart governance contributes to sustainability is sparse and mixed. Finally, the chapter explains why there is more evidence for the process effects of smart governance and that context matters.

1.1.6 Part VI: Designing inclusive smart cities policy instruments

Part VI discusses another key challenge faced by the smart city concept: To ensure the engagement of all groups of citizens and to promote inclusivity of policy instruments. Smart city policy applications should be inclusive in terms of providing opportunities for all and ensuring that particular groups are neither left out of positive impacts nor disproportionately affected by any societal costs that may be imposed. For instance, the needs of vulnerable groups such as women, the elderly, and persons with disabilities should be integrated into smart city policy strategies. The need to ensure inclusivity for such vulnerable groups is especially important because there is a significant possibility that they will lack the skills to use smart city policy applications or that their livelihoods may be the most affected by smart city policy applications. Promoting participatory governance in cities is a prerequisite for developing inclusive policies for smart cities. For instance:

Chapter 23, "Provide Affordable Smart Infrastructure Policy Applications for Informal Sectors," focuses on how smart infrastructure provides the inspiration for all of the key themes associated with a smart city, as well as good quality, good economy, good living, good governance, and good atmosphere. The core characteristic that underlies most of those elements is that they are connected—which generates knowledge, and can be used to show intelligence, optimum use of resources, and improved performance. This chapter also introduces some key elements of smart city infrastructure, by including a requirement for an integrated approach in handling such infrastructure. It also covers a large number of variables and improved modeling specifications, by incorporating dummy variables for global crisis and/or decomposing the analysis for specific sectors of the infrastructure (Transport, Energy, Telecom, etc. ...). The chapter also determines how there are some attempts that could be done to extend the analysis pertaining to the estimation of determinants of Public—Private Partnerships in India. Finally, the chapter focuses on two kinds of sensors that have been used: light sensor and photo electric sensor.
Chapter 24, "Customizable Service Level Agreement (SLA) Generator Platform Using FCAPS Management to Enhance Quality of Experience (QoE) on Internet of Things (IoT)," discusses the capture and use of information in the SLA provisioning process. The proposal presented in this chapter recognizes the different manners in which customers may wish to engage with a network and online services, as either application users, or more typically of IoT technologies, as data providers or data consumers. The information collected here from a customer is used to influence the SLA provisioning process, which is driven by an ontology. The proposed ontology accommodates the needs of all scenarios anticipated in the IoT.
Chapter 25, "Effective Value Capturing of Mobility and Public Transportation Through Efficient Urban Planning," examines the role of connectivity, mobility, and public transportation in our neighborhoods, within the context of urban development and the future of sustainable cities. When viewed from both, the design and planning standpoint, as well as the financial perspective, it reinforces the importance of transit-oriented development in urban environments.

1.1.7 Part VII: Smart infrastructure design principles and financial policy approaches

Part VII discusses smart city projects worldwide, as well as some key principles that may guide smart infrastructure projects that may be distilled, as follows: People-centered and inclusive financial policies; resilient and sustainable; interoperable and flexible; and risk-mitigating and safe. For instance:

Chapter 26, "Urban Design Strategies and the Smart City Paradigm," explores the complicated and still relatively unexplored interrelationship between the strategies associated with the practice of urban design, and the application of the so-called "smart city" digital technologies in urban areas, within the smart city paradigm. Key concepts of the smart city movement and its critical applications and outputs are defined, and common criticisms of the smart city paradigm are outlined, so as to ground the various technologies associated with the smart city movement within the context of the urban realm. Following that framing, urban design as a discipline for shaping and understanding cities and their relative components are described and contextualized. The opportunities and obstacles for the intersection of urban design with smart city digital technologies in cities are then explored. Finally, the future design of cities, and the ways that smart cities might integrate urban design (for the well-being of all citizens), is considered.

Chapter 27, "Smart City Technologies Plus Nature-Based Solutions: Viable and Valuable Resources for Urban Resilience," presents the possible and desirable networking of Smart City Technologies (SCT) with the emerging resource of Nature-Based Solutions (NBS), for cities to efficiently and effectively achieve resilience from threats, hazards, and disturbances, occasioned by an unsustainable development model.

Chapter 28, "Financial and Technical Planning of Decarbonized Hybrid Energy Systems for Modern Cities," addresses an overview of Nuclear-Renewable (N-R) hybrid energy systems and carries out a case study with real data. Sensitivity analysis is also conducted to validate the key findings of the case study.

Chapter 29, "Reactive Power Compensation for Voltage Variation Mitigation in Grid with Renewable Energy," proposes the planning of static and dynamic reactors (TCR), to overcome the problems of overvoltages in the transmission system of the Rajasthan state of India. This network has complex network transmission lines and grid substations (GSS) rated at voltage levels of 132 kV, 220 kV, 400 kV, and 765 kV—where the conventional and RE generators are integrated to this network. The study of reactive power compensation is performed using the MiPower software, where the transmission network of the Rajasthan State of India is modeled.

Chapter 30, "DC Smart Micro Grid Protection System," examines the main problems faced on DC micro grid protection systems from many features such as the characteristics of the fault current, grounding system, and fault locating system. Also, recommendations on the DC micro grid protection system are clearly examined.

Chapter 31, "Identification and Classification of Faults Using Stockwell Transform and Decision Rule," focuses on the design of an algorithm based on the processing of current signals using the Stockwell transform for identification of faults on the power transmission line.

Chapter 32, "Evaluation of Fuel Production Technologies by Using Renewable Energy for Smart Cities," investigates the latest fuel production technologies providing energy outputs from renewable energy resources for their usage in smart cities. The chapter looks at the role of renewable resources under the current dynamics of the global energy market situation, as well as the effects of fuel production technologies to increase the efficiency and sustainability for the future of smart cities. Specifically, solar and wind production technologies, as well as future forecasts, compared with the conventional fuel resources, are reviewed comprehensively. In other words, this chapter aims to shed light on the current status of fuel production technologies in renewable energy resources and their usage in smart cities.

Chapter 33, "Solving Traffic Congestion and Parking Scarcity with Narrow Track Vehicles," looks at how Narrow Track Vehicles (NTVs) offer significant benefits to both commuters and municipalities. These benefits can also be applied to the general public.

Chapter 34, "Modeling and Simulation for Connected and Automated Vehicle (CAV) Deployment and Performance Evaluation," presents modeling and simulation of Connected and Autonomous Vehicles (CAV) toward their deployment in smart cities. Different use cases are analyzed and used as the basis for modeling and simulation of CAV to evaluate their performance in different scenarios and conditions. Control strategies with deep learning are explained and applied on CAV. Communications and cybersecurity layers are illustrated to enable the smooth integration with information infrastructures. In order to assess CAV functions, test platform is discussed to enable cities and regions to evaluate CAV functions prior to deployment. Robot-assisted CAV systems are discussed to show benefits of adding robots to CAV. Finally, a detailed analysis of CAV deployment strategies is presented toward practical implementation in smart cities.

Chapter 35, "VUCA: Volatility, Uncertainty, Complexity and Ambiguity in Our Smart City and Tech Finance Future," explores the lived experiences of corporate executives, about their VUCA business environment, and the strategies they employed for VUCA readiness and success within a manufacturing sector. The chapter also looks at how chaos theory and complexity leadership theory could be used to frame their VUCA environment. Finally, it also looks at how successful practices and strategies play a role in leading systems-wide change in a technology-rich VUCA ecosystem.

John R. Vacca
john2164@windstream.net
www.johnvacca.com

Acknowledgments

There are many people whose efforts on this book have contributed to its successful completion. I owe each a debt of gratitude and want to take this opportunity to offer my sincere thanks.

A very special thanks to my Senior Acquisitions Editor, Brian Romer, without whose continued interest and support would not have made this book possible. And, a very special thanks to Senior Editorial Project Manager, Barbara Makinster, who provided staunch support and encouragement when it was most needed. Thanks to my Project Manager, Punithavathy Govindaradjane, and Copyrights Coordinator, Indhumathi Mani, whose fine editorial work has been invaluable. Finally, thanks to all of the other people at Elsevier Science and Technology Books, whose many talents and skills are essential to a finished book.

Thanks to my wife, Bee Vacca, for her love, her help, and her understanding of my long work hours. Also, a very very special thanks to David Lee, for writing the foreword. Finally, I wish to thank all the following authors who contributed chapters that were necessary for the completion of this book: Amir Hooshang Fakhimi, Mojtaba Nazarzadeh Dehbozorgi, Javad Majrouhi Sardroud, Louis Burns, Michela Longo, Wahiba Yaici, Evan Evangelopoulos, Chris Castro, R. Bruce Stephenson, Doaa M. El-Sherif, Mizanur Rahman, Steven Polunsky, Steven Jones, Ali Mohammad Mirzaee, Sheila Boudreau, Gareth Gransaull, Nina-Marie Lister, Gary Pritchard, Neena Pahuja, Tuncay Ercan, Mahir Kutay, Essam E. Khalil, Bern Grush, Alanna Coombes, Sabrina De Capitani di Vimercati, Sara Foresti, Giovanni Livraga, Pierangela Samarati, Prasun Kamal Agrawal, Bipin Pradeep Kumar, Lidia Mierzejewska, Magdalena Wdowicka, Carles Agustí i Hernàndez, V. Padmavathi, K. Aruna, Cathryn Peoples, Adrian Moore, Mohammad Zoualfaghari, Parag Kulkarni, Mamun Abu-Tair, Ritu Mohanty-Padora, Mallory B.E. Baches, Ernesto Philibert Petit, Hossam A. Gabbar, Muhammad R. Abdussami, Md Ibrahim Adham, Om Prakash Mahela, Ashok Gocher, Baseem Khan, Sunil Agarwal, Akhil Ranjan Garg, Hassan Haes Alhelou, Endeshaw Solomon, Esayas Gidey, Mesfin Fanuel, Vishnu Dutt Sharma, Celal Hakan Canbaz, Orhan Ekren, Banu Y. Ekren, Vikas Kumar, Richard W. Woodbury, Jing Ren, Patrick C. K. Hung, and John Sanphillippo.

Part I

Overview of urbanization policies and financial trends: a comprehensive introduction

Chapter 1

Introduction to industry 4.0: smart cities policies and financing

Amir Hooshang Fakhimi[1], **Mojtaba Nazarzadeh Dehbozorgi**[2] **and Javad Majrouhi Sardroud**[3]

[1]*Kashan University, Islamic Azad University, Kashan, Iran;* [2]*Central Tehran Branch, Islamic Azad University, Tehran, Iran;* [3]*Department of Civil Engineering, Central Tehran Branch, Islamic Azad University, Tehran, Iran*

1.1 Introduction

With the steady growth of the urban population in recent decades, it is estimated that by 2050, about 68% of the world's population, 85% of the European population, and 90% of the North American population will live in cities. It means in 2050, about 68% of the world's population will live in cities, while in 2018, about 55% of the world's population have lived in cities [1]. However, the increase in urban population on the one hand and the need for a higher quality of life, on the other hand, have increased urbanization problems. More energy consumption, low productivity, economic instability, poverty, inequality, and social conflicts are some of the issues that policymakers will face [2]. And to address these problems, utilizing policies based on novel knowledge and technologies along with required financing and ways to provide should put in the focus of attention.

Different points of view for the smart city have led to numerous definitions for it. Most of the definitions have been derived from two main perspectives: Smart sustainable development and ICT-based infrastructure, while a few of the investment, financing, and policy have been put in focus. One of these definitions states that a smart city is a city in which the interaction of people and machines is considered to meet the economic and social needs of the people by focusing on smart policies of various sectors such as energy, raw materials, services, and financing [3]. The smart city can be considered as a bidirectional synergism of citizens and the city in which smart means the use of smart-based technologies for the smart growth of people and governments to achieve the main goals of the smart city through specified policies. Novel technologies and the policy of using them in the smart city over time have introduced different generations of the smart city. In other words, the level of existing smart technology and the policy of utilizing it has played an essential role in the formation of different generations of smart cities. The emergence of different generations has been resulted from technological improvements, attempts to overcome the important challenges facing urbanization to achieve a higher quality of life, and the investment of governments and the private sectors.

Smart city policies, generations, and levels of used technologies, also financing of utilizing the novel technologies in the fourth smart city generation, named city 4.0, are discussed in this chapter. The main aim of this chapter is to discover industry 4.0 usage, benefits, and challenges in a city 4.0 with a focus on financing aspects.

1.2 Smart city policy

Policy and strategy are two words that are used separately and together. These words are sometimes not used in their proper place. To get a clear understanding of them, they are explained first. Most of the reliable dictionaries refer to the policy as a set of rules and regulations made by the organization for rational decision-making. Along with the policy, there must be an appropriate strategy to achieve the goals. The strategy could be considered as an opted plan for achieving the specified goals and objectives through predefined policies. Strategy characterizes an appropriate plan of action through specified policy in which principle of actions is characterized. The policy applies within a defined environment, while strategies also take into account external environmental factors. So, it could be said that policy clarifies the decision orientation and allows strategies to move toward organization goals through it. The policy is a set of principles and rules that the top

Smart Cities Policies and Financing. https://doi.org/10.1016/B978-0-12-819130-9.00013-9

management of the organization provides as a guide to direct the decisions and actions of the organization. An organization's policy is written for the long term and specifies what the organization is aiming for and how the organization should respond appropriately in different situations. Therefore, it is clear that the policy of organizations depends on different factors and is not the same. In other words, an organizational policy is not a common guideline, and it is a unique characteristic of the organization and is determined by the needs and goals of the organization.

While there is a strong focus on defining the concept of smart city and numerous definitions that have been provided for it, most of these definitions have lacked paying enough attention to smart city policy. Consequently, less attention has been paid to the economic effects of smart city policy and its potential for smart city growth and prosperity. Surprisingly, all this is happening at a time when an unbelievable budget is being spent on smart city researches. Smart city policy should specify what the smart city is established for and how the smart city should respond to its citizens in different situations. A smart city stands on three main platforms: physical infrastructure, society infrastructure, and digital structure [4]. Each platform consists of four dedicated domains and several components (Fig. 1.1). To achieve all the goals set for the smart city, smart city policy should cover them all. In other words, the smart city must be built on specific policies so that the selected strategies can lead the smart city toward the goals set for it. Smart city policies should include and address these features specifically. This policy should define the economic aspects of smart city policies and articulate their expected impact on the financial growth of the urban area. Despite the importance of smart city policy in achieving its goals, few cities have a clear policy in this regard [5]. Recent evidence suggests that smart city policies not only increase urban smartness but also increase the likelihood of success in smart city policies by empowering citizens. The general policy of smart cities is not the same in different countries, and they choose policies depending on their assets, the goal they pursue to create smart cities, and their readiness for it. In this regard, the United States has formed smart cities ecosystem led by private companies. Europe focuses on sustainability issues, mobility, energy, and high value-added industry through public–private partnerships. In Asia, governments have led smart cities to improve the competitiveness of cities and revitalize the economy [6].

Studies show that smart city policies do not cover not only all platforms and domains of smart cities but are not implemented to generate a smart city in principle. Table 1.1 presents four different models for smart city policy. The first model emphasizes an intelligent citizen-based approach, which maximizes the participation of the main stockholders of the smart city, the citizens, but minimizes the power and influence of governments to distribute services fairly. In this context, a dynamic economy is formed, and business owners play a more prominent role in economic and business innovation and therefore enjoy the benefits of building smart cities more widely. The second model focuses on assigning different roles to different

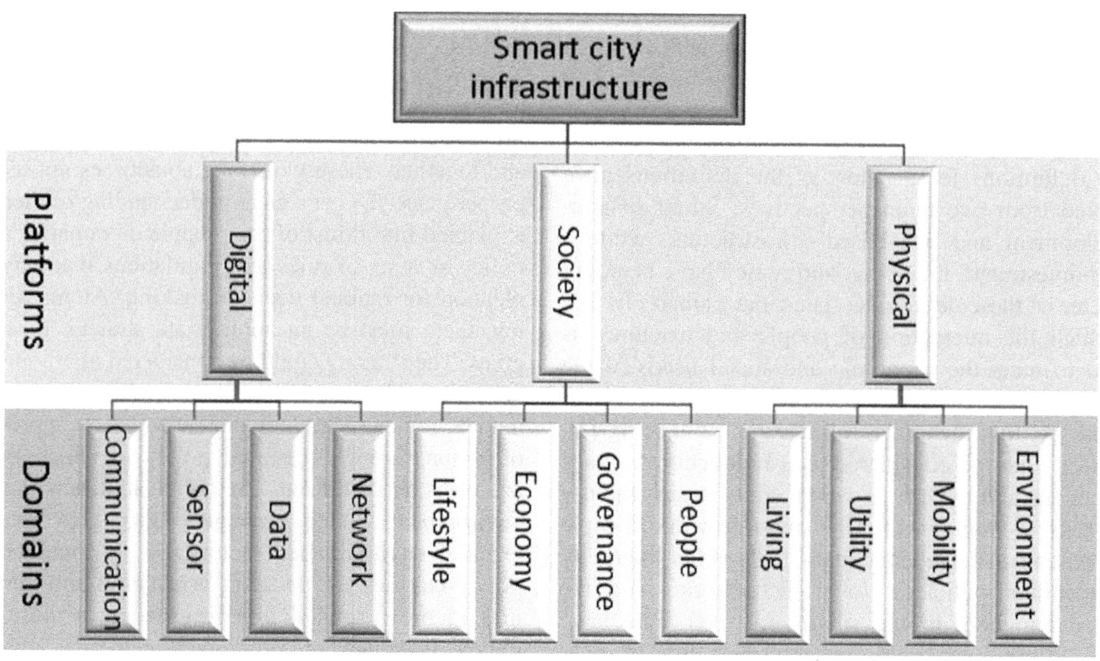

FIGURE 1.1 Structure of smart city infrastructures.

TABLE 1.1 Types of policy models in creating, developing, and managing the smart cities.

No.	Model description	Research fellow
1	Establishing a smart citizen-based approach instead of smart governance and focus on build and managing an intelligent community.	Castelnovo, W. et al. [10]
2	Focus on the role of all actors in a smart city such as political and governmental institutions, economic, civic institutions, also knowledge and technology production centers.	Fernandez Anez, V. et al. [11]
3	Control and domination of intelligent society by intelligent governance to redistribute resources and create social and economic justice.	Lara, A. P. et al. [12]
4	Policy and management of smart community ideas and innovations by smart governance to integrate different scopes of smart city as much as possible.	Komninos, N. [13]

actors in the structure of smart cities. Although this approach makes society more dynamic, it can be challenging to manage and establish communication between all these sectors. In this model, all elements of society, including public and private institutions and organizations and also knowledge production centers such as universities and science and technology parks, will be involved in the development of smart city infrastructure. In the third proposed model, smart governance is considered responsible for formulating and explaining smart city policies, also the task of constructing and developing infrastructure and monitoring the achievement of set goals will be done by government institutions. Although the concentration of power in the government and its affiliated institutions can be useful in the equitable redistribution of resources and interests, it increases the likelihood of discrimination and inequality due to political inclinations or profiteering of those in power. Finally, the fourth model emphasizes the role of governments as policymakers rather than executors of smart cities. In this model, governments are usually responsible for presenting strategic plans and play the role of coordinators of different sectors.

It should be noted that policymaking is a set of rules created by an institution, organization, or government in developed countries to achieve reasonable goals, and economy facilitates the achievement of the goals set in the adopted policy. More than 50% of smart city policy organizations are based in the United States (37%) and the United Kingdom (14%) [7]. The lack of indigenous smart city policy means that other countries tend to adopt policies that have been proposed and tested by developed countries. Obviously, these policies are not optimal for other countries and will not serve their national interests. These policies serve the interests of the countries that made and developed them rather than the national interests of the host countries [7]. This can be due to its infancy of the concept of smart city policy and most importantly the evolution of smart cities over time based on some specific needs and level of technology in each industrial revolution era. This means that smart city policy should be tailored to the desired level for the smart city and its intended goals. The positive impact of smart city policy can be examined from an economic perspective in various fields. These policies can increase the productivity of cities through widespread and online access to the services needed by the city stakeholders, increase the participation of citizens, which is the most important factor in creating a smart city, and increase the local and national economic growth by making the economy smarter. However, the financial impact of adopting smart city policies on urban growth and innovation is vague and should be put under more concentration by smart city researchers and policymakers [8]. Attention should be paid to discover the financial impact of smart city policy implementation on all platforms, domains, and components of smart cities.

As has been repeatedly stated, all policies and strategies in the field of smart urban planning must be in line with the goals of the smart city. Because the digital platform supports the physical and society platforms technologically, so before formulating any policy, must first specify the policies, dimensions, facilities, and needs in the digital field. For example, the issue of economic prosperity, which is one of the most important goals of the formation of smart cities, in combination with the issue of artificial intelligence and digital technologies, creates the term smart economy or digital economy, and this indicative of transregional economics and global. In this regard, the construction phase (smart city policy formulation); the connecting phase (connecting smart city platforms); and the enhancement phase (emphasis on creating a smart city ecology to gain expected financial benefits based on selected industry level) can be considered as three general steps to implement smart city policies [9]. Smart city policy should cover physical policy, society policy, and digital policy of the smart city.

1.3 Industry revolutions

The Industrial Revolution (IR) has had a great impact on the lifestyle of the people, and therefore it is called the most effective revolution in human history. The first IR began in the 18th century in England with the transition from agricultural and handicraft economy toward the industry-oriented economy. With the advent of the IR, mass production was introduced and the standard of living of the people changed in such a way that the comparison before and after is not imaginable. Technology, socioeconomic and culture are the main features of the IR. These characteristics are varied in different IR generations so that the first industrial revolution should not be confused with its subsequent ones. These characteristics are the driving force to go through different generations of the IR and have caused the IR to always continue the path of progress to meet the growing needs of human life. The IR focused on the cost advantage via increasing the level of output (scale of the economy). IR grew gross domestic product (GDP) per capita and living standards in parallel. The IR has started its movement from mechanization in 1765 (IR 1.0) and has reached industry 4.0 (IR 4.0) in 2011 by passing from mass production, assembly line and electricity (IR 2.0), and computer, automation, and automated production (IR 3.0). IR 4.0 can be defined as "the integration of complex physical machinery and devices with networked sensors and software, used to predict, control and plan for better business and societal outcomes" [14].

By autonomous decision-making, collaborative robots, on-demand manufacturing, individually unique products, flexibility and servitization, industry 4.0 overridden automation, industrial robots, demand forecasting, delimited variation, efficiency, and selling products of Industry 3.0 respectively [15]. IR 4.0 emerges to establish machine-to-machine (M2M) and machine-to-human (M2H) connections deeply through the Internet of things (IoT), cyber-physical systems (CPS), cognitive (cloud/fog/edge) computing, artificial intelligence (AI), machine learning (ML), digital twins (DT), and so on to create more value. Industry 4.0 has formed with the advancement of computers, the expansion of Internet communications, the development of the mobile communications industry and smartphone technology, the development of applications and communication platforms, and the creation of horizontal/vertical integration techniques. Industry 4.0 seeks to optimize all company parameters by making smart connections between all stakeholders while controlling them, using powerful processing tools. In general, the main idea of industry 4.0 is to create a comprehensive ecosystem for production processes, technologies, and applications through further automation and machine communications. By utilizing its technological tools (Fig. 1.2), industry 4.0 has been able to implement its policy (greater flexibility,

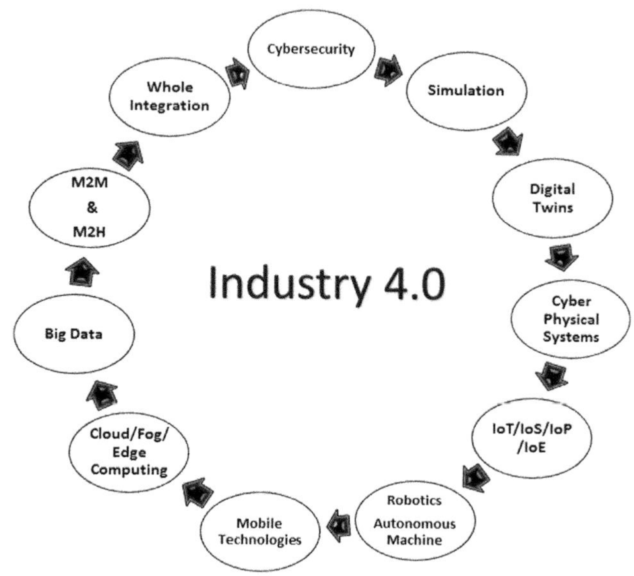

FIGURE 1.2 Industry 4.0 technologic tools.

higher quality standards, and greater efficiency) to create higher value. Despite these, many companies still refuse to move toward IR 4.0 due to a lack of specialized knowledge and especially uncertainty in its financial aspects [16]. Therefore, when considering the use of industry 4.0 in company policy, its financial aspects and its expected economic effects must be clarified.

1.4 Smart city generations development

As the urban population grows, cities become bigger and numerous. The increase in the size and number of urbanites has created new issues that cities have not faced before. These increase need for energy, roads, housing, infrastructure, etc. The increase of traffic, factories, and pollutants subsequently, on the one hand, and the lack of enough resources to meet the growing needs of urbanization, on the other hand, have caused traditional methods to lose their effectiveness for urban management. Urbanization had intertwined goals including quality of life, equity, resource efficiency, productivity, sustainability, economic growth, infrastructure resilience, and a few others. The intertwinement of these goals is due to their interaction with each other, which are in some cases synergistic and cooperative and in some other cases antagonistic and uncooperative. These goals, along with the amazing progress of the ICT industry and its applications, led city managers to put the policy of using ICT technology to manage urban areas on their agenda. In this regard, the era of the smart city began in 1994 in Amsterdam, the Netherlands, as the digital city. Digital city referred to the use of ICT to solve local problems within the urban space via ICT-based assets such as healthcare, public transportation, and administration services. Innovations in the ICT industry caused more penetration of this industry in all fields, and urban management was pushed to use it more in solving urban problems in the form of smart cities. Different generations of smart cities have been formed through more penetration of ICT physical infrastructure and social relations.

The first generation of smart cities (City 1.0) focused on the use of computer applications to facilitate city management and respond more quickly to clients. Since the formation of smart cities occurred in the third generation of the industrial revolution, the first and second generations of smart cities are not cofounded with industrial generations. City 1.0 uses fragmented local sensors to gather required data for special purposes such as climate change and parking occupancy. Diffusion of ICT-based devices is the main policy of city 1.0. In the next generation, city 2.0, city-led utilization of ICT policy emerged. In the context of city 2.0, city managers aimed to provide better services for the city's stakeholders through deeper and wider use of ICT. In this way sharing data and reply to clients within the specific organization via wired Internet (vertical integration) started. Simultaneously with the development of industry 3.0, city managers put the use of the related technologies on the agenda. Automation, industrial robots, interconnection of production processes, delimited variation, and computers are considered as main characteristics of city 3.0, and many problems are solved by utilizing them via the participation of citizens [17]. Wireless Internet and IoT are employed to establish a connection between fragmented platforms (horizontal integration) of city 3.0. There is a fundamental difference between horizontally integrated and interfaced platforms. Interfaced platforms are standalone platforms that are integrated vertically and at the same time share their data between themselves, while horizontally integrated platforms use all standalone created databases collectively. Sustainability, the collaboration of stakeholders, lack of integration across government systems, interoperability, availability, and compatibility of software and cybersecurity are the most important issues that City 3.0 failed to address [18]. This led to the use of industry 4.0 to overcome the mentioned challenges and to create the fourth generation of smart cities (city 4.0). City 4.0 is a smart city in which widespread use of industry 4.0 technologies including autonomous decision-making, collaborative robots, total interconnection, flexibility, servitization, CPS, IoT, Internet of People (IoP), Internet of Services (IoS), Big Data and Analytics, Augmented Reality (AR), autonomous robots, additive manufacturing (3D Printing), cloud/fog/edge computing, simulation, blockchain, digital twin are implemented. City 4.0 will be able to adopt the capabilities of industry 4.0 through smart city physical policy, smart city society policy, and smart city digital policy (Fig. 1.3).

In fact, transitioning from smart city 1.0 to smart city 4.0 is a journey from fragmentation to holistic integration of smart city platforms. Main implemented policies, characteristics, and focused areas of different generations of smart cities are presented in Table 1.2.

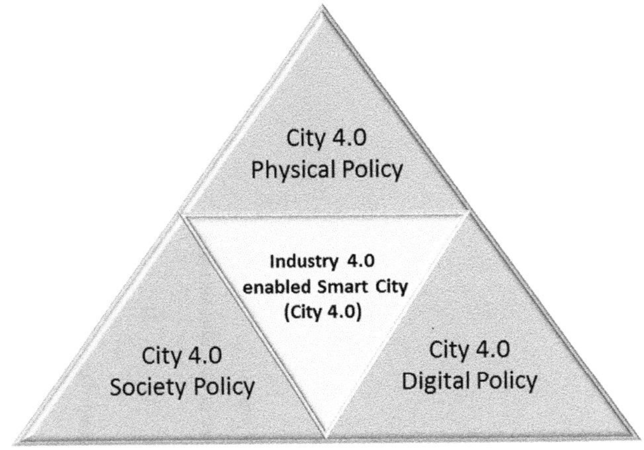

FIGURE 1.3 City 4.0 policies.

From an external perspective, it is understandable that the smart city and the economy have a two-way impact. The smart city supports the economic growth of the community at the local, regional, and national levels and interacts with the global economy simultaneously. Making cities smarter requires huge investment in infrastructure replacement with a newer one that should be provided by the government and private sectors separately or together. On the other hand, from an internal perspective, scoring e-economics include e-business and finance, smart tourism, smart culture, entrepreneurship, innovative economy, and smart marketing are crucial for smart city managers. In both perspectives, significant challenges arise that needed to consider by smart city managers. In this regard, city 4.0 policies can provide an appropriate infrastructure to pay attention to all raised issues. Interoperability, virtualization, decentralization, real-time capability, service orientation, modularity are six wonderful principles of industry 4.0 that support city 4.0 policies. Using these key principles, industry 4.0 has been able to implement its unlimited applications in most of the city 4.0 domains.

Tables 1.3–1.5 present most of industry 4.0-enabled technologies for society, physical, and digital platforms, respectively. Table 1.3 focused on society platform that is very important platform for the growth of citizens who are the main beneficiaries of smart cities. In this platform, the role of education and citizen participation as the human capital of the society as well as their economic and welfare issues such as entrepreneurship through the development of local and regional financial markets and even the development of tourism industry, to the desired lifestyle of society in terms of receiving fair health and social services, access to adequate housing as well as a safe work environment, with the help of tools available to smart governance in context of city 4.0 society policy are considered.

The physical platform of the smart city by focusing on sustainability is considered the objective manifestation of the growth and development of a dynamic system and includes environmental factors including energy, waste

TABLE 1.2 Smart city generations characteristics.

Smart city generation	Main policies	Main characteristics	Focused area	Ref.
City 1.0	Diffusion of ICT to: • optimization of infrastructures and services • preparation the interests of the demand side • new business opportunities incitement	• centralized and fragmented approach • passive citizens • ICT companies as a leader	City infrastructure and services	[6,19–21]
City 2.0	Greater utilization of ICT to improve: • social problems • citizens' wellbeing • public services, • quality of life	• decentralized approach • active citizen • more involvement of urban managers • ICT companies as a provider • establish vertical integration • city as a product approach	People, governance and policy	[6,19–21]
City 3.0	• improve quality of life • A city that thinks and acts • participatory urbanism	• improve vertical integration • holistic approach • establish horizontal integration	Physical, society and digital platforms	[20,21] [4] [6]
City 4.0	• Cocreator citizens (society policy) • sustainability—driven (physical policy) • industry 4.0 enabled (digital policy)	• end-to-end integration • stakeholders satisfaction • platform approach • self-organizing	Physical, society and digital platforms	[4] [6]

TABLE 1.3 Industry 4.0-enabled technologies in smart city society platform.

Platform	Domain	Main component	Industry 4.0-enabled technology
Smart society	**Smart people**	s-Education/digital education	e-Learning
			Mobile learning
		Human capital	E-skill/skilled workforce
			Cosmopolitanism/open-mindedness
			Creativity
			Social and ethnic plurality
			ICT-enabled
		Community engagement	Cocreate open platform
	Smart governance	E-government	Good administration
			Good policy
			Innovative organization and administration
		E-governance	Innovate governance network
			Transparent governance
			M-governance
		Citizenry engagement	Participatory decision-making/democracy
			Administration services to citizens/
			ICT based transparency and open data
		Distance services	Cocreated e-services
			E-services
	Smart economy	E-business and finance	Local and global interconnectedness
		s-Tourism	Touristic attractiveness
			Six A's (attractions, accessibility, amenities, available packages, activities, ancillary services)
		Entrepreneurship	Creative industry
			High-tech industry
		Innovative economy	ICT-enabled innovation
			Innovative spirit
			Smart banking/commerce
		Smart marketing	High productivity
			Flexibility of labor market
			International market connectivity
			Production in demand
	Smart lifestyle	Healthcare	E-health, M-health, home health
		Social services	Social care
			Public services/activities
			Recreation
		Smart surveillance	Environmental monitoring/alert
			Noise detection
		Social inclusion	Social cohesion
			Social capital
			Social media
			Social innovation
		Flexible work places	Distance working
		Smart home	Home automation

TABLE 1.4 Industry 4.0-enabled technologies in smart city physical platform.

Platform	Domain	Main component	Industry 4.0-enabled technology
Smart physical	**Smart environment**	s-Energy	Sustainable energy solutions
			Smart grid
			Green/Renewable energy sources
			Smart metering
			Resource use efficiency, reuse, and resource substitution
		s-Water	Water resource system
		s-Waste and sanitation	Waste management
			Drainage systems
		s-Pollution	Control and monitoring
		s-Building	Sustainable land use
		s-Relocation	
	Smart mobility	Intelligent public transport/e-mobility	Integrated mobility management (bus, train, metro, car sharing, bike sharing, pedestrian)
			Parking management
			Tolling and congestion charging
			Real time ticketing/e-ticketing
		s-Traffic	Dynamic traffic management
			Vehicle information services/real-time road navigation
			Travel guiding/real-time public transit info
		s-Departure (locomotion)	Smart airport
			Roads and highways
			Road bike
			Walkway
		s-Logistic	Demand-based logistics
	Smart utility	Urban facilities	Smart irrigation
			Street lighting
			Eco heating
	Smart living	E-connected	Digital payments
			Wi-Fi-connected city
			e-Identity
		Housing/shelter	Good-quality housing and accommodation
		Cultural facilities	Community biking
		Safety, security, and emergency	Crime or disaster prevention/management

recycling, pollution control, water resources and smart buildings, smart transportation, creating high-quality living conditions, welfare, safety, security, and emergency facilities/services. Table 1.4 presents some of industry 4.0 applications that support city 4.0 to accumulate its physical policy goals.

TABLE 1.5 Industry 4.0-enabled technologies in smart city digital platform.

Platform	Domain	Main component	Industry 4.0-enabled technology
Smart digital	Smart networks	Socially reliable networks	Broadband internet—wi-Fi
			International connectivity
		Smart city dashboard	Common operational framework via city simulation platforms
		Surveillance network	Wireless sensor networks
			Interoperable sensor networks
	Smart data	Data resource	Laser-scanned data (LiDAR)
			Real-time data
			Open data portal
			Data monitoring
			Interoperable data
			Big data
			Spatial data
		Data analysis	Cloud/fog/edge computing
			Artificial intelligence
	Smart sensors	Citizen sensors	Citizen as a sensor/on-body sensors
			Safety and security
			Clinical sensors
		Urban sensors	Public transportation sensors
			Sensors for parking
			Road/traffic sensors
			Building sensors
			Environmental sensors
			Digital signage (RFID)
	Smart communications	E-communication	High-speed communication
			Virtual reality/virtualization/digital twin
			ICTs/telecommunication
			Internet of things (IoT)
			Internet of services (IoS)
			Internet of people (IoP)
			Internet of energy (IoE)

Industry 4.0-enabled novel technologies for supporting digital platform of city 4.0 are presented in Table 1.5. Digital platform acts as a bridge between all city 4.0 platforms. Based on city 4.0 digital policy, it contains a large number of industry 4.0-enabled technologies for data collection, processing, and analyzing along with needed tools to widespread use of smart networks, smart sensors, and smart communication.

The experience of managing metropolitan and emerging big cities in the past decades has shown that if the issue of urban management is not given enough attention, it is not possible to overcome the problems ahead and future generations will face far greater problems. Sustainable development, environmental impacts, attention to renewable energy, improving the quality of life, attention to health, etc., are things that the expansion of cities has made it extremely necessary to address them. Implementing mentioned applications of the industry 4.0 and other applications that are not mentioned in Tables 1.3–1.5 or will be invented in the future requires a lot of investment, which will return with the economic prosperity created in the growth of income levels and state value-added GDP. A high rate for return on investment is estimated also in [22]. In other words, the implementation of industry 4.0 in smart cities indeed requires a huge amount of investment, but its positive economic effects will be such that it will pay back any investment at a favorable rate.

1.5 Summary

Economic instability, education, health, water and energy management, waste management, and urban area management are several of the many problems that have arisen due to the increase in urban population in the last two decades and decades ahead. A powerful tool to address these problems is the development of smart cities. In general, the smart city concept refers to a city that uses ICT, along with its effects on social and physical platforms and their related domains and components. The level and quality of this platforms, especially digital infrastructure, have undergone extensive changes in the last two decades. Nowadays smart cities have overridden industrial revolution 1.0 to 3.0 by industry 4.0-based smart cities (city 4.0), using technologies such as CPS, IoT, IoP, IoS, Big Data and Analytics, cloud/fog/edge computing, simulation, blockchain, and digital twin. To create or develop smart cities according to the requirements of city 4.0, four policy models have been identified, each of which has its strengths and weaknesses. Today, smart cities, given their wide financial dimensions, must have specific policies to be used in different dimensions of the industry, and these policies are adopted by governments for long periods, and with the strategies used, the smart city is moved toward achieving the predefined goals. Appropriate policymaking and strategy require the allocation of large budgets in this area, so many governments tend to benefit from the successful policies of developed countries. It should be emphasized that introduced policies by developed countries are not optimal for other countries, because the level of success and effectiveness of these policies following the infrastructures of each geographical area are variable and different. In addition, these policies will serve the interests of countries who have compiled them. It should be noted that the adoption of efficient policies (moreover the creation of appropriate and sufficient infrastructure) requires large investments; but, it should not be overlooked that with the implementation of the smart city platforms, investments can be made quickly and with high profitability rates of return. However, pushing this industry into an attractive and profitable scope for huge and long-term investments needs more research on financial aspects of smart city policy in industry 4.0 era.

Finally, let us move on to the real interactive part of this chapter: review questions/exercises, hands-on projects, case projects, and optional team case project. The answers and/or solutions by chapter can be found in Appendix G.

1.6 Chapter review questions/exercises

True/false

1. True or False? According to available statistics and forecasts, in the next four decades, we will see a significant decrease in urban population worldwide; this increase varies depending on the geographical location of the region.
2. True or False? With the steady decrease in growth of the urban population in recent decades, it is estimated that by 2060, about 79% of the world's population, 96% of the European population, and 91% of the North American population will live in cities.
3. True or False? Policy and strategy are two words that are used separately.
4. True or False? The Industrial Revolution (IR) has had a great impact on the lifestyle of the people, and therefore it is called the most effective revolution in human history.
5. True or False? As the urban population grows, cities become bigger and numerous.

Multiple choice

1. Different points of view for the smart city have led to:
   - **a.** Smart sustainable development
   - **b.** ICT-based infrastructure
   - **c.** Numerous definitions for it
   - **d.** Investment
   - **e.** Policy
2. While there is a strong focus on defining the concept of smart city and numerous definitions that have been provided for it, most of these definitions have lacked paying enough attention to:
   - **a.** Society infrastructure
   - **b.** Physical infrastructure
   - **c.** Smart city growth and prosperity
   - **d.** Economic effects
   - **e.** Smart city policy
3. Industrial Revolution 4.0 emerges to establish two connections through the Internet of Things (IoT):
   - **a.** Machine-to-machine (M2M)
   - **b.** Machine-to-human (M2H)
   - **c.** Cyber-physical systems (CPS)
   - **d.** Artificial intelligence (AI)
   - **e.** Digital twins (DT)

4. The first generation of smart cities (City 1.0) focused on the use of __________________ to facilitate city management and respond more quickly to clients.
   a. Holistic integration of smart city platforms
   b. Economic growth
   c. Making cities smarter
   d. Computer applications
   e. E-economics
5. In general, the smart city concept refers to a city that uses _____, along with its effects on social and physical platforms and their related domains and components.
   a. ICT
   b. CPS
   c. IoT
   d. IoP
   e. IoS

Exercise

Problem

Through industry 4.0 technology, why are the advantages of a smart city estimated to overcome the city's expenses with city platformization?

Hands-on projects

Project

Do research: Analyze the smart city concept. Identify key features of the smart city such as the smart economy (industry 4.0); smart population; smart government; smart mobility; smart environment; smart living; and intelligent infrastructure.

Case projects

Problem

What is the conjunction of the Smart City Initiative (SCI) and the concept of Industry 4.0?

Optional team case project

Problem

Describe the concepts of Industry 4.0; and, explain for clarity and completeness, some new technologies and challenges.

References

[1] United Nations. United Nations division for sustainable development goals. 2020 [Online]. Available: https://sdgs.un.org/goals.

[2] Angelidou M, Psaltoglou A, Komninos N, Kakderi C, Tsarchopoulos P, Panori A. Enhancing sustainable urban development through smart city applications. J Sci Technol Policy Manag 2018;9(2):146–69.

[3] Yatin Jog MC, Singhal TK, Barot F. Need & Gap analysis of converting a city into smart city. Int J Smart Home 2017;11(3):9–26.

[4] Fakhimi AH, Hossein Khani A, Sardroud JM. Smart-city infrastructure components. In: Vacca JR, editor. Solving urban infrastructure problems using smart city technologies. 1st ed. Elsevier; 2020. p. 16–54.

[5] Yigitcanlar T, Buys L, Ioppolo G, Sabatini-marques J, Moreira E, Joseph J. Understanding "smart cities": intertwining development drivers with desired outcomes in a multidimensional framework. Cities 2018;81:145–60.

[6] Yun Y, Lee M. Smart City 4.0 from the perspective of open innovation. J Open Innov 2019;5(4):92.

[7] Estevez E, Lopes NV, Janowski T. Smart sustainable cities. Reconnaissance study. United Nations University; 2016.

[8] B. Daniotti, M. Gianinetto, and S. Della Torre, Eds., Digital transformation of the design , construction and management processes of the built environment. Springer, Cham.

[9] Popkova EG, Sergi BS, editors. Artificial intelligence: anthropogenic nature vs. Social origin. Cham: Springer; 2020.

[10] Castelnovo W, Misuraca G, Savoldelli A. Smart cities Governance : the need for a holistic approach to assessing urban participatory policy making. Soc Sci Comput Rev 2016;34(6):724–39.

[11] Fernandez-Anez V, Fernández-Güell JM, Giffinger R. Smart City implementation and discourses: an integrated conceptual model. The case of Vienna. Cities 2018;78:4–16.

[12] Lara AP, Moreira E, Costa D, Furlani TZ, Yigitcanlar T. Smartness that matters : towards a comprehensive and human-centred characterisation of smart cities. J Open Innov 2016;2(2):8.

[13] Komninos N. Smart environments and smart growth: connecting innovation strategies and digital growth strategies. Int J Knowl Base Dev 2016;7(3):240–63.

[14] Alaloul WS, Liew MS, Amila N, Abdullah W, Mohammed BS. Industry revolution IR 4.0: future opportunities and challenges in construction industry. In: MATEC Web of Conferences. 203; 2018. p. 02010.

[15] Torn IAR, Vaneker THJ. Mass personalization with industry 4.0 by SMEs: a concept for collaborative networks. Procedia Manuf 2019;28:135–41.

[16] Mohamed M. Challenges and benefits of industry 4.0: an overview. Int J Supply Oper Manag 2018;5(3):256–65.

[17] Kumar A, Prakash SD, Shekhar S, Khant A. 'Smart sustainable Cities : a need for 21 century'. Int J Res Appl Sci Eng Technol 2016;4(Iv):464–74.

[18] Lim Y, Edelenbos J, Gianoli A. 'Identifying the results of smart city development : Findings from systematic literature review'. Cities 2019;95:102397. June.

[19] Appio FP, Lima M, Paroutis S. Understanding smart cities: innovation ecosystems, technological advancements, and societal challenges. Technol Forecast Soc Change 2019;142:1–14.

[20] Zawieska J, Pieriegud J. Smart city as a tool for sustainable mobility and transport decarbonisation. Transport Pol 2018;63:39–50. June 2017.

[21] C. Alexopoulos, G. V. Pereira, Y. Charalabidis, and L. Madrid, 'A taxonomy of smart cities initiatives', in Proceedings of the 12th international conference on theory and practice of electronic governance, pp. 281–290.

[22] Özüdoğru AG, Ergün E, Ammari D, Görener A. How industry 4. 0 changes Business : a commercial perspective. Int J Commer Financ 2018;4(1):84–95.

Part II

Implementing smart infrastructure: some key challenges and science, technology and innovation-driven policy and financial instruments

Chapter 2

The need to localize smart infrastructure policies and financing

Louis Burns
Urban Prosperity Network, Richardson, TX, United States

2.1 Introduction

Human history is full of stories featuring the struggle between individuals and families; between families and communities; between communities and states; and between local autonomy, regional economies of scale, and social cohesion. Generally speaking, the further removed a solution is from an individual, the less fit for the specific circumstances. Smart cities, infrastructure, policy, and financing are the same. They necessarily build on existing systems of political economy. Smart cities overlap the field of city planning, which was the original context for wicked problems. Finding solutions for these problems can draw from existing experience with design, public participation, and systems-based social change. The rate of adoption of smart city technology has likely been slower than our hopes and expectations. This could be partly due to the failure to recognize that solutions must be localized in their development and implementation. Like every other social and technological innovation, smart cities are not panaceas outside a municipal budgeting process.

This chapter will develop an essential framework for localizing smart city infrastructure policy and financing from first principles. We will review multiple approaches at each stage of conception and application. Rather than having one broad policy or financing recommendation, the reader will be equipped with the understanding to adapt smart city solutions for a local civic population and process. We will begin at the big-picture scale—context, culture, and economy. We then move through holistic problem-solving, local policy, and financing concepts and discuss specific policy and financing instruments.

Readers might be eager to skip ahead to the key civic policy and financing instruments section. The conceptual material in the first three sections is essential background context for understanding the key instruments. They are the why and how to the instruments' what. A skillful application requires the first principles' understanding found in the first three sections. Finally, being educated in, situated in, and therefore most familiar with the big cities of Texas, USA, many of the examples are drawn from those contexts.

2.2 Context, culture, and economics for local smart cities

Solutions that do not appreciate the local context, culture, and economy are no solutions at all. Picking technologies from a list of options in isolation has not yielded much progress in smart city implementations. Does every city need smart lighting or air quality sensors today? As with most important, complex problems, it depends. It depends on the local factors we will examine here. We cannot exhaustively define context, culture, economics, or even smart cities in a single chapter or text, but we can cover enough to substantially inform our decision-making process.

2.2.1 Context and smart cities

Smart cities have various definitions and scope far beyond the limited sense in which we are discussing infrastructure, policy, and finance here. For example, on the Wikipedia page [1], you will find frameworks related to technology, people, institutions, energy, and data management at the higher conceptual level. A comprehensive plan for a smart city would cover all those areas and more.

City planning was the original wicked problem [2]. The formulation of the wicked problem nearly equally applies to the development of smart city solutions [2]:

- There is no definitive formulation of a wicked problem.
- Wicked problems have no stopping rule.

Smart Cities Policies and Financing. https://doi.org/10.1016/B978-0-12-819130-9.00028-0

- Solutions to wicked problems are not true or false, but good or bad.
- There is no immediate and no ultimate test of a solution to a wicked problem.
- Every solution to a wicked problem is a "one-shot operation"; because there is no opportunity to learn by trial and error, every attempt counts significantly.
- Wicked problems do not have an enumerable (or an exhaustively describable) set of potential solutions, nor is there a well-described set of permissible operations that may be incorporated into the plan.
- Every wicked problem is essentially unique.
- Every wicked problem can be considered to be a symptom of another problem.
- The existence of a discrepancy representing a wicked problem can be explained in numerous ways. The choice of explanation determines the nature of the problem's resolution.
- The planner has no right to be wrong.

We understand that some of the preceding statements are based on the observation from regional science that cities are systems of systems [3] (see checklist: "An Agenda for Action for Taking The Appropriate Steps to Understand Systems of Systems").

An agenda for action for checking the appropriate steps to understand smart cities

A number of steps are required to understand a smart city system, which include the following key activities (check all tasks completed):

_____1. Model the stocks and flows (i.e., resource allocation) [4].

_____2. Balance exploration (solution seeking) and exploitation (solution implementation).

_____3. Balance value capture and value transfer through taxes and spending [5].

_____4. Balance *growth* that is measured through traditional economic indicators (GDP), with *prosperity* measured through various health, social, or community well-being factors.

_____5. Balance risk, reward, the scale of investment, and the political capital required for change and for maintaining the status quo.

_____6. Develop metrics for the progress or success of any solution within the system based on the local population's values. Disambiguate effects on different socioeconomic groups wherever possible.

_____7. Track the metrics at a spatial and temporal resolution that can yield meaningful insights while adapting to the local tolerance for privacy and surveillance [6].

_____8. Do all of the above in a publicly transparent and inclusive way.

Furthermore, smart city initiatives must compete with other budget requirements, including some already required by law. How important is a smart city initiative when compared to raising teacher salaries or improving police training?

In addition to city planning and regional science, the fields of urbanism, sociology, behavioral economics, user experience design, and architecture and engineering, among others, all have unique contributions to make to understanding cities in general and smart cities in particular. While it is not possible to consider all possible angles at once, we should be mindful of the selected areas of emphasis. Smart city solutions depend on context.

2.2.2 Culture and smart cities

While the local art scene or fine arts offerings are an important part of civic life, here we discuss city culture in the context of sociology, social psychology, and politics. The concept of tight and loose cultures is relevant here [7]. It gives us a spectrum of how local populations adhere to rules and social norms as part of their culture. It informs the possibility and likelihood of an eventual smart city initiative adoption. A tighter culture might be more amenable to more invasive surveillance if they feel it is for the public good. Other looser cultures might revolt against even noncontroversial, privately funded initiatives, such as the individuals that were burning down 5G towers in the United Kingdom based on misinformation [8]. Indeed, whether an initiative is publicly or privately funded will impact tight or loose cultures differently. Incorporating more transparent public discourse and human factors in the decision-making process can help [9].

How well the local population is informed regarding civic issues is important. During the 2021 Texas legislative session, a university and nonprofit conducted a poll to ask registered voting Texans what they believed the state legislative priorities should be. The pollsters were disconcerted to find that on an open-response question, Texans said the number one issue they wanted the State of Texas to work on was immigration—outranking the next three issues combined [10]. The disconcertment was that so many Texans would be unaware of the assumed obvious fact that immigration is handled at the federal level, not by the state.

The local political spectrum affects the range of possible solutions as well. Any mention of socialism in the United States could spell the death of a project before it begins. Arguments for initiatives often center on saving or conserving civic funds. The argument is usually that investment will spur additional investments. By contrast, stories of successes might gain wider spread in the United Kingdom when labeled "sensible socialism" [11]. Arguments for the public good might hold more weight there. Moreover, the conversation is different where the resources exist to build entire smart cities de novo.

Within localized cultures, conflicting subcultures exist between professions and the public. When asked to rank the following values for street design, engineers choose speed, volume, safety, and cost. Residents ranked those values as safety, cost, volume, and speed [12]. There is a clear disconnect between the professionals that value speed most and residents that value speed least. Any initiative would do well to factor in the intended audience for communication and involvement.

At the risk of stating the obvious, cultures change over time at variable rates. Polling and public participation can help us understand how much change has occurred as we prepare initiatives. For example, the City of Austin finally passed an initiative for rail service after multiple attempts over 20 years [13,14]. It took time for the local political economy to change and embrace rail and more public transit, whether through changing attitudes, net migration, or changing demographics.

Where available, psychographic studies can aid in understanding the local population [15]. Psychographics augment traditional demographics by developing personas or user stories to paint a clearer picture of the underlying statistics. The GIS services company, Esri, has a web application that allows users to look up and sample their psychographic product called Tapestry. In this author's zip code, the largest group representing 15.7% of the population is called "In Style," described as: "We are professional couples or singles with no kids and strong work ethics. We support the arts, travel, and extensive reading. We focus on home maintenance and improvement. We use our phones to check for the best prices and redeem both print and mobile coupons."

2.2.3 Economics and smart cities

The field of political economy connects the dots between culture and economics. It only makes sense that what we advocate for, what we do, and how we pay for it would be intertwined. Most smart city initiatives will pass through a city budget at some part of their journey. The local economic circumstances extend beyond the question of how we pay for it. Gross domestic product and per capita income are the broadest indicators. They do not measure many of the most important things to us captured in health outcome or well-being metrics.

Newer perspectives have augmented the traditional economic approaches. Neoclassical Economics, focusing on measuring supply and demand, was the dominant school of the 20th century and still frames most economic discussions [16]. A close contender is Keynesian Economics, which prescribes government intervention to manage aggregate demand to prevent the recurrence of events such as The Great Depression [17]. More recently, Modern Monetary Theory has gained in popularity. It can be characterized as saying governments that control their currencies are not constrained by revenues and can pay for the programs they want through printing new money [18]. How that might affect inflation or the special case of being a world reserve currency is an open question.

Beyond these more dominant perspectives, some of the emerging fields within economics include (alphabetically) Austrian Economics, Behavioral Economics, Complexity Economics, Cooperative Economics, Ecological Economics, Feminist Economics, Institutional Economics, and Marxist Economics [19]. The differences are in perspective and measurement emphasis.

For example, an Austrian economist might argue that government cannot solve any problem because it lacks free-market pricing signals. By contrast, a Marxist economist might argue that only the government can solve most problems because they lack the exploitative profit motive inherent and sometimes required in the free market. Like most issues, the truth is likely somewhere between the opposing views and local context matters.

Behavioral Economics' contribution is to supplement the caricature of the traditional assumptions of human actors with what we know about people from psychology. We can improve our model of the world when we understand better what motivates people [20] and how to subtly influence them to act in their own or the community's best interest [21].

Complexity Economics tells us to see the economy as a complex system or network as much as a system [22]. The difference is between complicated and complex—the former having many pieces and the latter having interrelated pieces that can result in emergent phenomena (not to mention wicked problems).

Cooperative Economics tells us that we have commons-based organizations that behave and contribute differently in the economy beyond the dichotomy of public and private. For example, the open-sourced community including many programming languages and Wikipedia are typically not counted in the GDP even though their contribution is substantial.

Ecological Economics considers the environment. Ecuador was the first country to grant nature constitutional rights [23] the way the United States has granted personhood to corporations.

Feminist Economics considers gender relations, including the implications of parenting a small child, accommodating feeding and bathrooms in public, running errands, and navigating the urban fabric with a stroller. Vienna, Austria, is perennially in the news as becoming a gender-neutral city (formerly city for women) for addressing these questions [24].

Institutional Economics considers institutions intertwined with the economy (e.g., big tech, hospitals, banking, higher education, HQ relocations, etc.) [19]. The analysis of the effects of a Walmart or Amazon on local retail falls into this category. Alternately, case studies where a major hospital focuses their purchasing and hiring power locally, benefitting the local area as we have seen in Cleveland fits here too [25].

Finally, while still emerging and somewhat politically contentious, race—either itself or as a socioeconomic proxy has recently emerged as another candidate perspective driving economics [26].

2.3 Holistic problem-solving for smart cities

There is a vast literature on problem-solving and design approaches. Here we will review some of the major approaches and end goals starting at the universal, global level and working our way to the local. Even though the concept of wicked problems specifies there is no end state possible, we can certainly formulate updatable milestones and targets. Furthermore, we can certainly broaden and focus our approach to better adapt to local conditions.

In the spirit of understanding bias, I share this quote: "The risk of becoming too steeped in any one framework is you start to be 'subject' to that framework, you can only look through its lens, not at the lens. I recommend trying to hold a handful of frameworks in your mind simultaneously in order to maintain flexibility" [27]. None of the frameworks that follow are complete in themselves and need periodic examination and updating.

2.3.1 Universal frameworks

The anthropic principle states that after information, time and space are foundational concepts [28]. Information was our context section. We will proceed through time and space here. A useful framework for beginning to assess local need and opportunity is the timeframe. The reader is undoubtedly familiar with the perils of short-term thinking and the drawbacks of relying on politicians who must face reelection. One conception of time is Pace Layers [29]. While without specific hard timeframe boundaries, the temporal levels from fastest to slowest are fashion/art, commerce, infrastructure, governance, culture, and nature. It makes sense to evaluate how an initiative could affect us over each time horizon in a systemic approach. At the local level, perhaps some of the slower than anticipated adoption of smart city technology has been due to the mismatch between corporations' timeframes and local governments. Corporations are typically on quarterly earnings schedules while cities move no faster than their annual budgets. To get on an annual budget might require an investment of multiple years, depending on the complexity and degree of change. Even then, Moore's Law tells us the rate of change is increasing and so technology might fall somewhere between fashion and commerce complicating the decision-making timeframe for smart city initiatives.

After temporal considerations, it makes sense to think about human dynamics situated in a particular space (i.e., location). Spiral Dynamics is a social psychology framework modeling individual and group dynamics [30,31]. It specifies a spiral to indicate that all levels are present rather than a directional hierarchy of needs. The memes—in the Richard Dawkins sense of human meta-expressions [32], not entirely unlike our current Internet memes that compress information—range between local and individual to global and group. The itemized memes are instinctive survival, clannish tribes, egocentric power, purposeful law, strategic prosperity, relativistic community, systems, and holism. The authors use arbitrary colors for visual appeal and can be found with a quick image search. The framework may have had a challenge in its widespread adoption due to its subsequent rebranding and popularization as Integral Theory by a prominent New Age teacher [33]. The underlying Spiral Dynamics was developed in line with the contemporary psychological literature. Its first principle of considering different perspectives is what is important here. For example, some local populations may emphasize a law-and-order focus while their neighbors may be more community-minded. However, another group in the same community might be most concerned with their day-to-day survival. The holistic understanding is that those different emphases can either coexist or conflict with each other. It would not do well for an "enlightened" systems or holistic perspective to arrive from the outside to "fix" the local community without their involvement. That is certainly part of the pushback against elites or experts coming into communities even when they only want to help with the best intentions and evidence.

2.3.2 Universal design principles

Once the temporal and social factors are better understood, we can draw from the best universal design frameworks. We draw from user experience (UX), which tells us to think about the end user in a multidimensional way. The UX process will typically include the development of personas and user stories. These are fictional narratives about individuals drawn from the psychographic groups. We build these characters whose motivations, hopes, and fears we can understand and around which we can design. UX necessarily is an incremental, iterative process where the end service or product develops with user testing and feedback. While it is not possible to extensively user-test something like an entire electrical grid or LED light coordination, the application is that a better practice is likely incremental development. Most places will not build an entirely new smart city to a finished state from scratch. Instead, we rely on pilot programs and whatever other evidence is available. More incremental applications, games, and simulations can fill in the gap where systems become more complex [34].

A good designer also understands that they are designing for the system. In many instances, the client is not the customer. The client's customer is the real customer. Suppose we were planning a smart city initiative. In that case, we understand that our immediate client may be the city but that the real customer is the city's residents and their social network, not just the city itself. It is incumbent on the designer to advocate for all affected parties.

A systems-level change process is described in Systems Thinking for Social Change [4]. They describe four stages of bringing about change [4]:

1. Build a foundation for change and affirm their readiness for change.
2. Clarify current reality at all levels of the iceberg (something akin to Spiral Dynamics memes) and accept their respective responsibilities for creating it.
3. Make an explicit choice in favor of the aspiration they espouse.
4. Begin to bridge the gap by focusing on high-leverage interventions, engaging additional stakeholders, and learning from experience.

To elaborate on point 3, the public is often given choices without understanding that the current status quo is a choice and will have its own set of consequences. This author recently witnessed a local city council vote down a zoning change because the neighborhoods opposed it. The proposal was to convert an old pig farm and processing plant (i.e., industrial zoning) to retail—and when that failed—to a very slight adjustment to their existing zoning to make it appropriate for warehousing (e.g., adding a few feet to the height allowance). The neighbors were quoted as saying they would only not oppose a city park. The developer evidently did not do enough engagement to explain to the neighborhoods that a city park was impossible. It would have required the city to purchase the land at a below-market rate, which would not have been in the city's budget. In all likelihood, the developer will get fed up and revert the property to the most profitable, most noxious use already allowable. The neighbors will not get a park, and they missed the opportunity to get something better than dirty industry. The point is that opposition matters, and the consequences of the status quo should be fully understood.

2.3.3 The need for civic engagement and involvement

Recently, an NPR article on raising helpful kids seemed an interesting parallel to many government initiatives [35]. The article found a substantial difference between whether kids were involved in the task with their siblings or not. Kids whose older siblings included them in the study tasks were generally more cooperative and more likely to want to engage with chores at home [36]. The cultural difference noted was between families that include children in chores versus those that delegate chores. The parallel to city initiatives is in the level of involvement. Do we involve the public in solution development, or are we forcing a set of limited choices? We may have a reasonably suitable smart city solution, but if we have neglected to involve the public, they may reject the solution for lack of understanding.

That exact situation happened when Massachusetts tried to pass a proposition for ranked-choice voting, which is widely popular, can cure various political ailments, and has as its only drawback that it needs to be explained [37]. With the pandemic raging, the promoters took for granted that everyone would side with them. It turned out not to be true in this instance, primarily due to their neglect to involve the public.

Residents are often aware of material facts that outside professionals might miss [38]. The planning and design communities have known people have mental maps about their city for ages [39]. Involvement is not solely for the residents' benefit to feel better about inclusion. Public participation is essential for an optimal design outcome. A motto could be, "The secret to getting people together is this: build your community *with* people, not for them" (author's emphasis) [40].

2.3.4 Local public participation processes

In addition to the framework and design literature, it serves our purpose to review the public participation literature and consider how we might derive local metrics for initiative success. Public participation can be messy. However, skipping it could be worse.

Public participation can span a wide range of resolutions—some federal legislation is open to public comment. Some public commenting consists of only a simple sentiment analysis noting whether it is in favor of or opposed. On the other end of the communication fidelity spectrum, we have local charrettes that bring available residents into a room to discuss, analyze problems, and formulate solutions as a group. Cities typically sponsor charrettes for planning purposes—to develop a comprehensive plan, for example. Even charrettes are not without fatal flaws. They tend to be poorly attended and primarily by older, wealthier residents who have the time to take off work or family obligations and self-select for an interest in planning or development. They are rarely representative of the underlying population. For the residents that do attend, their specific input can be limited. One group dominated by one individual might generate the majority of comments, diluting the wishes of others present. There are online, decentralized, and high-fidelity attempts at organizing the input, but this is still an emerging field [34].

The participation process continues beyond the formation and solution of local issues. Charrettes and communities need to discuss implementation. Top-down approaches sometimes feel heavy-handed, something done to us rather than with us. Do our communities agree so strongly that they want to enforce the policy with civil penalties? Or do they want a light touch or "nudge" that encourages people to default to prosocial behavior [20]? The cost of enforcement or compliance actions should be part of the discussion. What happens when someone inevitably games or circumvents the system?

In addition to charrettes, other relevant public participation processes could include civic hackathons [41], civic crowdfunding, crowdsourcing, surveys, or referendums. Of course, a vote is the least amount of public input possible, just above none. When a question reaches a vote, the question has been framed and settled, discussions have ceased, and alternatives are eliminated. It gives the perception that a matter is settled when we know there is no endpoint as a wicked problem. There is no point at which we can say "the people have spoken" with any clarity or finality through yes/no or single-option voting.

2.3.5 Local metrics for success

For all the reasons discussed thus far and more, the criteria for the success of any smart city initiative need to be locally developed. At this point, we will cover some of the categories worth consideration. Whatever the criteria, the initiatives eventually need to arrive on a city's budget and calendar so there must be time and resources available for any implementation. There are plenty of economic indicators to choose from—job growth, housing starts, new business formation, etc. These are fairly mainstream as success criteria. Corporations might look at the same population in terms of market penetration. With the possible exception of high-speed Internet, cities are less likely to agree with corporations on that metric.

What often gets overlooked is community well-being—mostly because it is more challenging to measure than the bottom line. These go beyond the socioeconomic indicators available in the Census and American Community Survey data. Community well-being consists of aesthetics, social offerings, and general openness [42]. These are subject to interpretation depending on the underlying population. Smart city initiatives can be a support or detriment to any of those. Across those areas, well-being spans through suffering, struggling, or thriving [42]. A straightforward approach to conducting some of a city's inventory asks where people are struggling (if not suffering)—both geographically and socially.

Another approach to developing success criteria is the public health angle. Quint Studer is a performance improvement consultant to both hospitals and cities. His dashboard for cities [43] includes health and wellness metrics such as preterm births, pre-K participation, kindergarten readiness, free and reduced lunch rate, cost of child care, high school graduation, college attendance, single-parent households, labor force participation, crime rate, middle-class households percentage, real per capita income, rent-burdened households, overweight and obesity rate, and voter turnout. The intention is to measure those over time and see improvements. A smart city initiative that demonstrates improvement in any of these has a stronger case to support it.

Local communities can discuss and select the well-being metrics that are most important to them through a public participation process. The goal is to ensure the selection process is open, transparent, and inclusive so that all of the best ideas make it to the final stages. Dashboards are a good starting point for aggregating decision criteria. The point of a holistic problem-solving process is to find the best solutions to the best problems. We waste too much effort solving problems no one has. Drawing from the best of the available frameworks, design research, and public participation gives us the best chance at achieving the best outcomes as well as knowing and agreeing on what we all mean by "best."

2.4 Locally relevant policies and financing concepts

While the previous sections tended toward more global concepts, here we start to hone in on local concepts. Central to all initiatives are public budgeting processes, which we will cover more in depth. This section could be considered the local strategy to the next section's local tactics. At the local level, policy, financing, and budgeting are intertwined to the degree that we will address them together.

I attended an event in 2017 in Dallas' West End living lab initiative (see https://www.dallasinnovationalliance.com/projects). I witnessed an exchange between the head of our university's GIS department and a local city planner. Of course, in academia, we were excited about smart city developments and always looking for cities or corporations to partner with for smart city projects. The planner said he loved smart cities and would be happy to recommend any projects for which we could find funding. Presumably, this is the crux of this entire book. We have run full steam ahead with the enthusiasm of technocratic idealists with the full encouragement of corporate tech infrastructure and service providers over the last 5–10 year timeframe [9]. Now we are rethinking how to pay for any of it. Most cities have not yet bought into funding smart city initiatives outright. Too many city officials have been burned by promises of projects that will pay for themselves and projects that cost too much for what they deliver in value.

2.4.1 Local balance sheets

Any conversation on policy and finance relies on the balance sheet—what are the assets and liabilities of the entity? Unbeknownst to many residents, most cities are completely upside down on their balance sheets without any idea or way to counter the trend. This state may be a surprise even to astute readers of this book. One primary reason for this is the often financially losing proposition of road maintenance. Every city in every country has potholes. Most roads are not built adjacent to financially valuable enough land to pay for their installation, maintenance, or eventual replacement. This condition is worse in auto-centric regions. Suffice it to say that few, if any, cities are well-positioned to maintain the roads, water, and sewer lines they currently have, and few residents are aware of this fact. Rather than reduce the liability, most infrastructure initiatives seek to expand it with new development. Any smart city initiative that can position itself to help remediate this situation might do well for itself.

Beyond the local infrastructure, cities might or might not be responsible for their energy grids, land owned by the city, machines or equipment owned by the city or local schools, parks, health facilities such as gyms or pools, city hall buildings, first responder facilities, libraries, and more. Additional liabilities typically include the salaries for the residents employed to maintain that infrastructure, pensions for retired employees, insurance against emergencies, and bond debt service. While accounting has its standards, it makes sense as a nonaccountant resident to think of assets as things that generate revenue while liabilities generate expenditures. That would mean that an accountant might consider a city park an asset on the city's balance sheet. However, for city budget considerations, they typically generate an expenditure, not a revenue. That is a situation that most people are fine with. They should be aware of it though when we begin to talk about budgets.

2.4.2 Local budgeting

The components of a budget are an aggregation of the revenue, expenditures, budgeting and decision-making process, balancing, execution and adoption, and implementation and control [5]. These components are not sequential, nor do they often have complete information with which to work.

There are multiple political orientations toward budgeting. Budgeting answers the fundamental question regarding what the scope of government should be. Some orientations include [5]:

- Reform—budgeting should focus on technical goals, not political outcomes.
- Incrementalist—focus on the public participation process for equity.
- Determinism—the most powerful groups get their way.
- Politics of process—focus on changing the rules to meet an outcome.
- Policymaking—policy-driven budgeting.

Difficulties can arise, for example, when the public believes the orientation is reform or incrementalist, but is actually deterministic behind the scenes. Accusations of corruption or other issues can result, according to Rubin (see checklist: "An Agenda For Action For Taking The Best Practices For Politicians To Pass A Budget And Avoid Being Voted Out Of Office").

An agenda for action for checking the appropriate steps for politicians to pass a budget and avoid being voted out of office

A number of steps are required for politicians to pass a budget and avoid being voted out of office, which include the following key activities (check all tasks completed):

_____1. "Make the extent of the revenue problem clear and credible."
_____2. Same with the situational consequences.
_____3. Possibly, make any tax increase temporary.
_____4. Optionally, use a referendum.
_____5. Itemize the needs and uses.
_____6. Preempt arguments of waste or mismanagement.
_____7. Show that the budget is currently well managed.
_____8. Make the increase as fair as possible.
_____9. The public needs to understand all the above.

The conversation prior to discussing the specifics of the budget needs to be how much government the population wants and how much it is willing to pay. Everyone wants nice things and assumes it is just a matter of making their voice heard. The reality of budgeting is more complex. The same is true for both individual residents and advocates of smart city initiatives. These are the waters into which smart cities wade. Awareness of the public budgeting process benefits all actors.

It is within this deficit of deferred infrastructure maintenance and required allocations such as pensions from which every budget begins. Some city budgets can cover their smaller maintenance jobs, but medium to large-scale jobs are outside the bounds of an annual budget. The principle to understand here for smart cities is that within the realm of expenditures, cities have *capital expenditures*, which are one-time upfront costs sometimes funded through bonds, and *operating costs* that are ongoing, usually included in the budget. Both are considerations for new initiatives.

2.4.3 Local project selection criteria

When we begin advocating for smart city initiatives for budget inclusion, we need to know how projects are typically selected. The simplest implementation method is substituting better parts during routine maintenance—switching to LED lights as bulbs are replaced, for example. Relying on maintenance upgrades does not give an initiative long-term security for future implementation but only requires a low-level, discretionary, nonpolitical decision. The most secure is through budget legislation, which relatively guarantees that the project will move forward.

Here we examine the type of criteria cities use in decision-making. We ignore circumstances where personal relationships bias the process since that is neither scalable nor equitable (and in some cases, it is illegal).

Typically, the first criterion is to ask whether this is something that this specific government entity should be doing. Should the local government spearhead 5G installations or make a deal with a private contractor and reach some final ownership and access arrangement? If the immediate need is police reform or pandemic mitigation, then the initiatives need to be relevant to the time and place.

The initiatives should fit into an overall plan for the city, such as the comprehensive plan or some other ideal, which could include being a walkable city, supporting historic preservation, an innovation district, or reducing crime or traffic fatalities. Moreover, hopefully, it goes without saying that the best approach is to work within the existing zoning or other relevant ordinances. It can be very costly to get those changed, not to mention the public might inexplicably turn out to oppose it vehemently. They rarely turn out in support of initiatives unless another group strongly opposes them.

Looking at comprehensive or zoning plans is the approach that looks backward or at the current state. Looking at what could be is often a specialty of tech companies and universities. It is worth considering the possibilities from a first-principles perspective. For what and whom are cities built? What would our city be like if we cared about every resident's well-being? Sometimes an idealized version is a good starting point—looking at smart cities built from scratch or other high quality-of-life "non-smart" cities. One such example is the 15-minute city paradigm that says most weekly trips should be able to be made within a 15-min walk [44].

A generalized systems design approach can be effective depending on the scope. For systems, we need to identify the stocks and flows. In network science, they would call them the nodes and links. Care must be taken to adequately describe the flows or links, which are the relationships holding everything together. Human perception research has found that humans have an object bias [45]. That means we tend to see everything as objects and sometimes overlook the underlying relationships.

Systems design could also incorporate a UX approach considering all the main actor types related to the project to assure that all the incentives are aligned. We can draw from gamification for this context [46]. Gamification would say to make sure that our interventions are congruent with the behavior we want without infringing on people's existing intrinsic motivation to do the right thing. For this criterion, we are checking to make sure everyone benefits and there are no losers to this change—including the city.

There are more formalized criteria. One is the Strong Towns approach [47]. Their basic criterion is that it either leaves the city in a financially better position or that the costs are known, and everyone agrees that it is worth paying in new taxes. They set a guideline of requiring $30 of private investment for every dollar of new infrastructure liability the city accepts. City budgets cannot cover much more expense than that of long term. And few people want higher taxes.

The last formal project evaluation framework suggested is the City Model Canvas [48], which was adapted from the Business Model Canvas to outline a business plan. It is a kind of systems analysis for a business idea. Items from the city plan version include:

1. Mission statement
2. Value proposition
3. Beneficiaries
4. Buy-in and support
5. Deployment
6. Key partnerships
7. Key activities
8. Key infrastructure and resources and key regulatory framework
9. Budget cost structure
10. Revenue streams
11. Environmental costs
12. Environmental benefits
13. Social risks
14. Social benefits

Writing out narratives for each of these factors will give everyone an excellent idea of what the initiative will do and its potential systemic impact.

As a closing comment, too many cities have shunned all criteria in favor of "build it, and they will come" and ribbon-cutting opportunities without enough thought to the long-term financial or well-being impacts. Large projects might be a temporary boon to a local economy but leave it less resilient should that project eventually lose relevance or be relocated in the future. An alternative approach follows.

2.4.4 Local incremental development

Given that cities are more than their local governments, we will address the impact of local real estate development on smart cities beyond any smart systems of individual buildings. The same kind thinking that incremental developers use can be applied to local economic development, social services, or other local issues.

We can define incremental development as seeking to fill the next missing piece of the urban ecology while minimizing the risk. It is akin to using the scientific method—form a hypothesis about what is needed next and test it out. Some might confuse incremental development with small to mid-sized projects. While in many cases, that is the appropriate increment, in a city of five-story buildings, the next increment is likely another five-story building as infill or at the edge of the existing ones.

The challenge is that most zoning does not allow that much flexibility. Single-family zoning is a key driver of the increasing crises of unaffordable housing currently facing the United States and many other countries. In many places, the next increment for single-family houses is residential buildings with two to four units or a single accessory dwelling unit behind. Rectifying this housing situation may be the "smartest" thing some cities could be doing right now. Furthermore, most single-family zoning excludes commercial uses aside from working from home. Using the personal residence—the home over the shop—has historically made the home more flexible and affordable by generating income onsite.

For our purposes here, this is to contrast the mega projects to which politicians often gravitate. Tests and small bets are less likely to lead to underutilized assets or even financial ruin in the future. If a project promises to pay for itself, we need to check the assumptions and calculations with a skeptical eye.

2.5 Key civic policy and financing instruments

As mentioned, this will be the section on tactics to the previous section's strategy. This section will be more specific solution-oriented. None of what follows is universal or should bypass a public participation or budgeting process. The intent of this chapter is the need to localize, not mindlessly parrot, plug, and play. It seemed that this chapter would best be organized almost as a glossary. Here is assembled a range of tactics and solutions alphabetically with their pros and cons or other considerations. Some were mentioned previously in the chapter. Most of these pertain to who owns and who pays as befitting policy and financing concerns. Specific technology solutions can be found in the first volume of this series, Solving Urban Infrastructure Problems Using Smart City Technologies [49].

2.5.1 Alternative engagement models to ensure the viability of IT initiatives

Forrester Research developed a report on new business models to support financing smart cities [50]. The contents of a summary table on engagement models are quoted here:

- External investment—Regional banks and investment funds such as the Asian Development Bank, Inter-American Development Bank, and the Brazilian Development Bank provide funding for public-sector IT initiatives.
- Revenue-generating (or cost-cutting initiatives)—Revenue-generating and cost-cutting initiatives, such as fee and tax collection or electronic government procurement, can become self-funding and prove appealing for budgetary and political reasons.
- Revenue-sharing and public–private partnerships—Partnerships with a vendor, service provider, systems integrator, or even real estate developer on a revenue-sharing basis can defray upfront costs and risks of a new initiative.
- Capacity reselling—Excess capacity from large municipal IT infrastructure or applications deployments can be provided to neighboring cities or organizations. The larger city IT department acts as a service provider or through a managed service provider.
- Multicity initiatives—Upfront agreements to pool resources and share infrastructure facilitate the launch of large IT initiatives.
- Leasing and financing—Traditional financing remains an alternative for purchasing IT infrastructure, particularly hardware and networking, and provides flexibility in case of budget shortfall or other political contingencies.
- Barter or in-kind exchange—Exchange of product testing or customer references for new technologies is a way of overcoming budget shortfalls, particularly for universities or research facilities with skilled developers and users.
- Data monetization—The use of primary data generated by instrumented infrastructure provides a potential revenue source for data owners.

2.5.2 Charrettes

Charrettes are a public participation method that typically brings the general public into a city facility for a workshop-style education and ideation session. Third-party consultants specializing in the format with event planning and architectural or planning backgrounds typically host these on behalf of cities. The time can vary. We usually get better results from multiple meetings spread over weeks or months than single meetings, depending on the situation's complexity. A minimum of two separate sessions plus an education session is a best practice so that the public can see and comment on how well the consultants were able to visualize their ideas. It is best to understand the opportunities and limitations of charrettes before taking bids. The most thorough, practical guide is The Charrette Handbook, published by the National Charrette Institute [51]. They also offer certification courses, which this author attended and can recommend.

2.5.3 Civic crowdfunding

Crowdfunding is when we amass funds through voluntary investments from the general public for a project. The civic variety is for civic projects. While it can be more complex to manage than budgetary or institutional funding, the benefit to civic engagement can be substantial. In crowdfunding, it is typical to offer investors (aka backers) a reward depending on their investment level. Equity crowdfunding offers the option of ownership as a possible reward. Aside from the products or artistic endeavors frequently featured on major crowdfunding sites such as Kickstarter or Indiegogo, we have examples of civic crowdfunding projects for smart city development [52] and place-making [53]. The six projects featured in Carè et al. all gave rewards to backers, accepted donations, and did not include any equity arrangements.

2.5.4 Civic hackathons

Hackathons are typically fixed-time contests for application developers to create a working piece of software or a minimally communicable prototype concept. These typically range from a weekend in-person to asynchronously over weeks. Sometimes they are data projects rather than application projects. Hackathons with even modest prizes can generate awareness, public interest, and solutions for specific problems at an affordable rate. Even cities on the tightest budget could offer prizes donated by corporations or even a set amount of time with local officials. We might expect that the winning solution could be promised a contract to work with a local corporation if needed to implement the solution for civic hackathons. Depending on the audience, we should take care that the promotions avoid confusion. In one instance, there was public opposition to the event because locals believed it had something to do with cybersecurity black-hat hacking, and that the city should not sponsor such a thing.

2.5.5 Commons-based organizations

In some contexts, neither public nor private solutions are optimal. The conditions best suited for commons management are when the scope is limited, boundaries are definable, monitoring and enforcement are possible, and externalities are limited [54]. Commons organizations require membership, rules, peer monitoring, and effective controls and consequences. These specific constraints avoid the so-called tragedy of the commons, which occurs when commons resources are completely unmanaged. The way I typically explain the ownership of commons to laypeople is:

- Private means specific people (or entities) own something exclusively.
- Public means no one owns something (i.e., in practice, no one has a negotiable, transferrable ownership interest in a public park).
- Commons mean a group or community owns something inclusively.

In this sense, exclusive means one typically needs a bill of sale to transfer ownership rights. Inclusive here means commons ownership is often membership-based, requiring a fee to join, and utilizing the resource. The commons ownership interest may or may not allow for being sold or transferred to a third party.

Traditional commons have managed fishing rights, park access, or riparian water rights. In another sense, mandatory homeowner associations could be considered a type of commons. More generally, commons-based organizations include community land trusts, cooperatives (aka co-ops), local stock markets, and public asset corporations, depending on their legal structures. These are all legal structures with varying degrees of and requirements for the spectrum of public and private ownership. Further research might shed insight into whether public infrastructure such as water, sewer, or power would serve people better if owned publicly, privately, or as co-ops or other commons entities as some are.

2.5.6 Community land trust

A community land trust is a trust owned by community members or municipalities that hold land or real estate to benefit the community. Sometimes the holding purpose can be as simple as maintaining the affordability of the housing that occupies the land or ensuring the commercial uses benefit the community [55]. More active land use management tends to become the domain of co-ops [56] or public asset corporations.

2.5.7 Co-ops

Co-ops (aka cooperatives) are an ownership model where the owners share ownership and usually actively participate in the success of the owned asset [56]. In addition to the profit-sharing aspect, co-ops often fill in market gaps where it might not otherwise be profitable to supply community needs. Typical co-ops include shared or multifamily housing or retail establishments (esp. grocery stores).

2.5.8 Crowdsourcing

In contrast to crowdfunding, crowdsourcing draws on the wisdom of the crowd without any requirement for the exchange of funds (i.e., backer rewards or prizes). It could be any project to which the community contributes time or expertise. Scale is not a defining characteristic as we can consider both the open-sourced software community and Wikipedia to be crowdsourced domains. We can classify charrettes, hackathons, or volunteerism as crowdsourcing with or without funding.

2.5.9 Dashboards

Dashboards are a type of data visualization. Dashboards on cars serve the same purpose as dashboards used by financial or data analysts—to give the user information about the system's state. In any project management scenario, we typically have an end goal in mind. Dashboards are fairly universally understood and adapted for smart city applications [57]. We discussed the public health dashboard as possible smart city initiative criteria. Dashboards can visualize emergency services or provide situational awareness for municipal services and reduce outages. Singapore is often cited as a leader in smart city applications. Payne et al. detail their dashboard that monitors their network of smart LED light posts that measure motion sensing, energy usage, manage conservation, can stream video, respond to panic button input, and adapt to ambient lighting [57]. Many more sensor types can be added or substituted depending on local political economy contexts.

2.5.10 Economic development corporations

These entities' mission is typically to promote the local economic development of their patron cities. Some get funding directly from the city. Others are privately funded or work through their allied organizations, such as a Chamber of Commerce. Their best use is likely local economic gardening initiatives, which seek to strengthen local businesses and create more jobs without relying on public funds [58]. More controversially, these organizations spend much effort trying to attract existing large businesses to relocate to their cities. It is not clear that the cost−benefit analysis works out long-term for cities, and this approach more closely resembles a race to the bottom. The approach does sometimes result in apparent "big wins" that gain publicity if not notoriety. Too often, the tax benefits and incentives used to bribe companies to relocate must be renewed to keep the company. If not, the company can get the same benefit from another city. Many of these relocations might have happened without any incentives at all. These efforts can undermine or discourage local businesses that already work within the city. Possibly peak corporate relocation mania was Amazon's second headquarters phenomena of 2017−18, which seemed a national horse race at the time [59]. In the end, cities should question whether they want corporate relocations that begin with an extractive or extortive relationship. Cities should take care that these organizations use their funds to benefit their local community transparently and sustainably.

2.5.11 Local currencies

Economic analysis usually finds a benefit to shopping locally. A primary benefit is that more of the profits circulate in the local economy. One study found that 32% of the funds stayed in the local economy for a local purchase compared with 16% for national chains [60]. Local business purchases are exactly twice as good for the local economy. We can further incentivize local shopping by introducing local currencies [61]. Beyond shopping, having a local currency solves the problem created when currency is a store of value—which it is by definition. As a store of value, people tend to take it out of service as a means of transaction—putting it in savings, investing it, or otherwise making it illiquid. When a currency is scarcer, the rate of transactions and its economy slows down.

Local currencies allow for demurrage, which penalizes the holder for holding it too long. Think of it as the fee that accrues if a gift card is left dormant too long. It is there to encourage the holder to put the funds back into productive use, not hoard them. The fees can be used for currency maintenance costs. Residents can be encouraged to adopt the currencies by offering discounts and appealing to their desire to improve their community. The currencies need to be coordinated between the local government and local businesses for widespread acceptance.

Local currencies can be physical [62] or digital [63]. Digital can be blockchain-based though this author is not aware of any full implementation yet. Any such digital implementation would need geographic restrictions and controls to avoid speculation. A stable coin pegged to the dollar might be more appropriate.

A local currency economy could be funded via crowdfunding, a loan guarantee, or a municipal bond. They need to be securitized one way or the other to hold value. The city could help maintain their value by accepting them for payment for city services.

2.5.12 Local stock exchanges

Local stock exchanges could be established to facilitate local investment in local businesses [64]. It would take crowdfunding to the next step—offering an equity backer reward. The details such as shares, rates, or residency requirements would be worked out with the local community and a securities attorney. This setup would be a system of local ownership rather than a single co-op or land trust. Local ownership also contributes to local patronage and social cohesion. We are much more likely to eat at a restaurant if we have an ownership interest there. We are more likely to back a smart city initiative if we have an ownership interest in its income stream.

2.5.13 Municipal annexation

Annexation is when a city acquires additional land to extend its boundaries. This is often politically or financially motivated. It could want the tax revenue from a commercial district. It could want a set of voters in a wealthy subdivision. Poor areas are often bypassed as financially precarious new service obligations (see Municipal Service Coverage below). The City of Austin, Texas, is a prime example. A quick look on Google Maps shows a serpentine and, in places, a nearly disconnected boundary that bypassed lower economic status areas. Annexation is a politically fraught proposition. It is mentioned here only for completeness.

2.5.14 Municipal bonds

Bonds are fixed-term loans governments use to pay for capital expenditures. Their debt service (i.e., payments) is a typical element of a budget. Some cities pass bond initiatives every 3–5 years to cover bigger projects.

2.5.15 Municipal Service Coverage

Service coverages are the areas served by cities with fire, police, schools, and any other service a city provides. This calculation is not as straightforward as it might initially seem. It is not a binary yes/no situation as it is whether there is electricity or not. The question starts to become one of response time and capacity. How many seconds is adequate coverage for first responders? How much reserve capacity does the city want to keep on hand before they have to ask for other levels of government such as their state or FEMA to intervene? What is the "fair" amount of coverage people expect?

Do we want more proactive coverage, such as the police or code enforcement actively looking for issues, or more reactive such as the fire department waiting for a call? Are the police incentivized to increase their budgets through citations? How often do we want residential trash service or recycling? Are multifamily properties included? How many hours should the library be open? Smart city initiatives can be of assistance in some of these instances.

2.5.16 Public asset corporation

In The New Localism, the authors profile a public asset corporation in Copenhagen [65]. The entity was structured as a publicly owned and privately managed corporation tasked with asset management of municipally owned lands. In the case of Copenhagen, the entity was 55% owned by the city and 45% by the National Ministry of Finance. The entity was empowered to borrow against underlying land assets for redevelopment funding, fast-tracked zoning proposals for developers, and recaptured value from transportation and infrastructure investments.

The public asset corporation structure results in a politically insulated entity with access to industry best practices, investment in the public good, and long-term thinking. This arrangement could rectify the situation many cities find themselves in where they are unaware of the land assets under control and make little effort toward improving their value. Confounding that process is that anything a city government does with their land becomes a political question. Professional land management can significantly increase the value and revenue. At the very least, land should not remain in a state of disuse for reasons of neglect. It would be worth considering making such a corporation a nonprofit, low-profit, or benefit corporation to avoid the pitfall of maximizing private wealth at the expense of the public.

2.5.17 Public participation methods

We discussed these extensively in the preceding sections. It is an umbrella term that includes charrettes, civic crowdfunding, civic hackathons, surveys, town hall meetings, and more. The goal is to engage with the public to solicit their input and gain their buy-in for initiatives.

2.5.18 Public–private partnerships (PPP)

These are arrangements whereby the city works with a private company to accomplish a task. Infrastructure or real estate developments are common targets for PPPs. These deals obligate the city beyond the typical service provider, such as what they might use to do their HR and accounting. The city is often responsible for additional funding, loan guarantees, or review processes using up city resources.

The benefit is that the city can contract out the expertise of private companies using their public procurement process. Disadvantages include loss of control, insufficient accountability, a changing political environment, insecure resident data [66], or conditions requiring contract renegotiation. Cruz and Sarmento [67] point out that, especially for smart city requirements, data and technology are a moving target, and monolithic service contracts often fail to cover a dynamically changing field (2017). They urge more flexible arrangements more akin to incremental development.

Another factor briefly touched on before is the inherent contradiction between the DNA of a company and a city. Especially for publicly owned corporations trading on a stock exchange, CEOs are under a legal obligation to maximize shareholder value, which often translates to squeezing their customers and employees. Cities should ensure that such driving factors are not influencing any of their agreements. All the incentives in their system should be aligned.

2.5.19 Tactical Urbanism

Tactical Urbanism is a subset of incremental development. It is a type of demonstration of minimum viable urbanism

[68,69]. They are low-cost and usually temporary installations to featurc how changes in urban form could improve the area. They could be anything from altering traffic flow or parking spaces to a weekend festival, adding pocket parks, street furniture, vendors, and activating the local community. Better Block in Dallas (betterblock.org) is an organization that specializes in these events. These might pair nicely with smart city pilot programs. They can scale from the block up to the city [70] though they do need a tipping point of participation to yield effect.

2.5.20 Tax increment financing (TIF)

Typically, a TIF is a geographic district within a taxing authority that the incremental tax gains from development into the district's infrastructure rather than into the general budget of the authority. TIFs incentivize developers to redevelop, knowing that the value of the improvement will go directly into further improvement of their properties.

As with any redistribution of wealth, there are issues with the TIF process. The Wikipedia page for TIFs lists nine specific unintended consequences [71]. Of those, it bears highlighting that TIFs can increase a city's service coverage demand without contributing to the budget for it. More schools, police, and waste services may be required, but those increased property tax revenues have already been allocated to other uses.

2.5.21 Upgrading to smart city infrastructure during routine maintenance

The most straightforward way to progress toward becoming a smarter city is to do it incrementally within the existing budget in a way that does not require additional permissions. As of this writing, the City of Dallas is replacing street lights with brighter, lower energy LED lights, saving money, and reducing crime [72]. Additionally, they are installing Wi-Fi routers to some of their traffic signals to expand Internet connectivity in areas where it is lacking [73].

As a graduate student working on a smart city project at university, this author was told the facilities maintenance group did not have a full map of the campus infrastructure on campus and relied on water leaks to locate their pipes. For their future reference, in addition to updating their maps, they placed sensible tape on any exposed pipes to be precisely located with sensing equipment in the future.

2.5.22 Zoning as a policy instrument

Many places in the developed world are facing housing affordability crises of varying degrees. A large part of that is constrained by zoning—land-use decisions that prohibit or allow certain uses. Whether they use technology to do it or not, a city would be smarter if fewer of its residents were rent-burdened, much less homeless. Code reform is one avenue that could benefit from smart city initiatives [74].

Single-family zoning is, by definition, exclusionary. It excludes a significant portion of the population not in its child-rearing age of life—students, single people, tech workers, digital nomads, blended families, multigenerational families, seniors, childless couples, singles living together, economically transitioning people, and excludes neighborhood commercial uses. Those people still live in the neighborhoods but adapt to the conditions.

Form-based codes are not a panacea per se (especially if they leave suburban parking requirements in place), but they remove the statutory exclusion. As-is, the excluded population is making the best of the SFH inventory—treating houses as dorms, home offices, short-term rentals, or struggling with accumulating deferred maintenance. As a student, this author was aware that the limited housing situation meant that it is common for seven students to live together in a two-bedroom apartment—two per room, two in the living room, and one in the hallway. Another student lived in a backyard shed without a bathroom or kitchen. Missing Middle Housing [75], where allowed, can start to include housing types more appropriate to the actual population.

Many cities are reforming their zoning to create innovation districts. These reforms need to consider their underlying population rather than expect dynamic tech companies to relocate when there are extensive, exclusive monocultures of single-family housing, business parks, or retail centers degrading the walkability and urban life.

2.6 Summary

In this chapter, we covered the deep need to localize. We covered the why, how, and what of localization. We moved from conceptual through strategic to tactical. Our journey traversed context, culture, economics, holistic problem solving, locally relevant policies and financing concepts, and key civic policy and financing instruments. We returned to systems and design thinking throughout and understanding the local political economy beyond the technical issues or specific solutions.

Fundamentally smart cities are simply cities trying to be better versions of themselves. As with all local political economies, we continue the conversation about who owns what and pays for what. Who is suffering, who is thriving, and how much struggle is okay? What should government do versus private industry or commons-based organizations? For how much government is the population willing to pay? What makes it into the budget and onto the calendar? These questions are necessarily local for the city. We have to stop, look, ask questions, and engage our local populations to build our best cities.

Finally, let us move on to the real interactive part of this chapter: review questions/exercises, hands-on projects, case projects, and optional team case project. The answers and/or solutions by chapter can be found in Appendix G.

2.7 Chapter review questions/exercises

True/false

1. True or False? Solutions that do not appreciate the local context, culture, and economy are solutions.
2. True or False? Smart city initiatives must compete with other budget requirements, including some already required by law.
3. True or False? How well the local population is informed regarding civic issues is not important.
4. True or False? Behavioral Economics' contribution is to supplement the caricature of the traditional assumptions of human actors with what we know about people from psychology.
5. True or False? The anthropic principle states that before information, time and space are foundational concepts.

Multiple choice

1. Once the temporal and social factors are better understood, we can draw from the best:
   a. User experience
   b. Smart city initiative
   c. Systems-level change process
   d. Proposal
   e. Universal design frameworks
2. Public participation can span a wide range of:
   a. Resolutions
   b. Sentiment analysis
   c. The communication fidelity spectrum
   d. Charrettes
   e. Local issues
3. The criteria for the success of any smart city initiative needs to be:
   a. Socioeconomic
   b. A public participation process
   c. Locally developed
   d. Upside down on their balance sheets
   e. Financially valuable
4. The components of a __________ are an aggregation of the revenue, expenditures, budgeting and decision-making process, balancing, execution and adoption, and implementation and control.
   a. Holistic integration of smart city platforms
   b. Political orientation
   c. Smart city initiative
   d. Computer application
   e. Budget
5. When we begin advocating for ______________ for budget inclusion, we need to know how projects are typically selected.
   a. Maintenance upgrades
   b. Smart city initiatives
   c. Budget legislation
   d. 5G installations
   e. Relevant ordinances

Exercise

Problem

Develop a methodological and conceptual framework to examine and document the localized hotspots of urban shrinkage that could be used as coworking spaces in a typical medium-sized city.

Hands-on projects

Project

Do research: Analyze why the concept of a smart city is highly context-specific.

Case projects

Problem

Why do Smart cities and infrastructure designs need to be people centered?

Optional team case project

Problem

Describe why an integrated approach should be adopted in developing smart cities and infrastructure.

References

[1] Smart city. In: Wikipedia; 2021. https://en.wikipedia.org/w/index.php?title=Smart_city&oldid=1020682863.

[2] Rittel HWJ, Webber MM. Dilemmas in a general theory of planning. Pol Sci 1973;4(2):155–69. https://doi.org/10.1007/BF01405730.

[3] Berry BJL. Cities as systems within systems of cities. Pap Reg Sci 1964;13(1):147–63. https://doi.org/10.1111/j.1435-5597.1964.tb01283.x.

[4] Stroh DP. Systems thinking for social change: a practical guide to solving complex problems, avoiding unintended consequences, and achieving lasting results. Chelsea Green Publishing; 2015.

[5] Rubin IS. The Politics of public budgeting: getting and spending, borrowing and balancing. 7th ed. CQ Press; 2013.

[6] Surveillance Issues in Smart Cities. In: Wikipedia; 2021. https://en.wikipedia.org/w/index.php?title=Surveillance_issues_in_smart_cities&oldid=1016214954.

[7] Gelfand MJ, Raver JL, Nishii L, Leslie LM, Lun J, Lim BC, Duan L, Almaliach A, Ang S, Arnadottir J, Aycan Z, Boehnke K, Boski P, Cabecinhas R, Chan D, Chhokar J, D'Amato A, Ferrer M, Fischlmayr IC, et al. Differences between tight and loose cultures: a 33-nation study. Science 2011;332(6033):1100–4. https://doi.org/10.1126/science.1197754.
[8] Reichert C. 5G coronavirus conspiracy theory leads to 77 mobile towers burned in UK, report says. CNET; 2020, May 7. https://www.cnet.com/health/5g-coronavirus-conspiracy-theory-sees-77-mobile-towers-burned-report-says/.
[9] AT&T. AT&T launches smart cities framework with new strategic alliances, spotlight cities, and integrated vertical solutions [Corporate]. AT&T; January 5, 2016. https://about.att.com/story/launches_smart_cities_framework.html.
[10] Ramsey R. UT/TT poll: Majority of Texans oppose permitless carry, would ban police chokeholds and taxpayer-funded lobbying. The Texas Tribune; May 3, 2021. https://www.texastribune.org/2021/05/03/texas-voters-legislature-poll/.
[11] Dennett P. Sensible socialism: the salford model. Tribune Magazine; January 19, 2021. https://tribunemag.co.uk/2021/01/sensible-socialism-the-salford-model.
[12] Marohn C. The ideology of traffic. Strong Towns; December 5, 2016. http://www.strongtowns.org/journal/2016/12/4/the-ideology-of-traffic.
[13] Henry T. Rail failed, but it's not the end of transit improvements in Austin [Radio News]. KUT: Austin's NPR Station; November 14, 2014. https://www.kut.org/post/rail-failed-its-not-end-transit-improvements-austin.
[14] Peters A, Peters A, Peters A. Austin, Texas, just voted to spend $7 billion on a transportation revolution. Fast Company; November 5, 2020. https://www.fastcompany.com/90572127/austin-texas-just-voted-for-a-transportation-revolution.
[15] DeMeritt M. Psychographics: market analysis moves beyond demographics. Esri; April 18, 2017. https://www.esri.com/about/newsroom/publications/wherenext/psychographics-market-analysis-moves-beyond-demographics/.
[16] Kenton W. Neoclassical economics. Investopedia. September 24, 2020. https://www.investopedia.com/terms/n/neoclassical.asp.
[17] Investopedia Staff. Keynesian economics definition. Investopedia; April 30, 2020. https://www.investopedia.com/terms/k/keynesianeconomics.asp.
[18] D'Souza D. Modern monetary theory (MMT) definition. Investopedia; February 23, 2021. https://www.investopedia.com/modern-monetary-theory-mmt-4588060.
[19] Fischer L, Hasell J, Proctor JC, Uwakwe D, Perkins ZW, Watson C, editors. Rethinking economics: an introduction to pluralist economics. 1 edition. Routledge; 2017.
[20] Pink DH. Drive: the surprising truth about what motivates us. Riverhead Books; 2011.
[21] Thaler RH, Sunstein CR. Nudge: improving decisions about health, wealth, and happiness. Revised and Expanded edition. Penguin Books; 2009.
[22] Farmer JD. Economics needs to treat the economy as a complex system. Institute for New Economic Thinking; 2012. https://www.ineteconomics.org/research/research-papers/economics-needs-to-treat-the-economy-as-a-complex-system.
[23] Global Alliance for the Rights of Nature. (n.d.). Ecuador adopts rights of nature in constitution. The rights of nature. Retrieved from https://www.therightsofnature.org/ecuador-rights/.28 May 2021.
[24] Illien N. How Vienna built a gender equal city. BBC Travel; May 25, 2021. https://www.bbc.com/travel/story/20210524-how-vienna-built-a-gender-equal-city.
[25] Shuman M. Local dollars, local sense: how to shift your money from wall street to main street and achieve real prosperity. Chelsea Green Publishing; 2012.
[26] Institute for New Economic Thinking. (n.d.). Economics of race. Institute for New Economic Thinking. Retrieved from https://www.ineteconomics.org/perspectives/collections/race-economics. May 28, 2021.
[27] Duncan, G. What's going on here, with this human? Graham Duncan blog. (n.d.). Retrieved from https://grahamduncan.blog/whats-going-on-here/.25 May 2021.
[28] John Archibald Wheeler. In: Wikipedia; 2021. https://en.wikipedia.org/w/index.php?title=John_Archibald_Wheeler&oldid=1022882543.
[29] Brand S. Pace layering: how complex systems learn and keep learning. J Design Sci 2018;(3). https://doi.org/10.21428/7f2e5f08.
[30] Beck DE, Cowan C. Spiral dynamics: mastering values, leadership and change. 2nd ed. Wiley-Blackwell; 2006.
[31] Beck PDE, Larsen TH, Solonin S, Viljoen DR, Johns TQ. Spiral dynamics in action: humanity's master code. 1 edition. Wiley; 2018.
[32] Dawkins R. The selfish gene. Oxford University Press; 1976.
[33] Wilber K. The integral vision: a very short introduction to the revolutionary integral approach to life, god, the universe, and everything. Shambhala; 2007.
[34] Burns LR. Social Gaming as a participatory urban design process. PhD Thesis. University of Texas at Dallas; 2017. https://utd-ir.tdl.org/bitstream/handle/10735.1/5651/ETD-5608-7374.70.pdf.
[35] Doucleff M. Are we raising unhelpful, bossy kids? Here's the fix. March 5, 2021. NPR.Org. https://www.npr.org/sections/goatsandsoda/2021/03/05/974069925/are-we-raising-unhelpful-bossy-kids-heres-the-fix.
[36] Alcalá L, Rogoff B, Fraire AL. Sophisticated collaboration is common among Mexican-heritage US children. Proc Natl Acad Sci USA 2018;115(45):11377–84. https://doi.org/10.1073/pnas.1805707115.
[37] The Boston Globe Editorial Board. A voting setback: Massachusetts rejects ranked choice. November 4, 2020. BostonGlobe.Com. https://www.bostonglobe.com/2020/11/04/opinion/voting-setback-massachusetts-rejects-ranked-choice/.
[38] Day C, Parnell R. Consensus design. Routledge; 2002.
[39] Lynch K. The image of the city. The MIT Press; 1960.
[40] Richardson B, Huynh K, Sotto KE. Get Together: how to build a community with your people. Stripe Press; 2019. https://www.amazon.com/dp/1732265194/.
[41] IBM. NYC BigApps 2014. 2014. https://smartercitieslaboratory.splashthat.com.
[42] Rath T, Harter J, Harter JK. Wellbeing: the five essential elements. Gallup Press; 2010.
[43] Studer Q. Building a vibrant community: how citizen-powered change is reshaping America. Be the Bulb Publishing; 2018.
[44] Andres Duany RS. Defining the 15-minute city [Text]. CNU; February 8, 2021. https://www.cnu.org/publicsquare/2021/02/08/defining-15-minute-city.

[45] Konkle T, Olivia A. Normative representation of objects: evidence for an ecological bias in object perception and memory. Proc Ann Meet Cogn Sci Soc 2007;29(29).
[46] Werbach K, Hunter D. For the win: how game thinking can revolutionize your business. Wharton Digital Press; 2012.
[47] Marohn C. Should we do this project? A strong towns answer. Strong Towns; March 1, 2021. https://www.strongtowns.org/journal/2021/2/27/should-we-do-this-project-a-strong-towns-answer.
[48] Timeus K, Vinaixa J, Pardo-Bosch F. Creating business models for smart cities: a practical framework. Publ Manag Rev 2020;22(5): 726–45. https://doi.org/10.1080/14719037.2020.1718187.
[49] Vacca J, editor. Solving urban infrastructure problems using smart city technologies. Elsevier; 2021. https://doi.org/10.1016/C2018-0-00326-6.
[50] Belissent J. Getting clever about smart cities: new opportunities require new business models. 2010. https://www.forrester.com/report/Getting+Clever+About+Smart+Cities+New+Opportunities+Require+New+Business+Models/-/E-RES56701.
[51] Lennertz B, Lutzenhiser A. The Charrette handbook: the essential guide to design-based public involvement. 2nd ed. edition. APA Planners Press; 2014.
[52] Carè S, Trotta A, Carè R, Rizzello A. Crowdfunding for the development of smart cities. Bus Horiz 2018;61(4):501–9. https://doi.org/10.1016/j.bushor.2017.12.001.
[53] Paynter B. Combining crowdfunding and place-making to help communities improve themselves. Fast Company; April 11, 2017. https://www.fastcompany.com/40404791/combining-crowdfunding-and-place-making-to-help-communities-improve-themselves.
[54] Ostrom E. Governing the commons. Reissue edition. Cambridge University Press; 2015.
[55] Solomon A. An atlanta neighborhood will finally own Property as a community. Next City; 2021, February 9. https://nextcity.org/daily/entry/an-atlanta-neighborhood-will-finally-own-property-as-a-community.
[56] Schneider B. How community land Trusts and Co-ops work: an explainer. CityLab; April 2019. https://www.citylab.com/equity/2019/04/home-ownership-ideas-housing-co-ops-shared-equity-land-trust/585658/.
[57] Payne B, Ling LO, Gorod A. Towards a governance dashboard for smart cities initiatives: a system of systems approach. In: 2020 IEEE 15th International Conference of System of Systems Engineering (SoSE); 2020. p. 587–92. https://doi.org/10.1109/SoSE50414.2020.9130542.
[58] National Center for Economic Gardening. National center for economic gardening. (n.d.). Retrieved from https://economicgardening.org/. 26 May 2021.
[59] Amazon HQ2. In: Wikipedia; 2021. https://en.wikipedia.org/w/index.php?title=Amazon_HQ2&oldid=1024330957.
[60] R. Stephanie B. Study: locally owned firms recirculate more money through the economy. NOLA.Com; September 22, 2009. https://www.nola.com/news/business/article_4c34f618-3db3-5240-82dc-0e36bb9b9187.html.
[61] Lietaer B, Dunne J. Rethinking money: how new currencies turn scarcity into prosperity. Berrett-Koehler Publishers; 2013.
[62] Waters M. Why a small town in Washington is printing its own currency during the pandemic. The Hustle; June 12, 2020. https://thehustle.co/covid19-local-currency-tenino-washington/.
[63] Bains. Why cities will soon choose digital currency over fiat money. CoinDesk; February 5, 2017. http://www.coindesk.com/cities-will-soon-chose-digital-currency-fiat-money/.
[64] Shuman M. Local stock exchanges and national stimulus. Federal Reserve Bank of San Fransisco 2009;5(2):81–4.
[65] Katz B, Nowak J. The new localism: how cities can thrive in the age of populism. Brookings Institution Press; 2018.
[66] Allam Z. The emergence of anti-privacy and control at the nexus between the concepts of safe city and smart city. Smart Cities 2019;2(1):96–105. https://doi.org/10.3390/smartcities2010007.
[67] Cruz CO, Sarmento JM. Reforming traditional PPP models to cope with the challenges of smart cities. Compet Regul Netw Ind 2017;18(1–2):94–114. https://doi.org/10.1177/1783591717734794.
[68] Lydon M, Garcia A, Duany A. Tactical urbanism: short-term action for long-term change. Island Press; 2015.
[69] Tactical Urbanism. In: Wikipedia; 2020. https://en.wikipedia.org/w/index.php?title=Tactical_urbanism&oldid=993512593.
[70] Steuteville R. Tactical Urbanism on a citywide scale [Text] PubSquare: A CNU J April 16, 2021. https://www.cnu.org/publicsquare/2021/04/16/tactical-urbanism-citywide-scale.
[71] Tax Increment Financing. In: Wikipedia; 2021. https://en.wikipedia.org/w/index.php?title=Tax_increment_financing&oldid=1023230889.
[72] Torralva K. One simple trick works to help reduce violence in high-crime areas, studies show. Dallas News: So Dallas is trying it; 2021, April 25. https://www.dallasnews.com/news/2021/04/25/one-simple-trick-works-to-help-lower-violence-in-high-crime-areas-studies-show-so-dallas-is-trying-it/.
[73] Cooper B. Dallas upgrading 1,000 street lights to LED, expanding Wi-Fi to needed areas [Local Newspaper]. Dallas Morning News; November 9, 2020. https://www.dallasnews.com/news/2020/11/09/dallas-upgrading-1000-street-lights-to-led-expanding-wi-fi-to-needed-areas/.
[74] Congress for the New Urbanism. The Project for code reform [Text]. CNU; March 20, 2017. https://www.cnu.org/our-projects/project-code-reform.
[75] Missing Middle Housing. Missing middle housing: responding to the demand for walkable urban living. (n.d.). Retrieved from http://missingmiddlehousing.com/.29 July 2015.

Chapter 3

Revolutionizing the localization of smart city infrastructure policies and financing

Michela Longo[1] and Wahiba Yaici[2]

[1]*Politecnico di Milano, Milan, Italy;* [2]*CanmetENERGY Research Centre, Ottawa, ON, Canada*

3.1 Introduction

Often referred to as a revolution, electric mobility is one of the biggest challenges in the transport sector. In fact, to counter the climate crisis, it is essential to change some habits, trying to reduce our impact on the environment as much as possible [1]. To do this, however, in addition to electric means of transport, charging infrastructures are needed: that is, electric columns, systems that allow you to recharge cars, bicycles, and motorcycles through the connection to the electric grid. These devices, which can be private, semipublic, or completely public, must be structured in an adequate and coherent way in a complex system such as the urban one that allows their use in a systematic way. The transformation of transport toward sustainable and ecological mobility will inevitably pass from the national diffusion of the latest generation of fast and ultrafast charging stations, without which the electric and hybrid car market will not be able to take off despite the general willingness of users to purchase. of zero-emission vehicles [2–5]. It should be emphasized, however, that charging stations for electric vehicles are infrastructures that are only apparently simple from a technological point of view. In fact, they must meet stringent requirements from the point of view of functionality, operational performance, reliability, and safety.

3.2 Mobility trend

The global crisis has not interrupted the process of development, innovation, and transformation already underway in mobility for some years now. The contingency has in fact certainly triggered a process of propulsion of some of the emerging and already developing Macro-Trends, also in anticipation of the macroeconomic growth estimated for 2021. Specifically, the main trends to be mapped include:

- Vehicle to everything—V2X
- Mobility as a Service (MaaS)
- Autonomous vehicle
- Electrification or the switch to electrical technology

3.2.1 Vehicle to everything—V2X

V2X is the communication between vehicles, of any type, and anything else, that is, infrastructures, other vehicles, the web, other people, and digital devices of all types. The transmission of enormous amounts of data that a vehicle is now able to carry out with the surrounding ecosystem is a frontier of research and development, especially in the context of improving the efficiency of the use of vehicles, especially electric ones, and safety on board, with an impact, to tend, even on the urban scenario [6–8].

3.2.2 Mobility as a service (MaaS)

MaaS refers to the gradual transition to the consumer's preference to use on-demand services to obtain overall cost advantages for the use of mobility solutions, time spent on end-to-end travel, and overall experience. In particular, the prospect of buying a vehicle is less attractive today, due to elements such as congestion of the roads and the onerousness of maintaining a vehicle, in terms of cost and time. Last but not least, global mobility is embracing mobility service solutions also to achieve the planet's climate sustainability objectives, reducing emissions and pollution and the number of total vehicles in circulation. The main development segments of this area are: multimodality of transport, understood as a set of smart solutions capable of integrating end-to-end public and private transport (up to the last mile) ensuring complete coverage for the movement of the person; subscription services (on demand mobility), which aim to offer on the market no longer just vehicles,

Smart Cities Policies and Financing. https://doi.org/10.1016/B978-0-12-819130-9.00016-4

but a set of services, such as maintenance, insurance, within the price of a service that can be stipulated for a short period, medium, or long according to the consumer's needs. Finally, the services that can be activated through "Subscription" represent the frontier of change for mobility players, especially among the traditional ones, as they allow to establish the creation of short-term revenue models, thus allowing to manage with the contingency of crisis, which traditionally undermines the liquidity of companies in all the main sectors, is more effective [9–11].

3.2.3 Autonomous vehicle

The autonomous vehicle is one of the main frontiers of transformation in the way of conceiving the movement of people and things. An autonomous vehicle is in fact a product that combines a system of sensors, cameras, GPS, radar, and artificial intelligence (AI) software that simplifies the driver's interaction with the vehicle and the automation of the driving process, with the futuristic goal of completely replacing the driver of the vehicle, although in this sense it is still difficult to break down the resistance and skepticism of the demand in terms of effective safety of the solutions offered and implementation times and presence on the market, at affordable prices. To classify the level of autonomy of the AI on board a car, the automotive sector uses a scale ranging from 0 to 5, where the minimum corresponds to a nonexistent level of automation, while the maximum corresponds to complete automation. At this moment, the technological effort, also in terms of investments, of the large car manufacturers is projected to increase the production of Level 4 vehicles, there are more than 10 large manufacturers who in fact have envisaged the release of new autonomous models within the next 5 years. Instead, we will have to wait until at least 2030 to see the first Level 5 prototypes on the market, which will be gradually reached through constant developments implemented on Level 4 models [12–14].

3.2.4 Electrification or the switch to electrical technology

The phenomenon of electrification, in strong progress, affects all mobility, involving the whole sector of private transport, two and four wheels, commercial transport, light and heavy, public transport, and finally that of micromobility (e.g., electric bikes and scooters). The path traced is that of finding solutions capable of maximizing the energy efficiency of electrical systems and a strong reduction in greenhouse gas (CO_2) emissions on the planet. From this point of view, the most important area of development concerns the lithium battery, and in particular the management of the so-called second life of the battery in terms of reuse of the part of the energy charge at the end of its first use and circularity of the materials. rare ones needed to manufacture them (e.g., lithium, cobalt) [15–18].

3.3 Smart mobility and important factors

The trend of some enabling factors will determine the overall development and speed of implementation of the innovative mobility solutions. Among these, there is the role of the public actor in terms of subsidies and regulations capable of encouraging the construction of a context that stimulates investments in this area and facilitates the grounding of an electrical and interconnected ecosystem [19,20]. From this point of view, the shift toward the promotion of the electrical switch in all the most developed areas of the planet is clear, to stimulate both demands, reducing the overall cost of electric vehicles and supply, with the preparation of the context. infrastructural necessary for the development of effective electric mobility, for example, with the construction of efficient charging stations widely present in the territory, and a regulation capable of bridging the current gaps that facilitate the rapid development of some segments of new mobility, such as management of the 5G connection, the use of micromobility in cities, and safety related to autonomous driving. Then there is the front of private investments necessary to stimulate the growth of new mobility, in particular in terms of development of the connection and interconnection, which will see a strong turning point with the large-scale introduction of 5G connection technology, as well as with the consolidation of 3G/4G technology. In fact, to reach the desired level of connectivity, security, and dynamism in the flow of data between digital devices, it is essential that we take the path of 5G technology to redefine the paradigm of the connectivity system. Finally, it is clear that the mobility ecosystem, as it presents itself today, must still witness a full consolidation of the new value chains and related supply chains, by virtue of the interconnection of numerous new players and industry up to a few years ago very distant from a technological and product point of view.

The study of the initiatives in these areas has shown how, although these are covered and considered among the country's priorities, a holistic approach to the development of smart mobility is needed that is not achieved through post initiatives, but which is developed through the launch of a true strategic plan of sustainable mobility for the country that defines not only the roadmap but clarifies the best governance to accelerate development. Governments and local administrations must set benchmarks in the application of digital technologies to urban activities and in the provision of mobility services. Similarly, economic logic suggests that the authorities must show the political will to outsource city management processes, in order to solve the technological capacity gap and make the process more effective and faster. It is necessary to regulate the use of smart mobility technologies, products, and solutions that are not only "intelligent" but also extremely sustainable for the community (e.g., electric scooters) [21,22]. Cities will need to define a more complex regulatory framework to govern the complex issues of liability, security, and privacy. Cities need to define new mobility standards in order to ensure effective use of technology and manage the most complex urban mobility problems. In fact, it is quite evident that extraurban mobility is more advanced in terms of intelligence and speed than urban mobility. Urban mobility standards should therefore be defined in a shared way between all the actors involved not only in individual urban areas but also between urban areas [21,23].

Cities will have to rely on partnership agreements between the public and private sectors to build and manage both physical and digital infrastructure [24,25]. This, in particular, is due to ever smaller public budgets and the competence and technological requirements that no single municipality could reach and adequately evolve. The development of truly smart mobility in cities is therefore a function of the correct and virtuous interaction between multiple public and private actors, supported and enabled by technology in the creation of a citizen-centric ecosystem.

3.4 Mobility drivers and trends

The transport of people is one of the great drivers of the energy transition: In the coming years, the electron will achieve primacy over the fossil molecule. At present, the world fleet of road passenger vehicles is made up of 97.5% internal combustion engines and 2.5% electric. By 2050, the fleet will transform into 73% electric motor and 27% internal combustion engine. The main drivers of electrical transformation:

- Policies to reduce local and global emissions.
- The decrease in costs for the purchase of electric vehicles.
- Improved battery performance.

3.4.1 Policies to reduce local and global emissions

According to experts, by 2100, the temperature could even rise by 1.8–4°, especially with the increasing use of fossil fuels. The time has therefore come to reduce man-made CO_2 emissions. With a 14% share of all global greenhouse gas emissions, the transport sector is particularly important for achieving the carbon neutral impact of the Paris Agreement by 2050. To date, the best and most efficient way to reducing emissions from the transport sector is precisely electric mobility.

3.4.2 The decrease in costs for the purchase of electric vehicles

The standardization of the production of electric cars and of all components will increasingly make it possible to reduce the prices of E-mobility. For example, the adoption of modular platforms to accommodate battery packs, the increase in the energy density of the cells, the design of more efficient cooling systems will allow the optimization of production in its entirety.

3.4.3 Improved battery performance

The development of better-performing batteries will allow greater autonomy of electric vehicles, as well as a reduction in costs. Batteries are the engine of electric cars, as well as the component on which studies focus most. And it is precisely here that the greatest progress is being made; new materials, better technologies, greater standardization are all elements that will allow not only greater autonomy, but also lower and competitive costs compared to classic cars, thanks to the greater technological availability.

Also, those who decide to buy an electric car can, therefore, expect:

- Less expensive and more performing vehicles.
- Possibility of selling energy on the grid (vehicle-to-grid—V2G).
- Reduction of emissions and fight against climate change.

3.4.4 Less expensive and more performing vehicles

Due to the drivers listed above, electric machines will be increasingly comparable, in terms of costs, to internal combustion machines. Forecasts indicate that batteries will decrease by up to 50%, significantly breaking down the economic barriers to purchase. Electric vehicles will also perform better than classic ones, as is already the case with the acceleration issue: the drive torque of some models is better than that of traditional cars, i.e., it goes from 0 to 100 in shorter times.

3.4.5 Possibility of selling energy on the grid (vehicle-to-grid—V2G)

On average, forecasts say 10% of vehicle batteries will be available as V2G storage. With BEVs (battery electric vehicles) dominating sales within the next decade, this will account for a significant share of networked storage. V2G technology comes into play when the electric vehicle cable is plugged into the charging socket. At that point the car battery can be rented to the electricity service manager, who can absorb the energy to balance its distribution network. In fact, in order to function at their best, electricity networks need storage systems that enable them to respond promptly to users. Experimental projects are already active in Italy and the prospects are encouraging.

3.4.6 Reduction of emissions and fight against climate change

The spread of electric cars will help reduce the amount of greenhouse gases. As we have mentioned, the transport of goods and people has the highest growth in CO_2 emissions of any industry. The United Nations conference in Paris in 2015 recognized the importance of building more sustainable, greener, and smarter transport systems. Electric cars are an important tool for reducing emissions: although several studies have shown that, to date, the production of electric cars in some countries can generate more CO_2 emissions than that of an internal combustion vehicle, once in e-mobility function ensures fewer emissions in the well-to-wheel cycle.

Sustainable mobility is the future and will be increasingly accessible and advantageous from all points of view.

3.5 New actions for to improve the e-mobility

The introduction of electric car sharing in cities can be the ideal solution to let people test electric cars on the road and to "touch" the convenience and advantages of battery-powered vehicles. Therefore, it represents an opportunity to educate on electric mobility and to accustom the user to the vehicle before buying it. Electric car sharing is also the most efficient solution for cities: it combines vehicle sharing with zero-emission technology, with positive effects both on traffic, thanks to the reduction of cars on the road, and on the environment, thanks to zero traffic emissions. Even traditional traction car sharing has had difficulties in establishing itself and finding a winning formula: initially, in fact, due to redelivery constraints in reserved parking lots (few and often illegally occupied by other vehicles) and the heavy penalties caused by noncompliance, the use of car sharing was not widespread. Thanks to computerization, the use of smartphones for booking, free parking, and the ability to enter the ZTL, the system has taken hold in many large cities. The electric car sharing initiatives, financed and supported by national and/or local public authorities, have also struggled initially to get off the ground and today have a good response among citizens.

However, car sharing could integrate into the business context and be the most suitable solution to meet the need for on-demand mobility, while relieving the company of the burden of ownership or long-term contracts when not needed: they reduced fixed costs and charges. Companies can no longer think in a standard way and need to adopt flexible models that combine different solutions, each suited to specific needs of use, even temporary ones.

The business context is very different from the urban one. The use of the vehicles of the company fleet is unfriendly: there are days with peaks of use and others in which the car fleet remains unused for some time. This normally occurs in the time slot from 18:00 to 9:00; in this slot it would therefore be possible to allow employees to use a company car to go home and make it available and ready to be used for business purposes, thus activating a sort of car sharing between employee and company. Today, rental or long-term leasing formulas are giving a strong boost to electric mobility, at least in the initial phase of the market. In fact, rental allows you to overcome the still high-cost obstacle, waiting for the prices of electric cars to drop as the cost of the battery, which weighs heavily on the final price list of the car, decreases.

However, the concessions should help reduce the price gap for electric vehicles compared to internal combustion vehicles and/or should help facilitate the installation of charging systems, possibly giving priority to electrified local public transport (or "collective electric mobility"), since the latter would pursue shared objectives also at European level (decarbonization and mobility sustainable), with an important plus of sustainability, making it possible to reduce emissions per passenger per km compared to individual mobility, thanks to the higher vehicle occupancy factor. It is believed that the concessions must be explicit, so that they can be more transparent, monitorable, and efficient, and it is also believed that they should not consist of exemptions from the application of tariff components (including those covering general system charges). In order, therefore, to stimulate the electrification of mobility, the proposals already presented that aim to reduce, in particular, the general system charges that weigh on the bills of all electricity users could be effective.

At the same time, with particular regard to the installation of recharging systems, the possibility of providing facilities for the installation of devices useful for optimal management of the electrical system could be considered. For example, contributions to reduce connection costs could be defined in the case of recharging infrastructures that allow intelligent management of recharging (smart charging), according to requirements that ensure optimal management of the available power, to be defined during technical standardization; promote the spread of integration technology between electric vehicles and the electricity grid. In this case, the concessions could consist in covering the additional standard costs for the installation of devices that allow dialogue between the charging infrastructure manager and its intermediary and/or network manager, for the purpose of providing services. ancillary costs, or in covering the higher standard costs (or part of them) that distinguish vehicle to grid solutions compared to traditional charging systems (which, not being equipped with bidirectional capacity, do not allow the entry into the network of electricity stored in vehicle batteries). To incentivize the market for a gradual but progressive transition to electric, the hybrid technology of PHEVs (plug-in hybrid electric vehicles) represents a possible solution to accompany speed of the change to a more sustainable growth of the market in the short and medium term. Hybrid models can in fact play a decisive role in the process of educating and raising awareness of users, accustoming them to make greater use of the electrical component of the engine and gradually abandoning traditional fuels, thanks to a driving range that can guarantee satisfactory performance even on average and long distances.

3.6 Summary

At least three are the key points to be addressed in order to have a widespread diffusion of the electric car. The first concerns the charging infrastructure and the adaptation of the electricity grid. The diffusion mass will inevitably require coordinated work between the car manufacturers, the company's electricity, and public administrations so that recharging, as well as in private, can be also carried out in the public area. From this comes a strong standardization and the need for solutions intelligent, as far as communication between vehicles and infrastructure is concerned, which are capable beyond that to identify the user to provide for the billing of consumption, also to manage the batteries in a perspective of use as a distributed storage, managing the flow of power in such a way bidirectional (V2G).

The second, the battery life: In recent years there has been great progress in this field through both laboratory and demonstration tests [26,27]. It is understandable that car manufacturers, before facing huge investments for mass production, want to have more and more consolidated certainties about performance and durability [28,29].

Finally, the economic aspect: If on the one hand the introduction of the electric car would lead, once fully operational, to the much sought-after environmental sustainability and to a reduction in costs for users, on the other hand it will certainly present an introductory phase of considerable difficulty, not only from the point of view in view of the distortion of the consolidated habits linked to old cars but above all from the economic point of view. In this sense, the community will have to bear the considerable costs for starting the market, also as regards the charging infrastructures. All these aspects must certainly not be underestimated, they are as important as the technological aspect.

Finally, let us move on to the real interactive part of this chapter: review questions/exercises, hands-on projects, case projects, and optional team case project. The answers and/or solutions by chapter can be found in Appendix G.

3.7 Chapter review questions/exercises

True/false

1. True or False? Often referred to as a revolution, electric mobility is one of the smallest challenges in the transport sector.
2. True or False? The global crisis has not interrupted the process of development, innovation, and transformation already underway in mobility for some years now.
3. True or False? V2X is the communication between vehicles, of any type, and anything else, that is, infrastructures, other vehicles, the web, other people, and digital devices of all types.
4. True or False? MaaS refers to the fast transition to the consumer's preference to use on-demand services to obtain overall cost advantages for the use of mobility solutions, time spent on end-to-end travel, and overall experience.
5. True or False? The autonomous vehicle is one of the main frontiers of transformation in the way of conceiving the movement of people and things.

Multiple choice

1. The phenomenon of ___________, in strong progress, affects all mobility, involving the whole sector of private transport, two and four wheels, commercial transport, light and heavy, public transport, and finally that of micromobility.
   a. Electrification
   b. Greenhouse gas
   c. Second life
   d. Interconnected ecosystem
   e. Electric vehicles
2. Mobility is needed that is not achieved through post initiatives, but which is developed through the launch of a true strategic plan of sustainable mobility for the country that defines not only the roadmap, but clarifies the best ___________ to accelerate development.
   a. Governance
   b. Physical infrastructure
   c. Smart city growth and prosperity
   d. Economic effects
   e. Smart city policy
3. The transport of people is one of the great drivers of the energy transition: In the coming years, the electron will achieve primacy over the:
   a. Fossil molecule
   b. Internal combustion engine
   c. Electric motor
   d. Electrical transformation
   e. Global emissions
4. According to experts, by 2100, the temperature could even rise by ___________, especially with the increasing use of fossil fuels.
   a. 2.8–5°
   b. 3.8–6°
   c. 1.8–4°
   d. 1.6–3°
   e. 1.4–2°

5. The standardization of the production of electric cars and of all components will increasingly make it possible to reduce the prices of:
   a. Electric vehicles
   b. Batteries
   c. E-mobility
   d. Electric machines
   e. Vehicle-to-grid storage

Exercise

Problem

In designing smart cities and infrastructure, what are the key design principles that need to be followed?

Hands-on projects

Project

Do research: Analyze why smart cities and infrastructure open new avenues, whereby local STI communities may actively contribute to sustainable urbanization.

Case projects

Problem

Why are existing national STI and ICT policies not adequate enough to integrate the needs of developing smart city initiatives?

Optional team case project

Problem

What are the broad array of tools that governments have at their disposal to develop smart cities?

References

[1] Slowik P, Kamakaté F. New Mobility: Today's Tecnology and policy landshape. Retrieved 30 April 2021 from: https://theicct.org/sites/default/files/publications/New-mobility-landscape_ICCT-white-paper_27072017_vF.pdf.

[2] Bastian A, Börjesson M. The city as a driver of new mobility patterns, cycling and gender equality: travel behaviour trends in Stockholm 1985–2015. Travel Behav Soc 2018;13:71–87. https://doi.org/10.1016/j.tbs.2018.06.003. ISSN 2214-367X.

[3] van Wee B. Peak car: the first signs of a shift towards ICT-based activities replacing travel? A discussion paper. Transp Pol 2015;42:1–3. https://doi.org/10.1016/j.tranpol.2015.04.002.

[4] Can A, L'Hostis A, Aumond P, Botteldooren D, Coelho MC, Guarnaccia C, Kang J. The future of urban sound environments: impacting mobility trends and insights for noise assessment and mitigation. Appl Acoust 2020;170:107518. https://doi.org/10.1016/j.apacoust.2020.107518. ISSN 0003-682X.

[5] Franco P. Emerging trends in transport demand modeling in the transition toward shared mobility and autonomy. In: Vickerman R, editor. International encyclopedia of transportation. Elsevier; 2021, ISBN 9780081026724. p. 331–9. https://doi.org/10.1016/B978-0-08-102671-7.10769-9.

[6] Li S, Gu C, Zeng X, Zhao P, Pei X, Cheng S. Vehicle-to-grid management for multi-time scale grid power balancing. Energy 2021;234:121201. https://doi.org/10.1016/j.energy.2021.121201. ISSN 0360-5442.

[7] Kavousi-Fard A, Abunasri A, Zare A, Hoseinzadeh R. Impact of plug-in hybrid electric vehicles charging demand on the optimal energy management of renewable micro-grids. Energy 2014/12/15/2014;78:904–15.

[8] Luo Y, Zhu T, Wan S, Zhang S, Li K. Optimal charging scheduling for large-scale EV (electric vehicle) deployment based on the interaction of the smart-grid and intelligent-transport systems. Energy 2016/02/15/2016;97:359–68.

[9] Lajas R, Macário R. "Public policy framework supporting "mobility-as-a-service" implementation. Res Transport Econ 2020;83:100905. https://doi.org/10.1016/j.retrec.2020.100905. ISSN 0739-8859.

[10] Sochor J, Karlsson ICM, Strömberg H. Trying out mobility as a service: experiences from a field trial and implications for understanding demand. Transport Res Rec 2016;2542:57–64. https://doi.org/10.3141/2542-07.

[11] Smith G, Sochor J, Karlsson IM. Mobility as a Service: implications for future mainstream public transport. In: International conference series on competition and ownership in land passenger transport; 2017. p. 15.

[12] Tian Z, Feng T, Timmermans HJP, Yao B. Using autonomous vehicles or shared cars? Results of a stated choice experiment. Transport Res C Emerg Technol 2021;128:103117. https://doi.org/10.1016/j.trc.2021.103117. ISSN 0968-090X.

[13] Acheampong RA, Cugurullo F, Gueriau M, Dusparic I. Can autonomous vehicles enable sustainable mobility in future cities? Insights and policy challenges from user preferences over different urban transport options. Cities 2021;112:103134. https://doi.org/10.1016/j.cities.2021.103134. ISSN 0264-2751.

[14] Pham M, Xiong K. A survey on security attacks and defense techniques for connected and autonomous vehicles. Comput Secur 2021;109:102269. https://doi.org/10.1016/j.cose.2021.102269. ISSN 0167-4048.

[15] Sharma DK, Rapaka GK, Pasupulla AP, Jaiswal S, Abadar K, Kaur H. A review on smart grid telecommunication system. Mater Today Proceed 2021. https://doi.org/10.1016/j.matpr.2021.05.581. ISSN 2214-7853.

[16] Sami BS. Prosumer in smart grids based on intelligent edge computing: A review on Artificial Intelligence Scheduling Techniques. Ain Shams Engi J 2021. https://doi.org/10.1016/j.asej.2021.05.018. ISSN 2090-4479.

[17] Fan D, Ren Y, Feng Q, Liu Y, Wang Z, Lin J. Restoration of smart grids: current status, challenges, and opportunities. Renew Sustain Energy Rev 2021;143:110909. https://doi.org/10.1016/j.rser.2021.110909. ISSN 1364-0321.

[18] Kumar Panda D, Das S. Smart grid architecture model for control, optimization and data analytics of future power networks with more renewable energy. J Clean Prod 2021;301:126877. https://doi.org/10.1016/j.jclepro.2021.126877. ISSN 0959-6526.

[19] van Oers L, de Hoop E, Jolivet E, Marvin S, Späth P, Raven R. The politics of smart expectations: interrogating the knowledge claims of smart mobility. Futures 2020;122:102604. https://doi.org/10.1016/j.futures.2020.102604. ISSN 0016-3287.
[20] Kamargianni M, Li W, Matyas M, Schafer A. A critical review of new mobility services for urban transport. Transport Res Procedia 2016;14:3294–303.
[21] Rahman I, Vasant PM, Singh BSM, Abdullah-Al-Wadud M, Adnan N. Review of recent trends in optimization techniques for plug-in hybrid, and electric vehicle charging infrastructures. Renew Sustain Energy Rev 2016;58:1039–47.
[22] How do electric car batteries work? Retrieved 30 April 2021 from: https://www.edfenergy.com/electric-cars/batteries.
[23] Shuai W, Maillé P, Pelov A. Charging electric vehicles in the smart city: a survey of economy-driven approaches. IEEE Trans Intell Transport Syst 2016;17:2089–106.
[24] Sanguesa JA, Torres-Sanz V, Garrido P, Martinez FJ, Marquez-Barja JM. A review on electric vehicles: technologies and challenges. Smart Cities 2021;4:372–404. https://doi.org/10.3390/smartcities4010022.
[25] Richardson DB. Electric vehicles and the electric grid: a review of modelling approaches, Impacts, and renewable energy integration. Renew Sustain Energy Rev 2013;19:247–54.
[26] Canals Casals L, Amante García B, Canal C. Second life batteries lifespan: rest of useful life and environmental analysis. J Environ Manag 2019;232:354–63. https://doi.org/10.1016/j.jenvman.2018.11.046. ISSN 0301-4797.
[27] Wood M, Alexander TH. Bradley Investigation of battery end-of-life conditions for plug-in hybrid electric vehicles. J Power Sources 2011;196:5147–54. https://doi.org/10.1016/j.jpowsour.2011.02.025.
[28] Swierczynski M, Stroe DI, Stan AI, Teodorescu R, Vikelgaard H. Selection and impedance based model of a lithium ion battery technology for integration with virtual power plant. In: Power electronics and applications (EPE), 2013 15th European conference on; 2013. https://doi.org/10.1109/EPE.2013.6634755.
[29] Mathews I, Xu B, He W, Barreto V, Buonassisi T, Peters IM. Technoeconomic model of second-life batteries for utility-scale solar considering calendar and cycle aging. Appl Energy 2020;269:115127. https://doi.org/10.1016/j.apenergy.2020.115127. ISSN 0306-2619.

Chapter 4

Harnessing the local innovation system: an overarching solution for policy and financing instruments for promoting the localization of smart infrastructure

Michela Longo[1] and Wahiba Yaici[2]
[1]*Politecnico di Milano, Milan, Italy;* [2]*CanmetENERGY Research Centre, Ottawa, ON, Canada*

4.1 Introduction

The electric mobility ecosystem benefits from a historic wave of innovation, from 5G connectivity, to increasingly green modes that are rapidly evolving [1]. Digital has increased the panorama of operators involved in the provision of transport services: a digital role is played, for example, by IT operators, whose presence is crucial in the offer of mobility services. Finally, digital has generated an opening of the market to competition: not only has the increase in digital accessibility to services reduced the information asymmetry between competitors on international markets, but also the barriers to entry typical of the market for transport operators have been drastically reduced since the advent of digital technology, to the extent that competition has shifted from the offer of mobility to the control of information useful for enabling new services. This raises the issue of data and process governance with increasing force, as well as the regulation of new mobility models.

4.2 Digital player

Designing mobility strategies today means identifying the priorities for the country's industrial policy. The digital player has in fact the effect of expanding the perimeter of the ecosystem of infrastructures, transport, and the territory, which by its nature is articulated and complex in terms of relations between public and private operators, on a local, national, and international scale.

Fig. 4.1 shows the different players necessary to consider for the new mobility models.

The opening of markets to international players, also favored by digital and by the large-scale use of data platforms for mobility, has led to a general expansion of mobility services, such as Car Sharing, Car pooling, and the introduction of new models for the management of mobility services thanks to the strong interrelationships with neighboring sectors, such as telecommunications and energy distribution networks. As far as the telecommunications sector is concerned, connectivity is a determining element in the provision and use of new mobility services, but it has a growing weight also in the more traditional transport segments, bringing significant added value in the generation of travel experiences seamless for passengers and the supply chain. The advent of 5G, currently being tested globally, promises to constitute a revolution in the telecommunications sector: the new mobile standard will have pervasive operational repercussions in all mobile-enabled sectors and that of infrastructure, transport, and mobility. The Smart Mobility market and its constant process of expansion of the stakeholders involved are based on four fundamental drivers: (1) information and communication technology that plays an enabling role in boosting the demand for mobility and allows the constant exchange of data between the various systems, facilitating decision-making processes aimed at improving the quality of life of citizens; (2) infrastructures and public transport; (3) the involvement of citizens both as users and as developers of the new Smart Cities; (4) the strengthening of the partnership between public and private to facilitate the development of projects aimed at making mobility increasingly smart.

Smart Cities Policies and Financing. https://doi.org/10.1016/B978-0-12-819130-9.00053-X

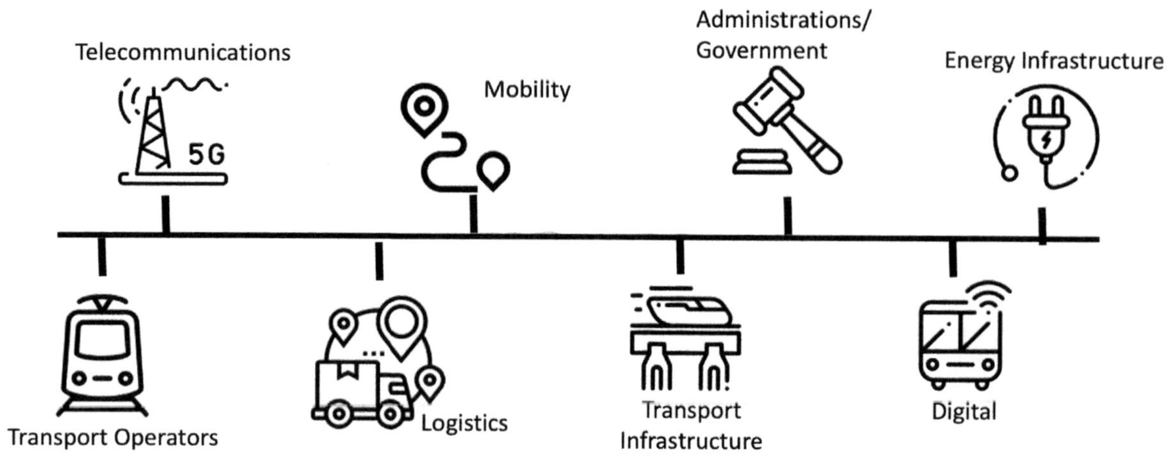

FIGURE 4.1 Different players for the new mobility models.

4.3 How mobility is changing

Technological and business model innovations are bringing about a profound change in the paradigms of mobility. Passenger habits and preferences are changing, but so are the needs of the production sector, which is increasingly inclined to satisfy the demands of a fast market based on the principle of same-day delivery. Mobility is conceived as a service: in the meaning of Mobility-as-a-Service (MaaS), the mobility experience must be as "seamless" as possible, without solution of continuity and at the same time satisfy the need for multimodal travel and access to "on demand" services [2,3].

Relevant in the passenger transport segment, MaaS consists in the integration of various types of transport services into a single mobility service, possibly accessible from a single platform with a unified payment channel [4,5]. Fig. 4.2 represents the new mode to live the mobility.

Directly connected to the idea of on-demand service is the concept of pay-per-use, which somehow replaces the traditional ownership of vehicles and integrates the traditional forms of payment based on hourly or kilometer rates: if the mobility service is calibrated exactly on the passenger's needs, the latter pays the amount due based on the actual use of the service or through inclusive subscription formulas. The transformation introduced by the MaaS has significant impacts on the business models of transport operators, but it also questions the policymaker on the need to favor the clustering of operators and to ensure uniform governance of services. In fact, if transport operators spontaneously tend to organize themselves according to market dynamics, the role of the public is more than ever necessary in guaranteeing the necessary integration of public and/or public utility services and in reducing the risk that the more peripheral areas remain excluded from the evolution of the sector and from access to new mobility services. However, in addition to this new way of experiencing mobility, the increasingly widespread use of electric vehicles must go hand in hand with the spread of charging infrastructure [6–9].

4.4 Charging infrastructure

When we are talking about charging stations, we are talking about devices on the road where it is possible to fill up the car's batteries with energy [10,11]. There is no standard form for these stations, and research is also improving induction systems that in the future may allow motorists to do without the charging cable. In fact, there are some pilot projects creating the Smart Road.

There are two types of electricity columns: public and private [12,13]. The former is installed by energy suppliers and network operators and can be located on public roads or in public car parks located at airports or railway stations, while the latter are those connected to the local network and present at homes, car parks, shops, restaurants, hotels, and so on. At the moment, the technology provides four different charging modes [14]:

- Slow charging only allowed in a private domestic environment. You can use a simple domestic socket or an industrial socket up to 32 A;
- Slow charging allowed in domestic and public environments. On the vehicle power cable there is a device called Control Box (PWM safety system), which guarantees the safety of operations during recharging. Usable sockets are domestic or industrial ones up to 32 A.
- Slow or medium-fast charging (Mode with PWM safety system), allowed in domestic and public environments. Charging must take place via a special power supply system equipped with specific connectors;
- Ultrafast charging in direct current up to 200 A, 400 V, only allowed in public environments. With this system it is possible to recharge the vehicles in a few minutes, the charger is external to the vehicle.

FIGURE 4.2 New user and new mobility.

Knowing in advance where the electric charging stations are is a fundamental aspect for a correct use of the car and eliminating the anxiety factor. However, the cars have integrated this type of application that allows the driver to accompany the driver for the entire duration of the journey, indicating the charging stations along the way.

Precisely on this aspect, the energy supply companies in collaboration also with the same administrations collaborate with each other in order to satisfy citizens' needs as much as possible by providing APPs or services that show the widespread diffusion of infrastructures.

4.5 Charging stations for public enviroment

To have a widespread diffusion of the infrastructure, administrations must consider different application scenarios and respond to people's needs. Attention obviously focuses on the public environment as for the private one, the user of the charging infrastructure is the sole responsible the environment in which the vehicle is charged, for example: box of a detached house, private garage of an apartment building, environment/place not accessible by third parties. The charging of the vehicle takes place under the responsibility of the owner of the private place. Certainly, thanks to incentives from the Government, this can allow a high diffusion of infrastructure. However, those who privately integrate this type of charging column mean that they own an electric vehicle [15,16].

To guarantee users the level of flexibility and the sense of security necessary to get the electric car market off the ground, the design of public charging systems and the choice of the related technical characteristics (e.g., installed power, charging time, number of available sockets, etc.) must be carefully evaluated in light of the real uses of these recharging points or in accordance with the installation scenarios [17]. It is therefore of great importance to match the recharging systems to the places of installation so that, ideally, the recharging times of the batteries are compatible with the activities that are carried out by users during recharging.

Additional design criteria are given by the costs of equipment and installation, but also by the consequences for the plants and electrical networks to which they are connected. To ensure interoperability between the various recharging infrastructures, the related management systems should be able to communicate with each other, in relation to authorizations and recharging data.

In Italy, for example, charging modes 1 and 2 are not allowed in public and private areas accessible to the public, which can be achieved by directly connecting the vehicle to the electrical network via common fixed sockets. Therefore, for all the scenarios indicated below, the use of charging modes 3 and 4 requires a dedicated power line and equipment that incorporates the relative control and communication circuits. Depending on the scenarios in which such charging systems are installed, three can be identified typical configurations.

4.5.1 Scenario 1: extended stop

The first scenario identified is characteristic of the places where vehicles stop for long periods. These areas generally correspond to parking lots at the workplace, single parking lots in garages, condominium or company fleet parking lots, parking lots along the sidewalk or in public structures, occupied by vehicles during the night or during the day for prolonged parking.

The recharging systems chosen offer a low-power recharge service ("normal" recharge about 3 kW in alternating current) for an extended period of time (typically 2–10 h); therefore, they must be able to charge several electric vehicles at the same time. This type of recharge is the one that—except for special needs—should be the most frequent. It can meet the user's so-called "primary" charging needs, "covering" the average daily distance of the vehicle during the day, with exceptions and needs of more intense use that can be served by the accelerated and fast charging systems referred to in subsequent Scenario 2 and Scenario 3. This solution can be supplied with multisocket systems to which several vehicles can be connected at the same time; or, with several independent systems, each equipped with one or more fixed recharging sockets to which users connect with their own cable, regardless of the connector available to the vehicle for low power charging:

- Type 2 and/or type 3C sockets for recharging points for passenger cars (i.e., vehicles not light),
- 3A type sockets for charging points for light vehicles (scooters, quadricycles, and the like).

For the purpose of counting the number of recharging points to be installed in consideration of the electric vehicles, it can be realistically estimated that, in the case of cars, every point can serve one vehicle a day.

4.5.2 Scenario 2: short stop

The second scenario is typical of transport hubs or commercial areas (cinemas, restaurants, shopping centers, etc.), where the duration of the stop is short (typically 30 min–2 h). This type of recharging, of the discontinuous type, is also suitable in company car parks where, alongside systems equipped with "normal" recharging (ref. Scenario 1), it can offer a complete recharge in the time typical of a short meeting or lunch break. Provided that the duration of the parking allows sufficient recovery of the range, this type of charging can satisfy the so-called "secondary" charging needs: additional mileage of the vehicle during the day, greater flexibility for the user.

The recharging systems chosen therefore mainly offer a medium power recharging service ("accelerated" recharging: over 3 kW and up to 22 kW, depending on the characteristics of the vehicle). As for scenario 1, this solution can be provided with multisocket systems to connect to several vehicles at the same time, or with several independent systems, each equipped with one or more charging sockets.

Unlike scenario 1, given the current diversity of technology vehicles (current alternating or direct) and for the connector in use for medium power charging, in order to be able to recharge the largest number of vehicles, it will therefore be appropriate to provide different types of charging points. For example, at least one AC connection point of 20–22 kW as well as a DC connection point of equal power. For the direct current connection points (mode 4) two systems can be envisaged, depending on the vehicles to be loaded, as indicated in the following Scenario 3.

For the connection points in alternating current (mode 3), you can choose between two possibilities, both able to charge vehicles suitable for recharging in alternating current at medium power:

- Fixed type 2 and/or type 3C sockets,
- Type 2 mobile connectors with cable fixed to the charging station.

In the first case, the same socket can also be used by vehicles equipped with a type 1 connector, using the cable supplied with the vehicle (but with these vehicles the power will not exceed 7 kW). The solution with fixed cable and type 2 connector, on the other hand, favors a more comfortable use of the charging station. For the purpose of counting the number of charging points to be installed in consideration of the electric vehicles in circulation, it can be realistically estimated that each point can serve 2–5 vehicles per day.

4.5.3 Scenario 3: stop for charging

Fast charging (43–50 kW) meets the needs of unexpected additional mileage and certain fleets of vehicles that necessarily require a recharge to be restored very quickly (for example, taxis and commercial vehicles for urban use). Offering this type of charging provides reassurance to the driver of the electric vehicle and ensures flexibility in the event of unforeseen user needs. Its use should generally remain sporadic, and tariffs applied to end users should reflect the insurance premium it represents to the user. Unlike Scenario 2, it will therefore be necessary to provide high-power charging points (43–50 kW).

Also, in scenario 3, the solution can be supplied with multisocket systems to which several vehicles can be connected at the same time, or with several independent systems, each equipped with one or more charging connectors. Given the current diversity of electric vehicles with recharging technologies (alternating or direct current) and different connectors, in order to be able to recharge the largest number of vehicles, it will be appropriate to provide recharging points with a power of 43–50 kW with following three different types of connectors:

- Type 2 mobile connectors for fast charging in mode 3 (AC) at 43 kW,
- Mobile connectors for fast charging in mode 4 (CC) Combined Charging System (Combo2),
- Mobile connectors for fast charging in mode 4 (CC) CHAdeMO system.

For the purpose of counting the number of recharging points to be installed in consideration of the Electric Vehicles, it can be realistically estimated that each point can serve 12 vehicles per day.

The spread of high-power recharging points, to ensure greater compatibility with electric vehicles on the road, should be divided into the three scenarios indicated. In this regard, it must be said that car manufacturers use both fast-alternating current charging based on the type 2 connector and direct current charging based on the Combo 2 connector. However, the CHAdeMO system already in use, supported by Japanese car manufacturers, will continue to be very popular in Europe due to the high market share of Japanese vehicles and European vehicles with Japanese technology. An interesting solution are the "multi-standard" charging systems, i.e., equipped with connectors in two or three of the existing systems, which allow to contain costs and space occupation, compared to solutions with three separate systems, having part of the shared components.

4.6 Cost of charging an electric vehicle

The spread of electric vehicles brings with it the need for a network of recharging points. The development of electric mobility will therefore have inevitable and important repercussions on the evolution of the national electricity system. It is therefore important to carefully evaluate the impact that this new demand may have on the electricity grids, in particular on the low and medium voltage distribution grids, where the withdrawals of the charging infrastructures will be attested. In this regard, it should be noted that the costs for the development, operation, and maintenance of the electricity grids are borne by all users of the electricity system, through the application of tariffs for grid services. However, depending on whether the column is public or private, the costs may vary.

To date, different access and billing systems coexist. In collective garages, electricity is metered by the property's common meter; and the amount billed is divided among the various tenants. In these cases, a separate meter can be installed for charging points or the flat rate billing method can be used. In shopping centers, restaurants, and corporate parking lots, it may be in the interest of the charging point provider to limit administrative costs and offer free EV charging to its customers. Models for defining costs and payment for charging energy are developing slowly and there are regulatory barriers. Some proposals favor billing by the energy provider directly to the vehicle owner's home electric utility. Others propose prepaid and billed services similar to mobile phone payment models.

Charging an electric car at the home socket can cost from 0 €/kWh, if the house is equipped with a photovoltaic system and you can only use its energy, at about 0.20 €/kWh, considering the cost average network current: given that these vehicles consume about 0.14 kWh/km, to travel 100 km, therefore, the cost of recharging from home would range from 0 to 2.8 €.

The minimum cost of electricity for public charging stations is € 0.40/kWh, double the domestic cost. And if it were only possible to recharge at the fast columns, the cost in proportion to their power could reach up to € 0.75/kWh. The public tariff consists of the presence of system charges and other parafiscal charges, in addition to taxes, which considerably reduce the margins for setting prices for operators. Costs in Europe are fairly in line with those in Italy. Perhaps the difference is that in European countries where the penetration of electric vehicles is more advanced, flat charging subscriptions are offered at controlled costs, also associated with the purchase of the car or as additions to the supply of domestic energy.

4.7 Summary

The use of electric vehicles to replace traditional ones is still the subject of many studies and research in various parts of the world, but the common attention to what would be a real environmental and technological revolution has never been so strong. It is perceived by the new way of experiencing the concept of mobility (Mobility-as-a-Service, Car Sharing, Car Polling).

There are many factors that push research toward the study of electric vehicles, first of all air pollution and the increasingly pressing need for sustainable development, i.e., a development that is able to meet current needs without compromising the possibilities of future generation to satisfy their own.

The economic aspect is perhaps the most delicate in order to spread the concept of green mobility. If on the one hand, the introduction of the electric car would lead, once fully operational, to the much sought-after environmental sustainability and also to a reduction in costs for users, on the other hand, it will certainly present an introductory phase of considerable difficulty, not only from the point of view in view of the distortion of the consolidated habits linked to old cars but above all from the economic point of view. In this sense, the community will have to bear extra costs for the launch of the market, also as regards the charging infrastructures. All these aspects must certainly not be underestimated, they are as important as the technological aspect, therefore it is necessary to create National Plans in favor of electric vehicles, as is happening in many countries. However, the incentive factor is a strong lever for the diffusion both for the purchase of an electric vehicle and for the integration of a private charging station. The goal is to create a critical mass of charging points so that companies guarantee mass production of cars at reasonable prices. A universal connector at EU level is an essential element for the spread of this type of mobility, trying to standardize the economic aspect as much as possible.

Finally, let us move on to the real interactive part of this chapter: review questions/exercises, hands-on projects, case projects, and optional team case project. The answers and/or solutions by chapter can be found in Appendix G.

4.8 Chapter review questions/exercises

True/false

1. True or False? The electric mobility ecosystem does not benefit from an historic wave of innovation—from 5G connectivity—to increasingly green modes that are rapidly evolving.
2. True or False? Designing mobility strategies today means identifying nonpriority items for the country's industrial policy.
3. True or False? Technological and business model innovations are bringing about a profound change in the paradigms of mobility.
4. True or False? When we are talking about charging stations, we are talking about devices on the road where it is possible to fill up the car's batteries with energy.
5. True or False? To have a widespread diffusion of the infrastructure, administrations must consider the same application scenarios and respond to people's needs.

Multiple choice

1. The recharging systems chosen offer a ________________ ("normal" recharge about 3 kW in alternating current) for an extended period of time (typically 2–10 h); therefore, they must be able to charge several electric vehicles at the same time.
   a. Very high-power recharge service
   b. Normal-power recharge service
   c. High-power recharge service
   d. Low-power recharge service
   e. Very low-power recharge service
2. Provided that the duration of the parking allows sufficient recovery of the range, this type of charging can satisfy the so-called ________ charging needs: additional mileage of the vehicle during the day, greater flexibility for the user.
   a. First
   b. Secondary
   c. Third
   d. Fourth
   e. Fifth
3. Fast charging (43–50 kW) meets the needs of unexpected additional mileage and certain fleets of vehicles that necessarily require a recharge to be restored very quickly (for example, taxis and commercial vehicles for urban use).
   a. 53–60 kW
   b. 63–70 kW
   c. 73–80 kW
   d. 43–50 kW
   e. 33–40 kW
4. The spread of electric vehicles brings with it the need for a network of:
   a. Recharging points
   b. Electric mobility
   c. The national electricity system
   d. Electricity grids
   e. Distribution grids
5. The use of ___________ to replace traditional ones is still the subject of many studies and research in various parts of the world, but the common attention to what would be a real environmental and technological revolution has never been so strong.
   a. Mobility-as-a-Service
   b. Car Sharing
   c. Electric vehicles
   d. Car Polling
   e. Old cars

Exercise

Problem

Why does smart infrastructure have the potential to promote inclusive development (including gender inclusiveness) in cities by helping to generate data on informal settlements and informal sectors and other marginalized groups in society (including women, the elderly, and persons with disabilities)?

Hands-on projects

Project

Do research: Analyze the significant scope for regional and cross-country collaboration and intercity learning with regard to the design, development, and management of smart cities and infrastructure.

Case projects

Problem

Do research: Collaborate with all relevant stakeholders to define a concept and vision of a smart city that is relevant for local contexts; that can help respond to specific local sustainable urban development needs; integrate smart city agendas into national STI and ICT policies; and have broader national development frameworks.

Optional team case project

Problem

Do research: Encourage national STI systems (including science parks and technology hubs) to make smart cities and infrastructure a priority theme by providing appropriate incentives and create an enabling policy framework in order for local innovation systems to generate and implement innovative smart infrastructure concepts that respond to specific local needs.

References

[1] Heineke K, Hornik T, Schwedhelm D, Szilvacsku I. Defining and seizing the mobility ecosystem opportunity. 2021. Retrieved 30 April 2021 from: https://www.mckinsey.com/industries/automotive-and-assembly/our-insights/defining-and-seizing-the-mobility-ecosystem-opportunity.

[2] MaaS Alliance. White paper. Guidelines and recommendations to create the foundations for a thriving MaaS ecosystem. 2017. Retrieved 30 April 2021 from:: https://www.drivesweden.eu/en/white-paper-maas-ecosystems.

[3] Kamargianni M, Matyas M. The business ecosystem of mobility-as-a-service. In: Conference: transportation research board; 2017. Retrieved 02 May 2021 from: https://www.researchgate.net/pub-lication/314760234_The_Business_Ecosystem_of_Mobility-as-a-Service.

[4] Matyas M, Kamargianni M. The potential of mobility as a service bundles as a mobility management tool. Transportation 2018;46:1–18. https://doi.org/10.1007/s11116-018-9913-4.

[5] Huhtala-Jenks K. Digital services and mobility as a service at ministry of transport and communication, Finland. 2016. Retrieved 30 April 2021 from: http://sunergos.ee/wp-content/up-loads/2018/10/From-Transport-to-Mobility-as-a-Service-Krista-Huhtala-Jenks.pdf.

[6] Hausler F, Crisostomi E, Schlote A, Radusch I, Shorten R. Stochastic park-and-charge balancing for fully electric and plug-in hybrid vehicles. IET Intell Transp Syst April 2014;15(2):895–901.

[7] Faria R, Moura P, Delgado J, de Almeida AT. Managing the charging of electrical vehicles: impacts on the electrical grid and on the environment. IEEE Intell Transp Syst Mag Fall 2014;6(3):54–65.

[8] Fox GH. Electric vehicle charging stations: are we prepared? IEEE Ind Appl Mag July/August 2013;19(4):32–8.

[9] Baouche F, Billot R, Trigui R, El Faouzi N-E. Efficient allocation of electric vehicles charging stations: optimization model and application to a dense urban network. IEEE Intell Transp Syst Mag Fall 2014;6(3):33–4.

[10] Timpner J, Wolf L. Design and evaluation of charging station scheduling strategies for electric vehicles. IET Intell Transp Syst April 2014;15(2):579–88.

[11] Singh S, Payne SR, Jennings PA. Toward a methodology for assessing electric vehicle exterior sounds. IEEE Intell Transp Syst Mag August 2014;15(4):1790–800.

[12] Bowermaster D, Gartner J, Hajjar N, Pronovost C. Overcoming barriers to deployment of plug-in electric vehicles. Washington, DC, USA: Nat. Res. Council Nat. Acad., Nat. Acad. Press; 2015, ISBN 978-0-309-37217-6 [Online]. Available: http://www.nap.edu.

[13] Shahan Z. Electric vehicle market share in 19 countries. Zurich, Switzerland ABB Conversation March 7, 2014.

[14] Falvo MC, Sbordone D, Bayram IS, Devetsikiotis M. EV charging stations and modes: international standards. In: Proc. IEEE SPEEDAM, Ischia, Italy, June 18–20; 2014. p. 1134–9.

[15] Sanchez-Miralles A, Gomez San Roman T, Fernandez IJ, Calvillo CF. Business models towards the effective integration of electric vehicles in the grid. IEEE Intell Transp Syst Mag Winter 2014;6(4):45–56.

[16] Alesiani F, Maslekar N. Optimization of charging stops for fleet of electric vehicles: a genetic approach. IEEE Intell Transp Syst Mag Fall 2014;6(3):10–21.

[17] Das R, Thirugnanam K, Kumar P, Lavudiya R, Singh M. Mathematical modeling for economic evaluation of electric vehicle to smart grid interaction. IEEE Smart Grid March 2014;5(2):712–21.

Chapter 5

Smart counties: technologies, considerations, characteristics, challenges, policies, and theoretical concerns

Evan Evangelopoulos
Pinal County Community Development Department, Florence, AZ, United States

5.1 Introduction

Smart is an interesting word and like any other word, it can be interpreted, stamped, stereotyped, and even misconstrued in myriad ways by different users. It is important to know, therefore, that when we all talk about smart, we more or less mean the same thing.

This is a fundamental principle in science, reasoning, law, or philosophy to make sure that we use any language terms—words or phrases—under the same meaning. So, let us identify smart as in smart cities. There are many attempts at a definition but in the most simple terms, smart could be defined as running a city by incorporating many different forms of informational and communication-based technologies (ICT) (adapted from Ref. [1] in Ref. [2]; and [3]. But, first, let us try to describe a nonsmart city.

A completely nonsmart or, for humor's sake, a stupid city, would be one that used not only no ICT technologies but no technology at all. So, maybe, if we took New York City of the early 1900s, then that would be a nonsmart city. It was a city of the Industrial Era, and although machines were used in many ways for the production of goods, most things were done by hand, even the cranking up of the machines themselves. Calculators of the time were often geared/hand-powered instruments with multiple cranks and levers to provide square roots, cosines, tangents, and other desired outcomes.

In the same era, in the urban planning field, when zoning and building ordinances were first being enacted, the plans were hand-drawn, the inspection was performed with pens and pencil markups, calculations were done by hand or those complex geared instruments I would often find at my grandfather's civil engineering office, while the notification that the plans were ready for pickup was sent via mail.

Although it is not clear when phone calls to architects started alerting them on the readiness of the reviewed plans, maybe that was the first form of technology use that elevated the field—to be humorous again—a step up from the "stupid level." Finally, after a turbulent 20th century, a new millennium, and many decades of technological innovation, which included satellites, Geographical Information Systems, sensors, the Internet, smartphones, and apps, the term "smart city" began to appear sometime between 2005 and 2008 [4].

Being smart, however, does not exist in a duality of smart versus nonsmart. There is a gray area in between, which probably represents where most cities or counties someday may lie in terms of smartification. This chapter will explore the technologies that enabled the terms smart cities, counties, or regions, will present the different parts of the smartification process, the major considerations, the specific applications to counties, and finally will explore the nature of the challenges along with the theoretical implications of smart technology applications. Smartify and smartification will be terms used throughout this text to describe the process of smart technology applications in cities/counties/regions.

5.2 Smart technologies

A smart city, county, or region is knowledge-based. Similar terms encountered are informational city, knowledge city, or creative city. A smart city relies on an informational or

Smart Cities Policies and Financing. https://doi.org/10.1016/B978-0-12-819130-9.00039-5

knowledge infrastructure that ensures the easy availability and dissemination of knowledge among its constituents and, of course, between government and constituents. Simply put, things become more accessible "through the internalization of smart technology" [3]. Transportation, utilities, energy savings, public transit, infrastructure, monitoring of the environment, and service exchanges between local governments and their respective communities become the beneficiaries of this sophisticated knowledge dissemination and infrastructure. "'Smart' urban management consists of the use of ICT-based technologies 'to deliver more effective and efficient public services that improve living and working conditions and create more sustainable urban environments.'" [1] in Ref. [2].

To get a coherent picture of city/county/region-smartness, it may be important to identify the technology aspects that contribute to such smartness. These are many and others will keep piling up in the future. For now, we will explore some of the most contributing technological aspects and concepts that have enabled the smart city or county. These are:

1. The Internet
2. Sensors and the Internet of Things (IoT)
3. User Generated Content Capabilities
4. Smartphones and Smartphone Applications
5. Cloud Computing
6. GIS
7. Analytics
8. Augmented and virtual Reality, and
9. Artificial Intelligence

5.2.1 The internet

The calculator, the personal computer, and finally the Internet began to smartify cities and raise them slowly higher and higher above the nonsmart level. Through personal experience, by the late 1990s, many cities and counties already had their zoning and parcel maps available online to anyone who wanted to identify information on parcels. The same jurisdictions had a website as well providing information on applications and procedures and identifying ways of accessing information and application forms.

As a side note, one of the problems that continues to this day was that although professionals were catching up and even demanding such services, many community members were still illiterate or inexperienced with the Internet, which made the use of electronic data unavailable to many. This has been the so-called Digital Divide between people who can afford and are adept to technology and those who are not.

The development of the Internet directly impacted the planning profession. The Internet provided direct access to documents, digital filing, facilitated project access, searching for information, accessing data bases, increased workers' productivity, and improved customer service efficiency and responsiveness. Time spent in the past looking for a three-year-old project in paper archives, now, with a good file organization, it could be achieved within minutes if not seconds.

5.2.2 Sensors and the internet of things

Sensors and the IoT are responsible for the term "smart city," which began to appear sometime between 2005 and 2008 with the proliferation of these devices. Sensors as smart devices and equipment able to understand, record, and disseminate information about the status of an underlying machine, infrastructure, or service, presented a plethora of opportunities to the digital landscape [4].

5.2.2.1 Sensors

Sensor Technology has deep roots but started to conceptualize in the 1960s. Some of the first sensors in the 1970s and 1980s were pressure and temperature sensors, artillery shell distance sensors, and tire pressure sensors. Sensor developing companies, however, started to flourish in Silicon Valley in the 1990s [5].

The opportunities started to skyrocket when the evolution of sensors was combined with the evolution of device-to-device wireless communication, expansion of satellite use, smart phones, and user-input information to help collect, monitor, and transmit data. The *interoperability* of devices was expanded eventually to allow a direct exchange of information and communication with other products and devices operating under different systems, without a necessary a need for mediation [6].

Sensors are an integral part of a smart city landscape, and according to Ref. [7], a city packed with censors is the city of the future attractive to people, companies, and business alike. With the proliferation of sensors, vast amounts of information, not available in the past, become available and new sources of insights become accessible [2], which can translate to opportunities for a more efficient city or county management.

5.2.2.2 IoT

As a result of the proliferation of sensors, the IoT started to materialize. When all things such as vehicles, TV screens, appliances, watches, streetlights, park meters, or trash receptacles are connected to and through the Internet, with abilities to communicate with each other, with servers, phones, and management centers, then a vast network of interconnectivity materializes [3,8,9].

It is really incomprehensible to imagine a reality of information and knowledge exchange among everything created by a human. In essence, in a nonsmart city such as New York in the early 1900s, humans were the only "things" communicating. Now humans are simply a slice of the existing communication ecosystem, and we have to learn to communicate with the things we have created.

Since anything can be connected to anything and everything exchanges information, it becomes necessary to find ways to assimilate this vast information and simplify operations. This creates new opportunities for faster and more efficient management of our milieu regardless of geographic extents whether it is a county, city, region, state, or a whole country [10].

This IoT relies on sensors that play the critical role of detecting, gathering, and communicating data, in addition to guiding microcontrollers and mechanisms to open, close, or adjust devices. The IoT also relies on the wireless connectivity of the sensors, a communication system, which is currently represented by Bluetooth or Wi-Fi Low-Energy.

The sensors command devices such as controllers and actuators, which are also part of this physical technological infrastructure. There are two additional levels, however, the management and the analysis of the information gathered. An effective management system such as SpeakThing can do this job while Blockchain or other ledger systems can securely track and record all exchanges. In the end, use of the data can be automatic but relies on analysts and managers as well to make use of it when needed.

Put it in a few words, the IoT forms the basis of a smart city or county. Gathering information, communicating that information to other devices, citizens, and government systems, and implementing relevant adjustment and reforms improve city operations and services and thus bring better management of cities or counties [11,12].

5.2.3 User-generated content capabilities

With the evolution of the World Wide Web, accessing information was not the only option anymore. User-generated content could be uploaded almost automatically and become part of the available online information. This was a characteristic of what was called "Web 2" [13].

The implication of such content became immediately apparent with the development of social media. In governments, such abilities allow any community member to digitally input opinions, estimates, questions, and directly participate in meetings. Although up to now software applications were being built for the desktop, now such applications were being built directly for the web where citizens could directly generate input. Wikipedia, Facebook, online questionnaires can all be seen as outcomes of Web 2 where the reader can actively participate and alter the webpage [13]. A useful comparison is that of Wikipedia with Encyclopedia Britannica: The latter relies on contributions from experts, while the former relies on contributions from community members to constantly write and edit.

5.2.4 Cloud computing

Another evolution that assists the operation of gathering and managing data has been cloud computing which relates to the concept of interoperability since it also allows the absence of an active mediator in collecting data. It exists independent of personal or organization servers, and everything and everyone can access and share informational resources. Available 24 hour a day, every day, it helps the continuous flow of data management which for cities or counties can be important for infrastructure and disaster management. Specifically in business and city planning, it helps professionals of a variety of specializations to work in teams independently of location, access and collaborate on the same information [14]. Without cloud computing, the development of smart cities would be impossible since sensors stream information constantly, updating it continuously in cloud locations accessible by different users [15].

5.2.5 Smartphones and smartphone applications

The advent of the smart phone in the 2000s, when it became available to everyone and not just business, presented a whole new step to the evolution of city smartness. An immense amount of knowledge suddenly became even more available to everyone. Instead of heading for the desktop, or opening the laptop, now someone could access this vast human library of knowledge anywhere there was phone reception.

The plethora of satellites circumnavigating the Earth, the multiplication of cell towers, and the immense increase in the speed of data traversing the atmosphere and reaching our smart phones have stretched the possibilities of a smart city or county and, of course, created a cultural transformation we are still grappling to understand its tentacles, possibilities, imaginable limits, and any possible potential interweaving with the pre-Internet culture. Everything around is coming together in the smart phone and its apps, almost like seeing life being assimilated within our mobile devices [3]. This life assimilation is happening because our smartphones bring us in closer contact with people and devices. Gathering, accessing, exchanging, and disseminating information has been revolutionized, and along with it opportunities to expand knowledge and inform increased exponentially, simply by the easy accessibility to the smart phone [16].

Communications and computing were just the beginning of a convergence, explained long ago by D. Messerschmitt [17]. Only the future will tell how far this convergence will go and how it will coexist with pre-convergence modes of interaction. But maybe, the real challenge and focus will be to make cities livable for the communities they represent and always use smartness as a way to improve quality of life.

Access to everything via an Internet search and direct access to a plethora of destinations via applications are becoming one of the basal structures of a smart city or county. Along with access comes an "effortless exposure" to things, thus offering a new avenue of knowledge beyond that of the era of the desktop/laptop-only Internet access [18].

Applications serve as the agent between the person and the destination. By becoming a shortcut directly to the site, they cut search time required for Internet searches; thus, in some ways, they are a replacement for the shortcuts used in computers. In addition, the ability to directly input information and interact with the source/destination services results in time savings in travel time, interactions, and in bypassing limitations imposed by regular business hours. Examples of personalized uses include exercise tracking apps, personal services such as health and finances, direct accessing of services such as tickets, direct access to information sources such as maps, and direct access to services and devices, which in the past required another person involved to access and use.

Apps create an interactive back and forth communication between destination and users, whether the destination is a machine or another person. This back-and-forth communication can be seen also in online computer games. Just grasping the increased volume of interactions occurring every second of time is inconceivable. If you think that in the past, telephone, mail, telegraph, and personal presence were the only ways of direct contact and compare that to the abilities/possibilities provided by smart phones and smart phone apps, one can clearly see the exponentially higher amount and types of interactive communications today. To this, we have to add the abilities of things to interact with each other, facilitating life as it happens in the background and releasing time for tasks previously requiring a certain human effort.

Life became more convenient because our normal daily activities were impacted by the new technologies. "The way we purchase things, to how we use energy within our homes, to the way we drive," all are impacted [19]. Communication, learning, entertainment, traveling, ordering, bureaucracy, and daily chores have exponentially benefited by the presence of apps on smart phones [20]. Solutions in life switched from the impossible or luxurious to the necessary.

For cities or counties, this means a direct access to meetings without in-person presence—such as city-council meetings—direct access to services such as application forms, payments, and submittals, available at any time and any location, access to GIS maps, the zoning code, complaints, and real-time reporting of infrastructure failures and maintenance.

5.2.6 GIS

A discussion about smart cities or counties would be insufficient without a discussion about GIS mapping—Geographical Information Systems mapping. Although paper maps and atlases are no longer necessities, maps have become an even more important aspect of today's world. With a geocoded influx of endless and diverse data from sensors and smartphone users, anything can be laid out on a map.

Although GISs started developing in the 1960s, it was 1981 when ESRI released ARC/INFO, "the first commercial GIS product" [21]. The ability to layer information and thus discover relationships between data was a concept mainly initiated by Ian MacHarg in his seminal "Design with Nature" of 1969 [22]. ESRI took the concept to the digital level with the help of computers aiming at using the concept of geographically laid data in layers to visualize relationships and patterns, "uncover meaning and insights," and solve problems by assisting in decision-making [21]. In addition, the geo-enabled data from a diversity of sources is a reminder of the importance of the concept of interoperability mentioned on the Internet of Things section, which allows different systems to recognize each other and exchange information without the need of a mediator system.

Since its inception, however, GIS has evolved. Spatial analysis evolved to help "uncover hidden patterns," and "improve predictive modeling" [21] by analyzing the geographic and geometric components in data, understanding the degree of dependency between data, the variables present, relationships among variables, the exploration of potential variables, and gravitational attractions among variables such as movements between cities of people, jobs, resources, and economic sectors. Endless more examples are conceivable. Analyzing the number of households, jobs, retail, and commerce an area could support, and incorporating the data into regional and city planning strategies to lay out future growth patterns, is one of the most instrumental urban planning applications of spatial analysis and analytics that GIS assists.

There is a prominent visual component in GIS mapping. Because the data is laid out on map layers, they literally become visible. Words, names, numbers, complicated series and sequences of numbers, and all sorts of data from nominal, ordinal, and interval to ratio all are closer to revealing their patterns and relationships when visible, and thus, eventually, they offer insights, solutions, and predictions.

The visual representation technologies have evolved over time to be more interactive; thus you can reveal, hide, and alter relationships between elements. The maps, through the superimposing of layers, the constant update of information, and the interactive environment, can reveal where things happen, where might happen, why they happen, or where they happen, conceptualized as location intelligence. As an example, locating stores, such as supermarkets or other chain stores, is possible based on population, traffic, and urban trends. Facilitating port traffic and operations is another example while with cloud-based software data collection, sorting out, and analysis happen 24 h a day.

From a simple 2D visualization, GIS has also evolved and continues to evolve into a 3D visualization, which makes exploration of data a very interesting endeavor, able to reveal more relationships, represent reality more efficiently, better acclimatize users to the relevant environment, and improve communication and sharing of data. Especially when it comes to land use and zoning, being able to transform zoning standards—numbers and words—into three-dimensional models can build scenarios and make the effects of policy decisions apparent and the impact of new projects obvious.

Ability for real-time information input from sensors and users is becoming an extremely useful tool for cities or counties. For example, in emergency management, real-time information enables efficient maintenance of infrastructure while in disaster management, knowing where and when infrastructure incidents are occurring or could occur can provide better assistance as needed. Similarly, in health care, mapping data of infections helps identify hot spots of disease and real-time connection with patients increases possibilities of positive outcomes. In transportation management, mapping the busiest intersection can prioritize projects.

Basically, everything is data. Every object, from a tree, to a trash bin, a building, a store, an empty lot, a rock, a road, a person walking, a bottle, a mug of coffee, a culvert, a channel, a river, a lake, or an intersection, are all able to provide real-time data. This explosion of data offers data storage, processing, and management challenges, but the analysis of data offers huge opportunities, as well, "to answer business questions, discover relationships, predict unknown outcomes and automate decisions" [23]. The field of Analytics is the next big item in smart cities or counties.

5.2.7 Analytics

In very simple terms, analytics is data-driven planning and could be described as a merger of GIS mapping, data input, analysis of data, and modeling and communication of the results. It is a step beyond the simple layering of information in GIS as it is "used to find meaningful patterns in data and uncover new knowledge based on applied mathematics, statistics, predictive modeling, and machine learning techniques" [23]. Communication of data is part of analytics [24], and once data is communicated, it can be used in decision-making. Urban analytics is a component of smart cities, and overall analytics help us understand urban systems by using "big data as exploratory material" [25]. In urban planning, understanding existing conditions and scenario buildup seem to be two major contributions of urban analytics to the profession.

Understanding existing conditions on a variety of urban planning topics results from the ability to merge mapping with the countless influx of data cities or counties can now obtain. On the other hand, like a soundboard, the scenario planning process modifies existing parameters, adds new, analyzes their interactions, and lays out options for future strategies. Scenario buildup is part of data interpretation. Throughout this process, conversations and strategies backed by analytics can make outcomes clear by quantifying and illustrating the potential impacts of policies [26], allowing to look at multiple outcomes simultaneously and, according to Peter Calthorpe (in Ref. [27], this "is the healthiest and the best way to think about possible outcomes."

Both the understanding of existing conditions and the scenario buildup offer insights that lead to policy and strategy development. Applications include multiple facets of urban planning such as:

- Assessment of potential natural hazard impacts
- Land consumption
- Future growth strategies
- Conservation
- Risk and resilience to shocks and stresses
- Transportation
- Environmental health
- Water use
- Energy consumption
- Community health
- Judicial records

Key planning parameters that scenario planning can manipulate are land use, zoning, traffic, pedestrian accessibility, transit accessibility, vehicle miles traveled, emissions, number of households, number of jobs, types of homes, household costs, and many more [26]. During this process, and due to the large amount of data available and the new possible ways of manipulating that data, new and more complex questions may arise with results portraying urban dynamics in more detail [28]. A new light rail, a reduction in single-family allocated zoning districts, potential pedestrian districts, a road realignment, a decrease in lane widths, an increase in the number of stories along an urban corridor, the number and size of water reservoirs all are policy and design decisions that can be tackled by urban analytics.

Visualization is a significant part of this process of analyzing and manipulating data. The goal is to efficiently communicate information to the decision-making process. There are many urban analytics tools that deliver the manipulations and visualization of data such as Urban Footprint, City Viz, or ArcGIS Urban, which can offer useful presentations to policy and decision-makers.

From the preceding, one could say that the IoT and analytics are the two most important aspects of a smart city. The IoT results in better management and responsiveness opportunities, while urban analytics, with data assessment and scenario buildup, helps assess the impact of existing and future policy and design decisions. The two represent two facets in all smart cities or counties: the fast, data producing landscape of sensors, and the slow landscape of city management and policymaking [2]. One can add the in-between landscape, the one of exploration, sorting, communication, analysis, discovery, dissemination, and manipulation of data, where analytics lies, and thus, we can have the smartification equation:

Sensors → data dissemination → exploration, sorting, analysis, communication, and manipulation (urban analytics) → city/county management and policymaking.

The communication of the data represents a very significant aspect of the process and is the place where the reality—virtuality continuum unfolds.

5.2.8 Augmented and virtual reality

There is a spectrum of visualizations of present and future information about an environment that starts with the representation of the existing physical environment, which is then superimposed to various degrees with existing information and future concepts, and eventually are supplanted by a complete virtual representation of both present and future suggestions. Reality, in other words, is extended or else expanded with additional information:

Physical reality → augmented reality → mixed reality → Virtual reality.

With augmented reality, information is placed visually on an existing environment, and one could say that virtual reality is layered on top of the physical environment. Layering virtual reality into the real world can be extremely helpful in exploring the present conditions of an environment and in visual simulations of future changes. Such applications can be helpful in urban planning education by providing immersive experiences to students, in navigation, disaster response, health services, or police training [29].

However, it seems that the major contributions are in advance urban planning, by first providing innovative ways of visualizing potential scenarios and second in public participation, where community members could actually walk along a street and see how the different scenarios would modify the spatial layout of the city and thus provide more effective input on potential projects and efforts [30—32].

5.2.9 Artificial intelligence (AI)

With artificial intelligence (AI), machines develop the capacity to respond to the environment the way humans would and can use reasoning with the most appropriate available data to provide answers or solutions by learning from past experience and assessing problems the way humans do [33—35].

5.2.9.1 *Artificial intelligence applications*

Smartphones, already equipped with GPS tracking, camera, and microphone, create an ideal platform for AI applications [36]. Under the concept of enhancing audience engagement, smartphone apps will be able themselves to interact with users, understand previous interactions, reason ways they can be helpful, suggest solutions to inputs, solve problems, and make autonomous decisions. Voice recognition devices such as Alexa or Siri can be a replacement of apps, an agent between the app and the person, or the agents between the person the Internet and the IoT. Suggestions for personal items such as clothing are based on previous patterns of behavior, trends, and familiarity with a person's personality.

In urban planning, AI could enable water reservoirs to self-regulate the inflow or outflow of water based on weather, past weather experiences, could empower traffic lights to self-operate and regulate traffic by combining traffic volumes with time of the day, holidays, unemployment levels, and similar past experiences, without the need for input. In zoning circumstances, an artificial intelligent map would be able to provide suggestions for future zoning and land-use adjustments based on real estate demand and direct community input on the map. A voice recognition device could potentially fill in a requested application, once it knows personal background, find answers to zoning questions, explore areas that enable specific development patterns, relieve community members from the cumbersome unknowns of the interaction with government entities, and make informed decisions regarding the best approach to development in a specific zoning district. Imagination will be the only limit for AI applications in urban planning.

AI closely relates to analytics because it can visualize and create scenarios and forecasts using the abundant data the IoT produces. According to Adie Tomer [35], although AI is still in its infancy, it is already around us, "when your video conference shifts the microphone to pick up the speaker's voice, when your smartphone automatically reroutes you around traffic, when your thermostat automatically lowers the air conditioning on a cool day, that's all AI in action." At the same time, "AI-informed water infrastructure would also switch into manual mode, potentially requiring extra workers to manage systems" "In Montreal, the regional public transportation agency and Transit, the maker of a well-subscribed smartphone application, are using machine learning to better predict future bus arrivals" "In New Orleans, the city's Office of Performance and Accountability used machine learning and public data to predict where fire-related deaths were most likely to occur, helping the fire department better target operations".

AI seems the last big element to explore here in the development of smart cities, since by definition—when and if it ever arrives at its full extent—it could replace us. It is a scary thought but one to keep in mind for all the possible ramifications in our world.

5.2.9.2 Drones and self-driving cars

Both drones and self-driving cars use sensors and AI to reproduce an understanding of the surrounding environment, respond to road conditions, and thus navigate their environment.

Driverless cars will change the way we park, how garages are conceived, and will impact parks and open spaces. The impact may be significant to replace surface parking lots with parks and plazas, replace roadside parking with drop-off lanes, and will impact the local transportation network. On the negative side there may contribute to sprawl. Still, many of these questions are being explored and answer will slowly begin to emerge as autonomous vehicles become more prevalent [37,38].

In planning practice, unmanned aerial vehicles (UAVs) or drones can be used in site analysis and visualization, local transportation network, photosimulations, project management and development monitoring, transportation and traffic applications, disaster planning and recovery, community engagement, emergency management, code enforcement and environmental protection, environmental assessment, infrastructure inspection, and surveying. All these applications can be valuable, prove pivotal, and thus become essential to planning practice [39] and as mentioned above, especially in counties where geography contributes to challenges of distance and remoteness.

UAVs were developed decades ago, but smart technologies have immensely expanded their capabilities and can be extremely useful in expansive county lands across the country. Drones can reach remote areas to map, investigate, analyze, visualize, report, and document existing conditions. They can also bring an added dimension to AI and augmented and virtual reality applications by making virtual visits to sites and presenting to distant audiences the potentials of a site. This can be especially true in cultural and archaeological sites [40–42] paving the way on how such technologies can be applied to remote rural communities.

5.3 The smartification equation: from collection to application

Considering that the IoT is at the basis of a smart city or county, [9] offers a simple lay out of a smart city data processing process that consists of three action layers:

1. The Perception Layer, which is the actual layout of the sensors to perceive the environment,
2. The Network Layer, which disseminates the information, and
3. The Application Layer, which refers to smart technology applications in different sectors of a jurisdiction.

Data however need to be sorted, collected, analyzed, and communicated, so a much more inclusive description of a smart city or county data processing may be one that reveals the analysis where data is processed before it is used for management and policymaking. Thus, inserting an analysis of platform layer between the network and application layers provides a more complete picture of the "architecture of the smart city" [43]:

1. The perception Layer, (the layout of the sensors to perceive the environment)
2. The Network Layer, which disseminates the information,
3. Analysis Layer (data exploration, sorting, storing, analysis, communication, and manipulation) and
4. The Application Layer, which refers to smart technology applications in different sectors of a jurisdiction

We can lay out the four action layers in the following smartification equation that includes:

Perception Layer (sensors) → Network Layer (dissemination and communication of information) → Analysis Layer (data exploration, sorting, storing, analysis, communication, and manipulation → Application Layer (city or county departments, management, and policymaking).

The last layer represents the many objectives in the application of smart technologies, which can include assessment of potential natural hazard monitoring, land use scenarios, future growth strategies, natural resource conservation, risk and resilience to shocks and stresses, transportation applications such as infrastructure assessment and monitoring, traffic monitoring and regulation, environmental pollution monitoring, health monitoring and intervention, water use, energy consumption, interface of citizens and government, and community participation [26].

All these applications include smart technology implementation with a specific outcome in mind, which often is to improve the function of something by understanding, cataloging, analyzing, and concluding the nuances of its operation, which result in improved decision-making. Such an approach represents the instrumental rationality approach where we have a specific goal that can be achieved step by step, following a specific process [44,45].

On the other hand, there are other considerations with a more abstract goal such as quality of life and place, and when one creates smart traffic signals to improve the flow of traffic, the goal is not to improve the flow of traffic but to improve commuting times and as a result, reduce stress, and thus improve quality of life. These are the value-based considerations according to philosophers Max Weber and John Rawls [46,47], which are implied by Refs. [44] and [45]; the latter referencing Jane Jacobs and her critique of the rational city. Value-based considerations would be important to keep in mind as a higher order of considerations in order to keep the aim on the community instead of simply aiming at the implementation of smart technologies. Such higher aims can potentially create a more selective approach to smart technologies with more focused aims. As a result of such an approach, the smartification considerations can be divided in two categories discussed in the next section:

- Instrumental considerations and
- Value-based considerations

5.4 Smartification considerations

Smartification technologies fall under the umbrella of information technology and have revolutionized the gathering, storing, and accessing of information. When implemented, local governments can then be labeled smart, within a range of smartness, depending on the extent of the applied modifications. Smart technologies can be considered under several general categories, distinguished in two types, instrumental and value-based as described below.

5.4.1 Overall considerations of smartification applications

Thinking of smart technology considerations broadly, we can accept the notion that smartification technologies offer value in the following three broad categories (adapted from Refs. [35,48,49]:

- Environmental,
- Social, and
- Economic

While a breakdown of applications to a more specific level is necessary, such a diagram of broad relations can be useful in keeping a big picture of smart technology impacts. The three categories overlap, and they seem to encompass much of the spectrum of issues affected by smart technology within a city or county.

It may also be important to keep in mind the difference between instrumental (process-based) and value-based types of applications of smart technology. These two types set different goals, although the means may overlap. Value-based considerations introduce more abstract concepts such as:

- Quality of life and
- Place

Such considerations (See Fig. 5.1) will help ensure that community and the spaces where the community operates remain the focus of smart technologies so we do not return to a rationalistic pre-Jane Jacobs era of planning [50,51].

As Christine [44] stated, uncertainty, complexity, and surprise still rule the city, therefore, "a concept of instrumental rationality can never provide a comprehensive image of the city." Therefore, value-based considerations can help focus a proposed set of policies toward the health of the community (See Fig. 5.1) and avoid the application of smart technologies simply for the sake of technology or to be labeled "smart." Indeed, introducing quality of life and place into the diagram of broad applications addresses a more complete set of possible smart technology considerations by introducing and maintaining the focus on two classic concepts in urban design and planning [52–57].

5.4.2 Instrumental considerations in smartification applications

Being smart should help a city or county improve several aspects of the interaction between government and community. The management of a city or county will change since it will involve a whole new array of equipment, technologies, information gathering, information analysis, which will ensue a transformation of operations, public services, direct services, communication linkages, cultural adaptations, opportunities, and conveniences [58]. Policies need to be set in place, however, to both facilitate and address these changes, which in the end, as already mentioned, will introduce significant transformations in the relationship between government and community [59].

At the instrumental level of considerations, specific functions of day-to-day government operations will improve. Applying sensors, for example, to water distribution pipes and transmission lines will enable to better detect defects and manage failures. Using smart traffic signals that can detect and predict traffic counts, the county can improve traffic flows and daily commute times. Using sensors in trash bins can help limit the number of trips to

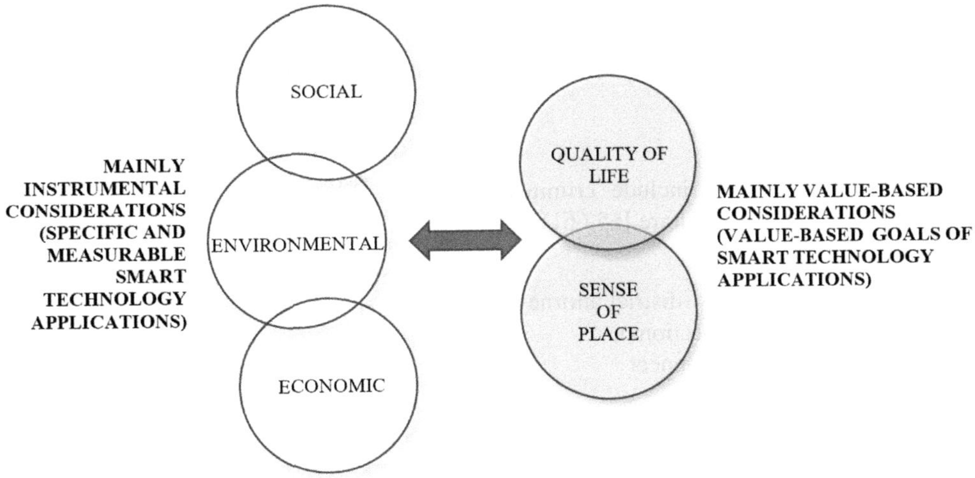

FIGURE 5.1 Broad level of smart technology considerations in a municipality.

empty them, to only when they are full. In addition, using advance 3-D analytics models will help recognize the most applicable zoning changes and minimize experimentation, while predictive analytics models can help with judicial applications and decisions, so unique to county services.

5.4.3 Value-based considerations in smartification applications

It is really important to consider here the impacts of smartification decisions on quality of life and Place. Smart technologies positively improve quality of life by impacting education, economy, mobility, public safety, sustainability, and quality of life [60]. Quality of life in urban form, however, relates to the presence of a neighborhood with a center and an identifiable edge, proximity of uses, and options in mobility [61]. Proximity of uses and options in mobility contribute to the 20-minute neighborhood, a major goal of smart growth, which aims at reducing commuting times, offering destinations at a walking distance, availability of jobs, and proximity to otherwise unreachable occupational and nutritional opportunities for underprivileged segments of the population. These and more are important to keep in mind as general guidelines when planning for smart technology implementation.

Place is a more complicated concept because it involves community and personal feelings, meanings, memories, and attachments to a location, besides a set of unique perceived characteristics [52,53]. In such terms, the impact of smartification to the unique social, economic, and environmental qualities of a place should be considered.

Since in every county there are a number of departments to respond to a variety of services, we can sort smart technology applications by county department to facilitate prioritization and organization of policy and management decisions. It is important, however, to discuss first the special characteristics of counties and the unique challenges they face.

5.5 Smartification in counties

Whether city or county, there are bureaucratic processes in both types of government that create similar difficulties on the way to policy and law adoption. Counties, however, are distinct from cities in some unique characteristics that can significantly affect the smartification process.

5.5.1 The nature of county–city differences—unique county characteristics

Counties and cities across the country are government entities that interact with their residents directly and daily on a multitude of everyday life issues such as property, health, justice, safety, utilities, elections, and infrastructure [62,63]. Although both are local government entities, counties are responsible for implementing state policy at the regional level and perform some unique services. Some of the basic differences between cities and counties result from the following characteristics [64]:

- Geography,
- Services provided,
- Operational provisions, and
- Structure

Geography, services, and operational provisions create some distinct characteristics for counties that can significantly affect the smartification process such as the logistical issues presented by geography for the provision of services. As already mentioned, however, experience shows that bureaucratic processes in the structure of both types of government present some parallel navigational difficulties on the road to policy and law adoption.

5.5.2 Unique county services

The unique services counties provide include criminal justice, recording of deeds, health, and welfare [65,66] and are typically:

- Judicial and legal offices and services (district attorney, sheriff, courts, public defenders, corrections),
- Transportation and infrastructure to connect communities,
- Public utilities such as water and sewage systems, and often gas and electricity,
- Tax collection,
- Election overseeing, monitoring, and certification of results, and
- Property tax evaluations.

According to Ref. [66]; "counties invest $106.3 billion in building infrastructure and maintaining and operating public works annually." Counties also operate hospitals and provide health services and testing sites as in the recent Covid pandemic, and invest up to $69.7 billion annually for such services [66]. One of the basic functions of counties, the provision of legal and judicial services, includes a district attorney, sheriff, courts, public defenders, and corrections to keep communities safe. Thus "most counties have a courthouse that houses county officials, such as the sheriff, the county clerk, the assessor, the treasurer, and the coroner" [64]. In addition, counties provide record keeping, tax assessments, 911 call centers, and organize and monitor elections and polling places coordinating more than 704,000 poll workers [66,67].

5.5.3 County departments

All these county services, unique to counties or common with cities, are provided via a certain number of county departments/divisions/offices. In each of the approximately 3006 counties across the United States, a very basic and necessary assortment of departments becomes apparent, especially if one examines low-budget counties. Such counties have fewer departments and present a good example of the very basic functions every county has to serve. High-budget, more affluent counties, tend to have numerous other departments, divisions, and offices that in some counties can exceed 50, as in Maricopa County, AZ, with the ability to provide a plethora of services. Some examples of small and low-budget counties in the country are Zieback County SD, Zapata County, TX, Benton County, OR, Whatcom County, WA, or Riley County, KS [68].

Zieback County S. Dakota, one of the poorest in the United States, has a budget of about $2.5 million [69] and about 13 offices/departments [70]:

1. Commissioners
2. Register of Deeds
3. Assessor
4. Auditor
5. Treasurer
6. Ziebach County 4h
7. Voter Information
8. Pesticide Testing
9. Nurse
10. Emergency Management
11. Sheriff
12. Clerk of Courts and
13. States Attorney

These 13 departments of Zieback County reflect some of the basic functions a county in the United States must provide that address economic, administrative, election, property, agriculture, health, emergency, and judicial services.

Additional departments are added as county budgets become larger and types of services increase. As an example, Riley County in Kansas with a budget of $65,501,116 seems to rest somewhere in the middle with about 21 departments [71,72], while Sonoma County in California, with a budget of almost two billion, has 43 departments/offices [73]. County affluence, however, does not always correlate to number of departments. As an example, Sacramento County in California, with a budget of almost six billion, has about 38 departments [74], while Pinal County in Arizona with a budget of about half a billion has about 48 [75]. Besides budget, such differences may also reflect different needs in the provision of services, or simply organizational dereferences. Department names also vary as, for example, Pesticide Testing in Zieback County is often generalized as Department of Agriculture in other counties.

After examining the websites of several counties, 10 from the west (CA: LA, San Bernardino, Orange, San Luis Obispo, Sonoma, Sacramento; AZ: Maricopa, Pinal; OR: Benton; WA: Whatcom) two from the Midwest, (Riley, KS and Zieback, SD) one from Texas (Zapata), two from the northeast (Albany, NY, Hancock, ME), and two from the south (Bartow, GA, Beaufort, SC), the following represent the typical departments present in those counties and probably to most counties across the country:

1. Community Development and Planning
2. Public Works
3. Building and Safety
4. Transportation
5. Facilities Management
6. Emergency Management
7. Fire

8. Agriculture
9. Environment, Parks, and Open Space
10. Education
11. Economy
12. Information Technology
13. Health Services
14. Social Services
15. Human Resources
16. Judicial and Legal offices and Services (district attorney, sheriff, courts, public defenders, corrections)
17. Tax Collection
18. County Clerk
19. County Treasurer
20. Assessor
21. Board of supervisors and commissioners' offices.

Some of the above departments such as the Assessor, Sheriff, and County Treasurer are unique to counties while many others such as Community Development, Planning, Transportation, and Information Technology are present in cities as well.

5.5.4 Smart technology applications by county department

Just like cities, counties can apply smart technologies to a variety of service areas spanning the full range of county departments, each applying smart technologies to improve the relevant services they provide (see sidebar: "**Variety Of Service Areas That Smart Technologies Address**").

Variety of service areas that smart technologies address

The variety of service areas that smart technologies address can be assigned by county department as follows (adapted from Refs. [3,9,11,21,58,63,76]:

1. Community development and planning (includes both advance and current planning):
   a. Smart entitlement process with direct citizen access to interactive property information including zoning minimized interaction between employees and citizens, and smart governance and government services (smart codes, smart customer service, smart ordinances, smart citizen interacting on property maps).
   b. Smart advance planning models, digital modeling, evaluation of impacts, data-driven design, reduce risk with digital twins evaluation of planning impacts, acceleration of policymaking).
   c. Smart citizen participation and interaction, (smart engaging of the public/easier citizen engagement, smart citizen input, smart advocacy, citizen advisory boards).
2. Public works:
   a. Smart infrastructure (smart transportation, smart/real-time public input on infrastructure issues)
   b. Smart utility infrastructure (smart sewers, smart drainage and storm water systems, smart reservoirs, smart water pipes, smart electricity infrastructure, smart energy savings).
3. Building and safety:
   a. Smart utilities/energy (smart meters)
   b. Smart buildings (energy efficient buildings, smart energy savings).
4. Transportation:
   a. Smart transportation (smart traffic management, smart traffic lights, smart traffic info, travel times, route choices),
   b. Smart mobility (smart parking monitoring, emergency services).
5. Facilities Management:
   a. Smart utilities/energy (smart meters, infrastructure),
   b. Smart buildings (energy-efficient county buildings, county building automation software).
6. Emergency Management:
   a. Smart disaster planning (modeling shocks and stresses),
   b. Smart disaster response (preparing and testing policy responses to shocks and stresses, emergency services).
7. FIRE:
   a. Broadband prioritization for firefighters "during large-scale events, incidents, and natural disasters,"
   b. More effective fire detection, notification, and extinguishment systems,
   c. Wearable technologies for firefighters,
   d. Use of drones [77].
8. Agriculture:
   a. Smart farming by monitoring of soil moisture, nutrition levels, salinity, harvest data, and use of satellites, drones, and remote sensors to provide detailed information about crops, use of robots for harvesting [78–82].
9. Environment, parks, and open space:
   a. Smart environment/nature (smart management of air quality management, noise, water levels, precipitation),
   b. Smart waste level and content management,
   c. Smart parks, using sensor technology to track animals, vehicles, people, electric fence protection, access control, traffic speed monitoring, vehicle location tracking, auto watering and mowing, conserving energy resources [83–86].
10. Education:
   a. Smart education (online classes, smart participation, user-generated content, virtual learning environments).
11. Economic development:
   a. Smart business support,
   b. Promote smart technology innovation,
   c. Assessing impacts on local economy.

Continued

Variety of service areas that smart technologies address—cont'd

12. Information Technology: Critical in a smart city or county with:
    a. Technology planning and policy development,
    b. Close collaboration with county strategic development,
    c. Strategy, planning, and implementation of the perception and network layers of a smart city or county and possibly providing the framework for the analysis layer,
    d. Smartification organization,
    e. Smart seamless management of smart systems,
    f. Smart surveillance/security.
13. Health services:
    a. Smart health care (making health care more mobile, distant elderly care, distant assessments and patient monitoring, smart home services, intelligent and connected medical devices).
14. Social services:
    a. Urban inclusion by empowering disadvantaged groups, to improve programming and provision of social services through digital platforms and service delivery platforms.
15. Other county essential departments affected by the smartification process are:
    - human resources,
    - judicial and legal offices and services (district attorney, sheriff, courts, public defenders, corrections),
    - tax collection,
    - county clerk,
    - county treasurer,
    - assessor,
    - board of supervisors and commissioners offices.

The list of departments presents an understanding of the breadth of impacts smart technologies can have and the extents of smart technology implementation. Of course, the Information Technology Department in any governmental unit is of primary importance in the smartification process, while Community Development, Public Works, and Transportation may be the most relevant departments in the urban and regional planning realm since they deploy a large amount of sensors, and utilize mapping, visualizations, and modeling. Any set of issues unique to counties that may involve other departments should be identified for the deployment of the necessary smartification policies. Finally, in the decision and policymaking processes of smartification, the Board of Supervisors and the Planning and Zoning Commission offices will be instrumental in finalizing and adopting relevant policies.

5.5.5 County characteristics, challenges, smartification, and departments

County departments will face the unique issues described earlier that result from:

- Geography,
- Services necessary locally or required by the state,
- Operational provisions, and
- Structure [64].

Counties, however, provide services to a variety of county residents including those in both unincorporated and incorporated areas within the county and "are part of a statewide system that delivers certain kinds of programs and services" [63]. Therefore, some issues, such as community participation or quantifying and illustrating the potential impacts of policies using GIS and Analytics, will be common in both counties and cities.

The preceding characteristics result in 13 unique challenges and issues when planning and implementing smart technologies (adapted from Ref. [87]. These challenges include:

1. Large territory
2. Unincorporated and incorporated communities
3. Small and remote communities
4. Rural communities
5. Agriculture
6. Natural disasters
7. Sprawl that results from geography
8. State funding
9. State provisions
10. State mandates
11. Coordination with municipalities
12. Annexations
13. Lack of visibility that results from the necessary services required by the state, operational provisions, and structure.

5.5.5.1 Large territory

Counties serve larger geographical areas than municipalities and must provide direct services to distant unincorporated areas within their territory [88]. This distinction implies large distances for infrastructure layout but also remoteness of communities from the county seat. Thus, often, it is almost financially and timely impractical to serve distant communities with all necessary services while infrastructure repairs are diagnosed and arrive late, sometimes resulting in extensive repairs. This is one of the areas where smart technologies can provide huge benefits to county residents and help counties save money by fast identifying infrastructure needs.

At the same time, large county distances might demand the installation of a large volume of sensors to implement a full capacity smart county, which can be both intimidating and mind-boggling. In Santander, Spain, more than 12,000 sensors have been employed in downtown [89]. Such a number can be daunting if not practically and fiscally prohibitive for tight county budgets and unprepared smart technology management system. As a result, in order to conserve valuable resources, identifying, prioritizing, and implementing smart technology in areas of highest need become paramount. County Departments to address such geographic issues will be Public Works, Transportation, and Community Development/Planning.

5.5.5.2 Unincorporated and incorporated communities

Counties include both rural and urban populations in the form of unincorporated suburbs of local municipalities and communities that decided not to incorporate. So, while cities are mainly faced with urban-related issues, counties often have to address both urban and rural issues and as a result a county might have to implement smart technologies to both a downtown and agricultural lands, which intensifies the complexity of the smartification process. County Departments to address such geographic issues will be Public Works, Transportation, and Community Development/Planning, Emergency Management.

5.5.5.3 Small and remote communities

The low population and remoteness of some small communities might constitute the costs of smart technology applications prohibitive to justify investment and prioritization of smart infrastructure in such areas. County Departments to address community remoteness challenges will be mostly Public Works, Transportation, and Emergency Management, and Community Development/ Planning.

5.5.5.4 Rural communities

According to the 2018 Farm Bill, "rural communities cover 72% of the country and include 14% of the US population" [60]. This is impressive as it constitutes the majority of the area of the country, which is mostly managed by counties. It is important to note that rural communities are often affected by a lack of economic opportunities that lead to population loss and reduction of human capital. These represent the first two types of rural communities, the chronically poor, and the ones transitioning to a new economic base. The third type is the resource-rich rural communities [60]. All three types can be found in the same or diverse counties, but these are unique problems for counties to address with the help of state legislatures.

Implementing a broadband network infrastructure will provide the basis for the implementation of smart technologies while at the same time will safeguard the reliance on online systems for remote working capabilities. During the times of the COVID-19 pandemic, working from home became very common, something that may continue even after the pandemic. Providing a solid and reliable broadband network infrastructure in small rural communities is a major priority to ensure many families can work from a distance and reside in rural areas, while it will increase accessibility to government services [8]. A reliable grid, broadband infrastructure, sensors, and even connectivity represent the first part of the smartification equation presented earlier and form the foundation of smart rural communities [60]. County Departments to address rural community smart technology issues will be Community Development/Planning, Public Works, and Transportation, Emergency Management.

5.5.5.5 Agriculture

Agriculture is prevalent within county borders. Since there is a desire to preserve agricultural lands and minimize losses to subdivision development, prioritizing agriculture in the smartification process may be a policy goal to keep in mind. Smart technologies with drones, precision measuring of soil and air conditions, AI robots monitoring crops and conditioning the soil are changing agricultural practices toward becoming more precise in terms of efficiency, cost, yields, nutritional value of crops, or any other target the farmer sets [60,90,91]. Analytics can play a vital role in sorting and modeling the vast amount of data as part of a research component of smart farming. It is important for counties to identify what role the agriculture departments can play and what policies to develop to promote and assist the development of smart farming. Such efforts can yield up to $65 billion annually "in gross benefit to the US economy" [60]. In this case environmental AI drones can play a significant role in monitoring environmental conditions [81,82]. Other County Departments to assist in smart precision farming besides Agriculture are Environment, Parks and Open Space, Community Development/Planning, Transportation, Public Works, Economic Development, along with County Clerk, County Treasurer, Board of Supervisors, and Commissioners offices.

5.5.5.6 Natural disasters

When it comes to natural disasters, local (municipal and county) governments are "on the front line of dealing with emergencies that occur within their jurisdictions" [92]. Although there is ambivalence in the literature on the long-term financial impacts of natural disasters, [93] and [94], the unpredictability of natural disasters raises the possibility of unexpected and sudden impacts in the near future. In such an instance, at least in the short term, natural disasters may add expenditures and limit funding available for smart technologies if there is a diversion of funds from other programs to help address the situation [95]. The large expanses of some counties and the remoteness of some populations may add some extra logistical and financial burdens to counties in relation to cities when responding to natural disasters. Possible exacerbation of natural disasters due to climate change is something else to consider, an issue that might increase the possibility of unexpected and sudden impacts and uncertainty on county finances in the near future [92]. County Departments to address such issues will be Emergency Management, Environment, Parks, and Open Space, Community Development/Planning, Transportation, Public Works, Economic Development, along with County Clerk, County Treasurer, Board of Supervisors, and Commissioners offices.

5.5.5.7 Sprawl

The rise of urban sprawl during the 20th century created continuous pressures on county lands that besides eliminating productive agricultural land, created peculiar living conditions that consumed land, increased commuting distances, pollution, infrastructure costs, traffic jams, needs for more and bigger roads, reduction of housing options, reduction of route options, elimination of biking or walking as a means of transportation, and a decrease in the richness of the urban environment available to the community. Sprawl is a main feature in many counties that border cities. Since infrastructure layouts are lengthier and densities lower, technology investments will have to sprawl over larger distances, thus possibly increasing the number of sensors required, installation and maintenance travels, while due to lower densities, devices may service fewer people. It is an area with little available research but smart growth in relation to smartification is a very relevant topic for counties that need further investigation. Sprawl pressure, for example, may constitute obsolete already installed sensors and smart technology in areas taken over by new subdivision developments. County Departments to address such issues will be Community Development/Planning, Transportation, Public Works, Economic Development, along with County Clerk, County Treasurer, Board of Supervisors, and Commissioners offices.

5.5.5.8 State funding

Sate funding is created by the state government and thus funding can be limited or overly controlled by the state. As a consequence, funding for smart technologies and applications may be harder to obtain [96].

5.5.5.9 State provisions

Typically operate under provisions set out in the state constitution, which also is another indication of less control over the county limits than a city has over the city limits [96]. County Departments to address such funding issues will be Economic Development, along with County Clerk, County Treasurer, Board of Supervisors, and Commissioners offices.

5.5.5.10 State mandates

As such, counties can be seen as administrative units of the state, and as a result, what are called state mandates are some of the most important aspects of the relation between states and counties. States mandate counties to implement certain programs and services without always providing the necessary funding. The unexpected nature of such requirements or the delay or absence of reimbursement can sap county funding and postpone prioritization and implementation of other programs such as physical and IT infrastructure and smart technologies [88,96–98]. County Departments to address such funding issues will be Economic Development, along with County Clerk, County Treasurer, Board of Supervisors, and Commissioners offices.

5.5.5.11 Coordination with municipalities

Counties must also coordinate resources and programs with incorporated municipalities under their jurisdiction, which places added complications to any smart technology implementations. Universities such as Arizona State University with efforts such as "the Connective," the "Smart Region Summit," and the collaborative Cox Connected Environments Collaboratory at ASU can be catalysts in research and organization of regional collaborations, thus helping both counties and municipalities work together to address smart technologies [8,99]. County Departments to address such issues will be Community Development/Planning, Transportation, Public Works, Economic Development, along with County Clerk, County Treasurer, Board of Supervisors, and Commissioners offices.

5.5.5.12 Annexations

Annexations of county lands by municipalities pose another hurtle to smart technology implementation since moneys spent on smart infrastructure, such as transportation, might soon be under the jurisdiction of an incorporated city and thus the county can be left bereft of its investment. Unincorporated areas within a city's Planning Area are still under the jurisdiction of a county, but why would a county finance smartification of areas soon to be annexed by neighboring municipalities? Reimbursement to counties and careful planning on the transference of services from county to municipality must be planned carefully. Therefore, a close collaboration and agreements of counties with adjacent cities seem necessary to address such issues [100]. Since smart technologies often involve management and maintenance of the technology by corporations, any agreements between county and city will have to involve the relevant corporations, an issue common in annexations since preexisting contractual agreements between counties and private interests may exist [101]. County Departments to address such issues will be Community Development/Planning, Transportation, Public Works, Economic Development, along with County Clerk, County Treasurer, Board of Supervisors, and Commissioners offices.

5.5.5.13 Lack of visibility

An additional unique characteristic of county governments is a relative lack of visibility in national or local politics, which results in a "deficit in knowledge and trust about the operations at a county level" among the community who is often unaware of the substantial services counties provide [88]. Such characteristics may add additional burdens to counties trying to gather support from the community on smart infrastructure programs. County Departments to address such issues will be Community Development/Planning, Transportation, Public Works, Economic Development, along with County Clerk, County Treasurer, Board of Supervisors, and Commissioners offices.

5.5.6 County planning department, priorities, applications, and considerations

Although the application of sensors and a fully deployed IoT can significantly improve transportation, infrastructure, agriculture, emergency functions and responses, smart technologies are not all about sensors. In the planning department the most important applications would relate to the improvement of services between the community and the department, opportunities for community participation input and decision-making, analytics, and methodologies of assessing the impact of various future development schemes and zoning impacts, buildout estimation, and visualizations of new projects and development proposals. However, there are four themes that will significantly impact the planning department/office [102]:

1. Demographic change
2. Technology as a disruptive force
3. 21st century problems and 20th century planning practices
4. Economics 101: the value of planning

These themes will increase the need for technology to monitor demographics, understand and address different cultural groups, promote transparency and communication with communities, ensure understanding of real estate economics, understanding economic impacts of projects, promote visualization as a means to evaluate alternative visions, forecast trends, and make policy recommendations. All these impacts can be distributed or commonly addressed in the current planning and advance planning divisions of a planning department.

5.5.6.1 Current planning

Improvement of services between the community and the department, visualizations of projects, and visualizations of development proposals belong to the Current Planning realm that deals with the everyday processing of applications, answering questions, and the entitlement process. Smart technologies involved would be a fast and reliable broadband infrastructure to provide access to all possible communities, smartphone, and tablet applications to easily access planning department information, interactive public engagement applications in phones and tablets, interactive GIS maps, project visualization, predictive analytics to assess project impacts, and a necessary smart technology collaboration with the Information Technology Department to provide for the above.

5.5.6.2 Advance or long-range planning

In Advance or Long-Range Planning, community engagement smart applications, analytics technologies, augmented and virtual reality, other visualization technologies, and AI drones will provide the necessary means to assess and communicate impacts of land use changes, zoning changes, density implications, forecasts, application of new ideas, and as a consequence better evaluate General Plan and Area Plan options and make policy recommendations. These types of applications in counties do not seem to differ from Planning Department smart applications in cities. The same types of smart applications would be used in city planning departments as well. The difference will be the context. County Planning Departments deal more with undeveloped lands, agriculture, suburbia, and small or isolated communities. In contrast, overall, Planning departments in cities address more developed neighborhoods, downtowns, greenfields, higher city densities, higher density developments, and different types of land use patterns. In other words, Planning Departments in counties will face their own challenges by utilizing the same smart technologies.

It is important to emphasize analytics again. As discussed earlier, analytics present one of the most sophisticated applications of smart technologies in planning departments whether they belong to a municipality or a county. With analytics, the plethora of data will be sorted out and provide opportunities for the modeling of different development options across a county. While similar to city planning departments, in counties, will be paramount to assist in the development of specific plans and special area plans for different parts of the county, thus facilitating the development of rural communities and remote areas.

5.5.7 County priorities

The application categories seem endless, and indeed they can encompass every aspect of a city or county and any of the departments listed. Considering the enormous cost of sensors and relevant technologies required to smartify every service area possible, jurisdictions need to set priorities in order to best allocate financial and human resources (see checklist: "An Agenda For Action For Taking Steps To Implement Policy Priority Areas In Counties Around The Country").

An agenda for action for taking steps to implement policy priority areas in counties around the country

According to Ref. [66], a number of steps are required to implement the following policy priority areas in counties around the country (check all tasks completed):

_____**1.** Advocate for additional federal COVID-19 relief in the form of direct, flexible aid to counties of all sizes.
_____**2.** Promote mental health and substance use treatment and address essential criminal justice reforms.
_____**3.** Restore the balance of federalism and optimize intergovernmental partnerships.
_____**4.** Promote county infrastructure priorities
_____**5.** Enhance community resilience through regional and local disaster preparedness.
_____**6.** Boost advanced broadband deployment and accessibility while preserving local decision-making.
_____**7.** Support full funding for payments in lieu of Taxes and the secure rural schools program.
_____**8.** Strengthen election security and safety.
_____**9.** Promote county priorities in future US Environmental Protection Agency (EPA) and other federal rulemaking.
_____**10.** Promote workforce opportunities and supportive services for county residents in changing economies.

Comparing the NACo priorities of 2021 to those of 2016 [76] and [103], county priorities across the country have remained similar with a strong focus on **health**—especially with the Covid Pandemic in 2020—**infrastructure** maintenance and development, **economic** opportunity, and **workforce** development especially for poor rural areas. The main difference between 2016 and 2021 is a current emphasis on **broadband development**, **strengthening of election security**, and emphasis on **disaster preparedness**.

The above clearly show a national shift toward smartification by emphasizing the importance of internet access for all residents and an emphasis on disaster preparedness and elections. The importance of health and infrastructure however has not diminished. Although this is positive news for Current Planning Departments since any investment on broadband development and Internet access will probably improve citizen services, the emphasis on infrastructure and health clearly shows a prioritization of investment toward the IoT in the areas of infrastructure and disaster preparedness in addition to smart health investments in smartphone applications, response centers, and other applications. Broadband development is closely linked to both health and infrastructure and as mentioned will positively impact Current planning.

Facing these priorities, the Advance Planning division of a planning department will have to advocate for rural community development and the importance of long-range planning, general plan, and special area plans, in the economic development of such areas in order to compete with limited financial resources and work closely with economic development departments. Infrastructure and health will probably be the first areas to receive funding for smart technologies. Every county though will have to juggle these priorities and concentrate on the topics more applicable and allocate funding accordingly.

5.5.8 Policies for smart counties

Adoption of policies will set the direction of action on decided topics. There is a general process that leads to policy adoption whether this is about smart technologies, General Plans, or any other topic and whether it is a municipality or a county. In any policy adoption process, however, it is very important to first set a vision. If the vision is not the right one, the ensuing policies will not be effective and an excellent example in urban planning is form-based codes, where the importance of capturing the right vision is paramount for the crafting of the right policies and zoning standards. Similarly, in smart technology adoption, setting the right vision for the county is a first and very important step in the smartification process.

5.5.8.1 The policy process

Once the decision is made to proceed to smart technology applications, the county needs to create a vision and a mission statement followed by a set of policies, objectives, and implementation strategies. These steps often vary between organizations but the process is often described as VMOSA after Vision, Mission, Objectives, Strategies, and Action plans [104]. There are variations to the steps of this process such as the goals, policies, and implementation measures, or goals, policies, and objectives [65]. The process is also used for setting a vision, policies, goals, objectives, and strategies in a General Plan. Once the vision is set, it will be very important to:

- Determine the necessary changes,
- Decide what is realistic,
- Set priorities, and
- Proceed with a set of objectives, goals, strategies, and action plans [65,104,105].

Since smartification sets a new focus for the county government, this vision process will describe the ways to focus the county's efforts, give structure and direction, set guidelines, priorities, provide tangible checkmarks for the accomplishment of the vision, take advantage, or define appropriate resources and opportunities, better respond to barriers and resistance, and make a more efficient use of time and resources.

5.5.8.2 Four schemes

Here it is important to note that smart technologies will put pressure on local governments to evolve and four schemes will possibly evolve, one that simply promotes smart initiatives and it continues the existing structure, a second one that restructures decision-making, a third one that restructures the government to an extent and creates a smart administration, and a fourth one that is based on smart collaboration among the various governmental departments [106].

Out of these four schemes, the fourth one, although it requires a significant restructuring of governmental operations, seems the most relevant to counties, since many smart technology decisions, applications, developments, management, and operations require not only a multidepartmental approach but also a multigovernmental approach that necessarily includes large corporations, holders of smart technologies.

5.5.8.3 Exploring a smartification vision for counties

The specific challenges stated for counties will create the need for some tailored visions, objectives, strategies, and action plans. The smartification process will demand a specific set of responses to every challenge and the county can facilitate the process by examining first the broad considerations of economy, community, environment, quality of life, and place. Value-based considerations of quality of life and place need to always be part of the discussion to avoid overextending budgets and focusing too much on the notion of smartification instead of the benefits to the community.

Critical to this process is engagement of the community and all stakeholders, creation of public–private partnerships, establishment of management practices, ensure the protection of the environment for both people and nature, protection of community and social capital, studies of best practices, and establishment of smart regulations and standards [107]. However, close coordination between departments will be paramount since smartification does not involve just the Information Technology Department [48]. Comparable to the general vision process described above, according to the Pinal County Association of Governments [59], the process to ensure smart community actions, benefit the community, and make prudent use of limited resources includes three steps:

1. Identify regional needs or issues.
2. Identify strategies for how data and technology potentially can address that need or issue.
3. Develop action-oriented initiatives tied to regional plans that use data and technology to address the identified need.

In addition to the above, however, it is also important to ask [44]:

4. Who should be in control of smart cities, corporations, government, or citizens,
5. Should *it be* comprehensive *city wide or should it be small scale, in little projects?*

Regional associations of government such as PAG will be instrumental in the smartification process as coordinators creating task forces, regional strategies, and in identifying local priorities, establishing regional innovation centers, and pilot programs [59].

5.5.8.4 Instrumental versus value-based vision explorations

There must be a balance, however, between instrumental and value-based considerations (see sidebar: "**Process To Ensure A Balance Between Instrumental And Value-Based Considerations**"). Questions must be asked, therefore that will set the foundation for establishing a smartification Vision and Mission.

Process to ensure a balance between instrumental and value-based considerations

The questions to guide the vision process within a county can be organized by the economic, social, environmental, quality of life, and place application considerations as follows:

Economic considerations:
- What areas will have the biggest financial return for the county?
- What sectors of the local economy will be mostly affected?
- Which regions will benefit financially the most from the smartification?

Social considerations:
- What populations are in most need of services?
- What actions/applications will have the widest impact?
- What smart technologies will reduce commute time?
- What smart technologies will improve community participation?
- Will smart technologies help address health concerns and responsiveness in a more effective way?

Environmental considerations:
- Will smart technologies reduce the vehicle miles travelled?
- Will smart technologies improve pollution generating traffic flows?
- Will smart technologies improve responsiveness to environmental pollutants and disaster events?
- Will smart technologies improve visitors' experience of parks and recreation areas?
- Will smart technologies improve sustainability indicators across the county?

Quality of life considerations:
- How will smart technologies impact the overall satisfaction of people living here?
- What are the things people consider more urgent for the improvement of their daily lives?
- Do all the proposed applications improve the daily and long-term routines of the residents?

Place considerations:
- 'What is the impact of smart technologies on the daily life of local neighborhoods?
- What is the impact of smart technologies on the local places where people hangout, on "third places"?
- What is the impact of smart technologies on people's perception of their neighborhood/town/region?
- What is the impact of smart technologies on the community identity?
- Will smart technologies create a sense of pride in the residents?

  Since place and locality matter, and each county is identified by its own unique idiosyncrasies, local needs, standards, and implementation strategies will differ and have to be locally adapted. As a result, the following are also very important place-based questions to ask according to public health law and policy [105].
- Does the policy reflect the community's vision? Was a broad and diverse set of stakeholders involved in identifying needs the policy aims to address?
- Does the policy include locally relevant health data? Benchmarks (existing conditions) as well as standards and targets (goals for future conditions) can reflect health and built environment data.
- How will we implement the policy?
- Does it identify actions and programs (such as updating zoning codes or securing funding for an infrastructure investment) that will guide us from vision to action?
- Does it name implementation partners so that local government agencies and the community know who will play what role?
- How do we know if the policy has been successful?
- Are there identified standards or targets that can be measured to highlight where policies have achieved a vision and where they may need to be reworked?
- Does the policy make progress toward eliminating health disparities?

Counties differ, however, and questions per county idiosyncrasy would help set clear Objectives, Strategies, and Action plans and complete the vision process (VMOSA). This way, each of the challenges described will get a chance to receive special attention and as with any significant effort, hearing the voices of the community and all the stakeholders would be important in these questions. As mentioned in the City of Austin's effort [108], creating an ecosystem of partners is an important part of the process.

5.5.8.5 Adapting the vision process to a County's unique characteristics

Smart cities and counties need the implementation of a framework of smart technology that includes sensors, connectivity, energy infrastructure, [109]. Factors such as geography, services unique and required by the state, and unique operational provisions [64] contribute to the county challenges discussed. These are challenges of distance, variety of populations, variety of governments, natural ecosystem diversity, dependencies, unexpected events, and technological intricacies that would require, as mentioned, creating a strong ecosystem of partners that reflects the people and culture of the locality [108,110].

The issues need to be aligned with department specific plans, provide a prioritizing set of initiatives for a period of 3–5 years, and "communicate progress on achieving measurable outcomes to all stakeholders" [110]. Such policies will require the involvement and coordination of all relevant departments as described earlier and not simply the Technology Information Department [111].

The first seven challenges mentioned are mostly a result of the geographic extent of counties. A set of exploratory questions (see sidebar: "**Setting Priorities, Planning, And Organizing Policy Development In Smart Technology**"), per challenge, can assist in setting priorities, planning, and organizing policy development in smart technology.

Setting priorities, planning, and organizing policy development in smart technology

Such questions can be part of a questionnaire to the community, among staff and most of them start from the mapping and identification of most prominent issues throughout the county as follows:

1. Large territory:
   - **a.** What are the most distant areas of the county?
   - **b.** What types of infrastructure are the most difficult to access?
   - **c.** What communities have the most difficulty reporting infrastructure failures?

2. Unincorporated and incorporated communities:
   - **a.** What type of coordination is required to address ROW and traffic smart technologies adjacent to incorporated communities?
   - **b.** Who are the main stakeholders?
   - **c.** How can we create an ecosystem of stakeholders to implement smart technologies?
   - **d.** What are some innovative public–private partnership models and what are the ways to ensure appropriate feedback loops?

3. Small and remote communities:
   - **a.** How many communities are the most difficult to access?
   - **b.** Which communities are most vulnerable to natural disasters?
   - **c.** Which communities are in the most need smart technologies such as broadband or hot spots, sensors, drones, and use of analytics?

4. & 5. Rural communities and agriculture [60]:
   - **a.** Where are the most extensive agricultural areas of the county?
   - **b.** What types of agriculture are in most need of smart technologies?
   - **c.** How can we simplify the process of applying for funding?
   - **d.** What are the program requirements?
   - **e.** How can we improve the accuracy of coverage maps?
   - **f.** What are other alternatives to broadband?
   - **g.** What are the ways to share costs and/or aggregate demand,
   - **h.** What are the community assets and needs?
   - **i.** What ancillary investments are required to fully realize the benefits of connectivity?
   - **j.** How will the county coordinate, share, and support the development of smart rural communities?

6. Natural disasters:
   - **a.** What are the most common and recurring natural disasters in the county?
   - **b.** What types of technologies can the county implement in those areas to address prevention, response, and remediation?
   - **c.** How can the county apply analytics and visual simulations to assist natural disaster prevention, response, and remediation?

7. Sprawl
   - **a.** What land-use policies will create land patterns financially and environmentally more compatible with smart technologies?
   - **b.** What smart technologies can be efficiently incorporated into existing suburbia?

The next three unique county challenges relate to funding and services necessary locally or required by the state, operational provisions, and structure of counties [64] and will provide another set of questions to structure the objectives, strategies, and action plans necessary to complete the vision process:

8. State funding,

9. State provisions, and

10. State mandates:
   - **a.** How can we provide the legal framework for the financial support of smart counties?
   - **b.** How can the county establish smartification promotion plans?
   - **c.** How can the county receive state funding and support for smart technologies?
   - **d.** What are the avenues to receive funding for construction projects, technology improvements, operation, and management of the smart technology implementation? [112]
   - **e.** What is the process for the county to ensure funding for state mandates?

Finally, the final three challenges involve the relation of counties to the internal or neighboring incorporated communities. Identifying the present landscape and ecosystem of stakeholders will be an important aspect of these challenges:

11. Coordination with municipalities,

12. Annexations, and

13. Lack of visibility:
   - **a.** What areas are most likely to undergo annexation by neighboring municipalities?
   - **b.** Which of those areas are critical in a smartification process?
   - **c.** Are neighboring municipalities interested in collaborating to develop a smartification program?
   - **d.** Who are the main public and private stakeholders and how can we create an ecosystem of stakeholders to implement smart technologies?

Continued

Setting priorities, planning, and organizing policy development in smart technology—cont'd

e. What are some innovative public–private partnership models and what are the ways to ensure appropriate feedback loops? [103,107,111,113]?

Finally, the final three challenges involve the relation of counties to the internal or neighboring incorporated communities. Identifying the present landscape and ecosystem of stakeholders will be an important aspect of these challenges:

14. Coordination with municipalities,
15. Annexations, and
16. Lack of visibility:
    f. What areas are most likely to undergo annexation by neighboring municipalities?
    G. Which of those areas are critical in a smartification process?
    h. Are neighboring municipalities interested in collaborating to develop a smartification program?
    i. Who are the main public and private stakeholders and how can we create an ecosystem of stakeholders to implement smart technologies?
    j. What are some innovative public–private partnership models and what are the ways to ensure appropriate feedback loops? [103,107,111,113].

5.6 General smartification concerns and benefits

Whether a city or county, many of the challenges appear common to both and often have theoretical tentacles that inescapably entangle urbanism and urban and regional planning thinking overall. These overarching concerns can be split into two categories, technology-related and major participant-related general concerns.

5.6.1 Technology-related general concerns

During the smartification process, a county or a city will have to address the following challenges (adapted from Ref. [77]:

1. Myriad Sensors: We mentioned earlier the sheer volume of sensors needed to implement a full capacity smart city or county, which can be intimidating and overwhelming. This is a common hurtle in all jurisdictions whether counties or cities. The big challenge is to sustain the cost of the myriads of sensors required, efficiently collect and analyze the amount of information continuously collected, and use it in a way that helps communities.
2. Security: Security can also be an issue since the whole network of sensors and devices communicating might be vulnerable to attacks and threats from external system attacks and hacking. The mere vastness of the landscape of sensors and communication networks will create cyber threats and security concerns requiring the necessary allocation of resources [12].
3. Device Communication: The overarching challenge will be the interdependency of these technologies. Ensuring all the disparate devices are able to communicate reliably and deliver reliable data is a major challenge. The vast expanses of some counties can exaggerate this issue, which may require some smart and financially feasible solutions.
4. Immense Amount OF Data: This leads us to another challenge, which is the large scale of data, and the challenge of creating a ledger of such data.
5. Digital Divide: Although making broadband networks available to everyone in the most remote section of a county can be achievable, if the residents do not have the smartphones or laptops to access information, what is called the digital divide becomes apparent and something to consider and overcome.
6. Comprehensive versus Incremental Approach: Another concern that should be incorporated into the policy vision is whether it should be comprehensive, city/county-wide, or small scale and incremental, in little projects [44].

5.6.2 Major participant-related general concerns

Besides these challenges, there are challenges that arise from the participation of many influential parties in the process [25], which, however, mainly focuses on the collaboration between three major groups: corporations, government, and citizens (See Fig. 5.2).

Whether it is in cities, counties, or regions, the mere presence of this triad of corporations, government, and citizens translates into potential internal conflicts [44] with the following concerns:

1. Large Technology Corporation Dependency: The first concern is the impact of the involvement of large corporations into smart city development: how much control that removes from local government and how much dependency that creates.
2. Type of Control: A question should also be posed about the type of the desired control of smart cities. Should it be top-down from corporations, government, or from the citizens [44].

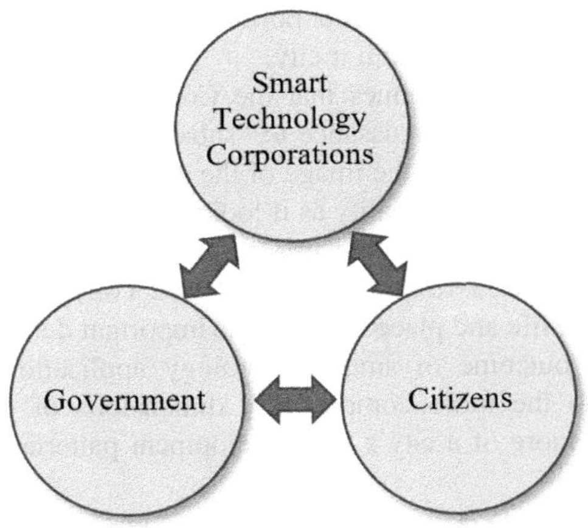

FIGURE 5.2 The three major players of the smartification process.

3. Data: The next issue relates to data. Code and data are always cooked, they are prepared toward certain goals, with algorithms and codes controlling the information input and action output that result in an abstract representation of the city. Both data collected and data ignored raise questions about political involvement and awareness since they relate to the management of the smart city or county [25]:
   - Are political decisions and policymaking converted into technical problems?
   - Are these technical problems created, resolved, and rely on experts in IT rooms of either corporations or government? [44,114,115].
   - Who controls the data information flows?
   - How do we keep the data information flows open?
   - How can the temporal nature of data help us understand a city better? [25].
4. Management Implications: The vast influx of data poses a huge challenge for budgets, administration, and management of this vast technological network of sensors.
5. Partnerships: Because of the interweaving of involvement between government and technology corporations, making wise decisions within such partnerships that are monopolizing these smart technologies can be a challenge for policymakers [44].
6. Surveillance and Freedom: There are concerns about these two issues, and [116] calls it the age of surveillance capitalism with two categories of citizens, the watchers and the watched. Along with this criticism falls the question of what data should remain private, what public, and what is the best avenue to ensure the privacy of citizens—part of this informational flow—and to what extent we can maintain a democratic society [44].
7. Citizen Dependency: The preceding triad poses a dependency of citizens on technology experts and exposure to data manipulations.
8. Accountability: Questions of accountability also arise within the triad of major participants:
   - Who is accountable for what within this triad?
   - To what extent is the corporation, the government, or the citizens accountable? [117] recommends smart city legislation, passing ordinances tied to specific surveillance technologies, or outlining a vision and developing guiding principles and smart practices, which can be typical for any new and major applications in government.

5.6.3 Spectrum of desired smartness and overall benefits

Because of these concerns, but also due to the challenges described earlier, smart technology applications may not be possible to apply in a blanket fashion throughout the city or county. An incremental approach may be more beneficial and as a result, smartness in a city or county will not be a simple duality of smart versus nonsmart. It seems almost impossible to replace every function and service in a jurisdiction with its smart equivalent. Therefore, it also seems important to try and understand smartness on a spectrum, the ***spectrum of desired smartness,*** which would most probably depend on the idiosyncrasies and resources of each locality.

"At any level of smartness, however, smart technology applications and 'an investment in ubiquitous, fast, reliable connectivity' involve positive impacts in **education**, **economy**, **mobility**, **public safety**, **sustainability**, and **quality of life**" [60]. To produce such positive results, making available the ability to connect to the Internet may be one of the first responsibilities of any county to its residents. In rural communities, broadband connectivity and Internet access have "cascading impacts" and improve access to health care, overall quality of life, enable community members to access distance learning programs, learn new skills, enable teleworking, "boost tourism and recreation in the area, enhance public safety, and offer more efficient transportation options" [60]. Obviously, for counties these would be very positive impacts to the overall quality of life of distant and rural residents. A similarly positive impact would have the improved responsiveness of a Public Works Department to infrastructure and utility maintenance in distant county locations.

5.7 Theoretical concerns

In one of the severe criticisms of smart cities, Professor in Practice of Architecture and Urban Design at the Graduate School of Design at Harvard University [118] asserts that the protagonists of the smart city movement create an apocalyptic scenario of problems cities have to resolve with the application of smart technologies while forgetting the notion of space and that it will "lead us to cities shackled by algorithmic conformity" [119]. Major theoretical implications of smart systems seem to exist in this duality between instrumental and value-based approaches to smartification but also in the ability to produce new understandings through the interpretation of big data collection [25]. However, the production of new urban theories based on a long-term interpretation of data is an instrumental rationality approach to understanding urban systems, with the goal of producing data-driven policies but is also an approach that increases the influence of technical experts [25].

Moving away from the instrumental understanding of urban systems, as described in the vision exploration section, the first step of creating urban policy requires a definition and categorization of major local needs followed by an assignment of smartness priority tasks to undertake. Thus, a city or county can plan, build, and manage high-priority smart systems and fulfill community-based desired outcomes. In the end, however, it is important to recognize that smartness may never resolve homelessness in LA or San Diego, which probably brings the most important issue in relation to smartness:

How can smartification deliver economic, social, and environmental value? [35].

As a result, sorting out needs, prioritizing interventions, and setting outcomes are a major process that can guide city and county smartification toward the benefit of the people [120]. acknowledge the same issue, the importance of the community needs. In other words, they argue for the AI to serve the needs of the community, improve existing conditions by focusing on such needs, for community participation, and in shaping smart systems to contribute to community trust.

Bailey and Coleman [121] argues a similar point and adds on the importance of an incremental approach to smartification, one that allows for an initial experimentation, and pilot program developing, involving all the stakeholders, in a way that the approach is tested with the public and all the involved parties, "massive upfront investment" is avoided and allows for the necessary technology and methodology adjustments. In addition, the authors argue for the importance of targeting a specific challenge and specifying the end beneficiary to generate measurable impacts from the technology.

Boyer [44] poses the same questions and fears regarding the smartification craze across the globe. She criticizes and asserts that the Smart City is the rejuvenation of the "Concept City," or "Dream City," an attempt to rationalize the city and see it from above as a two-dimensional project with issues that can be addressed with the implementation of one idea only, instead of realizing that uncertainty, complexity, and surprise rule the city. Within this notion of a smart city, no amount of data can represent all the economical, technical, and political aspects and negotiations that happen within a city.

The author continues that the Concept City has functioned since the Renaissance to "substitute, abstract, rationalize, and totalize" the image of the real city, and has kept planners aloof from reality as if looking down at the city on a two-dimensional graphic, erasing all other life dimensions of a city. As a result, the value-based considerations of quality of life and place can become important determinants for the outcome of smart technology applications, as a guide to the more complex life dimensions of cities to capture more of a city's full development patterns.

5.8 Smartification and smart growth

Some of the positive smartification impacts in education, economy, mobility, public safety, sustainability, and quality of life [60], in some respects, can be marginal. For example, an exclusive gated suburb is still an exclusive gated suburb whether smart or not. Similarly, a downtown devoid of city life after 5 p.m. is still devoid of city life after 5 p.m., whether the traffic signals are smart, the trash bins send signals when full, or the traffic flows are monitored by the latest technologies. Along the same line, smart technologies will not increase the number of intersections in suburbia or make intersections pedestrian friendly by reducing the distance between the mega strip malls across the four intersection corners.

Smart technologies will improve life comforts in a downtown until 5p.m. and will improve life for the gated community residents but how will smartification impact the spatial layout of a downtown or a new subdivision? How will it increase pedestrian interest on the first-floor fenestration of walkable areas? We can deduce that smartification creates a more comfortable life for what exists already or what is already spatially arranged, with sensors and the latest technologies. The spatial layout of the city, what leads to the 20-minute neighborhood and to a successful chrono-urbanism, the third places full of community members, the packed plazas and sidewalks full of cafes, bicycles, and scooters can be facilitated but cannot be created by smart technologies.

As Jane Jacobs articulated, in Death and Life of Great American Cities [122], urban life is complex and governed by a multitude of factors that contribute to quality of life so that no two-word utopian thinking can encapsulate or create. And, as [123] eloquently mentions, urban history is full of utopian experiments that failed to capture the complexity of urban life and only saw themselves collapse a few years later, under the weight of a missing urban life.

A variety of city layouts can implement smart technologies, but they may be dependent on a denser city layout. The physical structure of a city or community depends on the historical circumstances and decision-makers who created the layout. For example, the Public Land Survey System (PLSS), or else known as the Jeffersonian Grid, is a result of decision-making that brought a structural change to the layout of the country west of Ohio [124]. Would it be laid out differently if Thomas Jefferson had smart technologies available at the time? Considering Jefferson's admiration for the Roman equivalent of "Centuriation" and love of the classical world [125], that he was steeped in the Age of Reasoning and the rationalism of the Enlightenment intending to place reason upon the land, I would argue probably not; the system would be the same although more accurate and more sophisticated.

The reasons that initiated the historic idea of the PLSS seem unrelated to smart technologies and indeed [44] emphasized that, within the informational landscape of a smart city, the physical and architectural city and city space become invisible. Real-time data are transferred to control rooms and indicate the performance of different aspects of cities, but the notions of space and place are absent. Within this absence are the anecdotal data and stories that shape attachment, belonging, physical space and create place. This is one of the most significant criticisms of the smart city and according to Rem [118]:

> *Architecture used to be about the creation of community and making the best effort at symbolizing that community. Since the triumph of the market economy in the late 1970s, architecture no longer expresses public values but instead the values of the private sector.*

The preceding is a criticism that exemplifies a different functioning of smart technologies from the creation of architectural space. The same criticism points to a questionable relation between smart growth and smart cities if we discard "urban intelligence accumulated over centuries" [118].

However, it may be that we have to see the two smart processes as separate and value them as such. The most important of the two seems to be smart growth because it is the process that sets the physical structure of the city, the one that creates architectural and urban space. In itself is a smartification process since it reduces commute time, improves access to destinations, offers a variety of mobility options, reduces vehicle speeds, reduces intersection width, enables populations of different economic income to live more comfortable lives by reducing the distance between home and work, provides housing options, values public spaces as third places, and also values streets as public spaces to be shared by a variety of mobility modes and pedestrians alike.

I would argue that Smart Growth offers the foundation for quality of life exactly because it is based on "urban intelligence accumulated over centuries." Smart Growth with a garnish of sensors has the potential to improve quality of life even more and bring the results touted by Neal, Miller, Bansal, Fishman & Chew [60]. So, although both smart growth and smart cities claim similar results, a smart city that is just subdivisions, without a center, without a sense of community, devoid of third places, with streets dominated by the automobile, with pedestrians nowhere to walk, devoid of urban experiences, yet full of sensors is still a commute suburb no matter how many sensors it has, no matter how more efficient traffic has become because of the presence of the IoT.

In this case, smart technologies will improve the existing quality of life by shortening commute time, improving the operation of traffic signals, improving traffic flows, and the timely service of city trash bins in the strip malls. However, a resident will still have to use the automobile to go from one store to the next in the same shopping center, use the automobile to go the store across the street, navigate the gargantuan Black Friday parking lot, waste more money on gas, pollute the environment due to the extra trips the suburban layout generates, suffer the health consequences of increased pollution, and continue to live a life behind the steering wheel, scrambling to find some time personal or with the family or more money to spend on the car. Although this time, smart technologies will have saved the resident a mere half an hour of their day. It seems extremely important for the quality of life of the communities around us to not forget smart growth within this fervor of smartification.

5.9 Summary

The goal of this chapter has been to explore the smartification process in counties and address many different facets of such a process. They address technologies, inherent smartification processes, unique county characteristics and adaptations, policy creation, unique benefits, general concerns, and finally, theoretical implications.

5.9.1 Technologies

Several types of technologies have contributed to the appearance and evolution of the smart city/county concepts. Although 100 years ago no city could call itself smart, the turbulent decades of the 20th Century brought many technological innovations that lead to what today we call a smart city or a smart county. The development of sensor technology was probably the major contributing factor and what led to the term smart city in the late 2000s. However, without the development of the Internet nothing would be possible, because in a world blanketed with sensors, everything communicates with everything creating the IoT. Other important innovations that have contributed and continue to contribute to the development of the smart city/county concepts are user-generated content capabilities, smartphones, smartphone applications, cloud computing, GIS, analytics, augmented and virtual reality, and AI.

5.9.2 Process

When all smart technologies have been laid out, we can begin to discern the four basic parts of the smartification process. These include the following layers, which represent the different stages of the process:

1. Perception Layer (sensors),
2. Network Layer (dissemination and communication of information),
3. Analysis Layer (data exploration, sorting, storing, analysis, communication, and manipulation), and
4. Application Layer (city or county departments, management, and policymaking).

The last layer reflects the end goals of smart technologies, which can include land use scenarios, future growth strategies, natural hazard monitoring, natural resource conservation, risk and resilience to shocks and stresses, traffic monitoring and regulation, environmental pollution monitoring, health monitoring and intervention, water use, energy consumption, interface of citizens with government, and community engagement.

These end goals of smart technology applications are the instrumental type discerning them from value-based end goals, which reflect broader concepts or values in society such as quality of life and sense of place. In other words, two types of end goals can guide the smartification process:

- Instrumental and
- Value-based

Value-based goals are extremely important as an overall guide of the smartification process so the focus stays on the community.

5.9.3 In counties

Counties are an integral part of the American political landscape with some unique characteristics that distinguish them from cities. Differences from cities stem from geography, services provided, operational provisions, and structure of counties and while geography relates to the generally larger expanse of counties, states determine to a certain extent services, operational provisions, and structure with a resulting dependency on the state. Counties are required to provide some unique services compared to cities, such as judicial, transportation, and infrastructure to connect communities, public utilities, tax collection, election overseeing, monitoring, and certification of results, and property tax evaluations. Health services, hospitals and libraries, are usually operated by counties. To provide all these services counties, like cities, assign them to various departments that in some affluent counties may exceed 50 while in less affluent can be as few as 13.

Most county departments will benefit from the application of smart technologies and each department can utilize smartification accordingly using various tools for their own unique goals. Agriculture departments, for example, will aim toward smart agricultural monitoring and techniques, transportation departments will aim toward smart traffic monitoring and parking, while Planning departments will aim toward smart public engagement, predictive and spatial analytics, and immersive visualization methods. The extent of applications is immense and of course the information technology department will always play a leading role.

The unique county characteristics of geography, services provided, operational provisions, and structure result in 13 unique challenges for counties when implementing smart technologies. Geography creates challenges that relate to the larger county territories, unincorporated and incorporated communities, small and remote communities, rural communities, agriculture, natural disasters, and sprawl. The necessary services required by states, operational provisions, and county structure result in challenges that relate to state funding, state provisions, state mandates, coordination with municipalities, annexations, and lack of county visibility in relation to cities. Each of these challenges requires political and technological adaptations to address.

When prioritizing for smart technologies, it is very important to make Internet availability a priority to county residents. Next will be to understand the most urgent issues, decide what is realistic, prioritize the way smart technologies could help and, finally, set objectives, goals, strategies, and an action plan for implementation.

5.9.4 Benefits, concerns, and theory

Regardless of whether it is a county or city, smartification provides benefits in the areas of education, economy, mobility, public safety, sustainability, and quality of life. It poses though some challenges that relate to the interplay between the three major participants: technology corporations, government, and citizens. Such challenges include data and operations input and control, financing, the large volume of new operations, administration, management, surveillance, and cyber security issues.

Theoretical challenges stem from the interplay between instrumental and value-based end goals. Since the Renaissance, there has been a focus on creating the ideal city with all the ramifications of such one-sided approaches that forget the multidimensional character of human communities. Smartification often seems a similar desire to create a utopia forgetting that technology can only go so far in terms of improving human environments. There is an accumulation of knowledge over the centuries of what creates good urbanism, which is important to remember. It is also important to recognize that Smart Growth is a very different process and concept than smart cities with the ability to create meaningful architectural and urban spaces and places. Each of the two smart processes approaches quality of life from a very different perspective, can contribute positive results, but smart technologies lack the ability to create a foundation of good urbanism.

Finally, let us move on to the real interactive part of this chapter: review questions/exercises, hands-on projects, case projects, and optional team case project. The answers and/or solutions by chapter can be found in Appendix G.

5.10 Chapter review questions/exercises

True/false

1. True or False? A smart city, county, or region is knowledge-based.
2. True or False? The calculator, the personal computer, and finally, the Internet began to smartify cities and raise them slowly lower and lower below the smart level.
3. True or False? Sensors and the Internet of Things are responsible for the term "smart city," which began to appear sometime between 2005 and 2008 with the proliferation of these devices.
4. True or False? The opportunities started to lower when the evolution of sensors was combined with the evolution of device-to-device wireless communication, expansion of satellite use, smart phones, and user-input information to help collect, monitor, and transmit data.
5. True or False? As a result of the proliferation of sensors, the Internet of Things (IoT) started to materialize.

Multiple choice

1. With the evolution of the ___________, accessing information was not the only option anymore.
   a. Web 2
   b. World Wide Web
   c. Cloud computing
   d. Smart phone
   e. Computing
2. A discussion about smart cities or counties would be insufficient without a discussion about:
   a. 3D visualization
   b. 2D visualization
   c. GIS mapping
   d. Analytics
   e. Urban analytics
3. There is a __________________ of present and future information about an environment that starts with the representation of the existing physical environment, which is then superimposed to various degrees with existing information and future concepts, and eventually are supplanted by a complete virtual representation of both present and future suggestions.
   a. Reality
   b. Spectrum of visualizations
   c. Augmented reality of information
   d. Artificial Intelligence (AI)
   e. GPS tracking
4. Smartphones, already equipped with GPS tracking, camera, and microphone, create an ideal platform for:
   a. Holistic integration of smart city platforms
   b. Economic growth
   c. Making cities smarter
   d. AI applications
   e. E-economics
5. In general, __________ will change the way we park, how garages are conceived, and will impact parks and open spaces.
   a. Driverless cars
   b. Local transportation networks
   c. Unmanned aerial vehicles (UAVs) or drones
   d. Photosimulations
   e. Transportation and traffic applications

Exercise

Problem

How do you guide the vision process within a smart county?

Hands-on projects

Project

Do research: A set of exploratory questions must be answered as to why place and locality matter and why each county is identified by its own unique idiosyncrasies, local needs, standards, and implementation strategies that will differ and have to be locally adapted.

Case projects

Problem

Do research: A set of exploratory questions must be answered per challenge that can assist in setting priorities, planning, and organizing policy development in smart technology.

Optional team case project

Problem

Do research on the next three unique county challenges that relate to funding and services that are necessary locally or required by the state, operational provisions, and structure of counties [64].

References

[1] Menychtas A. European commission, ICT policy support programme. Online service delivery baseline and technical requirements report (D2.3 B). European Platform for Intelligent Cities (EPIC) Consortium; 2011.

[2] Rabari C, Storper M. The digital skin of cities: urban theory and research in the age of the sensored and metered city, ubiquitous computing, and big data. Camb J Reg Econ Soc 2014;8(1):27−42. https://doi.org/10.1093/cjres/rsu021.

[3] McDermott B. How the internet of things makes smart cities a reality. Dogtown Media; 2020. Retrieved 24 January 2021, from, https://www.dogtownmedia.com/how-the-internet-of-things-makes-smart-cities-a-reality/.

[4] Collier C. What a smart city is… and is not − smart cities connect. 2020. Smartcitiesconnect.org. Retrieved 19 December 2020, from, https://smartcitiesconnect.org/what-a-smart-city-is-and-is-not/#:~:text=Techopedia%3A%20%E2%80%9CA%20smart%20city%20is,consumption%2C%20wastage%20and%20overall%20costs.

[5] Sensor History. 2020. Allsensors.com. Retrieved 20 December 2020, from, https://allsensors.com/engineering-resources/sensor-history#:~:text=The%20person%20most%20responsible%20for,student%20in%20the%20late%20fifties.&text=In%201960%2C%20Art%20joined%20GE,on%20solid%20state%20motion%20transduction.

[6] O'Connor S. What is interoperability, and why is it important?. 2017. Adsc.com. Retrieved 7 January 2021, from, https://www.adsc.com/blog/what-is-interoperability-and-why-is-it-important.

[7] Newcombe T. The rise of the sensor-based smart city. 2014. Govtech.com. Retrieved 20 December 2020, from, https://www.govtech.com/data/The-Rise-of-the-Sensor-Based-City.html.

[8] National Conference of State Legislatures. Creating smart communities | A guide for state policymakers. 2020. Ncsl.org. Retrieved 24 May 2021, from, https://www.ncsl.org/research/energy/creating-smart-communities-a-gide-for-state-policymakers.aspx.

[9] Talari S, Shafie-khah M, Siano P, Loia V, Tommasetti A, Catalão J. A review of smart cities based on the internet of things concept. Energies 2017;10(4):421. https://doi.org/10.3390/en10040421.

[10] McCabe B. Creating smart cities with IoT - internet of things blog. IBM Business Operations Blog 2016. Retrieved 7 February 2021, from, https://www.ibm.com/blogs/internet-of-things/creating-smart-cities-iot/.

[11] Anttiroiko A. In: Roger W, editor. Encyclopedia of the city. Caves: Routledge; 2013.

[12] Baig Z, Szewczyk P, Valli C, Rabadia P, Hannay P, Chernyshev M, et al. Future challenges for smart cities: cyber-security and digital forensics. Digit Invest 2017;22:3−13. https://doi.org/10.1016/j.diin.2017.06.015.

[13] O'Reilly T. What is web 2.0. 2005. Oreilly.com. Retrieved 3 February 2021, from, https://www.oreilly.com/pub/a/web2/archive/what-is-web-20.html?page=1.

[14] Phillips A. How has the cloud affected the world of urban planning - CloudCow. CloudCow 2020. Retrieved 7 March 2021, from, https://cloudcow.com/content/how-has-the-cloud-affected-the-world-of-urban-planning/#.YL6NSvlKjdM.

[15] InsightsSuccess. How cities are getting smarter with cloud computing - InsightsSuccess. InsightsSuccess; 2021. Retrieved 6 March 2021, from, https://www.insightssuccess.com/cities-getting-smarter-cloud-computing/.

[16] Phillips C. How smartphones revolutionized society in less than a decade. Government Technology; 2014. Retrieved 13 January 2021, from, https://www.govtech.com/products/How-Smartphones-Revolutionized-Society-in-Less-than-a-Decade.html.

[17] Messerschmitt D. The convergence of telecommunications and computing: what are the implications today? Proc IEEE 1996;84(8):1167−86. https://doi.org/10.1109/5.533962.

[18] Keyideas. Impact of smartphones on society - use of mobile phones. 2017. Keyideasinfotech.com. Retrieved 16 January 2021, from, https://www.keyideasinfotech.com/blog/impact-of-smartphone-on-society/.

[19] Tiwari A. Impact of internet of things on mobile apps | mobile app development. JumpGrowth 2021. Retrieved 16 January 2021, from, https://jumpgrowth.com/impact-of-internet-of-things-on-mobile-apps/.

[20] Prescient Solutions Group. Psgincs.com; 2019. Retrieved 17 January 2021, from, https://www.psgincs.com/development-of-mobile-applications-its-positive-impact/.

[21] ESRI. History of GIS | early history and the future of GIS - Esri. 2021. Esri.com. Retrieved 7 February 2021, from, https://www.esri.com/en-us/what-is-gis/history-of-gis.

[22] MacHarg I. Design with nature. In: Ian L McHarg, editor; 1969 (Second printing). Published for the American Museum of Natural History (by) the Natural History Press.

[23] SAS Institute Inc. Analytics: what it is and why it matters. 2021. Sas.com. Retrieved 14 February 2021, from, https://www.sas.com/en_us/insights/analytics/what-is-analytics.html.

[24] Oracle. What is analytics?. Oracle. 2021. Oracle.com. Retrieved 15 February 2021, from, https://www.oracle.com/business-analytics/what-is-analytics/#analytics.

[25] Kandt J, Batty M. Smart cities, big data and urban policy: towards urban analytics for the long run. Cities 2021;109:102992. https://doi.org/10.1016/j.cities.2020.102992.

[26] Urban Footprint [Ebook]. In: Data driven planning for city resilience, climate adaptation, and recovery. Urban Footprint; 2021. Retrieved 17 February 2021, from, https://cdn2.hubspot.net/hubfs/2972972/UF_eBook_Risk-and-Resilience%20v3.pdf.

[27] Shieber J. Taking a page from SimCity, UrbanFootprint pitches new tools for urban development. 2021. Techcrunch.com. Retrieved 17 February 2021, from, https://techcrunch.com/2020/02/12/taking-a-page-from-simcity-urbanfootprint-pitches-new-tools-for-developers-and-urban-planners/.

[28] Murray P. Urban analytics: a 21st century introduction to spatial analytics. 2018. Carto.com. Retrieved 28 February 2021, from, https://carto.com/blog/urban-analytics-introduction-spatial-analytics/.

[29] Joshi N. Applications of immersive technologies in smart cities. Forbes 2019. Retrieved 14 March 2021, from, https://www.forbes.com/sites/cognitiveworld/2019/08/05/applications-of-immersive-technologies-in-smart-cities/.

[30] Meenar M, Kitson J. Using multi-sensory and multi-dimensional immersive virtual reality in participatory planning. Urban Science 2020;4(3):34. https://doi.org/10.3390/urbansci4030034.

[31] van Leeuwen J, Hermans K, Jylhä A, Nijman H, Quanjer A. Effectiveness of virtual reality in participatory urban planning: a

case study. In: MAB18: proceedings of the 4th media architecture biennale conference. New York NY United: States; Association for Computing Machinery; 2021. p. 128–36. Retrieved 14 March 2021, from, https://dl.acm.org/doi/10.1145/3284389.3284491.

[32] Yagol P, Ramos F, Trilles S, Torres-Sospedra J, Perales F. Erratum: Ramos, F.; Trilles, S.; Torres-Sospedra, J.; Perales, F.J. New trends in using augmented reality apps for smart city contexts. ISPRS Int. J. Geo-Inf. 2018, 7, 478 ISPRS Int J Geo-Inf 2020;9(9):537. https://doi.org/10.3390/ijgi9090537.

[33] Costa C. How AI can impact mobile app development and user experience. Medium 2019. Retrieved 17 January 2021, from, https://uxplanet.org/how-ai-can-impact-mobile-app-development-and-user-experience-e575c0f9068c.

[34] Sheladiya C. AI can impact mobile app development and user experience. ReadWrite 2019. Retrieved 17 January 2021, from, https://readwrite.com/2019/09/26/ai-can-impact-mobile-app-development-and-user-experience/#:~:text=Improving%20the%20App%20UX,the%20patterns%20of%20user%20behavior.&text=AI%20can%20also%20be%20utilized,and%20solve%20specific%20user%20problems.

[35] Tomer A. Artificial intelligence in America's digital city. Brookings; 2019. Retrieved 13 March 2021, from, https://www.brookings.edu/research/artificial-intelligence-in-americas-digital-city/.

[36] Clearbridge Mobile. How artificial intelligence is driving mobile app personalization. Clearbridge mobile. 2020. Retrieved 28 February 2021, from, https://clearbridgemobile.com/artificial-intelligence-driving-future-mobile-app-personalization/.

[37] Brazeal C. Could the next round of transportation innovations finally end the parking status quo? Planetizen - urban planning news, jobs, and education. 2017. Retrieved 10 March 2021, from, https://www.planetizen.com/node/92647/could-next-round-transportation-innovations-finally-end-parking-status-quo.

[38] H.O.K. Autonomous vehicles and the future of urban planning. HOK; 2021. Retrieved 10 March 2021, from, https://www.hok.com/ideas/research/autonomous-vehicles-urban-planning/.

[39] Stephens R, Dannenberg R, Kellington W, Sherman P [EBook]. Chapter 3. In: Using drones in planning practice. American Planning Association (APA) Planning Advisory Service; 2020. PAS 597. Retrieved 30 May 2021, from, https://www.planning.org/publications/report/9207028/.

[40] Chianese A, Picciallli F, Jung J. The internet of cultural things: towards a smart cultural heritage. Los Alamitos, CA, USA. In: 12th international conference on signal-image technology & internet-based systems (SITIS). IEEE Computer Society; 2016. Retrieved 30 May 2021, from, https://doi.ieeecomputersociety.org/10.1109/SITIS.2016.83.

[41] Silva M, Teixeira L. Developing an extended reality platform for immersive and interactive experiences for cultural heritage: serralves museum and Coa archeologic park. 2016 12th international conference on signal-image technology & internet-based systems (SITIS), Volume: 1. Los Alamitos, CA, USA: IEEE Computer Society; 2020. p. 300–2. Retrieved 30 May 2021, from, https://www.computer.org/csdl/proceedings-article/ismar-adjunct/2020/767500a300/1pBMgEgstH2.

[42] Unal M, Bostanci E, Sertalp E. Distant augmented reality: bringing a new dimension to user experience using drones. Digital Applications in Archaeology and Cultural Heritage 2020;17:e00140. https://doi.org/10.1016/j.daach.2020.e00140.

[43] Zhao Z, Zhang Y. Impact of smart city planning and construction on economic and social benefits based on big data analysis. Complexity 2020;2020:1–11. https://doi.org/10.1155/2020/8879132.

[44] Boyer M. M. Christine Boyer. Smart city and instrumental rationality in the 21st century [Video]. 2016. Retrieved 10 March 2021, from, https://youtu.be/rX34HiGVGvA.

[45] Callahan G, Ikeda S. Jane Jacobs' critique of rationalism in urban planning. Cosmos Taxis 2013;1(3):10–9. Retrieved 8 June 2021, from, https://cosmosandtaxis.files.wordpress.com/2014/11/ct_1_3_callahan_ikeda.pdf.

[46] Weber M. In: Wittich C, editor. Economy and society, guenther Roth. University of California Press; 1978. p. 24.

[47] Wenar L. John Rawls (Stanford encyclopedia of philosophy). 2021. Plato.stanford.edu. Retrieved 8 April 2021, from, https://plato.stanford.edu/entries/rawls/.

[48] Angelakoglou K, Nikolopoulos N, Giourka P, Svensson I, Tsarchopoulos P, Tryferidis A, Tzovaras D. A methodological framework for the selection of key performance indicators to assess smart city solutions. Smart Cities 2019;2(2):269–306. https://doi.org/10.3390/smartcities2020018.

[49] Ludlow D [EBook]. In: Tackling urban sprawl: towards a compact model of cities? University of the West of England; 2014. Retrieved 15 May 2021, from, https://ec.europa.eu/environment/land_use/pdf/ludlow.pdf.

[50] Callahan G, Ikeda S. Jane Jacobs' critique of rationalism in urban planning. Cosmos + Taxis 2014;1(3). Retrieved 23 April 2021, from, https://cosmosandtaxis.files.wordpress.com/2014/11/ct_1_3_callahan_ikeda.pdf.

[51] Neuman M. Does planning need the plan? J Am Plann Assoc 1998;64(2):208–20. https://doi.org/10.1080/01944369808975976.

[52] Evangelopoulos E. Landscape character dominant features in resort developments (Ph.D. Dissertation). University of Arizona; 2000.

[53] Najafi M, Shariff M. The concept of place and sense of place in architectural studies. Int J Hum Soc Sci 2011;5(8):1054–60. Retrieved 8 April 2021, from, https://publications.waset.org/14034/the-concept-of-place-and-sense-of-place-in-architectural-studies.

[54] Relph E. Place and placelessness. Sage; 1976.

[55] Seamon D. Life takes place: phenomenology, lifeworlds, and place making. Routledge; 2018.

[56] Seamon D, Sowers J. Place and placelessness, Edward Relph. In: Hubbard P, Kitchen R, Vallentine G, editors. Key texts in human geography. SAGE Publications; 2008. p. 43–51.

[57] Tuan Y. Space and place. University of Minnesota Press; 1977.

[58] Richman R. In: Caves RW, editor. Encyclopedia of the city. Routledge; 2013.

[59] PAG-EVAC [EBook]. In: Smart communities: regional issues statements, goals and potential strategies. Pima Association of Governments, Economic Vitality Advisory Committee (EVAC), Smart Communities Task Force; 2018. Retrieved 7 March 2021, from, https://pagregion.com/livability/.

[60] Neal C, Miller J, Bansal M, Fishman T, Chew B. The promise of smart rural communities. Deloitte Insights. 2021. Retrieved 15 May 2021, from, https://www2.deloitte.com/us/en/insights/industry/public-sector/the-promise-of-smart-rural-communities.html.

[61] Evangelopoulos E, Nuworsoo C. Form-based codes: an overview of the literature. Focus 2016;3(1). Article 17. Retrieved 10 March 2017, from, https://digitalcommons.calpoly.edu/focus/vol13/iss1/17.

[62] A.U.C. The American state and local government. 2021. TheUSAonline.com. Retrieved 18 May 2021, from, https://www.theusaonline.com/government/state-local-government.htm.

[63] I.L.G [EBook]. In: Types and responsibilities of local agencies. ILG; 2015. Retrieved 17 April 2021, from, https://www.ca-ilg.org/sites/main/files/file-attachments/types__responsibilities_of_local_agencies_0.pdf?1497852178.

[64] OpenStax. County and city government. 2021. Courses.lumenlearning.com. Retrieved 10 May 2021, from, https://courses.lumenlearning.com/os-government2e/chapter/county-and-city-government/.

[65] Feldstein L [EBook]. In: General plans and zoning: a toolkit for building healthy, vibrant communities. California Department of Health Services; 2007. Retrieved 29 May 2021, from, https://www.ca-ilg.org/sites/main/files/file-attachments/resources__finalbook_0.pdf?1441322939.

[66] NACo. Counties matter: stronger counties. Stronger America. NACo 2021. Retrieved 17 March 2021, from, https://www.naco.org/resources/featured/counties-matter.

[67] NACo. New NACo data release highlights the key role of counties in elections. NACo 2020. Retrieved 10 February 2021, from, https://www.naco.org/blog/new-naco-data-release-highlights-key-role-counties-elections.

[68] The United States Census Bureau. 2014–2018 poverty rate in the United States by county. The United States Census Bureau; 2019. Retrieved 9 March 2021, from, https://www.census.gov/library/visualizations/interactive/2014-2018-poverty-rate-by-county.html.

[69] West River Eagle [Ebook]. In: Meeting of the Ziebach county commissioners. West River Eagle; July 7, 2020. Retrieved 9 March 2021, from, https://newzgroup.com/SDLegals/2020/458183-2020-07-16_1002.pdf.

[70] Ziebach County. Welcome to the Ziebach county, SD website. Ziebach County, SD. 2021. Retrieved 9 March 2021, from, https://ziebachcounty.org/.

[71] Riley County. Riley county, KS official website. 2021. Rileycountyks.gov. Retrieved 9 March 2021, from, https://www.rileycountyks.gov/.

[72] Riley County. 2021 budget overview, Riley County, KS [EBook]. Riley County, KS; 2020. Retrieved 9 June 2021, from, https://www.rileycountyks.gov/ArchiveCenter/ViewFile/Item/2775.

[73] County of Sonoma. Departments & agencies. 2021. Sonomacounty.ca.gov. Retrieved 9 March 2021, from, https://sonomacounty.ca.gov/Your-Government/Departments-and-Agencies/.

[74] Sacramento County. Departments. 2021. Saccounty.net. Retrieved 9 March 2021, from, https://www.saccounty.net/CountyDepartments/Pages/default.aspx.

[75] Pinal County. Home - departments & offices of pinal county Arizona. 2021. Pinalcountyaz.gov. Retrieved 9 March 2021, from, https://www.pinalcountyaz.gov/Pages/Departments.aspx.

[76] NACo [EBook]. In: 2021 NACo policy priorities. National Association of Counties NACo; 2021. Retrieved 31 May 2021, from, https://www.naco.org/sites/default/files/documents/2021LegPrio_FINAL.pdf.

[77] Werner C. Fire technology - smart cities & the fire service - Charles Werner - firehouse. Firehouse 2021. Retrieved 8 May 2021, from, https://www.firehouse.com/tech-comm/article/21027703/fire-technology-smart-cities-the-fire-service-charles-werner-firehouse.

[78] Begemann B. Bayer BrandVoice: what smart cities are learning from smart farms. Forbes 2021. Retrieved 8 May 2021, from, https://www.forbes.com/sites/bayer/2019/11/27/what-smart-cities-are-learning-from-smart-farms/?sh=56db6214ac64.

[79] Emorphis Technologies. What is smart farming — Everything you want to know about it. Medium 2019. Retrieved 8 March 2021, from, https://medium.com/@emorphis.technologies/what-is-smart-farming-everything-you-want-to-know-about-it-28e0e5d8e142.

[80] Maayan G. How do AI-based drones work? Medium 2020. Retrieved 30 May 2021, from, https://heartbeat.fritz.ai/how-ai-based-drones-work-a94f20e62695.

[81] Michels M, von Hobe C, Musshoff O. A trans-theoretical model for the adoption of drones by large-scale German farmers. J Rural Stud 2020;75:80–8. https://doi.org/10.1016/j.jrurstud.2020.01.005.

[82] Rohi G, Ejofodomi O, Ofualagba G. Autonomous monitoring, analysis, and countering of air pollution using environmental drones. Heliyon 2020;6(1):e03252. https://doi.org/10.1016/j.heliyon.2020.e03252.

[83] Barnas A, Chabot D, Hodgson A, Johnston D, Bird D, Ellis-Felege S. A standardized protocol for reporting methods when using drones for wildlife research. J Unmanned Veh Syst 2020;8(2):89–98. https://doi.org/10.1139/juvs-2019-0011.

[84] Elsayed E, Ashrry A. A proposed model for measuring the performance of smart public parks. ERJ. Engi Res J 2020;43(3):245–60. https://doi.org/10.21608/erjm.2020.95301.

[85] Smart Parks [EBook]. In: Smart parks. ShadowView-Internet of Life; 2018. Retrieved 8 May 2021, from, https://www.sabineboogaard.nl/wp-content/uploads/2018/01/Smart-Parks-stichting-ShadowView.pdf.

[86] Smart Parks. Smart Parks - providing technology and services to protect wildlife. Smart Parks; 2021. Retrieved 8 May 2021, from, https://www.smartparks.org/.

[87] Krutz G, Waskiewicz S. County and city government - American government 2e. 2019. Openstax.org. Retrieved 10 February 2021, from, https://openstax.org/books/american-government-2e/pages/1-introduction.

[88] OpenGov [EBook]. In: 8 unique challenges for county governments and how to address them; 2020. OpenGov.com. Retrieved 10 February 2021, from, http://go.opengov.com/ebook-8-unique-challenges-for-counties.

[89] ESRI. Sensors for smart cities. 2015. Esri.in. Retrieved 12 May 2021, from, https://www.esri.in/esri-news/publication/vol9-issue1/articles/sensors-for-smart-cities.

[90] Begemann B. Bayer BrandVoice: what smart cities are learning from smart farms. Forbes 2019. Retrieved 16 May 2021, from, https://www.forbes.com/sites/bayer/2019/11/27/what-smart-cities-are-learning-from-smart-farms/?sh=308e2a88ac64.

[91] Emorphis Technologies. What is Smart Farming — Everything you want to know about it. Medium 2019. Retrieved 8 May 2021, from, https://medium.com/@emorphis.technologies/what-is-smart-farming-everything-you-want-to-know-about-it-28e0e5d8e142.

[92] Miao Q. The fiscal implications of managing natural disasters for national and subnational governments. In: Oxford research encyclopedia of natural hazard science; 2018. Retrieved 12 May 2021, from, https://www.researchgate.net/publication/323998267_The_Fiscal_Implications_of_Managing_Natural_Disasters_for_National_and_Subnational_Governments.

[93] Miao Q, Hou Y, Abrigo M. Measuring the financial shocks of natural disasters. Natl Tax J 2018;71(1):11–44. https://doi.org/10.17310/ntj.2018.1.01.

[94] Taulbee L. The local government financial response to natural disasters in Kentucky. 2019. Uknowledge.uky.edu. Retrieved 11 May 2021, from, https://uknowledge.uky.edu/cgi/viewcontent.cgi?article=1329&context=mpampp_etds.

[95] Benson C, Clay E [ebook]. (Disaster risk management series, No. 4). In: Understanding the economic and financial impacts of natural

disasters. World Bank; 2004. Retrieved 13 May 2021, from, https://openknowledge.worldbank.org/handle/10986/15025.

[96] I.L.G [EBook]. In: Understanding the basics of municipal revenues in California: cities, counties and special districts. Institute for Local Government; 2016. Retrieved 19 May 2021, from, https://www.ca-ilg.org/sites/main/files/file-attachments/basics_of_municipal_revenue_2016.pdf.

[97] Heaton S. State mandates | rural counties. 2021. Rcrcnet.org. Retrieved 2 April 2021, from, https://www.rcrcnet.org/state-mandates.

[98] Texas Conference of Urban Counties. Policy | urban counties. 2019. Cuc.org. Retrieved 9 May 2021, from, https://cuc.org/policy/.

[99] Busch K, Goldstein M. Arizona leads the smart region space by the founding "the connective". Green Living Mag 2020. Retrieved 6 April 2021, from, https://greenlivingmag.com/arizona-smart-region-space-founding-the-connective/.

[100] Yurko A. A practical perspective about annexation in Florida. Stetson Law Rev 1996;25:699—719. Retrieved 17 May 2021, from, https://www.stetson.edu/law/lawreview/media/a-practical-perspective-about-annexation-in-florida-25-3.pdf.

[101] Williams C. 5 Things about Cape coral: annexation from the county's point of view. 2018. News-press.com. Retrieved 6 May 2021, from, https://www.news-press.com/story/news/local/cape-coral/2018/10/28/cape-coral-annexation-county-pine-island-matlacha/1751835002/.

[102] Horwedel J, Costa F, Lindsey L, Poppel K, Silver M, Stoll G, Zelinka A [EBook]. In: Planning office of the future task Fordc. American Planning Association; 2015. Retrieved 10 April 2021, from, https://planning-org-uploaded-media.s3.amazonaws.com/legacy_resources/leadership/agendas/2015/spr/pdf/PlanningOfficeFutureFinal.pdf.

[103] NACo. NACo policy priorities. NACo 2016. Retrieved 10 December 2020, from, https://www.naco.org/resources/featured/2020-naco-policy-priorities.

[104] The University of Kansas. Developing a strategic plan | community tool box [Chapter 8]. 2021. Ctb.ku.edu. Retrieved 24 May 2021, from, https://ctb.ku.edu/en/table-of-contents/structure/strategic-planning.

[105] Public Health Law & Policy (PHLP) [EBook]. In: Healthy planning policies: a compendium from California general plans. Public Health Law & Policy (PHLP)-Planning for Healthy Places; 2009. Retrieved 29 May 2021, from, https://www.ca-ilg.org/sites/main/files/file-attachments/resources__healthy_planning_policies_compendium_final_web_090925_0.pdf?1441322939.

[106] Meijer A, Bolívar M. Governing the smart city: a review of the literature on smart urban governance. Int Rev Adm Sci 2015;82(2):392—408. https://doi.org/10.1177/0020852314564308.

[107] Menon A. 5 ways to make our cities smarter. World Economic Forum; 2015. Retrieved 19 May 2021, from, https://www.weforum.org/agenda/2015/10/5-ways-to-make-our-cities-smarter/.

[108] City of Austin, TX. In: Live from Austin, Texas: the smart city challenge; 2016 [EBook]. City of Austin, TX. Retrieved 30 May 2021, from, https://www.austintexas.gov/sites/default/files/files/Transportation/Austin_SCCFinal_Volume1_5.25.pdf.

[109] Lakamp B. What are the 10 key things that make a city smart? ReadWrite. 2017. Retrieved 22 February 2021, from, https://readwrite.com/2017/07/17/brian-lakamp-what-makes-cities-smart-cl1/.

[110] Fairfax County, VA. Countywide strategic plan | strategic plan. 2021. Fairfaxcounty.gov. Retrieved 30 May 2021, from, https://www.fairfaxcounty.gov/strategicplan/.

[111] Abbas M. Building smart cities with the smart citizen approach. IBM Business Operations Blog; 2021. Retrieved 31 May 2021, from, https://www.ibm.com/blogs/internet-of-things/building-smart-city-solutions/.

[112] OECD [EBook]. In: Smart cities and inclusive growth. Organisation for Economic Co-operation and Development; 2020. Retrieved 17 May 2021, from, https://www.oecd.org/cfe/cities/OECD_Policy_Paper_Smart_Cities_and_Inclusive_Growth.pdf.

[113] Rujan A. Risk versus reward: six considerations for smart cities | explore our thinking | plante moran. 2018. Plantemoran.com. Retrieved 31 May 2021, from, https://www.plantemoran.com/explore-our-thinking/insight/2018/08/risk-versus-reward-six-considerations-for-smart-cities.

[114] Boyer M. Dreaming the rational city. The MIT Press; 1983.

[115] Ellis C. Dreaming the rational city: the myth of American city planning by Christine Boyer. Berk Plann J 2012;2(1). https://doi.org/10.5070/bp32113206.

[116] Naughton J. 'The goal is to automate us': welcome to the age of surveillance capitalism. the Guardian; 2019. Retrieved 6 April 2021, from, https://www.theguardian.com/technology/2019/jan/20/shoshana-zuboff-age-of-surveillance-capitalism-google-facebook.

[117] Sunlight Foundation. Smart cities: best practices for transparency: sunlight foundation. 2020. Sunlightfoundation.com. Retrieved 6 April 2021, from, https://sunlightfoundation.com/our-work/open-cities/smart-cities/best-practices/.

[118] Koolhaas R. Rem koolhaas asks: are smart cities condemned to be stupid? ArchDaily. 2014. Retrieved 6 February 2021, from, https://www.archdaily.com/576480/rem-koolhaas-asks-are-smart-cities-condemned-to-be-stupid.

[119] Koolhaas R. The digital city will lead to "total conformity". New Perspect Q 2018;35(4):46—9. https://doi.org/10.1111/npqu.12177.

[120] Moreira S, Peixoto Almeida T. 7 practical applications of artificial intelligence in urban management. ArchDaily; 2020. Retrieved 13 March 2021, from, https://www.archdaily.com/937949/7-practical-applications-of-artificial-intelligence-in-urban-management.

[121] Bailey D, Coleman Y. Urban IoT and AI: how can cities successfully leverage this synergy? — AI business. AI business; 2018. Retrieved 14 March 2021, from, https://aibusiness.com/document.asp?doc_id=760580.

[122] Jacobs J. The death and life of great American cities. Jonathan Cape; 1961.

[123] Mora L, Deakin M. Moving beyond the smart city utopia. In: Mora L, Deakin M, editors. Untangling smart cities: from utopian dreams to innovation systems for a technology-enabled urban sustainability. Elsevier; 2019. p. 1—17. Retrieved 16 May 2021, from, https://www.sciencedirect.com/science/article/pii/B9780128154779000013.

[124] White C [Ebook]. In: A history of the rectangular survey system. U.S. Department of The Interior, Bureau of Land Management; 1983. Retrieved 13 April 2021, https://www.blm.gov/sites/blm.gov/files/histrect.pdf.

[125] Rashid M. The plan is the program: Thomas Jefferson's plan for the rectilinear Survey of 1784. In: 84TH ACSA annual meeting - regional papers. Association of collegiate schools of architecture (ACSA); 1996. Retrieved 13 June 2021, from, https://www.acsa-arch.org/chapter/the-plan-is-the-program-thomas-jeffersons-plan-for-the-rectilinear-survey-of-1784/.

Further reading

AFUL G. Definition: interoperability. 2021. Interoperability-definition. info. Retrieved 7 February 2021, from, http://interoperability-definition.info/en/.

Alshahadeh T. Smart cities, smarter management - smart city hub. Smart City Hub; 2018. Retrieved 3 February 2021, from, http://smartcityhub.com/governance-economy/smart-cities-smarter-management/.

Alshahadeh T, Marsap A. Smart cities, smarter management: developing a smart framework for smart cities management in Europe. GE-Int J Manage Res 2018;6(9):41–73. Retrieved 3 February 2021, from, https://www.researchgate.net/publication/327994661_Smart_Cities_Smarter_Management_Developing_a_Smart_Framework_for_Smart_Cities_Management_in_Europe.

Alshahadeh T, Marsap A. Smart cities, smarter management: developing a smart framework for smart cities management in Europe. GE-Int J Manage Res 2018;6(9). Retrieved 7 June 2021, from, https://www.researchgate.net/publication/327994661_Smart_Cities_Smarter_Management_Developing_a_Smart_Framework_for_Smart_Cities_Management_in_Europe.

Baek D. When smart cities become dumb cities. Business Korea; 2019. Retrieved 6 February 2021, from, http://www.businesskorea.co.kr/news/articleView.html?idxno=35247.

Betty Magome B. Beyond smart cities: the role of utilities in enabling smart regions. T&D World 2021. Retrieved 10 April 2021, from, https://www.tdworld.com/smart-utility/article/20972522/beyond-smart-cities-the-role-of-utilities-in-enabling-smart-regions.

Brasuell J. AI and the digital city. Planetizen - urban planning news, jobs, and education; 2019. Retrieved 13 February 2021, from, https://www.planetizen.com/news/2019/07/105509-ai-and-digital-city.

California State University Long Beach. Models of public policymaking. 2002. Web.csulb.edu. Retrieved 31 May 2021, from, https://web.csulb.edu/~msaintg/ppa590/models.htm.

Cohen B. The top 10 smart cities on the planet. Fast Company; 2012. Retrieved 18 October 2020, from, https://www.fastcompany.com/90186037/the-top-10-smart-cities-on-the-planet.

European Commission [EBook]. In: Analysing the potential for wide scale roll out of integrated smart cities and communities solutions SCC solution best practices. European Commission, Directorate-General for Energy; 2016. Retrieved 15 May 2021, from, http://icities4greengrowth.in/sites/default/files/casestudyfiles/2018-03/Smart%20MELIT%20Toyota.pdf.

European Commission b [EBook]. In: Analysing the potential for wide scale roll out of integrated smart cities and communities solutions final report. European Commission, Directorate-General for Energy; 2016. Retrieved 15 May 2021, from, https://ec.europa.eu/energy/sites/ener/files/documents/d2_final_report_v3.0_no_annex_iv.pdf.

Grossi G, Meijer A, Sargiacomo M. A public management perspective on smart cities: 'Urban auditing' for management, governance, and accountability. Publ Manag Rev 2020;22(5):633–47. https://doi.org/10.1080/14719037.2020.1733056.

Harrison L. Digital cities 2019: Winners gear up for a new decade. GovTech 2021. Retrieved 10 May 2021, https://www.govtech.com/dc/articles/digital-cities-2019.html.

Hitchcock D. Instrumental rationality. Toronto, ON, Canada. In: Argumentation in multi-agent systems-7th international workshop, ArgMAS 2010; 2010. Retrieved 26 March 2021, from, https://books.google.com/books?id=KnyhDgAAQBAJ&pg=PA535&lpg=PA535&dq=Hitchcock+D.+Instrumental+rationality.+Toronto,+ON,+Canada.+In:+Argumentation+in+multi-agent+systems-7th+international+workshop,+ArgMAS+2010;+2010&source=bl&ots=fhPJ2Ht-g2&sig=A.

Impact of Mobile apps on our daily life. Website designing company in Delhi. Web Development Company Delhi; 2021. Retrieved 18 January 2021, from https://www.unicodesolutions.com/impact-mobile-apps-daily-life.

Lancaster University. Report calls for National parks to get smart. Lancaster University; 2018. Lancaster.ac.uk. Retrieved 8 May 2021, from, https://www.lancaster.ac.uk/news/articles/2018/report-calls-for-national-parks-to-get-smart/.

Mainka A, Khveshchanka S, Stock W. Dimensions of informational city research. In: Presentation, conference presentation in: Digital cities 7 - real world experiences. Brisbane, Qld: State Library of Queensland; 2011.

Municipal Research & Services Center (MRSC) [EBook]. In: Local government policy-making process. Municipal Research & Services Center (MRSC); 2017. Retrieved 9 May 2021, from, https://mrsc.org/getmedia/e46223b6-f3ac-4afb-b7d9-b2362edf6890/Local-Government-Policy-Making-Process.pdf.aspx?ext=.pdf.

Northern Ontario School of Medicine. Policy development process with best practices and tools, NOSM. Nosm.ca. Retrieved 9 May 2021, from https://www.nosm.ca/about/policies-and-procedures/policy-development-and-process/.

Park E, del Pobil A, Kwon S. The role of internet of things (IoT) in smart cities: technology roadmap-oriented approaches. Sustainability 2018;10(5):1388. https://doi.org/10.3390/su10051388.

Riley County Commission. 2021 budget overview. 2021. Rileycountyks.gov. Retrieved 1 May 2021, from, https://www.rileycountyks.gov/ArchiveCenter/ViewFile/Item/2775.

Rodwell G [EBook]. In: The predominance of instrumental rationality in modern organizations; 2018. https://www.academia.edu. Retrieved 26 March 2021, from, https://www.academia.edu/36356309/The_Predominance_of_Instrumental_Rationality_in_Modern_Organizations.

Sarwar M, Soomro T. Impact of smartphone's on society. Eur J Sci Res 2013;98(2):216–26. Retrieved 16 January 2021, from, http://www.europeanjournalofscientificresearch.com.

Smartphones and how they impact our daily lives. Business matters. 2019. Retrieved 16 January 2021, from, https://bmmagazine.co.uk/business/smartphones-and-how-they-impact-our-daily-lives/#:~:text=The.

State Government of Victoria. Plan Melbourne - 20-minute neighbourhoods. 2016. Planmelbourne.vic.gov.au. Retrieved 8 June 2021, from, https://www.planmelbourne.vic.gov.au/current-projects/20-minute-neighbourhoods.

Sutton R [EBook]. In: The policy process: an overview. Chameleon Press Ltd, Overseas Development Institute; 1999. Retrieved 10 May 2021, from, https://www.files.ethz.ch/isn/104749/wp118.pdf.

Thornberg C, Kleinhenz R, Meux E [EBook]. In: Sonoma county complex fires: housing and fiscal impact report. Sonoma County Economic Development Board; 2018. Retrieved 11 May 2021, from, http://sonomaedb.org/WorkArea/DownloadAsset.aspx?id=2147564944.

Weiser M. The computer for the 21st century. 1991. Retrieved 21 December 2020, from https://web.archive.org/web/20141022035044/http://www.ubiq.com/hypertext/weiser/SciAmDraft3.html, Web.archive.org.

Weiser M, Brown J [EBook]. In: Designing calm technology; 1995. Retrieved 21 December 2020, from, https://people.csail.mit.edu/rudolph/Teaching/weiser.pdf.

Chapter 6

From community plan to clean energy infrastructure: policy, partnership, and investment

Chris Castro[1] and R. Bruce Stephenson[2]
[1]*Office of Sustainability and Resilience, Orlando, FL, United States;* [2]*Environmental Studies, Rollins College, Winter Park, FL, United States*

6.1 Introduction

With 40% of the most flood prone coastal land in the nation, Florida is confronting climate change and often referred to as "ground zero" for climate change impacts. In Miami Beach, $500 million was spent in securing a massive pumping system to keep the ocean at bay. This expenditure is symptomatic of the challenge facing metropolitan Miami, which has more than $400 billion in assets exposed to flooding, the highest total in the United States. At the same time, rising seas threaten the drinking water supply—the freshwater aquifer below the Everglades—with saltwater intrusion. Without significant investment, it is estimated that 2.5 million people in South Florida will be displaced due to sea-level rise and diminishing water supplies by the end of the century [1]. Chris Castro, a Miami native who heads the City of Orlando's urban sustainability and resiliency initiatives, is preparing for this influx. "You would have expected the federal government to be taking the lead, but the federal government seems to be backing away every day from the commitments they've made." [2].

> *The transition to a zero-carbon economy presents tremendous economic development opportunities for Orlando and the Central Florida region, some that we are already beginning to see stimulate our local economy, improve public health, reduce environmental impacts, and create meaningful high-wage jobs for our residents. In close collaboration with our hometown utility, OUC, and community partners, we remain committed to doing our part to advance a cleaner and more sustainable future for all.*
>
> Orlando May or Buddy Dyer [3]

The state government took a similar tact when Governor Rick Scott banned "climate change" as a topic of discussion in state business [4]. In Orlando, officials realized that deterring climate change was an environmental and economic necessity. Home to Disney World and Universal Studios, the fast-growing city of nearly 300,000 attracts a worldwide audience of more than 76 million visitors per year, and creating a green brand is good for business. Orlando Mayor Buddy Dyer believes that business, in general, has an attraction to sustainability. "The talented young people who are going to make the city successful want to work in cities that have sustainable practices." [5] In 2013, Dyer established a Sustainability Task Force and, at the first meeting, announced, "Our goal is to make Orlando the most sustainable city in the region, the Portland of the Southeast." [6].

The Task Force was charged with planning "an environmentally sustainable, socially inclusive and economically vibrant community." [7] Semantics helped sidestep the volatile issue of climate change, yet the Task Force's *Greenworks Action Plan* (see Fig. 6.2) was at heart a climate action plan. Modeled on Portland's climate action plan (see Fig. 6.1), its seven components, Livability, Energy/Green Buildings, Solid Waste, Water, Local Food, Transportation, and Green Economy (Green Economy was dropped in the 2018 update), paralleled those in the Portland's plan.

The Energy/Green Building was a priority given the potential of solar power in a city with the third highest average temperature in the United States. Solar power was the focal point of the *Greenworks Action Plan's* goal to supply 50% of energy from carbon-free sources by 2040 [10]. Four years later, the goal was updated to have a zero-carbon economy powered with 100% clean and renewable energy by 2050.

Smart Cities Policies and Financing. https://doi.org/10.1016/B978-0-12-819130-9.00010-3

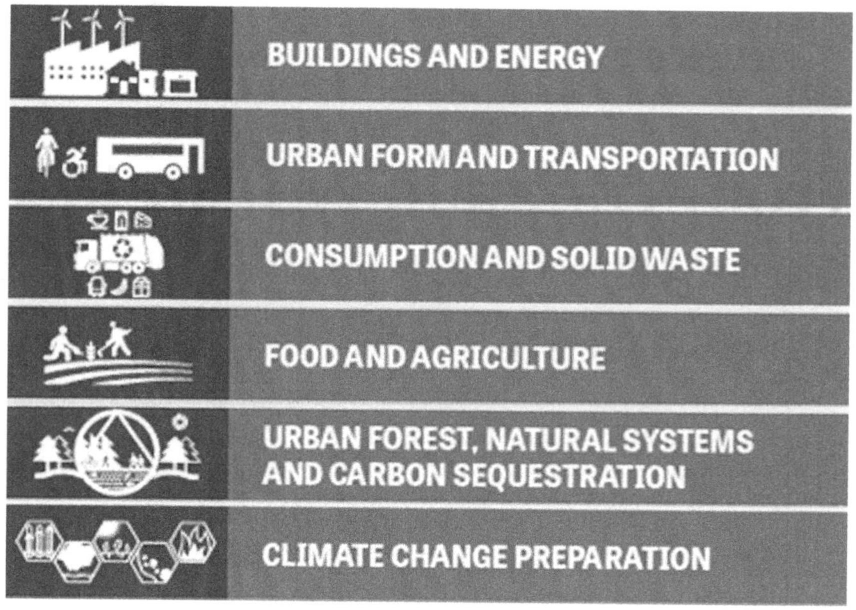

FIGURE 6.1 Portland climate action plan (Portland Sustainability and Planning Office, Public document) [8].

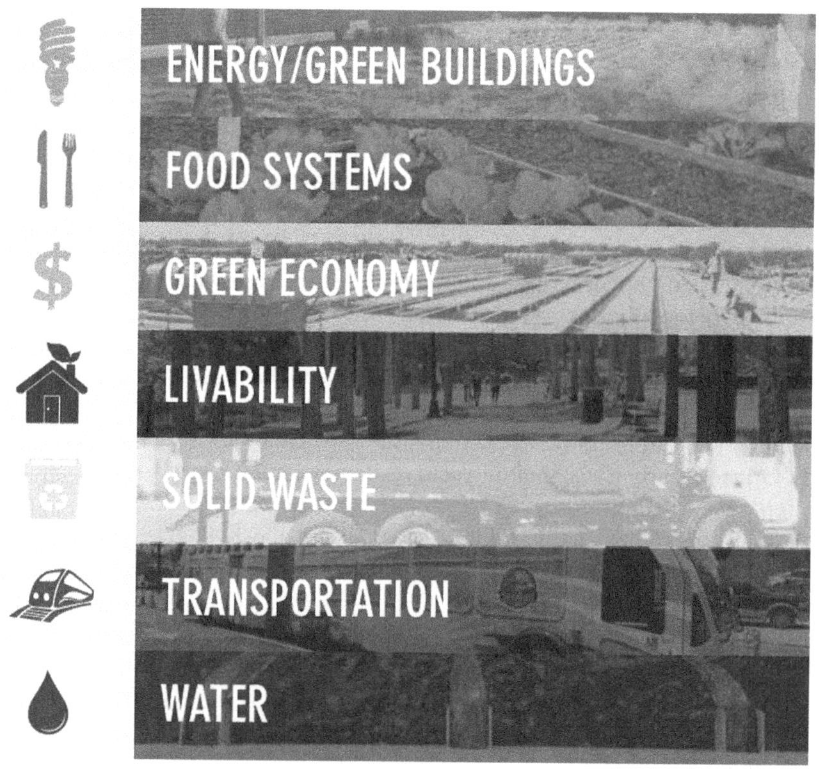

FIGURE 6.2 Orlando Greenworks Action Plan (Orlando Sustainability Office, Public document) [9].

6.2 Activist informed policy

Activist turned policy expert (with training in engineering), Chris Castro led the plan update. He is adept at building coalitions. In his early 20s, he founded IDEAS For Us, an accredited United Nation's nongovernmental organization that sponsors sustainability projects on campuses and in communities in more than a dozen countries. Having turned hopeful endeavors into ongoing programs, IDEAS for Us exemplifies Orlando's "Goldilocks condition," Castro states. "Orlando does not have the regimented power structure of older more established cities; it is a place where people realize they can make a difference." [11].

For a generation Orlando has been one of the nation's fastest growing cities, and the flux of change can be conducive to building community. In one high growth area, children from 200 different countries crowd into schools "finding new ways to learn, and new ways to live together," T.D. Allman writes in a *National Geographic* expose on Orlando [3]. In addition, the notion of the city as a laboratory has instilled the local culture ever since Walt Disney unveiled EPCOT, the Experimental Prototype Community of Tomorrow. Although Disney's vision for EPCOT was diluted into a theme park experience, during Michael Eisner's reign, the Disney Corporation built Celebration, a prototype New Urbanist community on a site planned for a theme park [12]. Celebration inspired Baldwin Park, a decommissioned navy base that Orlando's planning staff transformed into a pedestrian scaled neighborhood with 10,000 residents. Its compact form and mix of uses and housing types are a model for 25 walkable urban villages the *Greenworks Action Plan's* calls for by 2040 [13]. This same type of innovation and adaptation now drives the city's sustainability initiative that a new generation is championing.

Castro's success with IDEAS For Us, and the creation of a clean energy consulting firm Citizen Energy, translated into the young eco-entrepreneur being named Orlando's Director of Sustainability and Resilience before he turned 30. In 2017–18, he drew on a cohort of activists, experts, and concerned citizens in leading the 5-year update of the *Greenworks Action Plan*. Special care was taken to reach out to "communities of concern," primarily neighborhoods of color that were ignored in the past, to explain how the efficiencies of sustainability added cost savings to household budgets. As a result, environmental justice and social equity now inform the development of any program or policy change to enable sustainability [14].

Castro had a significant equity offering after brokering a partnership between utilities, academic institutions, and transportation agencies to secure more than $500 million in clean energy financing through the Property Assessed Clean Energy Program (PACE) and the Solar Energy Loan Fund (SELF). Both programs provide residents and business owners with access to clean energy capital at favorable rates. Low interest financing allows property owners to eliminate the high upfront costs for solar power. The investment is amortized over a five to 20-year period, which allows energy cost savings to help pay back the balance in conjunction with the property owners yearly tax bill. Solar co-ops are also an option for citizens interested in switching to solar to economize. By leveraging the power of bulk negotiation, this initiative has added 14 MW across the State of Florida, with roughly 33% to the Orlando Utility Corporation's (OUC) expanding renewable energy ledger [15]. The shift to solar power is good business for the municipally owned power company. "The biggest risk is not changing," Clint Bullock, the chief executive states [16].

Innovative engineering is also at play. In 2017, a massive landfill holding the ash refuse from a nearby coal producing electric plant was covered with 20,368 solar panels that produce 13 MW, enough electricity to power 2500 homes [17]. An additional 20 MW is generated from panels sited on the top of parking garages. Plans also call for two 600-acre solar farms to open by 2023 that will meet the demand of 27,000 homes. To counter the problem when solar power drops off due to clouds or sunsets, a $9 million investment from the federal government is turning hydrogen gas into electricity, what the industry calls "green hydrogen" because the electricity is coming from solar power [18]. Orlando's most inventive project is to cover stormwater retention ponds with floating grids of solar panels. Greater Orlando is pock marked with thousands of these manmade water bodies, and this breakthrough technology developed by the University of Central Florida has the potential to revolutionize solar power production. On pace to have 8% of its power fueled by solar in 2022, Orlando has also made significant investments in energy efficiency [19].

Castro's predecessor, Jonathan Ippel had assessed Portland's investment in green building, and he initiated a process that resulted in Orlando hiring Castro through the City Energy Project, and later leading to the city council adopting Florida's first Building Energy and Water Efficiency Strategy Ordinance. Designed to drive energy and water efficiency in Orlando's largest buildings, by 2017 more than 500 buildings (16% of the city's total) benchmarked energy and water use and reporting with the city. In total, $20 million in municipal green bonds funded LEED (Leadership in Energy and Environmental Design) certified "net zero" energy systems in 14 city buildings [20]. In 2020, Orlando was recognized as a "LEED Gold City," the first in the Southeastern United States. The award also identifies performance impact in moving toward the United Nations Sustainable Development Goals (UNSDG), and the city was ranked with Portland and other top sustainable cities in regard to clean energy.

This award signifies the gains Orlando has made since adopting the *Greenworks Action Plan*. In 2010, it held a lower tier spot on the City Clean Energy Scorecard computed by the American Council for an Energy Efficient Economy. In 2016, changes made to the city's land development code, and training permitting inspectors to streamline solar permitting guides, Orlando was named a "SolSmart Gold City," the award bestowed by the United States Solar Energy Office for cities reducing "soft costs." By 2017, Orlando ranked 20th on the City Clean Energy Scorecard and, 2 years later, it jumped five spots to 15th, the highest ranking among Southeastern cities [21].

Moving the city car fleet to electric vehicles is also reducing greenhouse gas (GHG) emissions. In addition, police motorcycles are electric, electric buses are operating in the downtown, and a prototype autonomous electric shuttle bus operates in the Lake Nona neighborhood taking residents from the neighborhoods to the Lake Nona town center. Electric cars are also available to rent at the Orlando International Airport through the Drive Electric Orlando program, with the incentive of receiving priority parking at the heavily visited theme parks. An international destination, Orlando is branding itself as a future-ready city at the forefront of sustainability and innovation; a harbinger of the future that celebrates sustainability as well as theme park fantasies [22].

6.3 The roadmap to carbon-free energy

The UNSDG informed the update of the *Greenworks Action Plan*, which set the goal of having 100% carbon-free energy sources by 2050. Mayor Dyer recognized this commitment when he signed the Under2 Coalition, a United Nations commitment to reach net-zero GHG emissions by 2050 and align with the Paris Agreement to limit global warming to less than 2°C above preindustrial levels. Interim emissions targets were set for 2030–50% and 2040–75% [23]. In April 2021, Orlando also joined the "Race to Zero" campaign as a reinstatement of its commitment to transition to a zero-carbon economy by 2050.

The agenda is ambitious. In 2020, energy use in Orlando relied on a mix of coal (38%), natural gas (52.9%), nuclear (5.9%), renewables (3.2%). Solar accounts for 45% of renewable energy and landfill gas converted to energy accounts for the remainder. After completing the *Greenworks Action Plan* update, the Sustainability and Resilience Office convened key partners and technical experts to construct a *Clean Energy Roadmap* to begin the transition to 100% renewable energy (see checklist: "An Agenda For Action For Taking Steps To Identify Action Items To Construct A Clean Energy Roadmap").

An agenda for action for taking steps to identify action items to construct a clean energy roadmap

Eight action items were identified, and by 2020, the Sustainability and Resilience office had responded to each of the following (check all tasks completed):

_____1. Conduct a municipal and community-wide forecast to determine rooftop solar energy potential.
_____2. Prepare a renewable energy and optimization analysis for 14 key municipal buildings.
_____3. Establish solar and electric vehicle design guidelines to inform future projects
_____4. Create a ranking matrix for solar siting.
_____5. Develop a *Clean Energy Strategy Toolkit* to transition to clean energy.
_____6. Utilize computer modeling, to determine the potential changes in the load forecast, by comparing 100% renewable energy versus existing trends.
_____7. Determine current and future load forecasts, which includes electric vehicle demand.
_____8. Invest in pilot projects [24].

6.3.1 Conduct a municipal and community-wide forecast

A study of solar energy and storage potential for rooftops assigned a solar radiance score for each building assessed in the study. The score was based on a building's yearly exposure to peak sunshine (hours where solar insolation achieves or exceeds 1 kW/m^2), and the amount of square feet available for solar panels. A solar kilowatt-hour production estimate was determined for each building, and solar radiance scores were ranked 1 through 5. Buildings with scores of 3 or higher were deemed most feasible, and it was determined that 748 averaging 755 h of peak sunlight hours could generate 35,519 kW of electricity. Bringing this energy online would more than double the supply to meet current electricity demand [25].

6.3.2 Prepare a renewable energy and optimization analysis

To determine the optimal production of renewable energy and energy storage for 14 municipal buildings, 15-min electric load profiles and utility rate structures were analyzed. The study included three water reclamation facilities, the emergency operations center, and several fire stations and community centers, which are used as makeshift hurricane shelters during Florida's storm season and a priority for energy resilience. The analysis found that installing solar (either with or without storage) would reduce total energy use by 74%. The result was a capital budget allocation of $3 million to start work on five existing buildings and five in-design buildings, with the potential to expand funding the 2021 fiscal year [26].

6.3.3 Establish solar and electric vehicle design guidelines

The Sustainability and Resilience Office tailored a set of municipal building code guidelines for solar photovoltaic storage and electric vehicle development based on the National Renewal Energy Lab's *Solar Ready Buildings Planning Guide*. The new operating procedure for solar readiness will first be tested on municipal buildings before creating new ordinances. Near-term plans are to accelerate renewable installations on city facilities and further investigate a utility scale solar array and battery storage facility at one of three water reclamation facilities selected for investment in the $3 million capital budget outlay [27].

6.3.4 Create a ranking matrix for solar siting

The matrix for solar siting went beyond traditional screening techniques: Evaluating the size of a project, its radiance score, and its load-bearing ability to support a rooftop solar installation, or the techno-economic analysis of project cost and financing options. First, a rooftop solar project is not always proportional to its impact. For instance, a small solar array may not contribute significantly to reaching a 100% renewable energy goal, but if it brings a building to net-zero, the installation would be viewed favorably. A second consideration is when a project has a less-preferable financing mechanism, but also contributes toward community resilience. For example, adding solar and storage to a community center so it can operate as a hurricane shelter is invaluable. Inclusion, in terms of city commissioner district representation and social equity considerations are also considered when weighing the costs and benefits of project siting. Finally, a new term, "Return on Visibility," describes how the visual impact a solar array illustrates Orlando's commitment to renewable energy. Visibility of solar on public sites is effective at encouraging private deployment, especially when linked to public incentives to install solar or when combined with programs such as civic-sponsored competitions between businesses, such as the PACE and SELF financing programs, and the annual solar coops to access preferred pricing [28].

6.3.5 Develop a clean energy strategy toolkit

In addition to technical information, a range of behavioral economic levers can accelerate the transition to solar technologies. Incorporating these elements drove the creation of the Sustainability and Resilience Office's *Clean Energy Strategy Toolkit*, which is, in essence, a "dictionary" of strategies to prioritize limited resources to meet the city's 100% renewable energy goal. After a comprehensive review of clean energy policies and program options, strategies were placed in five categories: financing, information, programmatic, regulatory, and technology, and ranked and ordered based on their feasibility and impact. The analysis revealed that high-feasibility measures generally corresponded with high-impact measures, such as the PACE program and SELF's green energy loan fund. Meanwhile, the most feasible strategies did not necessarily represent "low-hanging fruit," but were most often impactful. Additionally, several of the most impactful factors are currently employed, including a municipal energy manager, solar co-ops, PACE financing, a utility efficiency program for low-income residents, and a community solar program [29].

6.3.6 Utilize computer modeling

A modeling software, OACES, was used to produce future scenarios that revealed how changes in load forecasting could meet the 100% renewable energy goal. At the same time, this exercise gauged the cost of maintaining the status quo in a series of future scenarios. Utilizing hourly electricity demand and supply profiles, OACES calculates the price of clean and fossil-fuel based energy sources and the percentage of an energy source that is infused into the grid over time. In addition to assessing the potential for different clean energies to reach the 100% renewable goal, the modeling calculates the impact on public health. An analysis of a future scenario reveals the point where both power plant pollutant emissions will drop and shifts in air pollution disbursement patterns will occur. Identifying the synergy between clean energy and improved health over time is a key indicator for judging energy investments, and it will inform future decision-making.

The most important takeaway in the OACES modeling was how important energy efficiency is in meeting the renewable energy goal—energy efficiency represents over 60% of the savings in a reduced future load. It is key to meeting the 2030 and 2040 targets for emission reduction and mitigating detrimental effects to public health. In this scenario, utility costs also drop, while job creation expands. City officials and stakeholders will employ OACES to better understand the investments, benefits, and trade-offs. This process will also help procure more precise and equitable policy decisions as the city transitions to a sustainable future [30].

6.3.7 Determine current and future load forecasts

The market parallels between energy efficiency, solar installation costs, and electric vehicle charging rates are increasingly advantageous. New net-zero energy homes are more cost-effective than code homes, and older homes can be retrofitted cost effectively to reach net zero. Rooftop photovoltaic is less than half the cost than traditional sourcing and has a 12.7% internal rate of return in Florida, without more progressive state incentives. By 2025, the nation's customer-owned photovoltaic installations and energy storage will be equal to those installed by utilities and, by 2035, Florida is projected to reach 50% customer-owned rate. In 2030, a Florida net-zero home with 9.25 kW photovoltaics and 40 kW lithium ion battery will have a 10.6% internal rate of return.

The cost of electric vehicles is expected to cross over with the price of Internal Combustion Engine vehicles shortly after 2025. By 2030, electric vehicles will have 24 times the capacity of stationary storage, giving way for mobile storage opportunities where vehicles can supply backup power generation in times of emergency. Electric vehicles are projected to play a greater role in driving electricity demand over time and will account for two-thirds of passenger vehicles by 2050. Integrating utility and rooftop solar, energy storage, smart-charging "vehicle-to-grid" electric vehicles, building energy efficiency, and demand response to achieve the 2050 energy goal, but the returns are priceless, and costs will be less than the current network [31].

It is essential that both utilities and customers be empowered to achieve 100% renewables. Meeting this goal will require utilities to invest in energy efficiency on the supply and demand side and managing the electricity grid. At the same time, they must step away from making the grid a vertically integrated provider of production, transmission, distribution, and sale. Most important, the integration and optimization of solar energy, energy storage, green buildings, and electric transportation infrastructure generate must secure lower energy costs than fossil fuel [32]. Moving this agenda forward will have the added benefit of improving resilience in times of emergency, which will increase in frequency due to climate change.

6.3.8 Invest in pilot projects

To demonstrate the viability of the *Greenworks Action Plan's* goal of having municipal operations be 100% renewable-powered by 2030, the city invested in a series of pilot projects. During the Obama Administration, Orlando's Fleet and Facilities Division received grant money to build a solar and electric vehicle ready design for its 10-acre central compound. The 420 kW solar array was the city's flagship solar installation, and the cost savings let the city to float a $17.5 million bond to upgrade additional Fleet and Facilities structures. "Everything we've invested in had had dramatic effects on energy savings, water conservation, and better occupied space for people," David Dunn, manager for the Fleet and Facilities Division, explains [33]. In 2019, a partnership with the Florida Solar Energy Center and OUC produced a design to expand solar energy production on the central compound by scaling up lab-proven technology to meet the demand for 32 electric vehicles and fully offset the building's energy consumption [34]. The request for proposal went out in May 2021, and it requires installing bifacial photovoltaic panels that will produce 280 kW [35]. These solar modules allow energy to be generated on both sides of a panel, optimizing the production of solar power within a smaller footprint, reducing the balance of system costs, and increasing the ultraviolet light resistant durability of the system.

The solar array is connected to a large megawatt-scale battery that passes through service disconnects and the electrical panel before interconnection with the level two charging stations. Direct connection to the building was limited through an inverter via a service disconnect. The solar array will charge the battery that serves the charging stations. When a grid outage occurs, the battery supplements the energy to the existing building by a connection to the main disconnect panel. Once power is restored, the grid-energized panel signals the inverter to revert to normal operation. In an extended hurricane outage, a standby diesel generator would power normal operations. In this worst-case scenario, the solar storage system is "islanded" from the electric grid and will continue to charge the battery with solar energy serving the electric vehicle charging stations. As with other solar installations, the Facilities Engineering team deenergizes the solar network prior to each hurricane. After the storm passes, systems are inspected for damage before reenergizing [36].

Orlando's new Code Enforcement and Public Records building was constructed to model the type of LEED green building that will drive the city's transition to clean energy. A partnership between the Public Works Department and Facilities Engineering determined the viability of placing a 114 kW solar array on the white, flat membrane room covering the building. The tilt-wall building was erected while the electrical and mechanical engineers finalized system sizing and design. An energy model for the facility evaluated the practicality of attaining LEED certification, and it was concluded that a rooftop solar system was a good technical fit. The new building's $5.3 million price tag includes a $112,000 net savings over the building's lifetime and a measure of the city's path to sustainability [37].

6.4 The future takes shape: shutting down coal and envisioning equity

In 2020, the City of Orlando declared a climate emergency and urged residents, stakeholders, and partners to intensify efforts to become one of the nation's most sustainable cities. At the same time, OUC announced that its two monumental power plants on the city's eastern edge would cease burning coal by 2027. Two years before, OUC initiated a $1 million study in response to Mayor Dyer's call for the conversion to 100% renewable energy. Mayor Dyer noted that OUC's report represented the type of "forward thinking" the *Greenworks Action Plan* demands. Taking action shortly before the arrival of targeted dates would fail, he declared, and OUC's commitment illustrated the path to sustainability. Even if electric rates fluctuated, investing in renewable would be cheaper over time and the returns more beneficial [38].

OUC's decision marked a generational shift. Constructed respectively in 1987 and 1996, the Stanton Plant's giant cooling towers signaled the utility's original commitment to coal. Ending coal-fired power generation and moving to natural gas will be complete by 2027, and the plant will be entirely shut down by 2040. OUC first invested in natural gas in 2003, and it now sources 38% of the Stanton Plant's 940 MW. Renewables account for 3%, nuclear power 7%, and the remainder (52%) comes from coal [39]. Converting the Stanton units to natural gas will not be as efficient as a state of the art natural gas facility, but the conversion cost is minuscule in comparison to building a new plant [40]. "If we can build another couple 100 megawatts of solar, keep increasing our solar and renewable footprint at the same time, working way from fossil fuels, we're doing what's right for the environment, what's right for our customers, and as a municipal utility, absolutely what's right four our community," OUC representative Tim Trudell states [41].

"This is a long time in coming," Susannah Randolph, the senior Florida campaign representative for the Sierra Club's Beyond Coal Campaign, stated. "It's pretty momentous that they are ending coal usage. That's a big step forward." [42] In 1983, the Sierra Club brought a lawsuit challenging OUC's plan to build four coal fired plants on 3200 acres of pasture in eastern Orange County. In addition to securing a "reasonable balance between the need for the facility and the environmental impact from the construction and operation of the facility," the Sierra Club asked that "operation safeguards are technically sufficient for citizens welfare and protection." [43] Four decades later, the climate crisis has radically redefined the concept of the public welfare. In 1983, ending reliance on fossil fuel was unimaginable. Moreover, the Reagan Administration was pulling back from solar power—now the centerpiece of Orlando's energy future [44].

It is noteworthy that sustainability and resiliency match affordability and reliability in a survey of citizen priorities regarding energy production. This preference is also revealed in the advocacy for clean energy, which has moved beyond traditional environmental organizations. Moving away from coal "is a great step forward but we are still committed to an Orlando completely free of fossil fuels," Beverlye Colson Neal, the Orange County President of the National Association for the Advancement of Colored People, stated, "meaning no investment on toxic gas, and a strong commitment to increased investments in energy efficiency in our black and brown communities." At the same time, Orlando's Sustainability and Resilience Office is conducting an energy burden study with the data analytics firm, Greenlink Analytics, to address the glaring disparities that make equity an issue across the nation [45].

Societal imbalances instill current energy distribution but utilizing data-driven decisions can change this dynamic. The Sustainability and Resilience Office and its consultants are evaluating indicators of equity at a neighborhood/census tract level to delineate current positions and institute policies to advance equity [46]. A key indicator is utility burden—the percent of income spent on particular expenses (electricity, gas, and water)—which is unevenly shared across society. Low-income households spend a higher percentage of their income on electricity and gas bills than any other income group, and it is especially high in the South, rural America, and minority communities. Addressing utility burden is essential given that paying utility bills are a common reason for taking on predatory short-term loan services. Research demonstrates that improving utility equity leads to spillovers for economic well-being and quality-of-life benefits [47].

The city's initial utility burden map geospatially delineates both the average utility burden and the number of burdened households in a neighborhood. Additional equity indicators will be corelated with this data: evictions, foreclosures, households receiving Supplemental Nutrition Assistance Program, the distribution of housing types within neighborhoods, and the split between renters and owners [48]. This unfolding research is vital to quantify the multidimensional demands that link equity and sustainability.

These advances are essential to face the challenge facing metropolitan Orlando. In 2019, the Sustainable Development Solutions Network ranked Metro Orlando 77th among the nation's 105 metro areas in meeting the UNSDG. The metro area (Orange, Osceola, and Seminole Counties) was flagged for failing to meet Goal 11, Sustainable Communities, due to its poor access to active transportation (biking, walking, and transit) and the heavy rent burden for low wage workers in the dominant theme park economy [49]. In May 2021, a study conducted by *Bloomberg News* estimated the strain the pandemic placed on housing budgets by analyzing median wages and rents, and it found Orlando was less affordable than San Francisco [50]. A Smart Growth America report had a similar finding, ranking Metro Orlando last among the nation's 30 most populous metropolitan areas for equity when factoring the cost of transportation [51].

Central Florida is one of the nation's most automobile dependent regions, and car ownership is a financial burden—especially for low wage households. Since 1960, transportation costs have jumped from 10% to 20% of family income, and the culprit is car ownership. Eliminating one car from the typical households equates to a $9000 savings, the equivalent of an additional $100,000 in a mortgage [52]. With transportation emissions now exceeding power plants as the United States' largest source of GHG, forgoing automobile ownership is also one of the most effective individual actions to stymie climate change [53]. Walking and biking are the ultimate forms of clean energy and investing in walkable urbanism—as opposed to drivable suburbanism—is essential [54].

The Orlando *Greenworks Action Plan* recognized this fact: a shift from "heavy reliance on conventional private automobile" to active transportation "is an important component in Orlando becoming the greenest city in the Southeast." [55] Baldwin Park set the prototype and the new SunRail commuter line set the framework for creating 25 walkable villages to get to a majority of all trips being made by active transportation in 2040 [56].

Mayor Dyer had championed the construction of SunRail. In 2011, he drew the support of both the business community and Orlando's state and federal delegations to counteract Governor Rick Scott, after Scott put the 1.2 billion project on hold by refusing to comply with the federal formula for 25% state funding [57]. Three years later, SunRail was online and walkable urbanism was the default design near rail stops and an elemental component of affordable housing. In early 2020, Osceola County spent $8.9 million purchasing an 82-acre site next to SunRail. Originally slated for single family homes, the county commission rezoned the property for mixed-use affordable housing. This pedestrian scaled neighborhood will not only lower GHG emissions, it will also generate the informal face-to-face contacts that produce, the great urbanist Jane Jacobs wrote, the "small change" that forms "social capital." [58].

Pedestrian-scaled communities and clean energy are essential to ending dependence on fossil fuels. Moreover, Orlando's investment in sustainability illustrates how coupling a definitive vision and political will can empower citizens and initiate change. Challenges remain, but in a republic facing the specter of climate change integrating equity and sustainability must define the future if it is to be both livable and just.

6.5 Summary

Florida stands at the forefront in confronting the specter of climate change. Getting to the root cause of climate change defines a smart city, and Orlando's shift to clean energy highlights its investment in sustainability. The *Orlando Greenworks Action Plan* established policies, set targets, and identified best practices to shift to 100% renewable energy by 2050. The city sponsored pilot projects, then made strategic investments to earn a LEED Gold City ranking. At the same time, the Office of Sustainability and Resilience established a series of programs to mitigate the cost of investing in solar power for citizens and businesses. It also partnered with the Orlando Utility Commission (OUC) and the University of Central Florida to produce innovative strategies and technologies to generate solar power. These accomplishments inform the *Clean Energy Roadmap* that charts the path to 100% renewable future. Perhaps most important, the city's partnership with the OUC is quickening the move to solar power in historic fashion. Forty years ago, a lawsuit greeted OUC's plan to build the coal-fired Stanton Energy Complex because of its potential harm to public health. At the time, solar energy was an afterthought. In a shorter time than has elapsed since the lawsuit, the Stanton Energy Complex will be closed, Orlando will have a zero-carbon economy powered with 100% clean and renewable energy. Shifting from auto dependency is also a key to breaking free of fossil fuels. Investments in active transportation and walkable urbanism are setting the foundation for that day. Realizing the *Greenworks Action Plan* will not only make Orlando the "green capitol of the Southeast," its green brand will be an international moniker for the 76 million yearly visitors who find sustainability intertwined with their theme park experience.

Finally, let us move on to the real interactive part of this chapter: review questions/exercises, hands-on projects, case projects, and optional team case project. The answers and/or solutions by chapter can be found in Appendix G.

6.6 Chapter review questions/exercises

True/false

1. True or False? With 60% of the most flood prone coastal land in the nation, Florida is confronting climate change and often referred to as "ground zero" for climate change impacts.
2. True or False? To counter the problem when solar power drops off due to clouds or sunsets, a $9 million investment from the federal government is turning hydrogen gas into electricity, what the industry calls "blue hydrogen" because the electricity is coming from solar power.
3. True or False? The UNSDG informed the update of the *Greenworks Action Plan*, which set the goal of having 100% carbon-free energy sources by 2070.
4. True or False? A study of solar energy and storage potential for rooftops assigned a solar radiance score for each building assessed in the study.
5. True or False? To determine the optimal production of renewable energy and energy storage for 14 municipal buildings, 15-min electric load profiles and utility rate structures were analyzed.

Multiple choice

1. The Sustainability and Resilience Office tailored a set of municipal building code guidelines for solar photovoltaic storage and electric vehicle development based on the National Renewal Energy Lab's:
   a. Smart Sustainable Development Guide
   b. ICT-based Infrastructure Guide
   c. Solar-Ready Buildings Planning Guide
   d. Investment Guide
   e. Policy Guide
2. The matrix for __________ went beyond traditional screening techniques: Evaluating the size of a project, its radiance score, and its load-bearing ability to support a rooftop solar installation, or the technoeconomic analysis of project cost and financing options.
   a. Solar siting
   b. Physical infrastructure

c. Smart city growth and prosperity
d. Economic effects
e. Smart city policy

3. In addition to technical information, a range of behavioral economic levers can accelerate the transition to:
   a. Machine to machine (M2M)
   b. Machine to human (M2H)
   c. Solar technologies
   d. Artificial intelligence (AI)
   e. Digital twins (DT)
4. A modeling software, ________, was used to produce future scenarios that revealed how changes in load forecasting could meet the 100% renewable energy goal.
   a. HIOSC
   b. ECONG
   c. MAKCS
   d. OACES
   e. ECONO
5. The ___________ between energy efficiency, solar installation costs, and electric vehicle charging rates are increasingly advantageous.
   a. New net-zero energy
   b. Market parallels
   c. Rooftop photovoltaic
   d. Energy storage
   e. Smart-charging

Exercise

Problem

Why will investments in clean energy infrastructure need to be scaled up significantly in the coming years to support the broader development, economic, and climate agenda?

Hands-on projects

Project

Do research: Analyze why the key challenge for governments to catalyze investment flows in clean energy is to design and implement clear and predictable domestic policy frameworks.

Case projects

Problem

Discuss why a policy guidance for investment in clean energy infrastructure would not necessarily provide an exhaustive list of questions to consider for enhancing clean energy investment, nor would it follow a one-size-fits-all approach.

Optional team case project

Problem

Discuss why competition policy clean energy infrastructure investments often take place in a situation of imperfect competition, where a state-owned enterprise is the incumbent.

References

[1] Cox S, Cox P. Miami is sinking beneath the sea, but not without a fight. New Republ November 8, 2015. https://newrepublic.com/article/123216/miami-sinking-beneath-sea-not-without-fight.

[2] Castro quoted in Penn I. From rooftops to algae pools, Orlando's vision for carbon-free energy. N Y Times August 30, 2018. http://www.nytimes.com/2018/08/30/business/energy-environment/orland. www.nytimes.com/2018/08/30/business/energy-environment/orlando-carbon-climate.html.

[3] Orlando Sustainability Office (OSO). Greenworks Orlando: community action plan, vol. 3. Orlando: OSO; 2018.

[4] Korten T. In: Florida officials ban climate change, Miami Herald; March 11, 2015. https://www.miamiherald.com/news/state/florida/article12983720.html.

[5] Dyer quoted in, Kanowitz S. Simulation tools help Orlando deliver on environmental goals. GCN June 19, 2019. https://gcn.com/articles/2019/06/19/orlando-environmental-perfomance-tool.aspx.

[6] Stephenson B. Portland's good life: sustainability and hope in an American city Langham. MD: Lexington Books; 2021.

[7] OSO. Orlando greenworks action plan, vol. 3. Orlando: FL: OSO; 2014.

[8] Portland planning and sustainability office, Portland climate action plan https://www.portland.gov/sites/default/files/2019-07/cap-2015_june30-2015_web_0.pdf.

[9] OSO. Orlando, Greenworks action plan. 2014.

[10] OSO. Orlando Greenworks action plan. 2014. p. 11.

[11] Stephenson, Portland's good life, 188.

[12] Foglesong R. Married to the mouse: Disney World and Orlando Princeton, NJ. Princeton University Press; 2001.

[13] Stephenson B. The roots of new urbanism, john Nolen's garden city ethic. J Plan Hist 2002;1(2):99–123. 2002:101–126. Kim J, Larsen K. Can new urbanism infill development contribute to social sustainability? The Case of Orlando, Florida. Urban Studies; 2017;54(16):3843–3862.

[14] OSO. Greenworks plan. 2018. p. 7–10.

[15] City of Orlando PACE Program. www.cityoforlando.net/greenworks/pace/.

[16] Bullock quoted in, Penn, from rooftops to algae pools, Orlando's Vision for Carbon-Free Energy.

[17] Spear K. Orlando's new solar plant takes shape as Florida's solar energy erupts. Orl Sentin (FL); July 5, 2017. p. A1. https://www.orlandosentinel.com/news/os-ouc-solar-plant-compared-20170628-story.html.

[18] Spear K. Orlando utility to launch $9 million hydrogen system. Orl Sentin (FL); October 7, 2019. p. A1. https://www.orlandosentinel.com/news/os-ne-orlando-utility-hydrogen-solar-20191007-jajdl3upune orakhz4imd624pe-story.html.

[19] Penn, from rooftops to algae pools, Orlando's Vision for Carbon-Free Energy.
[20] OSO. Greenworks Plan. 2018. p. 14.
[21] American Council Energy Efficient Economy. 2019 city clean energy scorecard. July 2019. https://aceee.org/research-report/u1904.
[22] American Council Energy Efficient Economy, 2019 city clean energy scorecard, July 2019. https://aceee.org/research-report/u1904, 22.
[23] Orlando Mayor Buddy Dyer Joins International, Campaigns National. Reaffirming the city of Orlando's goal of using clean energy. May 31, 2017. wwwcityoforlando.net/news/2017/05/orlando-mayor.
[24] OSO. Orlando renewable and resilient road map. Orlando: OSO; 2021. p. 2–3.
[25] OSO. Orlando renewable and resilient road map. Orlando: OSO; 2021. p. 28–30.
[26] OSO. Orlando renewable and resilient road map. Orlando: OSO; 2021. p. 31–4.
[27] OSO. Orlando renewable and resilient road map. Orlando: OSO; 2021. p. 35.
[28] OSO. Orlando renewable and resilient road map. Orlando: OSO; 2021. p. 36–7.
[29] OSO. Orlando renewable and resilient road map. Orlando: OSO; 2021. p. 48–50.
[30] OSO. Orlando renewable and resilient road map. Orlando: OSO; 2021. p. 51–2.
[31] OSO. Orlando renewable and resilient road map. Office of Sustainability & Energy Orlando City Hall; December 19, 2017. https://www.bing.com/images/search?q=orlando+renewable+and+resilient+road+map&qpvt=Orlando+renewable+and+resilient+road+map&form=IGRE&first=1&tsc=ImageBasicHover.
[32] Goldsmith C. Renewable energy empowers consumers to take charge of their energy usage. N Econ October 29, 2018. https://www.theneweconomy.com/energy/renewable-energy-empowers-consumers-to-take-charge-of-their-energy-usage.
[33] Turner L. Orlando proves the power of solar arrays. May 5, 2021. https://www.government-fleet.com.
[34] OSO, Orlando renewable and resilient road map, vol. 39.
[35] Turner, O. Proves the power of solar arrays.
[36] OSO, Orlando renewable and resilient road map, 42–43.
[37] OSO, Orlando renewable and resilient road map, 44.
[38] Spear K. OUC to wipe out carbon emissions by switching from coal to natural gas, solar energy. Orl Sentin (FL); October 14, 2020. p. A1. https://www.orlandosentinel.com/news/environment/os-ouc-reveals-energy-future-20201014-g5azmsukrncc5kj34nds5t5cny-story.html.
[39] Gargotta J. Leaving coal behind: OUC switching to gas in as step-toward sustainability. Spectrum News February 16, 2021. https://www.mynews13.com/fl/orlando/news/2021/02/16/leaving-coal-behind–ouc-switching-plant-to-gas-a-step-toward-sustainability.
[40] Spear, OUC to wipe out carbon emissions by switching from coal to natural gas.
[41] Gargotta, Leaving coal behind: OUC switching to gas in as step toward sustainability.
[42] R. quoted in Spear, OUC to wipe out carbon emissions by switching from coal to natural gas.
[43] Fifth District District court of appel of Florida. August 18, 1983. 436 So. 2d 383, No. 83-77.
[44] Perlin J. Let it shine: the 6,000 Year history of solar power Novato. CA: New World Library; 2013.
[45] Brown MA, Soni A, Lapsa MV, Southworth K, Cox M. High energy burden and low-income energy affordability: conclusions fromliterature review. Progress in Energy. October 2, 2020;2(Number 4). https://iopscience.iop.org/volume/2516-083/2. https://iopscience.iop.org/issue/2516-1083/2/4. https://iopscience.iop.org/article/10.1088/2516-1083/abb954/meta.
[46] OSO, Orlando renewable and resilient road map, 56.
[47] Brown, S., Lapsa, S., and Cox, High energy burden and low-income energy affordability.
[48] OSO, Orlando renewable and resilient road map, 56-57.
[49] Lynch A, Anna LP, Fox C. Sustainability cities report: the 2019 US cities sustainable development report. New York: Sustainable Development Solutions Network; 2019. p. 17.
[50] Fox J. Covid has made Orlando less affordable than san Francisco. Bloomberg News; May 6, 2021.
[51] Leinberger C, Rodriguez M. Foot traffic ahead. Washington, DC: Smart Growth America; 2016. https://smartgrowthamerica.org/resources/foot-traffic-ahead-2016/.
[52] Garceau T, Atkinson-Palombo C, Garrick N, Outlaw J, McCahill C, Ahangari H. Evaluating selected costs of automobile-oriented transportation systems from a sustainability perspective. Res Trans Bus Manag 2013;7:43–53.
[53] Seth W, Nicholas KA. The climate mitigation gap: education and government recommendations miss the most effective individual actions. Environ Res Lett July 2017:24–40.
[54] The problem with cars is not that they do not get enough miles to the gallon, it is that they make it too easy for people to spread out, encouraging forms of development that are inherently wasteful and damaging. Cities dominated by drivable suburbanism intensify the consumption of land, energy, and water, the production of greenhouse gases and smog-related emissions, and the flow of stormwater pollution (due to extensive paving for roads and parking). In: addition, the economic outlay is staggering. Utility extensions are fiscal sinks, highway construction is cost prohibitive without charging tolls, and funds to maintain bridges and overpasses are increasingly scarce. The indirect costs are wasteful, deadly, and cold- hearted: more driving and pedestrian deaths, higher indices of road rage and obesity, less walking and biking, and the impairment of the human scaled public realm. "It is not possible to solve sustainability in cities without addressing automobile dependence," according to urban planning experts Peter Newman and Jeffrey Kenworthy. Newman and Kenworthy, Sustainability and Cities: overcoming Automobile Dependence. Washington, DC: Island Press; 1999. p. 334.
[55] OSO. Greenworks community action plan. 2014. p. 37.
[56] OSO. Greenworks community action plan. 2014. p. 53.
[57] Kassab B. Chamber turns to tough talk on Scott. *Orlando Sentinel*: Sunrail; April 19, 2011. p. B1.
[58] Jacobs J. The life and death of great American cities. New York: Random House; 1961. p. 138.

Chapter 7

Energy efficiency in urban planning for smart cities in the developing world

Doaa M. El-Sherif
Housing & Building National Research Center (HBRC), Ministry of Housing, Utilities & Urban Communities, Cairo, Egypt

7.1 Introduction

Currently, half of the world's total population (around three billion people) lives in urban settlements. Developing countries in particular are undergoing rapid change from rural to urban-based economies as they are transformed by their urbanizing populations. Over the next 30 years, population growth will be nearly entirely concentrated in urban areas in the developing world. Much of the current debate regarding sustainable cities focuses on the formidable problems for the world's largest urban agglomerations. However, smaller urban settlements are also growing rapidly and the majority of all urban dwellers reside in such smaller urban centers. While cities command an increasingly dominant role in the global economy as centers of both production and consumption, this rapid urban growth throughout the developing world is outstripping the capacity of most cities and urban centers to provide adequate services for their citizens [1–6]. While there is no "typical" city in terms of their energy needs and energy use, they do have a lot in common. Cities run on energy and require land—their burning of fossil fuels and their contribution to land clearing combine to contribute massively to the concentration of greenhouse gases in the atmosphere. They are also extremely vulnerable to energy scarcity, to energy price rises and health impacts of very poor air quality. In developing countries many urban citizens experience extreme energy poverty and their access to transport and so to urban goods is constrained.

7.1.1 The importance of efficient energy planning for cities

The use of energy, the types of energy used, and the lack of access to sufficient energy have far reaching implications for a city's economic development, its environmental health, and for the poor. The burning of fossil fuels to provide energy is the major contributor to excess carbon in the atmosphere, which is the cause of global warming. Cities that implement sustainable energy plans reduce their vulnerability to energy scarcity and to energy price rises, they have less traffic congestion and lower energy input costs, they have cleaner air and their low-carbon economies can afford them a competitive economic edge globally. And, specifically for cities in developing countries, a sustainable energy plan should consider the users' needs first—this means that poorer households and small energy users should be prioritized.

7.1.2 The sustainable energy efficient path for cities

The sustainable energy efficient path for cities should attempt to reduce carbon footprint, emissions, dependence on fossil fuels, and increased dependence on renewable energy. It should also include diversification of energy sources, support local and decentralized power supply, focus on energy efficiency, and provide support and information to users. In addition, it should include efficient resource use the basis of economic development, ensure that citizens have appropriate access to energy services and information on best energy use practices to reduce poverty. Planning for efficient spatial development and developing efficient accessible public transport using cleaner fuels are very important to energy-efficient path for cities [5].

7.2 Guidelines for energy efficiency in urban planning

In this section, the energy efficiency potential and why Energy Efficiency for Urban Planning are discussed. This will be followed by discussion of the principles of Energy Efficiency for Urban Planning.

Smart Cities Policies and Financing. https://doi.org/10.1016/B978-0-12-819130-9.00031-0

7.2.1 Energy efficiency potential in the building sector

Buildings account for a significant portion of primary energy consumption in south and east Mediterranean countries. By 2030, these countries are expecting a population increase of 40 million people and an increase of 24 million new housing units [1]). Growing populations and new housing stock increase energy demands of buildings, and cooling requires the largest energy load in these countries. Therefore, when planning for new urban communities to accommodate future populations, efficient use of energy resources should be considered. In order to be most effective and feasible, the reduction of energy consumption of buildings and the use of renewable energies should start with urban development. This covers a variety of scales, from regional and city levels, to district and neighborhood levels, as well as plot allocation levels. Affordable and sustainable solutions for communities are needed to secure energy supply and to reduce energy subsidies. New markets for energy-efficient products will be created when energy efficiency is introduced on urban planning level.

7.2.2 Why energy efficiency for urban planning

Urban planning is the basis for energy-efficient buildings. Planning new cities is a long and complicated process, which involves many stakeholders. Decisions made at the city or district level affect micro-climates in public space and influence the final energy consumption of buildings, their cooling, heating, or lighting loads. Planning for energy efficiency at an early stage includes context-specific issues such as site selection, type of regional connections, transportation structure and type, the main street layout, as well as the prevailing orientation of new developments. Energy-efficient urban planning offers opportunities to enhance the quality of urban life by providing spatial solutions, designed for a healthy and comfortable-built environment with minimum energy consumption. Passive methods that increase shade, cooling, and ventilation at all scales aim to reduce energy consumption, thus the focus of this brochure. Energy-efficient design and urban planning are the most cost-effective way to reduce energy consumption of buildings [6].

7.2.3 Principles of energy efficiency for urban planning

Energy-efficient architectural design is only possible when the site and climatic condition are considered appropriately at the urban planning level. So, passive design strategies are the most cost-effective approach to improve the energy efficiency of buildings. (see checklist: "An Agenda For Action For Taking Steps To Identify Principles That Form The Basis For Energy Efficient Urban Planning Guidelines").

An agenda for action for taking steps to identify principles that form the basis for energy-efficient urban planning guidelines

The following principles form the basis for the energy efficient urban planning guidelines at the city, neighborhood and plot level (check all tasks completed):

_____1. Make use of existing climate and site conditions to support sustainable passive design strategies for heating and cooling techniques and energy-efficient indoor and outdoor comfort.

_____2. Encourage energy-efficient transport modes such as (mass) public transit and non-motorized transportation (such as biking and walking).

_____3. Promote compact urban growth strategies and urban densities to sustain economic feasibility of public facilities and transit.

_____4. Integrate residential functions in mixed-use development, and allow for the distribution of services within short distances.

_____5. Provide amenities for a shaded comfortable outdoor environment and public space.

_____6. Accommodate renewable energy solutions through integration and flexibility in planning and design, to further minimize the carbon footprint.

7.3 Developing energy-efficient urban plan for cities

The Local Governments for Sustainability "ICLEI" [4], The United Nations Human Settlements Program (UN-HABITAT) and The United Nations Environment Program "UNEP" have developed a planning guide, which includes 10 steps as a framework for local action. This is of course not a linear process with public participation, building support and publicity as ongoing throughout the process. The main purpose of these planning steps is to assist people who are working in or with local government to develop sustainable energy and climate plans and implementation programs. There can be no single recipe for all cities—so it is up to each local government to develop its own innovative and appropriate plans based on local resources and needs. The energy-efficient urban planning steps in brief are as follows:

7.3.1 Step 1: designating a lead office, finding champions

Energy is vital to build support among staff across all of the city's responsibilities. The involvement of every city council will optimize the potential for successful planning and implementation at the local level. Leadership and direction need to come from one primary office (or department). "Champions"—people with charisma, commitment, and power—will play a vital role in capturing people's imagination. It is important to find both political and staff champions. The political champions need to build political will and lead the councilors, the staff champions are vital for bringing the departments on board and for building the sustainability of the program. A steering committee or task team that brings all the necessary players together will help build commitment and keep everyone on the same page.

7.3.2 Step 2: establishing partnerships

Partnerships bring expertise, resources, support, and information into the process. As energy impacts cut across all sectors, it is needed to start a partnership approach in order to address these challenges. It is recommended to check which NGOs, community groups, businesses, utilities, associations, funders, and government bodies can support the planning and implementation efforts. Which groups can help build in-house capacity and provide additional expertise? National and international energy programs can provide valuable support and so can other cities that already have sustainable energy action plans in place.

7.3.3 Step 3: finding the hooks

In addition to linking into national and international objectives, conventions, and laws, the sustainable energy plan must be clearly linked to the city's core goals and plans. This ensures that the city's energy action strategy is regarded as core city business and will ensure relevance and allocation of resources. Most of the city's goals and issues will have an energy component as energy is key to the delivery of all basic services. It is needed to find the "hooks" or links and make them clear. In addition, it is also needed to work across all the city's long-term and short-term plans and strategies: from economic development to environmental management, from land use planning to waste management, from housing to poverty alleviation and health programs. It is needed to start identifying energy issues during data collection and engagement in public participation. Then to filter these issues through local and national energy visions and policies so that it can be prioritized and develop energy goals that meet local needs. Focus group workshops are a good way of identifying important issues and links between energy and other primary local issues.

7.3.4 Step 4: conducting an energy and greenhouse gas (GHG) emissions audit

A vital step in developing the city's sustainable energy plan is to collect energy information about the city. This process is known as an "energy and greenhouse gas emissions audit." Use the review of the city's goals and policies to help decide what information should be prioritized.

7.3.5 Step 5: analyzing the collected data and developing a draft plan

After gathering a lot of information and developing relationships with many people, it is recommended that the following activities should take place:

- Firstly, identify the critical energy issues under each sector.
- Secondly, rank these issues according to the city's particular priorities.
- Thirdly, the draft plan could be developed on the basis of this list, identifying what to be done to promote a sustainable energy future for the city. The draft plan should list the possible measures (projects and programs), which can address these issues and identify those that will yield the greatest benefit.

7.3.6 Step 6: building public and internal support: the participation process

Developing and implementing a sustainable energy plan is, to many cities and local authorities, pioneering work. Changing the way people live and perceive reality is a difficult and long-term process and very often the most neglected and undervalued part of processes like this one. Building support is one of the most critical aspects of the planning effort and active engagement, community empowerment, ownership, and participation are all key to an effective plan. Achieving this is often challenging as people, businesses, and municipal departments may view energy policies as secondary to their core and everyday activities. It is very important, therefore, that the energy objectives share the concerns of the broader city and community.

7.3.6.1 Identifying stakeholders

There are internal and external people you will need to interact with in order to develop a good energy plan. It is needed to make a list of stakeholders who can provide information. Work with those local government agencies that should care about sustainability [3].

7.3.7 Step 7: finalizing the plan

Based on the participation process, a lot of input will be given on the draft plan and/or the different scenarios. In addition, a list of current and potential projects will be identified and their advantages and disadvantages will be analyzed. Then the city must decide on a set of reasonable objectives that can be implemented in the short and longer term. The next step is to identify priority projects that fit these objectives. Each project and program has both benefits and costs that are a mix of economic, social, political, and environmental impacts. The process of deciding which impacts to rank or which should be assigned more weight should be done as a staff/stakeholder process that is approved through political processes in order to adopt a long-term, sustainable plan. As soon as a high level of agreement is reached, it is recommended to take the sustainable energy plan to the city council for formal adoption and informing all the role players. It is also important to engage the media as soon as possible and arrange for opportunities for formal recognition of all who worked on the plan.

7.3.8 Step 8: implementing and financing the plan

It is all very well having a plan, but it is only meaningful if it follows through to implementation. It may found that there are severe capacity and institutional barriers to implementation. There may be financial hurdles to overcome. It is important to be fully aware of these constraints and build the city plan accordingly. It is needed to involve people to do the work, and it helps if the institution is structured in a way that enables it to assist. Energy should be dealt with as a cross-cutting sector, so the institutional structure needs to reflect this by bringing the different sectors together.

7.3.9 Step 9: monitoring and evaluating the plan

The strategic plan should be a living document with short-term plans being evaluated and updated every two or 3 years and long-term plans every 5 years. The energy field is such a rapidly developing and evolving area of work, so it is important to regularly check the plan with latest national, regional, and international progress and new developments. The evaluation process must compare the objectives against the results. It is needed to measure both the quantity and quality of the progress.

7.3.10 Step 10: publicizing the benefits

This step, like "Step 6 the participation process," needs to start from day 1. It is needed to tell everyone about the projects in order to build support for the sustainable energy goals and to change people's behavior. To build pride, enthusiasm, and a feeling of ownership and empowerment while building a more sustainable future. Education and publicity should be in mind all the time. This is an area that lends itself to many creative and innovative ways of ensuring public engagement and ownership.

7.4 Developing an strategic urban plan for cities in Egypt

The design of the Strategic Urban Planning for Small Cities Project in Egypt is built on a limited set of well-connected principles. The most important and far reaching is sustainable development. The Project seeks to strengthen political sustainability of leaders, social sustainability by including the vulnerable at every step of the process, economic sustainability of the projects, and environmental sustainability of resources used. To achieve sustainability, the project has structurally utilized participation and subsidiarity of the decision-making process; decentralized decision effectively takes advantage of local knowledge as a well-informed source of information about local situations. Local stakeholder not only embody the local knowledge to be accessed, but their participation presents an important entry point to the political decision-making needed for collecting differing viewpoints, interests, initiating negotiations leading to coordination if not cooperation. Risks of limited transparency, reluctance to attend public meetings, expectation to receive remuneration for participation, are recognized and need to be continuously addressed. Technical Teams (TTs) (experts and consultants) are required to balance their technical know-how with local priorities. The experts, besides providing technical analysis and insight, are required to assume more the role of facilitators in enabling local voices; their interpretations, whether in assessing the situation or formulating strategies, should be "hovering" over existing information.

7.5 Responsibilities

The responsibilities of the TT include detailed activities related to the following steps: (1) start-up, (2) Data collection, (3) Data analysis, (4) City consultation, (5) Strategy formulation, (6) Action planning, (7) Reporting, and (8) Evaluation. It is important to note that TT activities will move from one framework to another: First, the TT starts with a city wide and broad-based data collection activity, and second, as TT moves through activities, it increasingly focuses its activities into priority areas, thus conserving resources.

7.6 Different steps

The different steps are divided into the following items:

- Start-up
- Data Collection
- Data Analysis
- City Consultation
- Strategy Formulation
- Approval of Strategic Urban Plans
- Evaluation of Plans

7.6.1 Start-u

The TT is required to launch their city activities by meeting city executive/s, elected leaders, civil society, and natural leaders to open communication channels and:

- Brief stakeholders on project policies, approach, sectors areas, and record their views about the city;
- Identify stakeholders including those mandated to approve final outputs or provide development finance;
- Identify data sources available at city level;
- Assess stakeholders' resources and abilities; and
- Identify a development "champion."

It is imperative that the stakeholders identified include those representing the vulnerable, i.e., poor and women, and environmental concerns.

7.6.2 Data collection

The TT is required to use the data collection phase in the creation of synergies, build partnerships, and improve the institutional setting. This phase initiates the process of assessing development trends. Data collected should include secondary data, primary data through interviews, surveys, and access to data bases. Throughout the reporting on data found for the city, all sources of information, plans, and figures should be clearly mentioned. Interviews for each theme will reveal the multifaceted status and effect of policies, institutional setup and capacities, means of implementation, past and current projects, identifying priority issues, recommended action including beneficiaries, objectives, and visions as well as information related to cross-cutting sectors. In addition, it is needed to conduct visit sites to verify what has been revealed in interviews, e.g., slum areas and environmentally hazardous locations, and take snap shots and video recording to bring reality into foreseen public meetings.

7.6.3 Data Analysis

The objective of data analyses is to identify assets and issues for action as well as uncover sector trends and describe their properties individually and collectively, i.e., synergies, mismatches, and gaps. The TT is required to analyze both urban survey data and responses to interview question within each sector. At the end of the day, expert knowledge form indicators and analysis of the situation as well as local knowledge from stakeholder will be brought together and analyzed to reinforce and develop other projects. This is also underlined in the request to look at the sector as a whole once projects, specific in most cases, are identified.

7.6.4 City consultation

The function of the city consultation is for the city to review data collected and analysis results, reach consensus on the accuracy of city presentation, and identify priority issues and action. The city consultation will also discuss formulating a vision and related objectives for the city. In the consultation the TT will:

- Review the details of the city presentation in the city plenary meeting and stopping to correct any errors after a consensus is reached among audience.
- Manage stakeholders working groups according to stakeholder type and require each group separately to (1) read and reflect on identified issues and proposed actions.
- Collect prioritization results and consolidate these results across stakeholder groups indicating the level of priority for each project.
- Discuss with stakeholders the reason for their agreement on priority projects, the practical nature of these projects by probing into their readiness and ability to contribute to the project.

It is important to note that the priorities set in the city consultation shall define the focus of ensuing activities including strategy formulation and action plans. The strategic urban plan should not be considered an all-encompassing legal document, e.g., Master Plan, but rather a strategic plan that responds to what is important today and the near future and one that is expected to be readjusted as circumstance change.

7.6.5 Strategy formulation

The task of strategy formulation will require the TT to consolidate a stakeholder agreement on objectives and actions. The TT will also consolidate a stakeholder agreement on objectives and actions into a multisector urban planning strategy, with vision (goal and subgoal), development objectives, indicators, plans, and projects (see checklist: "An Agenda For Action For Taking Steps To Consolidate A Stakeholder Agreement On Objectives And Actions").

An agenda for action for taking steps to consolidate aa stakeholder agreement on objectives and actions

The TT will consolidate a stakeholder agreement on objectives and actions into a multisector urban planning strategy, with a vision (goal and subgoal) toward development objectives, indicators, plans and projects, as follows (check all tasks completed):

_____1. Recall vision and revisit end objectives.

_____2. Generate "horizontal" and "vertical" development alternatives for the city, for each sector, with an eye on related issues and without losing sight of an agreement on priority actions.

_____3. Undertake an overall vision workshop whereby stakeholders recall city strengths and opportunities and seeking what the city will, utilize its assets, become in the future. A realistic vision and objectives are required.

Continued

An agenda for action for taking steps to consolidate aa stakeholder agreement on objectives and actions—cont'd

_____4. Identify action properties including timing and duration needed to implement action, the location to implement the action, the financial resources and mechanisms, and the impact these actions have on environment and the vulnerable.
_____5. Prepare preliminary feasibility studies for alternative projects the city is committed to.
_____6. Prepare project profiles describing project objective, beneficiary, cost, etc.
_____7. Call land owners, local authority, and representative of local council to discuss the annexation of needed land to city limits. The TT should discuss the need for land in a competitive framework and find mechanisms to support agreed upon projects particularly those in support of the vulnerable.
_____8. Propose a city wide land use plan including the mechanisms for realizing the land use plan and prepare a preliminary land budget.
_____9. Generate development plan to locate agreed upon projects and identify local development action areas in the city

7.6.6 Approval of strategic urban plans

The TT, facilitated by the Governorate, will present the urban strategy to a public hearing in the Governorate Local Council to discuss and, possibly, approve or recommend changes to the strategic urban plan. Outputs expected for this step are Approving Authorities Reports and Decrees.

7.6.7 Evaluation of plans

The Evaluation Team (ET) reviews the work and issues evaluation reports to be delivered to the TT and GOPP. A meeting might be called by with the ET or the TT to discuss the evaluation reports. Table 7.1 describes the main steps and the subactivities worldwide and in Egypt.

7.7 Summary

In the following paragraph, the summary of this work is highlighted with emphasis on the more important findings:

- Participatory approaches are vital to the planning processes at the cities' level.
- Energy-efficient design and urban planning are the most cost effective way to reduce energy consumption of buildings.
- Energy aspects should be incorporated into the planning process as a cross-cutting issue, which intersects with all other aspects and sectors.
- The comparison between the Egyptian strategic urban planning process and the international energy efficient urban planning process shows that the Egyptian process has high inclusion of the participatory process and all local actors. However, it lacks the inclusion of the energy aspect as an important cross-sectorial aspect.

TABLE 7.1 Planning steps.

Planning steps			
Main steps	Sub-activities	International	Egypt
1. Designate a lead office	- Find a champion	✓	✓
2. Establish partnerships	- Identifying partnerships	✓	✓
3. Find the "hooks" in the vision	- Set goals, policies of the city - Identify issues	✓	✓
4. Conduct energy and greenhouse gas (GHG) emissions audit of city/local authority	- Conduct energy gas (GHG) emissions audit	✓	✓
5. Analyze the collected information	- Data analysis - Develop a draft plan	✓	✓
6. Build public and internal support	- Set participation process - Identifying stakeholders	✓	✓
7. Finalize the plan	- Identifying priority projects	✓	✓
8. Implement and finance the plan		✓	✓
9. Monitor and evaluate the plan		✓	✓
10. Publicize, communicate the benefits.		✓	✓

Finally, let us move on to the real interactive part of this chapter: review questions/exercises, hands-on projects, case projects, and optional team case project. The answers and/or solutions by chapter can be found in Appendix G.

7.8 Chapter review questions/exercises

True/false

1. True or False? Currently, half of the world's total population (around four billion people) lives in urban settlements.
2. True or False? The use of energy, the types of energy used, and easy access to sufficient energy have far reaching implications for a city's economic development, its environmental health, and for the poor.
3. True or False? The sustainable energy efficient path for cities should attempt to reduce carbon footprint, emissions, dependence on fossil fuels, and increased dependence on renewable energy.
4. True or False? Buildings account for a significant portion of primary energy consumption in south and east Mediterranean countries.
5. True or False? Urban planning is the basis for energy-efficient buildings.

Multiple choice

1. Energy-efficient architectural design is only possible when the site and climatic condition are considered appropriately at the urban planning level:
   a. Smart sustainable development level
   b. ICT-based infrastructure level
   c. Energy efficiency level
   d. Investment level
   e. Urban planning level
2. The Local Governments for Sustainability "ICLEI" [4], The United Nations Human Settlements Program (UN-HABITAT) and The United Nations Environment Program "UNEP" have developed a planning guide, which includes 10 steps as a framework for:
   a. Society infrastructure
   b. Public participation
   c. Local action
   d. Economic effects
   e. Smart city policies
3. Energy is vital to build support among staff across all of the city's:
   a. Leadership
   b. Direction
   c. Cyber-Physical Systems (CPS)
   d. Departments
   e. Responsibilities
4. Partnerships bring expertise, resources, support, and information into the:
   a. Holistic integration of smart city platforms
   b. Economic growth
   c. Associations
   d. Computer applications
   e. Process
5. In addition to linking into national and international objectives and conventions and laws, the sustainable energy plan must be clearly linked to the city's:
   a. Goals and issues
   b. Basic services
   c. Long-term and short-term plans and strategies
   d. Core goals and plans
   e. Sustainable energy plan

Exercise

Problem

Why is sustainability one of the strategic goals of smart cities?

Hands-on projects

Project

Do research: Analyze the environmental issues related to energy, efficiency, and sustainability.

Case projects

Problem

Provide an analysis of the determinants of economic efficiency and assess the prerequisites for the security of energy in smart cities.

Optional team case project

Problem

Describe what percentage of the world population will live in urban areas; what the rate will be in developed countries, by the year 2070; and why urbanization poses considerable challenges: including overpopulation, climate change, environmental quality, and access to energy.

References

[1] El-Sherif D. "Integrating urban-rural linkages in strategic plans of small cities -comprehensive approach to environmental preservation" 3rd international congress on environmental planning & management titled. In: "Climate changes and environmental planning", organized by the university of Seoul "Korea" & urbenviron international council "Brazil" – Seoul, Korea; July 2009. p. 26–30.

[2] El-Sherif D. "Environmental issues at the local level — an analytical review of five Egyptian cities" 4th international congress on environmental planning & management titled: "green cities — a path to sustainability", organized by the housing & building national research center "Egypt" & urbenviron. Egypt: International Council "Brazil"; 2011.

[3] Khalil EE, Dijk DV. ISO Focus+, the Magazine of the International Organization for Standardization. In: Future cities: building on energy efficiency, 2; May 2011. 5.

[4] ICLEI. UN-habitat, UNEP: "sustainable urban energy planning - a handbook for cities and towns in developing countries". ICLEI, UN-Habitat; 2009.

[5] Cohen B. Urbanization in developing countries: current trends, future projections and key challenges for sustainability. Technol Soc 2006;28(1−2):63−80.

[6] MED-ENEC. Energy efficiency urban planning, Guidelines for MENA region. 2013.

Chapter 8

Transportation policies for connected and automated mobility in smart cities

Mizanur Rahman, Steven Polunsky and Steven Jones
The University of Alabama, Tuscaloosa, AL, United States

8.1 Introduction

More than 1.2 million people die worldwide in traffic crashes, and some 35,000 in the United States alone. The majority of these crashes are attributable to some form of human behavior (e.g., speeding, driving under the influence of drugs or alcohol, failure to yield, and distracted driving). Considerable progress has been made in recent decades to deploy vehicle technologies to reduce and eliminate the possibility of human-related behaviors on crash occurrence and severity. These technologies have evolved from very specific applications such as antilock braking systems to more sophisticated advanced driving assistance systems (ADAS) and eventually to the various levels of automated driving [1]. As such, one of the primary motivating factors for the development of automated vehicles is the reduction of crashes. Additionally, vehicle-to-everything (V2X) connectivity enables additional safety and operational benefits from automated vehicles (AVs). Besides these benefits, automated vehicles can provide door-to-door mobility benefits to people with disabilities and aging populations. If people in a community cannot afford to have a vehicle because of high vehicle ownership costs or do not want to drive (or own) a vehicle, automated vehicles can be a solution.

Connected and automated mobility systems can revolutionize the movement of both people and goods. For example, if roadway traffic control signals are connected with vehicles and pedestrians via wireless communications, intersection safety and capacity can be substantially increased. Also, traffic volumes can be significantly increased while minimizing the likelihood of crashes with the use of dense platoons of connected and automated vehicles (CAVs) via optimal management of gap between a subject vehicle and a vehicle immediately in front of it and vehicle speed. Mobility services based on connected and automated technology (e.g., Uber and Lyft) are creating opportunities to replace the use of private vehicles with shared mobility. Also, improved tracking and tracing systems are greatly increasing the sophistication of door-to-door freight management.

Numerous studies call for the rapid adoption of CAV technologies that will be capable of supporting the unprecedented future demand. Several global smart city initiatives have shown that cities prosper when infrastructures are connected seamlessly. Interconnections of information systems are now occurring in ways never previously imagined, with potentially dramatic outcomes. A recent high-level report from the U.S. Department of Transportation (USDOT) noted that growing demand due to population and economic growth, coupled with a lack of adequate transportation policies and investment, are overwhelming transportation systems [2]. Without substantial technological progress supported by policy and investments, the transportation systems will face a "drifting toward gridlock" scenario along with unsafe operations of vehicles, as is already the case in many cities around the world.

This chapter is organized as follows. First, an overview of connected and automated mobility systems, including CAVs, transportation infrastructure, transportation data, few case studies, and curb management is presented in Section 8.2. Sections 8.3, 8.4, 8.5, 8.6, and 8.7 cover policies and guidance on safety, communication, cybersecurity and privacy, cooperative driving automation, and navigation systems. Sections 8.8 and 8.9 cover polices for connected and automated mobility services and parking and land use. Legislation related to connected and automated mobility is presented in Section 8.10. Section 8.11 presents public perception and education for improving user acceptance for new technologies such as connected and automated mobility systems.

Smart Cities Policies and Financing. https://doi.org/10.1016/B978-0-12-819130-9.00008-5

8.2 Connected and automated mobility

Modern mobility has many different components, such as wheelchair, micromobility devices, cars, trucks, motorcycles, and beyond. A person's chosen mobility methods can change given time, geography, ability, availability, affordability of acquisition and maintenance, fuel source, convenience, efficiency, and other factors. Technology is rapidly advancing on many fronts to make modes "smart," allowing communication among vehicles, people, and infrastructure. Communication among vehicles is expected to increase safety and efficiency for the traveling public. Communication between vehicles and people such as pedestrians should improve safety for people in proximity to motorways, whether they are walking, have visual or auditory impairments, or on scooters or bicycles. Communication between vehicles and infrastructure is hoped to improve the efficiency and safety of the infrastructure and may also have other benefits such as providing localized tourism or entertainment information, possibly including the use of augmented reality. Connected and automated mobility brings with it concerns for the technologies, data, and costs, as well as equity considerations.

8.2.1 Connected and automated vehicles

Technology has always been an important aspect of transportation. As technologies have evolved from the most basic forms of mobility (e.g., walking) to the complex multimodal transportation networks that serve modern society, so has the need to develop a wider, more interdisciplinary understanding of understanding of them and their potential impacts. Presently, the most advanced technologies defining modern transportation are those related to CAVs. CAVs may be buses (Fig. 8.1), commercial trucks, cars, and even motorcycles. The global CAV market has been projected to exceed $500 billion by 2026 [3], and the number of cities considering CAV policies has accordingly increased rapidly—from 6% of major US cities to 36% between 2016 and 2018 [4]. Citing over 100 cities worldwide currently integrating CAV planning or pilot projects into their transportation portfolios, the National League of Cities has acknowledged the growing awareness and importance of CAV technology to the future of cities [5].

It has been estimated that some 62 million vehicles worldwide will be capable of vehicle-to-vehicle communication by 2023, with 60% of new vehicles sold in the United States having that capability ([6]. and the global market for connected cars exceeding $160 billion by 2025 [7]. Further, the integration of technology into municipal infrastructure, referred to generally as the Smart City movement, presents both opportunities and challenges as to how CAVs will be affected. By June of 2019, one count noted 443 individual Smart City projects in 286 cities worldwide [8].

8.2.2 Connected and automated infrastructure

A brief list of services enabled by connected vehicles includes navigation, remote diagnostics, remote operation, integration with home services, vehicle management, and asset tracking. Vehicles may have hardware that facilitates connectivity such as intelligent antennas and telematics control units [7]. In order to function as part of an integrated system, CAVs require suitable infrastructure capable of collecting and communicating relevant data to and from vehicles. As of 2020, some 15,506 devices and 6182 infrastructure components were in use on operational connected vehicle projects [9]. Infrastructure that enables connected vehicles may also enable or enhance the performance of CAVs through the combination of technologies and the generation, collection, and analysis of relevant roadway and traffic data. In some cases, the infrastructure itself may perform automated functions such as: detection

FIGURE 8.1 Automated vehicle demonstration in transportation research board (TRB) annual meeting at Washington, DC, United States.

and enforcement of speeding, red-light running, inappropriate or illegal uses of special-purpose lanes, or even vehicles driving the wrong way on certain roadways. These detection devices are primarily photography-based (cameras) but may include radar or other sensor methods. A consortium of traffic safety organizations has developed an automated enforcement checklist to guide governments in implementing such programs [10]. The checklist is designed to have data drive the identification of problem roadways and intersections, place automated enforcement in the larger context of traffic safety strategies, and include engineering or signage changes to assist drivers with compliance [11].

The City of Austin, Texas, is actively looking at technologies that could be implemented to facilitate CAVs and, as such, has hosted several field tests of related technologies. One technology being evaluated is the public infrastructure network nodes (PINNS). PINNs are devices that combine technologies serving CAVs into single units that can be deployed across a geographic area, creating a mesh of PINNs. Potential PINN-based services include vehicle-to-infrastructure, atmospheric data collection, Internet service, and sound- and other sensor-based data collection [12]. At the same time, the Central Texas Mobility Authority, an independent government agency that owns and manages six Austin-area toll roads, is using CAV and roadway sensors coupled with artificial intelligence for detection and prediction of roadway incidents [13].

8.2.3 Transportation data

The rapid proliferation of CAVs brings with it a sharp increase in the amount of transportation-related data generated. Prior to CAVs, convectional vehicles were generating data that was limited to what could be used for performance diagnostics. Signals were received by radio for entertainment and primitive navigational purposes that were not retained or analyzed beyond their immediate use and discard. Driver-related data related to licensing and was maintained in agency mainframes. Even then there were policies addressing the access and resale of this information, for example, to vehicle insurance companies.

As capabilities for data sensing and collection were refined, more uses became apparent, making this an area of great growth in a relatively short period of time. Government agencies responsible for transportation now had easier access to information such as vehicle travel times and trip origins and destinations that used to require time-consuming and dangerous manual surveys of travelers. The rapid expansion in data available and collected as well as a technologically expanded variety of ways in which that data can be sensed raised a number of issues for policymakers, including whether data should be personally identifiable or anonymized, public use versus private use, whether data sharing should be opt-in or not, and the costs of data retention and analysis over time [14].

CAV infrastructure deployment tends to have a large federal component, usually funding, which defines their extent and use and may direct technologies and operation [15]. These arrangements may specify data ownership issues from a research perspective, leaving the larger societal issues untouched. Data resulting from signals received and sent, details of the physical environment and conditions and their effects on vehicle operational abilities, driver and cabin activity, and other uses have grown far beyond non-Internet based geolocational uses and now can be collected, aggregated, and associated with third-party Internet applications that already have significant data regarding the driver and passenger choices and preferences. There is growing demand for this data and the value it contains from data aggregators, infrastructure facility owner-operators, and automobile original equipment manufacturers, among others [16]. These issues will ultimately be sorted out through policy or, in the absence of it, through the courts.

8.3 Safety guidance for automated vehicles

Although automated vehicles will significantly increase safety benefits along with other benefits, new policies are required to ensure safety during pilot testing and on road operations. In the United States, all vehicle manufacturers must comply with the Federal Motor Vehicle Safety Standards (FMVSS) [17] for their manufactured vehicles, and after that, they will not have any barrier selling. However, the US National Highway Traffic Safety Administration (NHTSA) has recall, defect, and enforcement authority for any AV systems. The US Department of Transportation expects that manufacturer and other agencies will use NHTSA guidance for testing and deploying automated vehicles along with industry standards and best practices. In this way, all the automated vehicle testing and deployment will be safe under real-world conditions.

Fig. 8.2 presents the NTHSA 15-point checklist for automated vehicle safety for autonomy level 3 (i.e., conditional driving automation), level 4 (i.e., high driving automation), and level 5 (i.e., full driving automation), which are defined by the Society of Automotive Engineers (SAE). Some of them are applicable to all automation functions, and some of them applicable to each specific automation function of an automated vehicle. The following is the safety guidance that applies to all automation functions: data recording and sharing, privacy, system safety, cybersecurity, human–machine interface (HMI), crashworthiness, and consumer education and training. On the other hand, description of the operational design domain (ODD), object and event detection and

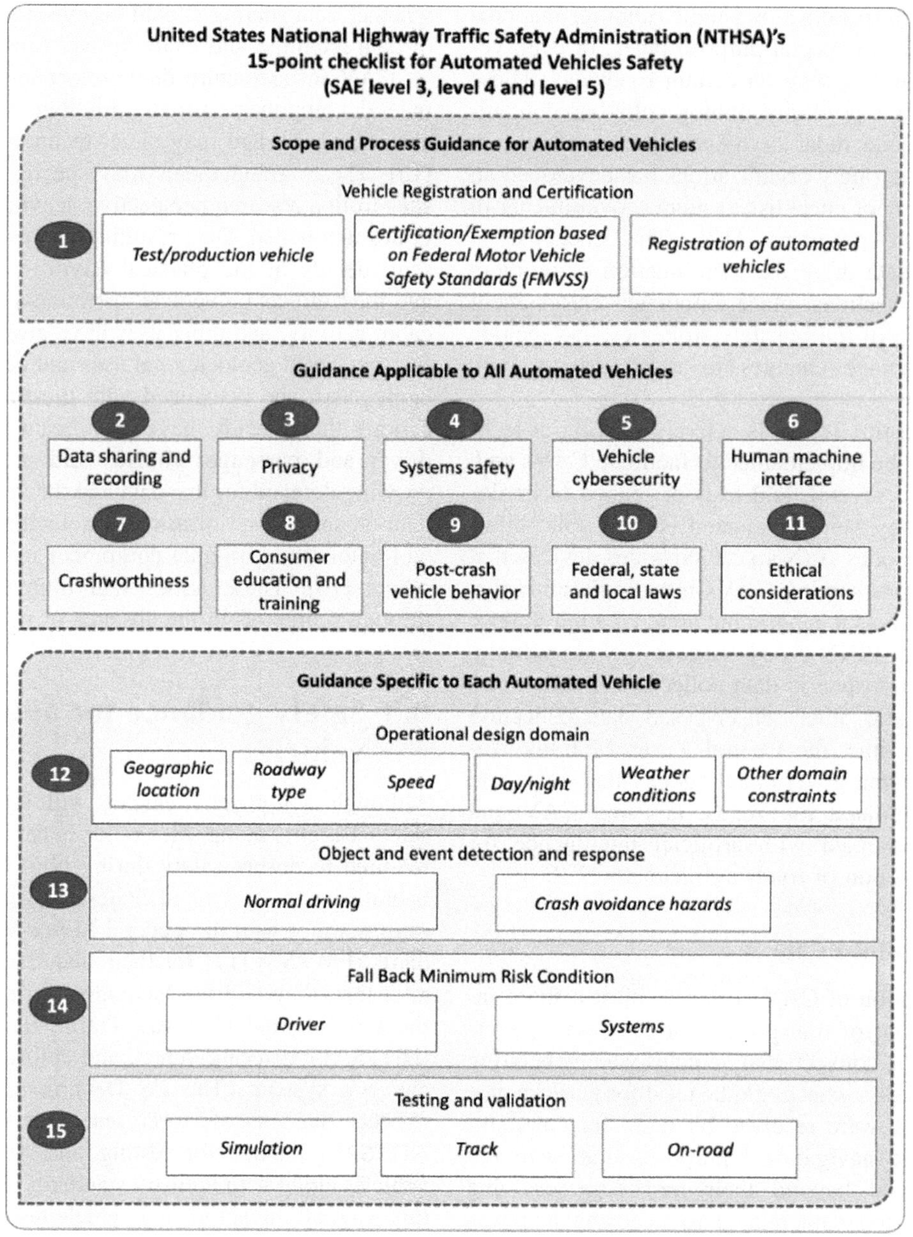

FIGURE 8.2 An example framework for automated vehicles safety performance guidance.

response (OEDR), and fall back minimum risk condition are applicable for specific automation functions. However, these functions should be clearly defined depending on the SAE automation levels. For example, the ODD needs to be defined based on the roadway types, geographical location, speed range, lighting conditions for operation (day and/or night), weather conditions, and other operational domain constraints. A manufacturer or other entities may request for an interpretation or exemption from NHTSA if needed. Note that the NHTSA guidance is applicable for both test and production vehicles. The manufacturer or other entities in the United States should then follow existing US Department of Transportation (DOT) identification/registration requirements (described in 49 Code of Federal Regulations (CFR) Parts 566 and 567).

On September 2nd, 2020, the U.S. Department of Transportation launched an online tool to track on road testing. It is available publicly at https://www.nhtsa.gov/automated-vehicle-test-tracking-tool. The purpose of this tool is to improve the safety and testing transparency of AV systems so that public can learn more about AV technology. At the beginning, it provides AV on road testing data for 17 cities across the United States. In this pilot initiative, nine states and nine companies participated in the automated vehicle pilot testing. On January 11th, 2021, the Automated Vehicle Transparency and Engagement for Safe

Testing (AV TEST) initiative became a full program from a pilot program. To date, 52 companies, governments, and associations are now participating.

KPMG International published a 2020 Autonomous Vehicle Readiness Index (AVRI) report to assess level of the preparedness of 30 countries and jurisdictions for AVs. The AVRI is determined by combining seven equally weighted measures as follows: (i) AV regulations; (ii) government-funded AV pilots; (iii) AV-focused agency; (iv) government readiness for change; (v) future orientation of government; (vi) efficiency of the legal system in challenging regulations; (vii) data-sharing environment. Note that Singapore and the United Kingdom are the top two countries worldwide in terms of policy and legislation related to AVRI. Following this AVRI, the safety rules for AV pilot testing of four countries are presented in Table 8.1.

8.4 Communication for connected mobility

In the coming years, vehicles will not only be controlled autonomously but will also be connected via V2X wireless communication networks [18]. V2X represents the communication between vehicles and everything, which includes other vehicles, traffic signals, roadside devices, pedestrians, etc. For example, vehicle-to-pedestrian (V2P) represents only communication between connected vehicles and connected pedestrians. Vehicle-to-Infrastructure (V2I) represents only communication between connected vehicles and connected infrastructures. Dedicated short-range communication (DSRC) and Cellular-V2X (C-V2X) are two examples of communication technologies for V2X communication. DSRC uses IEEE 802.11p standards and operates in the 5.850–5.925 GHz spectrum. On the other hand, C-V2X currently uses 3GPP standardized 4G cellular connectivity for V2X communication. Both technologies use the same message sets following SAE J2735 and J2945 standards [19,20].

Recently, the US Federal Communication Commission split the 75 MHz band, which was allocated before only for the DSRC spectrum at 5.850–5.925 GHz [21]. They allocated the lower 45 MHz of the band for unlicensed use and the upper 30 MHz at 5.850–5.925 GHz for connected mobility. It means that DSRC is no longer applicable for vehicular communication in the United States. However, C-V2X technology must be used for the upper 30 MHz, and it has several advantages compared to DSRC: (i) improved communication signal design when there is a high relative speed between vehicles; (ii) improved transmission structure in case of high vehicle densities; and (iii) allow utilization of Global Positioning Systems (GPS) timing for time synchronization when out of coverage. Although C-V2X uses 4G cellular connectivity, 5G will be used instead of 4G in the near future. 5G has several advantages over 4G: (i) increased spectrum allocation; (ii) availability of directional beamforming antennas used at both 5G enabled base stations and mobile users; (iii) increased capacity to aggregate numerous simultaneous users within the coverage area; (iv) highly increased bit rates within increased proportions of the 5G coverage areas; and (v) lower cost of communication infrastructure.

Although C-V2X uses a 5.850–5.925 GHz band, researchers are showing more interest than before on spectrums higher than 6 GHz as it can increase communication network reliability, throughput, and connectivity in terms of coverage area. There are several advantages for having spectrums above 6 GHz: (i) provide higher throughput for the increasing number of connected vehicles; and (ii) allows for increased data rates while reducing the communication latency compared to LTE. These advantages can support data-intensive safety and mobility application in a connected vehicle environment. Thus, the primary goal of telecommunication researchers from industry and academia is to improve spectral communication efficiency and increase the data rate. Table 8.2 presents the barriers for connected mobility-related applications and corresponding 5G Characteristics, which will help to overcome those barriers.

Since 1980, a new communication technology replaces an old technology every 10 years. Here is the list of technologies: (i) first-generation analog FM cellular systems in 1981; (ii) second-generation (2G) digital technology in 1992; (iii) 3G in 2001, and (iv) Long-Term Evolution–Advanced (LTE-A) in 2011 [22]. The 3GPP is creating 5G new radio (NR) standards, and they released a first complete set of 5G NR standards in 2017, which is "Release 15" [23]. Releases 16 and 17 will cover V2X application layer services. In addition, 3GPP standard committees have also started working on release 18. 5G has already been deployed in several major cities (e.g., Atlanta, Boston, Chicago, Houston, and San Francisco) in the United States [24]. Telecommunication companies, such as AT&T, Verizon, and Sprint, are leading these deployment efforts.

8.5 Cybersecurity and privacy

Connectivity and automation are fundamentally changing transportation systems and increasing the size of the surface, vulnerable to potential cyberattacks. It also allows malicious actors (potentially including malign individuals, criminal or terrorist organizations, and hostile nation-states) to exploit security vulnerabilities. In a connected and automated mobility system, cyber insecurity will affect every aspect of the transportation system. Connectivity and automation technologies will not only affect road transport but will also be deployed by ports and in air-traffic control systems, which will also be vulnerable.

Vehicles today contain high-end computers that process approximately 25 GB of data per hour [25]. In the future, this may increase to approximately 4 TB per hour for fully autonomous vehicles [26]. Today, each vehicle has multiple Electronic Control Units (ECUs), which are interconnected using the Control Area Network (CAN) bus [32].

TABLE 8.1 Safety rules for pilot testing.

Country name	Policy document	Safety rules for pilot testing
Singapore	Singapore Road Traffic Autonomous Motor Vehicle Rules, 2017 (January 2020 update) [27]	• For any automated vehicle trial, the land transportation authority (LTA), the traffic police and the center of excellence for Testing and Research of AVs-NTU (CETRAN) comprehensive safety assessment related to basic maneuvers and safe stop when the vehicle detects an obstacle. • A safety driver must be inside each vehicle as per LTA guidance, and the driver must have access to the steering wheel and/or emergency brake. However, an engineer and a safety driver were in each vehicle for most of the AV pilot trials so far.
United Kingdom	UK Code of Practice: Automated Vehicle trialing, 2019 (2015 version update) [28]	• There are three safety requirements for conducting any pilot trials: (i) a safety driver must be inside or outside each vehicle and ready to take over for any emergency situations; (ii) a vehicle must be compliant with roadway regulations and corresponding approval as needed; and (iii) insurance coverage. • Before any AV trials, british Standards institution (BSI) guidelines (BSI contains Connected and Autonomous Vehicle PAS Program including PAS 1880 guidelines for Developing and assessing Control Systems for Automated Vehicles, and 1881 assuring Safety for Autonomous Vehicles trials and Testing) is recommended to follow, and it is expected that safety features—i.e., information about the trial, driver and operator training, compliance, points of contact with related agencies, safety plans—will be shared publicly • BSI has published a PAS 1882 on July 17, 2020 related to data collection from any AV trials for supporting an incident investigation, and it is required to provide data for any investigation purpose. • The department for Transport and the center for the Protection of national Infrastructure (CPNI) have developed eight cybersecurity principles in 2017 for automated vehicle safety compliance. • It is a must to have contingency plans and corresponding agreements for each AV trail in conjunction with relevant authorities so that necessary actions must be taken in an unintended situation or accident.
Australia	Australian national guidelines for Automated Vehicle trials, 2017 [29]	• Every organization must have a safety management plan to identify and manage the following safety risks: (i) Security against hacking; (ii) risks to road infrastructure; (iii) appropriate transition processes for vehicles that can move between automated and human driving modes; (iv) risks to other road users; (v) system failure; (vi) elaborating on the management of system failures including hardware failures; (vii) software errors and human errors; and (vii) system redundancy and fallback options. • A safety driver must be present inside each vehicle for each AV as of October 2020. Otherwise, it is required to have a specific exemption or permit from relevant authorities • A safety driver must take the necessary training. • It is necessary to use an automated vehicle identification system so that other road users can recognize the vehicle as an AV.
United States: California	California Autonomous Vehicle Testing Regulations	• For any test trials, an AV must be evaluated for different operating design domain (ODD) in a controlled environment based on an indented operating environment and determine the ability of reasonable safe operation of an AV in the defined ODD scenarios. • A safety driver must be present inside an AV so that the driver can monitor and take control of the test vehicle if needed. • A human-machine interface must be available to engage and disengage an operator remotely to the autonomous technology and make sure the operator has completed required training. • There are two documents that are required to testing fully autonomous vehicles (driverless): (i) a Law enforcement interaction plan; and (ii) an assessment from manufacturers showing how safety is pursued.
United States: Arizona	All federal laws, Arizona state statutes, title 28 of the Arizona revised statutes, and Arizona department of Transportation (ADOT) [30,31]	An entity must follow the following safety protocols for pilot testing: • A communication procedure between a fleet support specialist and a vehicle operator during the operation of an automated vehicle • A safe removal procedure of an automated vehicle from a highway • A description of a city where the testing will be conducted • A recognition procedure of an automated vehicle while it is operating in autonomous mode and a safe towing procedure of an automated vehicle • A detailed guideline for any hazardous conditions and corresponding safety risks for the publics

TABLE 8.2 Barriers for connected vehicle applications and 5G characteristics.

Barriers for connected vehicle applications	Key characteristics of 5G to overcome barriers related to connected vehicle applications
Dynamic network topology of vehicular communication due to high vehicle mobility and frequent data flow disconnections	• **Enhanced mobile broadband (eMBB):** Provides high data rate mobile broadband services for seamless data access both indoors and outdoors.
Cross-channel interference and subsequent packet drop in adjacent channels	• **Ultra-Reliable low-latency communication (URLLC):** Enables stringent latency and reliability requirements in highly mobile vehicular communications to enable the CV network.
Higher channel access delay due to a high number of vehicles	• **Massive machine type communication (MMTC):** Supports a large number of devices, which generate a small amount of data in the irregular interval.

Examples of different ECUs are Engine Control Modules (ECM), Electronic Brake Control Modules (EBCM), and Transmission Control Modules (TCM). These ECUs communicate with each other via the CAN bus to perform different functions of the vehicle synergistically. The CAN bus and the ECUs are not well protected, as shown by multiple experiments that have simulated hacking. For example, in one experiment, the ECU responsible for processing wheel speed data was compromised, which could impact the EBCM and cause incorrect braking [33]. Such an attack could jeopardize the safety of the subject vehicle and other vehicles, as well as vulnerable road users in the vicinity, such as pedestrians.

As vehicles are moving toward connected and autonomous driving, and as cities are moving toward connected, adaptable, and automated systems, problems loom. With both vehicles and infrastructure communicating in real time, the extent of security threats is largely unknown. Thus, a cyberattack can cause catastrophic multivehicular accidents, which will lead to shutting down an entire city's transportation system. In sum, understanding what types of cyberattacks may occur, and developing new software and communication network technologies to protect against them, is of the highest importance.

The security requirements of applications related to CAVs can be classified as follows [34]: (i) confidentiality, (ii) integrity, (iii) availability, (iv) authentication, (v) accountability, and (vi) privacy (see Fig. 8.3). Any or some of these security requirements can be compromised because of a cyberattack by an attacker. Data breaches and data loss are two for data examples of cyberattacks, which are related to data. V2X communication networks can be compromised because of cyberattacks. Fig. 8.3 presents all these six security requirements and

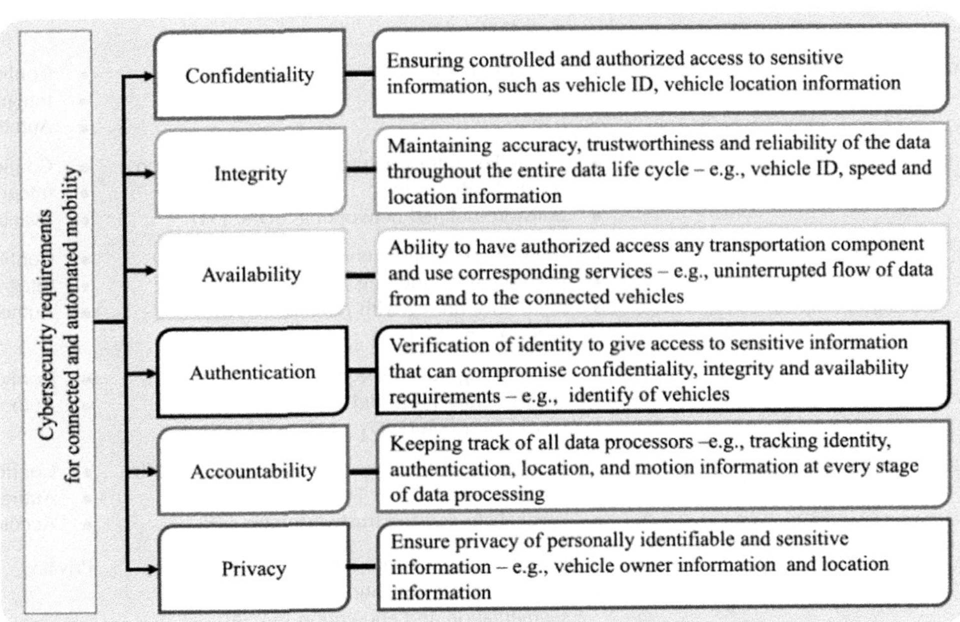

FIGURE 8.3 Cybersecurity requirements and their definitions.

corresponding definitions. Confidentiality refers to preventing unauthorized access and providing controlled and authorized access to sensitive CAV data. Vehicle ID and location are examples of sensitive data. Eavesdropping, account or service hijacking, spoofing, and Sybil are few examples of communication network-related cyberattacks.

Integrity refers to maintaining accuracy, trustworthiness, and reliability of the data throughout the entire data life cycle—e.g., vehicle ID, speed, and location information. Identity spoofing, repudiation, information disclosure, and Sybil are few examples of the attack that compromised data integrity. Availability means the ability to have authorized access any transportation component and use corresponding services—e.g., uninterrupted flow of data from and to the connected vehicles—whenever needed. For example, Denial of Service (DoS), Distributed Denial of Service (DDoS), malware attack, blackhole attack, replay attack, and jamming attack can compromise availability requirement. A DoS attack can flood a communication channel and disrupt connected vehicle application-related services. On the other hand, authentication means the verification of identity to give access to sensitive information that can compromise confidentiality, integrity and availability requirements—e.g., identifying a vehicle. Identity spoofing, information disclosure, brute force, and Sybil attacks can compromise authentication requirements.

Accountability refers to keep track of all data processors—e.g., tracking identity, authentication, location, and motion information at every stage of data processing. An attacker can duplicate, transfer, and use any vehicle-related data, and accountability requirements will be compromised. Privacy refers to ensuring the privacy of personally identifiable and sensitive information—e.g., vehicle owner information and location information. According to the US Fair Information Practices (FIP), there are five bases for defining the data protection and privacy laws: (i) data collection limitation, (ii) purpose specification, (iii) purpose use limitation, (iv) individual participation, and (v) visibility and transparency [35]. Table 8.3 presents the security issues related to connected automated mobility, example solutions, and corresponding security requirements.

The SAE J3016 defines common terminology related to security threats, vulnerability, and corresponding risk for connected and automated mobility as well as provides recommended practices [1]. These recommended practices are developed based on the automotive functional safety

TABLE 8.3 Security issues, corresponding security requirements, and examples of security solutions.

Security issues	Examples of security solutions	Corresponding security requirements
Authentication of connected vehicles	• Multifactor authentication [36]	• Integrity • Authentication
Establishing trust relationships with a high number of CVs	• Distributed agent-based trust establishment [37]; • Blockchain-based approach for establishing trust [38]	• Confidentiality • Accountability
Validation of connected vehicle's location	• Crowdsourcing [39]	• Confidentiality • Integrity • Authentication
Communication network security	• Public key infrastructure certificate revocation list (PKI CRL) [40]; • Software-defined networking (SDN) [41]	• Confidentiality • Integrity • Availability
Data security	• Multikey searchable encryption (MKSE) [42] • Data integrity auditing [43] and • Data outsourcing with public integrity verification [44,45]	• Confidentiality • Integrity • Authentication
Security of heterogenous communication networks and mobile nodes	• Establishing trust between CVs and infrastructure [46]; and • Blockchain [44,45,47];	• Confidentiality • Authentication
Scalability of security solutions	• Vehicular public-key infrastructure (VPKI) [48]; • Blockchain assisted trustworthiness [49]	• Confidentiality • Authentication • Accountability
Data privacy	• Data tracking, monitoring, and auditing [50]; • Data retention obligation [51,52]; data segmentation and encryption [53,54]	Privacy

standards, which is ISO 26,262. As the connected and automated mobility systems are changing rapidly, it is necessary to adapt these standards and develop required standards based on the system requirements.

8.6 Cooperative driving automation

Existing autonomous vehicle (AV) developments and real-world implementations have been conducted using in-vehicle sensors (e.g., radar, LIDAR, camera, and infrared), assuming that an AV can operate itself by detecting surroundings using these in-vehicle sensors. Only relying on in-vehicle sensors will limit the view of an AV, and it will not be capable of seeing what is happening around the corner of a road due to the line-of-sight constraint, behind an object, or further ahead of a road network beyond the sensor range, which will lead to unsafe and inefficient operation of CAVs. In addition, AVs require a high amount of power to run their onboard sensors and perform the processing functions to analyze the surrounding environment and make driving decisions. On the other hand, cooperative driving automation (CDA) can enable vehicle-related applications by sharing information between automated vehicles and other road users—e.g., other vehicles and pedestrians —as well as transportation infrastructures, such as traffic signal controller and roadside data infrastructure to improve roadway safety, efficiency, and reliability. The Society of Automotive Engineers (SAE) J3216 standard defines cooperative driving automation (CDA) as follows [55]:

Automation that uses machine-to-machine communication to enable cooperation among two or more entities with capable communications technology and is intended to facilitate the safer, more efficient movement of road users, including enhancing the performance of the dynamic driving task for a vehicle with driving automation feature(s) engaged.

CDA has the ability to enhance dynamic driving task (DDT), such as object and event detection (OEDR) and response, operational and tactical maneuvers, and situational awareness, such as signal phase and timing (SPaT), dynamic speed limits, and emergency response, of an AV. As per SAE J3216, there are four classes for CDA: (i) Class A: status-sharing cooperation; (ii) Class B: intent-sharing cooperation; (iii) Class C: agreement-seeking cooperation; and (iv) Class D: prescriptive cooperation. When a subject vehicle shares information about its location and perception information with surrounding vehicles, it can be called as status-sharing cooperation. The aim of this type of cooperation is to improve overall situational awareness. In case of an intent sharing cooperation, a subject vehicle shares it future actions with it close by vehicles, such as future states of a vehicle. Agreement-seeking cooperation intends to make collaborative, dynamic driving decisions by sharing a sequence of messages (may use status-sharing and intent-sharing cooperation information). When an action is prescribed for a subject vehicle based on the surrounding traffic states, it calls prescribed cooperation. Fig. 8.4 presents example scenarios for CDA based on different classes. This figure presents five different scenarios, and it also indicates the corresponding class of cooperation.

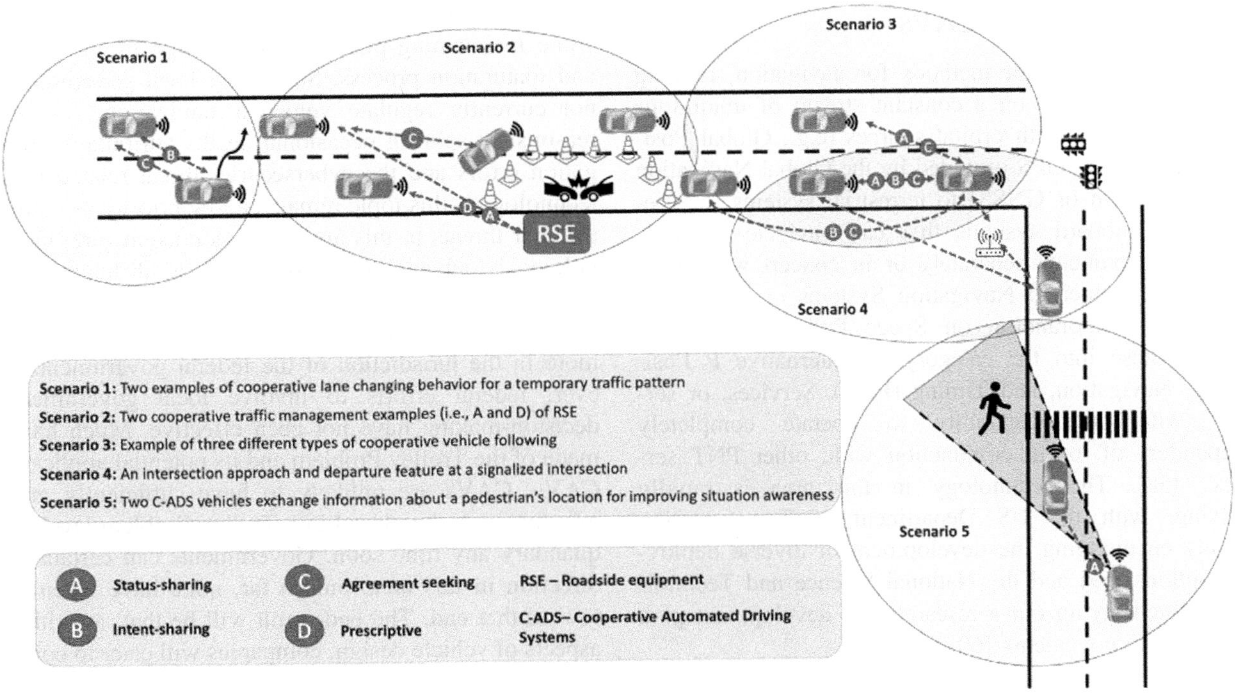

FIGURE 8.4 Example scenarios for cooperative driving automation (CDA).

An automated agent (e.g., an automated vehicle) may have different types of functionalities for completing its automation task. CDA depends on these functionalities. The National Institute of Standards and Technology (NIST) 4D-Real-time Control System (RCS) reference architecture defines three functionalities of an automated vehicle [56]: (i) situational awareness—e.g., sensor signal processing, object classification, world modeling, prediction, and other forms of knowledge; (ii) planning—e.g., behavior generation and value judgment; and (iii) action—e.g., vehicle motion control and state changes. Here, 4D represents four dimensions of hierarchical control structure, which includes sensor processing, world modeling, behavior generation, and time. Note that there is a relationship between CDA and the functionality of CDA agents, and all of these three functionalities can be influenced by any of the cooperation classes.

Table 8.4 presents relationship between CDA and different levels of automation. It indicates that types of cooperation can vary depending on the level of automation, which is described in the standard SAE J3016 [1]. However, there will be limited cooperation for levels 1 and 2 as a human driver is responsible for completing all the dynamic driving tasks. Cooperative driving automation is applicable for levels 3, 4, and 5 automation as automated driving systems is responsible to execute all the dynamic driving tasks. Table 8.5 presents different types of cooperation classes and corresponding sample message contents. Data exchange must be harmonized, and it can be achieved through appropriate data definitions. Data definitions must have the following characteristics: traceable, consistent, accurate, clear, complete, and concise.

8.7 Navigation of CAVs

CAVs use a variety of methods for navigation, ranging from systems reliant on a constant stream of ubiquitous information from earth orbital sources (e.g., Global Positioning System or GPS provided by the Global Navigation Satellite System or GNSS) to terrestrial systems (e.g., e-Loran) to onboard systems that can navigate in GPS-denied environments separately or in concert with other systems (e.g., Inertial Navigation Systems or INS) [57]. The 2021 Memorandum on Space Policy Directive 7 grouped these into the category of Alternative P Positioning, Navigation, and Timing (PNT). Services, or services with "the capability to operate completely independent of, or in conjunction with, other PNT services" [58]. The technology in this area is rapidly evolving, with the US Department of Transportation actively encouraging the development of diverse deployable options [59] and the National Science and Technology Council laying out a research and development plan for future improvements [60].

GPS is ubiquitous today due to its being provided without charge to the end user, its ready availability in urban and rural areas, its continuous maintenance by the US government, and its ability to be used by a variety of receivers. GPS provides sufficient data to calculate position, navigate, and attach timing information (PNT).

One CAV navigation technology has been around for a while—radar. Radar can perform well in a variety of weather conditions, has low power requirements, and an extensive body of literature and testing. Radar proponents believe that augmenting it with AI software can greatly increase its usefulness [61]. Other technologies include lidar and camera-based systems. Lidar excels in physical feature detection while cameras stand out in reading road signs and street markings and radar works in poor weather where other systems do not [62].

Any system or combination of systems intended for CAV navigation face a host of challenges. In addition to cybersecurity concerns, these include ever-changing environmental conditions, work zones, and other disruptions to known traffic patterns, route restrictions, and ethical dilemmas [62]. In addition, a seemingly infinite number of permutations of driving incidents continue to require attention. One example is the confusion experienced by an AV when it followed a truck hauling traffic signals, interpreting it as a continuing line of signals [63]. Competition for CAV-related resources such as the 5.9 GHz band [64] may impede CAV implementation and should be considered when planning.

In the absence of federal mandates regarding specific navigation technologies, it is likely that the private sector will drive the development and adoption of methods. For the policymaker, this means maintaining a regulatory atmosphere that allows private sector competition to thrive while maintaining public safety through the development and maturation process. States and local governments do not currently regulate vehicular navigation, for several reasons. In spite of occasional deaths attributable to navigation errors and the cybersecurity threat related to these technologies, this topic remains a low priority as compared to other threats in this area such as ransomware; the technology is advancing rapidly, and is difficult to keep informed about among competing and more urgent issues; and there is an expectation or assumption that the issue lies more in the jurisdiction of the federal government. However, federal efforts to involve local governments in decision-making have not been effective. Much has been made of the Trolley Problem and its potential application to CAVs. CAVs are unlikely to have sufficiently granular information to be placed in a Trolley Problem level ethical quandary any time soon. Governments can certainly give direction in this area, but so far, none have taken action toward that end. The end result will be that, as with other aspects of vehicle design, companies will cater to consumer

TABLE 8.4 Relationship between cooperative driving automation and levels of automation [55].

		Levels of automation defined by society of automotive engineers (SAE)					
		Level 0 (no driving automation)	**Level 1 (driver assistance)**	**Level 2 (partial driving automation)**	**Level 3 (conditional driving automation)**	**Level 4 (high driving automation)**	**Level 5 (full driving automation)**
Types of cooperative driving automation	**No cooperative automation**	Human is responsible for all driving tasks. Examples: Signage and traffic control devices)	Human driver is responsible to complete dynamic driving tasks and to supervise advanced driver assistance features in real-time.		• Automated driving systems is responsible for executing dynamic driving tasks, which varies based on automation level.		
	Status sharing	Human is responsible for all driving tasks. Examples: Brake lights and traffic signal	Limited cooperation: A human driver is responsible for driving and must supervise cooperative driving automation features.		• Cooperative automated driving systems (C-ADS) will be responsible for executing dynamic driving tasks. Situational awareness will be increased beyond in-vehicle sensing capabilities.		
	Intent-sharing	Human is responsible for all driving tasks. Examples: Turn signal and merge	Limited cooperation related to only lateral or longitudinal intent, which a human driver may take over.	Limited cooperation related to both lateral and longitudinal intent and it may be taken over by a human driver.	• Cooperative automated driving systems (C-ADS) will be responsible for executing dynamic driving tasks. • Situational awareness will be increased through prediction reliability and C-ADS plans of surrounding road users and road operators.		
	Agreement seeking	Human is responsible for all driving tasks. Examples: Hand signals and merge)	Not applicable	Not applicable	• Cooperative automated driving systems (C-ADS) will be responsible for executing dynamic driving tasks. • Performance of automated driving systems will be improved through accepting or suggesting actions based on mutual goals of C-ADS and transportation systems along with surrounding toad users and road operators.		
	Prescriptive	Human is responsible for all driving tasks. Examples: Hand signals, lane assignment by officials	Not applicable	Not applicable	• Cooperative automated driving systems (C-ADS) will be responsible for executing dynamic driving tasks. • C-ADS will except actions for which it is designed for and for which it allows prescriptive communication.		

TABLE 8.5 Cooperation classes and corresponding sample message content [55].

Cooperation classes		Sample message content
Status sharing	Object	Object classification—e.g., vehicle and pedestrian, and corresponding attributes—e.g., vehicle dimension and number of vehicle occupants
	Traffic participants	Object pose—e.g., position, speed and acceleration
Intent-sharing	Intent of C-ADS	Predicted future state and maneuvers
	Intents of other traffic participants	Traffic density, speed, and traffic volume
	Perception of intention	Events—e.g., traffic incident and work zone area
Agreement seeking	• Seeking and agreeing to engage in agreement-seeking cooperation • Proposed actions of relevant traffic participants • Acceptance of proposed actions, rejection of proposed actions, or alternative proposed actions	
Prescriptive	• Vehicle motion control • Traffic rules • State of traffic control device • evacuation orders	

preferences if those exist or can be determined or an aftermarket for products and services that alter the vehicle's factory settings will develop, similar to the rise of the "rolling coal" movement and its attendant environmental and ethical considerations.

8.8 Policies for connected and automated mobility services

The introduction of connected and automated mobility services has been marked an introductory period featuring a wide variety of vehicles and business models. These new private sector initiatives have entered the public sector under a regulatory scheme designed for a world where vehicles have drivers and a bicycle or scooter is ridden and secured by its owner. Also, while safety regulations may typically be found on the state level, business permitting of this nature is usually found on the local level, leading to great variation in if and how the general regulatory scheme addresses these new business models.

8.8.1 Connected and automated mobility

Every governmental body has a responsibility to address the safety and needs of its citizens and visitors. The evolving development of CAV laws and regulations at the federal level has led several policymakers to ask whether and how they might prepare for CAVs. The National Association of Counties created a toolkit/primer divided into four sections: technology, policy, deployment, and next steps [65]. The States of Connecticut and Florida, for example, used different approaches (see field reports below). In addition to bodies with legislative authority, other agencies with regulatory powers may also become involved. One example is the California Air Resources Board, which by rule set a clean-miles standard that requires 90% of ride-hailing miles to be in EVs by the year 2030 [66].

8.8.2 Field report: Connecticut Department of Transportation, Connecticut, US

Motivated by fast-moving technological advances and the potential for major impacts on the provision of mobility, the Connecticut Department of Transportation created a Strategic Plan for CAVs. The plan places an emphasis in the short term on policy coordination, workforce and infrastructure readiness assessments, and pilot projects. In addition to potential benefits, the plan notes risks and challenges such as public acceptance and planning uncertainties and suggests that legislative action may be required, primarily to provide funding and staffing resources for CAV research and deployment [67].

8.8.3 Field report: Florida, US

The State of Florida examined how CAVs in combination with other emerging technologies such as shared-use vehicles and electric vehicles might affect travel behavior. In spite of the uncertainties, researchers discovered that the rapid expansion of technologies and business models meant that innovations such as car sharing, ridesourcing, and CAVs are affecting travel attitudes and behavior. They concluded that these factors should be included in modeling for transportation planning, especially if a goal is to develop strategies to promote alternative mobility options [68].

8.8.4 Field report: case studies and pilot projects by university of Oregon, Oregon, US

The University of Oregon examined 11 case studies and 220 pilot projects to identify commonalities of successful efforts. Subject matter included micromobility, AVs, microtransit, and transportation network companies. Their report recommends using pilot projects if the goals and outcomes are well defined at the start of the process, if an easily understood policy framework is created for each pilot project, if relationships and trust are developed, and if a final report is produced that includes an evaluation of the project's outcomes [69].

8.8.5 Mobility as a service (MaaS)

Mobility as a Service (MaaS) is the technology-enabled provision of transportation-enhancing products and services that are new or may have been previously provided by the individual or employers. MaaS generally replaces current ownership/registration business models with subscription services [70] that seek to integrate disparate transportation services [71]. MaaS is intended to provide alternatives to private car usage that improve convenience, sustainability, and cost-effectiveness. It may involve subscription services, transportation on demand, and the creation of new markets [72]. Examples include taxis, TNCs, and apps that facilitate chaining trips involving multiple modes. One implementation in Australia used sensors to detect passenger train car weights and calculate how crowded they were, making that information available to people looking for less crowded cars to board [73]. The MaaS market is expected to exceed $100 billion by 2030 [74]. Examples include the Lime electric scooter company, which offers a monthly subscription service. This approach is in addition to their normal business model, in which anyone may use a Lime scooter upon paying a per-use access fee plus a per-minute fee. Subscription users may also be given other incentives such as longer reservation periods and promotional free access for a limited time [75].

8.8.6 Private sector involvement

Policymakers should consider the likely involvement in the private sector of future mobility efforts, especially where technology is employed. In early 2020, there were 44 startup companies valued over $1 billion each in the mobility and travel field, up from 22 in 2018 [74]. A variation on CAVs is the personal delivery device (PDD), a robot that delivers food, medicine, and supplies. These devices may use CAV technologies or may be remotely piloted. PDDs may be hampered by stairs, other sidewalk users, and the difference in time between arrival at the destination and receipt by the customer. The Center for Cities Solutions suggests there may be employment opportunities for curbside porters to handle the receipt and other tasks [76].

8.9 Parking and land use

Perhaps one of the most profound potential impacts of CAVs is the potential to fundamentally change the allocation of urban (and suburban) land use currently devoted to parking [77–79]. On the one hand, CAVs will change how vehicles and interact with, and indeed obviate much of the need for, on-street parking in urban areas. As such, urban planners and city officials will be able to repurpose this valuable urban space for other uses such as dedicated lanes for bicycles or buses, parking/storage of micromobility vehicles, MaaS pick-up/drop-off locations, outdoor dining, or even green spaces. Significant sections of land are devoted to temporarily storing cars (i.e., parking lots) in suburban and less dense urban environments. Typically, these spaces are adjacent to activity centers, such as shopping centers, apartment complexes, healthcare facilities, and places of employment. As with on-street parking in more dense urban areas, CAVs will reduce the need for extensive parking facilities (i.e., lots and decks) in less densely developed areas, thereby creating opportunities to repurpose land for more economically productive and socially supportive uses (e.g., enhanced accessibility for persons with disabilities). Additionally, it has been projected that widescale adoption of CAVs, especially ones devoted to shared mobility, will result in an overall decrease in the number of vehicles on the roads, a factor that will further reduce the amount of needed parking [80–83]. Indeed, some recent studies [84–86] have suggested that shared autonomous vehicles and ridesharing could reduce the number of required parking spaces by 80%–90%.

In order to accommodate the changes, local governments, planning organizations, etc., will need to revisit parking requirements (e.g., number of spaces per developed area) set out in building codes and zoning regulations. Such changes could not only reduce the amount of land assigned to parking, it could also impact the nature of land development projects. For illustrative purposes, Fig. 8.5A and B depict two different potential configurations for access to a shopping center. Both are directly accessible from the adjacent roadway and both have pedestrian access from the walkway. The configuration in Fig. 8.5B shows how an environment with CAVs might impact the development. It should be clear from the figure that the CAV scenario (see Fig. 8.5B) requires fewer overall parking spaces as it would be expected that some of the visitors to the shopping center

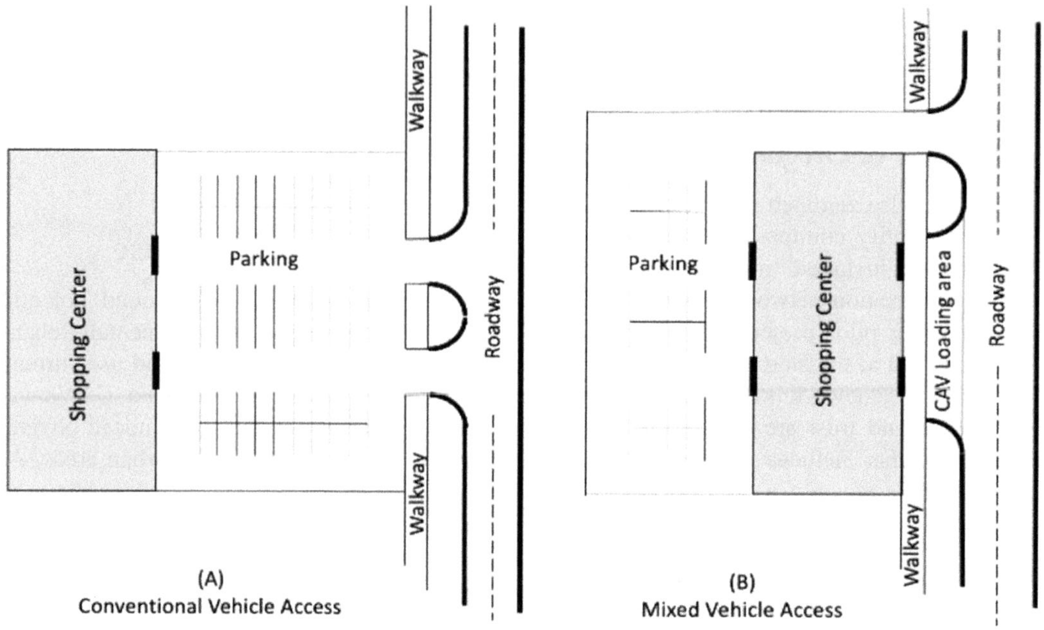

FIGURE 8.5 Potential impact of CAVs on parking and land use.

would be dropped off (and picked up) by an autonomous vehicle that could then proceed to its next destination such as a remote AV parking facility or, in the case of shared vehicles, to serve new customers. In addition to accommodating CAV traffic and reducing required parking spaces, thereby freeing up land for other purposes, the Fig. 8.5B scenario also better accommodates pedestrian access to and from the shopping center from the adjacent walkway, which has been shown to increase the attractiveness and thus monetary value of commercial land uses [87].

8.10 Legislation

Government, at all levels, has a responsibility to address the safety and mobility needs of its citizens and visitors. Governments have historically evaluated the effects of new technologies and, on occasion, taken action to regulate or prohibit such technologies, such as prohibitions on steam-powered vehicles in Chicago and Northampton, Massachusetts in the 19th century, current-day bans on scooters, or a recent Pennsylvania law that blocks cities from regulating certain drones [88]. The evolving development of CAV laws and regulations at the federal level has led many states and local governments to see gaps in needed direction for development, use, or implications of CAV activity in their jurisdictions, and the number of states considering CAV legislation is increasing. At least 41 states and the District of Columbia have considered AV legislation, with 29 states enacting legislation and governors of 11 states issuing AV-related executive orders (five states have both enacted laws and executive orders) [89]. While early legislative efforts centered on issues related to certification, testing, and liability [90], with much being made about their impact on insurance [91], the question of their very legality is not entirely settled (B) [92]. as the Federal Autonomous Vehicle Policy contains only guidance for states [90]. It is likely that states will continue to address CAVs legislatively, as well as federal and local governments, making it difficult for policymakers to provide a consistent context for vehicles sales and operation.

As it is, there is considerable leeway for how governments approach CAVs and therefore what kind of regulatory model is applied, if any. A governmental body may determine that no additional regulation is needed, or that prohibitions or geographical limitations may be necessary, or that existing regulatory models from other areas such as alcohol regulation may be serviceable. As with many public policy discussions, goals may be conflicting or difficult to reconcile. For example, stakeholders are calling for federal legislation that would greatly increase the number of CAVs while also calling for safety standards that could affect the market [93]. CAVs raise a variety of issues that policymakers may address. For AVs, these may include defining them, commercial versus passenger use, cybersecurity, infrastructure, insurance and liability, licensing and registration, operation on public roads, operator requirements, data including privacy and access, inspection, testing, and contribution toward infrastructure finance (Shinkle n.d.).

CAV manufacturing, distribution, and operation depend heavily on hardware and software and may at times become indistinguishable from aspects of the Internet of Things. Therefore, attention should be paid to legislation addressing cybersecurity and how it may affect CAVs, whether intentionally directed at this industry or having an effect as a result of unintended consequences. CAV cybersecurity concerns extend from Information Technology (IT) to Operational Technology (OT). IT concerns involve any data associated with a vehicle and its operation or the owners and users of the vehicle, including what data is collected, how it is stored, how long it is retained, who owns it, who has access to it, and whether and how that information can be released or shared. OT concerns have to do with processes and the mechanics of accessing, starting, stopping, and directing the vehicle, as OT systems are increasingly vulnerable to and victims of malicious compromise [94]. A survey of cybersecurity experts identified smart traffic signals to be among the riskiest of smart city technologies due to their vulnerability and potential for negative consequences (gridlock, crashes, misdirection of emergency vehicles) if attacked [95].

The arrival of CAVs raises issues that may or may not fall under existing laws, regulations, or practices. These include safety and crashes, mobility options for people unable to drive, effects on traffic congestion and land use, and energy and emissions implications [96]. They also include equity in access, safety, public input, and enforcement [97], and how the systems interact with pedestrians. Considerations from a governmental perspective may also extend to protections for the passengers or payloads; additional or waiving of regulations in emergencies; use by public agencies; and differentiations among vehicles by size or type, e.g., passenger, freight, low-speed vehicles, or usage—for example, policymakers may desire to regulate autonomous vehicles but not intend such regulation to extend to autonomous lawn mowers [98]. Governments may seek to overlay current processes such as public convenience and necessity schemes, alter existing laws and rules to accommodate these technologies such as redefining "vehicle operator," or establish wholly new provisions, which would require close attention to the definitions used for the technology and its applications, with sufficient latitude to accommodate changes over time. The recent proliferation of CAV laws suggests that policymaking bodies may wish to avoid confusion by standardizing their approaches regionally across cities or states. As CAVs acquire more miles in actual use, data will accumulate that may further inform the need and direction of policy and may suggest specific needs such as training for fire and police in the unique aspects of CAVs.

The introduction of new technologies from vehicle manufacturers also suggests there may be statutory or regulatory changes required for the vehicle aftermarket. The global vehicle aftermarket is expected to grow through 2030 with specialty areas such as advanced driver assistance systems and powertrain parts [99]. Policymakers may want to consider the potential for aftermarket electronic parts and software enhancements at the same time that cybersecurity issues are addressed in legislation affecting original equipment manufacturers.

8.11 Public perception

There have been myriad studies in recent years to assess the public acceptance of CAVs and, ultimately, to predict their adoption and transportation market penetration. Such studies include those aimed at identifying willingness-to-pay for certain CAV capabilities to those that ascertain the public perception of CAV safety (e.g., Bansal and Kockelman, 2018 [100,101]; Penmetsa et al., 2021). A 2019 study of global CAV awareness indicated that, as might be expected, that residents of countries with higher GDP were typically more informed. The study also reported, however, that these same respondents also exhibited a more pessimistic outlook toward the safety aspects of automated vehicles [100]. In investigating more specialized AV applications, [102] documented that roughly half of more than 2000 Michigan residents expressed hesitancy toward riding automated buses citing concerns over safety and a general distrust of the technology—young males expressed a higher willingness to ride automated buses while females, seniors, and persons with disabilities.

The overwhelming majority of evidence in the literature appears to suggest that there is a general curiosity toward CAVs. The way forward to a greater overall acceptance, however, is complicated. Some studies have suggested more "hands-on" opportunities for the public to interact and engage directly with these technologies. Others have suggested that continued media coverage may play an important role in shaping how CAVs are viewed. It nonetheless clear that continued education of the general public coupled with competent and transparent leadership from industry and government officials are essential to what many assume to be the inevitable transition to connected and automated mobility.

8.12 Summary

First, we present an overview on connected and automated mobility systems, including CAVs, transportation infrastructure, transportation data, and the history and political context for current and future considerations along with some case studies illustrating technologies, uses, and policy implications in Section 2. Then, we have presented policies, considerations, history, context, and guidance on safety, communication, cooperative driving automation, cybersecurity and privacy, and navigation systems (GPS/GNSS). After that, current and potential policies for connected and automated mobility services, intertwined with issues such as private sector involvement, new business models, and parking and land use are presented and discussed along with few field reports illustrating the wide variation in the drafting and implementation of policies in these areas. Finally, the history and context for legislation related to connected and automated mobility and public perception for improving user acceptance for new technologies such as connected and automated mobility systems are presented emphasizing the wide variation among state and local government approaches under broad federal guidance.

Finally, let us move on to the real interactive part of this chapter: review questions/exercises, hands-on projects, case projects, and optional team case project. The answers and/or solutions by chapter can be found in Appendix G.

8.13 Chapter review questions/exercises

True/false

1. True or False? The key advantage of automated vehicles is that it does not repeat the same mistake as human drivers.
2. True or False? The idea of fully automated vehicles was not first started with roadway safety.
3. True or False? Dedicated short range communication (DSRC) and Cellular-V2X (C-V2X) are two examples of communication technologies for V2X communication.
4. True or False? The SAE J3216 defines common terminology related to security threats, vulnerability, and corresponding risk for connected and automated mobility as well as provides recommended practices.
5. True or False? Mobility as a Service (MaaS) is the technology-enabled provision of transportation-enhancing products and services that are new or may have been previously provided by the individual or employers.

Multiple choice

1. What are the security requirements of connected and automated vehicles?
   a. Confidentiality
   b. Integrity
   c. Availability
   d. Authentication
   e. All of the above
2. ________________ refers to maintaining accuracy, trustworthiness, and reliability of the data throughout the entire data life cycle.
   a. Integrity
   b. Cybersecurity
   c. Privacy
   d. Confidentiality
   e. All of the above
3. As per SAE J3216, there are __________ classes for cooperative driving automation.
   a. Three
   b. Four
   c. Five
   d. Six
   e. Seven
4. When a subject vehicle shares information about its location and perception information with surrounding vehicles, it can be called as __________.
   a. Agreement-seeking cooperation
   b. Intent-sharing cooperation
   c. Prescriptive cooperation
   d. Status-sharing cooperation
   e. All of the above
5. The introduction of new technologies from vehicle manufacturers also suggests that there may be ______________ or regulatory changes required for the vehicle aftermarket.
   a. Cybersecurity
   b. Statutory
   c. Global
   d. Virtual
   e. Flexible

Exercise

Problem

Create report that covers key findings on smart city practices and implementation in the United States.

Hands-on projects

Project

Do research: Look at how an array of strategies can transform transportation in smart cities.

Case projects

Problem

Investigate the main smart city drivers with the potential to increase urban mobility intelligence, and then assign them a priority.

Optional team case project

Problem

Look at Artificial Intelligence (AI) transportation and the smart city and how this will affect urban futures.

References

[1] SAE. Taxonomy and definitions for terms related to driving automation systems for on-road motor vehicles (Surface Vehicle Recommended Practice (J3016_201806)): superseding J3016 Jan 2014). SAE International. June 15, 2018. https://www.sae.org/standards/content/j3016_201806/.

[2] Mendez VM, Monje CA, White V. Beyond traffic: trends and choices 2045—a national dialogue about future transportation opportunities and challenges. In: Disrupting mobility. Cham: Springer; 2017. p. 3–20.

[3] Jadhav A. Autonomous vehicle market by level of automation. May 2018. https://www.alliedmarketresearch.com/autonomous-vehicle-market. [Accessed 6 June 2021].

[4] Nyczepir D. City leaders envision future with driverless cars. June 4, 2018. https://www.route-fifty.com/smart-cities/2018/06/cities-envision-driverless-future/148712/. [Accessed 25 May 2021].

[5] National League of Cities. Autonomous vehicles: future scenarios. 2021. http://avfutures.nlc.org/. [Accessed 25 May 2021].

[6] Smith S. Vehicle to vehicle communication to be installed in 62M vehicles by 2023, as 5G disrupts established. December 11, 2018. https://www.juniperresearch.com/press/vehicle-to-vehicle-communication-installed-by-2023. [Accessed 26 May 2021].

[7] Markets and Markets. Connected car market - global forecast to 2025. July 2020. https://www.marketsandmarkets.com/pdfdownloadNew.asp?id=102580117. [Accessed 26 May 2021].

[8] Funicello-Paul L. Navigant research's smart city tracker 2Q19 highlights 443 projects spanning 286 cities around the world. June 20, 2019. https://www.bloomberg.com/press-releases/2019-06-20/navigant-research-s-smart-city-tracker-2q19-highlights-443-projects-spanning-286-cities-around-the-world. [Accessed 25 May 2021].

[9] National Academies of Sciences, Engineering, and Medicine. Business models to facilitate deployment of connected vevhicle infrastructure to support automated vehicle operations. Washington, D.C.: The National Academies Press; 2020.

[10] Young J. Safety groups issue automated enforcement guide. May 6, 2021. https://www.iihs.org/news/detail/safety-groups-create-automated-enforcement-checklist-to-encourage-well-designed-programs. [Accessed 4 June 2021].

[11] Automated Enforcement Program Checklist. AAA/GHSA/IIHS/NSC/Advocates for Highway safety. May 2021.

[12] Karmout K. Wonder when driverless cars could deliver your groceries? City of Austin works to pave the way. May 18, 2021. https://www.kxan.com/news/local/austin/wonder-when-driverless-cars-could-deliver-your-groceries-city-of-austin-works-to-pave-the-way/. [Accessed 27 May 2021].

[13] CTRMA. Central Texas using connected vehicle tech in traffic management operations. May 18, 2021. https://www.roadsbridges.com/central-texas-using-connected-vehicle-tech-traffic-management-operations. [Accessed 27 May 2021].

[14] Bibliography Stanley K, Wagner J. Data privacy considerations for automated and connected vehicles. College Station: Texas A&M Transportation Institute; 2015.

[15] Steadman M, Huntsman B. Connected vehicle infrastructure deployment and funding overview. College Station: Texas A&M Transportation Institute; 2018.

[16] Zmud J, Tooley M, Miller M. Data ownership issues in a connected car environment: implications for state and local agencies. College Station: Texas A&M Transportation Institute; 2016.

[17] Kahane CJ. Lives saved by vehicle safety technologies and associated federal motor vehicle safety standards, 1960 to 2012—Passenger cars and LTVs—with reviews of 26 FMVSS and the effectiveness of their associated safety technologies in reducing fatalities, injuries, and crashes. Report No. DOT HS, 812, 069. 2015.

[18] ARC-IT. National ITS architecture. 2021. http://local.iteris.com/arc-it/. [Accessed 25 July 2021].

[19] SAE International J2735D. Dedicated short range communications (DSRC) message set Dictionary™. PA: SAE International. Warrendale; 2016. https://www.sae.org/standards/content/j2735_201603/. [Accessed 24 May 2019].

[20] SAE J 2945. On-board system requirements for V2V safety communications. document SAE J2945/1. Mar. 2016.

[21] FCC. Available at: https://docs.fcc.gov/public/attachments/FCC-20-164A1_Rcd.pdf; 2020.

[22] Datta P, Kaushal S. March). Exploration and comparison of different 4G technologies implementations: a survey. In: 2014 recent advances in engineering and computational sciences (RAECS). IEEE; 2014. p. 1–6.

[23] Mastrosimone A, Panno D. A comparative analysis of mmWave vs LTE technology for 5G moving networks. In: 2015 IEEE 11th international conference on wireless and mobile computing, networking and communications (WiMob). IEEE; 2015, October. p. 422–9.

[24] Brake D. A US national strategy for 5G and future wireless innovation. Information Technology and Innovation Foundation; 2020.

[25] Deka L, Khan SM, Chowdhury M, Ayres N. Transportation cyber-physical system and its importance for future mobility. In: Transportation cyber-physical systems. Elsevier; 2018. p. 1–20.

[26] Upstream. Smart mobility cyber attacks repository. 2019. https://www.upstream.auto/research/automotive-cybersecurity/. [Accessed 3 June 2019].

[27] Singapore Statutes Online. Road traffic act (chapter 276), road traffic (autonomous motor vehicles) rules 2017. 2017. p. S 464. https://sso.agc.gov.sg/SL/RTA1961-S464-2017?DocDate=20170823.

[28] UK Code. Government of the UK, "Code of practice: automated vehicle trialling". February 2019. https://www.gov.uk/government/publications/trialling-automated-vehicle-technologies-in-public/code-of-practice-automated-vehicle-trialling.

[29] NTC. National transport commission (NTC), "National guidelines for automated vehicle trials, policy paper". May 2017. www.ntc.gov.au/sites/default/files/assets/files/NTC%20Policy%20Paper%20-%20National%20guidelines%20for%20 automated%20vehicle%20trials%20-%20May%202017.pdf.

[30] AZDOT. Available at: https://azdot.gov/motor-vehicles/professional-services/autonomous-vehicles-testing-and-operating-state-arizona Bansal; 2021.

[31] Title 28. Available at: https://www.azleg.gov/arsDetail/?title=28; 2021.

[32] Khan Z, Chowdhury M, Islam M, Huang CY, Rahman M. Long short-term memory neural network-based attack detection model for in-vehicle network security. IEEE Sens Lett 2020;4(6):1–4.

[33] Hunt, G. How much data does your car collect? More than you think. https://www.siliconrepublic.com/enterprise/data-privacy-connected-cars-infographic. Accessed Jun. 3, 2019.

[34] Jung R, Wernerus B. Cybersecurity for automated vehicles. ATZ worldwide 2021;123(1):38–43.

[35] Libaque-Sáenz CF, Wong SF, Chang Y, Bravo ER. The effect of fair information practices and data collection methods on

privacy-related behaviors: a study of mobile apps. Inf Manag 2021;58(1):103284.

[36] Jiang Q, Ni J, Ma J, Yang L, Shen X. Integrated authentication and key agreement framework for vehicular cloud computing. IEEE Network 2018;32(3):28–35.

[37] Yan G, Olariu S, Weigle MC. Providing VANET security through active position detection. Comput Commun 2008;31(12):2883–97.

[38] Singh M, Kim S. Trust bit: reward-based intelligent vehicle commination using blockchain paper. In: 2018 IEEE 4th world forum on internet of things (WF-IoT). IEEE; February 2018. p. 62–7.

[39] Agarwal Y, Jain K, Karabasoglu O. Smart vehicle monitoring and assistance using cloud computing in vehicular Ad Hoc networks. Int J Transp Sci Technol 2018;7(1):60–73.

[40] Hegde N, Manvi SS. A novel key management protocol for vehicular cloud security. Telkomnika 2019;17(2):857–65.

[41] Xie L, Ding Y, Yang H, Wang X. Blockchain-based secure and trustworthy internet of things in SDN-enabled 5G-VANETs. IEEE Access 2019;7:56656–66.

[42] Su Y, Wang J, Wang Y, Miao M. Efficient verifiable multi-key searchable encryption in cloud computing. IEEE Access 2019;7:141352–62.

[43] Garg N, Bawa S, Kumar N. An efficient data integrity auditing protocol for cloud computing. Future Generat Comput Syst 2020;109:306–16.

[44] Zhang X, Zhao J, Xu C, Wang H, Zhang Y. DOPIV: post-quantum secure identity-based data outsourcing with public integrity verification in cloud storage. IEEE Trans Serv Comput 2019. https://doi.org/10.1109/TSC.2019.2942297. https://ieeexplore.ieee.org/document/8844810.

[45] Zhang D, Yu FR, Yang R. Blockchain-based distributed software-defined vehicular networks: a dueling deep ${Q} $-Learning approach. IEEE Trans Cogn Commun Netw 2019;5(4):1086–100.

[46] Mudengudi SS, Kakkasageri MS. Agent based trust establishment between vehicle in vehicular cloud networks. Int J Comput Netw Inf Secur 2019;11(7).

[47] Lei A, Cruickshank H, Cao Y, Asuquo P, Ogah CPA, Sun Z. Blockchain-based dynamic key management for heterogeneous intelligent transportation systems. IEEE Intern Things J 2017;4(6):1832–43.

[48] Noroozi H, Khodaei M, Papadimitratos P. June). Vpkiaas: a highly-available and dynamically-scalable vehicular public-key infrastructure. In: Proceedings of the 11th ACM conference on security & privacy in wireless and mobile networks; 2018. p. 302–4.

[49] Wang C, Shen J, Lai JF, Liu J. B-TSCA: blockchain assisted trustworthiness scalable computation for V2I authentication in VANETs. IEEE Trans Emer Topics Comput 2021;9(3):1386–96. https://doi.org/10.1109/TETC.2020.2978866. https://ieeexplore.ieee.org/document/9037290.

[50] Brown J, Blough DM. Distributed enforcement of sticky policies with flexible trust. IEEE. In: 2015 IEEE 17th international conference on high performance computing and communications, 2015 IEEE 7th international symposium on cyberspace safety and security, and 2015 IEEE 12th international conference on embedded software and systems; August 2015. p. 1202–9. traffic-with-lidar-in-smart-cities.

[51] Kelbert F, Pretschner A. Data usage control enforcement in distributed systems. In: Proceedings of the third ACM conference on data and application security and privacy; February 2013. p. 71–82.

[52] Wüchner T, Pretschner A. Data loss prevention based on data-driven usage control. In: 2012 IEEE 23rd international symposium on software reliability engineering. IEEE; 2012, November. p. 151–60.

[53] La Manna M, Treccozzi L, Perazzo P, Saponara S, Dini G. Performance evaluation of attribute-based encryption in automotive embedded platform for secure software over-the-air update. Sensors 2021;21(2):515.

[54] Will MA, Ko RK, Witten IH. Privacy preserving computation by fragmenting individual bits and distributing gates. In: 2016 IEEE trustcom/BigDataSE/ISPA. IEEE; August 2016. p. 900–8.

[55] SAE J 3216. Taxonomy and definitions for terms related to cooperative driving automation for on-road motor vehicles J3216_202005. International, S.A.E., 2020. 2020.

[56] Albus JS, Huang HM, Messina ER, Murphy K, Juberts M, Lacaze A, Balakirsky SB, Shneier MO, Hong TH, Scott HA, Proctor FM. 4D/RCS Version 2.0: a reference model architecture for unmanned vehicle systems. 2002.

[57] Ali A, Chan B, Ahmed AS, Omr M, Krupity D, Wang Q, Al-Hamad A, Georgy J, Goodall C. Tight coupling between radar and INS/GNSS with AUTO software for accurate and reliable positioning for autonomous vehicles. In: Proceedings of the 33rd international technical meeting of the satellite division of the Institute of navigation (ION GNSS+ 2020); 2020. p. 1733–45.

[58] The White House. Memorandum on space policy directive 7. Washington, D.C.: The White House; 2021.

[59] Hansen A, Mackey S, Wassaf H, Shah V, Wallischeck E, Scarpone C, Barzach M, Baskerville E. Complementary PNT and GPS backup technologies demonstration report. Cambridge, MA: US DOT; 2021.

[60] National Science and Technology Council. National research and development plan for positioning, navigation, and timing resilience. Washington, D.C.: Executive Office of the President; 2021.

[61] Hong S. Why radar is the future of autonomous transportation. GPS World June 12, 2021. https://www.gpsworld.com/why-radar-is-the-future-of-autonomous-transportation/.

[62] Luccio M. GNSS in the fast lane. GPS World June 5, 2021:20–5.

[63] Robitzski D. Watch tesla autopilot get Bamboozled by a truck hauling traffic lights. June 2, 2021. https://futurism.com/the-byte/tesla-autopilot-bamboozled-truck-traffic-lights. [Accessed 5 June 2021].

[64] Johnson J, Yergin R, Pandya N. IoT update: the road Ahead for COnnected and automated vehicle Developments in Washington. December 22, 2020. https://www.lexology.com/library/detail.aspx?g=4780ac65-88de-4686-91bb-0ef7476514d0. [Accessed 25 May 2021].

[65] National Association of Counties. Connected and automated vehicles toolkit. 2019. https://www.naco.org/resources/featured/connected-autonomous-vehicles-toolkit. [Accessed 26 May 2021].

[66] Sumagaysay. Most uber and lyft trips will have to transition to electric vehicles in this decade, California declares. May 22, 2021. https://www.marketwatch.com/story/most-uber-and-lyft-drivers-will-have-to-transition-to-electric-vehicles-in-this-decade-california-declares-11621546765. [Accessed 28 May 2021].

[67] Connecticut Department of Transportation. Preparing for connected and automated vehicles strategic plan. CT: Newington; February 2021.

[68] Jin X, Rahimi A, Azimi G. The impacts of emerging mobility options and vehicle technologies on travel behavior, deliverable 6 - final report. Miami, FL: Florida International University; 2020.

[69] Urbanism Next Center. Perfecting policy with pilots: new mobility and AV urban delivery pilot project assessment. University of Oregon; 2020.

[70] Kaufman B, Leung A. We subscribe to movies and music, why not transport?. July 3, 2019. https://theconversation.com/we-subscribe-to-movies-and-music-why-not-transport-119538. [Accessed 1 June 2021].

[71] Alexiou G. Mobility as a service concept promises to revolutionize transport accessibility. May 23, 2021. https://www.forbes.com/sites/gusalexiou/2021/05/23/mobility-as-a-service-concept-promises-to-revolutionize-transport-accessibility/. [Accessed 28 May 2021].

[72] MaaS Alliance. n.d. What is MaaS? Accessed May 28, 2021. https://maas-alliance.eu/homepage/what-is-maas/.

[73] Alexiou G. Mobility as a service concept promises to revolutionize transport accessibility. https://www.forbes.com/sites/gusalexiou/2021/05/23/mobility-as-a-service-concept-promises-to-revolutionize-transport-accessibility/. [Accessed 4 June 2021].

[74] Sipe N. Billions are pouring into mobility technology - will the transport revolution live up to the hype?. February 23, 2020. https://theconversation.com/billions-are-pouring-into-mobility-technology-will-the-transport-revolution-live-up-to-the-hype-131154. [Accessed 1 June 2021].

[75] Hawkins A. Lime prime is the scooter company's new monthly subscription service. May 27, 2021. https://www.theverge.com/2021/5/27/22455553/lime-prime-scooter-subscription-price-unlock-fee. [Accessed 4 June 2021].

[76] National League of Cities. A human touch for robot delivery. 2021. http://avfutures.nlc.org/jobs-and-the-economy. [Accessed 6 June 2021].

[77] Bridgelall R, Stubbing E. Forecasting the effects of autonomous vehicles on land use. Technol Forecast Soc Change 2021;163:120444.

[78] Soteropoulos A, Berger M, Ciari F. Impacts of automated vehicles on travel behaviour and land use: an international review of modelling studies. Transport Rev 2019;39(1):29–49.

[79] Yigitcanlar T, Wilson M, Kamruzzaman M. Disruptive impacts of automated driving systems on the built environment and land use: an urban planner's perspective. J Open Innov Technol, Market, Complex 2019;5(2):24.

[80] Bischoff J, Maciejewski M. Simulation of city-wide replacement of private cars with autonomous taxis in Berlin. Procedia Computer Science 2016;83:237–44.

[81] Chen D, Kockelman KM. Management of a shared autonomous electric vehicle fleet. Transport Res Rec 2016;2572:37–46.

[82] Fagnant D, Kockelman KM, Bansal P. Operations of shared autonomous vehicle fleet for Austin, Texas, market. Transport Res Rec 2015;2536:98–106.

[83] Martinez LM, Viegas JM. Assessing the impacts of deploying a shared elf-driving urban mobility system: an agent-based model applied to the city of Lisbon, Portugal. Int J Transp Sci Technol 2017;6:13–7.

[84] ITF (International Transport Forum). Urban mobility system. Upgrade. How shared self-driving cars could change city traffic. Corporate partnership board report. OECD; 2015.

[85] Zhang W, Guhathakurta S. Parking spaces in the age of shared autonomous vehicles Transportation Research Record. J Transport Res Board 2017;2651:80–91.

[86] Zhang W, Guhathakurta S, Fang J. Exploring the impact of shared autonomous vehicles on urban parking demand: an agent-based simulation approach. Sustain Cities Soc 2015;19:34–45.

[87] Li C-N, Hsieh Y-K. Effect of pedestrian-oriented design benefits to land value. Wessex Institute of Technology Transactions on State of the Art in Science and Engineering 2014;78.

[88] Zipper D. When cities say no to new transportation technology. May 19, 2021. https://www.bloomberg.com/news/articles/2021-05-19/why-cities-resist-new-transportation-technology. [Accessed 27 May 2021].

[89] Autonomous Vehicles/Self-Driving Vehicles Enacted Legislation. February 18, 2020. https://www.ncsl.org/research/transportation/autonomous-vehicles-self-driving-vehicles-enacted-legislation.aspx. [Accessed 25 May 2021].

[90] Husch B, Teigen A. Regulating autonomous vehicles. April 2017. https://www.ncsl.org/research/transportation/regulating-autonomous-vehicles.aspx. [Accessed 25 May 2021].

[91] Wallace B. Autonomous vehicles and impact on auto insurance. February 16, 2021. https://gritdaily.com/autonomous-vehicles-and-impact-on-auto-insurance-infographic/. [Accessed 25 May 2021].

[92] Smith BW. Automated vehicles are probably legal in the United States. Tex. A&M L. Rev. 2013;1:411.

[93] Clifford Law Offices. The dangers of driverless cars. May 5, 2021. https://www.natlawreview.com/article/dangers-driverless-cars. [Accessed 6 June 2021].

[94] Lunden K, Zafra D, Brubaker N. Crimes of opportunity: increasingfrequency of low sophistication operational technology compromises. May 25, 2021. https://www.fireeye.com/blog/threat-research/2021/05/increasing-low-sophistication-operational-technology-compromises.html. [Accessed 25 May 2021].

[95] Frick K, Abreu G, Malkin N, Pan A, Post A. The cybersecurity risks of smart city technologies: what do the experts think? Berkeley, CA; UC Berkeley: Center for Long-Term Cybersecurity; 2021.

[96] Anderson J, Kalra N, Stanley K, Sorensen P, Samaras C, Oluwatola T. Autonomous vehicle technology: a guide for policymakers. Santa Monica, CA: RAND Corporation; 2016. Published Research, https://www.rand.org/pubs/research_reports/RR443-2.html.

[97] NACTO. #BlackLivesMatter: the responsibility of transportation agencies to make public places safer for people of all identities. May 25, 2021. https://nacto.org/2021/05/25/blacklivesmatter-the-responsibility-of-transportation-agencies/. [Accessed 25 May 2021].

[98] Cozzens T. Autonomous lawn mower hits the market this year. January 4, 2021. https://www.gpsworld.com/autonomous-lawn-mower-hits-the-market-this-year/. [Accessed 25 May 2021].

[99] Christof D, Hofmann J, Horner D, Kempf S. Making every part count: a component view on disruption in the automotive aftermarket for light vehicles until 2030. May 17, 2021. https://www.mckinsey.com/industries/automotive-and-assembly/our-insights/making-every-part-count-a-component-view-on-disruption-in-the-automotive-aftermarket-for-light-vehicles-until-2030. [Accessed 6 June 2021].

[100] Moody J, Bailey N, Zhao J. Public perceptions of autonomous vehicle safety: an international comparison. Saf Sci 2020;121:634–50.

[101] Penmetsa P, Adanu E, Wood D, Wang T, Jones S. Perceptions and expectations of autonomous vehicles – a snapshot of vulnerable road user opinion. Technol Develop Soc Change 2019;143:9–13.

[102] Kassens-Noor E, Kotval-Karamchandani Z, Cai M. Willingness to ride and perceptions of autonomous public transit. Transport Res Pol Pract 2020;138:92–104.

Further reading

AV 2.0. Automated driving systems 2.0: a vision for safety. National Highway Traffic Safety Administration; 2016.

Bansal P, Kockelman K. Are we ready to embrace connected and self-driving vehicles? A case study of Texans. Transportation 2018;45:641–75.

Brown P. Video: controlling traffic with lidar sensors in smart cities. May 11, 2021 (accessed 20 May 21, https://electronics360.globalspec.com/article/16702/video controlling-. [Accessed 27 May 2021].

California AV. https://www.dmv.ca.gov/portal/vehicle-industry-services/autonomous-vehicles/california-autonomous-vehicle-regulations/.

Scooters come to Hartford Thursday. 100 electric scooters deployed across downtown. April 22. Accessed June 4, 2021. https://www.courant.com/breaking-news/hc-br-downtown-hartford-link-scooter-rentals-20210422-fjt2ruim2zgddltji4hjnp75sq-story.html.

Hexagon. Automated driving systems (ADS) for rural America. 2021. https://autonomoustuff.com/products/case-studies/ads-rural-america.

Hong S. How connected cars will integrate with the smart city. https://www.thalesgroup.com/en/markets/digital-identity-and-security/iot/inspired/connected-cars/smart-cities. [Accessed 25 May 2021].

KPMG AVRI 2020. 2020 Autonomous Vehicles Readiness Index .Available at: https://home.kpmg/uk/en/home/insights/2020/07/2020-autonomous-vehicles-readiness-index.html.

NHTSA AV test Tool. Available at: https://www.nhtsa.gov/automated-vehicle-test-tracking-tool; 2021.

University of Iowa. ADS for rural America. 2021. https://adsforruralamerica.uiowa.edu/. [Accessed 4 June 2021].

Yan G, Wen D, Olariu S, Weigle MC. Security challenges in vehicular cloud computing. IEEE Trans Intell Transport Syst 2012;14(1):284–94.

Chapter 9

Public–private-partnerships (PPP) enabled smart city funding and financing

Ali Mohammad Mirzaee and Javad Majrouhi Sardroud
Department of Civil Engineering, Central Tehran Branch, Islamic Azad University, Tehran, Iran

9.1 Introduction

Although a smart city is a comprehensive concept to solve urbanism problems, its development needs financial investment and technological innovations [1]. On this basis, smart city policy has increasingly tended toward funding over recent years [2], has followed other feasible alternatives rather than only government budget. Statistics show that only 16% of cities are capable of governmental budget-based financing [3]. On the one hand, the urban population is growing, so smart cities should be developed consistent with such change. On the other hand, governmental financing constraints would postpone the development of smart cities [4]. Therefore, urban managers and other key stakeholders must prefer funding and financing approaches to smart cities considering the specifications and purposes of the smart city. The most substantial goals, in this case, include the engagement of key stakeholders, a specialized approach to the combination of sources, and actualization of sustainable development goals.

Key stakeholders, such as local government, citizens, jobs, living laboratories, and research and academic communities, play a vital role in making the city smart especially regarding fund attraction [5]. However, the mechanism of such involvement should be based on the smart city development-oriented approaches. For instance, a multi-sector partnership (MSP) can be used to overcome sophisticated challenges and uncertainty in urban development regarding the facilitation of participation and consolidation of resources in different fields [6]. Further, the actualization of sustainable development goals (SDGs) is highly related to governance, and Public–Private Partnership (PPP) is a distinct procedure for governance [7]. Hence, as a known theme of MSP and also as a distinct procedure of governance, PPP is the preferred option in funding and financing for the smart city.

As a new field in smart city projects, the PPP model has faced many problems. There is still a long way ahead of this model [8]. The problems are rooted in the use of traditional PPP models, and the long way is caused by innovative businesses and their different performance indices. For example, the basic approach of the traditional differs from the smart PPPs; in term of Value for Money (VFM) and Science, Technology, and Innovation (STI), respectively [9]. The abovementioned approaches must be understood and met by all key stakeholders. Further, the PPP model is assessed in various scopes, including contract, risk, funding and financing, management, and governance. Although financing mechanisms and funding sources can affect the decisions on project delivery systems (PDSs), there are few discussions on mechanisms of PPP projects regarding these financial aspects [4,5]. On the other hand, assessment of project risks, conditions, and financial sources is a necessity when selecting the most proper financing strategy for PPP [1]. These can be complementary or an introduction to anticipation measures to solve complicated construction problems [10]. These measures also can be used as a prerequisite to avoid disputes and to improve PPP project management [11]. Therefore, risk assessment of smart projects with a concentration on innovative aspects and SDGs along with potential financial sources available is known as the initial evaluation of possible funding and financing. Results of such evaluation in terms of financing and funding dimensions lead to a project delivery system, such as a PPP type.

Therefore, relevant subjects have been examined in specific order to make funding and financing of smart city projects through a simple and operational method using the

Smart Cities Policies and Financing. https://doi.org/10.1016/B978-0-12-819130-9.00011-5

PPP approach. First, the background of the smart city will be reviewed; this concept will be defined, and potential funding and financing options will be introduced. After PPP is defined, the theory and advantages of this approach to its funding and financing implementation will be described. In the next step, integrated funding and financing strategies and their components in specific PPP projects of the smart city will be discussed in detail. Finally, substantial points on new details of smart city funding and financing through the PPP technique will be mentioned. Inaccurate understanding of these strategies and their components and lack of knowledge on the connection between these components may make difficulty in having access to desirable performance in smart PPPs for key stakeholders.

9.2 Literature review

According to the structure designed in the previous section and concerns about choosing the best titles for the next sections, draft headings must be specified to facilitate understanding of the subject on choosing headings; in a way that is matched with both background and title of the section. In this lieu, previous studies have been described in form of tables.

In Table 9.1, definition or general-related approaches/methods for smart city, PPP, and funding/financing methods have been reported by focusing on the background/achievement of previous studies. For example, a smart city has been defined as an interdisciplinary concept

TABLE 9.1 Implication of previous studies regarding definition or general-related.

Basis for headings draft	Approach/ Method	Field of focus/Contribution (s)	Reference
Definition- or general-related	Smart cities (SCs)	ICT, people, and/or governance	[12]
		Diverse source disciplines	[13,14]
		Smart infrastructure	[9]
		Conceptual and/or development	[15]
		3-C Concept (competence, convenience, and cleverness) concept	[16]
		Structure, infrastructure, and superstructure	[17]
		Smart and sustainable urban regeneration ('smart-SUR')	[18]
		Decision making methods: Multi-criteria decision-making (MCDM), mathematical programming (MP), and artificial intelligence (AI)	[19]
		Distinguishing utopia from reality	[20]
	Public–private partnership (PPP)	Narrow and wide meaning; PPPs- and PPP-like contracts	[21]
		Overcoming modern challenges and developing new opportunities	[22]
		Forms and contracts	[23]
		Level of detail: General, project, and contract levels	[24]
		Combination of project finance theory and notion of public investments	[23]
		Sustainable development and stakeholder theory	[25]
		Main theories in public administration (PA) discipline	[26]
		Through-life management (TLM) benefit	[27]
		Capital expenditure (CAPEX) and operational expenditure (OPEX) reduction	[28]
		6 main benefits	[29]
	Funding/ Financing	Long-term or short-term debt, resource or mechanism, and repayment by tax and/or user charge	[30,31]

with specific dimensions and structures requiring concentration on potential smart aspects of infrastructures and decision-making methods. PPP has been introduced as an approach used to overcome modern challenges, which can create new opportunities within any semantic field based on the forms and contracts of the subset. Such attraction has been outlined as an advantage of this approach; however, the key factor of its actualization is based on the knowledge of theories pertained to this approach. Moreover, loan duration, mechanism, and repayment sources, as well as funding/financing methods have been differentiated.

According to the literature review, Project Company, capital structure, financial flows, project-based funding, business model, and performance can be named as approaches/methods of the PPP-related category of financial structure. The field/achievement of this category has been shown in Table 9.2. According to the literature, under the PPP structure, sponsors form a project company, in which is a contracting party of the government based on the project funding principles. These principles are underpinned based on both capital structure, including equity and debt, and no referral to sponsors only based on the projected revenue. In fact, PPP-related financial structure is a mechanism to achieve performance based on the business models.

Furthermore, the important field achievements in Table 9.3 point to the consideration of strategy-related mechanisms for funding and financing based on the literature. Funding and financing strategies, multisector partnership, main pillars of smart city, pilot projects' implementation, and performance improvement based on the approaches/methods of these mechanisms are in line with the literature. However, an integrated structure must be considered for these components under the conceptual framework regarding the goals of the smart city in addition to the specification of components of these strategies.

Regarding strategies-related funding and financing mechanisms based on the literature review, critical components have been identified in the form of approaches and methods (Table 9.4). These components include infrastructure, scale level of the pilot project, new PPP, PPPP, accessible funding matrix, sustainability performance-based evaluation model, and feasible level of private finance. Therefore, the literature review not only specified potential headings of the following sections but also provided an integrated framework based on the relevant strategies and components. This framework will bring an applied new PPP approach to smart city funding and financing.

9.3 Smart city definitions

In general, smart city refers to the identification and searching through smart solutions, which allow modern cities to improve the quality of services provided for citizens. These circumstances occur in a smart combination of financial supports and activities done by decision-maker, independent, and informed citizens [32]. Therefore, all aspects related to financial issues and the partnership of key stakeholders might be included in the smart city definition.

To provide the field for funding and construction of the smart city, there must be a link between smart cities and dimensions of smart infrastructure (SI). SI is a prerequisite for a smart city as a comprehensive system, which encompasses different components, including people, governance, environment, and economy [9]. Therefore, different definitions of a smart city can be provided regarding conceptual and infrastructural dimensions.

A smart city is where investment in human, social capital, modern and traditional information and communication technology (ICT) brings sustainable economic growth and high quality of life; these are achieved through

TABLE 9.2 Implications of previous research studies on PPP's financial-related structure.

Basis for headings draft	Approach/Method	Field of focus/Contribution (s)	Reference
PPP's financial-related structure	SPV/Concessionaire	Nonrecourse, joint venture	[21,33]
	Capital structure	Sustainability perspective, debt and equity ratio	[34]
	Financial flows	Inflows and outflows	[33]
	Project finance	The profitability and bankability of the energy efficiency initiative	[35]
	Bossiness models	Sustainable business model innovation	[36]
	Integrated ecosystem model	With the market, cash-flow of a project, and ecosystem around PPPs levels	[37]
	Performance	The group decision model brings both accurate and efficient, and reasonable for real-world uses	[38]

TABLE 9.3 Implications of previous research studies for strategies-related funding and financing.

Basis for headings draft	Approach/Method	Field of focus/Contribution (s)	Reference
Strategies-related funding and financing mechanisms	Funding and financing strategies	Constructive cash budget, choosing bank for relationship, possession of best financial leverage, technical knowledge of specialized finance institutions/agencies	[39]
		Financing strategies based on PPP low risk, high political risk, high financial risk, and high market risk	[40]
		Appropriate financing strategy requirements: project risks, project conditions, and financing sources	[41]
	Multisector partnership	Twelve key parameters in three models within the PPP value chain framework have been identified for the funding and financing; state, hybrid, and private models	[42]
		Conceptual framework of open strategy in multisector social partnerships	[43]
		Combining resources from different fields of urban development	[6]
		Collaboration innovation to address the problems of modern cities	[44]
	Main pillar of SC	Smart infrastructures components: Platforms and domains	[45]
	Project pilot implementation	Scaling up process	[46]
		Living lab	[47]
	Improving performance/Efficiency	Sustainability model	[48]
		Joint use of public and private finance	[49]
		Accessible funding matrix: Prioritize projects and design funding approach	[15]
		Generate a framework to better selection of investment and prioritization of SC initiatives in the various domains	[50]
		Cost–benefit analysis (CBAs) and feasibility analyses are suitable tools for evaluation of socially and economically acceptable smart projects and their implementation in real life.	[51]

intellectual natural resources management and government partnership [52]. Therefore, the smart city instrument can be considered, as a measure used to achieve sustainable economic growth goals and high quality of life.

A more detailed definition of a smart city can be presented by identifying issues and challenges occurring in the smart city. Seemingly, financing is the most important challenge. This stems from the fact that the smart city is forming the world future, where its financing certainly is facing challenges for the implementation [53]. The need for a smart city as a relative scope for investment decision-making has stemmed from increased urbanism, scarce resources, and a tendency toward comfort and welfare of people [54]. In this case, access to finance is a challenging experience. The reason is that several projects implemented by different financers with similar financing methods in the smart city that should be operated as integrated solutions through quick banking assessment [55].

The issues related to applied funding and financing models and their links to smart project implementation systems are challenging aspects of smart city financing. This is because dimensions of the smart city include description, business model, and smart city goal, where management, infrastructure finance, and financial sources have been hierarchically considered for the business model component [56]. Three main costly pillars of structure, infrastructure, and superstructure have been defined for the smart city concept [17].

TABLE 9.4 Indications of previous research studies for components-related funding and financing.

Basis for headings draft	Approach/Method	Field of focus/Contribution (s)	Reference
Components-related funding and financing mechanisms	Infrastructure components	Smart social, physical, and digital infrastructure	[45]
	Scaling up	Roll-out, expansion, and replication	[46]
	"New" PPP	Business-as-usual PPP, incremental PPP, and groundbreaking innovation PPP	[28]
		Coresearch PPP	[5]
	Public–private-people partnerships (PPPPs	A framework-based solution	[1]
		Development plans	[57]
		Collaboration innovation	[44]
	Accessible funding matrix	As the right funding approach to prioritize projects	[15]
	Sustainability performance-based evaluation model	Bring both promote the use of PPP for developing infrastructure and provide proper alpha as level of investment distribution between contracting parties	[48]
	Feasible level of private finance	Forms the contracting model for funding and financing of infrastructure	[49]

Therefore, presented definitions of the smart city indicate various aspects and definitions based on funding and financing considering the conceptual or developmental aspects, dimensions, or pillars. That is, financing is the millstone of the rolling wave planning of smart city development. Accordingly, business models should be determined to achieve financing for the smart infrastructural project. Business model steps lead to potential funding and/or financing options for the smart city.

9.4 The potential funding options for smart cities

There are various funding and financing instruments in the smart city based on the economic and legal conditions in any country. According to the facilities and limitations of such circumstances, governmental policymakers and urban managers will be encouraged to make decisions on infrastructure implementation systems within the smart infrastructure development process. To actualize such decisions, different funding/financing instruments and feasible alternative(s) should be chosen. After choosing these instruments, infrastructure implementation system will be determined to initiate contract signing between contractual parties.

Potential funding and/or financing instruments should be introduced at first because these instruments ought to be matched with the project procurement method [54]. There are various instruments in the smart city, such as taxes, charges, and bills, PPPs, land-based income, loans, and debt securities (bonds) [58]. Furthermore, concerns and viewpoints of key stakeholders of the smart city must be involved in decisions made about choosing the best instruments. At three levels, government, contract parties, and residents should be encouraged as funding resources suppliers to support projects [15]. Another important point in choosing instruments is a consideration of both economic and legal circumstances of countries, as well as the known and specialized theories, which can respond to challenges to smart city resources.

Infrastructure financing is under the authority of governments, especially corporate finance due to the high cost of private financing caused by wrong contract design. However, this method can meet a minor part of the total required infrastructure finance [59]. For example, financing sources in smart cities of India include funding options and government financial supports (including subsides, national and infrastructural investment); so that the major part of financing is provided by the private sector, and the central government just proposes legal support [60]. Moreover, these instruments that are used as financing mechanisms might be the disputes origin within the life cycle of smart infrastructure creation. In this case, financing mechanism and funding resources affect the project delivery system's issues and responsibility to design and construction of plan [4]. Therefore, contract parties should have close

TABLE 9.5 Smart city's funding and financing alternatives.

Potential financing/Funding options for smart cities	Reference
Government-level funding, local-level funding, community-focused funding, public–private partnerships (PPPs), loans and municipal bonds, private funding, and user charges and pay for performance.	[15]
Debt market • Banks • Pension funds, insurance funds, and infrastructure debt funds Bond market Multilateral financial institutions Equity investment • Venture capital • Alternate smart city vehicles • Real estate investment trust Financial models and tools for smart cities • Debt equity-based leveraged structure • PPP • Tax tool • Innovative business model	[61]
Funding (exchequer financing, government funding, and grant funding), debt (senior debt, bonds, and export credit), equity (multilateral finance, infrastructure funds, and sponsor equity), and hybrid options.	[62]

cooperation to avoid any dispute on choosing funding and/or financing instruments, as well as the operating procedure of these instruments in the smart city. Otherwise, smart infrastructure development might be stopped at any step of the life cycle, despite existing potential infrastructure funding, owing to any claim or dispute.

Researchers have named some of the smart city funding instruments with categorization or have categorized some overlapped instruments. Regardless of PPP application for ordinary or smart urban infrastructures, basic principles of project funding must be understood well. In addition, priority or latency of funding instrument and project procurement method requires specific measures, especially in the interdisciplinary concept of the smart city. Table 9.5 reports a summary of some relevant studies.

On the other hand, as shown in Table 9.5, some researchers consider PPP in smart cities as a subset of financial models considering new business models as a distinctive category. In the same reasoning, some others define PPP and mechanisms of payments to the investor as two separate instruments. To solve such conflicts, offtake contracts, availability-based contracts, and concession agreements must be considered based on project funding principles [21]. The mentioned models respectively provide some mechanisms to sell manufactured products to a buyer, provide the government with a service to use in exchange for a receipt, and provide a service to the public in exchange for receiving tolls. These mechanisms include various potentials, including hard infrastructure development, soft systems, and hard/soft aspects package regarding the domain of smart city infrastructures.

According to the literature review through critical and corrective approaches, categorizations of the debt market, nonprivate sector participation, and PPP and relevant factors bring some mechanisms for smart city funding and financing (Fig. 9.1). Therefore, this section of the study examines the case due to the high importance of the PPP method in smart city funding. After the PPP method is defined, as well as introducing its advantages, it is essential to complete and reveal other options since PPP grouping in Fig. 9.1 aims to bring initial transparency to the case. This requirement stems from the multidimensional specifications

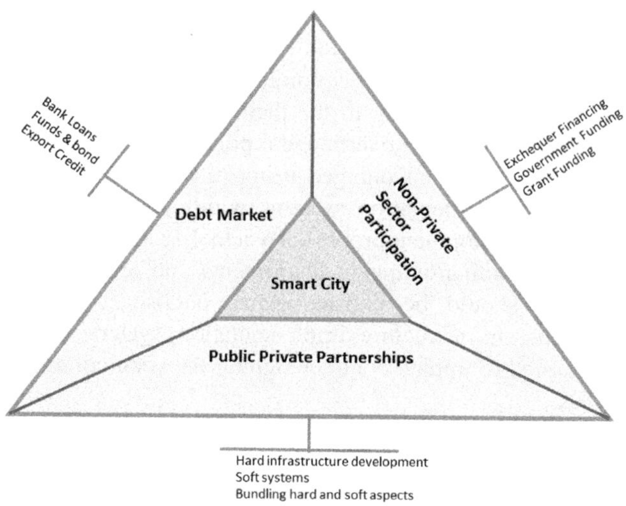

FIGURE 9.1 Potential funding and financing sources in smart city.

and the multisector nature of the smart city, which will be described in detail in the following parts of this chapter.

9.5 PPP definition, its theory, and advantages

There is not any fixed definition of PPP since the existing definitions are in the range of two different meanings, including a narrow meaning and a wide meaning. In narrowing meaning that covers concession and private finance initiative (PFI), PPP is considered as a long-term contract in which private sector performs the design, construction, finance, operation, and maintenance of the project. Other contracts of the public and private sector together are included in the wide meaning forming PPP-like contracts [21]. There is not any international consensus on the PPP definition because the available definitions are most incentive and country-specific. These definitions are ambiguous regarding terms and type, and partnership between public and private sectors is the only unchanged component in this definition. Furthermore, PPP is a popular method used to create synergy between public and private partners in order to overcome modern challenges and create new opportunities [22].

In terms of form and contract, PPPs can be categorized individually. In this classification, forms include BOT, DBO, DMB, DM, and BOOT. Contracts also consist of different types of concession, privatization, lease contract, and service contract [23]. Moreover, Project Delivery Systems (PDS) approach, which allocates and distributes the risk between public and private sectors, can be used to identify and define PPP. For example, a study has compared PPP with other PDSs by classifying factors at three levels. The general level revealed the difference between PPP and other types of PDSs while project and contract levels clarified the difference between various types of PPP [24]. Project ownership, performance quality, risk allocation, capital sources, and public sector control were considered as factors of the General Level. Nevertheless, in this classification, financing and service, and payment mechanisms indicating project level and contract level factors, respectively.

Theory assessment in PPP allows having accurate and rapid access to solutions for problems and concerns through an optimal technique. Furthermore, PPP is an interdisciplinary concept with different theories. To cover the shortage of social-sciences-related theories in PPP, project finance and public investment notion theories have been recommended [25]. The challenging point is about the combination of these two theories. Project financing is a debt-based method for large projects through financial engineering and is inclusively based on the loan against the liquidity flow of the project [21]. In project financing, the budget of PPP projects is often financed by equity and debt instruments with 10%–30% and 70%–90% ratios, respectively [41]. However, the risk model in PPP is caused by uncertainty in an inevitable issue based on measurable resources, indices, and parameters in the field of country, financial, revenue, project, and Project Company [63].

Most importantly, integration of concepts stakeholders, sustainable development, and PPP theory has been suggested as a new approach to urban development management [23]. In this sustainable development approach with smart measure, stakeholders have been divided into four types based on their direct and/or indirect involvement, as well as their financial and/or nonfinancial interests. Furthermore, specifications of power, legitimacy, and urgency include a mix of stakeholders through variable stock over time, which provide stakeholders with dynamism in long-term PPP for this urban development approach.

Regardless of traditional or smart type, PPP offers many advantages in developing infrastructure, which are implications for stakeholders regarding advantages taken from this infrastructural PDS. If private sectors are involved in this approach, capital expenditure (CAPEX) and operational expenditure (OPEX) will be reduced [28]. Through-life management (TLM) is another advantage of PPP since it shows the overall life cycle from idea to delivery of an infrastructure [27,29]. In summary, major advantages of PPP can be named as follows [29]: off-balance sheet debt, reduce infrastructure gap, achieve VFM, risk transfer to the private sector, public sector focus on strategies rather than operational tasks, and a single contract with one entity.

However, PPP in a smart city is beyond the PPP, such as people, participation, and politics [9]. Moreover, conventional contract-based concessions used in smart cities face groundbreaking innovation challenges, which lead to increased incompleteness level of contract and inventible condition for renegotiation. That is, renegotiation is indeed a claim instrument in PPP based on the agreement between contract parties to avoid dispute. However, the main challenge is about renegotiation over the mutual agreement on the overall claim of time, cost, quality, or scope because even a minor unsolved claim within renegotiation that is binding legally will cause dispute. Such dispute, in turn, postpones operation from the infrastructure. Furthermore, inappropriate funding and financing options in the smart city cause such disputes. Therefore, elements of these potential instruments in PPP and their relevant risks must be identified and described.

9.6 Components of PPP-Based funding and financing

Although funding and financing are used usually simultaneously, these two instruments can be employed individually or together to form or operate any PDS based on the considered purposes. This distinction is an issue related to monetary theory. In an investment project, short-term financing is used as a loan before completion of the project while long-term funding is the revenue obtained after completion of the project and covers assets used to meet financing costs. The funding source is mostly supplied from the tax revenues or user fees [30,33]. Therefore, the PPP option needs simultaneous availability of these two financial instruments in investment. To implement this PDS, public and private sectors should agree on the operational model of this option.

Funding and financing of PPP-based projects include two main bases: contractual and financial contexts. Under the contract, the private sector creates a project company with a commitment to result-based legal nature based on the principles of project finance. This project will be operated after signing a "concession agreement" with the public sector. Subsequently, the financial base includes financial flows of inputs and outputs as the potential financing and funding instruments, respectively. For example, there are some potential inflows (funds loans, capital, equity, and subsidies forms) and outflows (dividends, insurances, premiums, and debt service forms) in the life cycle of BOT projects [33]. The sum of these flows from the capital structure of the PPP project. This structure has more mature goals rather than financing concerns for infrastructure creation. This importance stems from the question of whether the public sector decides to create infrastructure with a certain goal and why it chooses PPP as the project implementation system. The answers to these questions can be found in access to the goal of creating integrated infrastructure in smart cities and innovative business models.

Business and planning models in the PPP projects of the smart city differ from the traditional PPP model. Moreover, these differences can be seen in the performance indices of these two PPP types. Multiobjective planning with qualitative analysis tends to access optimal and comprehensive capital structure in PPP projects using a sustainable development approach [34]. Innovative business models in various startups allow transferring and diversifying to additional businesses, converting business models to each other, conceptualizing and implementing a new business [36]. Procedures of these innovative business models will specify key activities, potential challenges, and available instruments. Therefore, innovative business models are preconditions used to determine funding and financing instruments if they are deeply identified and operated together with theoretical key concepts of smart-city-specified PPP.

9.7 Funding and financing strategies in PPP-Based smart cities

The higher access to finance does not necessarily ensure the higher success of a business model, but suitable financial funding and financing strategies should also be taken [39]. Suitable financial strategies play a vital role in complicated subjects, such as the smart city's infrastructure development. That is, specifications of conformity with relevant theories provide both specialization of issues and constructive measures simultaneously for the smart city. Policy-based management has gradually converted to multisector participation in codesign. The multisector partnership is used as an efficient approach to overcome complicated challenges and to remove uncertainty in urban development. This approach facilitates cooperation and combines the resource with different urban development fields [6]. Innovation-based collaboration between city and external actors (companies, third organizations, research institutes, and citizens) tends to present new solutions, as well as better services and products [44]. Therefore, complex challenges require new funding and financing solutions in the smart city regarding PPP projects for multisector innovative collaboration.

There are two other important strategies of the pilot and smart city pillars and optimization and distribution level in the capital structure. These strategies are related to multisector partnership strategy. The process of selecting capital structure includes equity and debt, which can be optimized based on the multiobjective program from viewpoint of sustainable development [34]. The importance of this optimization stems from the effect of distribution level of investment in the PPP's capital structure on the functional goals of the smart city [48]; that is, sharing the use of public–private financing along with optimization of capital structure will fund and finance the infrastructures [49]. Therefore, in a PPP project of smart city, financial flow between Project Company and government is similar to that between Project Company and private sponsors, including inflows and outflows considered as funding and financing instruments.

Smart social, physical, and digital infrastructures called smart city's pillars with the classification of platforms and domains point to the importance of examination at strategic, tactical, and operational levels [45]. Moreover, scaling dimensions of the smart city in pilot projects supported by municipalities and funded by subsidies must be considered [46]. These dimensions are known as the strategy of pilot smart city's pillars. It is critical to allay concerns of integrated development of PPP smart infrastructure and feasibility studies, which is possible by using living laboratories. This brings both design of the business model and develops potential funding and/or financing options.

According to the literature review and considering innovative business models, as well as challenges related to smart cities, the authors agree on three important strategies, including pillars and pilot, multisector partnership, and optimization and distribution of capital structure. Therefore, the multisector partnership was considered as a substantial strategy and source base required for the smart city

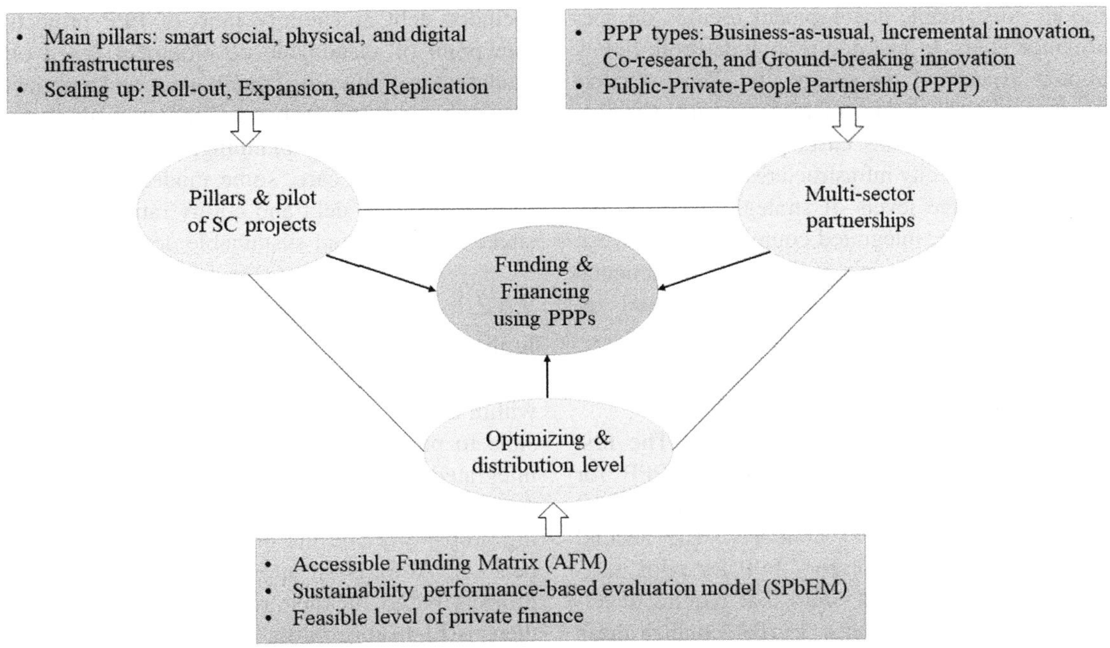

FIGURE 9.2 Underlying strategy with PPP-based projects funding and financing in smart city.

regarding the private sector partnership. Smart city pillar and pilot and optimization and distribution of capital structure strategies are considered as preconditions to actualize smart city funding and financing through PPP (Fig. 9.2).

Dimensions and factors of these strategies must be examined in detail. That is, the operation and completion process of these strategies require consideration of subsets including subjective relations and performance goals of smart city's PPP projects. The implementation process of strategies is complicated due to unknown key components or necessary measures for the actualization of any strategy. To solve these problems, key structural components of PPP that should be considered integrated have been described to actualize funding and financing for smart city projects.

9.8 Components of PPP-Based smart cities funding and financing

To actualize smart city funding and financing, the required preconditions must be hierarchically integrated at different levels. In this case, project finance and stakeholder theories play a vital role. Integration of PPP, stakeholder, and sustainable development concepts has been recognized as a new approach to urban management development. In this approach, public and private sectors face a conflict of interests. The public sector concentrates on the advantages of social, environmental, and economic dimensions while the private sector considers the actualization of the revenue predicted in the cash flow map [23]. As another substantial point, stakeholders and business models should be matched with integrated components of the smart city to create a specific PPP. In assessments taken by stakeholders, the actualization of preconditions for collaborative multisector and multicross aspects direct the specific PPP toward cases, which differ from project-based financing principles. However, project-based financing must be basically used regarding capital structure and expected performance of a smart city. Therefore, smart city funding and financing have their specific complications. Components of the relevant strategies should be identified and described to make them applicable.

9.8.1 Pillars and pilot of smart cities

As smart city components are not fixed but are continuously increasing, these components should be identified and integrated into the main pillars. This fundamental measure is two steps before PPP-based smart city funding and financing. In most cases, the pilot step is performed between the mentioned steps financed by the public sector. These measures are differently executed due to different structural dimensions of smart city projects, including single, cross, and overall dimensions [8]. Therefore, pilot components and requirements should be identified and implemented in infrastructures of the smart city before customization of PPP-based funding and financing.

9.8.1.1 Main pillars of smart cities

The smart city concept is interconnected with some concepts, such as infrastructure, sustainable development, and innovation. Of them, innovation highly pertains to the business model. Sustainable development focuses on specific performance goals. Infrastructure includes both policy and operational aspects of the smart city. These aspects should be determined by using an efficient and effective hierarchical mechanism. In this case, platforms, domains, and components of smart city infrastructures equal 3, 48, and 117, respectively at three levels of strategic, tactical, and operation [45]. Therefore, integrated compatibility between components of these three levels is the first fundamental precondition of smart social, physical, and digital infrastructures' funding and financing.

9.8.1.2 Scaling up in smart cities projects

Two PPP generations are used in smart cities. The first generation includes long-term under standard PPP for financing, design, construction, and operation in the framework of BOT and DBFO contracts. The second generation includes under exploratory PPP for pilot measures of initiative or conceptual ideas that require developmental and research components [1]. To actualize these components, the living lab will test the new business models [47]. Crucial point is that the dimensions and scope of these pilot projects must be clear and consistent with integrated components of smart cities infrastructure. Dimensions of scaling up include roll-out, expansion, and replication for smart city pilot projects supported by municipalities and financed with subsiding [46]. Smart city solutions for the customer or business-to-business, adding a partner or consumer, and replacing a tested solution in another part of the city are done to describe pilot scaling up dimensions for roll-out, expansion, and replication, respectively [46]. If results of the pilot under exploratory PPP are successful, then under standard PPP, funding and financing will be actualized. On the other hand, concentration on liquidity inflows and outflows of capital structure in smart cities projects that are affected by innovative business models requires some measures, including optimization of capital structure and multisector partnership.

9.8.2 Optimization and distribution level in capital structure

Recognition of priority or latency of categories, such as project company formation, capital structure, and financing management, is a critical point. Debt and equity, which form capital structure, should be determined before forming the project company. Management of the procedure is recognized as finance management. The future strategy of PPP-based infrastructures is used to comply with the smartness and sustainability of the public sector. This strategy also is used by the private sector to make a profit within a correct partnership with a joint ownership PPP entity [4]. The process of choosing the capital structure includes debt and equity ratio of PPP projects from the viewpoint of sustainable development based on the three main factors, including benefit, cost, and a project condition by using some key theories. The key theories comprise the game, multiobjective planning, and organizational principles [34]. In this case, some models and formulas can be used to measure debt and equity ratio to match them with smart city goals and sustainable development.

9.8.2.1 Accessible funding matrix (AFM)

In the smart city, three smart infrastructures, including social, physical, and digital categories, must be actualized within an integrated procedure. However, it seems impossible to provide funding and financing sources within an integrated and complicated procedure for all infrastructures. As a measure that determines the five-point scale of funding probability and the importance of smart cities projects benefits, AFM is an appropriate instrument used to rank projects based on their required funding [15]. Moreover, there is high uncertainty in smart city projects due to the link between the smart city concept and innovation. Hence, rolling wave planning should be considered as the benchmark planning in which AFM is applicable.

9.8.2.2 Sustainability performance-based evaluation model

Smart cities and sustainable development are inseparable concepts since performance of these cities includes all social, economic, and environmental dimensions. SPbEM model not only encourages using PPP but also specifies accurate capital level distribution between public and private sectors [48]. Therefore, at this optimal level, not only requirements of the new PPP model called private finance 2 (PF2) are actualized but also the economic, environmental, and social goals of smart cities are met.

9.8.2.3 Feasible level of private finance

The total financial amount is divided into two public and private sectors. According to indices of these categories, the debt and equity ratio of the private sector is considered in capital structure [49]. In fact, financial sources of both contract parties must be used commonly and feasibly for funding and financing.

9.8.3 Multisector partnerships

Cross-sector integrated planning regarding multistakeholder consensus on joint collaborative projects is a considerable challenge to sustainable development [64]. Unlike competitive technique, multisector partnership is cooperation-based, and this difference stems from the organizational strategy, particularly making strategic decisions [43]. Five multisector partnerships in urban development have been identified; Multistakeholder partnerships, community–organizational partnerships, end-user-oriented partnerships, PPPs, and public–private-people partnerships [6]. These five partnerships have been separated based on the engagement, relationship, barrier and driver, and influence. Regarding PPP-based smart cities funding and financing goals, as well as the role and importance of People and stakeholder theory, this part of the study examines PPP and PPPP as the core models. However, PPP in smart cities provides some specific requirements in terms of innovative business models rather than other conventional methods of this approach. Furthermore, due to the narrow development of PPPP both theoretically and academically, the operational framework of this method should be compiled as a supplement for PPP-based smart cities funding and financing.

9.8.3.1 New PPP approach

Different types of contract structure and capital structure of PPP are used in the development of infrastructures. This development is a fundamental concept with certain specifications regarding the required liquidity, complexity, uncertainty, and performance goals. The base case in each PPP is finalized by calculating CAPEX, COPEX, and Revenue [28]. PPP is the untraditional finance source with private-sector capacities, especially investment services finance. If the private sector does not find PPP feasible, the government will grant the project to support Viability Gap Funding (VGF) [65]. Financing is performed using MHT, HGH, land-based financing, and congestion change of new smart cities [66]. In this context, the PPP potential in using different innovation levels of technology will be assessed by categories, including business-as-usual, incremental, coresearch, and groundbreaking innovation [5, 28]. On the other hand, what is considered in smart cities may seem ordinary in another city [28]. Therefore, based on the results of the pilot test, smart cities' funding and financing can be undertake using flexibility in fundamental concepts of project finance under the category of New PPP.

9.8.3.1.1 Business-as-usual PPP

This type of PPP refers to ordinary BOT projects or other long-term concession agreements, which require a high level of private finance. Business-as-usual PPP is a standard PPP approach [28]. On the other hand, standard PPP does not mean that there is a lack of innovation in this kind of project. These projects have experienced innovation [28]. Therefore, this category has a high smart city funding and financing potential. Financial inflows, including debt through commercial loans and equity of stakeholders within the construction step, are smart cities financing instruments. These instruments are operated through financial contract arrangements. Moreover, financial outflows that are payment mechanisms of Project Company to sponsors are supplied from earnings given to end user, government service purchase, and/or end user–VGF combination.

9.8.3.1.2 Incremental PPP

This category of PPPs not only keeps the previous business models and structures but also promotes partial subsystems, such as ticketing systems and/or using electronic fleets. This category has no relation with the renovation of infrastructures [28]. On the other hand, since these PPPs pertain to the operation step and subsystems promotion does not need high cash, they are financed by just correcting the funding of the existing model of the concession agreement.

9.8.3.1.3 Coresearch PPP

The role of universities and research centers in innovative projects of smart cities is becoming more vital [5]. This kind of PPP financed by university budget is an instrument for smart cities financing. In fact, coresearch PPP is used to identify innovative potentials in new smart cities through purposive and step-by-step studies. PPPs financing is done to theoretically broaden the knowledge borders in the smart city. Therefore, financing is not used only for the construction and promotion of operation from smart infrastructures, while encompassing even research budgets of universities. Moreover, funding is not expected for this category of PPPs as funding and financing are not necessarily involved simultaneously in the financial flow of innovative business models of smart cities.

9.8.3.1.4 Groundbreaking innovation PPP

This kind of PPP is seldom used in smart cities but is highly employed for exploratory measures in pilot projects regarding "proof concept" to create a new business model or renovate an existing business model [28]. That is, the smart city is a platform for new ideas where a wide range of ideas and knowledge of stakeholders meet [5]. The first step of these PPPs begins with public financing within private-sector services and scaling up procedures of the pilot project. In the next step, the admitted results of the living lab will be converted to the business-as-usual category. It means finance mechanism changes from construction financing to financing and funding mechanisms in construction and operation phases.

9.8.3.2 Public–private-people partnerships (PPPPs)

PPPP is a strategy used to implement smart city initiatives by PPP [1]. Therefore, PPPP has been derived from PPP doctrine that their difference is not just in a "P". For instance, the 12th development plan of India concentrated on people to fill the gap of citizen's partnership and allay the concern on low performance of private sector [57]. PPPP approach aims to develop performance goals of sustainable development, as well as the participation of different sectors in new approaches based on innovation requirements of smart cities. That is, technical knowledge should be accessible through PPPP programs [44]. PPPP not only allows managing and controlling citizens within key decisions but also provides different PPP themes, including innovation in smart cities. Regarding funding and financing potentials, PPPPs are similar to standard PPPs based on long-term project financing. However, permits and controls on the people subset must be considered based on cross-sector and multisector frameworks. Therefore, permits and controls related to the people sector together with other conventional inflows and outflows of PPP's capital structure are final smart cities funding and financing with the new PPP approach.

9.9 Summary

Finance and smart city both point to conceptual space. These concepts must be designed in the form of funding/ financing and smart infrastructure options to visualize the operationally. These options can be done through general approaches, including nonprivate sector partnerships, debt market, and PPPs. These approaches are operated potential based on exchequer financing, corporate financing, and project finance. PPP is an approach to respond to smart cities' challenges and to make opportunities for innovation and finance. PPP is the preferred option for smart city funding and financing since this approach includes multi-sector partnerships and serves as an instrument for governance and sustainable development goals. Considering innovative business models and smart cities' challenges, three important funding and financing strategies (including pillar and pilots, multisector partnership, and optimization and distribution of capital structure) are needed.

The scale of pilot projects includes roll-out, expansion, and replication that must be matched with integrated pillars of smart cities' infrastructure. Pilot results of under exploratory PPP after acceptance of under standard PPP mechanism will actualize the funding and financing. AFM component is a prerequisite to rank projects based on their required funding. AFM is also a prerequisite for rolling wave planning for smart cities. SPbEM model evaluates the performance of smart cities regarding social, economic, and environmental dimensions. Feasible Level of Private Finance also suggests using sources provided by both partners of PPP through a shared and feasible approach.

As a most underlying strategy, the multisector partnership includes four components related to new PPPs and one component related to PPPP. Although some of these partnerships have funding and financing potentials, some others only provide one of these potentials. On the other hand, lack of attention paid to these strategies causes risk in the development process of smart infrastructures. That is, inappropriate funding and/or financing instruments in smart cities lead to theoretical problems of renegotiation and dispute. Besides conventional financial flows in PPP, partnership share permits and people control also should be outlined in financial flows existing in PPPP.

Findings will provide urban development managers and planners with an approach to correct defects of traditional PPPs regarding smart cities' funding and financing. These findings also will illustrate a roadmap for key stakeholders to find optimal performance indices for smart cities' PPPs.

Finally, let us move on to the real interactive part of this chapter: review questions/exercises, hands-on projects, case projects, and optional team case project. The answers and/ or solutions by chapter can be found in Appendix G.

9.10 Chapter review questions/ exercises

True/false

1. True or False? Key stakeholders, such as local government, citizens, jobs, living laboratories, and research and academic communities, play a nonvital role in making the city smart—especially fund attraction.
2. True or False? In general, a smart city refers to the identification and searching through smart solutions, which allow modern cities to improve the quality of services provided for citizens.
3. True or False? There are various funding and financing instruments in the smart city, based on the economic and illegal conditions in any country.
4. True or False? There is fixed definition of PPP, since the existing definitions are in the range of two different meanings, including a narrow meaning and a wide meaning.
5. True or False? Although funding and financing are used usually simultaneously, these two instruments can be employed individually or together to form or operate any PDS based on the considered purposes.

Multiple choice

1. Suitable financial strategies play a vital role in complicated subjects, such as the smart city's:
   a. Infrastructure development
   b. Specifications of conformity
   c. Policy-based management
   d. Multi-sector partnership
   e. Innovation-based collaboration

2. To actualize smart city funding and financing, the required preconditions must be hierarchically integrated at:
   a. The same levels
   b. Different levels
   c. The Integration of PPP
   d. Specific PPP
   e. Smart city policy
3. There is not any fixed definition of PPP since the existing definitions are in the range of two different meanings, including a:
   a. Private meaning
   b. Machine-to-human meaning
   c. Narrow meaning
   d. Wide meaning
   e. Design meaning
4. Although funding and financing are used usually simultaneously, these two instruments can be employed individually or together to form or operate any _____ based on the considered purposes.
   a. HIS
   b. ECG
   c. MCS
   d. COA
   e. PDS
5. The ______________ to finance does not necessarily ensure the higher success of a business model, but suitable financial funding and financing strategies should also be taken.
   a. Lower access
   b. Financial access
   c. Higher access
   d. Business access
   e. Secure access

Exercise

Problem

Why are stakeholders one of the main success factors of Public–Private Partnerships (PPPs) in smart infrastructure projects?

Hands-on projects

Project

Do research: Chose a project that corresponds to a smart city's strategy and thus, responds to real needs and demands.

Case projects

Problem

Why are feasibility studies essential in attracting the private sector on smart city projects, with unknown risks and a slow return on investment?

Optional team case project

Problem

Why are the development and implementation of Smart City projects (that require considerable investments) difficult to fund with traditional public finance?

References

[1] Liu T, Mostafa S, Mohamed S, Nguyen TS. Emerging themes of public-private partnership application in developing smart city projects: a conceptual framework. Built Environ Proj Asset Manag 2020. ISSN:2044–124X. https://doi.org/10.1108/bepam-12-2019-0142. http://hdl.handle.net/10072/399254.

[2] Caragliu A, Del Bo CF. Smart innovative cities: the impact of Smart City policies on urban innovation. Technol Forecast Soc Change 2019;142:373–83.

[3] Fishman TD, Flynn M. Using public-private partnerships to advance smart cities. London, UK: Deloitte Center for Government Insights; 2018.

[4] Alim S, Polak J. Public–private partnerships for future urban infrastructure. Proc. Inst. Civ. Eng.: Manag. Procure. Law 2016;169(4):150–8.

[5] Nguyen TS, Liu T, Mohamed S, Mostafa S. Public-private partnership in the context of smart cities: review of contemporary literature. 43rd AUBEA 2019:704–13.

[6] Xue Y, Temeljotov-Salaj A, Engebo A, Lohne J. Multi-sector partnerships in the urban development context: a scoping review. J Clean Prod 2020:122291.

[7] Anwar B, Xiao Z, Akter S, Rehman RU. Sustainable urbanization and development goals strategy through public–private partnerships in a South-Asian metropolis. Sustainability 2017;9(11):1940.

[8] Yang JQ, You JL. Research on the application of PPP model in smart city projects. IOP conference series: earth and environmental science, vol. 330. IOP Publishing; October 2019. p. 052039. 5.

[9] Selim AM, Yousef PH, Hagag MR. Smart infrastructure by (PPPs) within the concept of smart cities to achieve sustainable development. Int J Crit Infrastruct 2018;14(2):182–98.

[10] Mirzaee AM, Sardroud JM. Applications of artificial neural network in construction engineering and management: an academic literature review and direction for future research. J Construct Eng Technol Manag 2016;6(1):24–38.

[11] Mirzaee, A. M., Pourrostam, T., & Majrouhi Sardroud, J. Conceptual strategy of claims between stakeholders based on lifecycle in public-private partnership agreements (case study in Iran's highway projects). J Manage Improv, 13(1), 153-182.

[12] Arora RU. Financial sector development and smart cities: the Indian case. Sustain Cities Soc 2018;42:52–8.

[13] Chou JS, Tserng HP, Lin C, Huang WH. Strategic governance for modeling institutional framework of public–private partnerships. Cities 2015;42:204–11.

[14] Jayasena NS, Chan DW, Kumaraswamy M. A systematic literature review and analysis towards developing PPP models for delivering smart infrastructure. Built Environment Project and Asset Management; 2020.

[15] Galati SR. Funding a smart city: from concept to actuality. In: Smart cities. Cham: Springer; 2018. p. 17–39.

[16] Appio FP, Lima M, Paroutis S. Understanding Smart Cities: innovation ecosystems, technological advancements, and societal challenges. Technol Forecast Soc Change 2019;142:1–14.

[17] Atmojo, R. N. P., Kasih, T. P., & Chandra, Y. U. Alternative financing model for smart cities initiatives in Indonesia.

[18] Huston S, Rahimzad R, Parsa A. 'Smart'sustainable urban regeneration: institutions, quality and financial innovation. Cities 2015;48:66–75.

[19] Tran Thi Hoang G, Dupont L, Camargo M. Application of decision-making methods in smart city projects: a systematic literature review. Smart Cities 2019;2(3):433–52.

[20] Anthopoulos L. Smart utopia VS smart reality: learning by experience from 10 smart city cases. Cities 2017;63:128–48.

[21] Yescombe ER. Principles of project finance. Elsevier; 2013.

[22] Rybnicek R, Plakolm J, Baumgartner L. Risks in public–private partnerships: a systematic literature review of risk factors, their impact and risk mitigation strategies. Publ Perform Manag Rev 2020;43(5):1174–208.

[23] Wojewnik-Filipkowska A, Wegrzyn J. Understanding of public–private partnership stakeholders as a condition of sustainable development. Sustainability 2019;11(4):1194.

[24] Shang L. Achieving public agency Goals in public-private partnerships using innovative payment mechanisms (Doctoral dissertation). 2019.

[25] Uzunkaya M. Theory-based evaluation of public–private partnership projects and programmes. In: The emerald Handbook of public–private partnerships in developing and emerging economies. Emerald Publishing Limited; 2017.

[26] Wang H, Xiong W, Wu G, Zhu D. Public–private partnership in public administration discipline: a literature review. Publ Manag Rev 2018;20(2):293–316.

[27] Koskela L, Rooke J, Siriwardena M. Evaluation of the promotion of through-life management in public private partnerships for infrastructure. Sustainability 2016;8(6):552.

[28] Cruz CO, Sarmento JM. Reforming traditional PPP models to cope with the challenges of smart cities. Compet Regul Netw Ind 2017;18(1–2):94–114.

[29] Sarmento JM, Renneboog L. Anatomy of public-private partnerships: their creation, financing and renegotiations. Int J Manag Projects Bus 2016;9(1):94–122. https://doi.org/10.1108/IJMPB-03-2015-0023.

[30] Davidson P. Finance, funding, saving, and investment. J Post Keynes Econ 1986;9(1):101–10.

[31] O'Brien P, Pike A. 'Deal or no deal?'Governing urban infrastructure funding and financing in the UK City Deals. Urban Stud 2019;56(7):1448–76.

[32] Giffinger R, Pichler-Milanović N. Smart cities: ranking of European medium-sized cities. Centre of Regional Science, Vienna University of Technology; 2007.

[33] Aladağ H, Işik Z. Role of financial risks in BOT megatransportation projects in developing countries. J Manag Eng 2017;33(4):04017007.

[34] Du J, Wu H, Jin R. Capital structure of public–private partnership projects: a sustainability perspective. Sustainability 2019;11(13):3505.

[35] De Marco A, Mangano G, Michelucci FV, Zenezini G. Using the-private finance initiative for energy efficiency projects at the urbanscale. Int J Energy Sect Manag 2016;10(1):99–117. Emerald Publishing Limited. https://doi.org/10.1108/IJESM-12-2014-000. https://iris.polito.it/retrieve/handle/11583/2640841/110129/IJESM-12-2014-0005_PostPrint.pdf.

[36] Geissdoerfer M, Vladimirova D, Evans S. Sustainable business model innovation: a review. J Clean Prod 2018;198:401–16.

[37] Leviäkangas P, Ojala L, Töyli J. An integrated ecosystem model for understanding infrastructure PPPs. Util Pol 2016;42:10–9.

[38] Yuan J, Skibniewski MJ, Li Q, Zheng L. Performance objectives selection model in public-private partnership projects based on the perspective of stakeholders. J Manag Eng 2010;26(2):89–104.

[39] Akinola AO. Entrepreneurship: funding and financing strategies in Nigeria. Euro J Acc Audit Finan Res 2014;2(4):39–51.

[40] Schaufelberger JE, Wipadapisut I. Alternate financing strategies for build-operate-transfer projects. J Construct Eng Manag 2003;129(2):205–13.

[41] Kwak YH, Chih Y, Ibbs CW. Towards a comprehensive understanding of public private partnerships for infrastructure development. Calif Manag Rev 2009;51(2):51–78.

[42] Ruiters C, Matji MP. Public–private partnership conceptual framework and models for the funding and financing of water services infrastructure in municipalities from selected provinces in South Africa. WaterSA 2016;42(2):291–305.

[43] Pittz TG, Adler T. An exemplar of open strategy: decision-making within multi-sector collaborations. Manag Decis 2006;54(7):1595–614. https://doi.org/10.1108/MD-04-2015-0153.

[44] Ojasalo J, Kauppinen H. Collaborative innovation with external actors: an empirical study on open innovation platforms in smart cities. Technol Innov Manage Rev 2016;6(12).

[45] Fakhimi AH, Khani AH, Sardroud JM. Smart-city infrastructure components. In: Solving urban infrastructure problems using smart city technologies. Elsevier; 2021. p. 17–54.

[46] van Winden W, van den Buuse D. Smart city pilot projects: exploring the dimensions and conditions of scaling up. J Urban Technol 2017;24(4):51–72.

[47] Kwak YH, Lee J. June). Evolution of Korean smart city programs: challenges and opportunities. In: 2020 IEEE technology & engineering management conference (TEMSCON). IEEE; 2020. p. 1–5.

[48] Shen L, Tam VW, Gan L, Ye K, Zhao Z. Improving sustainability performance for public-private-partnership (PPP) projects. Sustainability 2016;8(3):289.

[49] Straub S, Fay M, Martimort D. Funding and financing infrastructure: the joint-use of public and private finance (No. 18–98). Institute for Advanced Study in Toulouse (IAST); 2018.

[50] Neirotti P, De Marco A, Cagliano AC, Mangano G, Scorrano F. Current trends in Smart City initiatives: some stylised facts. Cities 2014;38:25–36.

[51] Turečková K, Nevima J. The cost benefit analysis for the concept of a smart city: how to measure the efficiency of smart solutions? Sustainability 2020;12(7):2663.

[52] Caragliu A, Del Bo C, Nijkamp P. Smart cities in europe. J Urban Technol 2011;18(2):65–82.

[53] Shetty N, Renukappa S, Suresh S, Algahtani K. June). Smart city business models—A systematic literature review. In: 2019 3rd international conference on smart grid and smart cities (ICSGSC). IEEE; 2019. p. 22–5.

[54] Lam PT, Yang W. Factors influencing the consideration of Public-Private Partnerships (PPP) for smart city projects: evidence from Hong Kong. Cities 2020;99:102606.

[55] Kościelniak H, Górka A. Green Cities PPP as a method of financing sustainable urban development. Transport Res Proc 2016;16:227–35.

[56] Perboli G, De Marco A, Perfetti F, Marone M. A new taxonomy of smart city projects. Transport Res Proc 2014;3:470–8.

[57] Chattopadhyay S. Financing India's urban infrastructure: current practices and reform options. J Infrastruc Develop 2015;7(1):55–75.

[58] Merk O, Saussier S, Staropoli C, Slack E, Kim JH. Financing green urban infrastructure, OECD regional development working papers. OECD Publishing; 2012.

[59] Uppenberg K, Strauss H, Wagenvoort R. Financing infrastructure: a review of the 2010 EIB conference in economics and finance. Annual Economic Conference and Publication; 2011.

[60] Vadgama CV, Khutwad A, Damle M, Patil S. Smart funding options for developing smart cities: a proposal for India. Indian J Sci Technol 2015;8(34):1–12.

[61] Kavta K, Yadav PK. Indian smart cities and their financing: a first look. In: From poverty, inequality to smart city. Singapore: Springer; 2017. p. 123–41.

[62] Deloitte. Private sector participation in public sector financing, an introduction. 2018 [Online], Available: https://www2.deloitte.com/global/en/pages/public-sector/articles/smart-cities-funding-and-financing-strategies.html.

[63] De Marco A, Mangano G, Zou XY. Factors influencing the equity share of build-operate-transfer projects. Built Environment Project and Asset Management; 2012.

[64] Thabrew L, Wiek A, Ries R. Environmental decision making in multi-stakeholder contexts: applicability of life cycle thinking in development planning and implementation. J Clean Prod 2009;17(1):67–76.

[65] Alok VN, Vashist A. Financing smart cities in India. Indian J Publ Adm 2016;62(4):791–804.

[66] Mishra AK. Henry George and Mohring–Harwitz theorems: lessons for financing smart cities in developing countries. Environ Urbaniza ASIA 2019;10(1):13–30.

Further reading

Milenković M, Rašić M, Vojković G. May). Using public private partnership models in smart cities-proposal for Croatia. In: 2017 40th international convention on information and communication technology, electronics and microelectronics (MIPRO). IEEE; 2017. p. 1412–7.

Part III

Human resources skills gaps

Chapter 10

Preparing students for interdisciplinary work: green infrastructure curricula at Toronto Metropolitan University, Toronto, Ontario, Canada

Sheila Boudreau[1], Gareth Gransaull[2], Nina-Marie Lister[1] and Gary Pritchard[3,4]

[1]*Toronto Metropolitan University, Toronto, ON, Canada;* [2]*University of Western Ontario, Toronto, ON, Canada;* [3]*University of Guelph, Guelph, ON, Canada;* [4]*Carleton University, Ottawa, ON, Canada*

10.1 Introduction

To design a "smart city," a broad range of green infrastructure techniques is required that are designed to complement natural cycles, minimize environmental impacts, and optimize ecosystem services and associated social benefits. Given the complex environmental, asset management, and social issues inherent in this work, the need to collaborate effectively with other disciplines is fundamental to successful green infrastructure practice. This chapter explores a central question that has evolved through 2 decades of professional practice and teaching in this field:

How can students be better prepared for interdisciplinary practice to effectively address systemic challenges and to coproduce innovative solutions? The authors describe supportive social science theory, alternative ways of thinking and knowing, and collaborative approaches that help improve interdisciplinary outcomes. Curricula and extracurricular cases from Toronto Metropolitan University, Toronto, Ontario, Canada, are presented as examples that address this question, viewed through the lens of planning green infrastructure. Accompanying this material are descriptions of unique elements such as skills-based experiential approaches, storytelling, and the wisdom of Indigenous Knowledge and ancient wisdom.

10.2 Smart cities and green infrastructure

In Toronto, Ontario, smart city discussions are relatively recent and have focused primarily on controversies regarding the sharing and aggregating of personal digital data by private interests and the need for adequate protections and governance measures to protect the public from harm and loss [1]. Ideas regarding high-tech innovation and the integration of digital and social systems are generally welcomed by regulators and the public alike, in order to promote efficiency, sustainability, and livability. The need to combine both heavily engineered and nature-based designs for stormwater management is advocated by many experts (i.e., [2,3]) as a key instrument in the broader movement toward "ecological urbanism" [4,5].

"Green infrastructure" is defined by the Province of Ontario as (see Fig. 10.1):

> *... natural and human-made elements that provide ecological and hydrological functions and processes. Green infrastructure can include components such as natural heritage features and systems, parklands, stormwater management systems, street trees, urban forests, natural channels, permeable surfaces, and green roofs. [6]*

There is a mountain of evidence proving nature as being essential to human health, with the concept of "health" as holistic and encompassing mental, physical, emotional, and spiritual elements. Communities across the globe have experienced Covid-19 stay-at-home orders for over a year, and when people search for respite, they are visibly attracted to trees and green spaces. The result has been an overcrowding of local parks and streets and the overly hasty division and redesign of public spaces to maintain physical distancing requirements of local health agencies. Understanding the nuances of public realm use has become

Smart Cities Policies and Financing. https://doi.org/10.1016/B978-0-12-819130-9.00040-1

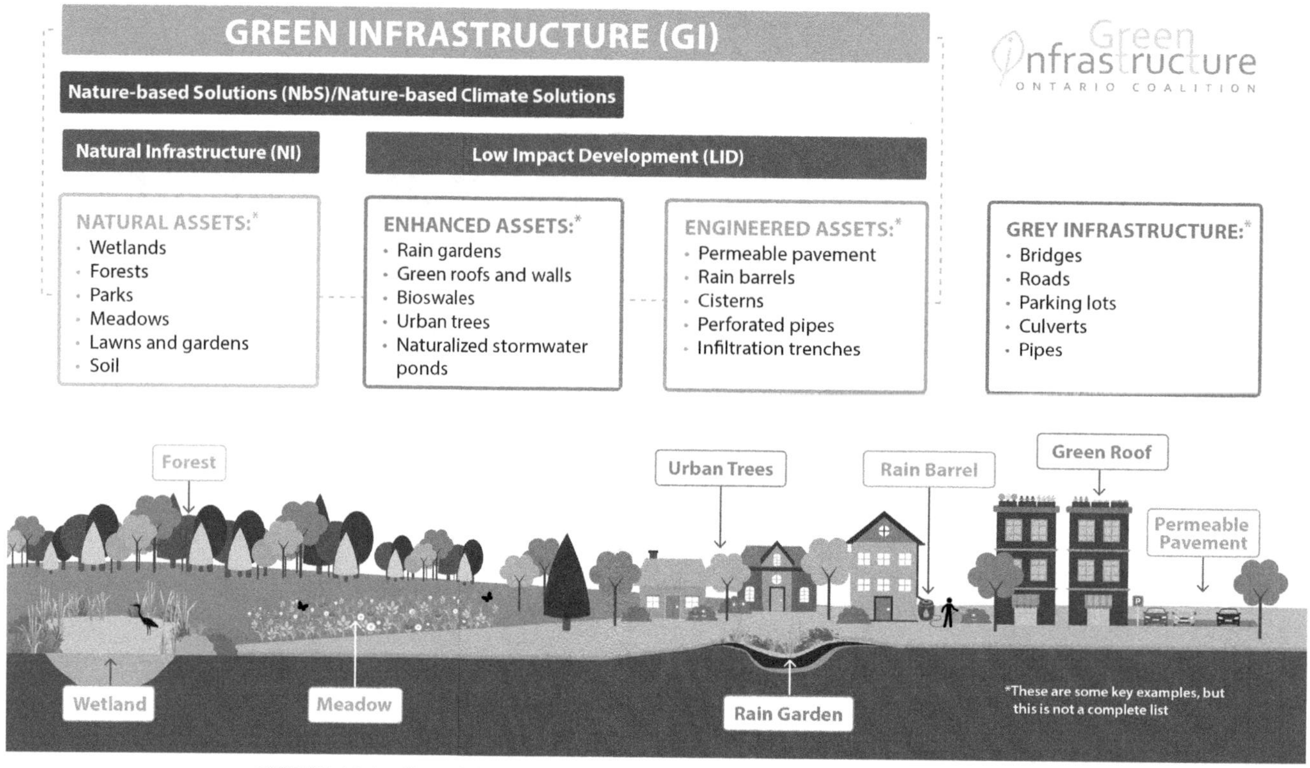

FIGURE 10.1 Green infrastructure types. *Source: Green Infrastructure Ontario Coalition.*

an urgent priority, even without the consideration of smart city elements such as automated vehicles, artificial intelligence, and other evolving technologies. Furthermore, it is clear that "the common unit driving many smart city initiatives is the evolving role of the citizen as not just a client, user, consumer, or recipient of city services, but additionally as a contributor to governance, either directly or indirectly" [7]; p.3). Planning for infrastructure must go beyond conventional single-discipline, siloed thinking, which has consistently failed to address complex and interrelated systems [8]. Adopting an integrated, systems thinking approach to problem-solving and conflict resolution is required to create novel solutions that result in mutual gains [9].

10.2.1 Considerations for practice

Green infrastructure is a complex field that works across disciplines, organizations, as well as political and geographic divides. While often seen as purely engineering due to the stormwater management performance objectives, disciplines such as landscape architecture, planning, urban forestry, and ecology, among others, are all areas of expertise that are essential in achieving the successful, long-term performance of green infrastructure. The recently published international standards for nature-based solutions [10] further expand on this understanding, denoting the interrelationships and challenges involved in this work (see Fig. 10.2).

To avoid the trap of siloed thinking, new interdisciplinary approaches are needed to facilitate effective collaboration between a myriad of disciplines and communities in addressing ecological challenges. Biodiversity goals, for example, must be a shared responsibility for those who oversee potable water service and sewer infrastructure asset management. Rain and snow melt can no longer be viewed exclusively as stormwater in need of removal, but as a critical part of local watersheds that replenishes groundwater systems, conserves waterways, and sustains aquatic habitats. More broadly, healthy environments must be seen as having an intrinsic connection to human health and well-being and as a means to promote physical exercise, psychological welfare, and spiritual or social connection.

For city plans to result in successful built form solutions, those responsible for the planning must also engage the end users of these environments. Research undertaken by the Design Council and Social Change UK on "Healthy Placemaking" demonstrates the effectiveness of user-focused design-driven innovation. The 2020 IUCN standards for nature-based solutions also note the need for careful collaboration in comprehensively mapping out potential scenarios and engaging with all parties involved (i.e., stakeholders, rights holders, and citizens) to identify

FIGURE 10.2 Societal challenges of nature-based solutions. *Source: Created by author, based on a similar IUCN graphic.*

and weigh the benefits and trade-offs involved in decision-making (see Fig. 10.3).

"Social prescribing," a successful client and community-centered healthcare initiative well underway in the United Kingdom and recently piloted in Canada, values health considerations for the whole person, rather than just their diagnostic medical needs. In this approach, public places that satisfy social and psychological needs (such as community gardens or healing spaces) play an important role. The Canadian Alliance for Healthier Communities's Rx: Community initiative involved 11 pilot projects that generated wide interest and resulted in plans "to scale social prescribing broadly to support a more integrated health system and build more connected communities." [11]. "Forest bathing" is another prominent example of how community-level public realm can be used to address health needs (see Fig. 10.4). In this approach, patients are recommended and even prescribed immersive walks in forests to receive direct medicinal benefits from the aerosol phytochemicals provided by specific trees (Kotte, D. et al., 2019).

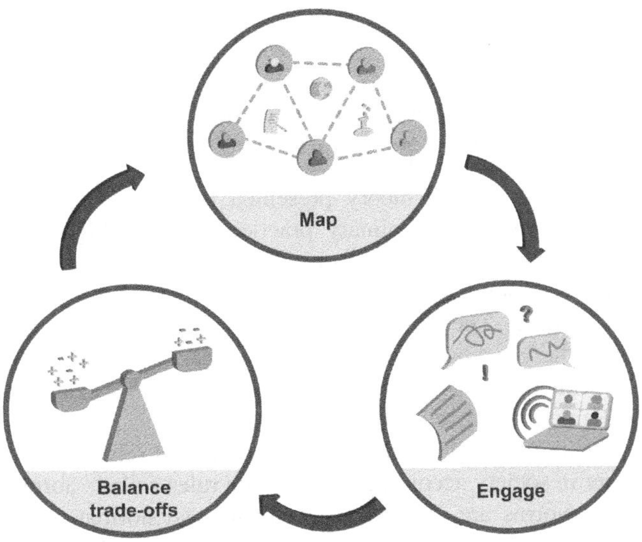

FIGURE 10.3 Collaboration for nature-based solutions. *Source: Created by author, based on a similar IUCN graphic.*

Rather than seeing green infrastructure as a purely technical problem, public realm practitioners must understand the implications of their work for human health and well-being. New research findings demonstrating the social benefits of nature are merely corroborating the traditional principles that Indigenous Peoples and ancient cultures have known since time immemorial [12]. Future considerations involving green infrastructure must reflect the need to go beyond "interdisciplinary" to that of "transdisciplinary," or "knowledge production that not only integrates knowledge from different disciplines (interdisciplinary) but includes values, knowledge, know-how, and expertise from non-scientific sources" [8].

The information contained in this chapter provides guidance as practice matures toward this transdisciplinary ideal.

10.3 Educating professionals

City building requires the involvement of numerous professions, which are regulated according to government legislation and regulations developed by their respective associations. Professionals are required to practice in a way that protects the public interest, as well as those of clients and colleagues. Codes of conduct and ethical principles related to specific professions are outlined clearly and made publicly available, as are the measures by which

FIGURE 10.4 Black Oak Savanna medicine walk led by Diana Beresford-Kroeger in High Park, Toronto. *Photo Credit: Sheila Boudreau.*

professionals are judged. In the education of new professionals, accreditation boards review curricula and evaluate schools to verify adherence to the above standards and to ensure that students are being taught the core competencies that are determined to be essential for work in their field.

Because of these restrictions, preparing students for interdisciplinary practice in the green infrastructure—which typically requires moving beyond the confines of a single field—is often quite challenging. Instructors face many challenges introducing knowledge that lies outside of their traditional domain; accreditation boards and university departments can often be skeptical that new knowledge will satisfy core curricula requirements, and funds are rarely budgeted for disseminating knowledge that is seen to lie outside of a profession's explicit mandate. Additionally, siloed structures and values within the academy itself, time constraints related to building interdisciplinary and cross-sectoral relationships, and the additional risk and complexity involved in dealing with new partners and initiatives all represent substantial obstacles. Without creative interventions, students will remain ill-equipped to address the inherent complexities of this work. It is important that educators tailor their instruction according to specific professional competencies that are fundamental to interdisciplinary success. For example, in Ontario, planning competencies are divided into "functional" skills, which focus on the central work of planning, and "enabling" competencies, which include the emotional and social skills necessary to perform interdisciplinary work (see Fig. 10.5).

For interdisciplinary practice, the most relevant enabling competencies are professional and ethical behavior, communication, interpersonal and critical thinking. Acting in ways that are in accordance with professional ethics is essential to building trust. Being able to communicate ideas and questions clearly, without professional jargon, leads to improved understandings and synergies. Having strong interpersonal skills such as high emotional intelligence, active listening, and empathy are crucial for building relationships. Across all of these, the ability to think critically helps with framing and following arguments of logic and opens up minds to unconventional ways of thinking. The following examples highlight approaches that the authors have found helpful in preparing students for green infrastructure work: social science theories for environmental sustainability, Indigenous Knowledge, and various systemic design techniques.

10.3.1 Social science theory for interdisciplinary practice

The social science theory presented below is helpful to consider in interdisciplinary practice. While a high level description of each theory is provided for each, the original sources should be consulted to gain a more in-depth understanding. The reader is encouraged to reflect on how each theory relates to the case studies provided in the subsequent section on supportive models or other examples presented and discussed in class.

Within "social norm theory," individuals relate to each other in society according to unwritten rules called "norms" [13]. Norms are culturally dependent, and normative expectations are often particular to specific workplace or disciplinary cultures. For people working across cultures, it is critical to learn about these norms in order to improve

FIGURE 10.5 Ontario Professional Planners institute competency Tree. *Source: Ontario professional planners institute.*

communication and understanding and avoid creating conflict, harm, or tension. Social norms can be "descriptive," referring to how individual behavior is perceived, or "prescriptive," which refers to others' expectation of how an individual should appropriately behave. These norms influence individual behaviors by reinforcing the benefits or consequences associated with particular actions that generate a pressure to conform. Indigenous Peoples refer to these norms in terms of "protocols" that are varied and community-specific. In this context, introductory meetings are dedicated to sharing personal experience, learning about others and their intentions, and also as an opportunity to explain community protocols so that they can be clearly understood and respected.

Extensive research from around the world shows that there are at least six moral foundations that are shared by all human cultures [14]. They include socially ingrained attitudes about care versus harm, fairness versus cheating, loyalty versus betrayal, authority versus subversion, sanctity versus degradation, and liberty versus oppression. In working with people from different cultural backgrounds, having an appreciation of shared moral foundations is necessary to find common ground, improve empathy, and drive collaboration.

Working with others requires sensitive consideration of how people view themselves, including their roles and social identities, and how they perceive others to view them. There is a greater need to appreciate "identity theory" and to avoid threats to identity, especially in situations where building trust is fundamental. To better understand identity, practitioners should take the time to research and learn about the values of others, which can also involve storytelling. Tools of "**framing**" are another set of techniques that are critical to interpersonal communication. "Interpretive frames" relate to personal beliefs that direct how we interpret our role in situations and how we should respond. Working with others requires independent research and deep listening in order to break free from the mental constraints we inadvertently employ when framing what we observe. "Communication frames" refers to the way we present shared information in terms of risks (e.g., avoiding losses), goals, and moral values.

Trust theory informs how interactions can be structured to promote trust, specifically through the "trust ecology" of a specific group context. Trust grows when individuals are well-liked or where protocols for interaction are coproduced and thus result in reduced risk. Conversely, prior experiences of broken trust make it difficult to create trust in the future. When working across cultures, it is important to consider how trust is influenced by values of independence, with a focus on individual needs and wants, as they relate to the common, collective good.

Developed by the Harvard Negotiation Project, "principled negotiation" is based on four principles: separating people from the problem; focusing on interests instead of positions; generating options for mutual gain (the "win-wins"); and developing shared criteria to evaluate outcomes [15]. This theory is useful in situations where differing beliefs or views are involved, and all participants are expected to agree and adhere to basic project principles.

"Team effectiveness theory" identifies the drivers that are responsible for team success. These are classified into four categories: inputs, such as traits, size, resources; processes, such as goal setting, structures, roles; emergent states, such cohesion and trust; and outcomes, including intangible qualities such as innovation and efficiency. In the development of team effectiveness theory, the inputs and processes most likely to produce positive outcomes were the traits of individual team members, including emotional intelligence, functional competence, willingness to work collaboratively, reliability, and flexibility. Additionally, a clear delineation of roles and a sense of agency and empowerment are also vital to success. All team processes require key elements such as leadership, working across divides, coordination, communication, monitoring, and evaluation. Team performance is tied directly to positive team culture, and if greater time is invested at the beginning of a project to identify norms for communication as well as common goals, positive outcomes are more likely. Effective leadership styles are typically empowering in nature, in which leaders promote a shared vision and give team members the autonomy to work toward it. Effective leaders support and respect the ideas of all team members and foster genuine concern for their well-being.

All teams both create and apply knowledge in the work they do. "Nonaka's Theory of Organizational Knowledge Creation" [16] refers to two forms of knowledge: "tacit" knowledge, including the implicit mental models and skills developed through experience; and "explicit" knowledge, the facts and figures that are explicitly and consciously remembered by the rational parts of our brain. Nonaka's theory claims that tacit knowledge related to experiences should be actively shared with others, through means such as reflection, observation, and conceptualization. To promote organizational learning, team members should have the opportunity to try their own approaches and should be given the means to experience meaningful tasks. Additionally, teams should create a culture of care and trust and incorporate adequate time to reflect on project outcomes.

Systems theories are particularly important in working with social-ecological systems, such as those found in green infrastructure. For example, in situations where unrelated efforts are made to address the same complex problem, "Collective Impact" theory offers support for the development of a shared vision and goals. Several key factors underlie the success of this work: long-term commitments, a common agenda, shared metrics, mutually reinforcing activities, continuous communication, and "backbone support organizations" to coordinate all actions. This theory offers a framework to guide collective impact initiatives:

- Initiating action
- Problem identification
- Team building
- Community involvement
- Organizing for impact:
- Build collective infrastructure
- Establish ground rules for participation
- Solidify a common agenda
- Develop an adaptive management plan
- Community involvement
- Implementation, adaptation and impact:
- Implementation
- Continuous communication
- Adaptive management
- Sustaining commitment and support
- Community involvement

Developing a clear framework that considers the needs of rights holders, stakeholders, and community members is vital for interdisciplinary work. This is the case for all public realm infrastructure, including green infrastructure, regardless of its scale, location, ownership arrangements, or technical innovation.

10.3.2 Supportive models for interdisciplinary practice

The following sections describe examples of approaches that the authors applied in teaching, as efforts to explore and introduce students to supportive models for interdisciplinary practice.

10.3.2.1 Indigenous Knowledge

The information shared below is offered to provide some context regarding why Indigenous Knowledge is included in this chapter. Teaching at a postsecondary institution in Canada mandates deeply reflecting on what it means to be an educator, following over 500 years of violence and oppression of Indigenous Peoples due to colonization. It is essential to acknowledge that this work is happening in an environment that resulted in the often unquestioned prioritization of Western science and ways of thinking and working, and that has benefited from the unspeakable losses of Indigenous Peoples. For guidance, educators are advised to turn to the Truth and Reconciliation Commission (TRC) of Canada 94 Calls to Action, developed as a 6 year undertaking across the country. The purpose of the TRC was to investigate and understand the legacy of the residential school system (created to separate First Nations children from their families and culture and to indoctrinate them into the dominant Euro-Christian society [17] operating for over 100 years), by listening closely to the tragic recounts of survivors and their families. The calls to Action were directed at settlers (non-Indigenous) as a means of advancing reconciliation, to begin to heal the trauma caused by the past, by offering tangible steps to begin to create improved and respectful relationships between Indigenous and non-Indigenous communities. Calls 62–65 include direct actions that are most relevant to postsecondary education in non-Indigenous institutions:

- 62: We call upon the federal, provincial, and territorial governments, in consultation and collaboration with Survivors, Aboriginal peoples, and educators, to:
- provide the necessary funding to postsecondary institutions to educate teachers on how to integrate Indigenous knowledge and teaching methods into classrooms.
- 63. We call upon the Council of Ministers of Education, Canada, to maintain an annual commitment to Aboriginal education issues, including the following:

Sharing information and best practices on teaching curriculum related to residential schools and Aboriginal history; building student capacity for intercultural understanding, empathy, and mutual respect; and identifying teacher-training needs relating to the above [18].

Understanding the real truth of the dark history of Canada, which was not taught in schools for decades, is a critical first step for all Canadians to undertake. However, a simple search online reveals countless reports, in both the traditional news media and across social media, that only a few of the Calls to Action have actually been implemented. In the work of "green infrastructure," which is intimately tied to land and water, improving biodiversity, and looking holistically at how infrastructure can be better planned and designed to work with nature, respected as Mother Earth by Indigenous Peoples, yet the inclusion of their voices in this work is rare. The reasons behind this situation are too complex to outline in this chapter, but the authors want to stress the importance of critically reflecting on what is shared above and taking the time to learn more about the Indigenous Peoples where you live and work, wherever you are located in the world. (It is also critical to note that at the time of writing, Toronto Metropolitan University, now also known as "X University," is currently undertaking a significant process that may lead to a name change due to its namesake, Egerton Ryerson, and his leading role as an architect of the residential school system in Canada, see: [19].)

The following text is written as a personal Indigenous person's account, in keeping with an Indigenous storytelling and knowledge sharing tradition. While this may seem out of place to the reader, the authors have intentionally changed the narrative to create space for Indigenous Knowledge to be shared in a way that honors ancestral methods used for teaching purposes since time immemorial.

It is very evident that Western science practitioners suffer from a term called "shifting baseline syndrome" (SBS). SBS has been gaining traction in the collection and implementation of Western science for quite some time in the environmental field. Essentially, this is the gradual change in the accepted normalities for the conditions present in the natural environment. This failure is easily achievable in Western science, which lacks transgenerational memory, compared to Indigenous Knowledge systems where this is central to understanding the health of the environment or landscape. To put it bluntly, landscapes that are considered healthy now would have been considered degraded to past generations. Indigenous knowledge systems are based on memory (shared oral history) and experience with past environmental conditions. Current generations of Western science practitioners cannot perceive how much their environment has degraded, because they are comparing it to their own "normal" baseline and not to a historical trend through time. Western science fails at defining ecological parameters or conditions by the following:

- "Scientists" have their own ideologies of environmental health;
- "Scientists" have their own methods, thus creating challenges on site performance or evaluation;

- Environmental conditions are evaluated in present time and not to presettler occupation.
- Western science has a difficult time qualifying and quantifying cumulative impacts that account for historical and future change of an area.
- Not truly defining or understanding the term "significant" from an Indigenous knowledge system or rights position.

This shifting of baseline ecological integrity from an Indigenous worldview makes it difficult for policymakers to make informed decisions that could ultimately impact Indigenous Rights. To Indigenous communities, policy around conservation and environmental protection are becoming less effective with each passing generation.

Defining the Term "Significant Harm": Practitioners and policies cannot agree on defining the term "significant harm" to the natural environmental and ecological features through a Western science SBS lens. When Western science attempts to define or identify with significant harm and relate that to the Indigenous community, they need to understand the following:

- Our belief of the natural world is that of a sacred commitment and bond to coexist together,
- People are just one element of this world, and we need it unharmed and whole for future generations.
- Indigenous people's, like all people's, health and wellbeing come from the environment, so that environment must be given the opportunity to provide and be stewarded.

Defining Meaningful Indigenous Engagement and Participation: Settler (non-Indigenous person) reconciliation with the Indigenous community needs to be at the center of all aspects to create a new relationship that is meaningful for all people in Canada. All settlers, not only governments, have a responsibility to uphold the treaty commitments. To foster a renewed relationship with all peoples, settler policymakers and practitioners need to hear that current methods of information sharing is only one step toward defining meaningful engagement and participation. Meaningful engagement aims to secure free, prior, and informed consent from actions or projects that affect the Indigenous way of life (see Fig. 10.6).

In-Class Application—Indigenous Voices: In order to introduce students to learning and thinking about the issues described above, Indigenous guests were invited into the classroom to share their Knowledge and experiences in an open discussion format. In one instance, after listening to a Traditional teaching, the class was involved in a hands-on activity where students held various "found" natural objects (e.g., stones, leaves, acorns, sticks) and reflected on their personal relationship with the natural world. Working in groups, they discussed ways to create something out of the objects together, while sharing stories and memories about nature.

In other instances, an Indigenous biologist (Gary Pritchard, coauthor) was invited to classes to talk about the Government of Canada's Indian Act and its unjust restrictions on First Nations, as well as working with Indigenous people, and how ecological conservation is understood differently from an Indigenous world view. Consistently, following these sessions, feedback from the students was overwhelmingly positive, but also sad and also angry, because for many of them, even at the graduate school level, this was the first time that they had been told this information. Some of the key messages shared included the need to think holistically and to reject the oversimplification of complex ecological systems by taking a singularly focused siloed approach. Also to think of nature respectfully as "All our Relations," where other creatures are seen as family, and that they give up their lives so that humans may live. And to question the valuation of nature as only being "resources" that we use. They also learned of the need to think seven generations ahead and not just the timeline of a human or the engineered infrastructure being planned for. For students interested in green infrastructure, where working with nature is central to its definition, appreciating these lessons shared from an Indigenous perspective is invaluable and resonated deeply.

10.3.2.2 User experience design (UX Design)

The concept of "User Experience Design" (UX Design) was developed by the software industry to improve human—computer interfaces. UX Design is also referred to as "interaction design," as it focuses on designing for the entire "interaction experience" [20]. In this approach, computer system designers have learned the critical importance of accounting for the emotional and psychological needs of users, including "self-identity, relatedness to others, and being satisfied with life" [20]. In particular, UX Design uses techniques to develop user "personas" or archetypes of typical users that identify their key traits and attributes. The Interaction Design Foundation (IDF) provides templates for creating user personas that interdisciplinary practitioners can use to empathize with the people they work for, especially early in the design phase:

"Personas are distilled essences of real users. In (UX Design) you use personas to build empathy with target users and focus on their world. You should always create personas from observations about real users (see checklist: 'An Agenda For Action For Taking Steps In The Creation Of User Personas'); because, you must map your users' needs to your design's functionality, you must first clearly define both the needs and the users" [21].

FIGURE 10.6 Indigenous art with clan symbols and medicine wheel at wandering spirit school, Toronto, Ontario. *Photo credit: Sheila Boudreau.*

An agenda for action for taking steps in the creation of user personas

The IDF recommends the following steps in the creation of user personas (check all tasks completed):

_____1. Collect extensive data on target users.
_____2. Determine the qualities of and differences between users.
_____3. Develop a hypothesis from the research, determining the qualities of and differences between users.
_____4. Ensure stakeholders agree on the hypothesis about the users.
_____5. Determine a number of personas, but focus especially on one.
_____6. Name and describe each persona in 1–2 pages, including: a picture, user values, interests, education, lifestyle, needs, attitudes, desires, limitations, goals, and behavior patterns; and details about the persona to make him/her more real and relevant and help build empathy.
_____7. Describe several situations/scenarios prompting the persona to use your product—put him/her in contexts with problems to overcome.
_____8. Include everyone involved in the project so they will accept the persona or advise revisions.
_____9. Send them the persona to use in their work.
_____10. Ensure everyone develops scenarios—these should expose the persona optimally to potential use cases.
_____11. Make continuous adjustments—revisit the persona; add new features; add required new personas; discard outdated personas [21].

The UX Design process is useful as a skills-building exercise for interdisciplinary practice. It is also useful as a comprehensive tool to build empathy, even in situations where the traits of future users remain unknown.

10.3.2.2.1 In-class application—user personas

The exercise of creating user "persona" was incorporated into a 2-day green infrastructure summer workshop for students entering the Master of Planning in Urban Development program at Toronto Metropolitan University. Students were asked to review the proposed masterplan for the university's campus (see Fig. 10.7), the information gathered through an Indigenous engagement workshop, and other information shared by Sidewalk Labs from their own Indigenous engagement process. Students were asked to imagine a day in the life of "Person X" by mentally walking around the proposed campus environments. The assignment outline was as follows:

- Create a persona in your online journal shared with the class, giving this person a name and description for others to understand what it would be like to be them. This person can be a student, staff, someone who lives in the community, or a visitor to the campus. To create this character, write down bullet point as answers to the following questions:
  - What is this person's story?
  - How will they interact with the campus?
  - Who and/or what influences them?
  - What are their goals and motivations?
- Explore the current campus plan using the maps provided, Google Streetview, images shared, as well as your memory, and record your thoughts about the green infrastructure you see and experience in your journal.
- Review the comments shared from the Indigenous engagement on the campus masterplan design, as well as the recommendations developed through the

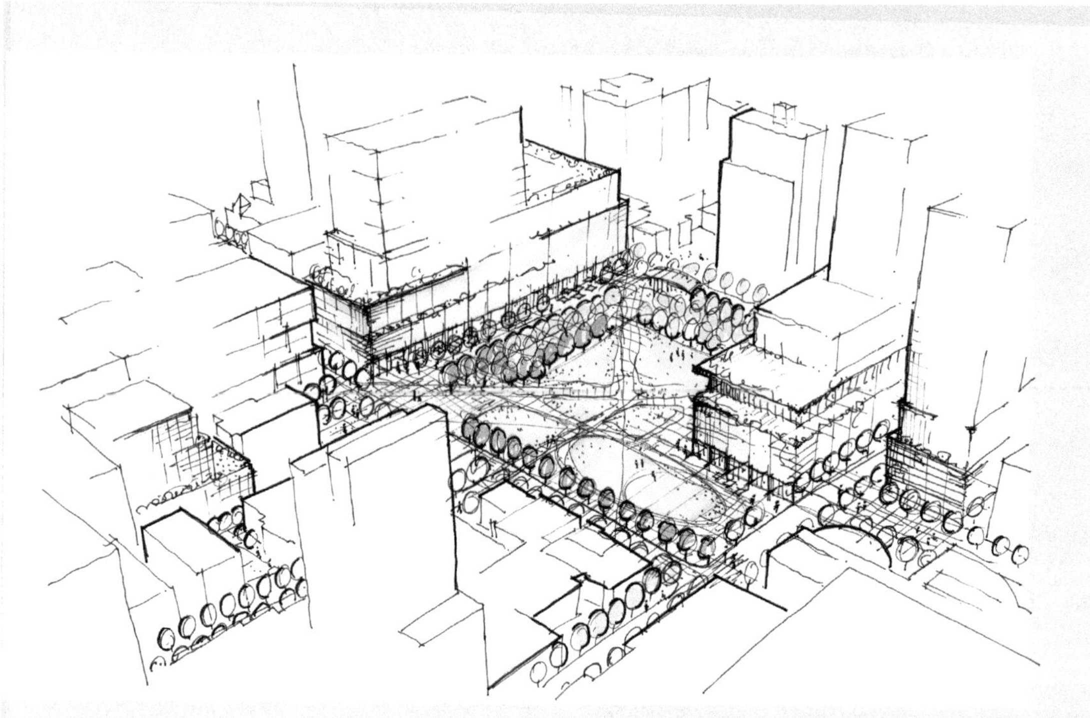

FIGURE 10.7 Toronto Metropolitan University masterplan sketch and location plan. *Source: Dialogue.*

Sidewalk Labs "Smart City" Indigenous engagement design workshop.
- Analyze the masterplan drawings and images, imagining how the green infrastructure changes would make you feel as "Person X," or as one of the Indigenous participants who contributed to the aforementioned engagement processes, and record your thoughts in your journal.
- Focus on 1–2 ideas and research them further. Record your ideas and relevant research, including images, links to videos, etc., in your journal, and be prepared to share them with the class.

The students involved with the workshop expressed that they found the activity interesting and engaging and that the experience made them think deeply about what it would be like to "walk in other people's shoes." For interdisciplinary practitioners, the ability to transcend one's own priorities, rethink preconceived assumptions, and closely empathize with the needs and interests of others is invaluable for design success.

10.3.2.3 Systems Thinking and Systems Design

Because most large-scale social and ecological problems are fundamentally integrated, socioecological systems challenges, or "wicked problems" (see Ref. [22], tackling them requires moving beyond the siloed, linear ways of thinking that are traditionally used to address governance challenges. Systems-level problems require us to see the inherent interconnections between the processes and actors that make up a particular system and understand how they are mutually interdependent in ways that transcend any single domain. This mode of thought is a holistic way of understanding the world that is known as "systems thinking." Peter Senge defines systems thinking as "a discipline for seeing wholes and a framework for seeing interrelationships rather than things, for seeing patterns of change rather than static snapshots" [23]. Systems thinkers see social and environmental systems as being more than the sum of their parts; they understand that the functioning of a system depends on the interdependence of all its components, none of which can be seen in isolation [24]. In this way, systems thinking moves beyond the reductionistic and mechanistic bias of Western science with a renewed appreciation for emergence, complexity, and nonlinear behavior. Systems thinking approaches rely on the tools of both systems science and complexity theory to understand how systems behave in dynamic (i.e., nonlinear) ways, often with unpredictable and chaotic results. Systems thinkers use diagrams to map entire systems and explore relationships between all stakeholders; these maps describe elements of a system such as negative and positive feedback loops, the stocks and flows of resources, and others [24].

Systems thinking, and the broader field of complex systems science, has a long and rich intellectual history. As an interdisciplinary discipline, systems thinking draws from the work of theorists from a wide variety of fields, including ecology, physics, cybernetics, management science, and more. Originating in Ludwig von Bertalanffy's work on General Systems Theory [25], tools from systems science became more widely applied to the study of ecological systems with the publication of H. T. Odum's 1983 text *Systems Ecology* [26]. These ideas subsequently made a deep impression in the planning world, where thinkers such as C. S. Holling argued for the application of ecological frameworks to the field of urban planning [27]. Complex systems science as a distinct field can be traced back to the pioneering work of thinkers such as Gregory Bateson and Nobel laureate Ilya Prigogine, and the explicit study of complex systems as a field coalesced during the 1970 and 1980s with the formation of the Santa Fe Institute. Building on these approaches, an influential community of researchers has since sought to bring an understanding of the complexity of social-ecological systems to the forefront of scientific literature, such as Holling and L. H. Gunderson in their text Panarchy: Understanding Transformations in Systems of Humans and Nature [28]. Simultaneously, theorists and design researchers such as Richard Buchanan, Chris Reed, Nina-Marie Lister, and Frances Westley have sought to merge systems thinking with design thinking approaches to popularize C. West Churchman's concept of "wicked problems" and influence professional designers address socioecological challenges through their work [29–31]. Easily accessible systems thinking tools have since been developed by scientists such as Donella Meadows, who identified 21 leverage points for intervening in systems [32], and Peter Senge, who elaborated on his concept of systems thinking for organizational learning in his influential text The Fifth Discipline [23].

Systems thinking is an indispensable approach when tackling interdisciplinary green infrastructure projects. Practitioners who are equipped with the tools of systems thinking are more likely to see outside of silos, look across disciplines, and borrow approaches from other fields, with the understanding that a multiplicity of perspectives is better than any single perspective. They are also more likely to pursue multistakeholder and cross-sectoral collaborations, knowing that systems-level problems can only be addressed through systems-level interventions. Most importantly, systems thinkers see past the borders of their own institution or organization to understand the broader social context and work to engage with all stakeholders and communities, rather than just individuals or firms. Many accessible and intuitive systems thinking tools have been developed in recent years, including Systems Thinking for Social Change, the Unschool of Disruptive Design, Symbiosis in Development (SiD), and the Systemic Design Toolkit, among others.

10.3.2.3.1 In-class application—systems thinking with systemic design toolkit

The Systemic Design Toolkit, introduced at the 2016 conference on Relating Systems Thinking and Design, is a set of tools and diagrams to help practitioners map a system and identify leverage points for intervention. By helping changemakers to holistically understand the systems in which they operate, the Toolkit allows actors to better achieve long-term, systems-level transformations. The Toolkit's methodology includes seven concrete steps:

10.3.2.3.2 Frame the system

Create a big-picture map of the system by setting the boundaries of the system in space and time. In addition, identify its component parts and their relationships.

10.3.2.3.3 Listen to the system

Interview actors within the system to better understand their concerns and expectations. Then, map their experiences over time.

10.3.2.3.4 Understand the system

Identify how variables within the system and interactions between actors influence the dynamics of the system. Next, identify leverage points for intervention.

10.3.2.3.5 Define the desired future and explore the possibility space

Help stakeholders within the system articulate a common desired future along social, psychological, economic, and ecological dimensions. Also, develop ideas for how to intervene at key leverage points by analyzing system characteristics such as feedback loops, information and resource flows, rules and limits, and other traits;

10.3.2.3.6 Design the intervention model and foster the transition

Define the engine for change and its variations, while identifying potential paradoxes. In addition, define how the interventions will mature, grow, and be adopted in the system.

10.3.2.4 Integrative thinking

The "integrative thinking" methodology is described by Ref. [9] and developed for use in business schools and applications. Concerned with the typical decision-making practice of choosing between unacceptable options and weighing trade-offs, integrative thinking is designed as a creative approach to problem definition and decision-making that questions assumptions and identifies unexplored alternatives (see checklist: "An Agenda For Action For Taking Steps In The Creation Of The Integrative Approach").

An agenda for action for taking steps in the creation of the integrative approach

The integrative approach outlines a series of steps that need to be taken as follows (check all tasks completed):

_____ 1. Develop opposing models (e.g., two extreme, alternative ideas).

_____ 2. Closely examine these models (using metacognition to critically reflect on personal held beliefs and also challenge common assumptions).

_____ 3. Generate alternative possibilities (searching for innovation or ideal conditions).

_____ 4. Prototype the preferred options (an interactive assessment of the prototype testing is followed to identify the optimal solution).

10.3.2.4.1 In-class application—urban and regional planning elective course

As the continuum of green infrastructure ranges from highly engineered interventions to the conservation of natural heritage systems, teaching students to be prepared for working in this sector necessarily requires skills that transcend disciplinary boundaries. Working closely with Ryerson Urban Water, an interdisciplinary consortium focused on urban water issues, two of the authors played a central role in planning and hosting a 3-day hackathon event for post-secondary students, initiated and funded in part by the City of Toronto (see Fig. 10.8). Students from across Ontario formed teams of about 5–7 students and dedicated themselves to learning about green infrastructure from the work of experts. Various experienced professionals challenged students to devise innovative solutions to complex problems and then to pitch a brief business plan.

Inspired by positive feedback from participants, one of the authors (Sheila Boudreau) developed an elective Urban and Regional Planning course titled "Green Infrastructure: Practice and Praxis." This course provided an overview of the relevant typologies, terminologies, environmental policy tools, and interdisciplinary design and lifecycle asset management techniques necessary for green infrastructure work. The course was based on the following learning objectives:

- To understand what green infrastructure is and to explore the complexities involved in planning, design, and implementation, related to:
  - the interrelationships of ecological, cultural/societal, and organizational systems;
  - Understanding people and place and the importance of storytelling in practice;
  - Respecting Indigenous Peoples, cultural awareness, and working in partnership;
  - Issues of equity and ethics in professional practice;
  - Balancing competing demands and interests, understanding the "triple-bottom line"; and
  - The need for advocacy, outreach and engagement, and strong leadership and support.
- To gain the knowledge and skills required to create and participate in effective planning processes for interdisciplinary, multistakeholder, and siloed environments, such as: inquiry and active listening; integrative thinking and collaboration; recognizing personal/others' abilities and biases; and knowing how to respond appropriately.

FIGURE 10.8 Green Infrastructure Youth Hackathon hosted by Ryerson Urban Water, Toronto Metropolitan University (Credit: Sheila Boudreau).

FIGURE 10.9 Dr. Kate Mulligan, speaking with students about how social prescribing relates to environmental planning and green infrastructure design. *Credit: Sheila Boudreau.*

- To appreciate the importance of creative research involving "reflecting in practice" to advancing the field of green infrastructure.

Visits from practicing professionals, Indigenous Knowledge Keepers, and others were held early in the term for students to begin to consider green infrastructure from a variety of different perspectives (see Fig. 10.9).

In-class projects included team collaboration to plan and undertake a formal interview with a green infrastructure professional or advocate. Termed "Practice Stories," student teams identified interview subjects, created a list of interview questions, and then documented the interview itself. Following this work, each team prepared a high level summary to share what they learned with the class, including reflections about the project and the process itself.

As a second, independent project, each student analyzed the interview and submitted a report written in the first person as a personal reflection on what they heard, and how this contributed to their understanding of green infrastructure and of their roles as future planning professionals. Student feedback on this exercise reflected the positive benefits of listening to others' stories, noting that the exercise promoted an in-depth, empathetic understanding of green infrastructure work and the people involved and impacted by it.

Another assignment focused on a proposed rooftop "Indigenous Healing Garden" in partnership with the Ted Rogers School of Management (TRSM) at Toronto Metropolitan University. Located in the heart of the downtown of Toronto, this proposed space will become a welcoming oasis that celebrates Indigenous cultures, entrepreneurship, and innovation. To generate interest and to begin to build a community around the project, TRSM hosted a series of conversations with Indigenous voices on a range of topics, including Traditional Medicines, urban agriculture, and creating a culturally appropriate biodiverse space. This initiative was the focus of another in-class project, where student teams were formed and introduced to the tool of "discourse analysis." Students listened carefully to these conversations in order to answer the following questions (See [33]):

- What was said;
- How was it said;
- Why it was said, and
- When it was said.

Using the software tool "Miro," an online whiteboard with graphic capabilities, the teams documented their discussion as they addressed each of these questions. The intent was to improve active listening, build empathy, and strive to appreciate all data and meaning relevant to the garden project. Following this exercise, the student teams also explored the university's anticipated "garden community," a network of faculty, students, and researchers, through the direct application of the Systemic Design Toolkit. Using the Miro software, each team went step by step through the Toolkit, recorded their ideas and questions, and summarized them in a graphic format.

Another in-class project involved learning about the integrative thinking methodology and working in small teams to discuss a specific green infrastructure problem. Using this method, teams developed alternative possibilities and then returned to the larger class to share what they had developed collectively. The final class discussion included a reflection on the need to be honest about "how we know what we know" and to remain open to new ways of thinking when working with complex issues, diverse communities, and in multistakeholder situations.

10.3.2.5 Ecological Design Lab

CoLaboratory praxis is an approach that involves transdisciplinary workshops for integrated planning and design of green infrastructure. Since 2010, the Ecological Design Lab (EDL) has been engaged in the cocreation of place-based solutions that (re)connect landscapes, people, and wildlife. Through design-based research and learning through collaborative praxis with-and-in communities, EDL has contributed to green infrastructure designs that physically and functionally facilitate the safe passage of people and wildlife across roads.

Why "Safe Passage"? Roads fragment landscapes, resulting in barriers to the safe movement of humans and animals. Canada's growing urban regions and expanding road networks are associated with increasing wildlife-vehicle collisions, higher wildlife mortality, and in the long term, landscape fragmentation that results in habitat degradation and the global decline in biodiversity. Where the construction of wildlife crossing infrastructure has emerged, it has been slow, sparse, and ad hoc. Yet resolving these issues is not solely a technical or research challenge; it is about working collaboratively and creatively across disciplines and building political, economic, and cultural consensus within communities of practice.

Reconnecting landscapes (between people, wildlife, and their habitats) is an interdisciplinary socioecological planning and design challenge that is not currently in the mandate of any single agency or even acknowledged within conventional systems of governance. Although the infrastructural design solutions to reducing fragmentation are known to work, there is an urgent and growing need for a coordinated, integrated approach to planning and design for widespread sustainable implementation of these infrastructures. This challenge cuts across scales and jurisdictions and requires a collaborative, transdisciplinary approach that connects researchers and practitioners to develop integrated, place-based solutions [34]. By engaging with these issues in an interdisciplinary and experiential learning format that links landscape design and road ecology with evidence-based policy and environmental planning, the EDL's CoLaboratory has created new opportunities for advanced design-based research and civic engagement.

CoLaboratories (or CoLabs) are a community–expert collaborative working forum for hands-on design and planning to solve a systemic problem that is complex, multisectoral, and interdisciplinary (see Figs. 10.10–10.13).

The CoLab praxis is grounded in theory, supported by research, and refined in practice in place and in community. However, the CoLab differs specifically in its emphasis on sequential iteration, shared learning, transdisciplinary knowledge generation, and the cocreation of outcomes. The CoLab is therefore not a self-contained activity but is grounded in a longer series of reflections, and the design-workshop or charrette may function as the culmination of a longer process of discussions and negotiations, from which a preferred physical form or structure emerges through consensus and coproduction.

Underlying the CoLab as method is design thinking, which supports an inter- and transdisciplinary mode of knowledge production and which can illuminate process development, making design techniques accessible to nondesigners. In this context, landscape-based

FIGURE 10.10 Montana CoLab. *Source: Ecological Design Lab.*

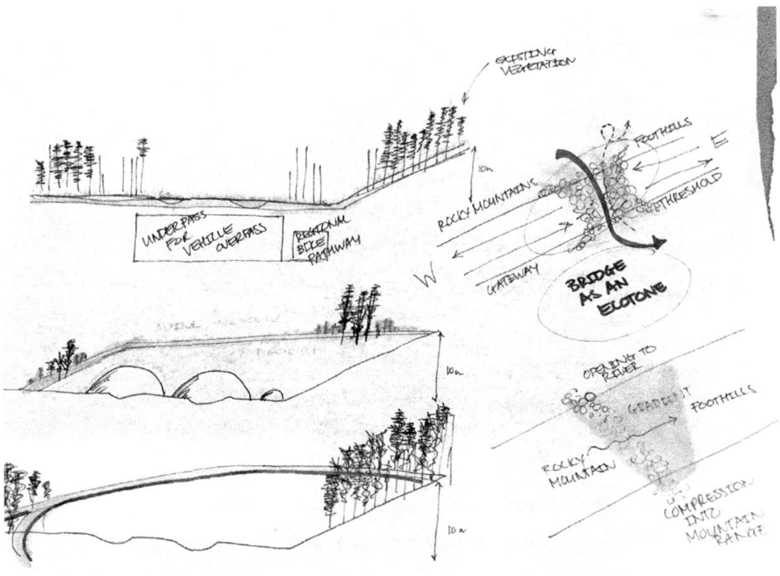

FIGURE 10.11 Calgary CoLab. *Source: Ecological Design Lab.*

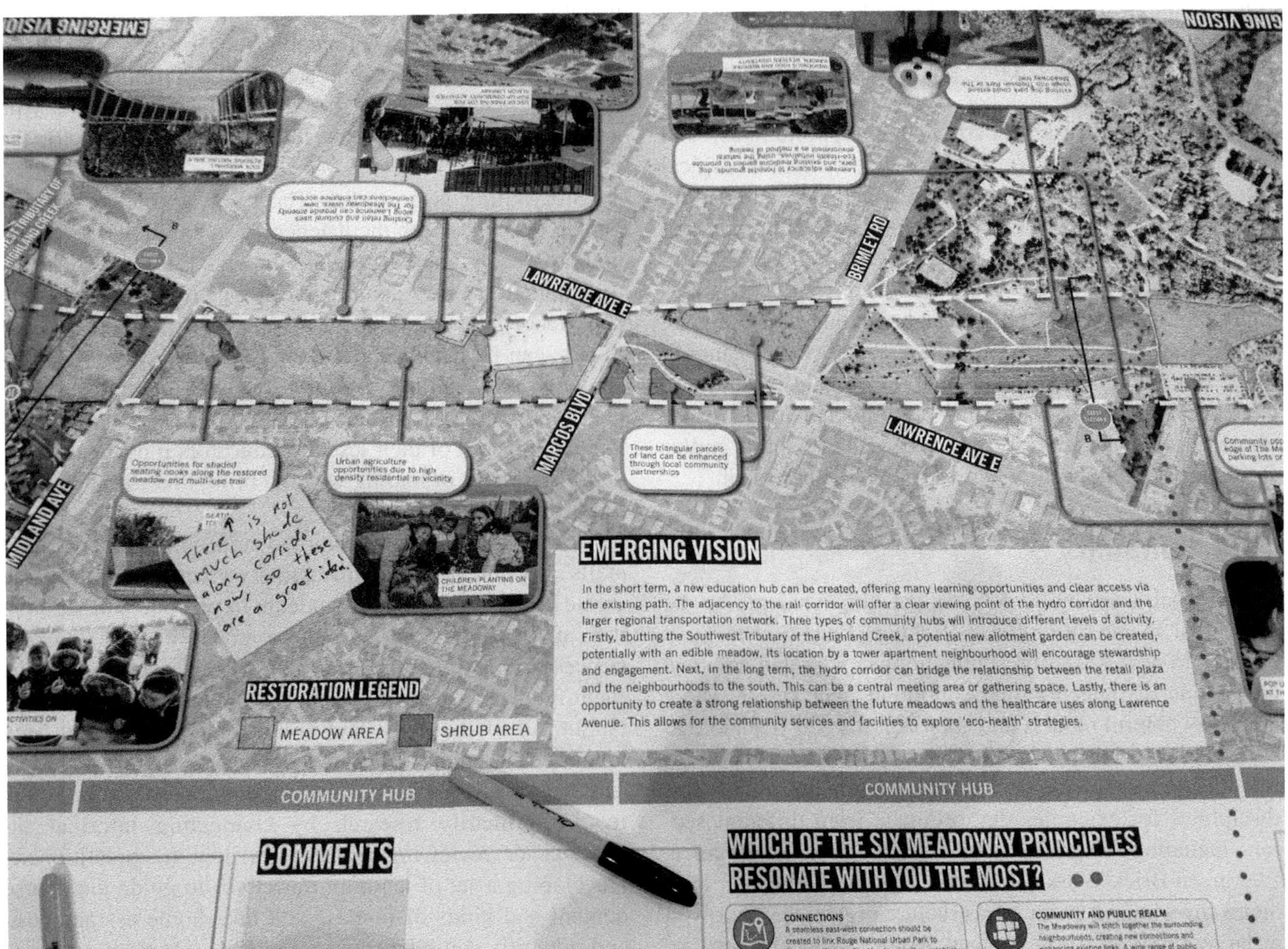

FIGURE 10.12 Meadoway CoLab. *Source: Ecological Design Lab.*

FIGURE 10.13 Liberty canyon CoLab. *Source: Ecological Design Lab.*

transdisciplinary challenges (such as how to move people and wildlife safely across roads) can successfully facilitate collaborative knowledge production across academic disciplines and with nonacademic practitioners, local actors, community members, and governing bodies.

The CoLaboratory puts planning, engineering, and design professionals together with community leaders to share knowledge, assess barriers and opportunities, and cocreate a resolution to this challenge. Methodological components of the CoLaboratories include: performing preliminary site and policy context research; compiling and analyzing technical data; mapping and visualization; cataloging existing and case-precedent decision-making strategies; analyzing policy options and planning tools for each site; soft-systems mapping of policy contexts and planning tools (including overlaps, boundaries, and constraints); developing preliminary design solutions and implementation strategies; ideation, programming, and exhibit and/or communications development for public engagement.

10.4 Summary

Experiential learning is key to teaching students to be successful in interdisciplinary practice. Practitioners must be open to new ways of teaching and learning and embrace an attitude of reflection and continuous adaptation. There are many helpful models and approaches freely available to educators that wish to teach students how to work in an interdisciplinary way. For example, looking outside of their disciplines for examples that originate from different sectors, including design, business, technology, or governance. Evergreen Brick Works, Toronto, Ontario, is an example of an award-winning green infrastructure project that involved a large team of professionals from a wide variety of disciplines, including those of one of the coauthors (see Fig. 10.14). Educating themselves about alternative approaches in collaborating with others in complex decision-making environments, and being open to experimenting with it as small projects integrated into the course curriculum. Working with colleagues from across disciplines and institutions to create new opportunities for student experiences. Asking for student feedback and if it is generally positive, adjusting the project as needed for the next iteration.

Storytelling is powerful. The truth as told in the first person resonates with the listener on an emotional level and is more deeply appreciated and understood. Inviting guests to tell their stories about their work, or how they have experienced the conditions that were created by the work of others, can be deeply moving and not easily forgotten. As students are learning from within the confines of the classroom, and also of the accreditation boards that govern the core content of their lessons, educators must be creative and resourceful in finding ways to bring diverse voices and ideas into the class.

"Soft skills" (human relations) are as important as "hard skills" (technical) in interdisciplinary practice. Find ways to create curricula that allow students to practice soft skills in the safety of the classroom environment, and create experiences and assignments that help to improve these skills. Identify the holistic requirements of the discipline being taught, all of the skills and competencies that have been identified as being critical by those working in this area. This information may not be found in one place and may require primarily research by conducting informal interviews or reviewing materials written by practitioners. Developing a set of learning objectives to guide the course content and kinds of in-class and take home assignments, and identify complementary optional extracurricular opportunities. Encouraging students to reflect on these

FIGURE 10.14 Evergreen Brick Works project, a global precedent in green infrastructure planning. *Credit: Toronto and Region Conservation Authority.*

experiences and writings and to discuss their thoughts and lessons learned with others in the context of the class for further discussion and shared learning.

10.4.1 Class materials: questions for class discussion

As part of the class materials, please see the following questions for discussion:

- What social science theories can help inform interdisciplinary practice, using a real-world example?
- What are potential scenarios that would help to demonstrate the social science theories described in this chapter?
- What skills and competencies that are considered critical to interdisciplinary work would be expected to play a major role in the case studies described in this chapter or in another real-world example?
- Review a case example that is well documented, and ask yourself: What were the disciplines that participated in the project?;
- What segments of the public were involved as the project evolved, and whose voices may still not have been heard?

10.4.1.1 Experiential learning activities: role playing

Have students work in teams to research and select a smart city initiative. Ask them to create a spreadsheet to chart those who were involved in the planning and design of the smart city project: government, property owners, consultants, organizations, community groups, etc. Students can then role play by selecting one of these agents, using their imagination to put themselves into this role, and creating a detailed persona for the person that they will be assuming. Working with their teammates, students are to select and discuss an issue that the team collectively decides is important to focus on. The task is to consider what challenges would be expected to arise in this working environment and to identify social science theories and methodologies that could help to improve creative problem-solving and decision-making.

10.4.1.2 Storytelling

Invite guest speakers to share their perspectives and experiences related to the course topic. Create assignments for students to encourage humility, self-reflection, and understanding, such as planning and conducting informal interviews with subject matter experts or users. Focus on improving skills of active listening and building a sense of empathy.

10.4.1.3 Coproducing

Create a project related to the course topic that requires students to collaborate to strategize, plan, and develop an implementation framework to realize their recommended solution to a complex "smart city" problem that is posed. Encourage students to use software programs and other tools that support multiuser brainstorming and project planning in real time, including those that can be borrowed from other areas of study (e.g., business programs, anthropology, etc.). Have student teams present their ideas and areas requiring additional exploration to the larger class, and describe the process that they used, for open discussion and to seek critical but helpful advice.

Finally, let us move on to the real interactive part of this chapter: review questions/exercises, hands-on projects, case projects, and optional team case project. The answers and/or solutions by chapter can be found in Appendix G.

10.5 Chapter review questions/exercises

True/false

1. True or False? To design a "smart city," a broad range of green infrastructure techniques is required that are designed to complement natural cycles, minimize environmental impacts, and optimize ecosystem services and associated social benefits.
2. True or False? In Toronto, Ontario, smart city discussions are relatively old and have focused primarily on controversies regarding the sharing and aggregating of personal digital data by private interests and the need for adequate protections and governance measures to protect the public from harm and loss.
3. True or False? Green infrastructure is a noncomplex field that works across disciplines, organizations, as well as political and geographic divides.
4. True or False? Smart city building requires the involvement of numerous professions, which are regulated according to government legislation and regulations developed by their respective associations.
5. True or False? The social science theory is not very helpful in considering interdisciplinary practice.

Multiple choice

1. Teaching at a postsecondary institution in Canada mandates deeply reflecting on what it means to be an educator, following over 500 years of violence and oppression of Indigenous Peoples due to:
   a. Racism
   b. Colonization
   c. Slavery
   d. Regulations
   e. Policies
2. The concept of ________________ was developed by the software industry to improve human–computer interfaces.
   a. Master of Planning in Urban Development
   b. Person X
   c. User Experience Design (UX Design)
   d. Transformations in Systems of Humans and Nature
   e. Interaction Design Foundation (IDF)
3. Systems-level problems require us to see the inherent interconnections between the processes and actors that make up a particular system and understand how they are mutually interdependent in ways that transcend any:
   a. Domain name
   b. Framework
   c. Diagram
   d. Resources
   e. Single domain
4. By helping changemakers to holistically understand the systems in which they operate, the Toolkit allows actors to better achieve long-term, systems-level:
   a. Transformations
   b. Toolkits
   c. Stakeholders
   d. Interventions
   e. Paradoxes
5. The ________ methodology is used in business schools and applications.
   a. Integrative approach
   b. Green infrastructure
   c. Ecological
   d. Integrative thinking
   e. Cultural/societal

Exercise

Problem

What are some lessons learned from the application of a program that seeks to educate and train students in low-impact development (LID) alternatives to traditional stormwater management through hands-on outdoor classroom activities?

Hands-on Projects

Project

Do research: Analyze the qualitative approach to explore the motivations and instructional design processes of interdisciplinary green infrastructure Contingency Planning Exchange (CPE) trainings.

Case projects

Problem

Do research: Look at smart city sustainability and sustainable design of the water infrastructure–urban hydrology.

Optional team case project

Problem

Discuss why solutions to smart city sustainability problems often require professionals with different training and backgrounds to work as a team.

References

[1] Vincent D. Sidewalk Labs' urban data trust is 'problematic,' says Ontario privacy commissioner. Toronto Star. 2019. Thurs. Sept. 26, 2019.

[2] Brown H, Stigge B. Infrastructural ecologies: alternative development models for emerging economies. Boston: MIT Press; 2017.

[3] Brown H. Next generation infrastructure: principles for post-industrial public works. Washington DC: Island Press; 2014.

[4] Lister N-M. Insurgent ecologies: (Re) claiming ground in landscape and urbanism. In: Doherty G, Mostafavi M, editors. Ecological urbanism. Basel, Switzerland: Lars Müller Publishers; 2010. p. 535–47.

[5] Lister N-M. Resilience and design: post-urban landscape infrastructure for the Anthropocene. The post-urban world: emergent transformation of cities and regions in the innovative global economy. NY: Routledge; 2018.

[6] Ministry of Municipal Affairs, Housing. Provincial policy statement for Ontario. Toronto: Queen's Printer for Ontario); 2020. p. 2020.

[7] Johnson PA, Robinson PJ, Philpot S. Type, tweet, tap, and pass: how smart city technology is creating a transactional citizen. Govern Inf Q 2020;37(1):101414. https://doi.org/10.1016/j.giq.2019.101414.

[8] Polk M, Kain J-H. Co-producing knowledges for sustainable urban futures. In: Co-producing Knowledge for sustainable cities. Merritt polk. NY: Routledge; 2015.

[9] Riel J, Martin RL. Creating great choices: a leader's guide to integrative thinking. Boston: Harvard Business Review Press; 2017.

[10] International Union for Conservation of Nature (IUCN). Global standard for nature-based solutions. A user-friendly framework for the verification, design and scaling up of NbS. 1st ed. Gland, Switzerland: IUCN; 2020.

[11] Mulligan B, Rayner, Hsiung. Social prescribing: creating pathways towards better health and wellness. J Am Geriatr Soc 2019;68(2).

[12] Beresford-Kroeger D. The global forest: 40 ways trees can save us. New York: Penguin Group; 2010.

[13] Parkinson G, Drislane R. Exploring society: pathways in sociology. ON: Nelson: Scarborough; 2003.

[14] Haidt J. The righteous mind: why good people are divided by politics and religion. New York: Vintage Books; 2012.

[15] Ury W, Fisher R, Patton B. Harvard Negotiation Project, Harvard Negotiation Project Archives – PON – Program on Negotiation at Harvard Law School. Cambridge: Massachusetts, USA; 1979. https://www.pon.harvard.edu/category/research_projects/harvard-negotiation-project/.

[16] Nonaka I. A dynamic theory of organizational knowledge creation. Organ Sci 1994;5(1):14–37.

[17] Truth and Reconciliation Commission of Canada, United Nations. National centre for truth and reconciliation. In: Truth and reconciliation commission of Canada honouring the truth, reconciling for the future: summary of the final report of the truth and reconciliation commission of Canada; 2015.

[18] Truth and Reconciliation Commission of Canada, United Nations. National centre for truth and reconciliation. In: Truth & reconciliation: calls to action; 2015.

[19] Pine L. Indigenous Toronto: stories that carry this place. Spacing Magazine 2021:77–8. May 3, 2021.

[20] Hartson HR, Pyla PS. The UX book: process and guidelines for ensuring a quality user experience. Morgan Kaufmann; 2018.

[21] Interaction Design Foundation. Personas. 2021. https://www.interaction-design.org/literature/topics/personas.

[22] Churchman CW. Wicked problems. Manag Sci December 1967;14(4). https://doi.org/10.1287/mnsc.14.4.B141. B-141–B-146.

[23] Senge P. The Fifth discipline, the Art and practice of the learning organization. New York, NY: Doubleday/Currency; 1990.

[24] Meadows D. Thinking in systems: a primer. White River Junction, VT: Chelsea Green Press; 2008.

[25] Bertalanffy L. General systems theory: foundations, development, applications. New York: George Braziller; 1968.

[26] Odum HT. Systems ecology: an introduction. New York: Wiley; 1983.

[27] Holling CS, Goldberg MA. Ecology and planning. J Am Inst Plan 1971;37(4):221–30. https://doi.org/10.1080/01944367108977962.

[28] Gunderson LH, Holling CS, editors. Panarchy: understanding transformations in systems of humans and nature. Washington, DC: Island Press; 2002.

[29] Buchanan R. Wicked problems in design thinking. Des Issues 1992;8(No. 2):5–21. https://doi.org/10.2307/1511637 (Spring, 1992).

[30] Reed C, Lister N-M. Ecological thinking, design practices. In: Reed C, Lister N-M, editors. Projective ecologies. Cambridge MA: Harvard University GSD and; 2014. p. 14–21. New York: ACTAR.

[31] Westley F, McGowan K. Design thinking, wicked problems, messy plans. In: Reed C, Lister N-M, editors. Projective ecologies. Cambridge MA: Harvard University GSD and New York: ACTAR; 2014. p. 290–311.

[32] Meadows D. Leverage points: places to intervene in a system. Hartland, VT: Sustainability Institute; 1999.

[33] Riley S, Wiggins S. Video tutorial on discourse analysis. Wales): Aberystwyth University; 2016.

[34] Lister N-M, Brocki M, Ament R. Integrated adaptive design for wildlife movement under climate change. Front Ecol Environ 2015;13(9):493–502. https://doi.org/10.1890/150080.

Further reading

Fey S, Bregendahl C, Flora C. The measurement of community capitals through research. Online J Rural Res Policy 2006;1(1):1–28.

Kaplan S, Kaplan R. Creating a larger role for environmental psychology: the reasonable person model as an integrative framework. J Environ Psychol 2009;29:329–39.

Maslow A. A theory of human motivation. Psychol Rev 1943;50(4):370–96.

Merriam SB, Tisdell EJ. Qualitative research: a guide to design and implementation. 4th ed. San Francisco: Jossey-Bass; 2016.

Ostrom E. Governing the commons: the evolution of institutions for collective action. New York: Cambridge University Press; 1990.

Chapter 11

Teaching sustainability in metro Orlando: the evolution of the pragmatic liberal arts at Rollins college

R. Bruce Stephenson
Environmental Studies, Rollins College, Winter Park, FL, United States

11.1 Introduction

Sustainability is the definitive ethic of our time. Its measure of humanity forces a moral response to the specter of climate change, species extinction, and resource depletion [1]. Sustainability, as put forth in the United Nations' landmark report, *Our Common Future*, imposes "limits to growth" to ensure development "meets the needs of the present without compromising the ability of future generations to meet their own needs." [2] With transportation emissions exceeding power plants as the United States' largest source of greenhouse gases, it is impossible to secure sustainability without addressing automobile dependence [3].

For half a century, Orlando has been one of fastest growing and most car-dependent metropolitan areas in the United States. In return, it suffers the nation's highest death rate for pedestrians and interstate motorists—determining factors the city being named the nation's most "sprawl threatened city" and "angriest city." [4] When factoring automobile dependence as a determinant of economic equity, Orlando ranked last among the nation's 30 most populous metropolitan areas [5]. Finally, this metric led the Sustainable Development Solutions Network to flag Metro Orlando for failing to meet the United Nations Sustainability Goal 11, Sustainable Communities [6]. Since the early 1990s, Rollins College students have prepared scores of projects to advance the tenets of sustainability. Active transportation (walking, biking, and transit) has been a particular emphasis, along with green infrastructure, ecological restoration, and walkable urbanism (as opposed to drivable suburbanism, a compact mixed-use development pattern makes the car an option) [7]. In 2014, sustainability became a watchword after the Orlando city council adopted a *Greenworks Action Plan* to make Orlando "the green capitol of the Southeast." [8] Rollins faculty partnered with the city's Sustainability and Resilience Office in directing student projects designed to advance the plan. The college's strategic planning took a similar path, and in 2021, Rollins is integrating sustainability into the college curriculum, an initiative inspired by the legacy of its iconoclastic president, Hamilton Holt.

11.2 Common sense education and an ideal campus: Hamilton Holt's legacy

Established in 1885, Rollins College raised civic awareness and cultural aspirations in the small town of Winter Park, Florida, located on the frontier's edge. The stripling liberal arts institution struggled, however, until the 1920s when the advent of middle-class tourism transformed a backwater state into the "Eden of the South," Lewis Mumford wrote, "the desire of the heart and the end of human aspirations." [9] In the midst of the greatest real estate boom in American history, acclaimed journalist Hamilton Holt accepted the presidency of Rollins, "because," he noted, "Florida is the one state where the spirit of progress most prevails and where results follow quickest from effort." [10].

Smart Cities Policies and Financing. https://doi.org/10.1016/B978-0-12-819130-9.00056-5

A consummate reformer, the longtime editor of *The Independent*, broke the staid academic formula by marrying pragmatism and the liberal arts. Leading the faculty on "an adventure in common sense education," he "put less emphasis upon information for its own sake, and more upon linking the subject studied with the life of the individual studying it." [11] After reforming the curriculum, in 1931, Holt named John Dewey to chair the Rollins College Colloquy on the Liberal Arts, which drew a distinguished assemblage to Winter Park. Dewey called on educators to extend the classroom and embrace "a citizenship of one earth." Analyzing the workings of nature and society from "outside the school" instilled the lesson that "all studies arise from aspects of the one earth and the one life lived upon it." [11].

Under Hamilton Holt, Rollins made its name merging pragmatism and the liberal arts. Poetry celebrated the Florida landscape, artifacts of Native American culture were cataloged and analyzed, and unique natural features were assessed. The college also hosted luminaries such as Sinclair Lewis, the first American to win the Noble Prize for literature, and presidents Herbert Hoover, Franklin Delano Roosevelt, and Harry Truman. The town and gown relationship prospered as Winter Park became a welcome partner in broadening the arts, cultural awareness, and eventually social reform.

In 1949, shortly before he retired, Hamilton Holt performed an act of humanism long in coming. An original member of the National Association for the Advancement of Colored People, he had wanted to honor Mary McCleod Bethune, the founder of Bethune Cookman College, at the 1947 graduation commencement, but the board of trustees thwarted him. Two years later, the trustees remained unmoved, but Holt forged ahead. Realizing it would be his last significant act as Rollins President, he bestowed Bethune with an honorary degree. *The Chicago Defender* called the act, "a milestone in human relations." It was the first time an African American had received such recognition from a white southern college [11].

Holt's reforming verve also extended to the physical campus. In fact, upon arriving in Winter Park, his first priority was to build an "ideal campus." [12] His aspiration was aided by the college's lake front setting in one of Florida's first planned suburbs. In 1881, Chicago businessmen Loring Chase and Oliver Chapman purchased 600 acres along a rail line to build a town modeled on Riverside, Illinois, the iconic suburb designed by Frederick Law Olmsted [13]. Their 1883 plan was a modified neoclassical grid with radiating concentric circles denoting 5-min walks from a train station in a central park. The town developed at a pedestrian scale and in a context-sensitive manner that employed a variety of building types, heights, and design approaches. Civic institutions occupied key locations, and like Riverside, Winter Park's compact downtown was aligned on a principal street, Park Avenue, that paralleled a central park [14].

Lying between the downtown's southern boundary and Lake Virginia, Rollins College imbued Winter Park with a priceless aesthetic. Holt was determined to build on this foundation and contacted prominent city planner John Nolen, to review his campus plan. Nolen had commissions from Davidson College, the University of North Carolina, Queens College, and the University of Wisconsin, but it was his pathbreaking work in designing the neo-Renaissance new town of Venice, Florida, that excited Holt. Holt was also inspired by Nolen confidant, George Merrick.

A former Rollins student and the developer of Coral Gables, Merrick drew on Winter Park in his plan for the nascent Miami suburb. Coral Gables drew national attention for its unmatched assemblage of Mediterranean Revival architecture. Open courtyards, tiled roofs, stucco exteriors, high ceilings, and arched windows and doorways combined practicality and aesthetics to meet the challenge of Florida's hot, humid climate. Holt spent a year examining Mediterranean Revival projects in Miami and St. Petersburg before hiring Richard Kiehnel, an Ecole de Beaux Arts trained architect who had designed noteworthy Mediterranean Revival buildings in both cities [15].

Venice and Coral Gables informed Kiehnel's campus plan that grouped 29 structures to form a neo-Renaissance academic village. It had an axial orientation with the theater and chapel set on a central square, but the plan's focal point was the library. Envisioned as a commanding 29-story structure, it was the terminus for the central axis that bisected a linear green and a common lawn. Kiehnel played to the lessons of the Italian Renaissance, which had emphasized a new more engaged urban culture where citizens spent much of their free time in public plazas and markets. Renaissance architect Leon Battista Alberti believed these public spaces helped divert young men and women from "the mischievousness and folly natural to their age," and the Rollins campus is a testament to this tradition [16].

Loggias, a staple of Florentine architecture, were paramount. Covered walkways linked the campus, while classrooms and dormitories were set on quadrangles and small courtyards to capture breezes and foster air circulation. "Breezy and cool" was how Holt described the scheme to create the first "open-air college" in the United States [15]. The library was never built to the plan's monumental dimensions, but Mills Library, a human-scaled three-story structure, did center the campus. Finally, intimate greens and small squares offered places of repose to activate thought and offer an escape from the swirl of campus life. A model of walkable urbanism, the campus was built to a pedestrian scale that respected the human form and celebrated intellectual inquiry.

When Holt retired in 1949, Rollins had taken shape on the lines he envisioned. The interplay between Winter Park and the college was significant. As one writer noted, "the town became a college." [17] But, Kiehnel's plan had a glaring omission—there were no parking lots. In 1927, the automobile was an option, but once it became a necessity, the campus's pedestrian orientation suffered. By the 1970s, the widening and realignment of Fairbanks Avenue severed the central axis running in conjunction with Interlachen Avenue, and parking lots soon covered sites designated for open space. This commitment to the automobile reflected a monomaniacal exercise in road building engineered by the Florida Department of Transportation.

After Walt Disney announced his plans for Disney World, the construction of a series of wide high-speed roads fueled a repetitive suburban pattern across metropolitan Orlando. This dogmatic approach peaked in 1998, when the Orange County Commission rejected federal dollars to build a light rail system (Charlotte garnered the funds and its $470 million investment in rail spurred a $1.8 billion return in walkable urbanism in its first 3 years). At the same time, the Orange County Expressway Authority was in the midst of constructing 153 miles of toll roads, which translates into the highest per capita concentration of toll roads in the nation. This investment gave the metropolitan area the shape of a disjointed mutant human body, where the intersections of the maze of toll roads with Interstate 4 (I-4) resemble the tentacles of a freakish octopus attached to the minuscule torso of the downtown [18].

Metro Orlando's fragmented mass of drivable suburbanism is the antithesis of sustainability. "The real problem with cars is not that they don't get enough miles to the gallon," David Owen, author of *Green Metropolis* writes. "It's that they make it too easy for people to spread out, encouraging forms of development that are inherently wasteful and damaging" [19]. Cities dominated by drivable suburbanism intensify the consumption of land, energy, and water, the production of greenhouse gases and smog-related emissions, and the flow of stormwater pollution (due to extensive paving for roads and parking). In addition, the economic outlay is staggering. Utility extensions are fiscal sinks, highway construction is cost-prohibitive without charging tolls, and funds to maintain bridges and overpasses are increasingly scarce. The indirect costs are wasteful, deadly, and cold-hearted: more driving and pedestrian deaths, higher indices of road rage and obesity, less walking and biking, and the impairment of the human-scaled public realm. In metropolitan Orlando, an eviscerated public realm was indifferent to human life as per capita pedestrian, motorist, and bicycle deaths rank among the highest in the nation [20].

Efforts to mitigate the problems of drivable suburbanism spawned and began to take hold in the early 1990s. Florida's counties and municipalities had adopted comprehensive growth management plans mandated by the 1985 Comprehensive Growth Management Act, and investments in sustainability proceeded on two fronts: building bike trails and acquiring natural lands. The passage of the 1992 Inter-Surface Transportation Efficiency Act allocated federal funds for bike trails, while Florida's Preservation 2000 program provided $300 million a year for the acquisition of natural lands [21]. Beginning in 1992, seniors in the Rollins College Environmental Studies program (ES) partnered with local government agencies and boards to support active transportation and the acquisition of natural lands. By decades end, they were preparing town plans set to walkable urbanism and informed by the Rollins campus and Winter Park's historic downtown.

These courses were predicated on project-based learning, the living embodiment of John Dewey's "learning by doing" approach to education. Shifting the emphasis from teacher-centered instruction to student-centered projects that are presented to a community partner creates a dynamic classroom focused on problem-solving and producing a quality product. Establishing connections to experts and local governing boards, students gain pragmatic and critical skills. In regard to sustainability, they must apply vision, idealism, and knowledge to steward resources and advance the well-being of citizens. Most important, when they present their work to experts versed in sustainability, they gain insight that extends beyond the academic community.

The presentation requires students to own their project. They must be knowledgeable and prepared for questions that require them to think on their feet. This real-world skill is a definitive achievement in an arduous process, and it yields a priceless commodity—intellectual self-confidence. In project-based learning, faculty assume the role of coach, facilitator, and colearner. To make this dynamic work, they spend more upfront time:

1. Assessing the scope of potential projects.
2. Plotting the logistics of meshing theory and analysis.
3. Determining the makeup of the deliverable product.
4. Securing a community partner.
5. Developing a strategy to procure quality presentations [22].

11.3 Bike trails, nature preserves, and green communities

In 1992, project-based learning centered on a capstone ES course where seniors prepared a comprehensive assessment of an abandoned rail corridor designated to accommodate metropolitan Orlando's first bike trail. Connecting Winter Park and Orlando, the Cady Way Trail drew the ire of adjacent property owners who feared it would attract

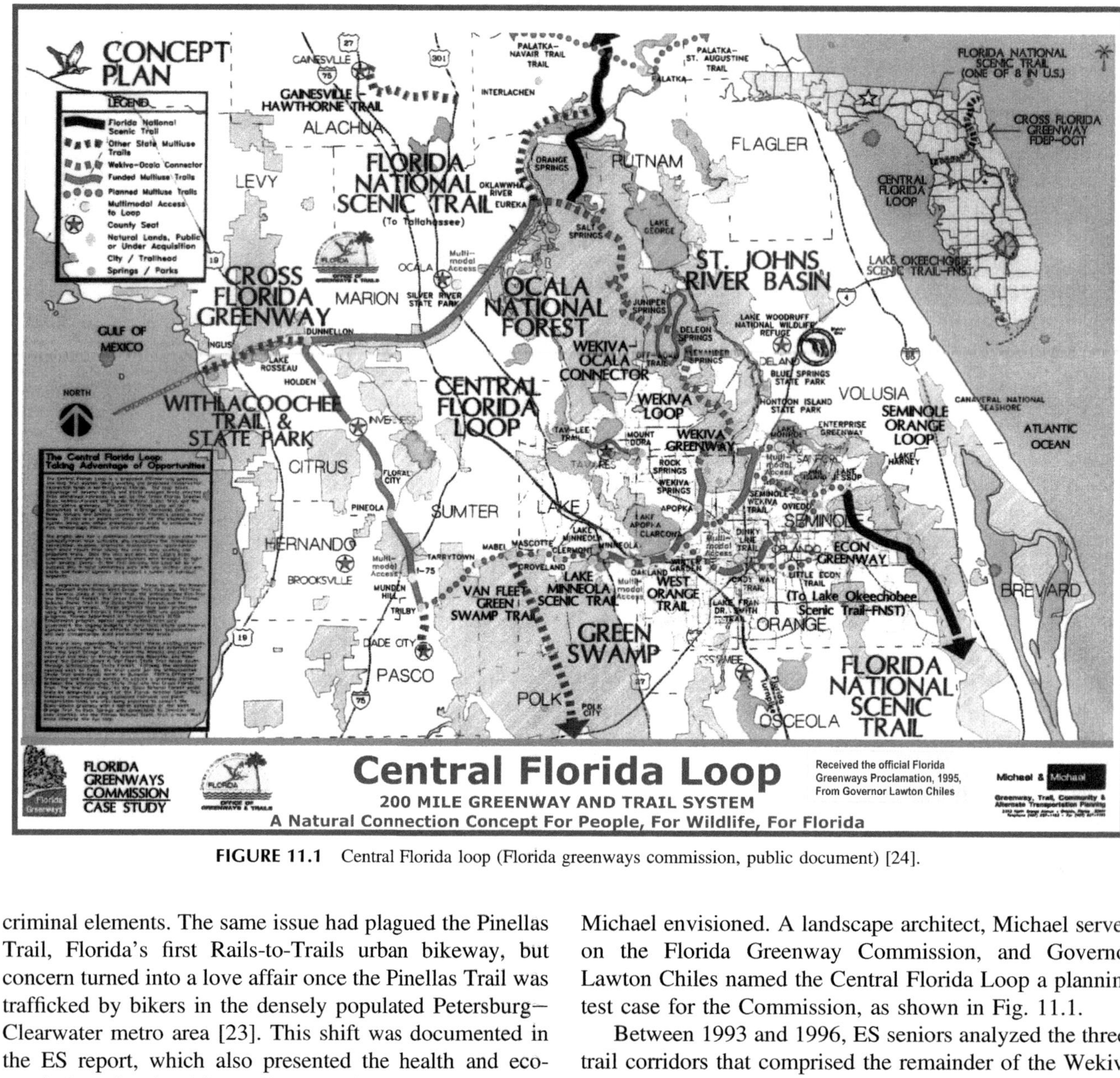

FIGURE 11.1 Central Florida loop (Florida greenways commission, public document) [24].

criminal elements. The same issue had plagued the Pinellas Trail, Florida's first Rails-to-Trails urban bikeway, but concern turned into a love affair once the Pinellas Trail was trafficked by bikers in the densely populated Petersburg–Clearwater metro area [23]. This shift was documented in the ES report, which also presented the health and economic benefits that bike trails engender, potential connections to parks and schools, and a native planting regimen (and costs) for the corridor. Students presented a summary of the 50-page study to the Orlando Parks Board and the project's planners. After the Cady Way Trail opened, opposition soon gave way and plans took form to create a trail network.

After the student presentation, Rollins faculty member Bruce Stephenson and Orlando Parks Board member Forest Michael partnered to plan the Wekiva Loop—a network of bike trails that would connect the Cady Way Trail to the Wekiva GeoPark—which encompasses a vital ecosystem centered on Wekiva Springs. Inspired by Portland's 40 Mile Loop, the Wekiva Loop was part of a larger Central Florida Loop—a 200-mile system of trails and greenways Michael envisioned. A landscape architect, Michael served on the Florida Greenway Commission, and Governor Lawton Chiles named the Central Florida Loop a planning test case for the Commission, as shown in Fig. 11.1.

Between 1993 and 1996, ES seniors analyzed the three-trail corridors that comprised the remainder of the Wekiva Loop: the Cross Seminole Trail, the Seminole Wekiva Trail, and a portion of the West Orange Trail. Their research extended the scope and detail of plans drawn by the county staffs and their consultants. Today, the Wekiva-Loop is long established and ties into the Cross Florida Trail connecting Titusville (on Florida's east coast) to the Pinellas Trail.

In 1997, ES seniors shifted their focus to prepare a conceptual management plan for High Oaks, a 240-acre priority acquisition for the Seminole County Natural Lands Program. In 1990, Seminole County, a mostly suburban enclave of 350,000, allocated \$20 million for the purchase of natural lands [25]. An advisory committee of field botanists ranked sites based on their corridor potential, prior land use, rarity of habitats, and vulnerability to development. Working with the state on a joint acquisition program, the county commission purchased over 4000 acres of mostly pristine river and lake front property by 1995. Of the 50-plus sites considered by the advisory committee, High Oaks received the highest ranking. On the county's rural edge, it was bounded by the Econlockhatchee (Econ) River, a designated outstanding Florida waterway that receives special protection. The property also had a definitive mix of xeric, mesic, and hydric habitats.

After analyzing the acquisition criteria that guided the advisory committee, students gained insight into essential ecological concepts such as corridors, biodiversity, and species extinction (see checklist: "An Agenda For Action For Taking Steps Establish Five Goals For The Project").

An agenda for action for taking steps to establish five goals for the project

Through a nominal group exercise, five goals were established for the project as follows (check all tasks completed):

_____ 1. Maintain the site's ecological processes.
_____ 2. Restore habitats to their "natural" state.
_____ 3. Protect endangered species.
_____ 4. Promote environmental education.
_____ 5. Promote outdoor experiences with minimal environmental impact.

Mapping, ground-truthing, and defining eight habitats keyed the plan. This work informed a proposed regimen of prescribed burning deemed crucial to maintaining the diversity and function of habitats, restoring degraded habitats, and ensuring the survival of the Gopher Tortoise, the site's keystone species. A series of transect analyses determined the locations and quality of Gopher Tortoise burrows, which are essential for a range of species to survive seasonal fires. This research also informed a fire regimen plan. For instance, an area with a mixture of mesic and scrubby flatwoods with a considerable oak scrub layer and high density of saw palmettos had a high number of abandoned burrows, which made managed fire a priority for this area.

A trail system was delineated to enhance education and highlight the site's premier natural features: pristine sandhill habitat, a grove of xeric Live Oaks, and a span of a large Bayhead swamp accessed by a boardwalk. The highlight was a sandy bluff overlooking the Econ River, where one of two benches designated for the property was sited. Finally, sampling tests of physical, chemical, and biological factors determined that the riparian ecosystem was healthy [26].

A quarter of a century later, the High Oaks property is called the Econ Wilderness Area—an important ecological steppingstone set within a growing matrix of suburban subdivisions. A fire regimen promotes habitat health, and there is a significant Gopher Tortoise population. Glimpses of bobcats, Bald Eagles, and Ospreys offer a hint of wildness for visitors [27]. Gabby Buendia, a 2019 ES graduate and college valedictorian, revealed the site's unique character in her essay, "The River that Raised Me," in *The Wilder Heart of Florida: More Writers Inspired by Florida Nature*.

The Econ Wilderness Area was Buendia's first nature experience, a short hike at the age of 16 stirred her senses and led to regular exploratory visits. Five years later, she set out on a final hike before leaving Central Florida. Departing from the sandy trailhead, she made her way to the bluff overlooking the Econ River that she had never experienced it in the summer, when the river runs high. The site was familiar yet different. "It was absolutely full—its banks felt as if they were swelling with freshwater and a new flow," she writes. The moment was accompanied by "a sudden flash of gratitude," she writes:

> *From my first hike ever to this ceremonial parting hike, the river helped me rise into adulthood and showed me the values of patience, exploration, and flexibility. It taught me that there is a wide diversity of benefits in getting to know the outdoors, and that everyone can find a sense of place and belonging in our natural landscapes [28].*

The idea of integrating nature into daily life infused the plan for the Winter Springs Town Center, as shown in Fig. 11.2. The ES analysis of the Seminole-Wekvia bike trail identified a 240-acre parcel, midway between the Econ Wilderness Area and Wekiva State Park, which was ideal for integrating active transportation and urbanism. The property was adjacent to Winter Springs High School, within walking distance of Lake Jessup, and it contained a mix of quality natural habitats. A suburb of 25,000 people, located north of Orlando, Winter Springs, Florida, lacked a focal civic point; and, in 1998, the city council hired Dover-Kohl, a New Urbanist planning firm; and, Michael Design Associates, a greenway design firm, to design a town center scaled to active transportation [29].

ES students supported the team. To prepare for the experience, they analyzed both the Rollins campus and Venice, Florida, to create a set of civic design principles. Focusing on spaces, connections, and buildings, the goal was to inform place-making that infuses civic awareness into the daily walk of life:

1. Set within a compact urban form, building height is regulated to provide a subtle but defined focus on educational or civic ideals.
2. Physical design secures natural benefits and uses nature to highlight urbanism.
3. A variety of uses are integrated in a proportional and balanced manner.

FIGURE 11.2 Winter springs town center plan (winter springs planning and zoning, public document) [30].

4. Spaces are aligned to enhance civic buildings and visual perspectives.
5. Beauty and function are synergized to celebrate learning and/or civic virtue.

A series of interconnected parks, natural areas, and trails defined the Winter Springs town center, as shown in Fig. 11.2. In contrast to surrounding developments, where parks are relegated to low quality, "left-over" land, sensitive upland habitat was preserved and linked to wetlands on either side of the site. Development was concentrated on a degraded pasture in a pedestrian oriented design. Neighborhood shopping, parks, and civic buildings were placed within a 10-min walk of residences. The plan provided the basis for a $4.98 million from grant from Preservation 2000 to purchase the proposed greenspace system. Usually, such grants are allocated for land lying within large natural systems, but the state agency felt the Winter Springs Town Center provided a model for the state. "The plan," the Florida Department of Environmental Protection contended, "is based on traditional town planning principles such as more compact urban form designed to the human scale—the pedestrian rather than the auto." [29].

The Winter Springs Town Center broke ground in January 2002 [31]. The urbanism did not take hold as planned, but the mix of parks linked to a bike trail offers a more sustainable pattern than the surrounding suburban landscape. While sustainability is multifaceted, it hinges on using resources in a more efficient manner. Protecting sensitive habitats and offering a mix of transportation modes is the first step. The obvious question is whether

FIGURE 11.3 Wekiva 2020: conceptual design principles for the Wekiva river protection area: seminole county, Florida (produced by author) [32].

Floridians will choose to live closer together with less private space and more public amenities and connections.

The Winter Springs Town Center informed a follow-up project, *Wekiva 2020: Conceptual Design Principles for the Wekiva River Protection Area: Seminole County, Florida*, as shown in Fig. 11.3.

In 1999, Governor Chiles issued a development moratorium for the Wekiva River Protection Area after the Seminole County Commission approved a series of projects that violated the Wekiva River Area Protection Act's mandate that projects have a "rural character" and be designed to limit "culminate ecological degradation." [33] Grants from the Florida Department of Environmental Protection and the Associated Colleges of the South Environmental Studied funded Wekiva 2020, which established design standards and applied them to a vacant 80-acre parcel to illustrate a sustainable pattern of rural development (see checklist: "An Agenda For Action For Taking Steps To Identify Six Design Principles").

An agenda for action for taking steps to identify six design principles

Building on the winter springs Town Center Plan and Randall Arendt's *Rural by Design*, the student team identified six design principles, as follows (check all tasks completed):

______1. Harmonize growth patterns with regional ecosystems.
______2. Maintain ecological integrity of rural forests.
______3. Maintain an interconnected system of green corridors.
______4. Minimize conflict between conservation and recreation on the seminole-wekiva trail.
______5. Establish edges to development and utilize vernacular architecture to maintain rural character.
______6. Cluster development to create greenway-oriented communities in the form of a rural village.

The final two points in the preceding checklist informed of a plan for a rural village employing vernacular architecture modeled on Haile Plantation Village, a New Urbanist neighborhood in Gainesville, Florida. A network of linear parks preserved sensitive natural areas and a provided a pedestrian-bike connection for three neighborhoods and a village center. Development was clustered on degraded pastureland with lots ranging from 40 feet by 100 feet in the village center to more secluded 70 foot by 100 foot "view lots" overlooking wetland buffers. A civic green highlighted the village center, which was a short walk from the Seminole-Wekiva Trail.

A 30-page report, *Wekiva 2020*, was the springboard for a Visioning Charrette the ES Department hosted to initiate discussion and aid in determining the path for future study and planning in the Wekiva River Protection Area. Free of the pressures of a public meeting, over 30 environmentalists, smart growth advocates, bike trail advocates, private developers, consultants, and representatives from the state, region, and local planning agencies initiated a robust discussion. After 4 hours, the concept of unifying bike trails, nature preserves, and traditional town planning drew a consensus, which read: "Land use policies which would require clustered multiuse development in harmony with the natural systems while providing linkages with existing development and habitat." [34].

Following the charrette, Seminole County hired Randall Arendt to help its planning staff draft policies and design guidelines for its rural–suburban interface. Both Seminole and Orange initiated small area studies within the Wekiva River Protection Area, which informed overlay districts that both county commissions adopted to meet the requirements of the Wekiva River Protection Area [35]. Clustering development within a network of greenways not only helps secure the region's water supply, this sustainable development pattern also informs the planning agendas for Florida's other springshed districts [36].

11.4 Winter park: sacrosanct space and active transportation

At the same time, the Vision Charette was meeting at Rollins, the college and Winter Park were championing walkable urbanism. Dover-Kohl was hired to redesign Park Avenue, and it was a significant project: utilities were buried, the avenue narrowed and bricked, sidewalks extended, and a street tree regimen added. By 2000, automobile traffic had slowed, pedestrian traffic had increased, sidewalk cafes appeared, and business vacancies dropped [37]. Rollins College commissioned Chael Cooper and Associates, in affiliation with Dover Kohl, to design the McKean Gateway and the Rinker Admissions Building at the entrance to the campus on Park Avenue. These architectural gems established a definitive terminus and signaled the willingness of both the college and city officials to restore a pedestrian-scaled environment based on historic plans [15].

In 2000, Winter Park took another step in this direction when the city commission hired Michael and Stephenson to draft a master plan for Central Park with the goal of keeping the historic park "sacrosanct." [38] Inspired by Frederick Law Olmsted, the design team was aided by ES students in preparing a plan that refashioned the historic park on traditional and sustainable lines. The results were significant: a two-acre parking lot was greened, a rose garden (with a mix of native plants) modeled on the work of Gertrude Jekyll was established, a native garden with a drought-resistant landscape took form, green sections that once defined Morse Boulevard were recreated to protect pedestrians, and a train station to serve the new SunRail commuter line was sited, built to historic style, and framed to a human scale.

After the city commission adopted the Winter Park Master Plan in 2002, the Elizabeth Morse Genius Foundation (EMGF) hired Stephenson to lead the ecological restoration of the Genius Preserve—a 50-acre conservation easement set between three lakes in the heart of historic Winter Park. An interdisciplinary team led ES students in inventorying and envisioning the preserve. The outcome was a management plan with the goal of "restoring and preserving a glimpse of an earlier generation's aesthetic ideal, while providing a working laboratory in ecological restoration." [39]

The management plan identified areas dominated by exotic canopies that became the focus of restoration work. Faculty led students in designing restoration plans (including plant acquisition costs) for designated sites. The EMGF oversaw the removal of exotics and students planted a regimen of native species. Once a site was restored, students monitored the area and devised management plans that evolved as conditions changed and information was collected. In 2006, an illustrated 70-page field guide was completed, and 2 years later, 6 acres of Mesic Oak Hammock and Hydric Hammock had been restored [39]. In 2008, 1000 Friends of Florida recognized Rollins College and the EMGF with its Community Betterment Award. "This is a well-planned and implemented effort to restore and protect an important piece of open space in an urban setting," Charles Pattison, President of 1000 Friends, noted, "It shows what a committed group of people can accomplish with vision and a workable plan." [40].

The EMGF provides stipends for ES students to manage the restored sites. The Genius Preserve's lessons, however, extend to a range of classes—art, philosophy, literature, and biology—outside ES. A prime ecological steppingstone in Metro Orlando for migrating species, the property is a repository for native flora with 195 plant species, including

124 native, 41 noninvasive exotics, 30 nonnative exotics, and three listed plants. Given that the majority of the nonnative species are planted cultivars that define the historic landscape, the flora biodiversity is unique for a site in an urban area [41].

In 2013, an Environmental Protection Agency Sustainability Curriculum Grant infused sustainability into the Rollins curriculum. A series of linked interdisciplinary classes informed the grant-funded study, *Sustainable Enterprise: Activating SunRail in Winter Park*, as shown in Fig. 11.4. In 2012, planning for the SunRail commuter rail line intensified after it received state and federal funding, and Winter Park was projected to be the prime destination. A richly illustrated report, *Sustainable Enterprise*, presented a series of plans and initiatives to improve pedestrian and bicycle access to the new SunRail Station.

It was presented at the Rollins SunRail Forum in 2014, where 250 citizens, activists, and experts gathered to share the benefits the new commuter rail would offer when it opened. "Before you know it, SunRail will be here, and that's why we're gathered here today – to make sure we're ready," declared Rollins professor Rick Foglesong, the program moderator [43].

In the ensuing 5 years, Winter Park invested over $2,000,000 in projects from the report: the redesign of Denning Drive for pedestrians and bicyclists, the SunTrail bikeway sited in conjunction with the rail right-of-way, and a bike share program housed at the Winter Park Library. A proposal to create a pedestrian underpass inspired by Frederick Law Olmsted's example at Central Park at the troublesome intersection at Fairbanks and Interlachen (the point where the Rollins graduation procession once crossed into Winter Park) was presented to the Winter Park Planning Commission, but the concept eventually stalled. In 2014, a gateway was built next to the project's campus terminus. Inspired by the Harvard Gates, the structure marked the end

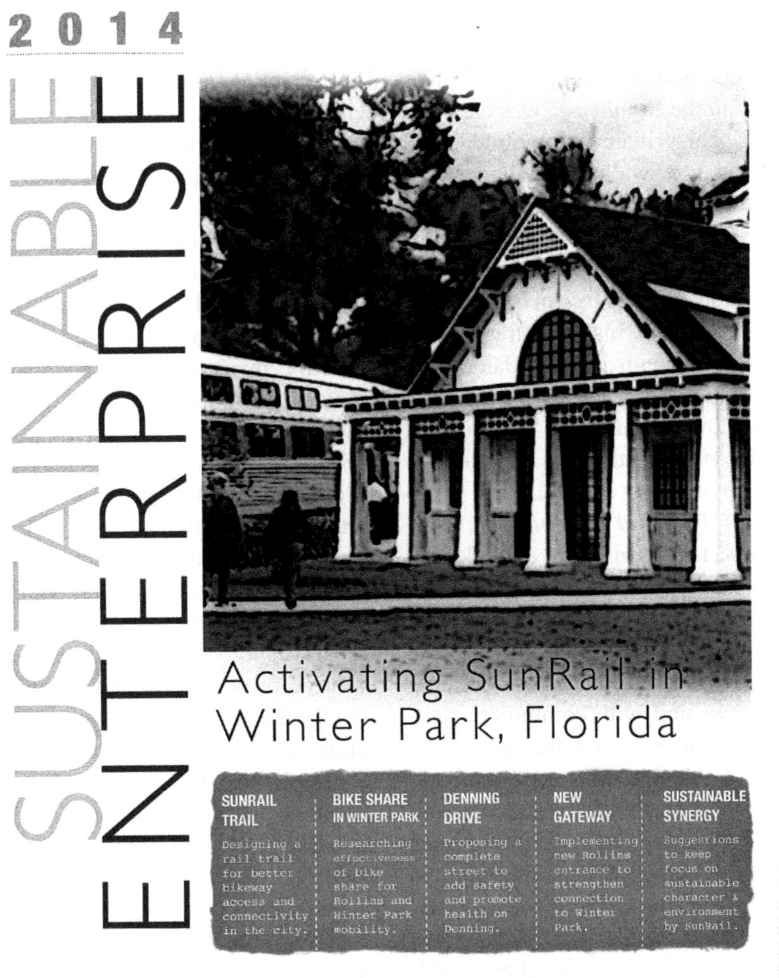

FIGURE 11.4 Sustainable enterprise: activating SunRail in winter park (produced by author) [42].

point of a new walkway that reestablished the historic central axis that centered on the Mills Library and defined Hamilton Holt's plan. Aligned with a new pedestrian crosswalk and traffic signalization system, the gateway enhanced the connection between the campus and community. At the same time, the Olmsted-inspired underpass is a reminder that Hamilton Holt's legacy still informs the vision of an ideal campus for the 21st century [44].

11.5 Greenworks action plan: making Orlando the greenest city in the Southeast

Sustainable Enterprise coincided with the drafting of the Orlando *Greenworks Action Plan*. Orlando Mayor Buddy Dyer had championed SunRail, and in 2013, he established a Sustainability Task Force. At the first meeting, he announced, "Our goal is to make Orlando the most sustainable city in the region, the Portland of the Southeast." [45] The Task Force's *Greenworks Action Plan* mirrored Portland's prototype climate action plan. Livability, Energy/Green Buildings, Solid Waste, Water, Local Food, and Transportation were the components. Key to Orlando becoming "the greenest city in the Southeast" was end "the heavy reliance on conventional private automobile," and called for creating 25 "quality walkable villages" to enhance Livability and meet the 2040 Transportation goal of active transportation being the mode for a majority of trips [45].

After the city council adopted the *Greenworks Action Plan*, the sustainability director and his staff traveled to Portland to study its accomplishments. A year later, Stephenson did the same, selling his car and taking up residence in the Pearl District (the focal point for the first streetcar system built in the United States in 50 years) to document the future the *Greenworks Action Plan* envisioned [46]. The result was *Portland's Good Life: Sustainability and Hope in an American City*, a book that helped secure a grant from the EMGF to advance sustainability both in the Rollins classroom and in the community.

The grant funded a 3-year faculty workshop, "Teaching Sustainability: Portland and Orlando," to promote teaching sustainability across disciplines and with community partners. A trip to examine Portland centered the workshop. Before the excursion, faculty attended three public forums led by experts who addressed a component of the *Greenworks Action Plan*. Next, workshop participants formulated a course project to advance a component of the plan. In Portland, they spent 5 days meeting with specialists and examining sustainability initiatives. In addition, Stephenson partnered with Orlando Sustainability and Resilience officer Chris Castro to model how to meld sustainability and project-based learning.

Building on the *Sustainability Enterprise* initiative, the ES senior capstone required students to prepare an illustrated project booklet that addressed a component of the *Greenworks Action Plan* and three United Nation's Sustainability Goals. Next, student teams identified three precedents to set the foundation for original research that included four original charts, drawings, and/or plans. In addition, the booklets included a set of findings, including projected costs, benefits, and an implementation strategy. After completing their booklet, student teams gave separate 7-min presentations that highlighted their original research to Castro and ES faculty.

The classroom operated like a consulting office, as the instructor worked with students to turn visions into illustrated project booklets. Student research informed the process, which led to concise proposals informed by precedents and data. By midterm students owned their projects, which were increasingly professional in look and substance. A week was devoted to preparing for presentations that had the students fine-tune their work. The presentations revealed when students mastered the material, as the give and take with their audience heightened confidence and empowered them to share their knowledge. Most importantly, students recognized the opportunity to foment change as Castro shared the top projects with Mayor Dyer and other department heads.

The ES sustainability projects covered a wide range and were often ingenious. They included identifying and modeling metrics for siting locally sourced food hubs in food deserts, detailed plans to invest in active transportation in Orlando's Main Street neighborhoods, turning a major arterial into a complete street, utilizing public health data to identify investments in pedestrian and bicycle improvements, developing a native landscaping ordinance and modeling it on a typical neighborhood lot, producing a feasibility analysis and strategy for a commercial recycling program in a Main Street District, intersecting a major arterial with a green street, reclaiming a vacant downtown lot by planting a fruiting orchard, turning a grassy expanse at City Hall into a rain garden, creating a sustainable moss lawn in a busy downtown park, and prioritizing neighborhoods for city subsidized tree planting. An especially inventive project proposed greening a section of Orlando's Bridge District, a plan to turn a linear five-block section underneath I-4 into a series of parks.

In 2015, the Florida Department of Transportation initiated Ultimate I-4, a $2.1 billion "makeover" of a 21-mile stretch of the nation's most dangerous interstate that runs through downtown Orlando. The Bridge District is ambitious, it would place soccer fields, basketball courts, tennis courts, playgrounds, a splash park, a bocce court, a dog walk, a civic plaza, and a retail district under a new elevated section of I-4. Officials recognize the proposal is highly conceptual. There is scant precedent for humans enjoying recreation under noisy, traffic-laden interstates, and some cities are instead opting to remove downtown freeways and replace them with surface-level boulevards. No one knows if the Bridge District will become a destination, but it could enhance livability and a space's vitality. A naturalistic environment designed for human movement fomented what Frederick Law Olmsted called "re-creation," the clearing of the mind from the psychic strain of urban life [47]. This principle informed a student project to green the Bridge District's eastern edge, the Parramore Gateway.

The plan proposed turning the 50-foot wide and 300-foot-long span of asphalt lying 26 feet below I-4 into a greenspace integrating water flow, native vegetation, and a bike-pedestrian path. Directing stormwater runoff from the overpass into a series of rivulets structured the alignment of a bike–pedestrian path that curved through a mix of shade-tolerant native hydric and mesic plant species. Creating this naturalistic environment would require excavating 12,000 square feet of impervious surface, laying out 36,000 cubic feet of soil, and planting 200 grasses, shrubs, and trees. In return, the network of green would filter approximately 325,440 gallons of water a year that would have flowed into sewer drains [48]. Greening the Parramore Gateway would not only be good for the environment, it would imbue a destitute impervious underpass with a beauty and tone conducive to contemplating art and recounting a forgotten history.

The Bridge District Plan calls for lining this section with murals that depict the history of Parramore, the historic African American neighborhood I-4 severed. Greening the Parramore Gateway could be a foundational piece in a network of green that would be a unifying thread in a community degraded by disinvestment and heavy-handed urban renewal [49]. Prioritizing equity is essential for sustainability, and in Portland, Rollins faculty experienced this reality.

11.6 Sustainability's challenge: affordable housing and homelessness

The workshop participants relied solely on active transportation during their 5-day stay. Immersed in walkable urbanism, they encountered complete streets centered on streetcar lines, the integration of affordable and market housing, and a model green infrastructure—restored wetlands, eco-roofs, bioswales defining green streets. In addition, the mix of uses—grocery stores, drug stores, medical offices, and eateries—placed everyday needs close at hand. "I understood the concept of sustainability, especially from a global perspective," one faculty member noted:

> *but the workshop was especially valuable in expanding my knowledge of local initiatives in Portland, from zoning laws that encourage density, to the design of public parks, to the creation of nature preserves, to innovative transportation infrastructure, to architectural design. It was fascinating to see Portland's progressive policies in all of these areas, which have successfully stimulated economic development and revitalized the city [50].*

The group also encountered concentrations of the homeless and soon learned that homelessness and affordable housing are intertwined. This issue has bedeviled Portland since October 2016, when Mayor Charles Hales issued a state of emergency to confront a "crisis of affordable housing and homelessness." In response, zoning codes were densified and voters approved a $258.4 million bond measure to support affordable. $20 million from the sale of housing bonds went to Right to Return, the first funded initiative in the United States to rectify displacement in African American communities. Portland also pioneered tax increment financing to spur urban renewal in the Pearl District, and this policy was extended to procure affordable housing in a range of underserved neighborhoods [51]. For Rollins faculty, Portland illustrated the rewards and challenges of sustainability, especially in regard to affordable housing, which is a pressing issue in Central Florida.

A 2019 study by the National Low Income Housing Coalition found Metro Orlando's provision of 13 affordable units per 100 families who need them to be the worst in the nation [52]. In response, Mark Brewer, the CEO of the Central Florida Foundation, announced the creation of a $100 million "affordable housing gap fund." [53] Parramore was a focal point. A Sustainable Communities Grant from the Obama Administration funded the *Parramore Comprehensive Neighborhood Plan*, which articulated an affordable housing strategy. In 2019, over 200 units came online, and a program to subsidize home ownership secured another 41 units. At the same time, the Central Florida Foundation formed the Central Florida Regional Housing Trust to build mixed-income housing in Parramore to set a prototype, where residents negate a significant real estate expense by only paying for the cost of housing [54].

Affordable housing was a constant in the Rollins sustainability courses, which art professor Josh Almond's Applied Design Solutions class exemplified. Almond's students examined sustainability through the lens of the tiny home movement. First, they designed their own unit, and then they collaboratively built a tiny home. The course's hands-on nature helped them learn skills and trades that are applicable outside the classroom. "This class allows us to discuss big concepts like homelessness and sustainability," one student stated, "while also teaching us how to hammer a nail into wood, how to downsize our material possessions, and how to digitally render buildings." [55] Tiny homes range from 96 to 400 square feet, and the Rollins' students' 96 foot square, "Tumbleweed Salsa Box," was designed to be mounted on an 8-by-12-foot trailer and features a kitchen, shower, compostable toilet, loft bed, couch/futon, and porch overhang with a living garden [55].

At the conclusion of the workshop, the faculty cohort moved to share their experience. Procuring a grant from the EMGF, they joined with the college administration to plan a symposium, "Equity the Key to Sustainability," for the fall of 2021. A workshop devoted to teaching sustainability will precede an expert panel, including Chris Castro and Mark Brewer, which will share their experience in a public forum. Finally, Cat Goughnour will deliver the keynote. A graduate of the London School of Economics, her sustainability-focused Right 2 Root initiative sought to readdress the root causes of gentrification in Portland and capacitate African Americans to leverage planning tools, public interest design, and community assets to create healthy, innovative places [56].

Sustainability is a new strategic initiative at Rollins, and the symposium will deliver criteria to infuse sustainability into the classroom aided by a network of mentoring and collaboration. The symposium will not only inform Rollins faculty and students, but a consortium of Central Florida colleges and universities that are committed to teaching sustainability [57]. Ideally, the Rollins Symposium will be the first in a series of yearly events hosted by the consortium's members.

The consortium is not merely an academic endeavor. Faculty participated in the East Central Florida Regional Planning Council's resilience collaborative. Integrating equity and sustainability defines the *East Central Florida Resiliency Action Plan*, which seeks to reduce the region's carbon footprint and vulnerabilities to climate change. Completed in May 2021, it parallels the *Portland Climate Action Plan* and the *Orlando Greenworks Action Plan*, and like its counterparts, the plan is visionary yet defined by research [58]. For a century, the pragmatic liberal arts have plied the same path. If the goals of sustainability are transcendent, reaching them is grounded in nature and reality. "If you have built castles in the air," Henry David Thoreau wrote, "that is where they should be. Now put the foundations." [59]

11.7 Summary

Sustainability is a watchword for smart cities and a buzzword on college campuses. John Dewey's community-based "learning my doing" brand of pragmatic education has a significant history, and it inspired Rollins College faculty teaching sustainability. Project-based learning requires that faculty share their expertise with both community partners and their students. Faculty research not provides a model for student projects, it is the basis of a common bond that turns the classroom into a laboratory for producing a quality product for the community partner. When students present their projects to a local governing board or agency, they gain invaluable real-world experience and provide, ideally, consultant quality work to their client. At Rollins College, initial project-based learning projects were relatively simple: rails-to-trails initiatives and management plans for natural lands. In the next phase, students analyzed walkable urbanism on the campus and in Winter Park to inform their plans for towns, villages, and neighborhoods incorporating active transportation. After Orlando adopted a sustainability plan, faculty and students worked in concert with the Sustainability and Resilience Office to illustrate investments in sustainability in the areas of livability, transportation, solid waste, local food, and water resources. Such partnerships are essential not only for educating the campus and the community but accelerating the pace for achieving sustainability. In 2021 Rollins made sustainability a strategic priority, and faculty versed in project-based learning are leading the effort to infuse sustainability into the curriculum. This initiative not only documents the evolution of project-based learning since John Dewey visited the campus 90 years ago, it also marks the generation that will envision and build smart cities.

Finally, let us move on to the real interactive part of this chapter: review questions/exercises, hands-on projects, case projects, and optional team case project. The answers and/or solutions by chapter can be found in Appendix G.

11.8 Chapter review questions/exercises

True/false

1. True or False? Sustainability is a measure of humanity's moral response to the specter of climate change, species extinction, and resource growth.
2. True or False? Efforts to mitigate the problems of drivable suburbanism spawned and began to take hold in the early 1980s.
3. True or False? In 1992, project-based learning centered on a capstone Environmental Studies (ES) course, where seniors prepared a comprehensive assessment of an abandoned rail corridor designated to accommodate metropolitan Orlando's first bike trail.
4. True or False? In 1997, Environmental Studies (ES) seniors shifted their focus to prepare a conceptual management plan for High Oaks–a 350-acre priority acquisition for the Seminole County Natural Lands Program.
5. True or False? After analyzing the acquisition criteria that guided the advisory committee, students gained insight into essential ecological concepts such as corridors, biodiversity, and species extinction.

Multiple choice

1. A __________ was delineated to enhance education and highlight the site's premier natural features: pristine sandhill habitat, a grove of xeric Live Oaks, and a span of a large Bayhead swamp accessed by a boardwalk
   a. Ecological system
   b. Habitat system
   c. Trail system
   d. Econ wilderness system
   e. Freshwater
2. The idea of integrating ________ into daily life infused the plan for the Winter Springs Town Center.
   a. Society
   b. Infrastructure
   c. Wilderness
   d. Nature
   e. Urbanism
3. A network of _________ can preserve sensitive natural areas and provide a pedestrian-bike connection for neighborhoods and a village center.
   a. Linear parks
   b. Vernacular architectures
   c. Wetland buffers
   d. Environmentalists
   e. Bike trail advocates
4. The hope is that _____________ will draw a consensus, which should read like: Land use policies that would require clustered multiuse development in harmony with the natural systems, while providing linkages with existing development and habitat.
   a. Holistic integration planning
   b. Traditional town planning
   c. Smart cities planning
   d. Smart counties planning
   e. Rural–suburban planning
5. Clustering development within a network of greenways not only helps secure the region's water supply, this sustainable development pattern also informs the planning agendas for Florida's other:
   a. Urban districts
   b. Significant projects
   c. Street tree regimens
   d. Business vacancies
   e. Springshed districts

Exercise

Problem

How can data skills be taught using a more unified and practical approach, which facilitates application of skills in genuine, smart city contexts?

Hands-on projects

Project

Do research: Design and validate a questionnaire for teachers on the knowledge and use of smart city concepts and their usefulness in online learning during the COVID-19 pandemic, based on a sustainable approach at different educational levels.

Case projects

Problem

How can a massive open online course be used to engage citizens in learning about smart cities and teach them how to cocreate a smart cities project in their community?

Optional team case project

Problem

Describe the relationship between Smart Cities and how the applicability of Machine Learning (ML) techniques is carried out with the aim of optimizing sustainability.

References

[1] McKibben B. Falter: Has the human game begun to place itself out New York: Henry Holt, 2019; Elizabeth Kolbert, the sixth extinction: an unnatural history. New York: Henry Holt; 2014.

[2] World commission on environment and development, our common future. New York: Oxford University Press; 1987.

[3] Newman P, Kenworthy J. Sustainability and cities: overcoming automobile dependence. Washington D.C.: Island Press; 1999. p. 334.

[4] Club S. America's most sprawl threatened cities. San Francisco: Sierra Club; 1999.

[5] Leinberger C, Rodriguez M. Foot traffic ahead. Washington D.C.: Smart Growth America; 2016.

[6] Lynch A, Anna LP, Fox C. Sustainability cities report: the 2019 US cities sustainable development report. New York: Sustainable Development Solutions Network; 2019.

[7] Leinberger C. The option of urbanism. Washington D.C.: Island Press; 2010.

[8] Orlando office of sustainability (OSO)vol. 4. Greenworks Orlando: Community Action Plan Orlando: OSO; 2014.

[9] Lane J. Rollins college centennial history: a story of perseverance, 1885-1985 Orlando, FL: story farm, 2017, pp. 1-324; Lewis Mumford. The intolerable city: must it keep on growing. National Humanities Center. Harper's Monthly. Sections: I-IV February, 1926:292.

[10] Hamilton H. Ideals for the development of Rollins college. Rollins College Archives and Special Collections (RCASC); June 1926.

[11] Stephenson B. Rollins college and winter park: exemplars of the American renaissance, precedents for the future. In: Maurice O'Sullivan Bruce Stephenson, editor. Florida's golden age,

1880–1930: the Rollins college Colloquy. Cocoa, Florida: Florida Historical Press; 2018. p. 71–86.
[12] Hamilton H, Letter to John Nolen, March 8, 1932, box 20, John Nolen papers, Cornell university manuscripts and special collections.
[13] Schuyler D. Riverside, Illinois. In: Mary CS, Isabelle G, Rob F, editors. Iconic planned communities and the challenge of change. Philadelphia: University of Pennsylvania Press; 2019. p. 40–60.
[14] Stephenson B. Rollins college and winter park: exemplars of the American renaissance. Classicist December 2020:14–21.
[15] Stephenson, Rollins college and winter park: exemplars of the American renaissance.
[16] Clark K. Civilisation. New York: Harpers & Row; 1969.
[17] Harris C. The Town that became a university winter. Park, FL: Rollins College; 1930 [RCASC].
[18] Stephenson B. Portland's good life: sustainability and hope for America's cities. Lanham, MD: Lexington Books; 2021.
[19] Owen D. Green metropolis. New York: Riverhead Books; 2010.
[20] Stephenson, Portland's good life, [xii-xv].
[21] Stephenson B. Visions of eden: environmentalism, urban planning, and city building in st. Petersburg, Florida, 1900–1995. Columbus: Ohio State Press; 1997.
[22] Lee JS, Blackwell S, Drake J, Moran KA. Taking a leap of faith: redefining teaching and learning in higher education through project-based learning. Interdiscip J Probl-based Learn 2014;8.
[23] Stephenson, Visions of eden, 174-177.
[24] Stephenson B. The roots of new urbanism: John Nolen's garden cityethic. J Plann Hist May 2002:101–26. Florida Greenways Commission, Central Florida Loop, 1995, 101–126.
[25] Seminole county natural lands program, https://www.seminolecountyfl.gov/departments-services/leisure-services/greenways-natural-lands/natural-ls/.
[26] Rollins college environmental studies department (RCES), A conceptual management plan for high oaks. Winter Park, FL: RCES; 1997.
[27] Seminole county natural lands, Econ wilderness area. https://www.seminolecountyfl.gov/locations/Econ-River-Wilderness-Area.stml.
[28] Buendia G. the river that raised me. In: Davis J, Poole L, editors. The wilder heart of Florida: more writers inspired by Florida nature. Gainesville: University Press of Florida; 2021. p. 44–9.
[29] Stephenson, The roots of new urbanism.
[30] Stephenson. The roots of new urbanism., winters Springs planning and zoning. Winter Springs Town Center Plan; 1998.
[31] Perez R. Winter springs to break ground on focal point for community. Orl Sentin (FL) January 18, 2002:B1.
[32] B. Stephenson, ed., Wekiva 2020: Conceptual design principles for the Wekiva river protection area: seminole county, Florida, Seminole County Water Atlas https://www.seminole.wataeratlas.usf.edu.
[33] Brooks M. New river homes may have to wait. Orl Sentin (FL) January 23, 1999:B1.
[34] Stephenson, ed., Wekiva 2020, seminole county water atlas.
[35] Orange County Planning Department. Rural settlement study. Orlando, FL: Orange County Planning Department; 2005.
[36] 1000 friends of Florida, protecting Florida's springs Tallahassee, FL: 1000 Friends of Florida.
[37] Dover V, Massengale J. Street design: the art and practice of making complete streets. New York: Wiley; 2013.
[38] Dover and Massengale, Street design.
[39] Stephenson. Ecological restoration: cultivating the "civilized landscape" in the liberal arts. Ecol Restor December 2006:290–1.
[40] June 20, 2008, 1000 friends of Florida Press release.
[41] Brown J. The vascular flora of the genius preserve and an analysis of its habitats. Winter Park, FL: RCES; 2021.
[42] Stephenson B, editor. Sustainable enterprise: activating SunRail in winter park. Winter Park Fl: RCES; 2014.
[43] Michael W. Freeman, winter park gears up for arrival of SunRail. Orl Sentin (FL) January 30, 2014.B1.
[44] Stephenson, Rollins college and winter park: exemplars of the American renaissance, precedents for the future.
[45] Stephenson, Portland's good life, xv.
[46] Stephenson, Portland's good life, xvi.
[47] Stephenson B. Planners could use highway makeover to plan a more humane city. Next City January 19, 2016. https://nextcity.org/daily/entry/orlando-bridge-district-design-planners-nature.
[48] Mucha E, Freedman H. Restoration of I-4 underpass. Winter Park: FL: RCES; 2018.
[49] Stephenson B. Creative village, UCF: the synergy to right a historic injustice. Orl Sentin (FL) April 24, 2015:A20.
[50] Stephenson B. Teaching sustainability workshop: report to the Elizabeth Morse genius foundation, winter park, FL. May 29, 2019.
[51] Stephenson B. Portland defines the next decade in planning. Planning January 2021. https://www.planning.org/planning/2021/winter/intersections-viewpoint/.
[52] Santich K. Orlando area now worst in the nation for shortage of extremely low-income housing. Orl Sentin (FL) March 14, 2019:A1.
[53] Lynch R. Non-profit, local business to create $100 million fund to aid affordable housing. Orl Bus J December 3, 2019. https://www.bizjournals.com/orlando/news/2019/12/03/nonprofit-localbusinesses-to-create-100m-fund-to.html.
[54] Stephenson, Portland's good life, 196-199.
[55] Humphreys R. Cool class: applied design solutions. Rollins News April 11, 2019. https://www.rollins.edu/college-of-liberal-arts/news/cool-class-applied-design-solutions.
[56] Stephenson, Portland's good life, 146.
[57] Institutions include the university of central Florida, Rollins College, Stetson University, Florida A&M College of Law, Barry University College of Law, Valencia College, Florida Institute of Technology, Bethune Cookman College, Florida Polytechnic University, Daytona State College, Full Sail, Lake-Sumter State College, Embry-Riddle.
[58] East Central Florida regional planning council, East Central Florida resiliency action plan. Orlando, FL: East Central Florida Regional Planning Council; 2021.
[59] Thoreau HD. Walden New York. Thomas Y. Crowell & Co.; 1910. p. 427.

Chapter 12

Partnering with technology firms to train smart city workforces

Neena Pahuja
ERNET India, Ministry of Electronics and Information Technology, New Delhi, Delhi, India

12.1 Introduction

There is little bit of Digital today in every business, every house, and every city. A Smart City uses Digital Technology, People and Processes to improve outcomes across every aspect of city operation and enhance the services it offers to its residents. With Digital Technology changing at an amazingly fast pace, the "City Processes" require redefinition and realignment with City Goals accordingly. The new-age processes of Smart City must be embedded in the city workflow, so that they are easily used. These workflows must be self-guiding for residents and also available as part of learning guides. The new-age learning and collaboration tools can be effectively used to complete the learning process, wherever required.

With Natural Language Processes (NLP), Artificial Intelligence (AI), Augmented Reality (AR), and Virtual Reality (VR), Social Media and Collaboration tools becoming easy to handle, transition to new processes and implementation of the new processes are becoming easier. Propagation of new processes to concerned person via commonly used tools, social media, and group collaboration tools are much easier and faster now.

Savvy cities can use Public–Private Partnerships (PPPs) to maximize their resources, go faster, and scale effectively on skilling and new best processes. A sustainable and well-functioning smart city is a tight orchestration of people, processes, policies, and technologies working together across the entire Smart City ecosystem.

In India and some other countries also, Skill Sector Councils have been created to support the PPP process for skilling, upskilling, and employment generation. The Sector Skill Councils also have representatives from the industry, which help in bringing in the changing landscape of skilling and upskilling requirements of the industries. Technology companies by partnering here for skilling and upskilling can help in learning about the "City Processes" and help in knowing the industry requirements so that training can be imparted on the employable skills.

Globally a lot of technology companies even attempted to create Smart City hubs, as the concept of smart cities was picking up. These Technology companies also attempted to create model for Smart Cities, so that the requirements can be understood better for replicating the model and bringing out the learning and reskilling requirements. This model city was also used as training ground for the workforce and citizens and helped in faster change management to the new city processes.

Role of e-trainings is reemphasized in papers [1]. During the outbreak of COVID-19 and the whole of 2020–21 and even 2021–22, the world realized the benefits of eLearning, tele-medicines, and collaboration tools. In fact, COVID proved to be the biggest technology disruptor of the century, where technology becomes part of life in every area. In various parts of world during lockdown, citizens became dependent on self-help for a lot of personal services and used technology to get the best practices in those areas. Even Work from Home (WFH) required different skills and learning. Remote care skills for patients also evolved during this period.

Responsibility of skilling, reskilling, and upskilling the workforce in the postpandemic world lies not just with governments but employees and companies as well. The WEF's Future of Jobs Report 2020 [2] states that 43% of the businesses surveyed are set to reduce their workforce, and 34% plan to expand their workforce due to technology integration. Experts need to examine how businesses and governments can work together to retrain workers and redeploy them toward jobs of the future.

The workforce is not just the backbone of a business but also of the economy of a city. It is said that if the workforce is highly skilled, it could significantly impact the growth of

Smart Cities Policies and Financing. https://doi.org/10.1016/B978-0-12-819130-9.00012-7

an economy and city. This factor is another opportunity for cities to transform into smart cities. Additionally, if businesses leverage the benefits of new-age technologies including AI for a competitive edge, they could gain a unique position in the marketplace. However, to incorporate AI into City business operations, it is important that the workforce is familiar with it. Hence, the first step to bringing AI to the business environment is by skilling the workforce, which will enable a cognitive collaboration between human and AI. While some believe that working along with AI could be daunting, some are positive toward the future of human–AI collaboration. To add more clarity to what AI can bring to businesses, we have discussed some of the visible benefits of AI that have been researched to be beneficial in area of Personalized Skilling.

This chapter brings out different needs for skilling, where some are required to be built as part of regular process for making cities more sustainable and/or create more job opportunities in the city, but there are others that may need to be acquired at a noticeably short interval, for example, in case of disaster or pandemic. The technology companies can play a major role by partnering to impart such skills. Another area where technology companies can partner is area of "Change Management" for every requirement of city together with ensuring behavior, Mental and Physical well-being of people in the city.

12.2 Why skill

With more and more persons living in cities, our cities are already facing growing environmental, societal, and economic challenges. The UN predicts that 68% of the world's population will live in urban areas by 2050. By making cities smarter, we can overcome some of these challenges and make cities better places to live. One report by McKinsey Global Institute found that smart city technology can improve key quality of life indicators, such as the daily commute, health issues, or crime incidents by 10%–30%.

Skilling and Reskilling must be a lifelong process for every economy to survive. With technology and environment changing the requirements so fast, need for new skills is becoming even more important. As per [3], reskilling costs $35,000 to an employer while removing him or her and hiring the other costs around $100,000. Globally, governments are also creating frameworks, in which incentives will have to be put in place to reward those who reskill employees in the required areas. Reskilling and upskilling are taken as an investment not cost.

Skill and Upskilling help in using the city resources more efficiently. Globally, mobile and broadband companies or communication hardware and platform companies have been investing heavily in smart city technology. In one example, sensors are attached to refuse containers to report, in real time, how full they are, which means refuse collectors do not have to waste time traveling to bins that are only half-full. It also means KPIs can be more closely tied to bottom-line impact, such as how many bins are close to overflowing and will not be emptied within the next few hours. This is far more meaningful than measuring abstract factors, such as how many waste collection trucks are out on the road. However, even to implement this and continuously using this, training will be required in handling control systems, managing sensors in bins, connectivity for sensors in the bin, and handling dashboard reporting.

In some countries, Education Loans are provided to fund the education, skilling, and upskilling, which have become quite expensive. Like incentives given for education loans, there could also be tax benefits that can be provided for spends on skilling and reskilling, to make it more attractive. Businesses and governments and individuals all must take part, and it has to be a lifelong process. Also, it will require huge finances. Companies, rather than laying off people, must focus on upskilling and reskilling should be their long-term strategic plan. Companies can also create resource pools to save costs, whereas resources could be tools and environments to provide training or even manpower. A concept of Virtual Laboratory may be used to further save resources.

A lot of technology companies have come forward and are helping in creating Digital Skills based on availability of manpower to be trained. Companies such as Microsoft are training factory workers on Digital skills at the time of their commute to factory. Even Google has created a group and is doing trainings in AI. The efficiency generated by digital technologies will create positive outcomes provided there is digital skilling. Starting from eCommerce to Remote Healthcare Deliveries to Industry 4.0, the knowledge of Digital Skills is expected to also create a new wave of new type of jobs.

Reskilling involves training employees on an entirely new set of skills to prepare them to take on a different role within the organization. This typically occurs when workers' previous tasks or responsibilities become irrelevant or can become irrelevant because of change in processes, which can be due to advances in technology. Organizations do this because an employee may fit well in a team and have marketplace or company knowledge that would be difficult to replace, "To keep that employee, the organization just needs to update their skills to match new systems and new capabilities." Reskilling may involve obtaining a new degree, certification, or education in a different field or area of expertise.

As opposed to reskilling's 180-degree pivot, upskilling occurs when workers improve upon existing skills and deepen their abilities and impact within their area of expertise. By expanding their knowledge, employees become better positioned for additional responsibilities and higher-level roles on a particular career track. For instance, a marketer will need to learn new digital tools and skills to better engage with their audience (and reach new ones) on emerging social platforms. If they do so successfully, they will simultaneously give their business a boost and set themselves up for future success in more advanced positions.

The demand for skilled talent is only going to continue to grow, and the skills gap will only widen as technology advancements and societal shifts disrupt the status quo. The new world of work requires people to continuously hone their skills to stay relevant and improve their employability. The term new skilling represents all types of continuous learning to help build high-demand skills, whether an individual is trying to upskill current capabilities, or they need complete reskilling to build entirely new capabilities.

A new skilling mindset keeps both a workforce and a company agile by ensuring that learning initiatives are relevant to future business objectives and tailored to the needs of learners. This is simply the new reality, and no business will survive for long without reskilling and upskilling initiatives driven by a new skilling strategy. By regularly identifying what skills will be needed in the future and which of those employees currently possess, organizations can build more thoughtful, continuous skilling programs to effectively develop those abilities in their workforce.

Another major area for skilling and upskilling is for improving life of persons with special needs in the city. This may be approached in two ways. One, training such persons with usage of technology tools and products, which can help their daily life. This also may cover basic employability skills to help them increase their self-confidence. Second, there could be trainings created for corporates, to help them earn their inclusivity goals. They can be trained on how to make work environment and city safe and inclusive for persons with special needs. A large number of technology companies including Microsoft are already doing a lot of work in this area.

In India, vocation education and skilling have been given a lot of importance as part of the latest National Education Policy 2020 [4]. Policy supports concept of starting life skills from the early stage of the school to provide a larger horizontal of education to the children.

12.3 Technology-based skilling verticals for a smart city

The Smart City leaders are realizing that smart-city strategies start with people and not only technology. "Smartness" is not just about installing digital interfaces in traditional infrastructure or streamlining city operations. It is also about using technology and data purposefully to make better decisions and deliver a better quality of life. This process requires training citizens and persons using and operating this.

Smartness of City uses many Technology components (hardware, software, processes), which also change over a period. Hence it is important to create processes and products for self-learning to ensure a faster and greater reach. Technology companies can also help in creating demonstration and practice centers for these city components, which can also act as training and innovation centers. This will not only help in learning and reskilling; it will also help locals to innovate in their areas. The demonstration centers can also help in Research and Development process and provide access to global expertise for problem solutions.

The trainings in these areas can also be started as part of hobbies or creation of Vocation skills at school level to create a good understanding in these areas and develop interest in these areas at an early age. Some countries are planning to include some of these courses in early studies as part of their education policies. The following sections cover some of the skill area for smart city where technology companies can support skilling and upskilling.

12.3.1 Telecommunication connectivity and infrastructure creation and maintenance

Connectivity is the base infrastructure requirement of any city to make it Smart. However, with technology evolving, the locals need to be skilled and reskilled to create and maintain such evolving infrastructure. Additionally, requirement for telecommunication connectivity is growing everyday with growing number of applications and areas it is supporting. Already there are some technology companies that are trying to create training platforms for regularly training the local manpower for ensuring connectivity. Most governments funding Smart Cities normally allocate funds for such trainings. The Technology companies can provide help in providing the right skills to the persons in the city to ensure high uptime and connected city environment. The training may be as simple as putting an "optical fiber network" to "home configuration of routers" to maintenance of towers for connectivity or setting up data centers, as required.

12.3.2 Traffic management

Traffic impacts city-dweller's life and technology offers some promising solutions to simplify it. For example, public transport routes can be adjusted in real-time according to demand, and intelligent traffic light systems can be used to improve congestion. Technology can be effectively used to keep the traffic moving. This will require creation and training on usage of technology-based tools, which are able to analyze based on current data and previous trends.

12.3.3 Data analytics and monetization

Data has value and can be monetized by both selling raw data or derived insights to other players in the market. City Data can then be used for Data Analysis and Decision-Making, which the local city persons can use for better decisions for improving their living conditions.

As AI is expected to touch every part of almost every business in the coming years, many Technology companies are already trying to train their own workforce and even persons working in other businesses on how AI can help them to improve their businesses and help in the required transformations. A large number of open-source training products with their testing modules have already emerged to help train persons in this required trend.

12.3.4 Healthcare

The need for healthcare transformation and methods for handling new diseases came as a big requirement during 2020–21. Working safely under the threat of the "Disease" will require new thinking and new ways of working across all aspects of processes so that everyone involved gets the right information and knowledge to play their role. Starting from need to train manpower in handling tele-medicines, remote and home healthcare to handling vaccination drive, healthcare skilling globally went through a major transformation. There can also be emergency requirement of training persons who can help in usage of emergency equipment such as ventilators and oxygen for patients and also repair such equipment. Hundreds of colleges, universities, and vocational schools in the United States offer online programs in a variety of healthcare fields, including nursing, medical assisting, healthcare administration, and dozens of other disciplines. NHS has already started online training program, where the medical professions can learn new skills at their own pace. Technology companies can come forward not only to meet the sudden need but also create training environment and modules to ensure that requirements of City are met.

12.3.5 Disaster management training

Globally, countries use online courses for disaster management, which have proved to be effective. The aim of these programs is to raise the level of awareness and preparedness as well as to enhance analytical **skills** and professional competencies of different key stakeholders in various areas of **Disaster** and Risk Management. Training may also need to be provided to the key the citizens on do's and don'ts during Disaster Management and even epidemics. Technology tools can be very effectively used for dissemination of the training to all.

12.3.6 Sales and marketing skills to promote local

One of the major requirements of a smart city is to promote local. It could be local handicrafts or skills or any product or even tourism. This may include usage of the local and Global Language skills to promote Globalization and local Tourism. Videography and creation of small products based of Video's, Audios, and AR can be effectively used for sales and marketing and promotion of "Local." This kind of training can help in creation of microentrepreneurs in a smart city. Technology tools such as Natural Language Processing (NLP) can also be effectively used to increase the reach.

The smart city persons may also be trained in carrying out Virtual Exhibition/fairs of their local products to increase its reach. They may be able to ride on hosted technology platforms for Virtual Fair to increase reach of selling their local products. Technology firms can help in creating such platforms and trainings and provide them both in controlled environment and open-source environment.

12.3.7 Smart asset management

A Smart City brings the requirements of Smart Asset Management, starting from handling Smart lighting, Smart streets, city equipment, buildings and infrastructure assets, or even capital goods used at homes. In the Smart Cities, since there are large type and number of products, it is important that product vendors come forward and support training the users on these products. These could again be provided using any of the tech-supported products.

12.3.8 Industry 4.0 and digital manufacturing

Smart Cities with manufacturing clusters need to be trained in area of Industry 4.0 to keep up with requirements of competitive manufacturing. Starting from training on basics of AI to Robotic arm, the training may cover concepts of digital manufacturing, control systems together with remote management of devices, and even fault detection and correction, if possible.

12.3.9 Usage of 3D printing for designing

Usage of 3D for designing various items is picking up. The item may be an automotive part or a shoe or a leather bag or a dress or furniture or a house. Tools for designing are used together with 3D printing to create a quick prototype model, which then can be replaced. This allows customized products to be created based on the need. This concept has also been used for jewelry designs or creating a sample house and designing new vehicles. The Technology companies can partner with locals and help them with this training, based on the local needs of the city. The artists can be given this training to increase creation of customized Handicrafts with new designs.

12.3.10 Smart energy: renewable energy and energy conservation

With Renewable energy getting the desired focus, the Technology companies can also provide training in handling of renewable devices and their maintenance. This will help in achieving the desired Carbon Footprint in the Smart City and give the desired focus to Green Energy.

Smart cities also use technology to help closely monitor real-time energy use and reduce energy consumption. For example, in many Smart Cities, homes are being provided with smart energy meters that are designed to incentivize reduced energy consumption. Also, some cities have upgraded their streetlights to LED technology, which allows the lights to be adjusted or dimmed based on real-time data of light and usage in that area. However, this again needs training, where the technology companies can partner to ensure that this knowledge is made available to the citizens.

12.3.11 eCommerce and supply chain management

One of the major learning in the pandemic has been a need to use eCommerce applications, digital payments and maintain a good supply chain of products and resources. Tracking of products and safe delivery processes came as a major requirement in all cities and countries. The employees of eCommerce companies and even the residents need to be trained in these safety processes. Smart City allows real-time access to data that can unleash tremendous value, driving smarter decision-making by Government, planners, community groups, and individual residents. With most governments working on Drone policy, the supply chain team may also be needed to be trained on usage and upkeep of Drones for maintaining contactless and immediate delivery of products and services.

12.3.12 Information security

With Smart City handling large number of Smart Objects, knowledge of Information Security becomes extremely important for the citizens of the city. Usage of digital payments and associated security and risk knowledge, etc., may also be covered as part of this. The Tech organizations can partner with collaboration platforms, media, TV, and other social media platform for dissemination of this knowledge. The citizens also may need to be trained on their data privacy rights as per the guidelines of the country.

12.3.13 Supporting specially enabled persons and elders

One very important component of Smart City is how the Specially Enabled Persons and Elders are treated in the city. With Technology changing every day, newer products are being launched to ease the life and provide safety. Trainings, social media, and collaborative tools can be used effectively for supporting and providing them this knowledge.

12.3.14 Entrepreneurial skill

For a city to be termed Smart, it is important that citizens are gainfully employed. This may require creating job opportunities and also training persons in different areas to create microentrepreneurs. Technology companies can also help in not only training such persons but also train the persons to set up their own businesses, carrying out the required legal and financial responsibilities.

12.3.15 Change management

Change Management is especially an important part of successful implementation of Smart Cities. With Technology becoming part of life of common citizen, usage of this technology requires constant knowledge upgrades. Smart Cities to become smarter must slowly bring-in the Change Management processes to be successful. Starting from usage of technology or process of wearing a mask, maintain a safe distance or even online education to remote working including remote healthcare, everything requires a Change Management. The Technology companies can create such training modules to help propagate the Change across. Tools such as social media tech tools or collaboration tools may also be effectively used for propagation of this knowledge.

12.4 Smart city training framework

For a Smart City to truly exist and thrive, it needs "smart" citizens who are engaged and actively taking advantage of new technologies. With any new city-wide tech project, part of the implementation process must involve educating the community on its benefits. This can be done through a series of in-person town hall-style meetings and email campaigns with voter registration, as well as an online education platform or collaboration platforms that keep citizens engaged and up to date.

When a community feels like it is playing a part in the overall decisions that affect daily life and is being communicated to in a clear and thoughtful manner, it is more apt to use the technology and encourage others to use it as well. This is key to a Smart City's success. The city needs collaboration with residents, entrepreneurs, large corporations, and start-ups to create a "city of tomorrow."

Despite the several obvious benefits of a smart city including digital urbanization, sustainability, quality of life, smart transportation, better governance, etc., a successful implementation needs several considerations. It needs a candid look at current operating models, architecture, planning orientations, resources, ROI and finance constraints, and process improvement. Some of the considerations for building a smart city are:

- Timely and strategic infrastructure upgrades
- Connectivity evolution, which is changing at a very fast pace
- Optimizing smart city communications
- Impact of open data, big data
- Secure, robust, and scalable architecture

There is also a need to establish radically new standards to ensure the effective use of technology to deliver services and manage complex citizen problems. It is important to benchmark the Smart cities against the very best cities globally, including cities using IoT and digital technology. City Policies may be extended by the city authorities for the use of smart technologies while safeguarding citizens. Issues around liability, security, and privacy need to be addressed through comprehensive regulatory frameworks. PPPs are key to build and operate both physical and digital infrastructure, especially given continually tightening civic budgets and enormous infrastructural requirements. Using the infrastructure, it should be possible to track the assets of the city to provide meaningful information. Most smart cities can have four layers for the training methodology framework as shown in Fig. 12.1.

Some of the latest tools required to build these trainings by the technology companies can be:

- Collaborative platform for one-to one or one-to-many interaction
- Videos/audios repositories together with website for dissemination of information
- Mobile applications
- Creation of virtual laboratories
- Augmented Reality (AR) and Virtual Reality (VR) tours
- Artificial Intelligence (AI) and Natural Language Process (NLP) can be used to bring in the desired personalization.

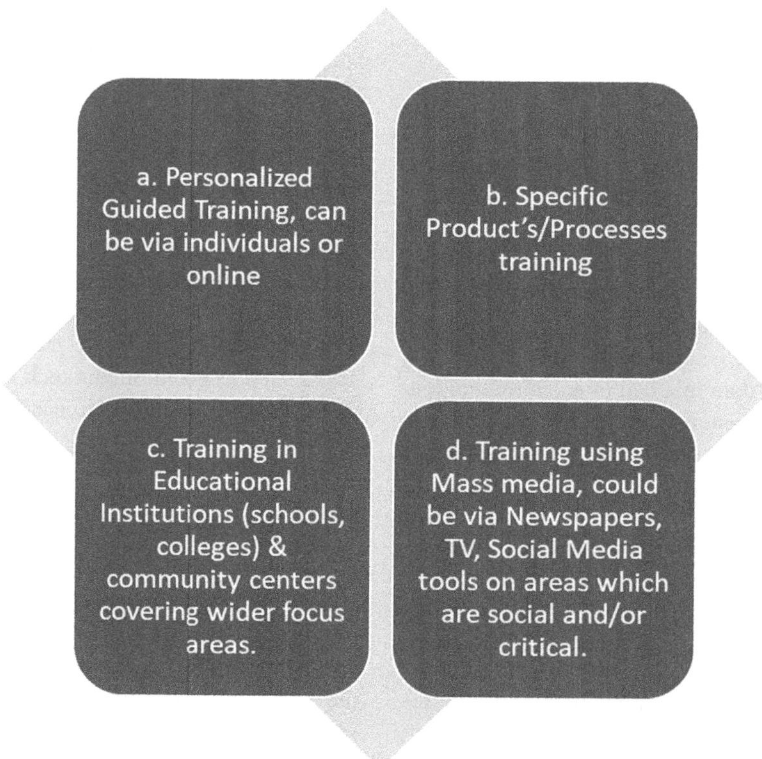

FIGURE 12.1 Smart city training framework.

12.5 Upskilling in creating a smart city

A precise and unanimously accepted definition of Smart city has remained ambiguous till date. There are multiple interpretations available, and a lot many amendments are being made to it every day, but there still remains a certain degree of imperfection in its definition. It is basically because of the wide range of components that come under the umbrella term—Smart City. And clearly, as "Smart" uses technology, "Upskilling" becomes a necessity for maintaining the city smart.

A Smart city is the one that supervises and fuses conditions of all its critical infrastructures, including roads, bridges, tunnels, rails, subways, airports, seaports, communication, water, power, even major buildings, with Information and communication technologies (ICT), to better optimize its resources, plan its preventive maintenance activities, and monitor safety and security aspects while maximizing services to its citizens. Construction Technology for a city has gone via a sea change in the last couple of years, and processes to reduce pollution during construction and cause minimum disruption are one of the major factors used in the newer methods of construction. Additionally, prefabricated components are used to speed the construction at the site. The locals must be trained to carry out such activities and also maintain them. Water harvesting is used wherever possible and different methods of renewal energy are used based on the location. This harvested energy ideally should meet the local requirement or even feed the grid to meet the larger energy goals of the city. Training products for the processes to be followed in the city may be created using AR/VR and other methodologies for training. These may also include training and processes on Disaster Management. In case of emergency situation of even power and connectivity disruption during the disaster, the city construction engineers must gear up for these eventualities and keep systems and processes ready to take care for such cases. Again, training modules may be created jointly by the technology companies and the specialists in that area.

Robots and 3D printing are already becoming part of most industries, and it is expected that we will see their usage increasing in the construction industry also in a big way in coming future. Usage of robots may even help in maintaining the desired hygiene standards of a city. Again, the workforce must be training to use these technologies. Additionally, even the city dwellers need to go via this change management process of using technology products.

12.6 Technology horizontals for skilling

Smart cities are cities where everything is connected to each other, and this is highly depended on technologies. Technological literacy is a key to turn a city into smart city, which is well connected, sustainable, and resilient, where information is not just available but also findable. It is not a new thing that smart city is all about providing smart services to its citizens, which can save their time and ease their lives. It is also about connecting them to the governance where they can give their feedback to the government as of how they want their city to be. And this aim cannot be turned into reality without technology. Using technology, the officials are able to gather city intelligence, and this intelligence when integrated with the operations makes the cities smarter and safer. The following seven horizontal technologies may be used to enhance the smartness of a city.

- Information and communication technology
- Internet of Things (IoT)
- Sensors
- Information security
- Geospatial technology
- Artificial intelligence
- Blockchain

A smart city is one that leverages technology to increase efficiencies and improve the quality of services and life for its residents. Smart city initiatives can cover anything from power distribution, transport systems, streetlights, and even waste collection. The idea is to use data and technology to make everyday life easier and better for the people who live and work in the city, while maximizing the use of resources. However, the preceding seven horizontals of training are a must as a background (see Fig. 12.2).

12.6.1 Information and communication technology

Creating a two-way communication channel is very important for a city to be smart, which is enabled using ICT. ICT builds a bridge between the citizens and the government where the citizens can interact with the government and in return, the government can build a city of choice of its citizens. ICT helps the government to analyze the demand pattern of the state and thus create a pool of resources to address the same online. The electronic medium of communication in a community helps in creating a collective intelligence, which can be deployed for resource optimization with the help of analytics and deep learning.

One of the important things about smart city technology is that it encourages residents to get more involved. Common examples include apps that allow citizens to report local issues more easily, or community networking platforms that allow neighbors to connect and share resources. In another example, a low-cost environmental testing kit or mobile application-based process may be used by the residents to collect local environmental data. This crowdsourcing methodology of collecting data helps lowering the

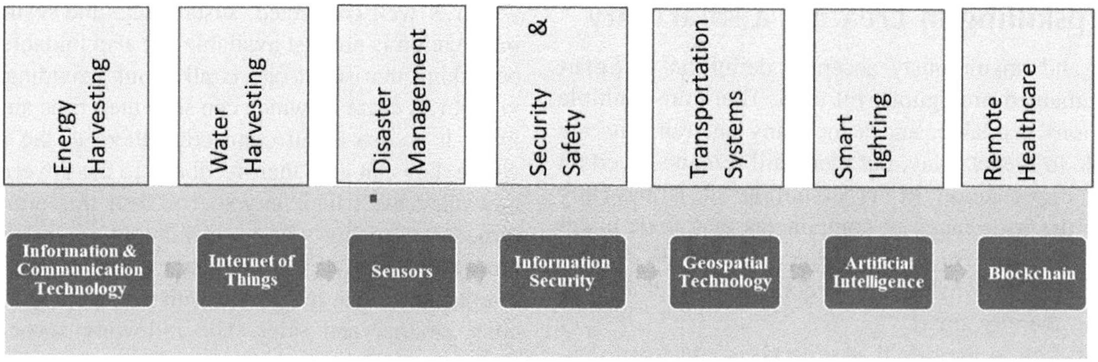

FIGURE 12.2 Smart city training architecture.

costs and speeds up decision-making processes for better governance. This process may be effectively used for collecting information on local environment, including air pollution and noise. The data is streamed to an online platform, effectively creating a crowdsourced map of data from all over the world.

12.6.2 Internet of things

IoT is like veins of the city that pick the pulse of city and are spread all across the city using the communication channels. Every device that is part of a smart city needs to be connected to each other so that they can talk among and can take decisions for themselves, which in return allows managing resources of a Smart City. IoT helps in providing the perfect solution of a set of communicating devices that provide smart solutions to everyday problems. All smart solutions in smart cities are based on "Internet of things" where they are connected and smart enough to decide their action and also provide the desired information to support the citizens.

12.6.3 Sensors

Sensors are hidden but ubiquitous components of the urban landscape. Sensors are a crucial component of any intelligent control system. A process can be improved based on its environment with a connected control system aware of its environment. A process may be typically fitted with an array of sensors, from which it collects the required data. This data together with the set algorithm helps in adjusting operations accordingly. The availability of a multitude of different sensors and continuously evolving connected technology enables applications that were infeasible in the past due to high costs and limited availability. Sensors are like converters, which convert parameters of a physical nature to an electronic signal, which can be interpreted by humans or can be fed into an autonomous system. These signals for conventional sensors, among others, include vibration, water level, light, pressure, temperature, humidity, moisture, and a variety of other parameters. Technology firms can provide the desired trainings on their usage.

12.6.4 Information security and safety

Information Security and safety training is one of the most important horizontals in trainings on Smart City. With so much of data/information flowing around, the citizens need to be trained on how to send their data to this information highway and also about their data privacy rights. With Wi-Fi connectivity, IoT technologies, and CCTV cameras, cities are able to harness technology to improve resident safety and boost incident response times. The real-time video data may be further analyzed to better track and allocate resources on the ground and improve public safety.

In a smart building, a camera detects the license plate of the car, matches it with the personnel records and the list of registered visitors, and raises the gate. For visitors, LED lights may point out the route to the parking place that has been reserved for that visitor. Technology companies need to interface in creation and maintenance of these smart systems.

12.6.5 Geospatial technology

Whatever is built in a smart city has to be right, and so to build a right and sustainable plan, it is required to know where it is coming from with the details the data. Geospatial technologies provide the underlying foundation and ultimately the fabric upon which solution can be built so that specific details on data come with its location. The location allows pinpointing exactly on the need so that better solution can be applied to it. Geospatial technology provides a necessary framework for collecting data and transforming observation in these collections to facilitate software-based solution around smart infrastructure. Technology-based solutions and products have been already created for enabling Geospatial solutions.

12.6.6 Artificial intelligence (AI)

Smart City becomes smart with the digital revolution generating a huge amount of data. This data is of no use until and unless it is processed and converted to useful information. This massive amount of data generation brings the role of AI that can make sense out of the data.

AI allows machine-to-machine interaction by processing the data and making senses out of that. To understand the interesting aspect of AI in the context of smart cities, let us take an example. In a system where energy spikes tend to happen, AI can analyze where they usually occur and under which circumstances, and this information can be used for better management of the power grid. Similarly, built-in intelligence with CCTV cameras can alert on the unexpected events at a particular location, making the process of decision and hence action faster. AI will push a new trend in businesses and enable workers to focus on the aspects that can benefit them and the customers on an everyday basis. Furthermore, it also gives opportunity and the freedom to the workforce to enhance their skills. This is because the tasks that are time-consuming and mundane will be handled by the built intelligence. From the business perspective, AI will help companies to become more productive. This plus point could be used in investing more time and money to engage with the workforce. It will usher in a dynamic company culture while keeping the retention rates high, helping workers to upskill, and also convert the unproductive hours into something productive. Investment in AI trainings can help at all levels.

This is the rising era of AI. And according to reports, if businesses leverage the benefits of AI for a competitive edge, they could gain a unique position in the marketplace. However, to incorporate AI into business operations and Smart Cities, it is important that the workforce is familiar with it. Hence, the first step to bringing AI to the business environment is by skilling the workforce to enable a cognitive collaboration between human and AI.

Taking the broader context in the view, skilling the workforce to work efficiently and smartly with AI is not just critical for businesses but also for the smart cities. Because smart cities are equipping the urban space with the intelligent technologies, providing people with a smart infrastructure that is ready to enhance their way of working, living, and relaxing. So even as citizens, we will need to learn to collaborate with AI in every walk of life in the near future.

12.6.7 Blockchain

Blockchain applications are a new to smart city concept. Blockchain technology secures data flow and ensures better data integrity. Its integration into smart cities could better connect all city services while boosting security and transparency. Some countries are already using it to manage Certificates, information about Blood banks/vaccine and its availability. It is also being effectively used as a tool to main land records and billing details. In some ways, blockchain is expected to influence cities through smart contracts, which help with billing, processing transactions, and handling facilities management. Smart contracts are self-executing contracts with the terms of the agreement between buyer and seller directly written into lines of code. They permit trusted transactions and agreements to be carried out among disparate parties without the need for a mediating third party, making the process safer, cheaper, and faster. Blockchain can be very effectively used in smart grids to facilitate energy sharing, a concept that is trending these days. Some Smart Cities are experimenting this concept of Blockchain for even better supply chain and logistics management. Technology companies can help in creating models and products and support the training.

12.7 Preparing for the future technologies

Technology companies can also help in preparing the citizens on future technologies to enable a smoother change management for the cities. For example, the Smart Cities powered by the latest technologies and 5G/6G connectivity are set change the city life and are also expected to benefit many technology companies globally in the coming couple of years. It will enable the creation of a faster infrastructure. As an example, a smart electrical grid will be able to monitor and analyze energy consumption, smart streetlights with gunshot detectors to improve police threat response, and connected traffic lights across the city to facilitate a seamless flow of traffic. Faster connectivity is also expected to fuel the growth of driverless cars and Drones. 5G/6G will be a key element in making smart cities a reality and, in the shorter term, it may even transform the way we travel, shop, learn, work, and even use City Resources. The technology companies can partner with the citizens to train the workforce in this upcoming technology and also prepare the citizens for the future.

12.7.1 Supporting Internet of transportation and logistics 4.0

There are several ways in which new technology including AR, image analysis, and faster connectivity using 5G/6G will change transportation, and they are all interdependent. The newer vehicles already are becoming more and more digital. First of all, 5G will help to create a network of interconnected vehicles and provide almost instant communication between these devices. To reach real-time communication, vehicles need to be receiving and sending data as fast as possible, which means sending it directly to other vehicles, rather than through a database. Such communication requires an out-of-vehicle connection as the technology used for in-vehicle connectivity does not require a constant mobile signal. In fact, today vehicles are processing the information they receive on board making them somewhat of a computer on wheels. As 5G/6G becomes widespread, all the processing will happen in the cloud, with a signal that will not bounce off buildings, vehicles, and other surfaces.

Self-driving cars will need to respond to surroundings instantaneously with built-in intelligence together with faster communication protocols. With the current 4G connection speed, they can do so in about 26 ms, but 5G will reduce this to just 10 ms. This has a lasting positive impact not only on autonomous decision-making and the smoothness of a commute, but also on the safety of drivers, pedestrians, and the entire city.

The newer technologies will enable smoother trips, which will eventually reduce exhaust emissions while allowing public transport to run more efficiently. With the 5G connection information flows at more than double the speed, making the decisions faster. For example, if there is an accident on the road ahead or cars are breaking sharply, the closest car could almost instantly communicate this information to the ones behind. In turn, they can start changing their routes to avoid creating congestion. This Internet of Transport can do more than just reduce traffic and make roads safer. It will help local governments to map the movement of all vehicles, allowing for real-time traffic management and visualization of the flow and capacity of a city. Connected vehicles will also contribute to the development of vehicle or ride sharing with precise real-time location tracking. However, implementation of these technologies will require upskilling the manpower on newer skills and the citizens on usage of this, and technology companies can provide the required support on it.

To reach this connected future, several improvements need to be made to infrastructure, which requires upskilling of manpower. The next generation of mobile technologies will require physical changes to the existing network architecture. At this point, hardware such as new antennae will need to be set up to allow for optimized capacity. With this in mind, it seems that the future of connected transport is in the hands of governments and providers. Technology companies will need to partner for skilling to make concepts of Internet of Transportation or Logistics 4.0 a reality. They may even start with small demonstration centers to introduce the citizens to this concept and expand it.

12.7.2 Connected city drones

Faster connectivity and newer technology will also pave way for City Drones, which will increase safety in the city and can also be used for faster transportation of equipment, medicines, or goods as the need be. Already, some cities are using Drones for policing and also for firefighting for high-rise buildings. **Drones** are playing a crucial role there. Most of the city police or private agencies use **drones** to keep an eye on the crowd during any event. **UAVs** are helping in smart policing.

Drones have flexibility in performing difficult tasks, supporting high-resolution imagery, and can easily cover the remote areas. However, usage of these requires training the work force. The related usage areas can be surveillance and security, Inspection and Detection, Surveying and Mapping, Emergency Transport and Delivery, Farming and Agriculture, and even firefighting. Drones are already getting used for Smarter Agriculture and in reducing weeds and also help in usage of right usage of pesticides and manage precision agriculture. A large number of technology companies are already working on using Drones, where they customize the drones for its required usage. The technology companies can help create, deploy, and train the Drones and users of drones to take real-time decisions.

12.7.3 Connected healthcare

New-age connectivity and availability of equipment and processes for remote management have given rise to the new business vertical of Connected Healthcare. Starting from connected ambulance, to connected patients, to connected healthcare assets and resources, and even connected pharmacies, Healthcare 4.0 aims at predictive processes for Wellness and Healthcare. This requires not only connectivity, but a technology connection of all of the components of the healthcare predictive processes, into a distributive network. This starts a methodology to approach healthcare in a new way and ensures that even citizens become more responsible for their health and wellness. Technology companies can play a very important role in this complete process of change and skilling.

12.8 Expectation from technology vendors

Citizens expect value-driven service from the technology companies. They need help and handholding also at times in their journey to adoption of new processes in Smart City. The technology companies must train and engage Citizens and local businesses to create Innovative Technologies that can help improve life in cities, make economical use of resources, and ensure stable economic growth. This also requires cities, technology vendors, and public funding institutions to work together to even develop the market.

12.8.1 Supporting innovation and digital city

The Smart City should create an environment of support innovation and create demonstration laboratories for the citizens to see and test their thoughts for their future innovations. Some of these lab environments will be helpful in providing an environment for learning, collaboration, and cocreation. Some of these cocreating centers and centers of excellence (COEs) may even be created by the technology companies in PPP mode with government and citizens. These kinds of models will also be able encourage new start-ups and innovators. These COEs can have best practices and use cases for "Smarter" processes. These COEs should also be able to provide the desired technology environment consisting of connectivity, sensors, 3D printers, testing devices, and others to be able to support the start-up culture so that newer products can be conceived, tested, demonstrated, and then operationalized. The technology companies can also create the cluster groups in different technology areas to enable collaboration.

12.8.2 Digital workplace and centers of excellence

The Digital Workplace and Centers of Excellence can function as a resource integrator of various ecosystem players, the local technology start-ups, system integrators, and global corporate enterprises, to cocreate solutions for smart cities, from ideation to testing to commercialization. This collaborative model provides a platform for local and overseas companies to exchange ideas, identify opportunities for new business models and innovations, as well as facilitate deployment of smart cities solutions.

The COE, particularly, can play a critical role in establishing the practice of automation into a habit and nurturing the automation culture to truly democratize automation within the organizations and cities. Usually, the automation programs run by COEs target large sets of cohesive manual processes and help innovators with space and connect to innovate. This also helps in creating a citizen developer community within the city. Over a time, this helps in increasing the pace of automation in the city. As the enabler of automation, the COE's role in developing the local community can be very crucial. It needs to work on a framework to expand awareness, organize training programs, and foster positive motivation.

An intelligent workplace is described by experts as a workplace that utilizes the increasing digital transformation of physical artifacts to provide innovative methods of functioning and increase the productivity of the workforce. During the outbreak of COVID, remote work became the preferred way of working, especially for large global enterprises. This makes the need to help employees do their jobs more in a secure and effective way, improve productivity, reduce costs, and provide an exceptional customer experience. In fact, these are common goals for organizations seeking sustainable, profitable growth, and digital technology enables it. Technology companies have come forward to provide such working environments.

12.9 Summary

Smart Cities for a sustained development must also have systems and processes to skill and upskill the manpower of city including the citizens. The skilling in some cases can even start at an early stage of education, where students can be introduced to different vocational and technology skills such as coding, robotics, information security, etc., and technology companies can help in creating eLearning modules and support in providing the desired infrastructure and tools for it. This will help in not only creating digitally savvy workforce, it will also help in creating the next generation of innovators.

Smart Cities present an opportunity for rendering Web of Things—enabled services. The anonymized data may be used for bringing out the required statistical reports for the city, for better planning and utilization of City Assets. Standardization of Infrastructure, Processes together with Open Data frameworks helps in faster scaling of applications across cities. Most Smart City technology implementations are supported by technology companies, researchers, and International Research Bodies. Stakeholder Consultation is a methodology to improve and add to the Smart City Operations and Facilities. While prioritizing the investment in applications and infrastructure, a social view of Smart Cities must be taken. This view may include improving ease of life to job creation and cover city sustainability goals.

Finally, let us move on to the real interactive part of this chapter: review questions/exercises, hands-on projects, case projects, and optional team case project. The answers and/or solutions by chapter can be found in Appendix G.

12.10 Chapter review questions/exercises

True/false

1. True or False? A Smart City uses Digital Technology, People, and Processes to improve outcomes across every aspect of city operation and enhance the services it offers to its residents.
2. True or False? With more and more persons living in cities, our cities are already facing growing environmental, societal, and economic challenges.
3. True or False? The Smart City leaders are realizing that smart-city strategies start with technology and not only people.
4. True or False? Connectivity is the base infrastructure requirement of any city to make it Smart.
5. True or False? Traffic impacts a city-dweller's life, and technology offers few solutions to make it simpler.

Multiple choice

1. Data has value and can be monetized by both selling raw data or derived insights to other players in the:
   a. Smart sustainable development
   b. ICT-based infrastructure
   c. Market
   d. Investment
   e. Policy
2. The need for healthcare transformation and methods for handling new diseases came as a big requirement during:
   a. 2017–2018
   b. 2018–2019
   c. 2019–2020
   d. 2021–2022
   e. 2020–2021
3. Globally, countries use __________ courses for Disaster Management, which have proved to be effective.
   a. Offline
   b. Virtual private
   c. Cyber-Physical Systems (CPS)
   d. Artificial Intelligence (AI)
   e. Online
4. One of the major requirements of a smart city is to promote.
   a. Outside
   b. Local
   c. Internationally
   d. Independently
   e. Within
5. A Smart City brings the requirements of _____________, starting from handling Smart lighting, Smart Streets, city equipment, buildings and infrastructure assets, or even capital goods used at homes.
   a. Manufacturing Clusters
   b. Artificial Intelligence
   c. Smart Asset Management
   d. Digital Manufacturing
   e. 3D Design Management

Exercise

Problem

Describe the challenges and suggest changes in content and delivery options that can be implemented in urban planning and public affairs programs.

Hands-on projects

Project

Do research: Concentrate on administration, people, and knowledge creation, for developing an organization's e-training platform; thus, helping in the building of smart cities with digital enterprise, administration, and people or smart citizens.

Case projects

Problem

Look at the future of smart work and how the network of education, smart workforce, and business professionals are working hard to broaden participation in the economy of tomorrow.

Optional team case project

Problem

Look at how the power of unified learning is considered a new approach for smart cities and utilities and how it delivers engaging custom and generic training programs for industrial operations.

References

[1] Garg S, Mittal SK, Sharma S. Role of e-trainings in building smart cities, 8th international conference on advances in information technology. 2017.

[2] World Economic Forum, reportFuture of jobs report. https://www.weforum.org/reports/the-future-of-jobs-report-2020.

[3] WEF 2021: 'Skilling workforce in post-pandemic world must to ensure global growth' https://www.businesstoday.in/latest/trends/wef-2021-skilling-workforce-in-post-pandemic-world-must-to-ensure-global-growth/story/429381.html.

[4] National Education Policy 2020, Ministry of Human Resource Development, Government of India, https://www.education.gov.in/sites/upload_files/mhrd/files/NEP_Final_English_0.pdf.

Chapter 13

Training professional smart city workforces through a partnership with technology firms

Tuncay Ercan and Mahir Kutay
Yasar University, Izmir, Turkey

13.1 Introduction

The term "smart city" is currently at the intersection of digital technologies, all transformative innovations, and urban living spaces. Smart cities, which are the development area of many new ideas due to their increasing population, are the first exciting places that many people think of to work and live. As the world is increasingly urbanized, large cities are also driving economic growth and becoming an attraction for different professional skills. In addition to this positive change, cities are constantly exposed to social problems. We can summarize these problems as follows:

- Unemployment
- Increases in crime rates
- Increasing energy need
- Population growth and its pressure on city infrastructure
- Increased responsibility of public institutions

The fact that technological innovation, which is the biggest driving force in the transformation into smart cities, has the potential to solve the problems of big metropolises increases the speed of use by all segments of the technology. The other most important content for smart cities is "data" which is the lifeblood of smart solutions. Being able to harness the power of all kinds of data collected for smart solutions that can address the real needs of the city people and that will be found meaningful by them is an important challenge. The third cornerstone of smart cities is smart people. Therefore, focusing on employment and winning the fight for the proper skills and talent is critical to sustainable economic growth [1,2].

Today, most city administrations that start smart city investments make significant investments in new technologies and see the positive effects of this on the economic environment and society. For this purpose, it is very important to have a large number of talented and well-trained employees for new business lines and new organizations established in cities. Since this need will increase exponentially in the coming years, there is always a need for an urgent solution. To become a smart and even a brand city, we should collect data about all utilities and systems in the city and be able to use the information obtained for our cities again. We need applicable systems where we can share the data to be able to comment on the data. Smart transportation systems that solve the traffic problem, security camera systems that reduce the crime rate, e-government applications, and the ability to keep citizen communication at a high level are the first smart city information systems coming to mind [3].

All cities in the world can imitate different smart city solutions with each other. However, the only difference that leads to success and achieving the goal is that local governments and the private sector must act together. Besides, both parties should explain all processes to their citizens and create a synergy for transformation throughout the city. By including educational institutions in this initiative, information such as crime zones, emergencies, construction, and excavation permits, mass demonstrations can be transferred to the citizens in real time with open data platforms that can be easily accessed by everyone in the city [3]. Thus, both technological applications and cooperation between public and private units can ensure that the city and the citizen remain in contact and increase public satisfaction to the highest level. The saying, "To be a smart city, data is needed, and technology companies are required to process this data" summarizes this issue very well [4].

Although there are many definitions of Smart Cities, cities that almost their all public services such as

Smart Cities Policies and Financing. https://doi.org/10.1016/B978-0-12-819130-9.00005-X

management, education, health, public safety, transportation are integrated by information and communication technologies and used efficiently and effectively, should come to mind [5]. Smart cities are also defined as cities that monitor the instantaneous status of all critical infrastructures such as roads, bridges, tunnels, rails, subways, airports, ports, communication, energy, and where all the resources used are better optimized [6]. Authors define the smart city as a city where information and communication technologies combine with traditional infrastructures and coordinate and integrate using new digital technologies. The smart city, which has many more definitions in the literature, can be defined as the application of technology to cities for a sustainable life and urbanization [7].

In the face of population growth, increasing rates of urbanization, and rapidly depleting resources, creating urban solutions, and maintaining this, regardless of the size of cities, made it necessary to develop smart approaches and created the idea of "smarting cities." The existence of information technology technologies, which are developing rapidly and have a great potential in terms of increasing the quality of life at every stage of urban processes, has made it necessary to develop smart approaches to create socially and environmentally sustainable cities, leading to the idea that it is possible to make cities smart in many ways, not one-sided. This approach, which targets not only the physical conditions of the cities but also the different development processes in the social field that concern their citizens, has been named the "Smart City" concept [8]. It will be possible to talk about a sustainable smart city when the applications in line with the concept, which is predominantly technology-oriented, are integrated with the social and social areas complementary [9].

The chapter is organized as follows. Section 2 makes statements about the new workforce qualifications and the training requirements in smart cities and also discusses the perspectives of the government and private sector on the subject. Section 3 lists the competencies required for the qualifications of the workforce in smart cities and provides information on policies and how to meet the needs in this regard. In Section 4, while giving information about the training and expertise areas for the required workforce, the impact of government, information technologies, and in-service training, the importance of internship and university–industry cooperation are explained in detail. Section 5 provides information on labor transitions, relations with technology companies, e-education, distance education, company collaborations, and workplace support. Section 6 concludes with a summary.

13.2 Smart city workforces

Technological developments and innovative approaches, which are constantly increasing in number and quality, have a strong transformative effect on city life. This transformation is expressed as a Smart City, by establishing strong networks and ensuring interoperability between the entities in the city in cooperation with all ecosystem stakeholders. It aims to enable the city to manage itself with future predictions based on data and expertise without the need for human intervention.

For this reason, Smart Cities attract the attention of countries, and countries make special efforts to understand the Smart City concept, to be prepared for changing conditions, to adapt and direct this dynamic process. Countries can prioritize a more effective and efficient Smart City Ecosystem policy by looking at their Smart City policies in the current conditions and all public services that need to be provided and implement them with the right projects and activities [10].

13.2.1 New features of workforce

It is becoming increasingly difficult for smart city managers to find trained employees for new business areas. The development of technology has accelerated so much that a significant part of the information taught in universities and vocational schools becomes useless in a short time. This makes methods that can be called in-service or lifelong education more useful. The resources that can be used to achieve this goal are summarized below.

The fact that the digitalization of the production phase in the labor market causes changes has brought the issue of technological unemployment back to the discussion. The extent of technological unemployment depends on the ability of the workforce to know new production techniques as well as the substitution of the machine for the workforce [11].

While the rapid increase in the processing power of computers brought "smart machines" to the surface, the sudden breakthrough of the autonomous industry led to the decrease in the labor force in the production and caused robot machines to be used more in production [12]. With the rapidly developing computer technologies in recent years, society has faced a new problem. This new problem is the dream of full employment falling out. The high levels of unemployment in economies in recent years are also among the structural features of today's new industrial economy. Today, machines are capable of performing most competencies and skills that the labor force can do. However, while offering all these competencies and skills, it can also perform faster and better than the labor force. As a result, many occupational groups owned by humans are becoming history with the rapid and dizzying development of automation [13].

The most important issue in the smartening process of a city is the Internet of Things (IoT). It is possible to define the IoT as an infrastructure that can connect physical and/or virtual objects based on information and communication technologies and use this system for all smart city services [14].

13.2.2 Training capabilities and opportunities

All concerns about the adaptation and effectiveness of work training in the labor market, which is necessary especially for smart cities, require further research at the national level. The training process in traditional business models is entirely based on the human factor. The smart factories of the future and the use of robots based on automation and the possible effects of all these on the labor market are one of the reasons why society has viewed this process in a threatening way. Robotization in production processes can be accepted as an indicator of the substitution of labor and capital. Repetitive works performed in assembly lines transferring it to automation can be considered as a factor reducing the employment of unqualified labor force in the short term. However, in addition to this, a highly qualified workforce will continue to work in interaction with the machine as a decision-making mechanism. This naturally means the need for a highly skilled workforce will increase.

Another point that should not be forgotten is that the replacement of the workforce with machines may have a decrease in employment rates and will also make it possible to minimize the error rates in production. It is the skill factor that is critical for people to be involved in the production and to be in the working workforce. This can deepen the low skill/low wage, high skill/high wage distinction [15]. For this reason, as a personal preference, in the automation period realized with the intensive use of technology in production, specialization in the workforce structure has come to the fore [16]. Of course, beyond such a personal preference, the society as smart city residents will be more pleased with the fact that the people working in the production and service sector around us are highly skilled and do their jobs well.

Change in production processes has led to the emergence of different areas of expertise. Reflections of technological change, artificial intelligence in factory usage, and developments in digital networks are developments that change the quality of the workforce and the way they do business [17]. Accordingly, the solutions that should be planned by the state or private sector are as follows [18]:

- Universities, continuing education centers, educators should focus on programs that will provide competence in the fields of digital technology and support postgraduate education.
- Workforce and industry should be brought together, cooperation among stakeholders should be increased, awareness of digital transformation should be spread to everyone in society.

13.2.3 Training capabilities and opportunities government trends

In today's world of constant change, public institutions have to be more intuitive to perceive and respond to emerging technology opportunities, social barriers, and citizen needs. Public institutions need to be more integrated to provide good service to their citizens. It is important for governments to seamlessly integrate and streamline data and process flows, find new solutions, increase security, and create personalized and engaging citizen experiences [19]. The Public Trends 2020 report focuses on nine of the most important trends that will affect public organizations today, aiming to assist public organization leaders who see the need for change and seek innovative ideas for the best way forward [19]:

- Innovation accelerators
- To act to be good
- Citizenship experience in the state
- Smart government
- The rise of data and artificial intelligence ethics
- Digital citizen
- Artificial-intelligence-powered government
- Cloud as an innovation starter

Governments all over the world are trying to create different business lines in the digitalized industrial environment and also work on collaborations with the private sector. Regardless of what the industrial sectors are, different technologies are transformed into different training opportunities by the state or private sector, creating opportunities for a new workforce. For now, the aim is to be able to change the working conditions as needed by smart cities and to share the advantage of all possibilities as a whole city. The rapid adoption of all relevant factors as listed above will enable both the state and private sectors to use them to their full capacity in a shorter time. Smarter government structures will provide solutions where different ideas are brought together, more informed decision structures, and where the goal is to continually improve society's living.

13.2.4 Status in the private sector

In smart cities, city dwellers feel high quality in almost every phase of their lives such as economy, mobility, living conditions, and education. The ability to achieve this depends naturally on the information technology infrastructure with the state and private sector facilities, as well as the individual and social capital that the target is people. The continuation of this depends on the fact that each component follows each other intelligently in a cycle such as a city, economy, people, state, mobility, environment, and life.

13.3 Skills gap

Another important point in this framework, which is expressed with the concepts of "smart life" and "smart person," is that ordinary people and their communities should have the skills to use information technologies to reach and benefit from urban services, innovations, and life quality enhancing activities to the maximum extent. The growth of smart cities created many jobs requiring new technological skills. Scientists and engineers develop technologies for smart cities, but the skills of the existing workforce lag behind them. A big group of technology professionals exist but not enough skilled for new technologies. In the first step, the necessary skills for smart cities should be identified. In the second step, policies to bring these skills to employees should be determined.

The number of vacancies is increasing with the job positions opened for smart cities. The reason for this is that professionals do not have the right education because existing education institutions failed to synchronize with smart city employers [20]. Most of the smart city employers demand skills that they could not acquire during their education life in job applicants.

In many smart cities, there is a tendency to find alternative training resources, which will be able to reduce the workforce shortage by providing much shorter pieces of training. For example, digital literacy training can be given to employees in all relevant departments. Courses can be designed and conducted with the help of technology firms following the needs specified. Nowadays, some technology companies are contributing to meet the workforce needs swiftly by establishing such partnerships with smart city administrations. Investing not only in smart technologies but also in the development of the skills of the workforce is vital for smart cities.

13.3.1 Skills analysis for smart city workforce

A skills gap can be defined as the difference between the skills that smart city employers need and the skills that the existing force's workforce provides. Skills needed can be correctly identified by conducting a skills gap analysis. Employee training schedules and recruitment programs can be prepared. The skillset needed for smart cities can be grouped into three main business groups as shown in Fig. 13.1.

- Technology skills
- Data skills
- Community skills

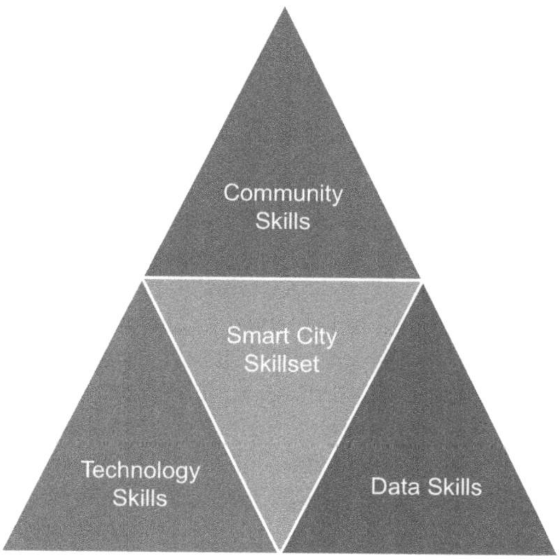

FIGURE 13.1 Skillset groups for smart cities [21].

Technology skills: Smart city technologies offer productive, efficient, and attractive solutions to the complex problems brought about by urbanization [22]. The building, operating, and development of the smart city can only be achieved with a skilled workforce. Technological business areas with the most labor shortage can be summarized as follows:

- Telecommunications
- Energy
- IoT
- Network Architectures
- Data Center Infrastructures
- Water and Environment
- Smart Buildings
- Health and Security
- Robotic Systems
- Logistics

Data Skills: In the modern smart city, there will be a need for IoT systems that create a huge network and massive information flow. The rapidly increasing trend of connected IoT devices in the world is given in Fig. 13.2 [23]. It will be inevitable to employ many employees with the following ICT qualifications.

- Cloud, Fog, and Edge Systems
- Data Analytics and Artificial Intelligence
- Software Developers and Web Designers
- Data and Network Security Specialists
- Database Specialists

Community skills: The smart city offers a lot of digital facilities to its citizens to participate and take advantage of new services to ease their lives. New digital services support both individual choices and communication/collaboration options of citizen groups. Occupational groups that need a workforce in smart cities are given in Fig. 13.3. Citizens also need training for efficient usage of digital

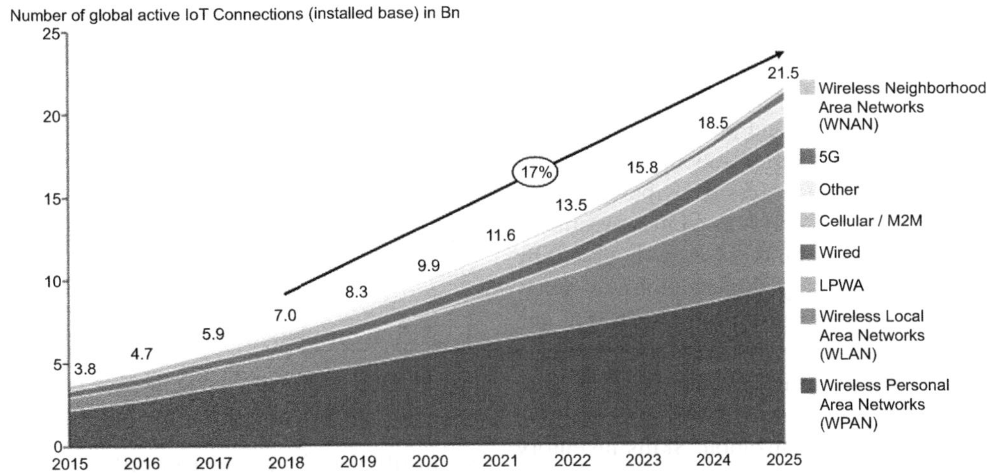

FIGURE 13.2 The rapidly increasing trend of connected IoT devices in the world [23].

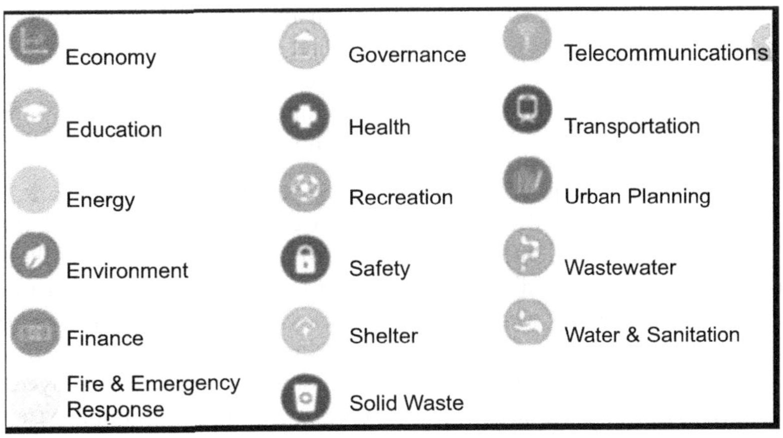

FIGURE 13.3 Occupational groups that need workforce in smart cities [25].

technologies depending on their education, experience, and their needs [24]. The new digital applications can be classified as follows:

- e-Information
- e-Business
- e-Marketing
- e-Participation
- e-Government
- e-Innovation
- e-Transport
- e-Health

13.3.2 Assessing the skills needed

Skills gap analysis can be conducted on three levels:

- Important skills identification
- Measuring existing skills of employees
- Measuring the levels of skills

Smart city managers can identify the skills needed in their cities simply by answering two questions:

- Which skills do we need to turn our city into a successful smart city?
- Which skills do our employees need to be able to do their jobs flawlessly and efficiently today and in the future?

Measuring current skills of employees: Employers can use the following methods to measure employees' abilities and levels.

- Surveys and skill assessments are conducted by professionals.
- Employee interviews.
- Feedbacks are given by managers evaluating performance.

Skill deficiency analyses can be conducted in a shorter time by using software tools such as Skills DB Pro and TrackStar [26,27].

13.3.3 Act on collected data and create a skillset inventory

A skills inventory is a detailed list of all the educational qualifications, work experiences, and professional skills of the existing employees of an organization. A skills inventory is prepared by the HR professionals based on the data collected using employees' surveys, skill assessments conducted by professionals, and the feedback given by managers. Skills inventories help organizations to determine whether the current employee skills are sufficient and where the skills gaps exist.

Skills inventories guide managers and the HR professionals of the organization to prepare workforce training and recruitment plans of the organization. Skills inventory also provides the data to calculate the cost of the investment for the workforce to reach the organizational goals.

13.3.4 Methods for skills development

There are two main ways to train the workforce to meet the proposed skillsets for smart cities:

A new education system should be introduced, which can be divided into two subclasses as graduate retraining and undergraduate education. Education programs offered by schools and courses for the training of different professions are relatively far from providing the current workforce needed by smart cities, and an interdisciplinary curriculum should be prepared for all students who need to focus on developing technologies [28]. Additionally, undergraduate students should learn how to resolve practical problems they will face in the future. Undergraduate education will play an important role in preparing young talents for careers in future smart cities.

Retraining of the graduate workforce may have been required to adapt to new technologies. Graduate workforce whose deficiencies are determined according to their areas of expertise can be trained in cooperation with academic institutions and companies. Training for those who are on a critical mission can be spread over time so that the operation is not interrupted.

The developing computer and communication technology now offers a new training method called e-learning or e-education. The most widely known advantage of e-learning is that it provides a very comfortable and flexible access opportunity for trainees. Training content can be accessed at any time, from any laptop, desktop computer, or mobile device. E-learning eliminates travel, food, space, and educational expenses and provides cost efficiency. In addition, hidden expenditures are eliminated.

As is widely known in the field of education, people do not learn in the same way. Some people learn by taking notes or reading, some with the help of animations and games, some with videos, and some by actually experimenting. The biggest advantage of e-learning in this area is that it can contain different types of content, so that content can be developed by different learning styles. Videos, documents, audio recordings, games, and many other types of content are supported by e-learning platforms. Trainees can instantly receive feedback from the system and have the opportunity to see their levels with quizzes and evaluation questionnaires. Components of smart training are given in Fig. 13.4.

13.3.5 Compensation of skill shortages

There may be many reasons for skill shortages. These reasons can be summarized as follows:

- Insufficient investment in skills improvement
- Weaknesses and incomplete planning in the education system
- A rapid change in the field of technology
- Inadequate wages and working environment for qualified personnel
- Lack of coordination between education institutions, companies, and the government also increases skills shortages

Skills shortage continues to be a major obstacle for smart city applications around the world. Smart cities that fail to fill positions with skilled workers will face both short-term and long-term detrimental consequences of the business. Of course, there are some viable ways to deal with this problem.

Training of current employees: Existing employees can receive on-the-job training based on monthly and annual programs. A knowledgeable employee or an experienced educator with valuable expertise may conduct these courses. This method is the most economical way of costing time and money. Another option is outside training in a training partner. Some cities are sending selected employees back to school and providing the training needed. This option is the most expensive time-consuming way but guarantees that the training quality of the person complies with the standards. Face-to-face or distance education supports with technology companies and/or private educational institutions can be obtained for shorter periods [30].

Shifting current workforce skills to the right positions: Nowadays, in many cities, talented people work in irrelevant jobs and are not productive. Quick solutions can be obtained in some skill shortage areas by reevaluating the employees according to their skills and assigning them to the right positions.

Reconsidering recruitment procedures: An accurate list is created to determine the competencies needed in the recruitment process. Everything written in these lists created for smart cities should be evaluated in terms of real

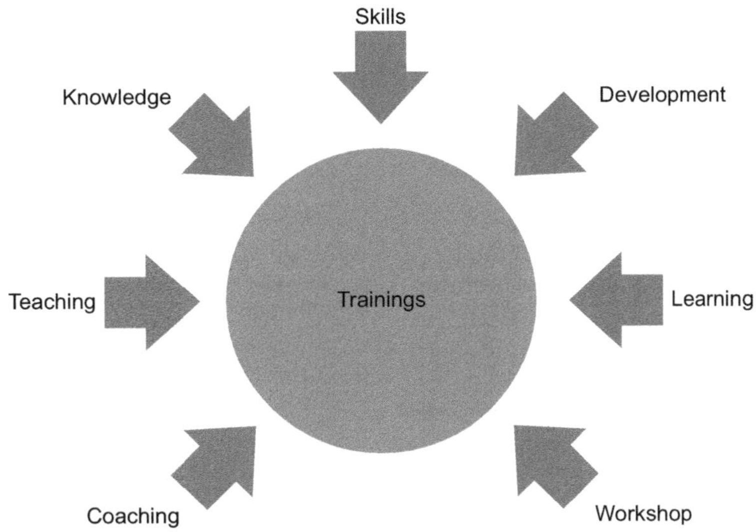

FIGURE 13.4 Components of smart training [29].

personnel requirements and cultural compatibility. Different internal and external channels may be needed to form the first qualified candidate group. Likewise, an effective prescreening process should be established for candidates with the highest probability of success [31].

13.3.6 Policymaking for skills gap

In smart cities, jobs are becoming more skills-intensive at all levels for both new and existing jobs. We can categorize the policy of raising the quality of the workforce for smart cities into three groups:

- Improving the education programs of universities for young graduates
- Training the existing workforce that lacks the skills to adopt new technologies
- Retraining the existing unskilled workforce in work areas where they can be useful

In many universities and colleges, vocational education is limited due to criteria such as the ratio of students per faculty member and the number of teachers who have completed their doctorate [32]. As cities are getting smarter, there will be a greater need for skilled employees for the smooth functioning of city services. Universities must review their education programs and keep up with the times. The main purpose of the university is to raise individuals who are smart, loving to learn, entrepreneurial, and honest. Universities should work in close collaboration with local governments and companies to develop and test new technologies. They must be able to transform their ideas into prototypes [33]. There is an almost infinite number of new technologies available today, so only the necessary training should be selected for the skill needs of the existing workforce. IT technology areas required for smart city applications are given in Fig. 13.5.

Employees may be afraid of unknown issues, so training should be kept simple at first and employees should be motivated. Training can be in-house training if there is company personnel with sufficient competence. Otherwise, it is necessary to benefit from external sources. These resources could be universities, partner companies, professional training companies, or e-learning resources. Training of staff without any skills is more difficult. These people should be directed to appropriate training according to their age, educational experience, and interests. These people will be more efficient if they receive their training under the supervision of an educator. Personnel training should not be done only once but should be repeated from time to time according to evolving needs.

13.4 Relations between education and areas of expertise

In recent years, it has been suggested that education should be restructured in line with the needs of the economy, both nationally and internationally, and to fulfill this, international Global arrangements are made under the leadership of organizations. Nowadays, this function is located in a central position and aims to shape education entirely vaccording to the economy. The importance of education in providing these qualifications is pointed out as new technologies require new qualifications. It is claimed that there is a new industrial revolution defined by concepts such as smart factories, the IoT, cyber-physical systems. Thus, it is claimed that, on the one hand, work, workforce qualifications, and organization of production will change, on the other hand, education should transform in this axis.

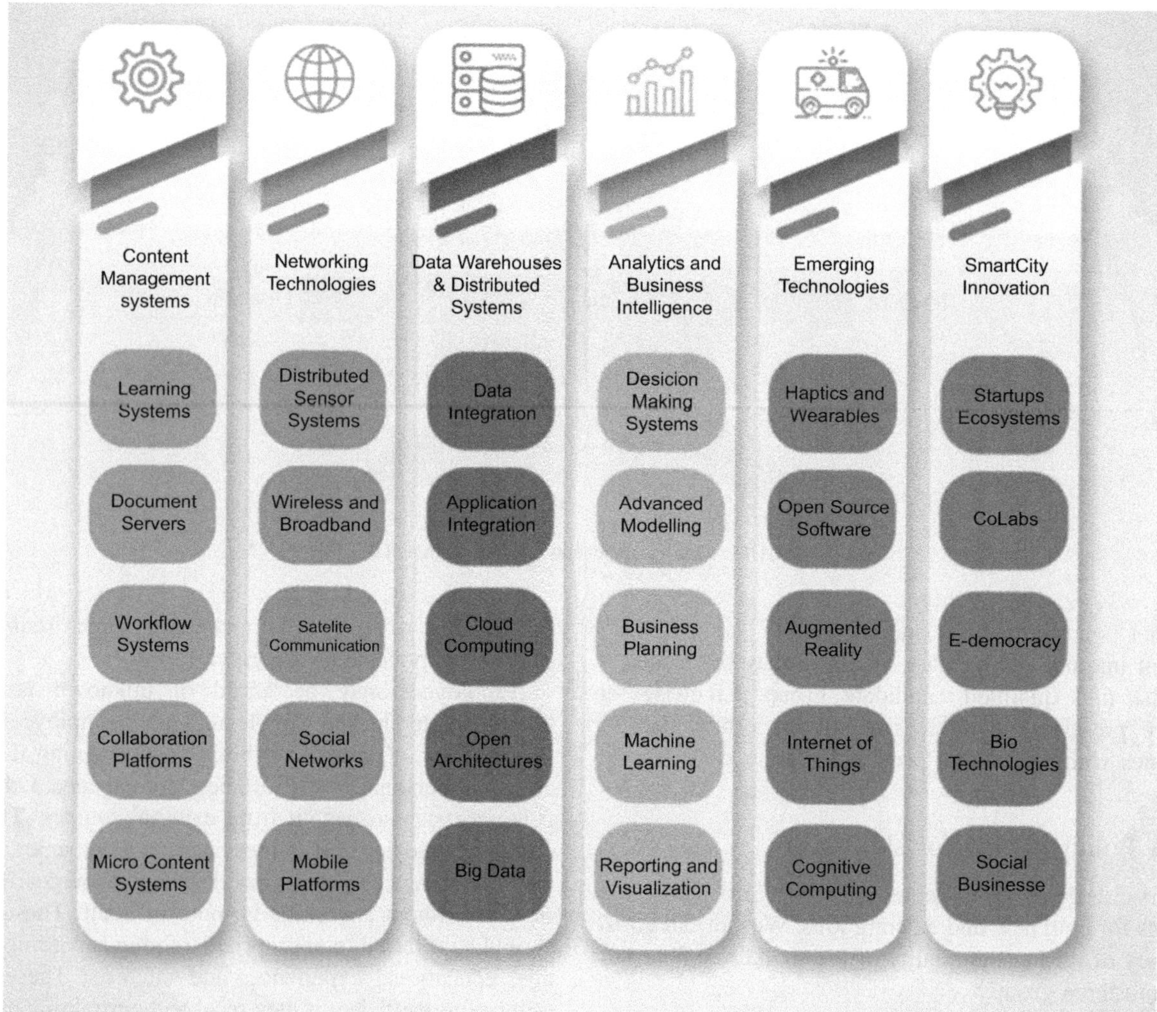

FIGURE 13.5 IT technology areas required for smart city applications [34].

Therefore, the concepts of qualification and employment, which occur in parallel with the changes experienced, should be understood by education institutions [35].

Qualified staff for the ever-expanding public administration cultivation has gained importance. In the process of social change and industrialization, it is one of the basic functions of public administration to adapt public officials to changing conditions and to enable them to serve their positions. In-service training, which is a tool to fulfill this function, is seen as an effective tool in ensuring organizational effectiveness by the officials, as well as in public institutions [36].

13.4.1 Partnering with technology firms to train smart city workforces

What is important for the term Smart City is the modernization effort of cities to use all their resources more effectively and to provide better service to the people living in the city. It must be admitted that cities are the center of attraction for people with the economic and social opportunities they provide. This situation makes cities more successful than rural areas in terms of labor mobility However, the lack of human resources in cities and the lack of awareness about the potential of smart applications have been identified as important obstacles to the spread of these solutions [37].

Cities can provide all public services at a high level for their people, if they can present information and communication technologies, which is one of the most important critical resources, in the most efficient way for both public and private sectors. Competency training stages to increase the skills of the workforce are given in Fig. 13.6.

13.4.2 Government effect

Digital technologies can transform the services and business models in smart cities. If governments' digital

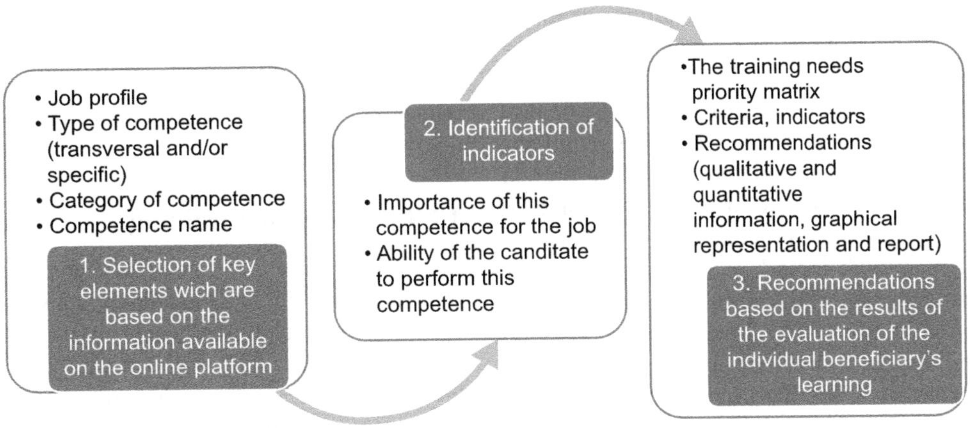

FIGURE 13.6 Competency training stages [38].

transformation support in private, public sectors in their services, smart city transformations will be faster. Government supports can be given for digitalization as shown in Fig. 13.7.

Digitizing government services such as business licensing, tax payment, passport application, etc., decreases the bureaucracy. Other contributions can be summarized as removal of risks, delays, savings of government expenditure, and more efficient usage of staff. Providing free digital pieces of training to the public increases the participation of citizens in digital services. Government can promote digitization projects of the public, educational, and business organizations by funding or subsidizing. The government must set rules and standards to address problems that may arise in the use of digital technologies.

13.4.3 IT effect in training

Technological structuring that develops with today's information society and dynamism has made it compulsory

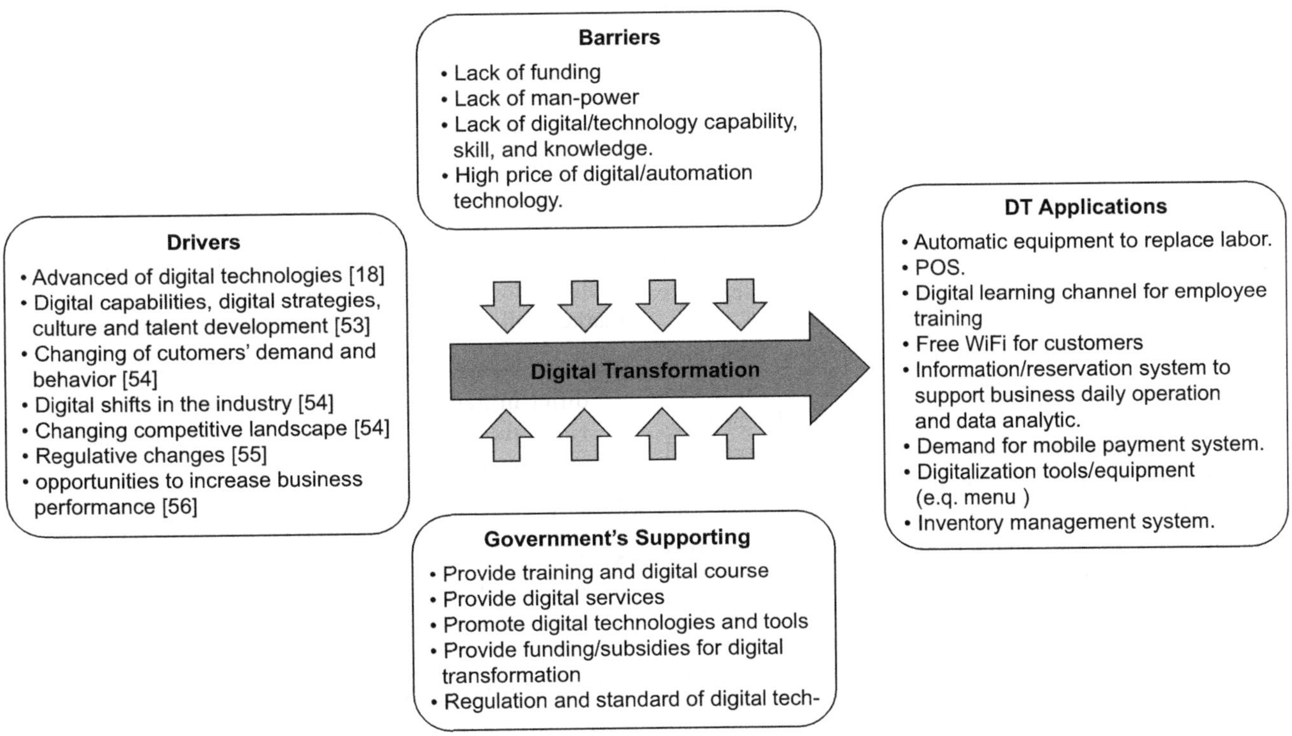

FIGURE 13.7 Government supports for digitalization [39].

for us to utilize technological tools and equipment in the field of science and education. The innovations that the information society and its extension, technological development bring to our lives have also entered the field of education as information technology tools [36]. Computers, which are the most important and most effective tool of information technologies, being used in the field of education have brought new training techniques together.

In addition to making their mark on the last decade, new teaching environments, in which digital technologies are used extensively, continue to develop further. Information technologies are not only a new tool but also a gateway to the world's resources. The importance of information technologies in educational environments can be listed as follows [40]:

- Make students sensitive to informatics
- Facilitate learning scientific concepts
- Develop cognitive ability in students
- Create new educational tools

13.4.4 Automation

Nowadays, urban planning takes advantage of the developments in information technologies. It seeks to plan, monitor, and manage as an intelligent system. Undoubtedly, in smart cities it is considered successful to the extent that developments in information technologies and data processing capabilities in cloud systems can be increased. Therefore, the first trend is that the developing technology will increase the workforce quality of automation systems that are used in production. Production automation is very important. For qualified products to emerge, production automation systems must be installed and monitored in the workshop. As a result of automation, production efficiency will increase and production costs will decrease. According to this approach, as the demand for qualified labor will increase as a result of developing technology, employment will also increase [35].

Automation systems that work in harmony with production tools and successfully fulfill their duties in human control also pave the way for globalization for businesses. Thanks to the automation systems that can be monitored regularly and work in a fast and coordinated manner, automation is in the first place among the indispensable requirements, especially in the field of production, in terms of the ability of manufacturing enterprises to carry out their business worldwide and to meet the demands of their customers or the company's goals in the market area.

One of the striking issues in future studies about the workforce is to reveal how much of the current or future jobs will be done by people or by machines. This distribution is likely to shift in favor of machines in the future. A fair intervention of the business management is required for such a change in the distribution of tasks within the workforce. It will be very important to redesign jobs and manage robot–human interaction as well. It is up to the managers to prepare their employees to cooperate with automated systems.

13.4.5 In-service training

In-service training, to ensure the predisposition of public officials in service, is educational activities that aim to increase their level of efficiency and increase their knowledge, experience, and skills so that they can better fulfill their future duties and responsibilities. The main purpose of in-service training is to ensure efficiency in the work of public officials. In an effective and efficient public administration, the personnel that have renewed and educated themselves according to the requirements of the age are needed. The required personnel can only be trained through in-service training programs that are carried out successfully and using the right tools and techniques [36]. It should not be forgotten that in-service training should be provided for any professional to educate himself or being educated for his profession from the first day he started the profession until the day he left the profession [41].

13.4.6 Internship

The internship is a training method that provides benefits such as improving their skills, working on future tasks that they will undertake, knowing the job conditions and learning the jobs, taking professional responsibility, and adapting to the job environment. In this method, which is very common in the public sector, the person doing the internship, depending on the internship officer, is aimed to gain some skills. The success of this method depends on monitoring and auditing the intern with the reports to be given [42]. The answer to the following questions to evaluate the status of young people entering the labor market will reveal the true value of the internship:

- How the young people participating in the programs differ in transition to the labor market?
- How the internship period affects the transition to employment?
- How first job times of internships and noninternships change?
- The status of the internship places
- Whether the compulsory internship application provides the expected benefit

13.4.7 University–industry cooperation

University–Industry Cooperation (UIL) has an important place in the training and supply of the workforce. Thus, the education model that can be provided can help the

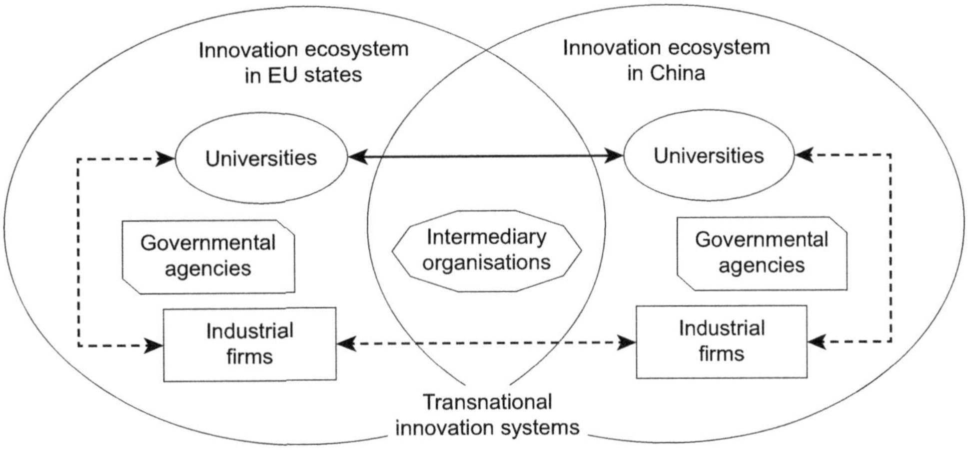

FIGURE 13.8 EU–China transnational innovation cooperation ecosystem [45].

country's industry, vocational and technical education by innovations and contributions. The university system, which is based on research and education, is going through important stages in the context of the developments that are described as the fourth industrial revolution in the world. In this context, trained manpower also occupies an important place. In terms of countries, universities will be the places that will train the qualified manpower that will be needed while adapting to Industry 4.0 [43,44]. As an example, the EU–China transnational innovation cooperation ecosystem is shown in Fig. 13.8.

Businesses generally have the following expectations from common relationships with public or private universities and research theories:

- Explanation about the research
- Access to the community of experts and scientific networks
- Analysis of the applicability of technological innovations
- Cost-effective project proposals that take into account the opinions and suggestions of external sources
- Information about scientific and technological trends
- Renewal of scientific and engineering capabilities for the industry.

The expectations of businesses from research centers and universities are not just to develop commercial products or processes. In addition, they want to voluntarily share their knowledge, technological and industrial infrastructure, workforce potential, education, and research funds with universities [46].

At the same time, region-based learning processes are very important, with the development of education programs that meet the regional requirements and the employment of students and graduates who know their region better. The goal is to create developments with high added value with the organization and strategy put forward by the universities, to increase RandD and PandD activities, to strengthen the competition with the product market, and to support development moves with a specialized structure. At this point, it will be important to come together with internal and external stakeholders, cooperate, and join forces.

13.5 Partnership with technology firms

The same effect is observed in domestic markets, as the technology sector is developing at a rapid pace globally. Not only the informatics sector but also the competencies that the employees should have in almost all positions that technology will enter vary depending on these developments. In developing economies, the technology sector has not reached the saturation point yet; therefore, qualified personnel and strong local capital are still needed in the eyes of employees. Continuing to provide attractive business opportunities for employees in different fields of activity such as e-commerce, telephone, and internet service providers, mobile gaming industry, research and development units in almost all sectors, industrial design companies, software industry, defense industry, consumer electronics, now and soon.

When all businesses operating in the field of technology are viewed from a global perspective, it can be easily seen that cooperation with newly established start-ups in the market is a method that becomes more evident every year for these companies to grow both faster and cheaper. These relationships have been a valuable approach as they increase the available business opportunities and capital accumulation that can be used. Because these mergers, which may cause mutual acceleration, not only add value to both companies and start-ups but also bring strong business connections.

Based on these determinations, to support the transformation of our cities into smart cities, a hybrid approach consisting of "top-down" and "bottom-up" policies has been adopted within the scope of the information society strategy. While guiding documents and case studies about smart cities are presented to the whole society by both the state and the private sector, the solutions offered especially at the local level are also supported by development agencies. Thus, it is aimed to create economic and social benefits by supporting the transformation of local creativity into smart applications, while the local governments are guided with central tools for the transformation into smart cities.

13.5.1 Drive technology and services

In the last 2 decades, advances in information technologies have brought the use of technology to the agenda in solving these chronic problems, especially in existing cities. When looking at the smart city approach in terms of existing technologies and their effects, each different technological approach should aim to increase the quality of life of the citizens as well as being applicable at the city scale. However, this time, the indispensable relationships of different technologies with each other can emerge:

- Transportation and energy together can increase the efficiency of urban systems.
- Reducing air and noise pollution increases the efficiency of services for citizens.
- Developing local economies increase the competitiveness of cities.

There are various approaches in the literature and practice regarding the planning of the transformation process into smart cities and monitoring this transformation process. Among these approaches, Cohen's "Smart Cities Wheel (SCW)" methodology stands out. According to this approach, which is also accepted by the European Union (EU), smart cities consist of six components. One of these components is smart people [37]. Within the scope of this component, it is aimed to create an inclusive society that encourages creativity and innovation by improving people's ability to use and produce information and communication technologies.

The "living laboratory" approach, which includes citizens in the innovation process, is critical in the development of knowledge-based and creative products and services needed during the transformation of our cities into smart cities. Living laboratory practices are organizations that enable the needs of the citizens living in the cities to be revealed, the design and development of the products suitable for the needs, and the testing of these products in real life.

13.5.2 Labor transition

Workers and employers are the two opposing social partners for the usual solutions in job changeovers. In this relationship, while trade unions represent workers and their interests at different levels, enterprises also participate in workplace negotiations as their representatives. These problem-solving meetings can be regional, sectoral, national, or international.

A fair job exchange between both parties ensures that environmental sustainability is maintained and decent work contracts are made legally. Indeed, this is what the Paris Agreement requires: Taking into account the imperatives of a just transition of the workforce and the creation of decent work and quality jobs by nationally defined development priorities—Paris Agreement (2015) [47]. It is an important future concern that people can find a job for themselves and their families and that they can work and lead their lives safely. The importance of dialogue between the employee and the employer should be considered as key, especially in the face of the unfair loss of a job by a former employee. Thus, the government can come together with businesses, trade unions, and civil society groups and cooperate as necessary for a fair transition. A fair transition pattern in labor mobility is possible with taking into account that it is imperative to create decent and quality jobs according to the national development priorities previously defined.

Negotiations, a bilateral process between trade unions and employers, or a tripartite process with their government participation are the essence of fair labor decisions. Therefore, these negotiations can be made in all countries, even in international environments, and workplaces of all sizes. Formal negotiations include consultation and information exchange, economic and social policies, and agreements. In addition, these negotiations play an active role in the creation and implementation of public investments required for labor exchange, labor rights, labor market, social protection policies, decent jobs, education and training, sustainable infrastructure, and community renewal at regional and national levels.

13.5.3 Relations with technology partners

The main problem in the management of technological capabilities and RandD studies is uncertainty. This situation affects all functions of the business. Businesses conduct research both to meet customer needs and to follow innovations. Business managers sometimes establish vertical business groups to encourage the sharing of experiences and the development of new ideas. The main purpose is to transfer market information and potential users to the technology research process. In this framework, there are two policies that business managers should maintain: (1) developing RandD capabilities by creating internal processes, (2) to implement processes that will allow the effective evaluation of RandD projects. By implementing these policies, businesses aim to follow technological developments, diversify their talent centers, and access resources, information, and technology that are not available within their structure. Nowadays, businesses direct their energies to developing talent groups that will best coordinate their strategies. Business managers use horizontal and/or cross-organizational structures that will bring together their resources and skills, control the degree of accomplishment of their strategic tasks, and increase innovation [47].

13.5.4 Role of e-training and distance education

In the rapidly digitalizing world and in a process called Industry 4.0, this change necessitates a serious transformation and change not only in the field of the industry but also in the field of education as in all segments of society. Education systems, which have a dynamic structure, have to integrate with other systems that make up societies. Industry 4.0 education is naturally the basic functions of education, and Industry 4.0 functions are somehow interrelated. Teachers and expert educator staff shape the teaching processes based on these functions. Among these functions, the main purpose of education in the context of smart cities can be defined as the training of qualified human resources by gaining the knowledge and skills that will contribute to the city life and economy with qualified and continuous education for the prosperity and strong economic driving force of the country. According to this definition, the importance of our teachers' being able to train their students according to the job requirements of tomorrow becomes clear for today and in the future.

The most efficient and effective application of information technologies in the in-service training of public personnel is the distance education model. This new model of education allows thousands of public officials to receive training on their own, whenever and wherever, they want. In this way, the institution reduces heavy-duty training costs and at the same time prevents job loss. Personnel who receive in-service training can easily get their training, always at hand, by logging into this online distance education system [36]. Individuals will be trained in an education system where they can receive theoretical knowledge outside the school with distance learning and practical knowledge through face-to-face education, where different tools supported by educational data mining will be used, independent of space and time, according to their understanding.

13.6 Conclusion

Smart cities are urban areas that can collect more data on any subject by their effective information technology infrastructures and manage their resources and services efficiently by using the information obtained from these data. Therefore, smart cities aim to meet the needs of the people following the possibilities of the digital age, establish effective city management, and increase the quality of life of their citizens. These cities have an environment that integrates all physical, digital, and human systems to offer their citizens a sustainable and rich future.

In the traditional governance model of the city, regardless of whether they are interrelated or not, different functionally oriented service providers have to come together for cooperation and innovation potential in all subjects that require data and expertise. For today, smart cities that are developing with a focus on production, innovation, and technology have aimed to constantly seek smarter searches and applications to increase efficiency and improve the quality of life in all public services. With the technological developments brought about by digital transformation, it is inevitable that in the coming years, many professions made with muscle power will be at risk and new professions that include the competencies required by digital transformation will come to the agenda.

It is becoming increasingly important that information processing systems are widely used, and individuals adopt information communication technologies, to obtain maximum benefit from every smart city application that we constantly experience not only in education but also in our daily life. To keep up with this transformation, it is necessary to implement a long-term employment policy by managing this change with a goal-oriented approach. To turn the threats presented by digital transformation into opportunities, first of all, these difficulties should be defined accurately and carefully, and studies should be initiated to solve these problems. Within the scope of this study, the professions that will be affected by digital transformation were examined, the new professions to be needed and the competencies that should be in the human resources performing these professions were analyzed, and an examination was made on private and general sector training supports for the efficient training of human resources with the required competencies.

13.6.1 Recommendations and solutions for the future

First of all, an approach that takes into account the change that will occur with the effect of digital transformation should be well adopted by both private and public sectors. It should be done urgently to plan how the training programs will be updated in the short, medium, and long term according to the new competencies needed to include teaching-learning, analysis, creativity, problem-solving, communication, and ethics.

It has become important to plan how the human resources, which are in danger of extinction due to the digital transformation of their professions today, can be brought back to the economy, which competencies will be gained by which method, and to determine the institution to be responsible and the activities to be carried out. It is necessary to implement a long-term employment policy by managing this change with a goal-oriented approach in order not to encounter the problem of increasing unemployment and to keep up with this transformation.

13.7 Summary

In the first part of this chapter, the definitions of smart cities are made, and the effects of expertise such as data, data collection, data analysis on digitalized cities, and job opportunities are mentioned. The importance of the cooperation of the state and private sector in all technological areas, smart application approaches, and education throughout smart cities was emphasized. Later, the workforce competencies needed and the training support required were mentioned. An evaluation has been made regarding the current and future possible jobs for smart cities and the search for new competencies that these will bring, and to what extent existing education and training contribute.

In the next fourth part, the relationship between experience and education in different industrial areas that are digitalized is reviewed, and the effects of regular education, intracompany training, internship periods, university—art cooperation, and industrial automation on workforce competencies are discussed and different technologies and services are mentioned. In the last section, an evaluation has been made on job differences and changes, changes in technological companies, the role of remote synchronous/asynchronous education, and firm support.

Finally, let us move on to the real interactive part of this chapter: review questions/exercises, hands-on projects, case projects, and optional team case project. The answers and/or solutions by chapter can be found in Appendix G.

13.8 Chapter review questions/exercises

True/false

1. True or False? It is necessary to change the training programs in technical schools and universities to train a highly qualified workforce, but the results of this can be seen in a long period.
2. True or False? Skills inventory is prepared by the technical managers based on the data collected by means of employees' surveys.
3. True or False? Skills gap analysis can be conducted on three levels:
   - Important skills identification
   - Measuring existing skills of employees
   - Measuring the levels of skills
4. True or False? The biggest disadvantage of e-learning is that it contains different types of content and it could not be developed easily.
5. True or False? A skills gap can be defined as the difference between the skills that smart city employers need and the skills of the existing forces workforce provide.

Multiple choice

1. The number of vacancies is increasing with the job positions opened for smart cities. The reason for this is that professionals do not have the right education because existing institutions failed to___________with smart city employers.
   a. Research and development, cooperate
   b. Educate and synchronize
   c. Government, cooperate
   d. Social, synchronize
   e. International standards, cooperate
2. Existing employees can receive _________ training based on monthly and annual programs.
   a. On-the-job
   b. Vocational school
   c. University
   d. Business
   e. All of the above
3. Skills needed can be correctly identified by conducting:
   a. An exam
   b. Job interview
   c. Survey
   d. Skills gap analysis
   e. All of the above
4. Skills inventory is a detailed list of:
   a. Educational qualifications
   b. Work experiences
   c. Professional skills
   d. Retrainings
   e. All of the above
5. The developing computer and communication technology now offer a new training method called:
   a. On-the-job training
   b. Hands-on training
   c. E-learning
   d. Self-learning
   e. All of the above

Exercise

Problem

Discuss why many cities and local communities are eyeing advances in technology as a way to increase efficiency, reduce costs, and improve quality of life for their residents—all in the face of shrinking budgets.

Hands-on projects

Project

Do research: Analyze why investing in smart-city infrastructure today, local governments will be laying the foundation for the sustainable societies of tomorrow.

Case projects

Problem

Develop a toolkit for government officials, urban planners, citizen groups, and others who want to implement successful smart cities projects.

Optional team case project

Problem

Describe how to equip communities with the know-how to build long-lasting partnerships that contribute to vibrant and sustainable smart cities.

References

[1] Deloitte, Smart Cities, <https://www2.deloitte.com/tr/tr/pages/public-sector/articles/smart-cities.html>.

[2] Version 1.0 Gov lab, Deloitte, smart cities, how rapid advances in technology are reshaping our economy and society. 2015. https://www2.deloitte.com/content/dam/Deloitte/tr/Documents/public-sector/deloitte-nl-ps-smart-cities-report.pdf.

[3] Kumar S. Top 10 best smart cities in the world—most impressive. 2015. https://www.worldblaze.in/top-10-smart-cities-in-the-world/.

[4] Akıllı Şehir Olmak için Ne Gerekiyor. 2020. https://www.endustri40.com/akilli-sehir-olmak-icin-ne-gerekiyor/.

[5] Bélissent J. Getting clever about smart cities: new opportunities require new business models. Forrester Research, Inc; 2020.

[6] Hall P. Urban renaissance/new urbanism. J Am Plann Assoc 2020;66(4):359—60.

[7] Batty M, Axhausen KW, Giannotti F, Pozdnoukhov A, Bazzani A, Wachowicz, Ouzounis G, Portugali Y. Smart cities of the future. Eur Phys J Spec Top 2012;214(1):481—518.

[8] Ateş M, Önder DE. The concept of 'smart city' and criticism within the context of its transforming meaning. Megaron Journal 2019;14(1):41—50.

[9] Mandl B, Zimmerman-Janschitz S. Smartes cities-ein modell Lebenswerter Städte. Wien: Real Corp Tagungsband; 2014. p. 611—8.

[10] 2020—2023 Ulusal Akıllı Şehirler Stratejisi ve Eylem Planı, T.C. Çevre ve sehircilik bakanlığı. 2019. https://www.akillisehirler.gov.tr/wp-content/uploads/EylemPlani.pdf.

[11] Bozdağlıoğlu UY. Türkiye'de İşsizliğin Özellikleri ve İşsizlikle Mücadele Politikaları. Sosyal Bilimler Dergisi 2008;20:45—65.

[12] Aydın N. Moore yasası ve geleceği. 2016. http://www.endustri40.com/mooreyasasi-ve-gelecegi/.

[13] Guerado E. Scientific societies and the third industrial revolution — the future role of the OTC. Injury J 2017;48(6):1—4.

[14] ITU-T: global standards for the internet of things. 2012. https://www.itu.int/en/ITU-T/techwatch/Pages/internetofthings.aspx.

[15] Schwab K. The fourth industrial revolution. World Economic Forum; 2016. https://www.weforum.org/about/the-fourth-industrial-revolution-by-klaus-schwab/.

[16] Görçün FÖ. Dördüncü sanayi devrimi endüstri 4.0, beta yayıncılık, İstanbul. 2016.

[17] Öztuna B. Endüstri 4.0 (dördüncü sanayi devrimi) İle çalışma yaşamının geleceği. Ankara: Gece Kitaplığı Yayınları; 2017.

[18] Doğru BN, Meçik O. The effects of industry 4.0 on the labor market in Turkey: firm predictions. Suleyman Demirel university. J Facul Eco Admin Sci 2018;(23):1581—606. Special Issue on Industry 4.0 and Organizational Change.

[19] Deloitte insights, government-trends-2020. 2020. https://www2.deloitte.com/content/dam/insights/us/articles/government-trends-2020/DI_Government-Trends-2020.pdf.

[20] Gerald P. Tackling the problematic labor skills gap [Infographic]. 2020. https://www.smartcitiesdive.com/ex/sustainablecitiescollective/tackling-problematic-labor-skills-gap/1096763/.

[21] David N, McNutt J. Building a workforce for smart city governance: challenges and opportunities for the planning and administrative profession. MDPI Informatics 2019;6(4):2—14.

[22] UN commission on science and technology for development nineteenth session. UN Nations Economic and Social Council; 2016. https://unctad.org/topic/commission-on-science-and-technology-for-development.

[23] Belli L, et al. IoT-enabled smart sustainable cities: challenges and approaches. MDPI Smart Cities 2020;3(3):1039—71.

[24] Stratigea A. The concept of 'smart cities'. Towards community development? Netcom J 2012;26—3(4):375—88.

[25] Iwańska KR, López ES. Chapter 10 - smart cities and the search for global talent- smart cities: issues and challenges. Elsevier; 2019. p. 171—84.

[26] Skills DB Pro: skills management software done right. 2021. https://skillsdbpro.com/.

[27] Trakstar - employee performance software. 2021. https://www.trakstar.com/.

[28] Velagapudi A, et al. Updating the smart grid workforce education and training process: an interdisciplinary approach. IEEE Power and Energy Society Innovative Smart Grid Technologies Conference (ISGT); 2019.

[29] Garg S, et al. Role of E-trainings in building smart cities. Proc Comp Sci 2017;111:24—30.

[30] Fournier J. 5 Ways to address the skills shortage in your industry, HCMWorks Insights. 2020. https://www.hcmworks.com/blog/5-ways-companies-can-address-a-skills-shortage.

[31] Mankiewicz K. The recruitment process: a guide to hiring top talent. 2020. https://www.jotform.com/recruitment-process/.

[32] Bhatia S. Sustainable smart universities for smart cities. J Eco, Manag Trade 2018;21(12):1—11.

[33] Dunkin M. How will higher education help to make our cities smart?. 2017. https://edtechnology.co.uk/comments/how-will-higher-education-help-to-make-our-cities-smart.

[34] Miltiadis DL, et al. Clustering smart city services: perceptions, expectations, responses. MDPI Sustain 2019;11(6).

[35] Buyruk H. Developing technologies, changing labour qualities and education. Int J Soc Res 2018;8(14):599—632.

[36] Can E. Türkiye'de kamu personelinin hizmetiçi eğitiminde bilişim teknolojilerinin rolü, Dokuz Eylül Üniversitesi, Sosyal Bilimler Enstitüsü, Kamu Yönetimi anabilim Dalı. Kamu Yönetimi Programı Yüksek Lisans Tezi; 2011.

[37] Akıllı Ş. İTÜ Vakfı Dergisi 2017;(77).
[38] Gabor MR, et al. Supporting employability by a skills assessment innovative tool—sustainable transnational insights from employers. MDPI Sustain 2019;11:3360.
[39] Chen C-L, et al. Role of government to enhance digital transformation in small service business. MDPI Sustain 2021;13(3):1028.
[40] Ozdemir N. Smart ports, gates of the digital future of the marine industry. 2021. https://www.ogunhaber.com/yazarlar/necmi-ozdemir/akilli-limanlar-101537m.html.
[41] Öztürk M, Sancak S. Hizmet İçi eğitimin çalışma hayatına etkileri. 2010. http://joy.yasar.edu.tr/makale/no7_vol2/10_ozturk.pdf.
[42] Dolgun U, et al. Ekin yayınevi, bursa. 2007.
[43] Dikmen MN, Baysal H. Research on faculty of technology of workplace training model in the scope of university-industry cooperation. Int J Sustain Engi Technol 2017;1(1):35–46.
[44] Etzkowitz H, Leydesdorff L. The dynamics of innovation: from national systems and 'mode 2' to a triple helix of university-industry-government relations. Res Pol 2000;29(2):109–23.
[45] Cai Y, et al. Building university-industry Co-innovation networks in transnational innovation ecosystems: towards a transdisciplinary approach of integrating social sciences and artificial intelligence. MDPI Sustain 2019;11(17):4633.
[46] Quelin B. Core competencies, RandD management and partnerships. Educ Sci 2006;31(141):71–82.
[47] Just transition, A report for the OECD. 2017. https://www.ituc-csi.org/just-transition-centre.

Further reading

Pekdağ B. Information and communication technologies in science education, balıkesir university. J Inst Sci 2005;7(2).

Chapter 14

Training smart city workforces with a wide range of electronic and digital technologies

Doaa M. El-Sherif[1] and Essam E. Khalil[2]

[1]*Housing & Building National Research Center (HBRC), Ministry of Housing, Utilities & Urban Communities, Cairo, Egypt;* [2]*Cairo University, Cairo, Egypt*

14.1 Introduction

Due to the breadth of technologies that have been implemented under the smart city label, it is difficult to distill a precise definition of a smart city. Deakin and Al Waer [1] list four factors that contribute to the definition of a smart city:

1. The application of a wide range of electronic and digital technologies to communities and cities.
2. The use of ICT to transform life and working environments within the region.
3. The embedding of such ICTs in government systems.
4. The territorialization of practices that brings ICTs and people together to enhance the innovation and knowledge that they offer.

Deakin defines the smart city as one that utilizes ICT to meet the demands of the market (the citizens of the city), and that community involvement in the process is necessary for a smart city [2]. A smart city would thus be a city that not only possesses ICT technology in particular areas, but has also implemented this technology in a manner that positively impacts the local community. Alternative definitions include:

- Giffinger et al.: "Regional competitiveness, transport and Information and Communication Technologies economics, natural resources, human and social capital, quality of life, and participation of citizens in the governance of cities." [3].
- Smart Cities Council [3]: "A smart city is one that has digital technology embedded across all city functions." [4].
- Caragliu and Nijkamp: "A city can be defined as 'smart' when investments in human and social capital and traditional (transport) and modern (ICT) communication infrastructure fuel sustainable economic development and a high quality of life, with a wise management of natural resources, through participatory action and engagement." [5].
- Frost and Sullivan: "We identified eight key aspects that define a smart city: smart governance, smart energy, smart building, smart mobility, smart infrastructure, smart technology, smart healthcare and smart citizen." [6].
- Institute of Electrical and Electronics Engineers smart Cities: "A smart city brings together technology, government and society to enable the following characteristics: smart cities, a smart economy, smart mobility, a smart environment, smart people, smart living, smart governance." [7].
- Business Dictionary: "A developed urban area that creates sustainable economic development and high quality of life by excelling in multiple key areas; economy, mobility, environment, people, living, and government. Excelling in these key areas can be done so through strong human capital, social capital, and/or ICT infrastructure." [8] Indian Government 2014: "Smart city offers sustainability in terms of economic activities and employment opportunities to a wide section of its residents, regardless of their level of education, skills or income levels." [9].
- Department for Business, Innovation and Skills, UK 2013: "The concept is not static, there is no absolute definition of a smart city, no end point, but rather a process, or series of steps, by which cities become more 'liveable' and resilient and, hence, able to respond more quickly to new challenges." [10].

Smart Cities Policies and Financing. https://doi.org/10.1016/B978-0-12-819130-9.00042-5

14.2 Characteristics

It has been suggested that a smart city (also community, business cluster, urban agglomeration, or region) uses information technologies to:

1. Make more efficient use of physical infrastructure (roads, built environment, and other physical assets) through artificial intelligence and data analytics in order to support a strong and healthy economic, social, cultural development [11].
2. Engage effectively with local governance officials by use of open innovation processes and e-participation, improving the collective intelligence of the city's institutions through e-governance [12], with emphasis placed on citizen participation and codesign [13,14].
3. Learn, adapt, and innovate and thereby respond more effectively and promptly to changing circumstances by improving the intelligence of the city [12,15].

They evolve toward a strong integration of all dimensions of human intelligence, collective intelligence, and also artificial intelligence within the city [16,17] The intelligence of cities "resides in the increasingly effective combination of digital telecommunication networks (the nerves), ubiquitously embedded intelligence (the brains), sensors and tags (the sensory organs), and software (the knowledge and cognitive competence)" [18]. These forms of intelligence in smart cities have been demonstrated in three ways:

1. **Orchestration intelligence** [12]: Where cities establish institutions and community-based problem-solving and collaborations, such as in Bletchley Park, where the Nazi Enigma cypher was decoded by a team led by Alan Turing. This has been referred to as the first example of a smart city or an intelligent community [19].
2. **Empowerment intelligence**: Cities provide open platforms, experimental facilities, and smart city infrastructure in order to cluster innovation in certain districts. These are seen in the Kista Science City in Stockholm and the Cyberport Zone in Hong Kong. Similar facilities have also been established in Melbourne [20].
3. **Instrumentation intelligence**: Where city infrastructure is made smart through real-time data collection, with analysis and predictive modeling across city districts. There is much controversy surrounding this, particularly with regard to surveillance issues in smart cities. Examples of instrumentation intelligence have been implemented in Amsterdam [21]. This is implemented through [12]:
   1. A common IP infrastructure that is open to researchers to develop applications.
   2. Wireless meters and devices transmit information at the point in time.
   3. A number of homes being provided with smart energy meters to become aware of energy consumption and reduce energy usage.
   4. Solar power garbage compactors, car recharging stations, and energy-saving lamps.

Some major fields of intelligent city activation are shown in Table 14.1.

According to David K. Owens, the former executive vice president of the Edison Electric Institute, two key elements that a smart city must have are an integrated communications platform and a "dynamic resilient grid." Both are large investments [22].

14.3 Frameworks

The creation, integration, and adoption of smart city capabilities require a unique set of frameworks to realize the focus areas of opportunity and innovation central to smart city projects. The frameworks can be divided into five main dimensions, which include numerous related categories of smart city development [23]:

14.3.1 Technology framework

A smart city relies heavily on the deployment of technology. Different combinations of technological infrastructure interact to form the array of smart city technologies with varying levels of interaction between human and technological systems, as follows [24]

- **Digital**: A service-oriented infrastructure is required to connect individuals and devices in a smart city. These include innovation services and communication infrastructure. Yovanof and Hazapis define a digital city as "a connected community that combines broadband communications infrastructure; a flexible, service-oriented computing infrastructure based on open industry standards; and, innovative services to meet the needs of governments and their employees, citizens and businesses." [25].
- **Intelligent:** Cognitive technologies, such as artificial intelligence and machine learning, can be trained on the data generated by connected city devices to identify patterns. The efficacy and impact of particular policy decisions can be quantified by cognitive systems studying the continuous interactions of humans with their urban surroundings [26].
- **Ubiquitous:** A ubiquitous city provides access to public services through any connected device. U-city is an extension of the digital city concept because of the facility in terms of accessibility to every infrastructure [27].
- **Wired**: The physical components of IT systems are crucial to early-stage smart city development. Wired infrastructure is required to support the IoT and wireless

TABLE 14.1 Sample intelligent cities.

Innovation economy	Urban infrastructure	Governance
Innovation in industries, clusters, districts of a city	Transport	Administration services to the citizen
Knowledge workforce: Education and employment	Energy/Utilities	Participatory and direct democracy
Creation of knowledge-intensive companies	Protection of the environment/Safety	Services to the citizen: Quality of life

technologies central to more interconnected living [28]. A wired city environment provides general access to continually updated digital and physical infrastructure. The latest in telecommunications, robotics, IoT, and various connected technologies can then be deployed to support human capital and productivity [29,30].

- **Hybrid**: A hybrid city is the combination of a physical conurbation and a virtual city related to the physical space. This relationship can be one of virtual design or the presence of a critical mass of virtual community participants in a physical urban space. Hybrid spaces can serve to actualize future-state projects for smart city services and integration [31].
- **Information city:** The multiplicity of interactive devices in a smart city generates a large quantity of data. How that information is interpreted and stored is critical to Smart city growth and security [32].

14.3.2 Human framework

Smart city initiatives have measurable positive impacts on the quality of life of its citizens and visitors [33]. The human framework of a smart city—its economy, knowledge networks, and human support systems—is an important indicator of its success, as follows [34]

- **Creativity**: Arts and culture initiatives are common focus areas in smart city planning [35,36]. Innovation is associated with intellectual curiosity and creativeness, and various projects have demonstrated that knowledge workers participate in a diverse mix of cultural and artistic activities [37,38].
- **Learning**: Since mobility is a key area of smart city development, building a capable workforce through education initiatives is necessary [34]. A city's learning capacity includes its education system, including available workforce training and support, and its cultural development and exchange [39].
- **Humanity:** Numerous smart city programs focus on soft infrastructure development, such as increasing access to voluntary organizations and designated safe zones [40]. This focus on social and relational capital means diversity, inclusion, and ubiquitous access to public services is worked in to city planning [30].
- **Knowledge:** The development of a knowledge economy is central to smart city projects [41]. Smart cities seeking to be hubs of economic activity in emerging tech and service sectors stress the value of innovation in city development [30].

14.3.3 Institutional framework

According to Moser, M.A., [39], since the 1990s, the smart communities' movement took shape as a strategy to broaden the base of users involved in IT. Members of these Communities are people that share their interest and work in a partnership with government and other institutional organizations to push the use of IT to improve the quality of daily life as a consequence of different worsening in daily actions. Eger [42] said that a smart community makes a conscious and agreed-upon decision to deploy technology as a catalyst to solving its social and business needs. It is very important to understand that this use of IT and the consequent improvement could be more demanding without the institutional help; indeed institutional involvement is essential to the success of smart community initiatives. Again, Moser [39] explained that "building and planning a smart community seeks for smart growth"; smart growth is essential for the partnership between citizen and institutional organizations to react to worsening trends in daily issues such as traffic congestion, school overcrowding, and air pollution. However, it is important to note that technological propagation is not an end in itself, but only a means to reinventing cities for a new economy and society. To sum up, it is possible to assert that any smart city initiatives necessitate the government's support for their success.

The importance of these three different dimensions is that only a link among them can make possible the development of a real smart city concept. According to the definition of smart city given by Caragliu, Del Bo, and Nijkamp [43], a city is smart when investments in human/social capital and IT infrastructure fuel sustainable growth and enhance quality of life, through participatory governance.

14.3.4 Energy framework

Smart cities use data and technology to create efficiencies, improve sustainability, create economic development, and enhance quality of life factors for people living and working in the city. It also means that the city has a smarter energy infrastructure. More formally, a smart city is: "... An urban area that has securely integrated technology across the information ... and Internet of Things (IoT) sectors to better manage a city's assets" [44]. Employment of smart technologies enables the more efficient application of integrated energy technologies in the city allowing the development of more self-sustaining areas or even positive-energy districts that produce more energy than they consume [45].

A smart city is powered by "smart connections" for various items such as street lighting, smart buildings, distributed energy resources (DER), data analytics, and smart transportation. Among these things, energy is paramount; this is why utility companies play a key role in smart cities. Electric companies, working partnership with city officials, technology companies, and a number of other institutions, are among the major players that helped accelerate the growth of America's smart cities [46].

14.3.5 Data management framework

Smart cities employ a combination of data collection, processing, and disseminating technologies in conjunction with networking and computing technologies and data security and privacy measures encouraging the application of innovation to promote the overall quality of life for its citizens and covering dimensions that include: utilities, health, transportation, entertainment, and government services [47].

14.4 Criticism

There is a rise in the number of criticisms of smart cities, and one of the big criticisms is that they focus on technological development, rather than on the problems that are faced every day by citizens. The criticisms of smart cities revolve around:

- A bias in strategic interest may lead to ignoring alternative avenues of promising urban development.
- A smart city, as a scientifically planned city, would defy the fact that real development in cities is often haphazard. In that line of criticism, the smart city is seen as unattractive for citizens as they "can deaden and stupefy the people who live in its all-efficient embrace". Instead, people would prefer cities they can participate to shape.
- The focus of the concept of smart city may lead to an underestimation of the possible negative effects of the development of the new technological and networked infrastructures needed for a city to be smart.
- As a globalized business model is based on capital mobility, following a business-oriented model may result in a losing long-term strategy: "The 'spatial fix' inevitably means that mobile capital can often 'write its own deals' to come to town, only to move on when it receives a better deal elsewhere." This is no less true for the smart city than it was for the industrial (or) manufacturing city
- The high level of big data collection and analytics has raised questions regarding surveillance in smart cities, particularly as it relates to predictive policing.
- As of August 2018, the discussion on smart cities centers around the usage and implementation of technology rather than on the inhabitants of the cities and how they can be involved in the process.
- Especially in low-income countries, smart cities are irrelevant to the majority of the urban population, which lives in poverty with limited access to basic services. A focus on smart cities may worsen inequality and marginalization.
- If a smart city strategy is not planned taking into account people with accessibility problems, such as persons with disabilities affecting mobility, vision, hearing, and cognitive function, the implementation of new technologies could create new barriers.

14.5 Summary

The present work defines a smart city as an urban area that uses different types of electronic methods and sensors to collect data. Insights gained from that data are used to manage assets, resources, and services efficiently; in return, that data is used to improve the operations across the city. This included data collected from citizens, devices, buildings and assets that is then processed and analyzed to monitor and manage traffic and transportation systems, power plants, utilities, water supply networks, waste, crime detection, information systems, schools, libraries, hospitals, and other community services. The smart city concept integrates ICT and various physical devices connected to the IoT network to optimize the efficiency of city operations and services and connect to citizens.

Finally, let us move on to the real interactive part of this chapter: review questions/exercises, hands-on projects, case projects, and optional team case project. The answers and/ or solutions by chapter can be found in Appendix G.

14.6 Chapter review questions/exercises

True/false

1. True or False? Due to the breadth of technologies that have been implemented under the smart city label, it is easy to distill a precise definition of a smart city.
2. True or False? The creation, integration, and adoption of smart city capabilities require a unique set of frameworks to realize the focus areas of opportunity and innovation central to smart city projects.
3. True or False? A smart city relies heavily on the deployment of technology.
4. True or False? Smart city initiatives have measurable negative impacts on the quality of life of its citizens and visitors.
5. True or False? Since the 1980s, the smart communities' movement took shape as a strategy to broaden the base of users involved in IT.

Multiple choice

1. Smart cities use data and technology to create efficiencies, improve sustainability, create economic development, and enhance quality of life factors for people living and working in the:
   a. County
   b. Country
   c. City
   d. State
   e. Parish
2. Smart cities employ a combination of data collection, processing, and disseminating technologies in conjunction with networking and computing technologies and data security and privacy measures encouraging the application of innovation to promote the overall quality of life for its citizens and covering dimensions that include:
   a. Utilities
   b. Health
   c. Transportation
   d. Entertainment
   e. All of the above
3. There is a rise in the number of criticisms of smart cities, and one of the big criticisms is that they focus on technological development, rather than on the problems that are faced every day by:
   a. Residents
   b. Humans
   c. Aliens
   d. Androids
   e. Citizens
4. A smart city as an urban area that uses different types of electronic methods and sensors to collect:
   a. Data
   b. Assets
   c. Bitcoin
   d. Meta data
   e. Devices
5. If a smart city strategy is not planned (taking into account people with accessibility problems, such as persons with disabilities affecting mobility, vision, hearing, and cognitive function), the implementation of new technologies could create new:
   a. Problems
   b. Usage
   c. Technology
   d. Barriers
   e. Policies

Exercise

Problem

Provide a perspective on smart cities and the various systems, platforms, sensors, computing devices, and network infrastructure that are required.

Hands-on projects

Project

Do research: Analyze smart city challenges, and make suggestions on changes in content and delivery options that can be implemented in urban planning and public affairs programs.

Case projects

Problem

Show how to build a smart city that incorporates successful public and private collaborations; open and flexible regulatory environment; ongoing investments in comprehensive planning for complete, compact and connected communities; and most of all, the diversity and creativity of people.

Optional team case project

Problem

Discuss how to build smart city transit access hubs, known as Smart Stations, that bring together a wide variety of mobility services, along with opportunities for other services useful to travelers, and can serve as centers for the deployment of autonomous and connected vehicles, urban freight logistics, and electric fleets.

References

[1] Deakin Mark, Al Waer H. From intelligent to smart cities. J Intel Build Int 2011;3(3):140–52. https://doi.org/10.1080/17508975.20.

[2] Deakin Mark. From intelligent to smart cities. In: Deakin M, editor. Smart cities: governing, modelling and analysing the transition, vol. 15. Taylor and Francis; August 22, 2013, ISBN 978-1135124144.

[3] Giffinger R, Fertner C, Kramar H, Kalasek R, Pichler-Milanovic N, Meijers E. In: "Smart cities – ranking of European medium-sized cities" (PDF). Smart Cities. Vienna: Centre of Regional Science; 2007. Archived (PDF) from the original on 9 April 2011. Retrieved 13 November 2009.

[4] Definitions and overviews. Smart Cities Council. Archived from the original on 3 November 2014. Retrieved 3 November 2014. The smart city sector is still in the "I know it when I see it" phase, without a universally agreed definition. The Council defines a smart city as one that has digital technology embedded across all city functions. https://na.smartcitiescouncil.com/smart-cities-information-center/definitions-and-overviews.

[5] Caragliu A, Del Bo C, Nijkamp P. Smart cities in Europe. Serie Research Memoranda 2009:0048. VU University Amsterdam, Faculty of Economics, Business Administration and Econometrics, Archived from the original on 3 November 2014. Retrieved 14 November 2015. https://ideas.repec.org/p/vua/wpaper/2009-48.html.

[6] Singh S. Smart cities – A$1.5 trillion market opportunity. Forbes June 19, 2014. Archived from the original on 3 November 2014. Retrieved 4 November 2014. https://www.forbes.com/sites/sarwantsingh/2014/06/19/smart-cities-a-1-5-trillion-market-opportunity/?sh=16a755e26053.

[7] About". IEEE Smart Cities. Archived from the original on 5 September 2015. Retrieved 3 November 2014.

[8] Smart City – Definition. BusinessDictionary.com. Archived from the original on 4 November 2014. Retrieved 3 November 2014.

[9] Draft concept note on smart city scheme (PDF). Government of India – Ministry of Urban Development. Archived from the original (PDF) on 3 February 2015.

[10] Smart cities background paper (PDF). UK Department for Business Innovation and Skills. Archived (PDF) from the original on 16 June 2018. Retrieved 11 August 2018.

[11] Hollands. 2008. p. 303–20.

[12] Komninos, N. (August 22, 2013). What makes cities intelligent? In Deakin, M. (editor.). Smart cities: governing, modelling and analysing the transition. vol. 77. Taylor and Francis. ISBN 978-1135124144.

[13] Deakin M. From city of bits to e-topia: taking the thesis on digitally-inclusive regeneration full circle. J Urban Technol 2007;14(3):131–43. Archived from the original on 18 March 2016. Retrieved 3 November 2014.

[14] Deakin M, Allwinkle S. Urban regeneration and sustainable communities: the role of networks, innovation and creativity in building successful partnerships. J Urban Technol 2007;14(1):77–91. https://doi.org/10.1080/10630730701260118. S2CID 153965022.

[15] Coe A, Paquet G, Roy J. "E-governance and smart communities: a social learning challenge" (PDF). Soc Sci Comput Rev 2001;19(1):80–93. https://doi.org/10.1177/089443930101900107. S2CID 53380562. Archived (PDF) from the original on 3 March 2016. Retrieved 3 November 2014.

[16] Komninos. 2008. p. 112–3.

[17] Atlee T, Pór G. Evolutionary Nexus: connecting communities for emergence. 2006. Archived from the original on 19 October 2015. Retrieved 6 November 2014.

[18] Mitchell W. Intelligent cities. E-J Know Soc 2007. uocpapers, issue 5 (2007) I ISSN 1885–1541. Archived from the original on 28 February 2017. Retrieved 1 February 2015. https://www.uoc.edu/uocpapers/5/dt/eng/mitchell.pdf.

[19] Komninos N. Intelligent cities: variable geometries of spatial intelligence. In: Deakin M, Al W, Husam, editors. From intelligent to smart cities. Journal of intelligent buildings international: from intelligent cities to smart cities, vol. 3; 2011. p. 140–52. https://doi.org/10.1080/17508975.2011.586671. S2CID 110580067.

[20] Department of Sustainability and Environment. Melbourne 2030. State Government of Victoria; 2005. Archived from the original on 30 May 2015. Retrieved 30 May 2015.

[21] Amsterdam Smart City. "Amsterdam smart city ~ projects". Archived from the original on 22 September 2012. Retrieved 30 May 2015.

[22] Riley Kim, November 16. America needs smart grid investments pronto, stakeholders say at NARUC event. Daily Energy Insider 2017. Archived from the original on August 28, 2019. Retrieved December 11, 2017.

[23] Sharifi A. A critical review of selected smart city assessment tools and indicator sets. J Clean Prod October 2019;233:1269–83. https://doi.org/10.1016/j.jclepro.2019.06.172.

[24] The World in 2019. World in 2019. economist.com. Archived from the original on 18 June 2019. Retrieved form 26 June 2019.

[25] Yovanof GS, Hazapis GN. An architectural framework and enabling wireless technologies for digital cities & intelligent urban environments. Wireless Pers Commun 19 March 2009;49(3):445–63. https://doi.org/10.1007/s11277-009-9693-4. ISSN 0929-6212. S2CID 207258536.

[26] Peris-Ortiz M, Bennett DR, Yábar DP-B. Sustainable smart cities: creating spaces for technological, social and business development. Springer; 2016, ISBN 9783319408958. Archived from the original on 30 October 2020. Retrieved 4 October 2020.

[27] Anthopoulos L, Fitsilis P. In: Sideridis AB, Patrikakis CZ, editors. Next generation society. Technological and legal issues. Lecture notes of the Institute for computer sciences, social informatics and telecommunications engineering. Springer Berlin Heidelberg; 23 September 2009, ISBN 9783642116292. p. 360–72. https://doi.org/10.1007/978-3-642-11631-5_33.

[28] Smart Cities World News Team, Smart cities are about people. Smart Cities World. Published May 20, 2018. Archived from the original on 29 June 2019. Retrieved 29 June 2019. https://www.smartcitiesworld.net/news/news/smart-cities-are-about-people-2932.

[29] Intelligent Cities: R&D offshoring, web 2.0 product development and globalization of innovation systems (PDF). Archived (PDF) from the original on 16 May 2018. Retrieved 20 December 2016.

[30] Nam, T.; Pardo, T.A>. "Conceptualizing smart city with dimensions of technology, people, and institutions" (PDF). Center for Technology in Government University at Albany, State University of New York, U.S. The Proceedings of the 12th Annual International Conference on Digital Government Research. Archived (PDF) from the original on 7 November 2018. Retrieved 29 June 2019 – via https://inta-aivn.org/images/cc/Urbanism/background%20documents/dgo_2011_smartcity.pdf.

[31] Koltun, N (Editor). Campaign Trail: why Uber built a virtual city to promote a product that doesn't exist yet. Marketing Dive. Archived from the original on 26 June 2019. Retrieved 26 June 2019. Published June 21, 2019. https://www.marketingdive.com/news/campaign-trail-why-uber-built-a-virtual-city-to-promote-a-product-that-doe/557301/.

[32] Nisenbaum, A. What's holding smart cities back? Scientific American Blog Netw. Archived from the original on 29 June 2019. Retrieved 29 June 2019. Published: March 18, 2019. https://blogs.scientificamerican.com/observations/whats-holding-smart-cities-back/.

[33] Smart city technology for a more liveable future | McKinsey. www.mckinsey.com. Archived from the original on 29 June 2019. Retrieved 29 June 2019.

[34] "United Smart Cities (USC)—United Nations Partnerships for SDGs platform". Sustain Times development.un.org. Archived from the original on 28 August 2019. Retrieved 29 June 2019. https://sustainabledevelopment.un.org/partnership/?p=10009.

[35] "engaging communities2005.org" (PDF). www.engagingcommunities2005.org. Archived from the original (PDF) on 27 February 2017. Retrieved 20 December 2016.

[36] Borda A, Bowen JP. Chapter 27: smart cities and digital culture: Models of innovation. In: Giannini T, Bowen JP, editors. Museums and digital culture: new perspectives and research. Series on Cultural Computing. Springer; 2019, ISBN 978-3-319-97456-9. p. 523—49. https://doi.org/10.1007/978-3-319-97457-6_27. ISSN 2195-9064.

[37] Creativity, ContributorZahn Professor of; Communications, Innovation is also the Van Deerlin Endowed Chair of; Policy, Public; University, Director of the Creative Economy Initiative at San Diego State Eger JM. Creativity in the smart city is what makes a city really smart. HuffPost; July 24, 2015. Archived from the original on 29 June 2019. Retrieved 29 June 2019.

[38] Malanga S. The curse of the creative class. City J 23 December 2015. Archived from the original on 11 August 2018. Retrieved 11 August 2018, Winter, 2004. https://www.city-journal.org/html/curse-creative-class-12491.html.

[39] Moser, M. A. "What is smart about the smart communities movement?". www.ucalgary.ca. Univ Calgary EJ vols. 10—11(1). Archived from the original on 10 February 2017. Retrieved 20 December 2016.

[40] Glaeser, Edward L.; Berry, Christopher R. "Why are smart places getting smarter?"(PDF). Harvard University. Archived (PDF) from the original on 28 August 2019. Retrieved 11 August 2018.

[41] Smarter cities for smarter growth: how cities can optimize their systems for the talent-based economy (PDF). www.zurich.ibm.com. Archived (PDF) from the original on 21 December 2016. Retrieved 20 December 2016.

[42] Eger JM. Smart growth, smart cities, and the crisis at the pump a worldwide phenomenon. Iways January 1, 2009;32(1):47—53. https://doi.org/10.3233/iwa-2009-0164. Archived from the original on 28 August 2019. Retrieved 11 August 2018.

[43] Caragliu A, Del Bo C, Nijkamp P. Smart cities in Europe. Serie Research Memoranda 0048, VU University Amsterdam, Faculty of Economics, Business Administration and Econometrics; 2009. Archived from the original on 11 August 2018. Retrieved 11 August 2018. https://ideas.repec.org/p/vua/wpaper/2009-48.html.

[44] Riley K. EEI's new board chairman cites smart-city opportunities as convention gets under way. Daily Energy Insider; June 12, 2017. Archived from the original on 28 August 2019. Retrieved 25 September 2017.

[45] Tuominen P. Yes to positive energy districts - but how to make it happen?. Archived 15 January 2021 at the Wayback Machine VTT Technical Research Centre of Finland; 2020.

[46] Riley K. Pittsburgh, San Diego city officials put utilities as major players in smart-city partnerships. Daily Energy Insider; 15 June 2017. Retrieved 25 September 2017.

[47] Gharaibeh A, Salahuddin MA, Hussini SJ, Khreishah A, Khalil I, Guizani M, Al-Fuqaha A. Smart cities: a Survey on data management, security, and enabling technologies. IEEE Comm Surv Tut 2017;19(4):2456—501. https://doi.org/10.1109/COMST.2017.2736886. S2CID 206578345.

Part IV

Lack of financing and well-developed business models

Chapter 15

Digitization, automation, operation, and monetization: the changing management of sidewalk and kerb 2000–25

Alanna Coombes[1] and Bern Grush[2]

[1]*University of Southern California, Los Angeles, CA, United States;* [2]*Harmonize Mobility Inc., Toronto, ON, Canada*

15.1 Introduction

The footway and kerb are the hottest venues in town. They have hosted centuries of competing actors vying for their use: as a thoroughfare; a vantage point to watch a Panathenaic Procession; the venue for Black Lives Matter (BLM) protests; a place to pay for parking; to meet your Uber driver; or the place to sell the best Kimchee dumplings in Toronto. Footways or pavements have set the stage for urban life for more than two millennia. With the expansion of the virtual world in the 21st century, it seems perhaps remarkable that footways and kerbs have continued to adapt to their audience to meet the needs of their many users. Not only have they continued to be used, but their purposes have diversified and expanded into the 21st century city.

15.1.1 A century of changes at the kerb

A fundamental change for footways and kerbs began on July 16th, 1935 when Park-O-Meter No. 1 was installed in Oklahoma City and the monetization of the kerb began. City free-for-all parking at the kerb began to ebb away as municipal management of kerb space across the globe started to get a grip on this increasingly important, valuable, and complicated city asset. Thus, the right to use the kerb moved decisively toward those with the ability to pay as the free, and democratic use of the kerb became more impractical in growing cities and their increasingly chaotic kerb space.

Monetized kerb space of course is attractive to city councils and not just to reduce local congestion. A municipality may perceive monetized kerb space as an opportunity to fund services or assets although there can be restrictions in practice. In the United Kingdom, for example, parking schemes must be self-financing, and the law does not allow local authorities to use parking enforcement schemes to raise revenue for other services [1]. As the 21st century has progressed, technology has brought greater sophistication to methods of balancing the variety of demands for kerb space and maximizing parking income.

Case study 1—SFPark, San Francisco

One of the more advanced parking systems is San Francisco's commercial parking, SFpark, which provides something of a blueprint for how technology can help to maximize the use of the varying value of kerbside parking across different locations, days, and different times of day. The system uses sensors to collect real-time parking space occupancy data to determine parking rates across the city's thousands of parking spaces. This data is used to determine future parking rates and to inform customers of block-by-block parking space availability.

SFpark points to a number of benefits of their system: increased trade for local businesses (on assessment, sales tax revenues rose over 30% compared to 20% in other parts of the city); a 43% reduction in search time for parking spaces; a 30% decrease in vehicle miles traveled because of less circling to find vacant spaces; a broad reduction of vacant parking spaces; fewer fines or citations for overstaying one's welcome in a parking space (made up for by increased revenue from legitimate parking); better air quality and improved safety [2]. The system uses gradual and regular (at least every 30 days) demand-responsive rate adjustments to find the lowest rate possible to achieve the availability target of one space free per block. The system increases parking prices when parking is hard to find and lowers where demand is low. A review of the system found that circling for parking spaces reduced by 50% [3].

Smart Cities Policies and Financing. https://doi.org/10.1016/B978-0-12-819130-9.00003-6

App-based automated payments are now common in North American and UK cities. With a ticket machine identification number linked to an app parking is easier than traditional meters parking and is swift and paperless. Shopper parking is less necessary as online shopping increases but if town and city centers move from retail provision toward more leisure provision (including restaurants, theaters, bars, or gyms) or service provision (including barbers or nail salons) parking demand by consumers can be expected to continue in city centers. The effect of the online shopping boom is increasing demand for delivery bays, rather than parking bays, including in residential areas or areas of residential and commercial mix.

This growth of online shopping—accelerated during the COVID-19 pandemic—has increased further use of the kerb and competition for space. Smart delivery bays—using sensors and apps—is enabling more efficient use of this increasingly precious resource because it means more vehicles can use it per hour. In some instances, it can also reduce footway clutter as there is less need for machines releasing space for other uses. In the United Kingdom, for example, Appyway has provided Pimlico Plumbers with a solution to the problem that this large plumbing business described as plumbers wasting as much as 20–30 min looking for a parking space in central London and receiving four or five parking fines a day due to lack of availability and having to guess the length of time required for a job—sometimes paying for 4–5 h when the job only took 1 h. This unnecessary time and cost put stress on the drivers and the company finances. Appyway has described their system:

"In July 2015 AppyWay teamed up with Vodafone drivexone to create the world's first 'One Click Parking' solution. Using a combination of M2M dongles and our mobile app we enabled drivers to find available spaces quickly and easily, start a parking session with one click and only pay for the minutes parked. We installed dongles into the diagnostics ports across the fleet, enabling the vehicles to become connected to the internet. The drivers use our app to start the session with one click. Once the drivers have completed their jobs, they simply hop in the car and drive away. The vehicles detect a key in the ignition and a certain distance traveled and the session ends automatically. Drivers then receive a notification with the details of their stay" [4].

Other infrastructure is already competing for the potential space created by redundant ticket machines. The significant increase in electric vehicles and the encouragement of it by governments are leading to a further use of the kerbside charging infrastructure. In the United Kingdom, the number of EV charge points per 100 km of road in the country has increased from 42 in 2011 to 570 in 2019 [5]; there are now 17,948 public charging points [6] in addition to owners' or users' hardware. This reflects the growth in number of electric vehicles, which in the United Kingdom has increased from just under 9000 at the end of March 2010 to 317,000 at the end of June 2020: an increase of 3427% [7]. Currently the more limited battery power limits long journeys, but this problem is resolving as recharging infrastructure across the country grows and battery capacity gets more powerful, enabling more miles to be covered on one charge. This could encourage electric vehicles to travel further and so lessen the need for charging at the kerb, which could provide more opportunities for competing needs for that space, such as pick-up and drop-off space for transport network companies (TNCs—such as Uber).

TNCs or any app-based taxi service has increased the need for pick-up–drop-off zones (PUDOs) in town and city centers. This has brought more activity at the kerb and reduced the availability of parking, which may be less attractive in the short term to a municipality than the revenue-generating parking space. This is changing however as technology now enables drivers to buy small slivers of time—enough for a pick-up or a drop-off or a package delivery. In turn this could provide an incentive for city councils to turn parking at the kerb into PUDO zones—small amounts of time quickly add up in a busy downtown. The first step toward getting the most from a kerbside asset in the short and longer term is to ensure there's a thorough understanding of the kerb space's short and long-term demands, opportunities, and risks, such as double parking. Washington DC took just such an approach to create effective PUDO zones.

Case study 2—Washington DC's PUDO zones

In 2019, Washington DC's Department of Transportation (DDoT) worked on a research project with curbFlow, a mobility company that coordinates commercial operator PUDO at the kerbside in real time. After removing parking spaces to create loading zones for commercial activities at nine locations across the District, the project sought to build on DDOT's efforts to improve the safety of PUDO activity. The results of the project are based on data collected from 6350 commercial drivers representing more than 900 companies who reserved space at the nine curbFlow PUDO locations across the District over 15,000 times during the research period. The zones were used by both commercial vehicles and private vehicles operating in a commercial capacity such as picking up for an online food delivery service or other online delivery platform. The District has reported the following highlights from the project:

- Incidents of double parking and illegal U-turns decreased by an estimated 64% in immediate proximity to the curbFlow PUDO zones.
- On-demand delivery (e.g., online food delivery services), freight, and parcel deliveries lasted an average of 7–11 min, while rideshare and taxi PUDO activity lasted less than two and a half minutes on average.
- On-demand deliveries were the most frequent users of the curbFlow PUDO zones, followed by freight and parcel deliveries [8].

Of course, TNCs and commercial drivers are not the only ones enabled by new technology; car-sharing has also brought new competitors for kerbspace. Cities across the United States have reserved kerbside car-sharing spaces working with companies including Zipcar and SHARE NOW, which is now a market leader with more than 20,000 vehicles and four million customers worldwide. Washington DC started its program in 2005 providing, initially, space free of charge to promote and maximize neighborhood access to these vehicles. As the scheme expanded and a more competitive market emerged in car-share, a market price charge was introduced. Meanwhile, demands from more active travelers have added to the competition for space at the kerb.

The first decade of the 21st century brought shared bike schemes to major cities often requiring docking space at the kerb or footway—with some difficulty in more narrow, European streets. The following decade saw an influx of new mobility competitors—both dockless bikes and scooters (see Fig. 15.1). The menace of abandoned scooters cluttering the footway—or dumped in waterways—emerged, and new mobility companies were forced to engage with local municipalities to stop the footway blight. The mighty Wandsworth Council in south west London simply deemed them pavement obstructions, rounded them up, and dumped them in their depot until the companies took seriously the problems they were causing pedestrians, wheelchair users, and people with strollers. While city councils have wanted to encourage cycling for health reasons and reduce vehicle use for air quality reasons in a competition between them and pedestrians, the footway has generally been prioritized for the latter.

FIGURE 15.1 Dockless in St Paul's: adapting medieval streets to 21st century London.

The focus on walkability, Walk Score [9], and pedestrian access has been a recurrent theme of global cities in the past couple of decades. As cities have acknowledged the importance of pedestrian and cycle movement to well-being, internationally competitive cities have given more priority to active travel. This has been demonstrated in the United Kingdom with the City of London's Transport Strategy of 2019 that prioritized walking and cycling in the Square Mile [10]. This reflects the surge in cycling across the broader London area—a growth of cycling of 229% between 2004 and 2014 [11]. Transport for London has reported that, in addition, during the pandemic, there was a 7% increase in inner London cycling and 22% in outer London (see Fig. 15.2). This is something forward-looking businesses have encouraged because TfL research shows people who walk, cycle, or use public transport to visit their local high street do so more frequently and spend up to 40% more than those who travel by car [12]. Planning authorities that insist on parking infrastructure as part of developments are better able to capitalize on these economic benefits for their city. The City of London's focus on the pedestrian experience is also seen to have important economic consequences as an international financial and business hub seeking to attract the very best in global talent.

While online shopping has reduced the number of shoppers on footways in downtowns increasing attention paid to urban design or placemaking has ushered in higher standards of public realm quality and better footways in cities. San Francisco has been an early pioneer of parklets, piloting its first in 2010. Scores of these mini parks taking up kerb space have since been introduced to the city providing seating and urban greening. Open to the public, these parklets are designed, created, and maintained at the expense of the applicant businesses, business improvement districts, retailers that usually live, work in, manage, or own the adjacent property. They can include tables, seating, bicycle parking, and green landscaping and have been found to encourage walking, cycling, and pedestrian flow by providing seating, cycle parking, landscaping, and public art [13]. The city also found that parklets provide an important buffer zone between traffic and the footway helping to create more serenity and comfort and a feeling of safety. As an outdoor gathering space, they are also seen as encouraging a sense of community, particularly in areas underserved by traditional parks [13]. Space for greening has not just emerged from kerb space but from pavements, sidewalks, or footpaths too.

FIGURE 15.2 Bike hire in the City of London, June 2021.

15.2 Twenty-first century footpaths, sidewalks, or pavements

Urban greening including street trees has helped to make a trip to downtown more pleasurable as well as functional and so enabling the businesses there to compete at least to some extent with online shopping. Importantly though such greening complements the development of town and city centers as places for leisure (see Fig. 15.3) as they move away from a primarily retail function in the face of online competition. The development of downtowns or town centers into centers for leisure (restaurants, barbers, coffee shops, and nail bars) rather than retail (shops) has encouraged a focus on the physical environment and need to upgrade paving, seating, and introduce street art and other means of delighting downtown visitors and shoppers. While all these new additions have brought a better experience for pedestrians, they have also brought pressures to accommodate the new street furniture or greening.

On footpaths, trees and greening can be viewed negatively by highways engineers (who might prefer unimpeded pedestrian flow) and safety officers (who might be understandably concerned with slip hazards of falling fruits or leaves). But there is a recognition of their importance for

FIGURE 15.3 Green wall at London Wall, City of London.

livability and for climate change adaptation (and indeed mitigation) as they help to combat the Urban Heat Island Effect, provide shelter from sunlight, and eat a little into carbon dioxide emissions. There are equity issues at play with greening too; city places subject to historic redlining, which has negatively impacted on the economic fortunes of generations of black Americans, have far less tree cover than city areas that were not, impacting on air quality, availability of shade, and well-being. A Duke University study found a 50% increase in tree cover in Durham, North Carolina, over the last 80 years for those areas not in redlined areas [14]. Such an absence of trees impacts on temperatures: a 2020 study of 108 urban areas in the United States by Portland State University found that 94% of studied areas displayed consistent city-scale patterns of elevated land surface temperatures in formerly redlined areas relative to their non-redlined neighbors by as much as 7°C, part of which is attributable to tree canopy cover [15]. Urban greening can be driven by environmental and equity objectives and perhaps nowhere is this more visible than the footpaths of downtowns, the heart of towns, and cities open to all.

On the central city footpaths, street vendors and the musicians continue to offer their services or talents as they did centuries ago. Today they may be licensed by the local municipality or vetted by the Business Improvement Area (BIA) or District or town center manager—an opportunity for municipalities to raise revenue from licenses while helping to promote a sense of vitality and interest to the street scene.

Encouraging a café culture is almost *de rigueur* in UK towns and cities; the environmentally unconscious might offer patio heaters to enable outside dining in the chill but the environmentally conscious, little blankets. What they take from the city in terms of space to walk they can more than make up for in terms of urban vitality, safety, and sensory interest (see Fig. 15.4).

FIGURE 15.4 Café culture in former road space in the City of London, UK.

As the footpaths have been seen to be more important to a city, the maintenance of footpath quality has also been put in the spotlight. Maintenance has been assisted by the increased ease of reporting problems such as broken streetlights or fly-tipping by ordinary city people. Apps that use GPS—such as fixmystreet.com—have enabled people to report problems sending photos and precise geographical locations allowing (if not delivering) quicker remediation. Similarly reporting faulty lamp columns has been made easier through unique QR codes for each to pinpoint the location and enable easier reporting. Such crowdsourced problem identification enables efficiency in public spending and public involvement in the quality of the street scene.

City centers footpaths—especially in major cities—may also need to provide space for bollards or other defensive measures (see Fig. 15.5). While bollards provide security against hostile vehicles in a terror attack, other measures (with appropriate reinforcement) can also be effective—such as kerb height, trees, benches, and other street furniture. Finding space for seating and other furniture can be hard though in older cities with narrow pavements, such as in the City of London.

The 2010s also saw the emergence of footway drones competing with other pavement users. Companies, such as Starship Technologies, began negotiating with universities and city governments to allow use of the footway for their robots to make milk, pizza, and other deliveries. By January 2021, Starship Technologies had completed a million deliveries 6 years after the company was founded; the appeal of a contactless delivery service during the pandemic heightening the appeal. Much more detail on this is set out in the accompanying chapter to this. In historic core areas such as the City of London, a planning authority might understandably baulk at having robots compete for footway space and question the utility of them when busy people, reading from their phones, in narrow pathways make for a complicated and potentially dangerous mix (see Fig. 15.6).

15.2.1 Heading toward the autonomous future

Technology is helping to ensure that footways and kerb space remain essential, into the 21st century. The concern is

FIGURE 15.5 Defensive bollards can be a necessity but eat into space at the kerb.

FIGURE 15.6 Awkward: The City of London's maze of passages and walkways where you do not want to meet a robot coming the other way—a not so "prudent passage."

how to manage this increased pressure. As described in Chapter 16, there is much evidence to suggest that sidewalk robots will be in regular use before autonomous vehicles, but urban and transport planners are already thinking ahead to street life with them. Predicting the future can be a futile business but with the many autonomous vehicle trials across the United Kingdom, United States, and elsewhere, there is much research already on how infrastructure or urban form might change as city centers adapt to their widespread introduction. The American Planning Association's Planning for Autonomous Mobility report, for example, provides excellent material [16]. There are also early indications from AV trials that can inform the ways in which cities can work to ensure that the local environment is shaped by the needs of its communities rather than it being shaped by the coming technology.

There seems to be some agreement on a few likely changes with AV introduction. First, that there will be an increase in the demand for pick-up and drop-off zones as there would be fewer reasons to exit a vehicle in a parking structure. All the people who otherwise would be parking their own vehicle would be vying for space to be dropped off may require a kerbside management system such as vehicle-to-infrastructure booking or differentiated pricing for drop-off depending on the popularity of the drop-off point.

We can expect autonomous vehicles to be electric, but advances in battery charging capacity, as described previously, are likely to lessen the need for clunky on-street equipment; trials of other forms of charging (such as embedded in road space) further question the form of future infrastructure needs. We can also see already that automated payment and instruction are likely to reduce the need for signage at the kerb, so we can expect with AVs' advanced technology that their vehicle-to-infrastructure capabilities will lead to a reduction in clutter at the kerb or on the footway. If all vehicles were autonomous, we could expect signage to become increasingly redundant with some exceptions such as place names.

It seems unlikely however that traditional vehicles will be dispensed with altogether given a country's car culture. Classic car enthusiasts or Harley riders may not wish to have their vehicles adapted for automation, and politically it would seem hard and perhaps foolish to compel them to do so. Some of the claimed benefits of autonomous vehicles such as dispensing with signage or releasing road space (because AVs do not need wide lanes because they do not weave within lanes)—would therefore not be realized. So, the touted wider footways, cycle lanes, or urban greening cannot be relied upon outcomes even if funding was secured. Whatever the outcome, we can assume that there would be a mixed fleet of AV and non-AVs for the foreseeable future.

Some city core parking will continue to be prized for the non-AV driver who will continue to want to leave their vehicle near to their destination. For the AV driver, there will be much less of a need for parking spaces in city cores whether or not they are owned or shared. Shared vehicles could be summoned in the same way as an Uber; privately owned could be sent to a cheaper parking structure away from the center or have it go home and return when collection is required. Of course, privately owned AVs in this scenario would be making double trips, impacting on congestion and the livability of city centers. We can expect that—as has always been the case—as technology becomes mainstreamed, the price reduces significantly, enabling the purchase of it for a wider section of the population.

A key issue with autonomous vehicles is whether they become commonly privately owned or whether they are shared mobility. This difference is what Robin Chase, founder of Zipcar, has described as the difference between the future "heaven or hell" on our roads [17]. This is important because it makes a big difference to the pressure on roads and at the kerb—and in city centers or downtowns more generally because, as described, a privately owned vehicle is more likely to drive empty to go home or to cheaper places to park or circle until its owner has got that email sent. All we can say for sure is that the future ownership structure is uncertain.

A key concern for the future of cities is whether AVs will lead to more vehicles and more Vehicle Miles Traveled (VMT), decreasing livability and adding to congestion. Research suggests we can expect a continued decline in heavy, fixed-route transit use—even without the negative effects of COVID-19. While AVs will enable people who currently cannot drive—the very young, those with disabilities, or those banned from driving—the increase in vehicles and VMT is a worrying prospect.

What is unclear so far is the long-term effects of the COVID-19 pandemic on travel preferences. Continued dislike and distrust of crowded transit are likely to encourage more cycling, walking, and private car use in cities especially if an investment in a bicycle, scooter, or car was made during the pandemic. The cumulative impact could be a less pleasant walking and cycling environment making footpaths and kerb space much less pleasant places to be and could add to pressures to pedestrianize streets.

With an increase in AVs and their VMT, we can expect further pressure on the kerb as they pick up, drop off, make deliveries, or pass through already congested towns and cities. For this to be managed, municipalities will need to decide who or what should have priority use of this precious resource.

15.2.2 Prioritizing space

To ensure efficient use of public space and thriving downtowns, it is important to manage this space thoughtfully and strategically. On the footpath we have seen the various competing uses but how should a municipality or BIA decide who gets priority? Or how do you balance the competing demand for space? For Blomley, his theory of pedestrianism—that the primary purpose of footways is the efficient flow of pedestrians [18]—casts food vendors, protesters, street trees—as impediments to footways' primary purpose. This can create a tension between a municipality's highway team (prioritizing the free flow of pedestrians) and its economic development team (prioritizing local business growth) or a municipal finance team (seeking to maximize its assets and gain a revenue stream through the licensing process). Blomley's theory of pedestrianism is practical while Annette Kim takes a more nuanced view—which is that laws, courts, and maps are social constructions and only have meaning if they are enforced [19]. So, the Christmas carollers, the ice cream van, or the merchandise displays outside the charity shop may well be overlooked in the interests of street vitality and community well-being even if they are obstructing part of the thoroughfare and bylaws require their removal.

Who or what makes use of pathway space is a matter of the rule maker and enforcer understanding and balancing competing needs that are in the best interests of the sustainability and livability of the city? Pathways need to contribute to the well-being of city people at large while providing for those less able. Space is limited; European streets have less scope because they are narrow; so, some sort of prioritization needs to be understood and enforced.

At the kerb, Seattle, WA, has tackled the issue of priority by defining the priority given to the competing users of kerb space (which they call the "Flex Zone") in three categories of districts. Their comprehensive planning process aims to balance competing demands, provide for efficient movement of people and goods, support local economies, and enhance the livability of communities [20]. Priorities differ depending on the surrounding land use whether it is residential, commercial, or industrial. So, for example, in a commercial or mixed-use district access for commerce takes precedence over access for people—while the reverse is true for a residential district. The priorities in all these districts however conform to the prioritized functions of the Modal Plan, the city's long-term plans for walking, biking, transit, and freight.

This defining of prioritization helps people to understand what is allowed and what is not. A similar process might be necessary on footways because that space too is increasingly under pressure. There are however ways in which that pressure can be alleviated through long-term planning, taking into account the context of footways.

Stretches of pathways should not be looked at in isolation; they should be considered in the totality of the surrounding space. Current and future strategies need to bring into play road space and private property. It may be that road narrowing is the solution to competing needs on the pathway to allow for its widening or that public access could be created by incorporating private space. A Torontonian may not think twice about walking straight through a pathway-level lobby to access a road on the other side—with or without their snowy boots—an idea that might be viewed with disdain in Munich. Such private space merging with public space is increasingly common and sometimes hard to know where one ends and the other begins. This can be mistaken for the privatization of public space but may well be a space that has been private for centuries but through negotiation with the local planning authority has agreed to provide it as public amenity. This "public-isation" of private space, as Matthew Carmona has called it [21], is apparent in the City of London.

Restricted by its narrow, medieval streets, the City of London seeks to carve out open space from ground floor developments to provide much needed seating, urban greening, and through access that is simply not possible otherwise; 122 Leadenhall Street is a good example of this (Fig. 15.7). Sensitivity to cultural norms and local needs and preferences need to be applied—but a broad, strategic view of city space—that takes into account private and public space—makes for better city places.

15.3 Rights to the footway and kerb

The question of who has the right to use space is perhaps a more fundamental question than that of priority. Municipalities and BIAs will have differing approaches to this. A much talked about concept is the "Rights to the City," a somewhat fuzzy concept that is most associated with Henri Lefebvre, who variously and vaguely defined it as, "the right to information, the rights to use of multiple services, the right of users to make known their ideas on the space and time of their activities in urban areas; it would also cover the right to use the center" [22]. To whom such rights belong has been described by Peter Marcuse as, "those who are excluded, the aspiration of those who are alienated; the cry is for the material necessities of life, the aspiration is for a broader right to what is necessary beyond the material to lead a satisfying life [23]." Such rights suggest that city people should not have to abide by municipal kerb rules because they are protected by more fundamental rights if they are one of the excluded many. In practice, applied to the kerb, it would mean access by a more limited number of people but enough to find that the value of kerb space could be destroyed over time.

15.3.1 Property rights

Public space, including the footway and kerb, can be seen as being subject to property rights. While legal ownership or title deeds can be seen as key to establishing such rights, other grounds for legitimate claims could be put forward depending on cultural contexts [19]. Economics is a strong claim: those businesses closest to it may assert stronger claims to its use; they may have duties (such as snow clearance or litter removal); will be more impacted by its use (such as if their customers can park there); and—if the business is a downtown asset—can increase its value because of the convenience of access. This right could be formalized or simply understood because of its proximity to a business or because a sign asks that only its customers park there. Increasingly though all city space at the kerb is formalized, monetized, and subject to enforcement.

FIGURE 15.7 The Leadenhall Building, City of London, where the ground level of the building footprint (framed by the metal pillars) provides public use of a previously closed private space.

The kerb can be seen as a victim of the *Tragedy of the Commons* [24]. A shared resource, in limited supply, if left unregulated, it would be overused and its usefulness potentially destroyed. For example, "free parking" in a town center may sound a very good idea to boost local businesses but if anyone could park there—including commuters who leave their car and get the train into the city—the parking space has absolutely no benefit to the town center and the vehicle merely adds to the already busy streets. For a week this is unhelpful, but not tragic; but over a period of a year, businesses can fail if there is no available parking because it is taken up by people not using the shops and businesses there.

As a resource, there are too many people wanting to use kerb space for parking for cooperation to work and make a Commons resource possible. There would seem to be more scope for a Commons approach to pathways given the amount of space. Indeed, many would consider that pathways are already for open, communal use; but that is not always the case. In the United States, we can see, for example, how to quell Socialist activists in Los Angeles the city amended an 1887 sidewalk obstruction ordinance forbidding, "meeting and public speaking debates or discussion in public streets" without a permit [25]. Other early sidewalk ordinances included restrictions on signage, parades, vending including such as food wagons, or other obstruction by people through loitering [26]. Today monetizing footpaths takes place such as through mechanisms such as licensing for pavement cafés, food kiosks, or markets. So, pavements, sidewalks, or footpaths are not quite a Commons either.

15.3.2 Kerb rights

In contrast to property rights, kerb rights remove the connection of the close business owner who instead would have no more right to that space than anyone else: instead, rights are gained because the kerb space has been rationed, regulated, and often monetized for those with the will and resources to lay claim to it. This is the approach of most cities as described in this chapter, but exploring basic philosophies is important in tailoring the priorities for each city. Such "Curb Rights" [27] are granted, for instance, to buses or jitneys to pull up to the kerb and pick up and drop off passenger while the footway played host to increasing amount of furniture to improve the user experience—bus shelters, seating, and live bus movement updates.

There is a tension between "kerb rights" and "property rights," but both can be at play at the same time. A business owner might be expected to clear snow but may have no preferential treatment as far as use of the pathway and kerb is concerned.

A community's interests as a whole are unlikely to be served if its city center businesses fail for lack of access to customers and residents' employment in (or indeed ownership of) those businesses end for the same reason. Similarly, the value of their property drops because there is no clearly available parking. Some sort of rationing of that space in the interests of the wider community in the longer-term is a necessary response.

15.3.3 Right to the smart city

In planning use of space at the kerb or sidewalk, issues of exclusion inevitably arise. Creating adaptations to enable wheelchair users, strollers, or those with mobility needs to enjoy the benefits of city centers is essential, of course, and usually demanded by statute. Care needs to be taken that those with additional needs or who are unable to make use of the necessary technology are not excluded from the smart city. The barriers may be physical or may be due to low income or other forms of exclusion. Those who cannot afford a phone with the capacity to book a space at the kerb—or who are unable to use such technology—need alternatives if they are to enjoy the benefits others do in their community.

Highly regulated environments, enforced through AI or other technology means, may preclude the willingness to overlook the behavior of those with extra needs. Technology that detects "loitering," for example, may lead to action against those who need time to rest for medical reasons or those perceived not to be sufficiently contributing to the local economy.

Open data and the encouragement of crowdsourced platforms are tools that local governments and BIAs can use to engage the wider community into the development of the downtown or city center. Augmented or virtual reality, for example, can be used to provide a clearer understanding of how imposing a new development would *feel* from the street level. But care would need to be taken to engage those without access to such technology. In part, the answer lies in providing public access to such technology or skills training for those who wish to have it or even recycling of unwanted equipment for those on low income. But there will still be people in danger of exclusion who would need alternative ways to pay for parking, book a parking spot for a home move, or a table at an outside café. Lefebvre's concept of the Right to the City is relevant here—citizens have the right to make use of public space and that space should be shaped around their needs [28].

15.3.4 Is smart, smart?

What can we learn from what has happened with kerb space that might be applied to footpaths as they come increasingly under pressure from sidewalk drones? Perhaps a key lesson is that just because the tech is advanced (such as enabling footpath drones) does not mean it is smart for the city to permit them. The long-term needs of the community are best served through a range of measures including those aimed at enhancing safety, livability, and local economic prosperity. Allowing one package to be delivered by drone for the benefit of one person is not necessarily in the interests of the local economy as a whole. Similarly, allowing one Amazon parcel to be delivered by motorcar with all the externalities including congestion or poor air quality that it brings is also not smart. Any policy that enables motor vehicles to dominate a city is not in the long-term interests of that city. Vehicles cumulatively create an unattractive scene, drawing the eye away from the possible beauty of the buildings, the landscape, or the people. Vehicle blight is not smart, whatever technology underpins it.

A truly smart city is not one that simply uses new, clever technology to monetize and enforce the kerb space and its rules but is able to take a strategic view of current and future needs that incorporates business, economic, social, community, and environmental needs. Such a calculation would be made with reference to an overarching strategy for the development of an area, which itself should be underpinned with appropriate data, consultation, and ethical and environmental commitments.

15.4 Summary

To thrive we need community, business, and political agreement on who has rights to the kerb and footways. In turn, these rights need to be turned into clearly defined priorities that meet the needs of citizens, including those traditionally excluded, and businesses. Public space, including the kerb and footways, should be designed for community and artistic expression and livability. These vital public spaces—like the city centers in which they exist—need to adapt to the needs of current and future generations, addressing their economic, social, community needs and their well-being.

Finally, let us move on to the real interactive part of this chapter: review questions/exercises, hands-on projects, case projects, and optional team case project. The answers and/or solutions by chapter can be found in Appendix G.

15.5 Chapter review questions/exercises

True/false

1. True or False? The footway and kerb are the hottest venues in town.
2. True or False? A fundamental change for footways and kerbs began on July 16th, 1925 when Park-O-Meter No. 1 was installed in Oklahoma City and the monetization of the kerb began.
3. True or False? Urban browning, including street trees, has helped to make a trip to downtown more pleasurable; as well as, functional, and so enabling the businesses there to compete at least to some extent with online shopping.
4. True or False? Technology is helping to ensure that footways and kerb space remain essential, into the 21st century.
5. True or False? To ensure efficient use of private space and thriving downtowns, it is important to manage this space thoughtfully and strategically.

Multiple choice

1. The question of who has the ______ to use space is perhaps a more fundamental question than that of priority.
   - **a.** Space
   - **b.** Infrastructure
   - **c.** Rules
   - **d.** Material
   - **e.** Right
2. Public space, including the footway and kerb, can be seen as being subject to:
   - **a.** Property rights
   - **b.** Legal ownership rights
   - **c.** Legitimate claims
   - **d.** Cultural contexts
   - **e.** Kerb rights
3. In contrast to property rights, kerb rights remove the connection of the close business owner who instead would have no more right to that space than anyone else: instead, rights are gained because the kerb space has been rationed, regulated, and often ______ for those with the will and resources to lay claim to it.
   - **a.** Regulated
   - **b.** Updated
   - **c.** Treated
   - **d.** Controlled
   - **e.** Monetized
4. In planning use of space at the kerb or sidewalk, issues of exclusion inevitably:
   - **a.** Arise
   - **b.** Decline
   - **c.** Improve
   - **d.** Engage
   - **e.** Conclude
5. The long-term needs of the community are best served through a range of measures including those aimed at enhancing safety, livability, and local economic:
   - **a.** Decline
   - **b.** Delay
   - **c.** Rise
   - **d.** Prosperity
   - **e.** Design

Exercise

Problem

Describe a ground-traffic control standard to address the loading and unloading of ground vehicles at curbs and the operation of robotic devices on sidewalks and at intersections.

Hands-on projects

Project

Do research: Look at why curb spaces have historically been a source of parking revenue for cities, but how the COVID-19 pandemic has accelerated the need for regulating a variety of other use cases, which most cities are not yet equipped for.

Case projects

Problem

How can new mobility contribute to making smart cities more sustainable and connected?

Optional team case project

Problem

Discuss how smart cities will able to manage the huge network of public curbside assets, a list that includes everything from parking signs and bike lanes to fire hydrants and street trees.

References

[1] Section 55, Road Traffic Regulation Act. 1984. https://www.legislation.gov.uk/ukpga/1984/27/section/55.
[2] https://www.sfmta.com/blog/san-francisco-adopts-demand-responsive-pricing-program-make-parking-easier.
[3] Millard-Ball A, Weinberger RR, Hampshire RC. Is the curb 80% full or 20% empty? Assessing the impacts of San Francisco's parking pricing experiment. Transport Res Pol Pract 2014;63:76–92.
[4] https://appyway.com/portfolio/connected-payments/.
[5] https://www.eafo.eu/countries/european-union/23640/infrastructure/electricity/compare.
[6] https://assets.publishing.service.gov.uk/government/uploads/system/uploads/attachment_data/file/891900/electric-vehicle-charging-device-statistics-april-2020.pdf.
[7] Paper Number CBP07480 Electric vehicles and infrastructure house of commons briefing. December 4, 2020.
[8] https://ddot.dc.gov/release/ddot-kerbflow-research-project-finds-high-demand-pickup-dropoff-zones.
[9] https://www.walkscore.com/.
[10] City streets: transport for a changing square mile. City of London; May 2019. https://www.cityoflondon.gov.uk/assets/Services-Environment/city-of-london-transport-strategy.pdf.
[11] https://content.tfl.gov.uk/travel-in-london-report-13.pdf Travel in London: report 13, (2020) Transport for London.
[12] https://tfl.gov.uk/info-for/media/press-releases/2021/january/outer-london-sees-22-per-cent-rise-in-cycling-as-new-data-further-highlights-vital-role-of-active-travel.
[13] http://groundplaysf.org/wp-content/uploads/San-Francisco-Parklet-Manual.pdf.
[14] Cooper G, Liberti A, Asch M. Replanting Durham's urban forest. Master's project. Duke University; 2016. Retrieved from, http://hdl.handle.net/10161/1189.
[15] Hoffman JS, Shandas V, Pendleton N. The effects of historical housing policies on resident exposure to intra-urban heat: a study of 108 US urban areas. Climate 2020;8(1):12. https://doi.org/10.3390/cli8010012.
[16] Crute J, Riggs W, Chapin TS, Stevens L. Planning for autonomous mobility PAS report 592. American Planning Association; 2018.
[17] https://www.bloomberg.com/news/articles/2014-04-03/will-a-world-of-driverless-cars-be-heaven-or-hell. Retrieved 09 June 2021.
[18] Blomley N. Rights of passage: sidewalks and the regulation of public flow. London: Routledge; 2011.
[19] Kim AM. The mixed-use sidewalk. J Am Plann Assoc 2012;78(3):225–38. https://doi.org/10.1080/01944363.2012.715504.
[20] https://www.seattle.gov/transportation/projects-and-programs/programs/parking-program/parking-regulations/flex-zone/kerb-use-priorities-in-seattle.
[21] Carmona M. The "public-isation" of private space – towards a charter of public space rights and responsibilities. J Urb Int Res Placemaking Urb Sustain 2021.
[22] Lefebvre H. "Les illusions de la moderite," Manieres de voir 13, Le Monde Diplomatique. In: Kofman E, Lebas E, editors. (1996) Henri Lefebvre: writings on cities, London: Blackwell; 1991. p. 34. Quoted in Marcuse, P., Whose right(s) to what city? In Brenner N, Marcuse P, Mayer M. Cities for People, Not for Profit: Critical Urban Theory and the Right to the City. Routledge; 2012.
[23] Marcuse P. Whose right(s) to what city? In: Brenner N, Marcuse P, Mayer M, editors. Cities for people, not for profit: critical urban theory and the right to the city. Routledge; 2012.
[24] Hardin G. The Tragedy of the commons. Science 1968.
[25] Loukaitou-Sideris A, Ehrenfeucht R. Sidewalks conflict and negotiation over public space. MIT Press; 2009. p. 102.
[26] Loukaitou-sideris A, Ehrenfeucht R. Sidewalks conflict and negotiation over public space. MIT Press; 2009. p. 30–1.
[27] Klein DB, Moore A, Reja B. Curb rights: a foundation for free enterprise in urban transit. Brookings Institution Press; 1997.
[28] Purcell M. Excavating Lefebvre: the right to the city and its urban politics of the inhabitant. Geojournal 2002;58(2/3):99–108. https://doi.org/10.1023/B:GEJO.0000010829.62237.8f.

Chapter 16

Digitization, automation, operation, and monetization: standardizing the management of sidewalk and kerb 2025–50

Bern Grush[1] and Alanna Coombes[2]
[1]*Harmonize Mobility Inc., Toronto, ON, Canada;* [2]*University of Southern California, Los Angeles, CA, United States*

16.1 Introduction

Dozens of companies are building and piloting small, electric delivery robots with a view to reduce the costs of delivering food and parcels over their final distance.[1] Amazon, FedEx, Postmates, and Starship are four among them (Fig. 16.1). Concurrently, cities are interested in reducing congestion and emissions associated with the use of trucks, vans, and cars for deliveries, which has more than tripled in the last decade. Growing expectations for instant delivery, high prices commanded by local delivery operators, and pressing urban traffic goals combine with accelerating digitalization to create demand for logistics-related innovation.

In addition to the four logistics operators just mentioned—and there are already many others—there will easily be several dozen more applications and manufacturers on a worldwide basis before a likely, late-decade market shake-out leaves only a few tens of significant players. These robots and their successors are expected to frequent pathways such as sidewalks and pavements over the next few years. As they become more capable, their adoption will become more pervasive. One of them, Starship, completed its first million commercial deliveries in January 2021. Within 4 months they completed another 500,000, and in mid-May of this year, they claimed 80,000 intersection crossings per day across the several locations in which they currently operate. Even if this accounted for 50% of the total number of commercial, robotic delivery trips worldwide, the other operators are not far behind. Once this technology matures, how can the express delivery operators such as DHL or UPS not add their machines to the mix? Seeing Swiss Post's success with its delivery robots, Canada Post, US Post Office, and many other national postal services can be expected to engage.

To date, each of the logistics companies using or trialing pathway robots operates in independent, closely constrained public spaces without spatial competition from other similar operators. Traditional logistics operators have long operated in common spaces—it is usual to see UPS and FedEx step-vans stopped in front of the same building—so we can expect to soon see multiple logistic-robot operators concurrently occupying the same pathways and crosswalks.

How can we prepare our cities for myriad robotic vehicles sharing public spaces with human-operated vehicles and human pedestrians? How will vehicles from various vendors operate simultaneously in these spaces? How can these vehicles be used to their greatest advantage within our cities? The chapter addresses these questions in three sections:

- Section 2 discusses reasons why mobile service robots will occupy the urban panoply of automated vehicles sooner and in greater numbers than will fully automated passenger vehicles.
- Section 3 outlines a draft, international standard designed to enable a mobility authority to manage, orchestrate, and monetize multiple, concurrent fleets of service robots. While this encompasses passenger and goods vehicles that use the roadway and kerbside (see Section 16.3.5), it predominately considers smaller

1. Pathway robots are being designed to perform several different services and are called by many names including sidewalk drones, delivery bots, and personal delivery devices (PDDs). In this chapter, they will be referred to pathway robots in general and as delivery robots in particular.

 https://doi.org/10.1016/B978-0-12-819130-9.00006-1

FIGURE 16.1 These are four of many pathway delivery robots in small-scale commercial use today. These range from 68 to 91 cm (length), 53–71 cm (width), 55–147 cm (height, without a flag), 33–136 kg (gross weight), and 5–24 km/h (max speed) [1]. Yet larger and faster delivery robots are planned for roadway use. Ambulatory robots are being developed to handle stairs. This is barely beginning. *Illustration commissioned by one of the authors.*

robots that will use pathways, footways, and active transport lanes (see Section 16.3.6).

- Section 4 describes one of the potential applications of this standard.

16.2 Delivery robot technology will out-disrupt the passenger robotaxi

Despite a decade and a half of investment, promotion, and anticipation about the coming of driverless taxis and shuttles, delivery robots will arrive sooner and in greater numbers than will robotaxis. These machines are essentially small containers on four or six wheels that use pathways such as footways, bikeways, crosswalks, or carriageway shoulders to move over modest distances (1–3 km), currently overseen by a human teleoperator (nonline of sight radio control) who can intervene if needed. These machines usually carry food or small express packages. The promise of their widespread deployment in driverless fleet operations is much closer to reality than is wide availability of passenger robotaxis. This is important because most cities are even less well prepared for robotic pathway systems than they are for driverless passenger vehicles on roadways.

There are a number of reasons tele-monitored delivery devices will become pervasive before driverless passenger vehicles. The barriers to deployment of delivery robot are far lower than they are for the robotaxi. Likewise, the accelerators driving development of delivery robots are more accessible to innovators, investors, and other participants. Several of these accelerators follow.

16.2.1 The safety barrier for delivery robots is much lower than for robotaxis

Delivery robots come in a variety of sizes and configurations. Smaller units for single deliveries are the size of a filing box and weigh less than 50 kg fully loaded. One of the most popular models, Starship, is a small cube less than 0.25 m. The top speed of these smaller robots is usually constrained to about 6 km/h, a hurried walk. Small and slow, they can stop quickly. Larger delivery robots—half the size and weight of a passenger sedan and perhaps traveling at 40 km/h—present greater safety challenges.

Considering only momentum, delivery robots pose less of a crash hazard than would a sedan-sized robotaxi. Because they carry only cargo, there would be no risk to human passengers. This, however, may not be entirely positive, as it could potentially affect unintended risk to pedestrians posed by algorithms that do not weigh passenger risk in their computations.

Like robotaxis, delivery robots are designed not to hit anything. If one of the smaller robots were to hit an adult human, it is far less likely the collision would be life-threatening. One exception to this is that a robot could precipitate a fatality in the same way that a pet running into the street might cause a vehicle to swerve and lose control. Or if a robot were struck by a bicycle, the cyclist could be seriously injured—or worse. Pedestrian-involved crashes with smaller, slower robots would be far less dangerous than crashes involving sedan-sized vehicles that may weigh 1400 kg and can travel at speeds exceeding 60 km/h.

The sheer variety of delivery robots presents challenges for protecting pedestrians and cyclists, and any regulations will need to account for a wide range of considerations. For example, smaller robots might best be kept off the roadway except when crossing at intersections, and the larger robots may need to be banned from pedestrian footways. While it is far too early to predict how this will play out, as of early 2021, approximately 20 US states have tabled legislation regarding Personal Delivery Devices (pathway delivery robots). At least 12 of these had been passed meaning there are already a number of somewhat variable regulations in place in that country [2].

16.2.2 The fear barrier for delivery robots is much lower than for robotaxis

According to the American Automobile Association, over half of people recently surveyed express fear of driverless vehicles [3]. This fear makes both makers and regulators sensibly conservative about removing the vehicle's safety driver. Notice that in all of the thousands of videos where a driverless-taxi safety driver is absent, the weather is especially clear, the roads are in excellent repair, and traffic is notably light.

Consumers may accept that the company using a delivery robot to deliver a meal or small package might delay for safety reasons until a downpour lets up or until a pathway is cleared of snow. They might not accept that from a passenger vehicle when they are late for an appointment. Fear of harm from crashes creates a much greater barrier for robotaxi acceptance and governance than it does for delivery robots.

16.2.3 Concerns for job loss from delivery robots are lower

Setting aside projections of driver shortages and arguments promoting "career retraining"—which are often not accepted by the people so employed—many workers and their families feel threatened by automation. In many cases, unions and associations can create effective, if limited, barriers to the deployment of larger, automated vehicles for passengers and goods.

Last-mile delivery—especially in the meal sector—generally provides temporary, part-time, or second jobs and employment for youths and gig workers. There are fewer coherent voices to speak out against automation of these jobs, implying that the union, social, and employment-equity barriers to the diffusion of pathway robots would be much lower than that for robotaxis.

16.2.4 Delivery robots will have fewer enemies and more friends than robotaxis

Standing against the robotaxi will be interests such as transit and taxi drivers and their agents and unions. Pushing against the delivery robot will be advocates for pedestrians, accessibility, and gig workers. These latter groups will have weaker voices than those potentially arrayed against the robotaxi.[2]

The delivery robot has the pedestrian pathway as a new space to exploit, and the exploiters of that space such as merchants, Amazon and FedEx, will have more power than any advocacy group that might wish to constrain the spread of these machines. The commercial weight of consumers demanding fast, cheap delivery that saves a trip to a shop or restaurant might outweigh pedestrian advocacy against the robots.

Until now, the footway has not been seen as a locus of employment as has been the case for the roadway and its kerbside. No municipality has monetized the footway as an entranceway or pathway to businesses as some have done with kerb parking. The footway has fewer powerful stakeholders as enemies of automation compared to the roadway, although it is possible that the coronavirus may have changed that.

2. This point is one of economic bias rather than social justice suasion.

16.2.5 Development and deployment costs for delivery robots is far lower

The investment required to build and prove delivery robots is far lower than that required to build and prove driverless passenger vehicles. The cost differential for a single robotaxi compared to a single delivery robot is currently in excess of an order of magnitude, exclusive of deployment and operations. While all these costs can be expected to drop over time, the relative differential will remain.

16.2.6 The regulatory barrier to delivery robot deployment is lower

In most countries, national and state/provincial governments consider regulations for automated passenger vehicles, mostly from a safety perspective. Regulations for robotaxi fleet deployment—which address issues that are quite different from matters of safety—generally receive little attention. To date, regulations for pathway robots appear to receive even less attention, although this may begin to change.

Pathway robots are generally seen as a municipal matter and that leads, as it did for ride-hailing and e-scooters, to regulatory outcomes that vary city by city. It is difficult to imagine this will not continue as the default. For that reason, opportunistic start-ups, which are currently a major source of innovation for pathway robots, will quickly target cities seeking smart-thinking reputations. These cities may turn a forgiving eye to the efforts of start-ups and innovators or even invite them to trial their devices in their municipalities.

Any laissez-faire attitude regarding regulations for pathway robots will shift rapidly once companies such as Amazon and FedEx deploy delivery robots monitored remotely by unseen human operators. The push to deploy, even at a modest scale, is likely to grow in response to congestion and environmental concerns driven by the demands of e-commerce. This will ensure closer attention from regulators.

16.2.7 The travel environment for pathway robots is a more complex

A robotaxi is often framed as "just a taxi with a silicon driver," and we are often told these machines will use the same roads and the same parking spaces as human-operated vehicles. This is generally expected to apply to automated goods-delivery vans, as well. While this assumption is only partially true, it is true that the physical infrastructure for road vehicles is already well developed. We will need to address automated loading and unloading rules (see Section 16.3.5), but we should expect to build very little new infrastructure if the technology is delivered as has been promised.

While this last point remains to be seen, the relative comparison with small pathway robots is germane. While delivery robots are expected to operate on existing infrastructure, there is a critical difference in that the rules governing the configuration, condition and certification of pedestrian clearways, and the systems to manage and broadcast information about any construction or changes thereon are neither as consistent nor as frequently complied with as they are for roadways.

Pathway robots will have to run a gauntlet of human legs, barking dogs, baby strollers, planter boxes, sandwich boards, tree roots, and uneven pavement—a much more disorderly environment than the more highly regulated roadway where robotaxis will operate.

Cities will have many more undigitized and nonconforming footways than they will have such streets. This constitutes a relative, but important barrier for operating delivery robots that exceeds that for robotaxis. This will need consideration in order to manage the arrival of these robots (see Section 16.3.7).

16.2.8 Psychotechnical barriers around full autonomy favor delivery robots

There is now a widespread understanding that the SAE Level 5, Fully Automated[3] vehicle has been overpromised by marketers, exaggerated by mass media, and misunderstood in the popular imagination. An object of engineering aspiration and popular fantasy, we are finally coming to understand what Professor Steven Shladover meant when he told us that the last stages of readiness for automated road vehicles—vehicles that can handle every driveable circumstance and mix with existing nonautomated vehicles on our roadways—are very difficult [5]. We see that while some robotaxis have begun to operate without a safety driver, they have so far been limited to operating design domains (ODD) where the road environment is relatively well organized, enjoys mild weather, and has low levels of traffic congestion.

3. The SAE "levels" of automation [4] span from 0 (no automation) to 5 (full automation). Level 4 (high automation) is generally thought to be suitable as a robotaxi, a vehicle that can be driverless in a defined area known as its "operational design domain." A pathway robot that requires a responsible teleoperator would be at SAE level 2 (partial automation) or level 3 (conditional automation).

Fear and negative perceptions evoked by fully autonomous vehicles create a significant barrier for widespread adoption of the robotaxi. Without full autonomy, fleets of these vehicles necessarily have limited operating domains, constraining their applicability, and reducing their profit potential. The delivery robot, also not fully autonomous, does not suffer to the same extent.

Robotaxis, with the recent exception of a minuscule number of driverless vehicles, require safety drivers in the vehicle who may intervene increasingly less frequently as these machines improve. Delivery robots also require the oversight of teleoperators, but these may be nonline of sight, situated several kilometers distant, with one person operating multiple robots at once.

As pathway robot technology improves, the number of robots that a single teleoperator manages will increase. With intelligent, collaborative, multiuser teleoperation systems, the ratio of machines to humans—now perhaps two or three per teleoperator in the more advanced instances—will reach 5 or 10 to 1 and eventually many more. In cities that are suitable and prepared, delivery robots will be able to diffuse sooner, in more places, and scale up much faster than will robotaxis.

The likely outcome is that none of these technologies will achieve *full* autonomy except in YouTube videos—humans will always be somehow in the loop. For this reason, delivery robots are better suited than robotaxis to overcome the constraints of working in environments with noninvolved humans and be able to operate effectively in a variety of settings without being fully autonomous.

16.2.9 Delivery robots evoke fewer perceived privacy issues

Will robotaxi trips be tracked, recorded, and remembered? Will data be searched, correlated, and sold? Will private conversations be recorded and passengers filmed? The capability to track, record, and film may be considered necessary to provide safe passage without a driver overseeing every part of the trip, but how can passengers know their data is secure and protected? Will it be destroyed at the earliest appropriate moment?

Similar questions can be asked of the purchases that are delivered by a robot. But that has not stopped e-commerce of all forms from growing dramatically. The greatest fear most people express about e-commerce is the fear of entering credit card information online—and that is a concern for property or financial security as opposed to personal privacy. The concern for privacy about what one eats, wears, or reads seems less significant than the concern for having one's trip behavior tracked, modeled, and sold. We do not need to debate the credibility or utility of these relative concerns. Differential perception is all that needs to be acknowledged.

16.2.10 Delivery robots gives rise to fewer perceived security issues

While imperfectly understood and still unexperienced, robotaxi security issues are clearly imaginable: vehicle hijack, passenger molestation, robbery, rape, or worse. If something is lost in the vehicle, will it be recoverable? Would parents be able to entrust the safety of young family members to a trip in one of these vehicles?

These concerns apply far less to the delivery of a pizza or a bag of vegetables in a robotic grocery cart. This means fewer, if any, psychological barriers to consumer acceptance of delivery robots compared to robotaxis depending on the demographic context of their deployment.

But there is a security concern that would more likely apply to delivery robots than to robotaxis. We might fear that a swarm of such robots could be commandeered for purposes of malfeasance. Systems for managing cybersecurity are currently being developed and standardized to address these concerns, but so far these efforts are incomplete, unproven, and unenforced. Worse, much of the current US state legislation permitting these devices provides no cyber guidance. This will need to be addressed for widespread adoption of service robots in public spaces.

16.2.11 The total risk equation for delivery robots is a magnitude lower

The issues above — cost, acceptance, liability, investment, ROI, privacy barriers, security concerns, and regulatory matters—combine to form a total risk picture. Because the payoff for products and services in the passenger transportation sector is projected to scale between US 7–10 trillion dollars annually, there is much more media, investment, and municipal focus on robotaxis and personal driverless vehicles than on delivery robots. But the first phase of automating mobility—using delivery robots for light, last-mile movement of goods—is a clear winner from the perspective of risk.

Governments, typically risk-averse, maintain an appropriately high barrier for driverless passenger vehicles. But small projects for teleoperated delivery robots are easily approved and may sometimes not even require approval, as was the case for one start-up in Toronto in 2019. And such delivery robot projects—given the comparatively little setup required—are far more easily decommissioned than are the more demanding driverless taxi or passenger shuttle trials.

16.2.12 The risk of nonreadiness to cities and towns

The greatest immediate risk facing cities is to ignore pathway robot technology until it is upon them. For cities that fail to prepare appropriately, the likelihood of repeating the chaotic introduction of ride hailing and e-scooters is high—and the likelihood of getting off easy this time is far lower.

Before these machines are available in volume, there are many aspects municipalities and retail communities (see Section 16.4) might consider in advance. In alphabetical order, some of these are:

- Business district opportunities
- Environmental goals
- Infrastructure readiness
- Licensing of fleets; registration of robots
- Pathway accessibility (i.e., relative to current legislation)
- Pathway maintenance (good order)
- Pathway pricing (similar to road or parking pricing)
- Safety
- Traffic goals

16.3 An international standard to manage pathway robots

The arrival of automated vehicle fleets in public urban spaces mixed with human-operated vehicles and pedestrians such as at the kerbside and on public pathways requires an overhaul of how public authorities operate and monetize these spaces. This challenge will involve increased digitalization, new communication technologies for operation and coordination, and highly coordinated collaborations between governing bodies and system providers.

16.3.1 ISO 4448

ISO Draft Technical Standard 4448: "Ground-based automated mobility systems" [6] is being prepared to create the requisite definitions for operating data and procedures that can impact ground-traffic management[4] and inform relevant vehicle operators, system makers, insurers, as well as community standards.

> ***Disclaimer***: *ISO 4448 is in an early draft state as of this writing. As a technical standard, 4448 will comprise data and procedural definitions covering critical elements of operations, security, and machine behaviors for automated passenger vehicles and service robots in public spaces shared with humans whether or not involved with the machines. The content of this section should be read as an*

4. The standard addresses loading and unloading at the kerbside as well as pathway reservation relative to service robots. It does not address machine control which is presumed to be with the vehicle operator and/or its software.

informal outline and implying no obligation on the part of the ISO to publish this as here described.

ISO 4448 sets out procedures and behavioral rules to make regulating, governing, and operating systems of automated fleets a coherent regional process. In addition to addressing automated data and procedures for loading and unloading goods and passengers, it provides the common data and procedures that would enable a robot *pathway reservation system* (PRS, see Section 16.3.6.5 in this chapter) and the monetization of its use for commercial purposes.

While the data and procedures within a PRS would be expressed using common definitions and forms, the local jurisdiction would set appropriate local behavioral constraints. An analogy is posted speed limits for automobiles—such signs use common sizes, shapes, and colors, as well as agreed metric units, but a local jurisdiction usually decides posted speeds. In an environment of digitalized, automated machines such as robotaxis and especially pathway robots, information is provided using telecommunications rather than analog signs. ISO 4448 is designed to enable:

- Makers to develop robots and their communication systems that can operate in multivendor environments,
- Planners to design deployments,
- Local jurisdictions to ensure appropriate infrastructure and governance guidelines,
- Insurance companies to provide coverage, and
- Transportation and logistics operators to proceed safety and profitably.

This section provides a preview of the draft, its intention, scope, and its critical components as currently configured.

16.3.2 The context for a standard for ground-based automated mobility

It is anticipated that in the very near future, many more urban and suburban jurisdictions will consider preparations for robotic cars, taxis, and trucks and other forms of robotic vehicles to move passengers and goods. At the same time and in the same places, service robots may be deployed for maintenance activities such as snow removal, trash pickup, sweeping, or surveillance. The location for these services will be in public spaces in towns and cities where kerb and pathway space are already under increasing pressure for access by a growing variety of uses, innovations, devices, businesses, and services.

Over the past decade, digitalization of mobility and commerce has brought rapid growth in new forms of taxi-class operations loading and unloading passengers at city kerbs as well as a dramatic rise in goods delivery from e-commerce systems. In some areas of larger cities, this change has been rapid and has already become unsustainable. Some of these are being addressed on a local and urgent basis often without consideration of future change, growth, or innovation. In addition, the rise in active transportation has added cycling, scooter, and e-bike lanes at the kerbside in many cities, as well as storage for these vehicles at the edge of the footway [7].

Since early 2020, the onset of the COVID-19 pandemic imposed yet more demands on these kerbside and footway spaces, including social distancing, an uptick in micromobility, and in some cases increased demand for outdoor dining space. This tended to create wider nonautomotive rights of way to accommodate the new demand. Additional width invites more variety and creates an even greater need for access management as social distancing continues, micromobility grows, walkability demand increases, and the need for cleaning, maintenance, and snow removal for these expanded and complex places grows.

To this mix, we expect to add the delivery of passengers and goods using driverless vehicles that load and unload at the kerb. And, as described above, we expect to add last-mile delivery of goods via pathway robots. Indeed, prior to 2020, such systems were already operating on a commercial basis, and they continue to do so.

All of this implies further increases in traffic volumes both at our kerbsides and our pathways. The introduction of automated vehicles without human accompaniment will necessitate highly automated (digitalized) management. Taken together this will change the nature of the interactions among these vehicles and their control systems—with each other, with the kerb, with payment systems, with active human mobility, and with our existing manual vehicles and devices.

The traffic and parking rules cities have relied on prior to 2020 represent governance that is already under stress, and their design and governance shortcomings have been made increasingly evident by the pandemic. Parking systems developed to date [7] are insufficient to support the loading and unloading of the anticipated automatic vehicle systems without additional data and procedures to support ground-traffic management systems.

Cities will need new operating guidelines as kerbsides and pedestrian spaces are joined by automated taxis and robots that arrive, stop, park, wait, and load under sensor, effector, and software control. Unaccompanied by human passengers or attendants, these machines will need to be prioritized, scheduled, queued, bumped, and placed in holding patterns, and all without blocking crosswalks, bicycle lanes, micromobility users, no-stopping areas, or transit stops that are common infractions by taxi and goods-delivery vehicles today. This must be done safely, mixed with human-operated vehicles, without inconveniencing active transportation users, pedestrian traffic, and those with accessibility challenges.

16.3.3 Five intentions for standardizing kerbside and pathway automation

Planning for the ISO draft technical standard 4448 was guided by several intentions for its use. These intentions span, safety and conflict avoidance, infrastructure planning, commercial use and activities, operational management, and legal, liability, and insurance matters.

16.3.3.1 Safety and conflict avoidance

As the number and variety of automated and nonautomated mobility vehicles and devices increase, so too does the potential for spatial and navigational conflicts involving vehicles arriving, stopping, parking, waiting, loading, passing, crossing, and overtaking. Spatial conflicts are already very common and cumbersome at many kerbsides and pathways. Machines that operate at kerbsides and on pathways must interact with each other and with human-operated vehicles and will be expected to operate without on-board human operators or even proximate human control and potentially without the spot- or lane-markings that often guide on-street vehicles. This requires a set of agreed-upon and tightly communicated behaviors and guidelines for real-time resolution. These guidelines require terminology, procedures, communications, and systems.

16.3.3.2 Planning

Projects to reformat and reorganize streets, kerbsides, or pathways will need to build and shape these spaces to be workable for vehicles and devices whose operating characteristics may be different, or differently constrained, than those vehicles and devices under human operational control. Such planning activities need guidelines and those guidelines need common data and systems. They will also need more detailed metrics and design parameter descriptions as more such spaces are prepared for automation.

16.3.3.3 Commercial

Some kerbsides and pathways can be expected to be used more heavily by commercial vehicles (taxis, shuttles, trucks, footway service robots, etc.), each with various automated capabilities. The use of automated (driverless) machines for loading and unloading passengers and goods requires forward planning for logistics. Such forward planning will need reservation systems updated in real time. The design and execution of such reservation systems require shared terminology, procedures, communications, and systems since we can expect multiple vehicle types, providers, and operators.

16.3.3.4 Operations

The kerbside and pathway comprise the spatial context for people who reside or trade in the buildings at or near such kerbsides or footways. People and goods that arrive or depart with the help of vehicles and devices, automated or not, expect to be able to arrive and depart in a timely manner without finding a pathway or loading facility blocked and without unexpected long waits. These spaces need to be managed in a reasonably smooth and coordinated fashion. This requires shared communications and systems.

16.3.3.5 Legal, liability, and insurance

Any kerbside or pathway is a public space shared by many types of users including local residents, vendors, visitors, and shoppers, whether able-bodied or not. Any conflict that causes injury, financial loss, or other harm or perceived harm may be subject to legal or claim action. Hence a common understanding and description for these spaces and the expected machine behaviors in those spaces are necessary to assign or determine liability. This shared understanding and description require common data, procedures, and system definitions.

16.3.4 Standard components

ISO DTS 4448 defines the data and communication systems needed to organize, expedite, and safeguard the flow of automated vehicular ground traffic relative to the loading and unloading of goods and passengers and the allocation and movement of service robots for delivery, garbage removal, sweeping, washing, snow removal, repair, food trucks, construction, etc., in public spaces such as kerbsides and pathways as shared with pedestrians and other automated or nonautomated vehicles. Such systems are intended to enable carefully defined and growing areas (operational design domains) of cities to manage any number of vehicles and vehicle varieties operated by any number of operators (public, commercial, private) for these various activities.

The remaining subsections of this chapter section look briefly at critical system components for managing public-system (urban) robotics. These roughly correspond to the current and planned parts of draft technical standard ISO 4448. Since this work is still in early draft stages, this outline may differ from its final form:

1. Robotic road vehicles for passengers or goods
2. Service robots for pathway and other public spaces
   a. Guiding principles for operation of robots in public spaces
   b. Guiding principles for governance of robots in public spaces
   c. Similarities between pathway robots and human accessibility devices
   d. Pathway robot access: surface conditions and path dimensions

  e. Service robot access permissions
  f. Service robot behavior
  g. Service robot social communication
  h. Integrating robotic kerbside and pathway access
  i. Robot cybersecurity
3. Certification for use
  a. Kerbside and pathway certification for automation
  b. Robot weather-worthiness

16.3.5 Robotic road vehicles for passengers or goods

Robotic ground transportation systems for passengers and goods require far more than just automated vehicles. Appropriate locations to load and unload require algorithms for prioritization, reservation, scheduling, and queueing. They also need recovery processes for exceptions that may occur.

An urban area that intends to permit or encourage the use of automated road vehicles will need to intermix a growing number of these complex, interacting and increasingly digitalized (fast, precise) components. As a system to operate among multiple vendors, this will be analogous to an air traffic control for numerous airplanes, flight operators, airports, and runways.

Current systems that match passengers to vehicles are plural, competitive, and disparate. Examples are taxi-dispatch and ride-matching services, which are individually workable, but taken together, suboptimal. We can often observe spatial conflicts for goods movement systems matching shippers to couriers; it is commonplace to see two or more stepvans from competing express delivery operators standing adjacent on the same street each blocking a bicycle lane while delivering just one or two packages each. That is suboptimal from a traffic, environmental, and total delivery-cost perspective.

Local or regional coordination will be required to create collaborative systems that match robotic vehicles with load/unload spaces, such as in publicly shared parking areas at the kerbside. In other words, a single, effective management system is required to coordinate loading/unloading of all passenger and goods vehicles, regardless of the number of taxis, shuttles, or logistics providers operating within a bounded region.

To load/unload passengers requires procedures for vehicles, or their operators, to reserve, queue, and access spaces at the kerbside or other controlled locations—i.e., mapped spots suitable to a passenger's start/end goals. A singular system is required within a given spatial domain to accommodate the complexities of admitting multiple passenger and goods transport operators dynamically sharing a large number of loading/unloading places. This is analogous to a computer operating system managing an arbitrary variety of programs and memory locations.

A system to manage loading/unloading of passengers is primarily concerned with trip terminus events and less with the routes between them. However, traffic flow or congestion along those routes naturally affects the real-time management of terminating events. Uncertainty in trip times will cause rescheduling, requeueing, and complexities of storage for queues, such as "circling the block," double-parking, waiting areas (oversupply of parking areas), or queueing in-motion (a process of having vehicles alter their travel speed to time of their arrival at a spot).

Flattening peak load/unload times would help this queueing process considerably. One way to accomplish this is through the use of variable pricing of loading/unloading privileges. Since a load/unload management system will require computation, IoT devices, oversight, maintenance, and spatial infrastructure for the vehicles, it will need to be funded. The best way to match a transportation system's expense with its management is through variable use pricing that is designed to flatten peaks.

Two critical elements related to both robotic passenger and goods movement are safety and accessibility. Safety considers passengers, pedestrians, as well as nearby vehicles and their passengers. Accessibility concerns are likewise threefold: passengers, nearby pedestrians with accessibility challenges, and the accessibility considerations of nonautomated vehicles and their passengers operating in the same space.

This road vehicle load/unload aspect of the standard needs a small set of data elements describing the location, dimensions, properties, permissions, and availability of load/unload spaces and a matching set of data describing the vehicles requesting those spaces. In addition, a set of rules, procedures, and processes are needed to request, prioritize, match, enqueue, dequeue, and manage the inventory of load/unload spaces. Methods to price loading/unloading activities according to jurisdictional requirements can be added readily since these processes require real-time location, scheduling, and monitoring. Loading/unloading goods has all of the same ground-traffic control issues as does passenger loading and unloading including: requesting, prioritizing, matching, queuing, and inventory (space) management, as well as additional considerations such as size, noise, emissions, and hazardous materials.

While the standard is largely agnostic to whether a ground vehicle is carrying passengers or goods, it admits distinctions so as to permit a jurisdiction to manage goods delivery schedules or locations differently from those of passenger systems. In this way, the standard can support separate loading areas for goods and passengers, dynamic loading areas that admit different vehicle purposes throughout the day, or even variable, on-demand mixing among modes without distinction in spatial allocation. This is done because it is not possible to predict the degree of segregation or mixing among passenger and goods systems.

Indeed, it is possible that passenger vehicles may also transport goods independently, either having the same vehicle perform different duties at different times (serial work assignment) or in parallel work assignments similar to the way that regional bus-passenger or air-passenger systems also transport goods (see Section 16.3.6.8).

16.3.6 Service robots on pedestrian pathways

Robotic vehicles intended for services such as personal deliveries, snow removal, sweeping, surveillance, or other light duties on footways, bike paths, road shoulders, or other urban pathways are a novel urban management problem. For centuries, cities have managed the loading and unloading of road-vehicles on or at the kerbside of roadways. Repurposing current kerbside-management practices for automated road vehicles is relatively straightforward to contemplate. Unfortunately, the ability of most cities has been sorely tested as they have had to manage a combination of high volumes of parked vehicles, dramatic growth in e-commerce, multiple active transportation modes, and now social distancing during the pandemic.

Considering these existing, and growing pressures, the management of even modest numbers of motorized, automated vehicles on pathways will be an even more daunting challenge. Worse, the current design and status of urban footways are already challenging for many pedestrians.

At base, the fundamental last-mile logistical concerns for automated vehicles at the kerbside and on pathways are analogous: match and schedule vehicles to use identified spaces. At the kerbside, spaces are loading or parking spots. For the pathway, the space is a city block-face or a segment of footway between two intersections or points.

But there are also critical differences. At the kerbside, vehicles queue while in motion to become stationary in order to load/unload. On the footway, crosswalk, or bike lane service robots queue to operate (move, navigate, work, and wait) in ways that are mixed with pedestrians of all abilities or active-mode users such as cyclists or scooterists. Pedestrians occupying this space walk pets, carry packages, push, drag, or ride in wheeled devices, chairs, scooters, or boards. They travel in small groups, meander slowly, stand in clusters such as at intersections or transit stops, and they window-shop, line-up, run, dash-across, or weave from one side of the pedestrian clearway to the other. Such normal pedestrian behaviors are at risk of becoming less safe or more difficult due to the presence and movement of robots among these existing activities.

Depending on the prevailing view of the governance of public space (more below), such pedestrian behaviors may be protected or curtailed by the introduction of service robots. The standard described here is agnostic to governance style or theory; it is designed to formalize communication and operation of any intended governance style. The next three subsections outline operational, governance, and accessibility principles for pathway robots.

16.3.6.1 Guiding principles for operation of robots in public spaces

To guide the development of a formal standard for robot behavior, a series of guiding principles are used. They are as follows:

1. Robots should grant **rights of way** to humans in close proximity, but rules of engagement may consider how to prevent a robot from being immobilized for an extended period in crowded circumstances. There may be explicit exceptions in the case of service robots as actors in emergency (police, fire, ambulance, evacuation) contexts.
2. Robots must be deployed to respect the cultural and contextual, interpedestrian distance normally observed when walking or standing in a public place, known as **shy distance**. This may be extended to **social distance** in the event that robots are identified as a disease vector.
3. Robots must **not harm or alarm** humans or animals on the pathway.
4. Robots must be **apparent** (visible and/or audible) to all humans on the pathway (flags, lights, sounds, gestures). This is not only to accommodate people who may have visual or auditory challenges but to avoid mishaps with distracted pedestrians. This is related to not harming, confusing, or alarming humans.
5. Robots must **signal** their presence, priority, and certain properties to other machines. This enables rights-of-way decisions and can help differentiate autonomous mobility devices from human-operated devices, humans, and nonmobility entities.
6. Robots must not diminish the **privacy** of humans or businesses using or residing near pathways. This implies constraints on the recording and retention of data.
7. Robots must not diminish the **security** of humans, businesses, or other machines on the pathway. This is also in regard to the security of humans residing and trading near such pathways. This includes both cyber and physical security.
8. Robot infrastructure must not be **intrusive**. Robots may be guided by localized infrastructure, high-resolution mapping, and other data or technologies, but any additional infrastructure cannot negatively affect (make more cluttered, riskier, more confusing, or less accessible) the use of this shared space by humans.
9. Robot **occupancy** within a defined area must be controllable to prevent unacceptable congestion on public pathways.
10. Robot **waiting and stand-aside** behaviors must not create obstacles for pedestrians. This impacts how robots may position themselves when pedestrians pass, wait at intersections, or travel at the edge of a pathway.

16.3.6.2 Guiding principles for governance of robots in public spaces

Kristen Thomasen outlines three views of public space that might guide a regulator of pathway robots: Communal Public Square, Regulated and Orderly Public Square, or State-Owned Property [8]. Depending on how these views influence relevant regulations, robots would be governed locally in more or less restricted ways.

An international standard, to the degree possible, must be agnostic to disparate legal theories. The primary goal of standardization is its role in equipment and system design, operation, as well as process design and certification. However, since the machines, systems, and processes being standardized operate in public spaces, in large numbers, for many purposes, and among many pedestrians, the deployment of a standard must necessarily impact, and be impacted by, jurisdictional governance in fact rather than in style.

Hence, it is critical to ensure necessary and sufficient operating data and procedures so that the respective sociolegal preferences can be supported in any country, state, or city by way of constructions that allow legislators to adapt the standard to their governance needs and be able to communicate relevant rules to makers, operators of automated devices, and their users (shippers, carriers, and receivers). Correspondingly, makers and operators of robots can anticipate and comply with the resultant rules.

In the simplest view of safe personal space for pedestrians, a clear space in the direction of travel must be open in order for a robot to proceed. The proximate, real-time issue resolves to whether the size and comfort of that clear space are such that pedestrians are not made worse off in terms of access, safety, convenience, or peaceful enjoyment of that public space.

Rules requiring robots to yield right of way and respect shy distance imply an optimal, clear space in this immediate real-time sense, but such rules do not prevent robots from entering a dynamic space that could, after a short period, develop into a circumstance that inconveniences or delays pedestrians or adds to pedestrian congestion potentially made worse as a consequence of the presence of the robot(s).

Robot navigational rules that operate by opportunistically moving into clear spaces as they open up (greedy algorithms) are essentially how humans navigate on busy pathways and cars operate in traffic. If such was the only local-decision approach employed by robots, then as these machines become more capable, nimble, and numerous, human pedestrians—especially those who are older or less nimble—would become increasingly disadvantaged as robots are technically enabled to dart opportunistically wherever possible. Average human skill as a pedestrian is unlikely to improve, especially as populations age. But over the next decade, robot skill will improve dramatically. In unregulated, congested circumstances, this could become deleterious to human comfort and rights of way.

Several instances of current US state legislation that have been enacted since 2017 indicate that robots (called personal delivery devices in these documents) must always give way to pedestrians [2]. This behavioral constraint is necessary but insufficient in the case of the use of greedy spatial algorithms.

For this reason, the standard provides data and procedures to regulate the ingress of robots to a block-face or pathway segment in real time so that their occupancy (count) at any one time can be limited. This reduces, but does not eliminate, the effect of greedy spatial and navigation algorithms.

Related to this, it is possible for a robot that must always give way to pedestrians and to maintain a shy distance to find itself temporarily constrained for unexpected or unintended periods of time especially in congested foot traffic ("robot trap"). Naturally, operators of such robots would like to avoid such circumstances, but this may not be possible on every occasion. It is likely that a logistics operator may not, or may not even be permitted, to know the travel plans of the robots of another operator. This is another reason to consider occupancy-count control according to pathway configurations and times of day so as to minimize the likelihood of such events and minimize robot extraction time when one does occur.

As robots become smarter, we can imagine that they might acquire, through machine learning, more foresight to further reduce the probability of being trapped among pedestrians or other robots. In the meantime, the standard provides a way to minimize the likelihood of this occurrence and provides a level of governance that acknowledges local contexts so that occupancy limits may act locally and dynamically.

16.3.6.3 Similarities between pathway robots and human accessibility devices

There are a number of useful comparisons between wheeled pathway robots and pedestrian accessibility devices such as wheelchairs or assistive scooters. As a vehicle, the wheeled (nonambulatory) pathway robot has characteristics similar to a wheelchair—it can easily travel faster or slower than the average human (walking) pedestrian, and it confronts issues of traversing uneven, damaged, steep, sloped, or potholed pavement, and ramps (kerb cuts). It cannot readily step aside as an ambulatory, abled pedestrian normally can, and it cannot streamline its width by turning sideways while walking as most pedestrians can. Basically, the wheeled pathway robot exhibits many of the rigid physical and motion characteristics of a pedestrian wheelchair. Depending on wheel diameter, number of wheels, and their suspension system, a robot may have somewhat different constraints compared to a wheelchair.

As a machine, the mobile service robot might be relegated fewer social rights or diminished rights of way compared to a pedestrian. Conversely, it may be an actor in an entitled social role, i.e., it may be performing a service critical to someone with special social rights. Perhaps some specially marked robots might inherit those rights in the way that a registered service dog inherits certain social rights of way from the human it is helping. Such a robot may be unable to cross certain path elements that an able-bodied pedestrian can readily traverse, it may be subject to vandalism or mischief in ways that are different or more frequent than those confronting a wheelchair user, or it might have a very much lower height profile compared to a wheelchair user, making it less apparent to pedestrians who are a short distance away, unless specially equipped in some way (flag, lights, sound, or beacon).

As an automated machine, the pathway robot would have no onboard or proximate human to provide or receive social signals. It may be programmed to send and receive social or directional signals and to exhibit more patience than does the average human. As a semiautomated machine, it might be teleoperated, but the ability of a teleoperator to engage in social signaling would likely be limited. An example of this might be a teleoperated micromobility device such as a self-standing e-scooter being guided back to a docking station. The eventual introduction of ambulatory robots will add still other considerations, relieving some constraints and adding others.

These three comparisons suggest that a standard for pathway robots should consider alignment with existing accessibility standards relative to wheelchair use. Such goal-congruent alignment provides opportunities to address pathway design and configuration to intentionally benefit accessibility goals while standardizing robot access and flow.

16.3.6.4 Service robot access: surface conditions and path dimensions

A ground-based robot is designed to effectively and safely operate with respect to a given set of surface conditions. Because a standard for mobile service robots cannot anticipate all possible robot designs in terms of weight, wheel diameter, wheel design, leggedness, or other physical properties related to roadworthiness, the standard defines a way to describe the surface properties of a pathway such that a logistics operator can make a decision—likely automated—regarding the relative roadworthiness of a vehicle to travel on a particular surface.

There are many aspects to surface condition and path dimensions that make up a particular set of conditions. These may be built, transient, temporary, or environmental, such as pavement width, garbage bins, construction, or ice, respectively. The standard specifies metrics such as roughness, firmness, stability, friction, and several other elements related to surface attributes. It also specifies metrics such as path width, height, and gradient, which taken together with several others form the basis of a navigability or accessibility calculation to be used for real-time routing and logistics decisions. Separate aspects of the standard address climate and weather features below.

Many of these metrics and their defaults have been gleaned from accessibility manuals related to wheelchair use. That the standard is drafted this way means it is biased for robots that are similar in configuration and dimensionality to commonly specified wheelchairs. This implies that any infrastructural preparation for automated vehicles on pedestrian pathways could easily benefit accessibility users at the same time. It is currently the case that very many pathways in our cities do not fully comply with the accessibility guidelines of their respective jurisdictions.

Nonetheless, the standard sets the information needed to perform a standardized accessibility calculation for machines with specific attributes known to their operator. It is the governing jurisdiction that sets and certifies pedestrian zones for accessibility by either humans or machines. The point of using the same metrics and parameters is to ensure that a designer of a shared pedestrian–robot space can be guided *by default* to address human accessibility certification at the same time.

16.3.6.5 Service robot access permissions

Access permissions differ from access conditions. In the case of access conditions, above, a jurisdiction is declaring information about the pathway, the majority of which is likely to be static. In the case of access permissions, the jurisdiction is expecting conformance from a robot for a particular task, time, and circumstance. Based on the dynamic nature of public mobility spaces, the description of this conformance can change in near realtime.

A governing jurisdiction may constrain access by restrictions regarding weight, width, height, length, noise, emissions, and schedule. It might stipulate requirements such as lights, sounds, flags, and registration ID display. A three-way match enables the assignment between the robot and its route among pathways within its operating domain:

1. The conditions a pathway offers including then-current pathway-occupancy constraints or weather conditions
2. What a robot declares about its physical properties
3. What a logistics operator requests to fulfill a delivery requirement, such as origin, destination, size, weight, and schedule

Access permissions would be affected by the purpose of each service robot. The route plan and permissions for a delivery robot would be very different from that for a snowplough robot. For this reason, a *pathway reservation system* (PRS) managing information about conditions, dimensions, and permissions would be required to offer real-time management in public spaces hosting the work of robots performing a variety of tasks from a multiplicity of robot fleets from a plurality of operators and designs. Such a PRS would grant trip permissions, or a *trip contract*, which in turn would require real-time monitoring.

Such traffic control systems have been developed to manage multivendor robot fleets on many shop and warehouse floors. Mass Robotics in Boston is in the process of standardizing such systems [9]. Another is under development in Israel for multifleet aerial drones [10].

The key operating problem faced by each of shop floor, air space, and public pathway is multivendor, proprietary designs of multiple fleets of robots sharing a common space safely and optimally. But there are also crucial differences among these three operating spaces:

1. The shop floor is an exclusive space shared with humans trained to work with these machines. These robots share a job purpose with these humans and with each other.
2. Airspace is 3D. Robots (drones) must manage a unique set of risks and complexities operating in a constrained space above cities among buildings and other infrastructure. This space may only seldom be shared with humans, but it is still complex and dynamic.
3. The urban pathway, established as a human mobility space over 11,000 years ago, is shared with humans of every age and ability, the majority of whom may move in hard-to-predict ways and have no involvement with the mission of any particular robot.

Today ground-based (pathway) robots are either used in experimental trials or controlled by dedicated human operators within a constrained, single-provider operating domain, while tele-monitoring route conditions via onboard cameras. That is, as of this writing, such pathway robot fleet operators deploy a small fleet of identical machines within a small geo-domain of a few square kilometers shared with no other competitive robot operator.

We can expect to soon reach a plurality of fleets, operators, and service purposes. When that happens, intensive human teleoperation will become untenable except for emergency oversight and resolution. Fully digitalized real-time schedule and flow coordination will be required. This will likely come from ground control systems such as a PRS and include distributed sensors and IoT networks.

16.3.6.6 Service robot behavior

Once a robot's route is determined and granted, the device may be expected to behave in particular ways. Such rules would most likely be mediated by software within the machine as governed by local settings and limits. These behaviors include speed, travel side, travel direction, shy distance, schedule, and several aspects regarding waiting, rights of way, and clustering. As an example, Fig. 16.2 illustrates some rules that might apply to robots queueing at an intersection. These behaviors comprise what are essentially "rules of the road" for service robots in public, shared places/spaces. In this regard, the standard could inform many of the elements of a jurisdiction's "robot traffic act."

Importantly, there would be a need for local and variable changes to settings and limits—perhaps delivery speed or street-sweeper access changes by time of day or current block-face traffic. These changes need to be reliably communicated to the machines in near real time, and to be effective, they must be ensured or enforced.

16.3.6.7 Service robot social communication

Some of the special aspects under development for the standard are uniform movement indicators and social communications. Because pedestrian traffic can be more chaotic than motor vehicles in traffic lanes, robots will need a bounded and precise vocabulary of lights and sounds.

Simple examples would be to signal a turn or to grant a right of way. Other examples include signals for apology, request, gratitude, and alarm to act as a machine replacement for the glances, gestures, vocalizations, and body language that pedestrians use now. These are being designed for the safety of both pedestrians and the robots and to increase the social acceptability of these robots.

Robots need to signal their intentions and moods in language- and culture-independent ways. Such signals will be matched triplets (lights, sounds, and radio signals) so as to be understood by pedestrians with auditory or visual challenges, as well as by proximate robots.

16.3.6.8 Integrating robotic kerbside and pathway access

One of the projected use cases for robotic goods delivery is a larger delivery van ("mother-ship") stopping at a kerb or other suitable location proximate to several deliveries, and releasing one or more "child-robots" to complete local deliveries.

To make this workable, a degree of coordination is needed between the load/unload reservation for the parent delivery van and the reservation (permission) needed by its child robot(s) to travel on the intended pathway(s). This is provided in the draft standard. Such an operational real-time coordination between kerbside and pathway is new and will be a mapping and data challenge for those larger cities for whom these domains are currently handled by separate city departments.

FIGURE 16.2 What rules should apply to robots at an intersection? ISO 4448 permits the managing jurisdiction to limit the number, to indicate shy distancing and where to wait. Still the jurisdiction must set these rules, as would differ from one city to another and possibly from one type of intersection to another. *Illustration commissioned by an author.*

16.3.6.9 Robot cybersecurity

The standard provides requirements and guidelines for secure application services data interfaces between vehicles and infrastructure. These are based on existing credentialing standards in ISO 21177 and ISO 5616.

16.3.7 Certification for use: kerbside and pathway certification for automation

A critical aspect of preparing for automated vehicles at the kerbside or pathway is to determine the readiness of a specific subset of such infrastructure or operating domain. This question can be asked in two ways: "Can a jurisdiction safely provide permission to deploy a described type of automated taxi or service robot at a particular kerbside or pathway?" or "What preparations must be made in order to safely attract deployment of a certain type of automated vehicle or service robot at this particular kerbside or pathway?"

Whether a jurisdiction is asked to permit these vehicles and devices or whether it, or a community association, seeks to attract them, a *gap analysis* is required based on a standardized readiness model. This involves considering multiple system and governance attributes for several classes of vehicle capabilities. Here are a few examples from a much longer list:

1. What must be done to ensure robotaxis are not loading or unloading in traffic or on bicycle lanes?
2. What human-readable signage is appropriate for a given level of automation to be permitted (or encouraged)?
3. What regulations should be in place for teleoperated robots? For fully autonomous devices?
4. What sounds, lights, signals, or markings should be regulated for these vehicles or devices to ensure compliance with accessibility guidelines?
5. When and how can city enforcement personnel (police) stop, examine, rescue, or seize a service robot?

Answers to such questions are dependent on the automation and IoT capabilities under consideration. Hence, the standard details multiply readiness attributes for each of several "maturity" classes for kerbside and pathway operating domains. These attributes and maturity classes define a readiness matrix to be used to gauge the automation readiness of a specific kerbside or pathway or a larger, contiguous domain comprising multiple kerbsides and pathways.

Kerbsides and pathways are independently assessed, so that a kerbside and any adjacent pathway may be categorized at different maturity levels. This has implications for automated logistics that may require integration between road vehicles and service robots.

16.3.7.1 Robot weather worthiness

Robots, especially smaller human-scale machines designed for pathway use at pedestrian speeds and weigh under 50 kg, may be less capable in extreme weather or climate conditions than would be larger robots intended for roadway use such as the cars or trucks we use today. Some of these conditions might disable such robots leaving them as pathway hazards. Severe weather conditions such as extreme winds might blow a robot into road traffic or cause one to become airborne and slam into a pedestrian, a shop window, or a car.

The standard identifies a body of weather-worthy and road-worthy criteria for matters of temperature, wind, rain, snow, ice, and sand. The standard describes criteria for certification of machines and conditions such that a jurisdiction or insurer can determine when various devices must suspend operations and return to a protected storage area.

16.3.8 Looking forward

While standards are critical, much more than standards are needed. Each of the *data elements* implied above needs to be parameterized by a governing jurisdiction. Updates to these parameters are sometimes required in near real time (e.g., occupancy, surface friction). Others require notice to allow logisticians to plan (e.g., construction, current maximum weight per location or conditions). Most, but not all, have tolerances (e.g., max height, $\pm$ 20 mm). All have update rules to guide their local maintenance.

If a PRS (see Section 16.3.6.5) is in place, there would be *procedures* for activities such as request, assign, enqueue, dequeue, yield, and reschedule. Many of these activities are precisely standardized; others such as impounding a robot are only suggested, and their specifics are not standardized. Hence, even with a comprehensive standard, much local thinking and preparation need to be carried out.

As of this writing, ISO/TS 4448 is slated to have approximately a dozen parts, three of which are in the working draft stage and the remainder are outlined. This work started in April 2020 and is expected to be published in stages for completion by 2024.

The importance of getting automated passenger and goods vehicles as well as service robots managed in a way that adds to our urban toolkit rather than its problem set cannot be overstated. We can only entreat cities to consider them seriously, to view them through an accessibility lens first, and to consider COVID recovery and global warming a close second.

16.4 Pathway robots and business improvement areas

As described in the two previous sections of the chapter, it is expected that in some places and times, and for a variety of objectives, multiple service robots will be permitted to operate on public footways, bikeways, and roadways. This section imagines ISO 4448 used within a Business Improvement Area (BAI) to maximize the economic value of these robots while mitigating their urban, societal, and personal risks.

> ***Disclaimer:*** *DTS ISO 4448 does not describe specific deployment aspects for any application. The material in this section is not part of a draft or description of any ISO-related matter. It is here for discussion purposes only.*

16.4.1 The context of a *business improvement district* for delivery robots

As described in section (see Section 16.2) of this chapter, unaccompanied pathway robots for personal delivery or maintenance activities are more likely to be ready for pervasive service in our cities and suburbs sooner than robotaxis will provide regular passenger service. Such robots are illustrated on the bottom of Fig. 16.3. The multiple barriers to deployment of these pathway robots are more easily surmounted than the equivalent barriers for the robotaxi.

Whether pathway robots deliver food and packages, keep the pavement free of debris and snow, or perform security and surveillance duties, a critical beneficiary of these machines will be the retailers whose shops line pedestrian footways. Depending on the nature of their businesses, the attraction of these robots for merchants will vary—some may become adversaries.

Many of these merchants will be members of local *BIAs (or Districts).* This section speculates on a constructive relationship between these organizations and pathway robots. This includes potential benefits, harms, synergies, and acceptance.

16.4.2 Using pathway robots for consumer deliveries

Assuming personal, direct-to-consumer deliveries as the primary application for pathway robots, these may be deployed in a number of ways. Several might be loaded

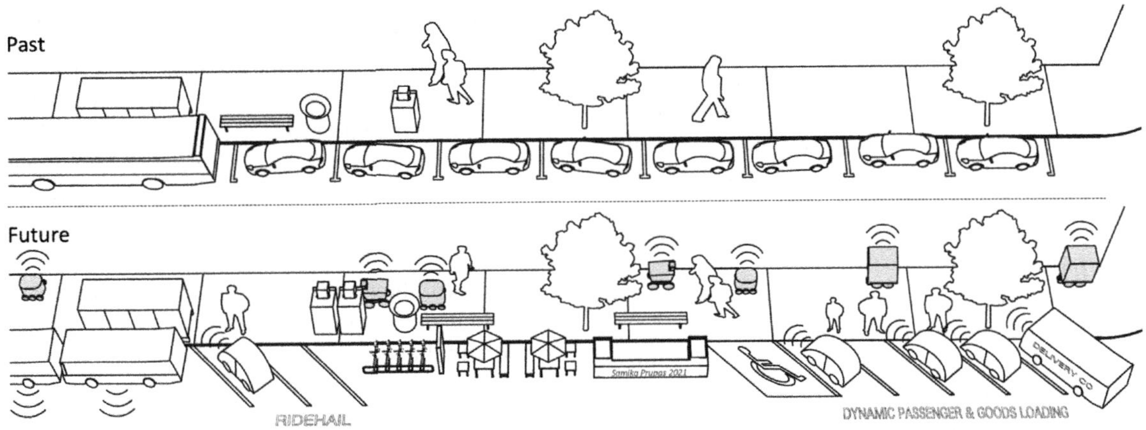

FIGURE 16.3 The evolution of preferred mobility in towns and cities (top) is expected to move away from personal vehicles toward multiple forms of automation and to make room for more social activities (bottom). COVID-19 moved us in both directions at the same time. It increased the relative use of private vehicles while boosting the use of delivery robots. This change will be a long process and may not always go smoothly. *Illustration commissioned by an author.*

inside a larger, specially fitted goods delivery van (sometimes called a "mothership"), brought to an area, then unloaded concurrently so that each delivers its packages within a radius of, say, 2 km. Some might be operated by individual retailers—perhaps a lunch shop or ghost kitchen—to make local meal deliveries. A larger number might be used by an e-commerce operator, such as Amazon or FedEx, to move goods from microwarehouses, strategically positioned staging areas, or from large, parked trailers to residents in a 2 or 3 km radius. Others could be used to deliver meals-on-wheels from a local charity kitchen, or be deployed as gig workers are now, moving on-demand from one delivery to the next, like a robotic *Uber-Delivers*. Still others might be used in an express-delivery service, such as DHL, FedEx, and UPS provide now when delivering packages for retailers and businesses, but in a strictly local environment suitable for personal-scaled deliveries, especially for residential deliveries, which have grown dramatically in the past few years. There are many dozens of such potential usage scenarios, each with varying advantages and disadvantages to merchants, residents, and pedestrians local to their operation.

What all of these types of delivery-service applications have in common is commercial trade. The more localized such trade—preferably in direct connection and without intermediary hand-offs between retailer and kitchen to consumer—the more this technology can act as an optimizer of time and cost. As well, pathway robots appear to fit well with ideas such as the 15-min city, the walkable neighborhood, and the car-free community in that they can help diminish the requirement for private automobile use and the contribution delivery stepvans make to congestion. Also central to the context of this section is whether the delivery robot can improve the post-COVID and eCommerce fortunes of BIAs and their communities.

16.4.3 Business Improvement Areas

The BIA, an innovation originating in 1970 in Toronto, Ontario, is a form of public–private partnership for local governance.[5] BIAs are important to the fabric of towns or cities, to their attractiveness, for the health of their commercial downtowns, and for their local communities. There are several thousand worldwide [11].

BIAs are "privately directed and publicly sanctioned organizations that supplement public services within geographically defined boundaries by generating multiyear revenue through a compulsory assessment on local property owners and/or businesses" [12]. Effective BIAs benefit both businesses and the community. "They can be an economic and social anchor to surrounding neighbourhoods and help to stabilize and add vitality to the local community … (they are an) important not-for-profit organizations which contribute to community development at a grassroots level [13]."

BIAs lobby for and channel local services on behalf of their members, according to their economic interests. Through various tax or business levies, these groups of private actors are able to execute local political decisions in public spaces. BIAs sometimes "intervene in strategic planning, e.g., land-use planning, that is a prerogative of local government [11]."

Each BIA focuses on their local programs, whether street furnishing, heritage preservations, reversing declines, attracting foot traffic, revitalization, managing panhandlers or graffiti, parking, and many others [14]. "Local businesses, working collectively as a BIA, become catalysts for civic improvements, ultimately enhancing the business

5. BIAs are known variously as *Business Improvement Districts* (BIDs) in the US and UK or Commercial Improvement District (CIDs) elsewhere. There are some 2250 of these in Canada, US and UK today.

climate and quality of life of the neighbourhood [15]." The 80-plus BIAs in Toronto are each a focal point of their respective communities taking on the characteristics of a local jurisdiction. They may be organized around income levels, ethnic groups, or sexual preferences, using their budgets for cosmetic improvements and branding. They generally work to attract visitors to their members' businesses and to locations for walking, gathering, shopping, and cultural pursuits.

But, as a matter of first order, BIAs seek to protect the interests of their members. One of those interests involve matters of local infrastructure such as trade-offs between street parking and bicycle lanes. Another is staying competitive with other channels of consumer supply such as malls or big-box stores—a matter that often turns on the same issue of sufficiency and location of parking facilities, invoking a fundamental conflict between our car culture and a desire for downtown community-oriented commercial areas (see sidebar: **Some merchants' concern for parking as critical to success may (or may not) be misplaced**).

Some merchants' concern for parking as critical to success may (or may not) be misplaced.

According to one study nearly half of local business owners estimated that more than a quarter of their customers arrive by car, when in fact the actual number turned out to be 4%. It was cyclists and pedestrians who were the majority customers. They were also the higher spending customers [16].

On the other hand, the recorded portion of 4% may reflect that those preferring to shop in cars simply shopped elsewhere. That some factors are causal and others are correlations is often misinterpreted.

What is certain is that more footfalls mean more shoppers and active modes such as cycling bring more footfalls per unit of parking space [17]. Maybe a smarter approach is to maximize downtown commercial spaces for active transportation and the kind of social, community and shopping environment that implies—and stop trying to compete with box-stores using underpriced parking.

Prior to the pandemic, a rising threat was eCommerce. COVID-19 has not only made that abundantly apparent, but most prognostication about the return to a new normal state indicates that eCommerce is unlikely to decline. Do online behemoths such as Amazon represent an existential threat to community merchants or will the Internet complement local sales to give these merchants a new resource for survival? Or does the cost and complexity to set up and manage an online presence mean this is a false promise—especially for the merchant with only one or a few outlets? Shopify may answer this for some, but it is too soon to be certain [18].

16.4.4 Are pathway robots a good idea?

The inventors and promoters of delivery robots promise that these machines will enable many advantages suitable to the BIA context. From a community perspective, they would be smaller, quieter, and cleaner (CO_2) than the delivery stepvans that are common today in many cities. They could deliver food, groceries, and other goods to seniors and the disabled. They could enhance meals-on-wheels programs. They would improve consumer reach by extending the effective radius of walkable neighborhoods while simultaneously reducing the demand for private car ownership.

From the perspective of the local retailer, they could help address the e-commerce crunch and compete against the "amazons" by lowering costs for local delivery. This could mitigate the high delivery costs for meals that were a hardship for so many restaurants during the pandemic.

From the perspective of jobs, these devices would generate direct, local employment (teleoperators, maintainers, managers, and handlers), but more importantly they add indirect employment to the degree that the local retailers prosper. Interestingly, some of the new jobs involving teleoperation can be performed from a worker's home including by employees with accessibility issues that might be confined to wheelchairs—people who would not have previously been able to engage in the logistics or delivery industry.

From the perspective of the urban environment, these devices would encourage better footway design and maintenance. John Kiru, the Executive Director of Toronto's TABIA, in a recent conversation with the first author of this chapter made it very clear that many local sidewalks are still too narrow and do not yet comply with applicable accessibility regulations. He also made clear the critical importance of uncluttered pedestrian clearways, planters, and the ability to promenade.

All of these spatial attributes are in competition with each other and may be further constrained by the migration of goods traffic from the kerbside and roadway onto the pedestrian pathway. Based on the aspects of draft ISO 4448 that address pathway robots, any infrastructure upgrades for robots must target guidelines for accessibility and pedestrian access. There is clearly an opportunity for a win-win for both commercial interests and pedestrian accessibility (see Section 16.4.7).

16.4.5 Are pathway robots a bad idea?

As just noted, many pathways are already difficult places for pedestrians with accessibility challenges. Sometimes they are complex for the abled to negotiate, for example, when they are crowded or when people are walking pets, pushing strollers, dragging carts, carrying bags, and using skateboards. If we add to this mix robots of varying sizes and speeds, fledgling pedestrian skills, and nonexistent social skills, we could threaten perceived safety and detract from pedestrian comfort and enjoyment.

Added to this are considerations of privacy and security. In the matter of privacy, these devices will capture image and other data in order to navigate and document their journey. While the data can be managed, guidelines are not fully mature and even when they are, will they be adequately enforced? In terms of security, physical mishaps might injure pedestrians or pets, or worse, a bad actor might intentionally employ a robot in a criminal enterprise.

We also need to consider that many people have worked very hard during the past decades to reclaim urban space for pedestrians, cyclists, and other active forms of transportation. For this reason, many people will be unwilling to consider pathway robots unless their advantages clearly outweigh any loss of pedestrian or cycling autonomy.

Lastly, I return to the question of jobs. While many new jobs will be created, automation also replaces and displaces human workers; even as it creates numerous opportunities, it causes dislocation, requires re-training, and leaves some behind. The jobs question is neither easy to answer or to project. It is naive to see this going only one way and impossible to dismiss. There will be shortages of both skills and jobs, requiring government intervention with programs to smooth this transition, while insisting on a degree of flexibility and learning of new skills.

I am often asked: "Do we really want these robots on our sidewalks?" My answer is always the same: Not the way many sidewalks are now, not without *fully attentive teleoperation* (i.e., we are not ready for full autonomy, SAE Level 4 or 5), not beyond a trial of devices from a single operator (we are unprepared for multioperator systems) and not without a secure, IoT-based, *pathway reservation system* under municipal governance. Clearly, there are many circumstances, times, and places where pathway robots—a least those currently available—would be a questionable idea.

16.4.6 Business change and societal issues

Done right, any technology that lowers local delivery costs could help restore the fortunes of local businesses and begin to heal the economic harm merchants have endured from the coronavirus pandemic. According to FedEx, this is because "[o]n average, more than 60% of merchants' customers live within 5 km of a store location … demonstrating the opportunity for on-demand, hyper-local delivery [19]." In the scenario FedEx is describing, an on-demand service would move goods and food directly from retail merchants to customers using small robotic machines, with the local merchant acting as a microwarehouse.

Similar to the way people are incentivised to shop where parking is free, community residents would prefer to order from merchants where delivery is near-free rather than inflating the cost of goods purchased. One projection claims that last-mile delivery costs could be driven as low as a US dollar [20] presumably for deliveries under about 3 km. As of this writing, in the United States Starship and Kiwibot charge US$2. Toronto's Tiny Mile charges $C3 ($US2.50).

There are two economic concerns in this scenario. Fast, convenient, and inexpensive delivery service would tend to change customers' goods delivery expectations from next day to next hour and for meal delivery from one hour to half-hour. This may be good for business, but not for congestion. And if the devices used are smaller and constrained to the footway, congestion would shift from the roadway and kerb onto the pathway. This would exchange one congestion problem for another—and could especially impact pedestrians.

Secondly, near-free delivery services risk replacing human couriers on bicycles and tricycles with robotic couriers. This unintended consequence of replacing a form of active transportation with automated transportation would have both health and employment consequences. Another potential impact would be to accelerate ongoing changes in the nature of food retail—grocery, restaurant, and fast food. In the past year, each of these has moved sharply toward pickup and delivery of online orders.

Although online grocery delivery has been around for at least 23 years, it ramped up dramatically during the pandemic. Ghost and virtual kitchens became a common way to sell prepared meals in 2020 for the same reason. In this form, order delivery has been mostly handled by gig workers and self-pickup. If delivery robots become viable, remote food preparation, coupled with robotic delivery, will become a permanent and growing fixture of the food economy.

Inexpensive sidewalk robots would disrupt several things at once: express delivery (van), bicycle couriers, average shopping radius, delivery-time expectations, e-commerce preferences, average total cost of goods purchased, size and frequency of purchases, and other structural buying habits. The net effect of all these disruptions would tend to increase consumption at the expense of sidewalk space, possibly with unintended negative impacts on livability.

16.4.7 Can local thinking sustain local growth?

In communities with a sufficient local population, pervasive local infrastructure such as wider and better designed footways, cycling systems, and slower streets, as well as local technology such as same-hour delivery systems are more likely to engender an increase in footfall and to sustain local commerce than would continuing the more car-oriented approach suitable for big-box stores. Can local deliveries using pathway robots help the post-COVID recovery of BIAs?

Given the extensive value BIAs represent and the range of business and community efforts they undertake (see Section 16.4.3), it will be important to consider the impacts of delivery robots on each of these efforts. Same-hour delivery robots would help some local merchants compete against eCommerce, but this would need to be done is a way that those same robots would not be a nuisance for customers of neighboring businesses such as those sitting at a near-by sidewalk café. If local delivery systems became pervasive, that would increase the number of local purchases in zero-sum competition with box stores. But in a world that prizes footway and parklet cafes and restaurants, pathway robots cannot be permitted to discourage footfall with its promenading and impulse purchases.

Renewed pedestrian infrastructure such as wider pedestrian clearways and improved pavement surface conditions would benefit all parties: residents, accessibility users, merchants, local shoppers, and robot delivery operators alike. Done right, this could benefit desired livability metrics. There are many thoughtful planning steps, anticipated by ISO 4448, that may be taken to prepare BIAs, towns, and cities for the turns this technology will take.

16.4.8 Potential municipal and BIA responses to pathway robots

There are a number of ways BIAs might ask their host municipalities to respond to pathway robots. These will range between two extremes. The business innovations to be enabled in gradual stages can be expected to adjust as change rolls forward. The technology and the standards that are being created to manage this are agnostic to the choices that cities will make and BIAs might prefer. The following list is not advice, but a menu of choices:

1. Severe constraints on the type, number, and schedule of robots permitted, including none permitted at all.
2. No constraints. Whoever operates these devices and however/whenever they are operated is acceptable, provided operations are within municipal guidelines that may apply to other mobility devices.
3. Local (BIA) override on municipally set restrictions such as weight, size, speed, occupancy constraints preferred by the local community. (In early 2021, a community in Pittsburgh, Pennsylvania, held an online town hall to discuss community preferences. A view toward some constraint was prevalent.)
4. Conform to ISO/4448 as configured by the municipality. Add an annual registration fee.
5. Conform to ISO/4448 as configured by the municipality. Add per-trip usage fees.
6. Conform to ISO/4448 as configured by the municipality. Add dynamic (variable) usage fees (this is enabled by ISO/4448).
7. Conform to ISO/4448 as configured by the municipality. Add variable permissions differing by block face and time of day and orchestrated by BIAs (this is enabled by ISO/4448).

16.5 Summary

The message of this Section 16.4 is that pathway robot technology is coming to shared, public spaces, bringing with it both opportunities and risks. Only by engaging with intention can BIAs tip the scale to their advantage.

External factors, whether box-store competition, changing approaches to urban planning, growth in e-commerce, a pandemic, or new technologies such as online storefronts, ghost retail, micromobility, or pathway robots, can challenge the financial and social viability of local retail. It follows that threats or advantages to a BIA can easily become threats or advantages to its local community.

Robotic deliveries, if they increase local purchasing, would help to restore financial viability and reduce retail vacancies. It is conceivable that the use of local, robotic deliveries may help sustain social viability as well as sales. If robots are purely and only for delivery and cost is driven out, this could encourage people to shop more from home. But if robots can enhance a district, keeping it cleaner, making it more secure, be used in novel ways acting, say, as BIA ambassadors—even interacting with visitors as such robots already do in some indoor environments—to encourage more community engagement, then these would be a good thing for both community and commerce. What might such a balance look like?[6]

A critical, net positive would be to reduce the need to use personal vehicles to get to nearly every activity we engage in. If local delivery takes away one more reason to drive, we may find more occasions to walk, more reasons to seek out 15-min cities, more times to buy locally, more opportunities to engage locally—and more community. Delivery robots can surely get us out of cars, but will they also get us off our couches and to prefer exploring our communities?

A negative would be that robots could reduce the need to leave home to shop, implying a drop in social activities,

6. It is worth reading online stories about the concurrent operation of well over 100 delivery robots in Milton Keynes [21]. Caution: at this point as of this writing such cases involve only a single provider, hence not requiring the higher level of coordination needed for the case multiple logistics operators sharing a common space.

community activities, active transportation, and the like. As the capability of robotic delivery grows, it makes sense for e-commerce logistics operators to operate out of small local warehouses or staging points simply acting as distributed box stores, which would have the opposite effect of reinvigorating local retail. Can this be avoided?

Digitalization and new mobility pull the local merchant in many directions. The need to be online [18], the threat from e-commerce, the introduction of the pathway robot, numerous changes in parking supply and pricing, as well as micromobility storage and charging stations at the kerb combine to rattle the retailers' world, demanding that they be responsive at every turn in order that their businesses survive. In a conversation with the first author, Professor Kenneth Jones of Ryerson University, said: "retail is the most flexible form of human commerce — it responds very quickly to change and opportunity." The current decade will test that assertion.

The influence BIAs have on shared public-commercial space (and vice-versa) implies that BIA leadership will seek to influence the regulation of automated devices on the pathways within the geographical boundaries of each BIA. There is no reason to assume BIAs would adopt common, uniform rules in terms of robot sizes, speeds, weights, and traffic volume—as implied in several bills introduced and passed in some US states [2]. Depending on the nature of local footways, local budgets, and local retail mixtures, some BIAs may welcome these robots, and others may wish to ban them. Surely, most of them will seek a say in their governance.

But there is already a concern here. As of late 2020, about 40% of US states have tabled or passed legislation permitting "personal delivery devices" on their cities' footways, bikeways, and roadways. This was done in favor of lobbyists representing companies making and deploying these robots (firms such as Amazon, FedEx, Starship). This was often done by these states without conferring with their constituent cities or their subconstituent BIDs [22]. Some BIAs may guard against this by lobbying for the ability to moderate such rules of use in ways appropriate to their community. Such lobbying could consider accessibility, community, economic, and safety factors not all of which have been considered in the US state bills tabled as of this writing.

16.5.1 Afterword

Taken together, the extraordinary potential and matching complexity of service robots in public spaces are likely to engender a difficult and ongoing social debate. If you are an advocate of this type of automation—and there are many of you—you should progress cautiously, understanding that there are very many unresolved issues before you will be able to reach substantial deployment. If you oppose this type of automation—and you are also not alone—you should be aware that there are many important opportunities that you may be inadvertently rejecting.

Each city and perhaps each local community may reach a different conclusion. It is possible that some communities will reject these robots while in other cases, neighboring communities will undertake different deployments. The impact of such variations might be hard to project and have surprising consequences. The best approach is to be open to research, trials, and public discussion—and above all to guard against hype.

Finally, let us move on to the real interactive part of this chapter: review questions/exercises, hands-on projects, case projects, and optional team case project. The answers and/or solutions by chapter can be found in Appendix G.

16.6 Chapter review questions/exercises

True/false

1. True or False? Dozens of companies are building and piloting large, electric delivery robots with a view to reducing the costs of delivering food and parcels over their final distance.
2. True or False? Despite a decade and a half of investment, promotion, and anticipation about the coming of driverless taxis and shuttles, delivery robots will arrive sooner and in greater numbers than will robotaxis.
3. True or False? Delivery robots come in a variety of sizes and configurations. Smaller units for single deliveries are the size of a filing box and weigh less than 100 kg fully loaded.
4. True or False? According to the American Automobile Association, over half of people recently surveyed express fear of driverless vehicles.
5. True or False? Setting aside projections of driver shortages and arguments promoting "career retraining"—which are often not accepted by the people so employed—many workers and their families do not feel threatened by automation.

Multiple choice

1. Standing against the robotaxi will be interests such as transit and taxi drivers and their agents and:
   a. Unions.
   b. Infrastructures
   c. Groups
   d. Investments
   e. Policies

2. The investment required to build and prove delivery robots is far lower than that required to build and prove driverless passenger:
   a. Planes
   b. Vehicles
   c. Boats
   d. Submarines
   e. Trains
3. In most countries, national and state/provincial governments consider regulations for automated passenger vehicles, mostly from a:
   a. Safety perspective
   b. Machine-to-Human perspective
   c. Cyber-Physical Systems perspective
   d. Artificial Intelligence perspective
   e. Digital Twins perspective
4. A robotaxi is often framed as "just a taxi with a silicon driver," and we are often told these machines will use the same roads and the same parking spaces as human-operated:
   a. Platforms
   b. Vehicles
   c. Vans
   d. Motorcycles
   e. Boats
5. Fear and negative perceptions evoked by fully autonomous vehicles create a significant barrier for widespread adoption of the:
   a. Delivery robot
   b. Autonomous vehicle
   c. Robotaxi
   d. Driverless vehicle
   e. Android

Exercise

Problem

Why is kerb space considered to be a prize for the increasing number of competitors that are bidding for their slice of it?

Hands-on projects

Project

Do research: Analyze why the kerb space is an area where many interactions take place, whether it is pickup, drop off, loading zones, bus stops, clearways, or parking spots.

Case projects

Problem

Who is in charge of the kerb?

Optional team case project

Problem

Why is monetization an outcome of management?

References

[1] https://www.dimensions.com/collection/autonomous-delivery-vehicles.
[2] Grush B. Are we doing justice to sidewalk robots?. 2021. https://www.linkedin.com/pulse/we-doing-justice-sidewalk-robots-bern-grush.
[3] Edmonds E. AAA: today's vehicle technology must walk so self-driving cars can run. 2021. https://newsroom.aaa.com/2021/02/aaa-todays-vehicle-technology-must-walk-so-self-driving-cars-can-run/.
[4] https://www.sae.org/automated-unmanned-vehicles/.
[5] Shladover S. What "self-driving" cars will really look like, scientific American (June). 2016.
[6] Grush B. The last block: towards an international standard to regulate and manage sidewalk robots. 2021. https://citm.ca/wp-content/uploads/2021/02/Harmonize-Mobility_The-Last-Block_21.02.01.pdf.
[7] Coombes A, Grush B. Digitization, automation, operation, and monetization: the changing management of sidewalk and kerb 2000-2025 (Chapter 12A in smart cities policies and financing handbook. Elsevier; 2021.
[8] Thomasen K. Robots, regulation, and the changing nature of public space. 2020. https://commons.allard.ubc.ca/fac_pubs/633/.
[9] Autonomous Mobile Robots (AMR) Interoperability Working Group. https://www.massrobotics.org/.
[10] Autonomous operations for multi-drone fleets. https://www.airwayz.co/.
[11] Peyroux E, Pütz R, Glasze G. Business Improvement Districts (BIDs): the internalization and contextualization of a 'travelling concept'. 2012.
[12] Hoyt L, Gopal-Agge D. The business improvement district model: a balanced review of contemporary debates. 2007.
[13] https://www.citywindsor.ca/business/Business-Improvement-Areas/Pages/Business-Improvement-Areas.aspx.
[14] Jamal A. From operational to aspirational? Business improvement areas (BIAs) in mid-sized cities. Plann Pract Res 2018;33(5):506−22. https://doi.org/10.1080/02697459.2018.1548214.
[15] https://www.toronto-bia.com/whats-a-bia/.
[16] Sorrel C. Why local businesses shouldn't worry about eliminating on street parking. 2017. https://www.fastcompany.com/3067515/why-local-businesses-shouldnt-worry-about-eliminating-on-street-parking.
[17] StrongTowns. How to talk to a sceptic about reducing parking requirements. 2018. https://www.strongtowns.org/journal/2018/11/23/how-to-talk-about-parking.
[18] Ovide S. Who wins the (online) corner store? New York Times, February 18. 2021. https://www.nytimes.com/2021/02/18/technology/who-wins-the-online-corner-store.html.
[19] http://cantruck.ca/fedex-begins-testing-autonomous-delivery-robot/.
[20] https://www.theguardian.com/sustainable-business/2017/may/31/delivery-robots-drones-san-francisco-public-safety-job-loss-fears-marble.
[21] https://www.buckinghamshirelive.com/news/buckinghamshire-news/starship-little-robots-milton-keynes-.
[22] Deto, R., (2020) Pennsylvania legalizes autonomous delivery robots, classifies them as pedestrians. https://www.pghcitypaper.com/pittsburgh/pennsylvania-legalizes-autonomous-delivery-robots-classifies-them-as-pedestrians/Content?oid=18482040.

Chapter 17

Innovative smart applications for solid waste management

Essam E. Khalil[1] and Doaa M. El-Sherif[2]

[1]Cairo University, Cairo, Egypt; [2]Housing & Building National Research Center (HBRC), Ministry of Housing, Utilities & Urban Communities, Cairo, Egypt

17.1 Introduction

There are multiple benefits to treating and managing waste. In this section, we will take a closer look at them.

17.1.1 Better environment

Probably the biggest advantage of managing waste is that it eventually leads to a better and fresher environment. Waste disposal units also contribute to the well-being of people by helping them become disease-free. The best part: all of this happens while the unnecessary is duly disposed of in a proper and sanitary manner. Multiple waste disposal units should be placed in tier-1 and tier-2 cities in a bid to prep up the process of waste disposal. This will also help implement remarkable safety measures in the long run.

17.1.1.1 Reduces pollution

When waste is managed the right way, it does not merely eliminate the subsequent waste but also reduces the impact and the intensity of harmful greenhouse gases like carbon dioxide, carbon monooxide, and methane that are often exuded from accumulated wastes in landfills. Managing waste reduces our reliance on landfills while also significantly cutting down the many factors that adversely impact our environment.

17.1.1.2 Conserves energy

Recycling is one of the biggest aspects of waste management, and over time, it helps conserve energy. One of the biggest instances of this advantage can be traced to the practice of recycling paper.

All of us are probably aware that thousands of trees are cut to produce paper. When a used paper is recycled to create new paper, the need of cutting trees is significantly minimized. This helps conserve energy while also reducing your carbon footprints.

17.1.1.3 Creates employment

The recycling industry alone creates hundreds of jobs. As more people adopt this eco-friendly practice, organizations creating and selling recycled products come to the forefront. This helps boost their business while also creating hundreds of jobs.

17.1.1.4 Helps make a difference

By managing waste, you are also making a difference to the society and the world in general. While none of us can completely get rid of garbage, we can always adopt eco-friendly practices of reducing and reusing waste. This way, you create an example for the people around you, who in turn are now motivated to embrace a sustainable approach. There are eight major categories of waste management, and each of them can be divided into numerous subcategories. The categories include source reduction and reuse, animal feeding, recycling, composting, fermentation, landfills, incineration, and land application [1,2]. Some of these methods such as reduction and reuse can be started from the comfort of your homes.

17.2 Various methods of waste disposal

There are many methods of disposing waste. In this section, let us take a look at some of the most commonly used methods that you should know about waste management.

17.2.1 Landfills

Throwing daily waste/garbage in the landfills is the most popularly used method of waste disposal used today. This process of waste disposal focuses attention on burying the waste in the land. Landfills are commonly found in developing countries.

Smart Cities Policies and Financing. https://doi.org/10.1016/B978-0-12-819130-9.00015-2

There is a process used that eliminates the odors and dangers of waste before it is placed into the ground. While it is true this is the most popular form of waste disposal, it is certainly far from the only procedure and one that may also bring with it an assortment of space.

This method is becoming less these days although, thanks to the lack of space available and the strong presence of methane and other landfill gases, both of which can cause numerous contamination problems. Landfills give rise to air and water pollution, which severely affects the environment and can prove fatal to the lives of humans and animals. Many areas are reconsidering the use of landfills.

17.2.2 Incineration/combustion

Incineration or combustion is a type disposal method in which municipal solid wastes are burned at high temperatures. The process eventually converts them into residues and gaseous products. The biggest advantage of this type of method is that it can reduce the volume of solid waste to 20%–30% of the original volume [3]. Additionally, it also decreases the space they take up while also reducing the stress on landfills. Incinerators are primarily used in thermal treatment where solid waste materials are converted to heat, gas, steam, and ash. Incineration is also widely popular in countries where landfill space is no longer available, such as the United States and Japan.

17.2.3 Recovery and recycling

Resource recovery is the process of taking useful discarded items for a specific next use. These discarded items are then processed to extract or recover materials and resources or convert them to energy in the form of useable heat, electricity, or fuel. Recycling is the process of converting waste products into new products to prevent energy usage and consumption of fresh raw materials. Recycling is the third component of Reduce, Reuse, and Recycle waste hierarchy. The idea behind recycling is to reduce energy usage, reduce the volume of landfills, reduce air and water pollution, reduce greenhouse gas emissions, and preserve natural resources for future use.

17.2.4 Plasma gasification

Plasma gasification is another form of waste management. Plasma is primarily an electrically charged or highly ionized gas. Lighting is one type of plasma that produces temperatures that exceed 12,600°F. With this method of waste disposal, a vessel uses characteristic plasma torches operating at +10,000°F, which is creating a gasification zone till 3000°F for the conversion of solid or liquid wastes into a syngas. During the treatment of solid waste by plasma gasification, the waste's molecular bonds are broken down as a result of the intense heat in the vessels and the elemental components. Thanks to this process, the destruction of waste and dangerous materials is found. This form of waste disposal provides renewable energy and an assortment of other fantastic benefits.

17.2.5 Composting

Composting is an easy and natural biodegradation process that takes organic wastes, i.e., remains of plants and garden and kitchen waste and turns into nutrient-rich food for your plants. Composting, normally used for organic farming, occurs by allowing organic materials to sit in one place for months until microbes decompose it. Note that composting is often deemed to be one of the best methods of waste disposal as it can turn unsafe organic products into safe compost. The process, however, has its downsides. Some people have found it to be slow, while others have observed that it takes a lot of space. But regardless of these issues, many people are still embracing home composting approaches to manage and reduce waste and turn it to.

17.2.6 Waste to energy (recover energy)

Waste-to-Energy, also widely recognized by its acronym WtE, is the generation of energy in the form of heat or electricity from waste. WtE process involves the conversion of nonrecyclable waste items into useable heat, electricity, or fuel through a variety of processes. This type of energy is a renewable energy source as nonrecyclable waste can be used over and over again to create it. WtE can also help reduce carbon emissions by offsetting the need for energy from fossil sources. Over time, this reduces global warming and makes our environment better.

17.2.7 Special waste disposal

There are certain waste types that are considered hazardous and cannot be disposed of without special handling, which will prevent contamination from occurring. Biomedical waste is one example of such a waste disposal method. It is primarily practiced in healthcare facilities and similar institutions. The special waste disposal system effectively disposes of hazardous biomedical waste.

17.2.8 Avoidance/waste minimization

The easier method of waste management is to reduce the creation of waste materials, thereby reducing the amount of waste going to landfills. Waste reduction can be done through recycling old materials such as jar, bags, repairing broken items instead of buying a new one, avoiding the use of disposable products such as plastic bags, reusing second-hand items, and buying items that use less designing.

17.3 The best methods of waste management

Recycling and composting are a couple of the best methods of waste management. Composting is so far only possible on a small scale, either by private individuals or in areas where waste can be mixed with farming soil or used for landscaping purposes.

Recycling, on the other hand, can be widely used around the world, with plastic, paper, and metal leading the list of the most recyclable items. Most material recycled is reused for its original purpose. In some instances, they may also be sold for generating profits.

17.3.1 Bottom line

As you can see there are plenty of important things that you should know about waste management and disposal in order to ensure that you and the environment around you are safe. While it may not be apparent, it is your choice that paves the way toward a better world and a healthier environment. That is why always be sustainable and make actionable efforts to manage and treat waste. Since we have listed multiple waste management methods, explore your options, before making a final choice.

Solid wastes include household refuse, agricultural remnants, food leftovers, plastic bags, tin cans, ash and packaging, such as cartons and used jonya sacks. They become waste once they have been discarded because they are no longer needed in their present form. Refuse, garbage, litter, and street sweepings are all terms used to describe solid wastes in various situations. In rural parts of Ethiopia, the type of solid waste generated will vary depending on the particular location and the socioeconomic and cultural conditions of the area.

In general, rural households produce wastes that are mostly organic in nature and result from agricultural production and processing activities. They include crop residues, food remnants, leaves and grass from clearing of sites, animal manures and dung, ashes, dead animal carcasses, etc. Other solid wastes could include glass, plastic containers, metal scraps, tin cans, plastic bags, condoms, and obsolete agricultural pesticides and insecticides. These solid wastes need to be managed properly in a way that avoids the potential risks to the environment and to human health.

The type of waste will determine the choice of possible disposal methods. Therefore, it is important to be able to classify solid waste according to its characteristics, Table 17.1. Solid waste can be classified based on its source as residential, agricultural, commercial, industrial, institutional, or healthcare waste. You were also introduced to the classification of waste as hazardous or nonhazardous and also as biodegradable or nonbiodegradable. Another word used in the context of biodegradable solid waste is putrescible, which means the waste decomposes (rots down) quite quickly. Waste can also be described as combustible or noncombustible depending on whether it will burn or not. Table 17.1 classifies waste using these different properties.

Putrescible wastes are generated by growing, handling, preparation, cooking, and consumption of food. These kinds of wastes tend to be more abundant during the summer (rainy) seasons. Nonputrescible wastes do not decompose easily; they may or may not be combustible. Because they do not break down, they persist in the environment and are often the cause of nuisance and aesthetic problems.

Plastic bags are commonly known as festal. They are widely used to carry goods from market to home in all areas because they are convenient, cheap, and easy to use. Plastic bags are usually nonbiodegradable and persist in the environment for a very long time. Moreover, when discarded indiscriminately, they pollute the land surface of your community, prevent rainwater from percolating into the soil, can easily be blown all over the place by wind, and create unsightly and nuisance conditions. The bags can also be easily swallowed by animals, which may block their digestive system and kill them, Fig. 17.1. In malaria-prone

TABLE 17.1 Classification of solid waste.

Main classification	Type	Short description	Examples
Hazardous waste	Solid or semisolid	Substances that are either ignitable, corrosive, reactive, infectious, or explosive	Some obsolete pesticides such as DDT, dieldrin, etc.
Nonhazardous waste	Putrescible	Easily decomposable/biodegradable solid waste	Food wastes
Nonputrescible	Nonbiodegradable solid waste, it includes combustible and noncombustible waste	Plastic bags, bottled water containers, tin cans, etc.	

FIGURE 17.1 Discarded plastic bags are a health hazard for grazing animals.

areas, there is also a potential to create breeding places for mosquitoes because the bags may trap small pools of water.

17.3.1.1 Hazardous waste

Hazardous wastes are treated as a separate category of waste because special management and disposal methods are required. The main sources of hazardous substances are agricultural offices/stores, health offices/stores, and possibly also private firms, homes, and retail shops. The toxic, corrosive, ignitable, explosive, and/or infectious characteristics of these wastes require careful and stringently controlled methods of handling, storing, and transportation [2]. Hazardous wastes, because of their particular environmental health hazard, should be identified, quantified, and reported to a higher level for further action.

The immediate health effects from hazardous wastes range from bad smells and simple irritation of eyes, skin, throat, and breathing (lungs), to serious health conditions that affect the nervous system and could cause paralysis of the functional body parts. Some hazardous solid wastes have teratogenic (birth defects) and carcinogenic effects.

The long-term effects of hazardous wastes can be devastating to community health and well-being. However, it is difficult to quantify the exposure level and predict the health consequences as a result of exposure because the types of hazardous waste are so variable. If you observe or get reports from others about the presence of any hazardous waste, you should report this to the environmental health office and/or the agricultural development agent in your locality.

If you know there are stores of hazardous materials in your community, you should monitor them regularly using observation checklists and by asking for information and then report your findings if you have any concerns. You should not attempt to bury, burn, or dispose of hazardous wastes. Hazardous solid wastes require specialized methods of handling, collection, transportation, and disposal. Some wastes, such as obsolete pesticides or radioactive material, may need to be transported out of the country for final disposal.

17.3.2 Functional elements of smart solid waste management

The functional elements (stages) of solid waste management can be classified as:

- Onsite handling, storage, and processing
- Collection
- Transfer and transport
- Resource recovery and processing
- Disposal

The complete set of functional elements will apply in urban centers managed by municipal authorities but not all of them will be relevant in rural areas.

17.3.2.1 Onsite handling, storage, and processing

Onsite means these functions are concerned with solid waste at the place where the waste is generated. For residential waste, this means at home in the household. Onsite handling is the very first step in waste management. It involves individual family members, households, and communities, all of whom need to know how to handle waste properly at this level. "Handling" means the separation of wastes into their different types so they can be dealt with in the most appropriate way, for example, separating putrescible waste for composting. The benefits of appropriate onsite handling include reducing the volume of waste for final disposal and recovering usable materials.

Onsite storage means the temporary collection of waste at the household level. It is important that waste is stored in proper containers. These could be baskets, preferably made from locally available materials, plastic buckets, or metal containers. Larger containers or dustbins, especially those used for food waste, should be leak-proof, have tight lids, and be long-lasting. The size of the container should be sufficient to hold at least the amount of solid waste that is generated per day at household level. Institutions and businesses should consider having onsite storage facilities with greater capacity. The proper location of storage containers and the frequency and time of emptying are important factors to be considered for efficient onsite storage. Some wastes will need some sort of onsite processing before the next steps, for example, in areas where false banana (enset) is used as a staple crop, the by-products should be chopped into pieces before composting to speed up the rate of decomposition.

17.3.2.2 Collection, transfer, and transport of solid waste

In urban centers, collection is a function that has its own process and services. Waste is collected and held at central transfer stations where waste is stored before it is transported to a final disposal site. In rural areas, waste is not normally collected in this way, and disposal is limited to onsite processing options, although sometimes there may be communal collection of solid waste using animal carts.

17.3.2.3 Resource recovery and processing

Resource recovery means finding a way to use the waste so it becomes a valuable resource, rather than just a disposal problem. This is a very important part of waste management. Resource recovery includes a range of processes for recycling materials or recovering resources from the waste, including composting and energy recovery. Converting the waste into a new product may require energy and equipment; therefore, there needs to be a careful assessment of inputs and outputs in case it is not economically sensible to do it. Resource recovery options are discussed further in the following sections.

17.3.2.4 Disposal of solid waste

Even after recycling and resource recovery, there will almost certainly be some residual waste that needs final disposal. Methods of disposal can be sanitary or unsanitary. Open field dumping is the most unsanitary method of refuse disposal and is most likely to cause a health hazard. Sanitary methods—including controlled tipping or controlled burial, incineration and sanitary landfill—are discussed later.

17.3.3 Integrated solid waste management

As shown in Fig. 17.2 (the waste hierarchy), waste management options are listed in order of desirability from most desirable at the top to least desirable at the bottom. The concept of integrated solid waste management (ISWM) mostly applies to municipal solid waste management in urban centers. The principles can, however, be applied to some extent in rural and periurban solid waste management.

It can be explained that an ISWM approach means considering not only the appropriate disposal of solid waste but integrating this with other management options such as minimizing waste production, recycling, composting and other waste recovery options. These different options can be ranked in order of their desirability as management options. At the top of the hierarchy is reduction, which means keeping the production of waste to a minimum. Next comes reuse, which simply means using something more than once. The third option in the hierarchy is recovery, which includes several separate processes that enable material or energy resources to be recovered from the waste. These include recycling, composting, and energy from waste. Finally, the least desirable waste management option is disposal, which includes landfill, tipping, and incineration (burning) without energy recovery.

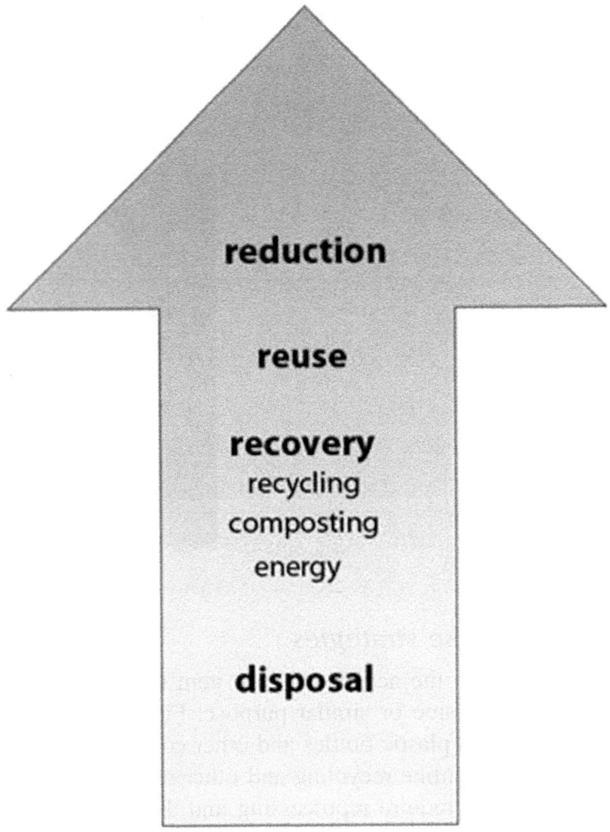

FIGURE 17.2 Waste hierachy.

17.3.3.1 Reduction strategies

Reductions strategies are the ways that a household or community may use to try to reduce or minimize the amount of solid waste they produce. This approach is generally more relevant in affluent homes and societies with a wasteful lifestyle. For example, people with more money may not worry about throwing household items away when they can afford to buy replacements. In a business context, using two-sided photocopying of a document reduces the paper used and also therefore the waste produced [3].

FIGURE 17.3 Plastic containers are frequently reused.

17.3.3.2 Reuse strategies

Reuse refers to the act of using an item more than once, either for the same or similar purpose. Fig. 17.3 shows a sample of used plastic bottles and other containers for sale to be reused. Unlike recycling and other recovery options, reuse does not require reprocessing and therefore requires less energy.

17.4 Recycling

Recycling is a process by which waste is processed in some way to be reformed into new or similar products. The principle is to make a usable product from the waste. Plastic bottles, newspapers, cardboard, and tin cans can all be reprocessed and made into new items. Plastic bags can also be recycled and used to make mats, carpets, and other products. Waste metal has a number of possible uses because it is relatively easy to reshape. Careful separation of the waste into its different types is important for the efficiency of recycling processes. Recycling not only reduces the quantity of waste but also saves money, so there is an economic, as well as an environmental, incentive to recycle.

17.4.1 Composting of organic solid wastes

Nonhazardous, putrescible solid wastes such as crop residues, leaves, grass, and animal manures can be managed onsite by composting. Composting is a controlled process in which this type of waste is collected in an open pit or heap and is decomposed by natural biological processes. The waste is broken down by the action of a variety of microscopic and other small organisms. The waste is converted into a stabilized material that can be used as fertilizer. Composting is an environmentally friendly way of recovering value from organic waste.

As we discussed in previous sessions, both human waste and organic household waste can be composted. However, the process is different. Human waste can be composted in alternating double pit latrines and in ecological sanitation systems. This process of composting is anaerobic. In the case of organic household waste composting, it is an aerobic process.

The pit for composting should be dug about 50 m away from a dwelling. The pit needs to be about 1 m deep and at least 1 m breadth and 1 m length. However, the size can vary with the amount of waste generated. The pit depth should be slightly less on one side, about 90 cm, to make a slope so that water does not collect at the bottom. To make the compost, organic matter such as grass, leaves, and kitchen/food waste should be thrown into the pit in a shallow layer. It is very important that only biodegradable material is added, so care is needed to sort the waste beforehand. No plastic should be included and bones should be avoided. The waste should then be covered with a thin layer of soil. Covering with soil encourages the composting process and prevents the breeding of flies and other vermin. Air must be allowed to mix with the compost so the contents of the pit need to be turned frequently by digging. The compost needs a small amount of water to keep it moist. The time for the compost to be ready will vary depending on the temperature and the mix of waste among other factors, but it should be ready within a few months. Composting is mostly practiced in rural communities [3]. In Ethiopia, it is becoming customary for households to prepare compost from their household organic wastes and you should encourage this practice.

17.5 Energy from incineration

To incinerate something is to burn it. In waste management terms, however, incineration means burning in a controlled and managed process—usually at high temperature. Incineration cannot be implemented at household level; it is mostly used for institutional waste management purposes. Different types of incinerator are used for burning waste. They differ by the temperature at which they operate the cost of construction, the method of operation, and the maintenance requirement. Incinerators can be used for disposal of wastes in health institutions/Health Posts and government and private institutions/offices/industries. Incineration can reduce the volume of refuse by up to 90%; the only remaining residual waste is ash. This significantly reduces the volume of material needing final disposal. Incineration is only classed as "recovery" in waste management if the energy (heat) that is produced is used in some way.

Sanitary landfill means the controlled filling of compacted layers of solid waste and soil into preprepared land. Large-scale landfill sites for municipal waste need to be designed to protect surface and groundwater from contamination by leachate, the liquid waste that may seep out into the ground underneath the layers of waste. Sanitary landfill sites are not just rubbish dumps for open field dumping. To be classed as sanitary, the site must be managed to minimize any negative environmental impact.

Controlled tipping or controlled burial is similar in principle to sanitary landfill but at a smaller scale that is appropriate in rural areas. In controlled tipping/burial, solid waste is disposed of into a dug pit and is regularly covered with soil to avoid attracting disease vectors such as flies and rodents. Covering the waste also stops it from being blown by the wind and avoids bad smells—hence "controlled."

Note that various similar terms may be used to describe different types of waste disposal pit. A refuse pit is a simple pit used to dispose of household refuse, which may or may not be used for controlled tipping (with soil). Some wastes will need to be buried under soil as soon as they are disposed of, in which case the pit may be referred to as a burial pit. When there is a need for preparing a refuse pit for households, you should advise them that sites for controlled tipping should be 10 m away from the house (preferably at the back of the house), at least 15 m and preferably 30–50 m away from water wells and at a lower ground level. At community level, a communal refuse pit should be 100 m away from houses, and they will also need to consider the direction of wind. The site should be easily accessible, with adequate space, and should be fenced so that it is not accessible to children and domestic animals. Care must be taken to avoid creating places that could harbor rats or encourage the breeding of flies and other insects. Waste from individual households should be taken to the site in suitable containers such as sacks, plastic bags, or buckets. For a community waste disposal pit, it should be a collective responsibility to keep communal areas clean.

Animal carcasses need to be disposed of carefully because they can encourage the breeding of flies and rodents, and attract scavenger animals. They can be disposed of in a common burial pit for the community.

Burning of waste is another possible, though less desirable, method of final disposal [3]. A burning site should be sited downwind and well away from houses. Noncombustible materials such as broken bottles, bones, etc., should be separated and buried at a safe location, not used by farming. Ashes that remain after burning can be used as fertilizer or, if mixed with mud, can be used for plastering of earth walls or floors.

17.6 Summary

The management of solid waste can have a significant impact on human and environmental health. You need to be able to survey the ways that solid waste is managed in your area so that you can assess the situation and identify possible areas for improvement. The list of questions below indicates some of the issues that should be covered in your survey; you may think of other questions to include:

- How many of the households in your tripe have a compost pit?
- Do households without compost pits have adequate space to make one?
- Are there other opportunities for recycling or reuse of waste material?
- Does the community have a communal refuse disposal pit?
- How many people use each refuse pit?
- Is there a need for another communal pit?
- Are the refuse pits fenced and properly managed?
- Is there adequate equipment in your tripe for managing solid waste?
- Is there any open dumping of waste in your tripe?

Answering the preceding questions will provide you with the baseline data you will need for planning improvements to the solid waste disposal facilities. If households do not manage their waste properly, you may want to try to find out the reasons for this. It may be lack of resources such as money, space, or labor, or there may be behavioral reasons that make people unwilling to use compost pits, for example. Understanding the reasons for the behavior will help you to give advice to the family and indicate the most effective ways to promote good waste management practices.

After analyzing your survey results, you may want to prepare an action plan. There are several possible measures that you could consider including in your plan to lessen the problems. You may want to promote general good practice by ensuring that all households have adequate waste containers at home. You may need to ensure that all pits are fenced and properly managed. You may need to mobilize the community to dig a new refuse pit, if your survey suggests that one is needed. You may want to focus on a specific problem such as plastic bags. You could advise people to collect their plastic bags in one place for recycling, and it may also be possible to encourage individuals and enterprises to undertake this profitable activity. You could encourage more reuse of plastic bags by suggesting that people take bags with them to the market or shop to be used again. You could promote other types of bag such as paper bags because they are made of biodegradable materials or permanent bags such as cloth bags. You could get the local school involved and initiate a campaign among the children to collect waste bags to clean up the local environment.

Keep in mind that any action plan should be drawn up with the involvement of your community because it will not succeed without their full cooperation [3]. You should include regular monitoring in your plan so you can assess if progress is being made.

Finally, let us move on to the real interactive part of this chapter: review questions/exercises, hands-on projects, case projects, and optional team case project. The answers and/or solutions by chapter can be found in Appendix G.

17.7 Chapter review questions/exercises

True/false

1. True or False? There are multiple benefits to treating and managing waste.
2. True or False? Probably the biggest advantage of managing waste is that it eventually leads to a better and fresher environment.
3. True or False? When waste is managed the right way, it does not merely eliminate the subsequent waste but also reduces the impact and the intensity of harmful greenhouse gases such as carbon dioxide, carbon monooxide, and methane that are often exuded from accumulated wastes in landfills.
4. True or False? Recycling is one of the biggest aspects of waste management, and over time, it helps conserve energy.
5. True or False? The recycling industry alone creates hundreds of jobs.

Multiple choice

1. By managing waste, you are also making a difference to the ________ and the world in general:
   a. People
   b. Society
   c. Environment
   d. Ecology
   e. Civilization
2. There are many methods of disposing:
   a. Landfills
   b. Contaminated water
   c. Waste
   d. Plastic
   e. Nuclear waste
3. Throwing daily waste/garbage in ________, is the most popular method of waste disposal used today.
   a. Rivers
   b. Landfills
   c. Lakes
   d. Oceans
   e. Containers
4. Incineration or combustion is a type disposal method in which municipal solid wastes are burned at:
   a. Medium temperatures
   b. Low temperatures
   c. High temperatures
   d. All of the above
   e. None of the above
5. The process of taking useful discarded items for a specific next use is called:
   a. Recycling
   b. Reduction
   c. Reuse
   d. Resource recovery
   e. Emissions

Exercise

Problem

How does waste in different forms such as solid waste, gaseous waste, and liquid waste (increases due to population increase, urbanization, and industrialization) affect the globe?

Hands-on projects

Project

Do research: Analyze how the current process of waste management starts with the waste being created by people in the cities and disposed in trash bins near its creation point.

Case projects

Problem

Propose a solution (hardware, software, and communications) that aims to optimize waste management and include a citizen in the process.

Optional team case project

Problem

Discuss the potential of smart cities and connected communities in facilitating waste management efforts.

References

[1] Waste 360. EREF study shows continued increase in average MSW landfill tip fees. 2018. August 1, https://www.waste360.com/landfill-operations/eref-study-shows-continued-increase-average-msw-landfill-tip-fees.

[2] US EPA. Recycling economic information report. 2020.

[3] Khalil EE, Khalil H. Energy efficiency in the urban development. CRC Press, Taylor and Francis; 2015.

Chapter 18

Digital infrastructure policies for data security and privacy in smart cities

Sabrina De Capitani di Vimercati, Sara Foresti, Giovanni Livraga and Pierangela Samarati
Università degli Studi di Milano, Milan, Italy

18.1 Introduction

In the vocabulary of our society, a smart city is a place where—to the benefits of its citizens and of the urban environment they live in—new services are created, and traditional services are made more efficient, through the use of Information and Communication Technologies (ICTs). At their heart, smart cities and smart services are based on the knowledge that can be extracted from the analysis of large and heterogeneous data streams and collections. For example, intelligent transportation can leverage mobility data coming from connected vehicles, images coming from surveillance cameras, information on accidents and plans for road works to support commuters in finding the best way to move from home to work and vice versa, with clear environmental benefits in terms of reduced traffic congestions and emissions. Similarly, intelligent healthcare can leverage medical records of patients, real-time sensing through wearable devices, and information on social contacts for providing citizens with smart healthcare, with benefits also for the environment in terms, for example, of better control of contagious diseases. It goes without saying that such a scenario, while unlocking new means for the common well-being and novel services that can be beneficial for all citizens, raises at the same time major concerns related to the security and privacy of the data used, analyzed, and produced [1,2]. The collected data can include users' identities, their locations and movements (e.g., collected for intelligent transportation), health status and medical conditions (e.g., collected for intelligent healthcare), personal habits and lifestyle (e.g., inferred from smart living applications or intelligent surveillance systems), social circles and activities (e.g., determined from sensing devices and online social media platforms), just as mere examples. Improperly accessing, analyzing, modifying, or sharing those data (be them raw collected data, associations derived from the combination of different sources, or the result of analysis) may result in major privacy and security violations, with negative impacts on both the involved citizens, and the overall system of the smart city that would appear vulnerable and could be considered a threat to its citizens.

It is interesting to underline that guaranteeing protection to data related to individuals is not simply a recommendation from privacy advocates and researchers, but a concrete requirement also coming from legislators and governmental agencies. For example, the European General Data Protection Regulation (GDPR), through its revolutionary requirements for data protection *by design* and *by default*, requires the data controller (i.e., the subject or entity that determines the purposes and means of the processing of personal data) to adopt appropriate technical and organizational measures designed to implement data protection principles, and to ensure that, by default, only personal data, which are necessary for the specific purpose of the processing, are elaborated.[1] Violating these requirements is severely sanctioned. To this end, ensuring proper protection to data is actually an enabling factor for permitting smarter cities that make use, in a controlled way, of citizens' data.

The security and privacy problems that may arise in such a complex scenario are multiple and diverse [3–7]. Data in smart cities are collected, transmitted over the network, stored, processed, and possibly shared. In any of these steps data could be violated, improperly accessed, tampered without permission, abused for purposes different from what was the original purpose. The scenario can be further complicated by the (almost inevitable) use of external platforms (e.g., in the cloud) necessary for storing and processing the huge amounts of data needed for the working of a smart city. Cloud providers are typically

1. http://data.europa.eu/eli/reg/2016/679/oj.

Smart Cities Policies and Financing. https://doi.org/10.1016/B978-0-12-819130-9.00007-3

considered honest-but-curious, meaning trusted for correctly managing data but not for accessing their contents, hence introducing another variable in the already complex scenario of data protection in smart cities. In addition, data used in a smart city can be under the authority of different entities, which may be entitled to specify access or usage restrictions and requirements. For instance, air quality measurements may be performed by a private company managing a sensor network, while healthcare data may be under the control of a public medical authority: combining and analyzing data coming from these two sources may be of interest for intelligent environmental monitoring, but inevitably requires to comply to all restrictions that those entities may have specified [8].

The scientific community has devoted major efforts toward the development of effective models and techniques for protecting data throughout the various steps of their lifecycle. Two of the main pillars on which more complex solutions, specifically tailored to different application scenarios, can be built are data *anonymization* and *encryption*. In the context of smart cities, anonymization can be adopted for different purposes, for instance, whenever data should be shared or undergo analysis that, while needing access to part of its informative content, do not need (and/or cannot access) the identities of the individuals to which the data pertain. For instance, consider a dataset containing health information of a set of citizens. To the purpose of intelligent medicine, such dataset can be analyzed by the local hospital, and knowing who are the individuals behind the analyzed data can permit, for instance, to link those data to real-time measurements coming from wearable devices for monitoring patients and promptly intervening whenever a situation arises. The same dataset could however also be of interest to the local municipality to determine relationships between pollutant emissions (collected through a sensor network) and respiratory diseases: in this case, the medical dataset could (and/or should, depending on the existing restrictions and regulations) be anonymized, so to allow the municipality to study the correlations without exposing the individuals' identities. Encryption can instead be adopted to protect data from unauthorized access, ensuring that only subjects knowing the encryption key can access a dataset. In the context of smart cities, this can prove extremely useful when leveraging the cloud, or more in general external providers, for storing or analyzing the data. In principle, encryption could be enforced before or after outsourcing to the cloud. In the first scenario, if the storage provider is not given access to the encryption key, data would be protected also against it.

Both data anonymization and encryption, while representing promising solutions for protecting data in different scenarios, still represent open research fields. In particular, anonymizing data is a complex problem, which has received attention by the research community for decades now and is still far from having a definitive solution. Similarly, encrypting data, while representing a simple approach for protecting them against unauthorized accesses, can represent an obstacle for performing operations and analysis on the data, which is however a primary goal in smart cities. In this chapter, we address the problem of protecting large data collections in the context of smart cities and focus on the issues of effectively anonymizing/encrypting them while permitting computations over them. We illustrate basic concepts and survey recent approaches, based on syntactic and semantic privacy requirements, for anonymizing data collections (Section 18.2), and discuss and overview recent approaches for data encryption permitting selective and fine-grained access without decryption (Section 18.3). We also overview the problem of allowing collaborative computations combining different data sources while obeying different access restrictions to the involved data (Section 18.4). We finally give our conclusions in Section 18.5.

18.2 Data anonymity

Ensuring data anonymity is a problem far from having a trivial solution. The straightforward solution of removing all identifying information (e.g., names or e-mail addresses) from a piece of data unfortunately does not offer any anonymity guarantee. Besides identifiers, there might in fact be other information, called *quasi-identifier* (QI), which may be used, possibly in combination with other data sources, to link deidentified (and only apparently anonymous) data to the individuals to whom such data pertain [9]. Examples of QI can include date of birth, gender, addresses, and other information whose combination may be peculiar to an individual. Completely removing or obfuscating also QI information along with the removal of identifiers, while possibly enhancing data privacy, may prevent useful analysis on the protected data, which is a key aspect in smart cities and must not be overlooked.

To illustrate how QI can be used to reidentify individuals, consider the dataset in Fig. 18.1, produced by the public hospital `Hosp` of a smart city for intelligent healthcare and reporting, for a set of citizens, the most recent diagnosis they had along with demographic information. Fig. 18.2A illustrates a deidentified version of the dataset in Fig. 18.1, where social security numbers and names of patients have been removed. Still, the deidentified dataset contains attributes BirthDate, Gender, and ZIP that may be linked to other non-deidentified data sources to restrict the uncertainty over some of the respondents (i.e., the individuals to whom data refer). Suppose that the deidentified dataset is shared by `Hosp` with private organization `Env`, in charge of maintaining a sensor network for intelligent mobility, and which has access to the smart city civil

SSN	Name	BirthDate	Gender	ZIP	Diagnosis
978-05-0111	Andrew	1955/07/05	M	12311	Stroke
978-06-0212	Barbara	1970/10/13	F	12333	Flu
978-04-0513	Carl	1955/07/10	M	12313	COVID-19
978-12-0114	Donna	1970/10/20	F	12331	Flu
978-07-0715	Elizabeth	1970/06/15	F	12332	Cancer
978-42-0116	Flynn	1955/07/20	M	12312	Asthma
978-56-0217	Gladys	1970/06/03	F	12324	Cancer
978-12-0118	Harriett	1970/06/28	F	12325	Dementia
978-00-0319	Isobel	1970/10/13	F	12326	Flu
978-63-0120	Janet	1970/12/01	F	12344	Depression

FIGURE 18.1 An example of a dataset including health information for a set of citizens.

SSN	Name	BirthDate	Gender	ZIP	Diagnosis
		1955/07/05	M	12311	Stroke
		1970/10/13	F	12333	Flu
		1955/07/10	M	12313	COVID-19
		1970/10/20	F	12331	Flu
		1970/06/15	F	12332	Cancer
		1955/07/20	M	12312	Asthma
		1970/06/03	F	12324	Cancer
		1970/06/28	F	12325	Dementia
		1970/10/13	F	12326	Flu
		1970/12/01	*F*	*12344*	Depression

(a)

Name	Address	City	ZIP	BirthDate	Gender
...	...	...	...	...	...
Janet	1100 Main Street	Portland	*12344*	*70/12/01*	*female*
...	...	...	...	...	...

(b)

FIGURE 18.2 A deidentified version of the dataset in Fig. 18.1 (A) and an example of publicly available non-deidentified (B) dataset.

registry. If the registry includes only a female, born on 1970/12/01 and living in 12,344, then Env (and any subject having access to the registry) can easily reidentify the last record of the deidentified dataset as pertaining to *Janet*, disclosing also the fact that *Janet*'s most recent diagnosis is depression. Such linking attack is unfortunately a concrete threat to the protection of individual's identities: a uniqueness analysis over the US 2000 Census data has estimated that more than the 60% of the entire US population is *uniquely identifiable* by the simple combination of their gender, ZIP, and full date of birth [10].

The scientific community has devoted major efforts toward the definition of effective approaches for anonymizing data while ensuring their utility, according to different privacy requirements and definitions. A first definition of privacy (*syntactic definition*) captures, with a numerical value, the degree with which a dataset is protected. An example of a syntactic privacy definition can state that each release of data should be indistinguishably related to no less that a threshold number of individuals in the overall population (Section 18.2.1). Syntactic privacy definitions are at the basis of different approaches aimed at protecting both the identities of the individuals represented in a dataset and their sensitive information. A second definition of privacy (*semantic definition*) captures the protection that should be ensured by the analysis carried out on a dataset that includes information to be protected. An example of a semantic privacy definition can state that the result of an analysis on a released dataset must be insensitive to the insertion or deletion of a record in the dataset (Section 18.2.2). Semantic privacy definitions are at the basis of different protection approaches aimed at hiding the actual original informative content of a dataset, while ensuring adequate statistical results.

SSN	Name	BirthDate	Gender	ZIP	Diagnosis
		1955/07/**	M	123**	Stroke
		1955/07/**	M	123**	COVID-19
		1955/07/**	M	123**	Asthma
		1970/10/**	F	123**	Flu
		1970/10/**	F	123**	Flu
		1970/10/**	F	123**	Flu
		1970/06/**	F	123**	Cancer
		1970/06/**	F	123**	Cancer
		1970/06/**	F	123**	Dementia

FIGURE 18.3 An example of 3-anonymous version of the dataset in Fig. 18.2A.

Protection approaches enforcing syntactic privacy definitions assume a precise definition of what information can operate as QI and of what information is sensitive. These approaches then typically satisfy the privacy desideratum by modifying the QI with nonperturbative techniques, in such a way to guarantee data truthfulness. On the other hand, protection approaches enforcing semantic privacy definitions do not need such strict classification of information as QI and sensitive. They typically perturb data and thus do not guarantee their truthfulness. In the remainder of this section, we overview some of the most well-known protection approaches pursuing both syntactic and semantic privacy definitions.

18.2.1 *k*-anonymity

k-Anonymity is an anonymization approach pursuing a privacy requirement that demands that no release of data can be related to less than a certain number *k* of individuals [11]. *k*-Anonymity has been specifically designed to counteract the improper disclosure of individuals' identities through the QI-based linking attack illustrated in Fig. 18.2. *k*-Anonymity is based on proper modifications to the QI (besides the complete removal of identifiers) to ensure that no piece of data in a dataset can be uniquely related to its respondent through her QI, and vice versa. Practically, *k*-anonymity ensures that each combination of QI values in a dataset appears with at least *k* occurrences. In this way, each individual in any external data source could be mapped to 0 or at least *k* records in the anonymized dataset. This guarantee can be obtained in different ways. The original proposal of *k*-anonymity adopts data generalization (to the QI) and data suppression. Data generalization is a data protection technique, which replaces data values with other more general values. For instance, an individual's complete date of birth ⟨`Year/month/day`⟩ can be generalized to ⟨`Year/month`⟩ or just to ⟨`Year`⟩. Since generalization (while maintaining data truthfulness) removes details from data, with reference to the QI-based linking attack in Fig. 18.2, it is easy to see that the more generalized the QI, the less probable the risk of finding unique correspondences with external data sources. Data suppression is adopted by *k*-anonymity to reduce the amount of generalization that would otherwise be needed. For example, the dataset in Fig. 18.3 represents a *k*-anonymous version of the dataset in Fig. 18.2A with $k = 3$. Note that the last record of the original dataset, pertaining to Janet, has been suppressed to reduce the amount of generalizaton adopted to ensure 3-anonymity: since Janet is the only respondent born in December 1970 in the original dataset, to obtain 3-anonymity (or, more precisely, *k*-anonymity with $k > 1$) while including her record, it would have been necessary to generalize the date of birth removing also the month, as illustrated in Fig. 18.4. Note also that such solution would collapse all the female records in a single equivalence class (i.e., the set of records sharing the same QI values). In these situations, it is better to remove a few outlier records if this can require less adoption of generalization.

The 3-anonymous dataset in Fig. 18.3 has been obtained adopting suppression at the level of entire records and generalization at the level of attributes (in such a way that all the values of a certain attribute are uniformly generalized). Clearly, both techniques could be adopted at different granularity levels also. Regardless of the granularity level, the more the generalization, the more the protection, but also the more the information loss caused by the release of imprecise and incomplete information. The problem of computing a *k*-anonymous version of a dataset minimizing information loss is NP-hard, and several approaches have been proposed [9]. Alternative proposals achieve the *k*-anonymity requirement by using, instead of generalization, microaggregation [12], which selectively replaces data items with new ones and hence, unlike generalization, does not preserve data truthfulness. Microaggregation-based *k*-anonymity clusters record in groups of size at least *k* based on their similarity and then replace their QI values with a representative one computed through an aggregation operator (e.g., mean or median).

While a *k*-anonymous dataset effectively protects the identities of data respondents and is less vulnerable to the

SSN	Name	BirthDate	Gender	ZIP	Diagnosis
		1955/**/**	M	123**	Stroke
		1955/**/**	M	123**	COVID-19
		1955/**/**	M	123**	Asthma
		1970/**/**	F	123**	Flu
		1970/**/**	F	123**	Flu
		1970/**/**	F	123**	Flu
		1970/**/**	F	123**	Cancer
		1970/**/**	F	123**	Cancer
		1970/**/**	F	123**	Dementia
		1970/**/**	F	123**	Depression

FIGURE 18.4 An example of 3-anonymous version of the dataset in Fig. 18.2A without suppression.

SSN	Name	BirthDate	Gender	ZIP	Diagnosis
		1955/**/**	M	1231*	Stroke
		1955/**/**	M	1231*	COVID-19
		1955/**/**	M	1231*	Asthma
		1970/**/**	F	1233*	Flu
		1970/**/**	F	1233*	Flu
		1970/**/**	F	1233*	Cancer
		1970/**/**	F	1232*	Flu
		1970/**/**	F	1232*	Cancer
		1970/**/**	F	1232*	Dementia

FIGURE 18.5 An example of 3-anonymous and 2-diverse version of the dataset in Fig. 18.2A.

linking attack in Fig. 18.2, it still may leak information not intended for disclosure on respondents' sensitive information. As an example, consider the 3-anonymous dataset in Fig. 18.3, and suppose that a recipient knows that a target respondent is born on October 1970. The recipient can therefore see that the target respondent belongs to the second equivalence class in Fig. 18.3, and that her most recent diagnosis is *flu*, even without discovering which is her actual record. This is due to the fact that all the records sharing QI value $\langle 1970/10/**\rangle$ have the same value for the sensitive attribute Diagnosis. As another example, consider the last equivalence class. Respondents in this class have around 0.7 probability of having cancer, since two records out of three assume this value. If the actual probability of suffering from cancer in the overall population is (significantly) lower than 0.7, the simple fact that a target respondent is included in the last equivalence class can signal to a recipient an increase in the likelihood that such respondent is suffering from cancer. To counteract similar issues, the original definition of k-anonymity has been extended to the consideration of the sensitive values when clustering records in the equivalence classes for ensuring the k-anonymity requirement. Examples of well-known extended approaches are ℓ-diversity [13], which demands that each equivalence class contains at least ℓ well-represented values for the sensitive attribute (therefore counteracting the first issue mentioned above) and t-closeness [14], which demands that the distribution of the values of the sensitive attribute in equivalence classes be close to the distribution of the attribute in the overall table (therefore counteracting the latter issue mentioned above). For example, the dataset in Fig. 18.5 is a 2-diverse (and 3-anonymous) version of the dataset in Fig. 18.2A, where each equivalence class counts at least $\ell = 2$ different values for the sensitive attribute Diagnosis. It is interesting to note that any approach devised for practically enforcing k-anonymity can be easily extended for ensuring ℓ-diversity and t-closeness, by considering the values assumed by the sensitive attribute.

The generalization-based approaches illustrated above ensure a certain degree of unlinkability between an individual and her identity and/or sensitive information by hiding each individual in a crowd of at least other $k-1$ individuals that could possibly have her identity. Such protection comes at the price of producing imprecise or incomplete QI information. A different strategy, pursuing the same rationale of breaking the correspondence between an individual and her information but without generalizing

SSN	Name	BirthDate	Gender	ZIP	ID
		1955/07/05	M	12311	1
		1955/07/10	M	12313	
		1955/07/20	M	12312	
		1970/10/13	F	12333	2
		1970/10/20	F	12331	
		1970/06/15	F	12332	
		1970/06/03	F	12324	3
		1970/06/28	F	12325	
		1970/10/13	F	12326	

ID	Diagnosis	Count
1	Stroke	1
	COVID-19	1
	Asthma	1
2	Flu	2
	Cancer	1
3	Cancer	1
	Dementia	1
	Flu	1

FIGURE 18.6 An example of a 3-diverse version of the dataset in Fig. 18.2A with data fragmentation.

QI, is represented by data fragmentation. With fragmentation, the original dataset is split in a set of vertical fragments defined over nonoverlapping subsets of the original attributes, to break the correspondence between data that should not be visible together, such as the QI and the sensitive data. The subrecords in the different fragments are then clustered, and information on the association among clusters is released, so to hide the original precise associations between subrecords in coarser-level associations among clusters. This approach can be effectively adopted for enforcing the protection approaches illustrated above without resorting to generalization. For instance, ℓ-diversity can be achieved by clustering records in groups containing at least ℓ well-represented sensitive values and then splitting the dataset in two fragments so that one contains the QI and the other the sensitive attribute. Each subrecord in the fragments is enriched with the identifier of the cluster to which it belongs, so to release associations among groups [15]. Fig. 18.6 illustrates an example of a fragmented version of the medical dataset in Fig. 18.2A (again, after the removal of the last record) that satisfies ℓ-diversity with $\ell = 2$: it is easy to see that each respondent in the left-hand side fragment can be associated with at least two different values for the sensitive attribute diagnosis. Note that in the figure, in line with the original fragmentation-based approach [15], sensitive attribute values are reported in their fragment only once, accompanied by their number of occurrences (attribute Count).

Data fragmentation has been investigated also for protecting, besides the association between respondents' identities and their information, generic associations among data items when such associations are considered sensitive [16–18]. Intuitively, the original dataset can be split in a set of fragments so to break all sensitive associations, while ensuring that fragments are unlinkable by unauthorized subjects (so to avoid the possibility of reconstructing the broken associations, either directly or indirectly). Subrecords in the fragments can then be clustered, and associations among the clusters can be released [19].

18.2.2 Differential privacy

Differential privacy is an example of a protection approach that pursues a semantic privacy definition [20]. Unlikely the generalization-based and fragmentation-based approaches illustrated in Section 18.2.1, differential privacy does not guarantee the truthfulness of the protected data, since its protection guarantees come from the perturbation of either the original data or the results of a computation. Differential privacy permits to perform computations over a dataset, without exposing data of the single individuals. The requirement pursued by differential privacy is that, given a dataset T and an analysis over it, the impact that the insertion of a record in T (and, similarly, of the removal of a record from T) can have on the result of the analysis is limited. More precisely, given two datasets T and T' differing only for one record, a randomized function $\mathscr{K}$ is said to satisfy *ε-differential privacy* if and only if $P(\mathscr{K}(T) \in S) \leq \exp(\varepsilon) \cdot P(\mathscr{K}(T') \in S)$, with S a subset of the outputs of $\mathscr{K}$ and ε a privacy parameter (clearly, the lower the value of ε, the greater the protection offered). This guarantees that the contribution of the specific data of an individual to the computation results remains negligible and, therefore, that the privacy of that individuals is not exposed by her contribution to the dataset: should she withdraw her data from the dataset, the results of the computations would not be altered significantly. In other words, differential privacy limits the information gain that a recipient can have on the specific data of an individual. Such protection is obtained through the controlled perturbation (e.g., controlled injection of noise) in the released data.

With respect to the moment in which protection is enforced (i.e., when data are distorted), differential privacy can be enforced in a centralized scenario, where individuals contribute their original (unprotected) data to a central entity in charge of data collection, and protection is then enforced, by the central entity, interactively (i.e., adding noise to the results of analysis carried out on the original dataset) or noninteractively (i.e., producing a sanitized version of the original dataset). This scenario assumes that the central entity be trusted for accessing the original and unprotected data, as also implicitly assumed also by the generalization-based and fragmentation-based approaches illustrated in Section 18.2.1. Differential privacy can however also be enforced in a local scenario, based on the randomization of the individual pieces of data directly at the respondents' side, that is, before being collected (e.g., Refs. [21,22]).

18.3 Data encryption

Data anonymization permits to sanitize data collections to be shared, publicly or selectively. Since the amounts of data that need managed in the context of smart cities can be extremely large, also storage can represent an issue. The availability of a multitude of specialized cloud providers, offering storage capabilities at competitive prices, represents a promising solution for data management. However, cloud providers are typically not considered fully trusted to access the content of the data they store, especially in the context of smart cities where (citizens') data can be sensitive, proprietary, or simply subject to access restrictions. For instance, health data collected and produced by `Hosp` for intelligent healthcare are sensitive and should not be improperly disclosed, even if outsourced to the cloud to relieve `Hosp` storage and management burden. A possible solution leverages owner-side encryption, by means of which a data collection is encrypted by its owner before being outsourced to a cloud provider. While effectively protecting data from improper observations by the cloud provider and by unauthorized subjects that do not know the encryption key, encryption can have an impact when the outsourced data are to be retrieved by authorized subjects. In the remainder of this section, we overview the problems of ensuring selective and fine-grained access to outsourced data.

18.3.1 Enabling selective access

The main problem of enforcing selective access to data that have been outsourced to the cloud is caused by the fact that neither the cloud provider (for security and/or trust reasons) nor the data owner (for performance/convenience reasons) can evaluate access requests. A possible solution consists in the outsourced data to *self-enforce* access restrictions, through *selective owner-side encryption* [23]. With selective owner-side encryption, different data items are encrypted by their owner (before being outsourced to the cloud) with different keys, which are then distributed to users in such a way that they can decrypt all and only the data items for which they are authorized. With such an approach, data are protected also against the cloud provider itself, which is not communicated any encryption key. To reduce the burden of key management for the users, selective owner-side encryption is typically accompanied by *key derivation* techniques, by means of which the value of an encryption key can be computed starting from the value of another encryption key and of a piece of public information [24].

The derivation of a key k_y from another key k_x is enabled by a public token $t_{x,y}$ computed as $k_y \oplus h(k_x, l_y)$, with $\oplus$ the bitwise `xor` operator, h a deterministic non-invertible cryptographic function, and l_y a public information associated with k_y. A careful adoption of key derivation permits each user to manage a single key only, from which she can then retrieve all the keys of all the data items she is authorized to access. The overall key derivation process, regulating which user can derive which keys, is controlled by the data owner through the definition of appropriate key derivation structures correctly modeling the authorization policies [25]. For example, consider two users Alice and Bob and two resources (e.g., two data items) r_1 and r_2, and an authorization policy such that Alice can access r_1, and Bob can access r_1 and r_2 (as illustrated by the access matrix in Fig. 18.7). With selective owner-side encryption and key derivation, a possible solution can require to encrypt r_1 and r_2 with keys k_1 and k_2, respectively, and to communicate to Alice and Bob encryption keys k_{Alice} and k_{Bob}, respectively. Fig. 18.8 illustrates an example of a key derivation structure along with the catalog $\mathcal{T}$ storing the tokens enabling a possible derivation for enforcing the authorization policy. Nodes in the structure correspond to keys, and edges to tokens. An edge from k_a to k_b models the existence of a token enabling the derivation of k_b from k_a. For example, with reference to the structure in Fig. 18.8, it is easy to see that Bob can derive both keys k_1 and k_2 for accessing the two resources he is authorized to access starting from his own key k_{Bob}, that is, managing a single key only.

Since access to a resource is granted with the knowledge of (or the possibility to derive) the key adopted for its encryption, changes in authorization policies should be reflected by a reencryption of the involved resources, so that encryption correctly reflects the updated policy. Since the cloud provider is typically not trusted to access plaintext data, such an operation would in principle require the data owner to locally perform the reencryption herself and to upload to the cloud the reencrypted resources. To reduce the burden for the owner, a possible solution is to adopt two independent layers of encryption over data items: one for enforcing the original policy, and a second one for enforcing updates to the policy [23]. If the cloud provider is

	r_1	r_2
Alice	1	0
Bob	1	1

FIGURE 18.7 An example of access matrix.

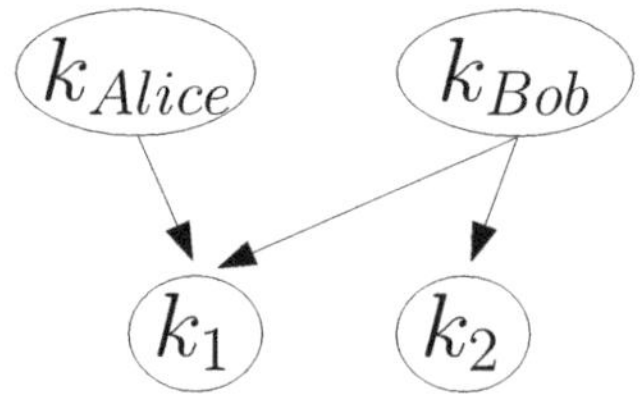

$\mathcal{T}$

From	To	Value
k_{Alice}	k_1	$k_1 \oplus h(k_{Alice}, l_1)$
k_{Bob}	k_1	$k_1 \oplus h(k_{Bob}, l_1)$
k_{Bob}	k_2	$k_2 \oplus h(k_{Bob}, l_2)$

FIGURE 18.8 An example of key derivation structure and token catalog.

trusted to correctly manage data (as commonly assumed in the *honest-but-curious* trust model), the management of such second layer of encryption can be delegated to the provider itself, which can then easily manage policy updates without accessing plaintext data (which remain always wrapped by the first layer of encryption). The token catalog is then updated to allow authorized users to derive the two keys used at the two levels for the resources they can access.

Selective owner-side encryption can also support dynamic scenarios of digital data markets, when data owners wish to make their data selectively available to others possibly receiving an incentive. Incentivizing owners to share, in a controlled way, their data is indeed relevant to smart cities, where analysis and computations of different data sources can fuel novel services to citizens. For examples, intelligent healthcare can benefit from more individuals sharing their health data and measurements. In these scenarios, selective encryption (adopted for protecting data against the storage provider as well as unauthorized external subjects) can be coupled with blockchain [26]. A possible solution is to publicly store on-chain the structure to be used for key derivation, in such a way to securely log the interactions among the subjects, and the possible incentives that the owners should receive for making their data selectively available [27].

The owner-side selective encryption approaches illustrated above typically use symmetric encryption. An alternative solution for enabling selective access with asymmetric encryption is represented by *attribute-based encryption* (e.g., Refs. [28–31]). Similarly to selective encryption, also with attribute-based encryption authorizations are enforced by permitting a user to decrypt different data items. In particular, attribute-based encryption is based on the definition of *attributes*, relevant for the authorizations, associated with users and data items. The decryption of a data item by a certain user is permitted if that user's attributes match the attributes of the data item. Encryption keys are then generated accordingly so to allow the correct enforcement of the authorization policy.

18.3.2 Enabling fine-grained access

When data are outsourced to a cloud provider in encrypted form, and the cloud provider is not given access to the encryption keys, enabling fine-grained access (e.g., retrieving the name and age of the patients whose most recent diagnosis is flu and who live in a certain area) can represent an issue. In fact, the cloud provider cannot access plaintext data to retrieve only those data items that satisfy the conditions in the access request. Clearly, permitting only the retrieval of the entire dataset and performing the filtering at the requesting side is not a viable solution, due to its unacceptable overhead. The research community has therefore investigated the problem and proposed different solutions for (partly) delegating the evaluation of fine-grained access conditions directly at the provider side.

A first family of solutions is based on the definition of *indexes*. Indexes are metadata that can be associated with data by their owner before being outsourced. In the context of relational databases, indexes are represented as additional attributes added to the outsourced relation. Indexes are defined for attributes expected to be involved in query evaluation and reflect the values assumed by the attributes for which they are defined without disclosing their values. Given a relation, the owner then outsources to the cloud an encrypted and indexed version. Fig. 18.9 represents an encrypted and indexed version of the medical dataset of Fig. 18.1. In this case, encryption is applied at the granularity level of records (i.e., each record is encrypted as a whole), and two indexes have been defined for attributes

encr_record	I_Z	I_D
j5as?$	ι	α
%grT6	λ	β
z12#	ι	γ
f*grTi_6	λ	β
lF=+	λ	ε
6#R_u	ι	ζ
1wp(yQ	κ	ε
yKu8$	κ	η
nfP*r;	κ	β
a%g_6	ι	ϑ

FIGURE 18.9 An example of an encrypted and indexed version of the dataset in Fig. 18.1.

ZIP (I_Z) and disease (I_D) Queries are translated for operating on indexes: since indexes are not sensitive, queries on them can be executed by the cloud provider itself, and the final user can possibly discard from the result only those records that have been included due to the imprecise evaluation over indexes (e.g., when more plaintext values are mapped to a same index value). Indexes can be defined in different ways and can support different kinds of operations (e.g., to support range conditions over indexes, it is necessary that the mapping between original and index values correctly reflects the original ordering) [32–34]. For example, index I_Z is defined as a *bucket-based* index (a type of index that splits the domain of an attribute in buckets and associates each bucket with a label), such that ZIP values from 12,310 to 12,319 are mapped to index value ι, from 12,320 to 12,329 are mapped to index value κ, and from 12,330 to 12,339 are mapped to index value λ. Index I_D is instead an encryption-based index, which associates each attribute value v with an index value $i(v)$ obtained by applying a deterministic encryption function to v.

Alternative solutions are represented by the adoption of *searcheable* or *homomorphic* encryption schemes. Searcheable encryption schemes support the search, over encrypted data, of data whose decrypted version satisfies a certain condition (e.g., Refs. [35,36]). To increase the efficiency of the search process, searcheable encryption schemes can tolerate a certain amount of information leakage [37,38]. Homomorphic encryption schemes support the evaluation of operations over encrypted data without the need of decrypting them (e.g., Ref. [39]). Homomorphic encryption schemes can be classified depending on whether they support only a specific kind of operation (*partially* homomorphic schemes, such as El-Gamal scheme for multiplications, or Paillier scheme for additions), generic computations over data an arbitrary number of times (*fully* homomorphic schemes), or generic computations for a limited number of times (*somewhat* homomorphic schemes) [40–42].

Different kinds of encryption may also be used in combination in an onion-like approach, where each data item is encrypted with different layers of encryption, and each layer supports a specific operation on data. Inner layers support more operations and are expected to provide less protection [43]. When an authorized user requires a given operation on a data item, then the encryption layers are removed (i.e., through decryption) until the one supporting the required operation is reached and the operation can be executed.

18.4 Collaborative computations

Combining different data sources and performing computations and analysis over them can generate knowledge. This is particularly relevant in the context of smart cities, characterized by the collection and production of huge amounts of heterogeneous and diverse data about citizens and their environments. For instance, health information collected through a variety of sensors may be usefully combined with mobility data and social interactions extracted from social media services to monitor and possibly control the diffusion of infectious diseases. Since computations involving different data sources can be intensive, a promising solution for mitigating their economic costs consists in delegating (parts of) the computation to cloud providers offering computation capabilities at competitive prices. However, the data involved in a computation can be sensitive or subject to access restrictions, and it is therefore necessary to ensure that the involvement of external subjects (i.e., the cloud providers) in the computation does not expose information not intended for disclosure. In the context of relational databases, several approaches have been proposed for enabling secure collaborative computations over data. Traditional solutions can be based on the definition of *views* (e.g., Refs. [44,45]), representing portions of a dataset for which different subjects are authorized, or on *access patterns* (e.g., Refs. [46,47]), restricting the data that can be accessed based on the input that is provided by a subject. Such solutions however only capture the information that is explicitly involved in the computation, while a computation may also leak information not explicitly visible. For instance, consider a relational query of the form SELECT Name, Surname FROM MedicalData WHERE Diagnosis = "Cancer." Its result, despite containing only patients' names and surnames, implicitly contains information on their diseases as well, in terms of the fact that all returned patients suffer from cancer. For this reason, a subject not authorized to access patients' diagnosis should not access the result of the query.

To permit collaborative computations with the selective involvement of nonfully trusted providers, while controlling both explicit and implicit information flows, recent proposals have built on the idea of keeping track, as the computation proceeds, of the operations that are being executed and of the implicit information flows they cause. This permits, for instance, to maintain information on the join conditions evaluated in the computation of a relation, which reveal the fact that the tuples surviving the join appeared in both the joined relations [48]. To enlarge the spectrum of the possible subjects that can collaborate for the computation including, for instance, providers that offer computational resources at competitive prices but are not fully trusted to access plaintext data, a recent proposal permits the data owners to specify different levels of visibility over their data: the classical yes/no visibility over a data item, and the *encrypted visibility*, meaning that the subject can only access an encrypted version of the data

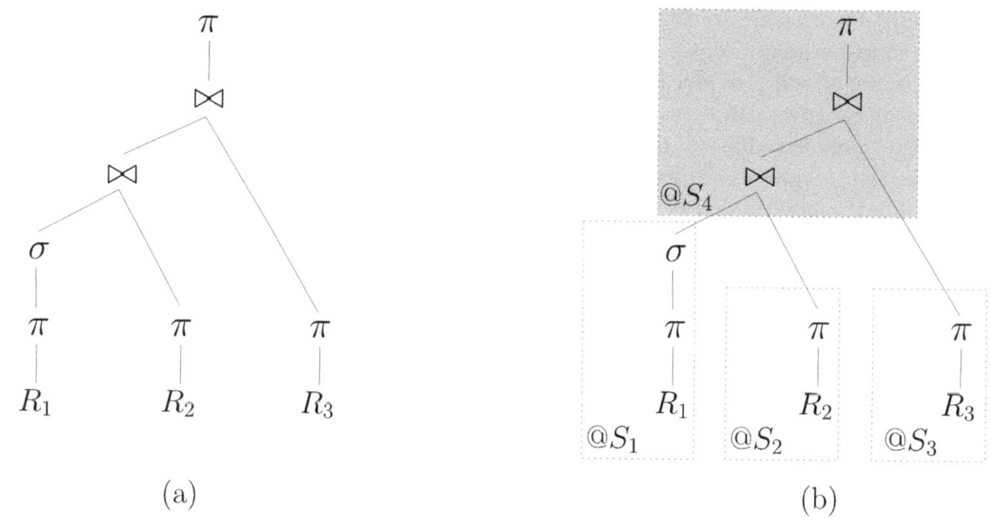

FIGURE 18.10 An example of a collaborative query execution with dynamic encryption.

item [49]. The proposal in Ref. [49] enforces authorizations regulating the visibility over data by dynamically injecting, in the computation, on-the-fly encryption and decryption operations. In this way, depending on the subjects to which the different parts of the computation are delegated, visibility over plaintext data is dynamically disabled and enabled. A computation can then be distributed to the most convenient providers, for instance, based on economic factors, while enforcing the authorization policies set by data owners. To counteract improper flows of information, the proposal in Ref. [49] evaluates the information flow enacted by the computation, considering for each step of the computation both its explicit and its implicit informative content, depending on the specific operations that have been executed until then. Authorizations are then enforced against both explicit and implicit information flows. Fig. 18.10A illustrates an example of an execution plan for a query over three relations R_1, R_2, and R_3 (owned by three subjects S_1, S_2, and S_3 respectively), where nodes represent traditional operators of relational algebra (projection π, selection σ, join $\bowtie$). Assume the existence of a computational providers S_4, which offers computational resources at competitive prices but cannot access all involved data in plaintext. Fig. 18.10B illustrates an example of assignment of query operations to the different subjects in the system (owners and the computational provider S_4), assuming that the computation of the joins and of the final projection is executed, by S_4, on encrypted data. This requires the injection of on-the-fly encryption operations on the input relations to the join operations. The encryption operations are performed by subjects S_1, S_2, and S_3 (which produce the relations to be encrypted). Note that if S_4 were not authorized to access the attribute(s) over which the selection over R_1 is executed, then it would not be allowed for the join computations, since such a selection would cause an implicit information flow leaking the value of the attribute(s) over which it is performed [49]. The proposal in Ref. [49] has been also extended to scenarios where the original relations are outsourced to nonfully trusted storage providers and hence stored in encrypted form (Section 18.3) [50].

Other aspects that have been investigated in distributed query evaluations include the possibility of permitting users to specify constraints on how the query should be evaluated, for instance, to specify that all join operations should be executed by a specific subject only (e.g., Ref. [51]), and cooperative executions of predetermined computations with the help of trusted hardware components (e.g., Ref. [52]).

18.5 Summary

Anonymization and encryption are two of the main pillars that can be adopted in smart cities for protecting the privacy and security of data whenever they are shared or outsourced for storage or computation. In this chapter, we have considered the problems of anonymizing datasets and of encrypting them while permitting computations over them, and we have surveyed possible approaches for addressing them. We also have presented recent approaches for collaborative computations that can be enabled by data encryption for protecting visibility over sensitive data.

Finally, let us move on to the real interactive part of this chapter: review questions/exercises, hands-on projects, case projects, and optional team case project. The answers and/or solutions by chapter can be found in Appendix G.

18.6 Chapter review questions/exercises

True/false

1. True or False? In the vocabulary of our society, a smart city is a place where—to the benefits of its citizens and of the urban environment they live in—new services are created, and traditional services are made less efficient, through the use of Information and Communication Technologies (ICTs.
2. True or False? Ensuring data anonymity is not a problem far from having a trivial solution.
3. True or False? k-Anonymity is an anonymization approach pursuing a privacy requirement that demands that release of data can be related to less than a certain number k of individuals.
4. True or False? Differential privacy is an example of a protection approach that does not pursue a semantic privacy definition.
5. True or False? Data anonymization permits to sanitize data collections to be shared privately or selectively.

Multiple choice

1. The main problem of enforcing selective access to data that have been outsourced to the cloud is caused by the fact that neither the cloud provider (for security and/or trust reasons) nor the data owner (for performance/convenience reasons) can evaluate:
   a. Smart sustainable development
   b. ICT-based infrastructure
   c. Access requests
   d. Investments
   e. Policies
2. When data are outsourced to a cloud provider in encrypted form, and the cloud provider is not given access to the encryption keys, enabling fine-grained access (e.g., retrieving the name and age of the patients whose most recent diagnosis is flu and who live in a certain area) can represent an:
   a. Additional attribute
   b. Indexed version
   c. Economic effect
   d. Issue
   e. Encrypted version
3. Combining different data sources and performing computations and analysis over them can generate:
   a. Machine learning
   b. Machine to human learning
   c. Knowledge
   d. Artificial intelligence (AI)
   e. Digital twins (DT)
4. Anonymization and encryption are two of the main pillars that can be adopted in smart cities for protecting the privacy and security of data whenever they are shared or outsourced for storage or.
   a. Holistic integration
   b. Economic growth
   c. Making cities smarter
   d. Computation
   e. E-economics
5. Consider the problems of ____________ datasets and of encrypting them while permitting computations over them, and survey possible approaches for addressing them.
   a. Sensing
   b. Providing
   c. Unlocking
   d. Accessing
   e. Anonymizing

Exercise

Problem

Discuss the privacy and security issues in current smart applications along with the corresponding requirements for building a stable and secure smart city.

Hands-on projects

Project

Do research: Analyze the key themes within smart cities research: Privacy and security of mobile devices and services; smart city infrastructure; power systems; healthcare; frameworks; algorithms and protocols to improve security and privacy; operational threats for smart cities; use and adoption of smart services by citizens; use of blockchain; and use of social media.

Case projects

Problem

Provide a stakeholder model for a security and privacy framework for secure and privacy-aware service provisioning in smart cities.

Optional team case project

Problem

As smart cities get smarter, why have citizens begun to express concerns about the use of their data, and whether their privacy is being sacrificed for the benefit of big data?

Acknowledgments

This work was supported in part by the EC within the H2020 Program under projects MOSAICrOWN and MARSAL, by the Italian Ministry of Research within the PRIN program under project HOPE, and by JPMorgan Chase & Co under project "k-anonymity for AR/VR and IoT/5G."

References

[1] Braun T, Fung BC, Iqbal F, Shah B. Security and privacy challenges in smart cities. Sustain Cities Soc 2018;39:499–507.
[2] Zhang K, Ni J, Yang K, Liang X, Ren J, Shen XS. Security and privacy in smart city applications: challenges and solutions. IEEE Commun Mag 2017;55(1):122–9.
[3] De Capitani di Vimercati S, Foresti S, Livraga G, Piuri V, Samarati P. A fuzzy-based brokering service for cloud plan selection. IEEE Syst J 2019;13(4):4101–9.
[4] De Capitani di Vimercati S, Foresti S, Livraga G, Piuri V, Samarati P. Supporting user requirements and preferences in cloud plan selection. IEEE Trans Serv Comput 2021;14(1):274–85.
[5] Jhawar R, Piuri V. Fault tolerance and resilience in cloud computing environments. In: Vacca J, editor. Computer and information security handbook. 2nd ed. Morgan Kaufmann; 2013.
[6] Jhawar R, Piuri V, Samarati P. Supporting security requirements for resource management in cloud computing. In: Proc. of CSE 2012, Paphos, Cyprus; December 2012.
[7] Jhawar R, Piuri V, Santambrogio M. Fault tolerance management in cloud computing: a system-level perspective. IEEE Syst J 2013;7(2):288–97.
[8] De Capitani di Vimercati S, Genovese A, Livraga G, Piuri V, Scotti F. Privacy and security in environmental monitoring systems: issues and solutions. In: Vacca J, editor. Computer and information security handbook. 2nd ed. Morgan Kaufmann; 2013.
[9] De Capitani di Vimercati S, Foresti S, Livraga G, Samarati P. Data privacy: definitions and techniques. Int J Uncertain Fuzziness Knowledge-Based Syst 2012;20(6):793–817.
[10] Golle P. Revisiting the uniqueness of simple demographics in the us population. In: Proc. of WPES 2006, Alexandria, VA, USA; October 2006.
[11] Samarati P. Protecting respondents' identities in microdata release. IEEE Trans Knowl Data Eng 2001;13(6):1010–27.
[12] Domingo-Ferrer J, Torra V. Ordinal, continuous and heterogeneous *k*-anonymity through microaggregation. Data Min Knowl Discov 2005;11(2):195–212.
[13] Machanavajjhala A, Kifer D, Gehrke J, Venkitasubramaniam M. ℓ-diversity: privacy beyond *k*-anonymity. ACM TKDD 2007;1(1):3:1–3:52.
[14] Li N, Li T, Venkatasubramanian S. *t*-Closeness: privacy beyond *k*-anonymity and ℓ-diversity. In: Proc. of ICDE 2007, Istanbul, Turkey; April 2007.
[15] Xiao X, Tao Y. Anatomy: simple and effective privacy preservation. In: Proc. of VLDB 2006, Seoul, Korea; September 2006.
[16] Aggarwal G, Bawa M, Ganesan P, Garcia-Molina H, Kenthapadi K, Motwani R, Srivastava U, Thomas D, Xu Y. Two can keep a secret: a distributed architecture for secure database services. In: Proc. of CIDR 2005, Asilomar, CA, USA; January 2005.
[17] Ciriani V, De Capitani di Vimercati S, Foresti S, Jajodia S, Paraboschi S, Samarati P. Combining fragmentation and encryption to protect privacy in data storage. ACM Trans Inf Syst Secur 2010;13(3):22:1–22:33.
[18] De Capitani di Vimercati S, Foresti S, Jajodia S, Livraga G, Paraboschi S, Samarati P. Fragmentation in presence of data dependencies. IEEE Trans Sec Depend Comput 2014;11(6):510–23.
[19] De Capitani di Vimercati S, Foresti S, Jajodia S, Livraga G, Paraboschi S, Samarati P. Loose associations to increase utility in data publishing. J Comput Secur 2015;23(1):59–88.
[20] Dwork C. Differential privacy. In: Proc. of ICALP 2006, Venice, Italy; July 2006.
[21] Bernau D, Robl J, Grassal PW, Schneider S, Kerschbaum F. Comparing local and central differential privacy using membership inference attacks. In: Proc. of DBSec 2021, Calgary, Canada; July 2021.
[22] Erlingsson Ú, Pihur V, Korolova A. RAPPOR: randomized aggregatable privacy-preserving ordinal response. In: Proc. of CCS 2014, Scottsdale, AZ, USA; November 2014.
[23] De Capitani di Vimercati S, Foresti S, Jajodia S, Paraboschi S, Samarati P. Encryption policies for regulating access to outsourced data. ACM Trans Database Syst 2010;35(2):12:1–12:46.
[24] Atallah M, Blanton M, Fazio N, Frikken K. Dynamic and efficient key management for access hierarchies. ACM Trans Inf Syst Secur 2009;12(3):18:1–18:43.
[25] De Capitani di Vimercati S, Foresti S, Livraga G, Samarati P. Practical techniques building on encryption for protecting and managing data in the cloud. In: Ryan P, Naccache D, Quisquater J-J, editors. The new codebreakers: essays dedicated to David Kahn on the occasion of his 85th birthday. Springer; 2016.
[26] De Capitani di Vimercati S, Foresti S, Livraga G, Samarati P. Empowering owners with control in digital data markets. In: Proc. of IEEE CLOUD 2019, Milan, Italy; July 2019.
[27] Foresti S, Livraga G. Selective owner-side encryption in digital data markets: strategies for key derivation. In: Proc. of SECRYPT 2021, virtual; July 2021.
[28] Goyal V, Pandey O, Sahai A, Waters B. Attribute-based encryption for fine-grained access control of encrypted data. In: Proc. of CCS 2006, Alexandria, VA, USA; October/November 2006.
[29] Lewko A, Waters B. Decentralizing attribute-based encryption. In: Paterson KG, editor. Proc. of EUROCRYPT 2011; May 2011. Tallin, Estonia.
[30] Li J, Zhang Y, Ning J, Huang X, Poh GS, Wang D. Attribute based encryption with privacy protection and accountability for cloud IoT. IEEE Trans Cloud Comput 2020 [pre-print].
[31] Waters B. Ciphertext-policy attribute-based encryption: an expressive, efficient, and provably secure realization. In: Proc. of PKC 2011, Taormina, Italy; March 2011.
[32] De Capitani di Vimercati S, Foresti S, Jajodia S, Paraboschi S, Samarati P. On information leakage by indexes over data fragments. In: Proc. of PrivDB 2013, Brisbane, Australia; April 2013.
[33] Hacigümüs H, Iyer B, Mehrotra S, Li C. Executing SQL over encrypted data in the database-service-provider model. In: Proc. of SIGMOD 2002, Madison, WI, USA; June 2002.
[34] Wang H, Lakshmanan L. Efficient secure query evaluation over encrypted XML databases. In: Proc. of VLDB 2006, Seoul, Korea; September 2006.
[35] Demertzis I, Papadopoulos D, Papamanthou C. Searchable encryption with optimal locality: achieving sublogarithmic read efficiency. In: Proc. of CRYPTO 2018, Santa Barbara, CA, USA; August 2018.

[36] Li J, Huang Y, Wei Y, Lv S, Liu Z, Dong C, Lou W. Searchable symmetric encryption with forward search privacy. IEEE Trans Sec Depend Comput 2021;18(1):460–74.

[37] Bösch C, Hartel P, Jonker W, Peter A. A survey of provably secure searchable encryption. ACM Comput Surv 2014;47(2).

[38] Domingo-Ferrer J, Farràs O, Ribes-González J, Sánchez D. Privacy-preserving cloud computing on sensitive data: a survey of methods, products and challenges. Comput Commun 2019;140–141:38–60.

[39] Boneh D, Gentry C, Halevi S, Wang F, Wu DJ. Private database queries using somewhat homomorphic encryption. In: Proc. of ACNS 2013, Banff, AB, Canada; June 2013.

[40] Acar A, Aksu H, Uluagac AS, Conti M. A survey on homomorphic encryption schemes: theory and implementation. ACM Comput Surv 2018;51(4).

[41] Gentry C. Fully homomorphic encryption using ideal lattices. In: Proc. of STOC 2009, Bethesda, MA, USA; May 2009.

[42] Wood A, Najarian K, Kahrobaei D. Homomorphic encryption for machine learning in medicine and bioinformatics. ACM Comput Surv 2020;53(4).

[43] Popa R, Redfield C, Zeldovich N, Balakrishnan H. CryptDB: protecting confidentiality with encrypted query processing. In: Proc. of SOSP 2011, Cascais, Portugal; October 2011.

[44] Guarnieri M, Basin D. Optimal security-aware query processing. Proc VLDB Endow 2014;7(12):1307–18.

[45] Rizvi S, Mendelzon A, Sudarshan S, Roy P. Extending query rewriting techniques for fine-grained access control. In: Proc. of SIGMOD, Paris, France; June 2004.

[46] Amarilli A, Benedikt M. When can we answer queries using result-bounded data interfaces?. In: Proc. of PODS 2018, Houston, TX, USA; June 2018.

[47] Benedikt M, Leblay J, Tsamoura E. Querying with access patterns and integrity constraints. Proc VLDB Endow 2015;8(6):690–701.

[48] De Capitani di Vimercati S, Foresti S, Jajodia S, Paraboschi S, Samarati P. Authorization enforcement in distributed query evaluation. J Comput Secur 2011;19(4):751–94.

[49] De Capitani di Vimercati S, Foresti S, Jajodia S, Livraga G, Paraboschi S, Samarati P. An authorization model for multi-provider queries. Proc VLDB Endow 2017;11(3):256–68.

[50] De Capitani di Vimercati S, Foresti S, Jajodia S, Livraga G, Paraboschi S, Samarati P. Distributed query evaluation over encrypted data. In: Proc. of DBSec 2021, Calgary, Canada; July 2021.

[51] Farnan N, Lee A, Chrysanthis P, Yu T. PAQO: preference-aware query optimization for decentralized database systems. In: Proc. of ICDE 2014, Chicago, IL; March–April 2014.

[52] Dave A, Leung C, Popa RA, Gonzalez JE, Stoica I. Oblivious coopetitive analytics using hardware enclaves. In: Proc. of Euro Sys 2020, Heraklion, Greece; April 2020.

Chapter 19

Financing smart cities—an Indian case study

Prasun Kamal Agrawal[1] and Bipin Pradeep Kumar[2]
[1]India Japan Technology Leader, Mumbai, Maharashtra, India; [2]Systems Biology Institute, Tokyo, Japan

19.1 Introduction

A city is a social construct whose effective functioning is determined partly by the rules of the state, partly by the values of society, and partly by its economic structures. Cities require the right mix of social capital, economic capital, and human capital. With technology making giant leaps in multitudes of ways, a new starting point is a right balance between technology and the state, and this has shifted in recent decades with technology gaining in stature, influence, and acceptance. Technology is becoming the organizing framework, not only for new-age companies, but also increasingly seen as fundamental infrastructure that affects broader citizen interactions. It has gone mainstream with its reach extending well into civic and family life. It is within this context that the Government of India sought to revitalize its urban agglomerations. This newfound focus was long overdue, and these reforms are expected to unleash a new dynamism in its cities.

The genesis of Smart Cities in India was the report by the Planning Commission of India in its 12th and last Plan (2012–17), where it envisioned to create Smart Cities to address India's urbanization challenges [1]. The Smart Cities Mission (SCM) in India was launched on June 25, 2015 by the Ministry of Housing and Urban Affairs (MoHUA) of the Government of India (GOI). It sought to address unprecedented urbanization that has not been witnessed in the history of India. The SCM aims to strengthen the urban infrastructure through the application of technology and smart solutions. This will lead to a better quality of life for a growing populace of citizens and is thus one of the most transformational urban missions for India.

Hundred cities were chosen as a part of the Smart Cities project in India, with a total of 5151 smart projects in these cities. While fundamentally altering the cities by virtue of driving large investments, these, in turn, affect the social and cultural ethos of the citizens it touches. These projects proposed total investments equivalent to USD 28.09 Billion (INR 2,05,018 Crores, at an exchange rate of INR 72 per 1 US Dollar) [2]. A bird's-eye view breakup of the source of funds for this is depicted in Table 19.1.

The Indian smart cities have been at the forefront of leveraging technology to manage COVID19 crisis, with 69 [2] of them using the newly built Integrated Command and Control Centers (ICCC) as crisis management war rooms. These were built as a result of projects under SCM, as cities used the process to complete them at the earliest.

This essay describes a few aspects from the Agra Smart City as examples. The Agra Municipal Corporation has an annual budget of INR 660 Crores (about USD 90 Million) for the financial year 2020–21 [3]. Its budget outlay under the Smart City Mission is INR 2133 Crores (or about USD 296 Million) [4]. To put these numbers in perspective, the GDP of Mumbai, India's economic capital, in 2020 was pegged at USD 310 Billion [5]. The budget of its Municipal Corporation—Brihanmumbai Municipal Corporation (BMC) that manages city affairs was about USD 5 Billion for the period of 2020–21 [6]. Mumbai is not a part of India's SCM, though. Bengaluru, a large metro and one of the Smart Cities, had a GDP of USD 86 Billion in 2020. Its municipal budget in 2020–21 was INR 9287 Crores or about USD 1.2 Billion [7]. Bengaluru's Smart City Budget is INR 2090.67 Crores (or about USD 290 Million) [8].

To cater to growing urbanization needs in India, the 15th Finance Commission [9] for the years 2021–26, in its budget report, recommended a total of INR 1,21,055 Crores as grants for Urban Local Bodies (ULB) for the same period. This is to be distributed among states using a formula with a weightage of 90% on population and 10% on area of the ULB.

Smart Cities Policies and Financing. https://doi.org/10.1016/B978-0-12-819130-9.00049-8

TABLE 19.1 Breakup of the source of funds for India's 100 smart cities.

Sources of funds	Amount (INR crores)	Amount (USD billion)	Percentage of total
Central and state government	93,552	12.82	45.63
Convergence funding from other missions, programs	42,028	5.76	20.50
Funds from PPP	41,022	5.62	20.01
Loans/debt	9843	1.35	4.80
Own sources	2644	0.36	1.29
Other sources	15,929	2.18	7.77
Total	205,018	28.09	100.00

19.2 Convergence of missions

The SCM (see Fig. 19.1) has been championed and implemented under the aegis of the Ministry of Housing and Urban Affairs, the erstwhile Ministry of Urban Development, as part of the transformation of urban landscape goal pursued by the GOI. The SCM is one of the six missions in this direction. The other missions include:

- **Swachh Bharat Mission (Urban)**: The Clean India mission launched on October 2, 2014 for a five-year period aimed at achieving 100% Open Defecation Free (ODF) status and 100% solid waste management in all Urban Local Bodies (ULBs).
- **Pradhan Mantri Awas Yojana (Urban)**: launched on June 25, 2015, aims to ensure "Housing for All" in urban areas. The Mission provides central assistance to implementing agencies through States/Union Territories (UTs) and Central Nodal Agencies (CNAs) for providing houses to all eligible families/beneficiaries.
- **Deendayal Antyodaya Yojana—National Urban Livelihoods Mission (DAY-NULM)**: The Mission aims to
- reduce poverty and vulnerability of the urban poor households by providing self-employment and skilled wage employment opportunities.
- provide shelter equipped with essential services to the urban homeless.
- address the livelihood concerns of the urban street vendors.
- **Atal Mission for Rejuvenation and Urban Transformation (AMRUT)**: Launched on 25th June 2015 with the objective of universal water supply; improvement in sewerage network; developing children and Divyang (physically challenged) friendly green spaces and parks; improvement in stormwater drainage and Non-Motorized Transport (NMT) and implementing a reform agenda under which States/UTs are incentivized for implementation of reforms.

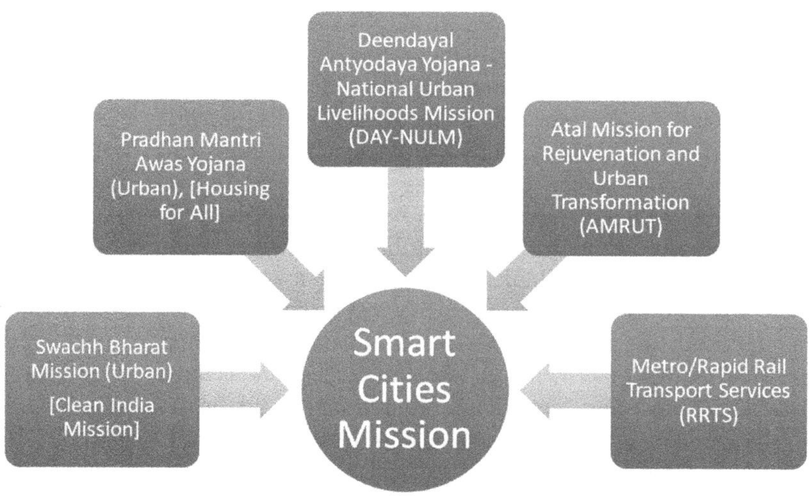

FIGURE 19.1 Convergence of different missions.

- **Urban Transport services:** Involving 900 kms of metro/Rapid Rail Transport Services (RRTS) construction in 27 cities and 700 kms of metro operationalized in 18 cities along with promoting multimodal integration of metro with city transport services (city bus, bus rapid transport services), regional transport services (rail, air, regional bus), NMT (pedestrian, cycle, rickshaw), intermediate para transit (auto-rickshaw, taxi), and water transport (ferry).

A three-level strategy to take advantage of the opportunities presented by the Indian cities was adopted by the Ministry of Housing and Urban Affairs. At the first level, for over 4300 Indian ULBs, poverty alleviation, affordable housing, and sanitation are the three biggest challenges. Deendayal Antyodaya Yojana—National Urban Livelihoods Mission (DAY-NULM), Pradhan Mantri Awas Yojana—Urban (PMAY-U) and Swachh Bharat Mission-Urban (SBM-U) are getting implemented in all the ULBs.

At the second level, for 500 cities, basic infrastructure such as water supply, sewage/septage projects, and green parks are the focus. These are implemented in towns with 100,000 and above population through AMRUT. This second level covers over 60% of urban population. Finally, at the third level, 100 cities are being developed under SCM to address the issue of ease of living by evolving new paradigms of urban governance with communities at the core and increased use of digital technology to improve the city infrastructure, services, and utilization of resources.

19.3 Six differentiators

The SCM was launched after detailed consultation with various stakeholders, including State and City Governments. It focused on targeted outcomes and better resource utilization through decentralizing project formulation, appraisal, and approval, thus empowering states/UT governments and cities. The SCM also promoted a healthy competitive spirit among cities and states through the smart cities selection process, thereby strengthening competitive and cooperative federalism. Periodic review of projects by the Center and State-level monitoring of the urban missions further ensured adherence to projects' timelines. The ULBs were encouraged to become self-sufficient and mobilize resources for attaining viability through instruments such as Municipal Bonds emphasizing improving urban governance for better service delivery, transparency, and accountability. In summary, the SCM created six differentiators of SCM. These are depicted in Fig. 19.2.

19.3.1 Competition

Cities are selected through a competitive mechanism. The GOI did not prescribe any particular model to be adopted by the Smart Cities. The approach was not a one-size-fits-all approach. Each city had to formulate its own concept, vision, mission, and plan (proposal) appropriate to its local context, resources and levels of ambition. Accordingly, they had to choose their Smart City model and answer the question: "What kind of Smart City do you want?" For this, cities prepared their Smart City Proposal (SCP) containing the plan for mobilization of resources and intended outcomes regarding terms of infrastructure upgradation and applications.

19.3.2 Different starting points

Not all the 100 smart cities were selected at the same time. The first 20 smart cities were selected on January 28, 2016 (see Fig. 19.3). The next 13 cities were selected as fast-

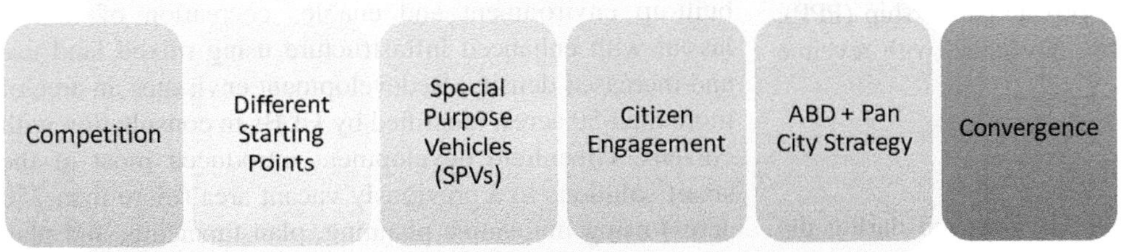

FIGURE 19.2 Six differentiators.

	Round 1	Fast Track Round	Round 2	Round 3	Round 4
Number of Selected Cities	20	13	27	30	10**
Selection Period	*Oct'15 to 28 Jan 2016	Feb'16 to 24 May 2016	May'16 to 20 Sept 2016	Mar'17 to 24 June 2017	Oct'17 to 19 Jan 2018

FIGURE 19.3 Roll out of smart cities. **Smart Cities Mission Launch Date was June 25, 2015. **Shillong was selected as the 100th Smart City in June 2018.*

track cities wherein states that did not find any of their cities selected in the first round nominated one city for the fast-track competition. Then, in round 2, the next 27 cities were selected. In round 3, another 30 cities were selected and, finally in round 4, the last 10 cities were selected, with the 100th smart city selected in June 2018. This phased approach of selecting smart cities allowed for cities to learn from each other's experiences of implementing smart city projects.

19.3.3 Special-Purpose Vehicles (SPVs)

Special-Purpose Vehicles (SPVs) are a new Administrative Mechanism to implement the smart city proposals. The implementation of the Mission at the city level has been done by an SPV with a responsibility to plan, appraise, approve, release funds, implement, manage, operate, monitor, and evaluate the Smart City development projects. Each smart city in India has an SPV, which is headed by a full-time CEO and has nominees from Central Government, State Government, and ULB on its Board. The States/ULBs ensure that:

- A dedicated and substantial revenue stream is available to the SPV to make it self-sustainable and evolve its own credit-worthiness for raising additional resources from the market.
- Government contribution for Smart City is used only to create infrastructure that has public benefit outcomes. The execution of projects has been done through joint ventures, subsidiaries, public–private partnership (PPP), turnkey contracts, etc., suitably dovetailed with revenue streams.

19.3.4 Citizen engagement

More than 150 million citizens were consulted during the smart city challenge process to prepare the smart city proposals. The SCM emphasized the development of the Smart City Proposal as a cocreation process, resulting in the evolution of a plan, which reflects citizen consensus and support for the vision and goals enshrined in the plan. This engagement focused on seeking citizen's responses to meet their aspirations on innovative solutions for improved livability and governance. Using the MyGov platform, residents of all the selected smart cities could post their ideas and suggestions pertaining to smart solutions in infrastructure services. This covers water supply, wastewater recycling and conservation, energy management, sanitation, including solid waste management. It also includes efficient urban mobility, robust IT connectivity and digitization, safety and security of citizen particularly women elderly, smart street lighting, public WiFi network, smart parking, development of nonmotorized transport, sustainable environment, intelligent traffic management, health and education services using IT-enabled smart applications, innovative use of open spaces, or any other feature related to city infrastructure and services identifying creative and innovative solutions to common urban challenges.

19.3.5 ABD (area-based development) + Pan City strategy

Area-based development in the SCM is city improvement (retrofitting), city renewal (redevelopment), and city extension (greenfield development) projects. Retrofitting is undertaken by introducing planning in an existing built-up area to achieve Smart City objectives, along with other objectives to make the existing area more efficient and livable. In retrofitting, an area consisting of more than 500 acres is identified by the city in consultation with citizens. Redevelopment strategy affects replacement of the existing built-up environment and enables cocreation of a new layout with enhanced infrastructure using mixed land use and increased density. Redevelopment envisages an area of more than 50 acres, identified by ULBs in consultation with citizens. Greenfield development introduced most of the smart solutions in a previously vacant area (more than 250 acres) using innovative planning, plan financing, and plan implementation tools (e.g., land pooling/land reconstitution) with provision for affordable housing, especially for the poor. Greenfield developments are primarily required around cities to address the needs of the expanding population. A Pan City initiative is one in which one solution is applied to areas that cover large parts of the city, if not the entire city. Pan City development envisages the application of selected smart solutions to existing citywide infrastructure. The application of Smart Solutions involves the use of technology, information, and data to make infrastructure and services better.

19.3.6 Convergence

Convergence has been one of the biggest differentiators in the implementation and financing of smart cities in India, both for sectoral and financial convergence. It is believed that comprehensive development occurs in areas by integrating the physical, institutional, social, and economic infrastructure. Many of the sectoral schemes of the Government converge on this goal, although the paths are different. There is a strong complementarity between the AMRUT and the SCM in achieving urban transformation. While AMRUT follows a project-based approach, the SCM follows an area-based strategy. Similarly, great benefits were derived by seeking convergence of other Central and State Government Programs/Schemes with the SCM. At the planning stage itself, cities sought convergence in the SCP with AMRUT, Swachh Bharat Mission (SBM), National Heritage City Development and Augmentation Yojana (HRIDAY), Digital India, Skill development, Housing for All, construction of museums funded by the Culture Department, and other programs connected to social infrastructures.

19.4 Various sources of finance for an smart city

Financing by the GOI plays a vital role in creating an enabling environment for implementation of smart city concepts and solutions for the benefit of the common man.

The SCM is operated as a Centrally Sponsored Scheme (CSS) and the Central Government has given financial support to the Mission to the extent of INR 48,000 Crores (about USD 7 Billion) over 5 years, i.e., on an average INR 100 Crores (USD 15 Million) per city per year. An equal amount, on a matching basis, has to be contributed by the State/ULB; therefore, nearly Rs 100,000 crores (USD 14 Billion) of Government/ULB funds have been made available for smart cities development.

The project cost of each Smart City proposal varies depending upon the level of ambition, model, and capacity to execute and repay. Both the Center and State government grants are also leveraged to attract funding from internal and external sources. The success of this endeavor depends upon the robustness of the SPV's revenue model and the comfort provided to lenders and investors. The complementary sources of financing for the 100 Indian smart cities, apart from their primary sources of finances, are as follows:

- The grant provided by the GOI to every city SPV in a phased manner to the tune of INR 500 crores (~75 Million USD)
- The matching contribution by states/ULBs to the tune of INR 500 crores (~75 Million USD)
- States/ULBs own resources from collection of user fees, beneficiary charges and impact fees, land monetization, debt, loans, and more.
- Innovative finance mechanisms such as municipal bonds with credit rating of ULBs, Pooled Finance Mechanism, Tax Increment Financing (TIF).
- Private sector through PPPs.
- Additional resources transferred due to acceptance of the recommendations of the 14th Finance Commission (FFC)
- Leverage borrowings from financial institutions, including bilateral and multilateral institutions, both domestic and external sources
- National Investment and Infrastructure Fund (NIIF)
- Convergence with other central Government schemes such as SBM, AMRUT, National Heritage City Development and Augmentation Yojana (HRIDAY)
- International donor agencies

19.5 The SPV—an Agra case study

The Smart City selection process begins when a city administration based on interdepartmental stakeholder discussions, studies, and citizen consultation creating the Smart City Proposal (SCP) [10]. This proposal contains all the projects with their cost estimates and the total budget. It is submitted to MoHUA for assessment against a predefined checklist. Based on a smart city challenge contest, MoHUA nominates the Smart City, and the funding follows. The SPV of that city is then formed as a public limited company. This SPV is tasked with responsibility and authority to coordinate with all concerned agencies and stakeholders to detail out and implement the projects. The SPV at this point hires a Project Management Consultant (PMC) to undertake further work. The PMC works with the SPV to create a Request For Proposal (RFP), which further details the proposal and brings a Master System Integrator (MSI). The MSI puts together all the solutions identified by the SPV. The MSI working with the RFP brings in relevant Solution Providers for the different solutions. This process is illustrated in Fig. 19.4. Detailed Project Reports (DPRs) would be created for further tendering and execution.

SPVs have not been used in the overall city administration in India until they were introduced as the implementing agency of SCM. So, in the area of performance, the efficacy of SPVs is yet to be proven. By its strict definition, it is a legal entity (usually a limited company or a limited partnership) created to fulfill a narrow, specific, or temporary objective [11]. The genesis of SPVs occurred with "Junk-Bond King" Michael Milken and his firm

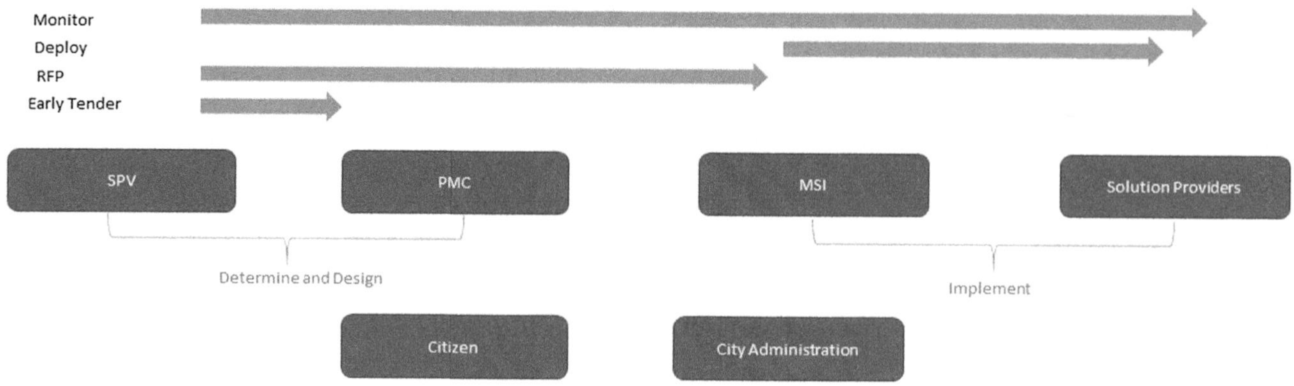

FIGURE 19.4 Process of implementation.

Drexel Burnham Lambert [12]. The success of an SPV in dealing with conflicts depends on two factors:

- The quality of the legal institutions and laws.
- The particulars of each relationship and contract affecting risk perceptions of debt holders. SPV for project planning and implementation focuses more on project finance and delivery [13].

At Agra, as required by the SCM's guidelines, the Special-Purpose Vehicle of Agra Smart City Limited (ASCL) was incorporated on January 9, 2017.

19.5.1 Structure of the SPV in ensuring financial control

The city level SPV was established as a Limited Company under the Companies Act, 2013. It was promoted by the State and the ULB jointly, both having 50:50 equity shareholding. This shareholding pattern is to be maintained at all times. Private sector or financial institutions can be considered for an equity stake in the SPV, provided the State and the ULB's share is equal to each other, and together the State and ULB have majority shareholding and control of the SPV. For example, the State/UT:ULB: Private sector shareholding can be in the ratio 40:40:20 or 30:30:40. Ratios such as 35:45:20 or 40:30:30 are not permitted since State/UT and ULB shares are not equal. Ratios such as 20:20:60 are also not permitted since the State/UT and the ULB together do not have majority shareholding. In addition to equity, the State/UT can provide its contribution to the SCM as a grant to fulfill the state governments' responsibility for ensuring availability of funds for the mission and for ensuring the financial sustainability of the SPV.

19.5.2 Structure of the SPV in ensuring accountability

From a financial management perspective, Central and State government money is transferred by their respective machineries to the SPVs account. The impact of the powers bestowed to an SPV in decision-making and utilization of funds were instantly visible to the authors who were members of the Project Management Consultant team (PMC) to ASCL. Personnel of the SPV were mandated to appraise their decisions to all the stakeholders that included members of the board from various government departments in the city as well as other representatives from the State and Central government.

The SPV structure also ensured greater accountability in the usage of the funds given to them. Characterized by the structure of the SPV and the possibility of an inquiry by central investigative agencies such as the Central Bureau of Investigation (CBI) and the Directorate of Enforcement (ED), it was ensured that the funds were not misused even though the SPV's had large government grants in their bank accounts. There was strict monitoring and control on the usage of funds by the management structure imposed by interdisciplinary stakeholders. This was further enhanced by monitoring of the Central Government of India through an online platform, which all the SPVs had to update every fortnight. On the one hand, there was strict rigor, and on the other, the SPV structure gave complete flexibility to the cities for implementing and managing the smart city projects.

19.6 ICCC project of Agra smart city limited

One of the primary objectives of Agra smart city was to enhance the safety and security of the citizens of Agra as well as of the tourists from across the world. These tourists visit Agra for the UNESCO World Heritage Monument, the Taj Mahal. Other objectives of the ICCC were to bring law enforcement and public awareness among the citizens of Agra on various aspects of Traffic Rules, Safe Driving, Solid Waste and Sanitation. All of this aimed at enhancing and improving the efficiency of municipal services and promoting a better quality of life for its residents.

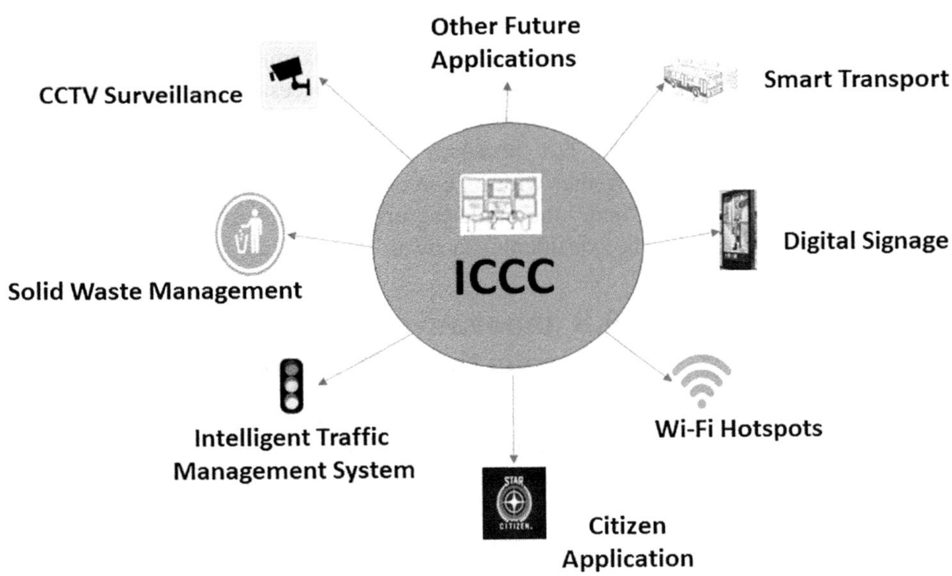

FIGURE 19.5 Integrated Command and Control Center.

Following a competitive bidding process to design, develop, implement, and maintain the Smart City System on a turnkey basis for 5 years, Agra Smart City Company Ltd identified an MSI. One of the mandates of the MSI was to develop an Integrated Command & Control Center (ICCC) [14]. Significant components of this were the CCTV camera system for City Surveillance, and Law Enforcement managed by the Police force. A unified dashboard of Solid Waste Management activities, providing Proactive Monitoring of disasters or catastrophic situations, were other areas where the ICCC was mandated to address. In addition, the ICCC managed VIP and Emergency Movement and controlled related utilities such as Automated Traffic Signal and Public Transport Movement. Fig. 19.5 provides a bird's-eye view of the solutions the ICCC was expected to control from a central facility.

Together, with the above initiatives, the ASCL ICCC also intended to embrace Artificial Intelligence (AI) and Deep Learning technologies to analyze the huge amount of data and metadata from various systems. Multiple surveillance cameras across several locations are saved as signatures and can be reproduced rapidly when time is of essence. Actionable intelligence derived from such technologies could be used to solve many local problems such as "Lapka" culture (culture of touts), locating and tracking a suspect's movement, traffic violations around critical infrastructures, monitoring activity around heritage sites, and others.

19.7 The tender evaluation process by cities

To issue projects, the Smart City SPV constitutes a tender evaluation committee that evaluates the bidders' responses. The tender evaluation committee evaluates the responses to the RFP and all supporting documents/documentary evidence. The steps involved in the evaluation process are as follows:

19.7.1 Stage 1: prequalification

Stage 1 of the evaluation process involves validating the documents required for participation in the bidding process. This is supported by Bid Security and the Earnest Money Deposit (EMD).

19.7.2 Stage 2: technical evaluation

The technical bid is evaluated only for the bidders who have made it through Stage 1. The committee reviews the technical bids of the short-listed bidders to determine whether they are substantially responsive. Bids that are not substantially responsive are liable to be disqualified at SPV's discretion. Technical solutions proposed in the bid document are evaluated as per requirements specified in RFP. Bidders submit their detailed Approach, Methodology and Solutions proposed as per the technical evaluation framework.

Each technical bid is assigned a score out of a maximum of 100 marks. Only the bidders who get an overall technical score of 70 or more, and minimum 50% in each section of the Technical Evaluation Framework qualify for the next stage—the Commercial Evaluation stage. Failure to secure minimum marks shall lead to rejection of the bid.

19.7.3 Stage 3: commercial evaluation

All technically qualified bidders are notified of their progress and are invited to participate in the opening of the

TABLE 19.2 Normalized scores in commercial bidding evaluation.

Bidders	Total cost of bid	Calculation	Normalized commercial score
Bidder-1	110	(110/110) * 100	100
Bidder-2	140	(110/140) * 100	78.57
Bidder-3	160	(110/160) * 100	68.75
Bidder-4	130	(110/130) * 100	84.61
Bidder-5	150	(110/150) * 100	73.33

TABLE 19.3 Arriving at QCBS scores.

Bidders	Technical score	Normalized commercial score	Final score (80:20)
Bidder-1	92.6	100	94.08
Bidder-2	94.7	78.57	91.47
Bidder-3	84.2	68.75	81.11
Bidder-4	100.0	84.61	96.92
Bidder-5	76.8	73.33	76.10

commercial bid. These commercial bids are now opened on the prescribed date and time and reviewed to determine whether the commercial bids are substantially responsive. Bids that are not substantially responsive are liable to be disqualified at ASCL's discretion. The method used to assign a score is as follows:

Normalized Commercial Score of a Bidder = [Lowest TCB/Bidders TCB] * 100

(adjusted to 2 decimals)

The bidder that has quoted that the lowest Total Price (Capex + Opex) shall be treated as L1. The bid price shall include all taxes and levies and shall be in Indian Rupees and mentioned separately. This is illustrated in Table 19.2.

19.7.4 Stage 4: final score calculation through QCBS

The final score is calculated through Quality and Cost-Based Selection (QCBS) method based on the following weightage:

Technical: 80%, Commercial: 20%

Final Score = (0.80 * technical score) + (0.20 * normalized commercial score).

The bidder with the highest final score is considered as the successful bidder. In the example shown in Table 19.3, Bidder 4 is considered a successful bidder. This is not because Bidder 4 provided the lowest commercial price (or L1), but because of better technical prowess proven together with its commercial value. In case final scores are "tied," the bidder securing the highest technical score will be considered for award of the Project.

19.8 Innovative quality cum based selection (QCBS) methodology

As observed from other similar Smart City tenders, the 80/20 QCBS method of evaluation that was employed in Agra was a game-changer in the Indian tendering process. This concept initially met with a lot of skepticism and even criticism. Several high-level discussions with the state government officials at Lucknow, the state capital, were held to prove the efficacy of such an approach. Once the capability of such a model was successfully demonstrated and adopted, the 80/20 QCBS methodology was made a norm across all the MSI RFP tenders in the state of Uttar Pradesh in India.

On the vendor side, a substantial number of vendors expressed their surprise that an Indian city was seeking to implement not only the latest technology and demanding future proof analytics for local city requirements, but they were also ready to recognize technical superiority and pay a premium for it. Vendors were typically habituated to the common L1 or Least Cost Selection Methodology of tendering. A100% EPC tender with world-class payment terms was both a huge surprise and a great business opportunity for vendors. ASCL received bids for the MSI tender from five world-class bidders that included Bharat Electronics Limited, L&T Limited, Tata Projects, Tech Mahindra and NEC Technologies.

19.9 Payment terms and schedule

Winning the bid is only one part of the process. Design, implementation, and validation and approval are the other vital parts of the process. Payments are made to the vendor following strict adherence to the payment terms. In the case of ASCL, this involves:

- A request for payment to be made to ASCL in writing, accompanied by invoices describing the services rendered. Required documents are to be submitted pursuant to general conditions stipulated in the contract.
- ASCL will make the due payment generally within sixty (60) days after submission of an invoice or request for payment by the MSI.
- The currency for payment to MSI under the Contract is in Indian Rupees (INR) only. All remittance charges shall be borne by MSI.

TABLE 19.4 Payment terms.

Sr. no.	Scope of work	Timelines	Payment
A	Phase I		
1.	Advance	On resource mobilisation = T	10% of contract value
2.	Upon finalization SRS, FRS, and SDD	T+2 Months	As completed 10% of contract value
B	Phase II		
3.	On supply, installation and commissioning of each component	T + 6 Months	As completed 20% of contract value of each component
C	Phase III		
4.	After 3 months' testing and go live of all components	T + 9 Months	As completed 20% of contract value of each component
D	Phase IV		
5.	Operations and maintenance of 60 months from the date of go live	T1 + 60 Months	40% of contract value in equal quarterly instalments to be paid over a period of 60 months

T is the date of signing of contract; T1 is the date of go live.

- In case of disputed items, the disputed amount shall be withheld and shall be paid only after the settlement of the dispute.
- Any penalties/liquidated damages as applicable for delay or nonperformance shall be deducted from the due payment of the respective milestones.
- Taxes as applicable shall be deducted as per prevalent rules and regulations

The schedule of payments to the MSI after successful completion of target milestones (including specified project deliverables) are made as shown in Table 19.4.

19.10 Conclusion

Multidisciplinary projects such as Smart Cities Urban projects may not necessarily have the only solution, but they do have a technologically and financially sounder alternative for providing a method for a way of life suitable to the city. The key to make our cities better is to exchange ideas and learn from diverse stakeholders who are well versed with their domain and the city. It naturally follows that an integrative approach like this is necessary for technologically appropriate, financially viable, socially inclusive, and ecologically sustainable models.

19.11 Summary

The innovative approach adopted by the Indian cities makes for substantial learning for other countries in the urban transformation of their cities. The strategic use of competition at the smart city proposal stage, citizen participation, SPV creation, enabling PPP, involving PMCs, and implementation of the smart solutions are all good instances of glocalization (adopting global best practices locally). However, a change in citizen and stakeholder mindset is the most crucial element to transforming the city to one that is smart and sustainable while retaining its unique identity.

The impact of smart city projects, such as the Integrated Command and Control Center, is seen in the manner that the cities leveraged them to manage the COVID19 crisis. In 69 cities, newly built ICCC infrastructure rose to the occasion as effective crisis management war rooms. These provided the much-needed data, dashboards of hotspots, applications for contact tracing, along with other services crucial to the pandemic management.

Citizen services, clean air, housing, safety, power, water, sanitation, mobility, green spaces, skill development and job creation, have emerged as core areas of smart cities. Healthy start-up support ecosystems, building resilience in the cities by keeping a check on crime, preventing flooding, and alleviating poverty are other critical aspects. Through financial support from the Center, the State, and with the help of convergence from key urban programs, the mission provided the much-needed impetus and catalysis for the transformation of Indian cities. In total, 5151 smart projects in 100 cities with investments of about USD 28.09 Billion (INR 205,018 Crores) [2] is a significant milestone in urban rejuvenation. When seen within the contexts that urban areas are expected to house over 500 million of India's population and contribute 75% to its GDP by 2030 [15], it is a mere beginning.

Finally, let us move on to the real interactive part of this chapter: review questions/exercises, hands-on projects, case projects, and optional team case project. The answers and/or solutions by chapter can be found in Appendix G.

19.12 Chapter review questions/exercises

True/false

1. True or False? A city is a social construct whose effective functioning is determined partly by the rules of the state, partly by the values of society, and partly by its economic structures.
2. True or False? The Smart Cities Mission has been championed and implemented under the aegis of the Ministry of Housing and Urban Affairs, the erstwhile Ministry of Urban Development, as part of the transformation of urban landscape goal pursued by the Government of Australia.
3. True or False? The Smart City Mission was launched after some consultation with various stakeholders, including State and City Governments.
4. True or False? Cities are selected through a competitive mechanism.
5. True or False? Not all of the 100 smart cities were selected at the same time.

Multiple choice

1. Special-Purpose Vehicles (SPV) are a new Administrative Mechanism to implement the smart city:
   a. Developments
   b. Proposals
   c. Projects
   d. Investment
   e. Policy
2. More than _____ million citizens were consulted during the smart city challenge process to prepare the smart city proposals.
   a. 250
   b. 350
   c. 150
   d. 450
   e. 550
3. Area-based developments in the Smart Cities Mission are city improvement (retrofitting), city renewal (redevelopment), and city extension (greenfield development):
   a. Density
   b. Envisages
   c. Bodies
   d. Citizens
   e. Projects
4. Convergence has been one of the biggest _________ in the implementation and financing of smart cities in India, both for sectoral and financial convergence.
   a. Differentiators
   b. Economic growth
   c. Infrastructure
   d. Computer applications
   e. E-economics
5. Financing by the ________ plays a vital role in creating an enabling environment for implementation of smart city concepts and solutions for the benefit of the common man.
   a. ICT
   b. GOI
   c. IoT
   d. IoP
   e. IoS

Exercise

Problem

Identify the reasoning behind the following conclusion: Government leaders, citizens, and business professionals across India understand that their infrastructure needs major reinvestment and modernization.

Hands-on projects

Project

Do research: Propose various sources of funding that could be available in the world economy and their feasibility, considering the Indian Economy, which can be adopted for a generation of required funds.

Case projects

Problem

Suggest a robust strategy for financing infrastructure in smart cities.

Optional team case project

Problem

Identify the barriers and potential solutions for the financing of smart city innovative technological solutions, in particular in the areas of low-carbon energy and transport, as well as ICT.

Acknowledgments

We are particularly grateful to Renuka and Srivenkatesa Manchiraju for their help in editing this essay and importantly the shot of enthusiasm at the right time. We also thank Sushil Kumar and Ritu Mohanty Padora for their support and appreciate information from Gaia Smart Cities in their role as PMC's of Agra Smart City Limited.

References

[1] Bholey M. India's urban challenges and smart cities: a contemporary study. Scholedge Int J Bus Pol Govern ISSN 2394–3351 2016;3(3):17–38. https://doi.org/10.19085/journal.sijbpg030301.

[2] Ministry of Housing and Urban Affairs, Government of India Data. http://mohua.gov.in/, Viewed on Feb 24, 2021.

[3] Agra Municipal Corporation to Table Rs 660 Crore Budget in Next Meet. https://timesofindia.indiatimes.com/city/agra/amc-to-table-rs-660-crore-budget-in-next-meet/articleshow/74698683.cms.

[4] 3.19 Project Cost Sheet, Agra Smart City Proposal, The Smart City Challenge Stage2. https://smartnet.niua.org/content/884491e6-2981-4a37-86fe-481cea313a58.

[5] Top 10 richest cities in India in 2020. https://www.indiatoday.in/information/story/top-10-richest-cities-in-india-in-2020-1726702-2020-09-29.

[6] BMC Budget 2021: How Much India's Richest Civic Body Plans to Spend This Year? https://www.hindustantimes.com/india-news/bmc-budget-2021-how-much-india-s-richest-civic-body-plans-to-spend-this-year-101612352349149.html.

[7] BBMP Presents Rs 9,287.81 Crore Budget for Bengaluru. https://www.deccanherald.com/city/top-bengaluru-stories/bbmp-presents-rs-928781-crore-budget-for-bengaluru-967171.html.

[8] SCP Financial Plan Summary, Bengaluru Smart City Proposal, Smart City Challenge Round 3. https://smartnet.niua.org/content/c8fdaabb-767c-4854-befc-9e9966c28282.

[9] 15th FC Report for 2021-26. https://fincomindia.nic.in/writereaddata/html_en_files/15thFcReportIndex.html.

[10] Smart City Proposal template of Agra Smart City Limited. https://smartnet.niua.org/sites/default/files/resources/Agra_SCP.pdf.

[11] UNECE. Chapter 4 -Special Purpose Entities. 2015. p. 39–65.

[12] Healy PM, Palepu KG. The fall of Enron. J Econ Perspect 2001;17(Dec 17):30–6. https://doi.org/10.1257/089533003765888403.

[13] The Rationale of SPV in Indian Smart City Development, Kranti Kumar Maurya and Arindam Biswas. www.researchgate.net/publication/333089308_The_Rationale_of_SPV_in_Indian_Smart_City_Development/link/601cf8e0a6fdcc37a8026765/download.

[14] Request for Proposal for Selection of Master System Integrator (MSI) for Implementation of Intelligent Traffic Management System, Adaptive Traffic Control System, CCTV and Surveillance System, Solid Waste Management and Integrated with Command and Control Centre (ICCC) for Agra Smart City Limited. https://smartnet.niua.org/content/596b0821-956c-4b78-bd9e-66a32b26b342.

[15] About Indian Economy Growth Rate & Statistics. https://www.ibef.org/economy/indian-economy-overview, Viewed on Jun 19, 2021.

Part V

Governance: balancing top-down and bottom-up policy approaches

Chapter 20

Policy instruments for facilitating smart city governance by promoting platforms for bottom-up participatory governance

Lidia Mierzejewska and Magdalena Wdowicka
Adam Mickiewicz University, Poznań, Poland

20.1 Introduction

Cities face enormous challenges today, related more broadly to the need to take actions for sustainable development and more narrowly to ensuring prosperity, safety, public health, and creating conditions for social integration [1,2]. Meeting these requirements in a traditional way turns out to be ineffective, which is indicative of the need to seek new, innovative solutions in this respect. Therefore, the transformation of cities into "smart cities" is becoming increasingly popular among urban authorities and specialists in various fields worldwide [3,4]. This concept refers to many different dimensions of urban development. It embraces, for instance, such issues as the need for social inclusion, importance of social capital and rural development [5,6], the role of the creative industry in urban economic growth [7–9], but also such aspects as safety, greenery, energy, transport, effectiveness, etc. [3]. This is because the "smart city" conception is about ensuring high living standards with the maximum efficiency, reliability, and integrity of technical infrastructure systems and tailoring urban services to inhabitants' requirements.

Smart city is often associated with the application of modern technology to daily urban life. This concerns both technology for the functioning of various urban systems and information-communication technologies (ICT). Urban systems equipped with detector networks provide data and information, which, adequately processed, may enable the optimization of adopted and implemented urban policies (transport, parking, energy, space, education, housing, etc.). To make a "smart city," however, it is not enough to fill the urban space with modern technology. Its overriding aim is to serve residents, not to keep up appearances of innovativeness [10]. It is the needs and expectations of users, not available technologies, that should be the basis for designing infrastructure of urban services [11–14]. As no one knows the needs better than the residents themselves, they should be involved in urban governance, and new technologies ought to make it possible [15]. This is the best way to improve adaptation of policies and urban services to the needs of recipients, because the inhabitants themselves could take part in this process. A "smart city" requires a new method of governance, where the city "works internally and together with outside partners and citizens" [16]. The growing influence of technology in the functioning of urban systems and communication forces local authorities to rethink its role in a "smart city" [3]. Their role is most often determined as "smart governance."

This chapter indicates policies that should be implemented for smart city governance and platforms serving to support bottom-up participatory. It will include numerous references to cities adopting specific policies on smart governance, in which various instruments for supporting participation and bottom-up initiatives of inhabitants are applied.

The chapter has been divided into five sections. The first includes the explanation of the general idea of a "smart city" and the presentation of the scope of the research. The second section explains "smart city governance," its basic elements, and the role of ICT in it. The next one indicates three essential policies serving to implement smart city governance, which can be briefly described as increasing public participation with particular emphasis on its bottom-up forms, improvement in ICT technologies and the open data policy. The penultimate section discusses selected platforms dedicated to bottom-up participatory as exemplified by particular cities and measures they adopted. The fifth part concludes with the summary of the study conducted.

Smart Cities Policies and Financing. https://doi.org/10.1016/B978-0-12-819130-9.00037-1

20.2 Smart city governance—understanding, aspects and the role of ICT

The term "smart governance" denotes a complex process of institutional changes, and it refers most often to intelligent use of modern technologies in order to facilitate a decision-making process and create conditions for better cooperation between urban authorities and residents [15,17–19]. Thus, this is a new model of governance, based on cooperation, involvement, and public participation, which is supposed to replace red tape and centralized, keeping inhabitants at bay, power structures [20]. The different aspects of smart governance are emphasized in various approaches to the smart city, and they include the following [15]:

- The process of gathering different kinds of data and information related to public governance, for example, by means of various types of detectors monitoring the functioning of urban services [21];
- Public participation in decision-making processes, transparency of measures as well as the quality and accessibility of public services [3];
- Inclusion of inhabitants in the development of public services [5];
- Governance, which allows organizing and integrating measures within the "smart city" idea better [3,5].

Thus, smart city governance is based on the availability of data about the city, cooperation, responsibility, openness, transparency, and it also allows residents to influence decisions regarding the city they live in and the services they use in the widest possible participatory process.

The use of ICT plays a significant role in smart city governance. It should be understood as technologies that make it possible to collect, process, and send electronic information [22]. Free access to ICT is conducive to smart city governance through:

1. The development of new relation networks between public and private actors as well as inhabitants for cooperation, cocreation, and coimplementation of new public services and new initiatives;
2. The implementation of public tasks in a more effective, personalized, and participatory way; and also
3. Increasing public participation in the formation of development policies, their implementation and the improvement in the quality of services provided by the public sector [23–26]. Moreover, it provides the inhabitants with better access to information about the city, fosters the development of ICT competences, and intensifies social interactions making it possible for residents to be better prepared and more interested in engaging in common, also bottom-up, smart city governance.

The employment of some ICT solution can open new opportunities for citizens, actively create the city's future by introducing new forms of residents' activity, social inclusion, and greater accessibility of services for the disabled; it can also reduce infrastructural barriers and help share resources and information [27]. Therefore, ICT should not be treated as an element of a "smart city," but as a tool that is to increase the effectiveness of conducted activities and provided services in its area as well as to serve to achieve the basic aim, which is an active, innovative, tolerant, enterprising, open, and flexible society, prepared to face new challenges of the contemporary world [28,29].

It does not mean that traditional forms of communication are not important. There are numerous occasions when face-to-face interaction brings better results. In this context it is not surprising that despite a continuous increase in online engagement, there are still many citizen–public authority interactions taking place either by direct contact or by phone, letter, or by the representatives of a local community [19]. This way of communication is often preferred by residents who find it difficult to navigate the Internet and obtain relevant data and information. Naturally, the provision of an alternative, traditional method of engagement is also needed when Internet access is not universal [30]. Cohen [31] indicates three approaches of cities to the ICT use depending on entities generating activities in this respect, i.e., (1) companies dealing with the new technology sector, (2) local authorities, and (3) residents, describing three phases of smart city evolution as 1.0, 2.0, and 3.0.

Smart City 1.0 is the earliest stage of creating intelligent cities. Initiators of using technologies were large ICT corporations, which sold the newest technological solutions to urban authorities, aiming at effective city governance. Cohen highlights, however, that at this phase, city authorities were neither aware yet nor had appropriate qualifications to make full use of the acquired technologies and also did not know how to choose them to fit in with the needs of a specific city.

Smart City 2.0 is a stage in which urban authorities become initiators of the ICT use. In order to improve the living standards, they search for specific technological solutions for the problems diagnosed in their cities. At this stage, the effect of the implementation of advanced technologies is of key importance. According to Cohen, most cities introducing smart city projects today belong to the 2.0 generation. They can be distinguished by a great number of projects and programs related to the application of intelligent technological solutions, starting from public WiFi networks to intelligent traffic management, the use of big data to the promotion of electrical transport. The weak point of Smart City 2.0 is an excessive technocracy of cities, where citizens are not the first priority.

Smart City 3.0 is a phase in which urban authorities are open to a proactive attitude of inhabitants toward development. The role of local authorities at this stage is creating possibilities for using citizens' different potential. This is related to both encouraging residents to apply modern technologies and enabling them to create their own technological solutions (e.g., by open data). In Smart City 3.0, despite using innovative technologies to improve the standard of living in cities, there appear social, equality, educational, and ecological issues. Smart City 3.0 fits in with an increasingly popular sharing economy. This requires a change in the attitude of municipal authorities, which have to accept a growing position of residents. The change must take place not only at the mental level (authority—citizens), but first of all in communication, hence the dominant role of dialog, mediation, and deliberation. It is this phase that can be determined smart city governance.

The main smart city conceptions hold that the residents should be the architects of changes in a city. Thanks to high qualifications, competences, creativity, and ability to cooperate and backed by ICT, they will continuously try to improve the quality of city life. Social involvement of citizens is clearly visible especially in the third generation of Smart City distinguished by Cohen. Local governments search for new ways of increasing the engagement of inhabitants in the implementation of projects for sustainable development. Thus, there is a tendency to create favorable conditions for using a local community's potential, e.g., by:

1. Access to urban data collected by a local administration with their possible use on mobile applications;
2. Access to data and public information increasing transparency of the functioning of local authorities and municipal property management (e.g., by open data);
3. Spending public means, among other things, based on the participatory budget, where the self-government empowers the community to take certain decisions or implement policies; and
4. Encouragement to use modern technologies by educational projects, organized, e.g., for the digitally excluded persons [29].

20.3 Basic policies for smart city governance

The implementation of the assumptions of smart city governance requires that urban authorities adopt specific policies and use relevant instruments. This involves mainly policies aiming at (1) increasing public participation, (2) improving access to technologies, primarily ICT, creating greater opportunities for participation, and (3) sharing data and information on the city, which, in the participation process and with the use of proper technologies, can be consequently processed and used to achieve smart city objectives. It is worth emphasizing that these policies are not independent, but related to one another.

20.3.1 Public participation

The most important element of the smart city conception is the proactive participation of a local community that should be involved in the planning and implementation of the urban development policy. Public participation is a very broad notion, embracing both management processes and particular behavior and methods. In broad terms, participation is associated with democracy in general, and in narrow terms, it is limited to specific forms of engaging in the life of democratic society [32]. It is usually understood as the active participation of residents in making decisions, which directly influence the functioning of the community they belong to. It ought to be emphasized that public participation should be treated as a cycle of mutually related activities and not only as a single, one-time action.

Local actors (residents, entrepreneurs, organizations, associations) may take on different roles in the governance of urban development, starting from a passive attitude to active participation. A passive style means that residents are recipients of information provided by authorities who consult about the actions they intend to take. In the active style, on the other hand, a local community can make suggestions on the solutions proposed by self-governments, then it is given specific powers and may carry out assigned tasks. In the next stage, it becomes a partner the self-government cooperates with and finally, at the last level, it can initiate local transformations [33,34].

Nonetheless, for residents to take an active part in the participatory process, they need to be interested first in what is happening in their local environment [19]. The starting point is then (1) social engagement, which provides an opportunity for active (2) participation, (3) coproduction, and (4) cocreation of value, which should be recognized as the final goal of a smarter city [19,35–38].

Engagement can be conceptualized in many different ways, allowing for such aspects as participants, objectives, timing, or the degree to which inhabitants are involved in urban governance. What is taken into account in most approaches, however, is the last of the listed criteria, which allows distinguishing the following [39]:

1. Voice, when the public opinion may openly present the view on a specific solution (e.g., by questionnaires, focus groups, etc.);
2. Representation, when the public opinion is presented indirectly by representatives (e.g., councilors) or directly (by other agents, such as associations, foundations, etc.);
3. Choice, when particular individuals may directly influence the decisions taken or services they use, e.g., by coproduction.

However, this is not a question of identifying one approach toward engagement, which seems to be better than the other ones, but rather of applying the most appropriate approach in a given situation under specific circumstances [40]. Engagement should be understood as a certain continuum, which starts from top-down, technocratic measures, controlled by authorities, consisting in providing information through consultations and empowering citizens, and ends with coproduction, delegation of powers, and public control over decisions taken in which there is room for bottom-up activities (Table 20.1). Ultimately, this is about real public participation understood as redistribution of powers, aimed at the inclusion of those excluded from political and economic decision-making processes.

One of the best-known typologies of participation is the public participation ladder prepared by Sherry Arnstein [41] for implementation of planning processes. It includes:

1. Lack of participation (with manipulation and therapy as part of it),
2. Simulated participation (embracing informing, consultations, and appeasement), and
3. Socialization of powers (comprising partnership, delegation, and civic control).

Arnstein treats participation as a synonym for the civic authority. However, as the number of models and typologies of participation increases over time, traditional forms of participation, consisting of technocratic urban planning, including information or public hearing, have now been joined by new ones that are more focused on using the freedom and creativity of the inhabitants and therefore suitable for smart city governance. This involves, among other things, codesigning or crowdsourcing, included in collaborative urban planning or BarCamp, civic hacking, or urban hacktivism proving self-organized urban planning (Fig. 20.1).

TABLE 20.1 Engagement continuum.

Level of engagement	Outcome sought
Inform	To provide information to the public
Consult	To obtain feedback from the public on an issue or decision
Involve	To work together with the public to explore an issue or decision
Collaborate	To work in partnership with the public, sharing decision-making power
Empower	To place final decision-making power in the hands of the public

Reproduced with permission from the author. Andersson E. Engagement in health: roles for the public and patients. In Smart governance for health and well-being: the evidences. World Health Organization. The Regional Office for Europe; 2014. pp. 34–48.

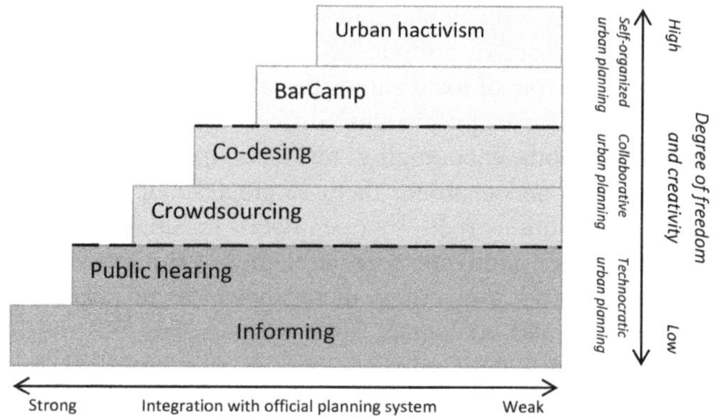

FIGURE 20.1 Degree of integration and creativity in urban planning. *Reproduced with permission from the author. Anttiroiko AV. City-as-a platform: the Rise of participatory innovation platforms in Finnish cities. Sustainability 2016;8(9): article 922. https://doi.org/10.3390/su8090922.*

This is an analogous approach to the degree a user is engaged in the preparation of the final product involving (1) designing for users, (2) designing with users, and (3) designing by users [43]. Designing for users may be associated to a certain degree with technocratic and more like top-down planning. We are dealing with bottom-up participation only in the two remaining cases. Designing "with users" can be linked to collaboration and coproduction, and "by users" with self-organization and cocreation of value.

Although coproduction and cocreation of value are often treated as synonyms in literature, they seem to be somewhat different. In the case of coproduction, an initiative usually lies with urban authorities, but ultimately, it is citizens who influence the decisions taken (policies adopted, urban services provided, etc.). In cocreation of value, on the other hand, a citizen acts as an initiator and the government as an actor that follows.

Another aspect related to participation is an analysis at which stage the society may enter the decision-making process. It is possible at the phase of the problem identification or making decisions about the way to solve it or the implementation of a specific solution (e.g., in the form of public or social services). It will greatly affect the role the participants of the process can play [30]. In smart city governance it would be desirable for residents to participate in all the abovementioned stages, although at each of them the participation forms will be probably slightly different.

Many studies show that processes engaging citizens to a greater degree, including stimulating bottom-up initiatives, such as cooperation, coproduction, or cocreation of value, give better results than traditional, technocratic consultations [44–46]. At the same time, participation turned out to be difficult to implement, causing cities to investigate the ITC potential as a way to boost social engagement. These technologies are often perceived as the most obvious and the best way to increase participation.

Thanks to ICT, the participation of residents in the planning process and city governance, carried out traditionally in the form of public meetings or consultations, has been extended to include new forms of participation conducted with online IT tools [47–50]. In effect, traditional expert planning loses in significance to bottom-up planning [49]. What gains in importance, however, are new communication channels such as social media, Internet platforms, applications, and other innovations by means of which urban authorities can actively engage residents in city matters and allow bottom-up participatory [51]. This is so, because ICT facilitates consultations for a large group making it possible to express opinions on a given subject and engage people who have little free time, work irregularly, or have mobility problems [40].

Participation, especially with the use of ICT, can be beneficial for smart city governance. It is favorable for [52,53]: (1) a more democratic decision-making process, (2) greater democratic legitimacy, (3) ensuring greater transparency of governance, (4) taking more intelligent political decisions based on more precise information, (5) increasing the sense of joint responsibility of residents for creating the development policy, (6) greater efficiency and the quality of public services, and (7) capability of local authorities to respond quickly to social needs [45,46,50,54,55].

Unfortunately, authorities of many cities often perceive the public as common information recipients, neither allowing for different views and proposals put forward by residents in the decision-making process nor providing feedback. This attitude results mainly from the underestimation of skills and knowledge of different actors in a city and also from the unwillingness of local authorities to share power with citizens, thus giving up part of the autonomy [50,56,57].

Therefore, it is not surprising that the research outcomes indicate that, for the time being, government initiatives, even those supported by ICT, bring modest results in terms of the participation level. This participation is limited mainly to such activities as informing, searching for political support and consultations. In effect, e-participation of residents in local decisions is practically nonexistent [14,50]. In conclusion, digitization itself does not mean extended and deeper civic involvement. Participation is constrained because residents are unwilling to get involved in public affairs and do not trust representatives of political leaders [24,50]. Thus, the success of smart city governance depends to a great extent on mutual reciprocity and trust [19].

20.3.2 Technology

The ICT popularization in smart city governance requires creating proper conditions. The most important include [40]:

- Access to the Internet, including Wi-Fi, in public places, such as railway stations, airports, buses, trams, etc.;
- Access to software to improve the functioning of urban systems and service availability as well as the functioning of inhabitants and visitors;
- Having appropriate devices enabling potential users to access the Internet and use available applications (computer, tablet, mobile phone etc.); and
- Ability to use these devices and applications.

The basic factor is, naturally, the development of appropriate technical infrastructure (mainly broadband networks), but an important role is also played by software (usually in the form of various platform types and Internet applications). There are many examples of platforms, applications, and other IT tools that can be used in smart city governance. Even more are expected with the development of new information and communication technology, nanotechnology, and new models of management and business solutions [58]. However, the possibility of public participation considering these devices is quite diversified.

At present, many new channels used for the communication with residents and available by ICT, such as "e-administration," "e-governance," or "e-democracy," provide mainly the mechanisms for commenting or voting, giving the residents an opportunity to choose options proposed by urban authorities (one-way communication). Gathering opinions obtained from these channels, urban authorities can test the level of social acceptance for their proposals, which goes slightly beyond conventional electronic voting tools and collecting petitions online, but it is still a top-down participatory form. It can be exemplified by Internet portals (including e-mail service within) and different types of visualization (e.g., Virtual Reality), which are aimed at building support for policies planned by authorities. Numerous applications that collect geographical data are also used exclusively to receive information by means of sensors placed in urban infrastructure networks, enabling joint mapping, monitoring the emission of greenhouse gases, testing the level of power consumption, etc. [13,59]. These tools play an important role in city governance; however, they do not support bottom-up initiatives of residents and give them no real influence on political decisions. They also do not use their competences, skills, creativity, local and professional knowledge, their awareness of local problems, and ideas to solve them, that is, generally speaking, the potential of a local community. They do not make it possible for residents to cocreate services, values, and open innovations, which could be used while formulating and implementing urban policies [49,60].

In smart city governance, what is most desirable are applications for two-way communication processes in which participants can cooperate and interact with one another. It can be illustrated by social media, the use of which not only improves the free flow of information but is conducive to obtaining a variety of opinions, a sociopolitical debate, and free speech [12,14,50]. The proposals, suggestions, and initiatives of residents can be helpful in making more informed political decisions and can ensure the higher level of services provided within the city area [50]. The existing participatory approaches often lack such feedback.

It should be considered, however, that despite growing access to Internet, it is not always possible for all groups of participants to take part in these processes due to time, geographical, or technological constraints. Unfortunately, sometimes urban authorities mistakenly assume that the mere presence of internet channels will automatically lead to greater participation in urban development processes. As the use of ICT tools limits the participation to those with access to digital devices, what is often mentioned is the problem of digital exclusion [50,61].

In the case of Ghent, local authorities recognized the priority of the issue of digital exclusion. The solution to this problem was found in the Ghent 2020 strategy by distinguishing the digital inclusion policy and the adoption of programs for the implementation of this policy, namely: Digitaal.talent, Train the Trainer, and Digikriebels. Under these programs, selected families were given computers with software, free Wi-Fi access throughout a year, and the opportunity to take part in ICT training held in the local points of the city. This project, dedicated primarily for the poor, brought not only an increase in the income of these residents, but also contributed to the improvement in their life conditions and the economic competitiveness of local communities. Therefore, new technologies may also lead to the reduction of divisions and social differences, provided the residents can have not only access to ICT but also necessary digital education. In the era of information society and the development of city applications enabling participation in making decisions related to urban development, depriving someone of the right to participate in the life of a local community means social exclusion the Ghent authorities want to minimize.

The problem of digital exclusion can be solved to some degree by smartphones increasingly often used also in participation processes, provided the available applications are dedicated to mobile devices as well. Smartphones enable access to various Internet platforms and information as well as applications put on them, ensuring connection with Internet resources and flexibility in voicing opinions on different issues. It is expected that smartphones help to eliminate the digital chasm and poverty, which is especially important in developing regions. This communication form is not only more comfortable but also more attractive than traditional forms of participation, particularly among young people [62,63]. It simultaneously promotes grassroots open governance, which enables citizens to cocreate, codesign, and be more deeply involved in planning processes.

In addition to the relevant technological infrastructure, the available Internet tools need to be transparent and user-friendly, and residents and civil servants alike must be able to improve their knowledge and skills regarding ICT [50,64]. What are often identified are insufficient competences in this respect on the side of a local government, unfavorable for participation, and urban coproduction. Therefore, before the implementation and formulation of smart governance, self-governments must become organizations that learn [65–67].

20.3.3 Policy of open government data (OGD)

Today's cities are drivers of a new urban economy, called a data-based economy. This means that a more important role in their economic and social development is played by a various kind of data, which are treated as a valuable resource. These data come, among other things, from the networks of sensors placed in different urban systems and many new digital services generating millions of interactions and transactions carried out every day in cities. The possession of Big Data is not particularly important if this knowledge cannot be uncovered and used. However, the analysis of this data by means of manual methods, such as spreadsheets or simple database queries, would not be possible. Only the dynamic development of processing power of computers, artificial intelligence, and related fields enabled intelligent and automatic exploration of Big Data volumes. Data from urban systems can be used to improve the quality of city services, the functioning of a city as a whole, and the transformation of cities into more sustainable smart places. Therefore, the capability to collect, analyze, and share gathered data and information is becoming the basis for modern, innovative city governance [20]. The policy of open data is aimed primarily at creating conditions for their innovative application and inspiring entrepreneurship and creativity, which means grassroots initiatives. There are two key aspects of the implementation of the open data policy [68]:

1. Digital transformation, specifically city governance with the use of digital technology, making it possible to collect urban data and share them, which requires a fundamental change in relations between the public sector and the private one as well as residents; and
2. Provision of a relevant open-source platform, which will ensure technological functions necessary to implement the objectives of digital transformation and smart city requirements, enabling the adoption of smart city governance.

It happens often, however, that cities pursuing data-based urban projects grapple with various kinds of problems. One of them is related to the outsourcing of key city services to large providers (e.g., smart parking, smart street lighting, etc.). In such circumstances, control over all the data that can be obtained is limited. What is more, data are gathered by means of various technologies and are dispersed among different departments of the Town Hall. The integration of these data poses a challenge in terms of both technology and governance. They should be widely shared in order to be analyzed, processed, and used effectively in the decision-making process. Both formal (from offices) and informal data (including private resources) often lost and not used during the preparation of urban policies can be significant for city governance [55].

Smart city governance also involves encouraging external users (residents, programmers, other companies, and organizations) to use open data and share them, design digital services, and introduce digital innovations, naturally in compliance with the principles of transparency, participation, and cooperation [20]. What can be helpful in effective city governance are data on public safety, health, energy and water use, waste production, road traffic, the movement of residents, education, computer and cell phone use, social media interaction, e-mails, financial and retail transactions, sustainable development attitudes, etc.

Providing residents with city data and background information (e.g., geographical coordinates) linked with the relevant technology creates opportunities for the analysis of these data in real time, their processing and as a result, generates new knowledge, ideas, services, and projects. Thus, it makes it possible to include residents in the active preparation of policies and coproduction of urban services embracing those services that could not exist without ICT (data-based and electronic services).

The policy of open data was introduced, among other cities, in Ghent (Belgium), where Open Data Portal was developed and shared. This portal is an Internet website that contains a lot of spatial and textual data on various topics, which are available both directly through a web browser and in the form of downloadable files in the formats of KML, CSV, DWG. It enables Ghent's residents to learn in a simple and transparent way about the data resources obtained and processed by the city. The available data are arranged in the following eight categories: (1) civil matters, (2) culture, sports and entertainment, (3) mobility, (4) education and child care, (5) city, politics and administration, (6) social welfare and health, (7) work, and (8) others. The largest possible data collection, used not only by NGOs or companies, but also by programming enthusiasts—freelancers, is available free of charge. This is so, because it has been assumed that the more actors will generate ideas about applying these data, the more valuable urban applications will be created.

The implementation of the open data policy in Ghent was related to the costs incurred for the IT infrastructure development, the arrangement of the data collected, protection of personal information, and the project promotion. However, the expenses proved to be disproportionately low compared to the multiple benefits it has brought, on both sides. What is beneficial is not only access of urban dwellers and other entities to data on public services and administration, which is a value in itself, but also the fact that these data can be analyzed, processed, and used when preparing different kinds of applications facilitating the application of urban services. This results in the dissemination of data-based tools and platforms designed to introduce more "responsive" (able to react quickly on receiving a user signal) and thus better tailored to the needs of users of urban services. The advantage also includes greater public participation and involvement of residents and other entities in city governance and development of many useful applications, facilitating the functioning of the city and its inhabitants.

Moreover, this policy relieves public entities, which are not usually keen on and do not have means, knowledge, and often also ideas how to use available data and prepare various applications on their basis. This is also a good step toward supporting innovative companies, commonly called start-ups, which create many valuable ideas but do not have money to implement them.

City Hall employees in Ghent derive benefits from the adoption of the open data policy as well. The publication of urban data via the portal reduces the need to submit requests for access to public information, which translates into fewer applications to be processed, especially from students and research companies.

20.4 Platforms supporting bottom-up participatory governance

The term platform denotes a physical, technological, or social base on which economic and sociotechnical processes are built [20]. This concept is not clearly defined, however, and is often treated as a synonym of such notions as portal or application. ICT platforms, which can be provided with data from many reliable sources, including the Big Data platform, play a special role in a smart city. Different organizations (e.g., start-ups) may activate their applications there. Such a platform ensures access to data and enables their processing, transmission, storage, and often also visualization as well as the preparation and implementation of new services for a smart city. Some of these platforms (including FIWARE) operate successfully in over 100 cities worldwide [68]. Nevertheless, no ICT platform in itself is a smart city yet [69]. ICT platforms facilitate planning, governance, urban policy and stimulate innovative processes and bottom-up initiatives. They assist in meetings, public debates, and institutionalized procedures of spatial planning, including revitalization. Open participatory platforms, usually organized by public authorities, are primarily dedicated to these types of functions.

Open participatory platforms fulfill various functions for the implementation of the assumptions of smart city governance. The most important are [49,70,71] as follows:

- Providing open access and encouraging interested parties to use their resources;
- Increasing individual, group, and community creativity;
- Facilitating open dialog and sharing ideas;
- supporting joint decision-making and integration of policies;
- Improving communication; and
- Better understanding and presentation of proposed spatial changes.

They increase the ability of citizens to cooperate and participate effectively in urban decision-making [49]. There are thousands of such platforms. The platform thinking gains significance because it enables using collective intelligence, which stems from the diversity of local communities and other interested parties involved in local participatory processes [42].

An important role in open platforms is played by open-source software. It is a nonrivalry system, the value of which grows with the number of persons that start to use it. The large number of users guarantees constant testing and contributes to the improvement in its quality (detecting faults, suggesting improvements, testing performance in various conditions, etc.). Goods of the highest quality in this environment are time, skills, creativity, and intellect of people engaged in the development of programs, usually in the form of different applications, but also other tools, prepared together with residents or by residents alone.

The further part of the study will present selected platforms, tools, and techniques (mainly applications), which are part of (1) coproduction and (2) cocreation of value. This is because it has been assumed that activities enabling a real influence of city-dwellers on urban decision-making may stimulate bottom-up participatory governance. The designation of specific instruments supporting grassroots initiatives for the mentioned participatory forms is for various reasons problematic, making it not fully detachable and can be treated only as an example.

20.4.1 Coproduction

Coproduction takes place when inhabitants cocreate products or services, being their recipients at the same time. They usually communicate with the city authorities by various types of Internet platforms and applications placed there and also other instruments and techniques, such as geoparticipation, crowdsourcing, crowdfunding, hackathons, Living Labs, or gamification.

A special role in coproduction is played by applications in the development of which residents participate themselves, sharing their opinions with authorities as well as observations and ideas about how to solve different kinds of problems, also those related to the functioning of a given tool. These applications, called horizontal, are targeted at the large number of users with different knowledge and skills [68]; there are many of them in cities aspiring to the name of smart city. This can be illustrated by the following available smartphone applications "Glasgow Cycling App" and "Glasgow Walking App," which were developed in cooperation with Glasgow residents.

The abovementioned applications ("Glasgow Cycling App" and "Glasgow Walking App") together with the dedicated educational tool are targeted at bikers and pedestrians [19]. Owing to various benefits of walking on foot and cycling, what are particularly promoted in a smart city are active traveling forms, whether for transport, leisure, or sports purposes. However, the promotion of this type of activity requires not only measures providing proper technical infrastructure for walking or cycling, but also those encouraging residents to use it. In Glasgow, it was inclusion of the interested residents, social groups, organizations, and cycling associations in the preparation of the both mentioned applications. Moreover, these organizations promoted applications by their networks and actively cooperated with an entity responsible for its development, passing on users' recommendations concerning the sections of the path requiring modernization and also proposals of building new ones. The marketing of the ready-made tool (application) was carried out via a peer-to-peer network and social media [19]. The inclusion of residents in the actions described above took place in various scopes and at various stages of the project implementation [19], namely during:

- The online stocktaking of the existing infrastructure—encouraging residents to communicate with actors responsible for the application development by Internet tools (in Glasgow, Internet proved to be effective in involving residents in the implementation of this task);
- The development of the application—at this stage residents could actively participate in designing new bicycle and pedestrian paths and indicating those sections that require modernization, providing information via the Internet (their online insights and opinions had a real impact on specific investment decisions in terms of bicycle and pedestrian infrastructure);
- The creation of administration portal—this is a tool that enables transmitting and editing the content dedicated for the application (at this stage health organizations, universities, schools, and passenger transport bodies were involved);
- The integration of the online stocktaking tool with the prepared applications—engagement of citizens at the development phase of the technologies involved;
- The establishment of the administration platform for innovative app development—this platform enables

local companies to contact residents and other entities operating in a city for the development and implementation of innovative solutions, contributing to a digital economy at the same time.

What is important, the transfer of knowledge acquired by the preparation of the said applications makes it possible for local communities to create their own applications and can help promote a healthy and active lifestyle. The urban development is also more sustainable.

20.4.1.1 Geoparticipation

An important role in supporting grassroots initiatives and participation in city planning may be performed by platforms applying modern geoinformation technologies in the participation process. These technologies enable different actors to interact with the system in order to create new spatial solutions with the use of 3D city models. Offering specific functions, these types of platforms increase the capability of citizens to cocreate, cooperate, and participate in urban decision-making. These functions should be linked to precise and contextual spatial data and information, which can enhance the efficiency of citizens in participatory planning processes (e.g., by obtaining automatic feedback on the changes in development proposed by residents).

The Smarticipate platform, based on the open data portal, can illustrate the possibilities of using 3D models. This is a program offering both the top-down and bottom-up participation of citizens, ensuring such functions as 2D and 3D visualization; remarks about the solutions proposed on the platform and receiving automatic feedback; social dialog; communication with residents; possibility of choosing preferences and submitting alternative proposals of city planning. This is mainly what decides about the innovativeness and uniqueness of this platform [49]. The organizational pattern of this process is presented in Fig. 20.2.

The Smarticipate platform was shared with Hamburg residents giving them an opportunity to propose a tree planting site in their immediate neighborhood. It enables inhabitants not only to propose the location for newly planted trees, but also receive feedback on the following questions:

1. Is it possible to plant trees in a selected location?
2. If not, seeking explanation why it is not possible and receiving an alternative proposal of an acceptable location?
3. What is the budget and cost of planting trees?
4. What is the expected influence of planting trees in a given location on the environment, i.e., CO_2 reduction, etc.?

On the basis of the preceding knowledge, residents can consult their proposals with other dwellers, obtain specific information and assessments of the expected social results. It helps those interested in making decisions related to planting trees, in sharing opinions with others in this respect, and even raising funds for this purpose (i.e., crowdfunding). The Smarticipate platform makes it possible for residents to use OGD, visualize tree plantings, obtain feedback on the planned trees, and also communicate with the city administration and other residents to get their opinion [49]. These possibilities are shown in Fig. 20.3.

3D modeling allows users not only to visualize proposals, but also to better understand the benefits of specific solutions. Three-dimensional spatial models help to guarantee a spatial context of the proposals brought forward by city authorities or other entities.

20.4.1.2 Crowdsourcing/citizensourcing

In the simplest terms, crowdsourcing is an attempt to solve different kinds of problems employing the "wisdom of the crowd," not a narrow group of specialists. Crowdsourcing

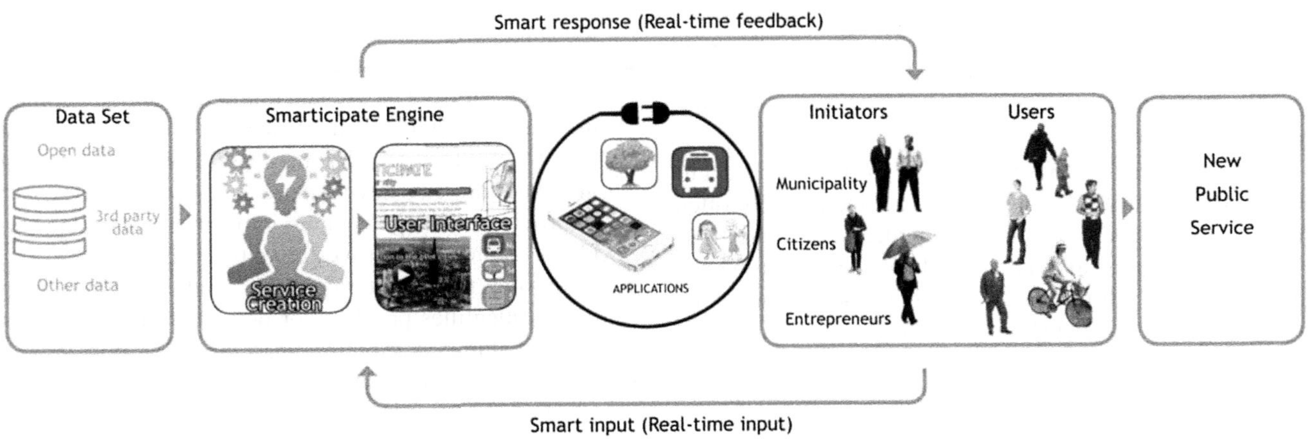

FIGURE 20.2 Pattern of the functioning of the Smarticipate platform. *Reproduced with permission from the author. https://www.smarticipate.eu/platform/.*

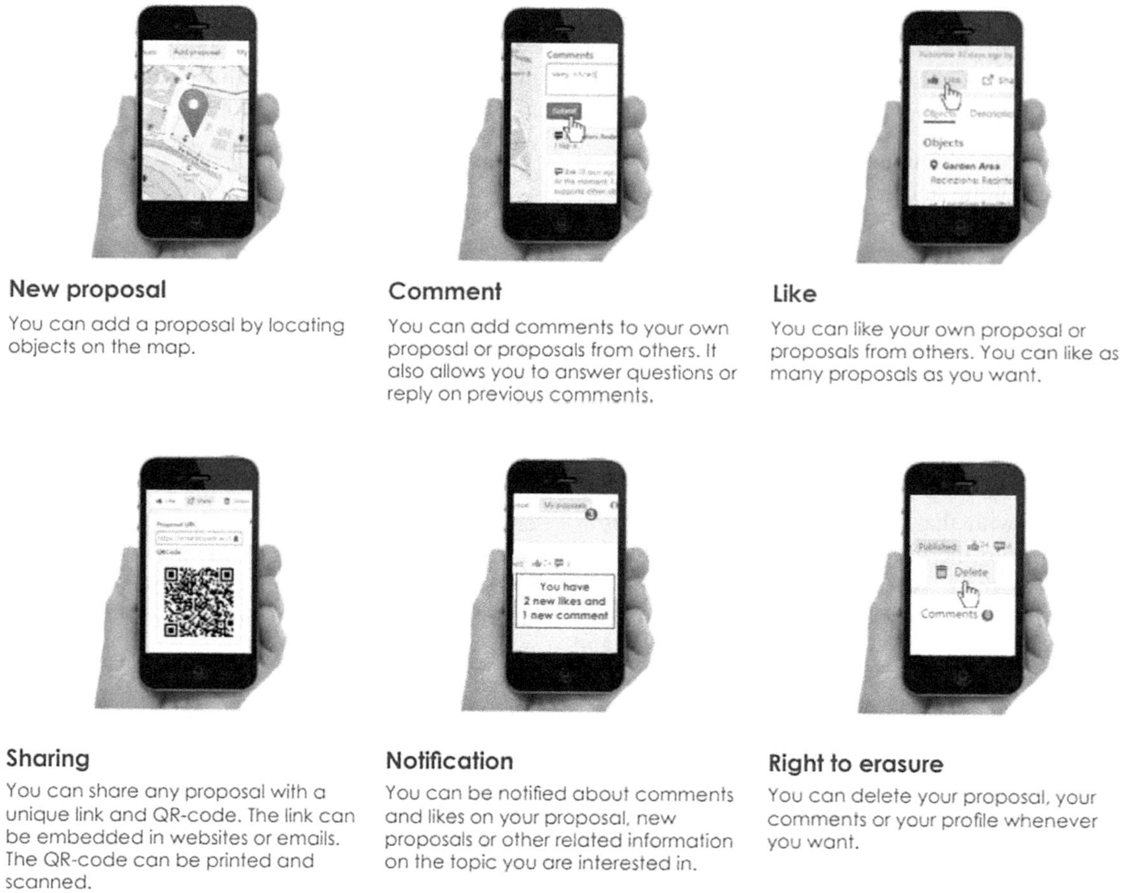

FIGURE 20.3 Features available for users of the Smarticipate platform. *Reproduced with permission from the author. https://www.smarticipate.eu/wp-content/uploads/20181022_SMARTICIPATE_App-Making-Manual_masterfile_Upload_web.pdf.*

enables looking at many different viewpoints on a given problem and collecting various ideas to solve it. The engagement of a large group of people (the crowd) in this technique is possible thanks to the Internet, social media, or the creation of specially prepared platforms for exchanging ideas. Opinions and ideas gathered in this way are then arranged and the best of them is selected. Such a choice is made by one person, a group, or asking other crowdsourcing participants for their opinion. Sharing thoughts often takes the form of a competition and individuals who produced the best ideas that are to be used and put into practice are rewarded.

Drawing on the "wisdom of the crowd," called citizensourcing when referring to a city, is increasingly often used by public institutions to pick up ideas and suggestions of urbanites related to planning and solving city problems. Thus, hermetic institutions closed so far, such as the public administration, are transformed becoming more open for the cooperation with residents and NGOs. Inhabitants, on the other hand, engaging in specific tasks, have the sense of a direct impact on shaping the city they live in.

The employment of citizensourcing has many advantages. The most important are as follows:

- Time-saving—the crowd in a relatively short time is able to demonstrate much more ideas than a narrow group of experts.
- Money saving—what needs to be financed is the platform and awards for the best ideas, which is much cheaper than outsourcing or hiring a group of professionals.
- Building social ties and local identity—residents feel that their viewpoints are taken into account when making decisions and have influence on what is going on in a city and that is why they are keener on getting involved into city governance.
- Improving an image—taking into consideration the residents' opinions improves the city image.
- Acquiring many options, ideas, and solutions—each person identifies problems or presented projects differently, making contribution that may prove priceless.
- Getting innovative ideas—as a result of brainstorming at a large scale, it is possible to get exceptionally interesting, often new ideas.

Citizensourcing can be an effective tool in smart city governance even more so, because it enables residents to participate at any time and at any place. The citizensourcing idea in city governance is used, for example, by the Ghent authorities. This can be illustrated by inricht-ding thanks to which urbanites can influence the development of the urban space. A user does not have to be a city planner or architect to create one's own project because the tools used by the application are friendly and intuitive. The program uses a 3D city model, created before. The available catalog, containing elements of small architecture, urban furniture, and plants, makes it possible to design space in a very simple way. Residents may decide about the construction of additional bicycle paths, pavement widening, a new bus stop, extra urban greenery, etc. Each suggestion, opinion, or remark goes automatically to a relevant division in the City Hall where the residents' ideas are analyzed and assessed. Inrich-ding allows Ghent's inhabitants to take part in spatial planning more effectively. Presently, the residents will be able to participate in the cocreation of each square and street in the city.

20.4.1.3 Crowdfunding

The term crowdfunding comes from the cluster of English words "crowd" and "funding." The existence of crowdfunding is closely related to open access to the Internet. It is based on the concept that an author of a given project asks Internauts for financial support offering specific benefits instead—e.g., a ready-made product or a discount on it when it goes on sale. In other words, this is a form of financing different kinds of projects by local or global communities with the use of advertisements and specialized Internet portals. Investors often make nominal small payments; their massive scale, however, can make the development of an innovative product or undertaking possible.

Crowdfunding takes place by a specific Internet platform. At present, the largest and most famous crowdfunding portal worldwide is Kickstarter [72]. Most portals take between 3% and 5% of the value of each payment collected by a user as a symbolic fee for using the website and for its development. In return, it offers the popularization of the project in social media and also supervises the timely achievement of the milestones of an individual project. A crucial factor behind the success of a crowdfunding project is its duration; the longer the time, the more likely it is to succeed [73].

The crowdfunding success can be understood in two aspects. The first can mean reaching the funding level, which is the threshold for taking action to achieve the project objectives, whereas the second denotes the general success of a project or its aims.

Such a platform, called crowdfunding.gent, operates in Ghent, enabling money to be raised in the first place for innovative and sustainable projects. Entities that can also receive support are nonprofit organizations and new companies starting their activity, so-called start-ups, distinguished, among other things, by the following features: higher risk than in the case of standard projects, low costs of starting the business, and a potentially higher return on investment in relation to a typical enterprise. Start-up companies and the like are most often linked to the Internet and new technologies. Adequate support for those meeting the abovementioned criteria is provided by the local authorities of Ghent, offering a high-quality, active, and easily available financing platform. However, some authors need not only money for a given project, but also an expert knowledge that can be also acquired through this platform. The city authorities allocate EUR 55,000 per year to cofinance nonprofit projects that use the crowdfunding website.

20.4.1.4 Hackathons

Hackathons are a productive method to support participatory governance, arouse interest in open data, boost creativity, innovativeness, and bottom-up designing of service solutions [19]. Urban hackathons are few days' events targeted at programmers, during which individuals related to developing software are to create urban applications, most often using open data. This is an opportunity to develop new applications for urban, raw data, and to find solutions that would bring specific benefits to city dwellers. Urban hackathons cover several stages: "brainstorming," designing, and testing ideas. They are specific incubators of ideas and forms of engaging the residents of a city (e.g., start-ups, students, and schoolchildren) in its improvement and development. Such events show urban authorities at the same time how urban data can be used in an innovative way and why they should be open to be widely applied.

Hackathons, also called designing marathons, are increasingly more popular in different cities. This can be illustrated by a cyclical, few days' hackathon Apps for Ghent, during which enthusiasts of new technologies may use satellite data shared by the city. Ghent's hackathons are aimed at designing mobile applications and Internet services that can improve the residents' life and help tourists visiting the place. It resulted in many interesting urban applications. One of them was the software Explore Gent. The program was created for tourists looking not only for interesting routes, but also for fun during sightseeing. The software navigates through a selected route, informs about the points visited, and provides attractive tasks to perform. Moreover, it encourages users to compete and meet other tourists.

Hackathons can be devoted to different topics. A good example is the 2-day online hackathon Hack the Crisis, organized by the president of Estonia in March 2020, who called on start-ups from all over the world to join in creating innovative ideas to combat the crisis caused by the coronavirus pandemic. The president emphasized the importance of cooperation between public and private sectors for seeking and finding new solutions aimed at minimizing socioeconomic effects of the coronavirus. The best five ideas were awarded with grants of up to EUR 5000. During the event there appeared numerous ideas for tools and applications combating the coronavirus and its negative socioeconomic results (designs of devices to detect risky gestures, such as face touching; software that allows people in need to connect quickly and effectively with volunteers willing to help; an application serving firms to transfer employees temporarily, e.g., persons hired in tourism could work in the commerce industry, etc.)

20.4.1.5 Living Lab (LL)

Living Lab is a concept, which functions in a narrow sense in technical and engineering sciences, whereas in a broader sense, it embraces technological and social dimensions. It is understood as an open, user-oriented environment of innovations with a clear territorial context (e.g., city, agglomeration, region), covering all types of research and development. In the Living Lab, users are included in the process of cocreating new, innovative services, products, and social infrastructure, developed within partnerships between public authority and the academic environment as well as between entrepreneurs and citizens. There are three main Living Lab conceptions [74]:

1. As a research environment for new technological solutions;
2. As a distributor and an active base of knowledge and tools; and
3. As a platform for the self-organization of groups and communities.

In the research on smart city, the Living Lab conception is defined in two separate ways: (1) as an early technological version (so-called beta version), requiring tests, verifications, and technical modification with the employment of users' experience; and (2) as a democratic social conception in which specific technical solutions shape bottom-up processes resulting from the needs and expectations of local communities.

In today's world, the search for new, technologically advanced solutions goes more and more often beyond the question of creating new tools and services concerning relations between a producer and an individual user (entity). The expectations grow as to the expansion of the scope of modernization of urban infrastructure systems as well as a greater socialization of the process of generating innovations. What is also changing is the perception of the role of users, from the traditional function of testers to inspirators in creating new solutions. In this context, the LL formula may become an effective link between urban participatory planning and creating a social environment conducive to generating and implementing technical improvements beneficial for strengthening local communities [75].

The Living Lab conception has been implemented, among other places, in the Helsinki region. The Living Lab Helsinki network was established in 2007. It provides a platform that serves to promote methods and tools dedicated for users to improve the functionality of the existing ones and the development of new products and services. The Helsinki Living Lab is a communication center as well, enabling companies and the public sector to cooperate with all actors operating within the Living Lab metropolitan area, becoming a kind of regional brand. This network works through various projects such as Living Virs Forum Virium, Arabianranta Living Labs, Laurea Living Labs, and Pasila Living Labs (Helsinki Living Lab on http://www.helsinkilivinglab.fi/). Residents are engaged in them in various ways. Some of the projects give the inhabitants the right to vote and others use their knowledge and experience as service users. Services are prepared and tested jointly by users and creators of applications. For instance, the Walk and Feel Helsinki project was initiated in the summer of 2011 to help cruise passengers to get to know Helsinki and ensure a new attractive way of visiting the city. Thus, the city's infrastructure and services develop by the implementation of various smart city projects, which serve smart mobility, improve security and other aspects of urban life, establishing conditions for the development of a really open city [42]. The examples of the ongoing platforms and projects within the Helsinki Living Lab are presented in Table 20.2.

The Helsinki LL brings benefits to different groups of actors. The authorities of cities use LL to promote economic growth by encouraging social, service, and government innovations. Higher education institutions have an opportunity to popularize methods and research results among creators and users. Innovative companies use them particularly to generate and test ideas as well as to develop products tailored to meet recipients' needs. Thanks to this diversity of actors involved, with the principles of openness and cocreation, Living Labs play an important role in shaping a smart city [42]. What is more, these types of initiatives make the residents more than merely city users or clients. However, it can only be achieved in the atmosphere of openness, democracy, and integration.

Living Lab was also set up in Ghent. This tool allows experimenting and engaging potential users at the stage of looking for solutions and ways of their implementation. The Ghent Living Lab is based on the partnership with numerous public and private actors (Ghent Living Lab on https://enoll.org/network/living-labs/?livinglab=ghent-living-lab). This

TABLE 20.2 Examples of platforms and projects within the Helsinki Living Lab.

Platform or project	Main aim
Smart Kalasatama	Model for building a sustainable smart neighborhood
Helsinki Region Infoshare	Public data for anyone to use in their applications
Open Ahjo	Interface that provides access to the documents of the city of Helsinki
Helsinki Loves Developers	Organises developer meet-ups and a portal (dev.hel.fi)
CitySDK	Open data interface project
D-CENT	Create new web-based tools and participatory platforms for citizen collaboration and open interaction between public authorities and citizens.
Walk and Feel Helsinki	Help cruise passengers to become acquainted with Helsinki and provide them with a new way to explore the city

Own study on the basis of Anttiroiko AV. City-as-a platform: the Rise of participatory innovation platforms in Finnish cities. Sustainability 2016;8(9): article 922. https://doi.org/10.3390/su8090922.

is a platform for creating, testing, and continuous modernization of innovative solutions in a user's real environment. The conducted usage tests verify not only the conception of the product itself, but also indicate the potential challenges or aspects of using it, which are difficult to predict. The testing process usually takes place in a structural, iterative way, that is from an idea to a prototype to its implementation. The activity of the Ghent Living Lab is focused on key areas, such as: (1) e-inclusion; (2) e-government and e-participation; (3) digital innovations for a smart city; (4) "green" digital development covering power efficiency in relation to ICT and specific initiatives such as Green Digital Charter.

The Ghent Living Lab presents an interdisciplinary approach to creating innovation, in which a user of innovative solutions is the main point of reference. The participation of users in this process may occur at the stage of [21]:

- Cocreation, i.e., designing a product or service jointly by users and producers;
- Exploration, i.e., finding the emerging applications, behavior, and market prospects;
- Experimentation, i.e., implementing real scenarios among the communities of users; and
- Evaluation, i.e., assessing concepts, products, and services in line with socioeconomic, socioergonomic, and sociocognitive criteria.

Ghent's Living Lab stimulates new ideas, provides a rich context for various challenges and R + D projects. It makes it possible to adjust new solutions to a user's needs and to identify the wrong answers as early as at the designing phase, not at the final stage of developing products or services. Therefore, technologies should be created for residents and by residents in their real domicile, place of work, or leisure, not in a hermetic laboratory; hence "living" should be treated as a catchword.

Undoubtedly, the employment of the Living Lab approach to create and test solutions together with residents as part of smart city projects goes beyond a traditional approach toward consulting decisions. Ghent is a city where the projects based on Living Lab methodology have been regularly implemented for years, making it possible to stimulate ideas for innovative projects, establish new companies, and introduce solutions thanks to which the quality of life improves.

20.4.1.6 Gamification: a gamified engagement tool

Gamification consists of using schemes and mechanisms known from games outside games, especially in different group activities in order to increase engagement of individuals performing specific tasks. These mechanisms and patterns include: tasks, challenges, achievement badges, levels of difficulty, progress bars, or other forms of showing how close a person is from completing a task or moving to the next level, etc. This technique is based on the pleasure gained by overcoming subsequent levels, facing challenges, competition, cooperation, etc. Gamification allows engaging people in tasks that are in line with the expectation of the author of a project, even if these activities are generally considered to be boring or routine.

Gamification was used, among other places, in Glasgow, where a tool for shaping a proper behavior pattern in power consumption was prepared. The introduction of new technologies and practices, aimed at reducing energy demand and the CO_2 emission, is a great challenge for smart cities and at the same time a concern for the public sector, individual citizens, and social organizations. These measures, however, must go hand in hand with the reduction of energy poverty. This poverty means that some households face a difficult choice between "eating" or "heating," which certainly fits neither in the smart city model nor the sustainable development idea. They should be also introduced along with the formation of appropriate, proecological attitudes of inhabitants, which can be achieved, e.g., by gamification. Gamification in Glasgow was dedicated particularly to school children. This is so, because it was assumed that the effect created would "trickle down" to their parents and family, leading to the reduction of energy consumption in households [19].

Another example of gamification is the preparation of a three-dimensional urban game 3D-Citygame-Ghent in Ghent. The local authorities engaged both professionals and amateurs, including residents, to create this game. The search was mainly for enthusiasts with a positive attitude and willingness to learn. In the game, one can play via the Internet or by using VR glasses; a 3D city model was used. Gamers think, build, and change the space in a virtual world, whereas clerks collect ample feedback from the residents (ideas, opinions, dreams). The game should be treated as an atypical, innovative tool for social consultations. It is aimed at developing civic attitudes and increasing residents' engagement in shaping the city space.

A different instance of gamification is the "ZWERM" project, which helped face the social problem of two neighborhoods in Ghent. The residents of Papegaai and Ekkergem did not know one another, were not integrated, and, in effect, did not create joint projects and did not feel that their mutual vicinity is a common good they should take care of together. Many actors were involved to solve this problem, including city dwellers, designers, engineers, sociologists, and urban servants. Residents were given out over 1400 magnetic cards ZWERM and the centers of both neighborhoods were equipped with card readers. After reading one's own card, a user got 10 points, but if at least two persons were registered at the same time and place, each received 50 points. Moreover, four boxes were installed in each neighborhood, which could be whistled in order to win points for a given district. Thus, in the urban game it was not only individuals that compete with one another but also neighborhoods. The game moderator sent tasks to users on mobile devices, consisting of joint meetings at a given event or looking for neglected parts of the estate and reporting them to relevant services. In total, 80% of the game participants confirmed that the project contributed to the integration of the local community, and 75% felt the sense of belonging to it [76].

The inclusion of city dwellers in urban problem-solving through gamification brings many benefits. One of the most important is the development of social capital by engaging citizens in taking on challenges faced by the city [19].

20.4.2 Cocreation of value

Cocreation of value refers to the process of designing a product or service in which users' input plays a key role from start to finish. It can be also mentioned when the recipients of the values created by residents (or other actors) are other residents. These processes can be illustrated by urban hacktivism or BarCamps.

20.4.2.1 Urban hacktivism

Hacktivism is the word combination of "hacking" and "activism." Although "hacking" is associated most often with the use of technology to launch malicious attacks on websites, it has also a different meaning. It can be understood as innovative problem-solving through joining new ideas and readily available materials (utilizing an old DVD as a drink stand) as well. Essentially, hacking means creative problem-solving, the example of which is hackathons. Hacktivism is the combination of computer hacking and sociopolitical activism. It means adopting information technology and networks to promote social and political objectives. It is understood then as a hacking activity for a good cause. This activity is undertaken on the assumption that the use of new technology may bring the same or even better effects than conventional protests, political and social activism, or civil disobedience.

Today, there are thousands of groups of hacktivists supporting various matters, publicizing their activities in social media. Their MO is often compared to traditional protest forms and embrace, e.g., from the defacement of Internet pages to graffiti or to blocking access to them. However, most of these methods are illegal, according to the law existing in a given country, and recognized as a form of cybercrime.

Urban hacktivism consists of creative engagement of citizens in urban planning and adopting the city space to the needs of its inhabitants. Urban and regional planning, as an institutionalized practice, is increasingly more often criticized for not being responsive to the needs of ordinary citizens. Urban hacktivism relates to solving problems by bottom-up actions of like-minded people who are able to oppose plans and decisions of authorities and institutions in pursuit of social objectives. This is all about objection to the measures taken by, e.g., self-governments or developers unfavorable for a local community.

Problems connected to urban planning are often shown during debates on urban space restructuring. Projects presented there are frequently perceived as unfavorable by locals, leading to the limitation of privacy, an increase in a traffic jam and noise, the destruction of favorite places, gardens, and the nature of the neighborhood. The related consultations are usually limited and aim to receive social approval for presented projects; the remarks made by residents are disregarded and the problems are underestimated. The protection of local interests by small groups of inhabitants is difficult in these circumstances. The bottom-up organization of civil groups, local planners, designers, ecologists, etc., and common actions to stop privatization of the public space and erosion of public functions have a better chance of success. These types of initiatives are undertaken mainly by social media.

The examples of hacktivism in urban planning can be found, among others, in Australia, where the initiative Gold Coast Hinterland and Environment Council (GECKO) undertaken by residents was aimed at "the hacking of planning." A social resistance was organized against the implementation of the "West Village" project in West End in

Brisbane within which seven 15-storys high residential towers and a large supermarket were to be built on the plot in the city center. Many inhabitants viewed such a proposal of urban development as wrong and causing serious problems. A group of residents and other interested parties began to create an alternative vision called the "Instead West Village" based on the existing social ties and the sense of local identity. It aimed to exert pressure on both the developer and local authorities and make them interested in the residents' objections to expectations about and aspirations toward the area under project. The inhabitants, using open, community-based social planning, developed a bottom-up alternative for the main urban scheme, taking into account models and development visions discussed during various meetings. Public debates and workshops also enabled the input of other neighborhoods in the final outcome of the jointly prepared project. Thus, local communities begin "hacking" urban planning. An alternative vision is still in the finalization phase, but the cooperation and the hacktivism spirit underlying it go beyond a regular, alternative master plan. Hacktivism restored the inhabitants' right to the city, ensuring planning that is fairer, more sustainable, and more conducive to social inclusion [77].

20.4.2.2 BarCamp

BarCamps are informal meetings, serving to openly share practical knowledge, experience, and ideas and integrating different environments, mostly Internet and marketing circles. However, BarCamps gathering experts in public relations, HR as well as ecologists or culture activists are also very popular. They are usually open for everybody, regardless of the background and discipline, thus ensuring interactive and interdisciplinary discourse on an equal footing. In contrast to scientific conferences, the participation fee is often low, there is no top-down agenda or an artificial division into lecturers and audience, and it is the participants themselves who decide about the program. There are no celebrities, large budgets, thick volumes of conference materials, or rigid rules. BarCamps are held in a relaxed, friendly atmosphere with plenty of time for discussion and questions from participants. Because of that, they are often called "unconferences" or "anti-conferences" [78].

The first BarCamp was organized between 19 and August 21, 2015 in California, in the offices of the Socialtext company. It was attended by 200 people. It turned out quickly that BarCamps are alternatives to often boring and expensive conferences where the time to exchange ideas between participants is generally limited because of the need to stick to the program. Since then, BarCamps have been attracting participants all over the world. As people usually turn up in their free time, their motivation for active engagement and benefitting from participation is very high [78].

Most BarCamps have specific themes (e.g., tourism, start-ups, etc.), attracting different individuals regardless of their subject of interest [78]. It does not mean, however, that other issues are not discussed. The cleverly chosen subject creates a greater opportunity for establishing contact with partners needed for cooperation and for free transfer of knowledge.

BarCamps have many advantages. They include [79]:

- Free transfer of knowledge—they make it possible to communicate openly, voluntarily, in a motivated, informal, and democratic way, encouraging every participant to take active part in the event;
- Focus on recipients—they offer greater possibilities of open discussion and creating knowledge than traditionally understood conferences (presentations and questions);
- Access to current information—they acknowledge current topics and works, which are not always ready to be sent a few months before a traditional conference;
- Cocreation—they provide room for changes in the discussion topics based on the participants' interests and ideas generated during the event itself.

Owing to numerous advantages, the notion to incorporate BarCamps, Unconferences and other open session formats, which concentrate on discussions initiated by participants into traditional conferences, is increasingly popular [80]. Originally, this took place at programmers' conferences, but it is also spreading into other areas of interest [79]. BarCamps and unconferences organized during classical meetings are treated as a way of engaging people, making contact, creativity, achieving objectives together, and establishing a valuable platform for the interactive transfer of knowledge [79,81].

20.5 Summary

Smart city can be understood as a well-functioning city of the future, based on proactive behavior of conscious, independent, and decisive citizens. It can be therefore concluded that it is not urban systems filled with modern technology, but a city properly governed, thus smart city governance, that is key to being smart.

Governance in a smart style consists of strengthening the importance of a local community in the city governance process through encouraging social participation, taking into account remarks and ideas of residents in making decisions concerning them, and supporting bottom-up initiatives. What is helpful in this respect is a wide use of ICT and various types of platforms, especially mobile applications, which extend traditional communication channels with residents with new ones and enable social involvement regardless of time and place.

Smart city governance is something much more than e-administration or even e-democracy, because it involves real social participation, which means readiness for joint design (coproduction) of solutions without the hierarchical submission of partners and even the implementation of fully bottom-up initiatives (cocreating of value). This participation is understood as a permanent relationship between local authorities including structures they supervise and active groups of citizens as well as other organizations and actors. The application of participation rules requires adopting an arduous process of arrangements, debates, consultations, but it happens to be less costly financially and politically than the results of arbitrary decisions taken by authorities, which can be protested in progress. The advantage of this process is the fact that all parties involved in planning changes identify with those changes, which makes them more easily accepted by a local community. Moreover, solutions developed with the participation of their users are usually much better than top-down ones, because they solve daily problems of residents, contributing to the improvement in the functioning of an entire unit. A city insensitive to the need for participation is viewed as exceptionally unfriendly and even hostile to inhabitants.

Participation has several levels—from a passive one (consisting in, e.g., informing residents unilaterally or monitoring their satisfaction with services) to active (demonstrated, among other things, by coproduction and cocreation of value). Active participation is supported by ICT development, which requires access to technology (Internet, computer or phone, ability to handle it) and the availability of data, which can be processed in order to develop solutions facilitating and optimizing the functioning of urban systems and residents in a city. The satisfaction of these three conditions (participation, access to technology, and open data) gives hope for greater cooperation and for launching a bottom-up participatory platform.

The combination of sociotechnological structures to achieve coproduction between local authorities and communities is of key importance for creating a smart city. This is so, because it opens up an opportunity to implement new, intelligent services and to introduce technology-based governance mechanisms [19].

In this process, however, smart city governance faces many problems and challenges. The most important are: (1) the lack of clarity as to the nature and purpose of participation, which may weaken the engagement, eliminate key interested parties, and create the impression that participation is a waste of time [82]; (2) the often observed public cynicism, understood as simulated actions, pretending that citizens' engagement plays an important role, when in fact, it is different [30]; (3) risk, related to the strengthening of those entities that have already been active and which influence the local decisions to a greater extend [83]. The use of ICT in the participation process brings also problems related to data safety and ensuring privacy of residents. However, the elimination of digital exclusion and the development of permanent mechanisms for residents' involvement in the process of smart city governance seem to be one of the major challenges.

20.6 Chapter review questions/exercises

True/false

1. True or false? Smart city governance uses modern ICT technologies.
2. True or false? Smart city governance is aimed at improving the living standards of residents.
3. True or false? The highest level of social participation is visible in the phase of Smart City 1.0.
4. True or false? The key principle of bottom-up participatory governance is the inclusion of the disabled persons in the planning process and city governance.
5. True or false? Coproduction enables residents to cocreate urban services.

Multiple choice

1. An underlying assumption of smart city governance is:
   a. Equipping the city with the newest technological systems.
   b. Engaging residents in shaping the urban living environment and public services.
   c. Implementing ecological solution for protection of greenery in a city.
   d. All of the above.
   e. None of the above.
2. In the phase of Smart City 3.0, urban authorities focus on:
   (a) Providing an opportunity to use intellectual potential of residents and their active attitude to create city development
   (b) Purchasing the newest technological solutions enabling effective city governance
   (c) Using big data in city governance
   (d) All of the above.
   (e) None of the above
3. In smart city governance urban authorities look, first of all, for new________________in/for the implementation of projects for sustainable urban development.
   (a) Financial means
   (b) Ways of boosting residents' engagement
   (c) Business partners included
   (d) All of the above.
   (e) None of the above

4. The highest form of public participation within smart city governance is:
   (a) Engagement
   (b) Coproduction
   (c) Cocreation
   (d) All of the above
   (e) None of the above
5. Bottom-up participation can be strengthened primarily by:
   (a) Traditional expert planning
   (b) Active cooperation between city authorities and the economic environment
   (c) New channels of bilateral communication (social media, open Internet platforms, innovative applications)
   (d) All of the above
   (e) None of the above

Exercises

Problem

Does providing residents with access to relevant ICT infrastructure guarantee the activation of bottom-up participatory governance?

Hands-on projects

How to effectively use OGD in smart city governance?

Case projects

Problem

Provide a specific urban planning perspective on smart governance, which represents a move away from the technocratic way of governing cities often found in smart cities.

Optional team case project

Problem

Present a systematic review of smart governance, defined as technology-enabled collaboration between citizens and local governments to advance sustainable development.

References

[1] Barber B. If mayors ruled the world: dysfunctional nations, rising cities. New Haven, CT: Yale University Press; 2013.
[2] Landry C. The art of city making. London, England: Routledge; 2006.
[3] Giffinger R, Fertner C, Kramar H, Kalasek R, Pichler-Milanović N, Meijers E. Smart cities, ranking of European medium-sized cities. Vienna UT: Centre of Regional Science; 2007.
[4] Tranos E, Gertner D. Smart networked cities? Innovat Eur J Soc Sci Res 2012;25:175–90.
[5] Caragliu A, Bo del C, Nijkamp P. Smart cities in Europe. In: 3rd central European conference in regional science, CERS, Kosice, Slovak Republic; 2009. p. 45–59.
[6] Kourtit K, Nijkamp P, Arribas D. Smart cities in perspective—a comparative European study by means of self-organizing maps. Innovat Eur J Soc Sci Res 2012;25:229–46.
[7] Hoon Lee J, Phaal R, Lee SH. An integrated service-device-technology roadmap for smart city development. Technol Forecast Soc Change 2013;80:286–306.
[8] Lombardi P, Giordano S, Farouh H, Yousef W. Modelling the smart city performance. Innovat Eur J Soc Sci Res 2012;25:137–49.
[9] Shapiro JM. Smart cities: quality of life, productivity, and the growth effects of human capital. Rev Econ Stat 2006;88:324–35.
[10] Sobol A. Inteligentne miasta versus zrównoważone miasta (Intelligent cities vs sustainable cities). Studia Ekonomiczne. Zeszyty Naukowe Uniwersytetu Ekonomicznego w Katowicach 2017;320:75–86.
[11] Allwinkle S, Cruickshank P. Creating smart-er cities – an overview. J Urban Technol 2011;18(2):1–16.
[12] Chiabai A, Paskaleva K, Lombardi P. E-participation model for sustainable cultural tourism management: a bottom-up approach. Int J Tourism Res 2013;15(1):35–51.
[13] Laspidou C. ICT and stakeholder participation for improved urban water management in the cities of the future. Water Util J 2014;8:79–85.
[14] Royo S, Yetano A. 'Crowdsourcing' as a tool for E-participation: two experiences regarding CO_2 emissions at municipal level. Electron Commer Res 2015;15(3):323–48.
[15] Bolivar MPR, Meijer AJ. Smart governance: using a literature review and empirical analysis to build a research model. Soc Sci Comput Rev 2015;1–20.
[16] Arup, Livable Cities, UCL, Smart City Expo. Delivering the smart city: governing cities in the digital age. London: Arup; 2014.
[17] Ferro E, Caroleo B, Leo M, Osella M, Pautasso E. The role of ICT in smart cities governance. In: CeDEM13 conference for EDemocracy and open government. Krems, Austria: Danube University; 2013. p. 133–45.
[18] Almirall E, Wareham J, Ratti C, Conesa P, Bria F, Gaviria A, Edmondson A. Smart cities at the crossroads: new tensions in city transformation. Calif Manag Rev 2016;59:141–52.
[19] Leleux C, Webster W. Delivering smart governance in a future city: the case of Glasgow. Media Commun 2018;6(4):163–74.
[20] Barns S. Smart cities and urban data platforms: designing interfaces for smart governance. City Cult Soc 2018;12:5–12.
[21] Schuurman D, Baccarne B, De Marez L, Mechant P. Smart ideas for smart cities: investigating crowdsourcing for generating and selecting ideas for ICT innovation in a city context. J Theor Appl Electron Comm Res 2012;7(3):49–62.
[22] Arendt Ł, Kukulak-Dolata I. Społeczeństwo informacyjne w mieście (Informative society in a city). In: EkoMiasto#Społeczeństwo. Zrównoważony, inteligentny i partycypacyjny rozwój miasta (Eco-City#Society. Sustainable, intelligent and participatory urban development). Wydawnictwo Uniwersytetu Łódzkiego, Łódź; 2016. p. 181–204.
[23] Hollands R. Will the real smart city please stand up? Intelligent, progressive, orentrepreneurial? City Anal Urban Trends Cult Theor Pol Action 2008;12(3):303–20.

[24] Yetano A, Royo S. Keeping citizens engaged A comparison between online and offline participants. Adm Soc 2015;49(3):394–422.
[25] Cimander R. Citizen panels on climate targets: ecological impact at individual level. In: Evaluating e-participation. Frameworks, practice, evidence. Basel: Springer International; 2016. p. 219–41.
[26] Sørensen E, Torfing J. Co-initiation of collaborative innovation in urban spaces. Urban Aff Rev 2018;54(2):388–418.
[27] Manual S. Using information and comunication technologies for smart and connected cities. A guide for sustainable urban development in the 21st century. 2012. http://www.un.org.
[28] Gotlibowska K. Propozycja modelu miasta inteligentnego (Smart City) opartego na zastosowaniu technologii informacyjno-komunikacyjnych w jego rozwoju (Proposal of the Smart City model based on the ICT application in its development). Rozwój Regionalny i Polityka Regionalna 2018;42:67–80.
[29] Szaja M. Rola lokalnej społeczności w zrównoważonym rozwoju miast funkcjonujących w myśl koncepcji smart city (The role of a local community in the sustainable development of cities functioning in the smart city conception). Prace Naukowe Uniwersytetu Ekonomicznego we Wrocławiu 2018;532:346–54.
[30] Andersson E. Engagement in health: roles for the public and patients. In: Smart governance for health and well-being: the evidences. World Health Organization. The Regional Office for Europe; 2014. p. 34–48.
[31] Cohen B. The top 10 smart cities on the planet. 2014. http://www.fastcoexist.com.
[32] Siemiński W. Cele i zasady partycypacji społecznej w planowaniu przestrzennym— przegląd literatury (Objectives and principles of social participation in spatial planning—literature review). Człowiek i środowisko 2007;31:37–59.
[33] Swianiewicz A, Klimska U, Mielczarek A. *Nierówne koalicje – liderzy miejscy w poszukiwaniu nowego modelu zarządzania rozwojem* (Unequal coalitions—urban leaders in search of a new model of development management). Warszawa: Wydawnictwo Naukowe Scholar; 2004.
[34] Szaja M. Partycypacja lokalnej społeczności w kształtowaniu procesu rozwojowego w gminie (Participation of a local community in shaping the development process in a commune). Zeszyty Naukowe Uniwersytetu Szczecińskiego, Ekonomiczne Problemy Usług 2015;118:281–96.
[35] Chourabi H, Nam T, Walker S, Gil-Garcia JR, Mellouli S, Nahon K, Scholl HJ. Understanding smart cities: an integrative framework. In: 45th Hawaii international conference on system sciences; 2012. p. 2289–97.
[36] Gabrys J. Programming environments: environmentality and citizen sensing in the smart city. Environ Plann Soc Space 2014;32(1):30–48.
[37] Alford J, Yates S. Co-production of public services in Australia: the roles of government organisations and co-producers. Aust J Publ Adm 2015;75(2):159–75.
[38] Osborne SP, Radnor Z, Strokosch K. Coproduction and the co-creation of value in public services: a suitable case for treatment? Publ Manag Rev 2016;18(5):639–53.
[39] Jakubowski E, Liste G. Ninth futures forum on health systems governance and public participation. Copenhagen: WHO Regional Office for Europe; 2006. http://www.euro.who.int/__data/assets/pdf_fi le/0006/97836/E89766.pdf. [Accessed 30 October 2013].
[40] Kickbusch I, Gleicher D. Smart governance for health and well-being: the evidence. WHO Regional Office for Europe; 2014. https://apps.who.int/iris/handle/10665/131952.
[41] Arnstein S. A ladder of citizen participation. J Am Inst Plan 1969;35:216–24.
[42] Anttiroiko AV. City-as-a platform: the Rise of participatory innovation platforms in Finnish cities. Sustainability 2016;8(9). https://doi.org/10.3390/su8090922. article 922.
[43] Kaulio MA. Customer, consumer and user involvement in product development: a framework and a review of selected methods. Total Qual Manag 1998;9:141–9.
[44] Simces Z. Exploring the link between public involvement/citizen engagement and quality health care – a review and analysis of the current literature. 2003. Ottawa, Health Canada, http://www.hc-sc.gc.ca/hcs-sss/alt_formats/hpb-dgps/pdf/pubs/2003-qualsimces/2003-qual-simces-eng.pdf. [Accessed 30 October 2013].
[45] Barabas J. How deliberation affects policy opinions. Am Polit Sci Rev 2004;98:687–701.
[46] Coulter A. Shared decision-making: the debate continues. Health Expect 2005;8:95–6.
[47] Khan Z, Ludlow D, Loibl W, Soomro K. ICT enabled participatory urban planning and policy development: *the UrbanAPI project*. Tran. Gov People Proc Policy 2014;8:205–29.
[48] Dambruch J, Krämer M. Leveraging public participation in urban planning with 3D web technology. In: Proceedings of the 19th international ACM conference on 3D web technologies, Vancouver, BC, Canada, 8–10 August 2014. New York, NY, USA: ACM; 2014. p. 117–24.
[49] Khan Z, Dambruch J, Peters-Anders J, Sackl A, Strasser A, Fröhlich P, Templer S, Soomro K. Developing knowledge-based citizen participation platform to support smart city decision making: the smarticipate case study. Information 2017;8(47):1–24.
[50] Tomor Z, Meijer A, Michels A, Geertman S. Smart governance for sustainable cities: findings from a systematic literature review. J Urban Technol 2019;26(4):3–27.
[51] Boes K, Buhalis D, Inversini A. Conceptualising smart tourism destination dimensions. In: ENTER 2015 proceedings, Lugano. Wien: Springer-Verlag; 2015. p. 391–404.
[52] Tenbensel T. Virtual special issue introduction: public participation in health policy in high income countries – a review of why, who, what, which, and where? Soc Sci Med 2010;71(9):1537–40.
[53] Fishkin JS. When the people speak. Oxford: Oxford University Press; 2009.
[54] Voorberg H, Bekkers VJJM, Tummers LG. A systematic review of co-creation and co-production: embarking on the social innovation journey. Public Manag Rev 2014;17(9):1333–57.
[55] Chirisa I, Matamanda AR. Smart cities in Sub-Saharan Africa: opportunities and challenges. In: Industrial and urban growth policies at the sub-national, national, and global levels. Hershey PA: IGI Global; 2019.
[56] Bifulco F, Tregua M, Amitrano CC, D'Auria A. ICT and sustainability in smart cities management. Int J Public Sect Manag 2016;29(2):132–47.
[57] Janowski T. Digital government evolution: from transformation to contextualization. Govern Inf Q 2015;32:221–36.
[58] Gotlib D, Olszewski R, editors. Smart City. Informacja przestrzenna w zarządzaniu inteligentnym miastem (Smart City. Spatial information in managing smart city). PWN; 2016.

[59] Wehn U, McCarthy S, Lanfranchi V, Tapsell S. Citizen observatories as facilitators of change in water governance? Experiences from three European cases. Environ Eng Manag J 2015;14(9):2073–86.

[60] Semanjski I, Bellens R, Gautama S, Witlox F. Integrating big data into a sustainable mobility policy 2.0 planning support system. Sustainability 2016;8:1142.

[61] Yigitcanlar T, Lee SH. Korean ubiquitous-eco-city: a smart-sustainable urban form or a branding hoax? Technol Forecast Soc Change 2014;89(2014):100–14.

[62] Ertiö TP. Participatory apps for urban planning: space for improvement. Plann Pract Res 2015;30(3):303–21.

[63] Zheng Y. Explaining citizens' E-participation usage functionality of E-participation applications. Adm Soc 2017;49(3):423–42.

[64] Lee JH, Gong Hancock M, Hu MC. Towards an effective framework for building smart cities: lessons from Seoul and san Francisco. Technol Forecast Soc Change 2014;89:80–99.

[65] Coleman S, Blumler JG. The internet and democratic citizenship: theory, practice and policy. Cambridge: Cambridge University Press; 2009.

[66] Molinari F, Ferro E. Framing web 2.0 in the process of public sector innovation: going down the participation ladder. Eur J ePractice 2009;9(1):20–34.

[67] Charalabidis Y, Koussouris S, Lampathaki F, Misuraca G. ICT for governance and policy modelling: visionary directions and research paths. In: Empowering open and collaborative governance. Berlin Heidelberg: Springer; 2012. p. 263–82.

[68] Achilleos A, Markides C, Konstantinidis A, Giorkas I, Kapitsaki M, Mettouris C, Papadopoulos GA. Adopting an open smart city platform: a survey. In: Fifth IEEE annual international smart cities conference (ISC2 2019), Casablanca, Morocco, 14–17 October 2019. IEEE Computer Society Press; 2019. p. 384–9. http://www2.cs.ucy.ac.cy/~george/files/ISC219.pdf.

[69] Chichernea V. Smart cities communities and smart ICT platform. J Inf Syst Oper Manag 2015;9(1):1–11.

[70] Wachhaus TA. Governance as a framework to support informatics. The Innov J Publ Sec Innov J 2011;16. article 5.

[71] Koliba C, Zia A, Lee BHY. Governance informatics: managing the performance of inter-organizational governance networks. Innov J TPub Sec Innov J 2011;16(1). article 3, Available online: http://innovation.cc/scholarly-style/koliba_governance_informaticsv16i1a3.pdf.

[72] Liebert F, Trzeciak M. Portal crowdfundingowy jako narzedzie wspierające realizacje małych projektów. (Crowdfunding portal as a tool supporting the implementation of small projects). Zeszyty Naukowe Politechniki Śląskiej, Seria: Organizacja i Zarządzanie 2017;114:255–71.

[73] Cordova A, Dolci J, Gianfranco G. The determinants of crowdfunding success: evidence from technology projects. Proced Soc Behav Sci 2015;181:115–24. Elsevier.

[74] Wallin S. APRILab: guidelines to define and establish an urban living Lab. University of Amsterdam, Amsterdam Institute of Social Science Research; 2014.

[75] Cardullo P, Kitchin R, Di Feliciantonio C. Living Labs and vacancy in the Neoliberal city. Cities 2018;73:44–50. https://www.sciencedirect.com/science/article/pii/S026427511730330X, dostep 15.12.2018.

[76] Coenen T, Mechant P, Laureyssens T, Claeys L, Criel J. ZWERM: stimulating urban neighborhood self-organization through gamification. In: Using ICT, social media and mobile technologies to foster self-organisation in urban and neighbourhood governance, Delft, The Netherlands; 2013. https://core.ac.uk/download/pdf/55828071.pdf dostep. 6.05.2020.

[77] https://theconversation.com/urban-hacktivism-getting-creative-about-involving-citizens-in-city-planning-62277.

[78] Dennerlein S. Assessing Barcamps: Incentives for participation in ad-hoc conferences and the role of social media. i-KNOW'13. 2013.

[79] Topps D, Dennerlein S, Treasure-Jones T. Raising the BarCamp: international reflections. MedEdPublish 2017;6(3):8. https://doi.org/10.15694/mep.2017.00012 [online: Available at:.

[80] Duffin K. Mob rule learning: camps, unconferences, and trashing the talking head by Michelle Boule. Publ Serv Q 2012;8(2):150–1.

[81] Budd A, Dinkel H, Corpas M, Fuller JC, Rubinat L, Devos DP, Wood NT. Ten simple rules for organizing an unconference. PLoS Comput Biol 2015;11(1).

[82] Anderson G, Gijón C, Whalley J. Internet access in Glasgow's deprived areas (Report 2015). Edinburgh: Citizens Advice Scotland; 2015. https://www.cas.org.uk/system/files/publications/internet_access_in_glasgow_2015_final_-_aug_15.

[83] Greenhalgh T, Humphrey C, Woodard F. Where next for user involvement?. In: User involvement in health care. Chichester: Wiley–Blackwell; 2010.

Chapter 21

Promoting platforms for bottom-up participatory governance: a policy instrument approach through the facilitation of strategic smart city governance

Carles Agustí i Hernàndez
Open Government Area of Presidency, Barcelona, Spain

21.1 Introduction

The arrival of citizen participation and its subsequent evolution and integration into the logics of Open Government, Smart Cities, governance, and knowledge management have led to the emergence of different terms that, before entering to analyze them and see their possible implementation in Smart Cities, we need to investigate in order to understand them from the point of view of their own definition.

21.1.1 Politics

The classic definition of politics would be that it is the decision-making process in human groups, a definition similar to that of governance. It is usually linked to governments and public administration, but it is a definition applicable to any human group and refers to the decisions and strategies implemented.

If we go to the field of Smart Cities and talk about "Smart Cities politics," we would be referring to what the decision-making process is and what decisions and strategies we implement in them. If we go to the field of Smart Cities policy in governance, we will be referring to the decisions and strategies we will carry out in Smart Cities governance. Later we will see the definition of the term "governance" and its variables and derivatives [1].

If we talk about general meaning of the term "politics," the subconscious leads us to talk about governance logics of governments and political parties. In this sense of "politics," we go to a more swampy terrain, and we must also take into account if we are working cities and, therefore, if we are engaged in public administration. Here we cannot ignore the "Crisis of Politics" concept.

The crisis of politics would be determined by a distance and mistrust present for years and which has not been reduced, between citizenship and politics, a definition in which citizens fit everything they consider public: governments, parties, civil servants, public administration.

Probably the explanation for this mistrust is found at a different rate in the evolution of society and public administration. Society advances much more rapidly than public sector and parties, thus causing distance. The analysis is certainly different around the world, but if we analyze the evolution of, especially, Western societies we can see that we have evolved a lot in the last 40 or 50 years in democratic terms, from dictatorships to consolidated democracy. From passive societies, which voted every 4 years and nothing more, to active societies that want to "vote," in a symbolic sense, every day, in the sense of giving an opinion and influencing. Above all, it has evolved into much more trained and prepared societies and much better integrated, by generational evolution itself, into technological governance and in general with the technological tools we enjoy today.

On the contrary, governments, public administrations, political parties, probably immersed in everyday life and without much capacity to rethink, innovate, and evolve,

Smart Cities Policies and Financing. https://doi.org/10.1016/B978-0-12-819130-9.00043-7

they have not evolved at the same speed. The techniques of transformation and adaptation to new realities, such as Open Government, have arrived but let us say that, at least for the time being, they have not had an implementation that has profoundly transformed the way politics are done and especially its relationship to the citizens and these with the public administration.

All this affects not only the governance of Smart Cities but also the Smart Cities themselves. Since, on the one hand, there is this crisis and distrust of citizens. On the other, the very concept of Smart City and its innovation and transformation can be a solution or solutions to the crisis of politics.

21.1.2 Smart cities

What are Smart Cities? Beyond the very meaning of the term, the issue deserves a deeper landing and analysis as its arrival has significantly affected our cities, the policy applied to them, and their design and evolution. When the term Smart Cities appeared, more than 10 years ago, beyond the very meaning of the term, the definition was nebulous and confusing. Nowadays it is not clear how it will affect cities and what it will mean. It floats in academia, but it does not land in the world of real urban governance.

Later, we can see Smart Cities identified in Technology. Smart Cities are technology. Therefore, we have technology applied to our cities, but how? Doing what? At the moment, the definition was no longer advancing, and we were in an era where, perhaps, some companies were beginning to notice it but, in general, citizens did not notice any impact.

Finally, we come to a definition that for me is the most accurate. "Smart Cities are strategy and technological solutions for the improvement of the daily lives of citizens." Here the citizen is already incorporated into the equation. He appears in it. He is already the main protagonist. He receives directly the benefits of Smart Cities: strategy, and they are also concrete solutions, like the ones we see on the street, and we can enjoy in our day to day in the city or through technological tools such as our own smartphones.

We can see how the citizen becomes part of this governance of Smart Cities as the final receiver. Not only as this, and here the intention of this chapter, but as the protagonist, taking advantage of their opinion, contribution, and knowledge.

21.1.3 Governance

Governance is a concept that refers to the model in how a government, association, or company is managed internally and in relation to its workers and citizens. This is a definition of governance which I work with and feel most comfortable with. First of all, we see two very remarkable issues. It is not just about governments, and it is not just about workers.

The term governance, curiously, does not exist as such in many languages, and language academies are working to incorporate it or have recently incorporated it. Classical governance was a concept basically attributed to governments and the public sector, but in this modern and up-to-date definition, we can see these two distinctive details:

It is not just about governments and the public sector, but governance, as we see in "politics," refers to any human community. Therefore, transcending governments and the public sector to go to business, association, institution, or any human collective.

The other big news is that it not only goes from public servants, in the case of public sector, or workers of a company in the case of private sector. It transcends these areas, especially in public sector, to go also to the citizen and incorporating them in the governance as they require the new models of management responding to the social demands.

Governance is also, like Smart City, a space with various aspects or families that has been incorporated with the evolution of policies, especially public ones. Governance, initially, was a term that could be equivalent to management or almost a synonym meaning of management of the term "politics." In other words, with relation to citizens, governance was practical and unique. Later on, the emergence of open government, incorporated what is known as "citizen participation," in the logistics of open government, in conjunction with transparency and open data. Eventually, the emergence of Smart Cities gives rise to what we know as Smart Governance or Smart Citizens, which would be the incorporation of governance in Smart Cities or Smart Cities in Governance as you look at it. Above all it involves the arrival of technological solutions and tools to governance, both internal and, as far as the case is concerned, especially in relation to the relationship with citizens. Finally, but this would be for another chapter, the Sustainable Development Goals (SDGs) are also incorporated into this world of governance, as, in addition to being a guide to sustainability, are also a guide for governance.

21.1.4 Platforms

Technology platforms are virtual spaces for interconnection between all the protagonists of a project, including citizens, in order to create, share, and enrich them. They are, in themselves, a Smart City product. A technology that allows a large number of people to be easily contacted, monitored, shared and, in fact, be a reference tool for the deployment of projects and public policies.

21.1.5 Bottom up

Continuing with the terms of the title of the chapter as a guide to advance in the definitions of the key concepts of this chapter, we bump into the concept "Bottom Up" Beyond a definition, a meaning, a different current through which ideas must flow, it is a philosophy that we must internalize if we want, not just innovate, but to incorporate workers and citizens into the governance models of Smart Cities.

Bottom-Up means not only the flow but also the effective incorporation of citizens and workers in the design, monitoring, and improvement of projects at all levels and also in the global project of strategic design of our cities. It is a principle to be incorporated in the design of global strategies, but also in each of the projects to be developed. It is a philosophy that recognizes the value and knowledge of any human community, whether citizens as a whole or a territory, concrete, or of the workers of a public administration or a company.

21.1.6 Participation

Participation is the right and the duty of the citizenry to be informed and to contribute with opinions and ideas to governance; initially understood as public, that is, governments. As we have seen throughout this chapter, however, where these issues evolve, are to transcend to the public sphere and then to go to the private sector; and, to consider that all human groups have that knowledge. In other words, the right and duty to participate, goes far beyond a government's citizenry, to also include public workers, companies' employees or associations.

Applied to Smart Cities, participation or participatory governance means recognizing the role that citizens can play in these projects. A topic that was extensively addressed in the previous volume of this series of books and where the citizen can participate from the ideation phase and going through all the phases: implementation (construction if we talk about an infrastructure), evaluation, maintenance, and improvement of the specific strategy, policy, or infrastructure to which we refer. Throughout the chapter we will see how, from a theoretical but also practical point of view, this integration of citizens in Smart Cities design models can be carried out.

21.2 The cities of the future

If we want to discuss and talk about participatory governance in Smart Cities, I think it is essential that we talk about what these cities will be like. Beyond the fact that we establish Smart strategies with the consequent solutions for the day-to-day improvement of their citizens, our cities are already undergoing an evolution, a change of model, which is marked by sustainability and accelerated by Smart Cities.

Where are cities evolving? In addition to going into detail on four key areas, such as mobility, urban planning, green spaces, and governance, we can also do a more generic analysis. We could say that our cities are evolving toward a model of more tranquility, less noise, less cars, more green spaces, and more prominence of the citizen in every way. Let us see it in parts:

21.2.1 Mobility

Modern cities were designed for exclusive vehicle mobility. It was like this in an innovative way, when the vehicle was residual, already foreseeing the car like the natural vehicle of intracity displacement. To this effect practically all the streets were designed with the classic and innovative format at the time. Two sidewalks and a central road for the circulation of vehicles.

Even today, this system has not been questioned. In fact, a majority of citizens do not yet question it, but there are some factors that are putting it in question and that are slowly transforming our cities.

First of all, the sustainable factor—pollution. Cities have become saturated with vehicles and for years they have been emitting an excess of gases causing an index of pollution above those tolerable for human and natural well-being. The car is one of the main causes of this increase in pollution.

On the other hand, beyond the evolution that public transport has undergone in recent decades, a whole series of vehicles have appeared. Some of them still in the experimental phase, which all at the same time and in addition to saturating or hindering the adaptation of the city's mobility to new uses, are an opportunity for intracity, ecological, cheap, and easy travel, which can greatly help reduce vehicle use in cities and improve pollution rates.

Finally, there is a growing pressure for the pedestrian recovery of public spaces, not only for walking but also for enjoyment, which has led to an increase in the size of sidewalks and pedestrian spaces in detriment of vehicles. All these are factors that lead to a decrease in the use of car and its travel as the vehicle par excellence in intracity travel, giving way to the other systems we have discussed. We could say that, in the cities of the future, the presence of vehicles, at least private ones, will be quite testimonial, gaining the city well-being in the form of more spaces, less noise, and less pollution. It is all these aspects that we must inevitably consider when designing Smart cities strategies, which have a challenge in providing solutions to this complex and new mobility, while providing information and facilitation tools in every way for citizens in these cities where they will become protagonists in front of the car.

21.2.2 Urbanism

Urban planning in cities, beyond the obvious effects we saw in the previous point, has increased the commitment to design and harmonize with the city. We move from functional urbanism to one of more design and image, which naturally must go hand in hand with the city and be one of its symbols and attractions.

It is also affected by large infrastructures, from newly designed parks to all kinds of infrastructures for public use that have undergone an important evolution, also toward design, in recent years or decades. The role of the citizen here is decisive, both from the point of view of the receiver and user of this urbanism and from the point of view of the design and contribution to the maintenance and improvement of urbanism and each of the infrastructures. It is the citizens who must provide us with data on the assessment, use, and improvement of our cities and their urban planning. What city we want from this point of view can only be said by its citizens. Within the Smart City ecosystem, we could place Smart Urbanism, or following the definition of Smart Cities, in the application of strategies and technological solutions for urban planning in cities, which makes it easier and improves the day to day of its inhabitants.

21.2.3 Green spaces

As we have already mentioned, due to a sustainable factor and to the growing prominence of citizens in our cities, they are and will increase their presence and their central position in citizen urbanism. The cities of the future will be greener and with more space for enjoyment and citizen walks.

Smart City strategies must consider this aspect and see which solutions to provide both from the strategy and from more specific aspects to technological solutions. Aspects such as air quality, weather, allergies, safety, or the design of green routes are growing needs in cities and susceptible to smart cities strategy and solutions.

Beyond green spaces, entering sustainability and more specifically in the field of Smart Green, smart cities strategies, must not only take into account the Sustainable Development Goals (SDGs), but must be fully aligned with them. The SDGs have a lot to do with participatory governance, as they have been conceived, and hence their success, not only as a guide to sustainability but also to governance, makes them perfectly identifiable by the entire population, far from being 17 purely technical goals as some believe. Some person once told me that if he had to form a government, he would take the 17 SDGs and make 17 ministries. So that be it.

21.3 Governance of smart cities

Analyzing the contents, definitions, and evolution of cities and governance, we can enter to examine in more detail the governance of Smart cities from a point of view of the role of the citizens or from a perspective and model of participatory governance of Smart Cities. We understand the governance of smart cities as the management of smart cities, especially in the case of the study, with respect to its citizens. In other words, defining what is or can be the role of citizens in the governance of smart cities, an aspect that was the subject of extensive analysis in the previous volume of this series of books on Smart Cities.

In summary, we could say that the citizen is the protagonist of Smart Cities and, therefore, with a prominent role in the design of their governance. First of all, the cities and the policies that are implemented in the cities, in this case in the Smart Cities, are made for the citizen, which is the last reason to be of these. We look for their comfort, facilities, and well-being; therefore, something citizens will have to say in all this debate.

The role of citizens in Smart Cities could be said to be transversal to the entire governance model, as they play a key role in detecting the needs of the city. They must play a role in designing solutions, but its role does not end here. Once these solutions have been implemented, there is no better expert than the user. The citizen who will make use of the city and the technological solutions that are provided, so you can assess. We must be able in the governance model to incorporate this opinion, an assessment that citizens make of it. And we must be able to go even further. The citizen as a user can be a very good source to determine the improvements to the implemented solutions and also an ally in the maintenance of the infrastructures, acting as a detector of those deteriorations or damages that there may have.

Citizen is no longer just the protagonist of cities, but is an ally for its degree of expertise and information in the model of governance and in all its phases. At this point, it is important to situate the evolution that citizen participation has undergone, which undoubtedly affects the participatory governance of Smart Cities. As we mentioned before, that isolated citizen participation, preopen government, noncentral, and nontransversal, gave way to Open Government being integrated into a logic of more global and transversal transformation, but has also evolved the management of these citizen contributions that arrive to improve projects. Moving, we might say, from citizen participation to knowledge management.

That citizen participation that in some cases took more into account quantity than quality; showing that you listened more than the real will to listen and internalize what citizens think and that it was isolated far from a cross-cutting influence on institutions, has evolved. The techniques of collective intelligence and knowledge management have arrived, based on the fact and recognition that knowledge resides in each of the citizens, each with their own abilities. Every human community: a business, a town, a city, a neighborhood, or even a group of friends or family, has knowledge, they hold a significant amount of knowledge.

If we take as an example a neighborhood or a government with hundreds or thousands of workers, or all the workers of a City Council or of a company, there is a very important knowledge in the sum of these individuals. How many economists do we have in the neighborhood? And sustainability experts? And architects? And data experts? I'm sure there are many. But it is not just a matter of valuing enlightened knowledge, but "popular" knowledge is also a value, it is also knowledge. For example, taxi drivers with mobility or users of a service (Metro, train, a bridge), users of a technological solution ... become experts in this service. The same methods are applicable to Smart Cities, considering the citizen as a source of knowledge for himself, for his expertise in the design of Smart City strategies and policies, but also for his leading role as a user of this city and the services and solutions implemented there.

Collective intelligence means that where we find two people more ideas and knowledge arise than where there is just one, or where there are 10 or 1000 more than where there are two, for a sum effect of knowledge but also as a result of the synergies that are taking place. For all this knowledge to contribute, to work, to flow, it is necessary to connect it, to put a method into it. If we do not connect it, it is lost. It will only be rendered by the person or people who are in contact with it. If we connect and create the tools and mechanisms needed to capture it, we can flow an unimaginable amount of knowledge, ideas and opportunities, and behind it all also detect talent.

21.4 Instruments of governance and cultural change

Prior to the study of governance in Smart Cities, in particular if we are studying models of participatory governance applied to our cities [2], we need to analyze the need for cultural change that should be done at all levels, but especially in public administration, and more generally in the public and political world in general. We have previously talked about the crisis of politics motivated by the distance and distrust of society with respect to politics and the public sector in general. The demands of evolved society will give us clues as to where this cultural change should go.

21.4.1 Listening

The representative public and if we transfer it to the person in charge of Smart City strategies, should be willing to listen, but not as an obligation, not as a smokescreen selling and pretending to be heard when in reality are not interested in what the citizen says, but in a sincere, mentally prepared way. Recognizing that value of knowledge that society has and that we commented before.

21.4.2 Observing

The participatory attitude must be proactive. It is not just waiting but going, being concerned about what people think. Observing what is happening in the city, paying attention to the behavior of citizens and their opinion.

21.4.3 Capture

The opinion of the people and beyond this, the knowledge that we have seen that people have and that can contribute to the improvement of our projects, must be captured in an orderly manner, not just stop the hearing. Symbolically yes, but it is not just that, you need to create the necessary channels, platforms, spaces for dialogue, participatory systems .... to capture and order this knowledge. Every participatory process has three fundamental phases, without which, or if only one of them fails, we are no longer in a real participatory process.

21.4.4 The creation of channels

With the objective of listening to the population, we must have activated the necessary channels in order to be able to capture their knowledge. Plural channels, to reach the whole population, or rather, for the opinions of the people to reach. They must be plural because not all the population responds to the same characteristics. There are those who always send their opinion, those who do not have the capacity to do so, and we must create the necessary channels to capture this opinion and knowledge of all typologies.

21.4.5 Internalization

What arrives must be processed, treated, and ordered for the purpose of enriching the projects. We have to believe in it and try to internalize it, apart from the fact that on a more practical level, we create the systems to order the knowledge.

21.4.6 The answer

There is nothing more discouraging for a person who has spent a few minutes or a few hours making proposals or giving his opinion on any topic, than not having any kind of answer as to whether his opinion has been processed, has helped, has not helped ... The response process must be guaranteed in all public administration and therefore also in Smart City projects.

21.4.7 Share

It is the principle of the 21st century, which explains the phenomenon of social networks and the engine that drives the generation of knowledge. If we do not share, if we do not create synergies, knowledge is closed. It does not connect; it does not produce. We need to share information and knowledge that we think might be useful.

21.4.8 From the outside in and from the top down

Analysis and change in currents of thought are also critical for knowledge to arrive and flow. The currents of opinion and thought, in any organization, obviously must not only go from top to bottom, but also from bottom to top. The famous "bottom up" that we have already discussed, but also from outside to inside. The knowledge management must not only remain in the organization itself, but everything that can come from outside is sure to contribute and enrich.

21.4.9 Citizen participation

Citizen participation, as such, along its evolutions and variants, is obviously a usable tool for the governance of Smart Cities, intrinsic in a model of participatory governance.

Citizen participation, apart from being a right, is an essential tool for achieving citizen opinion, improving projects, monitoring them, and socializing the chosen strategy.

21.4.10 Transparency

Transparency is another of the fundamental branches of Open Government. Open Government cannot exist without transparency, real transparency. The cultural change that we discussed before is not enough to comply with the laws, requirements, and regulations of transparency, but it must be clear. It is not enough to make glass walls, we also need to transform and adapt our intern activity.

Transparency, in addition, apart from being a citizen right, because we are talking about institutions and consequently about public money too, it is also, the mechanism that allows citizens to have information about the projects. Without this information, you cannot receive an opinion with criteria. Furthermore, we can add that transparency is a mechanism of prevention and fight against corruption. Applied to Smart Cities, transparency is necessary to publish projects and strategies, in order to socialize with the population, for the purpose of making it more their own and providing more solid criteria to gather public opinion.

21.4.11 Open data

Open Data is the third leg of Open Government and also an essential governance tool. Data governance is a whole world that is revolutionizing both public and private management. The power of reused and well-treated data is just beginning to be seen.

The derivative public of big data is open data. They constitute a right, as they are public. As with transparency, as long as teaching them does not affect the strategy of the institution, they have to teach. Here the cultural change lies in overcoming the Gollum effect, in moving from considering data as "my treasure," to seeing that shared (sharing, remember, the beginning of the 21st century that has revolutionized information) can contribute so much to the organization as the citizenry.

Open data means complete the three phases of opening" (1) opening, (2) following what happens with the opened data, and (3) reuse it. We talk about information that, being public and therefore searchable, downloadable, and treatable, can be used as a basis for all types of studies and also business models.

Applied to Smart Cities, the data allows a better understanding of studies, projects, and strategies. Working with it and developing business models that complement the strategy and services, operate as a source of information for a better opinion, and work as a reception method of citizen knowledge allows us to strengthen the project.

21.4.12 Open government

Throughout this chapter, we have spoken several times of Open Government [3], which is also a philosophy in the sense of a new direction in the way of governing that responds to the demands of society. At the same time, is a successfully tested government formula, which groups in a single strategy citizen participation, transparency, and open data, creating a practical paradigm of transformation in the relationship with citizens, of these with the administration and also internal. As we have also seen, Open Government is a cultural change, in fact, a cultural change aimed at opening, listening, internalizing, dialogue, explaining, and capturing knowledge.

Open Government can be optimally applied to Smart Cities if we want them to have a participatory governance model. The design and implementation of smart city policies and strategies must be done with an open mind, transparency, sharing, listening from the beginning, integrating citizens into the governance model (as key actors, users ...), and valuing the knowledge that, as I have argued, any human community has.

21.4.13 Collective intelligence

Collective intelligence is the sum of the intelligences and knowledge of any human community, can be a family, a village, a neighborhood, a city, a company, or an association among others. The interconnected knowledge and intelligence of two individuals will always add more than one ground. Knowledge sharing and dialog generate more evolved proposals and ideas. The more individuals, the more knowledge and the more collective intelligence, although there is also more difficulty in capturing and managing it.

Beyond the generation and internalization of this knowledge, it is necessary to order it to create a method and to put in place a system that allows us to really use it. Perfectly implementable in Smart Cities, which cannot waste the knowledge, ideas, and proposals of human communities to whom they are not only addressed, but who will also be the main users. Smart cities do not make sense if they do not reach the citizens and improve their day to day.

21.4.14 Cocreation spaces

Cocreation spaces are a technique for managing knowledge and citizen participation that act precisely in the line of ordering and managing the internalization of this citizen knowledge. They are usually web spaces, in which a project, idea, or question is explained in a pedagogical way. Citizens can post solutions or contributions they consider appropriate with the support of documents, videos, or any other plugin they deem appropriate.

Over time and technological advancement, cocreation spaces have acquired more possibility of interaction between participants, increasing the aforementioned effects of collective intelligence by creating synergies and exchanges of information.

It makes perfect sense to generate cocreation spaces in the design of Smart Cities and especially when it is applied to some specific projects.

21.4.15 Participatory platforms

Participatory platforms, among which there would be cocreation spaces. They are a broader term that groups these and all other technological variants of contact and dialogue with citizens. They would respond to the same approach as cocreation spaces although they have more variation, such as participatory processes, participatory budgets, citizen consultations, among others. They are also considered perfectly valid techniques in Smart Cities, not only for capturing knowledge and gathering public opinion, but also for other citizen participation systems like the ones we have seen.

21.5 Summary

In summary, as closure and final conclusion of the chapter, participatory governance for Smarts Cities is not only possible, but necessary. The tools technology brings to democracy and dialog with citizens, open a world full of possibilities between connection, dialog, participation, and contributions, they surely suppose a big enrichment of the projects and a better socialization thereof in behalf of the citizens.

Finally, let us move on to the real interactive part of this chapter: review questions/exercises, hands-on projects, case projects, and optional team case project. The answers and/or solutions by chapter can be found in Appendix G.

21.6 Chapter review questions/exercises

True/false

1. True or False? The arrival of citizen participation and its subsequent evolution and integration into the logics of Closed Government, Smart Cities, governance, and knowledge management, has led to the emergence of different terms that, before entering to analyze them and see their possible implementation in Smart Cities, we need to investigate in order to understand them from the point of view of their own definition.
2. True or False? The classic definition of politics would be that it is the decision-making process in human groups, a definition similar to that of governance.
3. True or False? Governance is a concept that refers to the model in how a government, association, or company is managed externally and in relation to its workers and citizens.
4. True or False? Technology platforms are virtual spaces for interconnection between all the protagonists of a project, including citizens, in order to create, share, and enrich them.
5. True or False? Bottom-Up means not only the flow but also the effective incorporation of citizens and workers in the design, monitoring, and improvement of projects at all levels and also in the global project of strategic design of our cities.

Multiple choice

1. Participation is the right and the duty of the citizenry to be informed and to contribute with opinions and ideas to:
   a. Governance
   b. Participate
   c. Implement
   d. Invest
   e. Policy

2. Modern cities were designed for exclusive vehicle:
   a. Displacement
   b. Mobility
   c. Circulation
   d. Pollution
   e. Sustainability
3. Urban planning in cities has increased the commitment to design and harmonize with the:
   a. Machine
   b. Humans
   c. City
   d. State
   e. County
4. Beyond green spaces, entering _________ and more specifically in the field of Smart Green, smart cities strategies, must not only take into account the Sustainable Development Goals (SDGs), but must be fully aligned with them.
   a. Holistic integration
   b. Sustainability
   c. Smart city platforms
   d. Computer applications
   e. E-economics
5. Analyzing the contents, definitions, and evolution of cities and governance, we can enter to examine in more detail the governance of Smart cities from a point of view of the role of the citizens or from a perspective and model of participatory governance of:
   a. Smart governance
   b. Smart Cities
   c. Open Government
   d. City Council
   e. Collective intelligence

Exercise

Problem

Address strategies, models, and the motivation behind smart cities by analyzing smart city project cases in medium-sized cities.

Hands-on projects

Project

Do research: Analyze the forms and social implications of citizen engagement in government-sponsored participatory innovation platforms that are designed to serve urban economic renewal.

Case projects

Problem

Discuss the forms and social implications of citizen engagement in government-sponsored participatory innovation platforms that are designed to serve urban economic renewal.

Optional team case project

Problem

Compare the various definitions of smart cities, and integrate the current insights from the field of e-participation.

References

[1] Anthopoulos LG. Understanding Smart Cities. A tool for a smart government or an industrial trick. April 13, 2017. https://books.google.es/books?id=2_eqDgAAQBAJ&printsec=frontcover&dq=smart+cities+governance&hl=ca&sa=X&redir_esc=y#v=onepage&q=smart%20cities%20governance&f=false.

[2] Cardullo P. Citizens in the smart city. December 2020. https://books.google.es/books?id=G2r0DwAAQBAJ&printsec=frontcover&dq=smart+cities+governance&hl=ca&sa=X&redir_esc=y#v=onepage&q=smart%20cities%20governance&f=false.

[3] The White House. Open government directive. December 8, 2009. https://obamawhitehouse.archives.gov/open/documents/open-government-directive.

Chapter 22

Policy instruments for facilitating smart city governance

Doaa M. El-Sherif[1] and Essam E. Khalil[2]
[1]Housing & Building National Research Center (HBRC), Ministry of Housing, Utilities & Urban Communities, Cairo, Egypt; [2]Cairo University, Cairo, Egypt

22.1 Introduction

The bulk of smart city research focuses on conceptualization—on mapping the meanings, components, and goals of the smart city from the perspectives of particular fields of study, such as public administration, information science, and urban development [1–4]. Although these studies consider interactive governance to be an essential component of smart cities, they fail to examine this in any great depth [2,5,6]. This paucity of knowledge prompted the present attempt to contribute to smart city discussions by exploring smart governance at the interface of different study areas.

A systematic analysis of the literature offers a means to map what is known about smart governance and its effects and thus to provide a more realistic perspective on the value of this approach for cities. To that end, the present study is guided by the following research question: What relationships exist between ICT-enabled citizen–government collaboration and sustainable urban development [7] and how do contextual circumstances influence these?

The contribution of this study to the discussion is in the first place conceptual: we aim to identify the various components and dynamics of smart governance to develop a rich understanding of this ICT-supported collaboration mode. Secondly, this paper contributes to the literature by broadening our understanding of the relationship between smart governance and outcomes in terms of its contribution to urban sustainability. While the literature describes many opportunities and threats in this relation, we still know very little about the environmental, societal, and economic outcomes of ICT-supported collaborative governance [3,8–11]. Finally, this study enhances our understanding of the role of context. Governance dynamics are mediated through multiscalar variables and moderated by contextual features of the local urban setting. Contextual aspects include external place- or domain-specific circumstances, such as physical, political, institutional, societal, economic, and cultural conditions, as potential influences in digitalized governance processes. Although widely recognized, a systematic overview of the influence of contextual factors on smart governance processes is still lacking [12–15]. This study fills this gap by opening the black box of the context-reliant nature of smart governance.

In sum, the major objective of this study is to generate a factual basis for debate by providing an overview of what is known about the context-dependent contribution of ICT-enabled citizen–government collaboration to urban sustainability. The article is structured as follows: in the next section, we introduce our research methods, followed by the section wherein we examine the main building blocks of smart governance. Then we present what the literature reveals about the (sustainability) outcomes of smart governance, while the section after that highlights the major contextual factors. In the last section, we discuss policy implications and suggest a future research agenda.

22.2 Literature review: methods and the corpus

In order to address the research question formulated in the preceding, we conducted a systematic literature review. Cooper and Hedges [16] formulated six stages for a systematic literature review, i.e., problem formulation, literature search, data evaluation, data analysis, interpretation of results, and public presentation. Our problem formulation has already been presented in the introduction, and this paper represents the presentation of our review to the public. The remaining steps can be grouped into two main

Smart Cities Policies and Financing. https://doi.org/10.1016/B978-0-12-819130-9.00052-8

categories: the literature search and the data evaluation, analysis, and interpretation. We will elaborate on these two phases below. For our literature search, we established a set of rules describing the study and report eligibility criteria and the search strategy:

- *Topic and search terms*: To cover all dimensions of our research topic in an integrated way, we used the following combinations of search terms: e-governance participation, e-governance sustainability, citizen e-participation urban sustainability, smart city participation sustainability, city participation ICT sustainability, governance e-collaboration citizen sustainability, smart city citizen sustainability, e-participation cocreation sustainability and collaborative e-governance. Our aim was to find articles dealing with the topic in its entirety and addressing, in particular, the relationship between smart governance and sustainability, rather than its distinct elements separately (such as collaboration, governance, innovation, e-government, participation, sustainability, etc.).
- *Study and research design*: We included all types of research designs (case studies, questionnaires, experiments, literature review, comparative research) with a goal of covering the full breadth of the research field to compensate the identified shortcomings in scope and depth in the scientific literature.
- *Year, language, and publications status*: Studies were retrieved that were published in the period 2006–2016, as smart cities and the use of ICTs for participatory/collaborative governance are recent phenomena (Harrison et al. [17]. Only publications written in English language were considered. In order to assure the quality of our literature review, we only included international, peer-reviewed (conference, proceedings) papers and books and book chapters from well-known academic publishers.

We started the search process by using established scientific databases, such as Scopus, PiCarta, and Web of Science. In the first round, we used the terminologies "e-governance" AND "participation and e-governance" AND "sustainability" as titles, abstracts, and key words in the query entry. Based on the above detailed search and eligibility criteria and their accessibility, we could identify only 11 relevant publications. To expand the results and to build up a robust body of relevant literature, we then used Google Scholar. While we are aware that the algorithm of Google Scholar is not transparent, we opted for its use as this yielded many more relevant publications. Since the more general search terms (such as "e-governance") on Google Scholar generated a huge number of irrelevant articles, we formulated specific combinations of terminologies, targeting our research scope (as listed above). Each combined search term delivered more than 40,000 hits. The selection took place by identifying appropriate articles from page one by reading their titles, abstracts, and introductions. Beyond the defined search criteria, the records were also assessed according to additional quality indications (journal, edited book by renowned scholars) and impact (number of citations). Each search term was assessed to about page 10 or 12, up to a point where either irrelevant or previously identified papers were exposed. This resulted in an additional 83 articles.

Our search through digital databases produced a total of 94 articles. In addition, we included further studies in the analysis from other sources, namely recommendations from academics and peer reviewers. Systematic literature procedures, such as the PRISMA method, allow the use of documents from sources other than database searching Liberati et al. [18]. The extensions we used provided additional perspectives, both as regards updating and the inclusion of other fields of study (public administration, urban issues, etc.) important for our theme. This resulted in 20 additional papers.

Almost half of the examined studies originate from specialized areas that deal with the application of (information) technologies in human-related interactions by connecting issues of government, technologies, collaboration, citizen participation, and sustainable development. At the same time, the number of articles that precisely match our topic—namely, the effect of technology-enabled collaboration between citizens and local governments on sustainable development—is small. This confirms that our review is a valuable contribution to understanding how technologies affect and are affected by urban environments.

For our data evaluation, analysis, and interpretation, we worked with a structured spreadsheet. For each main variable (i.e., smart governance, outcomes, and context), we identified subvariables as a starting point for the analysis since the main variables are often not explicitly mentioned in the papers and because these broad concepts consist of multiple components.

As subvariables of smart governance, we identified governance actors, organizational characteristic, types of ICT use, citizen participation, and the (spatial) scale of collaboration. The subvariables of sustainability outcomes were defined as the desired and adverse social, economic, and ecological outcomes and their impact on various societal and spatial levels. The third group of subvariables referred to the context and entailed culture and democratic tradition, policy domain, trust, socioeconomic characteristics, demographic characteristics, spatial characteristics, and Internet penetration. Given the explorative nature of the review and the preliminary character of these components, we iteratively further developed these subvariables throughout the research.

Based on the demarcation of the main and the initiatory subvariables, we created an Excel factsheet for encoding the articles. Three researchers encoded each of the articles independently from one another. In order to guarantee intercoder reliability, the three researchers discussed the results and the work methods and refined the analytical factsheet. The analysis and interpretation of the results were performed qualitatively, since the perspectives on the topics as presented in the various papers varied considerably in scope, nature, and depth. A follow-up assessment of evidence for certain relations was not possible at this stage, due to the low number of empirical evaluations of smart governance. To arrive at an overview picture, we aimed to map this variation and identify the various conceptualizations and mechanisms.

22.3 The components of smart governance

To this end, we examined the main components of smart governance. These components are in line with our definition of governmental organization, citizen participation (and, consequently, government–citizen collaboration), and the use of technology.

22.3.1 Governmental organization

The first building block of smart governance is the organization of government. This term entails a whole range of subfacets such as motivation, vision and strategies, attitudes, decision-making, process coordination, and roles and responsibilities, as well as the provision of financial, regulative, technological means and human resources, knowledge management and organizational culture, etc. [19]. It emerged from our analysis that three organizational characteristics are critical for smart governance: (1) commitment, (2) responsiveness, and (3) operational management.

The first organizational characteristic, commitment, refers to the extent to which local government is motivated to engage in sustainable development through ICT-supported urban collaboration. According to a large number of papers, smart sustainability governance will not ensue without the cooperation of local governments [20–23].

The second organizational characteristic regards the responsiveness of government. A long-standing relationship between government and citizens requires governments to become a receptive partner toward inhabitants [20,24–28]. In order to continuously improve the effectiveness of e-participation programs, government leaders should report on the level of e-participation to major stakeholders and solicit feedback. This aspect is important in sustaining fruitful online collaboration since the degree of satisfaction of e-participants with the responsiveness of their government has a direct and positive association with their perceptions of individual development, their perceived influence on decision-making, and their trust in government [15,29,30].

The third organizational characteristic is operational management. This classic organizational feature provides the operational basis to equip technology-supported collaborative governance arrangements with whatever is needed for their running and covers all facilitative types of measures [31], actions, and preconditions. Many articles emphasize that this requires an integrated approach combining infrastructural, technological, social, and political systems, as well as cross-sectoral bridging between policy domains and urban priorities [22,32,33].

Process management and coordinated leadership as well as a well-crafted reiterative participation process with clear guidelines for the interlaced elements of strategic planning, design, and implementation phases are vital. Likewise, flexible institutional designs based on networking among interdependent public, private, and civil actors will support ICT-supported collaborative forms of coinitiation, codesign, and coimplementation, which may lead to creating novel public services [23,29,30,34,35].

A large strand of literature identifies a plethora of other organizational aspects for stimulating smart governance between citizens and their local administrations, such as financial resources and funding; time; the provision of technological training and guidance; the evaluative monitoring of processes and outcomes; supportive and protective (privacy) legislation [36–39]. Hence, what is needed is clear. However, the literature tells us that the ambitions often remain within the realm of rhetorical phrasings. Despite the verbal adulation of sustainability and technology-enabled stakeholder participation, this does not often result in the implementation of significant changes in these areas [8,40–43]. Sustainability policies, therefore, often tend more to resemble "greenwashing" than actual commitment with digital technologies primarily being applied for administrative and transactional efficiency instead of, for instance, in support of dialog [35,44–46].

Urban governments still regard the public as mere recipients of information, often failing to consider citizens' various perspectives in decision-making or to provide feedback on ICT-supported participation processes [15,29,30,41,47–50]. This is mainly due to the municipalities' lack of understanding, capacity, and expertise. In addition, many articles point to the reluctance of local governments to share power with citizens and to give up their autonomy [1,42,44].

22.3.2 Citizen participation

The second building block of smart governance, and one according to the literature that offers huge potential, is citizen participation. Citizens can offer useful and helpful suggestions for government agencies to arrive at better informed policy decisions [36,51–53]. Relevant components are (1) the degree to which these forms of citizen participation in urban governance are interactive [54], (2) the representativeness of the participating population, and (3) motives for citizens to participate.

Concerning the level of citizen participation, the findings indicate that technology-supported governance initiatives have modest outcomes. The literature mainly referred to nonparticipative and nondeliberative activities such as political support-seeking, consultation, or one-way information provision [30,33,35,41,55]. The review shows that authentic e-participation, leading to policy changes through citizens' authority in decision-making, is nearly absent in practice [37,43,56]. According to some, better governance through digitally facilitated cooperation between citizens and governments is a myth, since very few empirical studies could offer evidence of improvement in this area [20,57]. In that respect, many papers cast doubt on the value of ICT tools in participation by arguing that digitization has not resulted in broader and deeper civic involvement.

The findings are more diverse regarding citizens' capacity and willingness to engage in smart governance. In this regard, questions on the issue of the representativeness of citizen participation in smart governance abound. Numerous papers stated that preexisting participation patterns, including socioeconomic and technological segregation, have been strengthened as a result of applying new technologies in government–citizen relations [22,58,59]. The reason for this is twofold: on the one hand, the technology has enabled the emergence of a group of very active participants who are usually better educated, more affluent, technologically more competent, and who know more about urban policy issues [43,53,58,60–62]. For instance, the US cities most noted for environmentalism, and a sustainability ethic is characterized by a youthful, idealistic, and prosperous citizenry that have strongly influenced municipal sustainability policies. On the other hand, the emerging technology also created a group of "unplugged" citizens who do not participate, typically made up of people with low incomes, insufficient schooling, and/or a marginalized status (immigrants, disabled, elderly), or living in isolated/rural locations [30,38,41,44,59,63,64]. Although members of the older generations are usually more engaged and interested in politics and public affairs, they are digitally less skilled and prefer traditional (face-to-face) participation [30,56,65,66]. Young people have considerable experience in using computers, the Internet, and smart phones, but they predominantly tend to lack the eagerness to engage in policymaking processes or urban development [67].

The motivations to engage in smart governance received far less attention in the literature than the other two components. What information there was suggested that intrinsic motivation and intangible rewards were far more decisive factors in the decision to participate in collaborative technology-supported governance than extrinsic motives (e.g., monetary reward). Civil actors were found often to be motivated for reasons of solidarity, altruism, and the felt need for problem-solving in environmental and community issues [29,30,68]. However, the examined studies also confirm that despite civic voluntarism by knowledgeable and active inhabitants, citizens' interest and involvement in public affairs exhibit a general decline [20,23,43,69].

22.3.3 Use of technology

The final building block of smart governance is the use of technology, in particular ICTs. We mapped how digital technologies are applied in participatory governance processes aiming at sustainable urban development. We identified the types of technologies used and the aims they are applied for, as well as their limitations.

The literature showed a growing diversity of devices, tools, and technologies that are deployed for diverse engagement [26,27,30,40,51]. One intention technologies serve is one-way communication with examples as web portals (including e-mail service) and different kinds of visualizations (Augmented/Virtual Reality), which aim at building support for intended policies. Openly available tools serve to lower the barrier to entry into the policy realm for the average citizen, allowing him/her to make connections and enhancements to often available, but not easy to leverage information from municipality websites [70] By soliciting ideas and gauging opinions through these channels, municipalities can test the level of public agreement on their proposals beyond the more conventional e-voting and e-petition tools. Other applications commonly employed for information receive-only purposes are the collection of geo-data by sensors and collaborative mapping, monitoring greenhouse gas emissions, or localized diversity in energy use [32,43]. Mobile phones, growing more than other types of communication, are in particular frequently deployed in these kinds of participatory sensing that enables citizens to collect a wide array of data in situ [22,43,52,67,71]. Mobile phones were seen as becoming increasingly important in developing regions, as they grant a viable alternative to computers and hence access to resources that might be a step toward bridging the digital divide and eradicating poverty [52,61,62]. The use of mobile phones is also expected to attract more youngsters to participatory processes, as these tools are much more appealing to them [15,67].

Technological applications, such as discussion forums, electronic town hall meetings, wikis, and blogs, are also employed in two-way communication processes, wherein participants can productively interact with each other [53,56,72,73]. The increased use of Web 2.0 and social media tools favors citizen-created content, which not only enhances the free flow of information but fosters diversity of opinions, sociopolitical debate, and freedom of expression while creating an environment conducive to crowdsourcing initiatives [20,30].

The literature also found that data platforms and software tools that allow for the retrieval, storage, modeling, analysis, and visualization of data play an increasingly important role. These applications are often used jointly with the aforementioned information gathering and communication support technologies and entail instruments such as Public Participatory Geographic Information Systems (PPGIS), Planning Support Systems (PSS), and Decision Support Systems (DSS) [74].

Although these technological applications play an increasingly important part in governance processes, they also have limitations in use, mainly due to a lack of technological infrastructure and/or ICT knowledge on the part of both public officers and citizens.According to the literature, the level of technology intensity is inversely related to the intensity of civic participation: projects with less dependence on technologies and data deliver more involvement of participants and vice versa, This is because ICT-facilitated collaborative projects with a strong technological and data-driven nature hinder citizens' understanding of complex issues and policies [37,47,64,69,75].

In this aspect, it is not surprising that the reviewed articles also indicate that, despite the ongoing increase of online involvement, much citizen–government cooperation still happens through face-to-face interactions, since this is what citizens prefer [30,37]. Citizens choose rationally between different participation methods. Whether they opt for offline or online engagement depends on which is more convenient, efficient, and cheaper. The functionality of e-participation applications affects a user's choice: websites that are difficult for citizens to navigate or from which it is difficult to extract information, discourage participation [12,15]. Compared to citizens' offline public activities, digital participation showed lower engagement levels and higher drop-out rates. Online cooperation projects are easily abandoned due to the referred causes, including low entry, transaction, and opportunity cost. The wide net of online activities of many people breeds shallow attention, time shortage, and transitory involvement [1,15,29,30].

Critiques revealed in the literature highlight other limitations, for example, how apps-based participatory sensing transforms citizens into operative units for data collection [44,76]. Also, various studies claim that the technologies available, rather than the users' needs and expectations, guide the design of online service infrastructures [20,30,32,44].

22.3.4 What are the outcomes of smart governance?

The aspired substantive outcome of smart governance is sustainable urban development integrating social, economic, and environmental values. These aspects are discussed in this section to provide an answer to the second guiding question: what are the outcomes of smart governance?

A first observation is that the effects of smart governance on sustainable urban development have remained strongly understudied. This is confirmed by the strand of the investigated corpus arguing that the way smart governance contributes to a more sustainable society is largely unknown [8,13,77]. Although both academics and practitioners commonly associate the potential of smart governance with creating greener, healthier, more equitable, economically and culturally thriving communities [1,78], verification of this was difficult to find in the examined literature. Despite the profusion of promises and expectations, the extent to which these were actually realized remained largely unexplored. Some papers initially stated an intention to examine in depth the actual contribution of smart governance and the use of ICT tools in collaboration in terms of public values and sustainable development. However, they tended not to venture beyond the discussion of either public service transactions and related administration efficiency or the presentation of collaboration and project objectives [1,52].

A second observation is that the studies were mixed regarding the effects on sustainable urban development. The few papers (listed and discussed in the following two paragraphs) that addressed the effects of ICT-supported cooperation on sustainable development in its integrative [79] concept including economy, social issues, and environment offered varying perspectives. We found skeptics/pessimists, and optimists. The optimists—who were, granted, limited in number—did report observing sustainability benefits from ICT-enabled cooperation projects [10,27,61]. Estevez et al. [27] analyzed EGovernance for Sustainable Development (EGOV4SD) initiatives and showed that these contributed to a range of sustainable development goals. The authors mentioned that the most common sustainability problems, effectively addressed in these projects, were empowerment (social), business [80] opportunities (economic), man-made activity (environment), and capacity building (institutional).

The skeptics and pessimists held that smart governance has not yet yielded long-term sustainability outcomes [52,64,69]. They mostly rejected the idea that digital cocreation could have value from the perspectives of political and social impact, scalability, and sustainability [77,81]. They pointed to the growing technoeconomic divide between tech haves and tech have-nots as a negative effect, as well as to a range of unfavorable ecological outcomes. The South Korean smart eco-cities of Songdo and Incheon, for example, were established on precious wetlands, destroying the habitat of some of the rarest species on the planet [64].

A third observation is that studies frequently focus on sole elements of sustainability such as only the social or only the environmental outcomes of ICT-supported state–citizen collaboration. In view of the environmental outcomes, researchers found that smart governance could produce decreasing emissions of carbon dioxide on both the individual (household) and the collective level. They emphasize that only a multitude of multidisciplinary actions, activating citizens on different scales, can trigger behavior changes and fossil energy reductions for environmental protection [34,82]. However, adverse effects have also been found such as those partially produced in a European [83] smart governance project targeting the reduction of CO_2 emissions. While significant savings were achieved in respect of heating energy and electricity, in some cases increased emissions occurred in the fields of nutrition and consumer goods. The mobility domains of private and public transportation and flights also exhibited heterogeneous tendencies [34]. The social aspect of sustainability, which was covered by studies, partly refers to citizen participation that has been detailed in the previous section "The Components of Smart Governance." Citizen participation here may also be termed democratic governance when considering the specific goals targeted by citizen participation: enhanced transparency, democracy, improved legitimacy, and efficiency of public service delivery [26,46,67,84–86]. However, whether these participation goals have been achieved is not examined in the articles selected. Rather, the various characteristics of citizen participation (e.g., the types, roles, and number of participants, the level and intensity of involvement, citizens' satisfaction with the project) are considered by the majority of the articles as the main outcomes of ICT-supported collaborative governance [20,37,43]. This solitary focus in the bulk of the articles suggests that civic engagement was assessed for its own merit and much less for its contribution to urban sustainable development through democratic governance [4,8].

Another social aspect of sustainability, which was addressed in the articles, refers to participants' learning capacity about which the literature is more hopeful. Much is made of the beneficial features of online citizen participation in terms of learning. Citizens' experiences in the policymaking process are thought to be able to serve as a "school of democracy," since it may help participants become more informed citizens. In addition, information technologies can positively affect citizen participation by enabling participants to become more knowledgeable about governmental and public affairs and sustainability challenges and also become more skilled in communication. These interactions make citizens more prepared and interested in engaging in collective policymaking and urban development [12,22,30,32,51,58,61,73].

The examined articles provide examples. For instance, learning and sharing experiences between living labs—user-centered, experimental areas in a realistic territorial context—have triggered additional expressions of interest from potential user organizations and a significant growth in demand for related e-services [72,87]. Similarly, hackathons—time-limited, venue-based events, where ICT professionals, civil servants, interested citizens, and the private sector develop tech-based solutions for societal challenges—engendered the bridging of physical, cognitive, and social boundaries between stakeholders from wide-ranging domains (e.g., agrarians, politicians, civil partners, etc.). App feedbacks have advanced the identification of shared values and joint problem-solving, resulting in pilots by government agencies as well as by public and private stakeholders [71,73,88]. Furthermore, the collective use of ICT instruments in spatial planning processes resulted in learning as the most crucial added value.

Learning in these cases positively influenced the governance process and the achievement of the objectives by different mechanisms. Learning—by becoming more conscious, informed, and aware—fostered the removal of mutual prejudices, the increase of partners' reciprocal understanding and trust, and a deeper knowledge of the issues at stake. Stakeholders' enhanced interest in the specific project and policy domain not only enabled the initiation of collaboration, but also allowed this to be sustained. Intellectual capital has been found to be a critical success factor in ICT-led government projects through multiparty collaboration. It refers to collective knowledge, largely in tacit form, embedded in governance actors' shared experiences during problem-solving [23,43,71].

These encouraging effects notwithstanding, a recurrent theme in many articles was the insufficient learning capacity of local government and the negative effects this has on participation and urban coproduction [35,41,48,50]. To increase the outcome effectiveness of their policies and of smart cocreation, city governments must themselves become learning organizations before formulating and implementing smart governance [37,51,61,73]. However, in practice, less priority is given to drawing up a knowledge agenda of governance arrangements that could help advance process management [4,53].

22.3.4.1 Contextual factors influencing smart governance

Finally, we explored the literature to learn about (expected) major contextual factors and how these influence ICT-supported, citizen–government governance aiming at sustainable urban development. The analysis resulted in a context-sensitive operational framework and a list of expectations that can be tested in further empirical research.

22.3.4.2 Policy domain

The core themes tackled and the characteristics of the problems within a policy domain are considered to be decisive for technology-enabled, citizen–government governance for sustainable urban improvement [20,27,43,46,53,85]. The salience, urgency, the sociopolitical intensity, or sensitivity of topics may influence (enhance) the commitment of both governments and citizens to smart governance. For instance, the generally acknowledged importance of environmental protection, climate change, or the safeguarding of cultural heritage has heightened smart policy agendas and joint online or offline activities across the world [20,60,89]. These strategic, collaborative agendas are frequently incited by external political pressure on governments to collectively find solutions.

Also, certain policy domains were seen to be more conducive than others to collectively taking sustainability measures and reaching the objectives through the support of digital technologies. This was illustrated by online participation projects such as the smart governance project aimed at reducing CO_2 emissions: while significant reductions were achieved in respect to heating energy and electricity, emissions failed to decrease or were even augmented in the fields of nutrition, consumer goods, and mobility [34,82].

Finally, some articles stressed that the complexity level of policy issues is critical for citizen–government relationships [37,67,82]. Complicated issues may be expected to have a negative effect on both online and offline citizen engagement, as tasks that are too specific and demand too much expertise can hamper citizens in the development of an informed opinion and discourage continued involvement. Simplicity also stimulates governments to facilitate ICT-supported collaboration, as it does not drastically increase workload [30,44].

22.3.4.3 Trust

Although trust can be an individual characteristic, the literature reveals that the general condition of trust within society is a vital factor in digital cooperation between public and civil actors [13,33,37,43,62,67]. Trust relates to the image people have of their government regarding existing policies, what they can expect from governments, and their own perceived influence in technology-facilitated decision-making.

Many scholars refer to the positive influence of trust, as it increases the probability that citizens will invest their resources, time, and knowledge in ICT-based participatory alliances [35]. These scholars consider trust as a reflection of citizens' willingness to comply, cooperate, adopt, and support government policies and innovative programs. By contrast, others state the exact reverse: that trusting citizens are less likely to engage in ICT-enabled, government–citizen participation, as trust decreases citizens' motivations to monitor government. It may well be that, although a lack of trust in politics motivates citizens to start collective actions, maintaining such initiatives necessitates the presence of trust. A third group of researchers argue that citizens' trust or distrust in mainstream representatives does not make them alienated from politics and their life communities. In that respect, political disappointment or satisfaction shows no effect on people's willingness to engage in smart governance practices [49,66,90].

These paradoxical findings display that technology-facilitated citizen participation can be triggered by either trust or distrust in governmental policy solutions though the endurance of citizen engagement necessitates trust in the procedural fairness of government [23,29,63,70].

22.3.4.4 Political and institutional environment

It further emerged from the literature that both the formal and informal facets of the wider political and institutional environment play a major role in smart governance. A rather obvious factor here is the strength of democracy. A strong democracy can be expected to result in more ICT-accommodated citizen engagement and to stimulate the top-down digitalization of citizen participation [42].

Many articles pointed to the existing political system as a conditioning factor in the outcome of smart governance arrangements [42,70,91]. Berry and Portney [60] underline that sustainability policies mostly prevail in cities that are politically liberal (progressive), while Tikka and Sassi [56] trace the relations between political liberties, electoral rights, and (online) political participation.

Several authors referred to the innovative atmosphere of a place as an influence on digitally supported open government and cooperative governance [24,25,77]. Lee et al. [62] mention the typical San Francisco culture of creative participation for software applications development and discovering new service areas. Another example is the municipality of Tartu, the adoption-initiator of online participatory budgeting in Estonia. The innovative and creative characteristics of Tartu, along with its reputation as "the city of good thoughts" and "the intellectual capital," drove other municipalities to follow and to implement their own public innovation programs.

The literature also suggested that the political legacy of a country may influence ICT-empowered citizen involvement. Various studies, Santos et al. [28]; Santos and Tonelli [42]; Cornwall [92], demonstrated how colonial exploitation and oppression, long-term slavery, military repression, and a patrimonial, centralized government long discouraged grassroots movements.

Inherited institutional frameworks also influence the evolution of smart governance between citizens and governments [33,93,94]. Administrative culture, in particular, rooted as this is in political traditions and societal value orientation would seem to be an important factor. More formal, hierarchical cultures and highly centralized public administration may therefore be expected to result in less either face-to-face or technology-driven citizen engagement [12,28,87].

The literature also contained references to informal institutional dimensions that might shape the evolution of e-government, as well as the propensity for electronic coproduction between civil and governmental stakeholders and the forms this could take [13,25,36,40,56,85]. These factors cover issues of customs, traditions, religion, and routines within a given society or community, although their effects are not clear.

22.3.4.5 Internet reach and use

The literature was unanimous on the fact that the Internet and the pervasiveness of digital infrastructures are transforming traditional governance and stimulating online public–citizen collaboration to reach more urban sustainability. A higher intensity and developmental level of online participation was shown to be strongly linked to countries with widespread Internet penetration, broadband availability, and high technological development. Researchers have also found evidence of a correlation between the availability of ICTs, open-source technologies, and the proliferation of technology-based participatory programs, including civil efforts to reverse existing geometries of power [36,39,47,55,61,90,93].

Many authors, Charalabidis et al. [47]; Santos et al. [28]; Stratigea et al. [53], stress that Internet dissemination and widespread ICT use have changed citizens' expectations regarding governmental actions. By gaining online access to a large number of resources, the public has become used to taking part in online or offline decision-making. It is a cumulative technological effect—a process in which attitudes are gradually being adjusted to the available means [65]. As the level of Internet infiltration increases, municipalities are feeling an even greater push from citizens and from societal and business organizations to provide (environmental) information, implement sustainability measures, and include multiple stakeholders in policy processes and online services [13,36,40,44,64,72,93,95].

In developing countries and in disadvantaged areas in developed regions, the absence of or limited access to the Internet is a significant obstacle in achieving civic empowerment, public–civil collaboration, and sustainability [36]. Moreover, the digital sociocultural and economic gap impedes the migration of public services and citizen–government cooperation to the World Wide Web [41,58]. The lack of accessibility to various ICT applications hampers collaborative, technology-based urban planning as well [53].

22.3.4.6 Sociospatial characteristics

The (human-) geographical features of a city can influence electronic public services and ICT-supported governance practices. For instance, the topography may render specific locations vulnerable to natural disasters (flooding, earthquakes). Such threats may invoke societal pressure and highlight the vital need for collective public–civil actions by making use of new technological applications [13,43,56,58].

It was unclear from the articles we examined whether and to what extent the size of an urban area affects the evolution of smart governance. Although big cities are not always more innovative, they have more staff and management resources, facilitating the development of novel instruments, service delivery options, and online interactions with citizens [90]. The efficiencies associated with larger geographical size, together with greater numbers of constituents, can thus provide local authorities with the motivation to employ technologies [35,96]. On the other hand, citizens in smaller or rural communities have been found to have higher frequencies of ICT-enabled participation, regardless of their relatively poor Internet connectivity [34,35,58]. Neirotti et al. [58] also argue that smart city initiatives [97] are not correlated with the size of a city in terms of population, but rather with the population density.

Finally, social cohesion is expected to be positively related to digitally enabled citizen engagement. Citizens' local embeddedness and place and community identity are significant for mobilizing resources to organize collaborative (online) activities in the public domain. In line with this, several studies demonstrated that in order to establish successful urban development, smart governance planning processes should fit the distinct community context [40,68,98].

22.4 Summary

The highly optimistic accounts of sustainability outcomes of technology-enabled government–citizen interactions that, however, were generally not based on empirical evidence triggered us to conduct this literature review. Filling these knowledge gaps was the main goal of our analysis as we sought an answer to our central research question: What relationships exist between ICT-enabled citizen–government collaboration and sustainable urban development and how do contextual circumstances influence these?

The first conclusion of our review is that smart governance, in the sense of ICT-enabled government–citizen collaboration to advance urban sustainability, is still rare. Despite the increasing variety of collaboration-based digital instruments, the literature reveals the dominance of a one-way information supply in citizen–government interactions. Although governments promote online and offline citizen engagement and civic empowerment, in practice they do not encourage deliberation or any broad-based public–civil interactions. Therefore, ICT-supported government–citizen cooperation for collectively shaping public matters seldom occurs. The reason for this lies in the lack of capacity and willingness to genuinely engage in smart governance for urban sustainability, both on the part of government and that of citizens. Old structures, patterns, and routines still dominate. Evidently, the mere availability of technological infrastructure is no guarantee that any radical attitudinal change will occur in public administration and the civil sphere regarding the development of cocreative collaboration to create more sustainable cities.

The second conclusion is that the evidence that smart governance contributes to sustainability is sparse and mixed. Notwithstanding all expectations in this direction, the literature fails to elucidate whether smart governance activities lead to more livable cities, i.e., cities with less social deprivation, more ecological diversity, and enhanced economic prosperity. The limited number of articles dealing with these aspects delivers mixed results. There are some indications for positive sustainability effects, although a significant number of adverse consequences also emerged. Smart governance often increases the gap between the haves and have-nots rather than generating more equal societal structures. It may also boost public consumption or jeopardize ecological values, rather than contribute to a city with a balanced program of development along economic, social, and environmental dimensions.

The third conclusion is that there is more evidence for the process effects of smart governance. This is because the majority of the examined articles focuses on solitary, particularly the social, aspects instead of assessing sustainability in its integrative—economic, social, and environmental—outcomes. Hence, scholarly attention centers on activities such as civil engagement and learning as primary outcomes of smart governance. Citizen participation and learning can indeed be (achieved) goals as social aspects of sustainable development. However, many studies seem to treat ICT-facilitated citizen participation as a merit in itself—engagement for the sake of engagement without questioning its benefits for the society. The review also revealed that, while the broad range of online collaborative experiments were often seen to contribute to social learning, surprisingly little attention was paid to their contribution to sustainable urban development.

The fourth conclusion is that context matters. We identified specific contextual factors such as the policy domain, political-institutional (e.g., democracy, innovation, administration styles), societal (e.g., Internet reach and use in the society, trust), and sociospatial (e.g., topography of the city, local-specific social cohesion) dimensions. These contextual factors influence the distinct components of smart governance (the role of governments and citizens as well as ICT use) and define how local governments and citizens collaborate through new electronic resources, which, in turn, determine the potential of advancing urban sustainability. Based on this literature review, we may expect specific governance approaches only to work in similar contexts. Our list of context-related expectations will support future empirical research on this domain.

These outcomes sketch the contours of some essential scope conditions for successful smart governance activities. This success can be understood in terms of the initiation and endurance of broad-based collaboration by means of technologies, which contributes to urban sustainability. For this, the major scope conditions lie at the interface between the human factor and the specific contextual circumstances. While the human factor implies both governmental and civil competences such as commitment, willingness, and capacities, the context conducive to effective smart governance refers to the circumstances summarized above.

Finally, let us move on to the real interactive part of this chapter: review questions/exercises, hands-on projects, case projects, and optional team case project. The answers and/or solutions by chapter can be found in Appendix G.

22.5 Chapter review questions/exercises

True/false

1. True or False? The bulk of smart city research focuses on conceptualization—on mapping the meanings, components, and goals of the smart city from the perspectives of particular fields of study, such as public administration, information science, and urban development.
2. True or False? The second building block of smart governance is the organization of government.
3. True or False? The second building block of smart governance is citizen participation.

4. True or False? The final building block of smart governance is the use of technology, in particular ICTs.
5. True or False? The aspired substantive outcome of smart governance is unsustainable urban development integrating social, economic, and environmental values.

Multiple choice

1. The core themes tackled and the characteristics of the problems within a __________________ are considered to be decisive for technology-enabled, citizen–government governance for sustainable urban improvement.
   a. ICT-based infrastructure
   b. Policy domain
   c. Cultural heritage
   d. Smart governance
   e. Policy
2. Although trust can be an individual characteristic, the general condition of trust within society is a vital factor in digital cooperation between:
   a. Infrastructure actors
   b. Physical actors
   c. Public actors
   d. Economic actors
   e. Civil actors
3. It further emerged from the literature that both the formal and informal facets of the wider political and institutional environment play a major role in:
   a. Political liberties
   b. Electoral rights
   c. Political participation
   d. Smart governance
   e. Cooperative governance
4. The Internet and the pervasiveness of digital infrastructures are transforming traditional governance and stimulating online public–citizen collaboration to reach more:
   a. ICTs
   b. Participatory programs
   c. Urban sustainability
   d. Resources
   e. E-economics
5. The human geographical features of a city can influence electronic public services and ______-supported governance practices.
   a. WEP
   b. CPS
   c. ICT
   d. IoP
   e. IoS

Exercise

Problem

Analyze administrative practices in relation to the emergence of novel technologies.

Hands-on projects

Project

Do research: Analyze the conceptualization of a smart city data plan; a data governance policy instrument intended to connect the production; and the use of urban data in a comprehensive and evolutionary long-term strategy aligned with city development goals.

Case projects

Problem

Look at how the smart city agenda is promoted and formulated by a diverse set of actors, ranging from multilateral development agencies, foreign countries, central and local governments, as well as industry players and academic experts.

Optional team case project

Problem

Discuss why the lack of a smart city regulatory framework is one of the major obstacles for a wider adoption of smart city services in practice.

References

[1] Bifulco F, Tregua M, Amitrano CC, D'Auria A. ICT and sustainability in smart cities management. Int J Public Sect Manag 2016;29(2):132–47.
[2] Pardo A, Nam T, editors. Smarter as the new urban agenda. Basel: Springer International; 2016. p. 49–71.
[3] Hollands RG. Critical interventions into the corporate smart city. Camb J Reg Econ Soc 2015;8:61–77.
[4] Meijer A, Thaens M. Urban technological innovation developing and testing a sociotechnical framework for studying smart city projects. Urban Aff Rev 2016. https://doi.org/10.1177/1078087416670274. [Accessed 19 February 2017].
[5] Hamza K. Smart city implementation framework for developing countries: the case of Egypt. In: Smarter new urban agenda; 2016. p. 171–87. Cham: Springer.
[6] Lin Y, Geertman S. Smart governance, collaborative planning and planning support systems: a fruitful triangle? Plan Supp Syst Smart Cities 2015:261–77. Cham: Springer.

[7] Gil-Garcia R, Pardo TA, Nam T, editors. Smarter as the new urban agenda. Basel: Springer International; 2016. p. 129–48.
[8] Aichholzer G, Kubicek H, Torres L, editors. Evaluating e-participation. Frameworks, practice, evidence. Basel: Springer International; 2016.
[9] Lazaroiu GC, Roscia M. Definition methodology for the smart cities model. Energy 2012;47(1):326–32.
[10] Scott T. Does collaboration make any difference? Linking collaborative governance to environmental outcomes. J Pol Anal Manag 2015;34(3):537–66.
[11] Winters JV. Why are smart cities growing? Who moves and who stays. J Reg Sci 2011;51(2):253–70.
[12] Hong S. Citizen participation in budgeting: a trade-off between knowledge and inclusiveness? Publ Adm Rev 2015;75(4):572–82.
[13] Meijer A. Smart city governance: a local emergent perspective. In: Gil-Garcia R, Pardo TA, Nam T, editors. Smarter as the new urban agenda. Basel: Springer International; 2016. p. 73–85.
[14] Sampson RJ. Urban sustainability in an age of enduring inequalities: advancing theory and ecometrics for the 21st century city. Proc Natl Acad Sci USA 2017;114(34):8957–62.
[15] Zheng Y. Explaining citizens' E-participation usage functionality of E-participation applications. Adm Soc 2015. https://doi.org/10.1177/0095399715593313. [Accessed 16 April 2017].
[16] Cooper H, Hedges V. Research synthesis as a scientific process. In: Cooper H, Hedges LV, Valentine JC, editors. Research synthesis as A scientific process. The handbook of research synthesis and meta-analysis. New York: Russell Sage Foundation; 2009. p. 3–18 [Google Scholar].
[17] Harrison C, Eckman B, Hamilton R, Hartswick P, Kalagnanam J, Paraszczak J, Williams P. Foundations for smarter cities. IBM J Res Dev 2010;54(4):1–16.
[18] Liberati A, Altman DG, Tetzlaff J, Mulrow C, Gøtzsche PC, Ioannidis JP, Moher D. The PRISMA statement for reporting systematic reviews and meta-analyses of studies that evaluate health care interventions: explanation and elaboration. Ann Intern Med 2009;151(4). W-65.
[19] Przeybilovicz E, Cunha MA, Tomor Z. Identifying essential organizational characteristics for smart urban governance. In: Paper presented at the proceedings of the 18th annual international conference on digital government research; 2017. p. 416–25. https://doi.org/10.1145/3085228.3085281.
[20] Chiabai A, Paskaleva K, Lombardi P. E-participation model for sustainable cultural tourism management: a bottom-up approach. Int J Tourism Res 2013;15(1):35–51.
[21] Hollands RG. Will the real smart city please stand up? Intelligent, progressive or entrepreneurial? City 2008;12(3):303–20.
[22] Nam T, Pardo TA. Smart city as urban innovation: focusing on management, policy, and context. In: Paper presented at the 5th international conference on theory and practice of electronic governance (Tallinn, September 26–29, 2011); 2011.
[23] Sørensen E, Torfing J. Co-initiation of collaborative innovation in urban spaces. Urban Aff Rev 2016. https://doi.org/10.1177/1078087416651936. [Accessed 20 April 2017].
[24] Alusi A, Eccles RG, Edmondson AC, Zuzul T. Sustainable cities: oxymoron or the shape of the future? Harv Bus School Organiz Behav Unit Working Paper 2011:11–062.
[25] Correljé A, Cuppen E, Dignum M, Pesch U, Taebi B. Responsible innovation in energy projects: values in the design of technologies, Institutions and stakeholder interactions. In: Koops B-J, Oosterlaken I, Romijn H, Swiertstra T, van den Hoeven J, editors. Responsible innovation 2. Basel: Springer International; 2015. p. 183–200.
[26] Engelken-Jorge M, Moreno J, Keune H. Developing citizens' observatories for environmental monitoring and citizen empowerment: challenges and future scenarios. In: Paper presented at the conference for E-democracy and open governement (Krems, May 21–23); 2014.
[27] Estevez E, Janowski T, Dzhusupova Z. Electronic governance for sustainable development: how EGOV solutions contribute to SD goals?. In: Paper presented at the 14th annual international conference on digital government research (Quebec City, June 17–20); 2013.
[28] Santos HR, Tonelli DF, Bermejo PH. Sociopolitical digital interactions' maturity: analyzing the Brazilian states. Int J Electron Govern Res 2014;10(4):76–93.
[29] Lee J, Kim S. Active citizen E-participation in local governance: do individual social capital and E-participation management matter?. In: Paper presented at the 47th Hawaii international conference on system sciences (Hawaii, January 6–9, 2014); 2014.
[30] Yetano A, Royo S. Keeping citizens engaged a comparison between online and offline participants. Adm Soc 2015;49(3). https://doi.org/10.1177/0095399715581625. [Accessed 25 March 2016].
[31] Misuraca G, Rossel P. Measuring and meta-measuring: in search of new pathways for modelling impacts of ICT-enabled services on the information society. In: Paper presented at the eGovernment conference (Delft, August 28–September 2, 2011); 2011.
[32] Laspidou C. ICT and stakeholder participation for improved urban water management in the cities of the future. Water Util J 2014;8:79–85.
[33] Molinari F. On sustainable eParticipation. In: Tambouris E, Macintosh A, Glassey O, editors. International conference on electronic participation. Berlin Heidelberg: Springer; 2010. p. 126–39.
[34] Cimander R. Citizen panels on climate targets: ecological impact at individual level. In: Aichholzer G, Kubicek H, Torres L, editors. Evaluating e-participation. Frameworks, practice, evidence. Basel: Springer International; 2016. p. 219–41.
[35] Cruickshank P, Ryan B, Smith CF. 'Hyperlocal E-participation'? Evaluating online activity by Scottish Community Councils. In: Presented at the conference for E-democracy and open government (Krems, 21–23 May 2014); 2014.
[36] Al Hujran O, Aloudat A, Altarawne I. Factors influencing citizen adoption of E-government in developing countries: the case of Jordan. Int J Technol Hum Interact 2013;9(2):1–19.
[37] Capra CF. The smart city and its citizens: governance and citizen participation in Amsterdam smart city. Int J E Plann Res 2016;5(1):20–38.
[38] Kingston R. Public participation in local policy decision-making: the role of web-based mapping. Cartogr J 2007;44(2):138–44.
[39] Schröder C. Some reflections on possible Uses of E-participation for the local level. In: Paper presented at the conference for E-democracy and open governement (Krems, May 21–23 2014); 2014.
[40] Álvarez-Crespo B. Co-creation of innovative experiences in the Spanish E-administration. Respons Sustain 2014;2(2):31–40.
[41] Molinari F, Ferro E. Framing web 2.0 in the process of public sector innovation: going down the participation ladder. Eur J ePractice 2009;9(1):20–34.

[42] Santos HR, Tonelli DF. Possibilities and limits of E-participation: a systematic review of E-democracy. In: Paper presented at the 38th meeting of ANPAD (Rio de Janeiro, 13–17 September, 2014); 2014.
[43] Wehn U, McCarthy S, Lanfranchi V, Tapsell S. Citizen observatories as facilitators of change in water governance? Experiences from three European cases. Environ Eng Manag Journal 2015;14(9):2073–86.
[44] Gabrys J. Programming environments: environmentality and citizen sensing in the smart city. Environ Plann Soc Space 2014;32:30–48.
[45] Loorbach D, Wittmayer JM, Shiroyama H, Fujino J, Mizuguchi S. Governance of urban sustainability transitions: European and Asian experiences. Tokyo: Springer; 2016.
[46] Wolfram M. Deconstructing smart cities: an intertextual reading of concepts and practices for integrated urban and ICT development. In: Paper presented at the proceedings tagungsband (Schwechat, May 14–16, 2012); 2012.
[47] Charalabidis Y, Koussouris S, Lampathaki F, Misuraca G. ICT for governance and policy modelling: visionary directions and research paths. In: Charalabidis Y, Koussouris S, editors. Empowering open and collaborative governance. Berlin Heidelberg: Springer; 2012. p. 263–82.
[48] Coleman S, Blumler JG. The Internet and democratic citizenship: theory, practice and policy. Cambridge: Cambridge University Press; 2009.
[49] Edelmann N, Höchtl J, Sachs M. Collaboration for open innovation processes in public administrations. In: Charalabidis Y, Koussouris S, editors. Empowering open and collaborative governance. Berlin Heidelberg: Springer; 2012. p. 21–37.
[50] Verdegem P, Verleye G. User-centered E-government in practice: a comprehensive model for measuring user satisfaction. Govern Inf Q 2009;26(3):487–97.
[51] Anttiroiko AV, Valkama P, Bailey SJ. Smart cities in the new service economy: building platforms for smart services. AI Soc 2014;29(3):323–34.
[52] Singh Kalsi N, Kiran R. E-governance success factors: an analysis of E-governance initiatives of ten major states of India. Int J Public Sect Manag 2013;26(4):320–36.
[53] Stratigea A, Papadopoulou CA, Panagiotopoulou M. Tools and technologies for planning the development of smart cities. J Urban Technol 2015;22(2):43–62.
[54] Arnstein SR. A ladder of citizen participation. J Am Inst Plan 1969;35(4):216–24.
[55] Edelmann N, Hoechtl J, Parycek P. Eparticipation for adolescent citizens (in Austria). In: Macintosh A, Tambouris E, editors. International conference on electronic participation. Berlin Heidelberg: Springer; 2009. p. 163–74.
[56] Tikka I, Sassi S. Immigrant inclusion and eparticpation. In: Paper presented at the electronic participation conference (Delft, August 29–September 1, 2011); 2011.
[57] Bekkers V, Homburg V. The myths of E-government: looking beyond the assumptions of a new and better government. Inf Soc 2007;23(5):373–82.
[58] Neirotti P, De Marco A, Cagliano AC, Mangano G, Scorrano F. Current trends in smart city initiatives: some stylised facts. Cities 2014;38:25–36.
[59] Neubauer T, Vuga T, Ilc B. ICT as the facilitator of postmodern and empowered forms of citizenship: myth or reality? In: Charalabidis Y, Koussouris S, editors. Empowering open and collaborative governance. Berlin Heidelberg: Springer; 2012. p. 223–40.
[60] Berry JM, Portney KE. Sustainability and interest group participation in city politics. Sustainability 2013;5(5):2077–97.
[61] Hanna NK. Grassroots innovation for the information society. In: Hanna NK, editor. Transforming government and building the information society. New York: Springer; 2010. p. 199–226.
[62] Lee JH, Gong Hancock M, Hu MC. Towards an effective framework for building smart cities: lessons from Seoul and San Francisco. Technol Forecast Soc Change 2014;89:80–99.
[63] Sun Y, Nakata K. The decomposed UTAUT model for analysing people's motivation in participatory sensing. In: Paper presented at the electronic participation conference (Delft, August 29–September 1, 2011); 2011.
[64] Yigitcanlar T, Lee SH. Korean ubiquitous-eco-city: a smart-sustainable urban form or a branding hoax? Technol Forecast Soc Change 2014;89:100–14.
[65] Åström J, Granberg M. Urban planners, wired for change? Understanding elite support for E-participation. J Inf Technol Polit 2008;4(2):63–77.
[66] Wijnhoven F, Ehrenhard M, Kuhn J. Open government objectives and participation motivations. Govern Inf Q 2015;32(1):30–42.
[67] Ertiö TP. Participatory apps for urban planning: space for improvement. Plann Pract Res 2015;30(3):303–21.
[68] Zait A. Exploring the role of civilizational competences for smart cities' development. Transform Gov People Process Pol 2017;11(3):377–92.
[69] Deakin M, Allwinkle S. Urban regeneration and sustainable communities: the role of networks, innovation, and creativity in building successful partnerships. J Urban Technol 2007;14(1):77–91.
[70] Gano G. The new face of government: publics and policymaking online. J Publ Adm Res Theor 2013;23(3):755–8.
[71] Das Aundhe M, Narasimhan R. Public private partnership (PPP) outcomes in E-government: a social capital explanation. Int J Public Sect Manag 2016;29(7):638–58.
[72] Cleland B, Mulvenna M, Galbraith B, Wallace JG, Martin S. Innovation of eparticipation strategies using living labs as intermediaries. Electron J eGovernment 2012;10(2):120–32.
[73] Termeer CJAM, Bruinsma A. ICT-enabled boundary spanning arrangements in collaborative sustainability governance. Curr Opin Environ Sustain 2016;18:91–8.
[74] Macintosh A. Characterizing E-participation in policy-making. In: Paper presented at the 37th annual Hawaii international conference on system sciences (Hawaii, January 5–8, 2004); 2004.
[75] Evans AM, Campos A. Open government initiatives: challenges of citizen participation. J Pol Anal Manag 2013;32(1):172–85.
[76] Yigitcanlar T. Smart cities: an effective urban development and management model? Aust Plan 2015;52(1):27–34.
[77] Voorberg WH, Bekkers VJJM, Tummers LG. A systematic review of Co-creation and Co-production: embarking on the social innovation journey. Publ Manag Rev 2015;17(9):1333–57. D. Walters, "Smart Cities, Smart Places, Smart Democracy: Form-Based Codes, Electronic Governance and the Role of Place in Making Smart Cities," Intelligent Buildings International 3: 3 (2011)198–218.
[78] Meijer A, Bolívar MPR. Governing the smart city: a review of the literature on smart governance. Int Rev Adm Sci 2016;82(2):392–408.
[79] Chourabi H, Nam T, Walker S, Gil-Garcia JR, Mellouli S, Nahon K, Pardo TA, Scholl HJ. Understanding smart cities: an integrative framework. In: Paper presented at the 45th Hawaii international conference on system sciences (Hawaii, January 4–7, 2012); 2012.
[80] Elkington J. Partnerships from cannibals with forks: the triple bottom line of 21st century business? Environ Qual Manag 1998;8(1):37–51.

[81] Prieto-Martín P, de Marcos L, Martínez JJ. A critical analysis of EU-funded eparticipation. In: Charalabidis Y, Koussouris S, editors. Empowering open and collaborative governance. Berlin Heidelberg: Springer; 2012. p. 241–62.
[82] Cimander R, Royo S, Yetano A. Citizen panels on climate targets: ecological impact at collective level. In: Aichholzer G, Kubicek H, Torres L, editors. Evaluating e-participation. Frameworks, practice, evidence. Basel: Springer International; 2016. p. 243–64.
[83] Caragliu A, Del Bo C, Nijkamp P. Smart cities in Europe. J Urban Technol 2011;18(2):65–82.
[84] Curwell S, Deakin M, Cooper I, Paskaleva-Shapira K, Ravetz J, Babicki D. Citizens' expectations of information cities: implications for urban planning and design. Build Res Inf 2005;33(1):55–66.
[85] de Araujo RM, Taher Y. Refining IT requirements for government-citizen Co-participation support in public service design and delivery. In: paper presented at the conference for E-democracy and open governement (Krems, May 21–23); 2014.
[86] Zavadskas EK, Kaklauskas A, Banaitis A. Application of E-technologies for regional development: the case of Vilnius city. J Bus Econ Manag 2010;11(3):415–27.
[87] Schuurman D, Baccarne B, De Marez L, Mechant P. Smart ideas for smart cities: investigating crowdsourcing for generating and selecting ideas for ICT innovation in a city context. J Theor Appl Electron Comm Res 2012;7(3):49–62.
[88] Wang S, Feeney MK. Determinants of information and communication technology adoption in municipalities. Am Rev Publ Adm 2016;46(3):292–313. V. Washington, "Local Government, the Internet, and Sustainability," Public Administration Review 74: 1 (2014) 99–100.
[89] Vanolo A. Smartmentality: the smart city as disciplinary strategy. Urban Stud 2013;1(6). https://doi.org/10.1177/0042098013494427. [Accessed 15 March 2016].
[90] Torres L, Pina V, Acerete B. E-governance developments in European union cities: reshaping government's relationship with citizens. Governance 2006;19(2):277–302.
[91] Sieber R. Public participation geographic information systems: a literature review and framework. Ann Assoc Am Geogr 2006;96(3):491–507.
[92] Cornwall A. Locating citizen participation. IDS Bull 2002;33(2):i–x.
[93] Azad B, Faraj S, Goh JM, Feghali T. What shapes global diffusion of E-government: comparing the influence of National Governance Institutions. J Global Inf Manag 2010;18(2):85–104.
[94] Ricciardi F, Lombardi P. Widening the disciplinary scope of eparticipation: reflections after a research on tourism and cultural heritage. In: Tambouris E, Macintosh A, Glassey O, editors. International conference on electronic participation. Berlin Heidelberg: Springer; 2010. p. 140–50.
[95] Twinomurinzi H, Phahlamohlaka J, Byrne E. The small group subtlety of using ICT for participatory governance: a South African experience. Govern Inf Q 2012;29(2):203–11.
[96] Saglie J, Vabo SI. Size and E-democracy: online participation in Norwegian local politics. Scand Polit Stud 2009;32(4):382–401.
[97] Bakıcı T, Almirall E, Wareham J. A smart city initiative: the case of Barcelona. J Knowl Econ 2013;4(2):135–48.
[98] Castelnovo W, Misuraca G, Savoldelli A. Smart cities governance the need for a holistic approach to assessing urban participatory policy making. Soc Sci Comput Rev 2015;34(6):724–39.

Further reading

Rodriquez Bolívar MP, editor. Smart technologies for smart governance: transparency, efficiency and organizational issues. Basel: Springer International; 2018.
World Commission on Environment Development (WCED). Our common future 17 special working session. 1987.

Chapter 23

Provide affordable smart infrastructure policy applications for informal sectors

V. Padmavathi and K. Aruna
A.V.C. College of Engineering, Mayiladuthurai, Tamilnadu, India

23.1 Introduction

The prototype, which is made with three prototypes that will gather the environmental data and store it in the Blynk cloud, communicates to the actuate sensors to complete the task. Tasks such as switching on and off of light, motion checking, and alert are done locally or online automatically.

Internet of Things (IoT) [1] can be defined as a network of physical objects or people called "things" that are embedded with software, electronics, networks, and sensors, which allows these objects to connect and exchange the data. It extends the Internet connectivity beyond the traditional devices such as desktop and laptop computers, smartphones, and tablets to a diverse range of devices and everyday things that utilize embedded technology to communicate and interact with the external environment. A device may have various types of sensors, which perform multiple tasks apart from sensing.

23.1.1 Connectivity

All of the collected data is sent to a cloud infrastructure. The sensors are connected to the cloud using various media of communication. These communication media include Bluetooth, Wi-Fi, WAN, etc.

23.1.2 Data processing

Once the data is collected and sent to the cloud, the software performs processing on gathered data. Data manipulation by a computer is known as data processing. It entails the transformation of raw data into machine-readable form, data flow through the CPU and memory to output devices, and output formatting or transformation. Data processing encompasses the use of computers to execute prescribed operations on data.

23.1.3 User interface

The information needs to be available to the end user in some way. which can be achieved by triggering alarms on their mobile phones or sending them notification through certain aspects, which impact the quality of infrastructure sectors across Indian states, such as the identity of operators, the nature of the regulatory framework, and the nature of the political economy and ultimately drive the public infrastructure investment that can be considered for future scope of work. In the Indian context, there is a huge research gap in this respect:

- One avenue for further research would be to impose sign restriction vector and carry out the analysis on estimating the impact of different modes of infrastructure financing on economic growth.
- Further research can be focused in terms of establishing a causal link between public debt and economic growth, via the use of an instrumental variable, which will have a direct impact on debt but no direct effect on economic growth.
- By estimating the threshold value of debt, we have tried to show that we can convert economic problems into machine learning problems and find out interesting relationships, which is relatively difficult to do while using statistics alone.

23.2 Smart cities and infrastructure

The intelligent infrastructure of the city is the preliminary stage to the entire structure of the intelligent city. Intelligent infrastructure might be characterized as a cyber-physical system that integrally manages all the parts it includes by using various technological instruments to assemble and analyzes data to achieve efficiency, sustainability, productivity, and security objectives.

Smart Cities Policies and Financing. https://doi.org/10.1016/B978-0-12-819130-9.00036-X

23.2.1 Smart city

A smart sustainable city [2] is an innovative city that uses information and communication technologies (ICTs) and other means to improve quality of life, efficiency of urban operation and services, and competitiveness, while ensuring that it meets the needs of present and future generations with respect to economic, social, environmental as well as cultural aspects.

23.2.2 Smart infrastructure

Smart City focuses on basic infrastructure (see Fig. 23.1), uses "smart" solutions to upgrade infrastructure and amenities, and depends on area-based growth. The true "smartness" of "Smart City" is to use IoT/M2M communications in a structured, safe, secure, accountable, and harmonized way to meet the dreams and expectations of its people, with regard to:

- Smart buildings
- Smart mobility
- Smart energy
- Smart water
- Smart waste management
- Smart health

23.3 Energy

Energy [3] is part of our everyday lives. Access to energy is required to sustain a variety of social and economic activities, including transport, connectivity, and processing of goods. Education and health systems cannot run optimally without energy, which could have an effect on healthcare and human resource growth. For cost-effective 100% renewable energy systems, the Smart Energy Grid [4] principle is crucial. To create energy grid resilience, reap synergies by using both infrastructures and lower energy storage costs; the definition requires an emphasis on energy efficiency, end-use savings, and business integration.

In order to define acceptable energy grid designs and operating plans, the smart energy system strategy involves the whole energy sector. Focusing exclusively on the smart griddle also means that the only way ahead in incorporating fluctuating renewable energy is to identify transmission lines, versatile power needs, and the storages of electricity. However, given the existence of wind power and related outlets, these steps are neither extremely effective nor cost-effective. When the electrical industry is paired with the heating, cooling, and/or transport sector, the most efficient and cheapest technologies can be found. In addition, the combination of electricity and gas facilities will play an important role when planning future renewable energy systems and heating and transport electrification—likely by means of electric fuels—can play a critical role in providing stability and ensuring convergence in all industries in renewable energy sources.

23.3.1 Key principles

It is a 100% renewable energy system and consumes a sustainable [5] level of bioenergy. It utilizes the synergies in the energy system to maximize efficiency and reduce costs. That ensures that the cost of electricity in contrast with a fossil-fuel energy [6] system does not increase significantly (sometimes it can reduce the cost and maximum increases of up to 10–15% are expected).

23.4 Transport

IoT has crucial applications in the transportation system. IoT plays an important role in all the field of transportation as air transportation, water transportation, and land transportation. All the components of these transportation fields are built with smart devices (sensors, processors) and interconnected through cloud server or different servers that transmit data to networks.

IoT in transportation is not only for traveling from one place to another, but it also makes safer, greener, and more

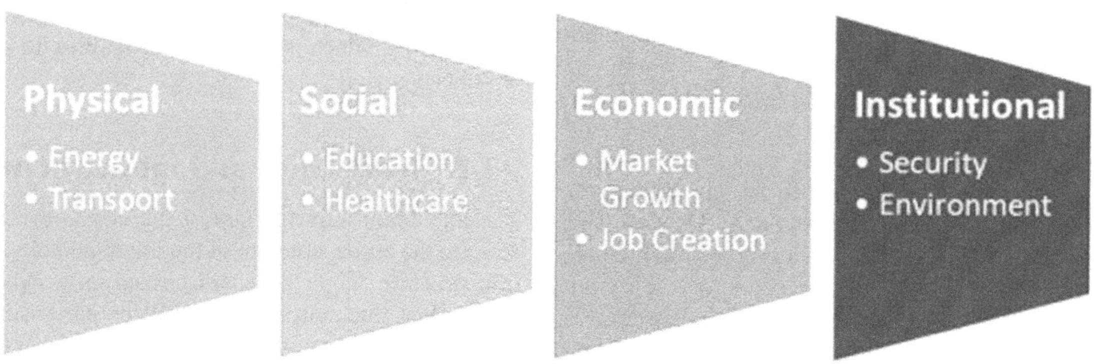

FIGURE 23.1 Smart infrastructure.

convenient. For example, a smart car performs work simultaneously such as navigation, communication, entertainment, efficient, more reliable travel. IoT facilitates travelers to remain seamlessly connected to every means of travel. The vehicle is connected with a variety of wireless standards to the Internet such as Bluetooth, Wi-Fi, 3G, 4G, intelligent traffic system, and even to other vehicles.

To make transport management more effective, several cities across the world are trying to build intelligence into existing systems. Transport management systems and software [7] tools have been effective to curtail traffic woes around some mega cities of the world. A smart traffic system helped Stockholm cut gridlock by 20%, reduce emissions by 12%, and increase public transportation use dramatically.

23.5 Traffic management

Reading is by far the biggest segment within transportation when it comes to the adoption of IoT technologies, and this is expected to grow as we head toward 2023. Within cities, data can be collected from CCTV feeds, which transmit vehicle-related data to traffic management centers. Applications using IoT technology include the following:

- Smart parking
- Traffic lights
- Smart accident assistance
- Toll and ticketing

Conventional toll systems are becoming rapidly outdated. With the increase in vehicles on the roads, queues at toll booths have become a common sight, not to mention the manpower needed to operate toll booths on busy highways. While automated tolls, using an RFID tag, have improved the flow of traffic, further improvements have been made possible by the use of IoT technology.

Many of today's modern vehicles are equipped with IoT connectivity. A vehicle can be detected up to a kilometer away from a tolling station, correctly identified, and the barrier lifted for the vehicle to pass through. Alternatively, for older vehicles, a registered smartphone could serve the same purpose, taking automatic payment from the digital wallet linked on the phone.

23.5.1 Connected cars

As mentioned in the preceding, cars today rely on connectivity, and a key part of that is many new cars are now equipped with Internet connectivity, sensors, and actuators, all monitoring a wide range of applications from brakes and engine to the control of tire pressure and exhaust gas composition.

23.5.2 Vehicle tracking systems

Vehicle tracking systems are typically used within the freight segment to help companies manage their fleets effectively. They also help to monitor driver behavior and can collect data, which informs on idling time and driving style. Examples of IoT-powered functionality include the following:

- Trip scheduling
- Fleet tracking
- Driving times and driver rest break scheduling
- Alerts for speeding, harsh cornering, acceleration, or braking
- Monitoring of vehicle load
- Distance traveled and fuel consumption

23.5.3 Public transport management

One key area in which NEC has been operating is smart transportation, with a focus on the public transport segment. IoT technologies are already in wide use in this segment, and our solutions, including integrated ticketing and automated fare collection, passenger information systems, passenger information display systems, and advanced vehicle Logistics solutions, all utilize IoT technology to help solve social and economic issues such as traffic congestion in public transport.

23.5.4 IoT technology for connected public transport systems

Real-time vehicle tracking—this helps public transport agencies better communicate with customers and provide accurate arrival times through both mobile devices and passenger information displays at transit stops and stations. Data analysis and real-time management—the technology allows transit agencies to monitor progress in real time and make adjustments for unpredicted incidents such as accidents, road works, emergencies, etc., helping to reroute and make journeys more efficient. Personalized travel information—transit agencies can track and monitor commuter behavior and travel patterns and deliver personalized information direct to their smartphone on key changes such as delays, station closures, or rerouting.

23.6 Health care

IoT applications can turn reactive medical-based systems into proactive wellness-based systems. The resources that current medical research uses lack critical real-world information. It mostly uses leftover data, controlled environments, and volunteers for medical examination. IoT opens ways to a sea of valuable data through analysis, real-time field data, and testing. The IoT also improves the current devices in power, precision, and availability. IoT focuses on creating systems rather than just equipment. Here is how an IoT-enabled care device works as shown in Fig. 23.2.

23.6.1 How an IoT-enabled care device works

IoT technology brings numerous applications in health care, from remote monitoring to smart sensors to medical device integration. It keeps the patients safe and healthy as well as improves the delivery of care by physician toward the patients. Healthcare devices collect diverse data from a large set of real-world cases that increases the accuracy and the size of medical data. A new paradigm, known as the Internet of Things (IoT), has an extensive applicability in numerous areas, including health care. It provides the support for the following:

1. Simultaneous reporting and monitoring
2. End-to-end connectivity and affordability
3. Data assortment and analysis
4. Tracking and alerts
5. Remote medical assistance

23.6.1.1 Simultaneous reporting and monitoring

Real-time monitoring via connected devices can save lives in the event of a medical emergency such as heart failure, diabetes, asthma attacks, etc. With real-time monitoring of the condition in place by means of a smart medical device connected to a smartphone app, connected devices can collect medical and other required health data and use the data connection of the smartphone to transfer collected information to a physician. The IoT device collects and transfers health data: blood pressure, oxygen and blood sugar levels, weight, and ECGs. These data are stored in the cloud and can be shared with an authorized person, who could be a physician, your insurance company, a participating health firm, or an external consultant, to allow them to look at the collected data regardless of their place, time, or device.

23.6.1.2 End-to-end connectivity and affordability

IoT can automate patient care workflow with the help healthcare mobility solution and other new technologies and next-generation healthcare facilities. IoT in health care enables interoperability, machine-to-machine communication, information exchange, and data movement that make healthcare service delivery effective. Connectivity protocols: Bluetooth LE, Wi-Fi, Z-wave, ZigBee, and other modern protocols, healthcare personnel can change the way they spot illness and ailments in patients and can also innovate revolutionary ways of treatment. Consequently, technology-driven setup brings down the cost, by cutting down unnecessary visits, utilizing better quality resources, and improving the allocation and planning.

23.6.1.3 Data assortment and analysis

Huge amount of data that a healthcare device sends in a very short time owing to their real-time application is hard to store and manage if the access to cloud is unavailable. Even for healthcare providers to acquire data originating from multiple devices and sources and analyze it manually is a tough bet. IoT devices can collect report and analyze the data in real time and cut the need to store the raw data. This all can happen over cloud with the providers only getting access to final reports with graphs. Moreover,

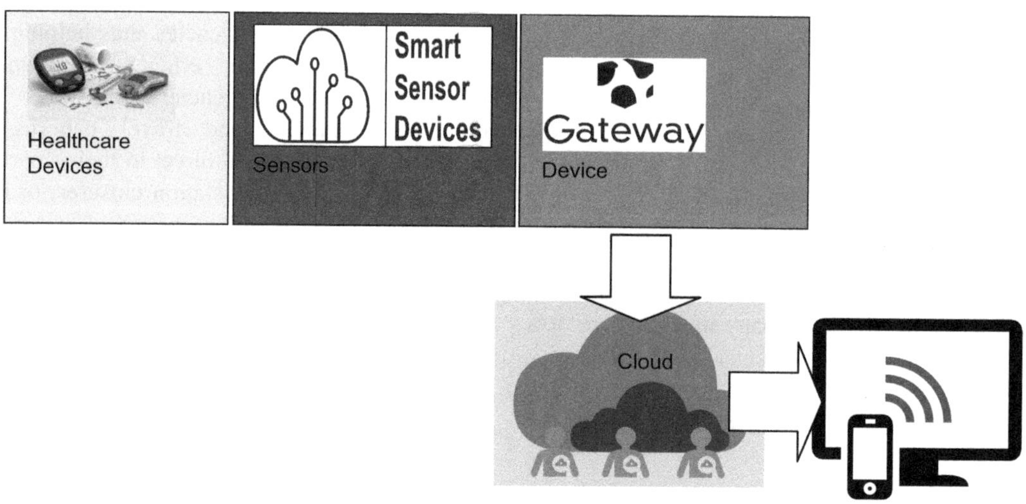

FIGURE 23.2 Health care using IoT.

healthcare operations allow organizations to get vital healthcare analytics and data-driven insights, which speed up decision-making and are less prone to errors.

23.6.1.4 Tracking and alerts

On-time alert is critical in the event of life-threatening circumstances. Medical IoT devices gather vital data and transfer that data to doctors for real-time tracking, while dropping notifications to people about critical parts via mobile apps and other linked devices. Reports and alerts give a firm opinion about a patient's condition, irrespective of place and time. It also helps make well-versed decisions and provide on-time treatment. Thus, IoT enables real-time alerting, tracking, and monitoring, which permits hands-on treatments, better accuracy, and apt intervention by doctors and improve complete patient care delivery results.

23.6.1.5 Remote medical assistance

In the event of an emergency, patients can contact a doctor who is many kilometers away with some smart mobile apps. With mobility solutions in health care, the medics can instantly check the patients and identify the ailments on the go. Also, numerous healthcare delivery chains that are forecasting to build machines can distribute drugs on the basis of patient's prescription and ailment-related data available via linked devices. IoT will improve the patient's care in hospital. This in turn will cut on people's expense on health care.

23.6.2 Applications of IoT in healthcare

The rise of IoT is exciting for everybody due to its different scope of use in various sectors. In health care, it has several applications. IoT in health care helps in:

- Reducing emergency room wait time
- Tracking patients, staff, and inventory
- Enhancing drug management
- Ensuring availability of critical hardware.

23.7 Market growth

The IoT market will grow because existing IT devices will need to be linked to the IoT. The IoT platform market is driven by the large-scale benefits of using IoT devices (see Fig. 23.3). IoT integrates technologies such as data communication, data storage, hardware design, and mining and is the latest trend in the global technology arena. Due to its potential applications and numerous advantages including high convenience, improved decision-making, improved handling of emergencies, cost benefits, improved quality of life, remote access, and improvements in safety and security, several organizations, industries, and technologists are adopting IoT since it empowers the end user to make informed decisions.

APAC was the largest IoT platform market in 2019, and the region will offer several growth opportunities to market vendors during the forecast period. This is attributed to factors such as the increased investments in infrastructure development by governments in APAC and the local business modernization by small and medium-sized enterprises (SMEs).

The global IoT platform market is fragmented. Alphabet Inc., Amazon.com Inc., AT&T Inc., Cisco Systems Inc., Huawei Investment & Holding Co. Ltd., International Business Machines Corp., Intel Corp., Microsoft Corp., Nokia Corp., Samsung Electronics Co. Ltd., and Siemens AG. To help clients improve their market position, this IoT platform market forecast report provides a detailed analysis of the market leaders.

23.7.1 Industrial internet revolution will be a key market trend

The industrial Internet revolution is a key trend in the market. Industrial Internet of Things (IIoT) is a combination of big data analytics [8] and IoT, which offers a number of opportunities for organizations across sectors such as agriculture, health care, manufacturing, mining, oil and gas, and transportation. To encourage the digitalization of industries, large companies and the governments of

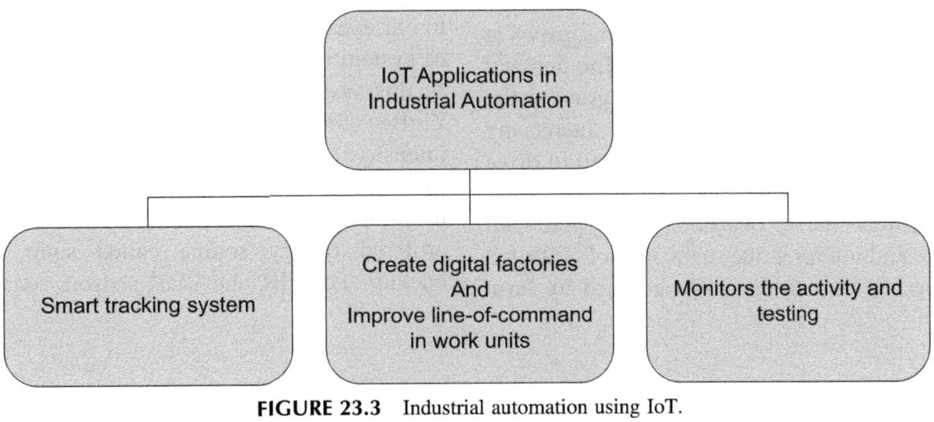

FIGURE 23.3 Industrial automation using IoT.

numerous countries across the globe have already adopted several industrial Internet initiatives. This in turn will help them create new revenue segments, enhance efficiencies, and improve the quality of public life. The IoT technology stack has advanced over the past 5 years in the meantime, each layer holds significant market growth opportunities. Device-enablement platforms have an especially strategic advantage of enabling related IoT growth while still in their own growth phase.

23.7.2 Networking technology to dominate the IIoT market

Networking technologies accounted for the largest market share in 2019. Networking technologies, especially wireless technology, are the backbone of IIoT. Numerous machines and sensor nodes are connected to the Internet using these technologies to facilitate machine-to-machine communication. These technologies also enable the collection of large volumes of data for improved decision-making. An extensive range of wireless communication technologies are available based on connectivity range, power requirements, and data transfer rates. The industrial sector is relying increasingly on cellular technologies for machine-to-machine communication.

23.7.3 Industrial IoT

Primarily devices work with one another without human interaction or intervention (M2M) robustness, controlled failure modes, security become primary issues. "Consumer" items can quickly cross over, as a phone app can control your home or even your car or monitor life-critical health issues. The main markets building automation industrial automation lightings are as follows:

- Commercial transportation/fleet management
- Enterprise asset management
- Smart cars
- Test and measurement
- Energy grid [9] (smart grid)

The IIoT market of farm management systems grows at the highest CAGR during the forecast period. The demand for farm management software is expected to grow at the highest CAGR during the forecast period. Increasing adoption of precision farming globally is expected to drive the growth of the farm management systems. Precision farming helps farmers make decisions that were not possible to analyze and interpret due to the lack of collected data without using effective tools. Software used by farm management systems allows cultivators to use the data collected over the years to create formula-based prescriptions, analyze the average yield data, create profit maps, and import soil test results.

The major device [10] and technology include networking technologies, industrial robotics, sensors, condition monitoring, and DCS. Decreasing cost of storing and processing of data and increasing convergence of Internet and industrial networks are the prime factors driving the growth of the IIoT market globally, which will lead to market opportunity of USD 110.6 billion by 2025.

Increasing use of cloud computing platforms, support from governments [11] of different countries for R&D activities related to IIoT, and decreased costs of automation are major driving factors. Predictive maintenance of machinery, increased number of data centers, and requirement of innovative cross-domain expertise, and collaborations for the development of comprehensive systems and solutions act as expected to create new opportunities IIoT [12] market. It focused on the necessity of the automated street light system and the peculiar way of implementation with embedded system tools. In this system, the piezoelectric sensor is used to detect the movement of the object on the street instead of using IR sensor. A microcontroller msp430 is a brain to control the process involved.

To detect the vehicle movement on the highways to switch ON only a block of the street light ahead of it and switch OFF the trailing light to save energy. During the night all the lights on the highways remain ON for the vehicle, but lot of energy is wasted when there is no vehicle movement on the highways. In this paper, two kinds of sensors have been used, which are light sensor and photoelectric sensor.

Automatic street light control system is not only the easiest but also the powerful technique. Relay uses as an automatic switch in this system. It releases the manual work at most up to 100%. As soon as the sunlight goes under the visible region of our eyes, this system automatically switches ON lights. Light-dependent resistor (LDR) is a type of sensor that actually does this work and senses the light as our eyes do. As soon as the sunlight comes, visible to our eyes it automatically switches OFF lights. Such type of system is also useful for reducing energy consumption. In this system, the system with LDR sensor, PIR sensor, ZigBee are used to intimate the status of humans' use, light intensity, and street light ON/OFF status to the EB section to avoid wastage of energy by glowing street lights in unwanted areas. The whole system is operated by using artificial energy source called solar and with battery backup. The PIR and LDR sensors sense the persons and light intensity of a particular place and transmit the data in wireless to the EB section with ZigBee. Depending upon the data received, the controller will turn ON/OFF the street light in wireless communication. This system is appropriate for street lighting in remote urban and rural areas where the traffic is low at times.

Streetlight control using ZigBee [13] wireless module includes LDR, microcontroller, and a transmission module. ZigBee allows wireless communication with the lamp module. The system consists of two LDR sensors to examine the day–night variations and lamp health conditions. The results from the LDR are transferred to the microcontroller after processing the data and further into the transmission module.

An automated streetlight monitoring system uses GSM technology [14], which the system consists of a server microcontroller and sensors such as smoke sensor, noise sensor, light sensor, etc., to measure various parameters. This system can detect surrounding temperatures, noise intensities, and alert the system for corrective measures. The problem is that the GSM modem has to be implemented in each streetlight, which makes the system very costly. It also has some connectivity issues. This model uses a lot of hardware to control and monitor the system so it is costlier.

A traffic-flow-based streetlight control system is powered by solar panels. They used 8052 series microcontroller and replaced normal bulbs with LED light, and energy consumption was decreased by three times. Sensors [15] are provided at either side of the roads to detect the vehicle movements and to instruct the microcontroller to switch on and off the lights accordingly. Here the lights are turned on only in the presence or motion of vehicles. Otherwise, all the lights are turned off even it is night time or dark weather conditions.

23.8 Implementation

With the help of sensors, the parameters motion, light intensity, and air quality are detected. In motion sensor (PIR sensor), the term PIR is the short form of the passive infrared. The term "passive" indicates that the sensor does not actively take part in the process, which means it does not emit the referred IR signals itself, rather passively detects the infrared radiations coming from the human body in the surrounding area (See Fig. 23.4).

The detected radiations are converted into an electrical charge, which is proportional to the detected level of the radiation. Then this charge is further improved by a built-in FET and fed to the output pin of the device, which becomes applicable to an external circuit for further triggering and amplification of the alarm stages. The PIR sensor (See Fig. 23.5) ranges up to 10 m at an angle of $+15°$ or $-15°$.

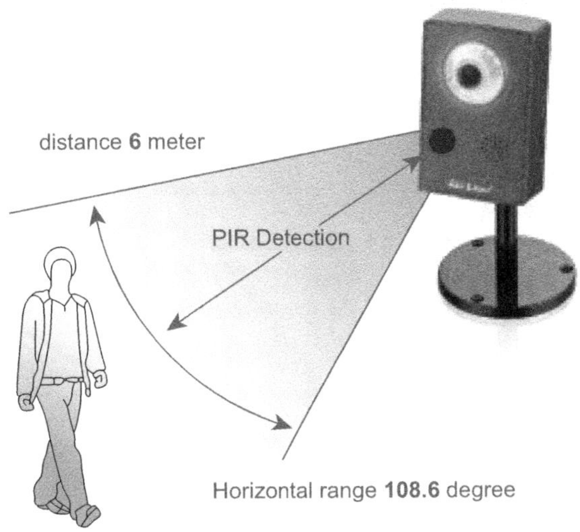

FIGURE 23.4 PIR sensor working.

The image below (see Fig. 23.6) shows a typical pin configuration of the PIR sensor, which is quite simple to understand the pinouts, and one may easily arrange them into a working circuit. The PIR sensors consist of three pins as indicated in the diagram shown below:

- Pin1 corresponds to the drain terminal of the device, which should be connected to the positive supply 5V DC.
- Pin2 corresponds to the source terminal of the device, which should be connected to the ground terminal via a 100 or 47K resistor. The Pin2 is the output pin of the sensor, and the detected IR signal is carried forward to an amplifier from the Pin2 of the sensor.
- Pin3 of the sensor is connected to the ground.

FIGURE 23.5 PIR sensor.

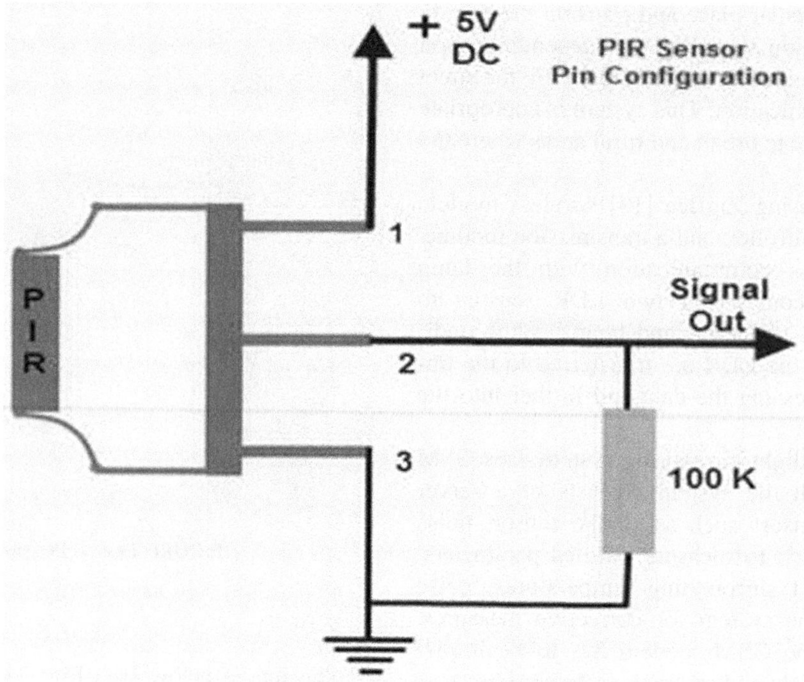

FIGURE 23.6 PIR sensor pin configuration.

23.8.1 PIR sensor's working principle

The PIR sensors are more complicated than the other sensors as they consist of two slots. These slots are made of a special material, which is sensitive to IR. The Fresnel lens is used to see that the two slots of the PIR can see out past some distance. When the sensor is inactive, then the two slots sense the same amount of IR. The ambient amount radiates from the outdoors, walls, or room, etc. When a human body or any animal passes by, then it intercepts the first slot of the PIR sensor. This causes a positive differential change between the two bisects.

When a human body leaves the sensing area, the sensor [16] generates a negative differential change between the two bisects (See Fig. 23.7). The infrared sensor itself is housed in a hermetically sealed metal to improve humidity/temperature/noise/immunity. There is a window, which is made of typically coated silicon material to protect the sensing element.

23.8.2 LDR sensor

An LDR or a photo resistor is a device that is made up of high-resistance semiconductor [17] material (See Fig. 23.8). The construction of an LDR includes a light-sensitive material that is placed on an insulating substrate such as ceramic. The material is placed in a zigzag shape in order to get the required power rating and resistance.

The area of zigzag separates the metal placed areas into two regions, where the ohmic contacts are made either on the sides of the area. The resistances of the contacts must be as less as possible to make sure that the resistance mainly varies due to the light effect only. The use of lead and cadmium materials is avoided as they are injurious to the environment.

23.8.3 Working principle of light dependent resistor

The working principle [18] of an LDR is photo conductivity that is nothing but an optical phenomenon. When the light is absorbed by the material, then the conductivity of the material reduces. When the light falls on the LDR,

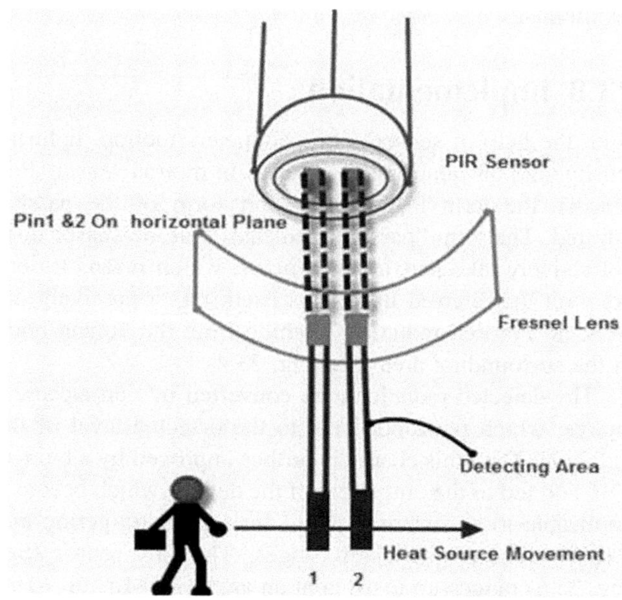

FIGURE 23.7 PIR sensor working principle.

FIGURE 23.8 LDR sensor.

then the electrons in the valence band of the material are jump to the conduction band. But, the photons in the incident light must have energy superior than the bandgap of the material to make the electrons jump from one band to another band. Hence, when the light has ample energy, more electrons are excited to the conduction band, which grades in a large number of charge carriers. With the effect of this process and more current starting to flow, the resistance of the device decreases.

23.9 Results and discussions

The LDR sensor checks for darkness and fault among the street lights and then notifies in the mobile application through a notification alert. The air quality sensor first detects the gas level and also checks for the threshold with giving notification in mobile application. System is also connected with motion sensor near the crossing of the road where it checks for the movements to the threshold level to alert locally with a display in road for vehicles to slow down and also notifies in the application. Finally, the system also detects the darkening of the area, or it will check for mode (day or night). Accordingly, it will send signal [19] for switching the light during night mode and switches off the unused light during day time with the help of LDR sensor.

23.9.1 Circuit connections

Fig. 23.9 shows the whole circuit connections with each sensors and Arduino. The suggested system consists of parameter sensing, data storage in the cloud, dashboard

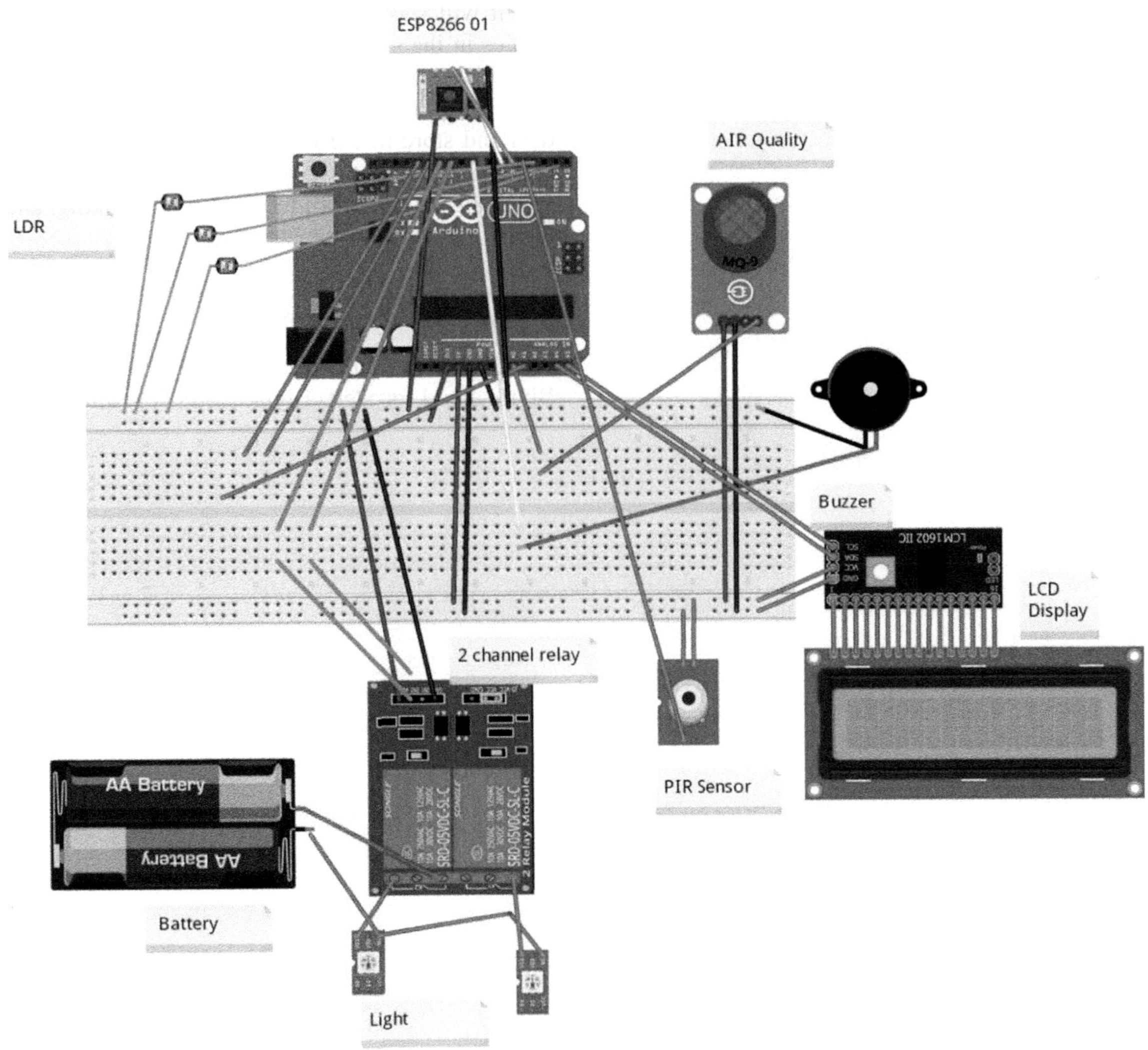

FIGURE 23.9 Circuit diagram of the proposed system.

display, and actuation control among electrical appliances to avoid any accidents or power switching.

23.9.2 Prototype model front view

Fig. 23.10 shows that the prototype model consists of sensors and actuators with all together attached with sensors. The prototype, which is made up of two parts, will collect environmental data [20], store it in the Blynk cloud, and connect with the actuation sensors in order to perform the operation. It automatically does tasks such as turning on and off lights, checking for motion, and issuing alerts locally or online.

23.9.3 Arduino IDE output

Fig. 23.11 represents the working of code in the IDE with initialization the connected Blynk server. The Arduino environment [21] may be expanded by using libraries.

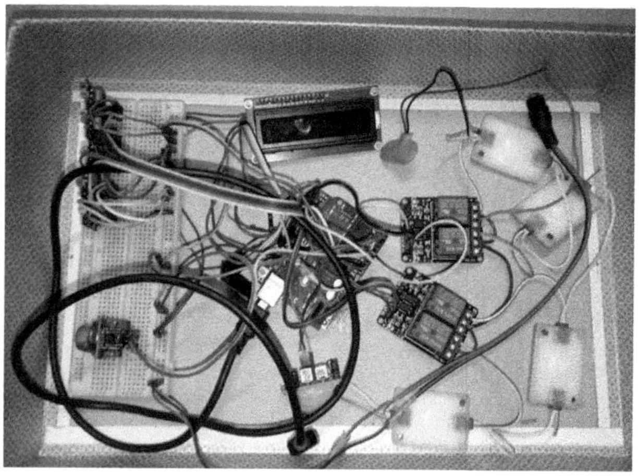

FIGURE 23.10 Prototype model front view.

Libraries add further functionality to sketches, such as the ability to work with hardware or manipulate data. Select a library from Sketch > import Library to utilize it in a sketch. The IDE comes with a number of libraries preinstalled.

Fig. 23.12 represents the motion detection parameters values in the serial output of IDE. The LDR sensor [22] checks for darkness and fault among the street lights and then notifies in the mobile application through a notification alert. The air quality sensor first detects the gas level and also checks for the threshold with giving notification in mobile application.

23.9.4 Blynk cloud representation

Fig. 23.13 represents all the parameter status in the Blynk application. MQTT protocol plays a vital role in this system in means of communication of all the commands instantaneously through Blynk cloud. This cloud [23] is helpful in storing the data and other threshold values to act accordingly. The safety environment can be monitored online in the cloud itself and also viewed in browser of mobile live, and it will show the live usage and currently running activities in the system. The control is one who is the admin that can handle or change respectively. The prototype is made with two prototype that will gather the environmental data and store in the Blynk cloud and communicate to the actuate sensors to complete the task. Tasks include switching on and off of light, motion checking, and alerting locally or online automatically.

23.9.5 Threshold alert

Fig. 23.14 represents parameter threshold value alert in the Blynk cloud for alert. Blynk cloud is helpful in storing the

FIGURE 23.11 Serial output of the proposed system in IDE.

FIGURE 23.12 Serial output of all the motion detection in IDE.

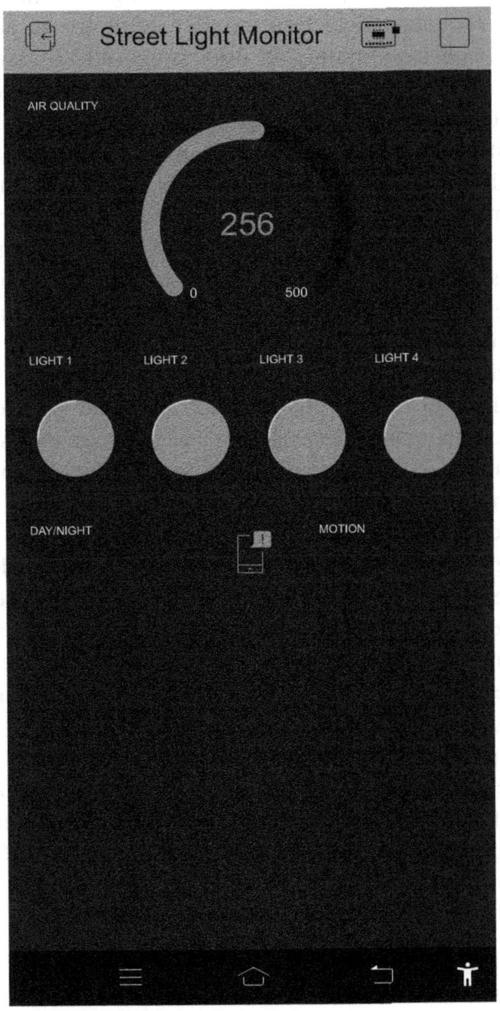

FIGURE 23.13 Live status of the system.

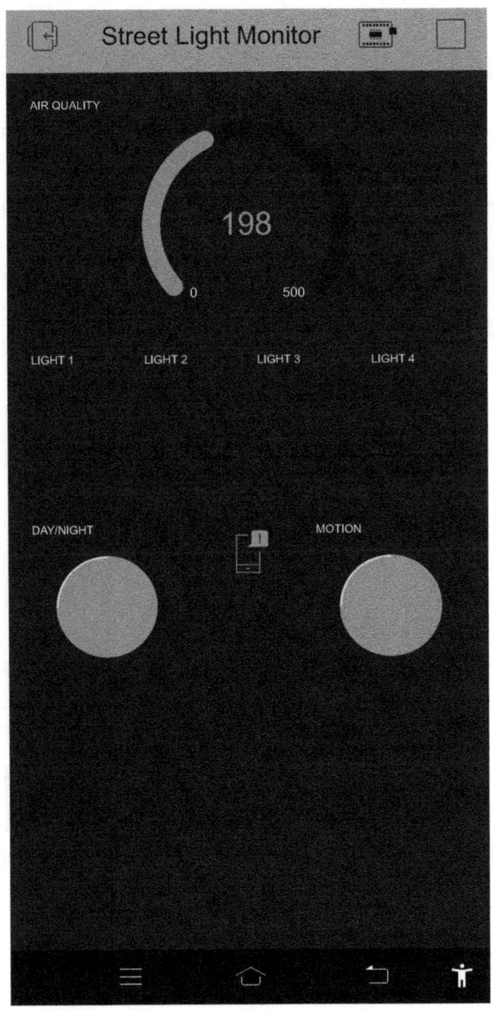

FIGURE 23.14 Threshold alert of the system.

data and other threshold values to act accordingly. The safety environment can be monitored online in the cloud itself and also viewed in browser of mobile live; it will show the live usage and currently running activities in the system. The control is one who is the admin that can handle or change respectively.

23.9.6 Motion reorganization

Fig. 23.15 represents movement detection through alert in Blynk cloud. The system is linked to a motion sensor near the road crossing, which monitors for movements beyond a threshold level and alerts locally with a display on the road for vehicles to slow down, as well as notifying the application. Finally, the system detects the darkening of the region or performs a mode check (day or night). As a result, it will transmit a signal to turn on the light.

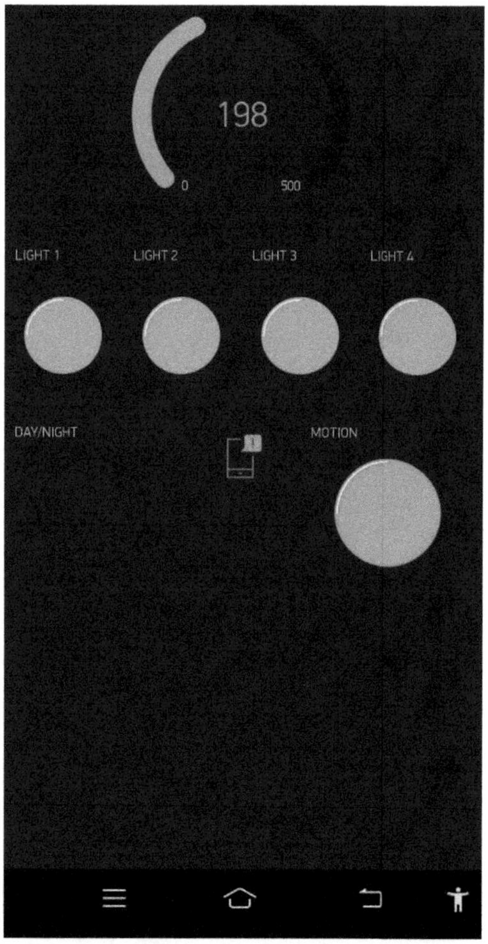

FIGURE 23.15 Motion reorganization on the Blynk app.

23.10 Summary

An informal economy is the part of any economy that is neither taxed nor monitored by any form of government. Although the informal sector makes up a significant portion of the economies in developing countries, it is sometimes stigmatized as troublesome and unmanageable. However, the informal sector provides critical economic opportunities for the poor and has been expanding rapidly since the 1960s. Integrating the informal economy into the formal sector is an important policy challenge.

In many cases, unlike the formal economy, activities of the informal economy are not included in a country's gross national product. However, Italy has included estimates of informal activity in their gross domestic product (GDP) calculations since 1987, which swells their GDP by an estimated 18%, and in 2014, a number of European countries formally changed their GDP calculations to include prostitution and narcotics sales in their official GDP statistics, in line with international accounting standards, prompting an increase between 3% and 7%. The informal sector can be described as a gray market in labor.

This system employs motion sensing, as well as light and air quality detection, to detect motions, air ppm levels, and low lighting in order to avoid accidents and losses. The data from the sensors is continuously scanned to record readings and check for faults or low light, and then it is delivered online. The Internet functionality is provided by the Wi-Fi module, to create a monitoring system to ensure that the degree of safety around street roads is maintained. This model will monitor tough conditions and send an instant alarm to the user. This prototype is separated into two components, one for actively detecting and sending data to the cloud via the Blynk App, and the other for actuating based on threshold values provided via cloud values, which it would get from the cloud. This connection takes place between the Arduino and the cloud via the MQTT protocol, which is included into the prototype.

Pedestrians and drivers can be protected in this system by signal alarm warning to neighboring suggested architecture in the event of an emergency, such as an accident between them. The data stored in the cloud can be studied further in the future to regulate street lights in a predictive manner. As a result, the projected data can be used to manage power utilization in this system.

Finally, let us move on to the real interactive part of this chapter: review questions/exercises, hands-on projects, case projects, and optional team case project. The answers and/or solutions by chapter can be found in Appendix G.

23.11 Chapter review questions/exercises

True/false

1. True or False? The prototype, which is made with three prototype that will gather the environmental data and store it in the Blynk cloud, communicates to the actuate sensors to complete the task.
2. True or False? All of the collected data is sent to a cloud infrastructure.
3. True or False? Once the data is collected and sent to the cloud, the software performs processing on gathered data.
4. True or False? The information needs to be unavailable to the end user in some way.
5. True or False? The intelligent infrastructure of the city is the preliminary stage to the entire structure of the intelligent city.

Multiple choice

1. Which of the following statements is/are correct about Smart Cities Mission?
   a. 500 cities and towns, with a population of 1 lakh and above each accounts for 73% of urban population is target of smart cities.
   b. It has a central outlay of Rs.48,000 cr for the next 5 years.
      a. Only a
      b. Only b
      c. Both
      d. None
      e. None of the above.
2. The center announced the names of cities and towns to be developed as smart cities on August 27, 2015. Which state gets the maximum number of aspirant smart cities?
   a. Maharashtra
   b. Tamil Nadu
   c. Uttar Pradesh
   d. Madhya Pradesh
   e. None of the above
3. Which of the following city got first Smart City nomination?
   a. Hubballi-Dharwad
   b. Surat
   c. Pune
   d. Puducherry
   e. None of the above
4. The core element of architecture of smart city is ________.
   a. Mobile unified service
   b. Urban application platform
   c. Management center
   d. Integrated information provider
   e. None of the above
5. Which one of the following protocols is lightweight?
   a. IP
   b. HTTP
   c. MQTT
   d. CoAP
   e. None of the above

Exercise

Problem

How is the concept of smart cities disrupting existing business models and creating new ones that have been positively impacting informal enterprise sectors?

Hands-on projects

Project

Do research: How does the applicability of smart infrastructure achieve sustainable urban development in cities?

Case projects

Problem

Examine the state of smart city development in developing countries, which includes understanding the conceptualizations, motivations, and unique drivers behind smart city development.

Optional team case project

Problem

Provide additional clarity and guidance on smart cities, considering that there is a significant amount of available material that may increase the complexity and confusion for practitioners and decision-makers in putting this information to best use.

References

[1] Al-Hasnawi A, Mohammed I, Al-Gburi A. Performance evaluation of the policy enforcement fog module for protecting privacy of IoT data. In: 2018 IEEE international conference on electro/information technology (EIT); 2018. p. 0951−7. https://doi.org/10.1109/EIT.2018.8500157.
[2] Dryjanski M, Buczkowski M, Ould-Cheikh-Mouhamedou Y, Kliks A. Adoption of smart cities with a Practical smart building implementation. IEEE Int Things Mag 2020;3(1):58−63. https://doi.org/10.1109/IOTM.0001.1900050.
[3] Das U, Dev VC, Namboodiri V. On the effectiveness of data aggregation to manage network congestion in smart grid AMI. In: 2018 IEEE power energy society innovative smart grid technologies

conference (ISGT); 2018. p. 1–5. https://doi.org/10.1109/ISGT.2018.8403329.

[4] Das U, Namboodiri V. A quality-aware multi-level data aggregation approach to manage smart grid AMI traffic. IEEE Trans Parallel Distr Syst 2019;30(2):245–56. https://doi.org/10.1109/TPDS.2018.2865937.

[5] Elghonaimy IH, Mahmoud WE. Sustainability and G.I.S. application in operating smart campuses in the smart cities. In: 2nd smart cities symposium (SCS 2019); 2019. p. 1–6. https://doi.org/10.1049/cp.2019.0208.

[6] Hrnjic T, Huseinagic I, Đonlagic T. Software architecture and communication protocols for integration of renewables in distribution smart grids. In: 2016 XI international symposium on telecommunications (BIHTEL); 2016. p. 1–6. https://doi.org/10.1109/BIHTEL.2016.7775730.

[7] Kardos P, Suter B, Mullican D, Nicol J, Kline M, York E, Salman A. Smart infrastructure: solutions to improve privacy and security. In: 2020 systems and information engineering design symposium (SIEDS); 2020. p. 1–6. https://doi.org/10.1109/SIEDS49339.2020.9106650.

[8] Marlen A, Maxim A, Ukaegbu IA, Kumar Nunna HSVS. Application of big data in smart grids: energy analytics. In: 2019 21st international conference on advanced communication technology (ICACT); 2019. p. 402–7. https://doi.org/10.23919/ICACT.2019.8701973.

[9] Karimi B, Namboodiri V, Jadliwala M. Scalable meter data collection in smart grids through message concatenation. IEEE Trans Smart Grid 2015;6(4):1697–706. https://doi.org/10.1109/TSG.2015.2426020.

[10] Kazmi A, Serrano M, Lenis A. Smart governance of heterogeneous internet of things for smart cities. In: 2018 12th international conference on sensing technology (ICST); 2018. p. 58–64. https://doi.org/10.1109/ICSensT.2018.8603657.

[11] Laicane I, Blumberga A, Rosa M, Blumberga D, Bariss U. The effect of the flows of information on residential electricity consumption: feasibility study of smart metering pilot in Latvia. In: Smart SysTech 2013; European Conference on Smart Objects, Systems and Technologies, 1–9; 2013.

[12] Liu N, Chen J, Zhu L, Zhang J, He Y. A key management scheme for secure communications of advanced metering infrastructure in smart grid. IEEE Trans Ind Electron 2013;60(10):4746–56. https://doi.org/10.1109/TIE.2012.2216237.

[13] Padmavathi V, Saminathan R, Selvamuthukumaran S. An adaptive QoS supportive approach for user based services using Krill Herd Approach over Internet of Things (KHAI). Int J Know Based Intell Eng Syst 2021;25(1):109–17. https://doi.org/10.3233/KES-210056.

[14] Piska S, Shetty M. A Java Card based approach for smart meter gateway security. In: 2013 IEEE innovative smart grid technologies-Asia (ISGT Asia); 2013. p. 1–5. https://doi.org/10.1109/ISGT-Asia.2013.6698780.

[15] Ramakrishnan R, Gaur L. Smart electricity distribution in residential areas: internet of Things (IoT) based advanced metering infrastructure and cloud analytics. In: 2016 international conference on internet of things and applications (IOTA); 2016. p. 46–51. https://doi.org/10.1109/IOTA.2016.7562693.

[16] Sun J, Zhang Y, Fan J. SmartAgents: a scalable infrastructure for smart car. In: 2011 12th international conference on parallel and distributed computing, applications and technologies; 2011. p. 99–103. https://doi.org/10.1109/PDCAT.2011.75.

[17] Sun L, Zhou K, Zhang X, Yang S. Outlier data treatment methods toward smart grid applications. IEEE Access 2018;6:39849–59. https://doi.org/10.1109/ACCESS.2018.2852759.

[18] Vance TR, Vance A. Cybersecurity in the blockchain era: a survey on examining critical infrastructure protection with blockchain-based technology. In: 2019 IEEE international scientific-practical conference problems of infocommunications, science and technology (PICST); 2019. p. 107–12.

[19] Yang J, Kwon Y, Kim D. Regional smart city development focus: the south Korean national strategic smart city program. IEEE Access 2021;9:7193–210. https://doi.org/10.1109/ACCESS.2020.3047139.

[20] Yau K-LA, Lau SL, Lin MWQ, Rahman MSA. Towards a smart city: the case of greater Kuala Lumpur in Malaysia. In: International conference on frontiers of communications, networks and applications (ICFCNA 2014 - Malaysia); 2014. p. 1–6. https://doi.org/10.1049/cp.2014.1393.

[21] Yeo W, Won D, Hahn H, Lee B, Lim J, Kim K, Kim S. Designing national RD planning framework in Korea government: focusing on smart grid. In: 2018 7th international conference on renewable energy research and applications (ICRERA); 2018. p. 1441–4. https://doi.org/10.1109/ICRERA.2018.8566708.

[22] Yu C, Ye B, Lin C, Wu YJ. Can smart city development promote residents' emotional well-being? evidence from China. IEEE Access 2020;8:116024–40. https://doi.org/10.1109/ACCESS.2020.3004367.

[23] Zhou Q, Natarajan S, Simmhan Y, Prasanna V. Semantic information modeling for emerging applications in smart grid. In: 2012 ninth international conference on information technology - new generations; 2012. p. 775–82. https://doi.org/10.1109/ITNG.2012.150.

Part VI

Designing inclusive smart cities policy instruments

Chapter 24

Customizable service level agreement (SLA) generator platform using FCAPS management to enhance Quality of Experience (QoE) on Internet of Things (IoT)

Cathryn Peoples[1], Adrian Moore[2], Mohammad Zoualfaghari[3], Parag Kulkarni[4] and Mamun Abu-Tair[2]

[1]Ulster University, Londonderry, United Kingdom; [2]Ulster University, Antrim, United Kingdom; [3]BT Group Plc, Suffolk, United Kingdom; [4]United Arab Emirates University, Al Ain, Abu Dhabi Emirate, United Arab Emirates

24.1 Introduction

Currently, customers pay service providers a monthly fee to connect to their online platforms, store data, access online services, and communicate. A provider will reimburse a customer through service credit if the service received is less than the service to which the customer and provider had agreed before the Service Level Agreement (SLA) became active. SLAs are defined to protect a customer and guarantee a minimum service level in response to the fee they are paying. FCAPS mechanisms, which provide Fault, Configuration, Accounting, Performance and Security (FCAPS) management, control network operations so that a minimum service level can be guaranteed within a dynamic and competitive network. When an SLA is achieved as a consequence of FCAPS, Quality of Experience (QoE), which reflects the extent to which a customer is satisfied with the network service received, can be assumed to have been facilitated.

FCAPS management is traditionally deployed to provide Quality of Service (QoS). QoS refers to a standard of performance achieved by the network, which allows the needs of applications to be achieved. "Needs" in this context refers to latency and data rate, for example, with exemplar metrics defined in ITU-T Recommendation G.1010 [1]. As a result of receiving QoS, QoE is also assumed to be achieved. QoE is assumed to be a by-product of QoS. QoE is more difficult to quantify in a standardized way than QoS due to the variety of human factors that influence it; a similar document to Recommendation G.1010 from the ITU-T does not exist for QoE. However, it is recognized that QoE can be achieved with less than QoS from a network provisioning and performance perspective and therefore less than the SLA (Fig. 24.1); this takes into account both the agreed-upon service and the cost. To put this in context, QoS according to the SLA may be achieved with a platform availability of 99%. QoE, on the other hand, could be achieved with an availability of 85%. There is no "one size fits all" mechanism, which is complicated by the fact that, "Two different users experiencing the same system condition may rate the situation differently due to user diversity" [2]. For customer groups who are satisfied with a lower level of service, there is an opportunity for the service provided to respond to this. There will be cost benefits for both the service provider and the customer.

It is understood that these characteristics give an appreciation of the service needs and personal tolerances, which can be used to appreciate how QoE varies across the demographic groups, which are associated with a city. This detail takes into account a user's typical application needs, in addition to the frequency and intensity with which they use the network. In a city, for example, we might generalize that service needs include capability for banking and booking tickets. Moving out toward modern housing, service needs include VoIP and uploading/downloading software. However, of course this scheme makes

Smart Cities Policies and Financing. https://doi.org/10.1016/B978-0-12-819130-9.00032-2

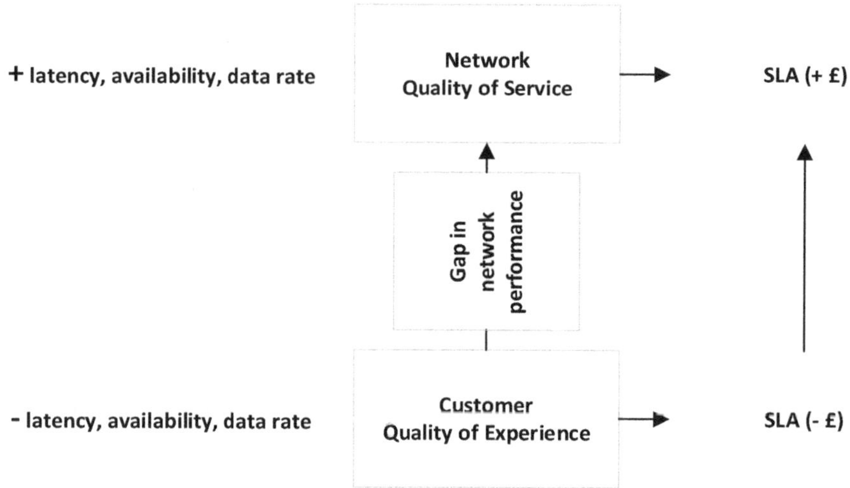

FIGURE 24.1 Relationship between QoS and QoE.

generalizations—while not typical of general assumed behavior, there may be users with specific operational needs. The process can accommodate this by separating the capture of user type information.

The extent to which this is true varies dependent on the user type. The notion of user types arises from customer classification guides, including the Acorn Guide [3], which is used as a point of reference in this work. The Acorn Guide has been designed as a tool to classify customers so that their characteristics may be exploited for optimized experiences from service providing perspectives and also from the point of view of the customer. This information can lead toward personalized service offerings.

SLAs are purchased from a set of tiers. A customer will agree to one that meets their potential maximum requirements, while one responding to their average requirements is typically sufficient in practice. Closure of this gap between an optimum and tolerable service helps to work toward the creation of personalized SLAs. This is something that can be exploited by service providers for more efficient operations, with efficiency measured by fewer redundant resource allocations, instead allowing resources to be provided for customers who need them. This is not to say, however, that a service provided using a free tier service is sufficient, i.e., basic service allowances provided without cost. In this scenario, the customer does not receive any service guarantees and is unable to negotiate their service terms with a provider.

Offering personalized SLAs requires either that the customer knows what they want or that the service provider knows what they want. However, understanding what customers want, such that an appropriate personalized SLA can be proposed, is complicated by the fact that, in 2019, "One in five [UK adults] … don't use the internet because it's too complicated …" [4]. This suggests that a number of (potential) customers will be unable to express what they want from a service and indicates that service autonomy is required. For customers who do not know what they want, we suggest that this process in managed using a schema similar to the Acorn Guide.

24.2 Research problem

Fulfillment of the SLA is made possible through monitoring the real-time network state. However, the ITU describes that "quality of service and quality of experience measuring is becoming more and more complex" [5]. We therefore seek to provide a solution by which QoE is fulfilled as a result of using the SLA to achieve QoS, with operational cost, which is optimized by recognizing the diversity in QoE expectations and satisfactions—not all users require the same level of operation from the network. The intention is not to capture QoE actively, but instead understand the tolerances through an awareness of the types of customers using the network. This is identified primarily through their fixed location.

To contextualize this: A customer may require 1 GB of bandwidth use per month, although they are purchasing 2 GB. There is therefore a redundancy in service provision, which can be allocated elsewhere. Another customer could be a data provider customer and needs capacity sufficient to provide one data reading per sensor at a minimum. The minimum SLA provisioned, however, may be in excess of this; therefore the customer is paying for and receiving more than they need. Again, there is an opportunity to refine the SLA to meet the customer's service needs more closely. Another SLA may provide service availability 99% of the time. The customer might be able to tolerate 80% service availability and can therefore be removed from the service provisioning process during constrained intervals, with capacity being allocated elsewhere to the customers who are less tolerant of disruption and need it.

This notion of tolerance is highly subjective and difficult to quantify from the perspective of a provider, so that the detail may fairly support SLA decisions. The ITU-T recognizes in Recommendation G.1080 [6] that, "cultural background, motivation, attention-related factors, [and] emotional state" affect perception of quality. They go on to state that, "Direct evaluations of QoE are designed to exclude these factors, since they are not generally under the control of a network operator and so do not contribute to network requirements." While it is agreed that these aspects are not controllable by a network operator, it is important to do something with this information given the role it plays in network (albeit perceived) service quality. It is this information that we accommodate in the service decisions made. Cultural backgrounds, motivation, attention-related factors, and emotional state are not aspects that a customer can be explicitly queried about. Indeed, a customer may not even have a good understanding of these personal qualities themselves. We can make parallels, which can be somewhat understood by where a person lives, however, between these elements and a customer's residential location, in relation to a city. Socioeconomic factors can be indicative of qualities such as motivation, for example, and emotional state. When these factors affect the QoE, they should be used to influence the SLA. It is the capture and use of this information in the SLA provisioning process that the proposal made in this chapter focuses on.

24.3 Prior art

The proposal presented in this chapter recognizes the different manners in which customers may wish to engage with a network and online services, as either application users, or more typically of IoT technologies, as data providers or data consumers. The information that is collected from a customer and used to influence the SLA provisioning process is driven by an ontology. The proposed ontology accommodates the needs of all scenarios anticipated in the IoT. Ontologies for the IoT in general do not accommodate all anticipated requirements, which can be encountered across the IoT. The proposal in Ref. [7], for example, contextualizes services in smart cities. In Ref. [8], the service is organized around needs within smart homes. IoT-Lite, as another example, captures metadata associated with a sensor built into a sensing device within a system [9]. The authors' intention is that the model can be applied across the IoT due to its simplicity; however, this simplicity leads to the fact that, "IoT-Lite does not intend to be a full ontology for the IoT", and that, "... different semantic modules can be added to provide additional domain and application specific concepts and relationships." Each of these solutions provides part solutions to specific use cases, but does not provision a solution, which accommodates all needs across the IoT. This is significant: interoperability across solutions is not achievable with standalone solutions. This results in a disjointed and ad hoc IoT landscape.

Monitoring the network for SLA fulfilment, and taking recovery action in the event that it is needed, is achieved using FCAPS mechanisms. FCAPS for the IoT generally applies to individual aspects of management problems, such as fault [10,11]. There is less evidence of solutions, however, which fulfill the needs of all aspects of FCAPS in a single, integrated solution. This is significant, as standalone solutions, which respond to individual service needs, are rolled out, the interoperability between IoT solutions becomes increasingly limited and restricted. Furthermore, management techniques apply to the network and the devices, but do not consider the needs and tolerances of customers who will be using the resources.

There is an opportunity to use the metadata collected to provision the SLA for network management objectives. Aazam et al. [12] present an approach to manage resources in line with customer SLA expectations, recognizing the opportunity to be exploited from diverse expectations in QoE. Their scheme requires that customers request a service, and the service provider fulfills this request. Customers are then required to feed back their satisfaction so that resource estimations can be made, and the service quality expectations can continue to be fulfilled. This approach, proposed as recently as 2019, requires ongoing input from customers as to whether or not they are satisfied with their service. This is an unnecessary overhead from the perspective of the customer.

In Ref. [13], the authors consider QoE metrics to include switching behavior and stall frequency, as a function of video streamed throughput. They then apply prediction within existing algorithms based on the DASH standard, with the aim of increasing QoE. This approach is less from the human perspective and more from the technical perspective.

An approach is defined in Ref. [14], which predicts perceived quality of IPTV services, with an understanding that the approach might be used for service planning. Input information includes bit rate, packet loss rate, and burstiness, in addition to information on the encoder/decoder settings, which include the number of slices per frame. This scheme does not take into consideration that different users will be satisfied with varying levels of service and is defined in a manner that is specific for IPTV.

It is described in Ref. [15] that, for many service providers, there has been a "race to the bottom" effect in terms of pricing. The authors acknowledge that this is not always to the benefit of service providers, "who see lower margins, or customers, who are stuck with whatever quality of experience the provider is able to deliver on this pricing model." We aim to avoid this approach with the proposal, by offering services and prices on a bespoke basis according to tolerance.

There are therefore several gaps to which this research proposal responds. Firstly, we propose an ontology to capture the needs of customers, which is presented in Ref. [16]. In this chapter, we demonstrate how information collected using the ontology is applied to influence the SLA. Customers are additionally included in autonomous decisions made to manage network resources through the information they have provided when the service was first established.

24.4 Research proposal

While the human behavioral characteristics described in ITU-T Recommendation G.1080 are generally not taken into account because they are not under operator control, this proposal advises that they are used to influence service provisioning decisions made. These characteristics are incorporated through a customer's socioeconomic status, which is inferred from their geographical location (excluding business customers). A customer will be resident in a location due to their socioeconomic status.

A personalized SLA can be provided on the same products for different customers, dependent on the customer's desire for certain performance levels and SLA options. This will be the case in situations where a customer is requesting the same application service, but has different tolerances to service levels and performance. The scheme is applied on a per household basis, and the household SLA will be an average of the SLAs offered for all service users within the household. The method of estimating QoE is invisible to the user. All they know is that there are certain questions they should respond to when setting up their service.

This proposal makes contributions to each stage of the SLA life cycle, from the SLA request to the SLA being terminated [16]. When a customer makes an SLA request to a service provider, several items of metadata are collected from the customer regarding their service needs and tolerances. This information is used to guide the SLA offered, and once it has been agreed upon by both the customer and the provider, and passed to an active state, the information is used to subsequently manage the network.

The Acorn Guide has been adapted into a reduced number of groupings to characterize customers according to socioeconomic status and their frequency and type of online activity. The Acorn Guide shows how this is primarily determined by customer location, which also reveals their type of home, the location of their home in relation to the closest city, the number of people they live with, cultural aspects, and disposable income. These have been categorized in this proposal according to the following groups:

- Less affluent nonfrequent low-intensity (user group 1)
- Less affluent frequent low-criticality (user group 2)
- Less affluent nonfrequent high-intensity (user group 3)
- Less affluent frequent high-criticality (user group 4)
- Affluent nonfrequent low-intensity (user group 5)
- Affluent, frequent low-criticality (user group 6)
- Affluent nonfrequent high-intensity (user group 7)
- Affluent, frequent high-criticality (user group 8)
- Business (user group 9)

Customers are categorized according to their ability to pay for a service, their online frequency, the intensity with which they use the network, and the importance of their traffic being successfully transmitted. One of the motivators of this proposal was marginalized groups, and the understanding that some are marginalized because of service cost. Financial ability, revealed through customer location, provides this detail. A customer will continue to belong to the same category until a point in time when they are able to move location, which will be triggered by a change in their socioeconomic status. This customer classification has been formed on the basis of customer mappings in Fig. 24.2. These categories are contextualized in relation to their network demands and service tolerances in Table 24.1.

While the concept behind this approach is feasible in a traditional and non-IoT network, the mechanics in this proposal focus on the IoT. This is in recognition of the fact that the IoT has unique requirements beyond traditional network environments, with the advent of a greater range of networked devices. It also drives the participation from a more diverse group of users than before due to the range of services offered, some of which are only available online. These devices generate data, which can be of interest to a broad user group. It can therefore be publicly hosted and results in customer types who may be considered to be data providers or data consumers. A data providing customer will contribute sensor data to a repository, which they or other consumers may consume.

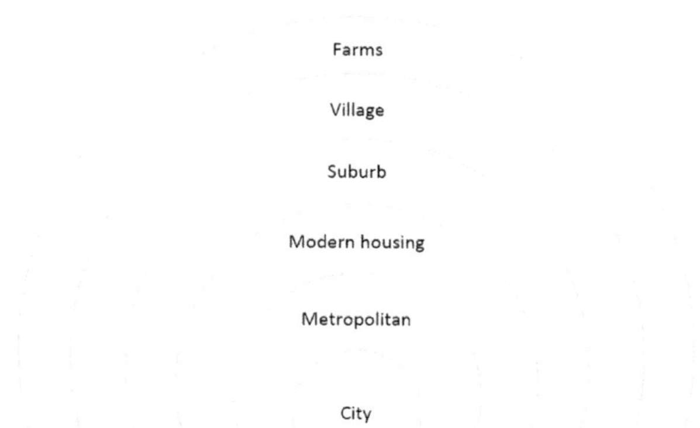

FIGURE 24.2 Customer classification.

TABLE 24.1 Customer network service needs and tolerances [3].

Acorn guide category	Classification	Characteristics	Online activity
Category 1 Group A Type 2	"**Metropolitan** Money"	Live in large apartments or town houses in London or other major cities.	Likely to use their tablet to access content relating to travel, news, weather and to play games and, less frequently, read blogs. More prolific internet users when organizing leisure, finances or household services. Use smartphones for email and to exchange pictures, or download and listen to music. Few are regular users of social media accounts.
Category 1 Group B Type 5	"Wealthy Countryside Commuters"	Live in **semi-rural areas, villages** and the fringes of small towns.	Mobiles and tablets are often used for practical purposes, accessing content relating to investments, and less likely to access entertainment content or social networking sites. They may regularly research financial products and services, property, cars, hotel reservations, airline tickets, consumer electronics, flowers and gardening equipment. They may read newspapers and download podcasts. They may not use social media particularly often.
Category 1 Group C Type 10	"Better-off **Villagers**"	Live in the larger, more expensive housing found in villages and the edge of small towns.	Occasional rather than regular users of the internet, shop online slightly more than average, research finances, cars, consumer electronics and property. Online purchases might be gardening equipment, wine and flowers.
Category 3 Group F Type 22	"Larger Families in **Rural Areas**"	Large detached houses or bungalows in more sparsely populated areas.	Use of social media is likely to be much lower than average. Many will find it convenient to bank online, monitor investments, read newspapers. Some will be active on auction sites such as eBay. They will also buy gardening equipment, electrical goods, cosmetics, tickets, travel, and holidays online.
Category 3 Group G Type 25	"Larger family homes, multi-ethnic areas"	Stable middle class neighborhoods often found in **suburban** areas of larger towns and cities.	Use VoIP/Skype more often than most, on both mobiles and PCs. Other frequent online activities might include uploading content and downloading software, music and phone apps. Some will read newspapers online.
Category 4 Group M Types 41, 42, 43, 33	"Striving Families"	Low income families typically living on traditional **low-rise estates**.	Their phone is less likely to have internet capabilities and with the possible exception of games consoles and TVs, these people are less likely to purchase the latest technological goods. Visiting the pub, computer games, DVDs, betting, bingo and the lottery are among the more common leisure activities.

Data providers typically operate in a best-effort approach in the IoT network, in that their operations and behaviors are not regulated, and they are not obligated to any certain behavior, i.e., a penalty is not applied if they do not do something that they did before, which was of benefit to others, such as contributing data. Data consumers relying on any data or services made available by data providers are therefore vulnerable [17,18], which contributes to a reduction in their trust of IoT services—in general, the activities and behaviors of customers are unpredictable. This proposal works to provide regulation in relation to the behavior of data-providing customers so that data-consuming customers may feel more secure. However, we recognize that customers may feel restricted with a fixed-term SLA and therefore seek to provide a flexible SLA, with flexibility considered from the perspective of the SLA being adaptable once it has moved into an active state. Relationships between the entities in this system are captured in Fig. 24.3.

In this scenario, a cloud service provider (CSP) provides services for one or more devices belonging to a customer. A customer can exist with other customers within a domain. Each customer will have one or more devices, for which the SLA has been set up. Data collected from each customer's device will populate a domain repository. Data may optionally be transferred to a centralized repository, depending on demand and need.

The platform setup is presented in Fig. 24.4. Fig. 24.4 is considered in more detail in terms of the components within each engine in Fig. 24.5.

The Domain Repository Manager (1) is applicable to a single domain and ensures that network requirements are fulfilled. This includes controlling the rate at which meta data is collected. The Domain Repository (2) will be populated so that terms in a dataset are specific to a domain. A single ontology template will support the needs of all domains. The Centralized Repository (3) is similarly populated using the ontology, but with information representative of multiple domains. The Centralized Repository accumulates a collection of data, which is frequently used by customers operating across and between domains. The Centralized Repository Manager (4) will play a similar role to the Domain Repository Manager (1), by controlling which context data is collected, the rate at which attributes are

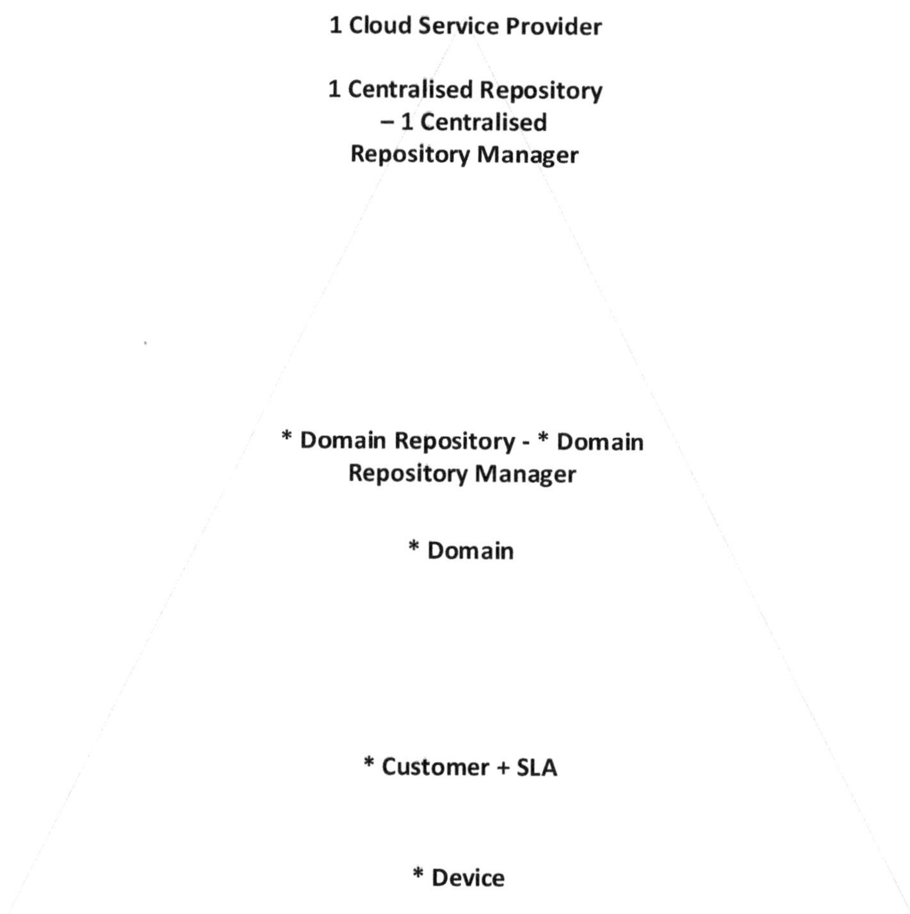

FIGURE 24.3 Entity relationship.

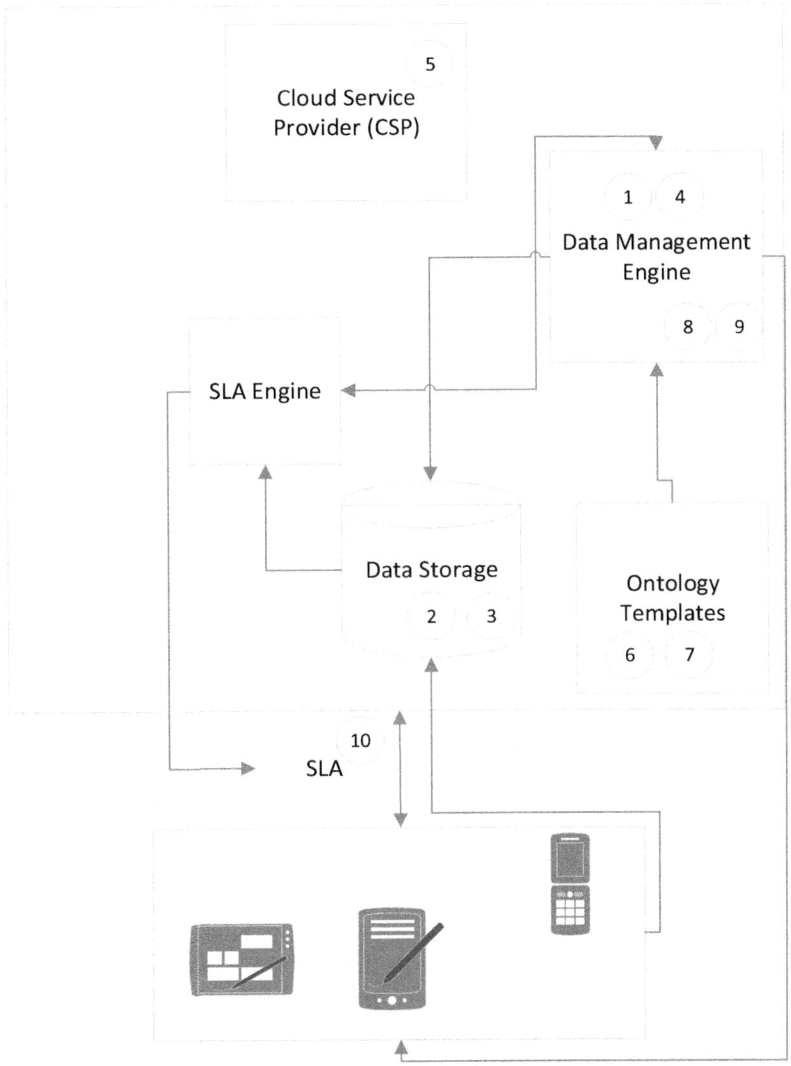

FIGURE 24.4 High-level platform setup.

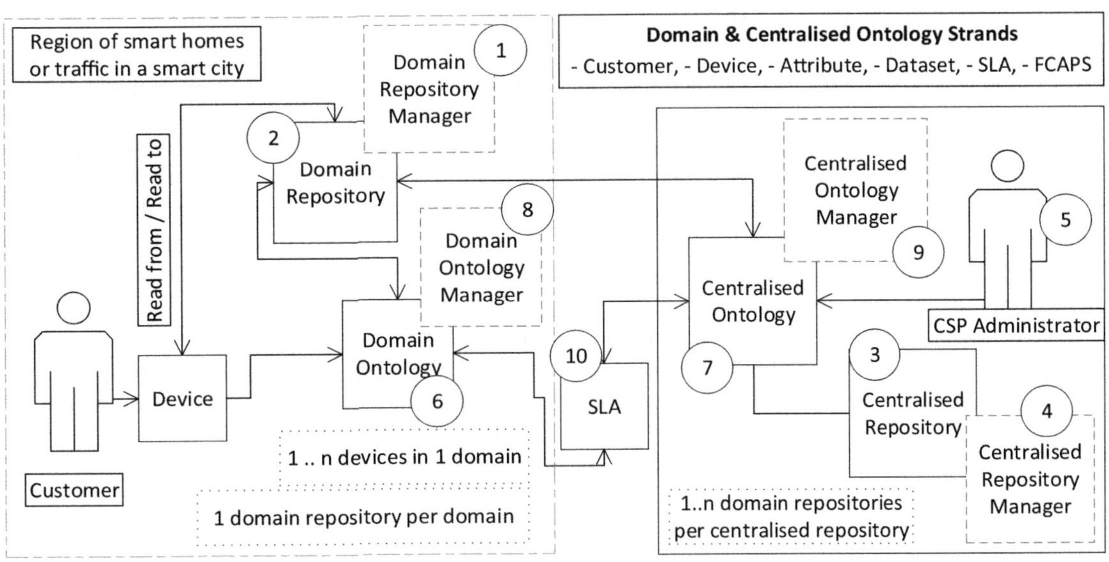

FIGURE 24.5 System platform architecture.

collected and updated, where they are stored, if they are transported to the Domain Repository, and when they are discarded. The Manager is also responsible for controlling which information is located in the Centralized Repository, by pulling detail from different domains, as required. The Administrator (5), resident on the server side of the network, will manage any manual configurations, such as ensuring that new domains or domain-specific attributes entered by customers are acceptably and accurately specified.

In relation to the ontology that governs the data collection process, the Domain Ontology (6) and Centralized Ontology (7) refer to the same standard template used to collect meta data. They vary in that the range of data collected in a domain might be less than the range of data retained centrally. Data retained in the Domain Repository is dependent on the Domain Ontology. Processing of data retained in both repositories will indicate if the SLA is being achieved as a result of management of the network. This will be the responsibility of the Domain Repository (8) and Centralized Repository (9) Managers. The ontologies are connected via the SLA (10). This relates to the fact that the SLA is dependent on information collected on a Customer, as controlled by the Domain Ontology. This will be combined with detail from the wider network to determine if the SLA is achieved from the perspective of the way in which the network is being managed. The Domain Ontology Manager and Centralized Ontology Manager are responsible for determining if a Customer SLA is being achieved by monitoring usage and performance achieved. The Ontology Manager will be responsible for filling in fields using information that has been collected using the ontology. In the case of determining the domain for example, this will be achieved by combining information on the device and sensor type; the Ontology Manager will be responsible for processing this information once received.

24.5 High-level ontology design

At the highest level of the ontology that supports the process, a device belongs to a customer, a customer signs an SLA, a device is associated with an attribute/application, and an attribute can contribute to a dataset. This ontology is presented in detail in Ref. [16]; however, a high-level consideration of the ontology design will be presented here for completeness. Each of the higher layer categories described (Device, Customer, SLA, Device, Attribute, and Dataset) is supported by several attributes at lower layers. In the first stage of the process, when an SLA is requested, a customer will be characterized. Guided by the ontology, a customer will first be asked their type or user group. Their type will be either a DP or a DC, or their user group will be a wearable e-health device user, gamer, or browser. A customer will be asked their ability to tolerate risk to service availability, and if they wish for the terms of their SLA to be renegotiable after the SLA becomes active. This gives consent to renegotiate when there is an opportunity to do so. The customer detail is processed alongside device detail. Device detail is collected for the purpose of ensuring that the SLA continues to be achieved throughout the service period, ensuring security, and providing opportunities for appropriate configurations in a dynamic environment. These decisions can be prioritized depending on the type of user, i.e., an elderly user might prioritize security while a younger professional might prioritize efficient memory consumption. Attribute information characterizes the properties of the data or application being communicated from the customer. These attributes allow the service requirements to be understood from an application/data traffic perspective in terms of, for example, the idea frequency of transmission and whether or not this attribute can be interrupted in its transmission. Detail retained about a customer's SLA defines the terms on which their SLA has been agreed, including the tolerance of the service being interrupted and the maximum disruption, which a customer is prepared to be exposed to. The SLA is also agreed on the basis of service allowances for a monthly cost, and an agreed reward for a breach in the event that one occurs. This information will be kept up-to-date in relation to the state in which the SLA exists at a point in time—states include definition, negotiation, active, renegotiation, recommendation, and termination.

Due to network dynamics within the IoT, configurations may be required, which are specific to the network performance achievable to ensure continued SLA fulfillment. For example, the communication of noncritical data from a sensor may be suspended during periods of network congestion to allow ability to support critical applications. Operating in this way allows the network to be managed to support SLA objectives—change may be applied to allow this to continue to be achieved. Furthermore, the approach recognizes that the change in one customer's SLA can have a subsequent impact on other customers' SLAs. In this case, the customer who is impacted will not necessarily have defined in their SLA that they desired the SLA to be renegotiable, but through conditions in the wider network, the way in which their service is being provided may need to be adjusted. The proposed SLA life cycle therefore incorporates a renegotiation and recommendation phase. Flexibility in SLAs is desirable so that customers are not restricted in their contractual obligations if personal needs change. The recommendation phase allows the service provider to make a service recommendation to a customer if situations beyond their control impact their QoE.

24.6 SLA assignment process

The SLA assignment process approach builds upon the mechanism captured in Ref. [16] and accommodates customers in one of two ways: In one approach, a customer can select their type from a predefined list. Alternatively, a customer can have their service tolerances characterized automatically according to their location. In both approaches, nonetheless, a customer's location will be accommodated in the service provisioning decisions made.

When we consider customers as DPs and DCs, the flow chart in Fig. 24.6 captures that particular domains and items of data are more popular than others, on the basis of research in Ref. [19]. The process first checks if a customer is contributing data to a popular domain. This will be encouraged, and a lower score is given for doing so. If a DC, on the other hand, is consuming data from a popular domain, this is weighted more highly in an attempt to regulate network activity in situations when congestion is possible. The scores applied are customizable and are represented in this embodiment with a higher value of 10 for alpha and a lower value of 1 for beta.

Three types of users are considered in a scenario to contextualize this process. In Table 24.2, Tables 24.3 and

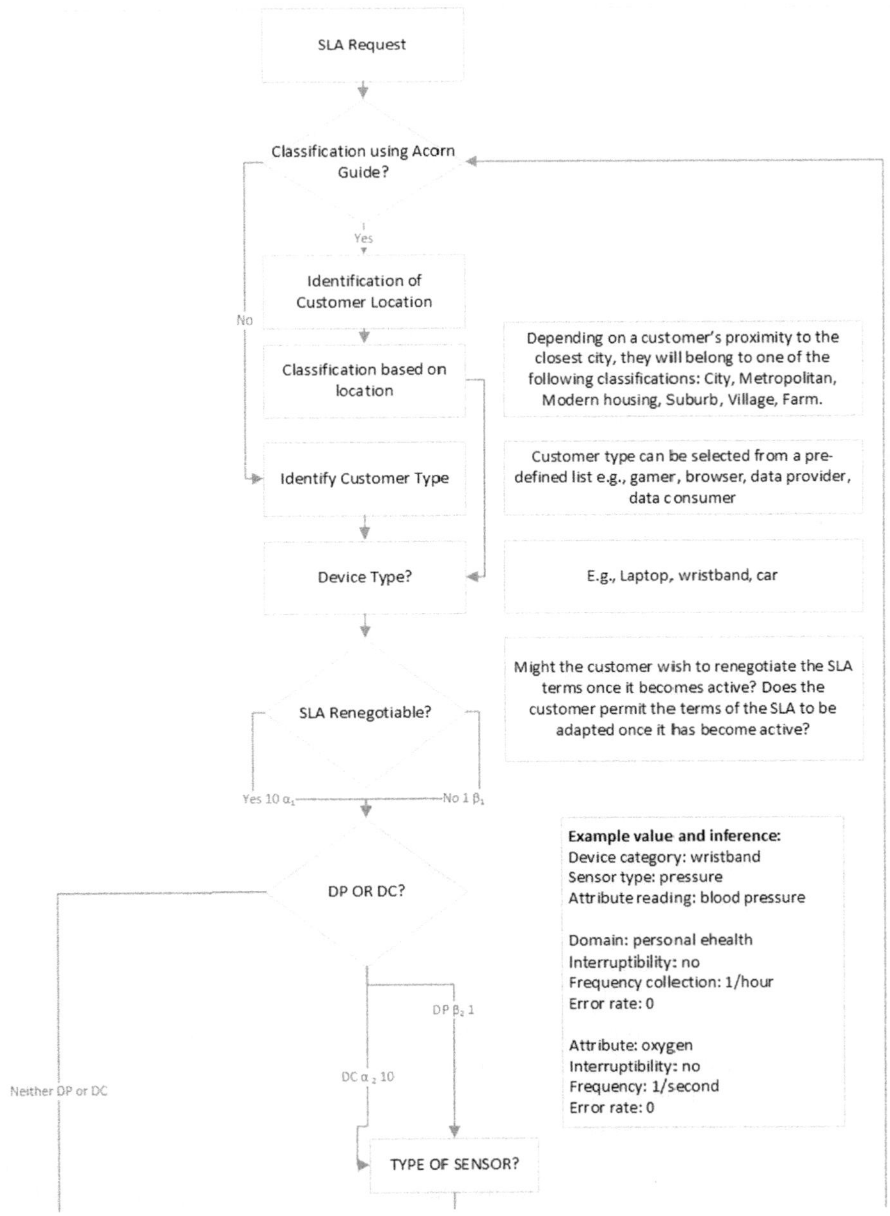

FIGURE 24.6 Flow chart of the SLA provisioning process.

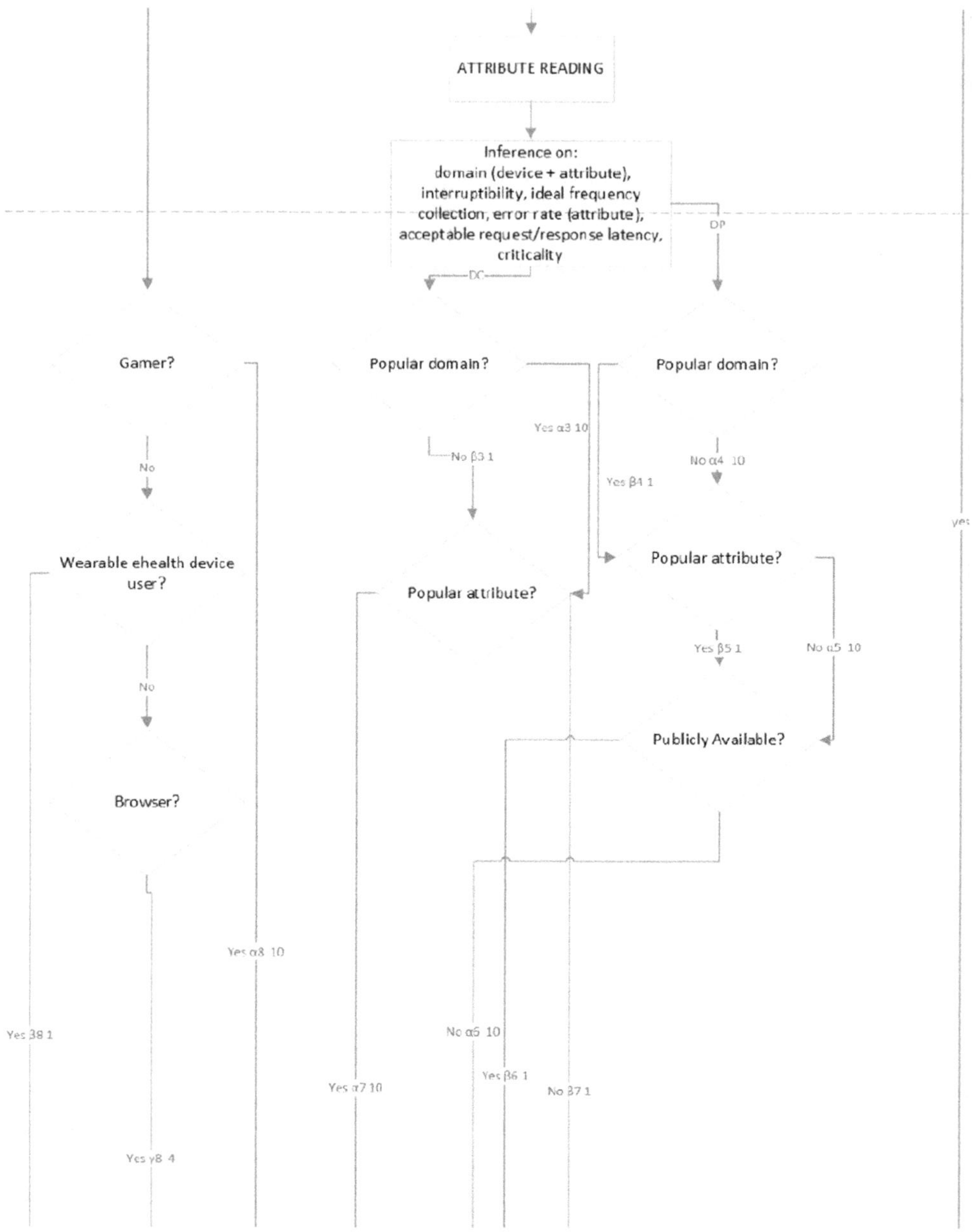

FIGURE 24.6 cont'd

24.4, the customers are flexible in their service needs and tolerant of the service provided.

Customer information is entered into the SLA provisioning process, which reflects the customer's tolerance of service interruption, risk to a change in their service, and desire to potentially renegotiate SLA terms after it becomes active. A score of 1 indicates that the customer can tolerate interruption and risk to service availability—a low score indicates tolerance. A score of 10 for the renegotiable attribute indicates that a customer wishes to have the flexibility in potentially renegotiating their SLA terms and will pay a premium for doing do.

Three customer types are considered: a data providing customer, a gamer, and a browser. These customer types are scored as a result of their application type:

DP	1
Gamer	10
Browser	4

These are arbitrary scores, and any value could be applied in practice depending on priorities.

DPs are subsequently scored according to the type of data that they are communicating to the repository, with a

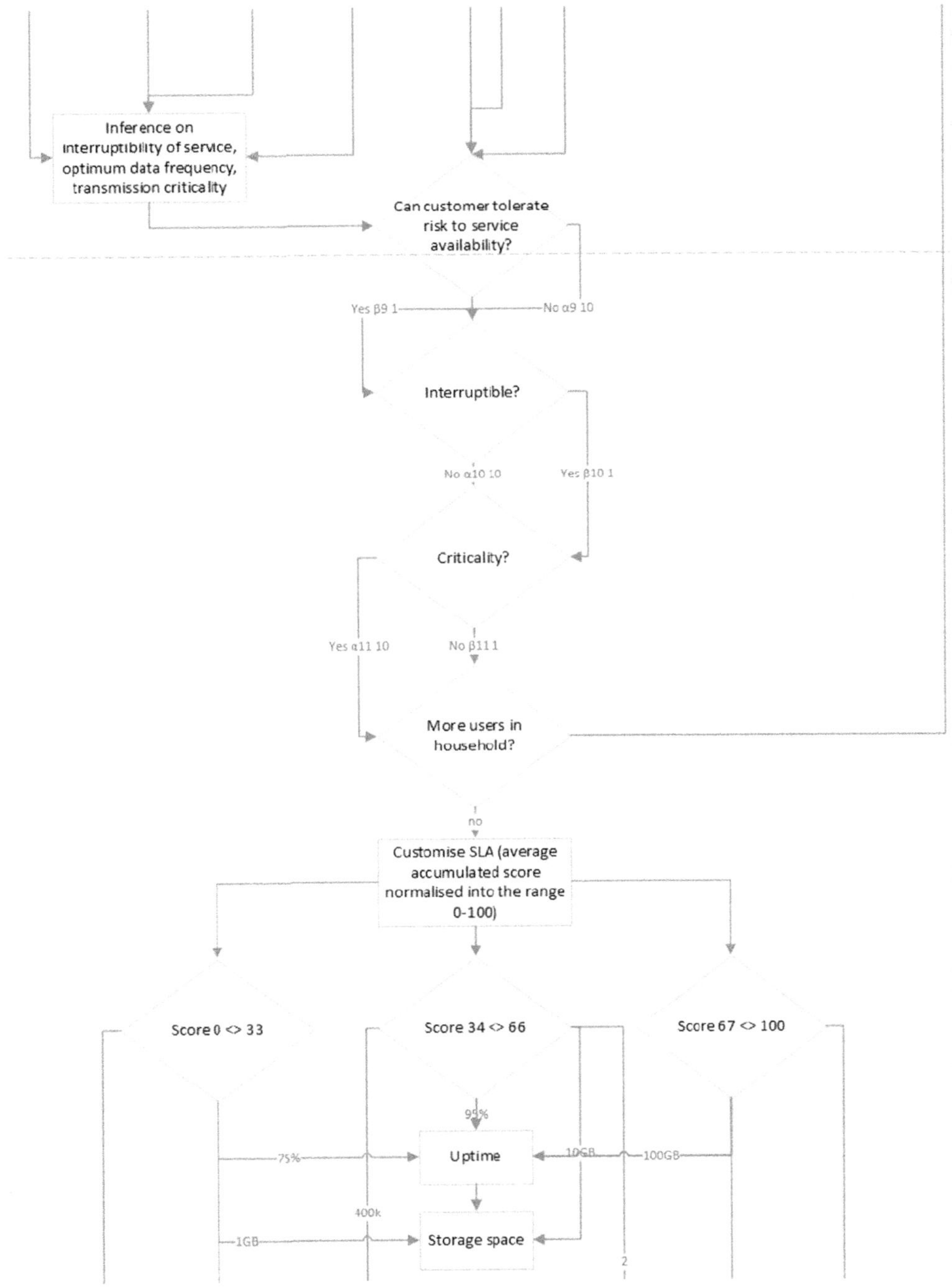

FIGURE 24.6 cont'd

lower score for data, which is more popular to encourage this type of activity in the network. A gamer is scored more highly (resulting in a higher SLA charge) in relation to the intensity of their traffic, and one due to the criticality of their traffic (it is not of critical importance, and this is traffic, which can cope with disruption). A browser, on the other hand, is given a low score for both traffic intensity and criticality. The customer will receive a service, which recognizes their lower service needs. The score based on application needs and customer tolerance is then weighted according to the location in which the customer resides in relation to their nearest city. The weighting increases as we move further from the city, taking into account typical customer service needs. An affluent frequent high-criticality data providing user is scored 21.7 while a less affluent nonfrequent high-intensity data providing user is scored 7.7. This scoring can be contrasted with another customer type, a tolerant gamer (see Table 24.3).

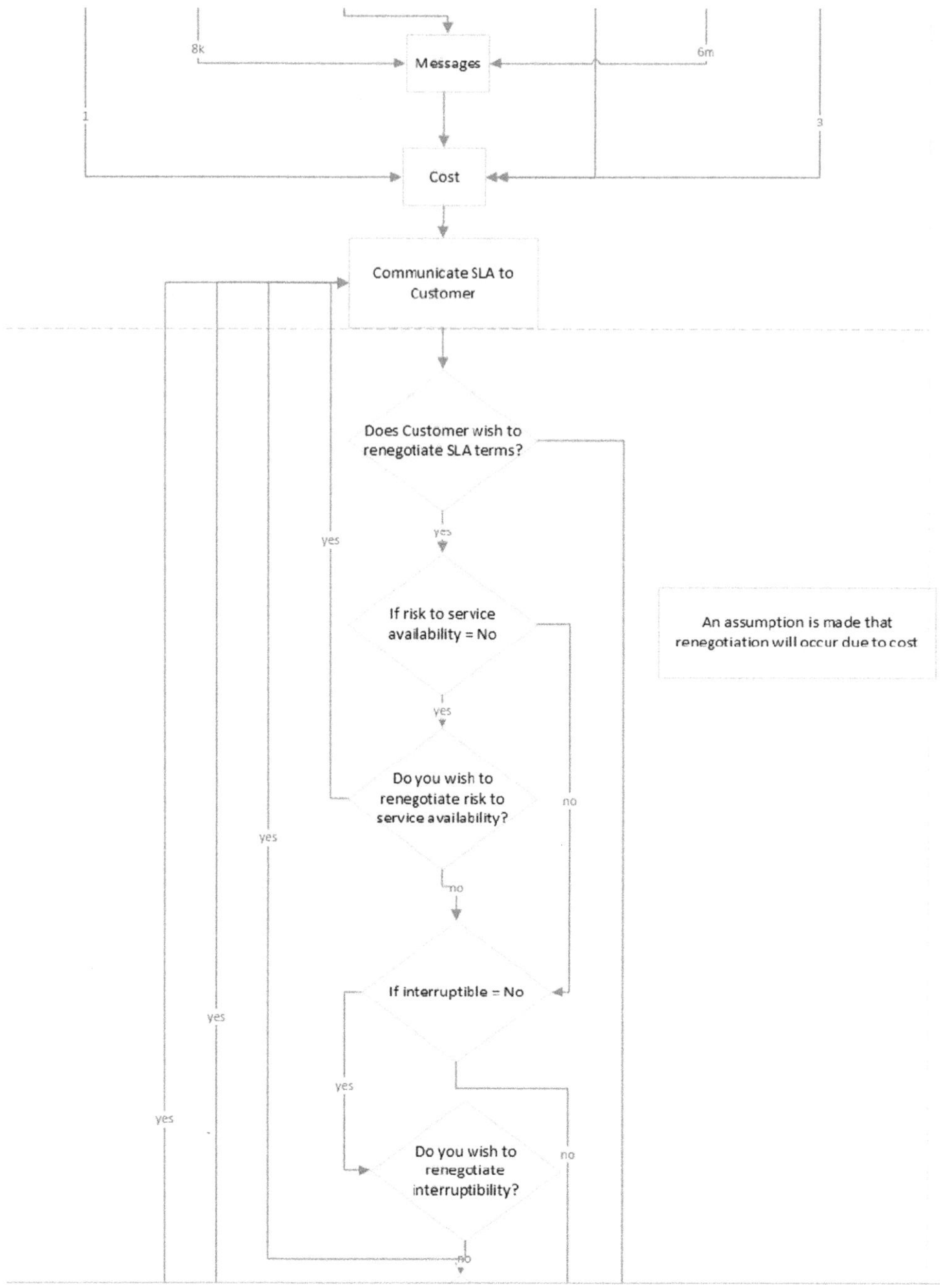

FIGURE 24.6 cont'd

In this scenario, the affluent frequent high-criticality user is scored 48.5, while the less affluent nonfrequent high-intensity user is scored 17.8. This customer type is scored more highly than the previous customer due to the greater intensity of application traffic. Another customer type is considered in Table 24.4, a tolerant browser:

The affluent frequent high-criticality user is scored 12.7, while the less affluent nonfrequent high-intensity user is scored 4.3. This is lower than for the gamer and the data providing customer. The customers in Tables 24.5 to 24.7, on the other hand, are less tolerant of the service received. In this setup, the customer cannot cope with interruption to service and is scored more highly with 10 in this embodiment to result in a higher service charge overall. The same weighting is applied for risk. The customers do not wish to have the ability to renegotiate their SLA terms; therefore renegotiable is given a lower score of 1. This results in a contrasting set of scores for this group of customers.

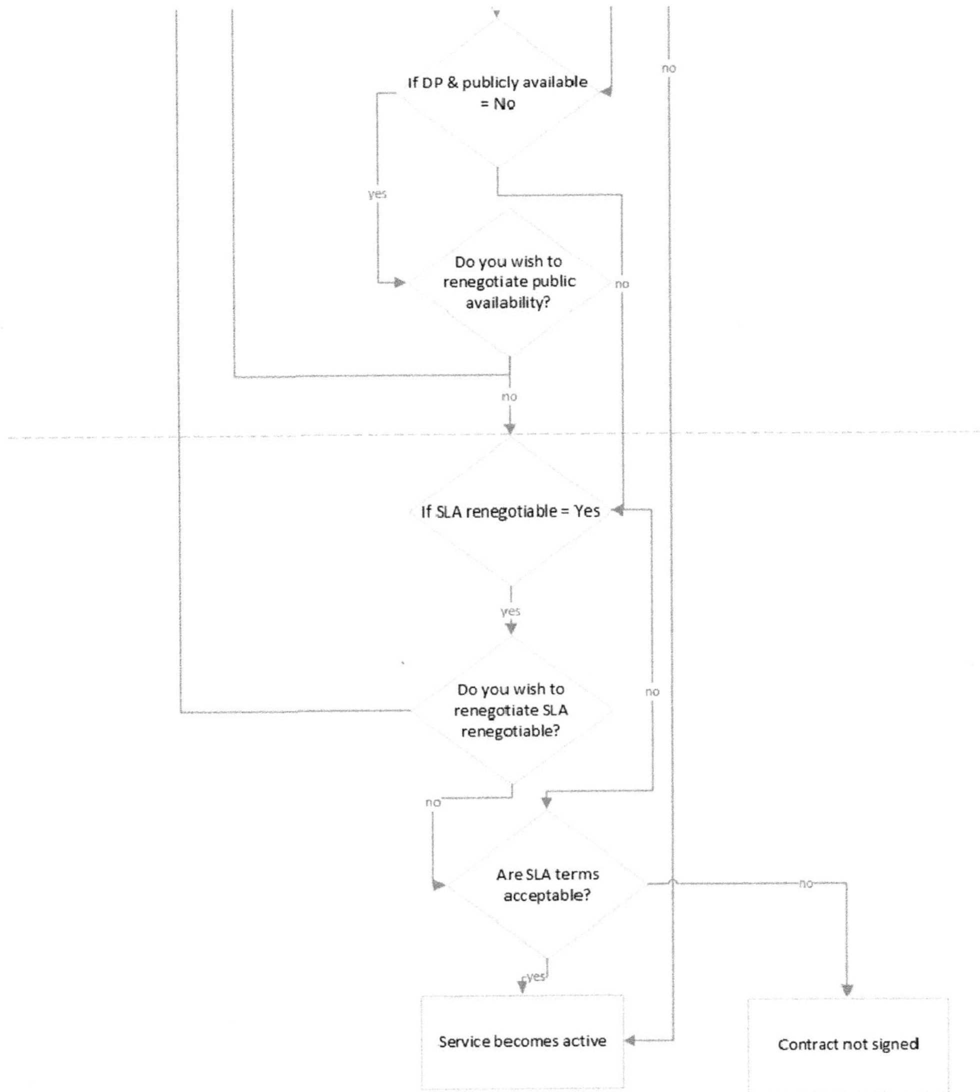

FIGURE 24.6 cont'd

TABLE 24.2 Tolerant data provider.

Service tolerances		Score	Customer classification	Weighted score	Normalized score
Interruptible	Yes	1			
Risk	Yes	1			
Renegotiable	Yes	10			
Popular domain	Yes	1			
Popular attribute	Yes	1			
DP	Yes	1			
Data provider	**Score**	**15**	Less affluent nonfrequent low-intensity (1) Less affluent frequent low-criticality (2) Less affluent nonfrequent high-intensity (3) Less affluent frequent high-criticality (4) Affluent nonfrequent low-intensity (5) Affluent, frequent low-criticality (6) Affluent nonfrequent high-intensity (7) Affluent, frequent high-criticality (8) Business (9)	**15** **30** **45** **60** **75** **90** **105** **120** **135**	**2.1** **4.9** **7.7** **10.5** **13.3** **16.1** **18.9** **21.7** **24.5**

TABLE 24.3 Tolerant gamer.

Service tolerances		Score	Customer classification	Weighted score	Normalized score
Interruptible	Yes	1			
Risk	Yes	1			
Renegotiable	Yes	10			
Application	Gamer	10			
	Intensity (high)	10			
	Criticality (low)	1			
Gamer	**Score**	**33**	Less affluent nonfrequent low-intensity (1)	**33**	**5.5**
			Less affluent frequent low-criticality (2)	**66**	**11.6**
			Less affluent nonfrequent high-intensity (3)	**99**	**17.8**
			Less affluent frequent high-criticality (4)	**132**	**23.9**
			Affluent nonfrequent low-intensity (5)	**165**	**30**
			Affluent, frequent low-criticality (6)	**198**	**36.2**
			Affluent nonfrequent high-intensity (7)	**231**	**42.4**
			Affluent, frequent high-criticality (8)	**264**	**48.5**
			Business (9)	**297**	**54.7**

TABLE 24.4 Tolerant browser.

Service tolerances		Score	Customer classification	Weighted score	Normalized score
Interruptible	Yes	1			
Risk	Yes	1			
Renegotiable	No	1			
Application	Browser	4			
	Intensity (low)	1			
	Criticality (low)	1			
Browser		**9**	Less affluent nonfrequent low-intensity (1)	**9**	**1**
			Less affluent frequent low-criticality (2)	**18**	**2.7**
			Less affluent nonfrequent high-intensity (3)	**27**	**4.3**
			Less affluent frequent high-criticality (4)	**36**	**6**
			Affluent nonfrequent low-intensity (5)	**45**	**7.7**
			Affluent, frequent low-criticality (6)	**54**	**9.4**
			Affluent nonfrequent high-intensity (7)	**63**	**11**
			Affluent, frequent high-criticality (8)	**72**	**12.7**
			Business (9)	**81**	**14.4**

TABLE 24.5 Low tolerance data provider.

Service tolerances		Score	Customer classification	Weighted score	Normalized score
Interruptible	No	10			
Risk	No	10			
Renegotiable	Yes	10			
Popular	No	10			
Popular	No	10			
DP	No	10			
Data provider		**60**	Less affluent nonfrequent low-intensity (1)	**60**	**10.5**
			Less affluent frequent low-criticality (2)	**120**	**21.7**
			Less affluent nonfrequent high-intensity (3)	**180**	**32.9**
			Less affluent frequent high-criticality (4)	**240**	**44**
			Affluent nonfrequent low-intensity (5)	**300**	**55.2**
			Affluent, frequent low-criticality (6)	**360**	**66.4**
			Affluent nonfrequent high-intensity (7)	**420**	**77.6**
			Affluent, frequent high-criticality (8)	**480**	**88.8**
			Business (9)	**540**	**100**

TABLE 24.6 Low tolerance gamer.

Service tolerances		Score	Customer classification	Weighted score	Normalized score
Interruptible	No	10			
Risk	No	10			
Renegotiable	No	10			
Application	Gamer	10			
	Intensity (high)	10			
	Criticality (low)	1			
Gamer		**51**	Less affluent nonfrequent low-intensity (1)	**51**	**8.8**
			Less affluent frequent low-criticality (2)	**102**	**18.3**
			Less affluent nonfrequent high-intensity (3)	**153**	**27.8**
			Less affluent frequent high-criticality (4)	**204**	**37.3**
			Affluent nonfrequent low-intensity (5)	**255**	**46.8**
			Affluent, frequent low-criticality (6)	**306**	**56.3**
			Affluent nonfrequent high-intensity (7)	**357**	**65.9**
			Affluent, frequent high-criticality (8)	**408**	**75.4**
			Business (9)	**459**	**84.9**

TABLE 24.7 Low tolerance browser.

Service tolerances		Score	Customer classification	Weighted score	Normalized score
Interruptible	No	10			
Risk	No	10			
Renegotiable	No	1			
	Browser	4			
	Intensity (low)	1			
	Criticality (low)	1			
Browser		**27**	Less affluent nonfrequent low-intensity (1)	**27**	**4.3**
			Less affluent frequent low-criticality (2)	**54**	**9.4**
			Less affluent nonfrequent high-intensity (3)	**81**	**14.4**
			Less affluent frequent high-criticality (4)	**108**	**19.4**
			Affluent nonfrequent low-intensity (5)	**135**	**24.5**
			Affluent, frequent low-criticality (6)	**162**	**29.5**
			Affluent nonfrequent high-intensity (7)	**189**	**34.5**
			Affluent, frequent high-criticality (8)	**216**	**39.6**
			Business (9)	**243**	**44.6**

Once the SLA has been calculated and communicated to the user, it can be renegotiated before the service becomes active. If the customer wants to adapt the cost, they can reconsider the interruptibility, renegotiability, and risk attributes.

24.7 Summary

Managing the IoT in a standardized way is extremely challenging due to the variety of applications and domains and the rate at which they evolve. The IoT is operated in an ad hoc manner at its back end, with solutions provisioned on an application and domain-specific (i.e., eHealth, gaming, browsing) basis. This is largely under the control of system service providers; however, it is unlikely to be a scalable approach, which will support IoT longevity. Furthermore, it is not a standardized technique. The IoT infrastructure is evolving and working to resolve this gap in solutions, ontologies are specified in an attempt to facilitate IoT service standardization. However, this has resulted in a landscape in which there are a number of ontologies proposed, which do not accommodate the requirements of all scenarios and domains that might be encountered across the IoT. One approach taken to unify the ad hoc ontology contributions is to integrate components from individual ontologies into a single schema. This is again, however, ad hoc and nonstandardized. We are therefore working toward contributing an ontology to support IoT operations in a more standardized approach. Furthermore, we additionally make a recommendation of the SLA provisioning process, which uses the relevant ontology attributes at this stage of the SLA management process. In this document, aspects of the proposed approach are defined.

Finally, let us move on to the real interactive part of this chapter: review questions/exercises, hands-on projects, case projects, and optional team case project. The answers and/or solutions by chapter can be found in Appendix G.

24.8 Chapter review questions/exercises

True/false

1. True or False? Currently, customers pay service providers a monthly fee to connect to their online platforms, store data, access online services, and communicate.
2. True or False? Fulfillment of the SLA is made possible through monitoring the real-time network state.
3. True or False? Monitoring the network for SLA fulfillment, and taking recovery action in the event that it is needed, is achieved using XCAPS mechanisms.
4. True or False? While the human behavioral characteristics described in ITU-T Recommendation G.1080 are generally not taken into account, because they are not under operator control, they are used however to influence service provisioning decisions that are made.
5. True or False? Customer information is entered into the SLA provisioning process, which reflects the customer's tolerance of service interruption, risk to a change in their service, and desire to potentially renegotiate SLA terms before it becomes active.

Multiple choice

1. Managing the _______ in a standardized way is extremely challenging due to the variety of applications and domains and the rate at which they evolve.
   a. SLA
   b. ICT
   c. IoT
   d. FCAPS
   e. QoE
2. FCAPS management is traditionally deployed to provide:
   a. ITU-T
   b. SLA
   c. Smart city growth and prosperity
   d. Quality of service (QoS)
   e. Smart city policy
3. Monitoring the network for SLA fulfillment, and taking recovery action in the event that it is needed, is achieved using:
   a. Machine to machine (M2M)
   b. Machine to human (M2H)
   c. Cyber-physical systems (CPS)
   d. Artificial intelligence (AI)
   e. FCAPS mechanisms
4. A personalized SLA can be provided on the same products for different customers, dependent on the customer's desire for certain performance levels and:
   a. Holistic integration of smart city platforms
   b. Economic growth
   c. Making cities smarter
   d. Computer applications
   e. SLA options
5. Due to network dynamics within the IoT, configurations may be required, which are specific to the network performance achievable to ensure continued:
   a. SLA fulfillment
   b. QoE objectives
   c. IoT fulfillment
   d. SLA fulfillment
   e. IoS objectives

Exercise

Problem

Discuss QoE inference and QoS improvement as forecasting techniques.

Hands-on projects

Project

Do research: How do you go about improving QoE metrics that reflect user-perceived performance?

Case projects

Problem

How can SLAs be used as an input to a cloud management system; increasing the efficiency of allocating resources, as well as that of infrastructure scaling?

Optional team case project

Problem

Describe service level agreement (SLA) predictions in multitenant environments.

References

[1] ITU. Recommendation G.1010: end-user multimedia QoS categories. 2001. Available: https://www.itu.int/rec/T-REC-G.1010-200111-I. [Accessed 14 June 2021].

[2] Hosfeld T, Heegaard PE, Skorin-Kapov L, Varela M. Fundamental relationships for deriving QoE in systems. In: Proc. of 11th int. conf. on quality of multimedia experience; June. 2019.

[3] CACI. The Acorn user guide. 2014. Available: https://acorn.caci.co.uk/downloads/Acorn-User-guide.pdf. [Accessed 14 June 2021].

[4] Ofcom. Access and inclusion in 2018: consumers' experiences in communications markets. January 2019. Available: https://www.ofcom.org.uk/__data/assets/pdf_file/0018/132912/Access-and-Inclusion-report-2018.pdf.

[5] ITU. Quality of service and quality of experience regulation. 2017. Available: https://www.itu.int/en/ITU-T/Workshops-and-Seminars/qos/201803/Documents/Badiel.pdf. [Accessed 14 June 2021].

[6] ITU-T. Quality of experience requirements for IPTV services. Recommend. G. Dec. 2008;1080.

[7] Ryu M, Kim J, Yun J. Integrated semantics service platform for the Internet of Things: a case study of a smart office. Sensors 2015:2137–60.

[8] Ning H, Shi F, Zhu T, Li Q, Chen L. A novel ontology consistent with acknowledged standards in smart homes. Comput Network Jan. 2019;148:101–7.

[9] Bermudez-Edo M, Elsaleh T, Barnaghi P, et al. IoT-lite: a lightweight semantic model for the internet of Things. In: Proc. of IEEE int. conf. on ubiquitous intelligence & computing; July 2016. https://doi.org/10.1109/UIC-ATC-ScalCom-CBDCom-IoP-SmartWorld.2016.0035.

[10] Li X, Ji H, Li Y. Layered fault management scheme for end-to-end transmission in internet of Things. In: 2011 6th international ICST conference on communications and networking in China (CHINACOM), Harbin; 2011. p. 1021–5. https://doi.org/10.1109/ChinaCom.2011.6158306.

[11] Gautam BP, Wasaki K, Sharma N. A novel approach of fault management and restoration of network services in IoT cluster to ensure disaster readiness. In: 2016 international conference on networking and network applications (NaNA), Hakodate; 2016. p. 423–8. https://doi.org/10.1109/NaNA.2016.46.

[12] Aazam M, Harras KA. Mapping QoE with resource estimation in IoT. In: Proc. of IEEE 5th world forum on internet of Things; 2019.

[13] Raca D, Zahran AH, Sreenan CJ, et al. Incorporating prediction into adaptive streaming algorithms: a QoE perspective. ACM NOSSDAV Jun. 2018:49–54.

[14] Garcia MN, Raake A. Parametric packet-layer video quality model for IPTV. In: Proc. of the 10th international conference on information science. Signal Processing and their Applications; 2010. p. 349–52.

[15] Varela M, Zwickl P, Reichi P, et al. From service level agreements (SLA) to experience level agreements (ELA): the challenges of selling QoE to the user. In: Proc. of the IEEE international conference on communication workshop; Jun. 2015.

[16] Peoples C, Abu-Tair M, Wang B, Rabbani K, Moore A, Morrow P, Rafferty J, McClean S, Zoualfaghari M, Cui Z, Kulkarni P. A multi-level ontology to manage service level agreements in smart cities. In: Published by Springer in 'smart cities: a data analytics perspective; Jan. 2021.

[17] Sato H, Kanai A, Tanimoto S, Kobayashi T. Establishing trust in the emerging era of IoT. In: Proc. of IEEE symposium on service-oriented system engineering; March 2016. https://doi.org/10.1109/SOSE.2016.50.

[18] Teixeira FA, Pereira FMQ, Wong H-C, et al. SIoT: Securing Internet of Things through distributed systems analysis. Systems March 2019;92:1172–86. https://doi.org/10.1016/j.future.2017.08.010.

[19] Shemshadi A, Sheng QZ, Qin Y, Sun A, Zhang WE, Yao L. Searching for the Internet of Things: where it is and what it looks like. Personal Ubiquitous Comput Dec. 2017;21(6):1097–112. Springer London.

Chapter 25

Effective value capturing of mobility and public transportation through efficient urban planning

Ritu Mohanty-Padora[1] and Bipin Pradeep Kumar[2]

[1]Urban Planning and Design Consultant, Mumbai, Maharashtra, India; [2]Systems Biology Institute, Tokyo, Japan

25.1 Introduction

Comprehensive policies that aim to guarantee efficient, equitable, user-friendly, and environment-friendly mobility infrastructure in our cities are linked to regulatory and economic reform [1]. In the Indian context, financing infrastructure investment through leasing land by the government as well as typical Smart City options such as SPVs (Special-Purpose Vehicles), etc., are mechanisms of revenue generation for implementing large-scale development projects. With meeting demands for increasing urban growth, inadequate basic services, and environmental degradation, there is also an urgent need for green investments and green-blue infrastructure planning, especially if we are to transition to a zero-carbon economy. Sustainable mobility as a master plan framework for cities is one of the first steps toward strategic adaptation to climate change and requires serious public investment. The timeless framework elements for master planning and sustainable urban development include: (1) city centers and cores (both old and new), (2) connectivity, (3) green infrastructure, and (4) a public realm [2]. This essay draws on two case studies in city centers, both old and new. These are framed against the backdrop of climate change and its implication on future growth as well as ecology of the city. While both case studies embrace all the four framework elements, this essay discusses the urban design, planning, and financial tools for optimizing reduced carbon footprint primarily through the framework element of "connectivity."

The discourse around urban finance at the policy level in India recommends not only making cities and states financially self-reliant, but also incentivizing cities to invest in low emission and climate-resilient infrastructure. It complements the urban policy treatise focused on urban transportation and mobility that advocates a multimodal public transport backbone while discouraging private transport modes through fiscal tools. The two case studies demonstrate that effective value capturing of transit oriented-development requires more efficient urban planning tenets. Both projects reiterate that walkable communities provide the most affordable and equitable transportation. Connectivity, mobility, and transportation networks, as key components of the overall urban development, also support inclusive development.

The first example is Mumbai's new Smart Central Business District (CBD) master plan in which connectivity and transport networks drive urban form and influence the framework plan. With three monorail stations, two metro stations and a large intra and intercity bus station, the new CBD is a mixed-use urban core located near transit hubs. As a commitment toward an integrated and sustainable city, the CBD is complimented by a public transportation node of the 21st century. It harnesses the potential of being implemented in a Special Zone by Mumbai Metropolitan Regional Development Authority (MMRDA). By reforming the zoning code to encourage mixed-income, mixed-use construction alongside walkability, multimodal transit and smart amenities at its core this new CBD serve as an asset to a city already burdened with heavily congested roads, overcrowded trains, and short supply of land. The source of revenue for this is through various forms of land value capture.

The second example includes key projects of the Bangalore Smart City Proposal—mainly roads implemented under an earlier successful model named Tender SURE and redevelopment of multimodal transit hubs, in the old city center. It envisages sustainable mobility as a framework for its Area-Based Development (ABD). Revitalizing the

Smart Cities Policies and Financing. https://doi.org/10.1016/B978-0-12-819130-9.00023-1

historic core of the city with 50 Tender SURE roads and two multimodal hubs as a comprehensive approach alongside other key proposals for the smart city is the main scope of this particular project. The ABD and Pan City initiatives are both covered in the Smart City Proposal, and the primary collaborator with the Smart Cities SPV is the Bruhat Bengaluru Mahanagara Palike (BBMP)—the city corporation. The resources for financing the Bangalore Smart City Project are through joint ventures, subsidiaries, public–private partnerships (PPP), turnkey contracts, with well-fitted revenue streams. Within this total scenario, the sources of financing specifically for the connectivity and mobility aspect of the project are through the Karnataka State government, BBMP, and Bengaluru Metropolitan Transport Corporation (BMTC).

Both examples offer an opportunity for achieving UN Sustainable Development Goals (SDG) implicitly—especially through improved economic activity, public transit-based solutions, last mile connectivity, safety for pedestrians, and Non-Motorized Transport (NMT). These in turn offer superior city services, better infrastructure, and citizen well-being.

25.2 Urban policies in India

Policies relating to large-scale urban development projects and urban financing act as catalysts for integrating public transportation and mobility into the master plan, thereby adding to the sustainability parameters of a city. A summary of the urban policies at the national level to contextualize the big ideas and principles for the examples that we cite include efforts by NITI Aayog and the Ministry of Housing and Urban Affairs (MoHUA), Government of India. As a policy think tank of the central government, NITI Aayog's mission is to achieve the SDG by encouraging collaboration with the State Governments of India in the economic policymaking process, as well as by adopting a decentralized and bottom-up approach. Few recommendations relate directly to urban financing vis-à-vis large-scale urban development projects focused on connectivity, mobility, and public transportation.

The National Urban Policy Framework (NUPF) by MoHUA, published in 2018 in collaboration with NITI Aayog, discusses an integrated approach toward the future of urban planning in India [3]. The NUPF recommendations pertaining to two functional areas of urban space and management—(1) Transportation and Mobility and (2) Urban Finance, are particularly relevant to this essay. The NUPF guidelines for the "transportation and mobility" function area promote a public transport system that is accessible, affordable, and efficient for all in order to alleviate the current state of urban congestion and unequal levels of mobility across social groups in our cities. It places people, not vehicles, in the center of transport planning. The broad objective is to promote transport planning with appropriate provision for pedestrian and street infrastructure. It calls for internalizing urban transport with its various components as a key variable in urban planning exercises, as illustrated in Fig. 25.1. It also

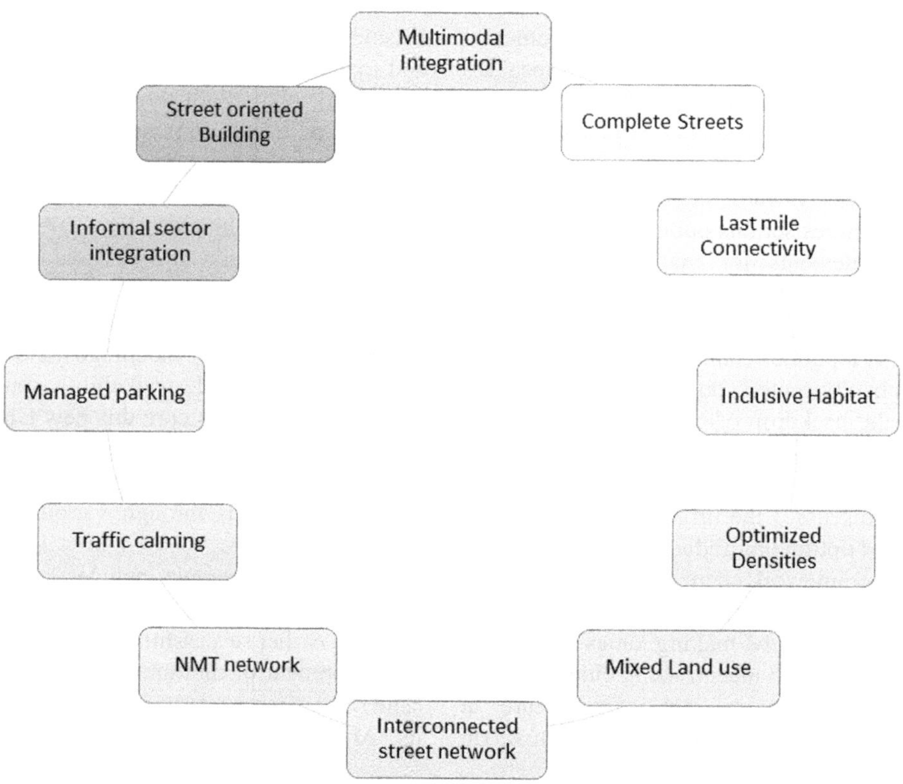

FIGURE 25.1 Components of mobility and transport.

discusses establishing institutional mechanisms and capacity building for better planning cum management of transport systems. Additionally, the NUPF guidelines for "urban finance" promote strengthening municipal finance in order to make cities self-reliant by stimulating innovative revenue mobilization techniques. The NUPF guidelines, as a supporting policy document for city planning, prioritize the sharing of common resources efficiently. More importantly, these guidelines lay emphasis on the local context and an approach that encourages conservation of heritage, revival, or reinvention of local identity, the importance of scalability and environmental sustainability—all of which differ from city to city.

The "Smart Cities for Urban Transformation" [4] proposed by Niti Aayog discusses strategies to leverage the Smart Cities Mission (SCM) in India by 2022–23, across the four paradigms of economy, equity, environment, and engagement. The directive for infrastructure and mobility by the central government included the following recommendations: (i) a Unified Metropolitan Transport Authority, (ii) spatial plans for integrating land-use and transport planning to support more mixed-use development, (iii) a focused attention to public transport, including existing intermediate and para-transit services, especially in smaller cities, (iv) a pooled green transport fund, and (v) a high-level interministerial electric vehicle (EV) mission.

Often, density is confused with overcrowding. However, a dense urban population does not imply low per-capita square footage. Low per-capita square footage, which results in unplanned squatter settlements and overcrowding, is largely due to poor urban policies [5]. There is widespread discussion among urban planners that overcrowding can be avoided, by firstly building robust cross-city public transportation and secondly enabling higher density in city centers [6]. This is still gathering momentum at the policy level in states/cities. It is the responsibility of the state government to manage its urban areas; however, India released a National Transit Oriented-Development (TOD) policy that serves as guidelines and is expected to play a catalytic role in formulating state-level TOD policies. The TOD policy typically applicable to cities (and metropolitan regions) comes into force, only after being accepted by the State Government. The state governments in the case of both our examples, Maharashtra government and Karnataka government, have yet to sanction their policies.

The main components based on the National TOD policy [7] are multimodal integration, mixed landuse, optimized densities, interconnected street network, last mile connectivity, NMT network, and traffic calming as demonstrated in the essay. The values of improving accessibility and creating walking communities are at the core of the developments. The National TOD policy also acknowledges certain supportive tools based on the objectives of the National Urban Transport Policy. The urban design and planning approach for both examples recognize key components of TOD such as density, diversity, and accessibility to destination; as well as certain supportive tools such as land value capture, high-quality transit system, technology integration, and universal accessibility. The examples have nodes of moderate-to-high density mixed land use within walking distance from mass transit stations. The mixed use in the proposed smart CBD, Mumbai, includes an array of housing, offices, retail, and recreation, along a network of walkable and bikeable streets within close proximity of the transit facilities; in the Bangalore smart city example, there is an attempt to integrate the development and built fabric of the city center with pedestrian, bicycle, feeder, and transit networks.

Developing effective planning strategies needs close collaboration and coordination not only between policy makers, but also public agencies, real estate developers, urban planning professionals, industry leaders, civil sector partners, community organizations, and citizens as stakeholders. The resilience and vitality of a city are much dependent on the policies and regulations that shape it [8].

25.3 Beginning with a strategic vision for urban planning and development

The creation of special zoning districts or Special Planning Authority (SPA) areas and SPV as two separate examples of management tools to implement innovative and efficient urban planning strategies are explained through the two case studies. SPA areas are used as an affirmative zoning technique, and SPVs are used as an overall mode of governance.

25.3.1 Case study 1: SPA for new smart business district in Mumbai

SPA areas or Special Zoning Districts are designated to achieve specific planning and urban design objectives in defined areas with unique characteristics. It stipulates zoning requirements as well as zoning incentives tailored to distinctive qualities that may not lend themselves to generalized zoning and standard city regulations. Mumbai's new Smart Business District in Wadala included in the SPA area is one such example. It does not fall under the city's development plan (DP) and development control regulation (DCR).

The total land area of the Wadala Notified Area—approximately 122 ha within the SPA boundary—is currently underutilized with predominantly a truck terminal and warehouses that will eventually be accommodated on land parcels in the outskirts of the city. The SPA area is planned as a mixed-use CBD with over four million sq m of offices, retail, leisure, hospitality, apartments, and civic facilities by MMRDA. The Wadala CBD has been designed along the lines of a large-scale TOD, with multimodal transportation uses. Creating a new CBD in conjunction with a node of international standard public transportation is a stimulus for the sustainable growth of the entire city.

On behalf of MMRDA, Edifice Consultants Pvt Ltd (ECPL), along with other key consultants, prepared a Detailed Master Plan for this new CBD in Wadala. The vision for the master plan is based on the following three principles:

1. **Connectivity**—recognizing the strategic value of the site location in Wadala—its proximity with the city's International airport and the neighboring business district of Bandra Kurla Complex (BKC) influenced the design process in reinforcing **connectivity** with the new CBD.
2. **Identity**—where a blend of uses to complement the commercial hub and job opportunities with valuable assets such as easy access to public transport as well as other amenities that reinforce the live–work–play community concept is incorporated. This ensures that it is not an insular development but plugged into the rest of the city with a unique **identity**. Mixed-use and transit-oriented planning thus became one of the main features of the new CBD.
3. **Sustainability**—walkability, pedestrian safety, conserving the natural environment, and aiming for a Smart CBD founded on inherent qualities of the master plan that are site (and context) responsive with ecological, social, and cultural values of **sustainability** became central to the plan.

Physical infrastructure including provision of utilities, social infrastructure, and participatory governance are focused upon as the main components for a self-sustaining Smart CBD. Hence, the overall approach and vision for the master plan of the Wadala Commercial Center are influenced by creating economic opportunities, providing a high quality of life, and ensuring sustainable development. Flexible planning, future proofing, building on existing infrastructure, maximizing land area potential, and pedestrian-oriented street network are some of the tenets of this Smart CBD.

The vision of the Master Plan is supported not only by the Zoning Plan at a building block level, but also areas allocated for traffic and transportation network, as well as the interconnected public spaces for pedestrian circulation and leisure. A set of complementary Urban Design Guidelines ensure that the entire development eventually strikes the right balance between a cohesive new urban district with strict building alignments, street setbacks, and building forms. At the same time it allows for enough diversity in the individual expression of the different buildings and precincts to create an inspiring urban environment. These guidelines align with the multicultural character of Mumbai. The Urban Design framework and guidelines recommend the following:

1. A network of attractive pedestrian public spaces in combination with an international standard public transport feeder network/system.
2. Parks and plazas—pedestrian public open spaces, whether developed as public recreational open spaces or as part of recreational open spaces on the private plots.
3. A cohesive street frontage with adequate retail at ground floor level to maintain safety as well as a lively, pedestrian-friendly environment.
4. Developing a range of residential units, serviced apartments, and hotels.
5. Allowing flexibility for individual developers to make their buildings more prominent and through a less restrictive DCR.

The mobility and public transportation vision in the urban design framework for the master plan are reinforced with a large network of wide pedestrian only streets, parks, and plazas lined with retail, leisure, and civic amenities, seamlessly connecting all commercial and residential towers to the public transit hubs. Prioritizing NMT via bicycle lanes along streets and secure bicycle parking at all stations, as well as ensuring last mile connectivity, have added to the sustainability goals for the Smart CBD.

25.3.2 Case study 2: SPV for smart city with Tender SURE and mobility in Bengaluru

SPVs in the context of the MoHUA's Smart Cities Mission are entities/companies having representation from central and state governments as well as municipal bodies to expedite the urban development in cities. The private sector or financial institutions may take equity stake in the SPV. The SPV is set up as a Public Limited Company with its own assets and liabilities. It is a separate legal entity that can circumvent certain regulatory restrictions and is capable of undertaking high financial risks. The Bengaluru Smart City Limited as an SPV was incorporated as the agency in charge of the ABD and Pan City projects under the SCM program. Jana Urban Space—an appointed consultant for this project—contributed toward the urban design and planning of the ABD and Pan City proposals included in the Bangalore Smart City Proposal by the Ministry of Urban Development.

As part of the Bangalore Smart City Plan, seven strategic redevelopment projects were proposed in the core city defined as the ABD covering approximately 2000 ha. The seven projects, as per the original proposal (few projects were altered in their scope at implementation stage), include:

1. Thirty (30) km of Tender SURE roads.
2. Creating an integrated mobility hub at Shivaji Nagar as well as the Kempegowda Bus Depot.
3. Redevelopment of KR Market and Russell Market.
4. Rejuvenation of Cubbon Park.
5. Cleanup of drainage system.
6. Increasing affordable housing stock.
7. Retrofitting a healthcare facility [9].

All seven redevelopment projects have been categorized into the following four smart initiatives proposed for the ABD project. The initiatives are aimed at establishing an urban form based on three sustainable planning principles of equity, economy, and the environment:

1. Smart and comfortable mobility for people (30 km of Tender SURE roads, redevelopment of two multimodal transit hubs).
2. Rejuvenation of public places (one park and two lakes).
3. Vibrant economic centers (two markets).
4. Improving quality of life for urban poor (one slum redevelopment).

The Pan City proposals incorporate citizen-centric e-governance and other inclusive features in order to facilitate public participation and enhance the city's infrastructure. However, only two out of the above listed seven strategic redevelopment projects of the ABD—(1) Tender SURE roads and (2) mobility hubs under "Smart mobility," will be discussed in this essay.

Thirty (30) km of roads selected in the historic core of the city were to be developed as per the Tender SURE street design guidelines. Improved last mile connectivity to public transport, safe continuous footpaths, designated cycle tracks, safe intersections, uniform travel lanes, planned parking as well as designed underground utilities contained within the footpath, improved vending, enhanced public space, and landscape addition are all part of the guidelines [10]. The NMT-friendly roads, once completed, are expected to connect with other finished Tender SURE roads, several transit hubs such as metro stations and bus stops, key institutions and landmarks, thereby providing the city center with high-quality TOD components.

One of the unique features of Tender SURE is that the street standards mandate the integration of networked services under the road. All the underground utilities are contained within the footpath, thereby ending cyclic expenditure of repairing and rebuilding the same roads to lay and fix lines with temporary upgrades. It is for this reason that the Tender SURE roads are also financially sustainable.

25.4 Capturing value by implementing unique financing mechanisms

Recognizing and leveraging the essential components of a city that enhance the quality of life of citizens, through establishing a key master planning framework of elements such as "mobility," help in capturing the greater value for new developments and are both tangible and intangible. Walkable neighborhoods and public transportation network services promote sustainable urban regeneration through actively adopting various smart technologies and harnessing emergent tools. Integrated transport entities and other TOD components as illustrated in Fig. 25.1 are not only low-cost tools for reducing emissions but also a planning approach for gaining greater land and energy efficiencies. In addition, high-quality transit system, land value capture, PPP, and technology integration as applied to TOD zones are supporting tools, for meeting large-scale developments' financing needs, as illustrated in Fig. 25.2.

As our cities continue to encounter the challenges of urbanization and climate change, a sustainable approach for accommodating the ever-increasing population in our city centers and cores is becoming imperative: a vibrant mix of uses with employment opportunities, housing options, transport choices, safe and shared streets, services and shops at walking distance, active recreational spaces, all serving a high density of population provide adaptable and efficient settings for a number of socioeconomic and environmental goals—including conservation of natural resources, technological innovation, and high quality of life [11].

25.4.1 Case study 1: Mumbai's new smart business district at Wadala

In the context of the new smart CBD in Wadala, Mumbai, a TOD approach and a combination of different land uses in close proximity to one another foster collaboration as well as reduce the development's carbon footprint. A higher density (appropriate to its context) and mixed-use planning with public transit facilities fortifies the immediacy and interaction between jobs, housing, and support services, thereby reducing vehicle miles traveled (VMT) and allowing for the day-to-day needs to be accessible with less energy and time investment. Therefore, the success of the new CBD as well as its surrounding depends on the

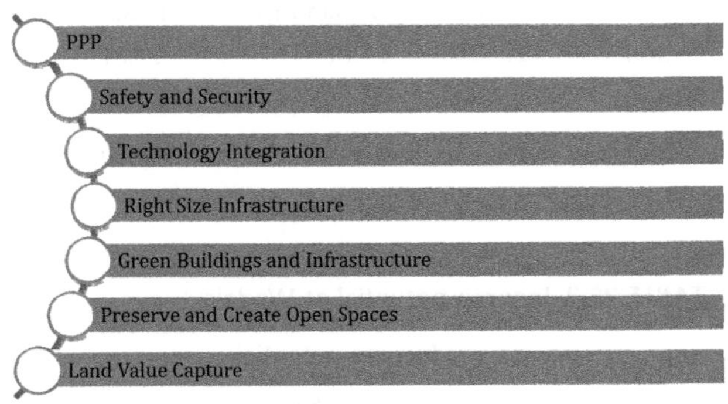

FIGURE 25.2 TOD support principles and tools.

appropriate land-use mix and density determined. Through adjustments to land-use balance and other measures such as providing adequate as well as inclusive housing supply, the total VMT may be reduced by a vast margin, resulting in lower levels of greenhouse gases emitted.

MMRDA is setting a new benchmark for international standard city planning in the Mumbai Metropolitan Region by creating a Conceptual Master Plan for the entire Wadala Commercial Center first, before developers and private stakeholders are invited to participate in the site's development. For revenue generation, MMRDA will auction land to developers and institutions; the revenues from long-term leasing of land to multiple stakeholders will finance the new Smart Business District.

25.4.1.1 Financing for Wadala CBD

Revenues are to be generated by the MMRDA through forms of land value captured in public auctions over the project life cycle. The estimated demand schedule from 2020 to 2053 is distributed over three phases of approximately 10 years each: (1) early phase, (2) growth phase, and (3) established phase, as illustrated in Table 25.1.

Even though the actual sale price will be established during the auction process, revenue potential of the plots was estimated based on the 2017 ready reckoner values, i.e., INR 65,900/sq m (FSI-1) [12]. The future rate will be an escalation on the 2017 ready reckoner average rate of open land rates, as shared by MMRDA.

Two scenarios as illustrated in Table 25.2, with different escalation rates—5% and 7.5%. have been assumed by the consultants, and the following data calculated as tentative income potential is part of their report.

TABLE 25.1 Introduction of supply at Wadala.

Introduction of supply at Wadala notified area			
Phase	Period	Estimated demand (%)	Cumulative demand (%)
Phase 1	2020–29	27.01	27.01
Phase 2	2030–39	30.84	57.85
Phase 3	2040–53	42.15	100.00

TABLE 25.2 Income potential at Wadala.

Income potential		
Escalation factor	INR	USD
Based on 5% escalation	40,000 crores	5.4 Bn
Based on 7.5% escalation	70,000 crores	9.6 Bn

Based on a 5% escalation the revenue potential for MMRDA over the entire project life cycle (until 2053) is computed as approximately INR 40,000 crores, and based on a 7.5% escalation, the revenue potential for MMRDA over the same project life cycle is computed as approximately INR 70,000 crores. In either of the escalation scenarios, the revenue from land is expected to cover the costs to be incurred by MMRDA for civil works, smart city related ICT costs, and provision of public infrastructure.

25.4.2 Case study 2: Tender SURE and mobility hubs at Bangalore Smart Cities

The Tender SURE street design guidelines as part of the Bangalore Smart Cities proposal for redeveloping 30 km of roads are a critical value addition in the historic heart of the city. It has revitalized the core into a walking, cycling, tree-lined, economically vibrant, environmentally sensitive and inclusive center with smart components such as intelligent traffic management, smart parking and e-rickshaw stands, smart bus shelters plus environmental sensors, e-bus charging points, solar/LED lighting, etc. The enhanced mobility networks through the Tender SURE roads between integrated transport hubs, civic amenities as well as social infrastructure, such as schools and markets, have transformed the core into a vibrant livelihood destination. Few prominent Tender SURE features are listed in Table 25.3.

The Tender SURE road development—known for its street and intersection design guidelines, has been branded as a best practice by Niti Aayog, and a recommendation has been made to emulate the scalable and financially sustainable model in cities and towns across India. Its scalability and influence on increase in public transport ridership have had a catalytic impact in transforming the Bangalore CBD

TABLE 25.3 Tender SURE salient features.

Tender SURE features	
Sl. no.	Feature
1	Improved last mile connectivity to public transport
2	Safe continuous footpaths
3	Designated cycle tracks
4	Safe intersections
5	Uniform travel lanes
6	Planned parking
7	Underground utilities contained within the footpath
8	Improved vending
9	Enhanced public space
10	Landscape addition

by attracting new high-rise, mixed-use buildings as well as transit-friendly investment. It has improved real estate values and spurred urban rejuvenation in the city center. Based on the Tender SURE Phase 1, executed independently of the smart city agenda between 2014 and 16, 30 km of Tender SURE roads were included in the Bangalore Smart Cities Proposal in 2017. The Tender SURE model, by itself, is an example of a pilot project that has resulted in a serious government (city, state, and center) fund allocation not only for road development projects but also to be part of the smart cities initiative, as discussed in case study 2. The SCM has taken this further in terms of the design principles and model guidelines with sizable fund allocation to roads development projects across the 100 selected cities. It has created opportunities for funding from bilateral and multilateral agencies. Institutionally, it has triggered a transformation in the way governments build and fund roads across the country for a better quality of life [13]. More importantly, by being inclusive and instilling civic pride, it has transformed the behavioral patterns of the citizens, and these intangible values are equally prized.

It is an example of how insights gained in smaller, prototypical projects can be translated into larger-scale development in terms of both processes and design methodologies, in order to impact a larger section of population and thereby achieve sustainable urban development.

25.4.2.1 Financing for Bangalore Tender SURE roads and mobility hubs as part of ABD

While the total budgetary allocation proposed for Pan City is INR 2250 crores approximately, the following items included in Table 25.4 are the proposed financing mechanisms for Tender SURE Roads and Mobility Hubs, as discussed in the Bengaluru Smart City Proposal. The Tender SURE proposal along with the multimodal transit hubs as one component of the Bengaluru Smart City Proposal achieves convergences with BBMP, Karnataka State funds, and BMTC in terms of human and financial resources, common activities and goals. The end goals and value addition from both case studies are designed to enhance the user experience, promote sustainability, and extract more effective value for the SPA and SPV.

25.5 Global lessons and way forward

There is global concern on sustainable development of cities leading to increased attention on compactness and density that can be achieved by good connectivity. This is made possible by a network of roads as well as mass transit systems. The urban renewal of city cores (old and new, both) presents an opportunity to meet the public housing demands as well as the growing need for adequate green infrastructure. Much like TOD projects, the prospect of these developments is measured through unique indicators optimizing public resources and maximizing the quality of life index (QLI). Table 25.5 lists the various indicators of a TOD.

The perils of gentrification such as unavailability of affordable housing or overpricing are offset by the introduction of mixed-use development and improved mobility. The question is whether the proposed mixed use justifies an equitable distribution of housing types, for not just the marginalized but also the middle class.

New York's Hudson Yards Redevelopment Project is one of the most prominent examples of the mutually beneficial relationship between transportation improvements and TOD tenets. The strategic partnership between transportation infrastructure improvements and real estate development has enabled Hudson Yards to create a new livable and workable district in a previously depressed area of Manhattan. This further catalyzed economic development of the area. However, while extremely successful with addressing connectivity, mobility, and public transportation; the Hudson Yards is currently saddled with a sizable number of empty high-end apartments in a city already saturated with unsold luxury dwelling units. Critics argue that as a TOD it promised affordability through rezoning, but it also delivered gentrification. These are lessons learnt for promoting more sustainable urban regeneration.

TABLE 25.4 Financing models.

Financing for Bengaluru Tender SURE and mobility			
Sl. no	**Scheme**	**Project**	**Amount**
1	Convergence with existing budget/schemes: Bruhat Bengaluru Mahanagara Palike (BBMP) and Karnataka state funds (Nagarothana scheme)	50 Tender SURE roads of 103 km (51.6 KM falls in ABD)[a]	INR 200 crores
2	PPP projects	1500 bus shelters	NR 2.5 lakhs each
3	BMRCL (Bangalore Metro Rail Corporation Ltd)	Construction of Shivajinagar metro station	NR 180 crores

[a]*This scope was altered at implementation.*

TABLE 25.5 Measures of ToD.

Measures and indicators of a ToD		
1	Land use diversity and choice	1. Land use entropy index
		2. Public transport hubs and amenities
		3. Mixed use of residential with office
2	Urban density	1. Population density
		2. Employment density
		3. Commercial density
3	Walkability	1. Walkable and cyclable roads
		2. Intersection density
		3. Live–work–play proximity
4	Economic development	1. Business development
		2. Tax earnings
		3. Employment levels

Prioritizing the development of underutilized sites within the city centers and cores, rather than opening up new Greenfield sites, is an opportunity to further optimize land use. Another lesson to be learnt from rapidly urbanizing cities is the need for conservation of ecological zones and natural resources to mitigate climate change. Furthermore, states and cities need to ensure that the relationship between urban form, balanced land use, and sustainability is tied to inclusive development, efficient energy management, and utilization of new technologies.

Alain Bertaud, in his book, "Order without design: How markets shape cities," recommends that urban planners should collaborate with urban economists to purposefully harness the power of market forces, toward an efficient and well-designed city. This resonates with the Wadala CBD's approach toward configuring its mixed-use development. The mix of land-use and floor space index (FSI) is calculated by applying the theories of urban economics to the practice of urban planning in order to increase both the productivity of the CBD and the welfare of urban citizens. Bertaud suggests that urban planners ought to concentrate on developing infrastructure and expanding land supply [14]. Along with urban planners, economists and financial planners were also key members of the team that prepared the master plan for MMRDA (case study 1). The economists focused on the financial advisory analysis based on market research and demand assessment to shape the master plan. This collaborative work reinforces Alain Bertaud's premise that design of infrastructure and market forces can complement one other.

The Tender SURE guidelines in Bangalore Smart Cities proposal (case study 2) unlock the potential of their streets as safe, accessible, and economically sustainable places. It also generates revenues through smarter use of existing public resources and acts as a catalyst for advancing sustainable real estate development. It sets a precedent for exploring the potential of implementing pilot projects that are self-sustainable across different scales. Such people-focused street design standards offer an urban mobility solution that can be replicated across cities in India. This also reinforces the "15-minute city" concept that Paris and Copenhagen have introduced as new urban models, becoming even more relevant since the outbreak of COVID. According to Carlos Moreno—scientist and professor, a "15-minute city" approach is based on four major principles: density, proximity, diversity, and ubiquity; and individual neighborhoods/areas within this city should be able to fulfill six social urban functions: living, working, supplying, caring, learning, and enjoying [15]. The mobility component of the Bangalore Smart Cities proposal much like the "15-minute city" idea, while increasing economic dynamism and innovation at the city center reduces pollution.

25.6 Conclusion

There is a need for an integrated approach to urban planning, financing, and economic growth. This collaborative approach reinforces the existing components of urban governance and urban finance that influence the development pattern of cities. In the face of the climate crisis, the critique of economic growth is gaining widespread attention but instead of deurbanization (or degrowth), building resilient cities is the need of the hour.

While economic growth may be essential to promote human development, it is certainly not the only parameter for determining the QLI of cities. Increasing unequal distribution of wealth (or income) leads to deteriorating human development for any given rate of economic growth. Tim Jackson, a professor of sustainable development, discusses how the economy of tomorrow may be transformed in ways that protect employment, facilitate social investment, reduce inequality, and deliver both ecological and financial stability, in his book "Prosperity Without Growth: Foundations for the Economy of Tomorrow." He advocates growth that is strong, sustainable, balanced, and inclusive as we are on the brink of a new economic era: one where growth is driven by the interaction between rapid technological innovation, sustainable infrastructure investment, and increased resource productivity [16].

The case studies included in this essay can be used as a reference by local planners and policymakers when adopting and supplementing existing regeneration plans. The impact of the mobility solution on Wadala's CBD and

Bangalore Smart Cities (ABD) proposal has resulted in revitalizing derelict public lands in the heart of the city and regenerating the city center, respectively. It addresses the new challenges of climate change, resilience, and liveability through initiatives to develop new policies. It is through the synergy of good urban design, planning, together with progressive economic and financial structures that long-term values of mobility-based development can be realized. A resilient approach to planning at the city, neighborhood, and street level balances economic development and equitable urbanization.

25.7 Summary

Due to the rising demands of climate change and overcrowding in cities, the master-planning decisions ought to be influenced by low-energy and time-efficient connectivity. India is struggling with an ever-growing urban crisis and inadequate public infrastructure. Sustainable mobility, multimodal connectivity, and accessibility to public transport as master plan framework accelerate the efficiency of land and energy for metropolitan cities. The organization of neighborhoods around well-composed cores and centers of cities is a cost-effective tool for reducing emissions as well as enhancing the quality of life of citizens. Cities can reinvent themselves for a future that does not rely on cars by nurturing communities with a vibrant mix of residential and commercial buildings, as well as easy access to public transit and NMT.

The essay recommends developing public transit efficiency alongside a broad range of mixed-use development offering high-value jobs and opportunities within the city cores. It establishes through case studies in two metropolitan cities of India that transportation and land-use planning are inherently tied to more effective value capturing for cities. Urban connectedness and urban compactness in the context of rapid urbanization need to be complemented with social and economic policies. The case studies corroborate that government actions are necessary to spur private investment and that synergy between urban planners and market forces is vital for realizing the vision of any sustainable development.

Finally, let us move on to the real interactive part of this chapter: review questions/exercises, hands-on projects, case projects, and optional team case project. The answers and/or solutions by chapter can be found in Appendix G.

25.8 Chapter review questions/exercises

True/false

1. True or False? Comprehensive policies that aim to guarantee efficient, equitable, user-friendly, and environment-friendly mobility infrastructure in our cities are linked to regulatory and economic reform.
2. True or False? Policies relating to large-scale urban development projects and urban financing act as catalysts for integrating public transportation and mobility into the master plan, thereby adding to the sustainability parameters of a city.
3. True or False? The creation of special zoning districts or Special Planning Authority (SPA) areas and Special-Purpose Vehicle (SPV), as three separate examples of management tools to implement innovative and efficient urban planning strategies, is explained through the three case studies.
4. True or False? SPA areas or Special Zoning Districts are designated to achieve general planning and urban design objectives in defined areas with unique characteristics.
5. True or False? SPVs in the context of the MoHUA's Smart Cities Mission are entities/companies having representation from central and state governments, as well as municipal bodies to expedite the urban development in cities.

Multiple choice

1. Recognizing and leveraging the essential components of a city that enhance the quality of life of citizens, through establishing a key master planning framework of elements such as "mobility," help in capturing the greater value for new developments and are both:
   a. Smart and sustainable
   b. Tangible and intangible
   c. Numerous and smart
   d. Harnessing and emergent
   e. None of the above
2. In the context of the new smart CBD in Wadala, Mumbai, a TOD approach and a combination of different land uses in close proximity to one another foster collaboration, as well as reduce the development's:
   a. Society infrastructure
   b. Physical infrastructure
   c. Smart city growth and prosperity
   d. Economic effects
   e. Carbon footprint
3. Revenues are to be generated by the MMRDA through forms of land value captured in public auctions over the project:
   a. Established phase
   b. Decline
   c. Life cycle
   d. Demand
   e. Domain
4. The Tender ________ street design guidelines as part of the Bangalore Smart Cities proposal for redeveloping 30 km of roads are a critical value addition in the historic heart of the city.

a. LED
b. SURE
c. MCS
d. CAC
e. CTU

5. There is global concern on sustainable development of cities leading to increased attention on compactness and density that can be achieved by:
a. Bad connectivity
b. Global connectivity
c. Good connectivity
d. Urban renewal
e. Unique indicators

Exercise

Problem

What is sustainable transport?

Hands-on projects

Project

Do research: Explore and document how aspiring smart cities are using data and models to inform mobility and energy initiatives within smart city programs and, in doing so, identify gaps in knowledge and processes that are guiding smart city mobility investment strategies, programs, projects, and pilots.

Case projects

Problem

Characterize and benchmark current smart city urban data, mobility models, and their evolutions.

Optional team case project

Problem

How do applications of big data support smart cities?

Acknowledgments

We thank Mumbai Metropolitan Region Development Authority (MMRDA), Edifice Consultants Pvt Ltd (ECPL), Mumbai, and Jana Urban Space, Bengaluru, for providing pertinent data related to the case studies and critical research, as well as supporting us by fact-checking from beginning to end.

References

[1] Chikarkane G. India 2030: rise of a Rajasic nation. Ebury Press; 2021.
[2] Forsyth A, Peiser R, editors. New towns for the twenty-first century: a guide to planned communities worldwide. University of Pennsylvania Press; 2021.
[3] National Urban Policy Framework. Ministry of housing and urban Affairs. 2018. Viewed on, https://smartnet.niua.org/sites/default/files/resources/nupf_final.pdf. [Accessed 15 April 2021].
[4] Strategy for new India @ 75. Niti Aayog. 2018. Viewed on, https://niti.gov.in/writereaddata/files/Strategy_for_New_India.pdf. [Accessed 16 April 2021].
[5] Abraham R, Tandel V. Don't blame dense cities. They have been resilient. IDFC Institute July 2020. Viewed on, https://www.idfcinstitute.org/knowledge/publications/op-eds/dont-blame-dense-cities-they-have-been-resilient/. [Accessed 17 April 2021].
[6] Patel B. Why Mumbai should embrace MCGM's new FSI policy. IDFC Institute March 2015. Viewed on, https://www.idfcinstitute.org/site/assets/files/13451/why_mumbai_should_embrace_mcgms_new_fsi_policy.pdf. [Accessed 18 April 2021].
[7] National Transit Oriented Development (TOD) policy. Ministry of Urban Development (MoUD). 2017. Viewed on, http://www.indiaenvironmentportal.org.in/files/file/National%20Transit%20Oriented%20Development%20(TOD)%20Policy.pdf. [Accessed 20 April 2021].
[8] 1000 Singapores- A model of the Compact city. DesignSingapore Council, Ministry of Communications and Information. 2015.
[9] Bengaluru Smart City Proposal. Jana urban space. 2018. Viewed on, https://www.janausp.org/portfolio/bengaluru-smart-city-proposal/. [Accessed 22 April 2021].
[10] The Story of Tender S.U.R.E. Jana urban space. 2020. Viewed on, https://www.janausp.org/2020/11/23/tender-sure/. [Accessed 22 April 2021].
[11] Chakrabarti V. "A country of cities: a Manifesto for an urban America". Metropolis Book; 2013.
[12] Internal confidential working report by ECPL et all. 2017.
[13] Bengaluru Smart Cities, Tender S.U.R.E. Jana urban space. 2020. Viewed on, https://www.janausp.org/portfolio/bengaluru-smart-cities-tender-sure/. [Accessed 20 April 2021].
[14] Bertaud A. Order without design: how markets shape cities. The MIT Press; 2018.
[15] Kim W. "Paris mayor unvelis'15-minute city' plan in re-election campaign". The Guardian; February 2020. Viewed on, https://www.theguardian.com/world/2020/feb/07/paris-mayor-unveils-15-minute-city-plan-in-re-election-campaign. [Accessed 25 April 2021].
[16] Cassidy J. "Can we have prosperiy without growth?". The New Yorker; February 2020. Viewed on, https://www.newyorker.com/magazine/2020/02/10/can-we-have-prosperity-without-growth. [Accessed 1 May 2021].

Part VII

Smart infrastructure design principles and financial policy approaches

Chapter 26

Urban design strategies and the smart city paradigm

Mallory B.E. Baches
Congress for the New Urbanism, Savannah, GA, United States

26.1 Introduction

Cities face an ever-increasing number of challenges: catastrophic climate impacts, overcrowding and housing affordability, global-scale economic impacts, and social unrest and racial reckoning, among the most pressing. Yet cities also provide the opportunity for efficiencies and opportunities that only the benefits of the population density, economic agglomeration, and systemic complexities can provide. Presuming that the planet continues on the current path of an increasing pace of global urbanization, with 55% of people currently living in urban areas, that number is predicted to increase to 68% by 2050 [1]. Based on nothing more than their increasing ubiquity, it is clear that cities are going to be the central landscape within which the global outcomes of our collective futures lie. One aspect of the debate around this topic, as this chapter will describe, are the tools that cities will use in best determining their own fate amidst those global challenges.

As a long-standing practice, urban design is a means by which cities can actively determine their spatial dimension. The discipline is one that leverages tools and technologies in order to inform understanding and improve the success of the built environment. Yet to date, urban design strategies have rarely been comingled with the growing breadth and depth of systemic and personal technological advances that characterize the 21st century. Utilizing the innovations that digital technology continues to evolve within the urban context, those supportive of advancing a smart city paradigm argue that the approach will provide us with cities that are more "responsive, efficient, sustainable, and safe" (Halegoua [2]; p. 1). Yet these innovations are being implemented, in many cases, without integration into the process of the design of cities.

Noted expert on smart cities, Anthony Townsend, describes smart cities as, "places where information technology is combined with infrastructure, architecture, everyday objects, and even our bodies to address social, economic, and environmental problems" (Townsend [3]; p. 15) and incredibly aspirational view of what cities might become if they were to truly be intelligent. This chapter will explore the paradigm he references when he describes the "smart city" and will review the details of urban design as a discipline for shaping cities. It will then review the ways in which these two fields intersect within the urban context and will propose ways in which the two might benefit from further integration, for the substantial benefit of cities overall.

26.2 Unpacking the smart city paradigm

The meanings of the terminology associated with the smart city paradigm can be elusive, evolving over time throughout the literature and providing more questions than answers, as to exactly what these technological advances mean within the smart city paradigm in the first place. Particularly over the past few decades, cities across the globe have been exploring and investing within the growing realm of information infrastructure and its requisite Information Communication Technologies (ICTs), smart city digitization, and big data. In an attempt to ground the terminology used in this chapter, the definition of various jargon is helpful.

By way of definition, *ICTs or information and communication technologies* is a catch-all phrase for those computerized systems and applications that network digitally through the use of radio, cellular, television, Internet, satellite, and other communication systems and structures. What began as the beginnings of a technological revolution with proliferation of various added means of communication is today a series of interconnected digital networks,

Smart Cities Policies and Financing. https://doi.org/10.1016/B978-0-12-819130-9.00048-6

interacting and communicating and surveilling our every move [2]. As a result, what were once mechanical or physical urban processes—traffic management or service response or civic participation—are now accessible digitally, perhaps without participants even knowing the process is taking place.

By comparison, the phrase *smart city* and its related digital intelligence systems more accurately reference the realm of urban data and system digitization as a whole. The phrase "smart city" has evolved from previous terminologies, to describe a city that utilizes digital technologies, either as the end itself insofar as the use of these infrastructures to monitor and respond within various systems is necessarily reflective of a more efficient and sustainable urban environment or as an intelligent and responsive framework serving as a means to the end of an improved quality of life for the community [4]. This distinction is important because it underscores the fact that the smart city paradigm, as discussed in this chapter, should not be considered inherently and absolutely "good," but that there is value in analyzing the implications of the use of these innovations in cities. Indeed, investigating these impacts gives us strong reason to rethink the relationship between these digital technologies and the built environment within which they are used [5]. Nonetheless, the domination of intelligent digital tools does seem to be a foregone conclusion of the urban future; what remains to be fully determined are the ways in which such tools are best utilized and how they will integrate with the urban design of cities.

Big data tends to be used to describe the collection and analysis of large datasets, though the precise volume of content is often less important than the use of the data to provide breadth of understanding [6]. In some ways, big data is simply a by-product of the acceleration of the smart city movement; it stands to reason that the more devices monitoring more data points across more urban areas there are, the greater the sheer volume of information that is gained and available for analysis [7]. In a report from 2018, the Royal Institution of Chartered Surveyors (RICS) describes the complexity inherent in big data by identifying that a high volume of information itself, a high velocity of ongoing generation of information, and a high variety of sources for information combine to form its key complex characteristics [8]. It is important to remember that there are power dynamics at play with data, particularly that of the magnitude (and implications on personal privacy) that big data carries within the urban context [4,9].

The *Internet of Things* (*IoT*) fits into discussion of the smart city paradigm, as it can be mistaken, and is sometimes purposefully used, as a proxy for the "city" in that it is a network of individual sources connected by a structure of interactions and common understandings. Proponents of the smart city paradigm will sometimes refer to this digital integrative framework of intelligent devices as replicative of the sociospatial interactions that people experience within the actual built environment. However, it is important not to lose sight of the fact that things are not people, and that the IoT is simply a technological tool for improving the efficiency, sustainability, and adaptability of cities rather than a "city" in and of itself [10]. With the advances of understandings that digital technologies provide comes a need to translate those understandings into improving the lived experiences of actual citizens within the urbanism of the city.

In all, the *smart city paradigm* is a set of systems of technological advancements, which both generally and more specifically apply to cities, but has also has proven to become a value system imposed on cities in some cases, either by decision-makers seeking a competitive edge in the global urban marketplace or by corporations eager to meet the extraordinary potential demand that might exist at the scale of an entire city worth of technological advancement. It is arguably reductive, to indiscriminately utilize digital representation in replacement of actual human actors and their agency, in determining the human interactions that constitute city life. A closer look at the criticism of the urban impact of the smart city paradigm helps illustrate this point further.

26.2.1 Criticism of the smart city paradigm and its impacts on urbanism

Too often, the label of "smart city" is used to describe a wide variety of advances in urban outcomes, without interrogating exactly what the phrase references or what impact the particular pejorative of a given "smart" change might have on the city itself [11]. Indeed, there is frequently an underlying assumption that the implementation of smart city technologies—whatever they may be—will necessarily have a positive impact on the urban form of the city, let alone the well-being of its inhabitants [12]. Various technologies can and do have an impact on urbanism; those impacts may or may not coincide with improved urban design or improvement to urban outcomes, whether social, economic, environmental, or otherwise.

When considering the smart city paradigm and the massive data collection that is often associated, it is important to remember that data is a resource, and that it cannot be extricated from the applications it is used to inform nor the context within which it is derived [4]. It is appealing to perceive data and the technology that provides it, as "neutral," yet doing so negates the human influence, and both the explicit as well as implicit biases that might exist, in producing that data. A variety of stakeholders are involved in the collection, analysis, and interpretation of smart-city-derived data, and any city or enterprise neglecting to acknowledge and thus account for the political nature of these technologies should be examined. Furthermore, criticism of the paradigm itself suggests its reliance on the very human instinct to idealize, while failing to acknowledge the financial and privacy vulnerabilities, chief among others, that do exist [11,13].

Cities are unique, and the individuals who populate cities are further unique, and so there is certainly room for criticism of technocratic top-down delivery of data-based systems that may or may not deliver benefit to the whole of the community. Still, it is possible that the smart city paradigm could have the capacity to deliver on the goal of a more responsive city, one that would reflect the evolutionary nature of cities and the advance of those who populate cities over time and generations [14,15]. Yet from the standpoint of implementation, much of the built examples of smart city development are unapologetically aimed at wealth creation, with the identified market being wealthy buyers and consumers of real estate and the goods and services found within. The argument for accepting this current inequity tends to focus around the need for experimentation, with the promise of broader reach once the smart city model is fully realized; the problem to this point has been that the deployment beyond upper economic segments of the community has rarely materialized [16].

Although criticism is widespread, consider that in 2018, McKinsey researchers noted that even those cities with the most advanced examples of the digital technologies most common to the smart city paradigm are still in the beginning stages of implementation [17]. The digital technologies that converge in cities will continue to expand and evolve, and to truly be "smart," cities will need to continue to assimilate these innovations into their built fabric, in ways that are environmentally sustainable, socially and racially equitable, and economically feasible. In consideration of the intersectionality of these demands, the discipline of urban design is a useful place to turn. The design of cities has been a multidimensional practice throughout human history and irrespective of the progress that cities have needed to adapt to, from engineering advances to industrial advances to technological advances.

The "smart growth" movement itself is worth considering briefly here, particularly since advocates for the smart city paradigm often tout the corresponding ICTs as a central component of "smart" urban growth strategies [2]. As defined by the Smart Growth Network, the national partnership of organizational stakeholders working in collaboration in order to encourage more sustainable development patterns, "smart growth" is an approach to land use and real estate development that implies diversity of use, housing, transportation, and activity options created within existing developed areas and with a priority toward community involvement [18]. Such a definition is generally shared by a variety of national organizations [19], and yet those references do not specifically implicate technologies, either as part of the operationalization of the agenda itself or as central to the methodology by which growth is determined to be "smart."

Yet critics of the smart city paradigm do address the subject of "smart growth" and its relevance to the relationship between the adoption of ICTs and related urban technologies and improvement to the built environment [2,12]. As a result, while the two movements may not be directly interrelated, the intention to promote what are perceived to be intelligent, efficient urban systems and strategies is shared by both. And certainly, the digital technologies of the smart city paradigm have had, and will continue to have, and impact on the urban form of cities [20]. The question then becomes: how can urban form benefit from that impact and/or mitigate the ill effects it may cause? To answer that question, it is important to begin with an exploration of the practice of urban design and the prevailing urban design strategies that cities are currently employing.

26.3 The role of urban design in shaping cities

There is no shortage of discord in identifying exactly what urban design is and what it pertains to within the context of the city [21,22]. Authors have provided a variety of definitions of the discipline of urban design, some lumping its provenance in with the origins of architecture and human settlement [23], and others aligning the initial practice of urban design along with that commonly assigned to urban planning, e.g., the postwar era [24–26].

Architecture is the design of buildings; urban design is the design of the places and contexts within which those buildings exist: the block, the neighborhood, and the town or city. It is the study of the public realm, of the spaces where people come together within an urban environment. Urban design utilizes understandings of architecture, applying them to the urban "rooms" of the city: spatial relationships, movement pathways, visual experiences, influence of light and air flow, impacts of noise, and so on. With this in mind, it is clear that cities face myriad challenges to which urban design might offer response. That said, recent practice as well as discourse has paid particular attention to urban design strategies that address equity and inclusion or the well-being of every citizen; strategies supporting housing access and affordability, urban walkability and livability, and community engagement and diversity are consistently elevated in the literature, as well as in policy and practice [27].

It is fair to say that the urban design of cities is a complex discipline, consisting of the consideration of that which can be described in scientific terms as well as that which require artistic context. It is both the static elements of a city and the transient characteristics, the structures as well as the people, the spaces as well as the sensory experiences those spaces provide. As Bahrainy and Bakhtiar detail thoroughly, there are as many definitions of urban design provided by writers and practitioners of the process, as there are interpretations of the successful application of design process upon the built realm [28]. While that may not provide perfect clarity to terminology, they go on to outline certain movements or eras of urban understanding—the City Beautiful, Garden City and New Town, Modernist, and New Urbanist movements, among others [28]—which, when investigated, offer a variety of theoretical approaches sharing the foundation of designing the city as a holistic system based on a set of principles.

In A New Theory of Urban Design, architect and author Christopher Alexander details the complexity that we experience, in an historic city built through generations of creation and replacement and alteration to the built form. That "wholeness" of spaces and places that Alexander observes as having been established through centuries of the ongoing adaptation of a city is a framework that can be strategically applied to cities today, specifically through the discipline of urban design [29]. By intentionally considering the city as a whole or a complete system, rather than simply the sum of its individual parts and participants, a more vibrant and responsive and sustainable urban realm can be cocreated through the process of urban design.

In practical use, urban design is then a process for understanding a place and providing responsive design, based on factors elevated in priority and data points by which they might be understood. As such, the smart city paradigm can and should contribute substantially to the urbanism that this process produces. So-called "smart streetlamps" or "smart crosswalks" or "smart stormwater systems" can provide invaluable data points to the urban designer and can enhance the user experience of the urban realm as well. However, it is important not to misinterpret the smart city paradigm as an urban design strategy or see digital technologies as urban design tools in and of themselves. Rather, ICTs and big data and the IoT are all technologies that can better inform urban design of the systems of the city and can better facilitate the successful functionality of the urban design of the city as a result.

26.3.1 Urban design as a framework for understanding cities

The shared experience of place is something that many thought leaders on urban design return to as a grounding point for their work, among them Kevin Lynch [25], Jane Jacobs [24], Aldo Rossi [30], Ali Madanipour [31], and many more. Where the smart city paradigm might correlate the city to an objective, functional machine, the urban thinkers of the present and recent past specifically criticize that notion, focusing instead on the experience of a city and the human interactions occurring within the fixed elements that make up the built environment. It is important to note the extensive literature supporting an understanding of place that extends beyond the physical elements—the buildings, streets, and open spaces—to include the lived experience of that place, in ways that data-driven devices cannot fully quantify [32,33]. Where the smart city paradigm might regard citizens as vehicles for digital technologies, urban design incorporates citizens as actors with agency in the production of space.

Urban design provides a mechanism for solving urban issues frequently cited as foundational to the need for the smart city paradigm in the first place, such as improved functionality and increased modality and optimized efficiency. Monitoring and analyzing the big data that ICTs and the IoT can provide does not automatically produce better urbanism. What is necessary to do so is a framework for response to the faster, broader understandings that smart city technologies can offer, across the systems and the spatiality of the city. The field of urban design addresses the experience of urban space at various scales; the smart city paradigm would do well to apply this approach. Smart city research would benefit from an agenda that balances the participants, e.g., the individuals participating in the daily life of that city, and the policymaking that thereby impacts these participants, e.g., most cities are not megacities yet smart city policy is not typically designed to adapt to urban scale [34].

An important tool within urban design practice is the creation of master plans, which provide a means for delivering a complex set of integrated solutions that correspond to the identified needs of the community. Urban master plans can take a variety of forms but are an intentional way to overlay a wide variety of understandings about the city into a coherent description, both of what currently exists and what might be possible within the built environment. Successful master plans both confirm context and, if adopted and implemented, provide predictability as to how the city will grow and change and evolve. Practitioners may describe master plans as "living documents," highlighting the way they both provide specific guidance while simultaneously allowing for flexibility, in instructing in the ongoing development of the city.

Interestingly, the literature can occasionally be found to tie the smart city paradigm specifically to the urban design movement known as New Urbanism [2], a relationship that is informal at best. According to the Charter the New Urbanism, the movement represents, "a broad-based citizenry, composed of public and private sector leaders, community activists, and multidisciplinary professionals" who are "committed to reestablishing the relationship between the art of building and the making of community, through citizen-based participatory planning and design" ([Talen 35]; p. vi). While it is true that there are certain shared goals and objectives between the smart city and new urbanist movements, namely around the improvement of the experience of the public realm and the overall sustainability of cities themselves, it would be misleading to imply that one has led to the other or vice versa. Rather, perhaps the alignment of the smart city and new urbanist movements is a by-product of the increasing development attention that cities are receiving overall.

It is possible that a solution to the inherent conflict of the quantitative nature of the data central to the smart city paradigm and the qualitative nature of the experientiality central to urban design strategies lies in the theoretical framework initiated by Brenner and Schmid, which describes a multidimensionality of the urban, whereby the digital ecosystem is simply a dimension of the city, interwoven amidst the physical and social and temporal intersections simultaneously occurring [36]. Furthermore, the efficiencies that are touted as a by-product of the smart city paradigm are extraordinarily valuable toward the goals many cities have for their urban development and overall sustainability. However, the distance between collecting and analyzing the data that digital technologies produces and delivering successful urban solutions to the understanding that the data provides is precisely where urban design aligns.

The digital city need not be mutually exclusive from the physical city, and certainly, urban design can be improved by the incorporation of digital technologies into its process. Indeed, there is a compelling need for urban designers to find successful ways to integrate the innovations of the smart city movement, designing around various tools and devices, and integrating adaptive infrastructure systems such as water management or lighting devices. In particular, the technologies of the smart city paradigm have the means for significantly contributing to the carbon reduction efforts that cities are finding themselves compelled to engage in Ref. [37]. The intersection of the smart city paradigm with the discipline of urban design is a necessary step in addressing the challenges that cities will continue to face, as we look toward an ever-increasing urban future.

26.4 Smart city technologies and urban design challenges

There is a difference between "smart" objects (e.g., the IoT platform and the extent of big data that it can produce) and "smart" places or spaces within the city, and while urban form offers an environment within which smart technologies can inhabit, they are not a replacement for the human experience of place. The information that these technologies collect can inform an "urban narrative" of a certain quantitative type [38]. However, it is reasonable to remain skeptical about claims, especially from private corporations selling products and services or even entire blank slate smart cities, such as Songdo City, South Korea; Masdar City, United Arab Emirates; or PlanIT Valley, Portugal [2,3,39]; indeed, considering the role of urban design (or lack thereof) in such "new" smart cities will be addressed later in this chapter.

As the reproduction of urban space continues in the era of the smart city, urban design has the potential to be a valuable agent within that change. As previously referenced, Madanipour does an excellent job of laying out the reasons for the increasing role that urban design is playing in the evolution of cities. He describes the manner in which urban design offers an added professional resource for cities, one that can provide strategy and stabilization to the urban development process [31]. Considering this perspective, the practice of urban design offers an empathetic framework for managing the myriad changes that advancing digital technologies—whether as part of "smart city" initiatives, or simply as a by-product of the ongoing technological revolution of the 21st century—impose on the city and on the actors within it, whether from the perspective of the municipal, private, or public point of view.

Understanding data points of the human experience of cities is not new with the emergence of the smart city paradigm, however. William H. (Holly) Whyte began his research into the performance of public spaces after joining the New York City Planning Commission in 1969, eventually publishing the results of The Street Life project and detailing the relative success of various city spaces [40]. His work reveals the significance of the opportunistic and unplanned interactions that spaces within the city are able to offer and the characteristics most likely to induce those sorts of interactions in passersby. Similarly, noted Danish architect and urban designer Jan Gehl first published his research on human behaviors within public open spaces in 1971, detailing the importance of designing cities in a way that is responsive to the fundamental human need for contact, desire to understand surroundings, and interest in activity [41]. His practice continues to explore the ways that the design of the built environment can promote an improved quality of life, with a focus on observation and analysis of data not unlike the methodologies central to the smart city paradigm, though perhaps with a more humanistic goal in mind.

One opportunity where smart city technologies and urban design strategies could be mutually beneficial is in regard to land use efficiency and the escalating housing crisis. Many cities have a dearth of vacant properties sitting idle and unproductive, either to municipal tax revenues or community benefit, and providing mapped identification of these properties could help in spurring their redevelopment [17]. Yet simply mapping these locations will not necessarily lead to development that will ensure their successful integration into existing neighborhoods; through participatory urban design processes, these opportunities for reinvestment can be calibrated to match the needs of the surrounding community, ensuring that they contribute to solving challenges rather than garnering more.

Smart technologies allow for the problem space of urban design—that is, the built environment—to be more efficiently managed in order to become more responsive to the activities within it, even becoming almost sentient in some cases [42]. With what is frequently reported as continually diminishing revenues and permanently limited capacity, it is no surprise that city governments might be highly attracted to (and indeed, might deeply benefit from) the types of efficiencies that the smart city paradigm stands to offer. In particular, the ways that digital technologies stand to improve the relationship between the city and the citizen and the ways in which ICTs can inform improvements to the urban design of the city for the benefit of the citizen are both central to that opportunity space.

26.4.1 Use of digital technologies in public engagement and participatory design

A central way in which digital technologies can serve to improve urban design processes is through the communication that they can facilitate. A city with a carefully developed digital technological ecosystem has the capacity to empower a more engaged community, both in the development process of smart city strategies themselves and in overall governance and participation in the urban design of the city, in so far as that ecosystem is equitably established and administered [43]. Where in recent decades citizen participation in local governance has diminished significantly, the use of digital technologies to broaden access to decision-making has the potential to reverse that trend [44].

Cities and their citizens may have become accustomed to the requirements of so-called "sunshine laws," community involvement, and public hearing processes now ubiquitous to municipal planning departments, but there is ample critique of the inclusiveness of many common practices, particularly with regard to those equity-seeking groups historically left out of decision-making processes. Many citizens are unable to attend or uncomfortable participating at in-person public meetings for a variety of valid reasons, an obstacle to participation for which digital technologies can serve as a useful tool. Nonetheless, to date many of the ways that the smart city paradigm has seen implementation have been by decree rather than consent; cities have partnered with corporations or test-driven solutions without first seeking the approval (or sometimes, even awareness) of the community [3]. Invoking a critical eye toward the equity, accessibility, and public compliance by which digital technologies are integrated into urban design processes will allow a city to leverage what will ideally result in an improved community engagement practice.

Furthermore, it is important to bear the critique made by Hollands in mind, that the prioritization of the smart city paradigm by municipal government can necessarily lead to a certain economic, social, or cultural polarization, brought on in part by the intentional attraction of an incoming subclass of citizens—in the case of smart cities, the attraction of those working in ICT industries and/or benefitting from the proliferation of urban ICT application [12]. In particular, personal digital technologies and tools are not equitably accessible, and likewise there is an inherent vulnerability to isolation caused by the increasing digitization of local governance, specifically for those members of the community who value the physical over the digital [43]. Cities seeking to leverage the smart city paradigm for their governance processes and procedures must do so with an eye toward ensuring social cohesion throughout the transition.

26.4.2 Urban design for smart(er) cities

Cities are more than just the sum of their interrelated parts. The assumption that we can digitize our urbanism to the point that we have removed the spontaneity and unpredictability of the humans inhabiting it (or that we might want to) flies in direct contrast to many of the most cited urbanism authors of our time. Whether it is Richard Florida discussing the structural impacts of the young, educated, and affluent returning to city centers [45], or Jane Jacobs describing the value of an active public realm in contributing to overall safety and security within the city [24], or going as far back as the early 1900s and to Ebenezer Howard describing the Garden City movement and the ways in which development might benefit the community as well as the individual [46] or Sir Patrick Geddes imploring everyday citizens to join in the "civics" necessary to improve the social condition of the city [3]; p. 96; [47], the role of human participation in the life of the city has remained an important theme. And too, there is something specifically cynical, in the idealization of the city as a computer rather than a community of human beings, that some smart city proponents, especially corporate actors such as Cisco, Siemens, IBM, and Google, who stand extraordinary financial benefit from that idealization, at times seem to hold.

The smart city has been described as an "upgrade" to existing urbanism, rather than a reunderstanding of urban form itself [3]. Yet the ever-increasing transition to "smart" technologies offers cities the opportunity to carry their existing urbanism and urban design practices and processes into the digital efficiencies that the smart city paradigm offers. Improving urban environments that already exist comes with its challenges, especially those involving integration with existing infrastructures, but likewise stands to benefit from the existing sense of place, and from the social and economic and cultural complexities already developed. By contrast, new cities are often critiqued for lacking these same qualities and characteristics, and newly built entire smart cities are no different.

Three of the most commonly referenced pilot projects of the smart city paradigm, PlanIT Valley in Portugal, Masdar in UAE, and Songdo in South Korea, are sometimes referred to as the "Empty Cities" due to the inactivity of their respective built realms [2,5]. What these newly designed cities provide in technological integration, they seem to lack in understanding and implementation of the urban design techniques that create spaces people are drawn to, encouraged to linger within, and inclined to activate. Further, the plans of these cities each lack a cohesive urban framework, one focused on buildings and the spaces between that are scaled to the human experience and reflect an understanding of social participation in urban life. Of those that have managed to see some level of build-out, a certain placelessness that from-scratch cities often display, particularly during early generations of implementation, is cited.

PlanIT Valley, designed to be a digitally immersive new city for 225,000 people, developed by an international smart-IT corporation for roughly $19 billion, was planned to serve as a living laboratory for those digital technologies, and so the success of its built environment is secondary to its functional intention as an urban R&D platform [2,48]. To date, that plan has not materialized into an actual city, with a unified investment team never coalescing to see the project through. As of current build-out, Masdar City abandons some of the traditional aspects of Arab urbanism in favor of a more vehicular-centric street network and a more Westernized hierarchy of open spaces [49]. This may lend to a disorientation from the sense of place inherent to the region and certainly informs a sense of modern intervention to the city, one that may reflect the technological and sustainable intentions of the design but could just as easily be described as forced or performative, an unsurprising by-product of the top-down planning policy and design approach [2,50].

Most notable of these new smart city examples may be Songdo, which was originally planned with the intention of attracting international investment and expatriate residents rather than local Koreans [48]. With a technocratic development approach, the top-down nature of the decision-making has yielded a quiet, wealthy residential enclave rather than the idealized city of the future that was originally envisioned [2,50]. True, some of this can be assigned to cultural differences, modern social behaviors, and the inevitably stark character of entirely newly built surroundings. Nonetheless, critiques of these newly developed and technologically integrated smart cities raise questions about the prioritization of digital connectiveness above an urban form that provides for a more fundamental social coherence.

Utilizing these technologies in a more collaborative manner, there are cities that have demonstrated a focus on community outcomes rather than strictly technological ones, avoiding the pitfalls of previously common technocratic top-down approaches to implementation by establishing a culture of cocreation with their citizens [5]. One such example is Amsterdam Smart City, a platform now over a decade in operation, which has developed a suite of tools and functions that leverage various digital technologies for the broad betterment of the community, including building opportunities for connection and networking, increasing individual mobility, and improving upon a circular economy [51]. While most of the innovations that the platform has provided have less of a direct influence on the design of the city, at least at this stage of implementation, the open-source data collection and availability that various smart city initiatives have accomplished (as example, documenting the 1248 public spaces in the city) may serve as the foundation for future urban design efforts that the city undertakes [52].

There are logistical implications to the ever-advancing technological revolution that cities are undergoing. More and more, the design of urban space must contend with the growing modalities that technology provides and must contend with a future where robotics will play a significant role in the activity of the public realm. Unless cities actively integrate their ongoing physical evolution with their simultaneous digital evolution, the public realm is likely to become an even more increasingly contested space. That said, digital advances also offer a potential competitive advantage to cities, serving as a resource for predicting and protecting against the fourth dimension of time, e.g., what cannot be known cannot be dealt with. In this way, digital advances are a critical resource to urban design as well, providing the ability to anticipate the needs of urban futures and the outcomes of urban interventions.

26.4.3 The practice of urban design with digital technologies

A growing body of practitioners are working at the intersection of urban design and digital technologies. Begun in the 1970s at The Bartlett School of Architecture at University College of London, the Space Syntax Laboratory has been at the forefront of integrating digital technologies with urban understandings. Today taking its applied research across the globe, Space Syntax leverages digital forecasting technologies to better understand the impacts that development might impose on the sustainability of a city [53]. Working directly with cities, their use of digital technologies to model outcomes of system and network designs provides performance metrics that can be incredibly helpful in better understanding the impact of urban design before the design is implemented, on strategies ranging from mobility networks to carbon-reduction master plans to urban growth approaches.

In the United States, the work of Urban3 explores a different aspect of applied data to urban design, by delivering land value and taxation analysis for the purpose of helping cities improve urban form and related policy decision-making [54]. In most cases, local governments make decisions about their land use without an accurate and/or complete understanding of the costs and revenues that those decisions will mean for the city, both in the present and in the years to come. With the "heat map" visualizations that Urban3 produces of the financial implications of development scenarios, particularly using geospatial mapping technology, cities are able to better understand the economic impact of physical conditions and envision their urban development while more accurately projecting the cost benefit of that vision.

Certainly, larger design and development firms with work around the globe are embracing the potential benefits that digital technologies can bring to their own design delivery. ARUP, leading global design and engineering firm with over 16,000 staff working in 143 countries, devotes an entire category of their website framing of perspective toward the Digital, and considerations for how cities can best utilize technologies for design, policy, implementation, and analysis [55]. Likewise, US-based design and delivery firm Stantec dedicates an entire division to Smart Cities, purportedly working at the intersection of technology and people with noted focus on mobility, utilities, public spaces, and buildings [56]. While it can be expected that big data and other smart city technologies will continue to be tools accessed by a growing proportion of urban design practitioners, the successful future of cities is dependent on the operationalization of those tools, for the betterment of urbanism itself.

26.5 The future design of cities

As referenced previously in this chapter, it is crucial to deeply question whether or not the operation of urban spaces and places as spatial "computers," as is often a default assumption within the smart city paradigm, is actually an ideal worth striving for. As Jane Jacobs describes, the "ballet of a good city sidewalk" is one of the particularly unique experiences of city life that cannot be quantified and should not be reduced down to some quantitative derivative simply for the sake of efficiency (Jacobs [24]; p. 65). The way in which the human convergences of the city prompt incidental interactions that would otherwise not occur is something that the smart city paradigm might be inclined toward reducing or even eliminating, in service of efficiency and which our human experience of cities would be the lesser for as a result [2]. To do so would provide an extraordinary loss to so many of the unexpected delights (if also, some of the undesired frustrations) that we experience, when in the close proximity of others in the way that the city provides.

We are only at the very beginning stages of smart technology integration into urban form and urban functionality. One promising way that cities are beginning to integrate urban design strategies with emerging digital information capabilities is through the creation of digital master plans, which serve to produce community-based visioning around the physical, social, and economic implications of the advancement of urban digital technologies [44]. Much like master plans as used within urban design practice, digital master plans provide a framework for growth and development, albeit that of technology-based elements and initiatives rather than those of the built realm. While digital master planning themes or approaches may vary widely, from one city to another, the goal of providing a predictable roadmap for stakeholders of the ways in which the city intends to adapt to digital technological advances is a worthy one. The opportunity for integration of digital planning with spatial planning is a natural evolution and one that would be a welcome advancement for both pursuits.

Digital master plans provide a way of synthesizing the digital dimensions of the city with the many other dimensions that already exist, something necessary if digital technologies are to prove successful on their promise to alter cities for the better [57]. Utilizing these sorts of strategic frameworks, there are certainly then myriad opportunities for ICTs and other smart technologies to be more meaningfully incorporated into cities, as they address future urban design challenges. Whether the challenge is climate change and climate migration, civic engagement and participatory decision-making, design impact forecasting and performance analysis, or other challenges not yet identified, the ability for cities to leverage a framework for digital integration that provides predictability yet allows for flexibility, so that a city can be spatially as well as technologically responsive, will be essential.

Unexpected threats lie on the horizon for any city, and as the Covid-19 global pandemic has shown, those cities with the greatest diversity of urban form, infrastructure and amenity, procedural and technological tools, and functional flexibility were able to show the highest resiliency [58]. Particularly as we look toward an unknown urban future, we must remember that the multidimensionality and multifunctionality of the urbanism we share can be significantly improved with the help of emerging digital technologies, but only insofar as those technologies are able to elevate citizens into actors with meaningful agency in the outcomes of their participation [59]. Adaptability will be key to addressing future urban challenges with improved urban design and improvement to urban outcomes, whether social, economic, environmental, or otherwise; smart city technologies offer advantage and opportunity to that proposition.

Perhaps rather than imposing a technocratic paradigm upon cities, it is more fitting to think of the smart city movement as reflecting an idea about cities: not implying the city as the computer itself, but instead responding to the metaphor of the city operating as does a highly advanced machine, one that should be a tool for improving the lives and well-being of the people that populate its spaces and structures. At their worst, ICTs and the uses of big data may become fully deterministic and remove the complexity that centrally contributes to the characteristics that have drawn humans to cities for millennia. However, at their best, smart technologies have the power to reveal critical knowledge about how cities are, can be, and should be designed, providing actors at all levels of participation with understandings that will improve and expand the value that the city can bring to their life, their work, and their lived experience.

26.6 Summary

As this chapter has described, there is a certain friction, yet ample opportunity for advancement, where the digital technologies of the smart city paradigm meet with the prevailing strategies of the practice of urban design. The integration of these separate but interrelated inquiries into the betterment of cities offers the ability to amplify the potential benefits that each might individually achieve. Still a relatively new realm of urban experimentation, it is important to remember that the impact of these transformations that the shift to the smart city paradigm brings has rarely been critically analyzed, and the political nature of the process of transitioning often goes unexamined [60]. The argument is made by some observing the smart city phenomenon that the integration of the physical with the digital city allows for a more complete, more personalized experience [2], yet so far there remains a lack of evidence of this outcome. More must be done to understand, advance, and coordinate the techniques and tools of smart city innovations, in ways that reflect and respond to the betterment of the built environment and the participants within.

Participation in the built environment is both a resource for the city as a system, one that can be understood and leveraged, as well as an experience for the individual as a citizen, one that can be designed for and improved upon. If a city is to truly be intelligent, one might argue that it is the community that would bestow such a distinction, more appropriately than a corporation or a government or a critic of some sort. As a result, such a distinction would necessarily involve the engagement and involvement of the community itself, elevating the participation of stakeholders in collaboration toward a "smarter" city [57]. It is through the lived experience of the community, within the context of the urban realm, that the identification of a city as truly "smart" must come.

Elevating the experience of people within the smart city paradigm, and doing so by means of successful urban design strategies, can provide individuals with the agency to better know their city and have their needs met within the city, yet still be able to enjoy the beneficial serendipity that the intensity of city life brings [2]. Urban design carries the power to fully elevate the spatial organization of a city, but only insofar as the practice can adequately interpret the needs of citizens and arrange the spaces and places of the city accordingly [31]. This contingency is key to how the smart city paradigm and the discipline of urban design might converge. As a species, cities have served as the central means of our communal nature as human beings, providing a platform for the cultural, economic, and social benefits that we derive from proximity within the urban realm [24]. As we face the challenges of our urban future, it will be through both design and technology, and the integration of the two, that our cities and their citizens will best find the means to thrive.

Finally, let us move on to the real interactive part of this chapter: review questions/exercises, hands-on projects, case projects, and optional team case project. The answers and/or solutions by chapter can be found in Appendix G.

26.7 Chapter review questions/exercises

True/false

1. True or False? Cities face an ever-decreasing number of challenges: catastrophic climate impacts, overcrowding and housing affordability, global-scale economic impacts, and social unrest and racial reckoning, among the most pressing.
2. True or False? The meanings of the terminology associated with the smart city paradigm can be elusive, evolving over time throughout the literature and providing less questions than answers, as to exactly what these technological advances mean within the smart city paradigm in the first place.
3. True or False? Too often, the label of "smart city" is used to describe a narrow variety of advances in urban outcomes, without interrogating exactly what the phrase references or what impact the particular pejorative of a given "smart" change might have on the city itself.
4. True or False? There is no shortage of discord in identifying exactly what urban design is and what it pertains to within the context of the city.
5. True or False? The shared experience of place is something that few thought leaders on urban design return to as a grounding point for their work.

Multiple choice

1. There is a difference between "smart" objects and "smart" places or spaces within the city, and while urban form offers an environment within which smart technologies can inhabit, they are not a replacement for the human experience of:
   a. Place
   b. Infrastructure
   c. Quantitative type
   d. Investment
   e. Policy
2. A central way in which digital technologies can serve to improve urban design processes is through the ____________ that they can facilitate.
   a. Infrastructure
   b. Physical infrastructure
   c. Smart city growth and prosperity
   d. Economic effects
   e. Communication
3. Cities are more than just the sum of their interrelated:
   a. Designs
   b. Impacts
   c. Systems
   d. Parts
   e. Twins
4. A growing body of practitioners are working at the intersection of urban design and digital:
   a. Platforms
   b. Economic growth
   c. Technologies
   d. Computer applications
   e. E-economics
5. It is crucial to deeply question whether or not the operation of urban spaces and places as spatial ____________, as is often a default assumption within the smart city paradigm, is actually an ideal worth striving for.
   a. Paradigms
   b. Sidewalks
   c. Interactions
   d. Computers
   e. Experiences

Exercise

Problem

In order to understand the flow of smart city research, conduct a central analysis (which is widely used in quantitative research), relating to social network analysis, and perform cluster analysis to identify relationships with related research.

Hands-on projects

Project

Do research: Investigate how the emerging data-driven smart city is being practiced and justified in terms of the development and implementation of its innovative applied solutions for sustainability.

Case projects

Problem

Explain how smart cities are a new paradigm for coping with sustainability objectives in urban environments.

Optional team case project

Problem

Describe how to further critical engagement, by distilling dimensions absent in current smart city urbanism.

References

[1] UN. World urbanization prospects: the 2018 revision. New York: United Nations, Department of Economic and Social Affairs, Population Division; 2019.
[2] Halegoua G. Smart cities. Cambridge MA: The MIT Press; 2020.
[3] Townsend A. Smart cities: big data, civic hackers, and the quest for a new utopia. New York: W W Norton & Company; 2013.
[4] Kitchin R. The real-time city? Big data and smart urbanism. Geojournal 2014;79:1−14.
[5] Biloria N. From smart to empathic cities. Front Architect Res 2021;10(1):3−16.
[6] boyd d, Crawford K. Critical questions for big data: provocations for a cultural, technological, and scholarly phenomenon. Inf Commun Soc 2012;15(5):662−79.
[7] Batty M. Smart cities, big data. Environ Plann Plann Des 2012;39:191−3.
[8] Mehdi N. Big data, smart cities, intelligent buildings - surveying in a digital world. London: Royal Institution of Chartered Surveyors; 2018.
[9] Bannister J, O'Sullivan A. Big data in the city. Urban Stud 2021;1:1−10. OnlineFirst.
[10] Kamal Boulos M, Al-Shorbaji N. On the internet of things, smart cities and the WHO healthy cities. Int J Health Geogr 2014;13:10. 10.
[11] Hollands R. Critical interventions into the corporate smart city. Camb J Reg Econ Soc 2015;8:61−77.
[12] Hollands R. Will the real smart city please stand up? City 2008;12(3):303−20.
[13] Grossi G, Pianezzi D. Smart cities: utopia or neoliberal ideology? Cities 2017;69:79−85.
[14] Brenner N. What is critical urban theory? City 2009;13(2):198−207.
[15] Parker S. Urban theory and the urban experience: encountering the city. London: Routledge; 2015.

[16] Clark J. Uneven innovation: the work of smart cities. New York: Columbia University Press; 2020.
[17] Woetzel J, Remes J, Boland B, Means J, Lv K, Sinha S, et al. Smart cities: digital solutions for a more livable future. New York: McKinsey Global Institute; 2018.
[18] Smart Growth Network. This is smart growth. Washington DC: Smart Growth Network; 2006.
[19] Ye L, Mandpe S, Meyer P. What is "smart growth?" -really? J Plann LitIncorporat CPL Bibliograp 2005;19(3):301–15.
[20] Graham S, Marvin S. Telecommunications and the city: Electronic spaces, urban places. London: Routledge; 1996.
[21] Cozzolino S, Polivka J, Fox-Kämper R, Reimer M, Kummel O. What is urban design? A proposal for a common understanding. J Urban Des 2020;25:1–15.
[22] Madanipour A. Ambiguities of urban design. In: Carmona M, Tiesdell S, editors. Urban design reader. Oxford: Routledge; 2007. p. 12–23.
[23] Kostof S. The city shaped: urban patterns and meanings through history. London: Thames and Hudson; 1991.
[24] Jacobs J. Death and life of great American cities. 1961 [London: Cape].
[25] Lynch K. The image of the city. Cambridge MA: MIT Press; 1960.
[26] Rowley A. Definitions of urban design: the nature and concerns of urban design. Plann Pract Res 1994;9(3):179–97.
[27] Talen E, Lee S. Design for social diversity. London: Routledge; 2018.
[28] Bahrainy H, Bakhtiar A. Toward an integrative theory of urban design. Switzerland: Springer; 2016.
[29] Alexander C. A new theory of urban design. New York: Oxford University Press; 1987.
[30] Rossi A, Eisenman P, Graham Foundation for Advanced Studies in the Fine Arts, & Institute for Architecture and Urban Studies. The architecture of the city. Cambridge MA: MIT Press; 1982.
[31] Madanipour A. Roles and challenges of urban design. J Urban Des 2006;11(2):173–93.
[32] Cresswell T. Place: a short introduction. Oxford: Blackwell; 2004.
[33] Dovey K. Becoming places: urbanism/architecture/identify/power. London: Routledge; 2010.
[34] Visvizi A, Lytras M. Rescaling and refocusing smart cities research: from mega cities to smart villages. J Sci Technol Policy Manag 2018;9(2):134–45.
[35] Talen E. Charter of the new urbanism. 2nd ed. McGraw-Hill Education; 2013.
[36] Brenner N, Schmid C. Towards a new epistemology of the urban? City 2015;19(2–3):151–82.
[37] Viitanen J, Kingston R. Smart cities and green growth: outsourcing democratic and environmental resilience to the global technology sector. Environ Plan A Econ Space 2014;46(4):803–19.
[38] Srivastava L, Vakali A. Towards a narrative-aware design framework for smart urban environments. The future internet. FIA 2012. Lect Notes Comput Sci 2012;7281:166–77. Berlin: Springer.
[39] Greenfield A. Against the smart city. The city is here for you to use Book 1. Do projects; December 20, 2013.
[40] Whyte W. The social life of small urban spaces. Washington DC: Conservation Foundation; 1980.
[41] Gehl J. Life between buildings: using public space. 6th ed. Washington DC: Island Press; 2011.
[42] Picon A. Smart cities: a spatialised intelligence. Chichester, England: Wiley; 2015.
[43] Angelidou M. Smart city policies: a spatial approach. Cities 2014;41(1):3–11.
[44] Townsend A, Lorimer S. Digital master planning: an emerging strategic practice in global cities. Working Paper #25. NYU Marron Institute of Urban Management; 2015.
[45] Florida R. The new urban crisis: how our cities are increasing inequality, deepening segregation, and failing the middle class – and what we can do about it. New York: Basic Books; 2017.
[46] Howard E. Garden cities of to-morrow. London: Swan Sonnenschein & Co; 1902.
[47] Geddes P. Cities in evolution: an introduction to the town planning movement and to the study of civics. London: Williams & Norgate; 1915.
[48] Carvalha L. Smart cities from scratch? A socio-technical perspective. Camb J Reg Econ Soc 2015;8(1):43–60.
[49] Hassan A, Lee H, Yoo U. From medieval Cairo to modern Masdar City: lessons learned through a comparative study. Architect Sci Rev 2015:39–52.
[50] Yigitcanlar T, Han H, Kamruzzaman M, Ioppolo G, Sabatini-Marques J. The making of smart cities: are Songdo, Masdar, Amsterdam, San Francisco and Brisbane the best we could build? Land Use Pol 2019;88.
[51] Amsterdam Smart City: About us. (June 1, 2021). Retrieved from Amsterdam Smart City: https://amsterdamsmartcity.com/.
[52] Gemeente Amsterdam: Data and information. (June 1, 2021). Retrieved from Gemeente Amsterdam: https://data.amsterdam.nl/.
[53] Space Syntax: About Us. (June 1, 2021). Retrieved from Space Syntax: https://spacesyntax.com/.
[54] Urban3: Services. (June 1, 2021). Retrieved from Urban3: https://www.urbanthree.com/services/.
[55] ARUP Perspectives: Digital. (June 1, 2021). Retrieved from ARUP: https://www.arup.com/perspectives/digital.
[56] Stantec: Smart Cities. (June 1, 2021). Retrieved from Stantec: https://www.arup.com/perspectives/digital.
[57] Komninos N, Tsarchopoulos P, Kakderi C. New services design for smart cities: a planning roadmap for user-driven innovation. Association for Computing Machinery; 2014. p. 29–38.
[58] Tabary Z. Cities reboot: urban life in the age of COVID-19. Thomson Reuters Foundation; October 27, 2020 [Inclusive Economies - Cities].
[59] Kitchen R. Toward a genuinely humanizing smart urbanism. In: Cardullo P, Di Feliciantonio C, Kitchen R, editors. The Right to the smart city. Bingley UK: Emerald Publishing; 2019. p. 193–204.
[60] Luque-Ayala A, Marvin S. Developing a critical understanding of smart urbanism? Urban Stud Int J Res Urban Stud 2015;52(12):2015–116.

Chapter 27

Smart City Technologies plus Nature-Based Solutions: viable and valuable resources for urban resilience

Ernesto Philibert Petit
Environmental Structure Research Group, Queretaro, Mexico

27.1 Introduction

This chapter is about the possible and desirable integration of urban networks and Smart City Technologies (SCT) with the emerging Nature-Based Solutions (NBS), for cities to efficiently and effectively achieve Resilience from threats, hazards, and disturbances occasioned by an unsustainable development model. A new appreciation of nature, innovation, technology, and an integrative vision is necessary to face global and local (urban) challenges on time.

My viewpoint or perspective while writing this chapter is that of a researcher, teacher, and consultant of architecture and urban design, but mostly as a promoter of nature-based and technologically resourceful integrated urban projects. This personal perspective is behind the text when written in first person, in the sense of propositions or assertions that express my opinion, like in an essay. But this chapter is part of a handbook, so you will hopefully find practical advices written in second person, directed to you, the reader, whom I think of as a potential promoter too. You will find in the text a good number of references intended for support, formality, and credibility, written in third person as in an academic paper, to recognize and give credit to individual authors, teams, or institutions that contribute with relevant knowledge to this chapter. You will also find what I call *timely tip boxes*, with direct recommendations of further reading or learning experiences on the subjects studied.

As a part of this Smart City Policies and Financing Handbook, in which I am honored to participate, the main purpose of this chapter is to point out the opportunity I foresee for SCT and for the Smart City approach to urban development, to become an important integral part of the construction of emerging models of urban development and urban retrofit, by networking with NBS, which are being increasingly applied to guide the design of resilient landscapes and cities to enable them to reach economic development goals with beneficial outcomes for the environment and society [1].

Through the research I made in preparing this chapter, I found out that in recent years, a surprising number of high quality academic resources have been published about how much ecosystem services; green and blue infrastructure; and, other resources [2–8] are covered under the umbrella of NBS [3]. In other words, the emerging Nature-Based Solutions (NBS), allows cities to efficiently and effectively achieve Resilience from threats, hazards, and disturbances at the urban-regional scale, which are also the main subject of the emerging Urban Resilience Policies [9].

I could also find in the literature that there has been some concern in the recent past for the separation of the concepts of "Sustainability" and "Smartness" of the city, a separation that brought heavy criticism to the Smart City concept [10], a situation that has been recently changing for good, when you can find interesting trends of integration of technology and ecology into *Technoecology* [11], the rising concept of *Smart Urban Sustainability* [12], an increased interdisciplinary collaboration that will span the conception, implementation, and enhancement phases of ideas, bridging the university, public, and private sectors [11], a call for better aligning smart city strategies to lead to concrete sustainable outcomes [10], and very timely explorations that consider whether and to what extent urban forests and other green spaces can be effectively integrated into smart city planning, to maximize green benefits for all city dwellers [13].

From my point of view and personal experience, SCT can enhance the benefits and reduce the costs of NBS and can even contribute with additional profits of adaptive

Smart Cities Policies and Financing. https://doi.org/10.1016/B978-0-12-819130-9.00001-2

management, competitiveness, governance, and social capital development for the cities. In the years to come, SCT + NBS, in the form or integrated urban projects [14], will have an extraordinary opportunity to be financed by complying with urban resilience policies, being more fundable, offering a better benefit—cost ratio over time, and providing a number of additional benefits for the cities and citizens.

My purpose for this chapter is to provide you, the reader, with information and useful references for imagining strategies to promote or at least support projects of SCT + NBS, to contribute to meeting goals of emergent policies of Urban Resilience. First, we make a brief exploration of the emergence or Resilience in the context of Sustainability, to establish a characterization of Urban Resilience Policies and what these policies are focused on, and the threats they are devised to face. Innovative urban resilience policies are allocating more financial resources worldwide for cities to recover from these threats and also to improve from the conditions that cause them.

The second part of this chapter is about NBS. We characterize NBS by defined concepts, the *nomenclature* of NBS. We then make an exploration of the ratios of benefits and costs of NBS found in different academic sources of the recent years. You will also find strong recommendations for implementation and assessment of NBS coming from very recent academic papers and reports from the EU.

From my own experience I have learned that SCT and approach can enhance the efficiency of NBS and even bring additional benefits to the formula. The third part of the chapter focuses on specific features and resources of SCT that may enhance efficiency, effectiveness, and governance of NBS and deliver additional benefits.

NBSs are financially viable per se, when understanding viability as a concurrence of policy compliance, fundability, and dynamic efficiency. The fourth part of this chapter refers to these three major elements of viability. SCTs definitely add value to NBS.

The fifth part of this chapter is a corollary of the four previous parts. It outlines how NBS can be even more valuable by incorporating SCTs in every step of cocreation, coimplementation, and comanagement of NBS, including effectiveness, stakeholder engagement, and synergism with other infrastructures and urban systems. The chapter finishes with propositions about a desirable and possible future for the integration of SCT + NBS to urban networks *by design*.

27.2 Urban resilience policies in the context of sustainability: the continuing battle for sustainability will be held in the cities

More than one-third of a century ago, the World Commission on Environment and Development published the Brundtland Report [15]. The Brundtland Report made a bold statement identifying the main causes of the world's environmental problems in the great inequality between the enormous poverty of the South and the unsustainable patterns of production and consumption of the North and made a sound call for an innovative strategy to address this problem, by fusing development and environment, describing it with the currently common term "Sustainable Development,"which was then defined as "development that meets the needs of the present without compromising the ability of future generations to meet their own needs" [15].

A new paradigm of integration of disciplines has emerged: development, along with its social end economic edges, has since been pondered as one with the environment. This emergent paradigm or way of thinking has been known since then roughly as *Sustainability*.

One of the main concerns in the Brundtland report in 1987 was that *the planet is passing through a period of dramatic growth and fundamental change. Our human world of 5 billion must make room in a finite environment for another human world. The population could stabilize at between 8 and 14 billion sometime next century, according to UN projections. More than 90% of the increase will occur in the poorest countries, and 90% of that growth in already bursting cities* [15].

In only 34 years, the world's population grew from 5 billion to almost 8 billion. According to the UN Habitat [16], urban areas are already home to 55% of the world's population, and that figure is expected to grow to 68% by 2050, when the current population estimates predict will reach almost 10 billion (10 thousand million inhabitants), an estimation that duplicates the population of the world when the Brundtland report was issued. Seven out of ten inhabitants of the planet will live in cities by 2050 [16].

One-third of a century ago, it was foreseeable that the battle for sustainability was to be held in cities, and today the goals of Sustainable Development: environmental preservation, social equity, and economic prosperity have not been reached already. With no doubt, the next quarter of the 21st century will be decisive in achieving Sustainability or not.

27.2.1 Resilience is the emergency room of sustainability

We must draw urgent attention to the decisiveness of the unsustainable way we have been building cities since the 1950s when the world's urban population barely reached 1 billion. Since then, the dominant urban model has been the diffuse or sprawled city. This means that from 1950 to 2020, we have built sprawled urban *tissue* for roughly 4 billion people, and we still need to build *new* city for 2 billion more by 2050. In order to meet the sustainability goals by that year, how are we supposed to build these new cities and to repair the sprawl we have built in these last 70 years?

The [17] World Cities Report has stressed the need to achieve sustainable urbanization in order to offer inclusive prosperity and opportunities for all as an economic value and to pursue inclusion and social equity. The report recognizes also that innovation and technology are an important edge of sustainable urbanization: the SCT as a means to support civic bottom-up participation to enhance governance and as a means for a better conception, operation, and administration of urban infrastructures. The fourth pillar of sustainable urbanization according to the report is the environmental value, and this pillar now incorporates the need to build *resilient* urban development. The report also remarks the importance of the local governments' institutional frameworks and presents the urban finance challenge to fund sustainable urbanization. **Resilience, technology, and funding** of sustainable urban development are recognized now as key players in the Sustainability agenda, and thus, they are central subjects of study in this chapter.

The Brundtland report introduced Sustainability at the end of the 20th century as a proactive way of thinking (and building) the city. The goals set by a sustainable urban development agenda were generally accepted by the developed and developing world and sustainability even became a trendy concept. However, a few years after, by the turn of the century, the effects of unsustainable development became so evident that an urgent call for action was released in order that we could be able to *recover* from such effects as a civilization. Recovery from diverse negative effects of growth and development is the main ethos of resilience. Resilience displaced Sustainability in the collective consciousness; however, the former being part of the latter and emerged as a priority.

Resilience is now the emergency room of the sustainability building and, since cities are the main stage where the battle for sustainability and resilience is being held, we need a solid defined concept of urban resilience. The team of Sarah Meerow, Joshua P. Newell, and Melissa Stults proposed this definition after a thorough literature review and aided by bibliometric analysis that we here take for the purpose of this work.

We can synthesize this even more by saying that urban resilience is the ability of an urban system to recover from a disturbance. The United Nations defines urban resilience as the ability of human settlements to recover quickly form any plausible hazards [17].

Sanchez at al. [9] have made a review of the urban resilience literature since the 1970s. In this publication they question dominant urban resilience conceptualizations and explain how different understandings of the concept of urban resilience can result in essentially different policies and governance interventions and outcomes. They identified seven types of urban resilience policies:

1. Disaster resilience policies, with focus on disaster hazard mitigation and preparedness.
2. Engineering resilience policies, with focus on quick recovery from shock events, short-term damage.
3. Ecological resilience policies, with focus on urban system relationships remaining functional.
4. Socioecological resilience policies, with focus on developing capacities to adapt and survive.
5. Evolutionary resilience policies, with focus on recovery as an opportunity for improvement.
6. Built-in resilience policies, with focus on proactive strategies to cope with dynamic change.
7. Climate change resilience policies, with focus on climate change mitigation and adaptation.

Because of the "emergency room" character of resilience, the correspondingly emergent urban resilience policies are conveying important financial resources for research, design, implementation, and assessment of "solutions" for resilience.

27.2.2 Urban resilience policies, their foci, and the threats they are devised to face

Considering that urban systems are complex systems far from equilibrium, we have to point out that the two kinds of policies that better respond to the complex and dynamic features inherent to cities include those with a focus on recovery as an opportunity for improvement and those with a focus on proactive strategies to cope with dynamic change and especially on the evident climate change adaptation and mitigation:

> *Urban resilience policy is a complex and evolving field characterized by significant challenges associated with urban governance systems, political pressures, uncertain and emergent nature of threats, speed of change, and the level of complexity of long-lived networks that form cities. Added to these issues, there are a number of resilience concepts that can potentially be used to develop such policies goals.*
>
> [9].

There are two major issues affecting the sustainability of cities (and civilization): environmental pollution and global warming. Urban resilience policies are in a vast majority, addressed to recover from, adapt to, and sometimes improve the conditions from the impacts that these two major issues produce in cities.

Pollution in cities is not a new problem. Cities have been struggling with pollution in great scale at least since the Industrial Revolution triggered a huge increase of anthropogenic carbon emissions by burning fossil fuels since more than 200 years ago. Besides environmental pollution, these anthropogenic carbon emissions upset the natural balance of emission and absorption of carbon dioxide or CO_2 and have majorly contributed to global warming. A publication by Ref. [18] points out that over 97% of climate experts are convinced that humans are changing global temperature (citing [19] and [20]), which include actions as burning fossil fuels, urbanization, deforestation, land use change, and others. Urbanization produces heat, whether by industries, buildings, households, and cars and is also responsible of artificial warming of the Earth environment [21] in Ref. [18]. Deforestation and land use change are the other human activities that contribute to global warming. Forests store massive amounts of carbon, and depleting them for urbanization, farming, or other uses, provokes a massive release of carbon to the atmosphere. The publication remarks that according to Ref. [22], about 25% of the world's greenhouse gas production is originated by deforestation alone.

To reduce or eliminate burning fossil fuels and the corresponding emissions, to diminish the negative externalities of the dominant patterns of (sprawled) urbanization and the excessive energy consumption and heat produced by buildings, to eradicate deforestation and to control land use change, have become the significant challenges to be addressed by urban resilience policies as a result from different approaches to these challenges and are intended for cities to recover from or adapt to the different threats, disturbances, or hazards that are relevant to identify, for the purpose of effectively promoting subsequent programs and projects that might also improve the existing conditions posed by the challenges themselves.

A review of [8,23–30] and [31] cited in chronological order helped us to identify 10 recurrent threats, hazards, or disturbances that cities suffer and have to be resilient of:

1. Severe heat waves with a huge impact in illnesses and mortality
2. Flooding by stormwater
3. Storminess and wind damage
4. Urban pollution health hazards
5. Urban heat island effect, which derives in an excess of energy consumption
6. Land subsidence by deficit of the underground water balance
7. Inequity by inaccessibility to social infrastructure
8. Environmental breakup by sprawl
9. Loss of economic competitiveness (disinvestment)
10. Sedentarism, obesity, and its health consequences.

All of preceding causes, challenges and threats, hazards, and disturbances identified are to be the major subjects of public policies concerning urban resilience nowadays and for the near future. Effective urban resilience policies ought to be designed with the goal of urban recovery from threatening conditions and also designed to obtain improvement from such conditions. They should provide proactive strategies (structured ways to achieve the goals) and be flexible in order to cope with dynamic change in the context they operate: the city.

27.3 Nature-Based Solutions, the ultimate resource for urban resilience

A conceptual dichotomy has existed for centuries between nature and city. Natural and artificial have been the two distinct categories of the environment in which we live, with the historic belief, since the 15th Century in the Western World that the artificial was meant to dominate the natural.

Since then, the world population has increased 23 times and the population in cities has increased 314 times. As we can deduce and observe, outcomes of this huge increment in population along with natural resources depletion and our excessive emissions of greenhouse gases to the atmosphere are also the causes for the concept of urban resilience to emerge.

27.3.1 Defined concepts of NBS

In a brief research on 21st Century's proactive human reactions to problems posed by natural depletion, specifically in the urban context, I found a good variety of defined concepts, such as building green and blue infrastructure, ecosystem services, urban forestry, green roofs, green spaces, urban green spaces, and others.

The publication of the methodological guidelines by the European Commission [32] includes the many emerging frameworks and methods emerging and will surely norm the requirements for funding projects, programs, and policymaking in the European Union. The EU has become a world leader in the complex fields of Sustainability and Resilience by launching, norming, and funding the concept of NBS, which is a very robust umbrella for other existing concepts, such as ecosystem-based adaptation (EBA), ecosystem-based disaster risk reduction (Eco-DRR), green infrastructure (GI), blue infrastructure (BI), green-blue infrastructure (GBI), urban forestry (UF), sustainable urban drainage systems (SuDS), ecological engineering (EE), best management practices (BMPs), low impact design (LID), water-sensitive urban design (WSUD), and ecosystem services (ESS). Ten statements on NBSs and their Ecosystem Services

1. The European Commission defined NBS as "solutions inspired and supported by nature, designed to address societal challenges which are cost-effective, simultaneously provide environmental, social and economic benefits, and help build resilience" [8,33]; in Ref. [3].
2. NBS were defined by the International Union for Conservation of Nature (IUCN) as "actions to protect, sustainably manage, and restore natural or modified ecosystems, that address societal challenges effectively and adaptively, simultaneously providing human well-being and biodiversity benefits" [34].
3. Nesshöver et al. [7] have referred to recently defined concepts in science, policy, and practice of the natural environment since the Brundtland report emerged in 1987: Biodiversity, natural capital, ecosystem services. They pointed out that the most recent entry of this discourse was "Nature-Based Solutions," a concept introduced specifically to promote nature as a means for providing solutions to climate mitigation and adaptation challenges, according with [34].
4. Lafortezza et al. [1] noted that the term NBS has largely evolved from previous concepts and/or principles (e.g., sustainability, resilience, ecosystem management, ecosystem services, coupled human and environment, and green (and blue) infrastructure), but with a central focus on the multiple cobenefits for the environment, economy, and society in urban (and to some degree rural) landscapes. The fundamental belief is that NBS can represent more efficient and cost-effective solutions than traditional approaches to development.
5. NBSs are increasingly applied to guide the design of resilient landscapes and cities to enable them to reach economic development goals with beneficial outcomes for the environment and society [1].
6. NBSs have emerged as a concept for integrating ecosystem-based approaches to address a range of societal challenges. NBSs directly address and contribute to increased urban resilience [35].
7. Niki [36] reported that NBSs are proliferating in European cities over the past years as viable solutions to urban challenges such as climate change, urban degeneration, and aging infrastructures. With evidence amounting about NBSs, there is a need to translate knowledge about NBSs to future policy and planning.
8. Dorst et al. [4] identified key NBS principles and compared them with those of EBA and GI. Key differences emerged: the NBS concept incorporates a broader array of interventions and a broader range of perspectives on what qualifies as "nature-based," and it is most explicitly oriented toward providing solutions to complex challenges.
9. One reason for the wide adoption of NBS is that the concept of nature providing solutions is simple in construct and logical for nonspecialist understanding. This has encouraged its uptake in policy, practice, and by the private sector [7].
10. Erik [37] documented that urban ecosystem services (provided by NBS) can play a critical role in reconnecting cities to the biosphere and reducing the ecological footprint and ecological debt of cities while enhancing resilience, health, and quality of life of their inhabitants. You may find in their *Classifying and valuing ecosystem services for urban planning* a useful classification and descriptions of ecosystem services (provided by NBS), based in turn on [38]. Table 27.1 is adapted from their classification:

27.3.1.1 Typology and examples of NBS

The recent publication by the European Commission, *Evaluating the Impact of Nature-Based Solutions/A Handbook for Practitioners* [32], provides a useful typology of NBS:

1. Type 1 NBS—minimal or no intervention in ecosystems, with objectives related to maintaining or improving delivery of ecosystem services within and beyond the protected ecosystems
2. Type 2 NBS—extensive or intensive management approaches seeking to develop sustainable, multifunctional ecosystems and landscapes in order to improve delivery of ecosystem services relative to conventional interventions
3. Type 3 NBS—characterized by highly intensive ecosystem management or creation of new ecosystems.

Castellar et al. [39] recently realized a comprehensive review of more than 250 NBS with the objective of setting a path of common understanding of NBS, resulting in taxonomy of 32 common NBS divided in two categories: NBS units for urban green spaces or green technologies and NBS Interventions when solutions are actions, measures, or techniques. Our main concern in this chapter are NBS units or NBSu, which are subdivided into NBS Spatial Units, when NBSs are urban green spaces and NBS Technological Units when solutions are a blue or green technology (see sidebar: "NBS Spatial Units, NBS urban green spaces and NBS Technological Units").

TABLE 27.1 Ecosystem services provided by Nature-Based Solutions.

Ecosystem services		Descriptions and references
Provisioning services	Food supply	Urban farming takes place in peri-urban fields, rooftops, backyards, and in community vegetable and fruit gardens [40] in Ref. [6].
	Water supply	Ecosystems play a fundamental role in providing cities with fresh water for drinking and other human uses and by securing storage and controlled release of water flows. Vegetation cover and forests in the city catchment influences the quantity of available water [6].
Regulating services	Urban temperature regulation	Urban blue and green space regulates local temperatures [41] in [6]
	Noise reduction	Urban soil and plants and trees can attenuate noise pollution through absorption, deviation, reflection, and refraction of sound waves [42–45] in [6]
	Air purification	Removal of pollutants operates through filtration of particulates through the leaves of trees and shrubs [46] in [6]
	Moderation of climate extremes	Ecosystems such as mangroves act as natural barriers that buffer cities from extreme climate events and hazards, including storms, waves, floods, hurricanes, and tsunamis [47–50] in [6]
	Stormwater runoff mitigation	By intercepting and retaining or slowing the flow of precipitation reaching the ground, trees (in conjunction with soils) can play an important role in urban hydrologic processes. They can reduce the rate and volume of stormwater runoff, flooding damage, stormwater treatment costs, and other problems related to water quality [26]
	Waste treatment	Ecosystems filter out, retain and decompose nutrients and organic wastes for urban effluents through dilution, assimilation and chemical re-composition [38] in [6]
	Biodiversity	Urban ecosystems are heterogeneous patchy mosaics of habitats where biodiversity in specific taxonomic groups can be surprisingly high [6]
	Climate regulation	Urban trees act as a sinks of CO_2 by storing excess carbon as biomass during photosynthesis [51–55] in Ref. [6]
Cultural services	Recreation	Green spaces in urban areas provide multiple opportunities for physical exercise, improved mental health, and cognitive development [56,57] in [6]
	Aesthetic benefits	
	Cognitive development	
	Place values and social cohesion	Because urban inhabitants develop affective links to the ecological sites of their cities, urban ecosystems also play an important role in sense of place [58] in [6]

NBS Spatial Units, NBS Urban Green Spaces And NBS Technological Units

The resulting list of 23 NBSu for an urban context is as follows:

1. Spatial arboreal units
   1. Street trees
   2. Urban forest
   3. Urban orchard
2. Spatial mixed vegetation units
   1. Large urban park
   2. Pocket garden/park
   3. Heritage garden
   4. Private garden
   5. Community garden
   6. Green corridors

NBS Technological Units:

3. Technological vertical units
   1. Green wall system
   2. Green façade
   3. Vertical mobile garden
   4. Planter green wall
   5. Vegetated pergola
4. Technological horizontal units
   1. Infiltration basin
   2. (Wet) Retention pond
   3. Rain garden
   4. Swale
   5. Constructed wetlands
   6. Extensive green roof
   7. Intensive green roof
   8. Semiintensive green roof
   9. Vegetated green pave

27.3.2 Benefits and costs of NBS

We will now refer to papers published by a number of research teams in the period between 2005 and 2017, regarding qualitative and quantitative benefits NBSs provide (and some of their associated costs). A scientific team made by Greg McPherson, James R. Simpson, Paula J. Peper, Scott E. Maco, and Qingfu Xiao made an analysis of the benefits (B) from energy use savings in buildings by tree shades, CO_2 sequestration by biomass, air quality improvement by pollutant deposition by trees (ozone, nitrogen dioxide, sulfur dioxide, and particular matter <10 micron diameter or PM_{10}), stormwater runoff benefits by estimating the amount of annual rainfall intercepted by trees and property value increase by analyzing the differences in sales prices of properties with and without trees, versus the costs (C) of planting, pruning, removal, and disposal of dead trees and other associated costs such as waste, infrastructural liabilities, administration, inspection, and other in five cities in the United States with populations ranging from 100,000 to 500,000 inhabitants. They found positive B/C ratios ranging from 1.37 to 3.09, with an average B/C ratio of 2.23 [25].

David J. Nowak and John F. Dwyer presented in 2007 their estimation of the associated monetary value of pollution removal by trees during nonprecipitation periods (dry deposition) in 14 cities of the United States and Puerto Rico. The pollutants removed included carbon monoxide (CO), nitrogen dioxide (NO_2), ozone (O_3), PM10, and sulfur dioxide (SO_2). They obtained absolute annual values ranging from 110,000 dollars in the smallest city to 8,071,000 dollars in the largest and relative values ($/ha) ranging from 378 to 663 $/ha with an average value of 504 $/ha per year [26].

In 2009, James C. Schwab edited the publication *Planning the Urban Forest. Ecology, Economy, and Community Development* for the American Planning Association.

The team made an analysis of (1) the environmental benefits of urban forests by treating stormwater runoff, shading, and cooling the urban heat island, reducing air pollution, storing and sequestering carbon, and providing wildlife habitat; (2) the social benefits of urban forests by contributing to health (the potential reduction of sedentarism and therefore obesity related costs, reduction of sun exposure and skin cancer, reduction of asthma incidence related to air pollution, and the influence of immediacy or direct view of green landscapes in patients recovery); by contributing to environmental justice, and community development, related to urban accessibility to green spaces; and (3) the economic value of the contributions urban forestry makes to real estate, both commercial and residential [28].

In 2015, a multidisciplinary team made by Thomas Elmqvist, Heikki Setala, Steven N. Handel, S. van der Ploeg, J. Aronson, J.N. Blignaut, E. Gomez-Baggethun, D.J. Nowak, J. Kronenberg, and R. de Groot published *The economic benefits and costs of trees in urban forest stewardship: A systematic review,* which is my recommendation of further reading in the subject of study. Their research resulted in the results of the empirical analyses and the resulting monetary benefits from urban ecosystem services in very diverse urban areas. The average obtained values in US$/ha/year of the following services provided by green spaces they estimated in urban areas are:

- 647 US$/ha/year for pollution and quality regulation;
- 395 US$/ha/year for carbon sequestration;
- 3125 US$/ha/year for carbon storage;
- 922 US$/ha/year for stormwater reduction;
- 1412 US$/ha/year for energy savings and temperature regulation;

- 6325 US$/ha/year for recreation and other amenity services, and
- 18,870 US$/ha/year for positive health effects [24], resulting in an average of 31,696 US$/ha/year.

Geoffrey H. Donovan wrote in 2017 the paper *Including public-health benefits of trees in urban-forestry decision making*. An important communication examining the relative importance of two biophysical benefits (reduced residential heating and cooling costs and reduced stormwater management costs) emphasizing on the nonbiophysical or social benefits urban trees provide, such as improvements in public health and crime reduction. Donovan found that the distribution of urban trees that would maximize the stormwater or energy-saving benefits of trees would not necessarily maximize the social benefits of trees. Therefore, considering social benefits could change the optimal location for tree planting and retention [23]. This information reminds us of the importance of careful and interdisciplinary design of NBS in cities. All these publications are evidence of how much NBS and their previous defined concepts can provide important qualitative and quantitative benefits as opposed to threats, hazards, and disturbances that cities are sustaining because climate change, pollution, etc., are a positive sign of their increasing value.

The observed values of NBS can attract more investing, thus improving the access of NBS projects, programs, and policies to more financial resources. Observed value ranges in $/ha/year increase in the measure that research advances and includes more potential benefit generators to the equation, such as public health and other social benefits. The average monetary values for pollution removal only, between the observations of Nowak & Dwyer in 2007 (504 $/ha/year) and the values obtained by Elmqvist et al. in 2015 (647 $/ha/year), are reasonably similar. We have to point out that by adding the benefits of carbon sequestration, carbon storage, stormwater runoff reduction, energy savings and recreation, and other amenity services, the value of the ecosystem services rises more than 25×, reaching 12,826 $/ha/year and much more (63×) when adding 18,870 $/ha/year by positive health effects (see Ref. [24]).

27.3.3 Implementation of NBS

We have to take a look ahead at the implementation of NBS, specifically in the urban realm—an emerging subject in theory and practice. For that purpose, you will find the recent work (2019) of a team integrated by Emmanuelle Cohen-Shacham, Angela Andrade, James Dalton, Nigel Dudley, Mike Jones, Chetan Kumar, Stewart Maginnis, Simone Maynard, Cara R. Nelson, Fabrice G. Renaud, Rebecca Welling, and Gretchen Walters: *Core principles for successfully implementing and upscaling Nature-based Solutions,* very useful. They argue that one reason for the wide adoption of the NBS concept in scientific literature and within governmental and nongovernmental policies and programs is that the concept of nature providing solutions is simple in construct and logical for nonspecialist understanding, a reason that will also facilitate its implementation [3].

They affirm that NBSs need clear and coordinated principles, on which evidence-based standards and guidelines for practitioners and decision-makers can be developed. In this paper, they present the definitions and principles underpinning the NBS framework:

1. NBS embrace nature conservation norms (and principles).
2. NBS can be implemented alone or in an integrated manner with other solutions to societal challenges (e.g., technological and engineering solutions).
3. NBSs are determined by site-specific natural and cultural contexts that include traditional, local, and scientific knowledge.
4. NBSs produce societal benefits in a fair and equitable way in a manner that promotes transparency and broad participation.
5. NBSs maintain biological and cultural diversity and the ability of ecosystems to evolve over time.
6. NBSs are applied at a landscape scale.
7. NBSs recognize and address the trade-offs between the production of a few immediate economic benefits for development and future options for the production of the full range of ecosystem services.
8. NBSs are an integral part of the overall design of policies and measures or actions, to address a specific challenge.

Recognition of nature conservation norms is a statement of respect for nature, as opposed to the *dominated* nature paradigm mentioned at the beginning of this chapter. Flexibility declared in Principle 2 is also very important, because almost any solution implemented in the city will have to be related to others. The holistic approach behind this principle is relevant for the integration with SCT (inter alia) and networking that we propose.

The frame that a site-specific definition provides in Principle 3 is essential to implement NBS, which have to be *adopted*, so to speak, by the natural and cultural contexts where they are to take place. Every city has a distinct set of complexities and contexts. There are no general solutions.

Principle 4 is very praiseworthy: NBSs are not only oriented to natural recovery, they require that the benefits are also for people, with a specific recommendation of fairness and equity (as in environmental justice), and moreover, advocating transparency and broad participation, features that are very helpful when implementing urban projects.

Principle 6, applying NBS at the landscape scale when applying them in the city, is quite appropriate because landscape-scale restoration includes large, contiguous, or fragmented areas (equal to or greater than several km^2) [59]; however, landscape scale is partially defined by the spatial scale (size and form, visual coherence) but also by the ecological scale and the functional scale (biophysical processes and human uses). Local landscapes are tangible to participants—in the field, in images, and in maps, and their visual character promotes the experience of local identity and sense of place. This inherent integration and *knowability* of the landscape scale ([60], in Ref. [61]) stimulates building common visions, shared decision-making, and collaborative invention of solutions ([62] in Ref. [61]), also employing landscapes as boundary objects can allow scientific concepts and terminology to be reframed in ways that are meaningful to multiple experts and stakeholders [61].

27.3.4 Assessment of NBS

The team integrated by Christopher M. Raymond, Niki Frantzeskaki, Nadja Kabisch, Pam Berry, Margaretha Breil, Mihai Razvan Nita, Davide Geneletti, and Carlo Calfapietra published *A framework for assessing and implementing the co-benefits of nature-based solutions in urban areas* in 2017 when the NBS concept was recently launched. They called attention to the fact that policy platforms were shifting their focus from ecosystem-based to NBS, broadly defined as solutions to societal challenges that are inspired and supported by nature and that NBSs result in the provision of cobenefits such as the improvement of place attractiveness, of health and quality of life, and creation of green jobs [8]. They proposed a framework for the assessment of NBS cobenefits that includes four dimensions:

1. Cobenefits for human health and well-being;
2. Integrated environmental performance;
3. Trade-offs and synergies to biodiversity, health or economy and
4. Potential for citizen's involvement in governance and monitoring.

27.3.4.1 Efficiency and effectiveness of NBS

When looking for concepts for evaluating projects, you generally find the concept of efficiency, which is the ratio of benefits and costs of a given project. Efficiency is a dominating feature for assessment. However, proposed concepts for assessment of NBS are also very much concerned on effectiveness, the actual performance of the project.

NBS effectiveness is defined as the degree to which objectives are achieved and the extent to which targeted problems are solved. In contrast to efficiency, effectiveness is determined without reference to costs [8].

27.3.4.2 Typology of societal challenges for assessment of impacts of NBS

The EINBS HB [32] presents a suite of recommended and additional indicators to evaluate NBS impact across the following 12 societal challenge areas:

1. Climate resilience
2. Water management
3. Natural and climate hazards
4. Green space management
5. Biodiversity
6. Air quality
7. Place regeneration
8. Knowledge and social capacity building for sustainable urban transformation
9. Participatory planning and governance
10. Social justice and cohesion
11. Health and well-being
12. New economic opportunities and green jobs

27.3.4.3 Evaluation principles for assessment of NBS

A proper assessment and evaluation of the targeted impacts are needed in a way that is relevant and useful: First, for immediate end users and second, to inform broader policy processes. Therefore, development of impact monitoring and evaluation plans should consider a few universal principles (see sidebar: "Nature Based Metropolitan Solutions MOOC by TU Delft"). Impact evaluation plans and its indicators must:

1. Be scientifically sound
2. Be practical and straight-forward
3. Use reference conditions and baseline assessment
4. Align with policy principles and reporting obligations
5. Be based on a transdisciplinary approach [32].

Nature-Based Metropolitan Solutions MOOC by TU Delft:

1. A Massive Online Open Course by the Technical University of Delft
2. This course establishes the priorities and highlights the direct values of including principles based on natural processes in urban planning and design.
3. This course will teach you about the design, construction, implementation, and monitoring of NBSs for urban ecosystems and the ecological coherence of sustainable cities.
4. Constructing smart cities and metropolitan regions with nature-based ecosystems will secure a fair distribution of benefits from the renewed urban ecology.

https://online-learning.tudelft.nl/courses/nature-based-metropolitan-solutions/

27.4 Smart City Technologies applied to NBS

My first experience with relating Urban Forests and ICT was some years ago as an official in a central sector of Mexico City. It was about developing a GIS of urban trees. The most recurrent kind of complaints the Urban Services municipal office received back then was related to nuisances from disservices of urban trees. The filing of complaints speed was faster than that of the correcting services the administration could offer, due in part to the very lean budget allocated for tree maintenance, but also because of the lack of information about the real size and kind of problems we were facing. Developing a GIS of trees taller than 1.8m in the sector allowed the administration to have an accurate diagnosis of the state of individual trees and a more precise view of the needs of maintenance of the urban forest. It was not difficult, once having the quantitative, qualitative, and location of trees information, to allocate enough funds to meet a policy of improvement of the urban forest, with a positive impact on the services and even on governance. I found out that cost-effectiveness and governance enhancement by social participation were at least two benefits that technology contributed so much to achieve.

Technology applied to NBSs has advanced very fast in less than two decades [11]. noted that increasingly complex research questions and global challenges (e.g., climate change and biodiversity loss) are driving rapid development, refinement, and uses of technology in ecology. They called this trend *technoecology* [11]. However, since we will be locating technology and ecology for resilience in the urban realm, with the implications of including the technological networks of a complex system (the city) to the formula, it seems more appropriate, at least in the context of this chapter and this handbook, to name this concept SCT.

This third section of the chapter is a brief exploration of current and emerging SCT applied to NBS with the vision that this formula (SCT + NBS) may further improve the NBS balance of efficiency, effectiveness, and other resilience and sustainability-related indicators. For this exploration, I found three main sources of updated reviews of technologies, here in chronological order:

1. In 2018, the research team of Blake Allan, Dale Nimmo, Daniel Ierodiaconou, Jeremy VanDerWal, Lian Pin Koh and Euan Ritchie, wrote in *Ecosphere*, *Futurecasting ecological research: the rise of technoecology*, where they highlight recent ground-breaking and transformative technological advances for studying species and environments, with the potential to revolutionize ecology by providing next generation data, and particularly by integrating with each other. They claim that this rise of technoecology will be fostering increased interdisciplinary collaboration [11], an idea in which I concur. They present a list of new technologies and its definitions in ecology and environmental science (in general), of which I have made a selection of those more adequate to NBS (in particular).
2. In 2019, a team integrated by Sophie Nitoslawski, Nadine Galle, Cecil Konijnendijk and James Steenberg, assembled a summary of state of the art technologies and corresponding definitions, focusing on the intersection between urban forest management and smart city planning and outlined how smart city trends and related technologies have been and can be applied to urban forestry and green space planning [13]. The team proposed a good number of technological trends and concepts to be studied and three turning points in the near future: the use of wireless sensor networks (WSN) and associated technologies such as cloud-based platforms and 5G, the recognition that urban forests are socio-ecological systems and the future self-regulation and self-organization of urban forests ecosystems enabled by smart technologies [13].
3. The European Commission published in May 2021, the *Evaluating the Impact of Nature-Based Solutions — A Handbook for Practitioners.* The European Union has taken a leadership in policies and financing of plans, programs and projects oriented to resilience from climate change. The very concept and framework of NBSs is welcomed as a consequence of this leadership. Now, with the publication of the handbook, I can foresee a normative order that could set the basis for a global standardized methodological understanding. In particular, Section 7 of the EINBS Handbook offers an extensive overview of the many type of data, data sources and data generation techniques for NBS monitoring and impact assessment [32].

27.4.1 A Smart City Technologies *plus* Nature-Based Solutions conceptual chart

The purpose of presenting a conceptual chart here (Fig. 27.1) is similar to that of the early charts of navigators exploring new territories: to have at least some orientation in the already quite complex and dynamic territory of technology applied to NBS in urban contexts. Because of the dynamics of the information in it, the graphic is more a chart than a table or an item list. You may track the route of any given project process starting from the type of data, to methods of collection, to emerging trends in data collection, to emerging technologies, and finally arriving to data integration, by drawing a line or as many lines as the complexity of the project requires. I suggest the use of the conceptual chart as an exercise for imagining the already traced routes and the ones to come in the near future.

Data required to assess a given NBS will most likely come from three large fields: environmental data, which can be mostly natural-environmental data; but, also built-environment data (buildings and infrastructures are closely related to urban NBS); and socioeconomic, demographic, and behavioral datasets, and data sources for

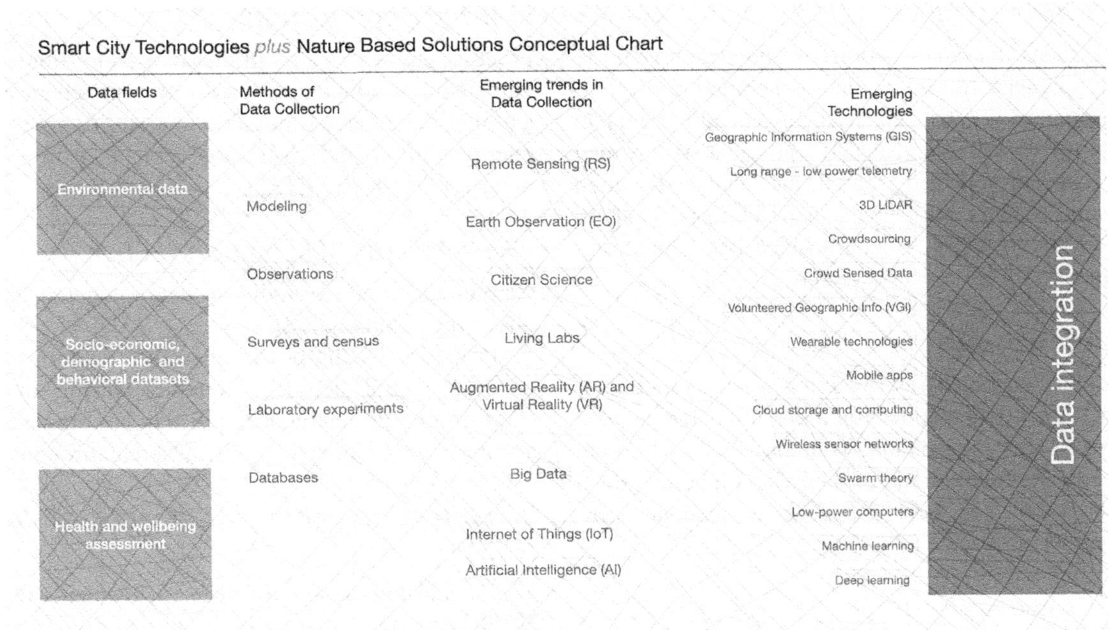

FIGURE 27.1 A Smart City Technologies *plus* Nature-Based Solutions conceptual chart.

the assessment of changes to health and well-being—for there is an increasing recognition of NBS cobenefits and influential determinants of human health and well-being [63–65]; all three in Ref. [32].

27.4.2 Glossary of defined concepts of Smart City Technologies

We are presenting next, a quick-reference glossary of defined concepts of emerging trends and technologies in alphabetical order.

27.4.2.1 Artificial intelligence (AI)

The simulation of human intelligence processes (such as learning and reasoning) by computer systems [13]. Artificial neural networks (see sidebar: "Intelligence for Ecosystem Services (ARIES)/probabilistic model") have been used to model and predict urban tree mortality [66,67]; both in Ref. [13].

Intelligence for Ecosystem Services (ARIES)/ probabilistic model

A networked software technology that redefines ecosystem service assessment and valuation for decision-making, maps out natural capital, natural processes, human beneficiaries, and service flows to society, as a new way to visualize, value, and manage the ecosystems on which the human economy and well-being depend on, and quantify the benefits that nature provides to society.

http://aries.integratedmodelling.org/

ARIES is meant to enable simple use of complex models through artificial intelligence [32].

27.4.2.2 Augmented reality (AR)

Augmented reality (AR) is where the real environment is enhanced with computer-generated, virtual objects, often for visualization purposes [68]; in Ref. [13]. To date, applications for urban forest and green space management have generally informed landscape design through public values elicitation.

Further applications of VR and AR could test physiological responses to varying urban green space designs. Researchers in Sweden have used VR to test stress responses and recovery (e.g., cortisol, heart rate) in virtual natural environments in order to elicit the impacts of nature sounds [69]; in Ref. [13], paving the way for management decisions informed by relationships between green spaces and human health.

27.4.2.3 Big data

Big Data are about the availability of data as a source of information and, ultimately, knowledge. The proliferation of information and communication technologies has led to a data surge. Big Data relate to the development of new tools and practices in order to collect, analyze, and work with this digital information.

The growing interest in big data has created a new community around digital pioneers, which represents a paradigmatic shift in how a diverse set of stakeholders interacts. Big data should therefore be understood as a phenomenon bringing together a large variety of stakeholders that individually or collectively engage in the processes that determine how data are collected and used for, among other things, policy goals [70].

27.4.2.4 Citizen science

A research focus that enables citizens and stakeholders to be actively engaged in science data generation and monitoring programs [32]. Citizen science generally refers to "partnerships between scientists and non-scientists in which authentic data are collected, shared, and analyzed", often with broader educational goals and impacts [71] in Ref. [13]. More recently, citizen science has gained significant traction in ecological research.

27.4.2.5 Crowdsourcing

Crowdsourcing encompasses obtaining a large amount of data from a crowd of people (or more often the general public) that shares information, voluntarily. This is often done through the Internet and/or using smartphones. Each single data supplied is then aggregated to generate a cumulative dataset [32].

27.4.2.6 Crowd-sensed data

Crowd-sensed data describes data. Data are specifically collected and shared by a large number of citizens through different types of devices, such as mobile phones, wearable sensors, or vehicles (e.g., sensors mounted on bicycles to measure air temperature or air quality parameters) [32].

27.4.2.7 Data integration

Data gathered from different sources will often have to be analyzed in combination and integrated together in order to provide valuable insights on the impacts and co-benefits of an NBS intervention in comparison to a baseline scenario [32]. Spatial modeling and spatial analysis may represent an effective strategy for the monitoring and/or planning of NBS, since it allows to integrate, analyze and visualize different data types. For example, using remote sensing data under a GIS environment, it is possible to provide geo-referenced information on the shape, size and distribution of different land-use classes of the urban environment [72] in Ref. [32]. This allows monitoring of urban growth (area change, structures, land consumption, soil sealing) and land cover/land-use changes (loss of agricultural area, wetland infringement, loss of areas important for biodiversity, spatial distribution of inner-urban green and open spaces, and natural areas) as well as mapping of various environmental parameters (data important for urban climate, access to and distribution of open space, calculation of sealed surfaces) [32]. In the EI NBS Handbook (2021) you may find (Table 7.10, page 333) a comprehensive review of examples of data integration for NBS projects, with technological approaches and their corresponding references.

27.4.2.8 Deep learning

Processes by which algorithms are used to simulate how the human brain adaptively learns. Recent advances in deep learning have made it possible to quantify urban metrics at fine resolution and over large extents using street-level images [73]; in Ref. [13].

27.4.2.9 Earth observation (EO)

Commonly used to indicate data gathered from Earth observing satellites [32].

Propelled by the powerful feature learning capabilities of deep neural networks, remote sensing image scene classification driven by deep learning has drawn remarkable attention and achieved significant breakthroughs [74].

27.4.2.10 Geographic information systems (GIS)

The spatial interactions of different elements of NBS and NBS with the environment are mostly managed by geographic information systems (GIS). GIS can be used to provide input variables required by simulation models and yield visualization and analysis of output data [32].

27.4.2.11 Internet of things (IoT)

IoT is a network of devices that can communicate with one another, transferring information and processing data [11]. In an IoT environment, devices can be aggregated according to their geographical position and also assessed by applying analyzing systems. Sensor services for gathering specific data are utilized with some ongoing projects regarding the monitoring of each cyclist, vehicle, parking lot, and so forth [75].

27.4.2.12 Light detection and reading (LiDAR)

Also known as 3D LiDAR, it is an accurate measurement of a 3D ecosystem structure [11]. LiDAR sensors, for example, allow gathering high-resolution elevation data, which can be applied for measuring the heights of trees or buildings [32]. Remote sensing technology (e.g., LiDAR, hyperspectral imagery), aided by machine learning, has mapped and assessed the species and structure of individual trees [76]; in Ref. [13].

27.4.2.13 Living labs

While a shared definition of urban living labs has not yet been agreed upon, they are generally understood to involve collaborative research and urban development activities undertaken alongside the intended end users, exploiting experimental platforms and/or approaches in real time [77].

27.4.2.14 Long-range, low-power telemetry and low-power computers

Long-range and low-power telemetry is a description of low-voltage, low-amperage transfer of data over several kilometers [11]. Also, small computers have the ability to connect an array of sensors and, in some cases, run algorithms and statistical analyses [11].

27.4.2.15 Mobile apps

Mobile technology facilitates participation in interactive platforms and projects. For citizen science, the deployment of mobile apps means that participants can contribute observations in real time. To report relevant sightings immediately not only increases the quality of the provided data (in terms of timeliness) but also improves the connection of the observer with the subject and its environment, as the observer is immersed in the field and is aware of the context of their observations [78].

27.4.2.16 Modeling

Modeling is a critical and often compulsory aspect of NBS impact assessment, it allows simulating the efficiency of one or more components of NBS, and to monitor and evaluate progress toward its goals [32]. Modeling NBS addresses the representation of processes that occur in the real world in space and time. The processes resulted or caused by NBS transform the environment through time and can be mostly described by dynamic models based on differential equations [32]. In the EI NBS Handbook (2021) you may find (Table 7.9, page 327), a comprehensive review of studies of the impacts of NBS, which show how numerical simulations and modeling can be applied [32].

27.4.2.17 Remote sensing (RS) and Swarm Theory

Remote sensing is a technique of observing and collecting information about an object or phenomenon from a distance, by means of sensors that are not in physical contact with the object of investigation [32]. Also, Swarm Theory is the autonomous, but coordinated use of multiple unmanned sensor platforms to complete ecological surveys or tasks without human intervention [11].

27.4.2.18 Volunteered geographic information (VGI)

VGI is a type of crowdsourced information where data have spatial information attached. The crowdsourced data are usually collected in, or converted to, a mapped form with spatial (and temporal) dimensions. Leading examples for this are OpenStreetMap (OSM) or the use of online mapping and social media such as Twitter to communicate information about natural disaster events (e.g., hurricanes and earthquakes) [32].

27.4.2.19 Wearable technologies and wireless sensor networks

IoT-enabled *wearables* are smart devices that can be worn as external accessories, embedded in clothing and garments, implanted in the body, or even adhered to or tattooed on the skin. These devices are able to connect to the Internet in order to collect, send data, and receive the information that can be used for smart decision-making. These wearables are becoming an increasingly important part of IoT technology and their development is moving from being simple accessories to more specialized and practical applications [79]. Wireless sensor networks have been deployed in greenhouse settings to measure and regulate environmental parameters [80,81]; both in Ref. [13].

27.5 Viability of NBS

NBSs are financially viable per se, when understanding **viability** as a concurrence of policy compliance, fundability, and dynamic efficiency. As instruments of public policies, concrete projects are subject to be funded according to their compliance with general or specific policies from local, state or federal governments; with their *fundability* and a positive benefit/cost ratio (efficiency). SCT may enhance NBS viability by (inter alia):

1. Informing programs or projects performance in real time to policy makers
2. Adding technical solvency to the project promoted
3. Providing more accuracy in the benefit and cost calculations
4. Reducing operational risks of projects

27.5.1 Policy compliance

It is not hard to think about a near future in the world where the axis of general policies will be to address resilience from global warming, including reduction or elimination of its main cause, burning fossil fuels to the atmosphere. I believe that a positive trend of developing local, national, and most importantly, global climate change resilience policies is on its way and will keep on growing if we are capable, as a civilization, to understand and address the decisive implications for our future. Urban projects that derive from urban resilience policies and projects that comply with urban resilience policies independently of which policy or set of policies they come from, will have a much better chance to be viable projects in the near future; however, they still have to prove to be fundable.

27.5.2 Fundability and solvency

In my own experience, the concept of *fundability*, i.e., the capability of a project to be funded, depends on a mix of three conditions: solvency of the project's promoter, access to financial resources and the acceptability of the project's risk. A promoter should prove to have economic and technical solvency to promote a given project. Promoters are in general, public entities or large (and solvent) private entities. A mixture of public and private entities can also play the role of promoter, but likewise, it should demonstrate economic and technical solvency. Technical solvency includes a reasonably interdisciplinary high professional quality of the teams promoting, including the teams designing, building, and communicating the project.

27.5.2.1 Access to financial resources

Accessibility to financial resources may come from a number of different sources. We can highlight taxes (general and labeled taxes such as carbon pricing), impact investments and bonds (general and labeled such as green bonds). I will here make a brief reference of three of the most emergent sources of funds for urban resilience projects, emphasizing on green bonds, which I believe are going to be the most recurrent source for *climate finance*:

27.5.2.1.1 Carbon pricing and impact investments

Most of the country and subnational carbon pricing programs take one of two forms:

1. A carbon tax or levy, which establishes a price on a measurable unit of greenhouse gas emissions, and taxes emitters a fixed amount per unit produced.
2. A cap-and-trade program, where the government sets a cap on allowable GHG emissions by each covered entity and allocates allowances equal to that cap that can be traded, through which a market price is established. Currently, nearly 40 countries either have a carbon tax or a cap and-trade model [82].

Impact investments are investments made with the intention to generate positive, measurable social and environmental impact alongside a financial return. Impact-investments can be made in both emerging and developed markets and target a range of returns from below-market to market-rate, depending on investors' strategic goals [82].

27.5.2.1.2 Green bonds

Like any other bond, are fixed income securities for raising capital to finance low carbon investments, or in general environmentally friendly investments [83]. Green bonds are debt instruments used to finance green projects that deliver environmental benefits [84].

Since 2013, the market for labeled green bonds has expanded dramatically almost doubling after each subsequent year: from \$11 billion in 2013 to more than triple, \$37 billion, in 2014; in 2015 the market experienced a plateau, with aggregate issuance just over \$40 billion; in 2016 it surpassed \$85 billion (more than 100% growth), and in 2017 reached \$155.5 billion (an 83% growth) [83]. Remarkable growth of the market continued with \$167.3 billion in 2018, \$255 billion in 2019, and in 2020, when reaching \$270 billion. This trend is expected to continue in the next years.

27.5.2.2 Risk acceptability criteria

A third important factor in the project's fundability is the plausible acceptance of the risk that the project poses. Financial risks include credit risk, liquidity risk, and operational risk. SCT + NBS will reduce at least the operational risk of projects. Economic risk from an NBS project will vary with the type of solution, targeted resilience outcome, level of investment, scale of actions, and the life span of the NBS. Performance measures of an NBS will vary with time and scale, leading to shifts in the level of resilience, and therefore risk mitigation of time [85].

27.5.3 Dynamic efficiency: a substantial positive benefit–cost ratio over time

The system-wide transformation needed to address the climate crisis will require substantial funding, but those actions will ultimately save much more money compared to business as usual [82]. A central part of the viability of a project depends on demonstrating a substantial positive benefit–cost ratio over time, generally by means of a cost–benefit analysis (CBA). CBAs are expected to be reliable, rigorous analytical tools for the decision-makers to back up a positive or negative verdict of whether or not funding a project.

CBA relates to any public decision that has an implication for the use of resources [86].

27.5.4 Efficiency and dynamic efficiency

Providing a positive (efficient) result from a reliable cost-benefit analysis is an important task in any project promotion, and moreover, regarding urban resilience projects, time is a very important factor in the analysis. Resilient urban projects offer results in the long term, thus the need of demonstrating efficiency over time. This is called dynamic efficiency.

The ThinkNature (https://platform.think-nature.eu/) project, recently edited a *Nature-Based Solutions Handbook* [85]. You will find in Section 6, *Financing and Business*, by Heather Elgar, Neil Coles, Juraj Jurik, Natasha

Mortimer, and Steven Banwart, a recommended reading regarding the economic opportunities of NBS, economic risks, and financial instruments.

27.6 Added value of SCT *plus* NBS

The fifth part of this chapter is a corollary of the four previous parts. It outlines how NBS can be even more valuable by incorporating SCTin every step of cocreation, coimplementation, and comanagement of NBS, including the engagement of stakeholders, assessment, planning, implementing, monitoring, evaluating, and most of all by enhancing NBS effectiveness in connecting nature to cities and cities to nature.

27.6.1 Effectiveness

NBS effectiveness (see sidebar: "A New Evaluation Framework For Nature-Based Solutions (NBS) Projects") is defined as the degree to which objectives are achieved and the extent to which targeted problems are solved. In contrast to efficiency, effectiveness is determined without reference to costs [8] in Ref. [32].

A new evaluation framework for Nature-Based Solutions (NBS) projects

Barbara Sowińska-Świerkosz and Joan García published recently (2021) *A new evaluation framework for nature-based solutions (NBS) projects based on the application of performance questions and indicators approach.* They have identified by means of a literature review of 43 papers related to NBS performance, seven main features that help us to get a clearer characterization of NBS effectiveness:

1. Stakeholders' participation
2. Policy and management capability
3. Economic efficiency
4. Synergies and trade-offs
5. Adaptation to local conditions
6. Performance in the long term
7. Adequate spatial size

[87].

In cases when NBS interventions combine solutions to achieve different impacts, it is important to ensure that the impacts and its cumulative effects are integrated throughout the process rather than simply synthesized at the end [32]. This makes the whole analysis of their effects and impacts complex, increasing uncertainty with respect to data collection, and precisely in this aspect is where SCT can prove to be most useful, not only by helping in the analysis of the NBS impacts but also by providing positive impacts themselves. SCT are useful tools for monitoring the performance of NBS against expected results.

27.6.2 Stakeholder engagement and social equity

Any societal problem or issue is likely to be affected by or have effects on several groups of stakeholders [7]. A list of NBS stakeholder groups, including but not limited to:

1. Policymakers, urban planner, and other public agents
2. Members of the scientific community
3. Businesses, nature-based enterprises, impact investors, and industries
4. Nongovernmental organizations and civil society at large [32].

The participative process may be more or less complex depending on the issues to be considered. In particular with NBS that would involve important trade-offs. A participatory process becomes critical, and can be especially pertinent, when their aim is to better connect people with their local natural resources. There is now much guidance and many ideas about how to facilitate collaborative efforts and engage relevant stakeholders [7].

27.6.2.1 Social equity

Among the actors to be involved in the decision-making process, citizens play a major role, and they should be informed, empowered, and eventually involved in NBS actions, planning, and implementation. Indeed, NBS initiatives focused on citizens can definitely contribute to social cohesion, replenishing the connection between citizens and nature, increasing their awareness of the multiple benefits of NBS, and raising a public request for a better environment [85]. SCTs have been useful in spatial planning for a number of years, addressing at least two of the three dimensions of social equity [88]: distributional and procedural. SCTs provide an ample variety of tools for promoting and managing citizen participation in plan development and in governance development and improvement.

27.6.3 Synergism with other urban infrastructures

Green and blue infrastructures are conceptually, part of NBS. However, it is important to note possible synergies of SCT + NBS with *other* infrastructures: mobility infrastructure, considering that a good number of NBS will take place in the public space in the city, most of it now taken for the purpose of mobility; gray infrastructure, which includes water, energy, and sewage infrastructure but also buildings are subjects of synergism with SCT + NBS. It is not difficult to think of a number of NBS as a set, connected by means of SCT, which in turn can connect to other infrastructures, providing important synergies to the city.

27.6.3.1 Mobility infrastructure

Occupying in many cases up to one half of the total urban space, the spaces of mobility, the thoroughfares of a city are the main spatial resource for developing NBS. There are three main kinds of mobility infrastructure that may have synergies with SCT + NBS:

27.6.3.1.1 Active mobility

NBS are devised to reduce the effects of global warming, among other hazards, disturbances, and threats, but they can also attack the cause directly, when in synergy with active mobility infrastructure. This kind of infrastructure has been gaining space in the urban thoroughfare. Bicycle lanes and better and wider sidewalks are now part of many plans, programs and projects all over. Active mobility makes us citizens less car-dependent, which reduces carbon emissions to the atmosphere. SCT + NBS embedded in active mobility infrastructure may, for instance, increase the canopy cover and the water absorbing capacity by means of urban trees, pocket parks, rain-gardens or swales, and the smart technologies used to monitor these NBS performances may also help in the stewardship of the natural solutions and the active mobility infrastructures at the same time.

27.6.3.1.2 Transit and private motorized vehicles

Technological networks to monitor NBS, linked to the spaces and corridors of transit mobility like street trees, green corridors, green walls, and roofs in the stations, and infiltration basins inter alia, can monitor and inform operators and managers of both infrastructures as well as transit users. The car as we know it, is going to have to adapt to new electric technologies, reducing and eventually stopping burning fossil fuels to the atmosphere. It will also lose its dominance in the public space, giving it up to NBS in thoroughfares such as street trees, pocket parks, corner planters, rain gardens, swales, and vegetated grid pavements inter alia. These *street* NBS will be connected to a technological network for management and monitoring performance that may as well serve the car or bicycle users, operators, and managers in emerging shared modalities (car-sharing and bicycle-sharing systems).

27.6.3.2 Gray infrastructure

SCT linked to NBS to monitor performance can also be linked to gray infrastructures via wireless sensor networks (for instance), with important savings in the operation of water networks in the city. IoT technologies will allow that NBS devised for water treatment to connect directly to water infrastructure. The same technological networks are capable of monitoring energy consumption in private buildings and public space. SCTs linked to NBS are capable to reduce CSOs or Combined Sewer Overflows.

27.6.3.3 Buildings

Private and public buildings can be connected to the network of NBS by the same SCT.

Buildings will have more active systems addressing reduction of energy consumption, energy and water production, water recycling, etc. However, passive systems in buildings such as trees, orchards, private gardens, green walls, green facades, vertical mobile gardens, planter green walls, vegetated pergolas, rain gardens, extensive and intensive green roofs, and other NBSs are more likely to interact or to be networked with other NBS in the public space of cities. Networking can be achieved by means of SCT such as wireless sensor networks, low-power computers, and other technologies.

27.6.4 Synergism with other urban systems

Synergies are significant in respect to sustainability objectives, as long as there is cooperation among diverse societal groups entailing different expertise and the assignment of roles to partners is flexible and complementary toward the best support of the adopted goals in each case (Frantzeskaki et al. [89] in [85]). The conceptual network of SCT plus NBS can also be connected to a number of other urban systems, providing important benefits to the city.

27.6.4.1 Urban revenue

A very important and convenient consequence of implementing NBS in cities is the rise in property values. As an example, Portland, Oregon had a 15.3M increase in tax revenues from a tree planting and maintenance project with a budget of 1.28M. A 12:1 ratio [90]. Surplus value capture from NBS projects can be supported by the use of SCT as well.

27.6.4.2 Urban security

Crime reduction is another convenient consequence of implementing NBS in cities [23]. Moreover, the network of NBS and SCT is capable of contributing to security surveillance in cities.

27.6.4.3 Urban services

Urban services such as urban lighting, waste management, and sewage management can also obtain operational and management benefits from the SCT + NBS network (see sidebar: "Evaluating the Impact of Nature-Based Solutions: A Handbook for Practitioners'), but especially of course, the green areas management services in cities, which can also be defined as *Smart Urban Forests*.

Evaluating the Impact of Nature-Based Solutions: A Handbook for Practitioners.

The Handbook aims to provide decision-makers with a comprehensive NBS impact assessment framework, and a robust set of indicators and methodologies to assess impacts of nature-based solutions across 12 societal challenge areas. Representatives of 17 individual EU-funded NBS projects and collaborating institutions have developed indicators collaboratively. They reflect the state of the art in current scientific research on impacts of NBSs and valid and standardized methods of assessment, as well as the state of play in urban implementation of evaluation frameworks.

https://ec.europa.eu/info/news/evaluating-impact-nature-based-solutions-handbook-practitioners-2021-may-06_en.

27.7 Propositions for an upcoming agenda: nature-based smart urban projects

NBSs, enhanced by SCTs are a viable and valuable resource for resilience. They will materialize by the promotion of Nature-Based Smart Urban Projects (NBSUP). NBSUP will multiply, because of the alignment of urban resilience policies and funding for research, planning, design, implementation, and assessment, as it is already happening in the European Union.

27.7.1 Network approach

NBSUP will materialize in the "space of places" and at the same time will be networked with the "space of flows," connected to both kinds of space by interfaces. A network approach is therefore needed to analyze, plan and design, implement and assess NBSUP. Complementarity constitutes a *certain coherence*, which makes that the whole is more than the sum of its parts [91].

27.7.2 Integration of disciplines

NBSUP needs to be integrated and promoted by interdisciplinary teams. They will overcome disciplinary fragmentation. *It follows that the fragmentation of knowledge and the isolation of bits of information can actually become a form of ignorance, unless they are integrated into a broader vision of reality* [92].

27.7.3 Integration to urban networks

The main interface of NBSUP with urban networks will materialize in the spaces of mobility, the thoroughfares, which presently occupy up to one half of the total urban space and are currently dominated by the automobile but where active mobility has started to gain terrain. Active mobility spaces will have important synergies with NBSUP; in fact, they will be effortlessly integrated. Private spaces such as building facades, walls, and gardens will also be part of this new emerging network of nature-based urban nodes and connections. *Integral urbanism considers means of integrating design with nature, center with periphery, local character with global forces, the various design/build professions, and people of different ethnicities, incomes, ages, and physical abilities* [14].

27.7.4 Research by design

The integration of SCT plus NBS with the urban networks is still a possible and desirable future. However, it will need a good amount of promotion to become a probable future. We must then undertake some action: exploring and promoting possibilities, making them probable.

How do we explore possibilities that are improbable? These must be designed. [93].

27.8 Summary

There is a global concern for achieving resilience from consequences of unsustainable development. Urban Resilience is nowadays positioned in the intersection of global and local policies, public and private funding, and urban complex networks, not only for recovering but also hopefully for improvement from the current conditions of environmental breakup, social exclusion, and economic inequality. In recent years, the European Union has led a robust effort of developing policies and providing funding for research, design, implementation and assessment of actions for urban resilience. The comprehensive concept of NBSs has emerged recently as a result of this effort.

The Smart City technological approach to urban development has also recently started to connect with sustainable and resilient aims and goals. SCTs operate in the *space of flows* of the complex urban networks and are starting to provide a *plus* to NBSs, as we could find out through the research for this chapter. Presently, SCTs are making NBSs more viable and valuable by enhancing NBS effectiveness, facilitating stakeholder engagement and providing synergism with other infrastructures and urban systems. Integrated, interdisciplinary teams will promote this convergence of Nature, Technology, and Urban Networks.

Finally, let us move on to the real interactive part of this chapter: review questions/exercises, hands-on projects, case projects, and optional team case project. The answers and/or solutions by chapter can be found in Appendix G.

27.9 Chapter review questions/exercises

True/false

1. True or False? A new appreciation of nature, innovation, technology, and an integrative vision is necessary to face global and local (urban) challenges on time.
2. True or False? An old paradigm of integration of disciplines has emerged: development, along with its social end economic edges, has since been pondered as one with the environment.
3. True or False? We must draw urgent attention to the decisiveness of the unsustainable way we have been building cities since the 1960s when the world's urban population barely reached 1 billion.
4. True or False? Considering that urban systems are complex systems far from equilibrium, we have to point out that the two kinds of policies that better respond to the complex and dynamic features inherent to cities include those with a focus on recovery as an opportunity for improvement and those with a focus on proactive strategies to cope with dynamic change and especially on the evident climate change adaptation and mitigation.
5. True or False? The observed values of NBS can attract less investing, thus improving the access of NBS projects, programs, and policies to more financial resources.

Multiple choice

1. We have to take a look ahead at the implementation of NBS, specifically in the urban realm—an emerging subject in theory and:
   a. Development
   b. Practice
   c. Work
   d. Investment
   e. Policy
2. When looking for concepts for evaluating projects, you generally find the concept of ____________, which is the ratio of benefits and costs of a given project.
   a. Infrastructure
   b. Efficiency
   c. Prosperity
   d. Economic effects
   e. Smart city policies
3. A proper assessment and evaluation of the targeted impacts are needed in a way that is relevant and useful: First, for immediate end users and second, to inform broader:
   a. Machine processes
   b. Machine to Human processes
   c. Universal processes
   d. Transdisciplinary processes
   e. Policy processes
4. Technology applied to Nature-Based Solutions has advanced very fast in less than two decades.
   a. Three decades
   b. Four decades
   c. Five decades
   d. Two decades
   e. Six decades
5. Data required to assess a given NBS will most likely come from three large fields: environmental data, which can be mostly natural-environmental data but also built-environment data (buildings and infrastructures are closely related to urban NBS) and socioeconomic, demographic, and behavioral datasets, and data sources for the assessment of changes to health and well-being—for there is an increasing recognition of NBS cobenefits and influential determinants of human health and well-being.
   a. Ecosystems
   b. Intelligence
   c. Benefits
   d. Services
   e. Well-being

Exercise

Problem

Look at an alignment between each of the NBS assessment schemes and a Sustainable Development Goal indicator framework that is normally mapped with a particular emphasis on consistency with city-level framework indicators for each Sustainable Development Goal subobjective.

Hands-on projects

Project

Do research: Analyze a classification into no-tech, low-tech, and high-tech green, which supports urban designers, to better utilize the ability of green elements to effectively manage water flows in different urban settings.

Case projects

Problem

How do smart city technologies change the nature and economics of infrastructure?

Optional team case project

Problem

Identify key NBS principles and compare them with those of the Ecosystem-Based Adaptation (EBA) and Green Infrastructure (GI).

References

[1] Lafortezza R, et al. Nature-based solutions for resilient landscapes and cities. Environmental Research, Academic Press; 6 Dec. 2017. www.sciencedirect.com/science/article/pii/S0013935117317115.

[2] Badiu DL, Nita A, Iojă CI, Niţă MR. Disentangling the connections: a network analysis of approaches to urban green infrastructure. Urban For Urban Green 2019;41:211–20. https://doi.org/10.1016/j.ufug.2019.04.013.

[3] Cohen-Shacham E, et al. Core principles for successfully implementing and upscaling nature-based solutions. Environmental Science & Policy, Elsevier; May 13, 2019. www.sciencedirect.com/science/article/pii/S1462901118306671.

[4] Dorst H, et al. Urban Greening through nature-based solutions – key Characteristics of an emerging concept. Sustainable cities and society. Elsevier; May 23, 2019. www.sciencedirect.com/science/article/abs/pii/S2210670719302550.

[5] Frantzeskaki N. Seven lessons for planning nature-based solutions in cities. Environmental science & policy. Elsevier; 3 Jan. 2019. www.sciencedirect.com/science/article/pii/S1462901118310888.

[6] Gómez-Baggethun E, et al. Urban ecosystem services. In: Elmqvist T, et al., editors. Urbanization, biodiversity and ecosystem services: challenges and opportunities: a global assessment; 2013. https://doi.org/10.1007/978-94-007-7088-1_11.

[7] Nesshöver C, et al. The science, policy and practice of nature-based solutions: an interdisciplinary perspective. Science of The Total Environment, Elsevier; 2 Dec. 2016. www.sciencedirect.com/science/article/pii/S0048969716325578.

[8] Raymond CM, Frantzeskaki N, Kabisch N, Berry P, Breil M, Nita MR, et al. A framework for assessing and implementing the co-benefits of nature-based solutions in urban areas. July 26, 2017. Retrieved from, https://www.sciencedirect.com/science/article/pii/S1462901117306317.

[9] Sanchez A, van der Heijden J, Osmond P. The city politics of an urban age: urban resilience conceptualisations and policies. Palgr Commun 2018;4:25. https://doi.org/10.1057/s41599-018-0074-z.

[10] Yigitcanlar T, Kamruzzaman M. Does smart city policy lead to sustainability of cities? Land Use Policy, Pergamon; February 21, 2018. www.sciencedirect.com/science/article/pii/S0264837717314667.

[11] Allan BM, Nimmo DG, Ierodiaconou D, VanDerWal J, Koh LP, Ritchie EG. Futurecasting ecological research: the rise of technoecology. Ecosphere 2018;9(5):e02163. https://doi.org/10.1002/ecs2.2163.

[12] Bibri SE, Krogstie J. Smart sustainable cities of the future: an extensive interdisciplinary literature review. Sustainable cities and society. Elsevier; March 6, 2017. www.sciencedirect.com/science/article/pii/S2210670716304073.

[13] Nitoslawski SA, et al. Smarter ecosystems for smarter cities? A review of trends, technologies, and turning points for smart urban forestry. Sustainable cities and society. Elsevier; 7 Aug. 2019. www.sciencedirect.com/science/article/pii/S2210670719307644.

[14] Ellin N. Integral urbanism: a context for urban design. In: Pickett, et al., editors. Resilience in ecology and urban design: linking theory and practice for sustainable cities; 2013.

[15] UN. Brundtland Report, Report of the World Commission on Environment and Development. 1987. https://www.are.admin.ch/are/en/home/media/publications/sustainable-development/brundtland-report.html.

[16] World Cities Report 2020. The value of sustainable urbanization: UN-habitat. 2021. Retrieved from, https://unhabitat.org/world-cities-report-2020-the-value-of-sustainable-urbanization.

[17] Resilience: UN-Habitat. (n.d.). Retrieved from https://unhabitat.org/resilience.

[18] Bayeh B, Alemayehu M. Scientific community Debates on causes and consequences of global warming. 2019. Retrieved from, https://journals.bdu.edu.et/index.php/jaes/article/view/201/301.

[19] Doran PT, Zimmerman MK. Examining the scientific consensus on climate change. Eos 2009;90(3):22–3. https://agupubs.onlinelibrary.wiley.com/doi/epdf/10.1029/2009EO030002.

[20] Anderegg W, Prall J, Harold J, Schneider S. Expert credibility in climate change. Proc Natl Acad Sci USA 2010;107(27):12107–9.

[21] Goel A, Bhatt R. Causes and consequences of global warming. Int J Life Sci Biotechnol Pharma Res 2012;1(1):27–31. https://www.researchgate.net/publication/265067277.

[22] Bennett L. Deforestation and climate change. A Publication of Climate Institute; 2017. p. 1400. http://climate.org/wp-content/uploads/2017/04/deforestation-final_r1.pdf.

[23] Donovan GH. Including public-health benefits of trees in urban-forestry decision making. February 16, 2017. Retrieved from, https://www.sciencedirect.com/science/article/pii/S1618866716303909.

[24] Elmqvist T, Setälä H, Handel S, Ploeg SV, Aronson J, Blignaut J, et al. Benefits of restoring ecosystem services in urban areas. May 22, 2015. Retrieved from, https://www.sciencedirect.com/science/article/pii/S1877343515000433.

[25] McPherson G, Simpson JR, Peper PJ, Maco SE, Xiao Q. Municipal forest benefits and costs in five US cities. 2005. Retrieved from, https://academic.oup.com/jof/article/103/8/411/4598681.

[26] Nowak, D. F., & Dwyer, J. F. (2007). Understanding the benefits and costs of urban forest. Retrieved from https://www.researchgate.net/profile/David-Nowak6/publication/226897077.

[27] Philibert Petit E. Connectivity-oriented urban projects. 2006. Retrieved from, https://repository.tudelft.nl/islandora/object/uuid:95545026-c11b-473b-983a-bfa28fa0cbc7?collection=research.

[28] Schwab JC. Planning the urban forest: ecology, economy, and community development. Chicago, IL: American Planning Association; 2009.

[29] Song XP, Tan PY, Edwards P, Richards D. The economic benefits and costs of trees in urban forest stewardship: A systematic review. Urban For Urban Green 2018. https://doi.org/10.1016/j.ufug.2017.11.017.

[30] Vallejo L, Mullan M. Climate-resilient infrastructure: Getting the policies right. 2017. Retrieved from, https://www.oecd-ilibrary.org/environment/climate-resilient-infrastructure_02f74d61-en.

[31] Ferranti EJS, MacKenzie AR, Levine JG, Ashworth K, Hewitt CN. First steps in urban air quality. 2nd ed. A Trees and Design Action Group; 2019.

[32] European Commission. Evaluating the impact of nature-based solutions : a handbook for practitioners. 2021. Retrieved from, https://op.europa.eu/en/publication-detail/-/publication/d7d496b5-ad4e-11eb-9767-01aa75ed71a1/language-en/format-PDF/source-206665393.

[33] European Commission. Horizon2020 work programme 2016-2017–12. Climate action, environment, resource efficiency and raw materials. Brussels, Belgium. 2016.

[34] Cohen-Shacham E, Walters G, Janzen C, Maginnis S, editors. Nature-based Solutions to address global societal challenges. Gland, Switzerland: International Union for the Conservation of Nature; 2016.

[35] Bush J, Doyon A. Building urban resilience with nature-based solutions: how can urban planning contribute? Cities. Pergamon; October 24, 2019. www.sciencedirect.com/science/article/pii/S0264275119313976.
[36] Frantzeskaki N, McPhearson T, Collier MJ, Kendal D, Bulkeley H, Dumitru A, Walsh C, Noble K, van Wyk E, Ordóñez C, Oke C, Pintér L. Nature-based solutions for urban climate change adaptation: linking science, policy, and practice communities for evidence-based decision-making. Bioscience 2019;69(6):455–66.
[37] Gómez-Baggethun E, Barton DN. Classifying and valuing ecosystem services for urban planning. Ecol Econ 2012. https://doi.org/10.1016/j.ecolecon.2012.08.019.
[38] TEEB. The economics of ecosystems and biodiversity: Mainstreaming the economics of nature: a synthesis of the approach, conclusions and recommendations of TEEB. 2010.
[39] Castellar J, Popartan L, Pueyo-Ros J, Atanasova N, Langergraber G, Säumel I, Acuña V. Nature-based solutions in the urban context: terminology, classification and scoring for urban challenges and ecosystem services. Sci Total Environ 2021;779:146237. https://doi.org/10.1016/j.scitotenv.2021.146237.
[40] Andersson E, Barthel S, Ahrné K. Measuring social-ecological dynamics behind the generation of ecosystem services. Ecol Appl 2007;17(5):1267–78.
[41] Hardin PJ, Jensen RR. The effect of urban leaf area on summertime urban surface kinetic temperatures: A Terre Haute case study. Urban For Urban Green 2007;6:63–72.
[42] Aylor D. Noise reduction by vegetation and ground. J Acoust Soc Am 1972;51:197–205.
[43] Ishii M. Measurement of road traffic noise reduced by the employment of low 848 physical barriers and potted vegetation. Inter-noise 1994;29-31:595–7.
[44] Kragh J. Road traffic noise attenuation by belts of trees. J Sound Vib 1981;74(2):235–41.
[45] Fang C-F, Ling D-L. Investigation of the noise reduction provided by tree belts. Landsc Urban Plan 2003;63(4):187–95.
[46] Nowak DJ. Notes: Estimating leaf area and leaf biomass of open-grown deciduous urban trees. For Sci 1996;42(4):504507.
[47] Costanza R, Mitsch WJ, Day JW. A new vision for New Orleans and the Mississippi delta: Applying ecological economics and ecological engineering. Front Ecol Environ 2006;4(9):465–72.
[48] Danielsen F, Sørensen MK, Olwig MF, et al. The Asian tsunami: A protective role for coastal vegetation. Science 2005;310(5748):643.
[49] Farber S. The value of coastal wetlands for protection of property against hurricane wind damage. J. Environ Econ Manage 1987;14:143–51.
[50] Kerr AM, Baird AH. Natural barriers to natural disasters. BioScience 2007;57(2):102–3.
[51] Birdsey RA. Carbon storage and accumulation in United States forest ecosystems. Washington, DC: U.S. Department of Agriculture, Forest Service; 1992.
[52] Jo H-K, McPherson GE. Carbon storage and flux in urban residential greenspace. J Environ Manag 1995;45(2):109–33.
[53] Scott KI, McPherson EG, Simpson JR. Air pollutant uptake by Sacramentos' urban forests. J arboric 1998;24:224–32.
[54] McPherson EG. Atmospheric carbon dioxide reduction by Sacramento's urban forest. J arboric 1998;24:215–23.
[55] Nowak DJ. Atmospheric carbon dioxide reduction by Chicago's urban forest. In: McPherson EG, et al., editors. Chicago's urban forest ecosystem: results of the Chicago urban forest climate project. Radnor: U.S. Dept. of Agriculture, Forest Service, Northeastern Forest Experiment Station; 1994. p. 83–94.
[56] Groening G. School garden and kleingaerten: For education and enhancing life quality. Acta Hortic 1995;391:53–64.
[57] Tyrväinen L, Pauleit S, Seeland K, et al. Benefits and uses of urban forests and trees. In: C. C. Konijnendijk, K. Nilsson, T. B. Randrup, et al. editors. Urban forests and trees: a reference book Heidelberg: Springer; 2005;81–114.
[58] Altman I, Low SM. Human behavior and environment, 12. New York: Plenum; 1992.
[59] Menz MH, Dixon KW, Hobbs RJ. Hurdles and opportunities for landscape-scale restoration. Science 2013;339(6119):526–7. https://doi.org/10.1126/science.1228334.
[60] Swaffield SR. Landscape as a way of knowing the world. In: Harvey S, Fieldhouse K, editors. The cultured landscape: designing the environment in the 21st century. New York: Routledge; 2005. p. 3–24.
[61] Opdam P, Nassauer JI, Wang Z, Albert C, Bentrup G, Castella J, et al. Science for action at the local landscape scale. Landsc Ecol 2013;28(8):1439–45. https://doi.org/10.1007/s10980-013-9925-6.
[62] Nassauer JI. Landscape as medium and method for synthesis in urban ecological design. Landsc Urban Plan 2012;106:221–9.
[63] Barton H, Grant M. A health map for the local human habitat. J Roy Soc Promot Health 2006;126:252–3. https://doi.org/10.1177/1466424006070466.
[64] Hartig T, Mitchell R, de Vries S, Frumkin H. Nature and health. Ann Rev Pub Health 2014;35:207–28.
[65] Kabisch N, van den Bosch M, Lafortezza R. The health benefits of nature-based solutions to urbanization challenges for children and the elderly – a systematic review. Environ Res 2017;159:362–73.
[66] Jutras P, Prasher SO, Mehuys GR. Prediction of street tree morphological parameters using artificial neural networks. Comput Electron Agr 2009;67:9–17.
[67] Steenberg JWN, Duinker PN, Nitoslawski SA. Ecosystem-based management revisited: updating the concepts for urban forests. Landsc Urban Plann 2019;186:24–35.
[68] Mirauda D, Erra U, Agatiello R, Cerverizzo M. Applications of mobile augmented reality to water resources management. Water 2017;9:699.
[69] Annerstedt M, Jönsson P, Wallergård M, Johansson G, Karlson B, Grahn P, et al. Inducing physiological stress recovery with sounds of nature in a virtual reality forest — results from a pilot study. Physiol Behav 2013;118:240–50.
[70] Mans U, Giest S, Baar T. Can big data Make a difference for urban management?1. Urban Planet; 2018. p. 218–38. https://doi.org/10.1017/9781316647554.013.
[71] Jordan RC, Ballard HL, Phillips TB. Key issues and new approaches for evaluating citizen science learning outcomes. Front Ecol Environ 2012;10:306–9.
[72] Herold M, Couclelis H, Clarke KC. The role of spatial metrics in the analysis and modeling of urban land use change. Comput Environ Urban Syst 2005;29(No 4):369–99.
[73] Cai BY, et al. Treepedia 2.0: applying deep learning for large-scale quantification of urban tree cover. In: 2018 IEEE international congress on big data (BigData congress); 2018. https://doi.org/10.1109/bigdatacongress.2018.00014.
[74] Cheng G, Xie X, Han J, Guo L, Xia G. Remote sensing image scene classification meets deep learning: challenges, methods, benchmarks,

and opportunities. IEEE J Selec Top Appl Earth Obser Remote Sens 2020;13:3735–56. https://doi.org/10.1109/jstars.2020.3005403.

[75] Talari S, Shafie-Khah M, Siano P, Loia V, Tommasetti A, Catalão J. A review of smart cities based on the Internet of things concept. Energies 2017;10(4):421. https://doi.org/10.3390/en10040421.

[76] Alonzo M, Bookhagen B, Roberts DA. Urban tree species mapping using hyperspectral and LiDAR data fusion. Remote Sens Environ 2014;148:70–83.

[77] Naumann S, Davis M, Moore M, Mccormick K. Utilizing urban living laboratories for social innovation. Urban Planet; 2018. p. 197–217. https://doi.org/10.1017/9781316647554.012.

[78] Lemmens R, et al. Citizen science in the digital world of apps. Sci Citiz Sci 2021:461–74. https://doi.org/10.1007/978-3-030-58278-4_23.

[79] Dian FJ, et al. Wearables and the Internet of things (IoT), applications, opportunities, and challenges: a survey. IEEE Access 2020;8:69200–11. https://doi.org/10.1109/access.2020.2986329.

[80] Mesas-Carrascosa FJ, Verd Santano D, Meroo JE, Snchez de la Orden M, García-Ferrer A. Open source hardware to monitor environmental parameters in precision agriculture. Biosyst Eng 2015;137:73–83.

[81] Srbinovska M, Gavrovski C, Dimcev V, Krkoleva A, Borozan V. Environmental parameters monitoring in precision agriculture using wireless sensor networks. J Clean Prod 2015;88:297–307.

[82] The Nature Conservacy. Playbook for climate finance. 2019. www.nature.org/content/dam/tnc/nature/en/documents/TNC_Playbook_ClimateFinance_01.pdf.

[83] Sartzetakis ES. Green bonds as an instrument to finance low carbon transition. Economic change and Restructuring. Springer US; Februray. 2020. https://doi.org/10.1007/s10644-020-09266-9. Retrieved from.

[84] OECD, et al. Green bonds: country experiences, barriers and options. OECD/ICMA/CBI/China GFC; 2016. Retrieved from, http://unepinquiry.org/wpcontent/uploads/2016/09/6_Green_Bonds_Country_Experiences_Barriers_and_Options.pdf.

[85] Somarakis G, Stavros S, Chrysoulakis N, editors. ThinkNature Nature-Based Solutions Handbook, ThinkNature project funded by the EU Horizon 2020 research and innovation programme under grant agreement No 730338; 2019. https://doi.org/10.26225/jerv-w202.

[86] Brent RJ. Applied cost-benefit analysis. Edward Elgar; 2010.

[87] Sowińska-Świerkosz B, García J. A new evaluation framework for nature-based solutions (NBS) projects based on the application of performance questions and indicators approach. May 07, 2021. Retrieved June 3, 2021, from, https://www.sciencedirect.com/science/article/pii/S0048969721026863.

[88] Meerow S, Pajouhesh P, Miller TR. Social equity in urban resilience planning. Local Environ 2019;24(9):793–808. https://doi.org/10.1080/13549839.2019.1645103.

[89] Frantzeskaki N, Wittmayer J, Loorbach D. The role of partnerships in "realising" urban sustainability in Rotterdam's City Ports Area, The Netherlands. J Clean Prod 2014;65:406–17. https://doi.org/10.1016/j.jclepro.2013.09.023.

[90] Speck J. Walkable City: how downtown can save America, one step at a time. Farrar, Straus and Giroux; 2012.

[91] Drewe P. The network city: a paradigm challenge for urban design and planning, the network society: the new context for planning. In: Thirdjoint congress ACSP – AESOP, July 8 – 12, 2003, Leuven; 2003.

[92] Francis P. Laudato Si': on care for our common home. Ideals Ideol. 2019. p. 503–10. https://doi.org/10.4324/9780429286827-79.

[93] De Jong, Taeke. Image archive and ways to study. In: Ouwekerk M, Rosemann J, editors. (2001) research by design conference proceedings DUP science. The Netherlands: Delft University Press; 2001.

Further reading

Atkinson G, Mourato S. Department of geography and environment and Grantham Research Institute on climate change and the environment. Environmental cost-benefit analysis. Annual reviews. 2008. www.annualreviews.org/doi/10.1146/annurev.environ.33.020107.112927.

Berkes F, Colding J, Folke C. Navigating social-ecological systems: building resilience for complexity and change. Cambridge, UK: Cambridge University Press; 2008.

Bibri SE. A foundational framework for smart sustainable city development: theoretical, disciplinary, and discursive dimensions and their synergies. Sustainable cities and society. Elsevier; December 28, 2017. www.sciencedirect.com/science/article/pii/S2210670717313069.

Boardman AE, et al. Cost-benefit analysis: concepts and practice. Cambridge University Press; 2018.

Caragliu A, Bo Cd, Nijkamp P. Smart cities in Europe. In: 3rd central European conference in regional science; 2009.

Castells M. The rise of the network society. Oxford: Blackwell; 2001.

Climate-resilient infrastructure. OECD environment working papers. 2017. https://doi.org/10.1787/02f74d61-en.

Coe A, Paquet G, Roy J. E-governance and smart communities: a social learning challenge. Soc Sci Comput Rev 2001;19:80–93.

Dennig F. Climate change and the Re-evaluation of cost-benefit analysis. Climatic change. Netherlands: Springer; August 30, 2017. https://doi.org/10.1007/s10584-017-2047-4. Retrieved from.

Dictionary by Merriam-Webster. Retrieved from https://www.merriam-webster.com/.

Eggermont H, Balian E, Azevedo JMN, Beumer V, Brodin T, Claudet J, Fady B, Grube M, Keune H, Lamarque P, Reuter K, Smith M, van Ham C, Weisser WW, Le Roux X. Nature-based solutions: new influence for environmental management and research in Europe. GAIA 2015;24(4):243–8.

Elmqvist T, Bai X, Frantzeskaki N, Griffith C, Maddox D, McPhearson T, et al., editors. Urban planet: knowledge towards sustainable cities. Cambridge: Cambridge University Press; 2018. https://doi.org/10.1017/9781316647554.

Elmqvist T, Andersson E, Frantzeskaki N, McPhearson T, Olsson P, Gaffney O, et al. Sustainability and resilience for transformation in the urban century. Nat Sustain 2019;2(4):267–73. https://doi.org/10.1038/s41893-019-0250-1.

Endreny T, et al. Implementing and managing urban forests: a much needed conservation strategy to increase ecosystem services and urban Wellbeing. Ecological modelling. Elsevier; July 30, 2017. www.sciencedirect.com/science/article/pii/S0304380017300960.

Ershad Sarabi S, Han Q, Romme AGL, de Vries B, Wendling L. Key enablers of and barriers to the uptake and implementation of nature-based solutions in urban settings: a review. Resources 2019;8(3). Art. no 121.

Falco E, Kleinhans R. Digital participatory platforms for co-production in urban development: a systematic review. Int J E Plann Res 2019;7(3): 52–79.

Flammer C. Green bonds: effectiveness and implications for public policy. NBER; June 17, 2019. www.nber.org/papers/w25950.

Foster A, Dunham IM, Kaylor C. Citizen science for urban forest management? Predicting the data density and richness of urban forest volunteered geographic information. Urban Sci 2017;1:30.

Gardiner J. In: Kindel P, Bothen A, editors. Why should We design and create green cities?; August 06, 2020. Retrieved May 18, 2021, from, https://www.the-possible.com/nature-in-cities/.

Gómez-Baggethun E, Ruiz-Pérez M. Economic valuation and the commodification of ecosystem services. SAGE Journals; 2011. https://doi.org/10.1177/0309133311421708.

Hollands RG. Will the real smart city please stand up. City. 2008. p. 12.

Ignatieva M, et al. Planning and design of ecological Networks in urban areas. Landscape and ecological engineering. Springer Japan; December 29, 2010. https://doi.org/10.1007/s11355-010-0143-y. Retrieved from.

IUCN (International Union for the Conservation of Nature). Resolution 69 on defining nature-based solutions (WCC-2016-Res-069). IUCN resolutions, recommendations and other decisions. 6-10 September 2016. USA: World Conservation Congress Honolulu, Hawai'i; 2016. https://portals.iucn.org/library/sites/library/files/resrecfiles/WCC_2016_RES_069_EN.pdf.

Janssen M, Charalabidis Y, Zuiderwijk A. Benefits, adoption barriers and myths of open data and open government. Inf Syst Manag 2012;29:258–68.

Knudsen C, et al. The value of sustainable urbanization: world cities report 2020. United Nations - Habitat; 2020.

Kumar P, Debele SE, Sahani J, Rawat N, Marti-Cardona B, Alfieri SM, et al. Nature-based solutions efficiency evaluation against natural hazards: modelling methods, advantages and limitations. April 15, 2021. Retrieved from, https://www.sciencedirect.com/science/article/pii/S0048969721021288.

Meerow S, Newell JP, Stults M. Defining urban resilience: a review. Landsc Urban Plann 2016;147:38–49. https://doi.org/10.1016/j.landurbplan.2015.11.011.

Møller MS, et al. Participation through place-based e-tools: a valuable resource for urban green infrastructure governance? Urban For Urban Green 2019;40:245–53. https://doi.org/10.1016/j.ufug.2018.09.003.

OECD. Mobilising bond markets for a low-carbon transition, green finance and investment. Paris: OECD Publishing; 2017. https://doi.org/10.1787/9789264272323-en. Available at:.

Pflieger G, Rozenblat C. Urban networks and network theory: the city as the connector of multiple networks. Urban Stud 2010;47(13):2723–35. https://doi.org/10.1177/0042098010377368.

Plan for climate change and environmental justice: Joe Biden. Joe Biden for President: official Campaign website, October 29, 2020, joebiden.com/climate-plan/.

Resilience in ecology and urban design: linking theory and practice for sustainable citiesPickett STA, et al., editors. Fut City 2013;3. https://doi.org/10.1007/978-94-007-5341-9_4.

Randrup TB, et al. Moving beyond the nature-based solutions discourse: introducing nature-based thinking. Urban Ecosyst 2020;23(4):919–26. https://doi.org/10.1007/s11252-020-00964-w.

Salingaros N. Principles of urban structure, Series design/science/planning. Amsterdam: Delft University Press, Delft/Techne Press; 2005.

Speranza P. [PDF] urban ecological interaction: air, water, light and new transit at the human scale of Barcelona's superilles: semantic scholar. September 2016. Retrieved May 17, 2021, from, https://cpb-us-e1.wpmucdn.com/blogs.uoregon.edu/dist/6/13503/files/2016/09/ICAUDT_SPERANZA_Barcelona-117wl17.pdf.

Stavrinidou E, Gabrielsson R, Nilsson K, Singh SK, Franco-Gonzalez JF, Volkov AV, Jonsson MP, et al. In vivo polymerization and manufacturing of wires and supercapacitors in plants. Proc Natl Acad Sci USA 2017;114:2807.

Stavrinidou E, Gabrielsson R, Gomez E, Crispin X, Nilsson O, Simon DT, et al. Electronic plants. Sci Adv 2015;1:e1501136.

Tietenberg TH, Lewis L. Environmental & natural resources economics. Pearson; 2018.

Tranos, Gertner. Smart networked cities? Innovation. Eur J Soc Sci Res 2012;25(2):175–90.

Trends in urbanisation and urban policies in OECD countries. 2010. https://doi.org/10.1787/9789264092259-en.

Ulmer JM, et al. Multiple health benefits of urban tree canopy: the mounting evidence for a green prescription. Health & Place, Pergamon; 14 Sept. 2016. www.sciencedirect.com/science/article/pii/S1353829216301332.

World Commission on Environment and Development. Our common future. Transmitted to the general assembly as an annex to document A/42/427 - development and international cooperation: environment. United Nations. 1987.

Yigitcanlar T, et al. Can cities become smart without being sustainable? A systematic review of the literature. Sustainable cities and society. Elsevier; November 29, 2018. www.sciencedirect.com/science/article/pii/S221067071831268X.

Zúñiga-Terán Adriana A, Staddon C, de Vito L, Gerlak AK, Ward S, Schoeman Y, Hart A, Booth G. Challenges of mainstreaming green infrastructure in built environment professions. J Environ Plann Manag 2020;63(4):710–32. https://doi.org/10.1080/09640568.2019.1605890.

Chapter 28

Financial and technical planning of decarbonized hybrid energy systems for modern cities

Hossam A. Gabbar[1], Muhammad R. Abdussami[2] and Md Ibrahim Adham[2]

[1]*Faculty of Energy Systems and Nuclear Science, Faculty of Engineering and Applied Science, Ontario Tech University, Oshawa, ON, Canada;* [2]*Faculty of Energy Systems and Nuclear Science, Ontario Tech University, Oshawa, ON, Canada*

28.1 Introduction

Electric demand for distant locations is typically served by Micro Energy Grid (MEG), since the electric grid footprint is not available. MEG refers to an energy network that links local Renewable Energy Sources (RESs) to support electrical and thermal demand. The generation side of an MEG involves fossil-fired internal combustion engines, microturbines, fuel cells, and RESs. The sizing of microgrids solely depends on assessment and understanding of system load and geographical characteristics; a specific sizing of MEG does not fit all types of system demand [1]. The "Combined Heat and Power (CHP)" and the "Distributed Generation (DG)" concepts are employed together in MEG to generate electric and thermal energy. An MEG can operate both grid-connected and off-grid mode. MEG allows high penetration of RESs, reduces transmission losses, enhances system reliability by operating in island mode during the fault, and improves energy management capability [2].

The CHP is also known as "Cogeneration"—concomitant generation of electric and thermal power from the same energy source. In a CHP facility, the waste heat is recovered and used as valuable thermal power for heating and cooling. Most of the electricity generation plant utilizes heat engines to produce electricity from heat. The theoretical heat-to-electricity conversion efficiency for modern combustion plants range from 30% to 60%; practical efficiency can be even less than the theoretical. CHP facility can improve the overall system efficiency of up to 90% and, CHP's benefits are widely recognized. Nevertheless, recovered waste heat is not usually considered for large-scale electricity generation plants since traditional generation plants are located far from the consumer end. Therefore, decentralized generation at the consumer end has a great potential to use waste heat by CHP units [3].

Microreactors can replace fossil fuel-based generators within MEG to reduce GHG emissions significantly and improve system operating conditions [4]. In traditional MEG, RESs and fossil fuel-based generators are regarded for energy production. It is imperative to introduce an energy source that will provide continuous electric and thermal energy supply for medium/large-scale demand. Usually, RESs cannot provide a constant supply for a longer duration. Hence, a nuclear microreactor, a continuous mode of power supply with essentially zero carbon emissions, is aimed to combine with RESs and energy storage systems. Nuclear-renewable integrated systems assure continuity, reliability, and resiliency for off-grid applications. An efficient energy management and control algorithm is also required to minimize the challenges of a nuclear-renewable integrated system.

28.2 Nuclear-Renewable Integrated Energy System (N-R IES)

A Nuclear-Renewable Integrated Energy System (N-R IES) is an energy hub combining nuclear reactors, energy storage systems, and all possible RESs, such as solar, wind, hydro, biomass, and geothermal. An N-R IES can operate grid-connected and off-grid mode. Electric and thermal demand can be met simultaneously within N-R IESs. Hybridization of an N-R IES can be carried out in multiple ways. However, three coupling methods have been proposed in Ref. [5].

Depending on applications and geographical scenarios, large-scale and small-scale N-R IES can be adopted to

Smart Cities Policies and Financing. https://doi.org/10.1016/B978-0-12-819130-9.00046-2

serve the system demand. In large-scale N-R IESs, traditional large-scale nuclear reactors are integrated with RESs. Conversely, Small Modular Reactors (SMRs) or microreactors are combined with available RESs in small-scale N-R IESs [6].

N-R IES can offer several secondary commodities. Fig. 28.1 represents a typical interconnected N-R IES with possible resources and applications [7]. An N-R IES can serve several applications, such as seawater desalination, electric transportation, district heating, and calcination. It is also economical to produce secondary commodities with N-R IES rather than only satisfying electrical and thermal demand [8].

28.3 A case study: modeling of system architecture

The Chalk River Laboratories, 286 Plant Road, Chalk River, ON K0J 1J0, Canada, is the selected location for the study here. Since the Global First Power has proposed the site readiness, construction, operation, and decommissioning of a 15 MWt microreactor [9], the authors consider this location an appropriate place for evaluating N-R IES. Among the six regions in Ontario, Canada, the Ottawa region's electric load data for 2019 are collected for this study. The wind speed data, solar irradiation data, and temperature data are obtained for the same location. No particular coupling method is applied in this case study.

Fig. 28.2 represents a typical small-scale N-R IES. The N-R IES consists of Wind Turbine (WT), Solar Photovoltaic (PV) panel, microreactor, Battery Energy Storage (BES), boiler, system converter, and Thermal Load Controller (TLC). The details of each system component are described later in this section. The project lifetime, nominal discount rate, and the inflation rate of the system are 60 years, 8%, and 2%, respectively, for this study.

28.3.1 Electric and thermal demand

The Ottawa region's electric load data for 2019 are collected and presented in Fig. 28.3 [10]. The annual energy consumption, average load demand, and peak demand are 24,444.2 kWh/day, 1018.5 kW, and 1623 kW, respectively. This baseload data is varied within a specific range in the simulation later to study the sensitivity analysis.

Due to the unavailability of the original thermal load demand (see Fig. 28.4) in the Ottawa region, two types (residential and commercial load) of typical thermal load are collected from the library of HOMER Pro software and studied here. Two types of thermal loads are considered to support the CHP criterion and assess the N-R IES accurately. The peak demands are different for the two kinds of thermal load, and the peak demands occur at a different time for both cases; this is the difference between residential and commercial thermal load. For the residential thermal load, the peak happens in the evening, whereas the

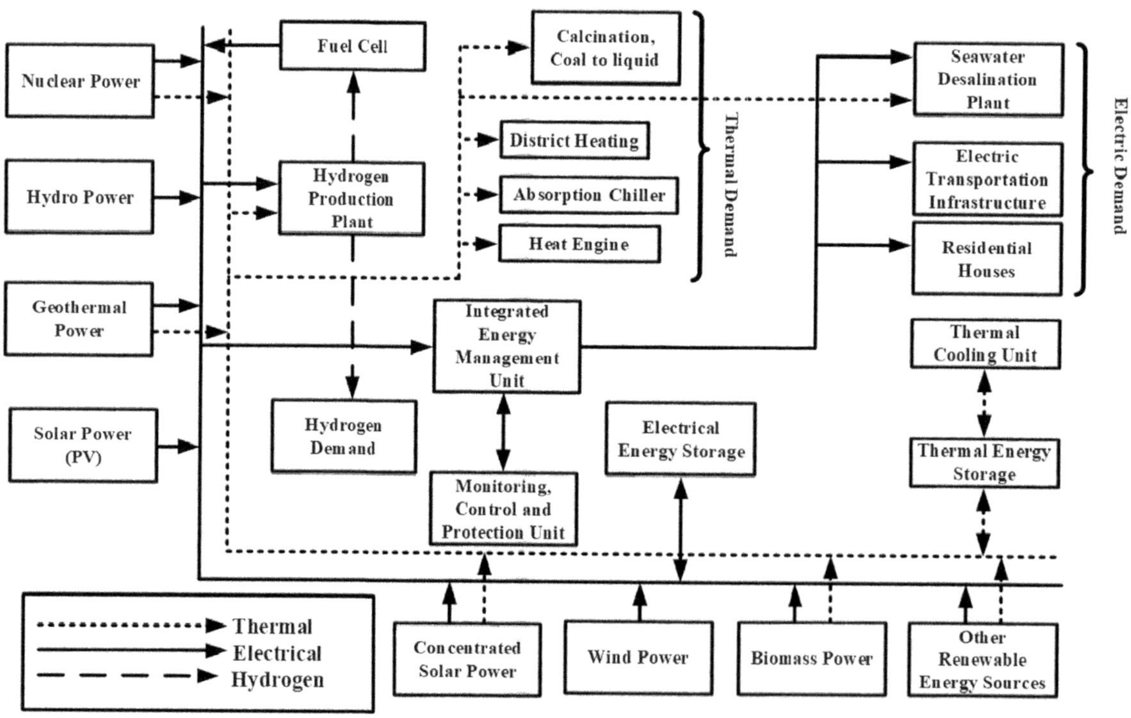

FIGURE 28.1 Schematic of an N-R IES.

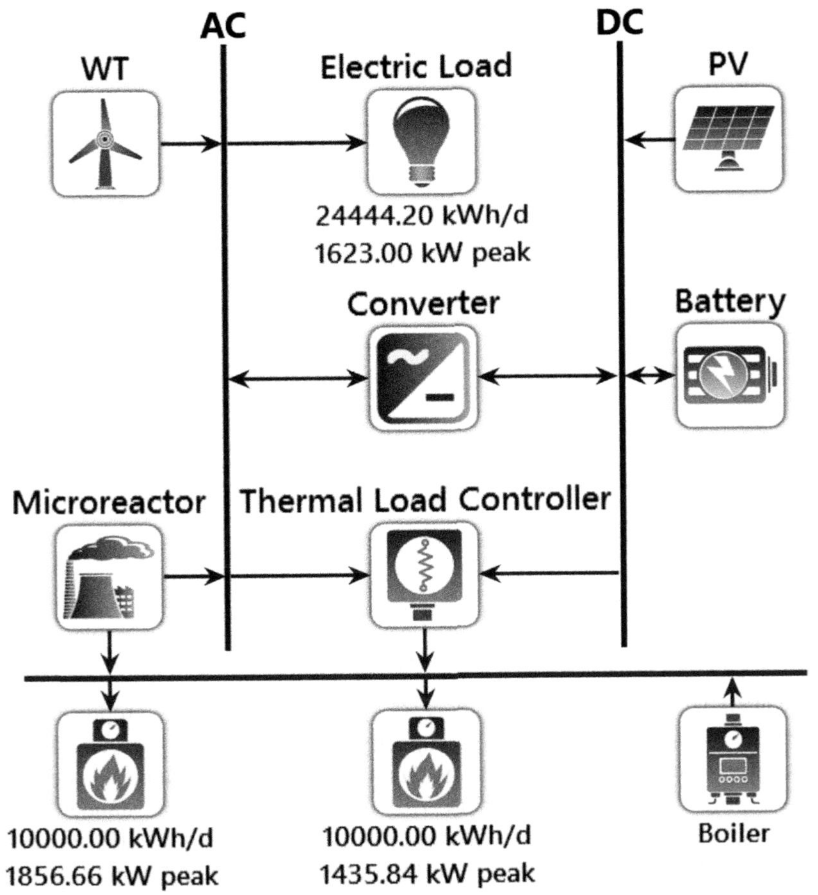

FIGURE 28.2 System architecture (case study).

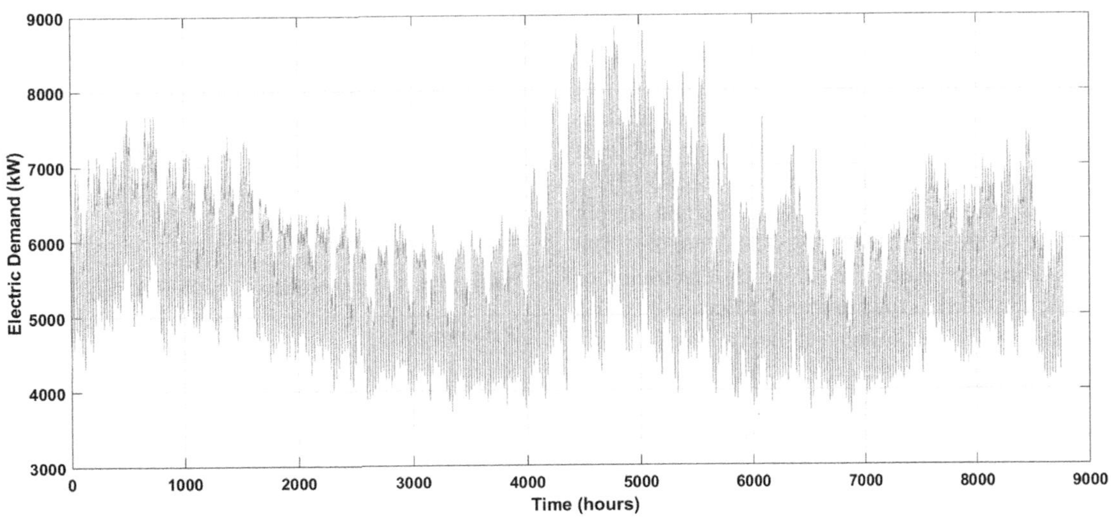

FIGURE 28.3 Electric load demand.

peak reaches at mid-day for the commercial load. The baseline data of both types of load are scaled-up and scaled-down in simulation to identify the optimal system configuration. The scaled values of the baseline load data do not change the load characteristics; the baseline data is only multiplied by a factor to emulate the high and low thermal demand. Both residential and commercial load changes from 1000 to 20,000 kWh/day in the simulation to characterize the optimal system arrangement for different scenarios.

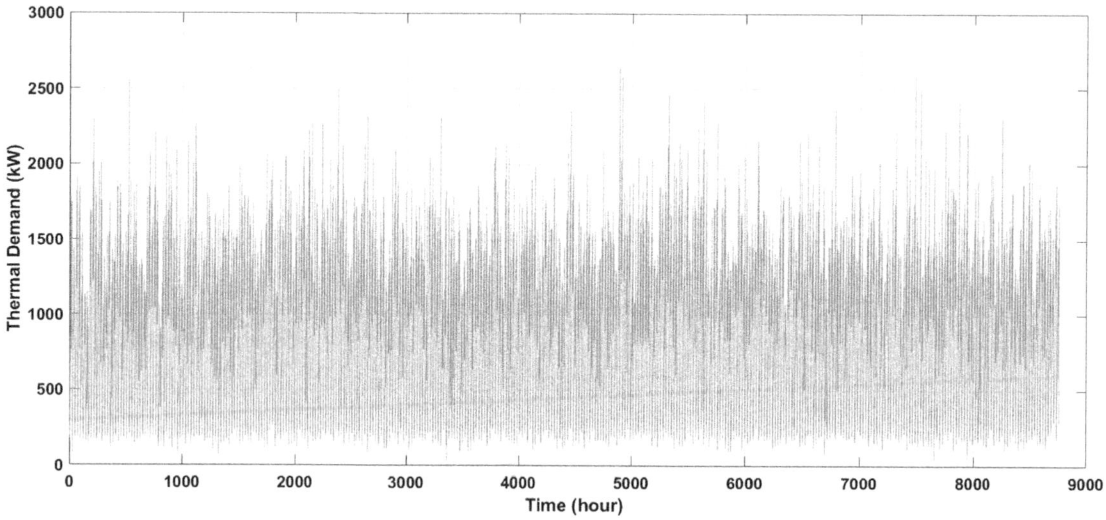

FIGURE 28.4 Thermal load demand.

28.3.2 Microreactors

Microreactors are a fourth-generation small-scale nuclear reactor that can produce electricity and thermal energy simultaneously by Combined Heat and Power (CHP). The main application of microreactors focuses on remote areas [9]. Several companies are progressing to develop the small/micro-scale nuclear reactor.

Microreactors involve several benefits compared to the conventional large-scale Nuclear Power Plant (NPP). It can save up to 50% of electricity and thermal energy cost. Microreactors are developed in the factory and delivered to the licensed site. They are sealed at the site, and it does not require refueling in its whole lifetime; this provides a unique advantage for remote location applications. After the fuel module's lifetime, the module is towed from the site and stored in a commissioned nuclear waste facility. Microreactors also have a lower development cost compared to typical Small Modular Reactor (SMR) [9].

Microreactors produce electricity at the site and supply the process heat to the adjacent plant of the facility. The nearby facility either converts the heat into electricity or utilizes the heat for thermal load demand. Microreactors do not have an active safety system; it has a passive heat removal system. Furthermore, since Microreactors are modular and scalable, the sufficient number of the module can be installed depending on the electric and thermal demand [11].

The details specification of microreactors considered in this study are listed in Table 28.1 [12].

28.3.3 Solar PV panels

The output power of a PV panel can be determined by the following equation [13]. The Location of solar data is Chalk River Laboratories.

$$P_{PV} = Y_{PV} f_{PV} \left(\frac{G_T}{G_{T,STC}}\right) \left[1 + \alpha_p (T_c - T_{c,STC})\right] \quad (28.1)$$

Where Y_{PV}, f_{PV}, G_T, $G_{T,STC}$, α_p, T_c, and $T_{c,STC}$ represent the PV array rated capacity under standard test conditions (kW), PV derating factor (%), Solar irradiance in the current time step (kW/m^2), solar irradiance at standard test condition (1 kW/m^2), temperature coefficient (%/°C), PV cell temperature in the current time step (°C), and PV cell temperature under standard test conditions (25°C), respectively.

A single-year (2019) solar irradiance data and temperature data for the selected location are collected and shown in Figs. 28.5 and 28.6, respectively. The chosen single PV panel has a nominal power rating of 1 kW. The capital cost, replacement cost, maintenance cost, and lifetime of a single PV panel are considered as 3000 USD/kW, 3000 USD/kW, 10 USD/year, and 25 years, respectively [14].

TABLE 28.1 Microreactor specifications.

Parameters	Value
Reactor size (kW$_e$)	500
Plant lifetime (Years)	40
Core lifetime (Years)	10
Nominal overnight capital cost (\$/kW$_e$)	15,000
Fixed O&M cost (\$/kW$_e$)	350
Fuel cost ($/MWh)	10
Decommissioning cost ($/MWh)	5
Refueling cost of fuel module ($)	20 million
Plant efficiency (%)	40

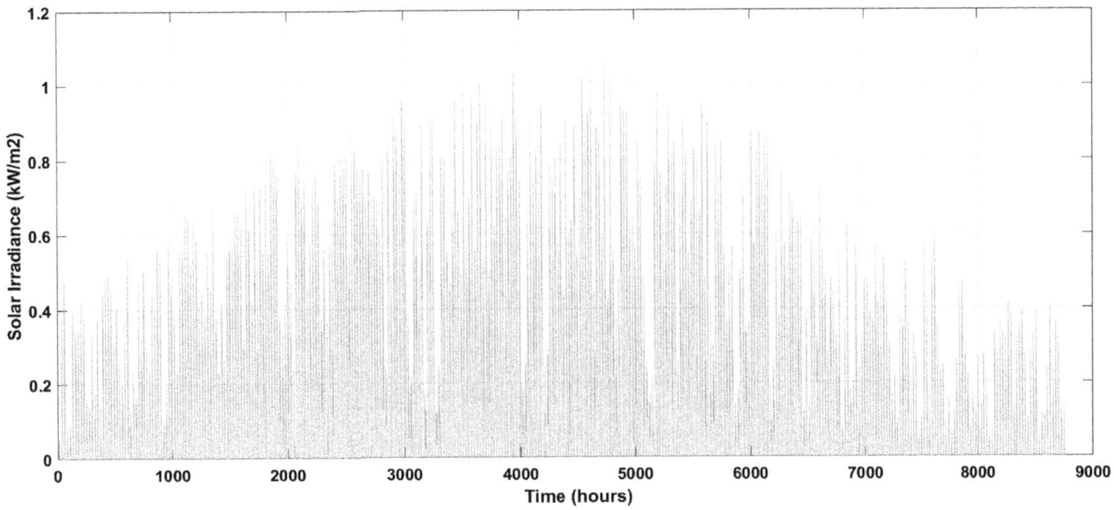

FIGURE 28.5 Solar irradiance.

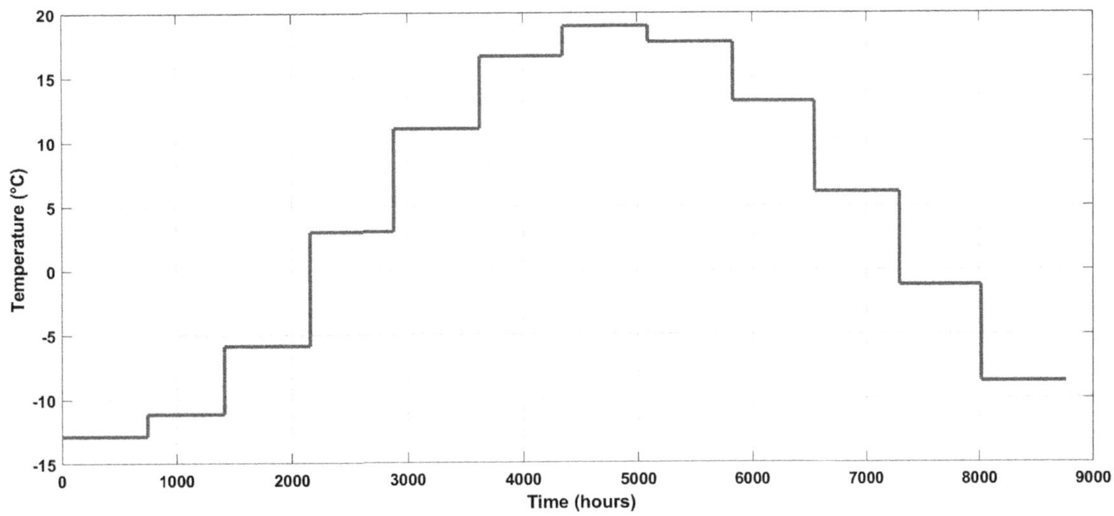

FIGURE 28.6 Temperature.

28.3.4 Wind turbines

In HOMER, WT power output is measured, mainly focusing on the WT power curve provided by the manufacturer [15]. Firstly, the wind speed is estimated from the hub height by using Eq. (28.2). Secondly, wind power is determined for the corresponding wind speed at standard air density from the WT power curve. Lastly, density correction is applied to calculate the real-world power for the WT by applying Eq. (28.3).

$$V_{hub} = V_a \frac{\ln\left(\frac{H_{hub}}{H_0}\right)}{\ln\left(\frac{H_A}{H_0}\right)} \tag{28.2}$$

$$P_{actual} = \left(\frac{\rho}{\rho_0}\right) P_{STP} \tag{28.3}$$

where V_{hub} is the wind speed at hub height (m/s), V_a is the wind speed at anemometer height (m), H_{hub} wind turbine hub height (m), H_0 is the surface roughness length (m), H_A is the anemometer height (m), P_{actual} is the actual power output (kW), ρ is the air density (kg/m^3), ρ_0 is the), ρ_0 air density at standard temperature and pressure (kg/m^3), and P_{STP} is the wind turbine power output at standard temperature and pressure (kW).

The wind speed data for the project location is shown in Fig. 28.7. A unit wind turbine rating for this study is 10 kW. The WT capital cost, replacement cost,

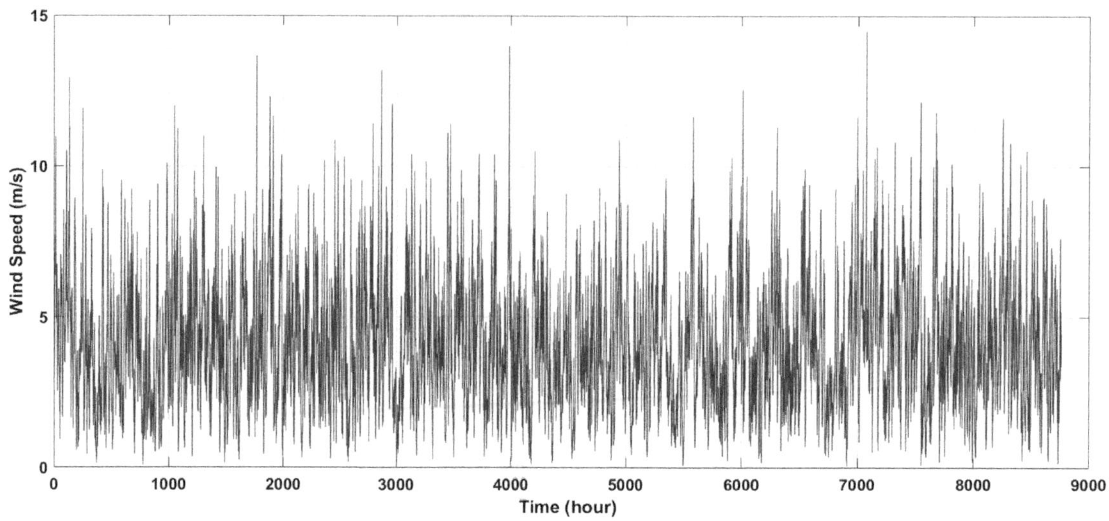

FIGURE 28.7 Wind speed.

maintenance cost, and lifetime are 1130 USD/kW [16], 1130 USD/kW [16], 48 USD/kW/Year [17], and 25 years [18], respectively.

28.3.5 Battery Energy Storage (BES)

The energy rating of a single battery cell for this study is 1 kWh. The capital cost (single cell), replacement cost, the maintenance cost, and a lifetime of the battery unit is 398 USD/kWh [19], 398 USD/kWh [19], 10 USD/kWh/year [20], and 5 years [19], respectively. The maximum and minimum State of Charge (SOC) of the BES are 30% and 100%, respectively.

28.3.6 Converter

The system converter is applied in between AC and DC bus bars. The size of the converter is optimized in each case in the simulation. The capital cost, efficiency, and lifetime of a unit converter (1 kW rating) are 300 USD, 95%, and 15 years, respectively.

28.3.7 Thermal Load Controller

A Thermal Load Controller (TLC) utilizes the excess electrical energy for serving the thermal load. The TLC is connected between electric and thermal power lines. The TLC persuades the surplus electric energy into thermal energy, which is eventually used to serve the thermal load demand. The initial cost, replacement cost, and lifetime of the TLC are 54 USD/kW, 54 USD/kW, and 20 years, respectively.

28.3.8 Boiler

In the HOMER simulator, it is assumed that the boiler is an existing infrastructure of the system. Hence, the capital cost is not included in the study. The diesel is utilized as the fuel for the boiler, and the fuel cost is considered in the simulation. HOMER assumes that the boiler has sufficient capacity to serve any amount of thermal load. The diesel fuel price is 0.719 USD/L [21].

28.3.9 Microgrid controller

An advanced microgrid controller, named predictive dispatch, has been employed in this study. The predictive dispatch has lower system operating costs compared to other microgrid controllers in HOMER Pro software. This control strategy can foretell future demand and resource data by utilizing 48 h of data stored within the system. Forecasting is used in this dispatch strategy to operate system components. It utilizes the energy storage system efficiently.

28.4 Optimization problem

A simulation is carried out in the HOMER Pro software. The NPC of the studied system can be written as follows: It should be noted that a carbon tax is included in this study. The value of the carbon tax is 40 $/ton [22].

$$minimization\ F_{NPC} = \sum_{i \in C} NPC_i \tag{28.4}$$

$$\begin{aligned} NPC_i = {} & E_{cap,i} + E_{rep,i} + E_{O\&M,i} + E_{fuel,i} + E_{refuel,i} \\ & + E_{dec,i} - E_{salv,i} \quad \forall i \in G \end{aligned} \tag{28.5}$$

where C, $E_{cap,i}$, $E_{rep,i}$, $E_{O\&M,i}$, $E_{fuel,i}$, $E_{refuel,i}$, $E_{dec,i}$, and $E_{salv,i}$ are set of component, capital cost, replacement cost, Operation and Maintenance (O&M) cost, fuel cost, refueling cost, decommissioning cost, and salvage value of ith component, respectively.

Capital cost of equipment is determined only once at the commencement of the project's lifetime. The following equation is used to calculate the capital cost of a system component.

$$E_{cap,i} = N_{com,i} \times E_{cap,unit(i)} \quad \forall i \in G \tag{28.6}$$

where $N_{com,i}$ and $E_{cap,unit(i)}$ are number of component and cost of unit component, respectively. Since microreactors are first-of-a-kind products, their capital cost is expected to reduce with lessons learned and experience gained. Eq. (28.7) calculates the capital cost of microreactors.

$$E_{cap,MR} = \sum_{n=1}^{N_{MR}} E_{cap,MR(1st)} \times (N_{MR})^{-R} \tag{28.7}$$

$$LR = 1 - 2^{-R} \tag{28.8}$$

where $E_{cap,MR(1st)}$ is capital cost of first microreactor module, N_{MR} is number of microreactor (MR), R is the rate of cost reduction, and LR is the learning rate (10%). Replacement cost of a component occurs after each component lifetime, and it continues throughout the entire project life span. Eqs. (28.9–28.12) determine the replacement cost of any system equipment.

$$E_{rep,i} = N_{com,i} \times \sum_{n=1}^{N_{rep}} E_{rep,unit(i)} \times \frac{1}{(1+r)^{j_{rep}}} \forall i \in G \tag{28.9}$$

$$N_{rep} = \left[\frac{T_{project}}{LT_{com,i}}\right] - 1 \tag{28.10}$$

$$r = \frac{i' - f}{1 + f} \tag{28.11}$$

$$j_{rep} = \sum_{n=1}^{N_{rep}} (n \times LT_{com,i}) \tag{28.12}$$

where $E_{rep,unit(i)}$, $T_{project}$, $LT_{com,i}$, i', and f are replacement cost of unit component, project lifetime, component lifetime, nominal discount rate, and inflation rate, respectively. Typically, O&M cost occurs at the end of each year. Eq. (28.13) calculates the O&M cost of a component.

$$E_{O\&M,i} = N_{com,i} \times E_{O\&M,annual(i)} \times \frac{\{(1+r)^{T_{project}}\} - 1}{r(1+r)^{T_{project}}} \quad \forall i \in G \tag{28.13}$$

where $E_{O\&M,annual(i)}$ is the O&M cost of unit component. Fuel cost is only calculated for microreactors. Renewable sources do not have any fuel cost. Fuel cost is also calculated at yearly interval for microreactors.

$$E_{fuel,i} = N_{com,i} \times E_{fuel,annual(i)} \times \frac{\{(1+r)^{T_{project}}\} - 1}{r(1+r)^{T_{project}}} \forall i \in G \tag{28.14}$$

where $E_{fuel,annual(i)}$ is fuel cost unit component. Refueling cost is also considered only for microreactors. Nuclear fuel modules are replaced at every 10 years intervals. The following equation determines the refueling cost of microreactors.

$$E_{refuel,MMR} = N_{MR} \times \sum_{n=1}^{N_{refuel}} E_{refueling,MR(unit)} \times \frac{1}{(1+r)^{j_{refuel}}} \tag{28.15}$$

$$N_{refuel} = \left[\frac{T_{project}}{LT_{fuel(MR)}}\right] - 1 \tag{28.16}$$

$$j_{refuel} = \sum_{n=1}^{N_{refuel}} \left(n \times LT_{fuel(MR)}\right) \tag{28.17}$$

where $E_{refueling,MMR(unit)}$ and $LT_{fuel(MR)}$ are refueling cost of single microreactor module and core lifetime of microreactor, respectively. Decommissioning Cost is collected uniformly starting in the first year, and it continues through the project life span. Since decommissioning is significant for nuclear industries, microreactor decommissioning cost is only considered in this study. Decommissioning cost for other system components is not considered here.

$$E_{dec,MMR} = N_{MR} \times E_{decom,MR(annual)} \times \frac{\{(1+r)^{T_{project}}\} - 1}{r(1+r)^{T_{project}}} \tag{28.18}$$

where $E_{decom,MR(annual)}$ is the decommissioning cost a microreactor module. Salvage value is the book value of a system component after its useful lifetime. HOMER considers linear depreciation to calculate the salvage value. The following equations are applied to calculate the salvage value of an equipment. If project life span is divided by a component lifetime and a remainder is not present, there will be no salvage value for that component.

$$E_{salv,i} = \begin{cases} 0, & \left(T_{project} \bmod LT_{com,i}\right) = 0 \\ x, & \left(T_{project} \bmod LT_{com,i}\right) \neq 0 \end{cases} \tag{28.19}$$

$$x = N_{com,i} \times E_{rep,unit(i)} \times \frac{LT_{com(i),rem}}{LT_{com,i}} \times \frac{1}{(1+r)^{T_{project}}} \forall i \in G \tag{28.20}$$

$$LT_{com(i),rem} = LT_{com,i} - \left(T_{project} - LT_{rep,i}\right) \tag{28.21}$$

$$LT_{rep,i} = LT_{com,i} \times \left[\frac{T_{project}}{LT_{com,i}}\right] \tag{28.22}$$

The optimization constraints for this problem are states as follows:

$$0 \leq MACS \leq MACS_{\max}$$

$$0 \leq MRF \leq MRF_{\max}$$

$$0 \leq OR(solar) \leq OR(solar)_{\max}$$

$$0 \leq OR(wind) \leq OR(wind)_{\max}$$

$$0 \leq OR(load\ in\ current\ time\ step)$$
$$\leq OR(load\ in\ current\ time\ step)_{\max}$$

$$0 \leq OR(annual\ peak\ load) \leq OR(annual\ peak\ load)_{\max}$$

where *MACS*, *MRF*, and *OR* denote Maximum Annual Capacity Shortage, Maximum Renewable Fraction, and Operating Reserve, respectively.

28.5 Simulation results

The objective function of the optimization problem for this study is to minimize the NPC. The NPC is the present value of all costs (e.g., capital cost, replacement cost, operation cost, and fuel cost) minus the current value of all revenues over the project lifetime. Microreactor will always provide constant power in this case study.

In this section, the simulations are carried out to determine the optimal system for the studied case. The three most influential system parameters—solar irradiance, temperature, and wind speed—are varied to handle the system's sensitivity analysis. In this investigation, it is shown how the system component operates depending on the contrast of solar irradiance, temperature, and wind speed while the electric demand thermal demand mare kept fixed. Later, the average annual electric and thermal demand are altered within a specific range to carry out another sensitivity analysis. At the same time, the other parameters (e.g., solar irradiance and wind speed) are kept constant at their original value. Besides, system equipment costs, such as microreactor capital cost, WT capital cost, and PV panel capital cost, have been varied by ±20% to assess the impact of equipment cost on system optimal configuration. This analysis represents the system equipment characterization and the requirement of different hybrid energy systems depending on the increased/decreased demand.

Table 28.2 represents the optimization results of the studied system. The optimal system includes PV panels, WTs, microreactors, battery banks, TLC, and converter within the system. The optimization tells that N-R IES meets the system demand economically.

Figs. 28.8 and 28.9 illustrate the generated electric and thermal energy distribution of the sources studied here. The results show that microreactor, wind, and solar fulfill all the electric load demand while the thermal demand is achieved by cogenerated heat from microreactor, excess electricity, and boiler.

Another technoeconomic analysis is conducted to compare the optimized winning system with an optimized base system. For this analysis, a ubiquitous system, a combination of PV panels, WTs, battery bank, converter, and boiler, is recognized as the base system. Fig. 28.10 presents the cumulative cash flow of the base case and winning system over the project lifetime. The comparison of technoeconomic parameters between the two systems is noted in Table 28.3. Though the initial cost is lower for the base case than the winning case, it results in a higher NPC and COE due to the increased operating cost. Moreover, due to reasonable IRR, ROI, and payback period, specified in Table 28.4, it is regarded as a good investment. So, it can be concluded that N-R IES could be an outstanding solution with lower cost and least GHG emission to satisfy the electric and thermal demand.

Figs. 28.11 and 28.12 show different system component costs and various cost contributing to system NPC over the

TABLE 28.2 Optimization results.

Architecture						Cost			
PV (kW)	WT	Microreactor (kW)	Battery	Thermal Load Controller (kW)	Converter (kW)	NPC ($)	COE ($)	Operating cost ($/yr)	Initial capital ($)
266	439	1,000	2,737	500	503	$68.0M	$0.425	$1.85M	$37.6M
		1,500		400		$68.1M	$0.424	$1.60M	$41.8M
0.0202		1,500		400	24.9	$68.1M	$0.424	$1.60M	$41.8M
	13	1,500		400		$68.1M	$0.424	$1.60M	$41.8M
		1,500				$68.2M	$0.425	$1.61M	$41.8M
0.0202		1,500			24.9	$68.2M	$0.425	$1.61M	$41.8M
	13	1,500				$68.2M	$0.425	$1.61M	$41.8M
		1,500	182	400	12.7	$68.4M	$0.426	$1.61M	$41.8M
		1,500	182		12.7	$68.5M	$0.427	$1.62M	$41.8M
	627	1,000	3,432	200	619	$68.6M	$0.430	$1.91M	$37.3M
147	686	1,000	2,862		525	$68.7M	$0.431	$1.90M	$37.5M
	17	1,500	320		4.29	$68.8M	$0.428	$1.63M	$41.9M
224	17	1,500		400	4.29	$69.0M	$0.430	$1.61M	$42.5M
224	17	1,500			4.29	$69.2M	$0.431	$1.62M	$42.4M

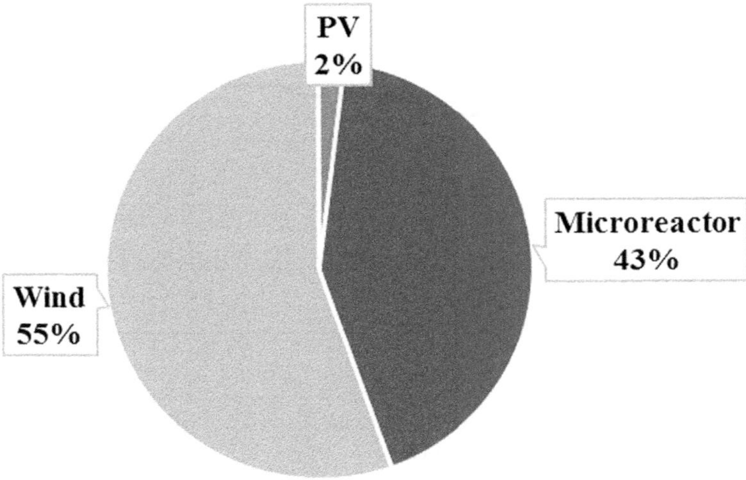

FIGURE 28.8 Production (electric) by sources.

project lifetime. Figs. 28.11 and 28.12, depict that capital cost and operating cost of microreactors have a significant share in total system cost. Figs. 28.13–28.19, represent power generation, conversion, and operating scenarios of different system components.

28.6 Sensitivity assessment

Three types of sensitivity analyses have been conducted in this study. In the first part, resources (e.g., wind speed, solar irradiance, and temperature) are varied to evaluate their impact on NPC. Fig. 28.20 shows that lower wind speed and solar irradiance suggest for stand-alone microreactor-based energy systems. On the other hand, higher wind speed and higher solar irradiance recommend only RES-based energy systems. N-R IESs are economical for the rest of the cases. Since the availability of RESs is site-specific, N-R IESs can be a potential solution to the energy crisis regardless of geographical location.

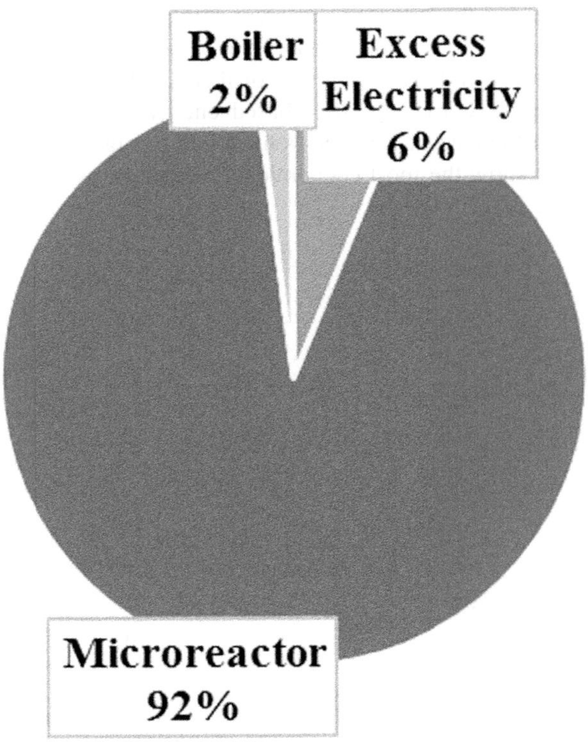

FIGURE 28.9 Production (thermal) by sources.

Fig. 28.21 tells that high temperature does not suggest including more PV panels since increased temperature degrades PV panel efficiency. However, PV panels can be combined with the systems if the wind speed is significantly high. WTs play an essential role in optimal system categorization and NPC due to lower cost and reasonable efficiency. Moderate wind speed always suggests N-R IES as a cost-effective system irrespective of temperature. Since solar irradiance and temperature are correlated, these are not varied and assessed in the same analysis.

In the second part of sensitivity analysis, different system costs, specifically WT capital cost, PV panel capital cost, and microreactors capital cost, have been altered to examine the results obtained in the earlier section. Fig. 28.22 shows that higher PV panels and higher WTs capital cost suggest stand-alone microreactor-based energy systems. N-R IESs are recommended for the rest of the scenarios for cost variations. Figs. 28.23 and 28.24 indicate that if microreactors' capital cost is degraded, stand-alone microreactors can serve the system demand economically. If microreactor capital cost is raised, only RES-based energy systems are the most reasonable solution. However, N-R IESs are still a feasible solution to serve the demand if both microreactors' capital cost and PV panel capital cost are high, as depicted in Fig. 28.23.

In the last part of the sensitivity analysis, average electric demand and average thermal demand are varied

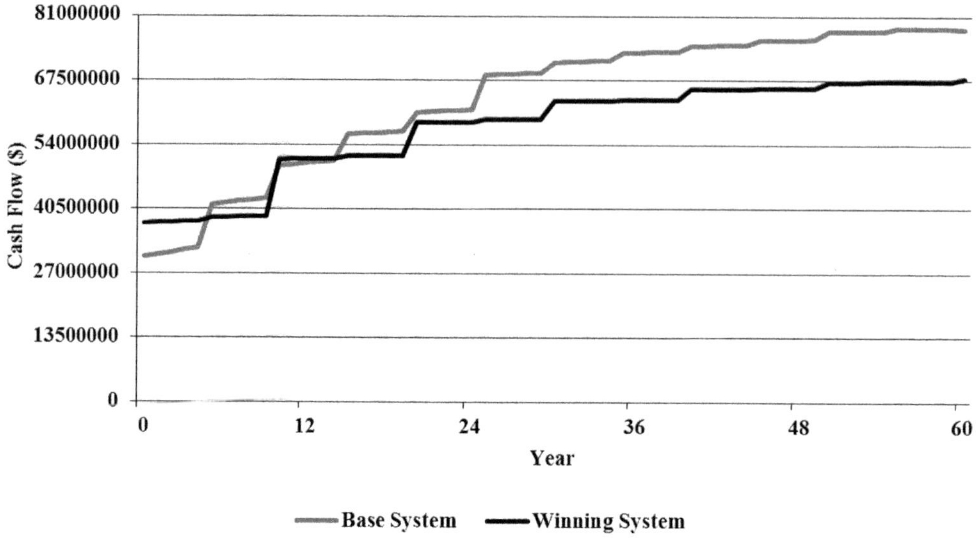

FIGURE 28.10 Cumulative cash flow over project lifetime.

TABLE 28.3 Comparison between optimized system and base system.

System	NPC (million $)	COE ($/kWh)	Initial cost (million $)	Operating cost (million $/year)
Base system	78.2	2.90	30.4	2.90
Winning system	68.0	0.497	37.6	1.85

TABLE 28.4 Economic metrics.

Parameters	Value
IRR (%)	16
ROI (%)	9.2
SPP (Year)	4.5

from 1000 to 30,000 kWh/day. Depending on the size of electric and thermal load, different combinations of system components serve the demand. Typically, lower electric demand, regardless of the size of thermal demand, can be accomplished by only RES-based energy systems. However, if the electric demand is moderate or high, N-R IESs always show the most economical investment, as indicated in Fig. 28.25.

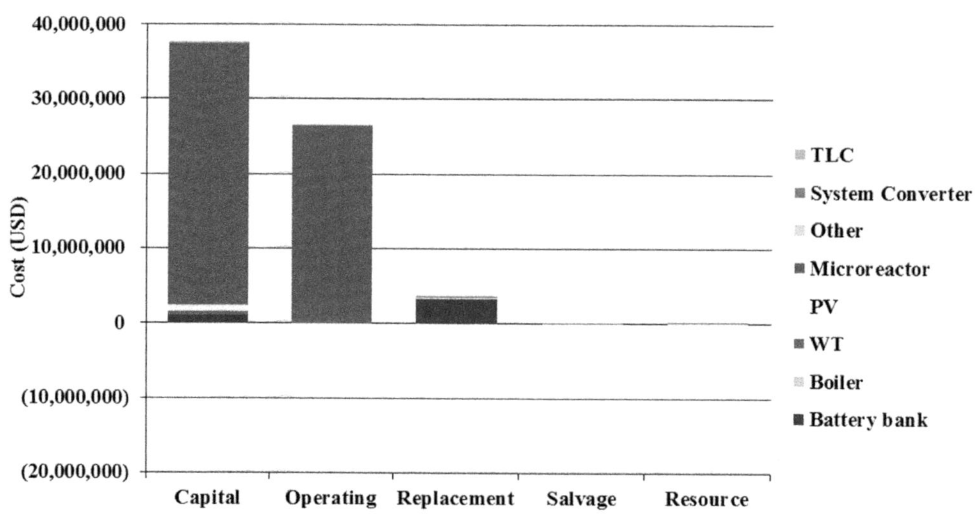

FIGURE 28.11 Cost summary.

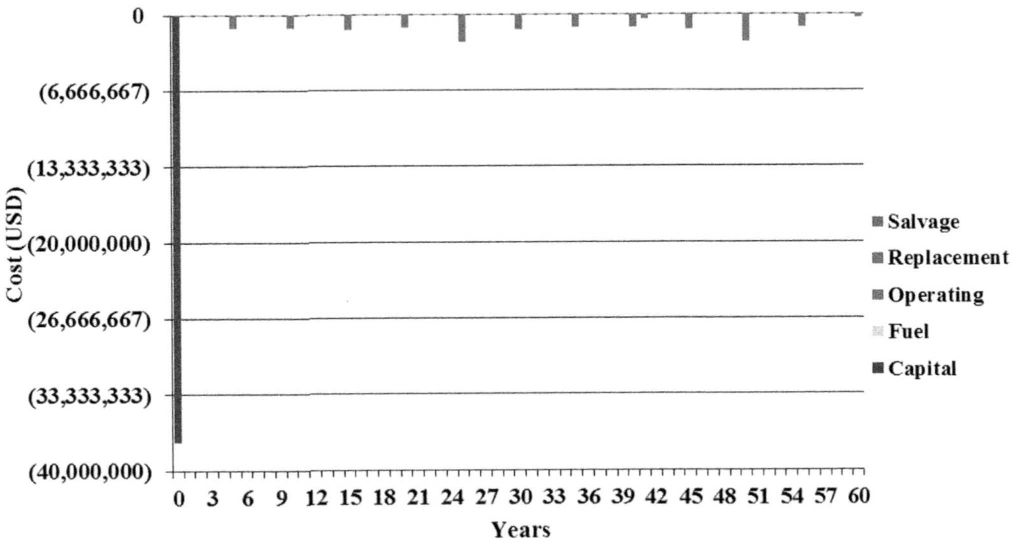

FIGURE 28.12 Cash flow.

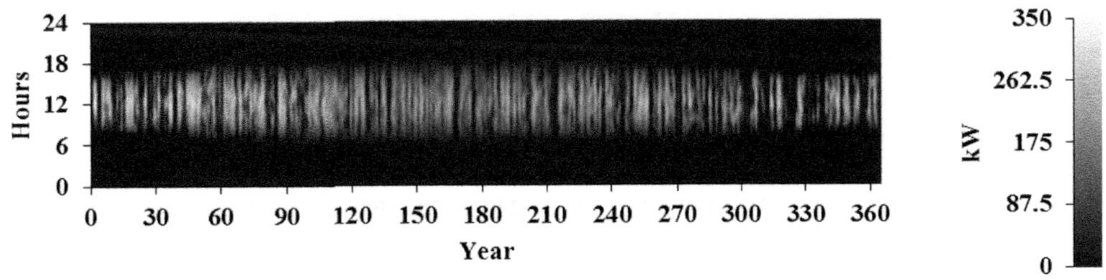

FIGURE 28.13 PV panels output.

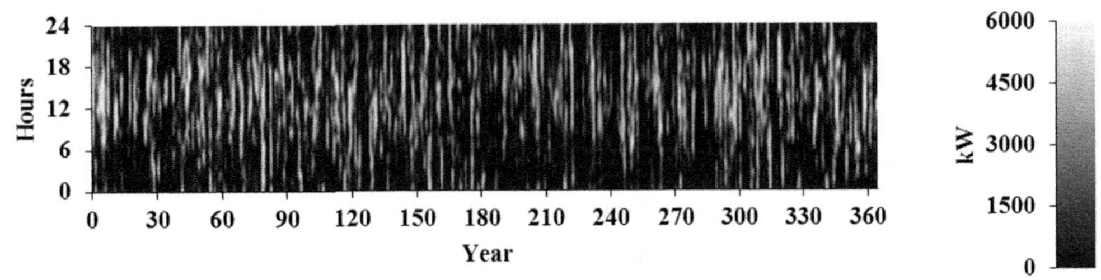

FIGURE 28.14 Wt output.

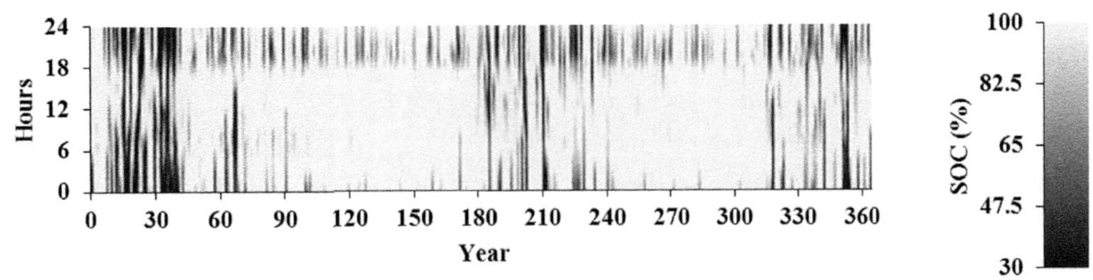

FIGURE 28.15 SOC of battery bank.

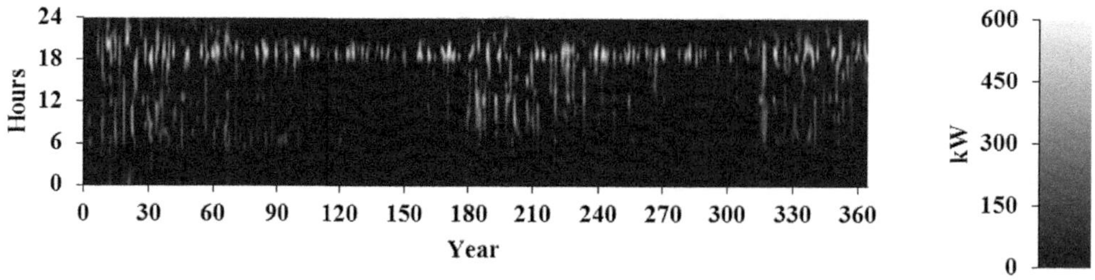

FIGURE 28.16 Inverter output.

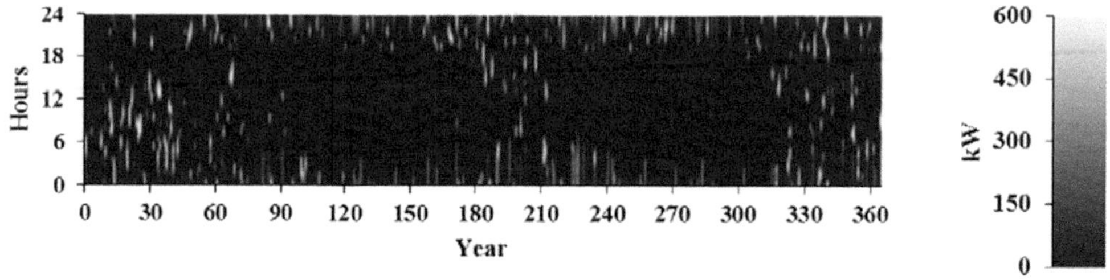

FIGURE 28.17 Rectifier output.

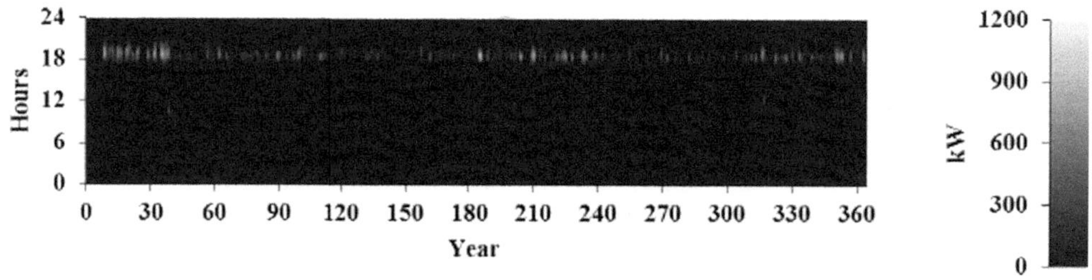

FIGURE 28.18 Boiler output.

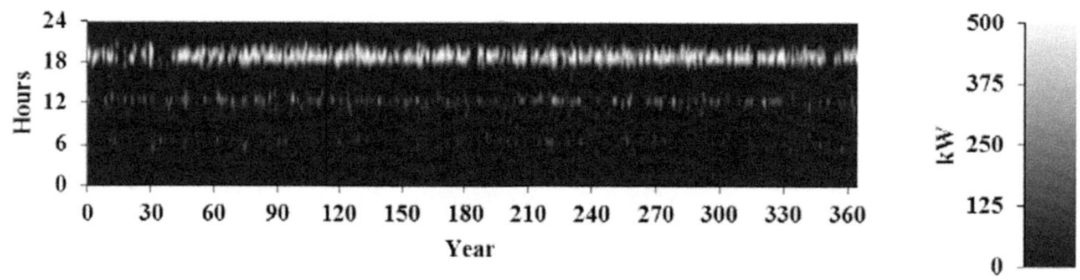

FIGURE 28.19 TLC output.

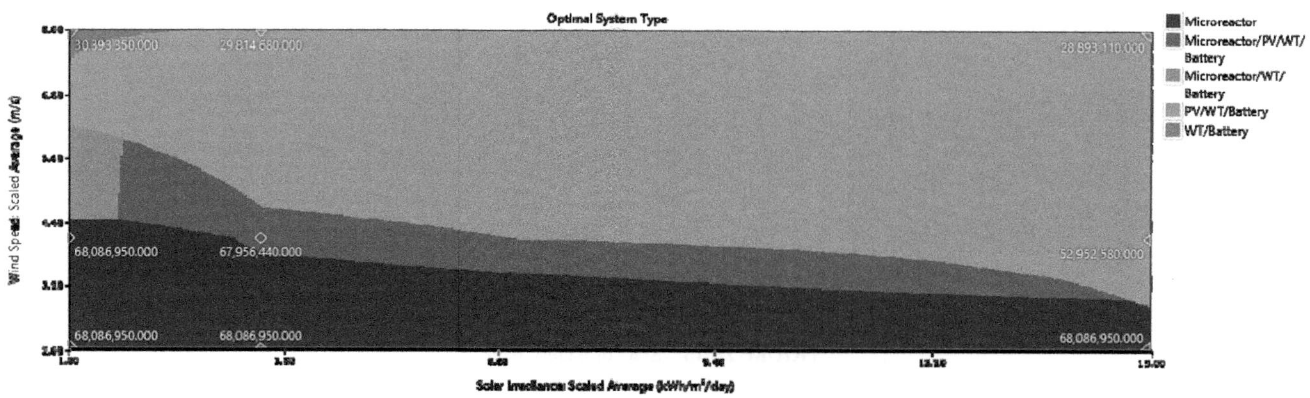

FIGURE 28.20 Sensitivity analysis (wind speed vs. solar irradiance).

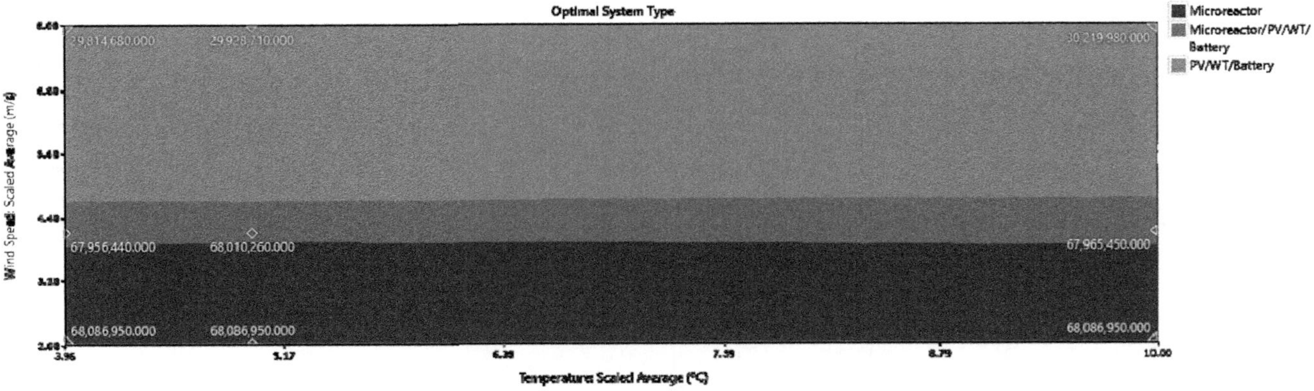

FIGURE 28.21 Sensitivity analysis (wind speed vs. temperature).

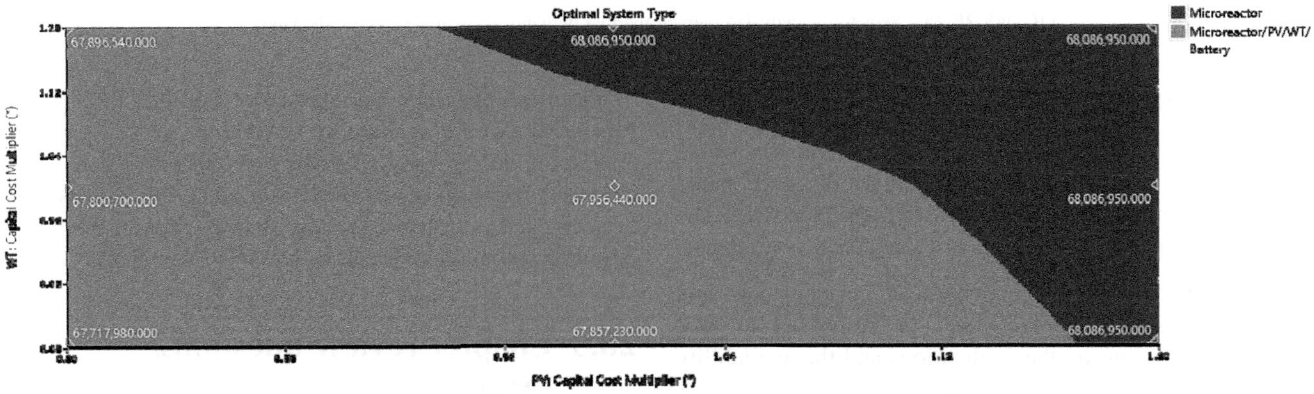

FIGURE 28.22 Sensitivity analysis (WT capital cost vs. PV panel capital cost).

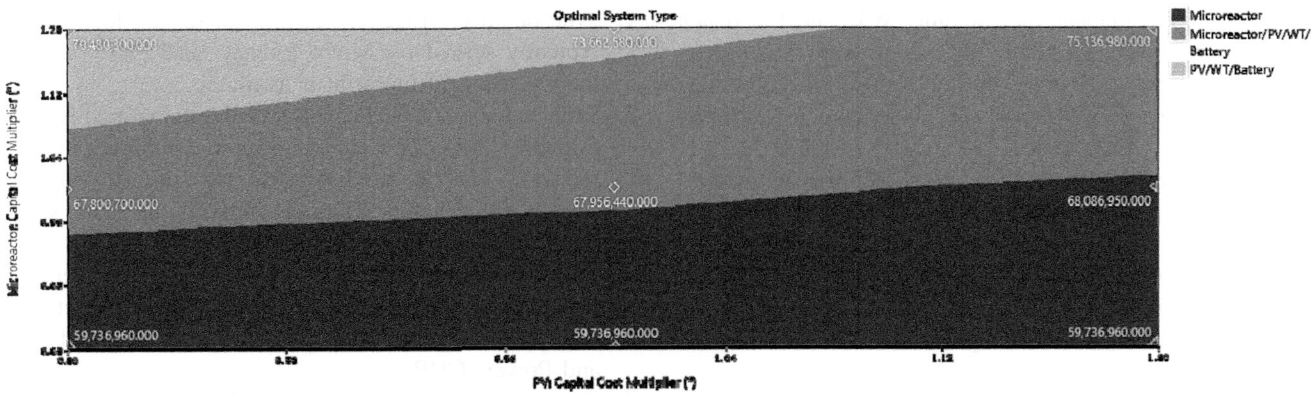

FIGURE 28.23 Sensitivity analysis (microreactor capital cost vs. PV panel capital cost).

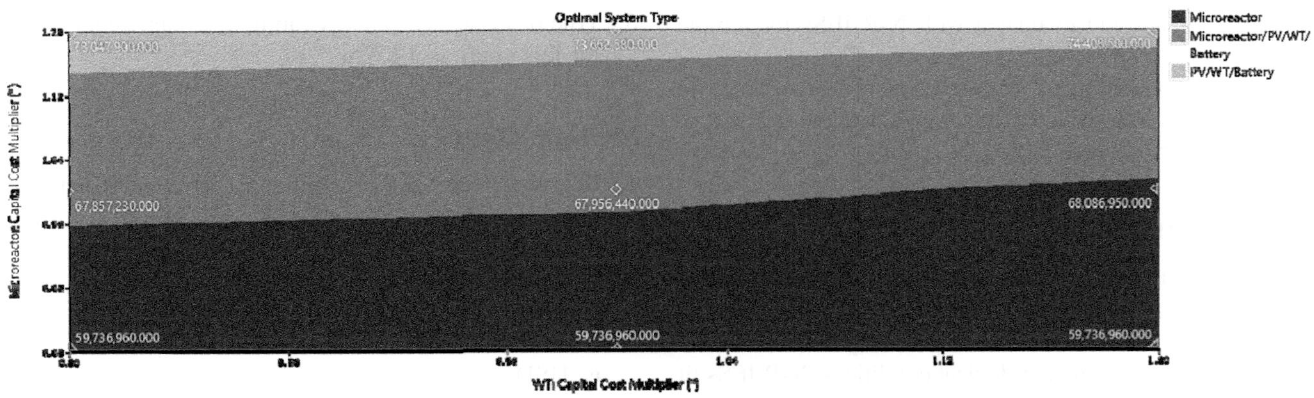

FIGURE 28.24 Sensitivity analysis (microreactor capital cost vs. WT capital cost).

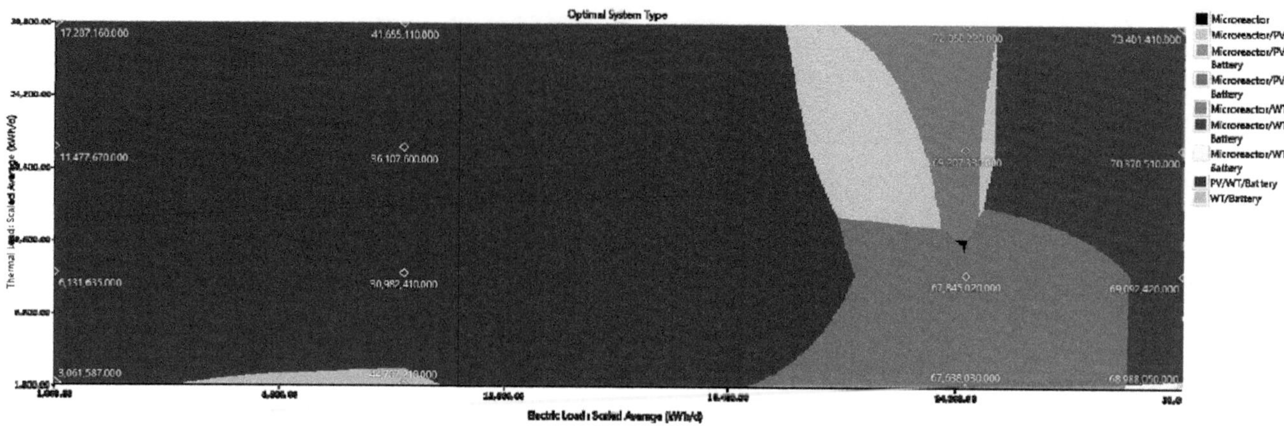

FIGURE 28.25 Sensitivity analysis (average thermal demand vs. average electric demand).

28.7 Future work

This study intends to consolidate nuclear and renewable sources to operate in tandem. Several works need to be done to determine the specific application of N-R IESs. The following can be proposed as future works:

- Different electric, electrochemical, mechanical, and thermal energy storage could be combined in N-R IESs depending on the applications and the availability to make the system more versatile.
- The cost associated with alternative energy generation sources that operate during equipment replacement, such as the nuclear fuel module, WT, and PV panel, should be incorporated in the future work. The energy management procedure during a specific component replacement period should also be documented in future research.
- The cost of real estate for nuclear facilities and renewable generation sources (where applicable) should be included in the future study if demanded to develop a realistic project outline.
- "Traceability Matrix" format can be adopted in the future to arrange the assumptions for extended research of this study.
- Electrification transportation is one of the most crucial issues at present, which has a massive impact on energy systems. Therefore, EV fast-charging stations could be incorporated and evaluated with N-R IESs for remote locations.
- Flexible operation of microreactors can be adopted in future research for N-R IESs to evaluate the technoeconomic performance analysis.

28.8 Summary

The case study indicates that N-R IESs could be a possible solution instead of only RES-based energy systems for off-grid applications. Since the availability of renewable sources depends on meteorological conditions, N-R IESs fit in almost all geographical regions for fulfilling the demand irrespective to the availability of RESs.

Finally, let us move on to the real interactive part of this chapter: review questions/exercises, hands-on projects, case projects, and optional team case project. The answers and/or solutions by chapter can be found in Appendix G.

28.9 Chapter review questions/exercises

True/false

1. True or False? Electric demand for distant locations is typically served by Micro Energy Grid (MEG), since the electric grid footprint is available.
2. True or False? A Nuclear-Renewable Integrated Energy Sources (N-R IES) is an energy hub combining nuclear reactors, energy storage systems, and all possible Renewable Energy Sources (RESs), such as solar, wind, hydro, biomass, and geothermal.
3. True or False? Microreactors are a fourth-generation small-scale nuclear reactor that can produce electricity and thermal energy simultaneously by Combined Heat and Power (CHP).
4. True or False? In HOMER, WT power input is measured, mainly focusing on the WT power curve provided by the manufacturer.
5. True or False? The energy rating of a single battery cell for this study is 1 kWh.

Multiple choice

1. The system converter is applied in between AC and DC bus:
   a. Cases
   b. Infrastructure
   c. Bars
   d. Unit converter
   e. USD

2. A Thermal Load Controller (TLC) utilizes the excess electrical energy for serving the thermal:
   a. Power
   b. Lines
   c. Load
   d. Energy
   e. Demand
3. In the HOMER simulator, it is assumed that the boiler is an existing infrastructure of the:
   a. System
   b. Machine
   c. Load
   d. Capacity
   e. Microgrid
4. Capital cost of equipment is determined only once at the commencement of the project's:
   a. Integration
   b. Growth
   c. System component
   d. Computer applications
   e. Lifetime
5. Depending on the size of electric and thermal load, different combinations of system components serve the:
   a. Demand
   b. Energy systems
   c. Renewable sources
   d. Applications
   e. System components

Exercise

Problem

Look at a mixed-integer linear optimization model to analyze the interdependencies between the electricity and gas distribution network operators' and the building owners' investment decisions during the transformation path.

Hands-on projects

Project

Do research: Analyze the scaling up of a hybrid system of energy generation.

Case projects

Problem

Look at the current developments in the smart energy sector, focusing on techniques in the main application areas, along with relevant implemented examples—while highlighting some of the key challenges currently faced and outlining future pathways for a sector.

Optional team case project

Problem

Describe how to coordinate clean energy generation technologies through integrated hybrid energy systems.

References

[1] Hirsch A, Parag Y, Guerrero J. Microgrids: a review of technologies, key drivers, and outstanding issues. Renew Sustain Energy Rev 2018;90:402−11. https://doi.org/10.1016/j.rser.2018.03.040.

[2] Gabbar HA, Zidan A. Optimal scheduling of interconnected micro energy grids with multiple fuel options. Sustain Energy Grids Netw 2016;7:80−9. https://doi.org/10.1016/j.segan.2016.06.006.

[3] Breeze P. "An introduction to combined heat and power," in combined heat and power. Elsevier; 2018. p. 1−11.

[4] Gabbar HA, Abdussami MR, Adham MI. Micro nuclear reactors: potential replacements for diesel gensets within micro energy grids. Energies 2020;13(19):5172. https://doi.org/10.3390/en13195172.

[5] Gabbar HA, Abdussami MR, Adham MI. Optimal planning of nuclear-renewable micro-hybrid energy system by particle swarm optimization. IEEE Access 2020;8:181049−73. https://doi.org/10.1109/ACCESS.2020.3027524.

[6] Gabbar HA, Abdussami MR, Adham MI. Techno-economic evaluation of interconnected nuclear-renewable micro hybrid energy systems with combined heat and power. Energies 2020;13(7):1642. https://doi.org/10.3390/en13071642.

[7] N. Honarmand, Key performance indicators modeling for optimized microgrid configuration, p. 141.

[8] Bragg-Sitton SM, et al. Nuclear-renewable hybrid energy systems: 2016 technology development program plan. 2016. https://doi.org/10.2172/1333006. INL/EXT-16-38165, 1333006.

[9] USNC. https://usnc.com/. (accessed December 10, 2019).

[10] Data directory. http://www.ieso.ca/en/Power-Data/Data-Directory. (accessed March 17, 2020).

[11] USNC. https://usnc.com/MMR.html. (accessed March 14, 2020).

[12] Cost competitiveness of micro-reactors for remote markets. Nuclear Energy Institute; April 15, 2019.

[13] How HOMER calculates the PV array power output. https://www.homerenergy.com/products/pro/docs/latest/how_homer_calculates_the_pv_array_power_output.html. (accessed November. 26, 2019).

[14] Huawei - building a fully connected, intelligent world Huawei. https://www.huawei.com/en/. (accessed December 22, 2020).

[15] How HOMER calculates wind turbine power output. https://www.homerenergy.com/products/pro/docs/latest/how_homer_calculates_wind_turbine_power_output.html. (accessed November 26, 2019).

[16] N. E. B. Government of Canada. Neb − market Snapshot: the cost to install wind and solar power in Canada is projected to significantly fall over the long term. February 13, 2020 (accessed July 24, 2020), https://www.cer-rec.gc.ca/nrg/ntgrtd/mrkt/snpsht/2018/11-03cstnstllwnd-eng.html.

[17] US wind O&M costs estimated at $48,000/MW; Falling costs create new industrial uses: IEA | new energy update. https://analysis.newenergyupdate.com/wind-energy-update/us-wind-om-costs-estimated-48000mw-falling-costs-create-new-industrial-uses-iea. (accessed April 18, 2020).

[18] Stehly T, Beiter P, Heimiller D, Scott G. 2017 cost of wind energy review. Renew Energy 2018:61.
[19] Kharel S, Shabani B. Hydrogen as a long-term large-scale energy storage solution to support renewables. Energies 2018;11(10):2825. https://doi.org/10.3390/en11102825.
[20] Johnson SC, et al. Selecting favorable energy storage technologies for nuclear power. In: Storage and hybridization of nuclear energy. Elsevier; 2019. p. 119–75.
[21] Canada diesel prices, 18-Nov-2019 | GlobalPetrolPrices.com. https://www.globalpetrolprices.com/Canada/diesel_prices/. (accessed November 26, 2019).
[22] HOMER - Hybrid renewable and distributed generation system design software. https://www.homerenergy.com/. (accessed January 03, 2020).

Chapter 29

Reactive power compensation for voltage variation mitigation in grid with renewable energy

Om Prakash Mahela[1], Ashok Gocher[2], Baseem Khan[3], Sunil Agarwal[2], Akhil Ranjan Garg[4] and Hassan Haes Alhelou[5]

[1]Power System Study Division, Rajasthan Rajya Vidyut Prasarn Nigam Ltd., Jaipur, Rajasthan, India; [2]Department of Electrical Engineering, Apex Institute of Engineering and Technology, Jaipur, Rajasthan, India; [3]Department of Electrical and Computer Engineering, Hawassa University, Hawassa, Ethiopia; [4]Department of Electrical Engineering, Faculty of Engineering and Architecture, J.N.V. University, Jodhpur, Rajasthan, India; [5]Tishreen University, Latakia, Syria

29.1 Introduction

Transmission system network in Rajasthan is constructed, operated, and maintained by the state transmission utility (STU). Rajasthan Raja Vidhyut Prasaran Nigam Ltd. (RVPN) performs the activities of STU in Rajasthan. RVPN maintains transmission network associated with voltage levels of 33, 132, 220, 400, and 765 kV. The generation of power is achieved using the thermal power plants, nuclear power plants, hydel power plants, and renewable energy (RE) sources. The RE potential (Solar and Wind) mainly is concentrated in the western part of Rajasthan. The generated energy is required to be transmitted for long distances to reach the load centers of Rajasthan, which are mainly in the central and eastern part of Rajasthan, and also to be transmitted to northern and western grids of India through Interstate Transmission System (ISTS). The government of India has ambitious plan of 175 GW of RE capacity by the end of year 2022. The predominance of this RE capacity shall be through solar plants for which the Rajasthan already has a strategic advantage. Government of Rajasthan has also stated to strengthen the transmission infrastructure to harness the full potential of renewable energy in Rajasthan. The existing installed capacity of RE power is 7384 MW. Further, the underexecution RE projects are owned by the state utilities having capacity of 10,000 MW. The RE projects under execution of the central transmission utility (CTU) have the capacity of 20,330 MW in the central sector projects category [1,2].

The intermittency of renewable energy, poor plant load factor for renewable generators, etc., pose various challenges in operating the transmission system. The integration level of existing RE (only considering state sector renewable generation) against existing handled peak load by transmission system is around 32%. At this integration level of renewable energy, the transmission system is experiencing high voltages on various EHV (extrahigh voltage) buses at substations, overloading of few transmission corridors, and constrained capability to transfer power toward load centers. Buses of 400 and 220 kV level are already experiencing the overvoltages in the range of 420–450 kV and 230–250 kV, respectively [3]. The abovementioned challenges create problems in the evacuation of RE power. However, this can be obviated by optimal restructuring of the existing transmission system by addition of the grid substation (GSS) and transmission lines of different voltage levels. A wide range of models has been reported for restructuring of power system network in India for meeting load requirements as well as requirements of policies of both the state and central governments. This is also done for meeting the needs of social as well as market of the state. This also helps to meet out the increased load demand and evacuation of increased generation either conventional or RE. The philosophy and approaches used for the purpose of planning as well as operation, which are well established during the era of past decades, are being changing. The need has also been arising that India should recognize and meet these forthcoming challenges. To make the market as competitive, there are

Smart Cities Policies and Financing. https://doi.org/10.1016/B978-0-12-819130-9.00004-8

various methods adopted for the purpose of restructuring the industry of power. Different types of reforms had been taken in the past looking to the setup of the organizations, financial condition of the utilities, control structure adopted, and coordination used [4]. Restructuring of power system network has been carried out in different countries of the world due to various reasons. Romanian government has taken a step toward restructuring of network of state utilities and assets of state-owned power generation system, which had been considered in 2010. This had been achieved with the help of creation of two companies at national level. These companies help the all power generation assets being owned by the state governments or by the two companies owned by the state and related to coal mining. This is also speed up by the three distribution companies owned by the state at regional electricity. Ministry of Economy, Trade and Business Environment has taken a decision in the Romania for restructuring of the majority assets of power generation in the Romania [5]. In Ref. [6], authors analyzed the welfare-related complicacies associated with implications of reforms of power sector during a condition where there is both the public-sector undertakings and utilities of private sector. This study is related to the restructuring of state power sector in West Bengal state of the India. Fifteen-year period has been considered for the review, which is associated with the prerestructuring period and postrestructuring period. Analysis of the panel data related to four customer categories gives the negative impact associated with the disintegration on welfare-related activities due to increased cost of the transmission and increased rates of the tariff, which results in decreased number of the power consumer and power producers. In Ref. [7], authors examined process adopted for the implementation of different governmental schemes based on the network perspective in Haoqiao, which is a poor village situated in northern China. Authors focused mainly on interactions between consumers and their impacts restructuring process in the rural areas.

29.1.1 Literature review

This section details the review of literature related to restructuring of transmission network to include the reactive power compensation and grid integration of RE generators. In Ref. [8], there are details the industry of electrical power has undergone restructuring in the United States of America (USA). This had started in the year 1990s, which had been aimed at improving the competitiveness of market subjected to effective allocation of the resources target at reduction in the cost. Firstly, restructuring had been implemented in the State of Delaware, which has now gained an experience of more than 20 years. In Ref. [9], it is essential for make appropriate modeling of the system and handle uncertain parameters during the power system studies. In this research work, sources of the uncertainties in networks of power systems have been described, and there are different methods for handling the uncertainties in power systems. Different methods have been reviewed in detail and merits as well as demerits of each method have been given. In Ref. [10], authors equipped the SDEA PO, which is based on the gradient of efficiency score, had been projected. Authors employed the indicators for gross domestic product (GDP) as well as carbon dioxide emissions (CO_2) level, which may be the output. This may be undesirable and population as well as clean energy consumption may be considered as the input. In the electrical power system networks, there may exist different diverse parameters, which are uncertain in nature such as the loads, prices of the electricity, wind energy power generation, and solar-energy-based photovoltaic power generation. In Ref. [11], authors investigated the efficiency as well as impacts of electricity market restructuring on the environment by examination of the changes that had taken place in the efficiency of fuel, cost of coal purchasing, and use in the coal-fired power generation plants with the help of panel data recorded from 1991 to 2005. This study is concentrated specifically on the coal-fired power generation plants. This has used panel data that covers different years after the implementation of restructuring has been started. In Ref. [12], authors presented detailed report on restructuring of electricity market in Iran. The different changes have undertaken in world electricity industry. Initial steps for restructuring of electricity industry in Iran had been considered in the year 1990. This had resulted in launching the wholesale electricity market of Iran in November 2003. Achievement of long-term security of electrical supply, clarification on prices of electricity and attraction of private investors are important key goals for the restructuring of the electricity industry in Iran. Privatization as well as breaking the monopoly of the governmental, had started in the year 2000 when the third development plan had started. This has resulted in 48.05% contribution from the side of private sector toward annual energy generation in the year 2015. A specific data envelopment analysis with a perfect object method (SDEA PO) method constituting of the restructuring approach has been reported in the scheme. This section details the review of literature related to grid integration of Renewable Energy. Grid integration of distributed generation (DG) sources such as solar photovoltaic (PV) system, wind energy, and fuel cell affects the quality of power, degrades the system reliability, and causes overvoltages in the network [13]. The power fluctuations at the output of wind and solar PV systems due to the type of turbines and solar PV systems, variations in the wind, and solar insolation over a period may lead to power quality (PQ) disturbances such as sag in voltage and voltage swell along with reactive power flow [14]. Recognition and localization of these PQ disturbances associated with outage event and grid synchronization event of solar PV plant in distribution power network are investigated in Ref. [15]. In Ref. [16], authors present a detailed study and current status of various applications of computational techniques such as conventional, optimization and artificial intelligence (AI). This technique has been used for assessment of impact of optimally placed and coordinated control of DGs and FACTS controllers for the improvement of power system performance parameters. In this we use analytical methods and heuristic optimization techniques. It is concluded that the analytical methods are simpler in implementation but have slow computations and poor convergence. The heuristic optimization techniques are fast and good in convergence. In Ref. [17], a three-phase three-wire distribution static compensator (DSTATCOM) along with PV system and battery system operated DC–DC boost converter has been proposed for the PQ improvement in the distribution system network. Various types of studies related to utility grid are reported in Refs. [18–35].

Based on the literature review presented in this section, it is observed that the voltage instability in certain part of the complex transmission system can be mitigated by optimal placing of reactors in the transmission network. However, this will depend on the complexity of the network and type of generations integrated to the system. Further, this type of restructuring should be in accordance with the regulatory frame work. In India, Central Electricity Authority (CEA) monitors all the activities related to addition, deletion, and restructuring of the network. In Rajasthan transmission system, large quantum of the RE power is expected to be integrated by 2023, which will be mainly concentrated in the western parts. Hence, it is pointed out that the planning of additional transmission system is needed to mitigate power evacuation constraints so that RE power can be easily and efficiently transported to the load centers. Further, optimal reactive power compensation is required to mitigate the voltage variations and over voltages.

29.1.2 Contribution of the research work

The research work detailed in this work is mainly focused on implementation of the additional reactive compensation both dynamic and static to mitigate voltage variations due to the RE penetration in the western parts of the Rajasthan to provide effective reactive compensation. The major contributions of this research work are as follows:

- To maintain the voltage profile of the transmission system by effective reactive power management using both the static and dynamic compensation.
- The proposed study is performed using the MiPower software using the load flow tools and short-circuit study tools.

29.1.3 Organization of the manuscript

The manuscript is organized as follows. Section 29.2 discussed the case study area. Section 29.3 discussed the methodology utilized in this work. Section 29.4 discussed the results and analysis. Conclusion and future scope are presented by the Section 29.5.

29.2 Overview of utility grid

Rajasthan Rajya Vidyut Prasaran Nigam Limited (RVPNL) is a power utility company registered under the provisions of Companies Act, 1956. This has been registered in the office of the Registrar of Companies as "Rajasthan Rajya Vidyut Prasaran Nigam Limited." This bears the registration number as No. 17-016485 of 2000–2001 and has Registered Office at Vidyut Bhawan, Jyoti Nagar, Jaipur – 302005. This company has been formed on 19 July 2000 by Govt. of Rajasthan after the implementation of Rajasthan Power Sector Reforms Act, 1999. This is the successor company of RSEB. Prime duty of RVPNL is related to the transmission of electrical power in the state of Rajasthan. It holds, develops, maintains, and operates transmission system rated at high voltage. This helps for keeping lights on and running and communities strong. RVPN is operating and maintaining about 34,688 circuit km of transmission lines all over Rajasthan with a total installed transformation capacity of about 69,456 MVA distributed in about 523 substations.

Rajasthan Rajya Vidyut Utpadan Nigam Limited (RVUN) is involved in the development of large-sized power plants owned by the state government. This is also responsible for operation and maintenance of state-owned power generating stations. Rajasthan RVUN has been constituted on 19th July 2000 by the provisions of the companies Act-1956. This company has been playing important and leading role for power generation in the Rajasthan State and rapid development of the state. Generating Stations of RVUN have the status of distinctive reputation all over the country due to efficient and economic generation of the electrical power.

29.3 Methodology

This section discusses the base case of transmission system and proposed methodology for planning for placement of dynamic and static reactors in western parts of Rajasthan for RE power evacuation. The transmission network of Rajasthan along with the network of central section and generation of Rajasthan state and central sector generation is used as base transmission system for proposed study of planning of transmission system for mitigation of RE evacuation constraints. Base transmission system is described in this chapter in detail. Important elements of Rajasthan power system such as conventional generation,

RE generation, and 765 kV Grid substation (GSS) are described in this chapter. Problem formulation is also presented in this chapter. The proposed approach used for the study of transmission system restructuring is also described in detail in this chapter.

29.3.1 Problem formulation

Voltages on the 765 and 400 kV network of the Rajasthan have recorded maximum voltages, which exceed the above the limits specified in Grid Code. Many transmission lines had been tripped or manually kept opened to curb high voltages in the system. Construction of large number of long EHV transmission lines for integration of large-scale RE power is the major reason for high voltages in Rajasthan. Further, the state transmission utility (STU) as well as central transmission utility (CTU) is constructing the additional transmission system in Rajasthan to evacuate RE power. This will result in further increasing of high voltages in Rajasthan power system. The high voltages result in failure of the equipment such as transformers, insulators, etc., and tripping of EHV transmission lines. Present practices include the manual operation of the Reactors and Capacitors to control the flow of reactive power in the transmission system. Hence, studies are required to introduce the additional reactors in the system and at critical location automatic switchable reactors, which can provide both the inductive and capacitive reactive powers as required. Hence, this study is focused to investigate the additional reactive compensation required in the transmission system of Rajasthan and location for the Thyristor-controlled reactors (TCR) to control the voltages.

29.3.2 Base transmission system

The existing practical network of the transmission system of the Rajasthan State Transmission Utility, including the system of voltage levels 132, 220, 400, and 765 kV, is considered for the study. The conventional generators such as thermal power plants, nuclear power plants, and hydro power plants are integrated to this system. RE is continuously being integrated to the existing network specifically in the western Rajasthan. The existing network consists of transmission lines and GSS rated at voltage levels such as 765, 400, 220, 132, and 33 kV. This system contains the power system network developed by the CTU and STU. In Rajasthan, the RVPN performs the responsibility of the STU for development of the transmission network. Power Grid Corporation of India (PGCIL) performs the responsibility of CTU to develop the high voltage level network in the India. Following are the important 765 kV GSS owned by the CTU and STU operating in the territory of Rajasthan:

TABLE 29.1 Grid substation statistics for lengths of transmission lines.

S. No.	Voltage level	Number of GSS	Circuit length of transmission lines (km)
1	765 kV	6	425.498
2	400 kV	27	7604.444
3	220 kV	124	15,443.394
4	132 kV	459	18,245.566
Total		616	41,718.902

- 765 kV GSS Anta owned by RVPN
- 765 kV GSS Phagi owned by RVPN
- 765 kV GSS Ajmer owned by CTU
- 765 kV GSS Chhitorgarh owned by CTU
- 765 kV GSS Bikaner owned by CTU and under construction
- 765 kV GSS Khetri owned by CTU and under construction
- 765 kV GSS Bhadla-I and II owned by CTU and under construction
- 765 kV GSS Fatehgarh-I under construction by tariff based competitive bidding (TBCB)
- 765 kV GSS Fatehgarh-II under construction and owned by CTU
- 765 kV GSS Kucheri under construction and owned by CTU
- 765 kV GSS Sikar under construction and owned by CTU

The existing GSS inside the Rajasthan and length of the total circuits of transmission lines as on March 31, 2020 are provided in Table 29.1. These include the 765 kV GSS, 400 kV GSS, 220 kV GSS, and 132 kV GSS. The 33 kV GSS and 11 kV GSS are owned by the distribution companies, and these were constructed according to the load demand. Extra high voltage (EHV) substations are constructed by RVPN, PGCIL, and under the scheme of public–private partnership (PPP). Table 29.2 presents the generation schedule of the Rajasthan state.

29.3.3 Planning of reactive compensation to mitigate voltage variations

The planning of the additional reactive compensation using the static reactors to mitigate the problem of voltage

TABLE 29.2 Generation schedule of Rajasthan.

S. No.	Power plants	Installed unit-wise capacity	Total installed capacity
1	Suratgarh power plant	6 × 250 + 2 × 660	2820
2	Chhabra power plant	4 × 250 + 2 × 660	2320
3	Kawai power plant	2 × 660	1320
4	KTPS power plant	2 × 110 + 3 × 210 + 2 × 195	1240
5	Kalisindh power plant	2 × 600	1200
6	GIRAL LTPS	2 × 125	250
7	MAHI HEP	2 × 25 + 2 × 45	140
8	RPS HEP	4 × 43	172
9	JS HEP	3 × 33	99
10	RAMGARH GTPS	1 × 35.5 + 2 × 37.5 + 1 × 50 + 1 × 110	270.50
11	DHOLPUR GTPS	3 × 110	330
12	Barsingsar LTPS	2 × 125	250
13	RAPP A	1 × 100 + 1 × 200	300
14	RAJWEST LTPS	8 × 135	1080
15	VSLP LTPS	135	135
16	RAMGARH GTPS (Stage IV)	1 × 50 + 1 × 110	160
17	ANTA GTPS	1 × 153.2 + 3 × 88.71	419.33
18	RAPP B	2 × 220	440
19	RAPP C	2 × 220	440
20	RAPP D	2 × 700	1400
21	SHRI CEMENT	2 × 150	300
	Total		27,727.795

variations in the Rajasthan Transmission system is also proposed in this work. This also includes the TCR.

29.3.3.1 Existing and additional transmission system in Rajasthan

Existing and under progress of installation of additional transformer capacity in Rajasthan state as on March 31, 2019 is detailed in Table 29.3. The CTU is installing additional transformation capacity of 24,000 MVA, 765/400 kV, and 10,500 MVA, 400/220 kV.

The details of the total installed capacity of shunt reactors rated as voltage levels of 765 and 400 kV including bus-type and line-type reactors planned by STU and CTU in Rajasthan are given in Table 29.4. Degree of inductive reactive power compensation is defined as ratio of sum of

TABLE 29.3 Existing transformer capacity existing transformer capacity.

S. No.	Voltage level	No. of GSS/MVA capacity	
		STU	CTU
1	765 kV GSS	2 (7500 MVA)	2 (6000 MVA)
2	400 kV GSS	18 (14,385 MVA)	9 (7855 MVA)
3	220 kV GSS	122 (30,295 MVA)	–
4	132 kV GSS	452 (31,850 MVA)	–
Total		594	11

TABLE 29.4 Reactive compensation in Rajasthan transmission system.

			STU		CTU	
S. No.	Particulars		Bus type (MVAR)	Line type (MVAR)	Bus type (MVAR)	Line type (MVAR)
1	765 kV GSS	Installed	240	1920	480	1440
		In progress	–	–	1680	4200
	Total		240	1920	2160	5640
2	400 kV GSS	Installed	1790	950	1480	1026
		In progress	125	160	875	–
Total			1915	1110	2355	1026

TABLE 29.5 Voltage limits.

Nominal voltage (kV)	Maximum voltage kV)	Minimum voltage (kV)
765	800	728
400	420	360

all bus-type and line-type reactors installed at the GSS to 50% of no load charging MVAR of directly connected Transmission lines at the GSS. In Present scenario, Degree of Inductive Reactive Power compensation of the GSS should be at least 100% to control flow of reactive power from one voltage level to another voltage level in order to curb high voltage and to reduce transmission losses. As per Indian Electricity Grid Code and CEA's planning criteria January 2013, 765 and 400 kV bus voltages should remain within the following limits as detailed in Table 29.5. Further, recorded voltage variation index (VVI) is observed as high as 5% in the Rajasthan Transmission System.

29.3.3.2 Additional requirement of static reactors and TCR

Based on the preceding recorded voltages at GSSs and to control 765 and 400 kV voltage within limits prescribed in the Indian Electricity Grid Code and CEA's Manual on Planning Criteria, Grid Substations have been identified for Static and Dynamic compensation at 765 and 400 KV levels. Installation of 240 MVAR, 765 kV bus reactor at 765 kV GSS Phagi has been planned by CTU. GSSs located in western Rajasthan have recorded large variations in the voltage and the variations. The following are the key issues for the additional reactive power compensation:

- 1495 MW capacity solar power projects are directly connected at 400 kV GSS Bhadla. Solar power generation connected to 220 kV GSS Kanasar (90 MW) and 132 kV GSS Khetusar (70 MW) is also evacuated through 400 kV GSS Bhadla. In addition to above, approximately 730 MW capacity solar projects are also envisaged to be connected at 400 kV GSS Bhadla, 220 kV GSS Kanasar, and 132 kV GSS Khetusar. To control voltage fluctuations, TCR has been proposed at 400 kV GSS Bhadla.
- 400 kV GSS Bikaner is directly connected to 400 kV GSS Bhadla through 400 kV D/C line. To control voltage fluctuations, TCR has been proposed at 400 kV GSS Bikaner.
- 400 kV GSS Merta is directly connected to 400 kV GSS Bhadla and 400 kV GSS Jodhpur through 400 kV lines. To control voltage fluctuations, TCR has been proposed at 400 kV GSS Merta.
- Approximately 2500 MW capacity solar projects and 300 MW capacity wind power projects are envisaged to be connected at 400 kV GSS Jaisalmer-2. 1985 MW capacity wind power projects are directly connected at 400 kV GSS Akal. 400 kV GSS Akal and 400 kV GSS Jaisalmer-2 are to directly connected at 400 and 220 kV voltage level. Therefore, to control voltage fluctuations, TCR has been proposed at 400 kV GSS Jaisalmer-2.
- As per load flow study results, 300 MVAR capacity TCRs have been planned at 400 kV GSS Bhadla, Bikaner, Merta, and Jaisalmer-2.
- Additional reactors suggested for installation are tabulated in Table 29.6. TCR-based reactors proposed to be installed are shown in Table 29.7.

29.4 Results and discussion

This section discusses the results computed by the proposed methodology on the case study system.

TABLE 29.6 Proposed static reactors.

S. No.	Name of GSS	Voltage level	Capacity of shunt reactor in MVAR
1	765 KV GSS Anta	765 kV	240
		420 kV	125
2	400 KV GSS Heerapura	420 kV	125
3	400 KV GSS Ajmer	420 kV	125
4	400 KV GSS Bhilwara	420 kV	125
5	400 KV GSS Babai	420 kV	125
6	400 KV GSS Chhitorgarh	420 kV	125
7	400 KV GSS Jodhpur(new)	420 kV	125
8	400 KV GSS Barmer	420 kV	125
Total capacity at 765 kV voltage		765 kV	240 MVAR
Total capacity at 400 kV voltage		420	1000 MVAR

TABLE 29.7 Proposed TCR.

S. No.	Name of 400 kV GSS	Capacity of TCR in MVAR
1	400 KV GSS Merta	300
2	400 KV GSS Bikaner	300
3	400 KV GSS Bhadla	300
4	400 KV GSS Jaisalmer-2	300
	Total	1200 MVAR (4 × 300 MVAR)

29.4.1 Load flow study to check feasibility of the additional reactive compensation

Load flow studies have been carried out for condition corresponding to 2022–23 for the light load system condition, namely 10,266 MW and low RE generation to simulate the effect of proposed shunt reactors on the Grid voltage. Following have been considered in the load flow study:

- In the load flow studies, net dispatch of Intra-state Solar/Wind generation is 3010 MW, which is 20% of the installed capacity of Solar and Wind generators.
- In the load flow studies, net dispatch of Inter-State Solar generation is 2446 MW, which is 20% of the 12,230 MW installed capacity of Solar and Wind generators.
- An additional load of 2608 MW has been reflected on ISTS 765 kV buses to simulate the conditions of export of Inter-state RE generation to outside the State.
- An additional Generation of 642 MW has been reflected on ISTS buses to simulate the conditions of import of power from Western Region to Rajasthan network.
- All shunt capacitors banks have been kept OFF.
- Load power factor is considered 0.982.
- Commissioned PGCIL 765 kV GSS, namely Chhitorgarh, Ajmer, and Bikaner and also 765 kV GSS Bhada-1, Bhadla-2, Fatehgarh-2, Khetri, and 400 kV GSS Fatehgarh-1 along with the associated transmission lines, have been considered in the load flow studies.

29.4.2 Proposed cases of study

As per load flow study, voltages in Case-1, Case-2, and Case-3 at the various Grid substations are tabulated in Tables 29.8–29.11. The following three cases have been simulated:

- Case-1: Load flow study for the condition corresponding to FY 2022-23 for total system load 10,266 MW and low RE Generation
- Case-2: Load flow study with Proposed Shunt Reactors as indicated in Table 29.8.
- Case-3: Load flow study with Proposed Shunt Reactors as indicated in Table 8 and proposed TCR as indicated in Table 8.

29.4.3 Observations of study

The following are the observations of results of load flow Studies performed for the feasibility of proposed reactors and thyristor Control Reactors:

- 765 KV Bus Voltage at 765 KV GSS Anta by 13 KV on the STU buses

TABLE 29.8 Voltage at GSS where BUS type shunt reactors are proposed.

		765 kV bus voltage			400 kV bus voltage		
S. No.	Name of substation	Case-1	Case-2	Case-3	Case-1	Case-2	Case-3
1	765 kV GSS Anta	791	780	778	413	409	408
2	400 KV GSS Heerapura	–	–	–	414	409	406
3	400 KV GSS Barmer	–	–	–	415	409	402
4	400 KV GSS Ajmer	–	–	–	419	413	410
5	400 KV GSS Bhilwara	–	–	–	416	409	407
6	400 KV GSS Babai	–	–	–	415	411	409
7	400 KV GSS Chhitorgarh	–	–	–	415	408	406
8	400 KV GSS Jodhpur	–	–	–	416	410	403

TABLE 29.9 Voltage at GSS where TCR is proposed.

		400 kV bus voltage		
S. No.	Name of substation	Case-1	Case-2	Case-3
1	400 KV GSS Bhadla	417	414	405
2	400 KV GSS Bikaner	417	414	406
3	400 KV GSS Jaisalmer-2	419	413	401
4	400 KV GSS Merta	416	413	404

TABLE 29.10 Voltage on 400 KV GSS of STU.

		765 kV bus voltage			400 kV bus voltage		
S. No.	Name of substation	Case- 1	Case- 2	Case- 3	Case- 1	Case-2	Case-3
1	765 kV GSS Phagi	796	789	785	415	411	408
2	400 KV GSS Ramgarh	–	–	–	415	411	401
3	400 KV GSS Jodhpur	–	–	–	415	410	403
4	400 KV GSS Ratangarh	–	–	–	414	412	408
5	400 KV GSS Akal	–	–	–	416	411	401
6	400 kV GSS Hindaun	–	–	–	403	399	397
7	400 KV GSS Alwar (PPP)	–	–	–	402	399	397
8	400 KV GSS Deedwana (PPP)	–	–	–	417	413	408

- 400 KV Bus Voltage has been reduced by 5–18 KV on the STU buses.
- 765KV Bus Voltage has been reduced by 6–16 KV on the CTU buses.
- 400KV Bus Voltage has been reduced by 3–10 KV on the CTU buses.

29.4.4 Effect of TCR on voltage under the condition of RE generation variation

The following three cases have been considered to simulate effect of TCR on bus voltages with variation of RE generation as detailed in Table 29.12. As per load flow study,

TABLE 29.11 Voltage on 765 and 400 KV GSS of CTU.

		765 kV bus voltage			400 kV bus voltage		
S. No.	Name of substation	Case-1	Case-2	Case-3	Case-1	Case-2	Case-3
1	765 KV GSS Ajmer	801	794	789	418	414	411
2	765 KV GSS Chittorgarh	799	791	788	416	410	408
3	765 KV GSS Bhadla-1	799	794	783	417	414	407
4	765 KV GSS Bhadla-2	800	795	785	417	415	408
5	765 KV GSS Fatehgarh	800	796	785	417	414	409
6	765 KV GSS Khetri	792	790	786	414	412	410
7	400 KV GSS Bhinmal	–	–	–	413	409	404
8	400 KV GSS Kankroli	–	–	–	410	406	404
9	400 KV GSS Kota	–	–	–	412	410	409
10	400 KV GSS Jaipur(South)	–	–	–	413	410	407
11	400 KV GSS Bassi	–	–	–	413	410	407
12	400 KV GSS Kotputli	–	–	–	412	409	407
13	400 KV GSS Neemrana	–	–	–	415	413	411
14	400 KV GSS Sikar	–	–	–	415	413	410
15	400 KV GSS Bhiwadi	–	–	–	415	414	412

TABLE 29.12 Different cases for simulating effect of TCR.

S. No.	Particulars	RE generation		Exhibits of power plots
1	Case-3	20% of IC	3010	Exhibit-3A and Exhibit-3B
2	Case-4	30% of IC	4515	Exhibit-4A and Exhibit-4B
3	Case-5	40% of IC	6021	Exhibit-5A and Exhibit-5B

TABLE 29.13 MVAR loading of thyristor-controlled reactors.

		Loading of TCR (MVAR)		
S. No.	Name of substation	Case-3	Case-4	Case-5
1	400 KV GSS Bhadla	284	271	247
2	400 KV GSS Bikaner	294	283	265
3	400 KV GSS Jaisalmer-2	259	245	218
4	400 KV GSS Merta	275	264	245

voltages in various cases at the GSSs where TCRs have been proposed are tabulated in Tables 29.13 and 29.14. From Load Flow Studies, it is observed that TCRs keep the voltage of GSS within limits with increase in RE Generation. Under scenario of increase in RE generation, MVAR absorbed by TCRs is reduced. To control the over voltage and reactive power, static compensation and TCR are proposed as detailed in Tables 29.15 and 29.16 respectively.

29.5 Summary

This research work presents a detailed study to propose optimized restructuring of Rajasthan transmission system to mitigate the constraints fro evacuation of RE power from western parts of the Rajasthan. The study is carried out for the base transmission system Rajasthan without considering the restructuring and loadings on the transmission elements. The overloaded elements are identified, and different schemes are designed to relieve such overloading. Different schemes are analyzed, and most feasible schemes are identified. To control the overvoltage and reactive power, static compensation and TCR are proposed.

Since the proposed planning of the power system network involves major changes in the existing network, it might affect the relay coordination system in the network.

TABLE 29.14 Voltage at GSS where TCR are proposed.

S. No.	Name of substation	400 kV bus voltage		
		Case-3	Case-4	Case-5
1	400 KV GSS Bhadla	405	403	400
2	400 KV GSS Bikaner	406	405	402
3	400 KV GSS Jaisalmer-2	401	399	395
4	400 KV GSS Merta	404	402	399

TABLE 29.15 Proposed static reactors.

S. No.	Name of GSS	Voltage level	Capacity of shunt reactor in MVAR
1	765 KV GSS Anta	765 kV	240
		420 kV	125
2	400 KV GSS Heerapura	420 kV	125
3	400 KV GSS Ajmer	420 kV	125
4	400 KV GSS Bhilwara	420 kV	125
5	400 KV GSS Babai	420 kV	125
6	400 KV GSS Chhitorgarh	420 kV	125
7	400 KV GSS Jodhpur(new)	420 kV	125
8	400 KV GSS Barmer	420 kV	125

TABLE 29.16 Proposed TCR.

S. No.	Name of 400 kV GSS	Capacity of TCR in MVAR
1	400 KV GSS Merta	300
2	400 KV GSS Bikaner	300
3	400 KV GSS Bhadla	300
4	400 KV GSS Jaisalmer-2	300
Total		1200 MVAR

To avoid the protection problem, a detailed related to relay coordination should be performed before charging the new elements. Furthermore, the high RE penetration level will affect the voltage profile; hence a detailed study should be carried out for the reactive power compensation.

Finally, let us move on to the real interactive part of this chapter: review questions/exercises, hands-on projects, case projects, and optional team case project. The answers and/or solutions by chapter can be found in Appendix G.

29.6 Chapter review questions/exercises

True/false

1. True or False? Transmission system network in Rajasthan is constructed, operated, and maintained by the system transmission utility (STU).
2. True or False? Rajasthan Rajya Vidyut Prasaran Nigam Limited (RVPNL) is a power utility company registered under the provisions of Companies Act, 1956.
3. True or False? Voltages on the 865 and 400 kV network of the Rajasthan have recorded maximum voltages, which exceed the preceding limits specified in the Grid Code.
4. True or False? The existing practical network of the transmission system of the Rajasthan State Transmission Utility, including the system of voltage levels 132, 220, 400, and 765 kV, is considered for the study.
5. True or False? The planning of the negative reactive compensation using the static reactors to mitigate the problem of voltage variations in the Rajasthan Transmission system is also proposed in this work, which includes the Thyristor-controlled reactor (TCR).

Multiple choice

1. The _______ is installing additional transformation capacity of 24,000 MVA, 765/400 kV, and 10,500 MVA, 400/220 kV.
   a. STU
   b. ICT
   c. GSS
   d. CTU
   e. MVA
2. Based on the preceding recorded voltages at __________________ and to control 765 and 400 kV voltage within limits prescribed in the Indian Electricity Grid Code and CEA's Manual on Planning Criteria, Grid Substations have been identified for Static and Dynamic compensation at 765 and 400 KV levels.
   a. CTU
   b. Grid Code
   c. Planning Criteria

d. Grid Substations
e. Smart city policies

3. Load flow studies have been carried out for condition corresponding to 2022–23 for the light load system condition, namely 10,266 MW and low RE generation to simulate the effect of proposed shunt reactors on the:
   a. Wind generators
   b. Grid voltage
   c. Solar generators
   d. RE generation
   e. Rajasthan network
4. Since the proposed planning of the power system network involves major changes in the existing network, it might affect the relay coordination system in the:
   a. Platforms
   b. Network
   c. Cities
   d. Applications
   e. E-economics
5. In general, the high RE penetration level will affect the voltage profile; hence a detailed study should be carried out for the reactive power:
   a. Planning
   b. Compensation
   c. Proposal
   d. Network
   e. System

Exercise

Problem

Look at how the greater integration of solar photovoltaic (PV) systems into low-voltage (LV) distribution networks has posed new challenges for the operation of power systems.

Hands-on projects

Project

Do research: Analyze environmental concerns with regard to how power system generation is shifting from traditional fossil-fuel resources to renewable energy, such as wind, solar, and geothermal.

Case projects

Problem

What are the emerging challenges and research opportunities for voltage control in smart grids?

Optional team case project

Problem

Describe the different types of Distributed Generation (DG) and investigate the newly emerging challenges arising in the presence of DG in electrical grids.

References

[1] https://energy.rajasthan.gov.in/content/raj/energy-department/rrecl/en/home.html#.

[2] https://niti.gov.in/writereaddata/files/175-GW-Renewable-Energy.

[3] https://energy.rajasthan.gov.in/rvpnl.

[4] Tripathi MM, Pandey AK, Chandra D. Power system restructuring models in the Indian context. Electr J 2016;29:22–7.

[5] Popovici V. 2010 power generation sector restructuring in Romania—a critical assessment. Energy Pol 2011;39:1845–56. Elsevier.

[6] Saha D, Bhattacharya RN. Analysis of the welfare implications of power-sector restructuring in West Bengal, India. Util Pol 2019;56:62–71. Elsevier.

[7] Chen C, Gao J, Chen J. Behavioral logics of local actors enrolled in the restructuring of rural China: a case study of Haoqiao Village in northern Jiangsu. J Rural Stud 2019. https://doi.org/10.1016/j.jrurstud.2019.01.021. ISSN: 0743-0167.

[8] Chen W-M. The U.S. electricity market twenty years after restructuring: a review experience in the state of Delaware. Util Pol 2019;57:24–32.

[9] Rezaee Jordehi A. How to deal with uncertainties in electric power systems? A review. Renew Sustain Energy Rev 2018;96:145–55.

[10] Vaninsky A. Energy-environmental efficiency and optimal restructuring of the global economy. Energy 2018;153:338–48.

[11] Ron Chan H, Fell H, Lange I, Li S. Efficiency and environmental impacts of electricity restructuring on coal-fired power plants. J Environ Econ Manag 2017;81:1–18.

[12] Reza Yousefi G, Kaviri SM, Latify MA, Rahmati I. Electricity industry restructuring in Iran. Energy Pol 2017;108:212–26.

[13] Mahela OP, Shaik AG. Power quality improvement in distribution network using DSTATCOM with battery energy storage system. Int J Electr Power Energy Syst 2016;83:229–40.

[14] Ray PK, Mohanty SR, Kishor N. Classification of power quality disturbances due to environmental characteristics in distributed generation system. IEEE Trans Sustain Energy April 2013;4(2):302–13.

[15] Mahela OP, Shaik AG. Detection of power quality events associated with grid integration of 100kw solar pv plant. In: Energy economics and environment (ICEEE), 2015 international conference; March 2015. p. 1–6.

[16] Singh B, Mukherjee V, Tiwari P. A survey on impact assessment of DG and FACTS controllers in power systems. Renew Sustain Energy Rev November 2015;42:846–82.

[17] Kamatchi Kannan V, Rengarajan N. Investigating the performance of photovoltaic based DSTATCOM using I cosU algorithm. Electr Power Energy Syst July 2014;54:376–86.

[18] Mahela OP, Khan B, Alhelou HH, Siano P. Power quality assessment and event detection in distribution network with wind energy penetration using stockwell transform and fuzzy clustering. IEEE Trans Indus Inform November 2020;16(11):6922–32.

[19] Rathore B, Mahela OP, Khan B, Alhelou HH, Siano P. Wavelet-alienation-neural-based protection scheme for STATCOM compensated transmission line. In: IEEE Transactions on Industrial Informatics April 2021;17(4):2557–65. https://doi.org/10.1109/TII.2020.3001063.

[20] Mahela OP, Khan B, Alhelou HH, Tanwar S. Assessment of power quality in the utility grid integrated with wind energy generation. IET Power Electronics; January 2020.

[21] Ola SR, Saraswat A, Goyal SK, Jhajharia SK, Rathore B, Mahela OP. Wigner distribution function and alienation coefficient based transmission line protection scheme. IET Gener, Transm Distrib May 22, 2020;14(10):1842–53.

[22] Agajie TF, Khan B, Alhelou HH, Mahela OP. Optimal expansion planning of distributed generation integrated distribution network by utilizing grid-based multi-objective harmony search algorithm. Comput Electr Eng 2020. https://doi.org/10.1016/j.compeleceng.2020.106823.

[23] Singh Chawda G, Shaik AG, Mahela OP, Padmanaban S, Holm-Nielsen JB. Comprehensive review of distributed FACTS control algorithms for power quality enhancement in utility grid with renewable energy penetration. IEEE Access 2020;8:107614–34.

[24] Singh Chawda G, Shaik AG, Shaik M, Sanjeevikumar P, Holm-Nielsen JB, Mahela OP, Palanisamy K. Comprehensive review on detection and classification of power quality disturbances in utility grid with renewable energy penetration. IEEE Power & Energy Society Section. IEEE Access 2020. https://doi.org/10.1109/ACCESS.2020.3014732.

[25] Mahela OP, Jaya S, Bipul K, Baseem K, Hassan HA. An algorithm for the protection of distribution feeder using Stockwell and Hilbert transforms supported features. CSEE J Power Energy Syst 2020. https://doi.org/10.17775/cseejpes.2020.00170.

[26] Gupta N, Khosravy M, Patel N, Dey N, Mahela OP. Mendelian evolutionary theory optimization algorithm. Soft Comput 2020. https://doi.org/10.1007/s00500-020-05239-2.

[27] Kulshrestha A, Mahela OP, Gupta MK, Gupta N, Patel N, Senjyu T, Shah Danish MS, Khosravy M. A hybrid protection scheme using Stockwell transform and Wigner distribution function for power system network with solar energy penetration. Energies 2020;13(14):3519.

[28] Yogee GS, Mahela OP, Kansal KD, Khan B, Mahla R, Alhelou HH, Siano P. An algorithm for recognition of fault conditions in the utility grid with renewable energy penetration. Energies 2020;13(9):2383. ISSN 1996-1073.

[29] Ola SR, Saraswat A, Goyal SK, Jhajharia SK, Khan B, Mahela OP, Alhelou HH, Siano P. A protection scheme for power system with solar energy penetration. Appl Sci 2020;10(4):1–22. Paper No. 1516.

[30] Ola SR, Saraswat A, Goyal SK, Sharma V, Khan B, Mahela OP, Alhelou HH, Siano P. Alienation coefficient and Wigner distribution function based protection scheme for hybrid power system network with renewable energy penetration. Energies 2020;13(5). Paper No. 1120, March 2020.

[31] Chawda GS, Mahela OP, Gupta N, Khosravy M, Senjyu T. Incremental conductance based particle swarm optimization algorithm for global maximum power tracking of solar-PV under nonuniform operating conditions. Appl Sci 2020;10(13):4575.

[32] Venu Madhav G, Obulesu YP. A control strategy for DFIM to minimize DC bus voltage fluctuations and torque oscillations. Indon J Electr Eng Comput Sci February 2017;5(2):277–89.

[33] Alhamrouni I, Hanafi FN, Salem M, Rahman NHA, Jusoh A, Sutikno T. Design of shunt hybrid active power filter for compensating harmonic currents and reactive power. TELKOMNIKA Telecommun Comput Electron Control August 2020;18(4):2148–57.

[34] Benali A, Khiat M, Denai M. Voltage profile and power quality improvement in photovoltaic farms integrated medium voltage grid using dynamic voltage restorer. Int J Power Electron Drive Syst September 2020;11(3):1481–90.

[35] Mohammed AB, Ariff MAM, Ramli SN. Power quality improvement using dynamic voltage restorer in electrical distribution system: an overview. Indon J Electr Eng Comput Sci January 2020;17(1):86–93.

Chapter 30

DC smart micro grid protection system

Endeshaw Solomon[1], Baseem Khan[1], Esayas Gidey[1], Mesfin Fanuel[1], Om Prakash Mahela[2] and Hassan Haes Alhelou[3]

[1]*Department of Electrical and Computer Engineering, Hawassa University, Hawassa, Ethiopia;* [2]*Power System Study Division, Rajasthan Rajya Vidyut Prasarn Nigam Ltd., Jaipur, Rajasthan, India;* [3]*Tishreen University, Latakia, Syria*

30.1 Introduction

Micro grids are used for distributed energy resources. Distributed energy resources can decrease the effect of pollutants on the surroundings, and also it increases the quality and consistency of power [1,2]. There are two types of micro grid AC and DC micro grid. AC micro grids are mostly examined by many researchers due to its similarities to AC power system. As compared with AC micro grid, DC micro grid needs few conversion steps and transmits high DC power. Likewise, DC systems are efficient without skin effect, corona effect, and reduce power losses in a transmission line [3,4]. DC micro grids are used for unique purposes such as low voltage up to 1.5 kV for data centers [5] and low-voltage DC distribution systems [6]; and medium voltage is up to 35 kv DC systems used for electric ships [7—10].

Due to good characteristics of DC micro grid, reliable fault protection system is the main problem. Earlier research studies show that the typical voltage source converters can accept two times converter full load current [3,11].

Presently, papers on DC micro grids (see Fig. 30.1) are mainly concentrated on the structure, control method, and energy management system. Designing of DC protection system for micro grid has not drawn high concentration.

30.2 DC micro grids

This section discusses the various concepts related to the DC micro grid design. It also includes control strategy.

30.2.1 DC micro grid design

A micro grid is a stand-alone system, which is composed of many distributed generators; energy storages and energy conversion devices; loads; monitors; and, protective devices. Micro grids can be either interconnected with national utility grid or stand-alone system [12]. Two types of micro grids are discussed AC and DC. A DC micro grid is mainly composed of sources, converters, storage devices and loads. Distributed generations in a DC micro grid are divided into AC and DC sources. DC sources include solar panel, which is directly connected to the DC bus through DC/DC converter or chopper. Wind generators are connected to the DC bus through AC/DC converter or rectifiers. So as compared to AC micro grid, energy conversion steps of DC micro grid are simple. DC micro grid has superior advantage over AC micro grid. Energy storage devices are necessary for renewable energy sources to improve power quality and reliability for emergence power system [3]. Most widely used energy storage devices are batteries, hydrogen storage, compressed air, supercapacitor, and flywheel [13]. Flywheel devices are connected to DC bus through machines and converters, whereas batteries and super capacitors can be connected directly. DC micro grids are more appropriate to give sources for sensitive loads.

30.2.2 DC micro grid control strategy

Efficient control methods are vital to stable and competent operation of Micro grids. A control strategy of DC micro grid includes centralized, decentralized, distributed, and sequential controls [14]. Decentralized control is used to optimize resources consumption and keep system stability [15]. It is appropriate for a DC micro grid with less capacity [16]. Droop control is the main representation of decentralized control. To improve the transient characteristics of DC micro grid resultant to fault, control method should be steadily developed. A dynamic droop method in Refs. [17—21] planned for hybrid energy storage system (HESS) had been checked to be more effective over the preceding droop control.

Smart Cities Policies and Financing. https://doi.org/10.1016/B978-0-12-819130-9.00018-8

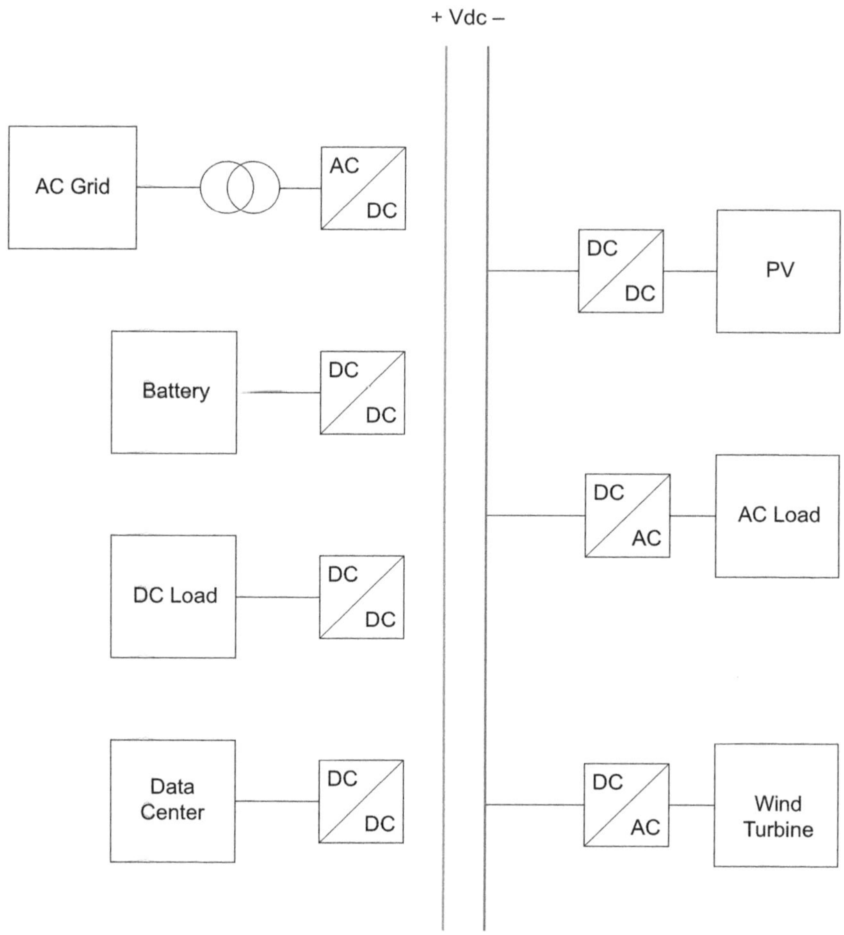

FIGURE 30.1 DC micro grid bus architectures.

30.3 DC protection system

This section deals with the various protection devices and Grounding Options utilized for the DC system.

30.3.1 Protection devices

Currently, the existing protective devices for DC micro grids are fuses and circuit breakers [22–25]. Interruption of current leads to the creation of arc that cannot be easily quenched in DC system without current crossing through zero. Fuses are used for both AC and DC systems. The rating of AC system is expressed in terms of RMS values. Circuit breakers are the main devices in DC systems, but these devices do not satisfy protection requirements. Conventional molded case circuit breaker (MCCB) is the most usual protection device in current DC systems. To attain sufficient voltage blocking capacity, the contactors can be connected in series in both cases [26]. To break down the restriction of usage of fuses and MCCBs, power-electronics-based protection tools were planned and improved. These are solid-state circuit breakers (SSCB) [27,28] and hybrid circuit breaker (CB) [29,30]. Semiconductor devices include IGBT, insulated gate commutated thyristors, and gate turn-off thyristors [31,32]. However, these devices have more power losses. Table 30.1 presents the comparison of different protection devices.

Hybrids of both mechanical and solid-state CBs are not used in practice. Active and passive resonance CBs are uniquely applied in CSC HVDC [33]. So, SSCBs are presented and in principle feasible for MVDC applications. Since the main disadvantages of SSCB are due to its high power loss, the selection of SSCB is the main design factor [34–39]. Hybrid solid-state DC CB with mechanical disconnector gives very rapid and well-ordered protection solution [40]. For now, the improvement of semiconductor technology forces hybrid DCCB to a new stage. Consequently the improvement of DCCBs for DC micro grid mainly focuses on topology structure and selection of power electronic devices.

30.3.2 Grounding options

Grounding is a significant concern of the DC micro grid protection system. There are many types of grounding options having different fault characteristics and affect the

TABLE 30.1 Comparison of protection devices.

Constraints	Full solid state circuit breakers	Hybrid state breaker with dis-connector	Solid circuit Hybrid mechanical solid state CB	Mechanical active or passive resonance CB
Disruption time	<1 ms	<2 ms	<5–30 ms	60 ms
Maximal rated voltage Un	<800 kv	120 kv 120 kv confirmed by test (up to 320 kv applicable)	AC CB > 500 kv Ultra fast switches <12 kv	<550 kv obtainable
Maximum DC breaking current In	<5 kA Predictable	9 kA Investigational confirmed (up to 16 kA expected)	~6–12 kA (predictable)	Up to 4 kA proven in operation (up to 8 kA)
Predicted power loss in comparison with VSC converter station	<30% (large forward voltage due to serial connection of solid state devices)	<1% (only few IGBTs in series in the main path)	<0.001% (metal contacts)	<0.001% (metal contacts)
Further development steps	Improvement in solid state device technology to reduce on-state forward voltage and number of modules in series	Field experience with prototype in a test grid; reduction of IGBT costs	Growth of ultra-fast mechanical drives to reduce commutation time	Optimization of DC arc for passive resonance to achieve higher current rating and to minimize time for current zero creation

arrangement and ordering of protection. The purpose of grounding design is to help the ground fault identification, minimize stray current, and ensure safety of equipment and personnel [41,42]. DC micro grids can be grounded with either low or high resistance, and even it can be ungrounded [22]. For low-voltage applications, ungrounded mode is highly advised [43,44]. For this issue, the common mode voltage can be high to cause an intimidation to devices and personnel safety [45–47]. For now the system operates constantly when single line to ground fault happens. Solid grounding has been rarely used due to the deterioration caused by stray current [42]. As we compare with low resistance, high-resistance grounding restricts the ground fault current. Similar to an AC power system, the IEC 60364 standards [48–60] decide three grounding methods including TN, IT, and TT. TN systems are low impedance grounded, and the detection of line fault is easy to execute, but the voltage may be very high. Whereas IT system needs high impedance; the fault current is low and increases the recognition difficulty [60–69]. In other case TT systems have many grounding options [44]. According to IEC 60364-1, DC grounding systems can be categorized into five types such as Terre Terre (TT), Terre Neutre (TN) with three subclasses, and IT [70]. The first letters T and I refer to direct connection of the earth and no connection to earth, respectively. The second letters T and N represent direct earthing of the exposed conductive parts and connection of exposed parts to earth neutral, respectively. Table 30.2 presents the comparison between the four grounding systems.

30.4 Principle of protection system

There are two types of protection systems; these are unit protection and nonunit protection techniques. The necessary characteristics used to design protection systems are reliability, selectivity, speed, performance, economics, and simplicity [26].

30.4.1 Nonunit protection system

For the moment, nonunit protection systems have inherent uses for integrating all the protection systems [4]. Nonunit protection system analyzes fault isolation in DC micro grid by analyzing current, voltage, di/dt, dv/dt. A variety of faults are simply quenched with the derivative of currents. But the selectivity of nonunit protection system is based on complex setting values and proper time delays. A resistance superconductive fault current limiter SFCL is composed of parallel resistance for limiting over voltage. The resistance SFCL could be operated in the superconductive mode.

30.4.2 Unit protection

Due to the extra cost, the vital communication and relay devices for the realization of the unit protection system are closely restricted. However, the expansion of smart micro grid will call for an increased investment of sensors and communication facilities [71,72]. Unit protection helps a brief zone and does not give response to an external circuit fault. When we compare with nonunit protection, unit protection does not

TABLE 30.2 Comparison between the four grounding system.

	TT	TN-S		TN-C
IT				
Safety of person	Good	Good	Good	Good
Safety of property	Good Fault current less than a few dozen amperes	Good Poor fault current around a 1 kA	Good Poor fault current around a 1 kA	Very good Fault current less than a few dozen mA
Continuity of service	Average	Average	Average	Excellent
EMC	• Good • Voltage imbalance • Equipotential problem • Require to manage devices with high leakage currents	• Excellent • Less equi-potential problem • Requires to manage devices with high leakage currents • High fault currents	• Poor • High fault current (transient disturbance)	• Poor • Risk of over Voltage

have secondary protection or backup protection to the adjacent elements in the system [73]. Therefore it is familiar that non-unit protection is used in net unit protection as a restricted protection. There are two types of unit protection: data-based and event-based protection system [74–79]. According to Ref. [76], event-based protection method does not call for high-speed communication and transfers less data than data-based protection method.

30.4.2.1 Data-based protection system

Besides the current differential protection system, an undervoltage protection system is usually installed as a backup [80]. Intelligent electric device (IED) combines the function of both detection and isolation. A high-speed fault location and discrimination that use the natural characteristics of DC difference are explained in Refs. [81,82].

30.4.2.2 Event-based protection system

Data-based protection systems differ from event-based protection systems, in that event-based protection systems enable each power unit to classify the type of fault immediately [76]. Like nonunit protection, artificial inductive line impedances (AILIs) are installed in each of the feeders to assist the fault isolation with the value of di/dt. Consequently, the last decision will be made, and the protection devices will quickly respond. Though, given the severity of bus fault, a decision will be made automatically without reconfirmation [83]. Fig. 30.2 presents the event-based

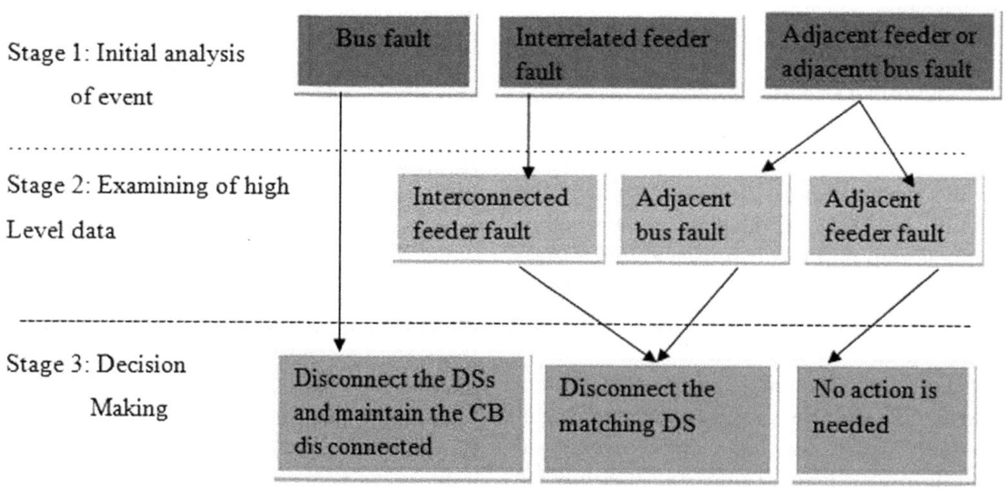

FIGURE 30.2 Event-based protection strategy.

protection strategy. Although, unit protection system works on communication delay or breakdown temporarily fault location and discrimination will become more difficult, and all the systems may be exposed to risk [84].

30.4.3 Other protection problems

There are various issues that affect the protection of the DC microgrid system. This section discusses some of these issues in detail.

30.4.3.1 Ground fault monitoring technique

Locating a high-resistance ground fault shows a tough and detecting a high-resistance grounding fault proves a tough and difficult task for safety of DC system. Conventionally, the strategy of AC injection and DC leakage are mainly used [85]. A ground fault monitoring method for extra low voltage direct current (ELVDC, up to a 60V DC) distribution system was discussed [86]. It is better explanation for ground fault location when DC leakage current is a low level and not obvious. However, the planned method realizes fast location and exact location of the ground fault. A novel high-resistance ground fault monitoring method for DC railway traction system was presented [87]. Voltage and current of two adjacent substations are applied in calculating the ground fault.

30.4.3.2 Effects of control system on protection

A single converter voltage control (SVC) method and distributed voltage control (DVC) method were engaged respectively. There are two aspects of fault response: the fast fault discrimination and post fault control of system recovery. Fast fault discriminations were autonomous from control method [88–92]. But the post fault operation was affected to some degree. As soon as SVC is applied, once the single VSC controlling the DC voltage was blocked after the fault. In DC system the system maintains the voltage control capability and came back to the set voltage operation state. Consequently DVC method supported the fault ride through its capability to some extent. Overcurrent protection is highly resistant to the voltage control method, by examining the voltage or dv/dt as undervoltage protection.

30.4.3.3 DC fault detection methods

DC micro grid line impedance is very low. Therefore, fault current variation is too high, and the fault current reaches 100 A with less than few milliseconds. Concerning developed sensors, communication, and control systems, protection systems should be quick, reliable, and precise. There are many types of protection methods such as overcurrent, current derivative, directional over current, distance, and differential protections, which are used to detect and isolate the fault section. The main features of the DC protection system are as follows [93]:

- **Speed:** The protective system, which identifies the fault as fast as possible to protect the equipment from breakdown/damage.
- **Selectivity:** Ability of protection system, which selects the fault part of the system.
- **Sensitivity**: Ability of protection system that detects all types of fault.
- **Reliability**: A protection system, which must isolate the fault part when primary protection fails to protect.

30.4.3.3.1 Overcurrent protection

Like AC overcurrent protection, a threshold is measured to decide the happening of fault. Time–current curves (TCCs) are composed of overload and instantaneous characteristics of upstream OCR. TCCs of downstream OCR must be below the TCC of the upstream with OCR with an adequate margin to ensure selectivity. Overcurrent protection is designed for a DC micro grid where AC/DC converter or rectifiers have an ability to limit the fault current [94,95]. But, installing this type of protection system is more complex DC micro grid. For a compact DC micro grid, the margin between upstream and downstream protection operation is small. Interconnection of unit-based protection has high sensitivity, speed, selectivity, and overcurrent protection to have a quick and effective operation and minimizes installation costs. The main disadvantage of overcurrent protection is its low sensitivity for high-impedance fault.

30.4.3.3.2 Current derivative protection

If the occurs, the current is increased from zero to a high value. This attribute can be concerned to specify a fault in very short time. But current derivative depends on the length of the cable, line loading, and fault impedance. Because of this, it is very hard to find an appropriate threshold. First- and second-order derivatives are measured to detect low and high fault impedance [4,96–98]. Sensors are operated at high sampling rates, which are used to measure current derivative.

30.4.3.3.3 Directional overcurrent protection

The direction of current can be from either side if a complex DC micro grid is used. Applying directional overcurrent protection could develop selectivity. Currently, the directional over current protection was planned for DC micro grid [99]. Based on the planned system, once the fault happens, the direction and magnitude of fault current will be altered and then the direction of all the branches is specified by assisting communication system, and it will help in identifying the fault line.

30.4.3.3.4 Distance protection

This is one of the protection system in which relay action depends up on the distance or impedance in between the point where relay is implemented and the point of fault. It works based on measuring of impedance from point of measuring (POM) to the fault point. If the measured value of impedance is within a specified distance value, a tripping signal will be sent to CB after specific time delay.

30.4.3.3.5 Differential protection

It is ideal method for transmission lines and is more costly to use. The two terminals of the line may cover several kilometers far apart then as comparison of the two currents, a costly pilot wire circuit is required. Differential protection is used to measure the amplitude of the current of each side of the specific element. Table 30.3 presents the comparison between five main DC systems.

30.4.3.4 The operation criterions and setting rules

There are various operating criterions and setting rules for the protection systems of the DC microgrid. Table 30.4 presents the summary of protection systems of DC micro grids.

30.5 Research prospect

In this chapter, the main protection systems and grounding systems have been presented. To calculate protection problems and to encourage the improvement of DC micro grid, it is essential to specify the main lines of the future paper. The future research direction lies on the expansion of protective devices derived from electronic technology, which provides loose of protection parameters and the development of protection systems appropriate for DC micro grids.

30.5.1 Grounding and protection system

A grounding system has not been determined for DC micro grids. The fast-acting grounding equipment to restrict fault current because of solid-state technology requires more concentration in advance. To understand quick and correct fault location and discrimination, a novel protection system based on fault characteristics is still required. So, the whole protective devices are required to incorporate into the overall protection system.

30.5.2 Control and protection of coordinated strategy

Micro grid combines many distributed generators, energy storage devices, energy conversion devices [100]. Splitting control and protection in future consideration ensure effective, protected, and economical operation of the system. Efficient control strategy takes the system back to normal after some form of disorder. In an abnormal situation, an efficient method decreases the harshness of the fault to some degree by monitoring the fault current and voltage. An organized control method makes the fault characteristics more important to assist fault location and prevention. The control method becomes primary protection system of DC micro grids and helps to understand the fault location and discrimination to guarantee system safety. Fig. 30.3 presents the protection and control strategy.

30.5.3 Comparison of optimal power flow methods

In this part, a comparison of the five main DC protection systems is presented in Table 30.3. Table 30.1 demonstrates the comparisons of the main protective devices. A review of the protection methods of DC micro grids for power systems with nonconventional energy resources is discussed in Table 30.4. Table 30.2 shows a comparative study of the four main grounding systems used for DC micro grids.

TABLE 30.3 Comparison between five main DC systems.

	Overcurrent protection	Current derivative protection	Directional overcurrent protection	Distance protection	Differential protection
Speed	Medium	High	Medium	Low	High
Selectivity	Medium	Low	High	High	High
Sensitivity	Low/medium	High	Low/medium	Low/medium	Low
Reliability	Low/medium	Low/medium	Medium/high	Medium/high	Low
Cost	Low	Low	Medium	Medium	High

TABLE 30.4 Summary of protection systems of DC micro grids.

	Nonunit protection			Unit protection	
Protection scheme	**Current based**	**Voltage based**	**Hybrid electrical quantities based**	**Differential**	**Event based**
Principle	Identify the fault through current, di/dt, and even d^2i/dt^2	Identify the fault through voltage, du/dt	Detect the fault through hybrid electrical quantities	Detect the internal and external fault with current differential	Judge fault type locally and connect with other units to make fault separation decision
Communication	No	No	No	Yes	Yes
Operation criterion	$i > i_{th}, \Delta_i > i'_{th}, \Delta_i^2 > i''_{th}$	$u > u_{th}, \Delta u, u'_{th}$	$> i_{th}, \Delta_i > i'_{th}, \Delta_i^2 > i''_{th}$ $u > u_{th}, \Delta u, u'_{th}$ $x < x_{th}$	$I_{diff} = /i_{in} + i_{out}/ > i_{d-th}$	Fault event is determined by contact with other integrated units
Threshold criteria	It is based on the current related values when direct short-circuit fault happens at the end of protected scale	It depends on voltage related values when direct short circuit fault occurs at the end of shielded scale	It depends on the hybrid electrical quantities related or fault location values when short circuit happens at the end of shielded scale	Threshold is mainly based on the overcurrent value of the power electronic devices, measurement inaccuracy and induced noise or transient of external faults	Fault type judgment threshold setting rules similar to nonunit protection schemes
Comments	Appropriate threshold is need; sensitive to fault resistance	Appropriate threshold is need; sensitive to fault resistance	Setting rules is complex to some degree; sensitive to fault resistance	Only the magnitudes of currents are needed to compare; insensitive to high di/dt and fault resistance	Achieving fault identification accurately; isolating fault within 30 ms

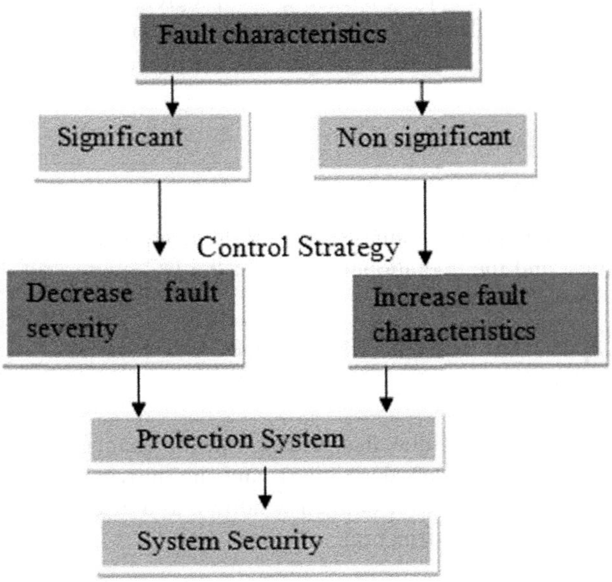

FIGURE 30.3 Protection and control strategy.

30.6 Summary

An appropriate protection method must be the main concern for the separate characteristics of DC micro grids during the fault situations. Important part of system design process such as grid topology, grounding, and connection between interrelated converters in case of fault events must be incorporated. This chapter discusses the present research condition on relative case of DC micro grids and summarizes the planned protection methods. In DC micro grids such as distributed generators, converters, storage batteries, and many loads are conducted. The fault current rapidly increases with large scale due to low fault impedance, in which fault location and discrimination should be carried out as soon as possible. Lack of efficient overall protection method limits the improvement and encouragement of DC micro grids. Novel protection systems are quick and exact fault location and isolation. Control system should be the primary purpose of DC micro grid protection methods. Generally integrated approach of control and protection can be more efficient than considering protection system by themselves. This chapter mainly focuses on DC fault current in three steps such as capacitor discharge, freewheeling diodes, and grid side current feeding steps. The kind of grounding system can also affect safety, capability to recognize a fault, and survivability of DC micro grid under fault condition. Based on these there are five grounding methods such as TT, TN-S, TN-C, T-N-C-S, and IT. This chapter also summarizes different methods of fault detection, discrimination, and location. There are five main DC fault detection methods; these are overcurrent protection, current derivative protection, directional overcurrent protection, differential protection, and distance protection.

Finally, let us move on to the real interactive part of this chapter: review questions/exercises, hands-on projects, case projects, and optional team case project. The answers and/or solutions by chapter can be found in Appendix G.

30.7 Chapter review questions/exercises

True/false

1. True or False? Micro grids are used for distributed energy resources.
2. True or False? A micro grid is a distributed system, which is composed of many distributed generators; energy storages and energy conversion devices, loads, monitors, and protective devices.
3. True or False? Efficient control methods are vital to a stable and competent operation of micro grids.
4. True or False? Currently, the existing protective devices for DC micro grids are buses and circuit breakers.
5. True or False? Grounding is not a significant concern of the DC micro grid protection system.

Multiple choice

1. There are two types of protection systems; these are unit protection and nonunit protection:
   a. Systems
   b. Techniques
   c. Definitions
   d. Investments
   e. Policies
2. Due to the extra cost, the vital communication and relay devices for the realization of the unit protection system are closely:
   a. Restricted
   b. Restrained
   c. Compared
   d. Protected
   e. Backed up
3. Besides the current differential protection system, an undervoltage protection system is usually installed as a:
   a. Machine
   b. Backup
   c. System
   d. Device
   e. Model
4. Data-based protection systems differ from event-based protection systems, in that event-based protection systems enable each power unit to classify the type of fault:
   a. ASAP
   b. Later
   c. Immediately.
   d. Soon
   e. Right away
5. There are various issues that affect the protection of the DC micro grid:
   a. Section
   b. Fault
   c. Monitor
   d. System
   e. IoS

Exercise

Problem

Clarify the importance of smart grids, in launching smart cities, by reviewing the advancement of micro/nano grids, applications of renewable energies, energy-storage technologies, and smart water grids in smart cities.

Hands-on projects

Project

Do research: Analyze the value of nano-grids in the launch of smart cities, by examining the progress of nano and microgrids; alternative energy storage technologies; energy applications; and smart gadgets in smart cities.

Case projects

Problem

Do a comparative analysis of different topologies, structures, and the operational modes of the DC micro grid.

Optional team case project

Problem

Propose a technique that uses a rate of change of fault current (first-order derivative) and its moving average (cumulative sum over defined window), as a discriminative feature for fault detection.

References

[1] Hatziargyriou N, Asano H, Iravani R, Marnay C. Microgrids: an overview of ongoing research, development, and demonstration projects. IEEE Power Energy Mag 2007;5(4):78–94.

[2] Ding G, Gao F, Zhang S, et al. Control of hybrid AC/DC microgrid under islanding operational conditions. J Mod Power Syst Clean Energy 2014;2:223–32.

[3] Salomonsson D, Soder L, Sannino A. Protection of low-voltage DC microgrids. IEEE Trans Power Deliv 2009;24(3):1045–53.

[4] Fletcher SDA, Norman PJ, Galloway SJ, Crolla P, Burt GM. Optimizing the roles of unit and non-unit protection methods within DC microgrids. IEEE Trans Smart Grid 2012;3(4):2079–87.

[5] AlLee G, Tschudi W. Edison Redux: 380 Vdc brings reliability and efficiency to sustainable data centers. IEEE Power Energy Mag 2012;10(6):50–9.

[6] Rodríguez-Díaz E, et al. Voltage-level selection of future two-level LVdc distribution grids: a compromise between grid compatibiliy, safety, and efficiency. IEEE Electr Mag 2016;4:20–8.

[7] Hooshyar A, Iravani R. Microgrid protection. Proc IEEE 2017;105(7):1332–53.

[8] Huang AQ, Crow ML, Heydt GT, Zheng JP, Dale SJ. The Future Renewable Electric Energy Delivery and Management (FREEDM) system: the energy internet. Proc IEEE 2011;99(1):133–48.

[9] Bifaretti S, Zanchetta P, Watson A, Tarisciotti L, Clare JC. Advanced power electronic conversion and control system for universal and flexible power management. IEEE Trans Smart Grid 2011;2(2):231–43.

[10] Liserre M. HEART- the highly efficient and reliable smart transformer. 2015.

[11] Park JD, Candelaria J, Ma L, et al. DC ring-bus micro grid fault protection and identification of fault location. IEEE Trans Power Deliv 2013.

[12] Li YW, Nejabatkhah F. Overview of control, integration and energy management of micro grids. J Mod Power Syst Clean Energy 2014.

[13] Kusko A, Dedad J. Stored energy—short-term and long term energy storage methods. IEEE Ind Appl Mag 2007.

[14] Meng L, Shafiee Q, Trecate GF, et al. Review on control of DC micro grids and multiple microgrid clusters. IEEE J Emerg Sel Top Power Electron 2017.

[15] Olivares DE, Mehrizi-Sani A, Etemadi AH, et al. Trends in micro grid control. IEEE Trans Smart Grid 2014.

[16] Jin C, Wang P, Xiao J, et al. Implementation of hierarchical control in DC micro grids. IEEE Trans Ind Electron 2014.

[17] Li J, Yang Q, Robinson F, et al. Design and test of a new droop control algorithm for a SMES/battery hybrid energy storage system. Energy 2017.

[18] Li J, Xiong R, Yang Q, et al. Design/test of a hybrid energy storage system for primary frequency control using a dynamic droop method in an isolated micro grid power system. Appl Energy 2016.

[19] Park JD, Candelaria J. Fault detection and isolation in low-voltage DC-bus micro grid system. IEEE Trans Power Deliv 2013.

[20] Fletcher SDA, Norman PJ, Galloway SJ, et al. Determination of protection system requirements for DC unmanned aerial vehicle electrical power networks for enhanced capability and survivability. IET Electr Syst Transp 2011.

[21] Yang J, Fletcher JE, O'Reilly J. Multi terminal DC wind farm collection grid internal fault analysis and protection design. IEEE Trans Power Deliv 2010.

[22] Dragičević T, Lu X, Vasquez JC, Guerrero JM. DC microgrids—Part II: a review of power architectures, applications, and standardization issues. IEEE Trans Power Electron 2016;31(5):3528–49.

[23] ETSI EN 300 132-3-1. Environmental engineering (EE), power supply interface at the input to telecommunications and datacom (ICT) equipment, part 3: operated by rectified current source, alternating current source or direct current source up to 400 V, subpart 1: direct current source up to 400 V. 2011.

[24] IEEE recommended practice for the design of DC auxiliary power systems for generating stations. IEEE Std; 1993.

[25] IEC—standardization management board—SG4 LVDC distribution systems up to 1500 V DC. 2009.

[26] Cuzner RM, Venkataramanan G. The status of DC micro grid protection. In: Proceedings of the 2008 IEEE industry applications society annual meeting, Edmonton, Alta, Canada, October 5–9, 2008; 2008. p. 1–8.

[27] Meyer C, Schroder S, De Doncker RW. Solid-state circuit breakers and current limiters for medium-voltage systems having distributed power systems. IEEE Trans Power Electron 2004.

[28] Lawes D, Ran L, Xu Z. Design of a solid-state D.C. circuit breaker for light rail transit power supply network. In: Proceedings of the 2014 IEEE energy conversion congress and exposition (ECCE), Pittsburgh, USA, September 14–18, 2014; 2014. p. 350–7.

[29] Zheng H, Lin H, Du Y. Designing and research of a novel current limiting DC hybrid circuit breaker with the combinatorial electronic switch. In: Proceedings of the 2015 IEEE innovative smart grid technologies—Asia (ISGT ASIA), Bangkok, Thailand, November 3–6, 2015; 2015. p. 1–5.

[30] Wen W, Huang Y, Sun Y, et al. Research on current commutation measures for hybrid DC circuit breakers. IEEE Trans Power Deliv 2016.

[31] Bucher MK, Walter MM, Pfeiffer M, et al. Options for ground fault clearance in HVDC offshore networks. In: Proceedings of the 2012 IEEE energy conversion congress and exposition (ECCE), Raleigh, USA, September 15–20, 2012; 2012.

[32] Heidemann M, Nikolic G, Schnettler A, et al. Circuit breakers for medium-voltage DC grids. In: Proceedings of the 2016 IEEE PES transmission and distribution conference and exposition-Latin America (PES T&D-LA), Morelia, Mexico, September 20–24, 2016; 2016.

[33] Mobarrez M, Kashani MG, Bhattacharya S, et al. Comparative study of DC circuit breakers using real-time simulations. In: : Proceedings of the 40th annual conference of the IEEE industrial electronics society (IECON 2014), Dallas, USA, October 29–November 1, 2014; 2014.

[34] Shen ZJ, Miao Z, Roshandeh AM. Solid state circuit breakers for DC micro grids, current status and future trends. In: Proceedings of the 2015 IEEE first international conference on DC microgrids (ICDCM), Atlanta, USA, June 7–10, 2015; 2015.

[35] Urciuoli DP, Veliadis V, Ha HC, et al. Demonstration of a 600 V, 60 A, bidirectional silicon carbide solid-state circuit breaker. In: Proceedings of the 2011 twenty-sixth annual IEEE applied power

electronics conference and exposition (APEC), Fort Worth, USA, March 6–11, 2011; 2011.
[36] Sato Y, Tanaka Y, Fukui A, et al. SiC-SIT circuit breakers with controllable interruption voltage for 400 V DC distribution system. IEEE Trans Power Electron 2014.
[37] Miao Z, Sabui G, Chen A, et al. A self-powered ultra-fast DC solid state circuit breaker using a normally-on SiC JFET. In: Proceedings of the 2015 IEEE applied power electronics conference and exposition (APEC), Charlotte, USA, March 15–19, 2015; 2015.
[38] Shen ZJ, Roshandeh AM, Miao Z, et al. Ultrafast autonomous solid state circuit breakers for shipboard DC power distribution. In: Proceedings of the 2015 IEEE electric ship technologies symposium (ESTS), Alexandria, USA, June 21–24, 2015; 2015.
[39] Urciuoli DP, Ibitayo D, Koebke G, et al. A compact 100 A, 850 V, silicon carbide solid-state DC circuit breaker. In: Proceedings of the 2016 IEEE energy conversion congress and exposition (ECCE), Milwaukee, USA, September 18–22, 2016; 2016.
[40] Peng C, Song X, Rezaei MA, et al. Development of medium voltage solid-state fault isolation devices for ultra-fast protection of distribution systems. In: Proceedings of the 40th annual conference of the IEEE industrial electronics society (IECON 2014), Dallas, USA, October 29–November 1, 2014; 2014.
[41] Peng C, Song X, Huang AQ, et al. A medium-voltage hybrid DC circuit breaker—part II: ultrafast mechanical switch. IEEE J Emerg Sel Top Power Electron 2017;5(1):289–96.
[42] Paul D. DC traction power system grounding. IEEE Trans Ind Appl 2002;38(3):818–24.
[43] Baran ME, Mahajan NR. DC distribution for industrial systems: opportunities and challenges. IEEE Trans Ind Appl 2002.
[44] Park J. Ground fault detection and location for ungrounded DC traction power systems. IEEE Trans Veh Technol 2015.
[45] Valdes M, Papallo T, Premerlani B. Finding fault—locating a ground fault in low voltage, high-resistance grounded systems via the single-processor concept for circuit protection. IEEE Ind Appl Mag 2007.
[46] Yang J, Fletcher JE, O'Reilly J. Short-circuit and ground fault analyses and location in VSC-based DC network cables. IEEE Trans Ind Electron 2012.
[47] Jacobson B, Walker J. Grounding considerations for DC and mixed DC and AC power systems. Naval Eng J 2007.
[48] IEC 60364-1 Low-voltage electrical installations—part 1: fundamental principles, assessment of general characteristics, definitions. IEC. https://webstore.iec.ch/publication/1865&preview=1.
[49] Lee CH, Lu CJ. Assessment of grounding schemes on rail potential and stray currents in a DC transit system. IEEE Trans Power Deliv 2006.
[50] Bam L, Jewell W. Review: power system analysis software tools. In: Proceedings of the IEEE power engineering society general meeting, San Francisco, USA, June 16, 2005; 2005.
[51] Power system analytical software tools. http://www.itee.uq.edu.au/pssl/drupal7_with_innTheme/?q=node/34.
[52] Venayagamoorthy GK. Comparison of power system simulation studies on different platforms-RSCAD, PSCAD/EMTDC, and SIMULINK Sim Power Systems. In: Proceedings of the international conference on power system operations and planning (ICPSOP), Praia, Cape Verde, May 22–26, 2005; 2005.
[53] A review of power system modeling platforms and capabilities. 2015. http://www.theiet.org/sectors/energy/documents/modelling-3.cfm?type=pdf.
[54] Xue S, Chen C, Jin Y, et al. Protection for DC distribution system with distributed generator. J Appl Math 2014;7:1–12.
[55] Meghwani A, Srivastava S, Chakrabarti S. A non-unit protection scheme for DC microgrid based on local measurements. IEEE Trans Power Deliv 2017.
[56] Baran ME, Mahajan NR. Over current protection on voltage-source-converter-based multiterminal DC distribution systems. IEEE Trans Power Deliv 2007.
[57] Meghwani A, Srivastava SC, Chakrabarti S. A new protection scheme for DC microgrid using line current derivative. In: Proceedings of the 2015 IEEE power and energysociety general meeting, Denver, USA, July 26–30, 2015; 2015.
[58] Mollet R. Over current protection of DC power plant equipment using modern high performance current limiting fuses. In: Proceedings of the 17th international telecommunications energy conference (INTELEC 95), The Hague, the Netherlands, October 29–November 1, 1995; 1995.
[59] Jin C, Dougal R. Current limiting technique based protection strategy for an industrial DC distribution system. In: Proceedings of the 2006 IEEE international symposium on industrial electronics, Montreal, Canada, July 9–13, 2006; 2006.
[60] Mahela OP, Khan B, Alhelou HH, Siano P. Power quality Assessment and event detection in distribution network with wind energy penetration using Stockwell transform and fuzzy clustering. IEEE Trans Ind Inform 2020;16(11):6922–32.
[61] Rathore B, Mahela OP, Khan B, Haes Alhelou H, Siano P. Wavelet-alienation-neural based protection scheme for STATCOM compensated transmission line. IEEE Trans Ind Inform 2021. https://doi.org/10.1109/TII.2020.3001063.
[62] Gangwar AK, Mahela OP, Rathore B, Khan B, Haes Alhelou H, Siano P. A novel K-means clustering and weighted K-NN regression based fast transmission line protection. IEEE Trans Ind Inform 2021. https://doi.org/10.1109/TII.2020.3037869.
[63] Rathore B, Mahela OP, Khan B, Padmanaban S. Protection scheme using wavelet-Alienation-Neural technique for UPFC compensated transmission line. IEEE Access 2021;9:13737–53. https://doi.org/10.1109/ACCESS.2021.3052315.
[64] Kaushik R, et al. Recognition of islanding and operational events in power system with renewable energy penetration using a Stockwell transform-based method. IEEE Syst J 2020. https://doi.org/10.1109/JSYST.2020.3020919.
[65] Gebru FM, Khan B, Alhelou HH. Analyzing low voltage ride through capability of doubly fed induction generator based wind turbine. Comput Electr Eng 2020;86:106727.
[66] Ram Ola S, Saraswat A, Goyal SK, Sharma V, Khan B, Mahela OP, Haes Alhelou H, Siano P. Alienation coefficient and Wigner distribution function based protection scheme for hybrid power system network with renewable energy penetration. Energies 2020;13:1120.
[67] Ram Ola S, Saraswat A, Goyal SK, Jhajharia SK, Khan B, Mahela OP, Haes Alhelou H, Siano P. A protection scheme for a power system with solar energy penetration. Appl Sci 2020;10:1516.
[68] Mahela OP, Sharma J, Kumar B, Khan B, Alhelou HH. An algorithm for the protection of distribution feeder using Stockwell and Hilbert transforms supported features. CSEE J Power Energy Syst 2020. https://doi.org/10.17775/CSEEJPES.2020.00170.
[69] Yogee GS, Mahela OP, Kansal KD, Khan B, Mahla R, Haes Alhelou H, Siano P. An algorithm for recognition of fault

conditions in the utility grid with renewable energy penetration. Energies 2020;13:2383.

[70] IEC. 60364-1. IEC 60364-1 low-voltage electrical installations Part 1: Fundamental principles, assessment of general characteristics, definition, vol. 11; 2005.

[71] Venkataramanan G, Marnay C. A larger role for micro grids. IEEE Power Energy Mag 2008.

[72] Sood VK, Fischer D, Eklund JM, et al. Developing a communication infrastructure for the smart grid. In: Proceedings of the 2009 IEEE electrical power and energy conference (EPEC), Montreal, USA, October 22–23, 2009; 2009.

[73] ALSTOM. Unit protection of feeders. Network protection and automation guide ALSTOM, vol. 10; 2011. https://zh.scribd.com/doc/120022911/Alstom-Network-Protection-and-Automation-Guide-2011.

[74] Monadi M, Koch-Ciobotaru C, Luna A, et al. A protection strategy for fault detection and location for multi-terminal MVDC distribution systems with renewable energy systems. In: Proceedings of the 2014 international conference on renewable energy research and application (ICRERA), Milwaukee, USA, October 19–22, 2014; 2014.

[75] Emhemed AAS, Burt GM. An advanced protection scheme for enabling an LVDC last mile distribution network. IEEE Trans Smart Grid 2014.

[76] Farhadi M, Mohammed OA. Event-based protection scheme for a multi terminal hybrid DC power system. IEEE Trans Smart Grid 2015.

[77] Pradeep Y, Khaparde SA, Joshi RK. High level event ontology for multi area power system. IEEE Trans Smart Grid 2012.

[78] Fariba L, Hossien S. An economic-based special protection system in a restructured environment. Electr Power Compon Syst 2013.

[79] Tang L, Ooi BT. Locating and isolating DC faults in multi-terminal DC systems. IEEE Trans Power Deliv 2007.

[80] Yu M, Wang Y, Zhang L, et al. DC short circuit fault analysis and protection of ring type DC micro grid. In: Proceedings of the 2016 IEEE 8th international power electronics and motion control conference (IPEMC-ECCE Asia), Hefei, China, May 22–26, 2016; 2016.

[81] Fletcher SDA, Norman PJ, Fong K, et al. High-speed differential protection for smart DC distribution systems. IEEE Trans Smart Grid 2014.

[82] Villamagna N, Crossley PA. A symmetrical component based GPS signal failure-detection algorithm for use in feeder current differential protection. IEEE Trans Power Deliv 2008.

[83] Farhadi M, Mohammed OA. A new protection scheme for multi-bus DC power systems using an event classification approach. IEEE Trans Ind Appl 2016.

[84] Yuan C, Haj-Ahmed MA, Illindala MS. Protection strategies for medium voltage direct current micro grid at a remote area mine site. IEEE Trans Ind Appl 2015.

[85] Li L, Ma P, Wang Y. A novel method for DC system grounding fault monitoring on-line and its realization. In: Proceedings of the 2008 IEEE international conference on automation and logistics, Qingdao, China, September 1–3, 2008; 2008.

[86] Chew SH, Tseng KJ. A line to ground fault detection technique for ELVDC distribution system in built environment. In: Proceedings of the 2016 IEEE 7th international symposium on power electronics for distributed generation systems (PEDG), Vancouver, Canada, June 27–30, 2016; 2016.

[87] Dong CY, He JH, Wang XK, et al. High-resistance grounding fault detection and location in DC rail ways system. In: Proceeding of the 11th IET international conference on the developments of power system protection, Birmingham, UK, April 23–26, 2012; 2012.

[88] Kontos E, Pinto RT, Bauer P. Effect of power flow control methods on the DC fault response of multi terminal DC networks. 2014.

[89] Li A, Cai Z, Sun Q, et al. Study on the dynamic performance characteristics of HVDC control and protections for HVDC line fault. 2009.

[90] Dhar S, Patnaik RK, Dash PK. Fault detection and location of photovoltaic based DC micro grid using differential protection strategy. IEEE Trans Smart Grid 2017. https://doi.org/10.1109/TSG.2017.2654267.

[91] Feng L, Gou R, Zhuo F, et al. Development of a 10 kV solid-state DC circuit breaker based on press-pack IGBT for VSC-HVDC system. In: Proceedings of the 2016 IEEE 8thinternational power electronics and motion control conference(IPEMC-ECCE Asia), Hefei, China, May 22–26, 2016; 2016.

[92] Kim JY, Choi SS, Song SM, et al. New DC solid state circuit breaker with reclosing and re-breaking capabilities. In: Proceedings of the 2016 IEEE 8th international power electronics and motion control conference (IPEMC-ECCE Asia), Hefei, China, May 22–26, 2016; 2016.

[93] Qi L, Antoniazzi A, Raciti L. DC distribution fault analysis, protection solutions, and example implementations. IEEE Trans Ind Appl 2018;54(4):3179–86.

[94] Salomonsson D, Member S, Söder L, Sannino A. Protection of low-voltage DC microgrids. IEEE Trans Power Deliv 2009;24(3):1045–53.

[95] Baran E, Mahajan R. Overcurrent protection on VoltageSource-converter- based multiterminal DC distribution systems. IEEE Trans Power Deliv 2007;22(1):406–12.

[96] Monadi M, Koch-ciobotaru C, Luna A, Candela JI, Rodriguez P. A protection strategy for fault detection and location for multi-terminal MVDC distribution systems with renewable energy systems. In: International Conference on Renewable Energy Research and Application (ICRERA); 2014. p. 496–501.

[97] Saleh KA, Hooshyar A, El-Saadany EF. Hybrid Passive Overcurrent relay for detection of faults in low-voltage DC grids. IEEE Trans Smart Grid 2015:1–10.

[98] Meghwani A, Srivastava SC, Chakrabarti S. A non-unit protection scheme for DC microgrid based on local measurements. IEEE Trans Power Deliv 2017;32(1):172–81.

[99] Emhemed AAS, Burt GM. An advanced protection scheme for enabling an LVDC last mile distribution network. IEEE Trans Smart Grid 2014;5(5):2602–9.

[100] Wang C, Li X, Guo L, et al. A nonlinear-disturbance observer-based DC-bus voltage control for a hybrid AC/DC micro grid. IEEE Trans Power Electron 2014.

Chapter 31

Identification and classification of faults using Stockwell transform and decision rule

Om Prakash Mahela[1], Vishnu Dutt Sharma[2], Baseem Khan[3], Sunil Agarwal[2] and Hassan Haes Alhelou[4]

[1]Power System Study Division, Rajasthan Rajya Vidyut Prasarn Nigam Ltd., Jaipur, Rajasthan, India; [2]Department of Electrical Engineering, Apex Institute of Engineering and Technology, Jaipur, Rajasthan, India; [3]Department of Electrical and Computer Engineering, Hawassa University, Hawassa, Ethiopia; [4]Tishreen University, Latakia, Syria

31.1 Introduction

The fault identification, classification, and localization on the power transmission lines are essential steps to be considered for initiating remedial and correcting measures in minimum time. With fast developments in the technology of digital computers, many techniques for recognition of faulty events have been developed. The fault estimation techniques reported in literature are divided in three groups. These are the techniques based on use of impedance measurement, based on the information extracted from the traveling waves, and the artificial intelligence or signal processing approach supported methods [1]. Research activities are reported that are aimed to develop the fast and accurate approaches for identification and to classify the faults on power transmission lines by the use of signal processing methods. Use of the combined features extracted by the use of two more these techniques might improve performance of the algorithm deployed for the detections of faults, which is considered as a main issue and for designing the primary as well as backup protection schemes. In Ref. [2], authors introduced an algorithm to identify faulty events incident on the transmission line. This is achieved by the use of pilot impedances and synchronized data. This resulted in the improvement of sensitivity as well as reliability of protection scheme. In Ref. [3], author proposed a method to identify, classify, and localize the faulty conditions on the both the overhead transmission line and underground cable. This method used the concept of entropy computed by hybrid combination of the fast discrete orthogonal ST (FDOST) to extract the features to identify the faults. Classification of these faulty events has been achieved using support vector machine (SVM) and support vector regression (SVR). In Ref. [4], authors introduced an algorithm that uses the wavelet transform and alienation coefficients to design a protection scheme for the multiterminal transmission line (MTTL). This is achieved by the application of wavelet transform (WT)-supported approximate coefficients of voltage and current signals computed over a time period of quarter cycle to identify, classify, and localize the faults on MTTL. In Ref. [5], authors introduced an approach for estimating the location of fault incident point on the power transmission lines. Voltage measurements performed using the Wide Area Measurement Systems (WAMS) have been used to implement the proposed protection scheme. A network bus admittance matrix is also computed to recognize the faults on transmission line. Recognition of faulted line and fault location is achieved with high accuracy in single step.

31.1.1 Related work

The review of literature focused on transmission line protection is detailed in this section. In Ref. [6], an approach for identification of faults that is fast and has low computational burden is introduced to design protection scheme for the transmission network. This is based on the calculation of mean error. To avoid false detection, a particle swarm optimization supported method with selection of optimal threshold is introduced by the authors. In Ref. [7], a technique is proposed that is based on estimation of Euclidean distance among successive samples for designing a fast and efficient algorithm for fault detection and classification in the network of electric power

Smart Cities Policies and Financing. https://doi.org/10.1016/B978-0-12-819130-9.00026-7

transmission. Magnitudes of fundamental components of currents were evaluated using the discrete Fourier transform considering as actuating signals for the proposed algorithm. In Ref. [8], a fault recognition approach that is effective to provide protection to the transmission line is proposed. This algorithm uses a combination of WT and linear discriminant analysis (LDA). Proposed algorithm used the current signals for identification of phases, which are faulty in nature. The zero sequence currents are utilized for identification of involvement ground in the fault. In Ref. [9], authors introduced a study for investigating effect of Multichamber Insulator Arresters (MCIA) on back flashover of lightning. This is obtained by transient modeling of the transmission line. Power system computer-aided design (PSCAD) software is implemented for 132 kV, Matugama to Kukule power transmission line. Simulation results are obtained with and without MCIA for evaluation of improvements in lightning back flashover due to the installation of MCIAs on the line. This increased the reliability of utility grid in the Sri Lanka. In Ref. [10], authors proposed an algorithm that uses discrete orthogonal Stockwell transform (DOST) to detect, classify, and locate (SFDCL) fault on the transmission systems, which has integrated multigenerators. Method is implemented with the help of synchronized current measurements from remote telemetry units (RTUs) available at both ends of transmission line. Energy coefficients computed from transients of current signals generated due to incidence of faults with the help of DOST are applied for identification and classification of faults. Fault location estimation is achieved using the support vector machine (SVM), which has considered the energy coefficients as inputs. In Ref. [11], fault detection and classification on a double circuit transmission line are proposed where one circuit is equipped with Static Synchronous compensator (STATCOM). Discrete WT (DWT)-based differential relaying protection scheme is developed for detection and classification of fault pattern on power transmission line. The scheme is based on use of current signals measured on two ends of power transmission line. In Ref. [12], authors presented an algorithm for identification and classification of the faults on transmission line using DWT as well as back-propagation neural network (BPNN) supported by transformation of Clarke. The current-based WT coefficients (WTC) as well as wavelet energy coefficient (WEC) are utilized for recognition of faults. Daubechies4 (Db4) is taken as mother wavelet for decomposition of high-frequency signal components. In Ref. [13], a technique is proposed to identify faulty section of the line commutated converter equipped with high-voltage direct current (HVDC) transmission line. The proposed method is implemented by extracting the current signal on rectifier end of the transmission line. Protection activity was achieved by the use of three separate fuzzy inference system modules. First fuzzy module detects faults on AC and DC sections. Second fuzzy module is aimed to discriminate the faulty section. Third fuzzy module identifies the faulty pole of the HVDC line. Proposed method found to be effective to achieve 100% reach setting, fault detection within a time less than one cycle, and elimination of requirement of communication links. In Ref. [14], a scheme based on resistance and power frequency component of differential current to identify faults on transmission line is proposed. Scheme is effective to avoid influence of the frequency-dependent feature of transmission line parameters. Proposed approach is also effective for reduction in the unbalanced differential current. This improves the selectivity of protection. Different techniques for recognition of faults have been reported in Refs. [15–29].

Technique using signal processing approaches for detection, classification, and localization of the faults on transmission lines is observed in literature. However, this method uses single features, which have their demerits. Recent focus is for designing smart protection schemes, which are fast, accurate, and reliable. This can be achieved by combining the several features. This has been taken as the key factor for the research undertaken in this thesis, which is mainly focused on developing an algorithm by the combination of different features of voltage and current signals for detecting and classifying the faults.

31.1.2 Contribution of the research work

Research work included in the work is aimed to design an algorithm that can be implemented for identification and classification of faults on power transmission lines using the current features. The major contributions of this work are as follows:

- This research work investigates to design an algorithm using processing of the current signals using Stockwell transform for identification of faults on power transmission line.
- The classification of the faults has been achieved using the rule-based decision tree.
- Proposed study is performed using MATLAB software in Simulink environment.

31.1.3 Organization of the chapter

Contents in the chapter are arranged in four sections. First section details the introduction of research work, related research work, and contribution of the chapter. Section 31.2 describes the methodology proposed for recognition of faults and test system used for the study. Results and discussion are included in Section 31.3, and finally at end of the chapter concluding remark is included in Section 31.4.

31.2 Methodology and test system

Test system implemented for implementation of the proposed study of fault estimation on the transmission line is described in this section. The data of the test system are also detailed in this section. Further, algorithm proposed for the estimation of faults using the Stockwell transforms and rule-based decision tree are detailed in this section. Brief descriptions of the Stockwell transforms and rule-based decision tree is also illustrated in this section.

31.2.1 Problem formulation

Presently the complexity level of the utility transmission network is continuously increasing. Transmission lines are up-rated in terms of the voltage levels as well as the power transmission capacities. Further, the capacities of the power transformers are also increasing day by day. The dynamic components such as thyristor controlled reactors (TCR), static synchronous compensator (STATCOM), etc., are also being continuously deployed in the network of the transmission. In recent scenario, the renewable energy (RE) is also integrated to the utility network in terms of the large-sized RE generators. The abovementioned dynamic changes have resulted in the following issues/problems in the power system network, which needs to be addressed:

- Protection systems deployed needs upgradation. The signal processing approaches may leads to the design of the advanced protection schemes.
- Stability of the power system network is at threat. Suitable practices and design parameters need to be deployed to maintain the stability of the utility grid.
- Power quality is deteriorating. Suitable techniques need to be designed for the recognition and mitigation of the PQ issues in the present-day power system network.

This study has considered the issues of the power system protection. A suitable technique has been designed, which can be used for the estimation of faults for designing the protection schemes for complex power system network.

31.2.2 Test system used for the study

Test network consisting of a transmission line with two terminals is used to perform the proposed study. This test transmission line is described in Fig. 31.1. Transmission line is integrated in between bus-1 and bus-2 as detailed in Fig. 31.1. A circuit breaker (CB) is placed near the bus-1, which is used to trip the transmission line during the faulty event. Generator sources 1 and 2 are used to simulate the large area utility networks. In the proposed study, bus-1 is treated as the sending terminal of power transmission line, and bus-2 is taken as the receiving terminal of transmission line. Parameters of test line and test system are included in Table 31.1 [29].

31.2.3 Current supported algorithm used for identification of fault conditions

The proposed current supported algorithm used for identification of the fault conditions on a test transmission line can be implemented (see checklist: "An Agenda For Action For The Proposed Current Supported Algorithm Used For The Identification Of The Fault Conditions On A Test Transmission Line").

An agenda for action for the proposed current supported algorithm used for the identification of the fault conditions on a test transmission line

The proposed current supported algorithm used for identification of the fault conditions on a test transmission line can be implemented using the detailed steps below (check all tasks completed):

_____1. Simulate a faulty condition on test line and measure currents on the sending end of test transmission line.

_____2. Process current or voltage signals using Stockwell transform to obtain output matrix and designate this as the STM, which is a complex valued matrix.

_____3. Compute absolute values of each element of the STM and designate it as the ASTM.

_____4. Calculate the median intermediate fault index (MIFI) from the ASTM matrix using the following MATLAB commands corresponding to all phases. The subscripts 1, 2, and 3 are used for the phases A, B, and C respectively:
A1=median(ASTM1);
A2=median(ASTM2);
A3=median(ASTM3);

_____5. Calculate the maximum value intermediate fault index (MVIFI) from the ASTM matrix using the following MATLAB commands corresponding to all phases. The subscripts 1, 2, and 3 are used for the phases A, B, and C respectively:
1. B1=max(ASTM1);
2. B2=max(ASTM2);
3. B3=max(ASTM3);

_____6. Calculate the summation intermediate fault index (SIFI) from the ASTM matrix using the following MATLAB commands corresponding to all phases. The subscripts 1, 2, and 3 are used for the phases A, B, and C respectively:
1. C1=sum(ASTM1);
2. C2=sum(ASTM2);
3. C3=sum(ASTM3);

_____7. Calculate the covariance fault index (CFI) from the ASTM matrix using the following MATLAB commands corresponding to all phases. The subscripts 1, 2, and 3 are used for the phases A, B, and C respectively:
D1=cov(max(ASTM1))
D2=cov(max(ASTM2))
D3=cov(max(ASTM2))

Continued

An agenda for action for the proposed current supported algorithm used for the identification of the fault conditions on a test transmission line—cont'd

_____8. Calculate the proposed hybrid fault index (HFI) using the following MATLAB commands corresponding to all phases. The subscripts 1, 2, and 3 are used for the phases A, B and C respectively:
FI1=A1.*B1.*C1.*D1;
FI2=A2.*B2.*C2.*D2;
FI3=A3.*B3.*C3.*D3;

_____9. Select a threshold value for the hybrid fault index (HFIT) equal to 200 for the proposed HFI to estimate the faulty condition from the healthy conditions. This HFIT is identified by testing algorithm for the different conditions such as different values of the fault location, fault impedance, fault incidence angle, reverse power flow, etc.

31.3 Result and discussion

Results for estimating the fault conditions on test transmission are detailed in this section. Discussion of the simulation results is included in detail. The results for estimation of fault condition such as phase-A to ground (AG), phase-A and Phase-B fault (AB), phase-A and Phase-B to ground fault (ABG), three-phase fault (ABC), and three phases to ground fault (ABCG) are described in this section. Further, the results of fault estimation during different operating scenarios have also been discussed in the section.

31.3.1 Faults on phase-A and involvement of ground

A phase-A to ground (AG) fault is simulated at the center of the test transmission line (at distance of 115 km from bus-1). Current associated with all phases is recorded on the bus-1 of test line. The currents associated with all phases are illustrated in Fig. 31.2A. This current is processed with the help Stockwell transform for obtaining STM matrix. Absolute value matrix (ASTM) is computed from the STM. The proposed MIFI, MVIFI, SIFI, and HFI are computed from the ASTM using the proposed algorithm and detailed in Fig. 31.2B–E in respective order.

Fig. 31.2A details that the current associated with the phase-A increased after incidence of AG faulty event, whereas values of currents associated with the phases-B and C follow the sinusoidal nature maintaining the same as that recorded in the healthy period. Fig. 31.2B details that the values of MIFI increase after incidence of fault condition for all phases. However, increase in the values of MIFI corresponding to the faulty phase-A is higher in comparison to healthy phases-B and C. Fig. 31.2C detailed that values of MVIFI associated with the faulty phase-A increase after incidence of fault condition. However, the values of MVIFI associated to healthy phases-B and C are the same in the postfault condition as that observed in the prefault condition. Fig. 31.2D details that the values of SIFI

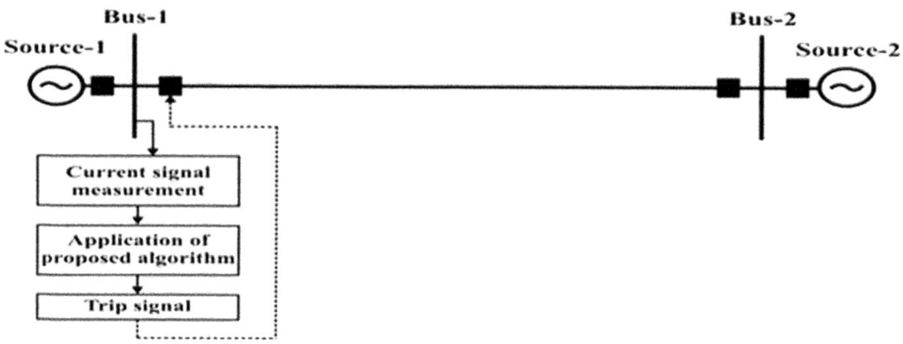

FIGURE 31.1 Test system with transmission line integrated between node-1 and node-2.

TABLE 31.1 Test system parameters.

Attribute	Value of parameter
Total length of line	230 km
Voltage of generator-1	500∠20 degrees kV
Voltage of generator-2	500∠0 degree kV
Equivalent impedance of generator-1	17.177 + j45.529 Ω
Equivalent impedance of generator-2	15.31 + j45.925 Ω
Positive sequence impedance of transmission line	4.983 + j117.83 Ω
Zero sequence impedance of transmission line	12.682 + j364.196 Ω
Positive sequence admittance of transmission line	j1.468 m℧
Zero sequence admittance of transmission line	j1.099 m℧
Transmission line capacity	433.63 + j294.52 MVA

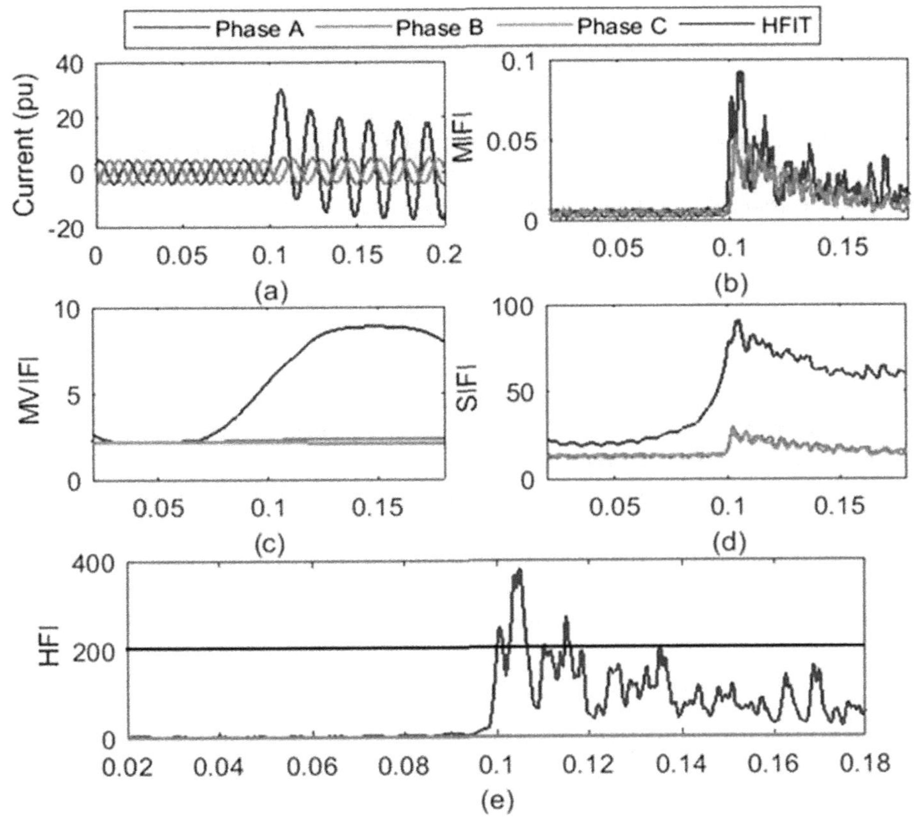

FIGURE 31.2 Current-based estimation of fault on phase-A and involvement of ground. (A) Currents of all phases, (B) median-based intermediate fault index, (C) maximum value-based intermediate fault index, (D) summation-based intermediate fault index, (E) proposed hybrid fault index.

increase after incidence of faulty condition for all phases. However, increase in the values of SIFI corresponding to the faulty phase-A is higher in comparison to healthy phases-B and C. Fig. 31.2E detailed that values of HFI associated with the faulty phase-A increase after incidence of the fault condition. However, the values of HFI for healthy phases-B and C are same in postfault condition as that observed in the prefault condition, which is zero. The values of HFI have crossed the threshold HFIT indicating that the phase-A is faulty. This index for healthy phase-B and C is below to HFIT. Hence, the phase-A to ground (AG) fault has been recognized efficiently.

The values of the covariance fault index (CFI) for all the phases during the event of the phase-A to ground faulty event are detailed in Table 31.2. It can be identified observed that CFI for the phase-A has high magnitude, whereas its value corresponding for phases-B and C is nearly equal to zero.

31.3.2 Fault on phases-A and B and without involvement of ground

A phase-A to phase-B (AG) fault is simulated at the center of the test transmission line (at distance of 115 km from node-1). Current associated with all phases is recorded on the node-1 of test line. The currents associated with all phases for event of phase-A to phase-B faulty event are illustrated in Fig. 31.3A. This current is processed using Stockwell transform for obtaining STM matrix. Absolute value matrix (ASTM) is computed from the STM. The proposed MIFI, MVIFI, SIFI, and HFI are computed from the ASTM using the proposed algorithm and detailed in Fig. 31.3B–E in respective order.

TABLE 31.2 Magnitude of current-based covariance fault index during faulty conditions.

	Covariance fault index		
Type of fault	Phase-A	Phase-B	Phase-C
AG	7.0959	0.0038	7.3309×10^{-4}
AB	14.7591	9.2744	5.3100×10^{-6}
ABG	13.6327	12.9619	7.7177×10^{-4}
ABC	16.5751	16.7919	16.5330
ABCG	16.5751	16.7919	16.5330

Fig. 31.3A details that the current associated with the phase-A increased just after incidence of AB fault, and values of currents associated with the phase-C follow the sinusoidal nature and maintain the same as that recorded in the healthy period. Fig. 31.3B details that the values of MIFI increase after the incidence of fault condition for the phases-A and B. However, magnitude of MIFI associated to healthy phase-C is low. Fig. 31.3C details that the values of MVIFI associated with the faulty phases-A and B increase after incidence of the fault condition. However, magnitude of MVIFI associated to the healthy phase-C is same in the postfault condition as that observed in the prefault condition. Fig. 31.3D details that the values of SIFI increase after the incidence of fault condition for the faulty phases-A and B. However, increase in the values of SIFI associated to healthy phase-C is the same in the prefault and postfault conditions. Fig. 31.3E details that the values of HFI associated with the faulty phases-A and B increased after incidence of fault condition. However, magnitude of HFI associated to healthy phase-C is the same in the postfault condition as that observed in the prefault condition, which is zero. The values of HFI have crossed the threshold HFIT for the phases-A and B and indicated that phases-A and B are faulty. This index for the healthy phase-C is below the HFIT. Hence, the phase-A to phase-B (AB) fault is recognized effectively. It is observed that the CFI for the phases-A and B is high, whereas its value associated to phase-C is close to zero.

31.3.3 Fault on phases-A and B and involvement of ground

Phases-A and B to ground (ABG) fault is simulated at the center of the test line (at distance of 115 km from node-1). Current associated with all phases is recorded on node-1 of test line. The currents associated with all phases in the event of phases-A and B to ground fault are detailed in Fig. 31.4A. This current is processed with the help of Stockwell transform for obtaining STM matrix. Absolute value matrix (ASTM) is computed from the STM. The

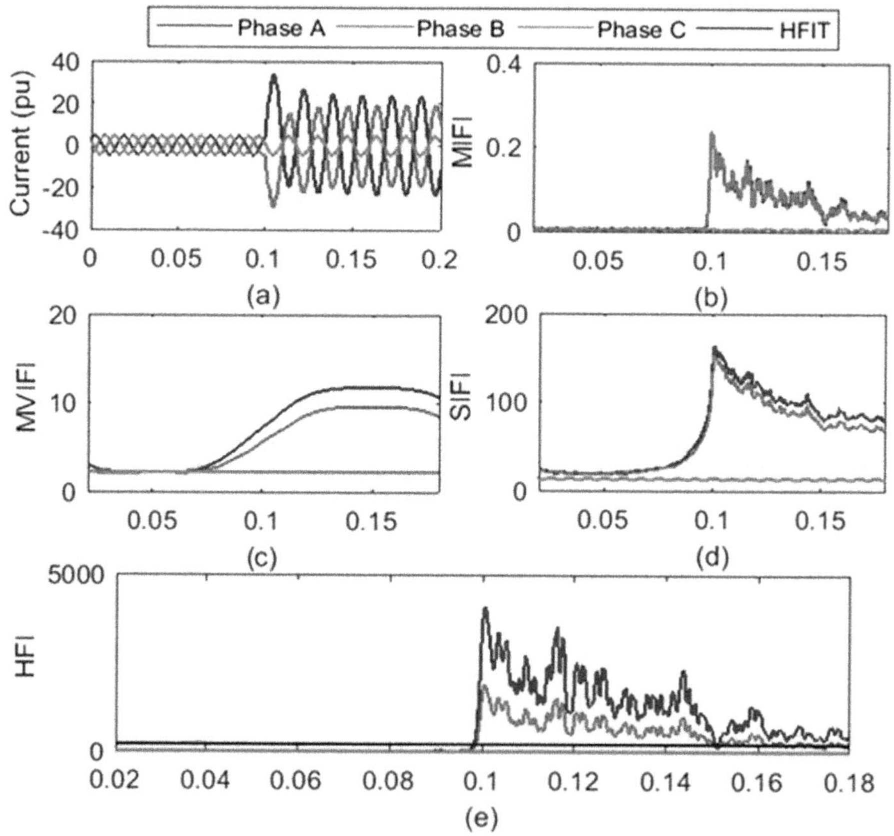

FIGURE 31.3 Current-based estimation of fault on Phases-A and B and without involvement of ground. (A) Currents of all phases, (B) median-based intermediate fault index, (C) maximum value-based intermediate fault index, (D) summation-based intermediate fault index, (E) proposed hybrid fault index.

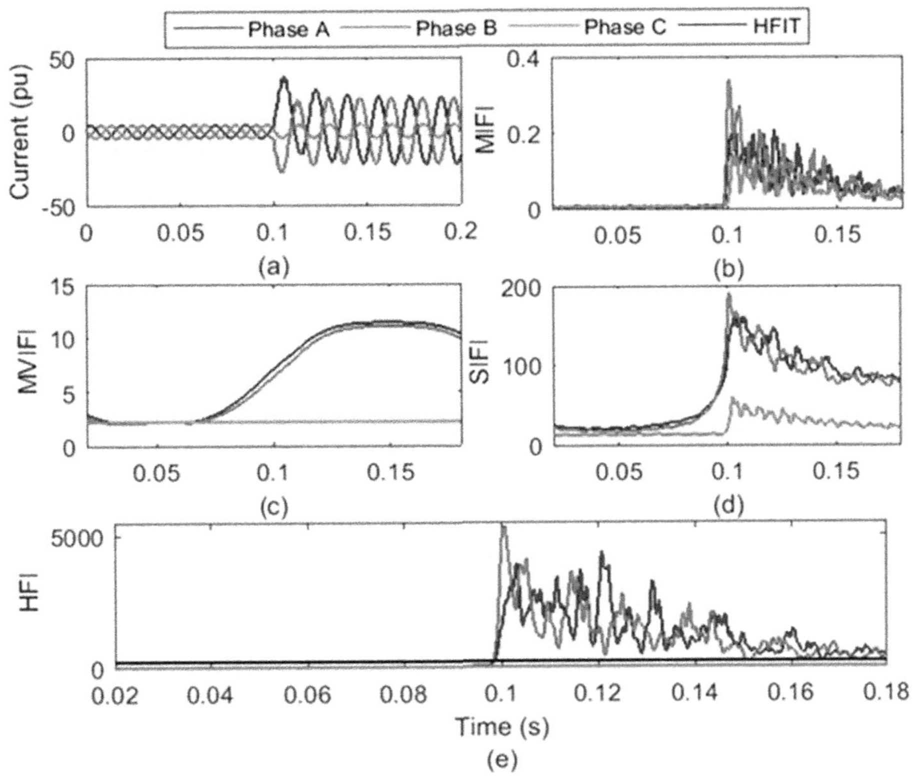

FIGURE 31.4 Current-based estimation fault on Phases-A and B and involvement of ground. (A) Currents of all phases, (B) median-based intermediate fault index, (C) maximum value-based intermediate fault index, (D) summation-based intermediate fault index, (E) proposed hybrid fault index.

proposed MIFI, MVIFI, SIFI, and HFI are computed from the ASTM using the proposed algorithm and detailed in Fig. 31.4B–E in respective order.

Fig. 31.4A details that the current associated with the phase-A increased just after incidence of ABG fault, and values of currents associated with the phase-C follow the sinusoidal nature and maintain the same as that recorded in the healthy period. Fig. 31.4B details that the values of MIFI increase after the incidence of fault condition for the phases-A and B. However, magnitude of MIFI associated to healthy phase-C is low. Fig. 31.4C details that the values of MVIFI associated with the faulty phases-A and B increase after incidence of the fault condition. However, values of MVIFI corresponding to the healthy phase-C are the same in the postfault condition as that observed in the prefault condition. Fig. 31.4D details that the values of SIFI increase after the incidence of fault condition for faulty phases-A and B. However, increase in magnitude of SIFI associated to healthy phase-C is the same in prefault and postfault condition. Fig. 31.4E details that the values of HFI associated with the faulty phases-A and B increased after incidence of fault condition. However, magnitude of HFI associated to healthy phase-C is the same in the postfault condition as that observed in the prefault condition, which is zero. The values of HFI have crossed the threshold HFIT for phases-A and B indicating that these phases are faulty. This index for the healthy phase-C is below the HFIT. Hence, the phases-A and B to ground (ABG) fault has been recognized effectively. It is found that CFI for phases-A and B has high values, whereas its value associated to phase-C is close to zero.

31.3.4 Fault on all the three phases and without involvement of ground

A fault condition involving all the three phases (ABC) is simulated at center of test transmission line (115 km from node-1). Current associated with all phases is recorded on node-1 of test line. The currents associated with all phases during event of ABC fault are illustrated in Fig. 31.5A. This current is processed by the use of Stockwell transform for obtaining STM matrix. Absolute value matrix (ASTM) is computed from the STM. The proposed MIFI, MVIFI, SIFI, and HFI are computed from the ASTM using the proposed algorithm and detailed in Fig. 31.5B–E in respective order.

Fig. 31.5A details that the current associated with all phases has increased just after incidence of ABC faulty event. Fig. 31.5B indicates that values of MIFI increase after the fault condition has incident for all phases after the

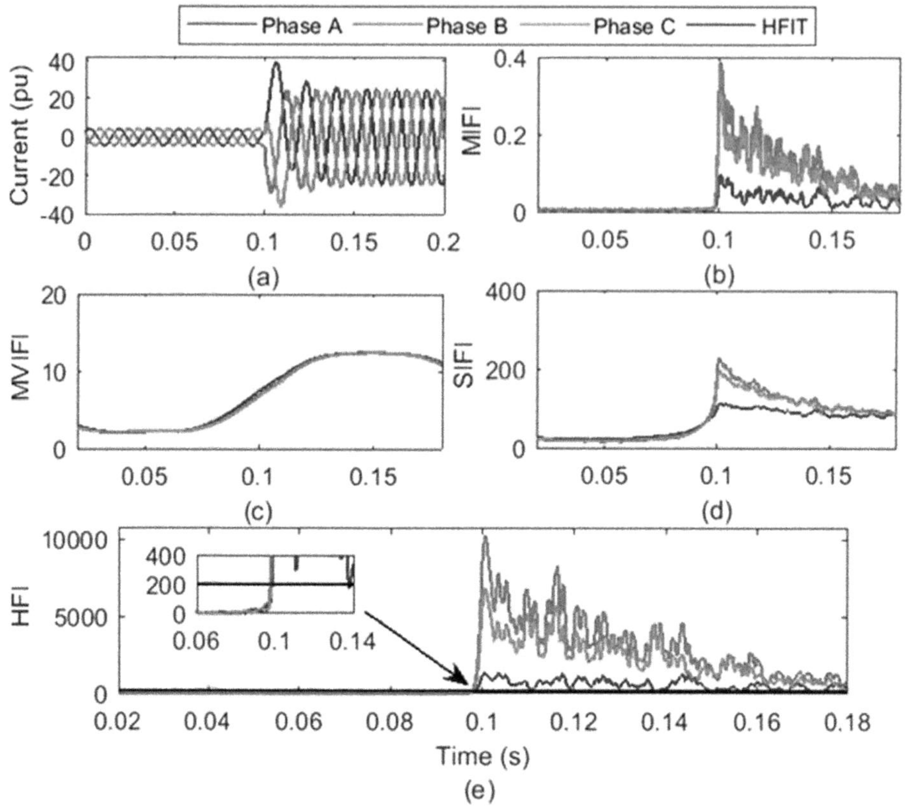

FIGURE 31.5 Current-based estimation fault on all the three phases and without involvement of ground. (A) Currents of all phases, (B) median-based intermediate fault index, (C) maximum value-based intermediate fault index, (D) summation-based intermediate fault index, (E) proposed hybrid fault index.

incidence ABC fault. Fig. 31.5C details that the values of MVIFI pertaining to all phases increase after incidence of the ABC fault condition. Fig. 31.5D details that the values of SIFI increase after incidence of ABC fault condition for all phases. Fig. 31.5E details that the values of HFI pertaining to all phases increase after incidence of the ABC fault condition. The values of HFI have crossed the threshold HFIT for all the phases indicating that all the phases are faulty. Hence, all phases (ABC) fault is recognized effectively. It is established that the CFI associated to all phases has high magnitudes.

31.3.5 Fault on all the three phases and involvement of ground

A fault condition involving all the three phases and ground (ABCG) is simulated at center of test transmission line (115 km from node-1). Current associated with all phases is recorded on node-1 of test line. The current signals associated with all phases during event of ABCG fault are detailed in Fig. 31.6A. This current is processed by the use of Stockwell transform for computing STM matrix. Absolute value matrix (ASTM) is computed from the STM. The proposed MIFI, MVIFI, SIFI, and HFI are computed from the ASTM using the proposed algorithm and detailed in Fig. 31.6B−E in respective order.

Fig. 31.6A details that the current associated with all phases has increased just after incidence of ABCG fault. Fig. 31.6B describes that magnitude of MIFI increases after incidence of fault condition for all phases after the incidence ABC fault. Fig. 31.6C details that the values of MVIFI pertaining all phases increase after incidence of the ABCG fault condition. Fig. 31.6D details that the values of SIFI increase after incidence of ABCG fault condition for all the phases. Fig. 31.6E details that the values of HFI pertaining tp all phases increased after incidence of the ABC fault condition. The values of HFI have crossed the threshold HFIT for all the phases indicating that all the phases are faulty. Hence, all phases to ground (ABCG) fault is recognized effectively. This can be observed that CFI for all phases has high values.

31.3.6 Impact of fault incidence angle

A phase-A to ground (AG) fault is simulated at center of test transmission line (115 km from node-1) with fault incidence angle equal to 45 degrees. Current associated with all phases is recorded on node-1 of test line. The current signals associated with all phases are described in Fig. 31.7A. This current is processed by the implementation of Stockwell transform to compute STM matrix. Absolute value matrix (ASTM) is computed from the STM.

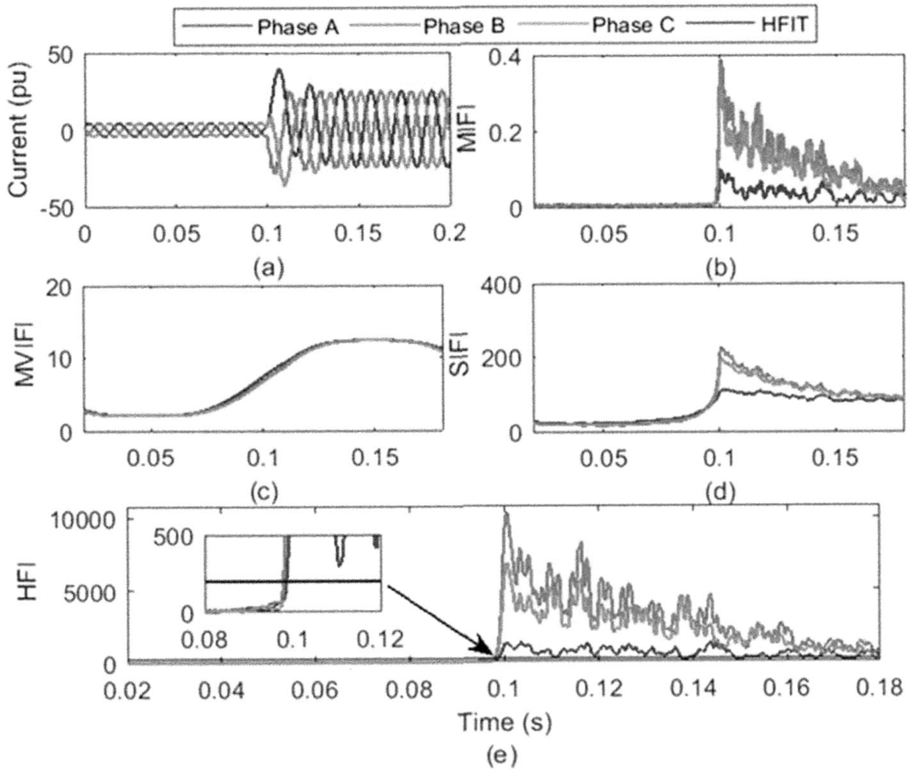

FIGURE 31.6 Current-based estimation fault on all the three phases and involvement of ground. (A) Currents of all phases, (B) median-based intermediate fault index, (C) maximum value-based intermediate fault index, (D) summation-based intermediate fault index, (E) proposed hybrid fault index.

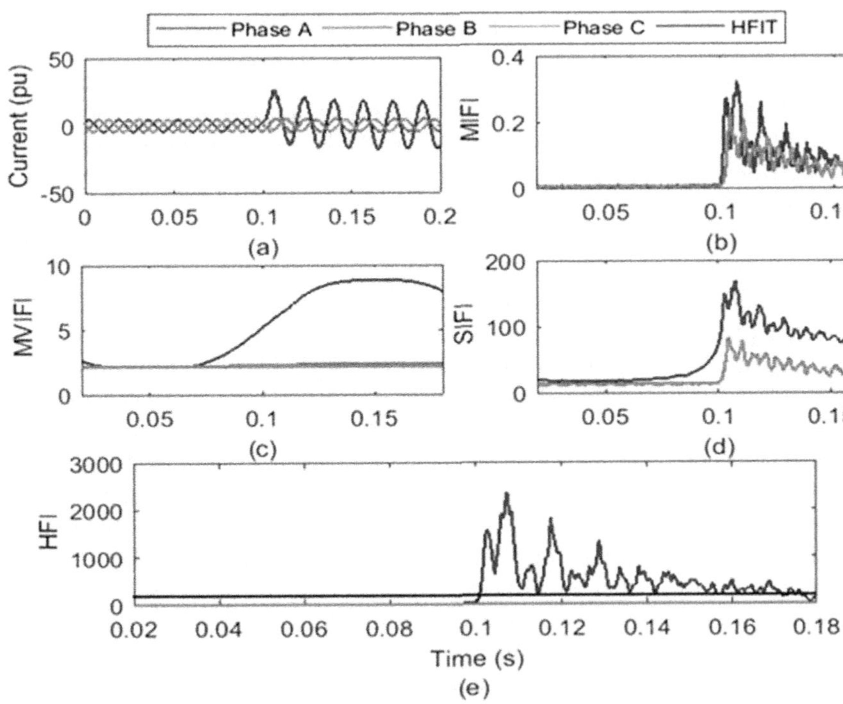

FIGURE 31.7 Current-based estimation fault on phase-A and involvement of ground with fault incidence angle of 45 degrees. (A) Currents of all phases, (B) median-based intermediate fault index, (C) maximum value-based intermediate fault index, (D) summation-based intermediate fault index, (E) proposed hybrid fault index.

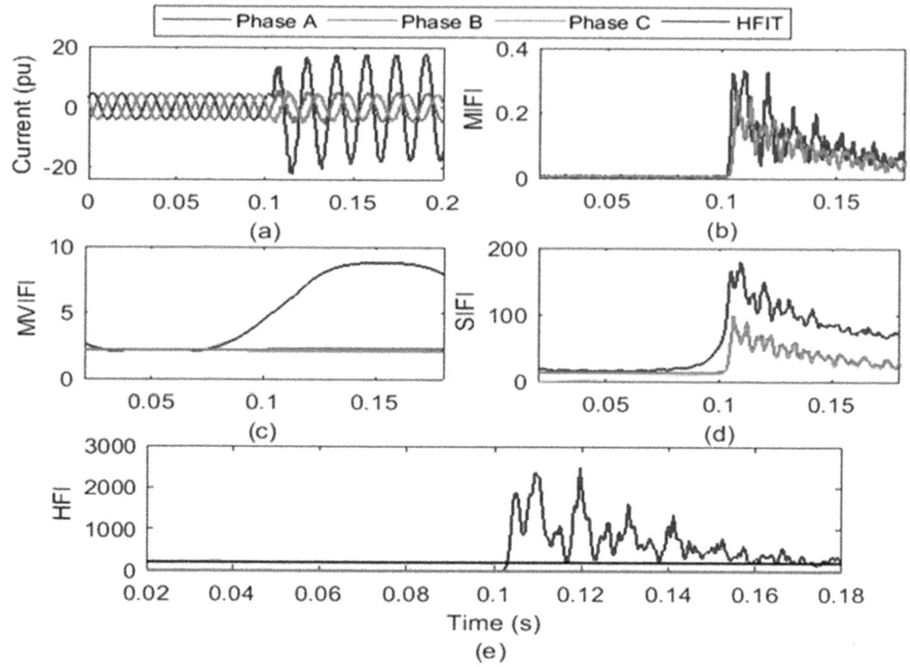

FIGURE 31.8 Current-based estimation fault on phase-A and involvement of ground with fault incidence angle of 90 degrees. (A) Currents of all phases, (B) median-based intermediate fault index, (C) maximum value-based intermediate fault index, (D) summation-based intermediate fault index, (E) proposed hybrid fault index.

The proposed MIFI, MVIFI, SIFI, and HFI are computed from the ASTM using the proposed algorithm and detailed in Fig. 31.7B–E in respective order.

Fig. 31.7A details that the current associated with the phase-A increases after the incidence of AG fault with fault incidence angle equal to 45 degrees, whereas the values of currents pertaining to phases-B and C follow the sinusoidal nature maintaining the same as that recorded in the healthy period. Fig. 31.7B details that the values of MIFI increase after incidence of fault condition for all phases. However, increase in the values of MIFI corresponding to the faulty phase-A is higher in comparison to healthy phases-B and C. Fig. 31.7C describes that magnitudes of MVIFI pertaining to faulty phase-A increase after incidence of the fault condition. However, the values of MVIFI pertaining to healthy phases-B and C are the same in the postfault condition as that observed in the prefault condition. Fig. 31.7D details that the values of SIFI increase after the incidence of fault condition for all phases. However, increase in the values of SIFI corresponding to the faulty phase-A is higher in comparison to healthy phases-B and C. Fig. 31.7E describes that magnitude of HFI associated with the faulty phase-A increases after incidence of the fault condition. However, the values of HFI pertaining to healthy phases-B and C are the same in the postfault condition as that observed in the prefault condition, which is zero. The values of HFI have crossed the threshold HFIT indicating that the phase-A is faulty. This index for healthy phases-B and C is below the HFIT. Hence, the phase-A to ground (AG) with fault incident angle equal to 45 degrees fault has been recognized effectively.

The magnitudes of the CFI pertaining to all phases during event of the phase-A to ground fault at an incidence angle of 45 degrees are observed as 7.0765, 0.0037, and 7.3475×10^{-04}. Hence, the CFI for the phase-A has high magnitudes, whereas its value associated to phases-B and C is nearly equal to zero.

A phase-A to ground (AG) fault is simulated at center of test transmission line (115 km from node-1) with fault incidence angle equal to 90 degrees. Current associated with all phases is recorded on node-1 of test line. The current signals associated with all phases are detailed in Fig. 31.8A. This current is processed by the use of Stockwell transform to obtain STM matrix. Absolute value matrix (ASTM) is computed from the STM. The proposed MIFI, MVIFI, SIFI, and HFI are computed from the ASTM using the proposed algorithm and detailed in Fig. 31.8B–E in respective order.

Fig. 31.8A details that the current associated with the phase-A increases after the incidence of AG fault with fault incidence angle equal to 90 degrees, whereas the values of currents pertaining to phases-B and C follow the sinusoidal nature maintaining the same as that recorded in the healthy period. Fig. 31.8B details that the values of MIFI increase after incidence of faulty event for all phases. However, increase in the values of MIFI corresponding to the faulty phase-A is higher compared to the healthy phases-B and C. Fig. 31.8C describes that magnitudes of MVIFI associated with the faulty phase-A increase after incidence of the fault condition. However, the values of MVIFI associated with the healthy phases-B and C are the same in the postfault condition as that observed in the prefault condition.

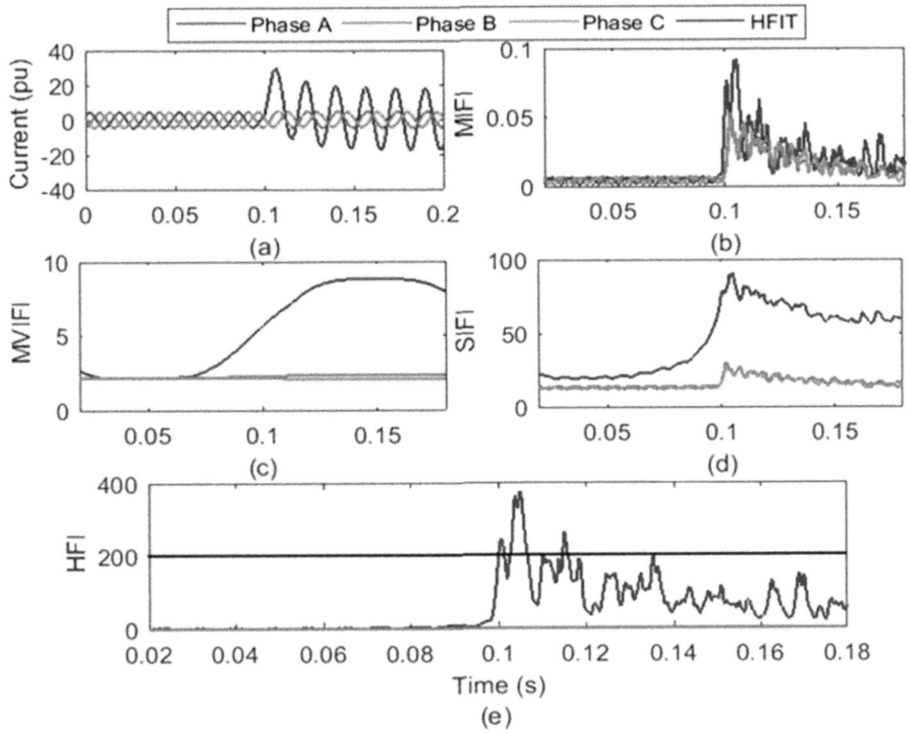

FIGURE 31.9 Current-based estimation fault on phase-A and involvement of ground with fault impedance equal to 10 Ω (A) Currents of all phases, (B) median-based intermediate fault index, (C) maximum value-based intermediate fault index, (D) summation-based intermediate fault index, (E) proposed hybrid fault index.

Fig. 31.8D details that the values of SIFI increase after incidence of fault condition for all phases. However, increase in the values of SIFI corresponding to the faulty phase-A is higher in comparison to healthy phases-B and C. Fig. 31.8E describes that magnitudes of HFI pertaining to faulty phase-A increase after incidence of the fault condition. However, the values of HFI corresponding to the healthy phases-B and C are the same in the postfault condition as that observed in the prefault condition, which is zero. The values of HFI have crossed the threshold HFIT indicating that the phase-A is faulty in nature. This index for the healthy phases-B and C is below the HFIT. Hence, the phase-A to ground (AG) with fault incidence angle equal to 45 degrees has been recognized effectively.

The magnitudes of CFI pertaining to all phases during event of phase-A to ground fault at and fault incidence angle of 90 degrees are observed as 7.512, 0.0041, and 7.563×10^{-04}. Hence, the CFI for the phase-A has high magnitude, whereas its value associated to phases-B and C is nearly equal to zero.

31.3.7 Impact of variations in the fault impedance

A phase-A to ground (AG) fault is simulated at center of test transmission line (115 km from node-1) with fault impedance equal to 10 Ω. The fault incidence angle is kept equal to 0 degree. Current associated with all phases is recorded on the node-1 of test line. The currents pertaining to all phases are described in Fig. 31.9A. This current is processed by the use of Stockwell transform for obtaining STM matrix. Absolute value matrix (ASTM) is computed from the STM. The proposed MIFI, MVIFI, SIFI, and HFI are computed from the ASTM using the proposed algorithm and detailed in Fig. 31.9B–E in respective order.

Fig. 31.9A details that the current associated with the phase-A increases after the incidence of AG fault with fault impedance equal to 10 Ω, whereas values of currents pertaining to phases-B and C follow the sinusoidal nature maintaining the same as that recorded in the healthy period. Fig. 31.9B details that the values of MIFI increase after incidence of faulty condition for all phases. However, increase in the values of MIFI corresponding to the faulty phase-A is higher in comparison to healthy phases-B and C. Fig. 31.9C describes that magnitudes of MVIFI associated with the faulty phase-A increase after incidence of the fault condition. However, the values of MVIFI associated with the healthy phases-B and C are the same in postfault condition as that observed in the prefault condition. Fig. 31.9D details that values of SIFI increase after the incidence of fault condition for all phases. However, increase in the values of SIFI corresponding to the faulty phase-A is higher in comparison to healthy phases-B and C. Fig. 31.9E describes that values of HFI associated with the faulty phase-

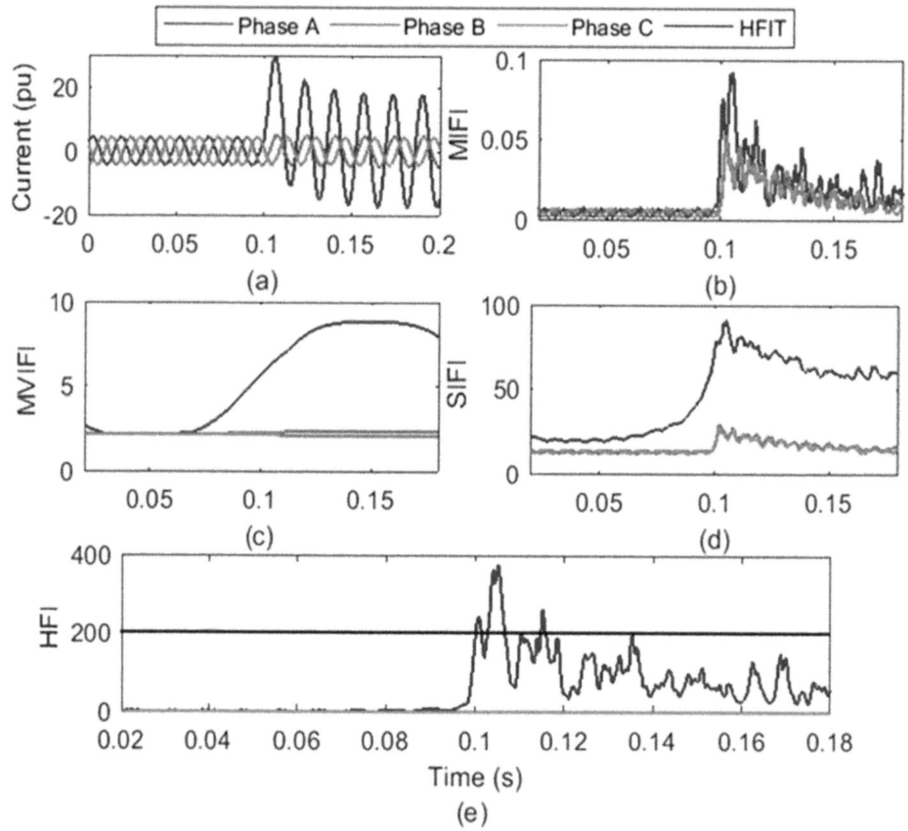

FIGURE 31.10 Current-based estimation fault on phase-A and involvement of ground with fault impedance equal to 5 Ω. (A) Currents of all phases, (B) median-based intermediate fault index, (C) maximum value-based intermediate fault index, (D) summation-based intermediate fault index, (E) proposed hybrid fault index.

A increase after incidence of the fault condition. However, the values of HFI pertaining to healthy phases-B and C are the same in postfault condition as that observed in the prefault condition, which is zero. The values of HFI have crossed the threshold HFIT indicating that the phase-A is faulty. This index for healthy phases-B and C is below HFIT. Hence, the phase-A to ground (AG) with fault impedance equal to 10 Ω is recognized effectively.

The magnitude of CFI for all phases during event of the phase-A to ground fault with fault impedance equal to 10 Ω is observed as 7.1158, 0.0038, and 7.6704×10^{-04}. Hence, the CFI for the phase-A has high magnitudes, whereas its value associated to phases-B and C is nearly equal to zero.

A phase-A to ground (AG) fault is simulated at center of test transmission line (115 km from the bus-1) with fault impedance equal to 5 Ω. The fault incidence angle is kept equal to 0 degree. Current associated with all phases is recorded on the bus-1 of test line. The currents associated with all phases are described in Fig. 31.10A. This current is processed by the use of Stockwell transform for obtaining STM matrix. Absolute value matrix (ASTM) is computed from the STM. The proposed MIFI, MVIFI, SIFI, and HFI are computed from the ASTM using the proposed algorithm and detailed in Fig. 31.10B–E in respective order.

Fig. 31.10A details that the current associated with the phase-A increases after the incidence of AG fault with fault impedance equal to 5 Ω, whereas values of currents pertaining to phases-B and C follow the sinusoidal nature maintaining the same as that recorded in the healthy period. Fig. 31.10B details that the values of MIFI increase after incidence of faulty condition for all phases. However, increase in the values of MIFI corresponding to the faulty phase-A is higher compared to the healthy phases-B and C. Fig. 31.10C details that the values of MVIFI associated with the faulty phase-A increase after incidence of the fault condition. However, the values of MVIFI pertaining to healthy phases-B and C are the same in the postfault condition as that observed in the prefault condition. Fig. 31.10D details that the values of SIFI increase after incidence of faulty condition for all phases. However, increase in the values of SIFI corresponding to the faulty phase-A is higher in comparison to healthy phases-B and C. Fig. 31.10E describes that values of HFI associated with the faulty phase-A increase after incidence of the fault condition. However, the values of HFI corresponding to the healthy phases-B and C are the same in the postfault condition as that observed in the prefault condition, which is zero. The values of HFI have crossed the threshold HFIT indicating that the phase-A is faulty in nature. This index for the healthy phases-B and C is below the HFIT. Hence, the phase-A to ground (AG) with fault impedance equal to 5 Ω is recognized effectively.

Values of the CFI for all the phases during the event of the phase-A to ground fault having fault impedance equal to 5 Ω are observed as 7.1188, 0.0039, and 7.6708×10^{-04}. Hence, the CFI for the phase-A has high magnitude, whereas its value for the phases-B and C is nearly equal to zero.

31.3.8 Effect variations in fault location

A phase-A to ground (AG) fault has been simulated at a distance of 30 km from sending end of power transmission line. Current associated with all phases is recorded on the node-1 of test line. The currents associated with all phases are described in Fig. 31.11A. This current is processed by the use of Stockwell transform for obtaining STM matrix. The Absolute value matrix (ASTM) is computed from the STM. The proposed MIFI, MVIFI, SIFI, and HFI are computed from the ASTM using the proposed algorithm and detailed in Fig. 31.11B–E in respective order.

Fig. 31.11A details that the current associated with the phase-A increases after the incidence of AG fault with fault at distance of 30 km from sending terminal of transmission line, whereas values of currents pertaining to phases-B and C follow the sinusoidal nature maintaining the same as that recorded in the healthy period. Fig. 31.11B details that the values of MIFI increase after incidence of faulty condition for all phases. However, increase in magnitude of MIFI pertaining to faulty phase-A is higher in comparison to healthy phases-B and C.

Fig. 31.11C details that the values of MVIFI associated with the faulty phase-A increase after incidence of the fault condition. However, the values of MVIFI pertaining to the healthy phases-B and C are the same in postfault condition as that observed in the prefault condition. Fig. 31.11D details that the values of SIFI increase after incidence of faulty condition for all phases. However, increase in the values of SIFI corresponding to the faulty phase-A is higher in comparison to healthy phases-B and C. Fig. 31.11E details that magnitude of HFI associated with the faulty phase-A increases after incidence of the fault condition. However, the values of HFI associated to healthy phases-B and C are the same in the postfault condition as that observed in the prefault condition, which is zero. The values of HFI have crossed the threshold HFIT indicating that the phase-A is faulty. This index for the healthy phases-B and C is below the HFIT. Hence, the phase-A to ground (AG) at a distance of 30 km from sending terminal of transmission line is recognized effectively.

The magnitude of CFI for all phases during event of phase-A to ground fault at a distance of 30 km from sending terminal of transmission line is observed as 33.7446, 0.0682, and 0.0168 corresponding to phase-A, phase-B, and phase-C in respective order. Hence, the CFI

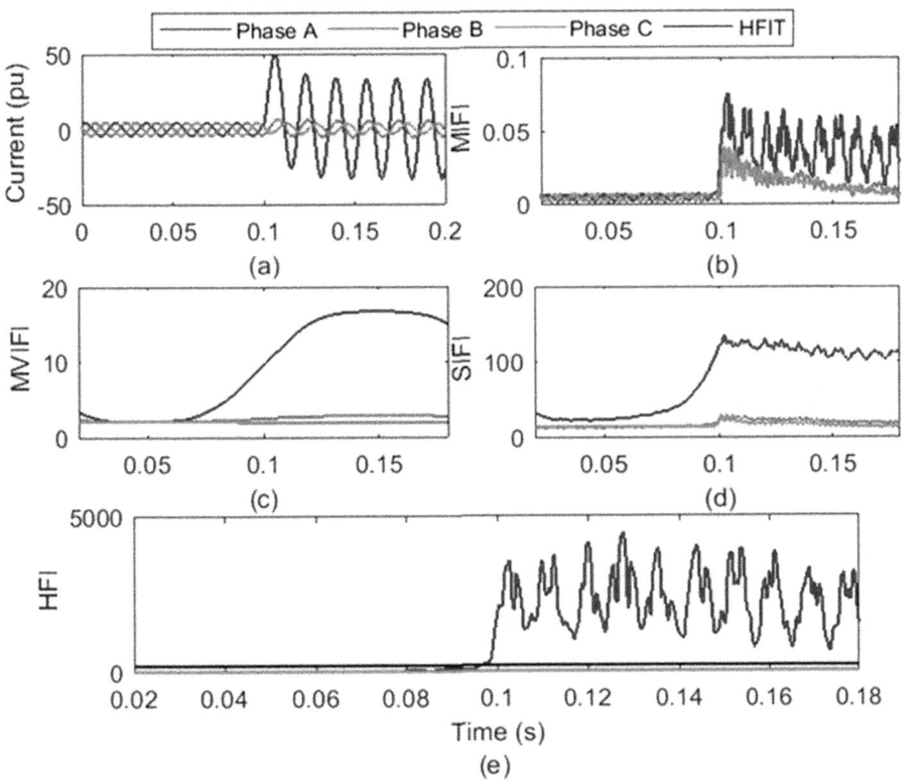

FIGURE 31.11 Current-based estimation fault on phase-A and involvement of ground incident at a distance of 30 km from sending terminal of transmission line. (A) Currents of all phases, (B) median-based intermediate fault index, (C) maximum value-based intermediate fault index, (D) summation-based intermediate fault index, (E) proposed hybrid fault index.

for the phase-A has high magnitudes, whereas its value associated to phases-B and C is nearly equal to zero.

A phase-A to ground (AG) fault is simulated at a distance of 200 km from sending terminal of transmission line. Current associated with all phases is recorded on the node-1 of test line. The currents associated with all the phases are illustrated in Fig. 31.12A. This current is processed by the use of Stockwell transform for computing STM matrix. Absolute value matrix (ASTM) is computed from the STM. The proposed MIFI, MVIFI, SIFI, and HFI are computed from the ASTM using the proposed algorithm and detailed in Fig. 31.12B–E in respective order.

Fig. 31.12A details that the current associated with the phase-A increases after the incidence of AG fault with fault at a distance of 200 km from sending terminal of transmission line, whereas the values of currents pertaining to the phases-B and C follow the sinusoidal nature maintaining the same as that recorded in the healthy period. Fig. 31.12B details that the values of MIFI increase after incidence of fault condition for all phases. However, increase in the values of MIFI corresponding to the faulty phase-A is higher in comparison to healthy phases-B and C. Fig. 31.12C describes that values of MVIFI associated with the faulty phase-A increase after incidence of the fault condition. However, the values of MVIFI pertaining to healthy phases-B and C are the same in the postfault condition as that observed in the prefault condition. Fig. 31.12D details that the values of SIFI increase after incidence of fault condition for all phases. However, increase in the values of SIFI corresponding to the faulty phase-A is higher in comparison to healthy phases-B and C. Fig. 31.12E describes that values of HFI associated with the faulty phase-A increase after incidence of the fault condition. However, the values of HFI pertaining to healthy phases-B and C are the same in the postfault condition as that observed in the prefault condition, which is zero. The values of HFI have crossed the threshold HFIT indicating that the phase-A is faulty in nature. This index for healthy phases-B and C is below the HFIT. Hence, the phase-A to ground (AG) at a distance of 200 km from sending terminal of transmission line has been recognized effectively.

Magnitudes of CFI for all phases in the event of the phase-A to ground fault located at 200 km from sending terminal of transmission line are observed as 7.7626, 0.0058, and 0.0012 corresponding to phase-A, phase-B, and phase-C in respective order. Hence, the CFI for the phase-A has high magnitude, whereas its value associated to the phases-B and C is nearly equal to zero.

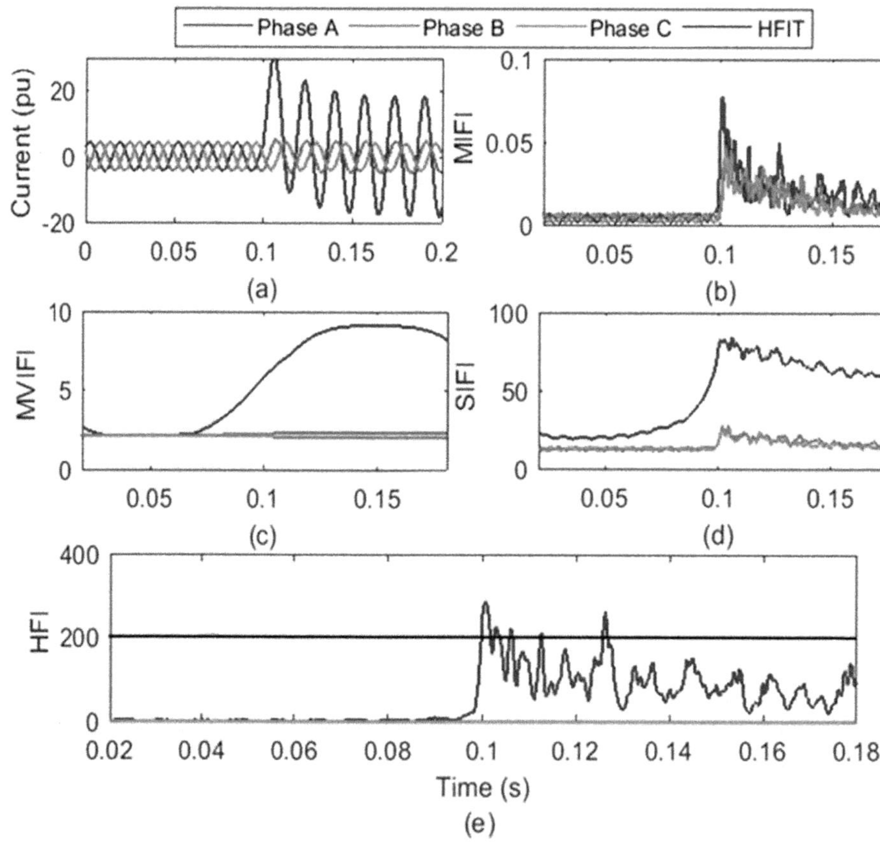

FIGURE 31.12 Current-based estimation fault on phase-A and involvement of ground incident at a distance of 200 km from sending terminal of transmission line. (A) Currents of all phases, (B) median-based intermediate fault index, (C) maximum value-based intermediate fault index, (D) summation-based intermediate fault index, (E) proposed hybrid fault index.

31.3.9 Fault classification

The classification of faults has been achieved using the decision supported by the rules. This rule-based decision tree algorithm classified different types of faults using the magnitudes of CFI.

The values of CFI detailed in Table 31.2 are used for classification of the faults. A threshold TH1 is used to classify the faults using number of faulty phases. Value of TH1 is set equal to 1 for classification of faults in the category based on the number of faulty phases. If CFI $>$ 1 for all the phases, then the fault is three-phase fault (ABC or ABCG). Furthermore, if CFI $>$ 1 for one phase only, then the type of fault is phase to ground fault (AG). For the two-phase fault, the values of CFI are greater than 1 for two phases. The threshold TH2 is used to classify the AB and ABG faults. A threshold value of 10 is set for the threshold TH2. If CFI $>$ TH2, then there is ABG fault, else AB fault. The classification of the different faults using the rule-based decision tree using the CFI is detailed in Fig. 31.13.

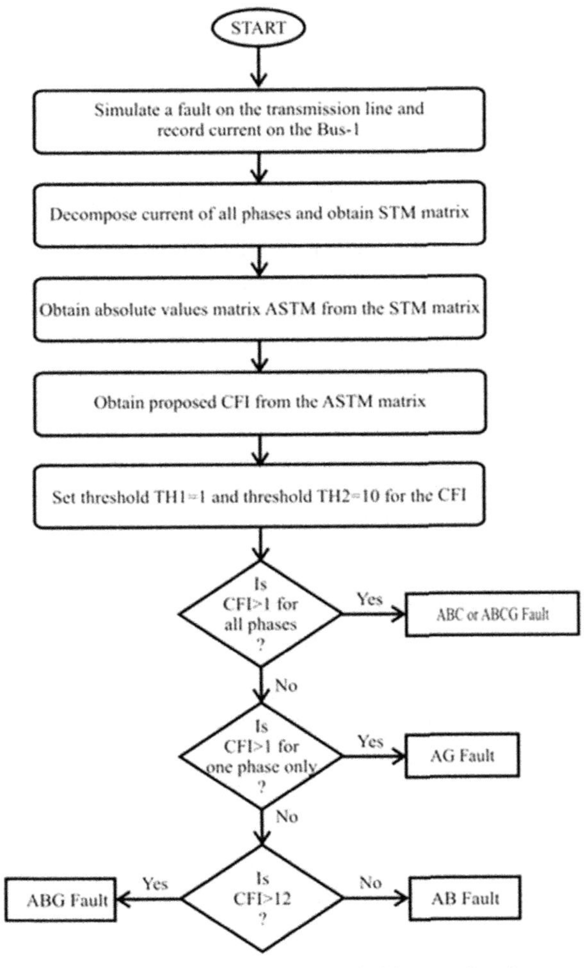

FIGURE 31.13 Fault classification using decision tree based on current features.

31.4 Summary

This research work presents an approach based on the current signals for recognition of the faulty events on the transmission line. Current signals recorded during the faulty condition are processed using the Stockwell transform to obtain output matrix. This matrix is used to compute the indices such as MIFI, MVIFI, SIFI, and CFI. A HFI is computed by multiplying the MIFI, MVIFI, SIFI, and CFI. A threshold value for the HFIT equal to 200 is decided for the proposed HFI to estimate the faulty condition from the healthy conditions. This HFIT is selected by the testing the algorithm for the different conditions such as different values of the fault location, fault impedance, fault incidence angle, reverse power flow, etc. The faults have been effectively classified using the rule-based decision tree. It is concluded that the proposed algorithm is found to be effective in the recognition of faults such as phase to ground, two phases fault, two phases to ground fault, all the three phases fault, and all three phases to ground fault. The algorithm is found to be effective in identification of the faults during the conditions such as the variations in fault incidence angle, variations in the fault impedance, variations in the fault location. The proposed algorithm has been tested for the two terminal transmission lines. This algorithm may also be tested for protection of the multiple transmission line as well as also in the presence of renewable energy.

Finally, let us move on to the real interactive part of this chapter: review questions/exercises, hands-on projects, case projects, and optional team case project. The answers and/or solutions by chapter can be found in Appendix G.

31.5 Chapter review questions/exercises

True/false

1. True or False? The fault identification, classification, and localization on the power transmission lines are essential steps to be considered for initiating remedial and correcting measures in maximum time.
2. True or False? Presently the complexity level of the utility transmission network is continuously increasing.
3. True or False? Test network consisting of a transmission line with three terminals is used to perform the proposed study.
4. True or False? The proposed current supported algorithm used for identification of the fault conditions on a test transmission line can not be implemented.
5. True or False? A phase-A to ground (AG) fault is simulated at the center of the test transmission line (at distance of 115 km from bus-1).

Multiple choice

1. A phase-A to phase-B (AG) fault is __________ at the center of the test transmission line (at distance of 115 km from node-1).
   a. Simulated
   b. Associated
   c. Recorded
   d. Invested
   e. Illustrated
2. The proposed MIFI, MVIFI, SIFI, and HFI are computed from the ASTM using the proposed:
   a. Infrastructure
   b. Algorithm
   c. Prosperity
   d. Economic effects
   e. Smart city policies
3. The magnitudes of the _____ pertaining to all phases during event of the phase-A to ground fault at an incidence angle of 45 degrees are observed as 7.0765, 0.0037, and 7.3475×10^{-04}.
   a. M2M
   b. M2H
   c. CFI
   d. AIS
   e. DTI
4. Values of the covariance fault index (CFI) for all the phases during the event of the phase-A to ground fault having fault impedance equal to 5 Ω are observed as 7.1188, 0.0039, and:
   a. 7.6748×10^{-04}
   b. 7.6508×10^{-04}
   c. 7.6608×10^{-04}
   d. 7.6708×10^{-04}
   e. 7.6808×10^{-04}
5. A phase-A to ground (AG) fault has been simulated at a distance of _________ from sending end of power transmission line.
   a. 20 km
   b. 30 km
   c. 40 km
   d. 50 km
   e. 60 km

Exercise

Problem

Present an algorithm for detecting and classifying faults on the transmission network of a smart city.

Hands-on projects

Project

Do research: Analyze an algorithm for detection and classification of different types of the power system faults occurring in smart cities on transmission lines that provide an effective protection to the line against these faults.

Case projects

Problem

Present a novel hybrid approach that combines S-transform (ST) and a feed forward neural network (FFNN) for the detection and classification of distribution grid faults in smart cities.

Optional team case project

Problem

Describe the concepts of S-Transform (ST)-based fault classification rules, in cases of overhead transmission lines in smart cities.

References

[1] Akmaz D, Mamis MS, Arkan M, Tăgluk ME. Transmission line fault location using travelling wave frequencies and extreme learning machine. Elec Power Syst Res 2018;155:1–7.

[2] Tong X, Wen H. A novel transmission line fault detection algorithm based on pilot impedance. Elec Power Syst Res 2020;179:106062.

[3] Patel B. A new FDOST entropy based intelligent digital relaying for detection, classification and localization of faults on the hybrid transmission line. Elec Power Syst Res 2018;157:39–47.

[4] Rathore B, Shaik AG. Wavelet-alienation based protection scheme formulti-terminal transmission line. Elec Power Syst Res 2018;161:8–16.

[5] Das S, Singh SP, Panigrahi BK. Transmission line fault detection and location using wide area measurements. Elec Power Syst Res 2017;151:96–105.

[6] Prasad CD, Nayak PK. Performance assessment of swarm-assisted mean error estimation based fault detection technique for transmission line protection. Comput Electr Eng 2018;71:115–28.

[7] Prasad CD, Nayak PK. A DFT-ED based approach for detection and classification of faults in electric power transmission networks. Ain Shams Eng J 2019;10:171–8.

[8] Yadav A, Swetapadma A. A novel transmission line relaying scheme for fault detection and classification using wavelet transform and linear discriminant analysis. Ain Shams Eng J 2015;6:199–209.

[9] Hemapala KTMU, Gnana Swathika OV, Dharmadasa KPRDSK. Techno-economic feasibility of lighting protection of overhead

transmission line with multi-chamber insulator arrestors. Dev Eng 2018;3:100–16.

[10] Jaya Bharata Reddy M, Gopakumar P, Mohanta DK. A novel transmission line protection using DOST and SVM. Eng Sci Technol Int J 2016;19:1027–39.

[11] Mishra SK, Tripathy LN, Swain SC. DWT approach based differential relaying scheme for single circuit and double circuit transmission line protection including STATCOM. Ain Shams Eng J 2019;10:93–102.

[12] Zin AAM, Saini M, Mustafa MW, Sultan AR, Rahimuddin. New algorithm for detection and fault classification on parallel transmission line using DWT and BPNN based on Clarke's transformation. Neurocomputing 2015;168:983–93.

[13] Agarwal S, Swetapadma A, Panigrahi C, Dasgupta A. A method for fault section identification in high voltage direct current transmission lines using one end measurements. Elec Power Syst Res 2019;172:140–51.

[14] Tang L, Dong X. A travelling wave differential protection scheme for half-wavelength transmission line. Int J Electr Power Energy Syst 2018;99:376–84.

[15] Mahela OP, Sharma J, Kumar B, Khan B, Alhelou HH. An algorithm for the protection of distribution feeder using Stockwell and Hilbert transforms supported features. CSEE J Power Energy Syst 2020. https://doi.org/10.17775/CSEEJPES.2020.00170.

[16] Kulshrestha A, Mahela OP, Gupta MK, Gupta N, Patel N, Senjyu T, Danish MSS, Khosravy M. A hybrid protection scheme using Stockwell transform and Wigner distribution function for power system network with solar energy penetration. Energies 2020;13(14):3519. https://doi.org/10.3390/en13143519.

[17] Yogee GS, Mahela OP, Kansal KD, Khan B, Mahla R, Alhelou HH, Siano P. An algorithm for recognition of fault conditions in the utility grid with renewable energy penetration. Energies 2020;13(9):2383. https://doi.org/10.3390/en13092383.

[18] Ola SR, Saraswat A, Goyal SK, Jhajharia SK, Khan B, Mahela OP, Alhelou HH, Siano P. A protection scheme for power system with solar energy penetration. Appl Sci 2020;10(4):1–22. https://doi.org/10.3390/app10041516. Paper No. 1516.

[19] Ola SR, Saraswat A, Goyal SK, Sharma V, Khan B, Mahela OP, Alhelou HH, Siano P. Alienation coefficient and Wigner distribution function based protection scheme for hybrid power system network with renewable energy penetration. Energies 2020;13(5). https://doi.org/10.3390/en13051120. Paper No. 1120.

[20] Ola SR, Saraswat A, Goyal SK, Jhajharia SK, Mahela OP. Detection and analysis of power system faults in the presence of wind power generation using Stockwell transform based median. Springer Lect Notes Electr Eng Ser 2019 ISSN: 1876-1100:319–29. https://link.springer.com/chapter/10.1007/978-981-15-0214-9_36.

[21] Gangwar AK, Rathore B, Mahela OP. K-means clustering and linear regression based protection scheme for transmission line. In: IEEE 9th Power India International Conference (PIICON 2020) from 28 Feb to 01 March, 2020. Murthal, India: Deenbandhu Chhotu Ram University of Science and Technology; 2020. https://doi.org/10.1109/PIICON49524.2020.9113038.

[22] Tailor N, Joshi S, Mahela OP. Transmission line protection schemes based on Wigner distribution function and discrete wavelet transform. In: IEEE 9th Power India International Conference (PIICON 2020) from 28 Feb to 01 March, 2020. Murthal, India: Deenbandhu Chhotu Ram University of Science and Technology; 2020. https://doi.org/10.1109/PIICON49524.2020.9113011.

[23] Sharma J, Kumar B, Mahela OP, Garg AR. Protection of distribution feeder using Stockwell transform supported voltage features. In: IEEE 9th Power India International Conference (PIICON 2020) from 28 Feb to 01 March, 2020. Murthal, India: Deenbandhu Chhotu Ram University of Science and Technology; 2020. https://doi.org/10.1109/PIICON49524.2020.9113014.

[24] Khoker MZ, Mahela OP, Ahmad G. A voltage algorithm using discrete wavelet transform and Hilbert transform for detection and classification of power system faults in the presence of solar energy. In: 2020 IEEE International Students' Conference on Electrical, Electronics and Computer Science (SCEECS 2020), MANIT Bhopal, India, February 22-23; 2020. https://doi.org/10.1109/SCEECS48394.2020.7.

[25] Khoker MZ, Mahela OP, Ahmad G. A current based hybrid algorithm using discrete wavelet transform and Hilbert transform for detection and classification of power system faults in the presence of solar energy. In: 2020 IEEE International Students' Conference on Electrical, Electronics and Computer Science (SCEECS 2020), MANIT Bhopal, India, February 22-23; 2020. https://doi.org/10.1109/SCEECS48394.2020.6.

[26] Karmakar S, Ahmad G, Mahela OP, Choudhary RR. Algorithm based on combined features of Stockwell transform and Hilbert transform for detection of transmission line faults with dynamic load. In: First IEEE International Conference on Power, Control and Computing Technologies (ICPC2T), NIT Raipur, India, January 3-5; 2020. https://doi.org/10.1109/ICPC2T48082.2020.9071516.

[27] Gupta D, Mahela OP, Ali S. Voltage based transmission line protection algorithm using signal processing techniques. In: 2020 IEEE International Students' Conference on Electrical, Electronics and Computer Science (SCEECS 2020), MANIT Bhopal, India, February 22-23; 2020. https://doi.org/10.1109/SCEECS48394.2020.15.

[28] Gupta D, Mahela OP, Ali S. Current based transmission line protection algorithm using signal processing techniques. In: 2020 IEEE International Students' Conference on Electrical, Electronics and Computer Science (SCEECS 2020), MANIT Bhopal, India, February 22-23; 2020. https://doi.org/10.1109/SCEECS48394.2020.14.

[29] Ola SR, Saraswat A, Goyal SK, Jhajharia SK, Rathore B, Mahela OP. Wigner distribution function and alienation coefficient based transmission line protection scheme. IET Gener, Transm Distrib 2020. https://doi.org/10.1049/iet-gtd.2019.1414. IF 3.229, Online ISSN 1751-8695, Print ISSN 1751-8687, (in press).

Further reading

Rathore B, Mahela OP, Khan B, Alhelou HH, Siano P. Wavelet-alienation-neural based protection scheme for STATCOM compensated transmission line. IEEE Trans Industr Inform June 2020. https://doi.org/10.1109/TII.2020.3001063. Electronic ISSN: 1941-0050, Print ISSN: 1551-3203, IF 9.112 (SCI), paper ID: TII-20-1327.

Chapter 32

Evaluation of fuel production technologies by using renewable energy for smart cities

Celal Hakan Canbaz[1], Orhan Ekren[1], Banu Y. Ekren[2] and Vikas Kumar[3]

[1]*Ege University, Izmir, Bornova, Turkey;* [2]*Yasar University, Izmir, Turkey;* [3]*University of the West of England, Bristol, United Kingdom*

32.1 Introduction

The "Smart City" concept is mainly an approach for urban planning that uses the latest technological resources. It uses various electronic tools and sensors to collect data within the city to provide efficient implementations also enabling sustainability in the whole system. On the other hand, a "Smart City" can be described as a localized geographical space that enables monitoring and managing the resources and wastes to make the city environmentally sustainable [1]. Herein, monitoring is based on the "Collected Information," which is used to have efficient asset management, usage of resources and services, and improve all the operations, which occur inside the city. Two crucial technological areas assisting the smart city concept are Information and Communication Technology (ICT) and the Internet of Things (IoT) concepts. The integration of ICT and IoT enables continuous monitoring and in-situ optimization of the parameters that affect sustainability. It also enables robust decision-making as well as enhances the quality and performance of the services by serving a more interactive approach. This also helps in reducing costs and the consumption of resources.

In smart cities, data are collected from all assets, buildings, devices, and even from citizens. By using those data, traffic, and transportation systems, utilities, all the networks such as water and electric supply, waste management, library and schools, hospitals, and all other community services are efficiently managed by taking the necessary actions. Therefore, it should be expected that a smart city can take simultaneous actions to generate new ideas by developing living laboratory environments by including the citizens of the city [2].

From this perspective, a smart city can be described as an integrated approach aiming to increase the quality of life of citizens rather than just an application of ICTs at the city scale. Hence, complementing each other, the following urban development axes come to the fore:

- Increasing the efficiency of urban systems, especially transportation and energy,
- Air and noise in living areas improvements such as reduction of pollution provision,
- Increasing efficiency in services provided to citizens,
- Development of local economies and increasing the competitiveness of cities.

In order to ensure the process of retransformation with smart cities, those functions become critical:

- Increase of economic and political efficiency and, social, cultural, and urban development communicating with each other to ensure the use of infrastructures.
- Creation of new job opportunities in the cities increasing socioeconomic development.
- Provision of social inclusion for the equal benefit of public services.
- Building a knowledge-driven and globalized economy attracting qualified human resources to cities and creating a constructive cultural environment.
- Adoption and learning of new technologies, contributing to creating new solutions for cities through structures such as living laboratories.

Smart Cities Policies and Financing. https://doi.org/10.1016/B978-0-12-819130-9.00038-3

- Ensuring social and environmental sustainability and efficient use of resources.

Based on the preceding information, an optimized smart city system requires to satisfy the following parameters [3]:

1. For energy and transportation, electricity and hydrogen energy should be utilized.
2. To prevent the seasonal changes of renewable resources, the city should use hydrogen energy for seasonal energy storage and also for road vehicle transportation.
3. An extensive electricity network, as well as hydrogen energy smart centers and hubs supplying renewable energy, should be installed.
4. The energy network must be integrated with all the buildings and the substructure of the smart city.
5. The pipeline network of hydrogen transportation has to be installed outside of the city, e.g., underground area.

32.1.1 Energy management and planning in smart cities

In a smart city concept, renewable energy resources become a critical issue to focus on. This is because these energy resources are infinite on earth, and they are environment-friendly. However, to provide uninterrupted energy source utilization, efficient and cost-effective methods should be developed.

The way of energy production in smart cities can be by conventional resources (e.g., fossil fuels) or by renewable energy sources. Although utilization of renewable energy resources may require initial investment costs for their system installations, it is well known that the alternatives (i.e., conventional resources) would cause environmental pollution and CO_2 emissions, which cannot be measured in terms of money costs. Therefore, it is worth pointing out that the smart city concept would definitely transform the situation into fully adopting renewable energy resources and green fuel production technologies [4]. The main subcategories of energy systems that could be considered in a smart city are given in Fig. 32.1.

In Fig. 32.1, the management and planning process of energy is divided into five categories: energy generation, energy storage, the infrastructure of energy systems, energy facilities, transportation of energy. These five steps are connected with each other by three different layers helping the sustainability of the system. These are intelligence, communication, and hardware.

As mentioned, energy generation for the future of smart cities by using renewable energy resources is one of the significant issues to focus on. For instance, photovoltaics can convert *solar energy* to electricity currents by using semiconductor materials. Thermal collectors can be used for absorbing the heat in solar energy utilization. Electricity generation is also possible by using the generated heat. However, this method is most effective in small-scale areas.

Geothermal energy is another electricity generation approach utilizing thermal energy on earth that can be considered to implement for smart cities. Although geothermal energy is cheaper compared to other energy sources, the location of the thermal source should exist close to the smart city. Besides that, *wind power* utilizing the power of wind energy can also be considered as an alternative method to generate electricity in smart cities. It is an advantage that wind turbine technologies are mature and cost-effective in large-scale applications. It can also be feasible in large-scale systems. Therefore, most of the smart city plans have been concentrating on providing energy to the city by using both solar and wind powers that are called the hybrid structure of renewable energy resources. For instance, Farahani et al. (2019) study a hybrid system design for hydrogen energy production for a smart city assumed to be built in Europe. Fig. 32.2 shows a figure for the workflow of hydrogen energy production by using hybrid energy resources in a smart city concept [3].

According to Fig. 32.2, the main constituents of the system are: a solar power system used for electricity and hydrogen production; upgraded buildings with smart electricity grid control mechanisms; hydrogen transportation

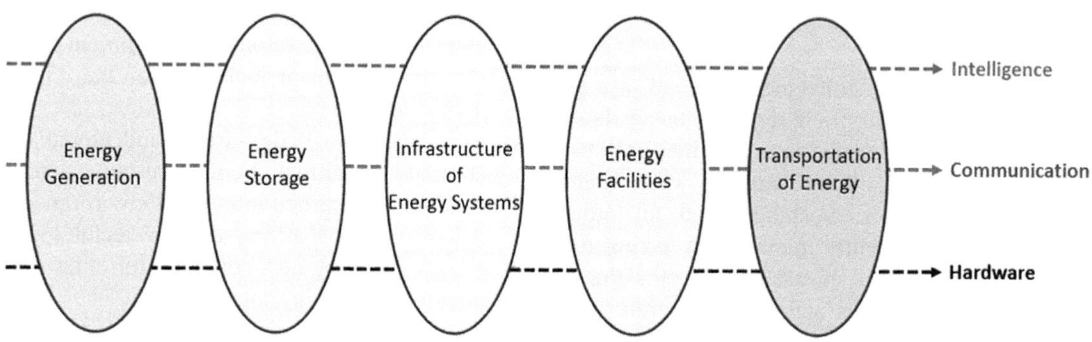

FIGURE 32.1 Subcategories of energy systems in a smart city. *Adopted from Calvillo CF, Sánchez-Miralles A, Villar J. Energy management and planning in smart cities. Renew Sustain Energy Rev 2016;55:273–87.*

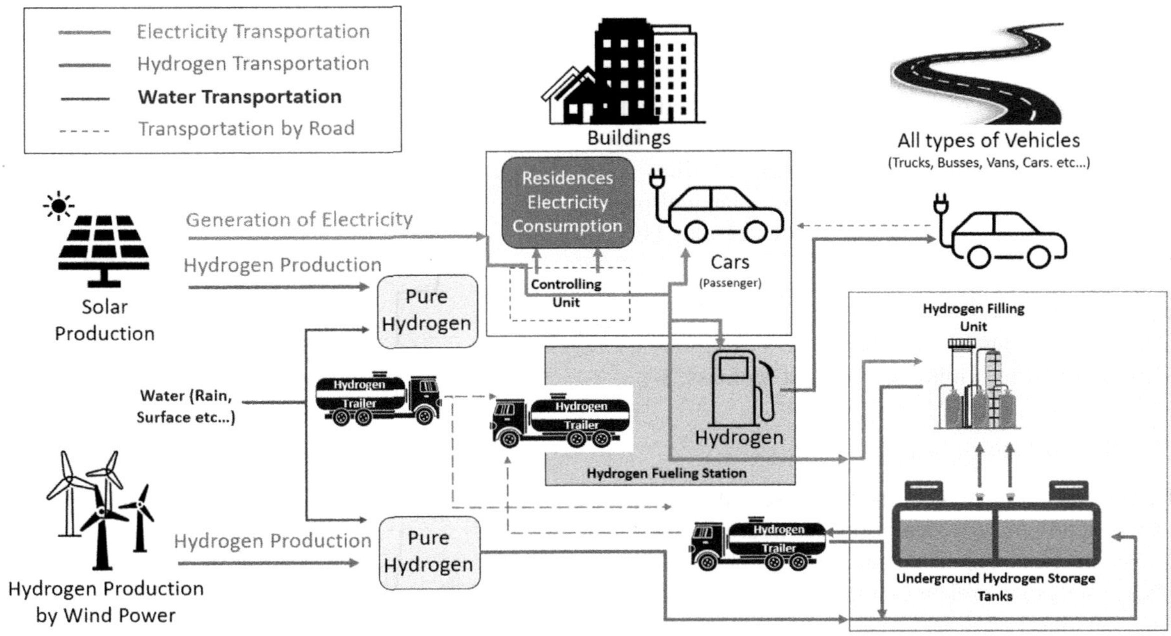

FIGURE 32.2 An example for hybrid (solar–wind) renewable energy usage for hydrogen energy generation in smart cities. *Adopted from Farahani SS, van der Veen R, Oldenbroek V, Alavi F, Lee EHP, van de Wouw, N, et al. A hydrogen-based integrated energy and transport system: the design and analysis of the car as power plant concept. IEEE Trans Syst Man Cybern Syst 2019;5(1):37–50.*

trailer fleet; fuel stations that provide hydrogen fuels; vehicles, cars, vans, buses using hydrogen energy; large-scale wind power systems located in a turbine park and producing compressed hydrogen fuel for the storing as seasonal hydrogen storage [3]. Here, the hydrogen energy can be produced in lot-for-lot order quantity and can be directly transported to the city or can be stored in underground reservoirs.

32.2 Renewable fuel production technologies

The word "fuel" is mainly described for the materials utilized for the production of energy. The first known fuel usage belongs to Homo Erectus using sticks and woods for combustion nearly two million years ago [5]. Animal- and plant-originated materials (including hydrocarbon-based fuels) are the only source that is used as fuel in the large time-scale of human history. For instance, charcoal has been used since at least 6000 B.C. for the aim of melting metals, and it was supplanted by coke (which is derived from coal) in the 18th century [6]. Hydrocarbons are also firstly discovered and transported in bamboo-made pipelines by the Chinese in 600 B.C. However, the efficient usage of hydrocarbon sources materials was started with the invention of kerosene that was procured by the distillation of crude oil by Persian chemists. Kerosene has been used as a primary energy resource for centuries. The popularity and common usage of coal start with the invention of the Steam Engine in the United Kingdom in 1869. Coal is used as energy resources for ships and trains, and it was even used for street lighting with the extraction of gas from coal in London in the 19th century. It is worth pointing out that coal is still providing 40% of the world's electrical power supply [7].

Fuel production materials and methods are continuously evolving with the development of new technologies. Environmental concerns and governmental regulations are accelerator factors for the utilization of clean energy sources in fuel production. Sourced by the growth of the population, fossil fuel consumption has reached the highest levels of all times, and it has directly increased the emissions of greenhouse gases in the atmosphere. Hence, that affected the energy usage trends. Today, even oil companies are starting to change their business plans, and they aim to increase fuel production from renewable energy sources such as solar, wind, and geothermal energy in the future [8].

United Nations Framework Convention on Climate exchange has signed the Paris Agreement with the participation of 174 countries. The agreement aims to limit the global warming increment level below 2°C by decreasing the amount of CO_2 emissions sourced by fossil fuel burning within a decade [9].

Besides, fuel production from hydrogen is a rising trend as renewable synthetic. Apart from that, biomass also has a great capacity in terms of producing environmentally friendly fuels.

32.3 Fuel production by using solar energy

The worldwide usage of solar energy has drastically increased due to the global climate targets in the last decade. This has also affected the research activities on solar power utilization for fuel production. Solar fuel production can be examined in four different categories [10]:

1. Thermolysis of water.
2. Thermochemical cycles.
3. Cracking of gaseous hydrocarbons.
4. Gasification and reforming of coal and biomass.

32.3.1 Thermolysis of water

The Thermolysis Of Water method uses the simple molecular structure of water and produces hydrogen from a water molecule by using solar energy. Eq. (32.1) shows how water molecule is broken to bonds by the endothermic process using solar radiation.

$$H_2O \rightarrow H_2 + 1/2O_2 \quad (32.1)$$

where, $\Delta H_{300K} = 286$ kj/mol.

The preceding process requires around 2225°C temperature to break all the bonds of a water molecule. It is possible to reach that temperature by using solar power. However, degradation of reactor materials might occur.

32.3.2 Thermochemical cycles

Metals are good at transportation and storage of solar energy. Therefore, metal oxides are utilized for the thermochemical cycle by using solar energy. In a thermochemical cycle with a metal oxide, an oxidation reaction comes after the step of reduction to produce a solar fuel such as hydrogen. The oxidation step requires lower temperatures compared with the reduction process. However, it is crucial to have the required temperature for the materials in the reduction step, their reaction kinetics as well as the dissociation to obtain the best efficiency.

Chueh et al. (2010) develop a system by utilizing purge gas, H_2O, CO_2, and a solar cavity receiver to produce fuel with the assistance of thermochemical dissociation. Fig. 32.3 illustrates the mechanism of a solar reactor that is used for thermochemical fuel production [11].

According to Fig. 32.3, the system includes a cavity receiver having a windowed aperture, and it allows the concentrated solar radiation to enter inside. The inner part ensures isolation with alumina, which is covered with porous ceria. Purge gas, H_2O, and CO_2 enter the system and meet with solar radiation. Thus, the thermochemical process, which will be resulted in the production of chemical fuel, starts. Thermochemical reactions inside the reactor can be defined by Eqs. (32.2–32.6).

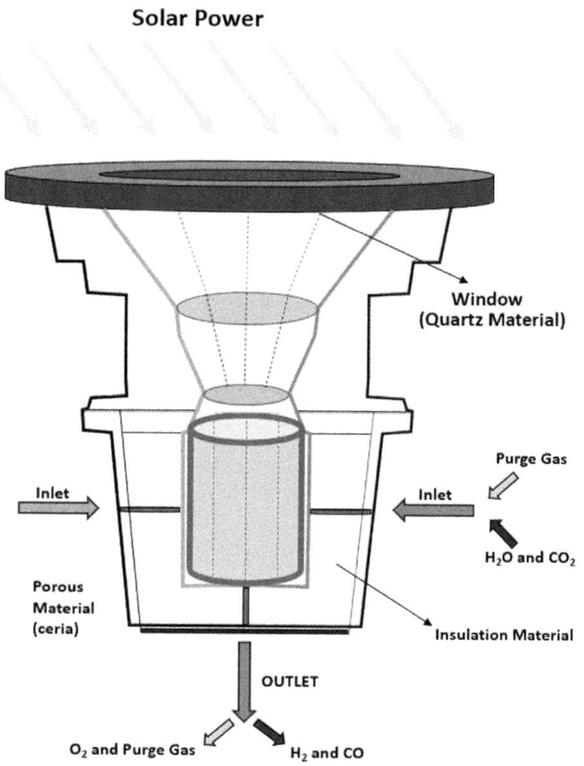

FIGURE 32.3 Solar reactor for solar-based thermochemical fuel production. *Adopted from Chueh WC, Falter C, Abbott M, Scipio D, Furler P, Haile SM, Steinfeld A. High-flux solar-driven thermochemical dissociation of CO_2 and H_2O using nonstoichiometric ceria. Science 2010;330(6012):1797–1801.*

$$^1/_\delta MO_2 \rightarrow {^1/_\delta} MO_{2-\delta} + 1/2O_2(g) \quad (32.2)$$

$$H_2O(g) + {^1/_\delta} MO_{2-\delta} \rightarrow {^1/_\delta} MO_2 + H_2(g) \quad (32.3)$$

$$CO_2(g) + {^1/_\delta} MO_{2-\delta} \rightarrow {^1/_\delta} MO_2 + CO(g) \quad (32.4)$$

$$H_2O(g) \rightarrow {^1\!/_2}O_2(g) + H_2(g) \quad (32.5)$$

$$CO_2(g) \rightarrow {^1\!/_2}O_2(g) + CO(g) \quad (32.6)$$

Here, M represents the metal Ce/combination of Ce and a dopant element. Eq. (32.2) is valid for higher temperature values (>800°C), Eqs. (32.3 and 32.4) are valid for lower temperatures (<800°C). Eq. (32.5) gives Net H_2O dissociation, and Eq. (32.6) gives the Net CO_2 dissociation.

Efficiency improvement is one of the key factors affecting the applicability and feasibility of solar-to-fuel systems. Davenport et al. (2016) study the ways of maximizing fuel production rates. They develop a numerical model that describes all possible scenarios for isothermal swing-cycling conditions [12]. Bork et al. (2020) increase the oxygen exchange capacity and its kinetics in solid oxides by using composite materials, and it resulted in improving the performances of many energy-related materials used in solar-to-fuel technology [13].

The thermochemical process by using solar energy can also be used in jet fuel production. Falter et al. [14] explore the climate impact and feasibility, geographical potential [15], water footprint, and land requirement [16] for efficient conversion of H_2O and CO_2 into the liquid hydrocarbon fuels for jet fuel production.

32.3.3 Gaseous hydrocarbon cracking

Gaseous hydrocarbon cracking is based on the mechanism of thermal decarbonization of hydrocarbon structures. Carbon, hydrogen, and other side products are formed by the cracking process. There is no CO_2 production during all the processes and that makes the whole production process environmentally friendly. The process is applicable at high temperatures (around 1225°C), and it can be reached by using concentrated solar energy [10]. Methane (CH_4) is preferred as the input material due to its short hydrocarbon chain and being the basic constituent of natural gas (Eq. 32.7). A basic diagram of the methane-cracking process is given in Fig. 32.4 [17].

$$CH_4(g) \rightarrow C_{(solid)} + 2H_2(g) \tag{32.7}$$

where, $\Delta H_{R,0} = 37.4\ \text{kj/mol}\ H_2$.

Another mechanism of the cracking process is given in Eq. (32.8) [10]:

$$\begin{aligned}2CH_4 &\rightarrow C_2H_6 + H_2 \rightarrow C_2H_4 + 2H_2 \rightarrow C_2H_2 \\ &+ 3H_2 \rightarrow 2C_{(solid)} + 4H_2\end{aligned} \tag{32.8}$$

Cracking is a good alternative process for hydrogen production with the thermal cracking of methane. The optimal temperature that needs to be applied by the solar cavity reactor is around 1225°C for the methane cracking process [18–20]. Thermochemical cycles and reaction steps for "High-Temperature Cycles" and "Low-Temperature Cycles" are given in Tables 32.1 and 32.2, respectively [10].

32.3.4 Gasification by solar energy

Gasification is a process in which the input materials benefit from the heating value. Carbon-based raw materials are transformed into fuels in the gas phase in the presence of a certain amount of steam and/or oxygen. Similar to the processes above, the temperature range varies between 1400 and 1700°C [9].

Solar gasification is a process that uses coal or other carbon-based materials and converts these into syngas (synthesis gas) by using concentrated solar power. Syngas mainly consists of CO, CO_2, water vapor, and H_2.

Solar thermal energy can also be integrated with the combined cycles of the gas turbines. In conventional systems, the steam powerplant part is repowered by a gas turbine to have better efficiency and increase of lifetime. Normally, the exhaust product gas of coal-fired plants is used as heat input for the steam cycle. In solar-modified systems, a solar-based thermal heat generation field is installed, and this produces steam for the steam plant (see Fig. 32.5) [21].

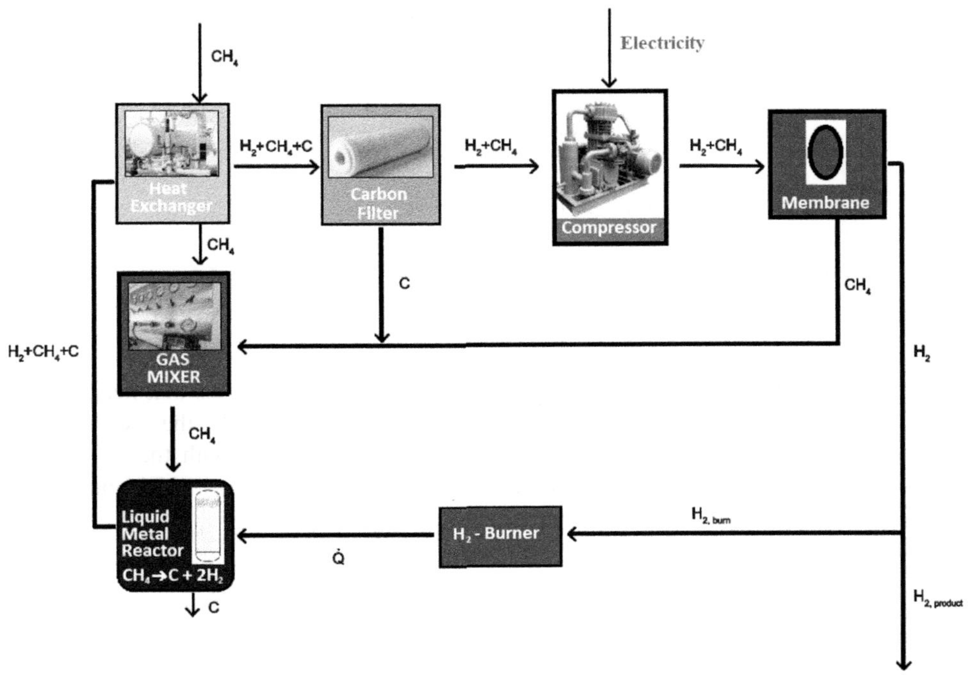

FIGURE 32.4 Block diagram of methane cracking process by applying liquid-metal technology. *Adapted from Postels S, Abánades A, von der Assen N, Rathnam RK, Stückrad S, Bardow A. Life cycle assessment of hydrogen production by thermal cracking of methane based on liquid-metal technology. Int J Hydrogen Energy 2016;41(48):23204–12.*

TABLE 32.1 High-temperature thermochemical cycles and reaction steps.

High-temperature cycles	
Cycle	FeO/Fe_3O_4
Reaction steps	$Fe_3O_4 \xrightarrow{2000-2300^\circ C} 3FeO + \frac{1}{2}O_2$ $3FeO + H_2O \xrightarrow{400^\circ C} Fe_3O_4 + H_2$
Cycle	Cadmium Carbonate
Reaction steps	$CdO \xrightarrow{1450-1500^\circ C} Cd + \frac{1}{2}O_2$ $Cd + H_2O + CO_2 \xrightarrow{350^\circ C} CdCO_3 + H_2$ $CdCO_3 \xrightarrow{500^\circ C} CdO + CO_2$
Cycle	Hybrid Cadmium
Reaction steps	$CdO \xrightarrow{1450-1500^\circ C} Cd + \frac{1}{2}O_2$ $Cd + 2H_2O \xrightarrow{25^\circ C(\text{electrochemical})} Cd(OH)_2 + H_2$ $Cd(OH)_2 \xrightarrow{375^\circ C} CdO + H_2O$
Cycle	Sodium Manganese
Reaction steps	$Mn_2O_3 \xrightarrow{1400-1600^\circ C} 2MnO + \frac{1}{2}O_2$ $2MnO + 2NaOH \xrightarrow{627^\circ C} 2NaMnO_2 + H_2$ $2NaMnO_2 + H_2O + \xrightarrow{25^\circ C} Mn_2O_3 + 2NaOH$
Cycle	M–Ferrite
Reaction steps	$Fe_{3-x}M_xO_4 \xrightarrow{1200-1400^\circ C} Fe_{3-x}M_xO_{4-y} + \frac{y}{2}O_2$ $Fe_{3-x}M_xO_{4-y} + yH_2O \xrightarrow{1000-1200^\circ C} Fe_{3-x}M_xO_4 + yH_2$

Adapted from Hall L. "Charcoal Briquette". How products are made. 2007. Retrieved 1 October 2007.

In Fig. 32.5, the power plant system has both high- and low-pressure modules. The system is designed to use solar energy to replace the heaters to maintain continuous energy production when any of these heaters are need to be stopped (e.g., due to maintenance-related issues, etc.)

Katayama and Tamaura (2005) develop a system utilizing both hydrocarbons and solar energy to produce methanol with almost zero CO_2 emission. The system uses solar energy for the endothermic process that needs an increase of enthalpy [18]. It converts the conventional hydrocarbon-based fuel to methanol with the coal gasification process, which is indicated in Eqs. (32.9 and 32.10).

$$C_2H + O_2 \rightarrow 2CO + 1/2H_2 \quad (32.9)$$

$$CO + 2H_2 \rightarrow CH_3OH \quad (32.10)$$

TABLE 32.2 Low-temperature thermochemical cycles and reaction steps.

Low-temperature cycles	
Cycle	Sulfur–Iodine
Reaction steps	$H_2SO_4 \xrightarrow{850^\circ C} SO_2 + H_2O + \frac{1}{2}O_2$ $I_2 + SO_2 + 2H_2O \xrightarrow{100^\circ C} 2HI + H_2SO_4$ $2HI \xrightarrow{300^\circ C} I_2 + H_2$
Cycle	Hybrid sulfur
Reaction steps	$H_2SO_4 \xrightarrow{850^\circ C} SO_2 + H_2O + \frac{1}{2}O_2$ $SO_2 + 2H_2O \xrightarrow{77^\circ C(\text{electrochemical})} H_2SO_4 + H_2$
Cycle	Hybrid Copper Chloride
Reaction steps	$Cu_2OCl_2 \xrightarrow{550^\circ C} 2CuCl + \frac{1}{2}O_2$ $2Cu + 2HCl \xrightarrow{425^\circ C} H_2 + 2CuCl$ $4CuCl \xrightarrow{25^\circ C(\text{electrochemical})} 2Cu + 2CuCl_2$ $2CuCl_2 + H_2O \xrightarrow{325^\circ C} Cu_2OCl_2 + 2HCl$

Adapted from Hall L. "Charcoal Briquette". How products are made. 2007. Retrieved 1 October 2007.

The study applies four different cases (3 Coal Gasification + 1 steam reforming of Natural Gas) successfully. These are:

Case 1: Coal + natural gas, solar heat utilization, and electrolysis of water by the power generation using solar heat.

Case 2: Coal + natural gas, solar heat utilization, and the cryogenic air separation unit using solar heat.

Case 3: Natural gas, solar heat utilization, and the cryogenic air separation unit using solar heat.

Case 4: Usual steam reforming of natural gas.

Although the system is applicable to generate clean fuel, it extremely increases the cost compared with conventional fuel consumption [22].

32.3.5 Solar energy fuel production in smart cities

Fuel production by using solar energy can be used in smart cities. Especially generating electricity for hydrogen production and designing the hydrogen refuelling stations that adapted to solar power usage can be realized in smart cities. Dispenzia et al. (2017) study the design of solar-powered stations, which are used for hydrogen refuelling [23]. A hydrogen minibus, an electric minivan, two bicycles, which are designed to work by electricity fueled by a solar plant, and the hydrogen refuelling unit are presented. Refuelling of the hydrogen bus, as well as the entire operations, is performed fully automated. The hydrogen production capacity of the refuelling unit is 8.5 kg at 350 bar in 630 min (with the peak load of 35 kW and the baseload of 10 kW). The results show that the refuelling of the hydrogen bus can be completed in 35 min where the temperature is reached 50°C max. The study was one of the pioneer studies that prove that solar energy can be used effectively for hydrogen refuelling [23]. On the other hand, Nambiar et al. (2018) mention that energy efficiency is possible with the electricity production of each building by using their own micro-grid solar PV generating emission-free electricity [24]. The study suggests assisting those micro-grid systems with a macro-grid PV system is significant when an extra energy supply is required. Alghamdi

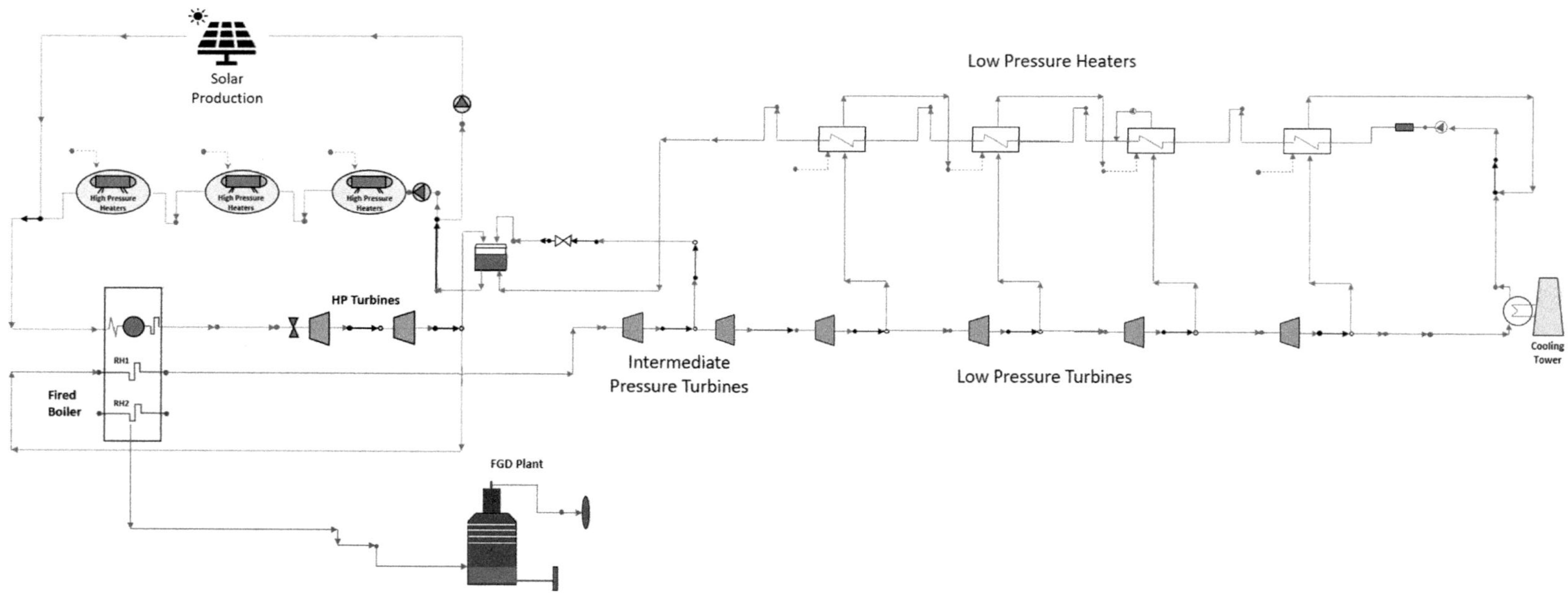

FIGURE 32.5 A diagram for a power plant that uses solar energy for repowering. *Adopted from Popov D. An option for solar thermal repowering of fossil fuel fired power plants. Sol Energy 2011;85(2):344–9.*

et al. (2020) study how to supply power for electric vehicles in a smart city and investigate the success of decentralized power production by using solar energy as the main power source [25].

Toh et al. (2020) investigate the advances in smart road applications in Chinese smart city projects [26]. The study gives the parameters of smart transportation and includes the details of smart roads. Roads that are capable to harvest and store energy, roads that can charge (electrified roads), road with smart street lights, roads that weight cars, roads that detect traffic violations are investigated. The usage of solar-powered smart systems is the most efficient way for the smart tools, which is being used for the above sections, and they can be used for activation (for electrified roads) or as a continuous power supply.

32.4 Fuel production by using wind energy

In this section, we discuss fuel production by the wind energy source. As mentioned, fuel production is also possible by using wind energy. Production of hydrogen by using water electrolysis and using wind energy as a power supply in the process has taken the attention of the energy industry in recent years. This new technique aims to decrease the consumption of hydrocarbons in hydrogen production and gather electricity with zero CO_2 emission and carbon footprint. In the system, hydrogen is obtained by using the electrolysis of water. Electricity required by the electrolysis process is generated by wind power as shown in Fig. 32.6 [27].

Hydrogen is obtained by using an inverse reaction. In most cases, an alkaline water electrolyzer, which gained recognition and considered the safest and reliable technology, is used. Eq. (32.11) shows how molecules of water are broken down while the water passes through the electricity, and it results in oxygen and hydrogen as products.

$$H_2O \rightarrow +H_2 + \frac{1}{2}O_2 \quad (32.11)$$

where, $E_0 = -1.23$ V (Standard Hydrogen Electrode).

The produced hydrogen can be used in different ways. It can be mixed with methane, transferred via pipelines for industry needs, used as a fuel directly, or used to enhance the wind turbine performances [28].

Several scientists have worked on the wind-hydrogen systems previously. Pioneer studies have started to generate electricity in the areas that the power lines could not reach. Ntziachristos et al. (2005) study the effectiveness of a wind power/fuel cell hybrid system for the Greek Cyclades. The system stores the energy that comes from the wind turbines in the form of hydrogen as shown in Fig. 32.7. The study shows that it is feasible to apply a wind power/fuel cell hybrid system instead of conventional stations [29].

Morgan et al. (2014) investigate ammonia production by using wind power for remote destinations such as isolated islands. The study demonstrates an analytical model to show the technical and economic advantages of local synthetic fuel production for the areas having huge fuel transportation costs [30]. Beccali et al. (2013) perform a size optimization for wind-hydrogen systems to decrease the operational problems such as dips in the voltage, variations in the frequency, power stability problems, and imbalances in the flow of the power [28]. Matzen et al. (2015) investigate the sustainability and technoeconomic analysis

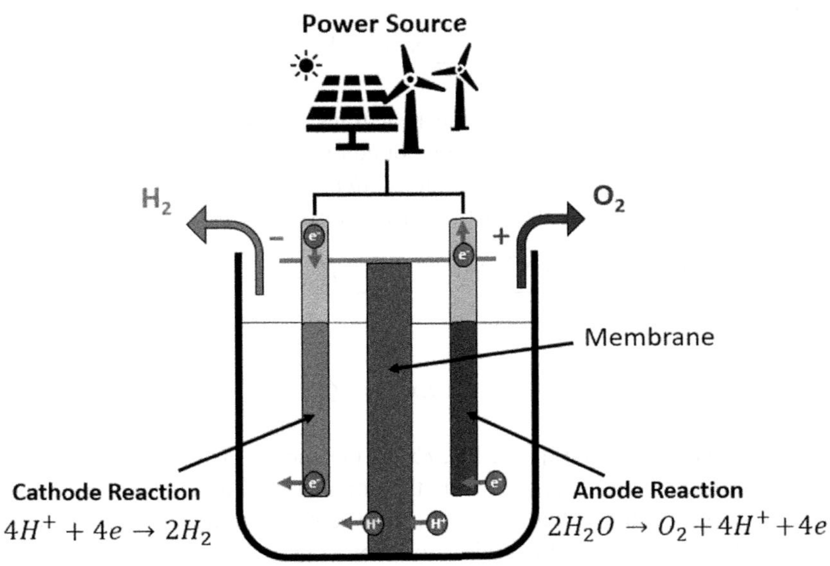

FIGURE 32.6 Hydrogen production by using wind energy for power supply. *Adopted from Hydrogen from wind energy and solar power, the future, REVE website, archieved from. [Accessed 08 June 2021].*

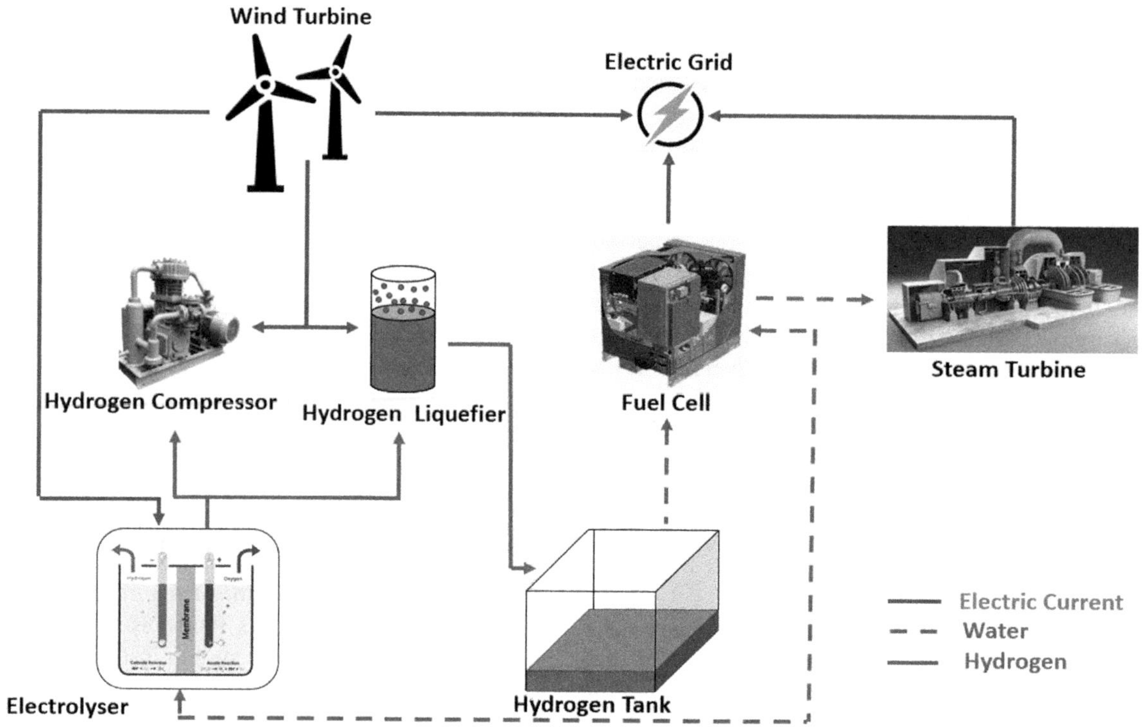

FIGURE 32.7 Workflow of fuel cell system powered by wind energy. *Adopted from Ntziachristos L, Kouridis C, Samaras Z, Pattas K. A wind-power fuel-cell hybrid system study on the non-interconnected Aegean islands grid. Renew Energy 2005;30(10):1471–87.*

of ammonia and methanol productions by using hydrogen production systems powered by wind [31].

Fasihi et al. (2016) performed a power-to-syngas study by using the hybrid PV-wind power plant with various storage options. Their proposed method uses two main steps for syngas production: (1) production of hydrogen by using water electrolysis and (2) hydrogenation of CO by using Reverse Water Gas Shift (RWGS) system. In the first part, the electrolysis plant uses water and electricity as inputs, hydrogen, and oxygen produced as a result of the endothermic process. The second part includes carbon monoxide production by using hydrogen and carbon dioxide that is captured from the ambient air with CO_2 capturing plants (Fig. 32.8) [32].

Muller et al. (2018) study diesel production by using biomass and wind energy. The process is similar to Fasihi's (2016) work till the syngas production. Following the syngas production, a laboratory-scale Fischer–Tropsch plant is used. Syngas is used as an input material at the plant, and following a series of process, wax, diesel, gasoline is produced as output materials [33].

Safari and Dincer (2018) perform the thermodynamic analysis of a hydrogen and methane production system that has been integrated with a wind energy system. They consider those assumptions in their study, ignore the heat losses from the boundaries of the system, consider steady-state conditions applying the power coefficient at the maximum value for the wind turbine, and ignore the pressure losses in the electrolyzer, pipelines, and methanation unit, etc. Additionally, the conditions of application environment are $T = 25°C$ and $P_0 = 101.325$ kPa [34]. By applying the first law of thermodynamics for each of the components involved in the system, the general energy balance equation can be written as;

$$\frac{dE_{cv}}{dt} = \dot{Q}_{cv} - \dot{W}_{cv} + \sum_i \dot{m}_i\left(h_i + V_i^2 + gZ_i\right) - \sum_e \dot{m}_i\left(h_e + V_e^2 + gZ_e\right) \tag{32.12}$$

Here, $\dot{Q}$ is the heat transfer and $\dot{W}$ is the work done crossing the system boundaries. As the process assumes the steady-state conditions, $dm_{cv}/dt = 0$, and $dE_{cv}/dt = 0$. Exergy destruction of a control volume also written as

$$\dot{E}x_d = \left[\sum \dot{m}_{\text{in}} ex_{\text{in}} - \sum \dot{m}_{\text{out}} ex_{\text{out}} - \sum Ex_w - \sum Ex_Q\right] \tag{32.13}$$

$$\dot{E}x_{Q,\,i} = Q_i\left(1 - \frac{T_0}{T_{s,i}}\right) \tag{32.14}$$

In Eq. (32.13), Ex_w is the energy of work and Ex_Q is the heat transfer inside the boundaries. The total exergy of the individual stream is the combination of chemical and physical exergies given by (Eqs. 32.15 and 32.16).

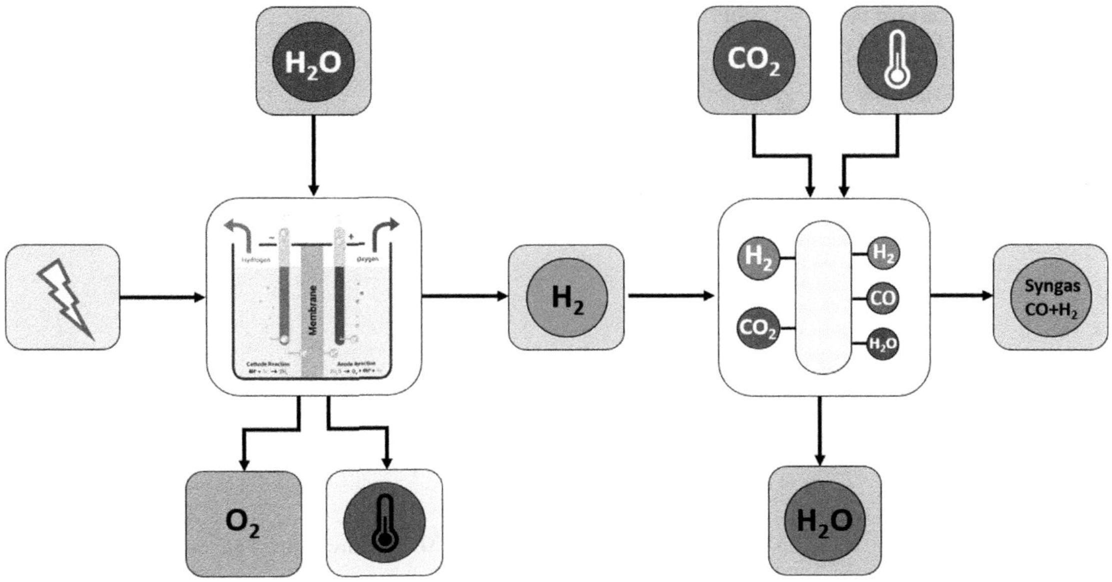

FIGURE 32.8 Electrolysis and Reverse Water Gas Shift (RWGS) processes. *Adopted from Fasihi M, Bogdanov D, Breyer C. Techno-economic assessment of power-to-liquids (PtL) fuels production and global trading based on hybrid PV-wind power plants. Energy Procedia 2016;99:243–68. 10th.*

$$ex_i = ex_{ch,\,i} + ex_{ph,\,i} \tag{32.15}$$

$$ex_{ph} = \sum_i [(h_i - h_0) - T_0(s_i - s_0)] \tag{32.16}$$

Chemical exergy of the gas mixture is given by the following Eq. (32.17):

$$\overline{ex}_{ch,\text{ gas mixture}} = \sum x_i ex^0_{ch,i} + RT_0 \sum x_i \ln(x_i) \tag{32.17}$$

Bicer et al. (2020) study ammonia production by using the wind power system and the electrochemical process and show that the ammonia produced by a renewable resource can be a good alternative for future energy production via hydrogen. The measured ammonia production amounts are given in Fig. 32.9.

Bicer et al. (2020) study for a pilot site in Qatar as the location that does not have strong seasonal wind changes. In their study, average ammonia production is around 3300 kg at an average wind speed.

32.4.1 Wind energy fuel production in smart cities

Wind power is becoming one of the most growing energy production methods due to its low production phase costs and environmentally friendly structure. Besides, it is a sustainable energy source and has a huge potential to be applied in smart cities directly or indirectly, for instance, for producing hydrogen fuels. The usage of hydrogen fuels in smart cities is direct fuel usage for vehicles and buildings; using a mixture of hydrogen and methane; transportation of hydrogen to users by using pipelines; and using hydrogen fuel to increase the wind turbine performances to supply the necessary energy amount [35]. In literature, several wind-hydrogen fuel production systems are investigated. For instance, Bernalagustin et al. (2008) compare different scenarios of wind-only and wind-hydrogen systems, and the results show that the wind-hydrogen systems are completely suitable for the management of energy generation [36]. A hybrid wind-powered fuel cell system application is performed by Nyziachristos et al. (2004) for Greek Islands, and the results show that it is crucial to select the appropriate sizes of the components of hydrogen storage, fuel cells, and electrolyzer [29]. The most feasible solution is described as the case that maximizes the electrolysis with a lower volume of the hydrogen tanks (Nyziachristos et al., 2004). Safari and Dincer (2018) have adapted the wind power system to hydrogen and methane production by using two separate units (Hydrogen fuel production and methanation units), which uses wind-powered electricity systems [34]. The study shows that the sustainability of the whole system is assisted by methane production, and by that the energy and exergy efficiencies are significantly increased.

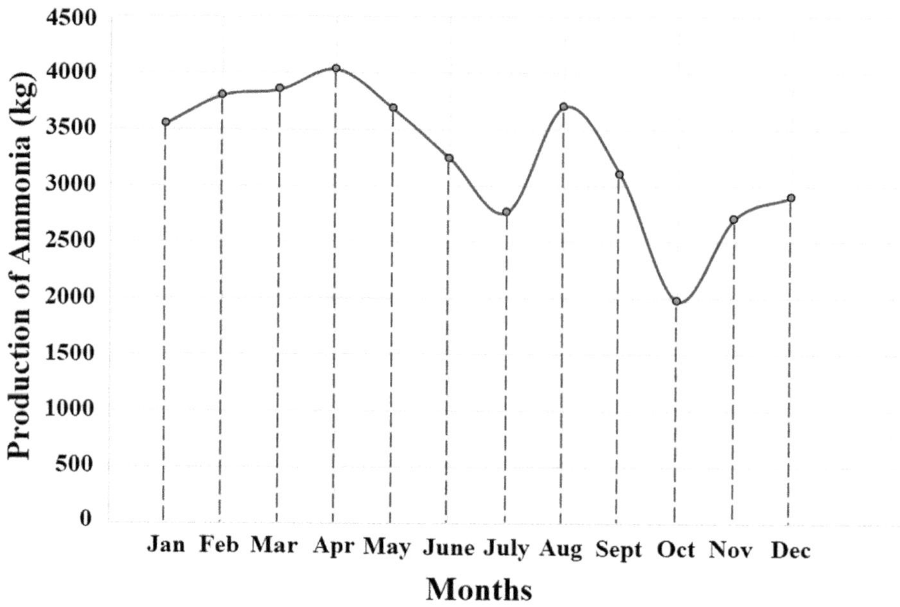

FIGURE 32.9 Amount of ammonia production by using 6 MW wind turbine. *Adopted from Bicer Y, Khalid F, Mohamed AM, Al-Breiki M, Ali MM. Electrochemical modelling of ammonia synthesis in molten salt medium for renewable fuel production using wind power. Int J Hydrogen Energy 2020;45(60):34938–48.*

32.5 Summary

In this work, we study the latest fuel production technologies for energy production from renewable energy resources for their usage in smart cities. This review includes a broad spectrum that spans from the technical energy resources from literature. By this work we aim to shed a light on the current status of fuel production technologies by renewable energy resources and their usage in smart cities.

Renewable energy production becomes crucial for smart cities due to minimizing carbon emission in the environment. Hence, we believe that fuel production from renewable resources would shape the future energy dynamics of smart cities. Technological developments open the gates of developing new methods of producing hydrogen and other fuel sources by using solar and wind energy resources as power supply. Besides, it is expected that the methodologies and workflows explained will be cost-effective processes that can compete with conventional hydrocarbon fuels in near future.

The latest developments of fuel production by using solar energy are given in four different categories: thermolysis of water, thermochemical cycles, cracking of gaseous hydrocarbons, and gasification and reforming of coal and biomass. Although the first three can compete with conventional fossil fuel usage, gasification-related processes need more technological developments to decrease the costs and can be applied to smart cities.

Wind energy is a good resource as a power supply for the production of raw energy materials used as input for the fuel production processes. The efficiency and production of hydrogen and ammonia by using wind power indicate that they will have the potential to compete with fossil fuels with technological developments in close future. Thus, it will pave the way for decreasing the market share of fuels, which increases the greenhouse levels in the atmosphere and will help to prevent global warming. Hence, it will become one of the common fuel production methods for the future of smart cities.

Finally, let us move on to the real interactive part of this chapter: review questions/exercises, hands-on projects, case projects, and optional team case project. The answers and/or solutions by chapter can be found in Appendix G.

32.6 Chapter review questions/exercises

True/false

1. True or False? The "Smart City" concept is mainly an approach for urban planning that uses the latest technological resources.
2. True or False? In a smart city concept, renewable energy resources become a noncritical issue to focus on.
3. True or False? The word "fuel" is mainly described for the materials utilized for the production of water.
4. True or False? The worldwide usage of solar energy has drastically increased due to the global climate targets in the last decade.
5. True or False? The Thermolysis Of Water method uses the simple molecular structure of water and produces hydrogen from a water molecule by using solar energy.

Multiple choice

1. Metals are good at transportation and storage of:
   a. Electrical energy
   b. Solar energy
   c. Nuclear energy
   d. Hydrogen energy
   e. Geothermal energy
2. Gaseous hydrocarbon cracking is based on the mechanism of thermal decarbonization of hydrocarbon:
   a. Infrastructures
   b. Physical infrastructures
   c. Structures
   d. Temperatures
   e. Products
3. Gasification is a process in which the input materials benefit from the:
   a. Machine value
   b. Human value
   c. Physical value
   d. Heating value:
   e. Carbon-based value
4. Fuel production by using solar energy can be used in:
   a. Holistic integration of smart city platforms
   b. Economic growth
   c. Making cities smarter
   d. Computer applications
   e. Smart cities
5. Wind power is becoming one of the most growing energy production methods due to its low production phase costs and environmentally friendly:
   a. Structure
   b. Sustainable energy
   c. Hydrogen fuel
   d. Hydrogen and methane
   e. Wind turbine

Exercise

Problem

Provide a comprehensive methodology for planning and assessing the development of smart energy systems, leading to complex energy provision technology networks, by using different on-site as well as off-site resources.

Hands-on projects

Project

Do research: Analyze the importance of smart grids in launching smart cities, by reviewing the advancement of micro/nano grids, applications of renewable energies, energy-storage technologies, and smart water grids in smart cities.

Case projects

Problem

What is the role of smart cities in climate change and environmental pollution worldwide?

Optional team case project

Problem

Provide an analysis of the determinants of economic efficiency, and assess the prerequisites for the energy security of smart cities.

References

[1] Eremia M, Toma L, Sanduleac M. The smart city concept in the 21st century. Proc Eng 2017;181:12–9.

[2] Bakıcı T, Almirall E, Wareham J. A smart city initiative: the case of Barcelona. J Knowl Econ 2013;4(2):135–48.

[3] Farahani SS, van der Veen R, Oldenbroek V, Alavi F, Lee EHP, van de Wouw N, et al. A hydrogen-based integrated energy and transport system: the design and analysis of the car as power plant concept. IEEE Trans Syst Man Cybern Syst 2019;5(1):37–50.

[4] Calvillo CF, Sánchez-Miralles A, Villar J. Energy management and planning in smart cities. Renew Sustain Energy Rev 2016;55:273–87.

[5] Leakey R. Origin of humankind. Basic Books; 1994, ISBN 978-0-465-03135-1.

[6] Hall L. "Charcoal Briquette". How products are made. 2007. Retrieved 1 October 2007.

[7] Archived from the original on 7 October 2006. Retrieved 10 August 2006 History of coal use. World Coal Institute; 2006.

[8] Pickl MJ. The renewable energy strategies of oil majors—From oil to energy? Energy Strat Rev 2019;26:100370.

[9] United Nations Climate Change. The Paris agreement. Adoption of the Paris agreement - Paris agreement text English. unfccc.int. [Accessed November 2020].

[10] Taylan O, Berberoglu H. Fuel production using concentrated solar energy. Appl Sol Energy 2013:33–67.

[11] Chueh WC, Falter C, Abbott M, Scipio D, Furler P, Haile SM, Steinfeld A. High-flux solar-driven thermochemical dissociation of CO_2 and H_2O using nonstoichiometric ceria. Science 2010;330(6012):1797–801.

[12] Davenport TC, Yang CK, Kucharczyk CJ, Ignatowich MJ, Haile SM. Maximizing fuel production rates in isothermal solar thermochemical fuel production. Appl Energy 2016;183:1098–111.

[13] Bork AH, Carrillo AJ, Hood ZD, Yildiz B, Rupp JL. Oxygen exchange in dual-phase La_0. $^{65}Sr_0$. $35MnO_3–CeO_2$ composites for solar thermochemical fuel production. ACS Appl Mater Interfaces 2020;12(29):32622–32.

[14] Falter C, Batteiger V, Sizmann A. Climate impact and economic feasibility of solar thermochemical jet fuel production. Environ Sci Technol 2016;50(1):470–7.

[15] Falter C, Scharfenberg N, Habersetzer A. Geographical potential of solar thermochemical jet fuel production. Energies 2020;13(4):802.

[16] Falter C, Pitz-Paal R. Water footprint and land requirement of solar thermochemical jet-fuel production. Environ Sci Technol 2017;51(21):12938–47.

[17] Postels S, Abánades A, von der Assen N, Rathnam RK, Stückrad S, Bardow A. Life cycle assessment of hydrogen production by thermal cracking of methane based on liquid-metal technology. Int J Hydrogen Energy 2016;41(48):23204–12.

[18] Rodat S, Abanades S, Coulié J, Flamant G. Kinetic modelling of methane decomposition in a tubular solar reactor. Chem Eng J 2009;146(1):120–7.

[19] Abanades S, Flamant G. Hydrogen production from solar thermal dissociation of methane in a high-temperature fluid-wall chemical reactor. Chem Eng Process 2008;47(3):490–8.

[20] Tou M, Michalsky R, Steinfeld A. Solar-driven thermochemical splitting of CO_2 and in situ separation of CO and O_2 across a ceria redox membrane reactor. Joule 2017;1(1):146–54.

[21] Popov D. An option for solar thermal repowering of fossil fuel fired power plants. Sol Energy 2011;85(2):344–9.

[22] Katayama Y, Tamaura Y. Development of new green-fuel production technology by combination of fossil fuel and renewable energy. Energy 2005;30(11–12):2179–85.

[23] Dispenza G, Sergi F, Napoli G, Randazzo N, Di Novo S, Micari S, et al. Development of a solar powered hydrogen fueling station in smart cities applications. Int J Hydrogen Energy 2017;42(46):27884–93.

[24] Nambiar R, Shroff R, Handy S. Smart cities: challenges and opportunities. In: 2018 10th International Conference on Communication Systems & Networks (COMSNETS). IEEE; 2018. p. 243–50.

[25] Alghamdi TG, Said D, Mouftah HT. Decentralized Game-Theoretic scheme for D-EVSE based on renewable energy in smart cities: a realistic scenario. IEEE Access 2020;8:48274–84.

[26] Toh CK, Sanguesa JA, Cano JC, Martinez FJ. Advances in smart roads for future smart cities. Proc R Soc A 2020;476(2233):20190439.

[27] Hydrogen from wind energy and solar power, the future. REVE website, archieved from. [Accessed June 8 2021].

[28] Beccali M, Brunone S, Finocchiaro P, Galletto JM. Method for size optimisation of large wind–hydrogen systems with high penetration on power grids. Appl Energy 2013;102:534–44.

[29] Ntziachristos L, Kouridis C, Samaras Z, Pattas K. A wind-power fuel-cell hybrid system study on the non-interconnected Aegean islands grid. Renew Energy 2005;30(10):1471–87.

[30] Morgan E, Manwell J, McGowan J. Wind-powered ammonia fuel production for remote islands: a case study. Renew Energy 2014;72:51–61.

[31] Matzen MJ, Alhajji MH, Demirel Y. Technoeconomics and sustainability of renewable methanol and ammonia productions using wind power-based hydrogen. 2015.

[32] Fasihi M, Bogdanov D, Breyer C. Techno-economic assessment of power-to-liquids (PtL) fuels production and global trading based on hybrid PV-wind power plants. Energy Procedia 2016;99:243–68. 10th.

[33] Müller S, Groß P, Rauch R, Zweiler R, Aichernig C, Fuchs M, Hofbauer H. Production of diesel from biomass and wind power–energy storage by the use of the Fischer-Tropsch process. Biomass Convers Biorefin 2018;8(2):275–82.

[34] Safari F, Dincer I. Assessment and optimization of an integrated wind power system for hydrogen and methane production. Energy Convers Manag 2018;177:693–703.

[35] Bicer Y, Khalid F, Mohamed AM, Al-Breiki M, Ali MM. Electrochemical modelling of ammonia synthesis in molten salt medium for renewable fuel production using wind power. Int J Hydrogen Energy 2020;45(60):34938–48.

[36] Bernal-Agustín JL, Dufo-López R. Hourly energy management for gridconnected wind–hydrogen systems. Int J Hydrogen Energy 2008;33:6401–13.

Chapter 33

Solving traffic congestion and parking scarcity with narrow track vehicles

Richard W. Woodbury
Commuter Cars Corporation, Spokane, WA, United States

33.1 Introduction: the traffic challenge

Sitting in a typical traffic jam on a Los Angeles area freeway was my inspiration for developing the Tango, a narrow-track vehicle (NTV)—defined as one that is less than 40 inches wide and 8.5 ft long—that can double lane capacity and virtually quadruple parking availability (see Fig. 33.1). As I looked around and noticed that most of the cars held only one person, the driver, I thought that there must be a better way. What if all those full-size cars were replaced by ones that were half as wide, designed to move one or two people instead of carrying the usual three or four empty seats most of those cars had? Thus, the Tango was born.

At the turn of the 20th century, most people still lived close to their work: above the shop, or at most, a short walk from the store or factory. Even small cities were densely packed with apartments, houses on small lots, stores on many blocks, churches, offices, and all the other necessities of life close by. The more well-to-do people lived a bit farther from the center of town and had horses and carriages to travel back and forth, but for the majority of people, that was a luxury that was out of reach.

Then came the automobile. Luxury items at first, with the advent of the Model-T they became affordable to the middle class. Soon people availed themselves of the mobility cars offered, many choosing to move to the outskirts of town where they could have houses with yards—the American dream. The suburbs were on their way to becoming the standard residential choice for Americans.

But jobs were still in the city, so they became commuters, driving to and from work—a pattern that prevails to this day. Today more than 107 million commuters drive to

FIGURE 33.1 A one-person wide NTV fits two to a lane and is able to travel between slow-moving or stopped cars on congested roads, more than doubling road capacity and improving traffic flow for all vehicles.

Smart Cities Policies and Financing. https://doi.org/10.1016/B978-0-12-819130-9.00014-0

and from their jobs by themselves in full-sized cars, vans, and pickups—nearly 80% of all workers. To accommodate them, governments have spent hundreds of billions of dollars building roads, parking facilities, and other infrastructure. The irony is that all those new roads have encouraged people to move consistently farther out of town, resulting in more demand for more roads—the construction of which destroys existing residences and forces people to move still farther away from city centers.

All this traffic congestion has costs. It results in loss of productive time as commuters spend hours a week stuck in traffic, waste large amounts of fuel as engines are idling rather than productively moving cars toward their destinations and contribute to air pollution. In 2002, a study at the Texas Transportation Institute at Texas A&M University [1] calculated that in 2000, traffic congestion in the nation's 75 large urban areas it studied wasted 5.7 billion gallons of fuel, an amount that would fill 570,000 gasoline trucks. If lined up bumper to bumper, they would stretch from New York to Las Vegas and back. Their study estimated lost time at 3.6 billion hours. By 2011, the lost time had climbed to 5.5 billion hours, and in 2017, the figure was 8.8 billion hours (see sidebar: "Texas A&M Transportation Institute Study").

Texas A&M Transportation Institute Study

According to the Texas A&M Transportation Institute study [2] (https://mobility.tamu.edu/umr/congestion-data/. [Accessed 22 April 2021]), annual costs due to congestion in 2017 included:

Metropolitan area	Hours of delay	Excess fuel consumed (gallons)
Los Angeles	119	256,931,000
San Francisco	103	95,037,000
Washington, DC	102	89,885,000
New York	92	323,712,000
Boston	80	74,143,000
Seattle	78	62,742,000
Chicago	73	144,987,000

For cities, the issues around commuting include congestion (and therefore unhappy voters/taxpayers); the cost for construction and maintenance of roads; and, frequently, the loss of the tax base, as residents move outside local taxing zones. In essence, this involves using the ever-growing infrastructure without contributing to its costs.

33.2 Potential solutions

Often the first thing that comes to mind when municipalities are confronted with commuter-related congestion is public transport. If we could only get people to ride buses or trains, goes the thinking, we could reduce the number of cars coming into the city and therefore reduce the crowding on roads as well as the need for extensive parking within the city core. There are several problems with this solution, however. First, the more spread out the urban area is, the farther apart the spokes in the public transit wheel are in the outlying areas. So someone wanting to take a commuter train, for example, might have to drive an extended distance, then leave a car in a suburban parking lot for the day. It does not solve the parking problem, it just moves it on down the road. While buses work well in densely populated areas, the fewer residents per square mile, the less demand there is, thus the less service that is economically feasible. Further, such services would have to be funded by the outlying communities or by an area-wide public transportation district.

Second, public transportation systems are expensive to operate and provide a poor return on investment. One of the high costs of public transportation systems for commuters is that the systems need to carry large numbers of passengers for short periods twice a day. Yet systems have to be built as if those loads were constant. Thus, they are greatly oversized for the bulk of their operating time. High amounts of capital are invested for erratic payback. On average, passenger fares fund 33% of public transit operations in the United States, with another 12% generated directly by the transit operator. Local and State sources fund 24% and 23%, respectively; Federal Government sources fund the remaining 8% according to the 2015 National Transit Summary and Trends, Office of Budget and Policy https://www.transit.dot.gov/sites/fta.dot.gov/files/docs/2015 NTST.pdf [3] [online report Accessed 5 March 2021].

Further, public transportation is unpopular with commuters—even before the COVID pandemic, and more so now that commuters are concerned about the spread of infectious diseases through close contact with other people. Unless public transportation routes provide quick and relatively direct service between one's home and work, they are often perceived as being uncomfortably time-consuming and limiting commuters' flexibility. The majority of commuters avoid public transportation whenever they can (see Fig. 33.2).

Carpooling or vansharing is another often-recommended option. This type of ridesharing has the advantage of being more flexible in both time and route than fixed-route scheduled public transport. It does have limitations, though. It can work if several people both live and work close to each other—and are able to commute on the same schedule. Too often, however, schedules do not align well enough to make carpooling practical. Again, the denser the community, the greater the likelihood that groups of commuters could be gathered together to take

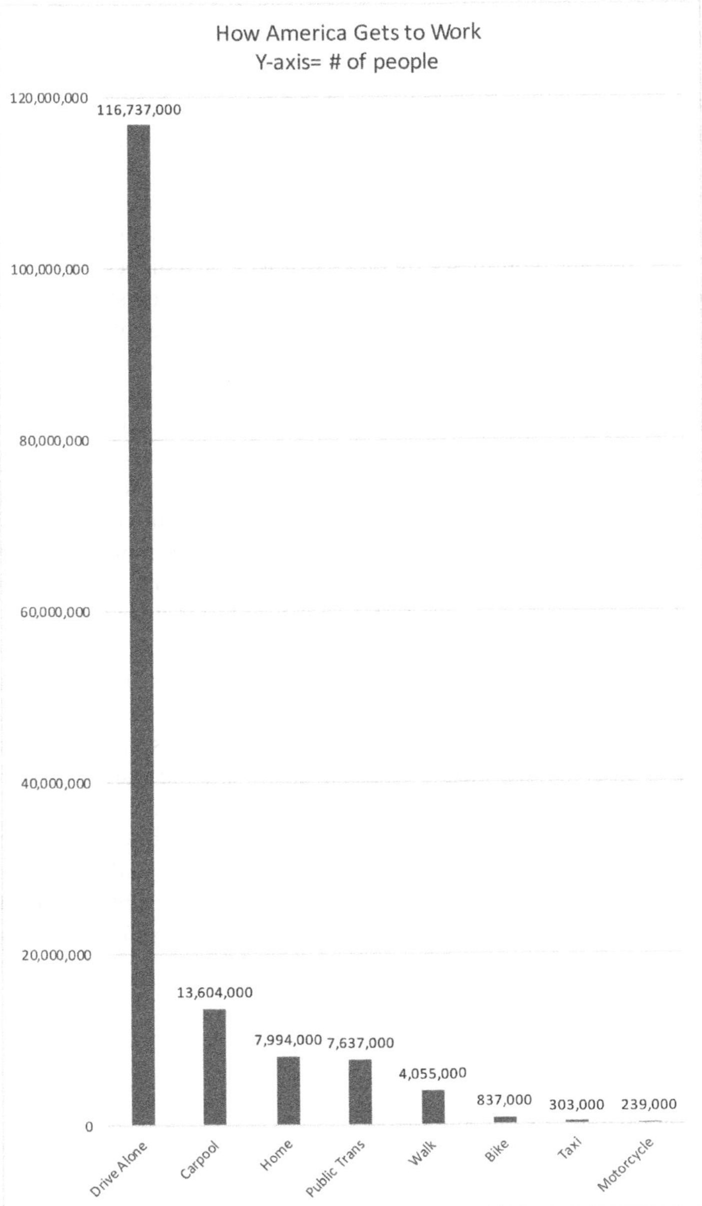

FIGURE 33.2 Commuters who drive alone vastly outnumber those who use all other transportation sources combined. Public transportation usage is less even than carpooling, the second-highest form of commuting. In fact, more people walk to work than take the bus.

The problem isn't the number of single-occupant drivers; it's the design of the cars they're driving.

They are carrying with them four or more empty seats, taking up an entire lane instead of half of one. Until now, the right tool for the job of commuting on freeways has not existed. Motorcycles are not even 0.2% of the commuter market due to their drawbacks.

Workers who can afford cars will use them, unless less convenient than other modes of travel. The vast majority of workers want the freedom to go where and when they want to, and want the security of a traveling locker.

Lane capacity can be increased from 2,000 vehicles per hour to 4,520 VPH with Narrow Track Vehicles (NTVs) according to a Booz-Allen-Hamilton / University of California, Berkeley study. They can meet all federal automotive regulations (FMVSS) while increasing safety beyond that of other cars due to the ability to avoid pedestrians, bicyclists and other vehicles.

advantage of carpooling. The more diluted (and therefore scattered) the population base is, the less likely it will be to find functional commuter matches.

Staggered start/end times can alleviate traffic congestion when great numbers of people work at large employment centers: manufacturing facilities, military bases, schools, etc. The constraint is having large enough employers who are willing to adjust their work schedules along these lines.

As we have seen during the Covid-19 pandemic, some people are able to successfully work remotely, but such electronic commuting works only for a few employees. The majority of workers in manufacturing, service jobs, and other hands-on work still need to be physically at their jobsites.

Finally, many solo commuters find these solutions unacceptable for personal reasons. Number one is convenience. A commuter who has to drop children off at daycare on the way to work and pick them up on the way home, someone who regularly runs errands such as grocery shopping or stopping at the dry cleaner as part of the commute trip, or someone who occasionally needs a car during the work day—for meetings or to visit alternate work sites, running office errands, or some other work-related reason—will find most of the abovementioned options impractical. Many others simply appreciate being able to drive their own cars, enjoy the transition time between home and work, listening to music or recorded books, or even just enjoying driving—even if traffic backups are involved. They want the comfort and flexibility that having their own cars provide. For many if not most Americans—and probably commuters throughout the world—their cars represent freedom, and they are not likely to give them up.

The following chart (see Table 33.1) indicates how people get to work. Note that over the last 28 years, single occupant driving has increased, while carpooling, despite all of the incentives, has decreased over the same time period from 11.8% to only 8.9% of all workers, a decrease of 25% despite taxpayer expense for HOV lanes of over $250 billion. Public transit in 28 years increased from 4.6% to 5.2% and then fell back to 5.0% over the last 2 years. At that trajectory, it would take 1750 years to achieve 25% of workers using public transit. Bicycling increased from 0.4% to 0.6% in 14 years despite the huge investment of capital and real estate for bicycle lanes, yet dropped back to 0.5% in the last year. At that trajectory it would take 3750 years to achieve 25% of the commuter market. This chart is based on data from the U.S. Department of Housing and Urban Development, American Housing Survey for the United States [4].

33.3 Another way of thinking

What if some way could be found to get drivers into ultranarrow cars that would still provide them the benefits they find so important in driving their own cars? This is where the Narrow Track Vehicle (NTV) comes into play. With a footprint the size of a motorcycle, such cars more than double lane capacity since they take up less than half the space of a typical private automobile, while offering the benefits of flexibility and of providing storage space for the things that get hauled around—briefcases, groceries, samples, library books, even a passenger—by the average commuter. In other words: Convenience. With the comfort of being fully enclosed, they provide shelter from the elements that keep many from riding motorcycles, while construction to federal safety compliance standards can offer at least as much protection as the average passenger car—another benefit motorcyclists do not enjoy. With four wheels and a low center of gravity resulting from the weight of batteries under the floor (about 2000 lbs. in the current design), an NTV has a rollover threshold comparable to that of a Porsche 911, just about the most stable car on the road. An NTV is capable of highway speeds, making its performance and usefulness comparable to the typical passenger car in everything except activities that require carrying several passengers and/or large amounts of material. In fact, it drives like a standard car, so there is no learning curve as there is in learning to ride motorcycles. Perhaps best of all, the narrow width gives NTVs a distinct advantage when it comes to accident avoidance. They can quickly maneuver between or to the side of cars and have much less width with which to collide with another vehicle, pedestrian, or object. All of these factors make NTVs potentially the safest vehicles available.

TABLE 33.1 A breakdown of how people commute.

	Principal means of transportation to work (thousands)							
	1989		2015		2016		2017	
	Number	Percent	Number	Percent	Number	Percent	Number	Percent
All workers	**106,630**	**100**	**148,324**	**100**	**150,377**	**100**	**152,803**	**100**
Automobile, total	93,943	88.1	126,924	85.6	128,348	85.4	130,341	85.3
Drives self	81,322	76.3	113,576	76.6	114,771	76.3	116,737	76.4
Carpool	12,621	11.8	13,348	9.0	13,577	9.0	13,604	8.9
Public transportation	4880	4.6	7761	5.2	7649	5.1	7637	5.0
Taxi	152	0.1	188	0.1	227	0.2	303	0.2
Bicycle	N/A	N/A	885	0.6	864	0.6	837	0.5
Motorcycle	795	0.7	266	0.2	251	0.2	239	0.2
Walks	3634	3.4	4114	2.8	4086	2.7	4055	2.7
Works at home	2736	2.6	6843	4.6	7592	5.0	7994	5.2

Based on data from 1989 to 2001: U.S. Department of Housing and Urban Development, American Housing Survey for the United States: 2005. Washington, DC: 2006: U.S. Department of Commerce, Bureau of Census, American Community Survey, table B08301, 1-Year estimates.

Because of their small size, there is a direct relationship between the number of NTVs on the roads in a given area and their impact on congestion. A study of the effects of motorcycles on congestion in Belgium (https://www.acem.eu/images/stories/conferences/2012/8.Conference_Leuven_study.pdf) [5], which was presented at the Association des Constructeurs Européens de Motocycles (ACEM) 2012 Conference in Brussels, found that if 10% of all private cars were replaced by motorcycles in the traffic flow of the test area, total time losses for all vehicles decreased by 40% and total emissions reduced by 6% (1% from the different traffic composition of more emission-reduced motorcycles and 5% from avoided traffic congestion). A 25% modal shift from cars to motorcycles was found to eliminate congestion entirely. It is significant that not all commuters would need to switch to NTVs for a clear benefit: not only would it speed up commute times, it would benefit all the other vehicles on the road—commercial and personal—as well.

A study in Auckland, New Zealand, suggested that traffic congestion problems are not because of the number of single occupied vehicles on the motorways at peak traffic times. It is because of the width of those vehicles. The solution, study author Auckland City Council member Toa Greening says, is to replace a specified number of the existing single-occupant vehicles with narrow cars that can travel side by side, offset, or staggered in a single lane as packs of motorcyclists do (http://www.projectmicrocar.co.nz) [6]. In California and much of Europe and the rest of the world, lane splitting or filtering (driving between slow or stopped lanes of traffic) is legal. This gives an added value to NTVs (see Fig. 33.3) because, like motorcycles, they can bypass long lines of stopped or slow-moving vehicles. Two additional states, Utah and Montana, now allow filtering.

As early as 1990, William L. Garrison and Mark E. Pitstick, in a study for the Institute of Transportation Studies at the University of California at Berkeley [7], analyzed the benefits of what they called a "lean machine"—a car with a nearly identical footprint to the Tango. It was a tiny three-wheeler that held only one person and required the driver to manually control the leaning like a motorcycle. Although this concept vehicle never took off, the key feature of this study is that it showed that with virtually no infrastructure increase, lane capacity would rise from 2000 vehicles per hour to 4520 vehicles per hour through the use of this size of vehicle—that is more than doubling lane capacity (see Fig. 33.4).

For municipalities, the benefits exceed simply reducing congestion. As the UC-Berkeley study pointed out, the increased roadway throughput could be achieved *with virtually no infrastructure increase*. For governmental agencies dependent on taxes, this is a huge benefit. As James Surowiecki pointed out in a 2017 *New Yorker* article [8], large-scale transport projects generally require cooperation and funding from multiple governing bodies and thus must satisfy a wide range of constituencies (*New Yorker*, January 23, 2017, p. 21). Surowiecki also points out the high costs of these construction projects, noting that most run over budget and behind schedule. He points to the "Second Avenue Subway" project in New York City, phase one of which cost approximately $1.7 billion for two miles of track. Boston's "Big Dig" tunnel is another example of the costs. According to https://www.roadtraffic-technology.com/projects/big_dig/ [9] [Accessed 8 March 2021] 7.8 miles of highway (161 lane miles) cost $24 billion (actual constructions costs $14.8 billion plus financing expenses) and took 17 years to complete. At the other end of Interstate 90, the Alaskan Way Viaduct Replacement project along the Seattle waterfront was estimated to cost $3.35 billion; in 2020, it received additional state funding according to the

FIGURE 33.3 NTVs can travel between lanes of slow or stopped vehicles, reducing congestion for all roadway users.

FIGURE 33.4 Two NTVs traveling side by side fit into the space of one standard passenger car.

Washington Department of Transportation [10] (https://wsdot.wa.gov/projects/sr99/alaskan-way-viaduct-replacement/home. [Accessed 8 March 2021]).

If, instead of these massive construction projects, NTVs were put into the traffic mix, congestion reduction could have been achieved at minuscule cost compared to the construction investment. According to Patrick Balducci of the Pacific Northwest National Laboratory [11], who did a study on the impact of NTVs on freeway performance in the Seattle area, if only 6% of commuters switched to NTV, reduced congestion would increase overall traffic flow by 6.6 mph—which does not sound like much unless you are sitting in traffic that is moving at a very slow rate. If traffic is moving at 5 mph, this increase would more than double the speed.

Further, the NTV provides the added benefit of being able to fit several into a typical on-street parallel-parking spot. This results not only in more efficient use of city space, but it also reduces congestion caused by drivers circling through areas waiting for parking spots to open up. In parking lots and structures, the NTV allows a 350% increase in space utilization, according to the above-mentioned California study and is also able to make use of those oddly shaped corner areas not accessible to larger vehicles (see Fig. 33.5). This results in reduced demand for parking facilities and allows more cost-effective use of expensive city real estate. In municipal terms, this means that instead of the relatively low tax return from a parking lot, the space could well be converted to an office building, apartment complex, retail stores, or other uses that would

FIGURE 33.5 NTVs parked perpendicular to the curb allow four to fit into a standard on-street parking space and, as shown here, are no longer than the width of a typical delivery truck. They also make use of otherwise wasted spots between other cars or in oddly shaped spaces on streets or in parking facilities.

be taxed at a higher rate. This would increase revenues without any additional expenditures on the part of the municipality: No new road construction and no major public transportation expenditures for construction or equipment. In essence, the major problem addressing transportation in and around cities could be solved with no new taxes or increased debt load.

33.4 Disrupting disruption

Although not always recognized as such, perhaps the greatest cost of commuting is the constant demand for new roadways. Construction projects cost the nation hundreds of billions of dollars. The website https://compass international.net/ [Accessed 5 March 2021] [12] lists major road constructions costs as running from $5.34 million per mile for a basic two-lane highway to $20.31 million per mile for a major four-lane freeway to over $71 million per mile for an elevated four-lane freeway. But perhaps the greatest cost is the disruption of the communities through which the new highways pass. Spokane, Washington, where I live, is still feeling the effects of the freeway construction of the 1960s that split the city, wiped out large sections of working-class housing, and left formerly thriving businesses bereft of the traffic that had sustained them. This pattern has been repeated in virtually every city that has a freeway through it, with low-income (frequently minority occupied) neighborhoods being the most impacted. Rural communities have been decimated by being bypassed by a highway that is now several miles away. Every new construction project, whether an entirely new road or addition of new lanes to existing ones, continues this pattern. From a municipality standpoint, the tax base of existing homes and businesses is wiped out. To survive, they are faced with raising taxes on existing residents, cutting services, or a combination—none of which are particularly popular with taxpayers/voters. And each new highway construction project forces (or allows) people to move farther away from the city center, thus increasing commute time in spite of the new roads.

33.5 What are the alternatives?

The simplest answer is *Do things differently*. Instead of building new roads and/or increasing expenditures on public transport designed to move large numbers of commuters over fixed routes, use NTVs. They give commuters what they truly want: flexibility, safety, comfort, independence. At the same time, they eliminate the need for those major investments by eliminating congestion naturally through the use of right-sized commuter vehicles.

The big question regarding implementing this solution is: how could we do it? There are several possible approaches, from incentivizing people to purchase and drive NTVs to penalizing the use of large cars, to creating a fleet of publicly owned NTVs that could supplement or replace existing public transit vehicles and services in areas where fixed-route service is not practical. This last option is probably the most practical. By taking a portion of the funds currently used for public transit and road construction and investing in NTVs, commuters could be incentivized to switch from their full-size vehicles to NTVs. Since the switch would be funded by government investment, commuters would have the benefit of an affordable alternative, one, in fact, that would probably save them money and improve their quality of life by speeding up their daily commutes.

33.6 Potential financial benefits to cities

Because this is a disruptive approach to solving traffic congestion, it will take time for the commuting public to fully accept the concept. While some might buy in quickly, it will be a while before enough people see the benefits to themselves to be willing to invest in an additional car just (or primarily) for commuting. Believe it or not, this actually provides an additional opportunity for municipalities and/or public transportation districts.

How? You ask? By purchasing and leasing the NTVs in quantities that will significantly alleviate congestion in key areas.

There are several advantages to this approach. First, government involvement in making NTVs available will encourage the immediate acceptance of the narrow vehicle concept. If NTVs are made available to commuters at a reasonable cost, people will be incentivized to switch, making an immediate impact on congestion.

Second, by providing the initial NTV lease option in key congested areas—say between suburb A and the downtown core area as a test program—their use will show the drastic congestion-reducing benefits of switching drivers to NTVs. For example, as the Belgian study referenced above suggests, replacing just 25% of full-sized vehicles with motorcycle-sized vehicles could entirely eliminate congestion. Saturating a commuter route with NTVs and eliminating congestion would encourage broad acceptance of NTVs by the commuting public elsewhere.

Third, instead of purchasing buses—the average cost of a diesel-powered bus being around $550,000 according to https://www.liveabout.com/bus-cost-to-purchase-and-operate-2798845 [Accessed 5 March 2021] [13], with the associated operating costs (labor, fuel, maintenance, etc.)—NTVs could be purchased and leased to commuters who would assume the maintenance/operating costs. In Toa Greening's Auckland study mentioned above, 15,000 NTVs was estimated to be the number necessary to eliminate congestion throughout the Auckland metropolitan area. His recommendation for solving Auckland's congestion is as follows:

- Purchase 15,000 NTVs at \$29,000 = \$435m;
- Amortize debt of \$435m at 7% over 10 years = \$660m;
- Lease out Tangos at \$55/week = \$2860/year (each) × 10 years = \$660m.

In other words, the purchasing agency/municipality would break even over 10 years on the purchase of the NTVs, meanwhile essentially eliminating traffic congestion and eliminating all the ancillary costs associated with operating an increased fleet of public transportation vehicles such as buses, with no significant construction costs.

Fourth, because the NTVs would also be available to commuters for other uses such as short trips near their residences, congestion in those areas would also be reduced. Imagine a shopping mall parking lot filled with NTVs rather than full-sized cars. It would require significantly less than half the space currently allocated to parking. As with downtown parking, these areas could be converted to structures that provide a higher tax base, meaning more income for the community.

Meanwhile, revenues deriving from automobiles such as licensing fees and parking meter income would not be negatively impacted. In fact, in the case of parking meter revenue, it could actually increase as more meters were installed in small spaces designed specifically for NTVs and motorcycles. Parked perpendicular to the curb, four NTVs will fit into a typical on-street parking space. With restriping, parking lot capacity may be increased by 350%.

For municipalities that are electricity providers, there are benefits from the fact that NTVs are electric. They would be charged primarily at night, thus evening out load on the distribution system and increasing revenue through increased demand. Beyond that, electric cars of all sorts open the potential for revenue from charging stations.

And then there are the less tangible, but no less valuable benefits. All those traffic jams caused by commuters result in wasted time. It is hard to put a dollar value on the time workers gain to spend with their families and on recreational activities, but the nonfinancial benefits are significant. Those traffic jams also contribute to air pollution. Again, it is hard to quantify, but the costs—both medical and quality of life—resulting from pollution are significant. At the same time NTVs, being electric, not only reduce pollution, they also reduce dependence on fossil fuels.

Because of their narrow footprint, NTVs will change the wear patterns on existing roads. They naturally straddle the ruts created by typical cars and trucks, evening out the usage of existing roadway surface areas. So not only will they reduce the requirement for new road construction, they can significantly lessen the frequency and therefore the cost of road maintenance.

33.7 Summary

In summary, NTVs offer significant benefits to both commuters and municipalities. These benefits can also be applied to the general public.

33.7.1 Commuters

With regard to flexibility, commuters can go when and where they need to without being limited by fixed-route public transportation service limitations. And because the NTV will hold a passenger as well as the driver, limited carpooling is still possible—reducing congestion even further.

With comfort, commuters have their own vehicles, and so are not subject to the space-sharing irritations experienced in public transportation. They can customize as they would any other car with everything from cushions to cupholders to GPS systems to child booster seats, while protected from weather.

Running errands, stopping for dinner on the way home—is not a problem, if you have independence. Dentist appointment mid-day? Try that with ride-sharing. No need to arrange alternative transportation because your own car is readily available.

Finally, time-saving is another important benefit. Apart from the obvious benefit of congestion being eliminated on main routes, having one's own vehicle allows the use of alternate routes/shortcuts on the trip.

33.7.2 Municipalities

Solving congestion problems without new construction is quite a challenge. Avoiding the major costs of road building, public transport facilities, or other capital expenditures, can probably be achieved without voter approval.

Maintain or increase the tax base: That is the question? Existing properties will remain, rather than being converted from taxpaying to public (roadway) uses. Reduced demand for parking has the potential to increase tax revenues by properties being converted to higher-tax-rate uses.

Reduce infrastructure investments, while maintaining or improving service levels. Roads and public transportation can be built to meet consistent rather than peak-demand needs. Current construction costs can be channeled to other programs, including maintenance needs.

Reduce emissions. The costs of emissions affect everything from building upkeep to public health. Reducing emissions reduces costs.

Recover the investment through a lease program. How many public expenditures actually end up being cost neutral? The leasing of NTVs should be considered at a rate that will return the investment over the life of the NTV.

FIGURE 33.6 NTVs are the happy way to commute.

Incentive programs may be self-limiting. By jump-starting the acceptance of NTVs through incentivizing programs on targeted routes, municipalities will show the benefits of NTVs to the majority of commuters. Acceptance will result in individuals who live in other areas becoming willing to purchase their own NTVs due to the benefits they have seen. Once this acceptance is more general, governments will be able to phase out these incentive programs and still enjoy the benefits NTVs provide.

33.7.3 Communities in general

Less road construction means less disruption and fewer delays in travel for all residents. Neighborhoods will not be "eaten up" by the demand for more and bigger roads.

Potentially lower taxes come into play, with fewer road construction projects. With less demand for major investments in public transportation facilities, combined with current tax bases being maintained and/or increased, residents could see lower tax rates.

A better quality of life can be had, by improved air quality and less congestion. This would benefit all residents, not just commuters. Elimination of massive parking lots in core areas of cities would make them more attractive to pedestrians, including shoppers and restaurant patrons.

With a more effective government, reduced focus on transportation-related expenditures would be at hand. This would free tax funds for other uses such as parks, police and fire services, innovative public service programs, and who knows what else.

Also, the implementation of NTV commuting programs has the potential to benefit everyone. This approach is a true win-win for everyone (see Fig. 33.6).

Finally, let us move on to the real interactive part of this chapter: review questions/exercises, hands-on projects, case projects, and optional team case project. The answers and/or solutions by chapter can be found in Appendix G.

33.8 Chapter review questions/exercises

True/false

1. True or False? For cities, the issues around commuting include congestion (and therefore happy voters/taxpayers); the cost for construction and maintenance of roads; and, frequently, the loss of the tax base, as residents move outside local taxing zones.
2. True or False? Often the first thing that comes to mind when municipalities are confronted with commuter-related congestion is public transport.
3. True or False? Because of their large size, there is a direct relationship between the number of NTVs on the roads in a given area and their impact on congestion.
4. True or False? Although not always recognized as such, perhaps the greatest cost of commuting is the constant demand for new roadways.
5. True or False? Instead of building old roads and/or increasing expenditures on public transport designed to move large numbers of commuters over fixed routes, use NTVs.

Multiple choice

1. Government involvement in making NTVs available will encourage the immediate acceptance of the ______________ concept.
   a. Smart vehicle
   b. Wide vehicle
   c. Narrow vehicle

   d. Suburban vehicle
   e. Electric vehicle
2. By providing the initial NTV lease option in key congested areas—say between suburb A and the downtown core area as a test program—their use will show the drastic congestion-reducing benefits of switching drivers to:
   a. SIVs
   b. PIVs
   c. SGPs
   d. NTVs
   e. SCPs
3. Instead of purchasing buses, NTVs could be purchased and leased to commuters who would assume the:
   a. Maintenance/operating costs
   b. Machine to human costs
   c. Cyber-physical systems costs
   d. Artificial intelligence costs
   e. Construction costs
4. A purchasing agency/municipality would break even over 10 years on the purchase of NTVs; thus, essentially eliminating traffic congestion and all the ancillary costs associated with operating an increased fleet of public transportation vehicles such as buses, with no significant:
   a. Smart city costs
   b. Construction costs
   c. Congestion costs
   d. Computer applications costs
   e. E-economics costs
5. Because the NTVs would also be available to commuters for other uses such as short trips near their residences, congestion in those areas would also be:
   a. Increased significantly
   b. Slightly increased
   c. Reduced
   d. Reduced significantly
   e. None of the above

Exercise

Problem

Conduct a comprehensive study and a comparison and extensive analysis of smart parking systems (SPSs) in terms of a technological approach, sensors utilized, networking technologies, user interface, computational approaches, and service provisions.

Hands-on projects

Project

Do research: Analyze double parking for autonomous cars, as a new approach to temporarily increasing parking capacity in locations in clear need for extra provisions, when best alternatives cannot be found.

Case projects

Problem

Describe the innovative design and development of new smart parking solutions that offer to reduce parking-related traffic congestion, enhance driver experience, and improve the overall parking scheme for smart cities.

Optional team case project

Problem

Describe a green IoT-based Pay-As-You-Go (PAYG) smart parking system, by utilizing unused garage parking spaces in smart cities.

References

[1] Texas A&M Transportation Institute, Mobility Division. Urban mobility report. 2002 [online report accessed 2003].

[2] Texas A&M Transportation Institute, Mobility Division. Urban mobility report. 2019 [online report accessed 22 April 2021].

[3] Office of budget and policy 2015 National transit summary and trends https://www.transit.dot.gov/sites/fta.dot.gov/files/docs/2015 NTST.pdf. [online report accessed 5 March 2021].

[4] 1989-2001: U.S. Department of Housing and Urban Development, American Housing Survey for the United States: 2005. Washington, DC: 2006: U.S. Department of Commerce, Bureau of Census, American Community Survey, table B08301, 1-Year estimates. http://www.census.gov/hhes/www/ahs.html. [Accessed October 2006]. http://factfinder.census.gov/faces/nav/jsf/pages/searchresults.xhtml?refresh=t. [Accessed 1 October 2018].

[5] Transport & mobility Leuven. https://www.acem.eu/images/stories/conferences/2012/8.Conference_Leuven_study.pdf. [last accessed 27 April 2021].

[6] Greening, Toa, project microcar (Microcar NZ), Aukland, NZ.

[7] Garrison WL, Pitstick M. Studies of road infrastructure requirements for small innovative vehicles. California Path Program. Berkeley: Institute of Transportation Studies University of California, UCB-ITS-PRR-93-16.

[8] Surowiecki J. The financial page: big ticket transport. New Yorker January 23, 2017.

[9] Road traffic technology, Boston big Dig, Central artery/tunnel project, Massachusetts, https://www.roadtraffic-technology.com/projects/big_dig/. [Accessed 8 March 2021].

[10] Washington Department of Transportation. https://wsdot.wa.gov/projects/sr99/alaskan-way-viaduct-replacement/home. [Accessed 8 March 2021].

[11] Balducci P. Commuter cars and congestion: the impact of the Tango F on freeway performance characteristics and parking capacity. Richland, WA: Pacific Northwest National Laboratory; July 23, 2003.

[12] Compass International, Inc. https://compassinternational.net/ [Accessed 5 March 2021].

[13] MacKechnie, C. Liveabout dotcom. https://www.liveabout.com/bus-cost-to-purchase-and-operate-2798845. [Accessed 5 March 2021].

Further reading

See www.commutercars.com [for more detailed information about the Tango NTV].

Chapter 34

Modeling and simulation for connected and automated vehicle (CAV) deployment and performance evaluation

Hossam A. Gabbar[1], Jing Ren[2] and Patrick C.K. Hung[3]

[1]*Faculty of Energy Systems and Nuclear Science, Faculty of Engineering and Applied Science, Ontario Tech University, Oshawa, ON, Canada;* [2]*Faculty of Engineering and Applied Science, Ontario Tech University, Oshawa, ON, Canada;* [3]*Faculty of Business and IT, Ontario Tech University, Oshawa, ON, Canada*

34.1 Introduction

As urbanization increases—an additional 2.5 billion people will live in cities by 2050—cities and suburbs will undergo significant transformations to create sustainable living conditions for their residents. Energy and Transportation are the twin pillars of these transformations, and both will require radical adaptation to meet the demographic and economic growth without increasing congestion and pollution.

Electric mobility is widely seen today to improve air quality and meet climate goals, and it is by nature integrated into a comprehensive vision for smarter cities. As Electric Vehicle (EV) becomes more affordable, some predict that they will constitute almost a third of new-car sales by the end of the next decade. Ride-sharing continues to surge, with estimates that by 2030, it will account for more than 25% of all miles driven globally, up from 4% today. Electrified Transportation can be used as a decentralized energy resource and provide new, controllable storage capacity and electricity supply that is useful for the stability of the energy system. In markets where regulation allows EVs to be used as a source of flexibility, energy players start betting on this vision, with cars working as "batteries on wheels."

These changes are just the first hints of what is to come as we will soon see Connected and Autonomous Vehicle (CAV) and commercial fleets of EVs integrated as parts of everyday life. In the future, CAVs will also cost significantly less per mile than vehicles with internal combustion engines for personal use—by as much as 40%—and could also reduce congestion and traffic incidents. CAV technology is expected to transform the nation's transportation system over the coming decades, with major implications for cities and regions' planning and design. Autonomous Vehicles (AVs), also known as driverless or self-driving cars, have been experimentally sharing city streets for several years. This technology moves very quickly, with the 11 largest automakers planning to have fully autonomous vehicles on highways between 2018 and 2021 (arriving somewhat later in urban driving conditions). AV technology, as defined by the International Society of Automotive Engineers (ISAE), ranges from a baseline of no automation up to five levels of increasing autonomy:

- Level 1, driver assistance (e.g., adaptive cruise control)
- Level 2, partial automation (e.g., Tesla's autopilot)
- Level 3, conditional automation (e.g., human drivers serve as a backup for an autonomous system that operates under certain conditions)
- Level 4, high automation (e.g., Google/Waymo test cars)
- Level 5, full automation (e.g., no steering wheel in the vehicle)

Researchers estimate that about two-thirds of vehicles in Canada currently have some connectivity (i.e., embedded telematics); by 2022, approximately 70%−95% of new cars in Canada will have a Vehicle-to-Infrastructure (V2I), Vehicle-to-Vehicle (V2V), and other telecommunications capabilities (e.g., vehicle to a smartphone). Currently, vehicles available to consumers are primarily Level 1 or 2 automation [1]. While a number of major manufacturers plan to launch autonomous passenger cars in the next year, the consensus in the literature is that deployment of Level 4

Smart Cities Policies and Financing. https://doi.org/10.1016/B978-0-12-819130-9.00035-8

or 5 vehicles on public roads will not be commonplace until the 2030s or 2040s. Major automanufacturers, including Ford and General Motor (GM), have indicated their first autonomous fleets will be dedicated to commercial operations.

AVs use artificial intelligence (AI) systems to interact with their surroundings and travel safely. One form of AI being used in fully autonomous cars is called fuzzy logic. Fuzzy logic allows the computer to use approximate reasoning instead of fixed reasoning to interpret situations that the AVs may encounter while traveling on public roadways. It means that the car interprets situations as something other than being entirely true or completely false. Another form of AI being used in fully AVs is called swarm intelligence. The definition of swarm intelligence is "The collective behavior of decentralized, self-organized systems, natural, or artificial." In common terms, swarm intelligence traces the velocity and time it takes groups of particles to return to the computer from which they were transmitted. Using a mathematical relationship, the computer can then determine its boundaries and other structures within the environment. Light Detection and Ranging (LIDAR) is another method used by the autonomous car to map out its surroundings instantaneously during travel. LIDAR operates the same way a radar system would but instead of transmitting radio waves, it transmits laser lights. Stefano Young from Nature Photonics says, "The system at the core of the LIDAR technology is a sensor called the HDL-64E that uses 64 spinning lasers and accumulates 1.3 million points per second in order to reconstruct a virtual map of its surroundings." These virtual maps created by LIDAR along with current GPS maps allow the autonomous car to construct extremely accurate depictions of its surrounding environment.

Significant changes to the transportation system are expected to occur as the development and adoption of CAVs continue. Notably, passenger safety and transport efficiency are expected to increase. According to the National U.S. Highway Traffic Safety Administration (NHTSA), over 94% of collisions are caused by human error. CAVs have the potential to almost eliminate the element of human error, likely resulting in increased overall passenger safety. Heightened V2V, V2I, and V2X connectivity may also decrease the risk of collision as vehicles become more aware of and reactive to their surroundings (e.g., other drivers' speed, traffic signals, and impending collisions). As the adoption of ride sourcing programs such as Uber and Lyft increases, in parallel with potential shifts toward platform-owned autonomous vehicle fleets, experts believe that the demand for personal vehicle ownership and potentially the number of vehicles on the road will decline. That said, the shift toward car sharing and ride sourcing may result in a range of scenarios. Other experts predict that the rise of car sharing/ride sourcing may increase road congestion as passengers who typically use modes of public transportation or are unable to drive (due to disability) begin using car sharing and ride sourcing programs. According to a study conducted by the Insurance Institute of Canada, these CV/AV trends will also significantly affect the commercial trucking/shipping sector, particularly as autonomous technologies will likely first be adopted by this sector.

There is a large number of vehicles on the road all over the world. However, a huge number of accidents have happened everywhere due to human driver errors [2]. According to the statistics in the United States, 539,000 two-lane traffic accidents occurred every year [3]. With the development of electronics, control systems, communication technology, etc., CAVs have been in the market for years. A CAV with fully equipped technology could significantly reduce these accidents due to human errors [4]. In addition, it is also able to increase the comfortability for traveling a long distance. Before autonomous vehicles are capable of driving on the road, a set of standards and testing methods should be proposed according to traffic control, public safety, etc. More importantly, virtual software simulation is utilized to test whether an algorithm is able to pass all the safety tests. This project aims to develop an integrated testing and training platform to support the design, development, operation, and evaluation of CAV using robotics and AI technologies.

34.2 CAV use cases

To study CAV and discuss planning, operation, and management, it is important to analyze how CAV is deployed in view of a number of possible use cases. Although there are many use cases presented in literature, the focus will be on the most common ones presented as examples in this section. The use cases are categorized to facilitate the analysis and utilization of the CAV.

34.2.1 Vehicle platooning

Vehicle platooning occurs when there is a group of vehicles in a single lane with close distance and the same velocity. The target speed is typically defined by the intervehicle distance and the speed of the leading vehicle. Different use cases for vehicle platooning are shown in Table 34.1.

As shown in Table 34.1, eight different use cases are categorized by three different factors: leading vehicle speed, intervehicle distance, and a number of vehicles in the platoon. Normally, intervehicle space defines the space policy, and the leading vehicle and road infrastructure (if available) decide the target speed of the whole platoon [5]. In addition, there are some benefits for grouping a number of vehicles into a platoon. For example, it is able to increase traffic flow and reduce traffic shock waves. More

TABLE 34.1 Use cases for vehicle platooning.

Use case #	Leading vehicle speed	Intervehicle distance	Number of vehicles in the platoon
1	Speeding up	Small (around 0.2 of vehicle length [6])	–
2	Speeding up	Small	–
3	Speeding up	Large	–
4	Speeding up	Large	–
5	Slowing down	Small	–
6	Slowing down	Small	–
7	Slowing down	Large	–
8	Slowing down	Large	–

importantly, it can reduce gas consumption and pollution emission. Besides all the reasons mentioned above, the main goal is to increase road safety. If all the vehicles on a highway are grouped into a platoon, it is possible to reach 8000 vehicles/hour/lane, which is about four times the original traffic flow [7].

34.2.2 Lane change

Changing lanes are common behaviors in daily driving. However, lane changing has a significant impact on transportation control. The decision will be made based on the vehicles within line of sight of on-board sensors. Table 34.2 shows different use cases in different situations. Discretionary Lane Change (DLC)/Mandatory Lane Change (MLC) are used to define these use cases.

Lane changes are common in a transportation system. It might impact the whole traffic flow and its oscillations, as well as public safety [8]. As shown in Table 34.2, there are 32 different use cases for a lane change. Five factors determine these use cases, such as DLC/MLC, turning right or left, other vehicle's position, and speed. The first factor is DLC/MLC, which indicates whether the vehicle performs lane change imposed by traffic rules or by the vehicle destination. The second factor that needs to know is that the vehicle performs a right lane change or left lane change. Also, other vehicles' position has to be determined to make an accurate decision while the vehicle performs lane change. In addition, knowing the vehicle's position is not enough to determine whether the vehicle can perform lane change or not. Thus, the speed of other vehicles is also an important factor for the decision.

34.2.3 Intersection traffic

The intersection is a special area that is shared by two or more roads. Although intersections represent only a small portion of the whole traffic system, two or more roads jointly use the same intersection, which is the heart of the entire system. However, there are too many different use cases for the case of an intersection. In this section, different tables are presented below based on different cars at the intersection, as shown in Tables 34.3–34.5.

Due to a large number of use cases presented in this section, they are categorized into three tables based on a different number of vehicles presented at the intersection. The first factor is whether the intersection is a controlled intersection or a noncontrolled intersection in each table. In addition to a controlled intersection, it is separated into signal light and stop signs installed at the intersection. Besides considering the ego vehicle, other vehicles and pedestrians' status are also two factors we need to consider.

34.2.4 Obstacle avoidance

Obstacle avoidance is the main goal for AV operating safely. If the obstacle is directly ahead of the vehicle, the controller has to decide what action to perform in order to avoid it. Different use cases are tableted in Table 34.6.

As shown in Table 34.6, there are 12 different use cases presented in this section. The first factor is whether the obstacle is moving or stationary. Also, the speed of the vehicle itself is another factor as well. In addition to vehicle itself, the obstacle's location and status are also somewhat important.

34.3 CAV integrated modeling and simulation

This section presents an integrated modeling and simulation framework for CAV. This includes functional modeling as shown in the following section.

TABLE 34.2 Use cases for lane change.

Use case #	DLC/MLC	Right/left lane change	A vehicle presented on left/right	A vehicle presented on front/behind	The vehicle is speeding up/slowing down
1	DLC	Right	Left	Front	Speeding up
2	DLC	Right	Left	Front	Slowing down
3	DLC	Right	Left	Behind	Speeding up
4	DLC	Right	Left	Behind	Slowing down
5	DLC	Right	Right	Front	Speeding up
6	DLC	Right	Right	Front	Slowing down
7	DLC	Right	Right	Behind	Speeding up
8	DLC	Right	Right	Behind	Slowing down
9	DLC	Left	Left	Front	Speeding up
10	DLC	Left	Left	Front	Slowing down
11	DLC	Left	Left	Behind	Speeding up
12	DLC	Left	Left	Behind	Slowing down
13	DLC	Left	Right	Front	Speeding up
14	DLC	Left	Right	Front	Slowing down
15	DLC	Left	Right	Behind	Speeding up
16	DLC	Left	Right	Behind	Slowing down
17	MLC	Right	Left	Front	Speeding up
18	MLC	Right	Left	Front	Slowing down
19	MLC	Right	Left	Behind	Speeding up
20	MLC	Right	Left	Behind	Slowing down
21	MLC	Right	Right	Front	Speeding up
22	MLC	Right	Right	Front	Slowing down
23	MLC	Right	Right	Behind	Speeding up
24	MLC	Right	Right	Behind	Slowing down
25	MLC	Left	Left	Front	Speeding up
26	MLC	Left	Left	Front	Slowing down
27	MLC	Left	Left	Behind	Speeding up
28	MLC	Left	Left	Behind	Slowing down
29	MLC	Left	Right	Front	Speeding up
30	MLC	Left	Right	Front	Slowing down
31	MLC	Left	Right	Behind	Speeding up
32	MLC	Left	Right	Behind	Slowing down

34.3.1 CAV functional modeling

In order to plan the effective deployment of CAV in different regions, it is important to analyze different functions of CAV and their interactions with infrastructures and other vehicles, such as V2X and V2I. Fig. 34.1 shows the functional modeling of CAV that will be used to model different interfaces with infrastructures and support CAV deployment.

TABLE 34.3 Use cases for intersections for two cars.

Use case #	Controlled/ noncontrolled intersection	Signal light/ stop sign	Another car is at left/ right	Another car approached the intersection first (yes/no/same)	Pedestrians present
1	Controlled	Signal light	Left	Yes	Yes
2	Controlled	Signal light	Left	Yes	No
3	Controlled	Signal light	Left	No	Yes
4	Controlled	Signal light	Left	No	No
5	Controlled	Signal light	Left	Same	Yes
6	Controlled	Signal light	Left	Same	No
7	Controlled	Signal light	Right	Yes	Yes
8	Controlled	Signal light	Right	Yes	No
9	Controlled	Signal light	Right	No	Yes
10	Controlled	Signal light	Right	No	No
11	Controlled	Signal light	Right	Same	Yes
12	Controlled	Signal light	Right	Same	No
13	Controlled	Stop sign	Left	Yes	Yes
14	Controlled	Stop sign	Left	Yes	No
15	Controlled	Stop sign	Left	No	Yes
16	Controlled	Stop sign	Left	No	No
17	Controlled	Stop sign	Left	Same	Yes
18	Controlled	Stop sign	Left	Same	No
19	Controlled	Stop sign	Right	Yes	Yes
20	Controlled	Stop sign	Right	Yes	No
21	Controlled	Stop sign	Right	No	Yes
22	Controlled	Stop sign	Right	No	No
23	Controlled	Stop sign	Right	Same	Yes
24	Controlled	Stop sign	Right	Same	No
25	Noncontrolled	–	Left	Yes	Yes
26	Noncontrolled	–	Left	Yes	No
27	Noncontrolled	–	Left	No	Yes
28	Noncontrolled	–	Left	No	No
29	Noncontrolled	–	Left	Same	Yes
30	Noncontrolled	–	Left	Same	No
31	Noncontrolled	–	Right	Yes	Yes
32	Noncontrolled	–	Right	Yes	No
33	Noncontrolled	–	Right	No	Yes
34	Noncontrolled	–	Right	No	No
35	Noncontrolled	–	Right	Same	Yes
36	Noncontrolled	–	Right	Same	No
37	Noncontrolled	–	Left	Yes	Yes

TABLE 34.4 Use cases for intersections for three cars.

Use case #	Controlled/noncontrolled intersection	Signal light/ stop sign	Other cars are at left and right/left and middle/ right and middle	Another car approached the intersection first (yes/no/same)	Pedestrians present
1	Controlled	Signal light	Left and right	Yes	Yes
2	Controlled	Signal light	Left and right	Yes	No
3	Controlled	Signal light	Left and right	No	Yes
4	Controlled	Signal light	Left and right	No	No
5	Controlled	Signal light	Left and right	Same	Yes
6	Controlled	Signal light	Left and right	Same	No
7	Controlled	Signal light	Left and middle	Yes	Yes
8	Controlled	Signal light	Left and middle	Yes	No
9	Controlled	Signal light	Left and middle	No	Yes
10	Controlled	Signal light	Left and middle	No	No
11	Controlled	Signal light	Left and middle	Same	Yes
12	Controlled	Signal light	Left and middle	Same	No
13	Controlled	Signal light	Right and middle	Yes	Yes
14	Controlled	Signal light	Right and middle	Yes	No
15	Controlled	Signal light	Right and middle	No	Yes
16	Controlled	Signal light	Right and middle	No	No
17	Controlled	Signal light	Right and middle	Same	Yes
18	Controlled	Signal light	Right and middle	Same	No
19	Controlled	Stop sign	Left and right	Yes	Yes
20	Controlled	Stop sign	Left and right	Yes	No
21	Controlled	Stop sign	Left and right	No	Yes
22	Controlled	Stop sign	Left and right	No	No
23	Controlled	Stop sign	Left and right	Same	Yes
24	Controlled	Stop sign	Left and right	Same	No
25	Controlled	Stop sign	Left and middle	Yes	Yes
26	Controlled	Stop sign	Left and middle	Yes	No
27	Controlled	Stop sign	Left and middle	No	Yes
28	Controlled	Stop sign	Left and middle	No	No
29	Controlled	Stop sign	Left and middle	Same	Yes
30	Controlled	Stop sign	Left and middle	Same	No
31	Controlled	Stop sign	Right and middle	Yes	Yes
32	Controlled	Stop sign	Right and middle	Yes	No
33	Controlled	Stop sign	Right and middle	No	Yes
34	Controlled	Stop sign	Right and middle	No	No
35	Controlled	Stop sign	Right and middle	Same	Yes
36	Controlled	Stop sign	Right and middle	Same	No

Continued

TABLE 34.4 Use cases for intersections for three cars.—cont'd

Use case #	Controlled/noncontrolled intersection	Signal light/ stop sign	Other cars are at left and right/left and middle/ right and middle	Another car approached the intersection first (yes/no/same)	Pedestrians present
37	Noncontrolled	–	Left and right	Yes	Yes
38	Noncontrolled	–	Left and right	Yes	No
39	Noncontrolled	–	Left and right	No	Yes
40	Noncontrolled	–	Left and right	No	No
41	Noncontrolled	–	Left and right	Same	Yes
42	Noncontrolled	–	Left and right	Same	No
43	Noncontrolled	–	Left and middle	Yes	Yes
44	Noncontrolled	–	Left and middle	Yes	No
45	Noncontrolled	–	Left and middle	No	Yes
46	Noncontrolled	–	Left and middle	No	No
47	Noncontrolled	–	Left and middle	Same	Yes
48	Noncontrolled	–	Left and middle	Same	No
49	Noncontrolled	–	Right and middle	Yes	Yes
50	Noncontrolled	–	Right and middle	Yes	No
51	Noncontrolled	–	Right and middle	No	Yes
52	Noncontrolled	–	Right and middle	No	No
53	Noncontrolled	–	Right and middle	Same	Yes
54	Noncontrolled	–	Right and middle	Same	No

34.3.2 Integrated CAV simulation

The use of an integrated CAV simulation environment should support the design, planning, operation, and safety of CAV deployment. Fig. 34.2 shows all aspects that should be incorporated in CAV integrated simulation. It is important to simulate all possible use cases and include all listed functions, as explained in Fig. 34.1. The integrated simulation environment should also evaluate human-to-vehicle, vehicle-to-vehicle, vehicle-to-infrastructure, robot-to-vehicle, drone-to-vehicle, and pet-to-vehicle.

There are a number of tools that can be used to simulate CAV functionalities. MATLAB is one of the useful tools to be used for CAV simulation, where Automated Driving Toolbox can be adopted. This toolbox is used to test automated driving algorithms using authored and synthetic detections from radar and camera sensor models.

34.3.2.1 Author driving scenarios

This part is used to generate a road network, actors, and sensors. A driving scenario design tool is used to test the vehicle vision system:

1. To create a road, the following parameters need to be defined to set it up properly:
   - Road center: two or more coordinates define the road on a map.
   - Road width: define total road width
   - Number of lanes: define lane numbers
   - Lane width: an array to define the width of different lanes.
   - Road elevation effects: the road is not flat here, detection error will increase.
2. Ego vehicle in the simulation:
   - The dimension of the vehicle: The dimension of the vehicle needs to be defined, such as length, width.

TABLE 34.5 Use cases for intersections for four cars.

Use case #	Controlled/non-controlled intersection	Signal light/ stop sign	Another car approached the intersection first (yes/no/same)	Pedestrians present
1	Controlled	Signal light	Yes	Yes
2	Controlled	Signal light	Yes	No
3	Controlled	Signal light	No	Yes
4	Controlled	Signal light	No	No
5	Controlled	Signal light	Same	Yes
6	Controlled	Signal light	Same	No
7	Controlled	Stop sign	Yes	Yes
8	Controlled	Stop sign	Yes	No
9	Controlled	Stop sign	No	Yes
10	Controlled	Stop sign	No	No
11	Controlled	Stop sign	Same	Yes
12	Controlled	Stop sign	Same	No
13	Noncontrolled	–	Yes	Yes
14	Noncontrolled	–	Yes	No
15	Noncontrolled	–	No	Yes
16	Noncontrolled	–	No	No
17	Noncontrolled	–	Same	Yes
18	Noncontrolled	–	Same	No

TABLE 34.6 Use cases for obstacle avoidance.

Use cases #	The object is moving/stationary	The vehicle is at high speed/ low speed	The object is at left/right/ middle	The object dimension
1	Moving	High speed	Left	–
2	Moving	High speed	Right	–
3	Moving	High speed	Middle	–
4	Moving	High speed	Left	–
5	Moving	High speed	Right	–
6	Moving	High speed	Middle	–
7	Moving	Low speed	Left	–
8	Stationary	Low speed	Right	–
9	Stationary	Low speed	Middle	–
10	Stationary	Low speed	Left	–
11	Stationary	Low speed	Right	–
12	Stationary	Low speed	Middle	–

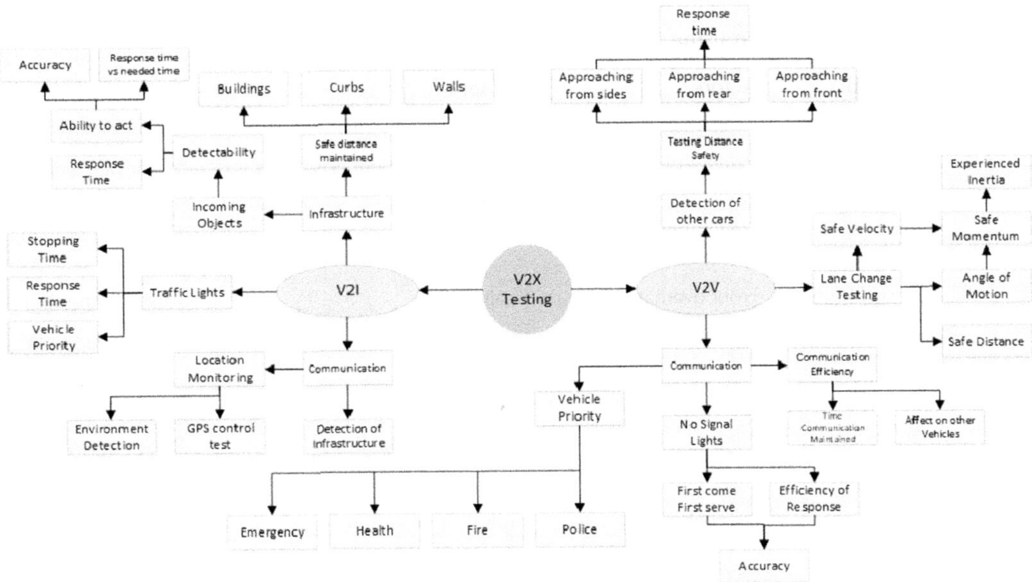

FIGURE 34.1 Functional modeling of CAV.

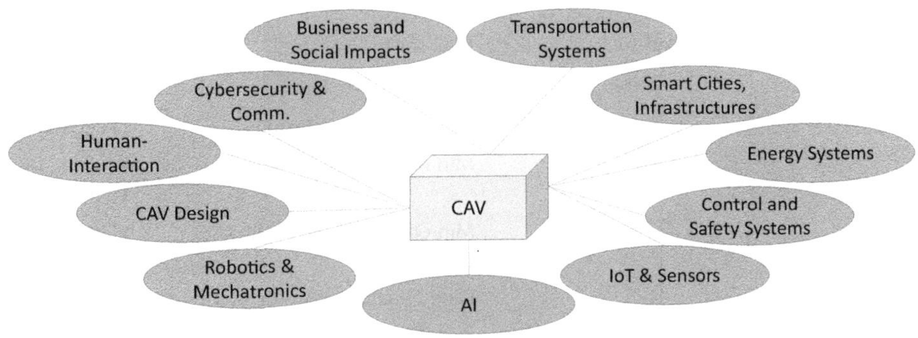

FIGURE 34.2 CAV integrated simulation.

- Waypoint: The way point is the point that the vehicle will follow in the simulation.
- Speed: To set up the vehicle speed in the simulation.

3. Other actors: To set up other actors, such as other vehicles and pedestrians, the same parameter (speed and waypoints) will be used here.
4. Vision detection sensor:
   - To set up cameras and radars, the following parameters in the simulation need to be defined.
   - Camera location on the actor: the location of the sensor on the vehicle needs to be known.
   - The properties of the camera need to be defined as well:
   - Focal length, Principal point, and image size.
   - The coverage area will be determined based on the information above.
5. Bird's eye plot: In order to see whether the logarithm is working properly, the bird's eye plot is needed.

After a brief introduction is presented in the preceding, all the detailed parameters are tableted in Table 34.7.

34.3.2.2 Detailed simulation

As the picture is shown in Fig. 34.3, it is an intersection scenario created using the software. In order to create the scenario, the detailed steps are introduced in the following paragraphs.

34.3.2.2.1 Road

First of all, in order to create a test scenario, the road network needs to be defined. As the picture is shown in Fig. 34.4, the rode center needs to be defined in order to locate where the road is on the map and the curvature of the road.

TABLE 34.7 Parameters for simulation using MATLAB.

Parameters	Road (unit)	Actors	Camera	Radar
Parameter 1	Width (m)	Length (m)	Update interval (ms)	Update interval (ms)
Parameter 2	Bank angle (deg)	Width (m)	Camera placement position	Radar placement position
Parameter 3	Number of lanes	Height (m)	Focal length	Detection probability
Parameter 4	Lane width (m)	Front overhang	Image width	False alarm rate
Parameter 5	Lane marking type	Rear overhang	Principal point	Field of view rate
Parameter 6	Road centers	Roll	Detection (road/object)	Max range
Parameter 7		Pitch	Detection probability	Range rate min
Parameter 8		Yaw	False position per image	Whether has elevation
Parameter 9		Azimuth angle (deg) [for radar cross section]	Limit # of detection	Whether has occlusion
Parameter 10		Elevation angle (deg) [for radar cross section]	Detection coordinates	Reference range
Parameter 11		Speed	Max speed	Reference RCS
Parameter 12		Waypoint	Max range	Limit # of detections
Parameter 13			Max allowed occlusion	Detection coordinates
Parameter 14			Min object image width	Azimuth (resolution and bias fraction)
Parameter 15			Min object image height	Elevation (resolution and bias fraction)
Parameter 16			Lane update interval (ms)	Range (resolution and bias fraction)
Parameter 17			Min lane image width	Range rate (resolution and bias fraction)
Parameter 18			Min lane image height	Whether it has noise
Parameter 19			Boundary accuracy	Whether it has false alarms
Parameter 20			Limit # of lanes	
Parameter 21			Bounding box accuracy	
Parameter 22			Process noise intensity(m/s^2)	

The next step is to give a name to the road created before. The road also needs to define how wide it is and the bank angle of the road, as shown in Fig. 34.5.

Finally, within each road created above, it needs to be separated into different lanes in order to have a regular traffic system. The picture is shown in Fig. 34.6 also shows that the lane width and marking type on the road need to be defined explicitly for the traffic system.

34.3.2.2.2 Actors

In order to create some moving objects in the traffic system, adding vehicles and pedestrians (refers to actors) is necessary. Firstly, an ego vehicle (the vehicle for testing) needs to be added to the system. The next step is to set how the vehicle goes in the road created before by setting the waypoints and speed shown in Fig. 34.7.

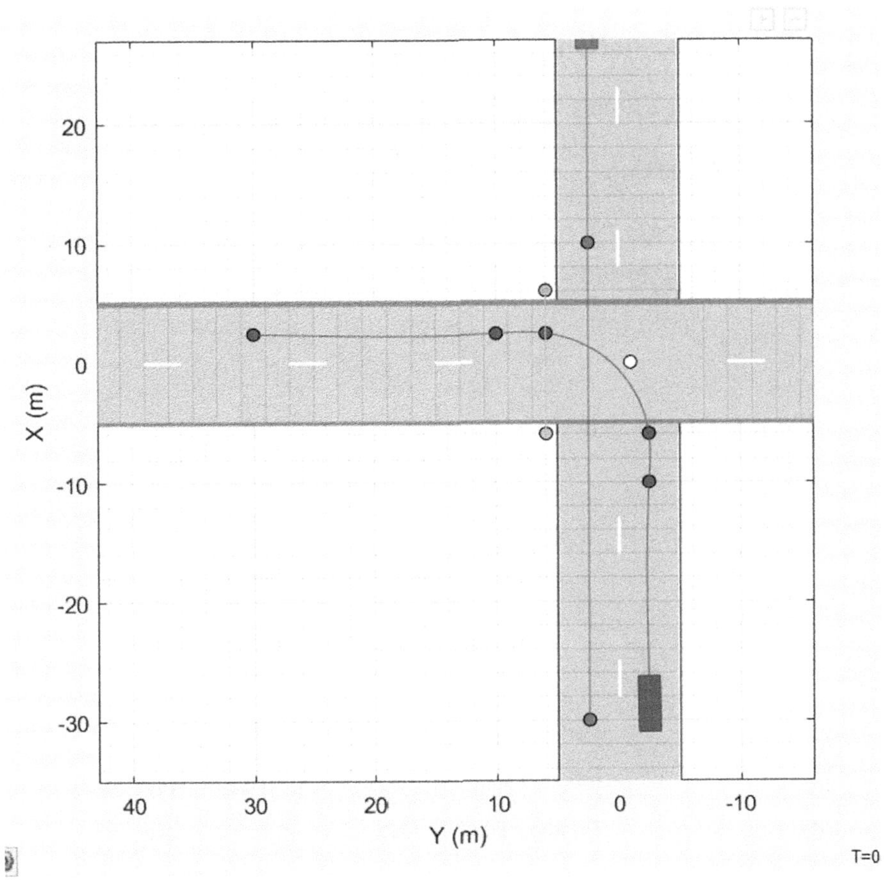

FIGURE 34.3 An example of using MATLAB.

	x (m)	y (m)	z (m)
1	0	99	0
2	0	-1	0
3	0	-101	0

FIGURE 34.4 Road center setup.

After the vehicle knows where it goes, defining the vehicle dimension is also necessary. Several parameters need to be defined for a vehicle in the system, such as vehicle length, width, height, front overhang, rear overhang, roll, pitch, and yaw, as shown in Fig. 34.8.

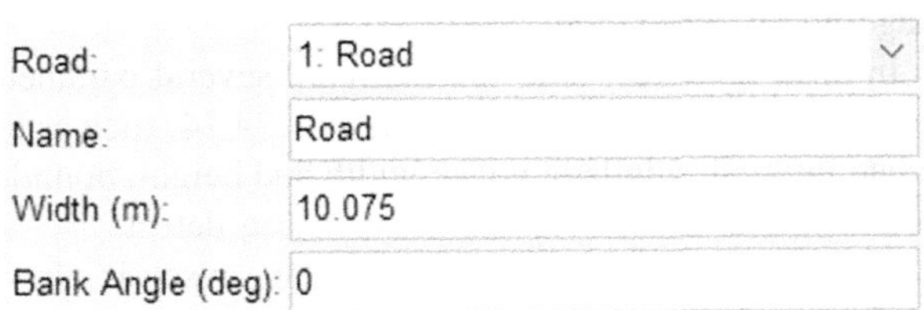

FIGURE 34.5 Road name, width, and bank angle.

Finally, the Radar Cross Section (RCS) needs to be defined by azimuth angle, elevation angle, and pattern in dBsm. RCS is the measure of a target's ability to reflect radar signals in the direction of the radar receiver in Fig. 34.9.

34.3.2.2.3 Camera

To make the vehicle see other objects, cameras are essential components for autonomous vehicles. To install a camera on the vehicle, the vehicle's camera location needs to install a camera. The camera location on the vehicle needs to be determined by the algorithm developed by engineers. Parameters relative to the vehicle need to be added into the system as shown in Fig. 34.10.

Once the camera's location on the vehicle is known, the camera's property is also important. Focal length, image width, and principal point need to set up the camera's basic functionality in Fig. 34.11.

Besides basic camera functionality, the detection parameters are also significant for the simulation in Fig. 34.12. The detection type defines whether the camera

FIGURE 34.6 Lane information.

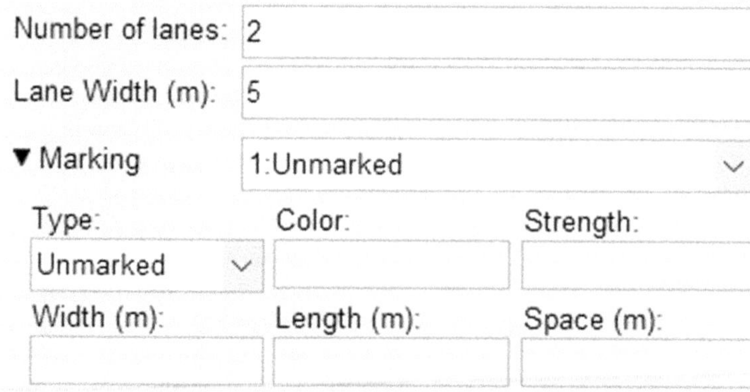

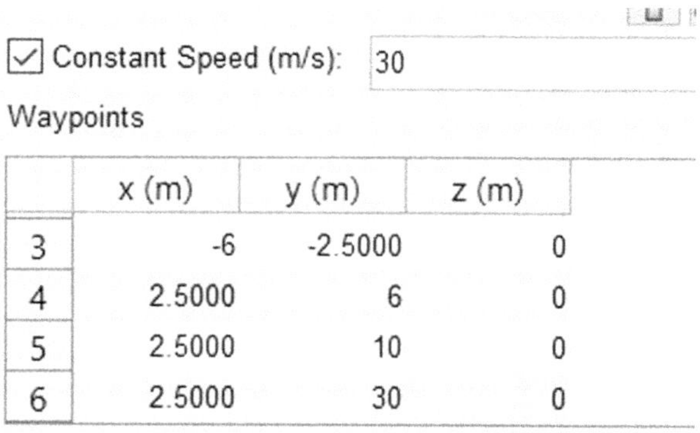

FIGURE 34.7 Waypoints of actors.

FIGURE 34.8 Vehicle dimensions.

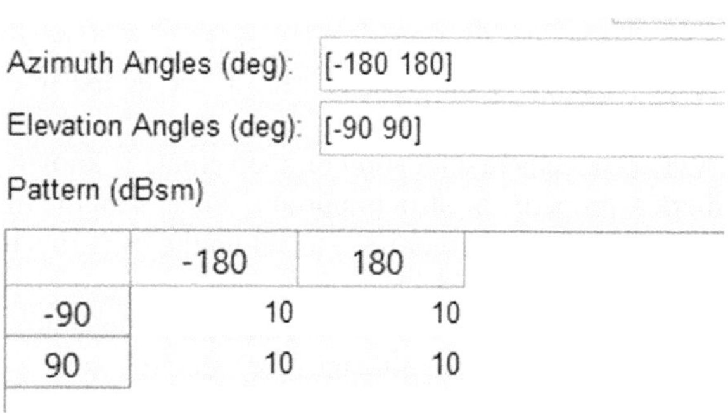

FIGURE 34.9 Vehicle RCS.

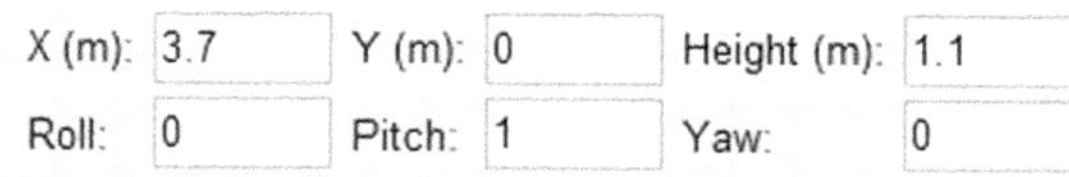

FIGURE 34.10 Relative position of sensors.

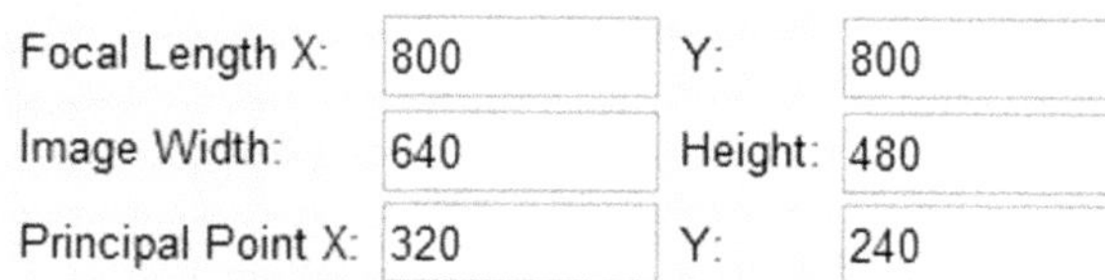

FIGURE 34.11 Properties of camera.

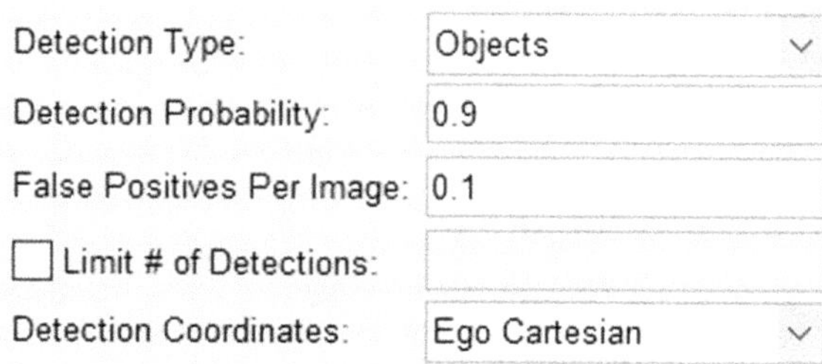

FIGURE 34.12 Detection parameters.

detects objects or the road lanes. Detection probability tells the camera the accuracy of the detection. Normally, the camera has a limit for the maximum number of objects it can detect. In Fig. 34.13, the camera has limit parameters, max speed, max range, max allowed occlusion, min object image width, and height.

In order to detect lanes on the road, several parameters need to set up for the lane as well in Fig. 34.14, such as lane update interval, min lane image width and height, boundary accuracy, and max number of lanes it can detect.

Finally, in real world, noise for the sensor is also an important parameter. Bounding box accuracy and process noise intensity, as is shown in Fig. 34.15.

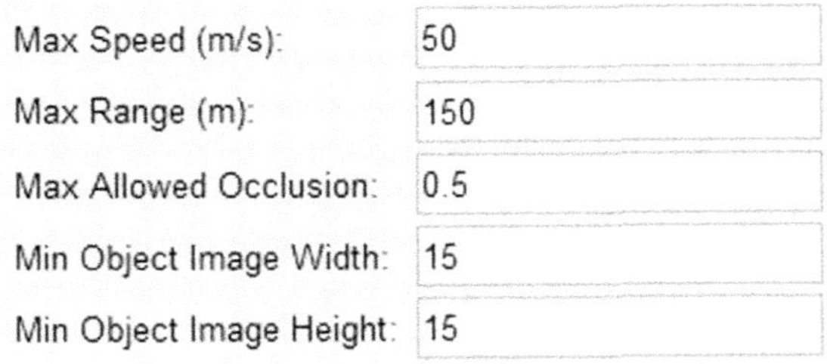

FIGURE 34.13 Sensor limitation.

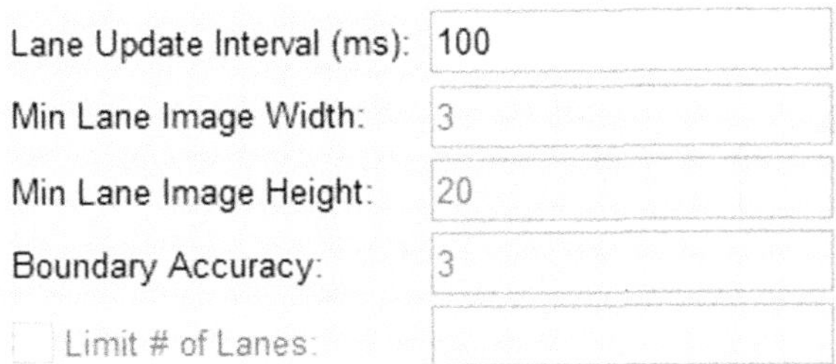

FIGURE 34.14 Lane parameters.

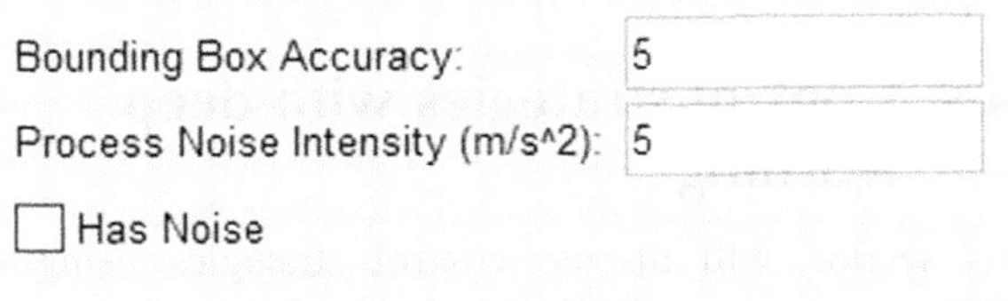

FIGURE 34.15 Sensor noise.

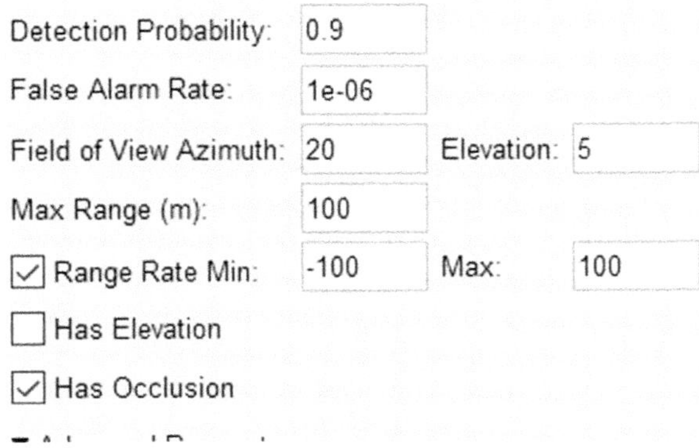

FIGURE 34.17 Radar detection parameters.

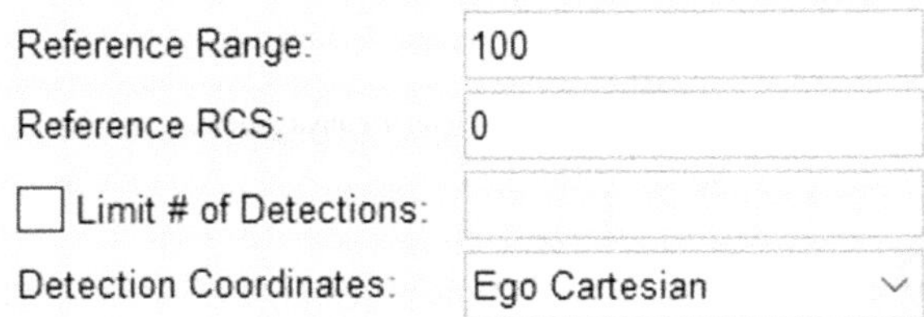

FIGURE 34.18 RCS setup.

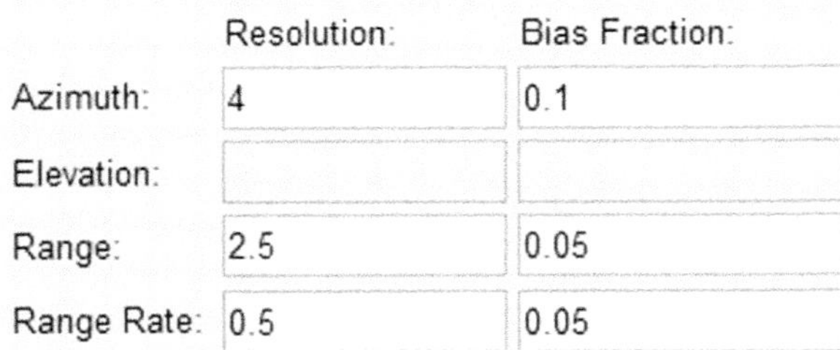

FIGURE 34.19 Accuracy parameters.

34.3.2.2.4 Radar

Radar has become another important part of autonomous vehicles. To simulate radar in the system, we need to define the radar location, like what did for cameras in Fig. 34.16.

However, the detection parameters are different from the camera setup. Several other parameters need to be defined, including detection probability, false alarm rate, a field of view azimuth and elevation, max range, and range rate from min to max, as shown in Fig. 34.17. For radar in Fig. 34.18, it also has a reference range, reference RCS, and a limited number of detections. Finally, the accuracy parameter is defined using resolution and bias fractions of azimuth, elevation, range, and range rate, as shown in Fig. 34.19.

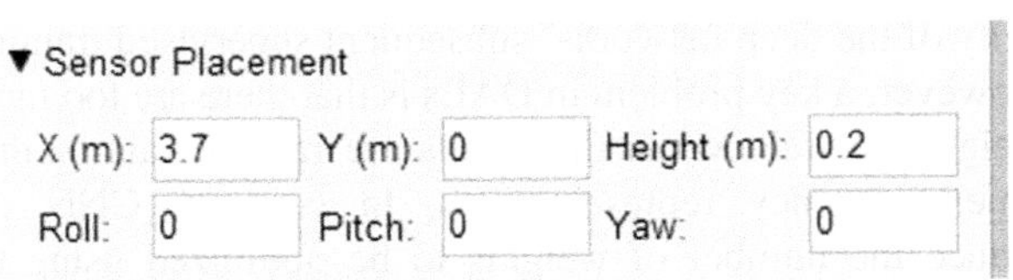

FIGURE 34.16 Relative position for radar.

34.3.2.2.5 Sensor canvas

As the picture is shown in Fig. 34.20, all the sensors installed on the car can be viewed using sensor canvas view. It uses the interesting vehicle as a reference to create a two-dimensional view to determine whether all the sensors are effectively detecting an object around the vehicle.

34.3.2.2.6 Egocentric view

To better view how the vehicle behaves in the scenario created before, the egocentric view is useful for users to see the view as the driver behind the car, as shown in Fig. 34.21.

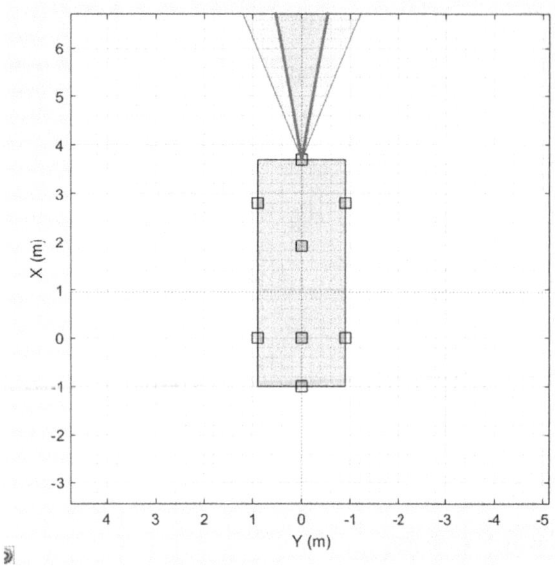

FIGURE 34.20 Sensor canvas view.

34.3.2.2.7 Bird's eye plot

To better view how sensors cover all the area around the ego vehicle, a bird's eye plot is necessary to determine whether all the sensors work effectively to locate the objects around the vehicle, while the vehicle moves as predefined waypoints, as shown in Fig. 34.22.

FIGURE 34.21 Egocentric view.

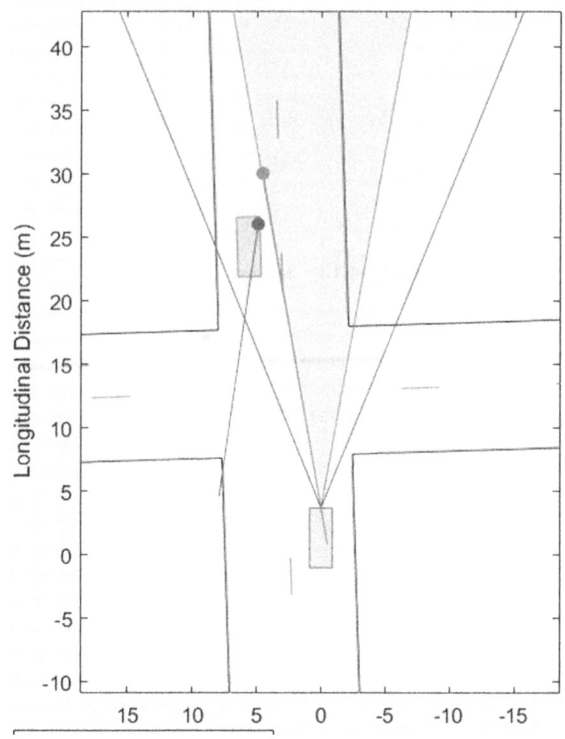

FIGURE 34.22 Bird's eye plot.

34.4 Control strategies with deep learning

This section will discuss control strategies using deep learning techniques. With the rapid advancement of deep learning techniques, they have find more and more successful applications in the field of autonomous vehicles.

34.4.1 Overview

Autonomous vehicles can be used for many applications where it may be inconvenient, dangerous, or impossible to have a human driver present. Deep learning, which uses multilayer transformations to approximate complex nonlinear relationships, has been developed to tackle problems in autonomous vehicles. There are many types of deep network architectures and learning paradigms. Several major types of deep neural networks include Convolutional Neural Network (CNN), Deep Autoencoder (DAE), and Deep Reinforcement Learning (DRL). DAEs can conduct unsupervised pretraining on the weights, easing the difficulty of the deep networks' subsequent supervised training. However, a key problem in DAEs is that there are too many weights to train when the inputs are raw signals or their time–frequency representations. In contrast, CNN can reduce the number of weights to be optimized using the local receptive field strategies and weight sharing, which can be effective for reducing computational burden during the training process. One disadvantage of CNN is the trap of local minima during the training phase. Deep Reinforcement Learning (DRL) uses Q-learning or policy gradient algorithms to gain the best rewards for the chosen actions. It can be used to achieve superhuman abilities through training without human intervention.

34.4.2 System designs and descriptions

Autonomous vehicles are controlled based on perception planning and control paradigm. Traditional control methods include optimal control, robust control, PID control, adaptive control, etc. However, the use of most of the above control techniques needs a mathematical model. On the other hand, deep learning, built upon the traditional shallow neural network, is a model-free, data-driven control technique applied to both discrete and continuous systems. Current deep learning solutions for autonomous driving provide state estimation or improve state estimation or replace the traditional controllers with end-to-end controllers based on deep neural networks.

In recent years, deep learning is used for state estimation for different controllers [9] or enhancing the quality of state estimation [10]. A dynamic model to identify the uncertainties and hidden states is the foundation of system control. However, model parameters are not easily identified by the traditional system identification method. In Ref. [11], a deep convolutional neural network–based dynamic identifier is proposed for the adaptive controller for an unmanned helicopter. It can robustly and accurately identify vehicles' dynamic behavior. In Ref. [9], the authors present a flight control method for autonomous helicopters. A deep learning network is used as the identifier for an adaptive control scheme. The complete model includes a first-principle-based dynamic model and a CNN-based model for modeling hidden states and uncertainties. All parameters and weights are trained with real flight data. In Ref. [10], a deep learning network with dropout technique is used to improve the attitude state estimation performance by Kalman filter. The network is trained to model the measurement noises, which is used to filter out the noise and enhance the quality of Kalman filter. In Ref. [12], deep learning is used to compensate for delays and measurement noises. The information extracted by a modular deep recurrent neural network is combined with sensory readings before feeding into the Kalman filter for state prediction and update. This deep learning network can detect the hidden states, which are generally difficult to be measured by sensors. In Ref. [13], a CNN and LSTM-based observer is presented. First, LSTM processes videos and adds a temporal dimension to the cost map. Then a particle filter is used for state estimation. The cost map generated by the deep learning network is combined with readings from IMU and wheel speed encoders in order to predicate and update the states for model predictive control.

Another application of deep learning in the control field is to replace the traditional controllers such as a Proportional Integral Derivative (PID) controller and generate continuous or discrete control actions as the output of the deep network [14] or combine the deep learning with traditional controller to form a hierarchical or better control scheme. In Ref. [14], a DRL and interactive Reinforcement Learning (RL) network is used for path following for underwater vehicles. Different from traditional PID controllers, this RL-based controller supports the vehicles' autonomy and can adapt to the dynamic or unknown environment. Both human reward and environmental reward are used to train the network. Through training, the network can learn optimal control policy from the human trainer and the environment. In Ref. [15], the authors investigated the low-level DRL-based control strategy for underwater vehicles. A deep reinforcement learning network is introduced with only the sensory signals serving as the input of the network. There is no need to have prior knowledge of vehicle dynamics. In Ref. [16], the deep learning network is used to analyze the environment in order to predict lateral and longitudinal control. They use two separate models for vehicle speed and steering. The inputs are road images, and the outputs are the speed and steering. In Ref. [17], the authors propose to use a CNN network as an end-to-end controller for driving, while two other CNN-based deep networks generating both the feature map and error map to help the controller better understand the scene. An attention model is used to identify the regions that affect the output most. In Ref. [18], the authors have explored the use of reinforcement learning for high-level decision-making in the context of a robotic game. In this hierarchy structure, the high-level DL is combined with low-level controllers to deliver a better control performance. The controller's design with multiple levels can accommodate the challenges in the game, such as action delay.

34.4.3 Case study

This case study will focus on the fault detection phase of fault tolerance control. we will show that the deep learning algorithm has the ability to perform fault detection tasks for a remotely operated multiwheel autonomous vehicle. The system dynamics model of the multiwheel autonomous vehicle is as follows:

$$Y(s)_{\text{Total}} = G_{R1}(s)\, u_{R1}(s) + G_{R2}(s)\, u_{R2}(s) + G_{L1}(s) u_{L1}(s) + G_{L2}(s)\, u_{L2}(s)$$

where the transfer functions are:

$$G_{R1}(s) = \frac{-14.97\, s^2 + 0.5626\, s - 13.09}{s^3 + 0.9098\, s^2 + 1.236\, s + 0.7621}$$

$$G_{R2}(s) = \frac{-149.6\, s^2 + 95.8\, s - 0.6111}{s^3 + 30.53\, s^2 + 9.796\, s + 7.421}$$

$$G_{L1}(s) = \frac{4.927\, s^2 + 0.1471\, s + 5.064}{s^3 + 0.4772\, s^2 + 0.7991\, s + 0.3155}$$

$$G_{L2}(s) = \frac{94.51\, s^2 + 93.14\, s + 86.09}{s^3 + 21.11\, s^2 + 2.626\, s + 22.96}$$

We use square waves as the input to the wheel motors. The output $y(t)$ was generated by feeding the system model with the input. $R1$ is the system without fault. The input (highlighted by green) and output (highlighted by red) signals are shown in Fig. 34.23. $R2$ is the system with fault, as shown in Fig. 34.24. The fault due to mechanical problems is reflected in the dynamic model.

We show a general framework using Continuous Wavelet Transform (CWT) and DNN network in Fig. 34.25. We will first generate simulated data and then convert one-dimensional signals to two-dimensional images through wavelet transform. We use three wavelet channels in order to extract complementary features in the input signals. These three channel 2D signals are fed to a deep neural network for classification. The output of the deep neural network indicates whether the signal has faults or no fault.

The loss function $J(w)$ is defined as follows:

$$\min_{w} J(w) = \sum_{n=1}^{N} \left(|y_n(I_n, w) - y_m|\right)^2$$

N is the number of training samples, w is all the deep neural network parameters, y_m is the ground truth label of normal class (0) or fault class (1). In this study, we need to detect the fault caused by a mechanical failure. We use a test dataset of 1000 images (500 normal and 500 defect

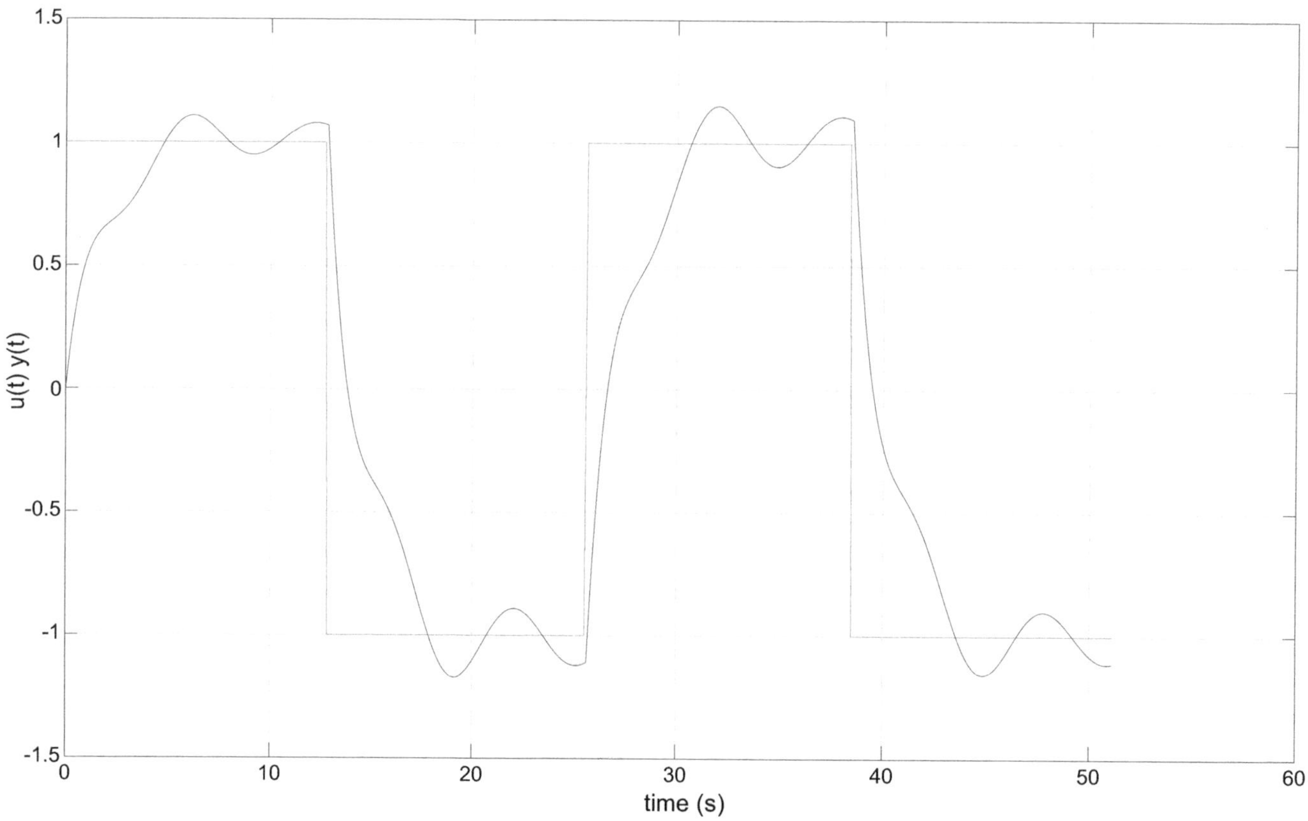

FIGURE 34.23 The input (green [light gray in print version]) and output (red [dark gray in print version]).

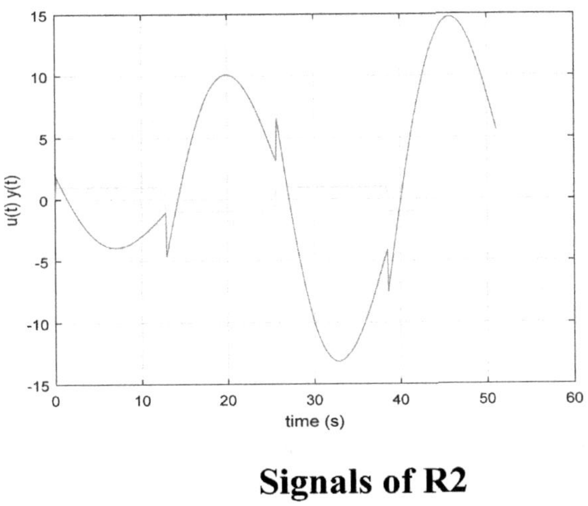

FIGURE 34.24 The input (green [light gray in print version]) and output (red [light gray in print version]).

images) to test the proposed DNN and wavelet transform method. In this study, we found that the proposed DNN wavelet combination can effectively detect the faults. The result is shown in Table 34.8.

This section has reviewed the recent advances in using deep learning for autonomous vehicles in the research areas of control strategies. We found that many types of deep neural networks have found successful applications in various design and operation aspects of autonomous vehicles. We can conclude that the deep learning technique has found its match in controlling autonomous vehicles, and more exciting research results will come to follow in the near future.

TABLE 34.8 Classification results.

Test data set	Detection correct rate (%)
500 normal signals	100
500 fault signals	100

34.5 Communication and cybersecurity layers for CAV systems

There are a large number of vehicles on the road all over the world. However, a huge number of accidents have happened everywhere due to human driver errors. According to the statistics in the United States, 539,000 two-lane traffic accidents occurred every year. During the autonomous drive, one of the most dangerous maneuvers is lane changing. Even with Advanced Driving Assistance Systems (ADAS), such as Adaptive Cruise Control (ACC), Autonomous Emergency Braking (AEB), Lane Keep Assistant (LKA), and other features, lane change is still complicated and potentially dangerous. With the development of electronics, control systems, communication technology, etc., CAVs with fully equipped sensory technology could significantly reduce these accidents due to

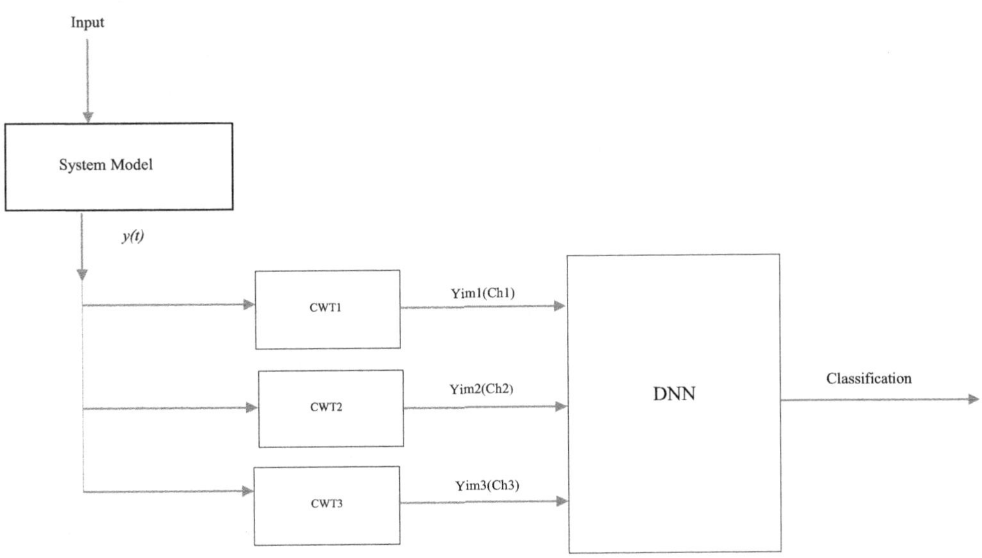

FIGURE 34.25 System framework.

cybersecurity attacks and human errors. Thus, a set of standards and testing methods should be proposed according to traffic control, public safety, etc., in view of the reliability of protection and mitigation layers [19]. For example, the working document ISO 21434 "Road Vehicles – Cyber Security Engineering" and ISO 26262 "Road vehicles – Functional safety" [20–24]. Safety Integrity Level (SIL) allocation should be fully addressed within the North American regulatory environment according to traffic control, public safety.

Trust in automated is a continuous concept rather than a binary variable of trust or not having trust. In general, trust is a dynamic factor that constantly continually changes based on trustee's performance and other factors that are independent of the trustee's performance [25]. Hancock et al. [26] categorize factors affecting trust in robots into three groups, including (a) machine performance, (b) user user-related factors, and (c) environment-related factors. The same classification can be used for factors affecting trust in the automated systems as well. Schaefer et al. [27] adopt the same classification for factors affecting trust in the automated systems. Gold et al. [28] show that age has a positive correlation with trust in automated vehicles. They showed that people above age 60 have a higher level of trust in automation than people below age 30. Sanchez et al. [29] investigate age's effect on the sensitivity of individuals' trust in the automated driving system when the system is 100%, 80%, and 60% reliable. Although they did not find any significant effect of age in trust and perceived reliability of the system, they found that older adults' perceived reliability of the system is more sensitive to a decrease in system reliability than younger adults. It shows the effect of age on shaping history-based trust in a dynamic form. The interpretation of system performance can affect trust. Jin et al. [30] investigate unsuccessful human interactions with ADAS and identified that 1338 of the complaints (0.6%) were related to ADAS issues. Unsuccessful ADAS interactions included situations where the user appeared unaware of the system's features, others were related to unintended consequences of regular ADAS operation, and some appear to relate to system errors. Privacy issues such as the disclosure of driver's personal information through the CAV network also affect trust [31,32].

CAV benefits are achieved by collecting relevant information from the CAV's context, such as geolocation, date, and time. Although such information is the source of CAV's benefits, they are also sensitive information because it can identify individuals. These are called Personally Identifiable Information (PII). The National Institute of Science and Technology (NIST) defines PII as any information about an individual maintained by an organization that can be used to distinguish or trace an individual's identity, including any information linkable to an identifiable individual. CAV can collect and process significant amounts of PII, such as his or her contextual data (e.g., geolocation, date, and time) and among other individual's attributes such as age, address, gender, and income. Therefore, CAV can infer an individual's interests, traits, beliefs, and intentions. At the same time that CAV offers benefits for their users and society, they may also open cybersecurity vulnerabilities such as, but not limited to, acquisition, storage, and unauthorized disclosure of collected PII. For this reason, the future of CAV is involved with security, privacy, and safety concerns [33,34].

CAV can be connected through external interfaces, such as Wi-Fi, Bluetooth, Geo-Positioning System (GPS), and Tire Pressure Monitoring System (TPMS). Internally, CAV works with a Controller Area Network (CAN) connecting different Electronic Control Units (ECU), as the engine itself. On the one hand, these connections are necessary to provide basic (e.g., driving) and advanced features (e.g., autonomous driving and entertainment) to involved persons, such as drivers, passengers, and pedestrians. On the other hand, those external interfaces can open several breaches for security attacks, leading to a series of privacy and safety issues. For instance, malicious data packages may be introduced via those connectivity interfaces into the CAN [33].

There are numerous safety issues related to security attacks. Those safety risks may bring implications for CAV's passengers, other vehicles, pedestrians, and road infrastructure. Human–Robot Interaction (HRI) researchers are making efforts to understand human aspects and perceptions toward CAV, such as trust [32], driving style [31], and physical safety of pedestrians and city infrastructure [35]. As a requirement to develop security solutions for CAV's issues, Thing and Wu [33] proposed an attack taxonomy for CAV and defense strategies for each type of security attack. Accordingly, attackers can be thieves, hacktivists, terrorists, or spies using remote or physical access to the available CAV interfaces. Attacks are motivated by distinct reasons, such as information collection, deception, financial gains, and aiming at different CAV targets (e.g., sensors, GPS, and infravehicular links). Potential consequences include disclosure of information, performance degradation, and loss of vehicle control or loos of the property itself. Suggested defenses cover preventive, passive, active defenses, and collaborative strategies such as using cloud computing services. Their proposal serves as the basis for developing other potential solutions for such issues.

Transport Canada is committed to working with all government, industry, and stakeholders across a range of disciplines [36]. They are also committed to compliance with international regulations to improve automotive cybersecurity in Canada for the context of CAV, as stated in the Cyber Security of Automated and Connected Vehicles Stakeholder Consultation Document in February 2019. TC approach for CAV cybersecurity follows best practices from both federal and international standards, such as compliance with the Transportation 2030: A Strategic Plan for the Future of Transportation in Canada. Transportation 2030 aims at building a safer and more secure transportation system that Canadians can trust. Besides, TC works under the Canada–US Regulatory Cooperation Council (RCC).

In Canada, CAV information collection must happen in accordance with the regulations of the Canadian Personal Information Protection and Electronic Documents Act (PIPEDA). PIPEDA regulates the collection, use, or disclosure of PII in the course of commercial activities. One of the principles of PIPEDA states that "an organization must make detailed information about its policies and practices relating to the management of personal information publicly and readily available." According to PIPEDA, CAV must also get customers" consent to collect their personal information and limit your collection and retention of it."

Privacy policy tools are usually composed of policy generation mechanisms and interfaces for policy comprehension, configuration, and system feedback [37]. They stand as a potential solution to mitigate CAV's security and privacy concerns. However, privacy policies are still long and complicated [38]. The usability of privacy policy interfaces for laypeople remains a great challenge in the field [39–41]. In CAV's context, each usability flaw may incur wrong decisions by the drivers and, as a consequence, security or privacy vulnerabilities. After that, it may also incur safety issues. For this reason, developing more useable interfaces and tools for privacy policy aiming lay drivers is a requirement to move forward on the topic.

Referring to Transport Canada, there are two important information flows in CAV described as: (1) Increased external communication between CAVs and the external environment: Vehicle-to-Infrastructure (V2I) and Vehicle-to-Internet of Things (V2IoT) communications; and (2) Increased internal communication among systems within the CAV, also known as intravehicular communication: CAVs have many ECUs interconnected via the CAN bus. The research challenge is how cybersecurity considerations are incorporated into this V2IoT experimental research with these standards: (i) ISO/TC 204 for standards catalog for intelligent transport systems; (ii) ISO 9241 for Human Centered Design (HCD) methods and processes as enabling, executing, and assessing HCD within organizations and usability inspections with experts to diagnose interface flaws and propose improvements; and (iii) ISO 14229 for unified diagnostics systems in CAV and ISO 15765 for standards on CAV communications.

34.6 CAV test platform

In order to evaluate different functionalities of CAV, an intelligent and integrated CAV test platform is required. The following section will discuss design requirements for CAV test platform.

34.6.1 CAVTE design requirements

CAVTE or Connected and Autonomous Vehicle Test Environment is developed with the research team at Ontario Tech University to support the testing of CAV technologies. The complete CAVTE platform will include the following modules: autonomous vehicle (selected technology, other autonomous vehicle technologies to be tested); energy system; sensor and monitoring system; communication system; a control system; human/robots; data analytics; and cybersecurity system. The figure below explains the main modules of CAVTE, which includes the design requirements of CAV and the communication layer with IoT and cybersecurity modules. An example of CAV test platform developed by the team at Ontario Tech University is shown in Fig. 34.26.

34.6.2 Test functions

CAV systems are based on sensor capabilities, and any failure in sensor functions will lead to high risks. It is important to test all sensors and the associated perception, including vehicle location, speed, surrounding objects, and humans. The analysis of sensor data will facilitate motion control and support the prediction of future movement in view of the available space and the associated constraints. It is vital to test detection, range, and communication sensor functionalities, including the following:

- Articulating Radars
- Long-Range Radars
- Short-Range Radars
- Lidars

Perception is the stage where the autonomous vehicle takes in raw information from sensors such as lidars, radars, and cameras to understand its environment—testing perception through simulated environments and scenarios in view of use cases. Sensor data can be integrated into a virtual reality environment to analyze vehicle actions and reactions and evaluate sensors and control functions' performance. The test functions will support CAV planning and operations, which are based on vehicle location, road infrastructure, pedestrian, road markings, and other external factors. Planning the operation of CAV identifies several paths per second and constantly chooses the best one. For example, suppose a vehicle aggressively changes lanes into a lane. In that case, the autonomous vehicle was planning

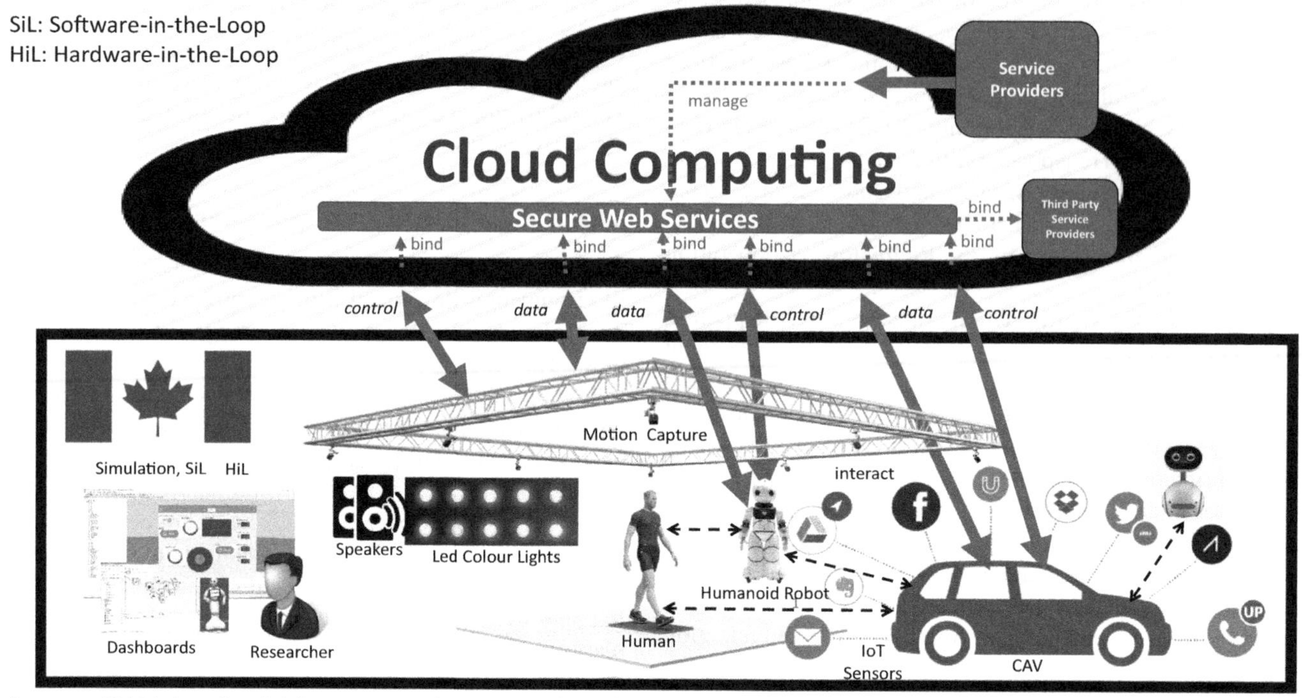

Human-to-Vehicle Interface (HVI), Human-to-Robot Interface (HRI), Robot-to-Robot Interface (RRI) and Vehicle-to-Robot Interface (VRI).

FIGURE 34.26 CAV test platform at Ontario Tech University.

on entering, and the autonomous vehicle will already have a backup plan and act upon that instead. Planning operations should be tested under different scenarios to test the vehicle's ability to adapt to road circumstances. Control functions convert instructions sent from planning and perception operations into actions carried out by the actuators. The functionality of communication through the process should be tested for accuracy and strength. Sensors of CAV will provide raw information regarding the environment. The vehicle's main sensors are the camera, lidar, GPS/Inertial Measurement Unit, and radar. Many autonomous vehicles use a process known as sensor fusion to combine multiple readings of different sensors to provide a more redundant performance. The following are examples of test functions:

- Sensor testing:
  - Lidar testing
    - Check light pulses → Performance evaluation
    - Check the capability of environment recognition
  - Basic Sensor testing
    - Readings, data transmission, and functionality
  - Radar testing
    - How well is radar detecting the speed of all objects
  - Camera testing
    - Feature recognition
    - Proper identification
  - Main Central Processing Unit (CPU) testing (Control unit)
- Mechanical component testing:
  - Breaking
    - Communication
    - Standard functionality
  - Motor
    - Communication motor control
    - Standard functionality
  - Battery
    - Communication
    - Power transmission and control
    - Safe voltage and current check
    - Temperature
    - Standard functionality
  - Steering
    - Accuracy → Input versus Output
    - Communication

V2X technology components V2V and V2I allow CAV to communicate by receiving information and talking to other systems in the environment. These environmental communicating systems can be other vehicles or even a smart city light that is changing signals, as follows:

1. Vehicle-to-Vehicle (V2V):
   - How far did the test vehicle stop a safe distance from an independent vehicle?
     - Approaching rear

- Approaching front
- Approaching from sides
- Does a vehicle safely execute lane changes?
  - Safe velocity and distance
  - The angle of lane change
  - Distance between the rear end and front end of a vehicle
- How fast was the response time of the test vehicle?
- Does the test vehicle maintain steady communication with an independent vehicle?
  - If not, how many segments of communications were made?
  - If not, where did communication fail, and how does this affect autonomous vehicles?
- Detectability and communication of all other vehicles on the road?
- How well does a test vehicle interact with an independent vehicle?
  - Do they communicate efficiently and properly at road intersections?
  - Do they communicate efficiently with each other when there are no signals?
  - Do they interact with unexpected objects collaboratively (e.g., animals, people, robots):
    - Accuracy
    - Distance and speed

2. Vehicle-to-Infrastructure (V2I):
   - How well do vehicles interact with signal lights?
     - Stopping time
     - Response time
     - Vehicle priority
     - Interaction between smart lights and regular lights
   - Do vehicles communicate safely with walls?
     - The safe distance on roads
   - Detectability and encountering of incoming objects:
     - Animals, falling ice, bicycle, etc.
     - How well an autonomous program detects these objects
     - How well it encounters them:
       - Response time
       - Distance
       - Safety to other vehicles
   - Was maneuver made safely?
     - Inertia test for passengers
     - Global environment detectability
     - Sense of location
     - GPS control test

34.6.3 Function test analysis

This section outlines the test parameters that were tested using the prototype 1 demo. Prototype 1 was designed for basic functionality testing of autonomous vehicles that could be introductory to complete complex test platform with CAVTE. Test parameters are tested and described in Tables 34.9–34.11 below.

34.6.4 CAVTE demonstration

The Test Platform will have a variety of test functionality, which will allow the user to specific testing.

A sample test scenario could be testing speed related to response time according to a gap between stopping a vehicle. In order to make the test platform test such precise data, there must be an appropriate structure between the mechanisms and algorithms. CAVTE is controlled by two separate main Raspberry Pi control boards. One will be for the test platform, and the other will be used for the Autonomous Electric Vehicle (AEV). CAVTE control board will have two main tasks, receiving input from a user then executing results such as turning platform, simulations on Liquid Crystal Display (LCD0 screen, moving objects on a platform, etc.). Testing will display on the GUI LCD touch screen, which will allow the user to select specific testing.

User input will be received via an LCD touch screen as part of the main control board. The main control board will execute functions. The main control board will be programmed accordingly to Graphical User Interface (GUI) app. Selecting a test function will execute a series of steps and recordings related to that function, as follows:

1. Selecting V2V test function:
   - Car stopping/approaching/changing lanes/vertical/horizontal/angle:

TABLE 34.9 V2V testing: vehicles communicating without signal lights.

Test	Recorded	Ideal	Performance (P)	Below standards	Average standards	Above standards
Response time between vehicles	0.060 s	0.070 s	85.7%	P < 84.0%	84.0%–95.0%	95.0% < P
Safe distance between rear of test vehicle	0.0048 m	0.0050 m	96.0%	P < 81.0%	81.0%–94.0%	94.0% < P
Speed prior to response	0.490 m/s	0.500 m/s	98.0%	P < 82.5.0%	82.5%–96.0%	96.0% < P

TABLE 34.10 V2I testing: vehicles communicating with signal lights.

Test	Recorded	Ideal	Performance (P)	Below standards	Average standards	Above standards
Response time between vehicle and signal	1.0 s	1.0 s	100%	P < 84.0%	84.0%–95.0%	95.0% < P
Vehicle priority	N/A	N/A	N/A	P < 84.0%	84.0%–95.0%	95.0% < P
Speed prior to response	0.490 m/s	0.520 m/s	92.3%	P < 82.5%	82.5%–96.0%	96.0% < P

TABLE 34.11 V2H: human detection.

Test	Recorded	Ideal	Performance (P)	Below standards	Average standards	Above standards
Response time of detection	0.060 s	0.070 s	85.7%	P < 84.0%	84.0%–95.0%	95.0% < P
Zenbo detection	N/A	N/A	N/A	P < 84.0%	84.0%–95.0%	95.0% < P
Safe distance	0.0048 m	0.0050 m	96.0%	P < 84.0%	84.0%–95.0%	95.0% < P

- From front/back/side/angle:
  - Test 1: Simulation is displayed of selected scenarios to observe AEV behavior:
    - Speed of tires is measured and recorded
    - Gap distance is measured and recorded.
    - Response time is measured and recorded.
    - App executes equation modeling to compare results to safety standards.
  - Test 2: Objects will be moved on tracks to observe AEV behavior.
    - Speed of tires is measured and recorded
    - Gap distance is measured and recorded.
    - Response time is measured and recorded.
    - App executes equation modeling to compare results to safety standards.
  - Test 3: Converging data from Test 1 and Test 2 and comparing it according to safety standards to pass/fail testing.

2. Selecting V2X test function:
   - AEV interacting with signal light/walking light/gate:
     - Test 1: Simulation is displayed to test AEV behavior:
       - Object recognition is testing.
       - Response time is recorded.
       - The distance at which AEV detects signal light.
       - Lighting scenarios.
     - Test 2: Real life objects are set into test platforms to test AEV behavior:
       - Object recognition is tested.
       - Response time is recorded.
       - Smart light communication if possible.
     - Test 3: Converging data from Test 1 and Test 2 and comparing it according to safety standards to pass/fail testing.
3. Selecting V2H test function:
   - AEV interacting with moving objects such as human/animal:
     - Test 1: Simulation shows a human attempting to crossroad several meters ahead:
       - Response time is compared to a decline in speed.
       - Testing detection of an object in scenarios day/evening/night.
       - The distance at which detected objects are tested.
     - Test 2: Real life human/animal objects are move on the test platform:
       - Response time is compared to a decline in speed.
       - The ability to respond to adverse movement is tested.
       - The distance at which detected objects are tested.
     - Test 3: Converging data from Test 1 and Test 2 and comparing it according to safety standards to pass/fail testing.

The first prototype of CAVTE is shown in Fig. 34.27 where it used to evaluate the set of CAV functions. Traffic signals are designed to test CAV as shown in Fig. 34.28. The second prototype of CAVTE with AEV was developed as shown in Figs. 34.29 and 34.30.

FIGURE 34.27 Prototype 1.

FIGURE 34.28 Prototype 1: test environment with test traffic signals.

FIGURE 34.29 Autonomous electric vehicle design.

FIGURE 34.30 CAVTE system design.

The developed prototypes I and II of CAVTE are designed and implemented at the Smart Energy Systems Lab, led by Dr. Hossam A.Gabbar, at Ontario Tech University.

34.7 Robot-assisted CAV systems

Human—Vehicle Interaction (HVI) concerns how a human performs tasks inside vehicles, and it has evolved dramatically over recent years. HVI interfaces developed from buttons to composite panels, including speech recognition functionalities, computer vision resources, visual displays, Augmented Reality (AR), until AI and robotics. Referring to the AI robot interfaces for SAV/CAV as In-Car Service Robots (ICSR), there are three essential components: sensing, cognitive, and execution. Thus, ICSR applies to personal or commercial SAV/CAV, in which AI functionalities provide human interaction, identification, and behavioral learning. It does so by collecting data such as image, voice, and GPS. ICSR acts on vehicle control, monitoring, and status, including privacy and security preventive defenses. ICSR solutions can work as driver assistance and for entertainment purposes. Often it can connect to third parties networked applications (e.g., music and video streaming, gaming, and social media). An example of an ICSR interface is the Affective Intelligent Driving Agent (AIDA) from the MIT lab [42]. AIDA companion robot interface helps drivers and serves as an entertainment facilitator for both drivers and passengers. The present research will address the HVI context by proposing a new companion robot interface for SAV/CAV that can provide better context awareness and improve the drivers' decision-making. At the same time that SAV/CAV offers benefits for their users and society, they may also open cybersecurity vulnerabilities. Vulnerabilities include but are not limited to acquisition, storage, and unauthorized disclosure of collected PII.

Companion or social robots can interact with humans by executing tasks adhering to specific social cues and rules [43]. Adopting companion robots as both assistance and entertainment for drivers and passengers is an emerging research topic. They offer human facial, voice, and emotion recognition, including adding human-like personality to the AI to achieve better human—machine communication. A companion robot may serve as the basis for developing more useable/accessible entertainment and safety tools for SAV/CAV. More functional and accessible HVI interfaces are essential and required to provide greater human control over SAV/CAV. Multiple tasks demanded by SAV/CAV may distract drivers' attention. If humans are needed to assume control, their lack of awareness can incur accidents [44]. In the automotive context, poor usability is a matter of consumer satisfaction and human safety [45]. Banks and Stanton [46] indicate significant usability issues in SAV/CAV context. It can impact the level of potential users' trust, which may lead to disuse or misuse. Usage issues include unexpected lane change, system initiation problems, perception of unsafe lane offering, and misunderstanding of the HVI interface. IVCR can overcome these issues by offering a real-time adaptive and context-aware interface for HVI.

The ultimate goal is to ensure that: (i) the supported person recognizes/accepts the robot as a trustworthy companion; and (ii) the information gathering by the robot complies with the security and privacy protection laws. To gain even more context of the user, the social robot may collect and retain historical data on the human such as previous movement patterns, to determine where the human is likely to be at certain times, if the human is traveling, or prior interactions with the service robot such as which content they had previously been interested in. In this context, the privacy requirements are modeled by a three-dimensional privacy infrastructure, i.e., user privacy, services privacy, and data privacy [47], and investigate the privacy requirements in information flow in between the human, CAV, and robot with privacy entities: purpose, recipient, obligation, and retention.

The four types of contextual knowledge model the CAV knowledge on services, contents, and device settings: (1) Environmental context describes the context knowledge derived by situational events on the surrounding environment such as traffic light and sound; (2) User preference knowledge describes each user's attributes and preferences on other types of knowledge; (3) Application-specific context represents available devices, such as sensors and actuators, and their settings and services; and (4) Data-centric context represents various tangible and intangible objects and their attributes [48].

For illustration, Fig. 34.31 introduces a framework of how the robot connects to CAV, based on the security and privacy requirements elicited in this CAV context. It shows the robot as a CAV Assistant Module, which has its robot library module. We aim to enhance the robot's features using a Security Vulnerability Exploration Module, such as Caring Caribou and CANtact, which are both open-source based on ISO 14229. To make the exploration module more robust, we aim to construct a V2IoT Interface Module for the CAV context, based on the service robot's attributes. Furthermore, all connections between the service robot and CAV Module will occur by ISO 15765, regulating CAN bus connections. According to the Driving Module in the CAV Module, the service robot reads and writes the Security and Privacy Module.

34.8 CAV deployment in smart cities

According to McKinsey and Company report published in January 2019, Canada will become one of the world's important autonomous vehicle markets, with revenue from sales of new cars and mobility services. Currently, companies are making efforts in developing technology concerning SAV/CAV that include 5G, lidar sensors, cameras, processors, and software, as well as maps and other GPS services. Together with CAVs have made a revolution in these sectors. CAV can excel numerous advantages for smart city citizens by offering them better and more effective transportation services, such as a dramatic reduction of car crashes and driver fatigue [34,49]. CAV offers benefits for private and public transportation. It includes vehicles as private and service (e.g., Uber), cars, buses for public transport (which provides for school buses), and trucks (e.g., garbage collector and agricultural trucks). The wide variety of CAV makes this topic of research particularly interesting, and its benefits are potentially available for different sections of society.

According to the Society of Automotive Engineers International (SAE), the level of automation of vehicles ranges from Level 0 to 5. Level 0 concerns vehicles that are

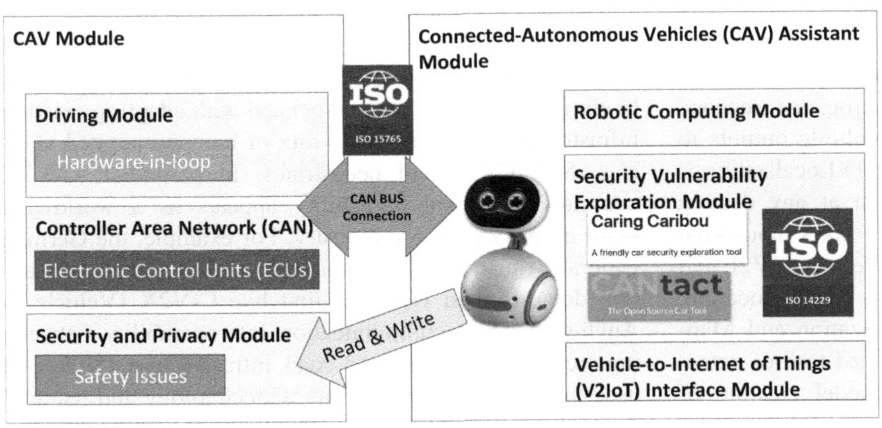

FIGURE 34.31 An Illustrative CAV robot framework.

entirely human-operated, while Level 5 vehicles are fully automated. For example, in Level 1 of automation, the vehicle may assist the driver with tasks such as steering or acceleration. It enables the driver to remain fully engaged with the driving task, but with gradual transfer of control from human to machine. Level 2 automation features include adaptive cruise control and automatic emergency braking. Most of the automated vehicles classify as Level 4 of automation in the industry, while the automotive companies have carefully explored Level 3. There might be a sharing of controlling the machine's driving task and the human under some circumstances. Such cases can be dangerous for both the driver and passengers due to the spare time between controlling exchange and human decision-making. Level 4 of autonomous capability means cars can be self-drive in most conditions without human intervention. However, there are many open design challenges to achieve it, including technical, ethical, and regulatory matters. A completely automated vehicle (Level 5) can perform all driving functions under all conditions. In this situation, humans are just passengers.

In general, CAVs have two main categories: hardware and software. Hardware governs over sensors, V2V, V2G, V2I, and V2X technology, and actuators. Software deals with processes of perception, planning, and control. V2X technology components V2V and V2I allow the vehicle to communicate by receiving information and talking to other systems in the environment. These environmental communicating systems can be other vehicles or even a traffic light that is changing signals. CAV benefits are achieved by collecting relevant information from the CAV's context, such as geolocation, date and time, and other individual attributes such as age, address, gender, and income. Another benefit of CAVs is to support people with disabilities, elderly people, children, and injured people.

CAVs have four functional components, and each component has multiple inputs from associated sensors, performs sensor fusion, and provides multiple outputs to associated functional blocks as follows. (1) Localization is the task of knowing the actual position at any driving condition. To perform this function, CAV takes input from its onboard GPS device, Telemetric service (map), V2I, and V2V commutation service to calculate its current location. They are termed as Simultaneous Localization and Mapping (SLAM). (2) Perception is a combined task of detection, classification, and a virtual 3D model creation of CAV's surroundings. In this task, CAV is assisted by many onboard sensors such as distributed cameras, RADAR, LIDAR, LASER, SONAR, etc. RADAR is a good device for detecting the distance of obstacles on the road in all weathers. However, it has a narrow field-of-view and low detection accuracy and resolution. LiDAR can provide adequate resolution but is affected by bad weather such as snow and rain. Cameras can give useful information about other vehicles and other obstacles on the road. However, its effectiveness will be greatly limited in the bad weather. Sensor fusion can integrate the data from different sensors and provide a better environment model to the decision-making module. Sensor fusion can be performed using AI-based techniques such as deep learning or some traditional techniques such as Kalman filters. Besides, the output from Localization is used in this step. (3) Path planning function comprises three separate functions: Mission Planning, Behavioral Planning, and Motion Planning. In recent years, AI-based algorithms have been developed to solve path planning problems. DRL is superior to the human brain in many applications when making optimal decisions. In contrast to traditional optimal control methods, which need models and a known environment, DRL-based approaches are model-free. They can be applied to an unknown environment and, at the same time, can fully exploit the recent advances in deep learning. A typical task of each function is outlined as follows. (a) The mission planner (or route planner): High-level decisions such as determining pickup destination locations and road selections achieve the target mission. (b) The behavioral planner (or decision-maker): Dynamic ad-hoc decisions such as lane change, intersection crossing, overtake, etc. (c) The motion planner (or local planning): collision avoidance with other vehicles, static and dynamic obstacle avoidance, alarm generation, etc. (4) Control takes inputs from Localization, Perception, and Path Planning process in addition to sensor inputs monitoring road conditions, passenger and driver behavior and comfort, vehicle dynamics, energy management systems, etc. Sensor and Information fusion from all these sources lead to optimal trajectory generation and tracking, path tracking, speed control, cooperative control for platooning, etc.

SAV/CAV's future directly involves the resolution of security, privacy, and safety concerns [33,34]. The goal is to eliminate the problems associated with adapting existing infrastructure to cope with a mix of human-operated vehicles, SAV/CAV, and pedestrians. In particular, C-V2X communication interoperability appears as a worldwide trend in the SAV/CAV industry. For example, the German Federal Ministry of Transportation and Digital Infrastructure demonstrated Europe's first live C-V2X (Vehicle to Anything) direct communication interoperability between motorcycles, vehicles, and road infrastructure [50,51]. It used the American Qualcomm 5G technology and featured Audi Q7 and Audi A4 vehicles and a Ducati Multistrada 1200 Enduro two-wheel vehicle.

34.9 Summary

This chapter presented modeling and simulation of Connected and Autonomous Vehicles (CAVs) toward their deployment in smart cities. Different use cases are analyzed and used as the basis for modeling and simulation of CAV to evaluate their performance in different scenarios and conditions. Control strategies with deep learning are explained and applied on CAV. Communications and cybersecurity layers are illustrated to enable the smooth integration with information infrastructures. In order to assess CAV functions, test platform is discussed to enable cities and regions to evaluate CAV functions prior to deployment. Robot-assisted CAV systems are discussed to show benefits of adding robots to CAV. Finally, detailed analysis of CAV deployment strategies is presented toward practical implementation in smart cities.

Finally, let us move on to the real interactive part of this chapter: review questions/exercises, hands-on projects, case projects, and optional team case project. The answers and/or solutions by chapter can be found in Appendix G.

34.10 Chapter review questions/exercises

True/false

1. True or False? Trust in automated is a continuous concept rather than being a binary variable of trust or not having trust.
2. True or False? As urbanization decreases—an additional 2.5 billion people will live in cities by 2050—cities and suburbs will undergo significant transformations to create sustainable living conditions for their residents.
3. True or False? To study CAV and discuss planning, operation, and management, it is not important to analyze how CAV is deployed in view of a number of possible use cases.
4. True or False? Vehicle platooning occurs when a group of vehicles is in a single lane with close distance and the same velocity.
5. True or False? Changing lanes are common behaviors in night driving.

Multiple choice

1. What is an example of Advanced Driving Assistance Systems (ADAS)?
   a. Adaptive Cruise Control
   b. Autonomous Emergency Braking
   c. Lane Keep Assistant
   d. All of the above
   e. None of the above
2. Referring to the In-Car Service Robots (ICSR), there are three essential components: ______, ______, and ______.
   a. Sensing
   b. Cognitive
   c. Execution
   d. All of the above
   e. None of the above
3. _____ is a good device for detecting the distance of obstacles on the road in all weathers. However, it has a narrow field-of-view and low detection accuracy and resolution. _____ can provide adequate resolution but is affected by bad weather such as snow and rain.
   a. RADAR
   b. Machine-to-human (M2H)
   c. Cyber-physical systems (CPS)
   d. LiDAR
   e. Digital twins (DT)
4. The intersection is a special area that is shared by:
   a. Holistic integration of smart city platforms
   b. Economic growth
   c. Making cities smarter
   d. Computer applications
   e. Two or more roads
5. Obstacle avoidance is the main goal for ______ operating safely.
   a. IC
   b. AV
   c. IT
   d. IP
   e. IS

Exercise

Problem

According to the Society of Automotive Engineers International (SAE), describe the level of automation.

Hands-on projects

Project

Do research: What is Simultaneous Localization and Mapping (SLAM)?

Case projects

Problem

Describe the broad classification of CAV applications (vehicle centric, infrastructure-centric, and traveler-centric).

Optional team case project

Problem

Describe the efforts that support CAV preparatory actions that include: Conducting a stakeholder survey to inform engagement activities to develop strategic partnerships in CAV deployment and conducting a crash analyses for deployment scenarios of CAV-based queue warning systems (QWSs).

Acknowledgment

Special thanks to Hassan Nauman and Chunhui Guo, who supported the work at Smart Energy Systems Lab and participated with Dr. Hossam A. Gabbar in the work related to Sections 34.2, 34.3, and 34.6. Special thanks to Andre de Lima Salgado, Anna Priscilla de Albuquerque Wheler, Manollo Guedes, and Kevin Liu for the research works at the Human Machine Lan with Dr. Patrick C. K Hung in Sections 34.5 and 34.7.

References

[1] Federal Trade Commission (FTC), Protecting Consumer Privacy and Security. Accessed on August 5th, 2019. [Online]. https://www.ftc.gov/news-events/media-resources/protecting-consumer-privacy-security.
[2] Critical reasons for crashes investigated in the national motor vehicle crash causation survey. Washington, DC, USA: Nat. Highway Traffic Safety Admin; February 2015. DOT HS 812 115.
[3] Sen B, Smith JD, Najm GW. Analysis of lane change crashes. Washington, DC, USA: Nat. Highway Traffic Saf. Admin; 2003. Tech. Rep. DOT HS 809 571.
[4] Rasekhipour Y, Khajepour A, Chen S, Litkouhi B. A potential field-based model predictive path-planning controller for autonomous road vehicles. IEEE Trans Intell Transport Syst May 2017;18(5):1255–67. https://doi.org/10.1109/TITS.2016.2604240.
[5] Xu L, Wang L, Yin G. Coordinated control and communication for enhanced safety of highway vehicle platoons. In: Proceedings of the 2013 international conference on connected vehicles and Expo (ICCVE), Las Vegas, NV, USA December 2–6; 2013.
[6] Klomp M, Jonasson M, Laine L, Henderson L, Regolin E, Schumi S. Trends in vehicle motion control for automated driving on public roads. Veh Syst Dyn 2019;57(7):1028–61.
[7] Ioannou PA. Automated highway systems. New York: Springer; 1997.
[8] Sasoh A, Ohara T. Shock wave relation containing lane change source term for two-lane traffic flow. J Phys Soc Japan 2002;71(9):2339–47.
[9] Maciel-Pearson BG, et al. Multi-task regression-based learning for autonomous unmanned aerial vehicle flight control within unstructured outdoor environment. IEEE Robot AutomLett 2019;4(4).
[10] Al-Sharman MK, et al. Deep-learning-based neural network training for state estimation enhancement: application to attitude estimation. IEEE Trans Instrum Measure 2020;69(1).
[11] Kang Y, et al. Deep convolutional identifier for dynamic modeling and adaptive control of unmanned helicopter. IEEE Trans Neur Netw Learn Syst 2019;30(2).
[12] Li Y, et al. Compensating delays and noises in motion control of autonomous electric vehicles by using deep learning and unscented kalman predictor. IEEE Trans Syst Man Cybern 2018;50:4326–38.
[13] Drews P, et al. Vision-based high-speed driving with a deep dynamic observer. IEEE Robot Autom Lett 2019;4(2).
[14] Zhang Q, et al. Deep interactive reinforcement learning for path following of autonomous underwater vehicle. IEEE Access; 2020.
[15] Carlucho I, et al. Adaptive low-level control of autonomous underwater vehicles using deep reinforcement learning. Robotics and Autonomous Systems; 2018.
[16] Sharma S, Tewolde G, Kwon J. Lateral and longitudinal motion control of autonomous vehicles using deep learning. In: IEEE International Conference on Electro Information Technology; 2019.
[17] Yang S, et al. Scene understanding in deep learning-based end-to-end controllers for autonomous vehicles. IEEE Trans Syst Man Cybern 2019;49(1).
[18] Zhu J, et al. Hierarchical decision and control for continuous multitarget problem: policy evaluation with action delay. IEEE Trans Neur Netw Learn Syst 2019;30(2).
[19] ISO 15007-1. Road vehicles — measurement of driver visual behaviour with respect to transport information and control systems — Part 1: definitions and parameters. 2014.
[20] ISO. ISO/TR 20545:2017(en). Intelligent transport systems — vehicle/roadway warning and control systems — report on standardization for vehicle automated driving systems (RoVAS)/Beyond driver assistance systems. 2017.
[21] ISO 14229-1:2013 Road vehicles – Unified diagnostic services (UDS) – Part 1: specification and requirements.
[22] ISO 15765-2:2016 road vehicles – diagnostic communication over controller area network (DoCAN) – Part 2: transport protocol and network layer services.
[23] ISO 9241-210:2019 ergonomics of human-system interaction – Part 210: human-centred design for interactive systems.
[24] ISO/TC 204. Intelligent transport systems.
[25] Heikoop DD, Winter JC Fd, Arem Bv, Stanton NA. Psychological constructs in driving automation: a consensus model and critical comment on construct proliferation. Theor Issues Ergon Sci 2016;17(3):284–303.
[26] Hancock PA, Billings DR, Schaefer KE. Can you trust your robot? Ergon Des 2011;19(3):24–9.
[27] Schaefer KE, Chen JYC, Szalma JL, Hancock PA. A meta-analysis of factors influencing the development of trust in automation: implications for understanding autonomy in future systems. J Hum Fact Ergonom Soc March 2016;58(3):377–400.
[28] Gold C, Körber M, Hohenberger C, Bengler K. Trust in automation – before and after the experience of take-over scenarios in a highly automated vehicle. Proced Manuf 2015;3:3025–32.
[29] Sanchez J, Fisk AD, Rogers WA. Reliability and age-related effects on trust and reliance of a decision support aid. In: Proceedings of the human factors and ergonomics society 48th annual meeting. California, USA: Santa Monica; 2004. p. 586–9.
[30] Jin L, Tefft BC, Horrey WJ. Mining consumer complaints to identify unsuccessful interactions with advanced driver assistance systems. In: Proceedings of the 11th international conference on automotive user interfaces and interactive vehicular applications; September 2009. p. 71–5.
[31] Basu C, Yang Q, Hungerman D, Sinahal M, Draqan AD. Do you want your autonomous car to drive like you?. In: Proceedings of the 12th ACM/IEEE international conference on human-robot interaction (HRI 2017); March 2017. p. 417–25.
[32] Lazanyi K, Maraczi G. Dispositional trust—do we trust autonomous cars?. In: Proceedings of the 2017 IEEE 15th international symposium on intelligent systems and informatics (SISY); September 2017. p. 135–40.
[33] Thing VL, Wu J. Autonomous vehicle security: a taxonomy of attacks and defences. In: Proceedings of the 2016 IEEE international

conference on internet of things (iThings) and IEEE green computing and communications (GreenCom) and IEEE cyber, physical and social computing (CPSCom) and IEEE smart data (SmartData); December 2016. p. 164–70.

[34] Bansal P, Kockelman KM. Forecasting Americans' long-term adoption of connected and autonomous vehicle technologies. Transport Res Pol Pract 2017;vol. 95:49–63.

[35] Millard-Ball A. Pedestrians, autonomous vehicles, and cities. J Plann Educ Res 2018;38(1):6–12.

[36] Ministry of Transportation of Ontario, Road safety: emergency vehicles. Accessed on April 30, 2020. [Online]. Available: http://www.mto.gov.on.ca/english/safety/emergencyvehicles.shtml.

[37] Paci F, Squicciarini A, Zannone N. Survey on access control for community-centered collaborative systems. ACM Comput Surv 2018;51(1):6.

[38] De Luca A, von Zezschwitz E. Usable privacy and security. It-Information Technology 2016;58(5):215–6.

[39] Oates M, Ahmadullah Y, Marsh A, Swoopes C, Zhang S, Balebako R, Cranor LF. Turtles, locks, and bathrooms: understanding mental models of privacy through illustration. Proc Pri Enhanc Technol 2018;4:5–32.

[40] Garfinkel S, Lipford HR. Usable security: history, themes, and challenges. Synth Lect Inform Secur Priv Trust 2014;5(2):1–124.

[41] Bertino E. Data security and privacy: concepts, approaches, and research directions. In: Proceedings of 2016 IEEE 40th annual computer software and applications conference (COMPSAC), vol. 1; 2016. p. 400–7.

[42] Williams KJ, Peters JC, Breazeal CL. Towards leveraging the driver's mobile device for an intelligent, sociable in-car robotic assistant. In: Proceedings of the 2013 IEEE intelligent vehicles symposium (IV); June 2013. p. 369–76.

[43] Yan H, Ang MH, Poo AN. A survey on perception methods for human-robot interaction in social robots. Int J Soc Robot 2014;6(1):85–119.

[44] Xu N, Guo G, Lai H, Chen H. Usability study of two in-vehicle information systems using finger tracking and facial expression recognition technology. Int J Hum Comput Interact 2018;34(11):1032–44.

[45] Harvey C, Stanton NA. Usability evaluation for in-vehicle systems. CRC Press; 2016.

[46] Banks VA, Stanton NA. Keep the driver in control: automating automobiles of the future. Appl Ergon 2016;53:389–95.

[47] Rezgui A, Ouzzani M, Bouguettaya A, Medjahed B. Preserving privacy in web services. In: Proceedings of the 4th international workshop on Web information and data management; 2002. p. 56–62.

[48] Rhee SK, Lee K, Kim H. Ontology-based context and preference model for personal service robot. In: Proceedings of the 2012 9th international conference on ubiquitous robots and ambient intelligence (URAI); 2012. p. 216–7.

[49] Johns M, Mok B, Sirkin D, Gowda N, Smith C, Talamonti W, Ju W. Exploring shared control in automated driving. In: Proceedings of the 2016 11th ACM/IEEE international conference on human-robot interaction (HRI); March 2016. p. 91–8.

[50] European Commission, Information Providers Guide: The EU Internet Handbook. Accessed on August 5th, 2019. [Online]. https://ec.europa.eu/ipg/standards/accessibility/index_en.htm.

[51] General Data Protection Regulation (GDPR). Accessed on August 5th, 2019. [Online]. https://gdpr-info.eu/.

Further reading

Wen-Xing Z, Li-Dong Z. A new car-following model for autonomous vehicles flow with mean expected velocity field. Phys Stat Mech Appl 2018;492:2154–65. https://doi.org/10.1016/j.physa.2017.11.133.

Rosolia U, De Bruyne S, Alleyne AG. Autonomous vehicle control: a nonconvex approach for obstacle avoidance. IEEE Trans Contr Syst Technol March 2017;25(2):469–84. https://doi.org/10.1109/TCST.2016.2569468.

Yoon J, Kim B. Vehicle position estimation using nonlinear tire model for autonomous vehicle. J Mech Sci Technol 2016;30(8):3461–8. https://doi.org/10.1007/s12206-016-0705-5.

Liu Y, Wang X, Li L, Cheng S, Chen Z. A novel lane change decision-making model of autonomous vehicle based on support vector machine. IEEE Access 2019;7:26543–50. https://doi.org/10.1109/ACCESS.2019.2900416.

Jardine PT, Givigi SN. A robust model-predictive guidance system for autonomous vehicles in cluttered environments. IEEE Syst J 2019. https://doi.org/10.1109/JSYST.2018.2806629.

Wu W, Liu Y, Xu Y, Wei Q, Zhang. Traffic control models based on cellular automata for at-grade intersections in autonomous vehicle environment. J Sens 2017:1–6. https://doi.org/10.1155/2017/9436054.

Sun C, Zhang X, Zhou Q, Tian Y. A model predictive controller with switched tracking error for autonomous vehicle path tracking. IEEE Access 2019;7:53103–14. https://doi.org/10.1109/ACCESS.2019.2912094.

Shim T, Adireddy G, Yuan H. Autonomous vehicle collision avoidance system using path planning and model-predictive-control-based active front steering and wheel torque control. Proc Inst Mech Eng - Part D J Automob Eng 2012;226(6):767–78. https://doi.org/10.1177/0954407011430275.

Moridpour S, Rose G, Sarvi M. Effect of surrounding traffic characteristics on lane changing behavior. J Trans Eng. 2010;136(11):973–85.

Eckermann E. World history of the automobile. Training 2011:4–20. 2011.

Adnan N, Vasant PM, Rahman I, Noor A. Adoption of plug-in hybrid electric vehicle among Malaysian consumers. Indus Eng.Manag 2016;5(185):316–2169.

Hashim HH, Omar MZ. Towards autonomous vehicle implementation: issues and opportunities. J Soc Autom Eng Malay 2017;1(2):111–23.

Litman T. Autonomous vehicle implementation predictions. Victoria Transport Policy Institute; 2017.

P. Gao, H. W. Kaas, D. Mohr, and D. Wee, Automotive revolution–perspective towards 2030 How the convergence of disruptive technology-driven trends could transform the auto industry, Advanced Industries, McKinsey & Company.

Heineke K, Kampshoff P, Mkrtchyan A, Shao E. Self-driving car technology: when will the robots hit the road? McKinsey Global Institute; 2017. http://www.Mckinsey.Com/Industries/Automotive-and-Assembly/Our-Insights/Self-Driving-Car-Technology-When-Will-the-Robots-Hit-the-Road.

Ioannou PA, Chien CC. IEEE Trans Veh Technol 1993;42:657.

Holland EN. Trans Res B 1998;32:141.

Hasebe K, Nakayama A, Sugiyama Y. Phys Rev E 2003;68:026102.

Bose A, Ioannou PA. IEEE Trans Intell Trans Syst 2003;4:173.

Jin WL, Recker WW. Trans Res B 2006;40:230.

https://cdn.sparkfun.com/datasheets/Sensors/Proximity/HCSR04.pdf.

https://cdn.sparkfun.com/datasheets/Dev/RaspberryPi/RPiCamMod2.pdf.

Asvadi A, Garrote L, Premebida C, Peixoto P, Nunes UJ. Multimodal vehicle detection: fusing 3D-LIDAR and color camera data. Pattern Recogn Lett 2018;115:20–9. https://doi.org/10.1016/j.patrec.2017.09.038.

https://cnib.ca/.

https://www.canada.ca/content/dam/esdcedsc/migration/documents/eng/disability/arc/reference_guide.pdf.

The state of autonomous legislation in Europe, Autovista Group. [Online]. Available: https://autovistagroup.com/news-and-insights/state-autonomous-legislation-europe. [Accessed: May 16, 2019].

https://autovistagroup.com/news-and-insights/state-autonomous-legislation-europe.

Accessibility must be at the heart of new transport tech, GOV.UK. [Online]. Available: https://www.gov.uk/government/news/accessibility-must-be-at-the-heart-of-new-transport-tech. [Accessed: May 16, 2019].

https://www.gov.uk/government/news/accessibility-must-be-at-the-heart-of-new-transport-tech.

D. for Transport. Trialling automated vehicle technologies in public. GOV.UK; February 6, 2019 [Online]. Available: https://www.gov.uk/government/publications/trialling-automated-vehicle-technologies-in-public. [Accessed 16 May 2019].

https://www.gov.uk/government/publications/trialling-automated-vehicle-technologies-in-public.

Huang S, Huang S. How the autonomous car works: a technology Overview. Medium; April 25, 2018 [Online]. Available: https://medium.com/@thewordofsam/how-the-autonomous-car-works-a-technology-overview-5c1ac468606f. [Accessed 16 May 2019].

https://medium.com/@thewordofsam/how-the-autonomous-car-works-a-technology-overview-5c1ac468606f.

Liu Y, Wang X, Li L, Cheng S, Chen Z. A novel lane change decision-making model of autonomous vehicle based on support vector machine. IEEE Access 2019;7:26543–50.

https://www.gm.com/content/dam/company/docs/us/en/gmcom/gmsafety-report.pdf.

Kulisch E. Reimagining federal safety standards. Automotive News; January 28, 2019 [Online]. Available: https://www.autonews.com/mobility-report/reimagining-federal-safety-standards. [Accessed 31 May 2019].

https://www.autonews.com/mobility-report/reimagining-federal-safety-standards.

Matthew.lynberg.ctr@dot.gov. Automated vehicles for safety. NHTSA; November 28, 2018 [Online]. Available: https://www.nhtsa.gov/technology-innovation/automated-vehicles-safety. [Accessed 31 May 2019].

https://www.nhtsa.gov/technology-innovation/automated-vehicles-safety.

How the steering system works, How a Car Works. [Online]. Available: https://www.howacarworks.com/basics/how-the-steering-system-works. [Accessed: May 31, 2019].

https://www.howacarworks.com/basics/how-the-steering-system-works.

Dastin J. Amazon's latest gadget: a self-driving toy car for coders. Reuters; November 29, 2018 [Online]. Available: https://www.reuters.com/article/us-amazon-com-selfdriving/amazons-latest-gadget-a-self-driving-toy-car-for-coders-idUSKCN1NX2LQ. [Accessed 12 June 2019].

https://www.reuters.com/article/us-amazon-com-selfdriving/amazons-latest-gadget-a-self-driving-toy-car-for-coders-idUSKCN1NX2LQ.

Maurer M, Gerdes JC, Lenz B, Winner H. Autonomous driving technical, legal and social aspects. Berlin: Springer Berlin; 2018.

https://www.researchgate.net/publication/303480945_Use_Cases_for_Autonomous_Driving.

Chapter 35

VUCA: volatility, uncertainty, complexity, and ambiguity in our smart city and tech finance future

John Sanphillippo
Granola Shotgun, San Francisco, CA, United States

35.1 Introduction

The smart city concept is first and foremost about creating and extracting value from the landscape. Any discussion on the topic needs to start there. What that value looks like and who does the creating and extracting is still an open question. A smart city is one in which things are tagged, tracked, analyzed, priced, and actively managed in real time in order to optimize specific outcomes in the most efficient manner. Such systems can be specifically designed for energy and water conservation, streamlined movement of vehicles, risk mitigation, effective distribution of products and services, revenue enhancement, or molding attitudes and behavior. As tech is deployed to wring out the value in all our systems, the built environment is going to shift in subtle but important ways. That process is likely to involve VUCA: volatility, uncertainty, complexity, and ambiguity [1].

35.2 What is new is old again

A close cousin of the smart city framework is digital finance. Banking, insurance, credit, and retail investment services are now available via mobile devices in a broadly accessible virtual environment. Point-of-sale transfers, contractual agreements, leases, fees, penalties, loans, tax collection, and so on can all be managed almost anywhere at any time in the great database ether. When coupled with smart city techniques, the ability to fine-tune and regulate the environment is amplified. The focus is always on drawing out value from each transaction. There is no incentive to invest in these elaborate systems if they do not somehow generate a return.

It is important to be clear about what tech is and what it is not. All our computerized bells and whistles are tertiary layers of organization superimposed over the existing underlying primary structures of civilization. Tech is not food. It is not fuel. Tech is not manufacturing per se. It is not a means of moving anything from point A to point B. Tech may enhance these things, but all our sophisticated interconnected machines are themselves dependent on lower levels of physical extraction, production, and distribution. And all that complexity has a metabolic cost above and beyond what was previously required to operate society.

A Tesla [2] is a slick incremental refinement of a Model T Ford [3] from a century ago. Google is an advertising company with a side hustle as a reference library [4]. Facebook is a low-grade tabloid newspaper populated by volunteers [5]. Amazon [6] is merely the latest version of the mail order Sears catalog circa 1897 [7]. Palantir [8] is a fresh take on Pinkerton's private investigators and strong men for hire from 1850 [9]. Apple makes stylish reinterpretations of telegraphs, televisions, cameras, telephones, typewriters, radios, and wrist watches [10]. They are cool, but they are not fundamentally different from their analog forebears. Taxis, home delivery of Chinese food and pizza, and renting spare rooms have all been around a lot longer than the mobile apps and Internet platforms that claim to be radically transformational. Technology is being used to consolidate market share, induce vertical integration, and capitalize on economies of scale. This is not a criticism of technology, just a clarification in case anyone has drifted too far away from external reality.

Digital Finance is the most recent version of how humans structure economic relations [11]. Online payment systems and cryptocurrencies are not intrinsically different from cowry shells, cuneiform tablets, silver coins, tally sticks, paper currency, or credit cards. These were all innovative and transformative in their day, but it is hubris to

Smart Cities Policies and Financing. https://doi.org/10.1016/B978-0-12-819130-9.00055-3

believe our present electronic accounting techniques are different. They are not. They all come down to tokens used to record and trade human promises—nothing more or less. Financial innovation can provide a reliable store of value, serve as a trusted intermediary between strangers, pull consumption forward with debt, cultivate reliable rents, facilitate tax collection, and generate interest payments on investment capital. There is nothing new about any of that. My apologies to Satoshi Nakamoto [12] acolytes, but everything else is just geeky fluff and speculative fervor.

The speed, power, and level of detail accompanying smart city protocols may allow society to do things it could not previously do. But that is a trap. Historically, our collective expectations rapidly adjusted upward to meet the new capabilities and fill the space that might otherwise have provided breathing room. For example, the miracle technologies of the so-called Green Revolution of the mid-20th century were intended to alleviate hunger and malnutrition [13]. Synthetic nitrogen fertilizers, pesticides, herbicides, hybrid seeds, mechanical irrigation, diesel tractors and combines, and so on all boosted food production to a remarkable degree. The resulting dramatic increase in global food supply drove an equally impressive increase in the human population. We grew from two and a half billion in 1950 to nearly eight billion just 70 years later, and that number is still rising [14].

Unfortunately, the hunger problem has not changed all that much. In 1970, when the tangible effects of the Green Revolution began to kick in, 34% of the world was chronically undernourished [15] while today it is just under 9% [16]. That sounds like a win by percentage, but the absolute number of malnourished people has remained roughly the same and that figure has been increasing year on year [17]. Technology improved, the problem of production was solved, but we took that immense amount of additional food and used it to make more people without actually feeding everyone. Meanwhile the new much larger population must continue to be fed with additional marginal production techniques. That is a consequence of human nature, not technology. Inventing more machines and chemicals will not change the way we use them.

Physical growth has been America's economic engine from the very beginning. Each era has had its own transformational technology and financial innovation to fund and manage it, each with an attendant boom and bust [18]. Poke around the existing landscape and there is plenty of evidence of previous infrastructure that shaped the built environment. Older towns have the vestiges of a river or sea port, a barge canal, and a rail line. These are often parallel to, or buried underneath, a modern highway. Sometimes all of the above are colocated along an old postal service route from the 18th century. Explore back alleys in forgotten neighborhoods, and you might spot the dregs of telegraph wires, landline public telephone booths, and C-Band satellite dishes the size of gazebos [19]. You will find them next to abandoned train stations and blacksmith shops or near a vacant video rental store at a dead mall. What these bygone enterprises and devices have in common is that they took cheap, low productivity land and added value to it, at least until the next new thing came along and displaced it.

Each new wave came with its own political shenanigans as well as often violent labor negotiations and pulses of immigration [20]. The details may be slightly different, but these are fundamentally the same issues we see in today's economic and political dynamics. The financial panic of 1837 was the result of speculative overinvestment in land, cotton, and slaves, much of it associated with canal construction into new territories [21]. The Panic of 1873 was the result of a protracted frenzy in railroad stocks, which was itself heavily intertwined with territorial westward expansion [22].

The annexation of Alaska [23] and Hawaii [24] in 1959 ended the official physical extension of the United States. Meanwhile, the US share of the global economy contracted from 40% in 1960 to 24% in 2019 [25] as the rest of the world grew and new powers emerged. The days of the Pax Americana [26] and de facto global empire are numbered. All our institutions are finely tuned to manage perpetual growth, but outward future expansion is becoming increasingly unlikely in a crowded world with aggressive competitors. America is having to turn inward now to find new value in what already exists, and it is going to be a hard squeeze. This is what the smart city is really all about.

Architecture firms and tech entrepreneurs love to present glossy graphics of green roofs, autonomous vehicles, solar panels, and healthy people in a safe, well-run futuristic city. It is one big happy Singapore or Copenhagen—a living breathing version of Richard Brautigan's poem, "All Watched Over By Machines of Loving Grace" [27]. What we are far more likely to get is the same old suburban strip mall, drive-thru, office park, and cul-de-sac terrain we have inherited from the late 20th century that has been wrung dry by tech for all sorts of uncoordinated parochial and proprietary reasons. Eventually these random systems will congeal into a novel, but deeply familiar set of interlocking organizations in a lumpy atavistic world of winners and losers.

In the American context, a smart city is unlikely to be implemented as part of a micromanaged top-down consolidated government plan. There is a long history of using corporate intermediaries to promote public policy indirectly with heavy subsidies and incentives in a clumsy, uneven manner [28]. American culture has conflicting values and impulses that are reflected in the general inability to compromise and resolve these differences coherently. So we do the easy thing by choosing not to resolve them. In the spread between all the political posturing and morality costume dramas, there is plenty of profitable wiggle room for private companies to insert themselves. That is a feature, not a bug. Tech companies are already following that pattern in search of yield.

Everything has a beginning, a middle, and an end. Smart cities will attempt to reconcile the underlying insolvencies of the aging post World War II development pattern as well as emerging new urban agglomerations through optimization protocols [29]. That is one way of describing rationing. The accumulated costs of perpetuating our physical infrastructure and associated soft obligations of salaries, pensions, and subsidies are increasingly burdensome. But the voting public has no interest in realistically addressing these profound funding gaps by actually paying for them [30]. Our present national landscape is both culturally sacred (the rugged individualist narrative) and structurally bankrupt (the cash flow is not there to maintain the basic support structures.) So technology will be deployed to perform triage in a manner that is socially acceptable and fits with the American ethos. What will that look like on the ground? Spoiler alert: this stuff is not sexy.

35.3 Flows of sewage, streams of revenue

Philadelphia had a problem. Its sewers are among the oldest in the nation and had an antiquated combined sewage and stormwater system [31]. Every time there was a heavy rain, the sewage treatment plants were overwhelmed and untreated effluent was released into the rivers [32]. Aside from the unpleasantness and health risks of the contamination itself, both state and federal regulators imposed fines and penalties for such overflows. The cost of bringing the entire system up to modern standards was exorbitant, and the political will associated with short-term election cycles was unable to resolve the issue. So a tech solution was implemented [33].

A private consulting firm digitized aerial images of the entire city [34], cross-referenced them with the city's property database, and used specialized algorithms to analyze each lot [35]. Every lot was then assessed a customized fee based on impervious surface area. Green fields that absorb stormwater were minimally impacted by the fee. Roofs, concrete, and asphalt paving were assessed a per square foot fee to generate the necessary revenue for stormwater remediation. This nuanced approach would not have been feasible in the past because manually observing and recording each of the thousands of individual city lots simply was not cost-effective. An algorithm did the job quickly and efficiently.

Many of the largest properties in the city are tax exempt such as schools, churches, and airports. Each of these entities has a substantial amount of hard surfaces that could not be taxed as property per se. So the stormwater surcharge is a user fee, not a tax [36]. This legal distinction was critical for the success of the program. An acre covered by a skyscraper paid the same stormwater surcharge as an acre of parking lot because they each had the same impact on water management. This rubbed a lot of people the wrong way. The first instinct was for business owners in Philadelphia to try and relocate to a suburban district that did not have this fee structure, but a significant number of outlying towns have since adopted similar programs. This prompted many property owners to dub it the "rain tax" [37]. People want flood control and clean water, but do not want to pay for it. The smart city approach bridges that gap.

The best way to mitigate the rain tax is to break up the concrete and asphalt on a property to allow stormwater to percolate into the earth. Alternatively, a low value parking lot might be converted into a high value building that generates more income to offset the fixed fee. It is a pro urban and pro rural fee structure. On the other hand, a low value yet highly impervious paved environment is the most heavily impacted by the fee since it is charged the most while having the least ability to create value to cover the cost of stormwater management.

That same basic smart city process is about to be used for any number of public safety and revenue enhancement efforts. Overhead images of a given municipality can be cross-referenced with building permits, inspections, and land use regulations. A computer-generated list of anomalies created by a private company might be shared with municipal officials. Was that addition on the back of the house on Apple Street registered with City Hall? Does that overly large shed on Cherry Lane meet code and conform to setback requirements? Does Mrs. Smith have more than the three legally permitted dogs on her property? If not there is a matter of penalties, fees, back taxes, and the need to bring these properties and residents up to standards. One possible business model might be for a private company to provide the analytic services up front and be paid a share of the revenue extracted from violations on the backend. What municipal administrator does not want free easy money while asserting that public safety has been improved and scofflaws have been brought to justice?

Thermal images [38] or metadata from cell phones and other devices [39] might detect the presence of people occupying illegal apartments in attics, garages, and basements or unauthorized enterprises that violate zoning regulations or other laws. Wastewater discharges can be monitored electronically to detect the presence of controlled substances. For example, the city of Tempe, Arizona, already monitors sewage citywide for Fentanyl, Oxycodone, Codeine, Heroin, Norfentanyl, and 6-Acetylmorphine [40]. People will undoubtedly cry foul at the Nanny State [41] and Surveillance Capitalism [42], but they will have been caught breaking the law. The "meta" part of these systems is about identifying broad trends without being so specific as to require a search warrant [43], but to be specific enough to establish patterns of behavior that achieve probable cause [44].

It is possible that private organizations such as Homeowners Associations (HOA) might be the most aggressive adopters of such methods since they are voluntary contractual entities rather than government agencies. The people who self-select into such arrangements are often looking for an age-restricted community or gated environment to filter out people and activities they do not approve of. They also frequently wish to escape so-called social engineering such as the imposition of inclusive affordable housing or homeless shelters, yet insist on rigorously enforcing highly detailed private codes in order to maintain property values. Smart city techniques are tailor-made to serve this massive private market, largely outside the legal confines of mere municipalities. The same demographic that might chafe at the government monitoring their personal behavior [45] may readily agree to private scrutiny of themselves and their neighbors with a chorus of "Think of the children!"

The question is not whether or not society will be engineered. The engineering is already here. It is about who decides on the parameters and end goals. It is mostly about the illusion of control and the need to be discerning about who is monitored and why. There are all sorts of mundane business opportunities to squeeze the built environment for a solid revenue stream while maintaining the desired cultural outcome. Will these tools be used intentionally to exclude "the wrong element" from select jurisdictions? Absolutely. We should expect many such programs to be implemented moving forward and for tech companies and governance structures alike to embrace them.

35.4 All drivers are above average

The current national development pattern is sometimes referred to as "drive till you qualify" [46]. In order to afford a newer larger home in a better neighborhood with lower taxes, the typical American will simply go a little farther down the highway. Businesses similarly migrate outward in search of lower cost property, cheaper labor, fewer regulations, lower taxes, and the ever-present economic development subsidies [47]. Meanwhile, many older towns quietly age and become more expensive to keep up and less desirable to live in as legacy costs accumulate and revenues stagnate or decline. So the American periphery is continually expanding leaving behind a semidisposable wake of insolvency. For each individual the extra time spent on the road is well worth the quest for the next shiny new place.

But it is going to be tricky to continue this practice once the algorithms begin to tighten things up. The promise of the smart city is that everything will be registered and priced in real time. That could subtly change the calculus of which places will be economical and desirable and which will be left behind. Metromile [48], Milewise [49], and Smartmiles [50] are all little pieces of the emerging smart city future that will have an unexpected influence on land use and property values. Today insurance companies offer a discount for drivers who voluntarily have their cars chipped and tracked. The early adopters are people who do not drive very much and believe themselves to be cautious and responsible. It will not be long before the average driver will be asked to pay extra to opt out of such plans. And eventually these billing systems will be mandatory for all drivers. People who attempt to drive without insurance, of which there are many, will be quickly identified and stopped, since other cars and roadside devices will communicate with each other, as well as the relevant authorities.

We all occasionally make illegal left hand turns. We all speed a little, or a lot. We are all more distracted more often than we realize. And many more of us are on the road when we should not be given our excess testosterone, advanced age, youthful inexperience, poor eyesight, drowsiness, level of intoxication, texting, and so on. At the moment, the only time this matters is after we have been noticed by the police or are involved in an accident. But those electronic trackers are not just measuring how many miles are driven. They are also keeping tabs on the quality of the driving as well as where and when the driving happens. Each little infraction will be noted and charged to the customer. Ding, ding, ding. The roads are about to get a whole lot safer. That is good news for society, and it boosts insurance industry profits by saving them a lot of money. Win win. But it is also going to take a lot of drivers out of the game. At the high end of the income distribution, a little extra insurance is of no consequence. At the lower end of the wage scale, it is the end of driving altogether.

The tech solution to human drivers is assumed to be the autonomous vehicle [51]. Luxury computerized chauffeurs and robotic tractor trailers are widely expected to take over the roadways over time. And perhaps they will. But the same systems that serve public safety and insurers will be pressed into service to monetize the use of public roads. America's municipalities are structurally insolvent [52]. Even prosperous places that have no trouble paying their bills at the moment have long-term obligations such as employee pensions and physical maintenance that simply cannot be met over the long term. Almost all municipal governments would experience extreme hardships if state and federal subsidies and transfer payments were withdrawn. This is equally true of big cities, suburbs, and rural outposts. The entire national administrative establishment has become dependent on a byzantine collection of interlocking and opaque programs that mask the true cost of everything and are themselves predicated on promises that Washington may not always be able to keep.

35.5 Repaving paradise, potholes in parking lots

The federal Interstate Highway System [53], which is separate from state and local road authorities, was established with a pay-as-you-go funding scheme in 1956 [54] using gasoline and diesel taxes, taxes from the sale of heavy trucks, and tire taxes. That was the theory at least, but the dirty little secret is that roads rarely ever really pay for themselves. Even when the amount of highway was modest and new and the revenue was strong, construction costs were a bit higher than incoming funds and the feds helped fudge the numbers year on year. There was a period of ticking along for a few decades that worked pretty well until the ever-expanding system began to age and deteriorate and the Interstate fund started to seriously dwindle just as costs spiked. Raising taxes to keep up with the need for maintenance was out of fashion by then. The last bump up on the federal gas tax was in 1993 [55], so in real terms the gas tax is now at 54% of its previous levels due to 28 years of inflation [56]. No one wants to pay more for gas, but everyone wants pristine roads.

Beginning in 2008 [57], money from the general federal budget was repeatedly transferred to prop up the Interstate system. No one seemed to notice because it was not a new tax, but rather a de facto increase in the national debt. This can continue for a while, but sooner or later there is going to be fiscal reckoning. We either need to pay for things for real, or we need to allow the roads to fail. Once again, tech will be pressed into service to make these hard choices.

Looking back to older technologies and infrastructure, there is a clear pattern. Today we continue to transport more tonnage of material by barge [58] and rail [59] than ever before. We just do so selectively on the limited bits of the old systems that have proven consistently productive. The remaining viable routes were preserved because they still cash flow. Raw timber, iron ore, coal, petroleum, gravel, grain, and other bulk commodities continue to travel most economically by inland waterways and rail. These happen to be items that are not in a hurry and do not need delicate care along the way.

In previous centuries, canals and railroads were so wildly overbuilt in a speculative flurry that the nonproductive spurs really needed to be let go. The cost of maintaining them exceeded their revenue and utility to society. Once the infrastructure failed, nearby property often lost its productive capacity and value too. We are about to see the same dynamic play out with our paved roadways, underground pipes, and overhead wires. The next shift is already upon us, and smart city technology is the mechanism that will perform the ruthless editing.

Enter the pay-per-mile model of public infrastructure accounting. Cell phone locations already provide detailed proxy data about how many vehicles pass over each little stretch of pavement at any given moment [60]. Many bridges and highways are already tolled, and most of those toll collection systems converted to automated transponders years ago. Some existing toll systems already use optical recognition software to charge vehicles based on the number of passengers in each vehicle [61]. And the Oregon Department of Transportation currently has a statewide system of per mile road usage fees [62]. It will not be long before each car and truck is charged a fee for each mile of every road it uses everywhere at all times. This is the smart city in action.

From a cultural perspective, one impetus for the pay-per-mile model will be the increase in electric vehicles that do not pay any gasoline tax. The percentage of electric vehicles is trivial at the moment, but public opinion against "freeloaders" will be levered in favor of universal road usage fees, possibly with a gas tax rebate such as Oregon's. But it will also make the full cost of driving immediately visible to drivers, which has not been true in the past. Soon, before any vehicle leaves the driveway or loading dock, there will be a display describing the cost in road fees for each potential journey. In addition, there may also be automated real-time feedback along the way from insurance providers and law enforcement. Again, this may have no impact at all on prosperous drivers, but it will hit the marginal middle and lower classes significantly enough to change their behavior.

This model for funding public infrastructure will also provide local officials with an accurate understanding of where their revenue is coming from and which segments of the system are worth maintaining with scarce resources. It will not matter if the cars are driven by humans or guided by automated systems. Driving is going to be more expensive for a long list of independent reasons that converge. Perhaps the cul-de-sacs that only ever see a dozen vehicles a day will not be given a million dollar resurfacing after all. These are really just collective driveways for a handful of homes. No one should be driving more than 10 or 15 miles an hour in these zones anyway, so gravel might just be the better surface. Property owners may object and demand fully paved side streets, and some may continue to receive them. But many others will not.

The frontrunners of this dynamic can be seen in any number of privately maintained areas. A sad mobile home park might have let the asphalt break up and turn to muddy ruts over the decades. Meanwhile, the owners of exclusive upscale equestrian-themed enclaves sometimes chose to keep their roads minimal and rustic with decorative pea gravel for both economic and aesthetic reasons. A gravel road can be a sure sign of poverty or a status symbol. It all depends on context and demographics. When private entities are tasked with paying the bills themselves, they often make rather pragmatic decisions.

The fire marshal will no doubt have something to say about all this since the minimum road width and lollipop shape of the average suburban cul-de-sac are dictated by the turn radius of the largest fire engine in service [63]. The compromise might be a mesh substrate within the gravel that provides a firm drivable surface for emergency vehicles while also allowing for natural drainage [64]. This will do double duty when the streets themselves are reclassified and assessed for stormwater utility fees, particularly in private gated subdivisions. All the little smart city pieces will come together haphazardly at first, but in time will relentlessly coalesce.

The majority of the American population will continue to live in diffused suburban environments because that is all that exists in most places, and we are now path-dependent on this model. Building anything new or wildly different from scratch is not an economical option and probably will not fly with a public that is deeply wedded to the American Dream. It takes years and an army of professionals to build much of anything within the existing institutional framework, even in the places that proclaim their progrowth business-friendly bonafides. But installing a few bits of interconnected technology here and there and spinning off some steady cash flow is relatively easy. Why attempt to have jackbooted thugs install spy cameras in every home when people willingly and freely install them themselves for $39 on sale with free delivery? The smart city is not about bold futuristic architecture and radical social change. Instead, it is highly recursive.

There is likely to be a process of incessant adaptation-in-place as the suburbs are used more intensely in some areas and depopulated in others. There is nothing new about this process as our empty farm towns and abandoned Rust Belt cities attest. Owning three or four vehicles per household will incrementally give way to the on-demand group minivan shuttle service because it gets the job done at a lower price point relative to insurance, road fees, fuel, and so on. The "on demand" part might not be as equally instantaneous in all locations once the system is optimized. Some places will be better positioned than others. These services will bridge the gap between the beloved private vehicle and detested public transit because the shuttle service will actively vet and filter passengers, thereby preserving class distinctions. The minivan shuttle will be the sweet spot for maximum use at minimum operating cost with every seat occupied for most trips in a segregated cultural milieu.

35.6 Complexity meets complications

In places with cyclical droughts, there are residents, farmers, and industries that protest the rationing of water and tiered rate structures that come at times of critically low supplies. In some instances, property owners have been legally guaranteed a particular amount of water at a fixed price in perpetuity, but that water may not actually physically exist. In other circumstances, it is illegal for utilities to charge more than the direct cost of providing water in the present moment, regardless of the longer-term consequences. Lawsuits over usage rates are a perennial occurrence in dry western states. Populations continue to grow in arid regions where the amplitude of climate variability is increasing with higher highs and lower lows. Even places with abundant water are hitting limits associated with the quality of their water and the cost of managing delivery systems. A huge opportunity is presenting itself for tech arbitrage. There is money on the table since no one can live without water.

This takes us to one of the other aspects of thc smart city that will assert itself in the coming years. Engineers are extraordinarily competent at creating hyper efficient systems based on established parameters. The current fashion is to use the waste stream of one system as a feedstock for another. For example, the 91st Avenue sewage treatment plant in Phoenix, Arizona [65], captures the methane gas created by the wastewater and sells it as natural gas on the open market. Then the treated effluent is pumped 45 miles west across the desert to the Palo Verde nuclear power station [66]. Palo Verde is the largest nuclear power facility in the country. When someone in Phoenix flushes the toilet or takes a shower, they are helping to cool the three local reactor cores. Palo Verde uses between 40,000 and 60,000 gallons of water per minute [66]. Phoenix needs electricity from the power station to run the sewage treatment plant. And the power station needs the water from the sewage plant to keep from melting down. On paper these synergies are gorgeous, but they are also fragile and vulnerable to catastrophic failure.

An unexpected cold snap in Texas [67] and massive forest fires in California [68] have disabled the electrical, natural gas, and water supply infrastructure in recent years. These were not isolated to single cities or counties, but afflicted entire states. They have also been coming in increasing and more intense waves over multiple years. Disruptions in the various grids created cascading failures in other systems that rely on uninterrupted supplies to perform as designed. We do not yet know which of our highly leveraged, tightly interconnected subsystems might tip and set off the dominoes elsewhere next time. A physical event might cause a financial or political crisis or vice versa. And since insurance companies are often on the hook for the resulting damage, the people at Swiss Re and Hannover Rück [69] might start to dictate some new terms moving forward. That is the dark side of the big tech squeeze. The smart city is a genius until the crops fail and the lights go out. Then we are all just isolated and alone with whatever resources happen to be on hand in each particular venue, frantically swiping our cell phones in search of sustenance.

35.7 Unintended consequences

Technologies frequently come with two parts. Part A is the mechanical fix. Part B involves a complimentary change in human nature to make the technical bits work as intended. But we are crafty creatures who have a long track record of enjoying the fast and dirty patch and spotting a fresh opportunity without making the concomitant accommodations. A classic example is the invention of the cotton gin in 1793 by Eli Whitney [70]. Northern abolitionists assumed that the mechanization of removing seeds from cotton bolls would eliminate the need for slaves to do this slow tedious work by hand. In reality, the cotton gin allowed marginal lower-quality upland cotton to become commercially viable, greatly expanded production, and increased the need for ever more field slaves. As the financial incentives of slave grown cotton increased, so did the concentration of wealth and power and the political will to preserve the institution. Southern culture embraced innovation and used it to become more like itself.

At the same time, Eli Whitney was a popularizer of standardized interchangeable machine parts that cultivated industrial efficiency. The South was slow to adopt such practices to the point where Southern railroads were often of completely different gauges and did not physically connect to one another [71]. Standardization and national connectivity were actively rejected by Southern plutocrats [72] as a threat to their regional sovereignty. Meanwhile, the North embraced interchangeable uniformed production techniques and collective standards [73]. Superior industrial production is generally acknowledged as the key to winning the Civil War, and the North had the upper hand. It can be said that Whitney's innovations were both a contributing cause of and definitive resolution to the Civil War on both sides [74].

Much of the smart city technology is already here, and we already know how it is being put to use. When any of us engages a customer call center we are usually categorized by an algorithm based on how valuable the company thinks we are. It uses our bank balances, credit scores, and past history to determine whether we reach an assistant immediately or we wait for half an hour as a recording assures us that "Your call is important to us." It is subtle and invisible by design. Likewise, products and services are increasingly sold using dynamic pricing. Airline tickets, rental cars, and hotel accommodations are priced differently at certain times of day, days of the week, and depending on geographic location [75]. People with good histories are offered discounts while those with blemished records are charged extra. And there are people who are simply denied service entirely.

Many doors open quite literally for some people, but not others. Office parks, apartment complexes, and various public accommodations are frequently hermetically sealed. Electronic badges and vehicle transmitters welcome authorized members while locking out the unidentified. Buildings have always had locks and guards, but tech is being used to create categories of people and places that are mutually exclusive based on all sorts of collated bits of data. For some the smart city will be an endless series of welcoming portals. For others it will be a continuous exercise in exclusion. The two will have fewer and fewer opportunities to mingle with greater fear on one side and alienation of the other. In other words, the smart city will be a lot like the world we already inhabit, only more so.

Landlords and property management companies increasingly use apps to screen prospective tenants. Companies such as TransUnion, Equifax, and TRW [76] already have everyone's detailed personal information, and predictive analytics give a thumb up or down. This dynamic is not new, although the highly automated version is likely to be more efficient, comprehensive, and ruthless. Our interconnected devices and algorithms do not care how we use them. Societies tend to deploy new tools in ways that promote and reinforce the dominant culture. Even after the most revolutionary moments of complete change, there is typically a cyclical reversion back to the mean. In the end the smart city is going to deliver a regressive version of America's long-standing historical tendencies, for better and worse.

Ubiquitous software already assesses job applicants based on credit scores, background checks, and peripheral metadata that are not directly associated with skills or qualifications. Human resource experts might suggest that people with a problem in one part of their lives tend to have problems in others as well so these are relevant considerations. A messy divorce, foreclosure, and a poor record with the car insurance agency might cost someone a job. Not having a job might close a lot of other doors. Being excommunicated by the algorithms makes it more likely that an individual struggles and faces new challenges. And the authorities tend to look at damaged people with extra scrutiny and prosecute them with less remorse. None of this is new. This sort of thing has played out repeatedly over the centuries, and we should expect more of the same moving forward.

35.8 The onerous ship

A popular meme emerged recently that suggested no one will own anything in the future. It was originally meant to be an optimistic vision of a bright future of abundance, but it was roundly interpreted as an ominous sign of global rentier elites plotting to confiscate everything of value. It might be worth asking who owns things right now. A "homeowner" only owns a home once the mortgage is paid off. Otherwise it is owned by whoever holds the debt. For most Americans, that is Fannie Mae and Freddie Mac [77], exactly halfway between a government agency and a private corporation.

Who escapes university or trade school without student loan debt—a debt that cannot be discharged even in bankruptcy? How many people have not finished paying off their own student debt when they begin to take on more student debt for their children? Who really owns their own time and effort when they have onerous debt to service for decades? How many people pay cash for their car or truck? As the economy tightens and the middle class contracts, auto loans have stretched out with longer terms to bring down the monthly payment and make vehicles more attainable. It is now common for 7-year-old used cars to be purchased with 7-year loans, all under the banner of freedom and mobility. The story is always about low interest rates, tax deductions, and savvy investors who know how to leverage their way around the system with clever accounting.

Most people go through life never really owning much of anything, even if the dominant narrative is one of self-determination and autonomy. The ominous vision of an ownerless society is already here for a significant proportion of the population. The house, the car, and the education are all owned by others, although it might look impressive from the facade. The future will probably look a lot like the past, only with a new advertising and marketing campaign to make it emotionally satisfying.

Value is created by a national infrastructure, much of it built by private enterprise with heavy government subsidies: canals, railroads, highways, copper wires, satellites, cell towers, and computers. These systems genuinely improve life for most people in all sorts of ways. But they are also monetized in the form of real estate speculation and steady rents of one sort or another. The profits are siphoned off to various centers of commerce and power through consolidation, vertical integration, and economies of scale. The pioneer log cabin in the woods, the homestead farm on the prairie, and the tract house in the suburbs were merely vectors for economic throughput halfway between a public federal agency and a private corporation. The places we think of as our homes and communities have been carefully crafted and aggressively marketed to appeal to rugged individualists, but they are just fodder. If the ongoing increase in productivity is greater than the cost of maintaining the supportive infrastructure these places can endure. If not, they are doomed to decline and abandonment once the cash flow dries up. This sort of thing has played out repeatedly over the centuries, and we should expect more of the same moving forward.

35.9 Summary

Changing the physical environment is expensive, it takes decades and requires armies of professionals to navigate multilayered regulatory and political obstacles. We probably are not going to build new Singapores or Copenhagens in America anytime soon [78]. But software and a few widgets can be installed in a day, spread quickly, establish committed constituencies, and become manifest in the world before the authorities have a chance to respond. Once these novel systems are in place and are busy spinning off profits and influence, officialdom tends to want their cut more than they want to put the genie back in the bottle.

The adventurous upstart disrupts the entrenched establishment and in due course becomes the establishment itself in what Vilfredo Pareto (1848–1923) described as a "circulation of elites" [79]. The more things change, the more they stay the same. The Mark Zuckerberg quote, "Move fast and break things" is a great tech slogan, but it is not so fun for those being broken [80]. The winners are generally tone deaf to the effects they have on others. The people on the losing end of these transitions tend not to want to fade away silently so they express their displeasure in unorthodox ways. So we are back to where we started. VUCA: volatility, uncertainty, complexity, and ambiguity.

Finally, let us move on to the real interactive part of this chapter: review questions/exercises, hands-on projects, case projects, and optional team case project. The answers and/or solutions by chapter can be found in Appendix G.

35.10 Chapter review questions/exercises

True/false

1. True or False? The smart city concept is second and foremost about creating and extracting value from the landscape.
2. True or False? A close cousin of the smart city framework is digital finance.
3. True or False? The best way to mitigate a rain tax is to break up the concrete and asphalt on a property to allow storm water to percolate into the earth.
4. True or False? The future national development pattern is sometimes referred to as "drive till you qualify."
5. True or False? The federal Interstate Highway System, which is separate from state and local road authorities, was established with a pay-as-you-go funding scheme in 1966.

Multiple choice

1. In places with cyclical droughts, there are residents, farmers, and industries that protest the rationing of water and tiered rate structures that come at times of critically:
   a. High supplies
   b. Medium supplies
   c. Normal supplies
   d. No supplies
   e. Low supplies

2. Technologies frequently come with:
   a. Three parts
   b. Four parts
   c. Five parts
   d. Six parts
   e. Two parts
3. A popular meme emerged recently that suggested no one will own anything in the:
   a. Present
   b. Future
   c. Past
   d. Future past
   e. None of the above
4. Changing the physical environment is expensive, it takes decades and requires armies of professionals to navigate multilayered regulatory and:
   a. Political obstacles
   b. Economic obstacles
   c. Smart obstacles
   d. Computer obstacles
   e. E-economics obstacles
5. Value is created by a ________________, much of it built by private enterprise with heavy government subsidies: canals, railroads, highways, copper wires, satellites, cell towers, and computers.
   a. Regional infrastructure
   b. International infrastructure
   c. National infrastructure
   d. All of the above
   e. None of the above

Exercise

Problem

How do the resilience, antifragility, and discontinuity of transformational, societal energy scenarios work?

Hands-on projects

Project

Do research: Analyze how a problem of inadequacy in leader preparedness could still lead to a win-win in a volatility, uncertainty, complexity, and ambiguity (VUCA) environment.

Case projects

Problem

Explore the VUCA framework in the context of projects; and make suggestions about how the VUCA approach can enhance a conventional risk assessment procedure.

Optional team case project

Problem

Conduct semistructured interviews to investigate the best practices and strategies employed by social entrepreneurs, who are leading change successfully within a VUCA context.

References

[1] Mackey Sr RH. Translating vision into reality: the role of the strategic leader. Carlisle Barracks, PA: US Army War College; 1992.

[2] McKenzie H. Insane mode: how Elon Musk's Tesla sparked an electric revolution to end the age of oil. New York, NY: Dutton; 2019.

[3] Casey RH. The model T: a centennial history. Baltimore, MD: Johns Hopkins University Press; 2008.

[4] Levy S. In the Plex: how Google thinks, works, and shapes our lives. New York, NY: Simon & Schuster; 2011.

[5] Levy S. Mark Zuckerberg, the hacker way and the art of the founder's letter. San Francisco, CA: Wired Magazine; February 3, 2012.

[6] Brandt RL. One click: Jeff Bezos and the rise of Amazon.com. New York, NY: Portfolio; 2012.

[7] Weil GL. Sears, Roebuck, U.S.A.: The Great American catalog store and how it grew. New York, NY: Stein and Day; 1977.

[8] Hassen A. From Pinkertons to Palantir: history and political economy of police surveillance. In: Hackers next door: an InfoSec conference, Tech Learning Collective, Brooklyn, NY December 15; 2019. https://hnd.techlearningcollective.com/2019/events/from-pinkertons-to-palantir-history-and-political-economy-of-police-surveillance.

[9] Mackay J. Allan Pinkerton: The first private eye. Hoboken, NJ: Wiley; 1997.

[10] Gibbs S. 40 Years of Apple - in pictures. London, UK: The Guardian; April 1, 2016.

[11] Reinhart CM. This time is different: eight centuries of financial folly. Princeton, NJ: Princeton University Press; 2011.

[12] Nakamoto S. The white paper. London, UK: Ignota Books; 2019.

[13] Mann CC. The Wizard and the Prophet: two remarkable scientists and their dueling visions to shape tomorrow's world. New York, NY: Vintage; 2019.

[14] United States Census Bureau. World Population:1950-2050. Washington, DC. May 17, 2016. https://www.census.gov/library/visualizations/2011/demo/world-population–1950-2050.html.

[15] Roser M, Ritchie H. Our world in data: hunger and undernourishment. Oxford, U.K: Martin School, University of Oxford. https://ourworldindata.org/hunger-and-undernourishment.

[16] United Nations. Sustainable development goals, "Goal 2, Zero Hunger." 2020. https://www.un.org/sustainabledevelopment/hunger/.

[17] FAO, IFAD, UNICEF, WFP, WHO. Safeguarding against economic slowdowns and downturns. The state of food security and nutrition in the world 2019. Rome. 2019. http://www.fao.org/3/ca5162en/ca5162en.pdf.

[18] Quinn W, Turner JD. Boom and bust: a global history of financial bubbles. Cambridge, UK: Cambridge University Press; August 2020.

[19] Segan S. What is C-band, and what does it mean for the future of 5G? New York, NY: PC Magazine; February 25, 2021. https://www.pcmag.com/news/what-is-c-band.

[20] Goodrich C. Government promotion of American canals and railroads, 1800−1890. New York: Columbia University Press; 1960.

[21] Campbell S. The Panic of 1837. The Economic Historian; November 12, 2020. https://economic-historian.com/2020/11/panic-of-1837/.

[22] U.S. Department of the Treasury. Financial Panic of 1873. Washington, D.C. 2021. https://home.treasury.gov/about/history/freedmans-bank-building/financial-panic-of-1873.

[23] U.S. Department of State, Office of the Historian. Purchase of Alaska, 1867. Washington, D.C. https://history.state.gov/milestones/1866-1898/alaska-purchase.

[24] National Archives, Hawaii Statehood, August 21, 1959. Washington, DC. https://www.archives.gov/legislative/features/hawaii.

[25] Govind Bhutada. The U.S. Share of the global economy over time. Visual Capitalist; January 14, 2021. https://www.visualcapitalist.com/u-s-share-of-global-economy-over-time/.

[26] Layne C. The end of Pax Americana: how western decline became inevitable. Washington, D.C: The Atlantic; April 26, 2012. https://www.theatlantic.com/international/archive/2012/04/the-end-of-pax-americana-how-western-decline-became-inevitable/256388/.

[27] Brautigan R. All watched over by machines of loving grace. San Francisco, CA: American Dust, Richard Brautigan's life and Writing; 1967. http://www.brautigan.net/machines.html.

[28] White AJ. Infrastructure policy: lessons from American history. Washington, D.C.: The New Atlantis; Spring 2012. https://www.thenewatlantis.com/publications/infrastructure-policy-lessons-from-american-history.

[29] White Paper, Orchestrating infrastructure for sustainable smart cities. The International Electrotechnical Commission (IEC) and The Centre for European Policy Studies (CEPS). Geneva, Switzerland. https://www.ceps.eu/wp-content/uploads/2015/01/iecWP-smartcities-LR-en.pdf.

[30] Kohut A. Debt and deficit: a public opinion dilemma. Washington, D.C: Pew Research Center; June 14, 2012. https://www.pewresearch.org/politics/2012/06/14/debt-and-deficit-a-public-opinion-dilemma/.

[31] Drago EB. How Philadelphia's water pollution problems shaped the city. Philadelphia, PA: Science History Institute, Distillations Podcast; January 7, 2020. https://www.sciencehistory.org/distillations/podcast/how-philadelphias-water-pollution-problems-shaped-the-city.

[32] Goldman TR. Why smart cities are turning themselves into sponges. Arlington, VA: Politico; April 20, 2017.

[33] Philadelphia Water Department. Stormwater management service charge. Philadelphia, PA. October 2020. https://www.phila.gov/water/PDF/scaa_manual.pdf.

[34] Azavea, Philadelphia, PA. https://www.azavea.com/.

[35] OpenDataPhilly. PWD Stormwater Billing Parcels. https://www.opendataphilly.org/dataset/pwd-stormwater-billing-parcels.

[36] WHYY. New stormwater fee structure explained, Philadelphia, PA; February 12, 2011. https://whyy.org/articles/philadelphias-new-stormwater-fee-plan/.

[37] Sauro S. The cost of rain: more towns ask property owners to pay for stormwater plans. Mechanicsburg PA: Pennsylvania Real-Time News; February 27, 2019. https://www.pennlive.com/news/2019/02/the-cost-of-rain-more-townships-boroughs-asking-property-owners-to-pay-for-stormwater-plans.html.

[38] Harvey B. Thermal imaging applications overview for law enforcement. Lexipol, Frisco, TX: Police1; July 24, 2007. https://www.police1.com/police-products/police-technology/thermal-imaging/articles/thermal-imaging-applications-overview-for-law-enforcement-ssC7sU4JA9onHojm/.

[39] They are watching you. ACLU of Washington, Seattle, WA, https://theyarewatching.org/technology/thermal-imaging-surveillance.

[40] Opioid wastewater collection data dashboard. Tempe data, City of Tempe, AZ. https://performance.tempe.gov/apps/69d996bc23dc461f82d01f47a5d70bfe/explore.

[41] Harsanyi D. Nanny state: how food fascists, Teetotaling Do-Gooders, Priggish Moralists, and other Boneheaded Bureaucrats are turning America into a nation of children. New York, NY: Broadway-Random House; September 18, 2007.

[42] Zuboff S. The age of surveillance capitalism: the fight for a human future at the new frontier of power. New York, NY: PublicAffairs; January 15, 2019.

[43] Search Warrant, Legal Information Institute Cornell Law School, Ithaca, NY. https://www.law.cornell.edu/wex/search_warrant.

[44] Probable Cause, Legal Information Institute Cornell Law School, Ithaca, NY. https://www.law.cornell.edu/wex/probable_cause.

[45] Farivar C. Surveillance Watchdog concludes metadata program is illegal, "should end. New York, NY: ARS Technica, Condé Nast; January 22, 2014. https://arstechnica.com/tech-policy/2014/01/surveillance-watchdog-concludes-metadata-program-is-illegal-should-end/.

[46] Dineen JK. Despite climate crisis, California continues to embrace exurban sprawl. San Francisco, CA: San Francisco Chronicle; September 20, 2019.

[47] Schwartz A. The realities of economic development subsidies. Washington, D.C: The Center for American Progress; November 1, 2018. https://www.americanprogress.org/issues/economy/reports/2018/11/01/457771/realities-economic-development-subsidies/.

[48] Metromile, National General Insurance Group, Tempe, AZ. https://www.metromile.com/.

[49] Milewise, Allstate Insurance Company, Northbrook, IL. https://www.allstate.com/auto-insurance/milewise.aspx.

[50] SmartMiles, Nationwide Insurance, San Francisco, CA. https://www.nationwide.com/personal/insurance/auto/discounts/smartmiles/.

[51] Metz C. The costly pursuit of self-driving cars continues on. And on.... New York, NY: New York Times; May 24, 2021. https://www.nytimes.com/2021/05/24/technology/self-driving-cars-wait.html.

[52] Marohn C. The Growth Ponzi scheme. Brainerd, MN: Strong Towns. https://www.strongtowns.org/the-growth-ponzi-scheme.

[53] US Department of Transportation. History of the interstate highway system. Washington, D.C.: Federal Highway Administration; June 2017. https://www.fhwa.dot.gov/interstate/history.cfm.

[54] Dutzik T, Davis B. Do roads pay for themselves? Setting the record straight on transportation funding. Denver, CO: U.S. PIRG Education Fund; January 2011. https://uspirg.org/sites/pirg/files/reports/Do-Roads-Pay-for-Themselves.pdf.

[55] It's been 28 years since we last raised the gas tax, and its purchasing power has eroded. New York, NY: Peter G.Peterson Foundation; March 16, 2021. https://www.pgpf.org/blog/2021/03/its-been-28-years-since-we-last-raised-the-gas-tax-and-its-purchasing-power-has-eroded.

[56] Webster I. Why a dollar today is worth only 54% of a dollar in 1993. San Mateo, CA: CPI Inflation Calculator; June 10, 2021. https://www.in2013dollars.com/us/inflation/1993.

[57] Jaffe E. The way congress has funded transportation since 2008 is not entirely legal. New York, NY: Bloomberg CityLab; December 3, 2014. https://www.bloomberg.com/news/articles/2014-12-03/the-way-congress-has-funded-transportation-since-2008-is-not-entirely-legal.

[58] US Army Corps of Engineers, Waterborne Commerce of the United States. Alexandria, Virginia: Institute for Water Resources; 2019. https://www.iwr.usace.army.mil/About/Technical-Centers/WCSC-Waterborne-Commerce-Statistics-Center-2/WCSC-Waterborne-Commerce/.

[59] US Department of Transportation, Federal Railroad Administration. Freight Rail Overview. https://railroads.dot.gov/rail-network-development/freight-rail-overview.

[60] Breyer N. Analysis of Travel Patterns from Cellular Network Data. Norrkoping, Sweden: Department of Science and Technology Linkoping University; 2019. https://liu.diva-portal.org/smash/get/diva2:1318916/FULLTEXT01.pdf.

[61] Metro Expresslanes, Torrance, CA. https://www.metroexpresslanes.net/.

[62] OreGo, Oregon Department of Transportation, Salem, OR. https://www.myorego.org/.

[63] Approved Fire Apparatus Turn-Around Minimum Dimensions. Vancouver Fire Marshal's Office. https://www.cityofvancouver.us/sites/default/files/fileattachments/fire_vfd/page/1404/turnarounddetail8-18-05.pdf.

[64] Permeable paving surfaces. Holmfirth, UK: ABG Creative Geosynthetic Engineering. https://www.abg-geosynthetics.com/applications/permeable-paving-surfaces.html.

[65] Lund S. Circle unbroken. Treatment Plant Operator Magazine; September 2019. https://www.tpomag.com/editorial/2019/09/an-arizona-plant-can-now-claim-full-capture-and-reuse-of-clean-water-plant-resources.

[66] Staten M. Palo Verde nuclear plant is cutting its wastewater use. Phoenix, AZ: Cronkite News; February 25, 2020. https://cronkitenews.azpbs.org/2020/02/25/palo-verde-nuclear-water-use/.

[67] Golding G, Kumar A, Mertens K. Cost of Texas' 2021 Deep freeze justifies weatherization. Dallas, TX: Federal Reserve Bank of Dallas; April 15, 2021. https://www.dallasfed.org/research/economics/2021/0415.aspx.

[68] McFadden C. California, battered by 2020, girds for more intense wildfires. Sacramento, CA: Capitol Weekly; June 14, 2021. https://capitolweekly.net/california-battered-by-2020-girds-for-more-intense-wildfires/.

[69] Buesnel C, et al. This time is different: six trends that will determine the future of non life reinsurance. London, UK: Deloitte; 2020. https://www2.deloitte.com/content/dam/Deloitte/global/Documents/Financial-Services/gx-fsi-future-of-reinsurance-report.pdf.

[70] The Eli Whitney Museum and Workshop, Hamden, CT. 2021. https://www.eliwhitney.org/7/museum/about-eli-whitney/inventor.

[71] Schafer M. Classic American railroads. Beverly, MA: Voyageur Press; November 8th 1996.

[72] Bateman F, Weiss T. Comparative regional development in Antebellum manufacturing. J Econ Hist 1975;35(1). The Tasks of Economic History (March 1975), pp. 182–208 (27 pages), Cambridge University Press, Cambridge, UK.

[73] Abrams M. Eli Whitney. New York, NY: The American Society of Mechanical Engineers; July 19, 2012. https://www.asme.org/topics-resources/content/eli-whitney.

[74] Harrison N. Did Eli Whitney cause the Civil war? Westminster, Maryland: Carroll County Times; March 18, 2012. https://www.baltimoresun.com/maryland/carroll/cct-arc-9f3452ac-e27e-5515-9cab-026e598cc17e-20120318-story.html.

[75] Sharda S. The extinction of the price tag: how dynamic pricing can save you. Potomac, MD: New Degree Press; April 25, 2018.

[76] Rosenblatt E. Credit data and scoring: the first triumph of big data and big algorithms. Amsterdam, Netherlands: Elsevier; January 7, 2020.

[77] Federal Housing Finance Agency, Washington, D.C. 2021. https://www.fhfa.gov/SupervisionRegulation/FannieMaeandFreddieMac/Pages/About-Fannie-Mae-Freddie-Mac.aspx.

[78] Schiller B. The U.S. is failing at making its communities more walkable. New York, NY: Fast Company; September 26, 2017.

[79] Sakthi. Vilfredo Pareto's circulation of elite: explanation and examples. Sociology Group. https://www.sociologygroup.com/circulation-elite/.

[80] Pearson T. Creative destruction: move fast and break things. Irvine, CA: Entrepreneur; November 29, 2016. https://www.entrepreneur.com/article/284609.

Part VIII

Appendices

Appendix A

List of top smart cities and critical infrastructure policies and financing implementation and deployment companies

John R. Vacca
TechWrite, Pomeroy, OH, United States

Organization	Description	URL
151 Advisors, LLC	Helps Smart Cities with building strategic execution, and focused Go-To-Market plans to generate profitable revenue streams from the deployment of projects.	http://www.151advisors.com
Asavie	Provides secure enterprise connectivity to everything; data controls; and, eliminates all Capital Expense costs.	http://www.asavie.com
Caves Valley Technologies	Provides core guidance and due-diligence support for the most efficient and optimum technology solutions achieved today, including Smart City transportation, infrastructure, and mobile solutions.	http://www.cavesvalleytechnologies.com
CITYWAY	Integrates transit, road, ticketing and pretty much all data necessary to provide the most complete service for mobility.	http://www.cityway.company/
Civic Resource Group	Provides a data integration and management platform; and, a suite of augmented reality-based mobile applications for Smart Cities.	http://www.civicresource.com/
Clever Devices	Provides innovative technology solutions for public transportation, including bus, bus rapid transit, para-transit and rail.	http://www.cleverdevices.com/
Driverless.global	A global directory established for all businesses involved with autonomous/driverless vehicles, and related products and services.	http://www.driverless.global
Eastern Research Group	Provides management, outreach, technical, and engineering support for the development and implementation of Smart City-related technologies, including vehicle electrification and infrastructure; ITS; and, goods and people movement.	http://www.erg.com
Eldorado Benson	A brand new development site that is planned to include a parallel autonomous roadway network within a proposed rights of way to construct, test and set the standard for Smart Cities of the future.	http://www.eldoradoholdings.net/
Electro-Optical Technologies, Inc.	Tenders expertise and consulting services in high-performance video imaging (video analytics; physical security and information management) for enhancing transit and transportation safety and security.	http://www.garlic.com/~biz

Organization	Description	URL
Genvict Technology	Focuses on vehicle-to-vehicle (V2V); vehicle-to- infrastructure (V2I); and, smart parking systems.	https://en.genvict.com/
HERE	Manages and enriches the data that connects travelers, vehicles, and infrastructure to improve safety, mobility, environmental and economic efficiency.	http://360.here.com
Inscape Data Corporation	Designs and manufactures total turnkey solutions for outdoor power over ethernet (PoE) switch, wireless, and fiber optics switch systems that allow interconnections of network devices installed in metropolitan or suburban environments.	http://https/www.inscapedata.com
Joyride	Specializes in downtown transit rides using a fleet of low speed neighborhood vehicles.	http://www.joyrideus.com
Laird	Provides systems, components and solutions that protect electronics from Electromagnetic interference (EMI) and heat; and, that enable connectivity in mission critical systems.	http://www.lairdtech.com/
Legion	Provides pedestrian modeling software that is used widely by transport operators, engineering consultants, smart cities and large scale events.	http://legion.com/
M2M Spectrum Networks	Provides comprehensive solutions for creating, connecting and managing machine-to-machine (M2M) communications for the Internet of Things (IoT).	http://www.m2mspectrum.com
Meridian Kiosks	Designs and manufactures interactive digital signage kiosks for indoor and outdoor installation to engage travelers and connect communities.	http://www.meridiankiosks.com/
Miovision	Explores innovative ways to use open data to advance the smart city agenda.	http://miovision.com/
Movatic	Builds a shared-mobility management platforms.	https://www.movatic.com/
Navigant Consulting	Creates flexible transportation systems and new mobility options.	http://www.navigant.com/
Nevana Designs	Provides UV powered glow for dark road/pathway markings.	http://www.nevanadesigns.com
Omnivex Corporation	Provides a common data and content repository that connects systems that otherwise don't communicate; and, then provides an intelligent visual presentation layer that allows this information to be automatically delivered to any visual medium.	http://www.omnivex.com/
Parsons Transportation Group	Provides the ability to assist clients with all phases of roads and highway projects: starting with planning and design; and, carrying through to operations and maintenance.	http://www.parsons.com/about-parso
Pulse Larsen Antennas	Provides solutions that reflect a far-reaching understanding of antennas and radio frequency (RF) technology.	http://www.pulseelectronics.com/
Rostra Precision Controls	Provides school bus driving aids called: student Detection Systems, that warn drivers of pedestrian activity or presence in and around the perimeter of the National Highway Traffic Safety Administration (NHTSA) danger zones.	http://www.rostrahd.com/
SAFECHECKportal	A mobile human resource management tool, with an analytic web dashboard for diverse, on demand contractual workers that promote an alcohol and substance free workplace; along with available telematics and ensuing data, with the purpose of reducing insurance claims activity.	http://www.safecheckportal.com/
SenseOps, Inc.	Delivers digital transformation as a service, by deploying well known operational data and processes on low-cost edge devices, and provide them as an easy way to consume service.	http://www.senseops.com/
Strategic Innovation Partners, LLC	Provides a partnership with public-sector clients to prepare internally for smart city transition with a particular focus on smart governance.	http://www.strategicinnovation-partners.com/
Strava Metro	Anonymizes and aggregates social network for runners and cyclists activities into a GIS format; and, partners with departments of transportation and planning groups to enable deep analyses of bicycling and pedestrian behavior, including activity counts by a specific time of day; and, day of week across the entire network.	http://www.metro.strava.com

Organization	Description	URL
T-Mobile US – Bellevue, WA	Redefines the way consumers and businesses buy wireless services through leading product and service innovation.	http://www.t-mobile.com/
Uptake	Delivers actionable intelligence that drives and protects growth.	http://www.uptake.com/
Vermont Energy Investment Corporation	Provides government, utilities, and businesses, with transportation demand management and electric vehicle program implementation and consulting services.	http://www.veic.org
Victor Stanley, Inc.	Modernizes waste and recycling collection by monitoring the fill levels and actual weights of all of their waste and recycling litter receptacles.	http://http/www.victorstanley.com
ZoomThru	A cloud platform that helps cities not only aggregate and visualize real time data from parking garages and spaces; but, then puts that data to work in a series of consumer facing applications that allow for better way finding.	http://www.zoomthru.com/
Axon Vibe	Specializes in enhancing the user experience by providing personalized services such as passive ticketing and end-to-end connected journey services.	http://www.axonvibe.com/
EV Charge Solutions	A national distributor of Electric Vehicle Supply Equipment (EVSE), offering the latest products and services.	http://www.evchargesolutions.com/
eX2 Technology	Provides swift and secure broadband and intelligent infrastructure networks to a variety of industries.	http://www.ex2technology.com/
Microcom Corporation	Specializes in thermal printing solutions for transportation applications.	http://www.microcomcorp.com/
Mindteck, Inc.	Provides smart city solutions that range from traffic management energy efficiency and smart healthcare, to e-governance, intelligent lighting systems and home and building efficiency systems.	http://www.mindteck.com/
Organic Transit	Provides urban mobility solutions: design, build, manufacture and solar hybrid designs.	http://www.organictransit.com/
Parking Sense USA	Develops and manufactures parking guidance hardware and software systems, with an integrated management reporting software and data analytics platform.	http://www.parkingsenseusa.com/
Placemeter	An urban intelligence platform that connects to a network of video cameras and analyzes pedestrian and vehicular movement, revealing hidden patterns and strategic opportunities.	http://www.placemeter.com/
Q Free TCS	Provider of full Information Technology Services (ITS) solutions (Parking, way finding, Tether Management System (TMS)).	http://www.q-free.com/
Rajant Corporation	Enables municipalities to build private wireless networks to support the data, voice, and video demands driving smart-city communications initiatives.	http://www.rajant.com/
skyTran	A patented, high-speed, low-cost, elevated Personal Rapid Transportation (PRT) system that moves passengers in a fast, safe, green, and economical manner.	http://www.skytran.com/
Socrata	Provides cloud-based solutions that help improve government through democratizing access to data.	http://www.socrata.com/
Triunity Engineering and Management, Inc.	Provides program, project and construction management services for large infrastructure projects nationwide.	http://www.triunityeng.com/
Truck Smart Parking Services, Inc. (TSPS)	Provides an Information Technology Services (ITS)-based platform and related services to improve parking for commercial vehicles at government-owned and private facilities on the highway network; as well as, at truck loading and unloading zones in urban centers.	http://www.trucksmartparkingservices.com/
way-CARe technologies ltd.	Forecasts road trouble to reduce congestion, road accidents and save lives.	http://www.waycare-smart-highway.com/
WIOMAX LLC	Focuses on real-time, smart and scalable modelling and optimization solutions, especially for sustainable multi-modal transportation, smart cities, and Internet of things (IoT), from basic research to industry transition and product development and support.	http://www.wiomax.com/

Organization	Description	URL
Bergkamp, Inc.	Provides a real-world pothole patching performance monitor and management system that helps governments stay informed and in control with reporting that captures data from each patching event and sends that data instantly.	http://www.bergkampinc.com/
BYD Heavy Industries	Manufacturer of electric buses.	http://www.byd.com/
Downtowner	Focuses on providing short-trip rides that are too far to walk and too short to drive.	http://www.ridedowntowner.com/
EZR Mobile	A smart signage company that provides a platform by way of Text, NFC, Tap or QR scans for customers at citywide parking lots to access platforms that are specific to the business attached to the parking structure or lot, airports, malls, etc.	http://www.ezrmobile.com/
Glocol Networks, Inc.	Provides the fabric for modular solution set for smart transportation and smart cities.	http://www.glocol.net/
Greenlots	Offers advanced electric vehicle fleet infrastructure and load management solutions.	http://www.greenlots.com/
Parkifi	Delivers real time occupancy data to cities and its residents.	http://www.parkifi.com/
R Systems	Provides software development for smart cCity IOT technology application development, by using some of the industry's highest quality certifications and standards, including CMMI Level 5, PCMM Level 5, ISO 27001:2013 and ISO 9001:2008.	http://www.rsystems.com/
Solar Power and Light	Provides EV charging stations; Solar carports and pergolas; and, LED lighting.	http://www.splsolar.com/
Traffic Technology Services, Inc.	A technology company and information provider for the automotive industry, commercial fleets, telematics integrators, and other transportation service providers.	http://www.traffictechservices.com/
ArqBravo Group	Helps solve urban environment transportation; energy and water problems; with a systemic vision, a disruptive business model and developments, that will revolutionize the way people are transported, data, goods, water and energy.	http://www.bravomotorcompany.com/
BRIDJ	A mass transit company that uses demand aggregation and real-time traffic information, to provide more efficient and accessible transportation; improve travel time reliability; and, capture valuable data to inform future decision making.	http://www.bridj.com/
Carmanah Technologies Corp.	Provides smart solar-powered LED traffic beacons and outdoor lighting.	http://www.carmanah.com/
Cisco	Provides secure connection to disparate intelligent transportation systems, in order to improve traffic flow, reduce roadside incidents, and provide a centralized view of highway systems.	http://www.cisco.com/
Multimer	Develops proprietary wearable sensors; deploys them on multitudes of people every day; and, then analyzes the data against existing geographic insights.	http://www.multimerdata.com/
EB Systems	Creates rich Internet of Things enabled environments to bridge the gap between the digital and physical web; where their technology is being used for smart city applications which includes Community Improvement Districts, Public Safety Organizations, Educational Institutions, and Transit.	http://www.ebsystems.com/
Michael Baker International	Provides intelligent transportation systems (ITS) master planning; scoping assessment studies; system definition analyses; agency coordination; ITS facilities design; and, operations and maintenance for a successful planning, development, and deployment of smart cities.	http://www.mbakerintl.com/
moovel	Focuses on technology that can solve complex problems and enhance mobility in the world's biggest cities.	https://www.moovel.com/en_us
Moovit	Provides commuters and tourists with a complete real-time snapshot of what their trip will be like and suggests the fastest, most comfortable routes.	http://www.moovitapp.com/

Organization	Description	URL
National Instruments	Manufacturer of instrumentation for autonomy, IoT, and Vehicle Electrification.	http://www.ni.com/
Overland ATS, LLC.	Focuses on the elimination of problems with regards to congestion, gridlock, and insufficient highway capacity.	http://www.overlandats.com/
Parkopedia	Provides drivers with the information on location, pricing and real-time availability of off-street and on-street parking; and, reduces the need for drivers to drive around the block looking for parking, thus causing unnecessary congestion and pollution.	http://www.parkopedia.com/
Q-Free	Global supplier and systems integrator for intelligent transportation systems, including tolling, traffic management, infomobility, and parking management.	http://www.q-free.com/
SICdrone	Offers drones for citywide use in easing traffic congestion, speeding up construction on roads and picking up trash.	http://www.sicdrone.com/
SiriusXM	Enhances and accelerates coast-to-coast deployment of connected vehicle pilots, as a means to dedicated short range radio communications (DSRC); as the foundation for a ubiquitous vehicle-to-vehicle and vehicle-to-infrastructure (V2X) system; leveraging the unique capabilities of existing network infrastructure, exclusively-licensed spectrum, and in-vehicle hardware/software platform.	http://www.siriusxm.com/
BestMile	Provides an agnostic fleet management platform designed to connect, operate and optimize any fleet of autonomous vehicles.	http://www.bestmile.com/
Giant Gray, Inc.	Provides an advanced, machine learning solution that teaches itself and continuously learns the expected operating conditions purely through observation.	http://http/giantgray.com/
Open Data Nation	Creates plug-and-play, productivity-based solutions that combine open, public data with data science techniques to increase transparency and productivity of public agencies.	http://www.opendatanation.com/
Place Global	Provides an ecosystem and platform that allows non-programmers to create, deliver and track proximity-based experiences which contain rich targeted content, audio, video and HTML.	http://www.placeglobal.com
Smart City Startup Consortium	Helps smart cities access a group of startup companies that work together to address the United States (US) Department of Transportation's (DOT's) 12 Vision Elements.	http://www.smartcitystartupconsortium.org/
Streetline	Delivers analytics and parking guidance to solve parking issues for smart cities.	http://www.streetline.com/
Think-Craft	A design studio that provides innovative design solutions that takes a human centered and research based approach to crafting spaces, strategies and content.	http://www.thinkcraft.com/
TouchSource, Inc.	Provides a green incentive based digital directory, signage, and announcement systems that display real time transit, shared ride, and bike information.	http://www.touchsource.com/
Veniam	Connects vehicles and their occupants to each other and to the Internet, delivering wireless Internet access to mobile devices inside a vehicle and collecting high-definition data from a fleet of vehicles	http://www.tveniam.com/
Black & Veatch	Offers Smart City Plan and Roadmap Development; Holistic Program Management; System Design and Integration; Infrastructure Deployment; Data Management; and, Analytics.	http://www.bv.com/
Brandbumps	Provides the infrastructure necessary for Smart City/Smart Transportation technology companies to securely and unobtrusively house the placement of their antennae and electronics that is pervasive, manageable and revenue producing.	http://www.brandbumps.com/
CitiSite Corporation	Creates seamless solutions and platforms, which are designed to enhance today's Smart Cities.	http://www.citisite.us/
CIVIQ Smartscapes, LLC	Provides design and deployment of high-performance public structures that integrate WiFi, mobile LTE and smart sensor technologies, by working together to create a seamless WiFi hotspot, that spans large corridors of a city, while the built-in sensors provide valuable data cities need for monitoring traffic, weather, and air pollution.	http://www.civiqsmartscapes.com/

Organization	Description	URL
Eno Center for Transportation	Focuses on the economic, environmental, safety, and health impacts of emerging advancements in metropolitan transportation.	https://www.enotrans.org/
InfraMarker by Berntsen	Combines RFID, magnetic locating, cloud data management, and geo-locating software to mark, locate, and manage infrastructure asset points.	http://www.inframarker.com/
Itron Idea Labs	Development and deployment of Smart Cities resource measurement systems that help Cities make infrastructure and asset decisions more easily.	https://www.itron.com/na/company/who-we-are/itron-idea-labs
New Eagle – New "Energy"	Supplies production control systems in a variety of applications: electric and hybrid system; alternative fuel engine control; hydraulics; autonomous systems; and, overall vehicle control, in order to develop EV City Transit Vehicles for customers and OEM's, by using proven industry validated components and vehicle controls.	http://www.neweagle.net/
Nimble Consulting Inc.	A program and change consultancy, that helps energy companies deploy transformational change initiatives.	http://www.nimbleconsulting.net
OPTIMYS	Deploys measurement technologies in the waste management industry to increase efficiency, improve recycling rates, reduce emissions and save city costs.	http://www.optimys.org
P3 Communications Inc.	Supports companies across the automotive, telecommunications and aerospace industries to bring their products from the innovation stage to full implementation.	http://www.p3-group.com/
PTV Group	Provides software and professional services to cities and regions for optimizing the movement of people and goods for a sustainable and smart future.	http://www.ptvgroup.com/
Realtime Technologies Inc.	Builds driving simulators for research needs.	http://http/www.simcreator.com/
Splunk, Inc.	Provides software to improve infrastructure and IT operations; application delivery; security and compliance; business analytics; and, to leverage the power of industrial data and the internet of things.	http://www.splunk.com/
Swiftly, Inc.	Creates data-driven software to improve urban mobility.	http://www.goswift.ly/
TEN – The Efficiency Network, Inc.	Provides turn key GIS/GPS asset inventory auditing; design and engineering; installation; project management; controls integration; advanced analytics; system commissioning; utility billing analysis; warranty management; project financing; and, utility rebate optimization.	http://www.tensaves.com/
Time Lab, Inc. (Kid CarPool)	Provides a mobile application that helps reduce unnecessary traffic by making carpooling fun and effective; saving parents time and money; while simultaneously decreasing emissions.	http://www.kidcarpool.com/
Wireless EDGE	Provides unique solutions to overcome difficult siting; a comprehensive array of services including turn-key antenna site development; materials and logistical support; and, construction management.	http://www.wirelessedge.com/
XLR8SUN	Provides energy storage and solar generation at vehicle to grid (V2G) bi-directional charging stations.	http://www.xlr8sun.com/
Zonar	Provides innovative technology solutions that are shaping the future of smart fleet management.	http://zonarsystems.com/downloads/spotlight/City-of-Denver-Spotlight.pdf
5D Robotics	Provides Autonomous Behaviors and Precise (+/−2cm) accuracy for Vehicle-to-everything (V2X) solutions (indoors, outdoors, air and ground vehicles).	http://www.5drobotics.com/
Air Expertise Colorado	Provides engineering and science expertise in applying U.S. EPA models to determine the most fuel-efficient and least carbon intensive transportation alternatives.	http://www.airexpertisecolorado.com/
BLERVAQUE Sprl	Provides support in facilitating cooperation with European stakeholders (cities, industry, research) for exchange of best practices and knowledge sharing.	https://www.companyweb.be/societe/blervaque/sprl/474847662

Organization	Description	URL
CIMCON Lighting, Inc.	Provides cities with an ideal on-ramp for creating a smart multi-services platform for driving energy efficiency, sustainability and a reduction in operating and maintenance costs.	http://www.cimconlighting.com/
CourthouseUSA, LLC	Provides a parcel-level affordable housing and traffic reduction model to aid in smart growth and sustainable development.	http://www.courthouseusa.com/
DatamanUSA	Provides IT Services to public sector organizations across the country.	http://www.datamanusa.com/
FLIR Systems	Provides a cloud based data analytics solution; fusing the traffic data with Wi-Fi sensor data; helping traffic engineers determine travel behavior; measures travel and delay times; and, monitors congestion and signal performance of the road network 24/7.	http://www.flir.com/traffic
Gemalto	Provides solutions that include an application enablement platform to simplify the deployment of machine to machine (M2M) cloud services; robust device connectivity and communication suite and security technology to protect everything from the data to the device to the entire network	http://www.gemalto.com/
Global Strategic Solutions, LLC	Provides advanced analytical health management solutions for use in defense aviation, commercial aviation, energy, and ground transportation systems.	http://www.gssllc.net/
GRIDSMART Technologies, Inc.	Provides cost effective systems for intersection control, data collection and situational awareness.	http://www.gridsmart.com/
Hill & Smith, Inc.	Provides cameras designed for the collection of data from moving traffic regardless of speed or illumination.	https://hillandsmith.com/
LQD	Provides Palo unified technology hubs that are street-smart, providing a network of public information, safety and emergency assistance.	http://www.lqdwifi.com/
Michelin Tire Corporation	Provides urbanization and the development of solutions for safer, cleaner, and more reliable transportation for future smart cities.	http://www.michelin.com/
MultimediaLED	Provides proprietary software which is GPS enabled that allows one the ability to quickly push out messages for public safety; such as route changes and closures, emergency weather alerts, amber alerts or other critical community information.	http://www.multimedialed.com/
NetSteady Communications, Ltd.	Provides distributed networks that deliver services where conventional networking techniques are not effective.	http://www.netsteady.com/
Prospect Silicon Valley	Works with technology and government innovators to launch smarter, cleaner solutions in energy, transportation and the built environment.	http://www.prospectsv.org/
Quicket Solutions	Provides Federal Bureau of Investigation (FBI) Criminal Justice Information Services Division (CJIS)-compliant cloud-based, mission critical solutions for law enforcement.	http://www.quicketsolutions.com/
RideLeads	Offers software that allows cities to offer Uber-style services, thereby investing in their own digital transportation infrastructure.	http://www.rideleads.com/
Siemens ITS	Provides Smart City solutions to customers related to the management of traffic, transport, logistics and intelligent street lighting.	https://www.mobility.siemens.com/global/en/portfolio/road/digital-ai-solutions/digital-lab.html
Navigant Consulting	Creates flexible transportation systems and new mobility options.	http://www.navigant.com
Nevana Designs	Creates UV powered glow in the dark road/pathway markings.	http://www.nevanadesigns.com
Omnivex Corporation	Provides a common data and content repository that connects systems that otherwise don't communicate; and, then provides an intelligent visual presentation layer that allows this information to be automatically delivered to any visual medium.	http://www.omnivex.com
Parsons Transportation Group	Assists clients with all phases of roads and highway projects, starting with planning and design and carrying through to operations and maintenance.	https://www.dnb.com/business-directory/company-profiles.parsons_transportation_group_inc.f5d5e7154ea769d4.html
Pulse Larsen Antennas	Provides solutions that for antennas and RF technology; and, antennas covering 2G/3G/4G LTE, WLAN (WiFi), Zigbee, Bluetooth, GPS/Glonass, Compass, ISM, VHF/UHF, NFC and custom applications.	https://www.pulselarsenantennas.com

Organization	Description	URL
Rostra Precision Controls	Warns drivers of pedestrians activity or presence in and around the perimeter NHTSA danger zones.	https://www.summitracing.com/search/brand/rostra-precision-controls
SAFECHECKportal	Mobile human resource management tool.	http://www.safecheckportal.com/
SenseOps, Inc.	Builds licenses and/or configures various software technologies to operate on edge-devices and interoperates with larger cloud based systems.	http://www.senseops.com/
Strategic Innovation Partners LLC	Prepares internally for smart city transition with particular focus on smart governance.	http://www.strategicinnovation-partners.com/
Strava Metro	Anonymizes and aggregates runners and cyclists activities into a GIS format; and, partners with departments of transportation and planning groups to enable deep analyses of bicycling and pedestrian behavior, including activity counts by specific time of day and day of week across the entire network.	http://www.metro.strava.com/
T-Mobile US – Bellevue, WA	Delivers wireless experiences to millions of customers who are unwilling to compromise on quality and value.	http://www.t-mobile.com/
Uptake	Delivers actionable intelligence that drives and protects growth.	http://www.uptake.com/
Vermont Energy Investment Corporation	Provides government, utilities, and businesses with transportation demand management and electric vehicle program implementation and consulting services.	http://www.veic.org/
Victor Stanley, Inc	Monitors the fill levels and actual weights of all waste and recycling litter receptacles.	http://www.victorstanley.com
ZoomThru	Helps cities not only TO aggregate and visualize real time data from parking garages and spaces; but, then puts that data to work in a series of consumer facing applications that allow for better wayfinding.	http://www.zoomthru.com
Axon Vibe	Facilitates explicit permission-based data sharing between companies and consumers.	http://www.axonvibe.com/
EV Charge Solutions	National distributor of Electric Vehicle Supply Equipment (EVSE).	http://www.evchargesolutions.com/
eX2 Technology	Provides secure broadband and intelligent infrastructure networks to a variety of industries.	http://www.ex2technology.com/
Microcom Corporation	Specializes in thermal printing solutions for transportation applications.	http://www.microcomcorp.com/
Mindteck, Inc.	Provides a comprehensive suite of smart city solutions that ranges from traffic management energy efficiency and smart healthcare to e-governance; intelligent lighting systems; and, home and building efficiency systems.	http://www.mindteck.com/
MotionLink	Technology partner with municipal public works departments.	http://www.motionlink.com/
Organic Transit	Provides Urban Mobility Solutions: design, build, manufacture and solar hybrid designs.	http://www.organictransit.com/
Parking Sense USA	Develops and manufactures parking guidance hardware and software systems with an integrated management reporting software and data analytics platform.	http://www.parkingsenseusa.com/
Placemeter	Connects to a network of video cameras; and, analyzes pedestrian and vehicular movement, revealing hidden patterns and strategic opportunities.	http://www.placemeter.com/
Q Free TCS	Provider of full ITS solutions (Parking, way finding, TMS).	http://www.qfree.com/
Rajant Corporation	Rajant Kinetic Mesh® technology enables municipalities to build private wireless networks to support the data, voice, and video demands driving smart-city communications initiatives.	http://www.rajant.com/
RouteMatch Software	Technology partner with over 600 transit agencies.	http://www.rajant.com/
skyTran	Moves passengers in a fast, safe, green, and economical manner.	http://www.skytran.com/
Smart Cities Summit Boston – Organized By Informa	Boosts infrastructure, increases investment, is sustainable, and improves the quality of life to meet the demands of the smart city.	http://www.smartcitiesevent.com/

Organization	Description	URL
Socrata	Provider of cloud-based solutions that help improve government through democratizing access to data.	http://www.socrata.com/
Triunity Engineering and Management, Inc.	Provides program, project and construction management services for large infrastructure projects nationwide.	http://www.triunityeng.com/
Truck Smart Parking Services, Inc. (TSPS)	Provides an ITS-based platform and related services to improve parking for commercial vehicles at government-owned and private facilities on the highway network; as well as, at truck loading and unloading zones in urban centers.	http://www.trucksmartparkingservices.com/
Velodyne LiDAR	3D Real-time LiDAR sensor provider.	http://www.velodynelidar.com/
way-CARe technologies ltd.	Forecasts road trouble to reduce congestion, road accidents and save lives.	http://www.waycare-smart-highway.com/
WIOMAX LLC	FocusES on real-time, smart and scalable modeling and optimization solutions; especially for sustainable multi-modal transportation, smart cities, and Internet of things (IoT), from basic research to industry transition and product development and support.	http://www.wiomax.com/
Bergkamp, Inc.	Pothole patching performance monitor and management system.	http://www.bergkampinc.com/
BYD Heavy Industries	Manufacturer of electric buses.	http://www.byd.com/
Downtowner	Utilizes a mobile phone application to dispatch rides using a fleet of fully electric low speed neighborhood electric vehicles.	http://www.ridedowntowner.com/
EZR Mobile	Provides information, schedules, and payment portals; as well as, a portal for directions back to the parked car.	http://www.ezrmobile.com/
Glocol Networks Inc.	Provides the fabric for a modular solution set for smart transportation and smart cities.	http://www.glocol.com/
Greenlots	Offers advanced electric vehicle fleet infrastructure and load management solutions; as well as, a global provider of open standards-based distributed energy resource solutions.	http://www.greenlots.com/
Parkifi	Delivers real time occupancy data to smart cities and its residents via a sensor.	http://www.parkifi.com/
R Systems	Provides software development for Smart City IOT technology application development, using some of the industry's highest quality certifications and standards, including CMMI Level 5, PCMM Level 5, ISO 27001:2013 and ISO 9001:2008.	http://www.rsystems.com/
Solar Power and Light	Provides EV charging stations, Solar carports and pergolas, and LED lighting.	http://www.splsolar.com/
Traffic Technology Services, Inc.	Provides data products for use in connected vehicle applications by utilizing existing public infrastructure and communication protocols.	http://www.traffictechservices.com/
ArqBravo Group	Solves urban environment transportation, energy and water problems, with a systemic vision, a disruptive business model and developments that will revolutionize the way people are transported by using data, goods, water and energy.	http://www.bravomotorcompany.com/
BRIDJ	Uses demand aggregation and real-time traffic information, to provide more efficient and accessible transportation; improve travel time reliability; and, capture valuable data for future decision making.	http://www.bridj.com/
Carmanah Technologies Corp.	Provides smart solar-powered LED traffic beacons and outdoor lighting.	http://www.carmanah.com/
Cisco	Securely connects disparate intelligent transportation systems to improve traffic flow, reduces roadside incidents, and provides a centralized view of highway systems.	http://www.cisco.com/
Multimer	Collects and analyzes millions of quantified self data points throughout a city; develops proprietary wearable sensors and deploys them on multitudes of people every day; and, then analyzes the data against existing geographic insights.	http://www.multimerdata.com/
EB Systems	Creates Internet of Things enabled environments to bridge the gap between the digital and physical web.	http://www.ebsystems.com/

Organization	Description	URL
Michael Baker International	Provides ITS master planning, scoping assessment studies, system definition analyses, agency coordination, ITS facilities design, and operations and maintenance, which are integral to the successful planning, development, and deployment of Smart Cities.	http://www.mbakerintl.com/
moovel	Looks at how technology can solve complex problems and enhance mobility in the world's biggest smart cities.	http://www.moovel.com/en_us
Moovit	Provides commuters and tourists a complete real-time snapshot of what their trip will be like and suggests the fastest, most comfortable routes; and, how to ride public transit smarter around the world.	http://www.moovitapp.com/
National Instruments	Manufacturer of instrumentation for autonomy, IoT, and vehicle electrification.	http://www.ni.com/
Overland ATS, LLC	Provides transportation infrastructure, which aims to eliminate the problems of congestion, gridlock, and insufficient highway capacity.	http://www.overlandats.com/
Parkopedia	Provides parking services to the automotive industry, by providing drivers with the information on location, pricing and real-time availability of off-street and on-street parking.	http://www.parkopedia.com/
Q-Free	Global supplier and systems integrator for intelligent transportation systems, including tolling, traffic management, infomobility, and parking management.	http://www.q-free.com/
SICdrone	Provides drones for citywide use in easing traffic congestion, speeding up construction on roads and picking up trash.	http://www.sicdrone.com/
SiriusXM	Enhances and accelerates coast-to-coast deployment of connected vehicle pilots; as a means to dedicated short range radio communications (DSRC); as the foundation for a ubiquitous vehicle-to-vehicle and vehicle-to-infrastructure (V2X) system; leveraging the unique capabilities of SiriusXM's existing network infrastructure exclusively-licensed spectrum; and, in-vehicle hardware and software platform.	http://www.siriusxm.com/
BestMile	Provides an agnostic fleet management platform designed to connect, operate and optimize any fleet of autonomous vehicles.	http://www.bestmile.com/
Giant Gray, Inc.	Provides a multi-sensor data fusion behavioral analytics platform that includes video, time phase numerical data (SCADA, building management data sources) and cyber.	http://www.giantgray.com/
Open Data Nation	Uses open data to predict outcomes, help prioritize resources, and establish data-driven performance metrics.	http://www.opendatanation.com/
Place Global	Provides non-programmers with the ability to create, deliver and track proximity-based experiences which contain rich targeted content, audio, video, HTML and more.	http://www.placeglobal.com/
Smart City Startup Consortium	Was created to help Smart Cities access a group of startup companies that work together to address the US DOT's 12 Vision Elements.	http://www.smartcitystartupconsortium.org/
Streetline	Provides analytics and parking guidance to solve parking issues for smart cities.	http://www.streetline.com/
Think-Craft	A design studio that takes a human centered and research based approach to crafting spaces, strategies and content.	http://www.thinkcraft.co/
TouchSource Inc.	A green incentive based digital directory, signage, and announcement system that displays real time transit, shared ride, and bike information.	http://www.touchsource.com/
Veniam	Enables vehicles with multi-network capabilities, allowing vehicle-to-vehicle (V2V) and vehicle-to-infrastructure (V2I) communication.	http://www.veniam.com/
Black & Veatch	Specializes in the deployment of energy, water, telecommunications, transportation and smart city infrastructure.	http://www.bv.com/
Brandbumps	Delivers the infrastructure necessary for Smart City/Smart Transportation technology companies to securely and unobtrusively house the placement of their antennae and electronics that is pervasive, manageable and revenue producing.	http://www.brandbumps.com/
CitiSite Corporation	Creates seamless solutions and platforms designed to enhance today's Smart Cities.	http://www.citisite.us/

Organization	Description	URL
CIVIQ Smartscapes LLC	Designs and deploys high-performance public structures that integrate WiFi, mobile LTE and smart sensor technologies.	http://www.civiqsmartscapes.com/
Eno Center for Transportation	Provides a multi-stage research and recommendation project, which is focused on the economic, environmental, safety, and health impacts of emerging advancements in metropolitan transportation.	http://www.enotrans.org/
InfraMarker by Berntsen	Combines RFID, magnetic locating, cloud data management, and geo-locating software to mark, locate, and manage infrastructure asset points.	http://www.inframarker.com/
Itron Idea Labs	Develops and deploys Smart Cities resource measurement systems that help Smart Cities make infrastructure and asset decisions more easily; as well as, bringing crucial gas, water, electricity and sensor data to provide real-time views of a Smart Cities resource utilization in co-operation with utilities and services departments.	https://www.itron.com/na/company/who-we-are/itron-idea-labs
New Eagle – New "Energy"	Supplies production control systems in a variety of applications: electric and hybrid systems; alternative fuel engine control; hydraulics; autonomous systems; and overall vehicle control.	http://www.neweagle.net/
Nimble Consulting Inc.	Program and change consultancy that helps energy companies deploy transformational change initiatives and services like: governance and stakeholder management; change management; training; facilitation; and, program management.	http://www.nimbleconsulting.net/
OPTIMYS	Deploys measurement technologies in the waste management industry to increase efficiency, improve recycling rates, reduce emissions and save smart city costs.	http://www.optimys.com/
P3 Communications Inc	A global consulting and engineering firm, that supports companies across the automotive, telecommunications and aerospace industries to bring their products from the innovation stage to full implementation.	http://www.p3-group.com/
PTV Group	Provides software and professional services to smart cities and regions for optimizing the movement of people and goods for a sustainable and smart future.	http://www.ptvgroup.com/
Realtime Technologies Inc.	Builds driving simulators for research needs.	http://www.simcreator.com/
Splunk, Inc.	Takes machine data and makes sense of it.	http://www.splunk.com/
Swiftly, Inc.	Creates data-driven software to improve urban mobility.	http://www.goswift.ly/
TEN – The Efficiency Network, Inc.	Provides GIS/GPS asset inventory auditing; design and engineering; installation; project management; controls integration; advanced analytics; system commissioning; utility billing analysis; warranty management; project financing; and, utility rebate optimization.	http://www.tensaves.com/
Time Lab, Inc. (Kid CarPool)	Reduces unnecessary traffic by making carpooling fun and effective; saving parents time and money; while simultaneously decreasing emissions.	http://www.kidcarpool.com/
Wireless EDGE	Develops and operates antenna facilities on Governmental properties.	http://www.wirelessedge.com/
XLR8SUN	Provides energy storage and solar generation at a V2G bidirectional charging station.	http://xlr8sun.com/
Zonar	Provides innovative technology solutions for the future of smart fleet management.	http://www.zonarsystems.com/
5D Robotics	Implements autonomous behaviors and precise (+/−2cm) accuracy for V2X solutions (indoors, outdoors, air and ground vehicles).	http://www.5drobotics.com/
Air Expertise Colorado	Provides expertise in applying US EPA models to determine the most fuel-efficient and least carbon intensive transportation alternatives.	http://www.airexpertisecolorado.com/
BLERVAQUE Sprl	Provides great support in facilitating cooperation with European stakeholders (cities, industry, research) for exchange of best practices and knowledge sharing.	https://data.be/en/company/Blervaque-SPRL-0474847662
Cimcon Lighting Inc.	Integrates intelligence, controls, wireless connectivity and smart sensors on street lighting infrastructures.	http://www.cimconlighting.com/

Organization	Description	URL
CourthouseUSA, LLC	Provides a parcel-level affordable housing and traffic reduction model to aid in smart growth and sustainable development.	http://www.courthouseusa.com/
DatamanUSA	Provides IT services to public sector organizations across the country.	http://www.datamanusa.com/
FLIR Systems	Provides cloud based data analytics solution for fusing traffic data with Wi-Fi sensor data; helping traffic engineers determine travel behavior; measuring travel and delay times; and, monitoring congestion and signal performance of the road network.	http://www.flir.com/traffic
Gemalto	Provides an application enablement platform that simplifies the deployment of M2M cloud services; robust device connectivity and communication suite; and, security technology to protect everything from the data to the device to the entire network.	http://www.gemalto.com/
Global Strategic Solutions LLC	Provides software technologies and develops advanced analytical health management solutions for use in defense aviation, commercial aviation, energy, and ground transportation systems.	http://www.gssllc.net/
GRIDSMART Technologies, Inc.	Manages congestion and safety at intersections by using technology that improves traffic signal operation, in order to reduce delays and optimize system efficiency.	http://www.gridsmart.com/
Hill & Smith Inc.	Provides cameras designed for the collection of data from moving traffic, regardless of speed or illumination.	https://hillandsmith.com/
Hortonworks	A big data platform that enables the Internet of things and smart technology solutions.	http://www.hortonworks.com/
LQD	Provides unified technology hubs, where anyone can engage with the community, the Internet and local services at any time; as well as, a network of public information, safety and emergency assistance.	http://www.lqdwifi.com/
Michelin Tire Corporation	Researches and investigates urbanization and the development of solutions for safer, cleaner, and more reliable transportation for future smart cities.	http://www.michelinman.com/
MultimediaLED	Provides proprietary software which is GPS enabled and allows you the ability to quickly push out messages for public safety; such as route changes and closures, emergency weather alerts, amber alerts or other critical community information.	http://www.multimedialed.com/
NetSteady Communications, Ltd.	Provides distributed networks that deliver services where conventional networking techniques are not effective.	http://www.netsteady.com/
Prospect Silicon Valley	Provides smarter, cleaner solutions in energy, transportation and the built environment.	http://www.prospectsv.org/
Quicket Solutions	Provides digitization of information at the source of an incident along with real-time data sharing, streamlined report generation and data analytic tools for real-time insight into operations.	http://www.quicketsolutions.com/
RideLeads	Provides software that allows cities to offer Uber-style services, thereby investing in their own digital transportation infrastructure.	http://www.rideleads.com/
Stellar Solutions, Inc.	Specializes in providing engineering talent to complex projects in the transportation, infrastructure, aerospace, defense, science, and commercial sectors.	http://www.stellarsolutions.com/
Teradata, Inc.	Provides end-to-end solutions and services in data warehousing, big data and analytics, and data science applications that enable cities to become data-driven.	http://www.teradata.com/
Traffic and Parking Control CO., Inc. (TAPCO)	Provides intelligent solar powered traffic warning devices.	http://www.tapconet.com/
Transit Labs, Inc.	A transportation analytics company that helps smart cities move people and goods.	http://www.transitlabs.com/index.html
Truck Smart Parking services, Inc. (TSPS)	Real time parking information services.	http://www.trucksmartparkingservices.com/

Organization	Description	URL
Center for Transportation and the Environment	Provides support to fleets to minimize the challenges associated with the operating characteristics and fueling requirements of medium- and heavy-duty zero-emission vehicles; helps organizations match propulsion and fuel technologies to the most appropriate fleet applications; and, provides independent, third party assessments to help organizations make informed zero-emission vehicle procurement and deployment decisions.	http://www.cte.com/
Kapsch TrafficCom	Provides intelligent mobility solutions and professional services focused on smart city requirements, such as vehicle miles traveled; urban congestion; integrated corridor management; and, connected vehicle and advanced traffic management systems.	http://www.kapsch.com/
Kinetic Metrics LLC	Assists clients in understanding and implementing U.S. safety compliance regulations; as well as, completing research into autonomous and electric vehicle solutions.	http://www.kineticmetrics.com/
McCain, Inc.	Provides Intelligent Transportation hardware and software.	http://www.mccain-inc.com/
MetroTech	Converts current camera infrastructure into traffic sensors; and, publishes real-time, lane-level data for improved signal timing, driver safety and transportation planning.	http://www.metrotech-net.com/
Mobilitie	Supports the US Department of Transportation and Smart Cities with connected and autonomous vehicles; rideshare and bike share programs; and, electric vehicle charging stations,	http://www.mobilitie.com/
PayBySky, Inc.	A parking payment service for commercial delivery and service vehicles.	http://www.paybysky.com/
Proterra, Inc.	Provides design and manufacturing of zero-emission, battery-electric transit vehicles, that enable bus fleet operators to significantly reduce operating costs, while delivering clean, quiet transportation to communities across North America.	http://www.proterra.com/
SGS Testcom	Provides inspection, verification, testing and certification on the use of information technology and data analytics in the transportation industry.	http://sgs.com/
Smarking, Inc.	Provides holistic data analytics platforms designed to get smart cities organized around their existing parking data, in order to implement smarter pricing; inventory allocation; planning decisions; and, seamless growth into the connected age.	http://www.smarking.net/
V5 Systems	Manufactures self-powered, portable computing systems for outdoor intelligent security and industrial internet applications.	http://v5systems.us/
American Council for an Energy-Efficient Economy	Quantifies the energy-related benefits of advanced transportation solutions; and, evaluates transportation programs at the state and local levels.	http://www.aceee.org/
Arup	Analyzes the future of smart cities and new mobility, in terms of travel demand; energy and emissions; policy; data; equity; economics; and, the built environment.	http://www.arup.com/
Carma	Provides real-time ridesharing, car sharing and carpooling.	http://www.gocarma.com/
Centerline Solutions LLC	Provides the design, building, modification and maintainability of wireless systems.	http://www.centerlinesolutions.com/
CMG	Provides a smart city planning framework and services that include the how to on development of strategies; use cases; business cases; processes; and, the use of best practices to create, deploy and manage the digitalization of energy, water, telecommunications, healthcare, transportation, education, public safety, public works, parks and recreation, economic development, legal, financial, and e-government services.	https://512cmg.com/
Connecthings	Deploys in the public space, contactless technologies (Bluetooth Low Energy, NFC, QR code, Wi-Fi) that turn urban, passive physical assets (street furniture, bus stops, public lighting) into smart, connected objects, that are able to interact in the smart city with mobile users, in bringing contextualized, high added value information.	http://www.connecthings.com/

Organization	Description	URL
Current powered by GE, Intelligent Cities	Enables smart cities to leverage their street lighting to be a data-collecting network that sees, hears and harnesses the data that is everywhere.	http://www.currentbyge.com/ideas/intelligent-cities
Econolite Group, Inc.	Provides a platform through which data is exchanged between vehicles and infrastructure, specifically signalized intersections and roadside equipment (RSE).	http://www.econolitegroup.com/
Engage Mobile Solutions	Develops applications to reduce friction for users in areas such as casual carpooling.	http://www.engagemobile.com/
FreeWave Technologies, Inc.	Provides long range wireless Machine-to-Machine communications devices; and, enables remote management and control of traffic control devices, video solutions and wi-fi access.	http://www.engagemobile.com/
Garageio	Is a WiFi-enabled garage controller that modernizes delivery logistics by turning garages into a personal delivery destination.	http://garageio.com/
Global Traffic Technologies, LLC	Provides solutions to move people through cities quicker, improves emergency vehicle safety and reduces transportation emissions.	http://http/www.gtt.com/
IBI Group	Designs every aspect of a truly integrated smart city for people to live, work, and play–from high-rises to industrial buildings; schools to state-of-the-art hospitals; transit stations to highways; airports to toll systems; and, bike lanes to parks,	http://www.ibigroup.com/
INRIX, Inc.	Provides traffic and parking information, connected car services, and smart city analytics.	http://www.inrix.com/
Iteris, Inc.	Provides transportation elements associated with Smart Cities, including intelligent, sensor-based infrastructure to collect and differentiate vehicle, pedestrian, and bicyclists at traffic signals; as well as, enabling urban automation with mobility data aggregation and real-time analyses to improve safety, enhance the value of regional mobility and local accessibility, and ready the roadways for connected vehicle transportation.	http://www.iteris.com/
LVL1 LLC	Provides support for smart cities and related solutions for Network Management, Content Development, Content Management, On-Site Technicians, and other services.	http://www.mylvl1.com/
moovee innovations Inc	Provides driverless vehicle platforms and technologies under centimeter level accuracy in positioning and navigation.	http://www.mooveeinnovations.com/
NRG EVgo	Provides drivers access to more than 1000 fast charging locations, along with individual charging stations at homes, schools, offices, multi-family communities and hospitals across the network.	http://https/www.nrgevgo.com/
Pangea Motors LLC	Provides the platform for application specific vehicles that can be tailored for specific use, operating environment and performance criteria.	http://www.pangeamotors.com/
Park Smart s.r.l.	Provides information on the availability of parking to mobile APPs for drivers; as well as, helping parking management companies to monitor their spaces efficiently.	http://www.buttiglieri.it/
Peek Traffic Corporation	Provides hardware, software and firmware to monitor and control ITS devices.	http://www.peektraffic.com/
Phantom Iot	Provides unified security and management for IoT systems.	http://www.phantomiot.com/
Pivotal Software, Inc	Provides predictions for the smart city and smart vehicle of the future and traffic data sets.	http://pivotal.io/
PLAT.ONE INC	Provides an application and data management platform that enables centralized monitoring and control of smart city services regardless of the underlying technologies, vendors, or devices.	http://www.plat.one/
Purple	Provides a multilingual service, monetizes the WiFi network, surface tourism, transport information and a better understanding of the footfall.	http://www.purple.ai/
RJ Vincent Enterprises LLC	Provides complicated multifaceted technical projects with an emphasis on design, development, and fielding of complex systems.	http://vincent-enterprises.com/blog/

Organization	Description	URL
Smart City Results	Provides vehicle electrification; electrification infrastructure; wireless communications; vehicle automation; new forms of shared mobility; cost-benefit analysis; performance measurement; transportation economics and finance; community development; team building; process analysis; and preparation of funding and project proposals.	http://www.smartcityresults.com/
Soofa	Provides residents and visitors with free charging for mobile devices, public Wi-Fi, and pedestrian analytics.	http://www.soofa.com/
TCS Digital Software and Solutions Group	Provides IoT enabled, integrated IUX platform for big data; pre-built IUX applications with domain-specific data models for transportation, water, and energy; and, a set of APIs for third party application development.	https://dss.tcs.com/
Teknotraffic	Provides smart city consulting services.	http://www.teknotraffic.com/
The Digit Group (TDG), Inc.	Provides master planning, design and development of smart cities.	https://www.thedigitgroupinc.com/
Transdev North America	Provides smart mobility solutions for electric vehicles, automated vehicles, ride-hailing, ride-splitting, first-last mile transit access solutions, shared ride, vehicle maintenance, taxi, paratransit, rail, and fixed-route bus operations.	http://transdevna.com/
Transmap Corporation	Provides pavement and roadway asset management.	http://www.transmap.com/
Velociti, Inc.	Manages the installation and deployment of a wide range of technology solutions for Transportation, Retail, Distribution, Manufacturing, Education, Healthcare, Government, Hospitality, Restaurants and Public Venues.	http://www.velociti.com/
Xaqt	Provides smart data applications that make smart cities and communities safer and more connected; as well as, open standards, frameworks and SDKs that bring new innovation to smart cities, and creates new markets for developers.	https://www.xaqt.com/
Acyclica	Provides high-resolution analytics and information at the arterial level.	http://www.acyclica.com/
Airbornway Corp.	Provides a transportation system that is like a tram; it rides on a pair of cables just like a tram, but that's where the similarities end–it drives itself along the pair of cables (self driven), where it can turn and travel any distance.	http://airbornway.com/
Bentley Systems	Provides users with a leverage on information mobility across many disciplines; and, infrastructure project lifecycles, to deliver better-performing projects and assets.	https://www.bentley.com/
ChargePoint	Provides charging stations, mobile apps, and the network that allow people to charge their cars everywhere they go; as well as, transforming the energy industry by developing intelligent energy management solutions.	http://www.chargepoint.com/
CivicSmart, Inc.	Provides a highly accurate, radar-based sensor that captures real-time, on-street parking occupancy and violation data; delivers guidance information to motorists; supports demand-based meter pricing; promotes parking program efficiency; reduces traffic congestion; and, increases smart city revenues.	http://www.civicsmart.com/
Grush Niles Strategic	Works with local governments, public transit agencies, taxi and transportation network companies, advanced vehicle developers, and transportation innovators, to prepare for the coming public and private markets for Transportation as a Service (TaaS); as well as, planning and preparing for the development, marketing and growth shared-vehicle fleets, including autonomous public transit fleets, designed to become massive and pervasive over time.	http://grushniles.com/
KIOSK Information Systems	Provides manufacturing, engineering, services and self-service Kiosks.	http://www.kiosk.com/
Louisville Metro – Advanced Planning	Provides real time traffic and bus data.	https://data.louisvilleky.gov/group/advanced-planning
MVMANT	Provides public transit on demand, where the displacement of vehicles is managed by predictive algorithms.	http://www.mvmant.com/

Organization	Description	URL
NEXT Future Transportation	Provides a modular, scalable and personalized smart city transportation system.	http://www.next-future-mobility.com/
PennGlobe/ PennSMART	Provides Smart IoT Lighting solutions, ranging from energy efficient LED platforms; discretely integrated security cameras; , sensors (gunshot, hazardous material, precipitation, etc); electric car charging stations; wifi; kiosks; and, back-end networking.	http://www.pennsmart.com/
Ridesharing Institute	Provides technologies such as autonomous and connected vehicles; sensor-based infrastructure; smart parking; urban analytics; smart grid and EV's; and, multi-modal transportation.	http://www.bosch.us/
Savari	Deploys communication ecosystems in smart cities that enable vehicle-to-vehicle, vehicle-to-smartphone, vehicle-to-traffic light, and vehicle-to-infrastructure communication.	http://www.savari.net/
StreetLight Data	Provides analytics describing key mobility behavior with smart cities and run routing; origin/destination; before/after comparisons; and, more in a matter of minutes, using billions of real-world trips sourced from mobile devices.	http://www.streetlightdata.com/
SYNC City LLC	Provides free Wi-Fi Kiosks to Smart Cities supported by advertising revenues.	http://www.sync.city/
URB-E	Links city dwellers to transportation hubs efficiently and effectively, while eliminating the need for parking, sitting in traffic, and paying for gas.	http://www.urb-e.com/
Vim and Vigor Development LLC	Provides a powerful, as well as, wired charging solution that is user-friendly, and safe to use to customers everywhere around the smart city.	http://www.vimvigordevelopment.com/
Vizonomy	Helps smart cities increase their resiliency against flooding and other natural hazards by using cloud computing, real-time mapping, and interactive graphics, which together save planners hundreds of hours in data collection, analysis, and visualization.	http://www.vizonomy.com/
iCabbi	Provides a dispatch system that monitors vehicle positions, availability, waiting times, advanced reservations and data relating to all trips (upcoming or completed)	http://www.icabbi.com/
Via Transportation, Inc.	Uses technology to move people from multiple origins to multiple destinations, to seamlessly share rides in professionally-driven vehicles – in real time.	http://platform.ridewithvia.com/
911Cellular	Provides a mass notification system; as well as, a safety app, which uses an indoor positioning system (IPS) to help responders more accurately locate distressed individuals.	http://www.911cellular.com/
Citilabs	Provides software and data platforms for use in smart management, operation and planning.	http://www.citilabs.com/
Commuter Cars Corp	Provides a car with the footprint of a motorcycle, and the weight of a midsize sedan with more side impact protection.	http://www.commutercars.com/
Connected Signals, Inc.	Uses existing infrastructure to connect traffic signals to vehicles.	http://www.connectedsignals.com/
Live Traffic Data (LTD), LLC	Provides ways to help smart cities provide their citizenry new and smart public benefits from their public agency data.	http://www.livetrafficdata.com/
NFC-Interactive	Specializes in mobility technologies and digital services to smart cities in several fields like mobile ticketing (lightweight solutions for public transportation); event management; construction site management (information for people, mobility tools for workers); and waste management.	http://http/www.nfc-interactive.fr/
OpenDataSoft	Provides a SaaS platform that serves as both an open data portal and a smart city data hub that is natively designed for real-time sensor data, and includes built-in API generation and data visualization.	https://www.opendatasoft.com/

Appendix B

List of smart cities and critical infrastructure policies and financing products

John R. Vacca
TechWrite, Pomeroy, OH, United States

Policy/product	Country/region	City	URL
Go-To-Market policies	USA	New York	https://zmaps.net/company/151-advisors-llc-9175215272
Software Defined Networking (SDN)	Ireland	Dublin	https://www.asavie.com/
CivicConnect Platform	USA	Los Angeles	http://www.civicresource.com/
HERE Open Location Platform	USA	Alpharetta	https://www.here.com/platform
MPower Smart City platform	USA	Phoenix	http://www.m2mspectrum.com
Mobile human resource management tool	USA	Miami	https://www.safecheckusa.com/
ZoomThru cloud platform	USA	New York	http://www.zoomthru.co
Urban Mobility Solutions	USA	Durham	http://www.organictransit.com
Placemeter urban intelligence platform	USA	New York	https://techcrunch.com/2016/05/24/placemeters-urban-intelligence-platform-gets-smarter/
Rajant Kinetic Mesh®	USA	Malvern	https://rajant.com/technology/rajant-kinetic-wireless-mesh-networks/
InPave	USA	Anaheim	https://www.bergkampinc.com/products/inpave-system-technology/
SKY Smart Charging charge management platform	USA	Los Angeles	https://chargedevs.com/newswire/sky-smart-charging-uses-open-standards-to-help-utilities-manage-ev-loads/
Microgrid with IoT	USA	Merced	https://www.f6s.com/arqbravogroup
Digital Transportation Solutions	USA	San Jose	https://www.cisco.com/c/en/us/solutions/industries/transportation.html
SaaS Decision Engines	USA	New York	http://dukodestudio.com/mindriderdata/

Policy/product	Country/region	City	URL
eBeacons mobile app, Bluetooth beacons, and Beacon Reader device	USA	Kansas City	https://ebsystems.co/
Smart City Challenge	USA	Portland	https://www.moovelus.com/
Moovit	Isreal	Ness Ziona	https://m.moovitapp.com/
Overland ATS	USA	Gainesville	https://www.overlandats.com/
Connected Vehicle pilots	USA	New York	https://www.siriusxm.com/
Graydient platform	USA	Houston	https://www.securityinfowatch.com/video-surveillance/company/10482504/giant-gray-formerly-brs-labs
Turnkey ecosystem and platform	USA	Denver	https://www.f6s.com/placeglobalPlace Global
Smart Panel	USA	Madison	http://www.brandbumps.com/about
ALMANAC	Italy	Torino	https://almanacproject.github.io/
Amsterdam Smart Light	Netherlands	Amsterdam	https://www.luciassociation.org/smart-lighting-for-amsterdams-hoekenrodeplein-square/
Automatic Urban and Rural Network (AURN)	UK	Not Available (N/A)	https://uk-air.defra.gov.uk/networks/network-info?view=aurn
Berliner Badewasserqualität	Germany	Berlin	http://daten.berlin.de/datensaetze/liste-der-badestellen-badegewäA4sserqualität
Bristol Air Quality Data	UK	Bristol	http://www.bristol.airqualitydata.com/
Bristol is Open	UK	Bristol	https://www.wired.co.uk/news/archive/2015-03/17/bristol-smart-city
Building Smarter and Safer Communities	USA	Charleston	https://smartcitiescouncil.com/resources/building-smarter-and-safer-communities
Chicago Array of Things	USA	Chicago	https://arrayofthings.github.io/
Cities Unlocked	UK	Not Available (N/A)	http://www.citiesunlocked.org.uk/
City2Mobil2	Europe	Not Available (N/A)	https://cordis.europa.eu/result/rcn/191677_en.html
City of Chicago Data Portal	USA	Chicago	https://data.cityofchicago.org/
Civic Hall NYC	USA	New York	http://civichall.org/
Connecting Cities	Global	N/A	http://connectingcities.net/
DIMIS	Germany	N/A	http://www.iis.fraunhofer.de/de/ff/lok/proj/dimis.html
Eeneind Street Lighting	Netherlands	Eeneind	http://www.tvilight.com/projects/city-nuenen/
EkoBus	Serbia	Belgrade, Pancevo	http://www.libelium.com/smart_city_environmental_parameters_public_transportation_waspmote/
EPIC European Platform for Intelligent Cities	Europe	N/A	http://www.epic-cities.eu/
Express Park Program	USA	Los Angeles	http://www.laexpresspark.org/about-la-expresspark/
Forum Virium Helsinki	Finland	Helsinki	http://www.forumvirium.fi/en
Glasgow Future City	UK	Glasgow	https://futurecity.glasgow.gov.uk/
Greater Manchester Data Synchronisation Programme	UK	Manchester	http://iyfsearch.com/?dn=gmdsp.org.uk&pid=9PO755G95
Hamburg Smart City	Germany	Hamburg	https://www.e-zigurat.com/blog/en/urban-development-the-smart-city-hamburg-project/
Hamburg smartPORT	Germany	Hamburg	https://newsroom.cisco.com/press-release-content?type=webcontent&articleId=1414144
Helsinki Region Infoshare (HRI)	Finland	Helsinki	http://www.hri.fi/en
iCity	Spain	Barcelona	http://www.icityproject.eu/
IssyGrid	France	Issy-Les-Moulineaux	http://imaginationforpeople.org/maintenance/maintenance.htm

Policy/product	Country/region	City	URL
Kalasatama Model District of Smart Urban Development	Finland	Helsinki	http://en.uuttahelsinkia.fi/
La Garrotxa	Spain	La Garrotxa	http://www.libelium.com/smart-city-urban-resilience-smart-environment/
LinkNYC	USA	New York	http://www.link.nyc/
London Data Store	UK	London	http://data.london.gov.uk/
Los Angeles Synchronised Traffic Lights	USA	Los Angeles	https://www.pcworld.com/article/2032914/los-angeles-synchronizes-all-its-traffic-lights-lets-a-computer-control-the-roads.html
Madrid smart parking meters	Spain	Madrid	https://www.theguardian.com/world/2014/apr/30/madrid-smart-parking-metres-polluting-cars
Milton Keynes' Internet of Things	UK	Milton Keynes	http://www.mksmart.org/
Monaco 3.0	Monaco	Monaco	http://blog.bosch-si.com/categories/projects/2013/12/smart-city-monaco-3-0-bosch-technology/
NAPA	Germany	N/A	http://www.iis.fraunhofer.de/de/ff/lok/proj/napa.html
Nijmegen Roadway Lighting	Netherlands	Nijmegen	http://www.tvilight.com/projects/intelligent-roadway-lighting/
Open Data Hamburg	Germany	Hamburg	http://transparenz.hamburg.de/open-data/
PlanIT Valley	Portugal	Porto	http://www.living-planit.com/
RESCATAME	Spain	Salamanca	http://www.libelium.com/smart_city_air_quality_urban_traffic_waspmote/
Sensing London	UK	N/A	https://futurecities.catapult.org.uk/project/sensing-london/
Smart Airport Experience	UK	London	http://www.airport-business.com/2015/06/london-city-investing-smart-airport-experience-expedite-passenger-journey/
Smart Blocks	Australia	Melbourne	https://communicationscollective.com.au/city-of-melbourne-smart-blocks/%20target=
Smart Cities Demo Aspern (SCDA)	Austria	Aspern	https://smartcity.wien.gv.at/site/en/
Smart City	Sweden	N/A	https://international.stockholm.se/city-development/the-smart-city/
Smart City Berlin	Germany	Berlin	http://www.berlin-partner.de/en/the-berlin-location/smart-city-berlin/
Smart City Commission	UK	Birmingham	https://birminghamsmartcity.wordpress.com/
Smart London Initiative/Datastore	UK	London	http://data.london.gov.uk/
Smart Santander	Spain	Santander	http://www.smartsantander.eu/
SMILE	Austria	Vienna	http://smile-einfachmobil.at/
Songdo International Business District	South Korea	Songdo	http://www.songdo.com/
Stockholm Royal Seaport Project	Sweden	Stockholm	http://www.environmentalleader.com/2012/07/11/smart-cities-stockholm-royal-seaport-project-update/
Streetmix	USA	N/A	http://streetmix.net/
Trash Track	USA	N/A	http://senseable.mit.edu/trashtrack/
UK Air Pollution	UK	N/A	http://uk-air.defra.gov.uk/air-pollution/
Whereabouts London	UK	London	http://whereaboutslondon.org/#/about
WindyGrid	USA	Chicago	http://www.cityofchicago.org/city/en/narr/foia/CityData.html
XALOC	Spain	Barcelona	http://www.uab.cat/web/latest-news/news-detail/new-system-helps-locate-car-park-spaces-1096476786473.html?noticiaid=1278568901108
Yokohama Smart City Project	Japan	Yokohama	https://www.toshiba.co.jp/about/press/2011_12/pr1601.htm

Appendix C

List of smart cities and critical infrastructure policies and financing standards

John R. Vacca
TechWrite, Pomeroy, OH, United States

Standard	Number	Type	URL
Services	03.080	International Classification for Standards (ICS)	https://www.standard.no/en/webshop/productcatalog/productpresentation/?ProductID=247499
Transport	03.220	ICS	https://www.standard.no/en/webshop/productcatalog/productpresentation/?ProductID=247524
Medical sciences and health care facilities in general	11.020	ICS	https://www.normadoc.com/english/ics/11-health-care-technology/11-020-medical-sciences-and-health-care-facilities-in-general.html
Aids for disabled or handicapped persons	11.180	ICS	https://www.iso.org/ics/11.180.html
Environmental protection	13.020	ICS	https://www.iso.org/ics/13.020.html
Wastes	13.030	ICS	https://www.iso.org/ics/13.030.html
Air quality	13.040	ICS	https://www.iso.org/ics/13.040.html
Water quality	13.060	ICS	https://www.iso.org/ics/13.060.html
Accident and disaster control	13.200	ICS	https://www.iso.org/ics/13.200.html
Alarm and warning systems	13.320	ICS	https://www.iso.org/ics/13.320.html
Energy and heat transfer engineering in general	27.010	ICS	https://www.iso.org/ics/27.010.html
Heat pumps	27.080	ICS	https://www.iso.org/ics/27.080.html
Solar energy engineering	27.160	ICS	https://www.iso.org/ics/27.160.html
Electrical accessories	29.120	ICS	https://www.iso.org/ics/29.120.html
Switch gear and control gear	29.130	ICS	https://www.iso.org/ics/29.130.html
Power transmission and distribution networks	29.240	ICS	https://www.iso.org/ics/29.240.html

Standard	Number	Type	URL
Electronic display devices	31.120	ICS	https://www.iso.org/ics/31.120.html
Electromechanical components for electronic and telecommunications equipment	31.220	ICS	https://www.iso.org/ics/31.220.html
Telecommunications in general (including infrastructure)	33.020	ICS	https://www.iso.org/ics/33.020.html
Telecommunication services. Applications (including supplementary services, service aspects and associated legal traceability aspects)	33.030	ICS	https://www.iso.org/ics/33.030.html
Telecommunication systems (including network (system) aspects)	33.040	ICS	https://www.iso.org/ics/33.040.html
Mobile services	33.070	ICS	https://www.iso.org/ics/33.070.html
Components and accessories for telecommunications equipment	33.120	ICS	https://www.iso.org/ics/33.120.html
Telecontrol. Telemetering	33.200	ICS	https://www.iso.org/ics/33.200.html
Information technology (IT) in general	35.020	ICS	https://www.iso.org/ics/35.020.html
Software	35.080	ICS	https://www.iso.org/ics/35.080.html
Networking	35.110	ICS	https://www.iso.org/ics/35.110.html
Interface and interconnection equipment	35.200	ICS	https://www.iso.org/ics/35.200.html
Applications of information technology	35.240	ICS	https://www.iso.org/ics/35.240.html
Road vehicles in general	43.020	ICS	https://www.iso.org/ics/43.020.html
Road vehicle systems	43.040	ICS	https://www.iso.org/ics/43.040.html
Electrical and electronic equipment and control systems	43.060.50	ICS	https://www.iso.org/ics/43.060.50.html
Commercial vehicles	43.080	ICS	https://www.iso.org/ics/43.080.html
Passenger cars. Caravans and light trailers	43.100	ICS	https://www.iso.org/ics/43.100.html
Electric road vehicles	43.120	ICS	https://www.iso.org/ics/43.120.html
Special purpose vehicles	43.160	ICS	https://www.iso.org/ics/43.160.html
Railway engineering in general	45.020	ICS	https://www.iso.org/ics/45.020.html
Materials and components for railway engineering	45.040	ICS	https://www.iso.org/ics/45.040.html
Rails and railway components	45.080	ICS	https://www.iso.org/ics/45.080.html
Freight distribution of goods	55.180	ICS	https://www.iso.org/ics/55.180.html
Distribution and vending machines	55.230	ICS	https://www.iso.org/ics/55.230.html
Construction industry	91.010	ICS	https://www.iso.org/ics/91.010.html
Physical planning. Town planning	91.020	ICS	https://www.iso.org/ics/91.020.html
Buildings	91.040	ICS	https://www.iso.org/ics/91.040.html
Installations in buildings	91.140	ICS	https://www.iso.org/ics/91.140.html
Lighting	91.160	ICS	https://www.iso.org/ics/91.160.html
Building accessories	91.190	ICS	https://www.iso.org/ics/91.190.html
Civil engineering in general	93.010	ICS	https://www.iso.org/ics/93.010.html
Road engineering	93.080	ICS	https://www.iso.org/ics/93.080.html
Construction of railways	93.100	ICS	https://www.iso.org/ics/93.100.html
Construction of airports	93.120	ICS	https://www.iso.org/ics/93.120.html

Standard	Number	Type	URL
Construction of waterways, ports and dykes	93.140	ICS	https://www.iso.org/ics/93.140.html
Home economics in general	97.020	ICS	https://www.iso.org/ics/97.020.html
Domestic electrical appliances in general	97.030	ICS	https://www.iso.org/ics/97.030.html
Kitchen equipment	97.040	ICS	https://www.iso.org/ics/97.040.html
Laundry appliances	97.060	ICS	https://www.iso.org/ics/97.060.html
Cleaning appliances	97.080	ICS	https://www.normadoc.com/english/ics/97-domestic-and-commercial-equipment-entertainment-sports/97-080-cleaning-ap-pliances.html
Domestic, commercial and industrial heating appliances	97.100	ICS	https://www.iso.org/ics/97.100.html
Automatic controls for household use	97.120	ICS	https://www.iso.org/ics/97.120.html
Shop fittings	97.130	ICS	https://www.iso.org/ics/97.130.html
Miscellaneous domestic and commercial equipment	97.180	ICS	https://www.iso.org/ics/97.180.html
Equipment for entertainment	97.200	ICS	https://www.iso.org/ics/97.200.html
Systems Evaluation Group on Smart Cities – Most of their activities seem to be working group reports	SEG 1	IEC	http://noderedguide.com/smart-city-standards-an-overview/
Sustainable cities and communities — Vocabulary	37100:2016	ISO	https://infostore.saiglobal.com/en-us/Standards/ISO-37100-2016-607033_SAIG_ISO_ISO_1391510/
Sustainable development and resilience of communities – Management systems – General principles and requirements	37,101	ISO	https://pecb.com/en/education-and-certification-for-individuals/iso-37101
Sustainable development in communities — Management system for sustainable development — Requirements with guidance for use	37101:2016	ISO	https://infostore.saiglobal.com/en-us/Standards/ISO-37101-2016-606594_SAIG_ISO_ISO_1390363/
Sustainable development and resilience of communities – Vocabulary	37,102	ISO	https://global.ihs.com/doc_detail.cfm?document_name=ISO%2037102&item_s_key=00674095
Sustainable cities and communities — Transforming our cities — Guidance for practical local implementation of ISO 37101	37104:2019	ISO	https://infostore.saiglobal.com/en-au/standards/iso-37104-2019-1148043_saig_iso_iso_2722831/
Sustainable cities and communities — Descriptive framework for cities and communities	37105:2019	ISO	https://infostore.saiglobal.com/en-au/standards/iso-fdis-37105-2019-1159394_saig_iso_iso_2754465/
Sustainable cities and communities — Guidance on establishing smart city operating models for sustainable communities	37106:2018	ISO	https://infostore.saiglobal.com/en-au/standards/iso-37106-2018-1132896_saig_iso_iso_2674837/
Sustainable cities and communities — Guidance on establishing smart city operating models for sustainable communities — Amendment 1	37106:2018/CD Amd 1	ISO	https://iso.ch/ru/standard/76545.html?browse=tc
Sustainable development of communities – Indicators for city services and quality of life	37120:2014: 2014	ISO	https://www.iso.org/obp/ui/#iso:std:iso:37120:ed-1:v1:en
Sustainable cities and communities — Business districts — Guidance for practical local implementation of ISO 37101	37,108	ISO/AWI	https://www.iso.org/standard/62067.html

Standard	Number	Type	URL
Sustainable development of communities — Management systems — Requirements with guidance for resilience and smartness	37,101	ISO/DIS	https://ec.europa.eu/eip/ageing/standards/city/smart-cities/isodis-37101_en
Inventory and review of existing indicators on sustainable development and resilience in cities	37,121	ISO/DTR	https://ec.europa.eu/eip/ageing/standards/city/smart-cities/isodtr-37121_en
Smart Cities	JTC 1	ISO IEC	https://www.iso.org/iso/smart_cities_report-jtc1.pdf
Sustainable development in communities — Indicators for smart cities	37,122	ISO/NP	https://ec.europa.eu/eip/ageing/standards/city/smart-cities/isonp-37122_en
Inventory and review of existing indicators on sustainable development and resilience in cities	37,121	ISO/TR	https://www.iso.org/obp/ui/#iso:std:iso:tr:37121:ed-1:v1:en:sec:5
Smart community infrastructures — Review of existing activities relevant to metrics	37,150	ISO/TR	https://standards.globalspec.com/std/1663072/iso-tr-37150
Smart community infrastructures — Common framework for development and operation	37,152	ISO/TR	https://standards.globalspec.com/std/10034071/iso-tr-37152
Sustainable cities and communities — Maturity model for smart sustainable communities	37107:2019	ISO/TS	https://shop.belgiss.by/en/mezhdunarodnye-standarty-iso/iso-ts-37107:2019
Smart community infrastructures — Principles and requirements for performance metrics	37151:2015: 2015	ISO/TS	https://www.iso.org/obp/ui/fr/#iso:std:iso:ts:37151:en
Sustainable development and communities — Practical guidance for project developers — Meeting ISO 37101 framework principles	37,109	ISO/WD	https://www.iso.org/standard/62068.html
Sustainable cities and communities — Management guidelines of open data for smart cities and communities — Part 1: Overview and general principles	37,110	ISO/WD	https://www.iso.org/standard/62069.html
Sustainable development of communities — Indicators for city services and quality of life	37,120	ISO/WD	https://ec.europa.eu/eip/ageing/standards/city/smart-cities/isowd-37120_en
Focus group on smart sustainable cities	SG5 FG-SSC	ITU-T	https://www.itu.int/en/ITU-T/focusgroups/ssc/Pages/default.aspx
Smart cities — Infrastructures — Universal accessibility	178,106	PNE	https://ec.europa.eu/eip/ageing/standards/city/smart-cities/pne-178106_en
Accessible mobility in smart cities	178,306	PNE	https://ec.europa.eu/eip/ageing/standards/city/smart-cities/pne-178306_en
Management system of smart tourist destinations — Requirements	178,501	PNE	https://ec.europa.eu/eip/ageing/standards/city/smart-cities/pne-178501_en
Smart Sustainable cities — Analysis of Definitions	0100-Rev 2	SSC	https://www.iso.org/iso/smart_cities_report-jtc1.pdf
Technical Report on Standardization Activities and Gaps for SSC and suggestion to SG5, ITU-T	0110	SSC	https://www.iso.org/iso/smart_cities_report-jtc1.pdf
Key performance indicators (KPIs) definitions for smart sustainable cities	162	SSC	https://www.iso.org/iso/smart_cities_report-jtc1.pdf

Standard	Number	Type	URL
Smart cities – Open Data	178301:2015: 2015	UNE	https://www.en-standard.eu/une-178301-2015-smart-cities-open-data/
Risk Management Framework for Information Systems and Organizations: A System Life Cycle Approach for Security and Privacy	Special Publication (SP) 800-37 Rev. 2	NIST	https://csrc.nist.gov/publications/detail/sp/800-37/rev-2/final
Managing Information Security Risk: Organization, Mission, and Information System View	SP 800-39	NIST	https://csrc.nist.gov/publications/detail/sp/800-39/final
Standards for Security Categorization of Federal Information and Information Systems	Federal Information Processing Standard (FIPS) 199	NIST	https://nvlpubs.nist.gov/nistpubs/FIPS/NIST.FIPS.199.pdf
Guide for Mapping Types of Information and Information Systems to Security Categories	SP 800-60 Rev. 1	NIST	https://csrc.nist.gov/publications/detail/sp/800-60/vol-1-rev-1/final
Minimum Security Requirements for Federal Information and Information Systems	FIPS 200	NIST	https://csrc.nist.gov/publications/detail/fips/200/final
Security and Privacy Controls for Information Systems and Organizations	SP 800-53 Rev. 5 (Draft)	NIST	https://csrc.nist.gov/publications/detail/sp/800-53a/rev-5/draft
Guide for Security-Focused Configuration Management of Information Systems	SP 800-128	NIST	https://csrc.nist.gov/publications/detail/sp/800-128/final
Contingency Planning Guide for Federal Information Systems	SP 800-34 Rev. 1	NIST	https://csrc.nist.gov/publications/detail/sp/800-34/rev-1/final
Computer Security Incident Handling Guide 21 of 65	SP 800-61 Rev. 2	NIST	https://csrc.nist.gov/publications/detail/sp/800-61/rev-2/final
Cybersecurity and Privacy Advisory Committee Guidebook	Global City Teams Challenge 2019: Smart and Secure Cities and Communities Challenge (SC3) GCTC-SC3	NIST	https://pages.nist.gov/GCTC/uploads/blueprints/2019_GCTC-SC3_Cybersecurity_and_Privacy_Advisory_Committee_Guidebook_July_2019.pdf
Assessing Security and Privacy Controls in Federal Information Systems and Organizations: Building Effective Assessment Plans	SP 800-53A Rev. 4	NIST	https://csrc.nist.gov/publications/detail/sp/800-53a/rev-4/final
Information Security Continuous Monitoring (ISCM) for Federal Information Systems and Organizations	SP 800-137	NIST	https://csrc.nist.gov/publications/detail/sp/800-137/final
Chain Risk Management Practices for Federal Information Systems and Organizations NIST Internal Report	SP 800-161	NIST	https://csrc.nist.gov/publications/detail/sp/800-161/final
An Introduction to Privacy Engineering and Risk Management in Federal Systems NIST Cybersecurity Framework v1.1 Framework for Improving Critical Infrastructure Cybersecurity	(NISTIR) 8062	NIST	https://csrc.nist.gov/publications/detail/nistir/8062/final
The Cybersecurity Framework: Implementation Guidance for Federal Agencies	NISTIR 8170 (Draft)	NIST	https://csrc.nist.gov/csrc/media/publications/nistir/8170/draft/documents/nistir8170-draft.pdf
Risk Management – Guidelines	31000:2018	ISO	https://www.iso.org/standard/65694.html

Standard	Number	Type	URL
Information Security Management Systems (ISMS) Standards Institute of Risk Management (IRM)/The Public Risk Management Association (Alarm)/The Association of Insurance and Risk Managers (AIRMIC) 2002 Risk Management Standard Committee of Sponsoring Organizations of the Treadway Commission	27,000	ISO/IEC	https://www.iso.org/obp/ui/#iso:std:iso-iec:27000:ed-4:v1:en
Enterprise Risk Management – Integrated Framework	2004	COSO	https://www.coso.org/documents/coso-erm-executive-summary.pdf
Red Book Governance, Risk, and Compliance (GRC) Capability Model	N/A	OCEG	https://www.oceg.org/standards/
A Business Framework for the Governance and Management of Enterprise IT Given the information/data- and technology-centric nature of Smart Cities	COBIT 5	ISACA	https://www.isaca.org/resources/isaca-journal/issues/2018/volume-5/using-cobit-5-to-get-and-give-board-support-for-revolutionizing-cities

Appendix D

List of miscellaneous smart cities and critical infrastructure policies and financing resources

John R. Vacca
TechWrite, Pomeroy, OH, United States

Resource	URL
OCIA – The Future of Smart Cities – Cyber-Physical Infrastructure Risk	https://www.us-cert.gov/sites/default/files/documents/OCIA%20-%20The%20Future%20of%20Smart%20Cities%20-%20Cyber-Physical%20Infrastructure%20Risk.pdf
Smart cities readiness guide	http://www.estudislocals.cat/wp-content/uploads/2016/11/SmartCitiesReadinessGuide.pdf
CSTEP Report Smart Cities Framework	https://india.smartcitiescouncil.com/system/tdf/india/public_resources/CSTEP%20REport%20Smart%20Cities%20Framework%20-%20Compendium%20of%20Resources.pdf?file=1&type
Digital Government Smart Cities and Communities	https://www.govloop.com/resources/
Protecting Critical Infrastructure and Roadways: How Smart Cities Create New Risks	https://blog.trendmicro.com/protecting-critical-infrastructure-and-roadways-how-smart-cities-create-new-risks/
Smart city technology: India joins global alliance	https://m.economictimes.com/news/economy/infrastructure/india-joins-global-alliance-on-responsible-use-of-smart-city-technologies/articleshow/71535525.cms
5 Reasons Why Securing Critical Infrastructure is Essential in Smart Cities	https://www.lanner-america.com/blog/5-reasons-securing-critical-infrastructure-essential-smart-cities/
Smart Cities – Arm	https://www.arm.com/solutions/smart-cities
Smart Cities and Communities Federal Strategic Plan – NITRD	https://www.nitrd.gov/drafts/scc_strategicplan_draft.pdf
Bridging the smart cities security divide	https://www.csoonline.com/article/3252844/bridging-the-smart-cities-security-divide.html
Smart City Indicators India – AIRSOFTNET	http://dqox.airsoftnet.nl/smart-city-indicators-india.html
Critical Infrastructure Security and Resilience	https://www.cisa.gov/critical-infrastructure-sectors
Cities policy responses – OECD	https://www.oecd.org/coronavirus/policy-responses/cities-policy-responses-fd1053ff/

Resource	URL
Smart cities readiness guide – Smart City Alliance	https://smartcityalliance.ca/site/assets/files/1465/readiness-guide-v2-8-24-2015.pdf
Cybersecurity: Critical Infrastructure Authoritative Reports and Resources	https://fas.org/sgp/crs/misc/R44410.pdf
Smart cities readiness guide – Estudis locals	http://www.estudislocals.cat/wp-content/uploads/2016/11/SmartCitiesReadinessGuide.pdf
Document OCR: National Security Archive	https://nsarchive.gwu.edu/media/23603/ocr
America's Global Infrastructure Opportunity: Three Recommendations to the New U.S. Development Finance Corporation. CSIS Briefs.	https://www.csis.org/analysis/americas-global-infrastructure-opportunity-three-recommendations-new-us-development-finance
Infrastructure deal could offer window of opportunity for $1.1B smart cities bill.	https://www.smartcitiesdive.com/news/infrastructure-deal-could-offer-window-of-opportunity-for-11b-smart-citie/602474/
Securing critical infrastructure in the smart city	https://www.smartcitiesdive.com/news/securing-critical-infrastructure-in-the-smart-city/579756/
Reconceptualising Smart Cities: Compendium	https://india.smartcitiescouncil.com/system/tdf/india/public_resources/CSTEP%20REport%20Smart%20Cities%20Framework%20-%20Compendium%20of%20Resources.pdf?file=1&type=node&id=2869&force=

Appendix E

Smart cities and critical infrastructure policies and financing frequently asked questions

John R. Vacca
TechWrite, Pomeroy, OH, United States

Question	Answer: URL
What are the techniques for managing municipal solid wastes for smart cities?	https://www.waste360.com/generators/how-smart-cities-are-managing-solid-waste
What does it mean that a city is smart?	https://internetofthingsagenda.techtarget.com/definition/smart-city
What does a smart city mean to you?	https://statescoop.com/what-does-smart-city-mean-in-2018/
What is the definition of smart data?	https://whatis.techtarget.com/definition/smart-data
Does the development of smart cities improve people's lives?	https://www.nlc.org/article/how-smart-cities-will-change-our-lives
What are Data Governance Frameworks for smart cities?	https://smartimpact-project.eu/app/uploads/2018/02/SmartImpact_Data-Gov-and-Intergration_A4_AW.pdf
What technologies dominate in the field of smart ecology sustainable development cities?	https://link.springer.com/article/10.1186/s40309-019-0157-0
What are the trends in smart cities?	https://www2.deloitte.com/us/en/insights/focus/smart-city.html
What is an active mode of transportation (walking/cycling) in smart cities?	https://www.railstotrails.org/media/847675/activetransport_2019-report_finalreduced.pdf
What are the current trends in smart city contracts?	https://cordis.europa.eu/docs/projects/cnect/6/270896/080/deliverables/001-D2221NewtrendsforSmartCities.pdf
What is a heterogeneous smart city dataset?	https://www.mdpi.com/1424-8220/18/9/2994/htm
What is the future of green buildings or smart buildings, or a combination of both practices?	https://idac.co/question/99/What-is-the-future-green-buildings-or-smart-buildings-or-a-combination-of-both-practices
Is there any existing method, technique, or sensors to control the flow of water in a smart pipeline?	http://www.libelium.com/smart_water_wsn_pipe_leakages/
What are the hot research areas in computer science for smart cities?	https://www.techrepublic.com/article/15-hot-tech-jobs-for-smart-cities-in-2018-and-beyond/
Which is the appropriate use of technology for urban interiors, is it intelligent or smart? Why?	https://www.askidac.com/question/434/Which-is-the-appropriate-use-of-technology-for-urban-interiors-is-it-intelligent-or-smart-Why

Question	Answer: URL
What could be the best road to start research in smart charging of electric vehicles?	https://www.sciencedirect.com/science/article/pii/S0968090X17301365
Are there any presentations of smart city projects?	http://smartcities.gov.in/content/presentation.php
What are the protocols and criteria for smart city multimodal traffic management?	https://www.diva-portal.org/smash/get/diva2:975199/FULLTEXT01.pdf
Will the development of autonomous cars be correlated with the development of electric cars?	https://www.quora.com/Will-the-development-of-autonomous-cars-be-correlated-with-the-development-of-electric-cars
How can Artificial Intelligence help make smart cities more competitive?	https://becominghuman.ai/artificial-intelligence-for-smart-cities-64e6774808f8
What are the known future visions of smart city technology development until 2070?	https://pittsburghpa.gov/domi/transport-vision-plan
What are the current examples of smart cities?	https://bipartisanpolicy.org/blog/five-innovative-examples-of-smart-cities-in-the-u-s/
Is it possible to build a fully autonomous energetical and ecological smart city?	https://www.plugandplaytechcenter.com/resources/9-smart-cities-startups-are-reshaping-cities-we-know-them/
What do we need to know to plan and design better smart cities?	https://medium.com/@anthonymobile/smart-cities-what-do-we-need-to-know-to-plan-and-design-them-better-b6d05e736ea1
Is 5G and Ethernet sufficient enough to connect, communicate and help to enable cyber security for 700 million vehicles in a smart city infrastructure?	https://arxiv.org/pdf/1909.08096
Clustering with intensive technology integration, how have smart cities impacted competition?	https://www.tandfonline.com/doi/full/10.1080/23311975.2018.1532777
What are smart city and IOT simulation tools?	https://datasmart.ash.harvard.edu/news/article/simcities-designing-smart-cities-through-data-driven-simulation-893
Is China the technological leader in the construction of environment-friendly metropolises developed in the smart city model?	https://www.tandfonline.com/doi/full/10.1080/13549839.2019.1628730
What technologies are used in the development of computerized smart city systems?	https://mobility.here.com/smart-city-technologies-role-and-applications-big-data-and-iot
Is a artificial intelligence based mathematical model possible to forecast smart cities growing requirements for increasing populations?	https://www.ncbi.nlm.nih.gov/pmc/articles/PMC6957485/
How do you apply the principles of smart cities to the operation of the smart campus?	https://www2.deloitte.com/us/en/pages/consulting/solutions/next-generation-smart-campus.html
What will a smart home look like in the future?	https://planetechusa.com/the-future-of-smart-homes/
Should smart cities be steered by IT technology alone or led by land use planning principles?	https://www.planning.org/planning/2019/mar/smartcities/
How can we connect between banal existent buildings (lacking originality, freshness, or novelty trite) in smart cities?	https://www.smartcitiesworld.net/opinions/opinions/why-smart-cities-need-smart-buildings
Why do we need to have smart cities and not the smart and social people?	https://www.forbes.com/sites/ellistalton/2019/07/09/smart-cities-are-built-by-smart-people-not-smart-things/
What actually is a smart city?	https://meetingoftheminds.org/exactly-smart-city-16098
What are the appropriate mobility models for smart city applications in IoT?	https://mobility.here.com/smart-city-mobility-7-major-cities-getting-it-right
Is smart city a utopia or has it become a reality under a methodological process?	https://lists.w3.org/Archives/Public/semantic-web/2018Jul/0090.html
How can artificial intelligence impact smart cities?	https://becominghuman.ai/artificial-intelligence-for-smart-cities-64e6774808f8
What are some of the most recent advances and future research directions in smart city transportation, by utilizing IoT infrastructures and cloud computing?	https://www.mdpi.com/2079-9292/7/11/309/htm
Where are the smart city monetization research boundaries?	https://thinkbigpartners.com/smart-city-data-monetization/

Question	Answer: URL
How are ideas implemented with regards to research, case studies and crowdsourcing for smart cities?	https://scielo.conicyt.cl/scielo.php?script=sci_arttext&pid=S0718-18762012000300006
What is the most common funding source cities use to finance their smart utility initiatives?	https://urbaninnovation.asu.edu
How does an investor finance a smart city?	https://urbaninnovation.asu.edu
What are common policy models to use for technology projects in public spaces?	https://smartcityconsultants.com
What combination of existing and new approaches exist to fund/finance smart cities?	https://www2.deloitte.com
Are municipal (project and framework) loans on-budget?	https://www.eib.org/
Can a smart city finance a mix investments and other urban investments?	https://www.eib.org/
Do smart city investments pay back on their own?	https://www.eib.org/
Can smart city investments be combined with grant funds?	https://www.eib.org/
Should funds be very bureaucratic?	https://www.eib.org/
Could smart city funding be counted as state aid?	https://www.eib.org/

Appendix F

List of smart cities and critical infrastructure policies and financing case studies

John R. Vacca
TechWrite, Pomeroy, OH, United States

Case studies	URL
Smart Cities: Case Studies – ResearchGate	https://www.researchgate.net/publication/310485601_Smart_Cities_Case_Studies
International Case Studies of Smart Cities – Publications	https://publications.iadb.org/publications/english/document/International-Case-Studies-of-Smart-Cities-Rio-de-Janeiro-Brazil.pdf
Examples and case studies – Smart Cities Council	https://smartcitiescouncil.com/smart-cities-information-center/examples-and-case-studies
Smart Cities and Infrastructure – UNCTAD Case Studies	https://unctad.org/meetings/en/Presentation/ecn162016p01_Wu_en.pdf
Smart City trends-and-case-studies – Deloitte	https://www2.deloitte.com/content/dam/Deloitte/us/Documents/process-and-operations/us-cons-deloitte-smart-city-trends-and-case-studies.pdf
Case study examples of smart city initiatives in environmental sustainability Case Studies	https://www.mdpi.com/2624-6511/2/2/19/htm
Smart Cities – What's In It For Citizens – Intel Newsroom: Case Studies	https://newsroom.intel.com/wp-content/uploads/sites/11/2018/03/smart-cities-whats-in-it-for-citizens.pdf
Smart Cities World: Case Studies	https://smartcitiesworld.net/AcuCustom/Sitename/DAM/012/Understanding_the_Challenges_and_Opportunities_of_Smart_Citi.pdf
Trends in Smart City Development – National League of Cities: Case Studies	https://www.nlc.org/sites/default/files/2017-01/Trends%20in%20Smart%20City%20Development.pdf
Bridging the smart cities security divide: Case Studies	https://www.csoonline.com/article/3252844/bridging-the-smart-cities-security-divide.html
OCIA – The Future of Smart Cities – Cyber-Physical Infrastructure Risk: Case Studies	https://www.us-cert.gov/sites/default/files/documents/OCIA%20-%20The%20Future%20of%20Smart%20Cities%20-%20Cyber-Physical%20Infrastructure%20Risk.pdf
Smart cities: Digital solutions for a more livable future – Case Studies	https://www.mckinsey.com/~/media/mckinsey/industries/capital%20projects%20and%20infrastructure/our%20insights/smart%20cities%20digital%20solutions%20for%20a%20more%.
The Smart City as Global Discourse: Case Studies	https://www.tandfonline.com/doi/full/10.1080/10630732.2018.1558387

Case studies	URL
Smart cities and internet of things: Journal of Information Technology Case and Application Research	https://www.tandfonline.com/doi/full/10.1080/15228053.2019.1587572
A Risk Management Approach to Smart City: Case Studies	https://pages.nist.gov/GCTC/uploads/blueprints/2019_GCTC-SC3_Cybersecurity_and_Privacy_Advisory_Committee_Guidebook_July_2019.pdf
Smart cities with big data: Reference models, challenges, and case studies	https://www.sciencedirect.com/science/article/abs/pii/S0264275117308545
5G security and privacy for smart cities: case studies	https://securelist.com/5g-security-for-smart-cities/95057/
A Guide to a Critical Infrastructure Security and Resilience – CISA: case studies	https://www.cisa.gov/sites/default/files/publications/Guide-Critical-Infrastructure-Security-Resilience-110819-508v2.pdf
Lack of Critical Infrastructure Cybersecurity Investments in Smart Cities: case studies	https://www.businesswire.com/news/home/20190820005103/en/Lack-Critical-Infrastructure-Cybersecurity-Investments-Smart-Cities
Global Innovators: International Case Studies on Smart Cities	https://assets.publishing.service.gov.uk/government/uploads/system/uploads/attachment_data/file/249397/bis-13-1216-global-innovators-international-smart-cities.pdf
Connected LED street lighting: Enabling smart cities: case studies	https://www.theclimategroup.org/sites/default/files/downloads/tcg_smart_cities_introduction.pdf
Smart Cities Seoul: a case study	https://www.itu.int/dms_pub/itu-t/oth/23/01/T23010000190001PDFE.pdf
Smart cities: five smart steps to cybersecurity: case studies	https://www.pwc.com/us/en/services/consulting/cybersecurity/library/broader-perspectives/smart-cities.html
Building Smarter Cities and Communities: case studies	https://www.comptia.org/content/research/building-smarter-cities-and-communities
ECSO Smart Cities Sector Report – European Cyber Security: case studies	http://www.ecs-org.eu/documents/uploads/smart-cities-sector-report-032018.pdf
Data Sharing in Infrastructure – National Infrastructure: case studies	https://www.nic.org.uk/wp-content/uploads/Data-sharing-in-infrastructure.pdf
Smart cities survey: Technologies, application domains and case studies	https://journals.sagepub.com/doi/10.1177/1550147719853984
Smart City Ahmedabad Development: Case Studies	https://www.nec.com/en/case/scadl/index.html
Case Studies in IoT for Smart Cities – Smart Cities India expo	http://www.smartcitiesindia.com/images/photo-gallery-2017/Room-B-Buildings-India/IoT_enabled_infrastructure_for_cities/Justin_Bean.pdf
Reconceptualising Smart Cities – NITI: case studies	https://niti.gov.in/writereaddata/files/document_publication/CSTEP%20Report%20Smart%20Cities%20Framework.pdf
Barriers to the Development of Smart Cities in Indian Context: case studies	https://link.springer.com/article/10.1007/s10796-018-9873-4
The Smart Cities Mission in Delhi: case studies	https://www.orfonline.org/research/the-smart-cities-mission-in-delhi-2015-2019-an-evaluation-60071/
Smart Cities in India – India Smart Grid Forum: case studies	http://indiasmartgrid.org/event2017/09-03-2017/1.%20Plenary%202-%20Smart%20Cities/6.%20Vishal%20Sharma.pdf
Smart Cities Mission in India: case studies	https://www.emerald.com/insight/content/doi/10.1108/SASBE-07-2018-0039/full/html
Quantitative analysis of the development of smart cities in India: case studies	https://www.emerald.com/insight/content/doi/10.1108/SASBE-06-2019-0076/full/html
Head Smart Cities in India: case studies	http://search.proquest.com/openview/b630eb47c05257bddc7c1121b65ebe2a/1?pq-origsite=gscholar&cbl=2029987
Smart and Sustainable Cities of the Future Key Initiatives: case studies	https://www2.deloitte.com/content/dam/Deloitte/in/Documents/public-sector/in-gps-CII-SmartCity_SustainableSmartCities.pdf
Smart City Emergence: case studies	https://www.sciencedirect.com/book/9780128161692/smart-city-emergence
Smart Cities and the Idea of Smartness: case studies	http://iopscience.iop.org/article/10.1088/1757-899X/245/8/082008/pdf

Case studies	URL
Trends in Smart City Development – National League of Cities: case studies	https://www.nlc.org/sites/default/files/2017-01/Trends%20in%20Smart%20City%20Development.pdf
Smart Cities – What's In It For Citizens: case studies	https://newsroom.intel.com/wp-content/uploads/sites/11/2018/03/smart-cities-whats-in-it-for-citizens.pdf
Insights into the incipient smart cities phenomena in india: case studies	https://smartech.gatech.edu/bitstream/handle/1853/58547/zahra_jeena_insights_into_the_incipient_smart_cities_phenomena_in_india.pdf
Analysing and Rating Smart City Development in India: case studies	https://www.krishisanskriti.org/vol_image/03Jul201503070817.pdf
India's "smart" cities mission: case studies	https://www.tandfonline.com/doi/full/10.1080/07352166.2018.1468221?src=recsys
Smart Cities Seoul: A Case Study	http://icities4greengrowth.in/casestudy/smart-cities-seoul-case-study
Why the Smart Cities Mission will miss its deadline – India Today: case studies	https://www.indiatoday.in/india-today-insight/story/why-the-smart-cities-mission-will-miss-its-deadline-1574728-2019-07-29
International Case Studies of Smart Cities – Publications	https://publications.iadb.org/publications/english/document/International-Case-Studies-of-Smart-Cities-Rio-de-Janeiro-Brazil.pdf
Smart City Success Stories and Events – City of Coral Gables	https://www.coralgables.com/media/IT/CGITSmartStories.pdf
Smart at Scale: Cities to Watch 25 Studies – WEF	http://www3.weforum.org/docs/WEF_Smart_at_Scale_Cities_to_Watch_25_Case_Studies_2020.pdf
Smart cities financing guide	https://urbaninnovation.asu.edu/sites/default/files/smartcitiescouncil_-_financing_guide-3_31_14.pdf
A Taxonomic Analysis of Smart City Projects in North America and Europe	https://www.mdpi.com/2071-1050/12/18/7813
China's Smart Cities Development	https://www.uscc.gov/research/chinas-smart-cities-development
Catalyzing private sector investments in climate smart cities	https://openknowledge.worldbank.org/handle/10986/35928?show=full
Issues Paper on Smart Cities and Infrastructure – UNCTAD	https://unctad.org/meetings/en/SessionalDocuments/CSTD_2015_Issuespaper_Theme1_SmartCitiesandInfra_en.pdf
Examples and case studies – Smart Cities Council	https://smartcitiescouncil.com/smart-cities-information-center/examples-and-case-studies
Trust in smart city systems – CISA	https://www.cisa.gov/sites/default/files/publications/Trust%20in%20Smart%20City%20Systems%20Report_0.pdf

Appendix G

Answers to review questions/exercises, hands-on projects, case projects, and optimal team case project by chapter

Chapter 1: Introduction to industry 4.0: smart cities policies and financing

Review questions/exercises

True/false

1. False
2. False
3. False
4. True
5. True

Multiple choice

1. C
2. E
3. A, B
4. D
5. A

Exercise

Solution

The following is a partial exercise solution. The students should be able to expand on the following:

Focus on why with the expansion of online to offline convergence, smart cities can grow without limitations on their size—where connectivity and innovation determine the inclination of the smart city's benefit–cost curve. As a city platform is responsible for connectivity, its value drastically increases through the Industry 4.0's O2O (online to offline convergence) platform. Finally, when a smart city reflects on its own as a Digital Twin (DT) in the Cloud and when complete information becomes accessible through a citizen's participation through smart phones, self-organization takes place—an ideal linkage between the city and its citizens.

Hands-on project

Solution

The following is a partial project solution. The students should be able to expand on the following:

Focus on the integration and interaction between new urbanization and the development of new technologies of Industry 4.0. Technologies of Industry 4.0 create fundamentally new infrastructure of the smart city. These new technologies allow the solving of problems of resources utilization and energy efficiency improvement; organize urban production and demographic changes in mega smart cities. Finally, the development of a smart city promotes new industrialization; creates new conditions for living, work, and education; accumulates social and human capital; and attracts financial resources for business development.

Case projects

Solution

The following is a partial project solution. The students should be able to expand on the following:

Focus on why the concept of Industry 4.0—the Internet of Things (IoT), should be used for the development of so-called smart products within smart cities. Subcomponents of the products are equipped with their own intelligence. Finally, added intelligence is used both during the manufacturing of a product and during subsequent handling: Up to the continuous monitoring of the product life cycle (smart processes).

Optional team case project

Solution

The following is a partial project solution. The students should be able to expand on the following:

Focus on the strategies and plans for Industry 4.0. Explain why Industry 4.0 is an imagined future, where it has a fundamental influence on smart cities and their environment and regions, thus given that their primary goal is to improve a country's competitiveness. Finally, demonstrate the relevant correlations between the concepts of disruptive innovation, smart city, and Industry 4.0.

Chapter 2: The need to localize smart infrastructure policies and financing

Review questions/exercises

True/false

1. False
2. True
3. False
4. True
5. False

Multiple choice

1. E
2. A
3. C
4. E
5. B

Exercise

Solution

The following is a partial exercise solution. The students should be able to expand on the following partial list:

Focus on the framework to identify the optimal localized hotspots in a medium-sized smart city. In other words, look at vacant commercial spaces to be considered as viable coworking spaces through an analysis of the respective spatial and quantitative data. Finally, these hotspots can become dynamic nodes of a network structure at a smart city scale contributing to the renaissance of its economic, social, and innovative profile.

Hands-on project

Solution

The following is a partial project solution. The students should be able to expand on the following:

Focus on why it is important for national and city governments to work together, with all relevant stakeholders to develop a common understanding of what a smart city entails in their specific national and local contexts.

Case projects

Solution

The following is a partial project solution. The students should be able to expand on the following:

Focus on the needs and challenges of specific urban systems and smart cities, and recognize the potential of technology as an enabler and at the same time understand its limitations.

Optional team case project

Solution

The following is a partial project solution. The students should be able to expand on the following:

Focus on why an integrated approach should be adopted that will help break down silos that may exist between existing infrastructure verticals and government departments that deal with related services in smart cities.

Chapter 3: Revolutionizing the localization of smart city infrastructure policies and financing

Review questions/exercises

True/false

1. False
2. True
3. True
4. False
5. True

Multiple choice

1. A
2. A
3. A
4. C
5. C

Exercise

Solution

The following is a partial exercise solution. The students should be able to expand on the following:

The key design principles that need to be followed in smart cities include inclusiveness, resilience, sustainability, interoperability, flexibility, risk mitigation, and safety.

Hands-on project

Solution

The following is a partial project solution. The students should be able to expand on the following:

Focus on why STI communities can play a key role in addressing some of the main challenges faced by smart cities and infrastructure projects.

Case projects

Solution

The following is a partial project solution. The students should be able to expand on the following:

Focus on why existing national STI and ICT policies may not be adequate enough to integrate the needs of developing smart city initiatives. Doing so would require the strengthening of a supportive ecosystem that enables the development of smart infrastructure, including human capabilities, legal frameworks, technology policies, institutional mechanisms and data use policies at both the Government level and in the workforce.

Optional team case project

Solution

The following is a partial project solution. The students should be able to expand on the following:

The broad array of tools include the following: inter alia output-based contracting; public–private partnerships; procurement policies; planning and development frameworks; social and entrepreneurial investment funds; research funds; and the provision of support services. Such tools can enable Governments to actively shape markets and correct market failures related to smart infrastructure in smart cities.

Chapter 4: Harnessing the local innovation system: an overarching solution for policy and financing instruments for promoting the localization of smart infrastructure

Review questions/exercises

True/false

1. False
2. False
3. True
4. True
5. False

Multiple choice

1. D
2. B
3. D
4. A
5. C

Exercise

Solution

The following is a partial exercise solution. The students should be able to expand on the following:

Focus on how data can be used to design infrastructure that specifically addresses the needs of marginalized groups in society in smart cities.

Hands-on project

Solution

The following is a partial project solution. The students should be able to expand on the following:

Focus on adopting a participatory approach to smart city development that actively engages citizens at all stages of development and ensures that cultural and livelihood factors of all sections of society are adequately integrated into the design of smart city plans.

Case projects

Solution

The following is a partial project solution. The students should be able to expand on the following:

According to the specific needs of smart cities, focus on strengthening the core ICT infrastructure required to support smart cities and facilitate the development of an ecosystem that supports the key technologies required to enable smart cities and infrastructure.

Optional team case project

Solution

The following is a partial project solution. The students should be able to expand on the following:

Focus on how to adopt an integrated approach to the conceptualization and design of smart cities and infrastructure to promote the more efficient use of resources and provision of public services.

Chapter 5: Smart counties: technologies, considerations, characteristics, challenges, policies, and theoretical concerns

Review questions/exercises

True/false

1. True
2. False
3. True
4. False
5. True

Multiple choice

1. B
2. C
3. B
4. D
5. A

Exercise

Solution

The following is a partial exercise solution. The students should be able to expand on the following:

In order to guide the vision process within a smart city in a county, you need to ask quite a number of questions, as they relate to economic, social, environment, quality of life, and place considerations. Please answer the following questions under each category:

Economic considerations:

- What areas will have the biggest financial return for the county?
- What sectors of the local economy will be mostly affected?
- Which regions will benefit financially the most from the smartification?

Social considerations:

- What populations are in most need of services?
- What actions/applications will have the widest impact?
- What smart technologies will reduce commute time?
- What smart technologies will improve community participation?
- Will smart technologies help address health concerns and responsiveness in a more effective way?

Environmental considerations:

- Will smart technologies reduce the vehicle miles traveled?
- Will smart technologies improve pollution generating traffic flows?
- Will smart technologies improve responsiveness to environmental pollutants and disaster events?
- Will smart technologies improve visitors' experience of parks and recreation areas?
- Will smart technologies improve sustainability indicators across the county?

Quality-of-life considerations:

- How will smart technologies impact the overall satisfaction of people living here?
- What are the things people consider more urgent for the improvement of their daily lives?
- Do all the proposed applications improve the daily and long-term routines of the residents?

Place considerations:

- What is the impact of smart technologies on the daily life of local neighborhoods?
- What is the impact of smart technologies on the local places where people hangout, on "third places?"
- What is the impact of smart technologies on people's perception of their neighborhood/town/region?
- What is the impact of smart technologies on the community identity?
- Will smart technologies create a sense of pride in the residents?

Hands-on project

Solution

The following is a partial project solution. The students should be able to expand on the following:

Please answer the following questions with regard to smart cities:

- Does the policy reflect the community's vision?
- Was a broad and diverse set of stakeholders involved in identifying needs the policy aims to address?
- Does the policy include locally relevant health data? Benchmarks (existing conditions) as well as standards and targets (goals for future conditions) can reflect health and built environment data.
- How will we implement the policy?
- Does the policy identify actions and programs (such as updating zoning codes or securing funding for an infrastructure investment) that will guide us from vision to action?
- Does the policy name implementation partners so that local government agencies and the community know who will play what role?

- How do we know if the policy has been successful?
- Are there identified standards or targets that can be measured to highlight where policies have achieved a vision and where they may need to be reworked?
- Does the policy make progress toward eliminating health disparities?

Case projects

Solution

The following is a partial project solution. The students should be able to expand on the following:

Please answer the following questions under each category, as part of a questionnaire to the community and among the staff, where most of them start from the mapping and identification of the most prominent issues throughout a smart city in the county:

1. Large territory:
  - **a.** What are the most distant areas of the county?
  - **b.** What types of infrastructure are the most difficult to access?
  - **c.** What communities have the most difficulty reporting infrastructure failures?

2. Unincorporated and incorporated communities:
  - **a.** What type of coordination is required to address ROW and traffic smart technologies adjacent to incorporated communities?
  - **b.** Who are the main stakeholders?
  - **c.** How can we create an ecosystem of stakeholders to implement smart technologies?
  - **d.** What are some innovative public–private partnership models, and what are the ways to ensure appropriate feedback loops?

3. Small and remote communities:
  - **a.** How many communities are the most difficult to access?
  - **b.** Which communities are most vulnerable to natural disasters?
  - **c.** Which communities are in the most need smart technologies such as broadband or hot spots, sensors, drones, and use of analytics?

4 & 5. Rural communities and agriculture:
  - **a.** Where are the most extensive agricultural areas of the county?
  - **b.** What types of agriculture are in most need of smart technologies?
  - **c.** How can we simplify the process of applying for funding?
  - **d.** What are the program requirements?
  - **e.** How can we improve the accuracy of coverage maps?
  - **f.** What are other alternatives to broadband?
  - **g.** What are the ways to share costs and/or aggregate demand?
  - **h.** What are the community assets and needs?
  - **i.** What ancillary investments are required to fully realize the benefits of connectivity?
  - **j.** How will the county coordinate, share, and support the development of smart rural communities?

6. Natural disasters:
  - **a.** What are the most common and recurring natural disasters in the county?
  - **b.** What types of technologies can the county implement in those areas to address prevention, response, and remediation?
  - **c.** How can the county apply analytics and visual simulations to assist natural disaster prevention, response, and remediation?

7. Sprawl
  - **a.** What land-use policies will create land patterns financially and environmentally more compatible with smart technologies?
  - **b.** What smart technologies can be efficiently incorporated into existing suburbia?

Optional team case project

Solution

The following is a partial project solution. The students should be able to expand on the following:

Three unique county challenges relate to funding and services that are necessary locally or required by the state, operational provisions, and structure of smart cities in counties and will provide another set of questions to structure the Objectives, Strategies, and Action plans that are necessary to complete the vision process. Please answer the following questions as they relate to State Mandates:

- **a.** How can we provide the legal framework for the financial support of smart counties?
- **b.** How can the county establish smartification promotion plans?
- **c.** How can the county receive state funding and support for smart technologies?
- **d.** What are the avenues to receive funding for construction projects, technology improvements, operation, and management of the smart technology implementation?
- **e.** What is the process for the county to ensure funding for state mandates?

Chapter 6: From community plan to clean energy infrastructure: policy, partnership, and investment

Review questions/exercises

True/false

1. False
2. False
3. False
4. True
5. False

Multiple choice

1. C
2. A
3. C
4. D
5. B

Exercise

Solution

The following is a partial exercise solution. The students should be able to expand on the following:

Focus on the following: Given the strains on public finances, engaging in private sector capital will be critical for smart cities. Several obstacles, resulting from market and government failures, including fossil, fuel subsidies, the lack of supportive policies, as well as outstanding barriers to international trade and investment, will hamper investment in renewable energy.

Hands-on project

Solution

The following is a partial project solution. The students should be able to expand on the following:

Focus on why a policy guidance for investment in clean energy infrastructure in smart cities would assist governments, including developed countries; emerging economies and developing countries; in identifying ways to engage private enterprises in financing and developing a clean energy infrastructure. Explain why a policy guidance for investment in clean energy infrastructure would provide policymakers with a list of issues and questions to consider for enhancing private investment in clean energy infrastructure. Finally, discuss the inclusion of electricity generation from renewable energy sources and improved energy efficiency in the electricity sector, particularly in the following areas: investment policy; investment promotion and facilitation; competition policy; financial market policies; and public governance.

Case projects

Solution

The following is a partial project solution. The students should be able to expand on the following:

Focus on why smart cities in countries at different stages of technological and economic development will find different issues more relevant to their specific situation. Every context is different; and a policy guidance will thus need to be adapted to the needs of each individual country. Finally, a policy guidance's ultimate goal would be to serve as a policy toolkit and as a basis for multistakeholder dialog on ways of increasing private investment in clean energy infrastructure, particularly for the purpose of broadening energy access.

Optional team case project

Solution

The following is a partial project solution. The students should be able to expand on the following:

Focus on why policymakers aiming to increase investment in clean energy infrastructure in smart cities will have to consider ways of creating a level playing field between independent power producers and state-owned enterprises, as well as between state-owned enterprises and other network operators. Also, explain why working with a state-owned energy utility to design and implement clean energy policy and technology solutions, including legislation for independent power provision, would be a good step in the right direction. Finally, explain why governments may consider embarking on a more extensive structural separation of the power sector, whereby multiple actors are encouraged to engage not only in power generation but also in transmission and distribution.

Chapter 7: Energy efficiency in urban planning for smart cities in the developing world

Review questions/exercises

True/false

1. False
2. False
3. True
4. True
5. True

Multiple choice

1. E
2. C
3. E
4. E
5. D

Exercise

Solution

The following is a partial exercise solution. The students should be able to expand on the following:

Focus on the influence of the smart sustainable cities concept on energy management from the labor market perspective. Do research about an energy manager's profession transformation, by considering the smart sustainable city concept. Finally, create an up-to-date holistic energy manager skill model with a focus on emerging technologies.

Hands-on project

Solution

The following is a partial project solution. The students should be able to expand on the following:

Focus on the vital environmental issue, when addressing urbanization with special focusing on the Arabian Gulf area, where the higher prices of oil are having their impact on the booming construction sector. Investigate how existing smart cities are planned, by considering energy efficiency in much older cities, as those in Egypt. Finally, look at the relationship between cities and energy consumption, in order to identify the factors having the most impact in planning for energy efficiency.

Case projects

Solution

The following is a partial project solution. The students should be able to expand on the following:

Focus on the economic analysis of infrastructure improvements that result in reducing the energy demand of the smart cities, represented by the intelligent light-emitting diode (LED) street lighting system. Do research on the estimation of net profit, net discounted saving, as well as the total savings, by using an example of an average smart city. Discuss why there should be a need for better management, including strong networks of leaders to drive smart city policies and investments and coverage of wider smart city areas with economically sustainable projects and plans.

Optional team case project

Solution

The following is a partial project solution. The students should be able to expand on the following:

Focus on how urbanization poses considerable challenges, including overpopulation, climate change, environmental quality, and access to energy. Discuss how urban planning must reassess how to supply the population in a sustainable way with basic services at an affordable cost. Show how the smart cities of tomorrow should be configured to meet two major challenges: climate change and worsening air quality.

Chapter 8: Transportation policies for connected and automated mobility in smart cities

Review questions/exercises

True/false

1. True
2. False
3. True
4. False
5. True

Multiple choice

1. E
2. A
3. B
4. D
5. B

Exercise

Solution

The following is a partial exercise solution. The students should be able to expand on the following:

Focus on the role of big data and the ways in which it is integral to smart cities. Also, look at the metrics for measuring the impacts of smart city implementation projects. Finally, look at how collaborators and innovators work together to advance smart energy and environment; smart transportation; smart governance; smart workforce; smart living; smart economy; and smart connections.

Hands-on project

Solution

The following is a partial project solution. The students should be able to expand on the following:

Focus on how to use a multimethod approach to do research. Develop a framework for smart cities that can be employed to: Understand what a smart city is and how to replicate smart city successes; the role of pilot projects, metrics, and evaluations to test, implement, and replicate strategies; and understand the role of shared micromobility, big data, and other key issues impacting communities. Finally, provide recommendations for policy and professional practice, as it relates to integrating transportation into smart cities.

Case projects

Solution

The following is a partial project solution. The students should be able to expand on the following:

Focus on how to do a systematic review and meta-analysis. Also, confirm why different smart city drivers are so important; and then assign them a priority level. Finally, consider why smart city drivers are so important; how they are related to city governance and technical solutions; and their ability to increase urban mobility intelligence.

Optional team case project

Solution

The following is a partial project solution. The students should be able to expand on the following:

Focus on key AI smart cities mobility initiatives: Connected and Autonomous Vehicles (CAVs); autonomous Personal and Unmanned Aerial Vehicles (PAVs and UAVs); and Mobility-as-a-Service (MaaS). Also, look at interventions that may work as enabling AI technologies for transport, such as the Internet of Things (IoT), Physical Internet (PI), and broader transformations such as Industry 4.0. Finally, discuss how using AI for researchers and city planners can provide clear and systematic definitions of the ambiguous smart mobility in terms of tomorrow; and describe their individual and collective roles underpinning the nexus in scope.

Chapter 9: Public–private partnerships (PPP) enabled smart city funding and financing

Review questions/exercises

True/false

1. False
2. True
3. False
4. False
5. True

Multiple choice

1. A
2. B
3. C, D
4. E
5. C

Exercise

Solution

The following is a partial exercise solution. The students should be able to expand on the following:

Focus on studying and analyzing the stakeholders role in smart cities infrastructure projects through the PPPs concept. Also, look at why the PPPs concept decreases opposition leading to a non or ineffective implementation of the project as a result of the gap between the expectations of different stakeholders involved in PPPs on the desired process or outcome of the project. Help establish successful PPPs for smart infrastructure project phases by presenting a clear image of stakeholders roles; the highlights of the PPPs challenges; characteristics; state's models; and agreements methods—especially, with the high cost of these projects. Finally, try and reach a set of recommendations that aid to improve efficiency, quality, and affordable costs to achieve sustainability, whether in retrofitting or in the new urban sprawl.

Hands-on project

Solution

The following is a partial project solution. The students should be able to expand on the following:

Focus on making sure that the national and local regulation is adapted to the development of a smart cities PPP. Also, check on data protection legislation and the flexibility of public procurement rules. Nest, make sure to have the necessary budget to finance the PPP process and the technical capacity to manage it. Local governments should not invest in too sophisticated tools that create digital dependency or incapacity to manage the project on the long run. Finally, take the necessary time to evaluate all aspects of the project; and in particular, carefully assess the risks and the return on investment beforehand.

Case projects

Solution

The following is a partial project solution. The students should be able to expand on the following:

Focus on how to ensure a competitive, fair, and transparent public procurement process. Next, look at how to ensure continuous communication and participation at all stages of the smart city PPP development toward the population, as well as municipal officials and the private operator. Finally, help build a relationship of trust with the private partners, by maintaining a close dialog and being transparent, which should help guarantee the success of the partnership.

Optional team case project

Solution

The following is a partial project solution. The students should be able to expand on the following:

Focus on why Public–Private Partnerships (PPP) appear to be suitable solutions to overcome the shortage of public finance and cuts on public spending. Look why the adoption of PPP forms for Smart City projects has not been fully explored and only experimentally applied so far. Discuss how to promote the usage of PPP to finance Smart City initiatives with some PPP financial instruments; and discuss the associated strengths and weaknesses. Finally, use the Project Finance, Revenue Sharing, and Social Impact Bonds, as sound alternatives and suitable sources of financing for Smart City projects.

Chapter 10: Preparing students for interdisciplinary work: green infrastructure curricula at Toronto Metropolitan University, Toronto, Ontario, Canada

Review questions/exercises

True/false

1. True
2. False
3. False
4. True
5. False

Multiple choice

1. B
2. C
3. E
4. A
5. D

Exercise

Solution

The following is a partial exercise solution. The students should be able to expand on the following:

Focus on why opportunities for interdisciplinary networking in smart cities; knowledge of the landscape construction process; and the ability to utilize scientific rationale for design decision making, all but increased the challenges of coordination efforts across disciplines. Also, why were maintenance issues the cause of overcoming unknown nomenclature specific to each field; delays due to unforeseen circumstances; and budgetary increases. Finally, discuss how collaboration between multidisciplinary professionals enables students to experience the professional design process and have a deeper understanding of the positive impacts of green infrastructure through interdisciplinary experiential learning.

Hands-on project

Solution

The following is a partial project solution. The students should be able to expand on the following:

Focus on how green infrastructure in smart cities providers could offer some perspectives from agencies, nonprofits, consulting firms, and academic institutions. Also, look at how provider organizations aim to increase worker competency and advance the field of green infrastructure by designing trainings that facilitate communication skills, enhance networking opportunities, and exploit social learning activities.

Case projects

Solution

The following is a partial project solution. The students should be able to expand on the following:

Focus on where urban and smart cities climate and evapotranspiration; water demand and stormwater runoff; water infrastructure; green infrastructure; extreme events; and integrative planning take place.

Optional team case project

Solution

The following is a partial project solution. The students should be able to expand on the following:

Show how a smart city sustainability curriculum should provide students with the skills needed to competently participate in an interdisciplinary team. Also, discuss a design/methodology/approach: Where instructors are drawn from different departments and divisions of a school, develop a pool of smart city sustainability-focused projects; and act as mentors to teams of students with diverse undergraduate backgrounds. Finally, describe how students who attend these workshops are provided training that is designed to participate in teamwork, research, and report preparation.

Chapter 11: Teaching sustainability in Metro Orlando: the evolution of the pragmatic liberal arts at Rollins College

Review questions/exercises

True/false

1. False
2. False
3. True
4. False
5. True

Multiple choice

1. C
2. D
3. A
4. B
5. E

Exercise

Solution

The following is a partial exercise solution. The students should be able to expand on the following:

Focus on why the use of urban and smart cities data games should be set to a context for learning, and demonstrating, practical application of skills for handling large complex data sets. Also, focus on participants who are tasked to design a novel application from real satellite data, after first learning and applying data skills to data about home energy consumption

Hands-on project

Solution

The following is a partial project solution. The students should be able to expand on the following:

Focus on why it is important to analyze the use of online education based on the construction of intelligent learning environments in smart cities, which favor the interaction between the environment and the student. Also, use a quasi-experimental methodology, to analyze the psychometric properties of a questionnaire, through structural equation modeling (SEM). Finally, discuss how a model can be adjusted through a multivariate regression analysis, to relate response patterns to a set of latent factors that cannot be directly observed, but exist in continuous dimensions of the people evaluated, and to create a valid and reliable instrument as a measurement tool

Case projects

Solution

The following is a partial project solution. The students should be able to expand on the following:

Focus on designing and delivering a smart cities massive open online course; a multidisciplinary course that is targeted at an international audience; and built upon the pedagogical approach of social learning. Use data to evaluate the motivations, demographics, online participation, and experiences of students. Finally, provide an analysis of social learning interactions; insight into citizens' views on smart cities; and identify how a massive open online course can be adapted to meet the learning needs of smart city initiatives and potential areas for future research.

Optional team case project

Solution

The following is a partial project solution. The students should be able to expand on the following:

Focus on why the machine learning models in smart cities, analyzed from the point of view of the models, techniques, and applications, are studied. Also, address the areas and dimensions of sustainability and how sustainable development goals are analyzed. Finally, provide an inclusive perspective on applicability; sustainability scopes and dimensions; sustainable development goals; tools; data types; and Machine Learning techniques.

Chapter 12: Partnering with technology firms to train smart city workforces

Review questions/exercises

True/false

1. True
2. True
3. False
4. True
5. False

Multiple choice

1. C
2. E
3. E
4. B
5. C

Exercise

Solution

The following is a partial exercise solution. The students should be able to expand on the following:

Focus on the growth of smart cities and how collateral movements offer new and exciting possibilities for the use of information and communication technologies (ICTs) for service delivery; civic engagement; and governance. Also, look at how the exponential growth of ICTs and their use in governance—both formally and informally—highlights the need for urban planners and public administrators who are trained on how to use ICTs to achieve the public interest, maximize the positive impacts of ICTs, and minimize the negative impacts of ICTs. Finally, discuss how all of the preceding presents a challenge for professional education to provide a supporting infrastructure that trains urban planners and public administrators for smart city governance in the 21st century.

Hands-on project

Solution

The following is a partial project solution. The students should be able to expand on the following:

Focus on the concept of designing an e-learning platform to provide instant training and education for shaping a new generation of smart citizens in smart cities.

Case projects

Solution

The following is a partial project solution. The students should be able to expand on the following:

Focus on why the systematic review of the smart workforce in smart cities will trend with new opportunities and challenges for the future.

Optional team case project

Solution

The following is a partial project solution. The students should be able to expand on the following:

Focus on the latest software technologies and proven instructional practices that are combined to deliver competency-based experiential learning in smart cities. Look at how the training system can be deployed to a specific physical location by using dedicated hardware; and how it can be deployed in the cloud for an enterprise-wide, multisite training solution. Finally, describe how cities and utilities can integrate a solution within their own existing learning management system; and how it can be deployed as a stand-alone learning management system to easily track and manage progress.

Chapter 13: Training professional smart city workforces through a partnership with technology firms

Review questions/exercises

True/false

1. True
2. False
3. True
4. False
5. True

Multiple choice

1. B
2. A
3. D
4. E
5. C

Exercise

Solution

The following is a partial exercise solution. The students should be able to expand on the following:

Focus on why many cities and local communities are seeking to become smart cities by embedding new digital technologies into municipal infrastructure. The range of smart-city solutions includes smart grids, intelligent transportation systems, connected street lighting, and remote healthcare, to name a few.

Hands-on project

Solution

The following is a partial project solution. The students should be able to expand on the following:

Focus on why many municipalities lack the resources and in-house expertise to develop, deploy, and operate large-scale technology projects in smart cities. Also, one way to meet these challenges is to harness the resources and strengths of private-sector stakeholders—innovators, businesses, anchor institutions, educators, and more. Finally, private-sector partners can be an important source of funding, technical knowledge, continuing innovation, and workforce development.

Case projects

Solution

The following is a partial project solution. The students should be able to expand on the following:

Draw from lessons learned. Provide a framework for getting the most out of public–private partnerships, including what to look for in a partner, assessing each partner's contribution, and guidance on how to structure the most fruitful partnership agreements in smart cities development.

Optional team case project

Solution

The following is a partial project solution. The students should be able to expand on the following:

Focus on why partnerships are the foundation of smart cities. Show how partnerships drive successful smart city initiatives, because each partner contributes unique knowledge and expertise to the project. Although local, state, or federal funding may be available to support smart cities projects, those funds rarely cover the entire cost. Finally, partners can provide additional resources that help defray capital expenditures.

Chapter 14: Training smart city workforces with a wide range of electronic and digital technologies

Review questions/exercises

True/false

1. False
2. True
3. True
4. False
5. False

Multiple choice

1. C
2. E
3. E
4. A
5. D

Exercise

Solution

The following is a partial exercise solution. The students should be able to expand on the following:

Focus on the smart grid and smart lighting systems to show some of the key challenges and headwinds that end users face when adopting smart city technologies. Also, explain why cyber security is often overlooked in smart city discussions. Finally, evaluate smart city technologies, which are based on their ability to provide business value and their impact on health, safety, and environment.

Hands-on project

Solution

The following is a partial project solution. The students should be able to expand on the following:

Focus on the growth of smart cities and how collateral movements offer new and exciting possibilities for the use of information and communication technologies (ICTs) for service delivery, civic engagement, and governance. Also, look at how the exponential growth of ICTs and their use in governance (both formally and informally) highlights the need for urban planners and public administrators (who are trained on how to use ICTs), to achieve the public interest, maximize the positive impacts of ICTs, and minimize the negative impacts of ICTs. Finally, present a challenge to professional education to provide a supporting infrastructure that trains urban planners and public administrators for smart city governance in the 21st century.

Case projects

Solution

The following is a partial project solution. The students should be able to expand on the following:

Focus on how to achieve the solutions for smart city values of equity, economic opportunity, and environmental stewardship; inform and provide testable hypotheses and proven pilots for other potential smart city communities; and build a foundation for partnerships for better urban mobility. Finally, look at potential smart city pilots and programs that continue to be on a successful track; and show how to adopt innovative technology, including: pilots, places, and services for people.

Optional team case project

Solution

The following is a partial project solution. The students should be able to expand on the following:

Focus on connected corridors that link Smart Stations in smart cities, with new transit services (including a deployment of electric bus rapid transit); dedicated transit lanes and signal priorities; better facilities for active transportation; and a sensor-rich environment that allows for rapid deployment of V2V and V2I connected-vehicle technology.

Chapter 15: Digitization, automation, operation, and monetization: the changing management of sidewalk and kerb 2000–25

Review questions/exercises

True/false

1. True
2. False
3. False
4. True
5. False

Multiple choice

1. E
2. A
3. E
4. A
5. D

Exercise

Solution

The following is a partial exercise solution. The students should be able to expand on the following:

Focus on why without such standards, innovators will have insufficient guidance to develop the operational details of their technical solutions in smart cities. Explain why fleet operators will be unable to collaborate effectively within tightly constrained urban spaces; and cities will be unable to express regulations in digital formats that will uniformly operate in the vehicle-to-everything (V2X) systems that control multiple fleets of large numbers of vehicles to move people and goods, as well as to provide services. Finally, discuss why the distance between the independent perspectives of municipal decision-makers and technology innovators is particularly problematic, about preparing for automated vehicles and devices.

Hands-on project

Solution

The following is a partial project solution. The students should be able to expand on the following:

Focus on how from an increase in food and parcel deliveries to outdoor streateries, the pandemic has changed the way smart cities are looking to organize and monetize their curb space. Also, look at how curb pricing models have changed very little since they were originally introduced in cities decades ago. Finally, discuss why metered parking is geared toward personal vehicles and does not allow many options for commercial operators.

Case projects

Solution

The following is a partial project solution. The students should be able to expand on the following:

Focus on investigating a methodology for designing the physical, virtual, urban realms in tandem, across a dynamic timeline. Also, look at how a methodology could contribute to ideas on redesigning public spaces to be responsive to real-time use and demand; and that are digitally connected. Finally, look at how a methodology could contribute to the design of a framework from which to visualize and understand the digital twin of the urban and smart cities realm.

Optional team case project

Solution

The following is a partial project solution. The students should be able to expand on the following:

Focus on how a smart city would know every asset's exact location and condition, and be able to coordinate projects across departments. Discuss, why it would be a very difficult task to know every asset's exact disposition, without a full digital inventory. Finally, describe how some smart cities have recognized the problem and launched programs to digitize their public street inventory.

Chapter 16: Digitization, automation, operation, and monetization: standardizing the management of sidewalk and kerb 2025–50

Review questions/exercises

True/false

1. False
2. True
3. False
4. True
5. False

Multiple choice

1. A
2. B
3. A
4. B
5. C

Exercise

Solution

The following is a partial exercise solution. The students should be able to expand on the following:

Focus on the obvious actors such as pedestrians; ambulant disabled people are also jostling for position with vehicles legally parked on the pavement; children on scooters (and an exponentially rising number of adults on scooters); parents with buggies; pushchairs and prams of varying widths; those in need of the aid of walking sticks; frames and crutches; and, postal workers and delivery personnel darting across the flow of foot traffic in smart cities.

Hands-on project

Solution

The following is a partial project solution. The students should be able to expand on the following:

Focus on the kerb as a fixed, two-dimensional asset, or liability in smart cities. Also, be able to look at the kerb in a 3D environment. Finally, understanding the kerb's vital position as a gateway, either from the delivery hinterland or beyond the kerb to the access hinterland, is integral to understanding the type of permission or user driven permission stack that will sit above the kerb.

Case projects

Solution

The following is a partial project solution. The students should be able to expand on the following:

Focus on the one question that affects every aspect of the era of kerbside digitization, with particular reference to responsibility, is: Who or what type of authority should oversee ground traffic control operations in smart cities?

Optional team case project

Solution

The following is a partial project solution. The students should be able to expand on the following:

Focus on reviewing and discussing why it may be that the amenity value from a parklet outweighs the monetization from a series of loading or parking permits in smart cities. However, there clearly needs to be a balance.

Chapter 17: Innovative smart applications for solid waste management

Review questions/exercises

True/false

1. True
2. True
3. True
4. True
5. True

Multiple choice

1. B
2. C
3. B
4. C
5. D

Exercise

Solution

The following is a partial exercise solution. The students should be able to expand on the following:

Focus on how waste management involves activities such as reuse, recycling, waste reduction generation, and other strategies, to combat the effect of waste generation due to increasing population and industrialization in smart cities. Also, look at how monitoring is one of the key functions of waste management, as it is needed to address the issues faced by waste management, which includes waste generation, waste collection, transportation of waste, waste treatment, and waste disposal processes. Finally, look at how the use of artificial intelligence (AI) can solve waste management, such as convolutional neural network for efficient classification and waste identification.

Hands-on project

Solution

The following is a partial project solution. The students should be able to expand on the following:

Focus on how the disposed trash is collected by a smart city municipality or private company and is then trucked at predefined times and then transferred to temporary collection centers. Then, describe how the trash at the collection centers is sent to recycling. Finally, explain how the recycling process in a smart city setting solves the waste problem partially, while it creates other problems.

Case projects

Solution

The following is a partial project solution. The students should be able to expand on the following:

Follow an IoT-based approach, where the discarded waste from a smart city bin is continuously monitored by sensors that inform the filling level of each compartment, in real time. Next, show how data are stored and processed in an IoT middleware, by providing information for collection with optimized routes; generating important statistical data for monitoring the waste collection accurately in terms of resource management; and the services for the community.

Then, create a prototype of a smart container; the development of a waste management application; and a real-scale experiment use case for evaluation, demonstration, and validation. Finally, show how the proposed system can efficiently change the way people deal with their garbage and how to optimize economic and material resources.

Optional team case project

Solution

The following is a partial project solution. The students should be able to expand on the following:

Focus on the shortcomings of existing waste management practices in smart cities; and a conceptual framework proposal for a centralized waste management system. Next, look at an infrastructure for the proper collection of product life cycle data to facilitate full visibility throughout the entire life span of a product. Then, design a set of new business models that rely on product life cycle data to prevent waste generation. Finally, look at an intelligent sensor-based infrastructure for proper upstream waste separation and on-time collection.

Chapter 18: Data security and privacy in smart cities

Review questions/exercises

True/false

1. False
2. False
3. False
4. False
5. False

Multiple choice

1. C
2. D
3. C
4. D
5. E

Exercise

Solution

The following is a partial exercise solution. The students should be able to expand on the following:

Focus on how smart cities are expected to improve the quality of daily life, promote sustainable development, and improve the functionality of urban systems. Next, discuss how many smart city systems have been implemented; and how security and privacy issues have become a major challenge that requires effective countermeasures. Finally, discuss why traditional cybersecurity protection strategies cannot be applied directly to intelligent applications, because of the heterogeneity, scalability, and dynamic characteristics of smart cities.

Hands-on project

Solution

The following is a partial project solution. The students should be able to expand on the following:

Focus on the security, privacy, and risks within smart cities, highlighting the threats relating to information security and challenges for smart city infrastructure in the management and processing of personal data.

Case projects

Solution

The following is a partial project solution. The students should be able to expand on the following:

Focus on why a framework should provide end-to-end security and privacy features for trustable data acquisition, transmission, processing, and legitimate service provisioning in smart cities. Then, as a proof of concept, test core functionalities of authentication protocols for data acquisition and service provisioning.

Optional team case project

Solution

The following is a partial project solution. The students should be able to expand on the following:

Explain why, ever since the concept of a smart city first emerged, it has been clear that technological innovation, intelligent systems, and big data would be the three key ingredients to a city's success. Next, discuss why very few urban residents complain about streamlined services and sustainable materials, while there are many who question the ethics behind the gathering, storing, and implementation of private data.

Chapter 19: Financing smart cities—an indian case study

Review questions/exercises

True/false

1. True
2. False
3. False
4. True
5. True

Multiple choice

1. B
2. C
3. E
4. A
5. B

Exercise

Solution

The following is a partial exercise solution. The students should be able to expand on the following:

Focus on how India's deferred maintenance challenges coexist with disruptive technologies—automated/autonomous vehicles; ride-sharing services; cloud-based management services for improving the delivery of municipal services; Internet of Things (IoT); and roads, water, and lighting systems with sensors embedded—which are reconfiguring how people live, work, and interact with smart cities. Also, look at how these technologies force city managers, planners, and financial and accounting managers to consider or rethink issues such as citizen data.

Hands-on project

Solution

The following is a partial project solution. The students should be able to expand on the following:

Focus on evaluating the various funding options and their feasibility with reference to the Indian perspective of smart cities.

Case projects

Solution

The following is a partial project solution. The students should be able to expand on the following:

Focus on the beneficiaries pay and congesters pay principles that are combined for smart cities in developing countries such as India, which can generate adequate revenues to service long-tenor debt incurred for core infrastructure facilities in smart cities.

Optional team case project

Solution

The following is a partial project solution. The students should be able to expand on the following:

Discuss the potential financial mechanisms and models that can be applied to overcome barriers such as debt and equity mechanisms and the use of new contractual models in smart cities. Also discuss how to integrate socioeconomic benefits in the project preparation process to estimate the value of the benefits to society that cannot be recovered by the project. Finally, estimate a justified and reasonable level of public subsidies that are required to implement a project, which would be not bankable otherwise.

Chapter 20: Policy instruments for facilitating smart city governance by promoting platforms for bottom-up participatory governance

Review questions/exercises

True/false

1. True
2. True
3. False
4. False
5. True

Multiple choice

1. B
2. A
3. B
4. C
5. C

Exercise

Solution

The following is a partial exercise solution. The students should be able to expand on the following:

Providing smart city residents with access to relevant ICT infrastructure does not guarantee the bottom-up participatory process in the city. ICT infrastructure should be treated as an important and necessary tool to increase the effectiveness of conducted activities and the quality of provided services in the smart city as well as to serve to achieve the basic aim, which is an active, innovative, tolerant, enterprising, open, and flexible society, prepared to face new challenges of the contemporary world. ICT infrastructure provides the inhabitants with better access to information about the smart city, fosters the development of ICT competences, and intensifies social interactions making it possible for residents to be better prepared and more interested in engaging in bottom-up, smart city governance. However, in order to activate the bottom-up social participatory, technical know-how is also necessary, indicating the possibilities of using ICT by residents in the cogovernance process, as well as appropriate education of the society and training aimed at shaping active social attitudes and a sense of social responsibility. An important issue is also the level of residents' trust in the smart city authorities, which is influenced, inter alia, by the attitude of municipal authorities toward residents and their growing position in city governance. It is not only about the mental layer (authority citizens), but first of all in communication layer, where dialog, mediation, and deliberation play an important role.

Hands-on project

Solution

The following is a partial project solution. The students should be able to expand on the following:

Focus on why students in teams as part of the workshop deal with a case study of their choice on the use of a specific group of OGD data in the smart city governance process in the city they live in, e.g., in terms of improving communication traffic, more effective management of electricity resources, water consumption, generation waste, public safety, health or the movement of residents, etc.

Case projects

Solution

The following is a partial project solution. The students should be able to expand on the following:

Focus on a framework of smart city urban governance, on the basis of three intertwined key components namely: spatial, institutional, and technological components. Provide a way to test the applicability of a framework. Finally, identify and discursively analyze two smart city projects, in order to illustrate how a framework works in practice.

Optional team case project

Solution

The following is a partial project solution. The students should be able to expand on the following:

Focus on the empirical evidence that alleges that sustainability benefits are sparse. In addition, look at how the emerging picture is ambiguous, in that it reports both positive and negative effects in respect to the sustainability achievements of smart city governance. Also, identify contextual conditions of smart city governance as crucial to understanding the preceding mixed outcomes. Finally, conduct more empirical work, and develop an agenda for researching the relationship between smart city governance and sustainability outcomes.

Chapter 21: Promoting platforms for bottom-up participatory governance: a policy instrument approach through the facilitation of strategic smart city governance

Review questions/exercises

True/false

1. False
2. True
3. False
4. True
5. True

Multiple choice

1. A
2. B
3. C
4. B
5. B

Exercise

Solution

The following is a partial exercise solution. The students should be able to expand on the following:

Focus on the internal process innovation through extensive employee training and education, regarding smart city concepts and emphasizing data-driven (rather than infrastructure-driven) policy decision-making. Then, explore smart city strategies, and how they result in distinctively different outcomes. Finally, look at how instilling a culture of innovation through the training of government managers and frontline workers is a critical component in achieving a holistic and sustainable smart city transformation that can survive leadership changes.

Hands-on project

Solution

The following is a partial project solution. The students should be able to expand on the following:

Focus on the forms and implications of citizen involvement in publicly supported participatory innovation platforms that facilitate urban economic development in the welfare society context. Then, review of the smart city discourse, which in the context of economic development policy translates into the cities' need to support innovativeness by creating smart environments. Finally, look at how a participatory innovation platform is a prime example of such an environment.

Case projects

Solution

The following is a partial project solution. The students should be able to expand on the following:

Focus on a review of smart city discourse, which in the context of economic development policy translates into the cities' need to support innovativeness by creating smart environments, including facilitated participatory innovation platforms. Then, look at how platform thinking is gaining prominence in economic renewal, as it enables the utilization of collective intelligence that emanates from the diversity of citizens and other stakeholders involved in local innovation processes. Finally, describe how enabler-driven innovation platforms and living labs involve different target groups in participatory innovation processes.

Optional team case project

Solution

The following is a partial project solution. The students should be able to expand on the following:

Focus on best-practice examples from over the world, which illustrate the various ways participation can be developed from various leadership perspectives with regard to smart cities. Also, look at a new conceptual framework: The Social Smart City framework and how it is derived from both the e-participation theory and best-practice examples. Finally, look at how the framework is comprised of a set of digital strategies for participatory governance in smart cities.

Chapter 22: Policy instruments for facilitating smart city governance

Review questions/exercises

True/false

1. True
2. False
3. True
4. True
5. False

Multiple choice

1. B
2. C
3. D
4. C
5. C

Exercise

Solution

The following is a partial exercise solution. The students should be able to expand on the following:

Focus on how administrative practices of governing create, delimit, and constitute smart city mobility as a governable object. Explain how this is done: By analyzing the policy instruments deployed by public administrations to obtain and disseminate knowledge in innovation processes, with the aim of controlling its development. Finally, analyze how public administrators utilize four categories of policy instruments: pilots, standards, scenarios, and collaboration.

Hands-on project

Solution

The following is a partial project solution. The students should be able to expand on the following:

Focus on how the concept of a smart city data plan has been elaborated by taking into account current issues related to privacy and manipulation of data in the smart city. Also, look at the methodological approach, which is adopted to define the nature of a smart city data plan, which is grounded on the conceptual and empirical parallelism with the corporate data governance plans and general urban plans, respectively; which are aimed to regulate decision-making powers and actions on data in enterprise contexts; and the interests of local stakeholders in the access and use of urban resources. Finally, look at the end result of an analytic process, which is a data governance policy instrument to support the iterative negotiation between the instances of data producers and data users for instantiating shared smart city visions.

Case projects

Solution

The following is a partial project solution. The students should be able to expand on the following:

Focus on how the smart city agenda has helped to create a ground for technology producers and policy entrepreneurs to propose and translate their solutions for smarter cities, including novel technologies such as integrated command centers or real-time problem-reporting systems.

Optional team case project

Solution

The following is a partial project solution. The students should be able to expand on the following:

Focus on why the smart city regulatory framework should be accompanied by examples of good practice, which stress the necessity of adopting interoperable smart city services. Also, look at why the development and deployment of smart city services can incur significant costs to cities, service providers, and sensor manufacturers; and thus, because of this, it is vital to adjust the national legislation to ensure legal certainty to all stakeholders, and at the same time, protect the interests of all citizens and the state. In addition, look at why a vast number of heterogeneous devices and smart city services (both existing and future) have had their interoperability become vital for service replicability and massive deployment, leading to a digital transformation of future smart cities. Finally, look at the classification of technical and regulatory characteristics of IoT services for smart cities, which are mapped to corresponding roles in the IoT value chain.

Chapter 23: Provide affordable smart infrastructure policy applications for informal sectors

Review questions/exercises

True/false

1. False
2. True
3. True
4. False
5. True

Multiple choice

1. B
2. C
3. D
4. D
5. C

Exercise

Solution

The following is a partial exercise solution. The students should be able to expand on the following:

Focus on the emerging phenomenon in the nexus between smart cities, new business models, and informal enterprises. Also, look at how the concept of smart cities is indeed facilitating new business models that are formalizing the informal sector.

Hands-on project

Solution

The following is a partial project solution. The students should be able to expand on the following:

Focus on the socioeconomic characteristics of the residents of smart cities, and assess the infrastructure, the building uses, and the environmental conditions of the smart cities areas in relation to smart principles and smart infrastructure application in the areas. Also, analyze the data by using inferential statistics; and show how the findings reflect that the building use is sprawl development in nature, with the informal settlement and infrastructure inadequacy as the major challenges. Finally, identify how smart infrastructure approaches are applicable for achieving sustainable development in areas with smart interventions in the socioeconomic status of the smart cities residents, the environment, the building uses, and the basic facilities and services in the areas.

Case projects

Solution

The following is a partial project solution. The students should be able to expand on the following:

Focus on why technology-enabled smart cities in developing countries can only be realized when concurrent socioeconomic, human, legal, and regulatory reforms are instituted.

Optional team case project

Solution

The following is a partial project solution. The students should be able to expand on the following:

Focus on an analytic framework to guide practitioners and decision-makers in assessing the status and devising strategies and actions to implement smart city initiatives. In other words, propose a smart city analytical framework for developing nations, which reflects their context to assess existing smart city programs, develop new smart city portfolios, and provide direct support for the planning and implementation of smart city solutions.

Chapter 24: Customisable Service Level Agreement (SLA) generator platform using FCAPS management to enhance Quality of Experience (QoE) on Internet of Things (IoT)

Review questions/exercises

True/false

1. True
2. True
3. False
4. True
5. False

Multiple choice

1. C
2. D
3. E
4. E
5. D

Exercise

Solution

The following is a partial exercise solution. The students should be able to expand on the following:

Focus on providing a QoE inference model in smart cities that estimates the mean opinion score with the root mean square error with additional traffic during probing. Then, evaluate the QoE inference model's generalization capacity when estimating QoE for videos that are different from the ones that are used for training, which can speed up the model's creation process. Next, explain why the network quality-of-service (QoS) does not always translate to user quality of experience (QoE). Also, discuss why the knowledge of user QoE is desirable in several scenarios that have traditionally operated on QoS information. Finally, to help address this problem, look at the offline generation of per-application models mapping application-independent QoS metrics to application-specific QoE metrics.

Hands-on project

Solution

The following is a partial project solution. The students should be able to expand on the following:

Focus on demonstrating how prioritizing traffic for latency-sensitive applications in smart cities can improve responsiveness and video frame rate. Next, show how a novel QoE-aware bandwidth allocation scheme for bandwidth-intensive applications can improve the average video bit rate for multiple users.

Case projects

Solution

The following is a partial project solution. The students should be able to expand on the following:

Focus on ways to present an extended SLA semantic model for modeling complex service dependencies in distributed applications in smart cities and for enabling automated cloud infrastructure management operations. Next, describe a multiobjective allocation algorithm for optimized resource allocation in infrastructure clouds. Finally, describe a method of discovering relations between the performance indicators of services belonging to distributed applications and then using those relations for building scaling rules that can be used for automated management.

Optional team case project

Solution

The following is a partial project solution. The students should be able to expand on the following:

Focus on a clustered-based multitenant database management system in smart cities. Next, look at how a dynamic proactive provisioning technique is built by using different prediction models. Finally, examine various experimental scenarios with different datasets that have been conducted to prove a promising solution in multitenant environments, where the tenants have irregular workload patterns.

Chapter 25: Effective value capturing of mobility and public transportation through efficient urban planning

Review questions/exercises

True/false

1. True
2. True
3. False
4. False
5. True

Multiple choice

1. B
2. E
3. C
4. B
5. C

Exercise

Solution

The following is a partial exercise solution. The students should be able to expand on the following:

Focus on some of the aspects of smart transport for in smart cities, such as cycling and the role of women in sustainable transport. Also, look into the idea of smart cities and sustainable transport and their global heterogeneous characteristics

Hands-on project

Solution

The following is a partial project solution. The students should be able to expand on the following:

Focus on the creation of an integrated data-sharing environment approach that ranges from establishing a smart city operating system; a data utility; a data archive; and an enterprise data management system. Then, discuss why most of these systems are being developed in parallel with multiple new data analysis tools, while regional metropolitan planning organizations continue to slowly evolve to consider emerging mobility technologies and services.

Case projects

Solution

The following is a partial project solution. The students should be able to expand on the following:

Focus on key gaps in data, knowledge, and methods required to advance energy-efficient smart city urban mobility innovation. Also, enable research and analysis into the collaboration between smart cities, requisite data resources, organization priorities, and objectives, as evidenced in smart city mobility goals.

Optional team case project

Solution

The following is a partial project solution. The students should be able to expand on the following:

Focus on discussing and comparing smart city and big data; and explore the opportunities, challenges, and benefits of incorporating big data applications for smart cities. In addition, identify the requirements that support the implementation of big data applications for smart city services. Finally, review opportunities that are available for utilizing big data in smart cities; but, keep in mind, that there are still many issues and challenges to be addressed to achieve better utilization of this technology.

Chapter 26: Urban design strategies and the smart city paradigm

Review questions/exercises

True/false

1. False
2. False
3. False
4. True
5. False

Multiple choice

1. A
2. E
3. D
4. C
5. D

Exercise

Solution

The following is a partial exercise solution. The students should be able to expand on the following:

Focus on why the results of a smart city analysis are represented in the form of network maps; and how the role of each result is clarified based on these network maps. In addition, explain the overall flow; change of flow; and the relationships with related fields as explained through cluster analysis. Finally, explain why the cluster analysis findings could serve as a basis for policymakers, urban managers, and researchers seeking a comprehensive understanding of the smart city concept in urban planning areas.

Hands-on project

Solution

The following is a partial project solution. The students should be able to expand on the following:

Focus on a qualitative research methodology that examines and compares smart cities as leading data-driven smart cities. Then, show how smart cities can have a high level of development of applied data-driven technologies; but, slightly differ in the level of implementation of such technologies in different city systems and domains with respect to sustainability areas. Finally, show how smart cities moderately differ in the degree of their readiness, as to the availability and development level of the competences and infrastructure needed to generate, transmit, process, and analyze large masses of data to extract useful knowledge for enhanced decision making; and deep insights pertaining to urban operational functioning, management, and planning in relation to sustainability.

Case projects

Solution

The following is a partial project solution. The students should be able to expand on the following:

Focus on how to critically link ongoing smart efforts with a development strategy of specific urban environments, by setting the context for selecting specific types of smart city applications.

Optional team case project

Solution

The following is a partial project solution. The students should be able to expand on the following:

Focus on the exploration of both the academic field of critical urbanism and smart urbanism, through the development of contributions to smart urbanism, from existing theoretical and conceptual approaches within critical urbanism. Next, look at the dimensions that require further development to facilitate a comprehensive analysis of what smart city policies mean for contemporary urban life: The acknowledgment that urban life is not confined to the administrative boundaries of a city; the importance of local socialeconomic, cultural-political, and environmental contingencies in analyzing the development, implementation, and effects of smart city policies; and the social-political construction of both the urban problems of how smart city policies aim to solve and the consider solutions; thus, there being a lack of consideration for the urbanism in smart urbanism.

Chapter 27: Smart city technologies plus nature-based solutions: viable and valuable resources for urban resilience

Review questions/exercises

True/false

1. True
2. False
3. False
4. True
5. False

Multiple choice

1. B
2. B
3. E
4. D
5. E

Exercise

Solution

The following is a partial exercise solution. The students should be able to expand on the following:

Focus on the results that facilitate an NBS assessment scheme selection that is based on an alignment between each analyzed assessment scheme and specific Sustainable Development Goal subobjectives. Also, look at how smart cities face multiple challenges amidst a complex hierarchy of legislative, regulatory, and other stakeholder obligations. Finally, look at a strategic NBS assessment scheme that closely aligns with one or more subobjectives within Sustainable Development Goal that can maximize operational efficiency by exploiting synergies between evaluation schemes.

Hands-on project

Solution

The following is a partial project solution. The students should be able to expand on the following:

Focus on no tech, where green is considered traditional urban green, by handling (rain) water like nature would. Also, look at low-tech green, which is suitable for dense urban settings with limited demand for water management and ecosystem services. Finally, discuss why specifying the demanded no/low/high-tech level of green infrastructure in urban design plans will help to yield the most of ecosystem services using appropriate levels of available technology in smart cities.

Case projects

Solution

The following is a partial project solution. The students should be able to expand on the following:

Focus on how smart city technologies reduce the cost of gathering information about usage patterns. Also, look at how smart city governments, employers, and residents can find new ways to optimize existing systems. Next, show how some smart city solutions respond to demand and involve the public in shaping it. Finally, look at how smart city technologies encourage people to use transit during off-hours; to change routes; to use less energy and water and to do so at different times of day; and to reduce strains on the healthcare system through preventive self-care.

Optional team case project

Solution

The following is a partial project solution. The students should be able to expand on the following:

Focus in part on the key differences that will emerge: The NBS concept that incorporates a broader array of interventions and a broader range of perspectives on what qualifies as nature-based; and how it is most explicitly oriented toward providing solutions to complex challenges. Next, look at how the NBS implementation could benefit from a more performance-based planning approach; a flexible approach to urban planning, which accommodates the integration of multiple land uses; and considers urban complexity. Finally, explain how the NBS concept has potential to unite currently segregated bodies of knowledge, generated as part of related approaches to urban greening;

and can therefore enable researchers and policymakers to more explicitly discuss the role of nature in addressing a broad range of sustainability challenges in smart cities.

Chapter 28: Financial and technical planning of decarbonized hybrid energy systems for modern cities

Review questions/exercises

True/false

1. False
2. False
3. True
4. False
5. True

Multiple choice

1. C
2. C
3. A
4. E
5. A

Exercise

Solution

The following is a partial exercise solution. The students should be able to expand on the following:

Focus on a real grid area, and how it is used to validate the simulation setup of a sensitivity analysis of building collectives and grid topologies, which provides a systematic insight into the interrelated system. Also, look at whether it is possible to identify building and grid configurations that increase the risk of a complete gas grid shutdown and those that should be operated as a flexibility option in a future renewable energy system in smart cities.

Hands-on project

Solution

The following is a partial project solution. The students should be able to expand on the following:

Focus on the simulation of hybrid urban renewable energy systems for technoeconomic feasibility analysis. Then, look at the past, present, and future trends of a typical hybrid energy system design, development, and implementation for an urban region, which can be later replicated in smart cities in other parts of the world.

Case projects

Solution

The following is a partial project solution. The students should be able to expand on the following:

Focus on a detailed overview of a framework developed for the EU H2020 funded Sharing Cities project. Also, illustrate the nature of the design stages encountered and control hierarchies required. Finally, look at the current state of computational intelligence in the field of smart city energy management, while providing insight into the ways in which current barriers can be overcome.

Optional team case project

Solution

The following is a partial project solution. The students should be able to expand on the following:

Focus in part on the potential to further revolutionize energy services at the system level, by coordinating the exchange of energy currency among the energy sectors in a manner that optimizes financial efficiency (including capital investments); maximizes thermodynamic efficiency (through best use of energy, which is the potential to use the available energy in producing energy services); reduces environmental impacts when clean energy inputs are maximized; and provides resources for grid management in smart cities.

Chapter 29: Reactive power compensation for voltage variation mitigation in grid with renewable energy

Review questions/exercises

True/false

1. False
2. True
3. False
4. True
5. False

Multiple choice

1. D
2. D
3. B
4. B
5. B

Exercise

Solution

The following is a partial exercise solution. The students should be able to expand on the following:

Focus on a typical LV distribution network to demonstrate the effectiveness of different reactive power control techniques in mitigating over voltage issues due to high PV integration. Then, analyze the performance of four reactive power control techniques: fixed power factor control, scheduled power factor control, power factor control as a function of injected active power, and voltage-dependent reactive power control. Next, look at how the four reactive power control techniques are analyzed and compared in terms of the number of customers with voltage violations, reactive power compensation, and network losses. Finally, look at the performance of three-phase, time-series, high-resolution power-flow simulations; investigate the potential overvoltage issues; and assess the performance of the adoption of reactive power controls in a network for a smart city.

Hands-on project

Solution

The following is a partial project solution. The students should be able to expand on the following:

Focus on strategies to mitigate negative impacts (like the inverter-interfaced nature of the PV systems significantly impacts the power system operation from protection, power flow, and stability perspectives.), so that PV penetration into the grid can continue. Next, thoroughly review such strategies from different research fields: such as communication, artificial intelligence, power electronics and electric vehicle charging coordination. Finally, recommend possible research directions for future work in smart cities.

Case projects

Solution

The following is a partial project solution. The students should be able to expand on the following:

Focus on transmission grids in smart cities; the voltage control for accommodating wind and solar power; fault-induced delayed voltage recovery (FIDVR); and measurement-based Thevenin equivalent for voltage stability analysis. Next, look at the distribution grids; how the impact of high penetration of distributed energy resources (DER) is analyzed; typical control strategies; and the challenges for a local inverter Volt-Var control. Finally, thoroughly discuss the motivation, state-of-the-art, and future directions of the coordination of transmission system operators (TSO) and distribution system operators (DSO).

Optional team case project

Solution

The following is a partial project solution. The students should be able to expand on the following:

Focus on the recent decades of the power system restructuring process in smart cities; and how centralized energy sources are being replaced with decentralized ones. Look at how this phenomenon has resulted in a novel concept in electric power systems—particularly in distribution systems, known as Distributed Generation (DG). Next, look at how utilizing DG is important for secure power generation and reducing power losses. Also, look at how the widespread use of such technologies introduces new challenges to power systems, such as their optimal location, protection devices' settings, voltage regulation, and Power Quality (PQ) issues. Finally, consider why specific DG technologies are based on Renewable Energy Sources (RESs)—such as wind and solar, due to their uncertain power generation.

Chapter 30: DC smart micro grid protection system

Review questions/exercises

True/false

1. True
2. False
3. True
4. False
5. False

Multiple choice

1. B
2. A
3. B
4. C
5. D

Exercise

Solution

The following is a partial exercise solution. The students should be able to expand on the following:

Focus on major European smart city projects. Determine why these projects will offer a wider vision for researchers in the operation, monitoring, control, and audit of smart-grid systems.

Hands-on project

Solution

The following is a partial project solution. The students should be able to expand on the following:

Focus on the implementation of challenges associated with a technical and economical point of view in smart cities. This will provide a wider vision for the researchers to focus on the monitoring and control of nano-grid technologies.

Case projects

Solution

The following is a partial project solution. The students should be able to expand on the following:

Focus on the global energy crisis and the increasingly atmospheric pollution in smart cities. Look at the distributed generation integration with renewable energy, as it is becoming a potential trend in technology. Finally, look at the attention that has been paid to the recent challenges to the DC micro-grid system.

Optional team case project

Solution

The following is a partial project solution. The students should be able to expand on the following:

Focus on the features, rate of change, and cumulative sum that are calculated, based on a local measurement. This means that there is no delay involved in the calculation—which means a real time system. Also, look at how a protection method detects the faults within a few microseconds of its occurrence. Finally, test a proposed method under different fault scenarios by using MATLAB simulations in smart cities.

Chapter 31: Identification and classification of faults using Stockwell transform and decision rule

Review questions/exercises

True/false

1. False
2. True
3. False
4. False
5. True

Multiple choice

1. A
2. B
3. C
4. D
5. B

Exercise

Solution

The following is a partial exercise solution. The students should be able to expand on the following:

Focus on how fault detection in smart cities is achieved by utilizing the fault index, which depends on a combination of characteristics extracted from the current signal by the application of the Stockwell transform and Wigner distribution function (WDF). Next, look at how various faults are categorized by using a quantity of phases with a faulty nature. Finally, investigate fault events such as phase to-ground (L-G), two phases (LL), two phases to-ground (LL-G), and three phases to-ground (LLL-G).

Hands-on project

Solution

The following is a partial project solution. The students should be able to expand on the following:

Focus on a technique that uses the Stockwell transform-based multiresolution analysis of current signals for detection of the faults for smart cities. Also, look at a rule-based decision tree for classification of faults in the power system network. Finally, investigate line to-ground (LG) fault, double line (LL) fault, double line to ground (LLG) fault, and three-phase fault with the involvement of the ground (LLLG).

Case projects

Solution

The following is a partial project solution. The students should be able to expand on the following:

Focus on how three-phase current signals are processed through ST with a view to extracting useful statistical features. Next, look at how the extracted features are forwarded to FFNN in order to detect and classify different types of faults in smart cities. Finally, implement two different test distribution grids, modeled and simulated in real time, by using a digital simulator and MATLAB/SIMULINK.

Optional team case project

Solution

The following is a partial project solution. The students should be able to expand on the following:

Focus on how the power system model is simulated in a MATLAB/Simulink environment, with unbalanced loading. Then, look at how this technique requires voltage and current signals to be extracted at the sending and receiving end of the network in smart cities. Finally, show how the current and voltage signals are processed, by using ST to produce complex S-matrices in smart cities.

Chapter 32: Evaluation of fuel production technologies by using renewable energy for smart cities

Review questions/exercises

True/false

1. True
2. False
3. False
4. True
5. True

Multiple choice

1. B
2. C
3. D
4. E
5. A

Exercise

Solution

The following is a partial exercise solution. The students should be able to expand on the following:

Focus on a dense settlement structures in smart cities that have high demands of energy. Show how these demands exceed the local resource availability. Finally, develop supply options to cover these demands; and show how they differ from place to place and vary within the boundaries of a smart city.

Hands-on project

Solution

The following is a partial project solution. The students should be able to expand on the following:

Focus on the intelligent application of smart devices, and how they stand as a key element in establishing smart cities. Next, look at the intelligent application of smart devices, which have been suggested as the solution to complicated future urbanization difficulties in coming years. Finally, consider the scarcity of traditional fossil fuels in the near future. Look beyond the ecological problems of traditional fossil fuels and the new smart grids that have demonstrated the potential to merge the nonrenewable and renewable energy resources into each other; thus, leading to the reduction of environmental problems and optimizing operating costs.

Case projects

Solution

The following is a partial project solution. The students should be able to expand on the following:

Focus on technology innovation, smart governance and main financing, and support programs in smart cities. Do an analysis of the most interesting initiatives at the international level that are being pursued by smart cities investigating the main areas of Green Buildings, Smart grid-Smart lighting, and Smart mobility. Finally, identify sustainable development plans and programs at the urban level, within the current legislative framework of your location.

Optional team case project

Solution

The following is a partial project solution. The students should be able to expand on the following:

Focus on an economic analysis of the infrastructure improvements that result in reducing the energy demand of the smart cities, which are represented by the intelligent light-emitting diode (LED) street lighting system. Discuss why smart LED streetlights are getting increasingly popular in the world's major metropolises, as one of the leading components of the smart city. Finally, compare the efficiency of LED street lighting that is used in smart cities, with the commonly used lighting systems that are based on sodium lamps.

Chapter 33: Solving traffic congestion and parking scarcity with narrow track vehicles

Review questions/exercises

True/false

1. False
2. True
3. False
4. True
5. False

Multiple choice

1. C
2. D
3. A
4. B
5. C

Exercise

Solution

The following is a partial exercise solution. The students should be able to expand on the following:

Focus on a clear insight into the suitability of SPSs in various environmental conditions, and highlight their advantages/disadvantages for smart cities. Also, look at how an extensive comparison among multiple aspects of SPSs would enable researchers, designers, and policymakers, to identify the best-suited SPSs; and understand the current trends in this sector.

Hands-on project

Solution

The following is a partial project solution. The students should be able to expand on the following:

Focus on curbside parking implementation, double parking operations, important advantages for drivers, the environment, and the smart city. Next, look at how autonomous car owners could (at least) double their probabilities of finding parking compared to traditional drivers, saving cruising time and emissions. Finally, look at the significant work and technological advances that are still needed, in order to make this feasible in the near future.

Case projects

Solution

The following is a partial project solution. The students should be able to expand on the following:

Focus on some of the challenges regarding the realization of smart parking systems solutions for smart cities.

Optional team case project

Solution

The following is a partial project solution. The students should be able to expand on the following:

Focus on an intelligent system that offers the most favorable prices to its users by matching private garages' pricing portfolio with a garage's current demand for a smart city. Look at how a pricing that is generated correctly forecasts the demand for a particular garage at a specific time of the day and year. Finally, look at how a PAYG smart parking system can effectively contribute to the macro solution to traffic congestion, by encouraging potential users to use the system's services; thus, reducing the orbiting time for parking.

Chapter 34: Modeling and simulation for Connected and Automated Vehicle (CAV) deployment and performance evaluation

Review questions/exercises

True/false

1. True
2. False
3. False
4. True
5. False

Multiple choice

1. D
2. D
3. A, D
4. E
5. B

Exercise

Solution

The following is a partial exercise solution. The students should be able to expand on the following:

The level of automation of vehicles ranges from Level 0 to 5. Level 0 concerns vehicles that are entirely human-operated, while Level 5 vehicles are fully automated. For example, in Level 1 of automation, the vehicle may assist the driver with tasks such as steering or acceleration. It enables the driver to remain fully engaged with the driving task, but with gradual transfer of control from human to machine. Level 2 automation features include adaptive cruise control and automatic emergency braking. Most of the automated vehicles classify as Level 4 of automation in the industry, while the automotive companies have carefully explored Level 3.

Hands-on project

Solution

The following is a partial project solution. The students should be able to expand on the following:

Localization is the task of knowing the actual position at any driving condition. To perform this function, CAV takes input from its onboard Global Positioning System (GPS) device, Telemetric service (map), V2I, and V2V commutation service to calculate its current location.

Case projects

Solution

The following is a partial project solution. The students should be able to expand on the following:

Focus on a number of typical CAV applications that have been examined in great detail, where a categorized analysis in terms of MOEs is performed. Then, draw some conclusions: including the identification of influential factors on system performance, and suggested approaches for obtaining cobenefits across different types of MOEs.

Optional team case project

Solution

The following is a partial project solution. The students should be able to expand on the following:

Focus on organizations that come under the context of CAV planning; and gauge their awareness, perception, and viewed importance of planning for CAV technologies.

Chapter 35: VUCA: Volatility, Uncertainty, Complexity, and Ambiguity in our smart city and tech finance future

Review questions/exercises

True/false

1. False
2. True
3. True
4. False
5. False

Multiple choice

1. E
2. E
3. B
4. A
5. C

Exercise

Solution

The following is a partial exercise solution. The students should be able to expand on the following:

Focus on the results of a futures clinique, where scenarios were tested and presented. Also, explore the implications of surprises for energy security, as the world increasingly seeks to move toward a renewable energy-based society.

Hands-on project

Solution

The following is a partial project solution. The students should be able to expand on the following:

Focus on a study that explores the lived experiences of corporate executives, about their VUCA business environment, and the strategies they employed for VUCA readiness and success within a manufacturing sector. Also, conduct research about the lived experiences of corporate executives in manufacturing companies, regarding VUCA readiness and success in a VUCA environment. Finally, look at how chaos theory and complexity leadership theory could be used to frame their VUCA environment.

Case projects

Solution

The following is a partial project solution. The students should be able to expand on the following:

Focus on an accessible diagnostic tool to assess the VUCA dimensions a project is facing, so that project managers can effectively isolate risk events. Also, discuss how an accessible diagnostic tool can be tested on complex projects in the manufacturing industry and its effectiveness.

Optional team case project

Solution

The following is a partial project solution. The students should be able to expand on the following:

Focus on the challenges that entrepreneurs encounter while leading; the ways in which they evaluate their success; the role that technology played day to day; and what recommendations they would make to future leaders of systems-wide change. Also, look at how successful practices and strategies played a role in leading systems-wide change in a technology-rich VUCA ecosystem.

Appendix H

Glossary

Accessibility The ability to reach desired goods, services, activities, and destinations (collectively called opportunities).

Acute shocks Sudden, sharp events that threaten a smart city.

Autonomous vehicles Autonomous vehicles can assume decision-making and operational tasks, enabling drivers to become passengers, entirely disengaged from the demands of driving.

Autonomy Automated Service Level Agreement (SLA) provision, negotiation, and renegotiation.

Ballast An electronic device for HPS light that increases the voltage of the supply power up to thousands of volts in order to produce the lighting.

Bicycle detection and actuation Sensors at regular traffic signals to alert the controller of bicycle crossing demand on a particular approach.

Carbon credits A permit that allows a country or organization to produce a certain amount of carbon emissions and can be traded if the full allowance is not used.

Cloud service provider The provider of the SLA, which is agreed with a customer, who provides a service and resources in exchange for a monthly fee.

Controls Electronic technology that interfaces power to a light or light output electronics to a sensor or timer or remote human interface.

Critical data Data on which human life depends.

Critical infrastructure Systems and assets, whether physical or virtual, so vital to the United States that the incapacity or destruction of such systems and assets would have a debilitating impact on security, national economic security, national public health or safety, or any combination of those matters.

Customer tolerance Customer ability to tolerate disruption to the service agreed with the provider and captured in the SLA.

Data Consumer (DC) Data Consumers access data that has been stored in a repository and is publicly available.

Data Provider (DP) Data Providers contribute data for hosting in the cloud, which may be consumed by themselves and, if publicly available, by other Data Consumers.

Dimming Reducing light output.

Disaster resilience A combination of a society's preparedness for a hazard; their ability to mitigate, plan, and respond immediately and effectively to it; and their ability to recover and regenerate from the event.

Domain Describes a collection of devices, which operate for a common application purpose (autonomous cars in a city or a group of smart homes).

Driver An electronic device that converts line electricity into the form that is needed to drive HPS or LED lighting.

Eco-smart city A human settlement modeled on the self-sustaining resilient structure and function of natural ecosystems.

Efficiency From the service provisioning perspective, efficiency is achieved through fewer redundant resource allocations across the network to support customer demand.

First and last mile Gaps in public transit, which require individuals to use other forms of transportation such as driving their car or riding their bike.

Green building An environmentally sustainable building, designed, constructed, and operated to minimize the total environmental impacts.

High albedo pavement A special type of concrete pavement that reflects more light than dark-colored materials due to its lighter color.

Household SLA A single Service Level Agreement would be assigned to each household.

HPS (High Pressure Sodium) lighting Lighting that uses an arc between electrodes in a sodium vapor environment to produce light from plasma.

Intelligent transportation system A system in which real-time data is gathered and used to inform automated decisions regarding the function of traffic-related infrastructure and hardware.

Intensive data High volume data transmissions.

Internet of Things (IoT) The interconnection via the Internet of computing devices embedded in everyday objects, enabling them to send and receive data.

LED (Light-Emitting Diode) A semiconductor device (electronics) that produces light directly from electricity.

Low carbon smart city A smart city that reduces its carbon footprint by focusing on renewable energy and mitigation measures.

Luminaire A device consisting of a light source, optics (reflectors, lenses), driver electronics (ballast, driver), and a housing to protect the components that is used to produce light from electricity.

Metadata Attributes that capture a customer's personal profile, their network service needs, and operational tolerances of the service provided by the network.

Mobility The movement of people or goods.

Modular bike share Is usually solar-powered, quick, and cheap to install; has the ability to alter and move the stations; and, typically does not require trenching, excavation, or other preparatory work.

Normalcy bias The difficulty to comprehend the fact that a disaster is occurring.

Ontology A template that governs the range of metadata collected about a customer service scenario.

Open bottom catch basin A box that is put into the ground near areas of standing water to help facilitate proper water drainage and avoid property damage.

Permeable pavement Can be asphalt, concrete, or pavers that lets storm water filter through and drain into the ground, instead of collecting on hard surfaces or draining into the sewer system.

Photocontrol An electronic circuit that senses the sun and turns off a luminaire.

Platform A group of management entities, data repositories, and ontology templates, which automate the SLA provisioning process, in addition to SLA management and network management for FCAPS responsibilities.

Positive train control A system of remote sensors and automated control devices primarily designed to stop or slow a train automatically to prevent dangerous situations.

Quality of Experience (QoE) The extent to which a customer is satisfied with the service received from the network, measured in a subjective and nonstandardized way.

Quality of Service (QoS) A level of network performance which satisfies application needs.

Repository A storage space populated with data collected from customers, with a storage allowance agreed in their SLA.

Resilient smart city A smart city that has developed capacities to help absorb future shocks and stresses to its social, economic, and technical systems and infrastructures, so as to still be able to maintain essentially the same functions, structures, systems, and identity.

Sensor An electronic device that senses a parameter such as presence of a vehicle; presence of a pedestrian(s); sunset or sunrise; WiFi; and, Bluetooth®, etc.

Service Level Agreement (SLA) A contract to ensure that a customer receives a level of service, which has been agreed with a service provider for a fixed length of time for an agreed monthly cost.

SLA active state The SLA is responsible for governing activities within the network in the active state, which involves ensuring that the terms to which the customer agreed are being fulfilled, and reimbursing the customer in the event that they are not.

SLA flexibility A SLA may be renegotiated with a service provider once it moves into the active state.

SLA life cycle The phases through which the SLA passes, from customer service request to SLA definition, SLA negotiation, SLA active, SLA renegotiation, SLA recommendation, and SLA termination.

Smart city An efficient city, a livable city; as well as, an economically, socially and environmentally sustainable city.

Smart connections A vital component of a smart city, incorporating transportation, online access, technology, and community.

Smart lighting Any lighting that turns on, dims, or turns off in response to a control.

Smart urban mobility Finding new ways to improve the flow of traffic in cities and encourage more efficient use of transport infrastructure.

Sustainable urban mobility Satisfies current mobility needs of smart cities without compromising the ability of future generations to meet their own needs.

Traffic Vehicle movement.

Transition smart city Refers to grassroot community projects that aim to increase self-sufficiency to reduce the potential effects of peak oil, climate destruction, and economic instability.

Transportation Moving stuff and people.

Urban Mobility Concept (UMC) An innovative mobility concept; characteristics of UMC include its suitability for short distance travel, for easy transport by train, and the possibility to use it on public roads.

Urban resilience The ability to adapt to changing conditions and withstand and rapidly recover from disruption due to emergencies.

Vehicle-to-Infrastructure Systems allow physical infrastructure to inform vehicles of their presence and provide additional data, and also allow vehicles to send information to the infrastructure.

Vehicle-to-Vehicle Vehicles that talk to one another to provide data about speed, location, and other information.

Vertical garden/farm An alternative form of farming stretching from the ground up, with the ability to be built in areas with limited soil space to grow crops.

Waste-to-energy Incineration of municipal solid waste that then uses the energy (in the form of heat) to produce electricity and/or steam for heating.

Zero-carbon smart city A smart city that runs entirely on renewable energy; it has no carbon footprint and will in this respect not cause harm to the planet.

Index

'*Note:* Page numbers followed by "f" indicate figures, "t" indicate tables and "b" indicate boxes.'

A

Accessible funding matrix (AFM), 126
Accountability, 104
Acorn Guide, 336, 338
Active mobility, 392
Activist informed policy, 80–82
Actors, 490–491
Adaptive Cruise Control (ACC), 497–498
Added value of SCT plus NBS, 391–393
  effectiveness, 391
  stakeholder engagement and social equity, 391
  synergism with
    other urban infrastructures, 391–392
    other urban systems, 392–393
Administrative culture, 312
Advanced driving assistance systems (ADAS), 97, 497–498
Affective Intelligent Driving Agent (AIDA), 504
Affordability, 324
Affordable housing, 165–166
Affordable smart infrastructure policy applications for informal sectors, 579
  review questions/exercises, 579
Agra, 267–268
Agra Municipal Corporation, 263
Agra Smart City Limited (ASCL), 268
  ICCC project of, 268–269
Agreement-seeking cooperation, 105
Agriculture, 61
Air pollution, 240
Alerts, tracking and, 325
Alexa (voice recognition devices), 54
Amazon, 221, 511
American Automobile Association, 220–221
Analytics, 53–54
Annexations, 27–28, 62–63
Anonymization, 249
Antilock braking systems, 97
APAC (IoT platform market), 325
Arabianranta Living Labs, 288
ArcGIS Urban, 53–54
Arduino, 329–330
  IDE output, 330
Area-based development (ABD), 266–267, 353–354
Artificial inductive line impedances (AILIs), 430–431
Artificial intelligence (AI), 6, 34, 54–55, 169, 177, 198, 269, 387, 482
  applications, 54–55
  drones and self-driving cars, 55
Assistive scooters, 228–229
Association des Constructeurs Européens de Motocycles (ACEM), 475
Atal Mission for Rejuvenation and Urban Transformation (AMRUT), 264
Attribute-based encryption, 256
Augmented reality (AR), 54, 169, 308, 387, 504
Austrian economics, 19
Automated agent, 106
Automated streetlight monitoring system, 327
Automated Vehicle Transparency and Engagement for Safe Testing (AV TEST), 100–101
Automated vehicles (AVs), 97
  safety guidance for, 99–101, 100f
Automatic street light control system, 326–327
Automation, 190, 213, 573–574
  review questions/exercises, 573
  systems, 190
Automobiles, 478
Autonomous Electric Vehicle (AEV), 501
Autonomous Emergency Braking (AEB), 497–498
Autonomous Vehicle Readiness Index (AVRI), 101
Autonomous vehicles (AVs), 34, 105, 212–213, 481, 494–495

B

Back-propagation neural network (BPNN), 439–440
Bandra Kurla Complex (BKC), 356
Bangalore Smart City Proposal, 353–354
BarCamp, 291
Barter, 25
Batteries, 35
Battery electric vehicles (BEVs), 35
Battery Energy Storage (BES), 400, 404
Behavioral economics, 19
Bengaluru Metropolitan Transport Corporation (BMTC), 353–354
Best management practices (BMPs), 380–381
Big data, 366, 387
Bike trails, 157–162
Biomedical waste, 240
Bird's eye plot, 494
Black Lives Matter (BLM), 207
Blackhole attack, 104
Blockchain, 177
  connected city drones, 178
  connected healthcare, 178
  preparing for future technologies, 177–178
  supporting internet of transportation and logistics 4. 0, 177–178
Blomley's theory of pedestrianism, 213
Blue infrastructure (BI), 380–381
Bluetooth, 321–323, 498
  LE, 324
Blynk cloud, 321
  representation, 330
Boiler, 404
Boston's "Big Dig" tunnel, 475–476
Bottom line, waste management, 241–242
  classification of solid waste, 241t
  discarded plastic bags, 242f
  hazardous waste, 242
Bottom-up participatory governance, 277, 299
  basic policies for smart city governance, 279–283
  platforms supporting, 283–291
    cocreation of value, 290–291
    coproduction, 284–290
  promoting platforms for, 577–578
    review questions/exercises, 577–578
  smart city governance, 278–279
Bridge District Plan, 165
Brihanmumbai Municipal Corporation (BMC), 263
Browser, 344–345
  low tolerance, 349t
  tolerant, 348t
Bucket-based index, 256–257
Budgeting, local, 23
Buildings, 392
Built environment, 365
Business Improvement Area (BIA), 211, 232–236
  business change and societal issues, 235
  context of, 232
  local thinking sustain local growth, 235–236
  merchants' concern, 234b
  potential municipal, 236
  responses to pathway robots, 236

Business Model Canvas, 24
Business-as-usual PPP, 127

C

Camera, 491–492
CANtact, 505
Capacity reselling, 25
Capital cost of equipment, 405
Capital expenditures (CAPEX), 23, 123
Capture, 301
Car pooling, 41
Car Sharing, 41
Carbon dioxide emissions (CO_2 emissions), 416–417
Carbon monoxide (CO), 383
Carbon pricing, 390
Carbon-free energy, 82–84
 conducting municipal and community-wide forecast, 82
 creating ranking matrix for solar siting, 82–83
 determining current and future load forecasts, 83–84
 developing clean energy strategy toolkit, 83
 establishing solar and electric vehicle design guidelines, 82
 investing in pilot projects, 84
 preparing renewable energy and optimization analysis, 82
 utilizing computer modeling, 83
Caring Caribou, 505
Carpooling, 472–473
CAV Assistant Module, 505
Cellular-V2X (C-V2X), 101
Centers of excellence (COEs), 178–179
Central Bureau of Investigation (CBI), 268
Central Business District (CBD), 353
Central Electricity Authority (CEA), 417
Central Nodal Agencies (CNAs), 264
Central transmission utility (CTU), 415, 418
Centralized Ontology, 342
Centralized Repository, 340–342
Centralized Repository Manager, 340–342
Centrally Sponsored Scheme (CSS), 267
CHAdeMO system, 45
Chalk River Laboratories, 400
Change management, 170, 173
Charging stations, 33
Charrettes, 25
Chicago Defender, The, 156
Circuit breakers (CB), 428, 441
Cities
 developing energy-efficient urban plan for, 90–92
  analyzing collected data and developing draft plan, 91
  building public and internal support, 91
  conducting energy and greenhouse gas emissions audit, 91
  designating a lead office, finding champions, 90–91
  establishing partnerships, 91
  finding hooks, 91
 developing strategic urban plan for cities in Egypt, 92
Citizen participation, 297, 300–302, 307–308
Citizen science, 388
City Model Canvas, 24
City of Austin, Texas, 99
"City of tomorrow", 174
City planning, 17–18
City Viz, 53–54
Civic crowdfunding, 26
Civic engagement and involvement, 21
Civic Hackathons, 26
Clean energy, 86, 413
 strategy toolkit, 83
Clean energy infrastructure. *See also* Green infrastructure
 activist informed policy, 80–82
 carbon-free energy, 82–84
 Orlando Greenworks Action Plan, 80f
 Portland climate action plan, 80f
 shutting down coal and envisioning equity, 84–86
Clean Energy Roadmap, 82
Climate change, 79
 reduction of emissions and fight against, 36
Cloud computing, 6, 51
Cloud service provider (CSP), 340
Cocreation
 spaces, 303
 of value, 290–291
  BarCamp, 291
  urban hacktivism, 290–291
Code of Federal Regulations (CFR), 99–100
Code reform, 29
Codes of conduct, 137–138
Cogeneration. *See* Combined Heat and Power (CHP)
Cognitive computing, 6
CoLaboratories (CoLabs), 148
Collaborative computations, 257–258, 258f
"Collective Impact" theory, 140
Collective intelligence, 301
Collective intelligence, 303
Combined Heat and Power (CHP), 399, 402
Combustion, 240
Commitment, 307
Commons-based organizations, 26
Communal Public Square, 228
Communication, 278
 for connected mobility, 101
 frames, 140
Community/communities, 479
 land trust, 26
 plan to clean energy infrastructure, 566
  review questions/exercises, 566
 skills, 184–185
 well-being, 22
Commuters, 478
Complexity economics, 19
Composting, 240
 of organic solid wastes, 244
Comprehensive approach, 68
Comprehensive Growth Management Act, 157
Concept City, 70
Congestion, 472
Connected and automated mobility systems, 97–99, 108
 CAVs, 98
 connected and automated infrastructure, 98–99
 field report
  case studies and pilot projects by university of Oregon, Oregon, US, 109
  Connecticut Department of Transportation, Connecticut, US, 108
  Florida, US, 108–109
  MaaS, 109
  private sector involvement, 109
 policies for connected and automated mobility services, 108–109
 transportation data, 99
Connected and automated vehicles (CAVs), 97–98, 481
 AV demonstration in transportation research board, 98f
 communication and cybersecurity layers for CAV systems, 497–499
 control strategies with deep learning, 494–497
 deployment in smart cities, 505–506
 functional modeling, 484
 integrated modeling and simulation, 483–494
 modeling and simulation for, 587–588
  review questions/exercises, 587–588
 navigation of, 106–108
 robot-assisted CAV systems, 504–505
 test platform, 499–504
  CAVTE demonstration, 501–504
  CAVTE design requirements, 499
  function test analysis, 501
  test functions, 499–501
 use cases, 482–483
Connected and Autonomous Vehicle Test Environment (CAVTE), 499
 demonstration, 501–504
 design requirements, 499
Connected cars, 323
Connected mobility, 98–99
 communication for, 101
Connected public transport systems, IoT technology for, 323
Connecticut Department of Transportation, Connecticut, US, 108
Connectivity, 321, 356
 protocols, 324
Continuous Wavelet Transform (CWT), 496
Control
 and protection of coordinated strategy, 432
 strategies with deep learning, 494–497
 system effects on protection, 431
Control Area Network (CAN), 101–103
Control Box, 42
Controller Area Network (CAN), 498
Convenience, 473–474

Conventional toll systems, 323
Convergence, 267
of missions, 264–265, 264f
Converter, 404
Convolutional Neural Network (CNN), 494–495
Cooperative driving automation (CDA), 105–106
Cooperatives (Co-ops), 26
economics, 19
Coordinated leadership, 307
Coproduction, 284–290
crowdfunding, 287
crowdsourcing/citizensourcing, 285–287
gamification, 289–290
geoparticipation, 285
Hackathons, 287–288
Living Lab, 288–289
Coresearch PPP, 127
Cost of charging electric vehicle, 45
Cost–benefit analysis (CBA), 390
Cotton gin, 517
County departments, 58–59
smart technology applications by, 59–60
Covariance fault index (CFI), 443
Cracking, 461
Crashworthiness, 99–100
Crisis of politics, 297
Critical infrastructure policies, 525, 541, 545, 551, 553, 557
Crowd-sensed data, 388
Crowdfunding, 26, 287
Crowdsourcing, 26, 388
crowdsourcing/citizensourcing, 285–287
methodology, 175–176
Cultural change, instruments of governance and, 301–303
Culture and smart cities, 18–19
"Curb Rights", 215
Current derivative protection, 431
Customers, 338
activities and behaviors of, 340
classification, 338f
entity relationship, 340f
network service needs and tolerances, 339t
pay service providers, 335
types, 344
Customizable service level agreement generator platform
prior art, 337–338
high-level ontology design, 342
relationship between QoS and QoE, 336f
research problem, 336–337
research proposal, 338–342
customers classification, 338f
customers network service needs and tolerances, 339t
entity relationship, 340f
high-level platform setup, 341f
system platform architecture, 341f
SLA assignment process, 343–349
Cyber-physical systems (CPS), 6, 187–188
Cyberattack, 103
Cybersecurity, 99–105
layers for CAV systems, 497–499
requirements and definitions, 103f

D

Dashboards, 27
Data
analysis, 323
analytics, 172
anonymity, 250–255
dataset including health information, 251f
deidentified version of dataset, 251f
differential privacy, 254–255
k-anonymity, 252–254
assortment and analysis, 324–325
consumers, 340
data-based protection system, 430
fragmentation, 254
generalization, 252
integration, 388
management framework, 200
monetization, 25, 172
privacy, 250
processing, 321
security and privacy in smart cities, 575
review questions/exercises, 575
Data encryption, 255–257
enabling fine-grained access, 256–257
example of encrypted and indexed version, 256f
key derivation structure and token catalog, 256f
enabling selective access, 255–256
example of access matrix, 255f
Data providers (DPs), 340, 344–345
low tolerance, 348t
tolerant, 347t
Daubechies4 (Db4), 439–440
DC fault detection methods, 431–432
DC micro grids, 427
bus architectures, 428f
control strategy, 427
design, 427
operation criterions and setting rules, 432
research prospect, 432
DC protection system, 428–429
grounding, 428–429
protection devices, 428, 429t
DC smart micro grid protection system, 584–585
review questions/exercises, 584–585
Decarbonized hybrid energy systems
future work, 412
modeling of system architecture, 400–404
BES, 404
boiler, 404
converter, 404
electric and thermal demand, 400–401
microgrid controller, 404
microreactors, 402
solar PV panels, 402
TLC, 404
wind turbines, 403–404
N-R IES, 399–400
optimization problem, 404–406
sensitivity assessment, 407–408
simulation results, 406–407
Decentralized control, 427
Decision Support Systems (DSS), 309
Dedicated short-range communication (DSRC), 101
Deendayal Antyodaya Yojana–National Urban Livelihoods Mission (DAY-NULM), 264–265
Deep Autoencoder (DAE), 494–495
Deep learning, 388
case study, 495–497
control strategies with, 494–497
system designs and descriptions, 495
Deep Reinforcement Learning (DRL), 494–495
Delivery robot technology, 220–223
concerns for job loss from, 221
development and deployment costs for, 221
fear barrier for, 220–221
privacy issues, 222
psychotechnical barriers, 222
regulatory barrier to, 221
risk of nonreadiness to cities and towns, 223
safety barrier for, 220
security issues, 223
total risk equation for, 223
travel environment for, 221–222
Democracy, 297
Denial of Service (DoS), 104
Development control regulation (DCR), 355
Development plan (DP), 355
Device communication, 68
Differential privacy, 254–255
Differential protection, 432
Digital city, 6–7
Digital divide, 50, 68
Digital exclusion, 282
Digital finance, 511–512
Digital infrastructure policies
collaborative computations, 257–258, 258f
data anonymity, 250–255
data encryption, 255–257
Digital manufacturing, 172
Digital technologies, 41, 169
practice of urban design with, 371–372
in public engagement and participatory design, 370
Digital twins (DT), 6
Digitization, 573–574
review questions/exercises, 573
Digitizing government services, 189
Directional overcurrent protection, 431–432
Directorate of Enforcement (ED), 268
Disaster management training, 172
Discrete orthogonal Stockwell transform (DOST), 439–440
Discrete WT (DWT), 439–440
Discretionary Lane Change (DLC), 483
Disposal of solid waste, 243
Distance education, 186, 193
Distance protection, 432
Distributed Denial of Service (DDoS), 104

Distributed energy resources (DER), 200
Distributed Generation (DG), 399, 416–417
Distributed voltage control method (DVC method), 431
Distribution static compensator (DSTATCOM), 416–417
Domain Ontology, 342
Domain Repository Manager, 340–342
Dream City, 70
Drive Electric Orlando program, 81–82
Driverless cars. *See* Autonomous vehicles (AVs)
Drones, 55, 178
Droop control, 427
Dynamic driving task (DDT), 105
Dynamic economy, 4–5
Dynamic efficiency, 390–391

E

E-administration, 281
E-democracy, 281
E-education, 186
E-governance, 281, 306
E-learning, 186
E-mobility, new actions for improving, 36–37
E-trainings, 169, 193
Earth observation (EO), 388
East Central Florida Resiliency Action Plan, 166
Ecological Design Lab (EDL), 148–150
Ecological economics, 19
Ecological engineering (EE), 380–381
Ecological restoration, 162
eCommerce, 173, 222
Econ Wilderness Area, 159
Economic(s)
  analysis, 27
  development corporations, 27
  and smart cities, 19–20
Ecosystem services (ESS), 380–381, 382t
Ecosystem-based adaptation (EBA), 380–381
Ecosystem-based disaster risk reduction (Eco-DRR), 380–381
Edge computing, 6
Edifice Consultants Pvt Ltd (ECPL), 356
Educating professionals, 137–150
Education
  rograms, 186
  system, 199
Effective value capturing of mobility and public transportation through efficient urban planning, 580–581
  review questions/exercises, 580–581
Efficiency, 200, 390–391
Egocentric view, 493
EGovernance for Sustainable Development (EGOV4SD), 309
Egypt, developing strategic urban plan for cities in, 92
Electric cars, 36, 81–82
Electric demand, 400–401
Electric mobility, 33, 41, 481
Electric vehicle (EV), 355, 481
  cost of charging, 45
  design guidelines, 82
Electrical technology, 34
Electricity columns, 42
Electrification, 34
Electrified Transportation, 481
Electronic and digital technologies, training smart city workforces with wide range of, 572
  review questions/exercises, 572
Electronic Brake Control Modules (EBCM), 101–103
Electronic Control Units (ECUs), 101–103, 498
Elizabeth Morse Genius Foundation (EMGF), 162
Empowerment intelligence, 198
Empty Cities, 371
Encrypted visibility, 257–258
Encryption, 250
End-to-end connectivity, 324
Energy, 322, 481
  distribution networks, 41
  Energy/Green Building, 79
  framework, 200
  key principles, 322
  management and planning in smart cities, 458–459
Energy efficiency
  developing energy-efficient urban plan for cities, 90–92
  guidelines for energy efficiency in urban planning, 89–90
  planning for cities, 89
  responsibilities, 92
  steps, 92–94, 94t
    approval of strategic urban plans, 94
    city consultation, 93
    data analysis, 93
    data collection, 93
    evaluation of plans, 94
    start-up, 93
    strategy formulation, 93–94
  sustainable energy efficient path for cities, 89
  in urban planning for smart cities in developing world, 566–567
    review questions/exercises, 566–567
Engine Control Modules (ECM), 101–103
Entrepreneurial skill, 173
Environmental justice, 81
Environmental Studies program (ES program), 157
Equifax, 517
Equity crowdfunding, 26
European Union (EU), 192
Evaluation team (ET), 94
Event-based protection system, 430–431
Experiential learning activities, 151
Experimental Prototype Community of Tomorrow (EPCOT), 81
Explicit knowledge, 140
External investment, 25
Extrahigh voltage (EHV), 415–416, 418

F

Face-to-face education, 186
Facebook, 511
Fair Information Practices (FIP), 104
Fall back minimum risk condition, 99–100
Fault, Configuration, Accounting, Performance and Security (FCAPS), 335
Fault(s), 439, 585–586
  characteristics, 428–429, 432
  classification, 453
  effect variations in fault location, 451–452
  impact of fault incidence angle, 446–449
  identification, 439
  index, 443
  methodology and test system, 441–442
    current supported algorithm used for identification of fault conditions, 441–442
    problem formulation, 441
    test system, 441
  on phase-A and involvement of ground, 442–443
  on phases and involvement of ground, 446
  on phases-A and B and involvement of ground, 444–445
  on phases-A and B and without involvement of ground, 443–444
  research work, 440
  results, 442–453
  review questions/exercises, 585–586
  on three phases and without involvement of ground, 445–446
  impact of variations in fault impedance, 449–451
  work, 439–440
Federal Interstate Highway System, 515
Federal Motor Vehicle Safety Standards (FMVSS), 99
FedEx, 221
Feminist economics, 19
Festal, 241–242
Finance Commission (FFC), 267
Financial and technical planning of decarbonized hybrid energy systems for modern cities, 583
  review questions/exercises, 583
Financial resources, access to, 390
Financing, 25
  for Bangalore Tender SURE roads and mobility hubs as part of ABD, 359
  case studies, 559
  frequently asked questions, 555
  implementation and deployment companies, 527
  products, 543
  resources, 553
  standards, 547
  strategies in PPP-based smart cities, 124–125

components, 125–128
for Wadala CBD, 358
Financing smart cities, 575–576
convergence of missions, 264–265
differentiators, 265–267, 265f
ABD + Pan City strategy, 266–267
citizen engagement, 266
competition, 265
convergence, 267
different starting points, 265–266, 266f
SPVs, 266
innovative QCBS methodology, 270
payment terms and schedule, 270–271
review questions/exercises, 575–576
sources of, 267
ICCC project of Agra smart city limited, 268–269
SPV–Agra case study, 267–268
tender evaluation process by cities, 269–270
FIP. *See* Fair Information Practices (FIP)
5G new radio (5G NR), 101
5G, 101
barriers for connected vehicle applications and 5G characteristics, 103t
technology, 34
Flows of sewage, streams of revenue, 513–514
Fog computing, 6
Footway, 207
rights to, 214–216
kerb rights, 215
property rights, 214–215
right to smart city, 215–216
Ford, 481–482
"Forest bathing", 137
Formal negotiations, 192
Fossil fuels, 89
4D–RCS, 106
4G, 322–323
Framing, 140
Fuel production technologies, 458, 586
energy management and planning in smart cities, 458–459
renewable fuel production technologies, 459
review questions/exercises, 586
solar energy, 460–465
wind energy, 465–467
Function test analysis, 501
Fundability, 390
Funding strategies in PPP-based smart cities, 124–125
components, 125–128
Fuzzy logic, 482

G

Gamer, 344–345
low tolerance, 349t
tolerant, 348t
Gamification, 24, 289–290
Gaseous hydrocarbon cracking, 461
Gasification by solar energy, 461–463
General Data Protection Regulation (GDPR), 249
General Motor (GM), 481–482
General Systems Theory, 144–145
Geo-Positioning System (GPS), 498
Geographical Information Systems (GIS), 52–53, 388
Geoparticipation, 285
Geospatial technology, 176
Geothermal energy, 458
Ghent Living Lab, 288–289
Global Navigation Satellite System (GNSS), 106
Global Positioning Systems (GPS), 101, 106
Gold Coast Hinterland and Environment Council (GECKO), 290–291
Goods delivery, 224, 230
Google, 511
Google Scholar, 306
Governance, 175, 298
instruments of governance and cultural change, 301–303
of smart cities, 300–301
Government of India (GOI), 263
Graphical User Interface (GUI), 501–502
Gray infrastructure, 392
Great depression, 19
Green bonds, 390
Green communities, 157–162
Green Energy, 173
Green infrastructure (GI), 135, 380–381
class materials, 151–152
coproducing, 151–152
experiential learning activities, 151
storytelling, 151
educating professionals, 137–150
EDL, 148–150
integrative thinking, 145–148
social science theory for interdisciplinary practice, 138–141
supportive models for interdisciplinary practice, 141–150
systems thinking and systems design, 144–145
smart cities and, 135–137
considerations for practice, 136–137
types, 136f
Green infrastructure curricula at Toronto Metropolitan University, Toronto, Ontario, Canada, 569
review questions/exercises, 569
Green Revolution, 512
Green spaces, 300
Green-blue infrastructure (GBI), 380–381
Greenhouse gas emissions (GHG emissions), 34, 81–82, 91
Greenworks Action Plan, 79, 81–82, 84–85, 155, 164–165
Grid substation (GSS), 415–418
Gross domestic product (GDP), 6, 416–417
Ground fault monitoring technique, 431
"Ground-based automated mobility systems", 223–224
Groundbreaking innovation PPP, 127
Grounding, 428–429, 430t
Grounding and protection system, 432
GSM technology, 327

H

Hackathons, 287–288
civic, 26
Hacktivism, urban, 290–291
Hamilton Holt's legacy, 155–157
Harvard Negotiation Project, 140
Hazardous waste, 242
HDL-64E, 482
Health care, 172, 323–325, 324f
IoT applications in, 325
IoT-enabled care device works, 324–325
Helsinki LL, 288, 289t
Heritage City Development and Augmentation Yojana (HRIDAY), 267
High-level ontology design, 342
High-voltage direct current (HVDC), 439–440
Home-owners Associations (HOA), 514
Homelessness, 165–166
HOMER Pro software, 404–405
Homomorphic encryption schemes, 257
Human accessibility devices, 228–229
Human Centered Design (HCD), 499
Human framework, 199
Humane–machine interface (HMI), 99–100
Humanity, 199
Human–Robot Interaction (HRI), 498
Human–Vehicle Interaction (HVI), 504
Hybrid city, 199
Hybrid energy storage system (HESS), 427
Hybrid energy systems, 406
Hydrocarbons, 459
Hydrogen, 465

I

Identity, 356
IEEE 802.11p standards, 101
Impact investments, 390
In-Car Service Robots (ICSR), 504
In-kind exchange, 25
In-service training, 190
Incineration, 240
energy from, 245
Incremental approach, 68
Incremental PPP, 127
India
smart cities in, 263
smart cities project in India, 263
Indiegogo (crowdfunding sites), 26
Indigenous knowledge, 141–142
Industrial Internet of Things (IIoT), 325–327
networking technology to dominating IIoT market, 326
Industrial internet revolution, 325–326
Industrial Revolution (IR), 6
IR 1.0, 6
IR 3.0, 6

Industry 4.0 (IR 4.0), 6, 172, 193
review questions/exercises, 561–562
smart city generations development, 6–12
smart city policy, 3–5
technologic tools, 6f
Inertial Navigation Systems (INS), 106
Information and communication technology (ICT), 41, 49, 119–120, 175–176, 197, 249, 277, 365, 457
employment of, 278
ICT-based infrastructure, 3
ICT-supported collaborative governance, 305
Information city, 199
Information security, 173
and safety, 176
Information Technology (IT), 110–111
effect in training, 189–190
Infrastructure
financing, 121–122
and public transport, 41
Innovative QCBS methodology, 270
Innovative smart applications for solid waste management, 574–575
review questions/exercises, 574–575
Institutional economics, 19–20
Institutional framework, 199
Instrumental based vision explorations, 65–66
Instrumentation intelligence, 198
Integrated CAV simulation, 487–494, 489f
author driving scenarios, 487–489
detailed simulation, 489–494, 490t
Integrated Command and Control Centers (ICCC), 263, 269
project of Agra smart city limited, 268–269
integrated command and control center, 269f
Integrated solid waste management (ISWM), 243–244
reduction strategies, 243
reuse strategies, 244
waste hierachy, 243f
Integrated urban projects, 377
Integrative thinking, 145–148
in-class application, 146–148
Integrity, 104
Intelligence for Ecosystem Services/probabilistic model, 387b
Intelligent city, 198
Intelligent electric device (IED), 430
Intelligent traffic system, 322–323
Interdisciplinary practice, 138
social science theory for, 138–141
supportive models for, 141–150
Internalization, 301
International Society of Automotive Engineers (ISAE), 481
International Union for Conservation of Nature (IUCN), 381
Internet, 50, 322–323
reach and use, 312
Internet of things (IoT), 6–7, 50–51, 175–176, 182–183, 187–188, 200, 321, 324, 366, 388, 457
applications in healthcare, 325
IoT-enabled care device works, 324–325
data assortment and analysis, 324–325
end-to-end connectivity and affordability, 324
remote medical assistance, 325
simultaneous reporting and monitoring, 324
tracking and alerts, 325
IoT-Lite, 337
technology for connected public transport systems, 323
in traffic management, 323
in transportation, 322–323
Internship, 190
Interoperability, 50
Interpretive frames, 140
Intersection traffic, 483, 485t–488t
Interstate Transmission System (ISTS), 415
Intravehicular communication, 499
ISO 4448 standard, 223–224

J

Jamming attack, 104

K

k-anonymity, 252–254
Kalman filter, 495
Kerb rights, 214–216
property rights, 214–215
right to smart city, 215–216
Kerbside and pathway automation
commercial, 225
five intentions for standardizing, 225
legal, liability, and insurance, 225
operations, 225
planning, 225
safety and conflict avoidance, 225
Kerosene, 459
Keynesian Economics, 19
Kickstarter (crowdfunding sites), 26

L

Labor transition, 192
Land use, parking and, 109–110
Landfills, 239–240
Lane change, 483, 484t
Lane Keep Assistant (LKA), 497–498
Laurea Living Labs, 288
Lean machine, 475
“Learning by doing” approach, 157
Leasing, 25
LEED Gold City, 81
Light detection and reading (LiDAR), 388, 482
Light-dependent resistor sensor (LDR sensor), 326–328, 329f
Arduino IDE output, 330
Blynk cloud representation, 330
circuit connections, 329–330
motion reorganization, 332
prototype model front view, 330
threshold alert, 330–332
working principle, 328–329
Lime electric scooter company, 109
Linear discriminant analysis (LDA), 439–440
Listening, 301
Living Lab (LL), 288–289
“Living laboratory” approach, 192
Living labs, 388
Living Virs Forum Virium, 288
Load flow study to check feasibility of additional reactive compensation, 421
Local actors, 279
Local balance sheets, 23
Local budgeting, 23
Local currencies, 27
Local innovation system, 563
charging infrastructure, 42–43
charging stations for public enviroment, 43–45
extended stop, 44
short stop, 44
stop for charging, 44–45
cost of charging electric vehicle, 45
digital player, 41, 42f
mobility changing, 42
review questions/exercises, 563
Local public participation processes, 21–22
Local stock exchanges, 27
Localization of smart city infrastructure
mobility, 33–34
autonomous vehicle, 34
drivers and trends, 35–36
electrification, 34
MaaS, 33–34
V2X, 33
new actions for improving e-mobility, 36–37
smart mobility and factors, 34–35
Logistics 4.0, 177–178
Long-range, low-power telemetry and low-power computers, 389
Long-Term Evolution–Advanced (LTE-A), 101
Low impact design (LID), 380–381
Lyft, 97, 482

M

Machine learning (ML), 6, 198
Machine-to-human (M2H), 6
Machine-to-machine (M2M), 6
Macro-Trends, 33
Malware attack, 104
Mandatory Lane Change (MLC), 483
Market growth and IoT, 325–327
industrial automation using IoT, 325f
industrial internet revolution, 325–326
industrial IoT, 326–327
networking technology to dominating IIoT market, 326

Marketing skills, 172
Marxist economics, 19
Master planning, 372
Master System Integrator (MSI), 267
Methane (CH_4), 461
Metromile, 514
Micro Energy Grid (MEG), 399
Micro grids, 427
Microaggregation-based k-anonymity, 252
Microgrid controller, 404
Microreactors, 399–400, 402
Milewise, 514
Ministry of Housing and Urban Affairs (MoHUA), 263, 354
Mobile apps, 389
Mobility, 33–34, 199, 299, 357
  autonomous vehicle, 34
  drivers and trends, 35–36
    decrease in costs for purchase of electric vehicles, 35
    improved battery performance, 35
    less expensive and more performing vehicles, 35
    policies to reduce local and global emissions, 35
    reduction of emissions and fight against climate change, 36
    selling energy possibility on grid, 35
  electrification, 34
  infrastructure, 392
  MaaS, 33–34
  services, 97
  standard for ground based automated, 224
  V2X, 33
Mobility as a service (MaaS), 33–34, 42, 109
Modeling, 389
Modern monetary theory, 19
Molded case circuit breaker (MCCB), 428
Monetization, 573–574
  of kerb, 207
  review questions/exercises, 573
Moore's Law, 20
Moral foundations theory, 139
Motion reorganization, 332
MQTT protocol, 330
Multichamber Insulator Arresters (MCIA), 439–440
Multicity initiatives, 25
Multisector partnerships (MSP), 117, 126–128
  new PPP approach, 127
    business-as-usual PPP, 127
    coresearch PPP, 127
    groundbreaking innovation PPP, 127
    incremental PPP, 127
  PPPPs, 127–128
Multiterminal transmission line (MTTL), 439
Mumbai Metropolitan Regional Development Authority (MMRDA), 353
Mumbai's new smart business district at Wadala, 357–358
Municipal annexation, 27–28
Municipal bonds, 28
Municipal service coverage, 28
Municipalities, 478–479

N

Narrow-track vehicles (NTV), 471, 474
  alternatives, 477
  disrupting disruption, 477
  potential financial benefits to cities, 477–478
  potential solutions, 472–474
  solving traffic congestion and parking scarcity with, 586–587
    review questions/exercises, 586–587
  traffic challenge, 471–472
National Institute of Standards and Technology (NIST), 106, 498
National Investment and Infrastructure Fund (NIIF), 267
National U.S. Highway Traffic Safety Administration (NHTSA), 482
National Urban Policy Framework (NUPF), 354–355
Natural disasters, 61–62
Natural Language Processes (NLP), 169
Nature preserves, 157–162
Nature-Based Smart Urban Projects (NBSUP), 393
  integration of disciplines, 393
  integration to urban networks, 393
  network approach, 393
  research by design, 393
Nature-Based Solutions (NBS), 377–378, 380–385
  assessment, 385
  benefits and costs, 383–384
  defined concepts, 380–383
  efficiency and effectiveness, 385
  evaluation principles for assessment of, 385
  implementation, 384–385
  Smart City Technologies applied to, 386–389
  typology and examples, 381–383
  typology of societal challenges for assessment of impacts of, 385
  viability, 389–391
Neoclassical Economics, 19
Networking technology to dominating IIoT market, 326
New urbanism, 368–369
Nitrogen dioxide (NO_2), 383
Non-Motorized Transport (NMT), 264, 354
Nonunit protection system, 429
Nuclear Power Plant (NPP), 402
Nuclear-Renewable Integrated Energy System (N-R IES), 399–400

O

OACES modeling software, 83
Object and event detection and response (OEDR), 99–100, 105
Observing, 301
Obstacle avoidance, 483, 488t
On-demand service, 42
On-time alert, 325
One Click Parking' solution, 208
"One size fits all" mechanism, 335
Online shopping, 208–209
Open data, 302
Open Defecation Free (ODF), 264
Open government, 302
Open government data policy (OGD policy), 282–283
OpenStreetMap (OSM), 389
Operating costs, 23
Operation, 573–574
  review questions/exercises, 573
Operation and Maintenance cost (O&M cost), 404–405
Operational design domain (ODD), 99–100, 222
Operational expenditure (OPEX), 123
Operational management, 307
Operational Technology (OT), 110–111
Optical fiber network, 171
Optimal power flow methods, comparison of, 432
Optimization
  analysis, 82
  problem, 404–406
Orchestration intelligence, 198
Organic solid wastes, composting of, 244
Orlando Greenworks Action Plan, 80f, 85, 166
Orlando Utility Corporation (OUC), 81
Overcurrent protection, 431
Owner-side selective encryption approaches, 256
Ozone (O_3), 383

P

Palantir, 511
Palo Verde nuclear power station, 516
Pan City
  initiatives, 353–354
  strategy, 266–267
Parking, 471
  and land use, 109–110
  rules cities, 224
Participation process, 91, 299
  finalizing plan, 91–92
  implementing and financing plan, 92
  monitoring and evaluating plan, 92
  publicizing benefits, 92
  stakeholders identification, 91
Participatory platforms, 303
Pasila Living Labs, 288
Passive design strategies, 90
Passive infrared sensor (PIR sensor), 327, 327f
  pin configuration, 328f
  working principle, 327f–328f, 328
Pathway reservation system (PRS), 224, 229–230
Pathway robots, 221–222, 232–236
  bad idea, 234–235
  BIA responses to, 236
  certification for use, 231–232

Pathway robots (*Continued*)
robot weather worthiness, 232
for consumer deliveries, 232–233
devices, 228–229
five intentions for, 225
good idea, 234
international standard to manage, 223–232
ISO 4448, 223–224
robotic road vehicles for passengers, 226–227
service robots on pedestrian pathways, 227–231
standard components, 225
standard for ground based automated mobility, 224
Pavement, 219, 229
Payment terms and schedule, 270–271, 271t
Pedestrian, 220
accessibility devices, 228–229
pedestrian-scaled communities, 86
Pedestrianism, 213
Perception, 499–500
Personal delivery device (PDD), 109
Personal Information Protection and Electronic Documents Act (PIPEDA), 499
Personalized SLAs, 336, 338
Personalized travel information–transit agencies, 323
Personally Identifiable Information (PII), 498
Phase-A to ground fault (AG fault), 442, 448–450, 452
Phases-A and B to ground fault (ABG fault), 444–445
PiCarta, 306
Pick-up–drop-off zones (PUDOs), 208
PlanIT Valley, 371
Planning Support Systems (PSS), 309
Plasma gasification, 240
Plastic bags, 241–242
Plug-in hybrid electric vehicles (PHEVs), 36–37
PM10, 383
Point of measuring (POM), 432
Policy compliance, 389
Policy instrument approach
cities of future, 299–300
green spaces, 300
mobility, 299
urbanism, 300
for facilitating smart city governance, 576–579
review questions/exercises, 576–579
governance, 298
governance of smart cities, 300–301
instruments of governance and cultural change
answer, 301
capture, 301
citizen participation, 302
cocreation spaces, 303
collective intelligence, 303
creation of channels, 301
internalization, 301
listening, 301
observing, 301
open data, 302
open government, 302
from outside in and from top down, 302
participatory platforms, 303
share, 302
transparency, 302
participation, 299
platforms, 298
politics, 297–298
smart cities, 298
Policy of open government data, 282–283
Policymaking, 5
Portland Climate Action Plan, 166
Positioning, Navigation, and Timing (PNT), 106
Power Grid Corporation of India (PGCIL), 418
Power quality (PQ), 416–417
Power system computer-aided design (PSCAD), 439–440
Power-to-syngas study, 466
Pradhan Mantri Awas Yojana, 264
Pradhan Mantri Awas Yojana–Urban (PMAY-U), 265
Pragmatic liberal arts, 166
Pragmatic liberal arts at Rollins College, 570
review questions/exercises, 570
PRISMA method, 306
Privacy, 101–105
Private finance, feasible level of, 126
Private finance initiative (PFI), 123
Private sector involvement, 109
Process management, 307
Project delivery systems (PDSs), 117, 123
Project Management Consultant (PMC), 267–268
Project-based learning, 157
Property Assessed Clean Energy Program (PACE), 81, 83
Property rights, 214–215
Proportional Integral Derivative controller (PID controller), 495
Protection system
nonunit protection system, 429
other protection problems, 431–432
principle of, 429–432
unit protection, 429–431
Prototype, 321
Psychographics, 19
Public affairs, 308
Public asset corporation, 28
Public infrastructure network nodes (PINNS), 99
Public Land Survey System (PLSS), 70–71
Public participation, 279–281
engagement continuum, 280t
methods, 28
Public Participatory Geographic Information Systems (PPGIS), 309
Public perception, 111
Public transport management, 323
Public transportation, 353
systems, 472
Public–private partnerships (PPP), 25, 28, 117, 169, 266, 353–354, 418, 568–569
components of PPP-based smart cities funding and financing, 125–128
multisector partnerships, 126–128
optimization and distribution level in capital structure, 126
pillars and pilot of smart cities, 125–126
funding and financing strategies in PPP-based smart cities, 124–125
implication of previous studies regarding definition or general-related structure, 118t
implications of previous research studies for components-related funding and financing, 121t
on PPP's financial-related structure, 119t
for strategies-related funding and financing, 120t
literature review, 118–119
PPP-based funding and financing components, 123–124
review questions/exercises, 568–569
smart city, 119–121
potential funding options for, 121–123
theory, and advantages, 123
Public–private-people partnerships (PPPPs), 127–128

Q

Qualified workforce, 183
Quality and Cost-Based Selection method (QCBS method), 270
final score calculation through, 270
Quality of Experience (QoE), 335
Quality of life, 71
Quality of Service (QoS), 335
Quasi-identifier (QI), 250

R

Radar, 106, 493
Radar Cross Section (RCS), 491
Rajasthan Rajya Vidyut Prasaran Nigam Limited (RVPNL), 415, 417
Rajasthan Rajya Vidyut Utpadan Nigam Limited (RVUN), 417
Ranking matrix for solar siting, 82–83
Rapid Rail Transport Services (RRTS), 265
Reactive power compensation for voltage variation mitigation, 416–417
in grid with renewable energy, 583–584
review questions/exercises, 583–584
literature review, 416–417
manuscript organization, 417
methodology, 417–420
base transmission system, 418
planning of reactive compensation to mitigate voltage variations, 418–420
problem formulation, 418

research work, 417
results, 420–423
cases of study, 421
load flow study to check feasibility of additional reactive compensation, 421
observations of study, 421–422
effect of TCR on voltage under condition of RE generation variation, 422–423
utility grid, 417
Real-time Control System (RCS), 106
Real-time management, 323
Real-time vehicle tracking, 323
Recovery, 240
Recycling, 240, 244
composting of organic solid wastes, 244
Reduction strategies, 243
Regional planning elective course, 146–148
Regulated and Orderly Public Square, 228
Reinforcement Learning (RL), 495
Remote medical assistance, 325
Remote sensing (RS), 389
Remote telemetry units (RTUs), 439–440
Renewable energy (RE), 82, 173, 415, 441, 458
Renewable Energy Sources (RESs), 399
Renewable fuel production technologies, 459
Replay attack, 104
Request For Proposal (RFP), 267
Resilience, 378–379
Reuse strategies, 244, 244f
Revenue-generating, 25
Revenue-sharing, 25
Reverse Water Gas Shift system (RWGS system), 466
"Rights to City", 214
Risk
acceptability criteria, 390
mitigation, 511
Road, 489–490
construction, 476–477
Robot-assisted CAV systems, 504–505
Robotaxi, 220–223
Robotic road vehicles for passengers or goods, 226–227
Robotics, 482
Robotization, 183
Robots, 175, 211
in public spaces, 227
Rollins SunRail Forum, 163
Royal Institution of Chartered Surveyors (RICS), 366

S

Safety guidance for automated vehicles, 99–101
Safety Integrity Level (SIL), 497–498
Sales skills, 172
Science, Technology, and Innovation (STI), 117
Scopus, 306
Searcheable encryption schemes, 257
"Second Avenue Subway" project in New York City, 475–476
Security, 68
Security Vulnerability Exploration Module, 505
Self-driving cars, 55
Selling energy possibility on grid, 35
Semantic privacy definitions, 251
Sense of place, 370–371
Sensible socialism, 18
Sensitivity assessment, 407–408
Sensors, 50, 176, 327
canvas, 493
Service level agreement (SLA), 335, 579–580
assignment process, 343–349
review questions/exercises, 579–580
Service robots on pedestrian pathways, 227–231
access, 229
permissions, 229–230
behavior, 230
governance of robots, 228
integrating robotic kerbside and pathway access, 230
operation of robots, 227
pathway robots and human accessibility devices, 228–229
robot cybersecurity, 231
social communication, 230, 231f
Shifting baseline syndrome (SBS), 141–142
Shy distance, 227
Sidewalk and Kerb management (2000–25)
century of changes at kerb, 207–209
rights to footway and kerb, 214–216
twenty-first century footpaths, sidewalks, or pavements, 210–214
heading toward autonomous future, 211–213
prioritizing space, 213–214
Sidewalk and kerb management (2025–50)
delivery robot technology, 220–223
international standard to manage pathway robots, 223–232
pathway delivery robots in small-scale commercial use, 220f
pathway robots and business improvement areas, 232–236
Signal phase and timing (SPaT), 105
Simultaneous Localization and Mapping (SLAM), 506
Single converter voltage control method (SVC method), 431
Single-family zoning, 25, 29
Siri (voice recognition devices), 54
Skill shortages, compensation of, 186–187
Skills analysis for smart city workforce, 184–185
Skills DB Pro, 185
Skills gap, 184–187
act on collected data and create skillset inventory, 186
assessing skills needed, 185
compensation of skill shortages, 186–187
methods for skills development, 186
policymaking for, 187
skills analysis for smart city workforce, 184–185
Small and medium-sized enterprises (SMEs), 325
Small Modular Reactors (SMRs), 399–400, 402
Smart asset management, 172
Smart Cities Council, 197
"Smart Cities for Urban Transformation", 355
Smart Cities Mission (SCM), 263, 355
Smart Cities Wheel methodology (SCW methodology), 192
Smart city, 525, 541, 545, 551, 553, 557t, 3, 117, 119–121, 135, 175, 197, 249, 277, 297–298, 322, 457, 511. *See also* Cities
agenda for action for checking appropriate steps to understanding, 18b
City 1.0, 7
City 2.0, 7
City 3.0, 7
City 4.0, 7, 7f
ecosystem, 169
energy management and planning in, 458–459
funding and financing alternatives, 122t
generations development, 6–12, 8t
governance of, 300–301
and green infrastructure, 135–137
in India, 263
industry 4.0-enabled technologies
in smart city digital platform, 11t
in smart city physical platform, 10t
in smart city society platform, 9t
paradigm, 366
criticism of smart city paradigm and impacts on urbanism, 366–367
policy, 3–5
infrastructures, 4f
policy model types, 5t
potential funding options for, 121–123
project in India, 263
research, 305
revolutionizing localization of smart city infrastructure policies and financing, 562–563
review questions/exercises, 562–563
right to, 215–216
Smart City 1.0, 278
Smart City 2.0, 278
Smart City 3.0, 279
solar energy fuel production in, 463–465
strategy, 200
technology, 17
training framework, 173–174, 174f
wind energy fuel production in, 467
Smart city governance, 277–279
basic policies for, 279–283
policy of open government data, 282–283
public participation, 279–281
technology, 281–282

Smart city governance (*Continued*)
  components, 307–312
    citizen participation, 307–308
    governmental organization, 307
    use of technology, 308–309
  literature review, 305–307
  outcomes, 309–312
    contextual factors influencing smart governance, 311
    internet reach and use, 312
    policy domain, 311
    political and institutional environment, 311–312
    sociospatial characteristics, 312
    trust, 311
Smart City Proposal (SCP), 265, 267
Smart City Technologies (SCT), 377–378
  applied to NBS, 386–389
    conceptual chart, 386–387
    defined concepts, 387–389
  plus nature-based solutions, 582–583
    review questions/exercises, 582–583
  practice of urban design with digital technologies, 371–372
  and urban design challenges, 369–372
    digital technologies in public engagement and participatory design, 370
    urban design for smart(er) cities, 370–371
Smart city workforces, 182–183
  new features of workforce, 182–183
  partnering with technology firms to train, 570–571
    review questions/exercises, 570–571
  partnership with technology firms, 191–193
  relations between education and areas of expertise, 187–191
    automation, 190
    government effect, 188–189
    in-service training, 190
    internship, 190
    IT effect in training, 189–190
    partnering with technology firms to train smart city workforces, 188
    university–industry cooperation, 190–191
  skills gap, 184–187
  status in private sector, 183
  training capabilities and opportunities, 183
    government trends, 183
Smart counties
  general smartification concerns and benefits, 68–69
  policies for, 64–68
    adapting vision process to county's unique characteristics, 66–68
    exploring smartification vision for counties, 65
    four schemes, 65
    instrumental *vs.* value-based vision explorations, 65–66
    policy process, 64–65
  review questions/exercises, 564–565
  smart technologies, 49–55
    analytics, 53–54
    artificial intelligence, 54–55
    augmented and virtual reality, 54
    cloud computing, 51
    GIS, 52–53
    internet, 50
    sensors and internet of things, 50–51
    smartphones and smartphone applications, 51–52
    user-generated content capabilities, 51
  smartification
    and smart growth, 70–71
    considerations, 56–57
    in counties, 57–68
    equation, 55–56
  theoretical concerns, 69–70
Smart energy, 173
Smart Energy Grid, 322
Smart government, 183
Smart growth, 199, 367
Smart infrastructure, 322, 322f. *See also* Green infrastructure
  holistic problem-solving for smart cities, 20–22
    local metrics for success, 22
    local public participation processes, 21–22
    need for civic engagement and involvement, 21
    universal design principles, 20–21
    universal frameworks, 20
  key civic policy and financing instruments, 25–29
    alternative engagement models to ensure viability of IT initiatives, 25
    Charrettes, 25
    civic crowdfunding, 26
    civic Hackathons, 26
    co-ops, 26
    commons-based organizations, 26
    community land trust, 26
    crowdsourcing, 26
    dashboards, 27
    economic development corporations, 27
    local currencies, 27
    local stock exchanges, 27
    municipal annexation, 27–28
    municipal bonds, 28
    municipal service coverage, 28
    PPP, 28
    public asset corporation, 28
    public participation methods, 28
    tactical urbanism, 28–29
    TIF, 29
    upgrading to smart city infrastructure during routine maintenance, 29
    zoning as policy instrument, 29
  local smart cities, 17–20
    context and, 17–18
    culture and, 18–19
    economics and, 19–20
  locally relevant policies and financing concepts, 22–25
    local balance sheets, 23
    local budgeting, 23
    local incremental development, 24–25
    local project selection criteria, 23–24
  policies and financing
    review questions/exercises, 562
  policy applications for informal sectors
    connectivity, 321
    data processing, 321
    energy, 322
    health care, 323–325
    implementation, 327–329
    market growth, 325–327
    results, 329–332
    traffic management, 323
    transport, 322–323
    user interface, 321
Smart mobility
  and factors, 34–35
  market, 41
Smart sustainable development, 3
Smarticipate platform, 285, 286f
Smartification
  considerations, 56–57
    instrumental considerations, 56–57
    overall considerations of smartification applications, 56
    value-based considerations, 57
  in counties, 57–68
    county departments, 58–59
    nature of county–city differences–unique county characteristics, 57–58
    smart technology applications by county department, 59–60
    unique county services, 58
  county characteristics, challenges, smartification, and departments, 60–63
    agriculture, 61
    annexations, 62–63
    coordination with municipalities, 62
    lack of visibility, 63
    large territory, 60–61
    natural disasters, 61–62
    rural communities, 61
    small and remote communities, 61
    sprawl, 62
    state funding, 62
    state mandates, 62
    state provisions, 62
    unincorporated and incorporated communities, 61
  county planning department, priorities, applications, and considerations, 63–64
    advance or long-range planning, 63–64
    county priorities, 64
    current planning, 63
  equation, 55–56
  general smartification concerns and benefits, 68–69
    major participant-related general concerns, 68–69
    spectrum of desired smartness and overall benefits, 69
    technology-related general concerns, 68
  policies for smart counties, 64–68

and smart growth, 70–71
Smartmiles, 514
Smartphones, 54
Social cohesion, 312
Social engineering, 514
Social equity, 81, 391
Social media, 309
Social norm theory, 138–139
Social prescribing, 137
Social science theory for interdisciplinary practice, 138–141
Socialism, 18
Society of Automotive Engineers (SAE), 99–100, 105, 505–506
Sociospatial characteristics of smart city governance, 312
Solar array, 84
Solar design guidelines, 82
Solar energy, 458. *See also* Wind energy
fuel production by using, 460–465
gaseous hydrocarbon cracking, 461
gasification by solar energy, 461–463
in smart cities, 463–465
thermochemical cycles, 460–461
thermolysis of water, 460
Solar Energy Loan Fund (SELF), 81
Solar gasification, 461
Solar photovoltaic (solar PV), 416–417
panels, 400, 402
system, 416–417
Solar-powered stations, 463–465
Solid waste management, 239
conserves energy, 239
creates employment, 239
difference, 239
energy from incineration, 245
recycling, 244
reduces pollution, 239
waste disposal methods, 239–240
waste management methods, 241–244
Solid-state circuit breakers (SSCB), 428
"SolSmart Gold City", 81
Solvency, 390
Special Planning Authority (SPA), 355
for new smart business district in Mumbai, 355–356
for smart city with Tender SURE and mobility in Bengaluru, 356–357
Special-Purpose Vehicles (SPVs), 266, 353
Agra case study, 267–268
ensuring accountability, 268
ensuring financial control, 268
process of implementation, 268f
Specific data envelopment analysis with a perfect object method (SDEA PO method), 416–417
Spectrum of desired smartness, 69
Spiral Dynamics, 20
Sprawl, 62
Stakeholders, 117
consultation, 179
engagement, 391
identification, 91
State funding, 62
State of Charge (SOC), 404
State provisions, 62
State transmission utility (STU), 415, 418
State-Owned Property, 228
Static Synchronous compensator (STATCOM), 439–441
Stockwell transform, 440
Strategic urban plan for cities in Egypt, 92
Strategic urban plans, approval of, 94
Strategic vision for urban planning and development, 355–357
SPA for new smart business district in Mumbai, 355–356
SPV for smart city with Tender SURE and mobility in Bengaluru, 356–357
Streetlight control, 327
Sulfur dioxide (SO_2), 383
Sunshine laws, 370
Supplemental Nutrition Assistance Program, 85
Supply chain management, 173
Support vector machine (SVM), 439–440
Support vector regression (SVR), 439
Supportive models for interdisciplinary practice, 141–150
indigenous knowledge, 141–142
UX Design, 142–144
Sustainability, 155, 197, 310, 356
performance-based evaluation model, 126
Sustainable energy
efficient path for cities, 89
plans, 89
Sustainable Enterprise, 164
Sustainable urban drainage systems (SuDS), 380–381
Swachh Bharat Mission (SBM), 264, 267
Swachh Bharat Mission-Urban (SBM-U), 265
Swarm intelligence, 482
Swarm theory, 389
Sybil, 104
Symbiosis in Development (SiD), 145
Synergism
with other urban infrastructures, 391–392
with other urban systems, 392–393
Systemic Design Toolkit, 145
Systems design, 144–145
in-class application, 145
Systems theories, 140
Systems thinking, 144–145
in-class application, 145
for social change, 21
Systems-level change process, 21

T

Tacit knowledge, 140
Tactical urbanism, 28–29
Tariff based competitive bidding (TBCB), 418
Tax increment financing (TIF), 29, 267
Teaching sustainability in Metro Orlando, 570
bike trails, nature preserves, and green communities, 157–162
Central Florida loop, 158f
common sense education and ideal campus, 155–157
Greenworks Action Plan, 164–165
review questions/exercises, 570
sustainability's challenge, 165–166
Winter park, 162–164
Team effectiveness theory, 140
Technical Teams (TTs), 92
Technology, 98, 281–282
framework, 198–199
horizontals for skilling, 175–177
AI, 177
geospatial technology, 176
information and communication technology, 175–176
information security and safety, 176
internet of things, 176
sensors, 176
smart city training architecture, 176f
platforms, 298
and smart governance, 308–309
technology-based skilling verticals for smart city, 171–173
blockchain, 177
change management, 173
data analytics and monetization, 172
disaster management training, 172
eCommerce and supply chain management, 173
entrepreneurial skill, 173
expectation from technology vendors, 178–179
healthcare, 172
industry 4.0 and digital manufacturing, 172
information security, 173
sales and marketing skills to promote local, 172
smart asset management, 172
smart city training framework, 173–174
smart energy, 173
supporting specially enabled persons and elders, 173
technology horizontals for skilling, 175–177
telecommunication connectivity and infrastructure creation and maintenance, 171
traffic management, 171–172
upskilling in creating smart city, 175
usage of 3D printing for designing, 172–173
technology-supported governance, 308
vendors, 178–179
digital workplace and centers of excellence, 179
supporting innovation and digital city, 178–179
Technology firms
partnership with, 191–193
drive technology and services, 192
labor transition, 192

Technology firms (*Continued*)
relations with technology partners, 192–193
role of e-training and distance education, 193
skill, 170–171
technology-based skilling verticals for smart city, 171–173
training professional smart city workforces through partnership with, 571–572
review questions/exercises, 571–572
Ted Rogers School of Management (TRSM), 147
Telecommunications, 41
connectivity, 171
Tender evaluation process by cities, 269–270
commercial evaluation, 269–270
arriving at QCBS scores, 270t
normalized scores, 270t
final score calculation through QCBS, 270
prequalification, 269
technical evaluation, 269
Tender SURE, 353–354
and mobility hubs at Bangalore Smart Cities, 358–359
Terre Neutre (TN), 428–429
Terre Terre (TT), 428–429
Tesla, 511
Texas A&M Transportation Institute Study, 472b
Theory of Organizational Knowledge Creation, 140
Thermal demand, 400–401
Thermal Load Controller (TLC), 400, 404
Thermochemical cycles, 460–461
Thermolysis of water, 460
ThinkNature project, 390–391
3D
3D-Citygame-Ghent, 290
LiDAR, 388
printing for designing, 172–173
3G, 322–323
Threshold alert, 330–332, 331f
Through life management (TLM), 123
Thyristor-controlled reactors (TCR), 418, 441
effect of TCR on voltage under condition of RE generation variation, 422–423
Time–current curves (TCCs), 431
Tire Pressure Monitoring System (TPMS), 498
Tracking and alerts, 325
TrackStar, 185
Traffic jams, 471, 478
Traffic management, 171–172, 323
connected cars, 323
IoT technology for connected public transport systems, 323
public transport management, 323
vehicle tracking systems, 323
Traffic volumes, 97
Traffic-flow-based streetlight control system, 327
Training capabilities and opportunities, 183
Training smart city workforces
characteristics, 198, 199t
criticism, 200
frameworks, 198–200
data management framework, 200
energy framework, 200
human framework, 199
institutional framework, 199
technology framework, 198–199
Transit and private motorized vehicles, 392
Transit Oriented-Development policy (TOD policy), 355
Transmission Control Modules (TCM), 101–103
Transparency, 302
Transport, 322–323
Transport Canada, 498–499
Transport management systems, 323
Transport network companies (TNCs), 208
Transportation, 481
Transportation policies
barriers for connected vehicle applications and 5G characteristics, 103t
communication for connected mobility, 101
connected and automated mobility, 98–99
for connected and automated mobility in smart cities, 567–568
review questions/exercises, 567–568
cooperative driving automation, 105–106
cybersecurity and privacy, 101–105
legislation, 110–111
navigation of CAVs, 106–108
parking and land use, 109–110
policies for connected and automated mobility services, 108–109
public perception, 111
safety guidance for automated vehicles, 99–101
safety rules for pilot testing, 102t
TransUnion, 517
Trust, 498
theory, 140
Truth and Reconciliation Commission (TRC), 141
TRW, 517

U

U.S. Department of Transportation (USDOT), 97
Uber, 97, 482
Ubiquitous city, 198
UN Sustainable Development Goals (SDG), 117, 300, 354
Undergraduate education, 186
Union Territories (UTs), 264
Unique county services, 58
Unit protection, 429–431
data-based protection system, 430
event-based protection system, 430–431
United Nations Environment Program (UNEP), 90
United Nations Human Settlements Program, 90
United Nations Sustainable Development Goals (UNSDG), 81
United States of America (USA), 416–417
University–industry cooperation, 190–191
Unmanned aerial vehicles (UAVs), 55
Uploading/downloading software, 335–336
Urban design, 365
as framework for understanding cities, 368–369
future design of cities, 372–373
in shaping cities, 367–369
smart city technologies and urban design challenges, 369–372
strategies and smart city paradigm, 581–582
review questions/exercises, 581–582
unpacking smart city paradigm, 365–367
Urban Footprint, 53–54
Urban forestry (UF), 380–381
Urban future, 366, 371
Urban governments, 307
Urban greening, 210
Urban hacktivism, 290–291
Urban Local Bodies (ULBs), 263–264
Urban planning
elective course, 146–148
guidelines for energy efficiency in, 89–90
energy efficiency potential in building sector, 90
principles of energy efficiency, 90
reason for energy efficiency, 90
Urban policies in India, 354–355
Urban population, 117
Urban resilience, 378
added value of SCT plus NBS, 391–393
NBS, 380–385
NBSUP, 393
policies in context of sustainability, 378–380
policies, foci, and threats, 379–380
resilience, 378–379
SCT applied to NBS, 386–389
viability of NBS, 389–391
Urban revenue, 392
Urban security, 392
Urban services, 392–393
Urban settlements, 89
Urban sustainability, 305
Urban systems, 277
Urban Transport services, 265
Urban3, 372
Urbanism, 300
Urbanization, 6–7, 181–182
US Department of Transportation (DOT), 99–100
US Federal Communication Commission, 101
US National Highway Traffic Safety Administration (NHTSA), 99
US Regulatory Cooperation Council (RCC), 498–499
User experience (UX), 20–21
design, 142–144

in-class application, 143–144
User interface, 321
User-generated content capabilities, 51
Utility
burden, 85
grid, 417

V

Value capturing by implementing financing mechanisms, 357–359
Mumbai's new smart business district at Wadala, 357–358
Tender SURE and mobility hubs at Bangalore Smart Cities, 358–359
Value for Money (VFM), 117
Value-based considerations in smartification applications, 57
Value-based vision explorations, 65–66
Vansharing, 472–473
Vehicle Miles Traveled (VMT), 213, 357–358
Vehicle platooning, 482–483, 483t
Vehicle tracking systems, 323
Vehicle-to-everything (V2X), 33, 97, 101
Vehicle-to-grid (V2G), 35
Vehicle-to-Infrastructure (V2I), 101, 481–482, 499, 501
Vehicle-to-Internet of Things (V2IoT), 499
Vehicle-to-pedestrian (V2P), 101
Vehicle-to-Vehicle (V2V), 481–482, 500–501
Viability gap funding (VGF), 127
Viability of NBS, 389–391
dynamic efficiency, 390–391
efficiency, 390–391
fundability and solvency, 390
policy compliance, 389
Virtual reality (VR), 54, 169, 308
Vision, Mission, Objectives, Strategies, and Action plans (VMOSA), 64–65
Visual representation technologies, 52–53
Visualization, 53–54
Voice recognition devices, 54
VoIP, 335–336
Volatility, uncertainty, complexity, and ambiguity (VUCA), 511, 588
complexity meets complications, 516
drivers, 514
flows of sewage, streams of revenue, 513–514
onerous ship, 517–518
repaving paradise, potholes in parking lots, 515–516
review questions/exercises, 588
unintended consequences, 517
Voltage variation index (VVI), 419–420
Volunteered geographic information (VGI), 389

W

Wadala CBD, 355
Walkable urbanism, 156
WAN, 321
Waste disposal methods, 239–240
avoidance/waste minimization, 240
composting, 240
incineration/combustion, 240
landfills, 239–240
plasma gasification, 240
recovery and recycling, 240
special waste disposal, 240
waste to energy, 240
Waste management, 91
bottom line, 241–242
classification of solid waste, 241t
discarded plastic bags, 242f
functional elements of smart solid waste management, 242–243
collection, transfer, and transport of solid waste, 243
disposal of solid waste, 243
onsite handling, storage, and processing, 242–243
resource recovery and processing, 243
ISWM, 243–244
methods, 241–244
Waste to energy process, 240
Water-sensitive urban design (WSUD), 380–381
Wavelet energy coefficient (WEC), 439–440
Wavelet transform (WT), 439
Wavelet transform coefficients (WTC), 439–440
Wearable technologies, 389
Web 2.0, 309
Web of Science, 306
Wekiva 2020, 162
Wekiva River Area Protection Act, 161
Wheelchairs, 228–229
Wi-Fi, 321–324, 498
Wicked problems, 144
Wide Area Measurement Systems (WAMS), 439
Wind energy
fuel production by using, 465–467
in smart cities, 467
Wind turbines (WT), 400, 403–404
Winter Springs Town Center, 160–161
Wired city, 198–199
Wireless internet, 7
Wireless sensor networks (WSN), 386, 389
Work from Home (WFH), 169

Z

Z-wave, 324
Zero-emission vehicles, 33
ZigBee, 324, 327
Zoning as policy instrument, 29
"ZWERM" project, 290

9780128191309